Ordered Pairs [Section 3.1]

A pair of numbers enclosed in parentheses and separated by a comma, such as $(-2, 1)$ is called an *ordered pair* of numbers. The first number in the pair is called the *x-coordinate* of the ordered pair, while the second number is called the *y-coordinate.* For the ordered pair $(-2, 1)$, the x-coordinate is -2 and the y-coordinate is 1.

The Rectangular Coordinate System [Section 3.1]

We graph ordered pairs on a rectangular coordinate system. A rectangular coordinate system is made by drawing two real number lines at right angles to each other. The two number lines, called *axes,* cross each other at 0. This point is called the *origin.* The horizontal number line is the *x-axis* and the vertical number line is the *y-axis.*

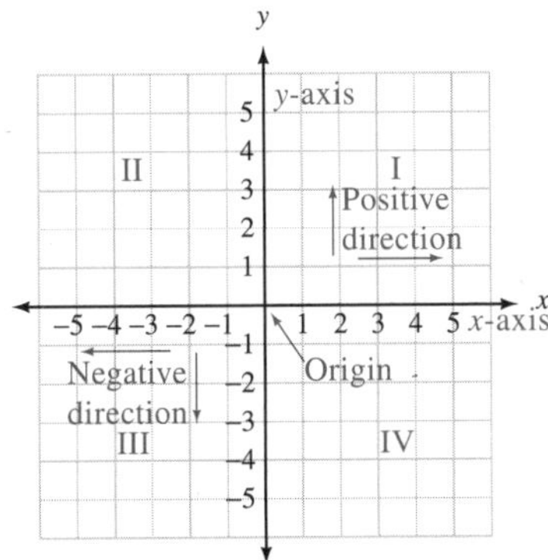

FIGURE 4 **The rectangular coordinate system.**

Graphing Ordered Pairs [Section 3.1]

To graph the ordered pair (a, b) on a rectangular coordinate system, we start at the origin and move a units right or left (right if a is positive and left if a is negative). We then move b units up or down (up if b is positive and down if b is negative). The point where we end up is the graph of the ordered pair (a, b). Figure 5 shows the graphs of the ordered pairs $(2,5)$, $(-2, 5)$, $(-2, -5)$, and $(2, -5)$.

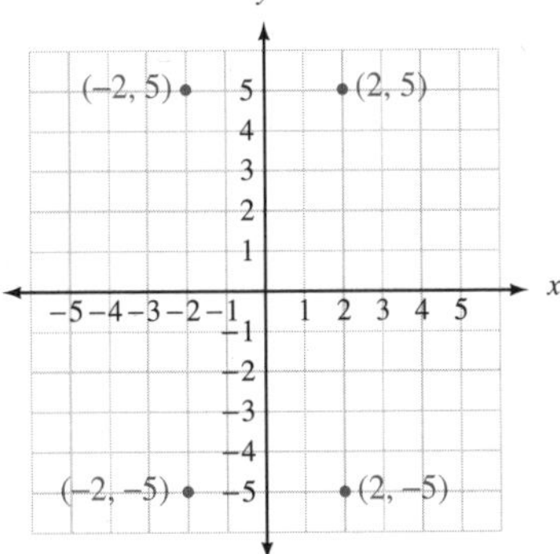

FIGURE 5 **The graphs of four ordered pairs.**

Graphing Functions [Section 3.5]

There are many ways to graph functions. Figure 6 is a graph constructed by substituting 0, 10, 20, 30, and 40 for x in the formula $f(x) = 7.5x$, and then using the formula to find the corresponding values of y. Each pair of numbers corresponds to a point in Figure 6. The points are connected with straight lines to form the graph.

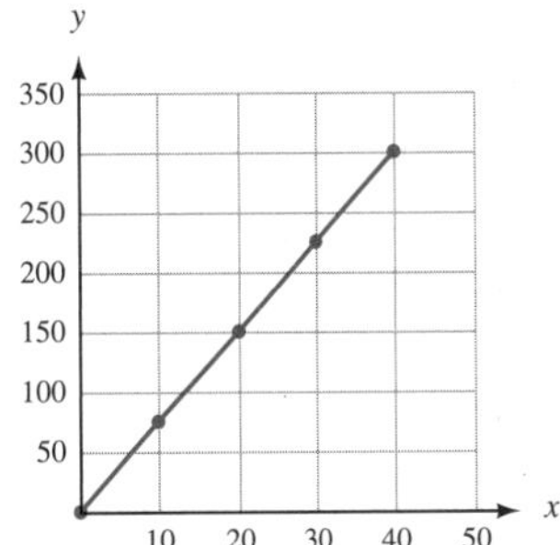

FIGURE 6 **The graph of $f(x) = 7.5x$, $0 \le x \le 40$.**

Function Notation [Section 3.6]

The notation $f(x)$ is used to denote elements in the range of a function. For example, if the rule for a function is given by $y = 7.5x$, then the rule can also be written as $f(x) = 7.5x$. If we ask for $f(20)$, we are asking for the value of y that comes from $x = 20$. That is

if $\quad f(x) = 7.5x$

then $\quad f(20) = 7.5(20) = 150$

Figure 7 is a diagram called a *function map* that gives a visual representation of this particular function:

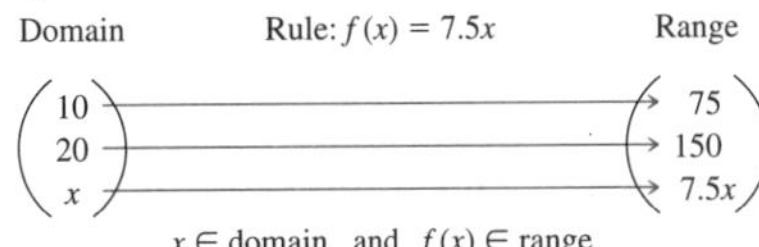

FIGURE 7 **A function map.**

A function can also be thought of as a machine. We put values of x into the machine which transforms them into values of $f(x)$, which are then output by the machine. The diagram is shown in Figure 8.

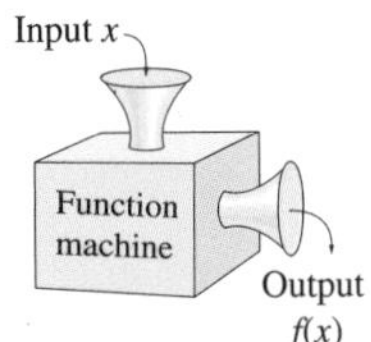

FIGURE 8 **A function machine.**

We can organize our work with paired data, graphing, and functions into a more standard form by using ordered pairs and the rectangular coordinate system.

(Continued on inside back cover)

ENHANCED WebAssign

The perfect instructional tool to help you bridge the gap!

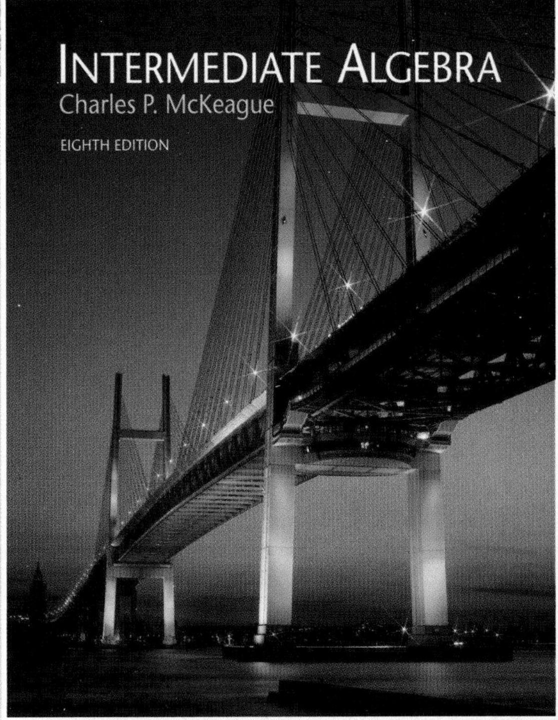

Intermediate Algebra is infused with Pat McKeague's passion for teaching mathematics. His attention to detail and exceptionally clear writing style move students through each new concept with ease. This Eighth Edition of ***Intermediate Algebra*** is enriched with new features and pedagogy that will help your students bridge the concepts. And now this best-selling text integrates **Enhanced WebAssign**—a dynamic homework management system that enables you and your students to work as partners in the teaching and learning process. McKeague's ***Intermediate Algebra*** is a bridge to student learning, and **Enhanced WebAssign** builds a bridge between the text's proven content and powerful learning resources!

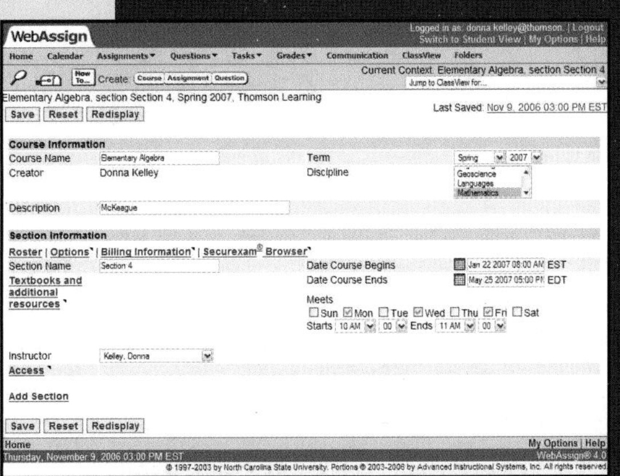

Open here to learn more about
Enhanced WebAssign!

Practice It

Enhanced WebAssign offers unlimited practice! Active examples create a bridge from homework to examples and additional practice. Students can work through these multi-step, fill-in-the-blank examples at their own pace and, when finished, return to the original homework problem. Plus, step-by-step tutorials with feedback guide students through solving a similar problem, which better prepares them to solve the question at hand.

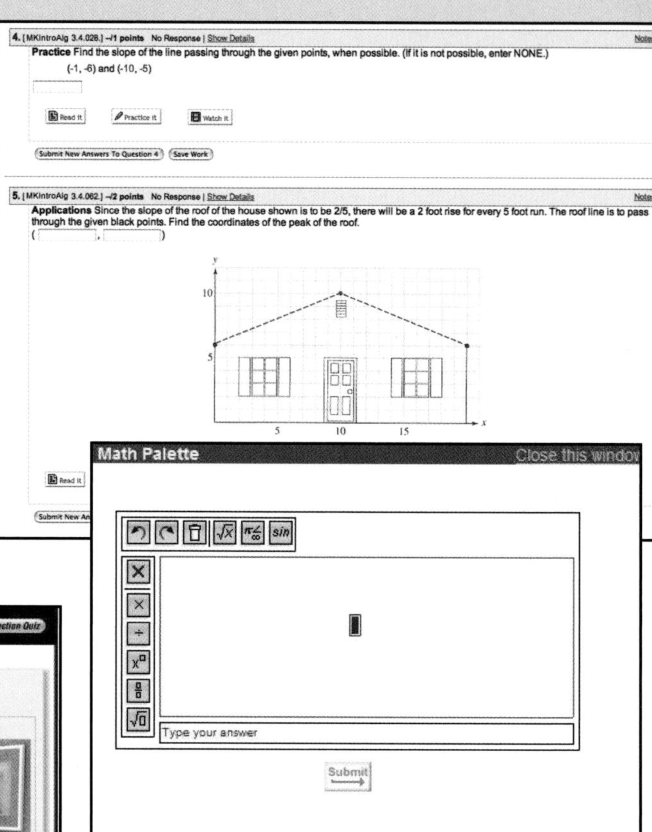

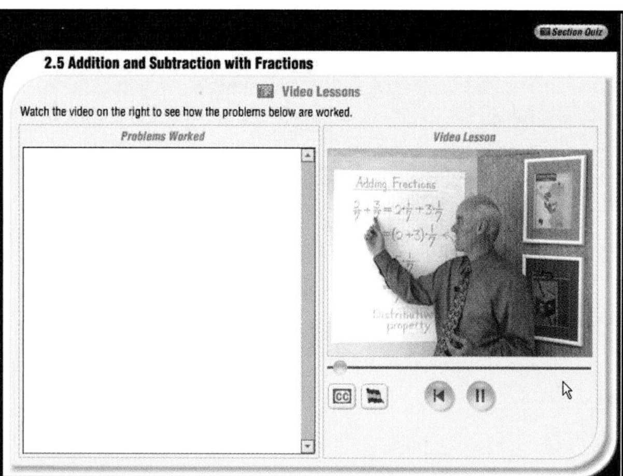

Watch It

Enhanced WebAssign features powerful visuals! The "Watch It" feature provides text material in multiple media formats to reinforce what students are learning in class and provide additional tutorial support. Students can see and hear additional instruction for each homework question via 1- to 4-minute problem-specific videos that correlate to the problems in the homework questions.

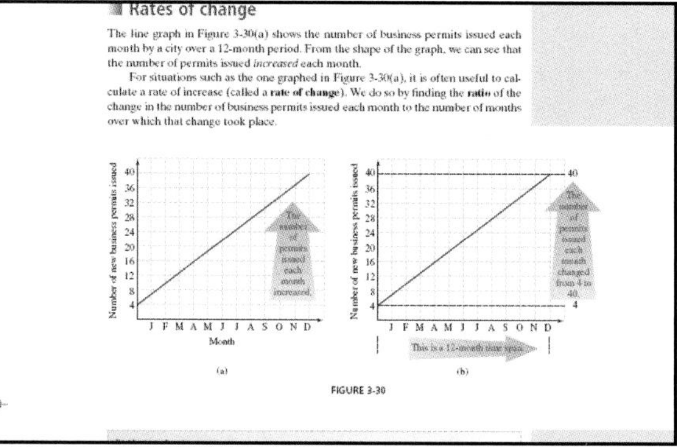

Rates of change

The line graph in Figure 3-30(a) shows the number of business permits issued each month by a city over a 12-month period. From the shape of the graph, we can see that the number of permits issued *increased* each month.

For situations such as the one graphed in Figure 3-30(a), it is often useful to calculate a rate of increase (called a **rate of change**). We do so by finding the **ratio** of the change in the number of business permits issued each month to the number of months over which that change took place.

FIGURE 3-30

Read It

Enhanced WebAssign builds a direct bridge to the text! Whenever students leave their book behind, they can always refer to important sections within their homework assignment by clicking the "PDF Read It" link, which refers students to PDF sections of the text.

Get your free 45-Day trial!

Visit **www.webassign.net/brookscole**

Enhanced WebAssign allows you to assign, collect, grade, and record homework assignments via the web. This proven homework system is **enhanced** to include videos, links to textbook sections, and problem-specific tutorials. **Enhanced WebAssign** is more than a homework system; it is a complete learning system for math students!

Enhanced WebAssign is perfect for:

Online homework ▪ Quizzes, tests, and exams ▪ Polls and surveys
Just-in-time teaching ▪ Coordinated teaching with a group of instructors

Grade It

Enhanced WebAssign saves you time! You can assign problems from the text and have them automatically graded. Problems match the end-of-section exercises and never deviate from the concepts covered in that section. As an instructor you have immediate access to all of your student grades. In addition, you can choose to enable your students to see their own quiz and test averages, grades, and homework scores as the semester progresses, as well as compare their scores against class averages.

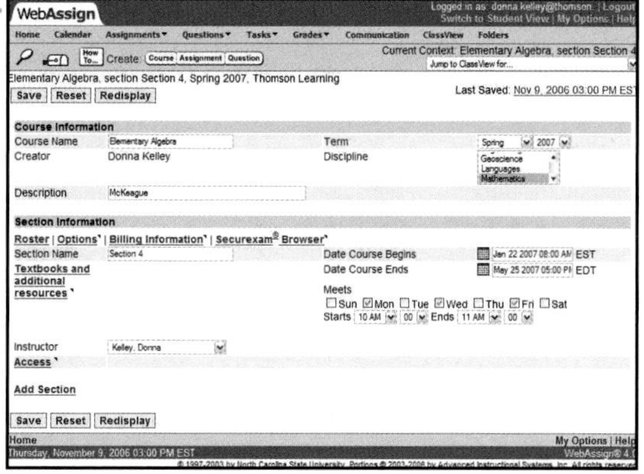

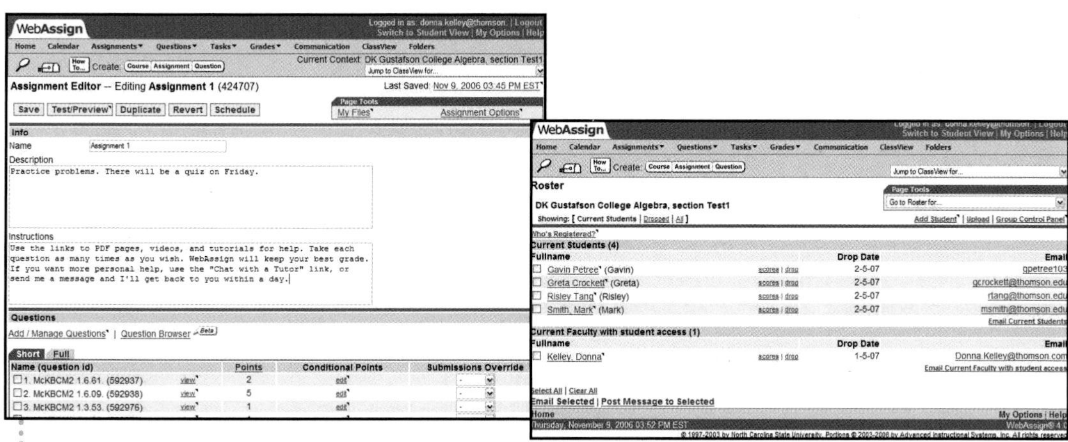

Solve It

Enhanced WebAssign includes algorithmic problems! Choose from approximately 1,500 text-specific, algorithmically generated problems including free-response, multiple-choice, and multi-step problem types. All problem sets use graphics and figures taken directly from the text.

INTERMEDIATE ALGEBRA

EIGHTH EDITION

Charles P. McKeague

CUESTA COLLEGE

Australia • Brazil • Canada • Mexico • Singapore • Spain
United Kingdom • United States

Executive Editor: *Charlie Van Wagner*
Development Editor: *Donald Gecewicz*
Assistant Editor: *Laura Localio*
Editorial Assistant: *Lisa Lee*
Technology Project Manager: *Rebecca Subity*
Marketing Manager: *Greta Kleinert*
Marketing Assistant: *Cassandra Cummings*
Marketing Communications Manager: *Darlene Amidon-Brent*
Project Manager, Editorial Production: *Cheryll Linthicum*
Creative Director: *Rob Hugel*
Art Director: *Vernon T. Boes*
Print Buyer: *Karen Hunt*
Permissions Editor: *Roberta Broyer*
Production Service: *Graphic World Publishing Services*
Text Designer: *Diane Beasley*
Photo Researcher: *Kathleen Olson*
Illustrator: *Graphic World Illustration Studio*
Cover Designer: *Diane Beasley*
Cover Image: Yokohama Bay Bridge at Sunset © 2006 Jupiterimages Corporation
Cover Printer: *R.R. Donnelley/Willard*
Compositor: *Graphic World Inc.*
Printer: *R.R. Donnelley/Willard*

Printed in the United States of America
1 2 3 4 5 6 7 11 10 09 08 07

Thomson Higher Education
10 Davis Drive
Belmont, CA 94002-3098
USA

Library of Congress Control Number: 2006937869

ISBN-13: 978-0-495-10840-5
ISBN-10: 0-495-10840-5

Brief Contents

Contents

CHAPTER 4 Systems of Linear Equations and Inequalities 209

CHAPTER 5 Exponents and Polynomials 269

CHAPTER 6 Rational Expressions and Rational Functions 337

CHAPTER 7 Rational Exponents and Roots 407

CHAPTER 8 Quadratic Functions 467

CHAPTER 9 Exponential and Logarithmic Functions 529

CHAPTER 10 Conic Sections 583

CHAPTER 11 Sequences and Series 615

Preface to the Instructor

I have a passion for teaching mathematics. That passion carries through to my textbooks. My goal is a textbook that is user-friendly for both students and instructors. For students, this book forms a bridge to college algebra with clear, concise writing; continuous review; and foreshadowing of topics to come. For the instructor, I build features into the text that reinforce the habits and study skills we know will bring success to our students.

The eighth edition of *Intermediate Algebra* builds upon these strengths. In this edition, renewal of the problem sets was my first priority. This renewal of the problem sets, along with the continued emphasis on foreshadowing of later topics, a program of continuous cumulative review, and the focus on applications, make this the best edition of *Intermediate Algebra* yet.

Renewal and Reorganization of Exercise Sets

This edition of *Intermediate Algebra* contains 1,800 new problems, roughly 30 percent of the problems in the reorganized problem sets. Most of these new problems have been used to shore up the midrange of exercises. Doing so gives our problem sets a better bridge from easy problems to more difficult problems. Many of these midrange of problems cover more than one concept or technique in a new and slightly more challenging way. In short, they start students thinking mathematically and working with algebra productively. This enhanced midrange of exercises also underscores our series' appeal to the middle level of rigor for the course. We think instructors will use many of these midrange problems as classroom examples, so we have labeled some of them as *chalkboard problems.*

As part of our revamping of the problem sets in this edition, we have also reordered the categories of problems to make a more logical bridge between sections.

A Better Progression of Categories of Problems

The categories in our problem sets now appear in the following order.

General (Undesignated) Exercises These "starter" exercises normally do not have labels. They involve a certain amount of the drill necessary to master basic techniques. These problems then progress in difficulty so that students can begin to put together more than one concept or idea. It is here that you will find the foreshadowing problems. Instead of drill for the sake of drill, we have students work the problems that they will need later in the course—hence the description *foreshadowing.* This category is also where you will find the midrange of problems discussed earlier. As in previous editions, we have kept the odd-even similarity of the problems in this part of the problem set.

Applying the Concepts Students are always curious about how the algebra they are learning can be applied, so we have included applied problems in most of the problem sets in the book and have labeled them to show students the array of uses of mathematics. These applied problems are written in an inviting way, to help students overcome some of the apprehension associated with application problems. We have a number of new applications under the heading *Improving Your Quantitative Literacy* that are particularly accessible and inviting.

Maintaining Your Skills One of the major themes of our book is continuous review. We strive to continuously hone techniques learned earlier by keeping the important concepts in the forefront of the course. The *Maintaining Your Skills* problems review material from the previous chapter, or they review problems that form the foundation of the course—the problems that you expect students to be able to solve when they get to the next course.

Getting Ready for the Next Section Many students think of mathematics as a collection of discrete, unrelated topics. Their instructors know that this is not the case. The new *Getting Ready for the Next Section* problems reinforce the cumulative, connected nature of this course by showing how the concepts and techniques flow one from another throughout the course. These problems review all of the material that students will need in order to be successful, forming a bridge to the next section.

Extending the Concepts Many of the problem sets end with a few problems under this heading. These problems extend the topics to a more challenging level than the previous problems in the problem set.

The Intermediate Course as a Bridge to Further Success

Intermediate algebra is a bridge course. Many students will go on to precalculus, calculus, and majors requiring quantitative literacy. The concepts in an intermediate-algebra course are bridges linked cumulatively to each other and to advanced quantitative concepts. The course and its syllabus, therefore, serve students who require some shoring up of their math skills, who have gaps in their mathematical preparation, or who need additional concepts to bring them to the level of ability to do quantitative work in their majors. Students in intermediate algebra can be motivated more readily if they see the "bridge" ahead to precalculus and their chosen course of study.

Our Proven Commitment to Student Success

After seven successful editions, we have developed several interlocking, proven features that will improve students' chances of success in the course. We placed practical, easily understood study skills in the first six chapters (look for them on the page after the chapter opener). Here are some of the other, important success features of the book.

Early Coverage of Functions Functions are introduced in Chapter 3 and then integrated into the rest of the text. This feature forms a bridge to college algebra by requiring students to work with functions and function notation throughout the course.

Getting Ready for Class Just before each problem set is a list of four questions under the heading *Getting Ready for Class.* These problems require written responses from students and are to be done before students come to class. The answers can be found by reading the preceding section. These questions reinforce the importance of reading the section before coming to class.

Linking Objectives and Examples At the end of each section we place a small box that shows which examples support which learning objectives. This feature helps students to understand how the section and its examples are built around objectives. We think that this feature helps to make the structure of exposition of concepts clearer.

Blueprint for Problem Solving Found in the main text, the *Blueprint for Problem Solving* is a detailed outline of steps needed to successfully attempt application problems. Intended as a guide to problem solving in general, the blueprint takes the student through the solution process to various kinds of applications.

Maintaining Your Skills We believe that students who consistently work review problems will be much better prepared for class than students who do not engage in continuous review. The *Maintaining Your Skills* problems cumulatively review the most important concepts from the previous chapter as well as concepts that form the foundation of the course.

Getting Ready for the Next Section At the ends of section problem sets, you will find *Getting Ready for the Next Section,* a category of problems that students can work to prepare themselves to navigate the next section successfully. These problems polish techniques and reinforce the idea that all topics in the course are built on previous topics.

End-of-Chapter Summary, Review, and Assessment

We have learned that students are more comfortable with a chapter that sums up what they have learned thoroughly and accessibly through a well-organized presentation that reinforces concepts and techniques well. To help students grasp concepts and get more practice, each chapter ends with the following features that together give a comprehensive reexamination of the chapter.

Chapter Summary The chapter summary recaps all main points from the chapter in a visually appealing grid. In a column next to each topic is an example that illustrates the type of problem associated with the topic being reviewed. Our way of summarizing shows students that concepts in mathematics do relate—and that mastering one concept is a bridge to the next. When students prepare for a test, they can use the chapter summary as a guide to the main concepts of the chapter.

Chapter Review Test Following the chapter summary in each chapter is the chapter review test. It contains an extensive set of problems that review all the main topics in the chapter. This feature can be used flexibly—as assigned review, as a recommended self-test for students as they prepare for examinations, or as an in-class quiz or test.

Chapter Projects Each chapter closes with a pair of projects. One is a group project, suitable for students to work on in class. Group projects list details about number of participants, equipment, and time, so that instructors can determine how well the project fits into their classroom. The second project is a research project for students to do outside of class and tends to be open ended.

Additional Features of the Book

Facts from Geometry Many of the important facts from geometry are listed under this heading. In most cases, an example or two accompanies each of the facts to give students a chance to see how topics from geometry are related to the algebra they are learning.

Using Technology Scattered throughout the book is material that shows how graphing calculators can be used to enhance the topics being covered.

Unit Analysis Chapter 6 contains problems requiring students to convert from one unit of measure to another. The method used to accomplish the conversions is the method they will use if they take a chemistry class. Since this method is similar to the method we use to multiply rational expressions, unit analysis is covered in Section 6.7 as an application of multiplying rational expressions.

Chapter Openings Each chapter opens with an introduction in which a real-world application is used to spark interest in the chapter. We expand on each of these opening applications later in the chapter.

Supplements

Test Bank (0495382655) Drawn from hundreds of text-specific questions, an instructor can easily create tests that target specific course objectives. The Test Bank includes multiple tests per chapter as well as final exams. The tests are made up of a combination of multiple-choice, free-response, true/false, and fill-in-the-blank questions.

Text-Specific Videos (0495382663) These text-specific DVD sets completed by Pat McKeague are available at no charge to qualified adopters of the text. They feature 10- to 20-minute problem-solving lessons that cover each section of every chapter.

Student Solutions Manual (0495382671) The *Student Solutions Manual* provides worked-out solutions to the odd-numbered problems in the text.

Annotated Instructor's Edition (049538271X) The Instructor's Edition provides the complete student text with answers next to each respective exercise.

Complete Solutions Manual (0495382701) The *Complete Solutions Manual* provides worked-out solutions to all of the problems in the text.

Printed Access Card (ThomsonNOW™) (0495393274) This printed access card provides entrance to all the content that accompanies McKeague's *Intermediate Algebra* within ThomsonNOW, a powerful and fully integrated teaching and learning system that provides instructors and students with unsurpassed control, variety, and all-in-one utility. ThomsonNOW ties together the fundamental learning activities: diagnostics, tutorials, homework, personalized study, quizzing, and testing. Personalized Study is a learning companion that helps students gauge their unique study needs and makes the most of their study time by building focused Personalized Study plans that reinforce key concepts. **Pre-Tests** give students an initial assessment of their knowledge. **Personalized Study** plans, based on the students' answers to the pre-test questions, outline key elements for review. **Post-Tests** assess student mastery of core chapter concepts. Results can even be e-mailed to the instructor!

JoinIn™ on TurningPoint® (0495383368) Thomson Brooks/Cole is pleased to offer book-specific JoinIn™ content from the Eighth Edition of McKeague's *Intermediate Algebra.* This content for student classroom response systems allows instructors to transform the classroom and assess students' progress with instant in-class quizzes and polls. Our agreement to offer TurningPoint® software lets an instructor pose book-specific questions and display students' answers seamlessly within the Microsoft® PowerPoint® slides of your own lecture, in conjunction with the "clicker" hardware of your choice. Enhance how your students interact with you, your lecture, and each other. For college and university adopters only. Contact your local Thomson representative to learn more.

Enhanced WebAssign (0495109630) Instant feedback and ease of use are just two reasons why WebAssign is the most widely used homework system in higher education. WebAssign's Homework Delivery System lets you deliver, collect, grade, and record assignments using the web. This proven system has been enhanced to include end-of-chapter problems from McKeague's *Intermediate Algebra,* Eighth Edition—incorporating figures, videos, examples, PDF pages of the text, and quizzes to promote active learning and provide the immediate, relevant feedback students want.

ThomsonNOW with Personalized Study (0495382728) ThomsonNOW, a powerful and fully integrated teaching and learning system, provides instructors and students with unsurpassed control, variety, and all-in-one utility. ThomsonNOW ties together the fundamental learning activities: diagnostics, tutorials, homework, personalized study, quizzing, and testing. Personalized Study is a learning companion that helps students gauge their unique study needs and makes the most of their study time by building focused Personalized Study plans that reinforce key concepts. **Pre-Tests** give students an initial assessment of their knowledge. **Personalized Study** plans, based on the students' answers to the pre-test questions, outline key elements for review. **Post-Tests** assess student mastery of core chapter concepts; results can be e-mailed to the instructor!

Blackboard ThomsonNOW Integration (0495385212) It is easier than ever to integrate electronic course-management tools with content from this text's rich companion website. This supplement is ready to use as soon as you log on, or you can customize WebTutor ToolBox with web links, images, and other resources.

WebCT ThomsonNOW Integration (0495385239) Another option is available for integrating easy-to-use course management tools with content from this text's rich companion website. Ready to use as soon as you log on, or you can customize WebTutor ToolBox with web links, images, and other resources.

Instant Access Code (ThomsonNOW) (0495393282) Instant access gives students without a new copy of the text one access code to all available technology associated with this textbook. ThomsonNOW, a powerful and fully integrated teaching and learning system, provides instructors and students with unsurpassed control, variety, and all-in-one utility. ThomsonNOW ties together the fundamental learning activities: diagnostics, tutorials, homework, personalized study, quizzing, and testing. Personalized Study is a learning companion that helps students gauge their unique study needs and makes the most of their study time by building focused Personalized Study plans that reinforce key concepts. **Pre-Tests** give students an initial assessment of their knowledge. **Personalized Study** plans, based on the students' answers to the pre-test questions, outline key elements for review. **Post-Tests** assess student mastery of core chapter concepts. Results can be e-mailed to the instructor!

ExamView® (0495383791) This computerized algorithmic test bank on CD allows instructors to create exams using a CD.

Website www.thomsonedu.com/math/mckeague
The book's website offers instant access to the Student Resource Center, a rich array of teaching and learning resources that offers videos, chapter-by-chapter online tutorial quizzes, a final exam, chapter outlines, chapter reviews, chapter-by-chapter web links, flash cards, and more interactive options.

Acknowledgments

I would like to thank my editor at Brooks/Cole, Charlie Van Wagner, for his help and encouragement with this project. Many thanks also to Don Gecewicz, my developmental editor, for his suggestions on content, his proofreading, and his availability for consulting. This is a better book because of Don. Patrick McKeague, Tammy Fisher-Vasta, and Devin Christ assisted me with all parts of this revision, from manuscript preparation to proofreading page proofs and preparing the index. They are a fantastic team to work with, and this project could not have been completed without them. Susan Caire and Jeff Brower did an excellent job of proofreading the entire book in page proofs. Mary Gentilucci, Shane Wilwand, and Annie Stephens assisted with error checking and proofreading. Thanks to Rebecca Subity and Laura Localio for handling the media and ancillary packages on this project. Cheryll Linthicum of Brooks/Cole and Carol O'Connell of Graphic World Publishing Services turned the manuscript into a book. Ross Rueger produced the excellent solutions manuals that accompany the book.

Thanks also to Diane McKeague and Amy Jacobs for their encouragement with all my writing endeavors.

Finally, I am grateful to the following instructors for their suggestions and comments: Jess L. Collins, McLennan Community College; Richard Drey, Northampton Community College; Peg Hovde, Grossmont College; Sarah Jackman, Richland College; Carol Juncker, Delgado Community College; Joanne Kendall, College of the Mainland; Harriet Kiser, Floyd College; Domíngo Javier Lítong, South Texas Community College; Cindy Lucas, College of the Mainland; Jan MacInnes, Florida Community College of Jacksonville; Rudolfo Maglio, Oakton Community College; Janice McFatter, Gulf Coast Community College; Nancy Olson, Johnson County Community College; John H. Pleasants, Orange County Community College; Barbara Jane Sparks, Camden County College; Jim Stewart, Jefferson Community College; David J. Walker, Hinds Community College; and Deborah Woods, University of Cincinnati.

Charles P. McKeague
January 2007

Basic Properties and Definitions

1

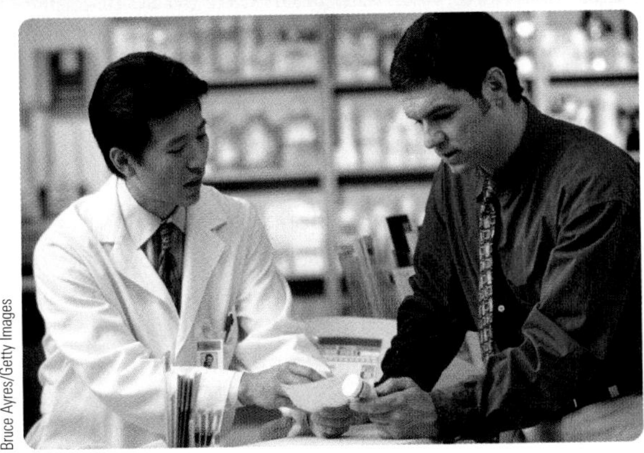

CHAPTER OUTLINE

The following table and diagram show how the concentration of a popular antidepressant changes over time once the patient stops taking it. In this particular case, the concentration in the patient's system is 80 ng/mL (nanograms per milliliter) when the patient stops taking the antidepressant, and the half-life of the antidepressant is 5 days.

Concentration of an Antidepressant

Days Since Discontinuing	Concentration (ng/mL)
0	80
5	40
10	20
15	10
20	5

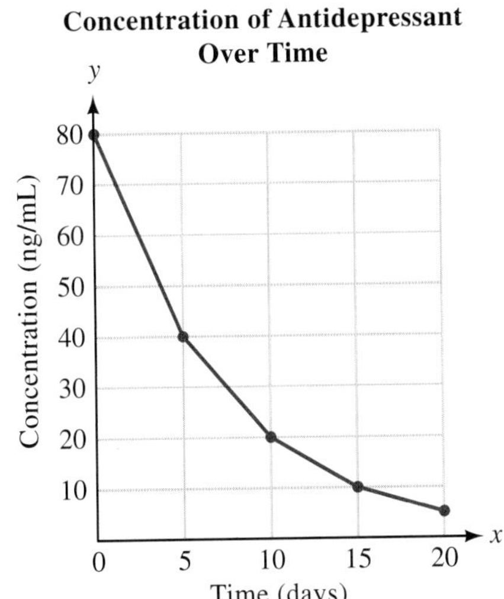

The half-life of a medication tells how quickly the medication is eliminated from a person's system: Medications with a long half-life are eliminated slowly, whereas those with a short half-life are more quickly eliminated. Half-life is the key to constructing the preceding table and graph. When you are finished with this chapter, you will be able to use the half-life of a medication to construct the table and graph.

Some of the students enrolled in my beginning algebra classes develop difficulties early in the course. Their difficulties are not associated with their ability to learn mathematics; they all have the potential to pass the course. Students who get off to a poor start do so because they have not developed the study skills necessary to be successful in algebra.

Here is a list of things you can do to begin to develop effective study skills.

1 Put Yourself on a Schedule

The general rule is that you spend 2 hours on homework for every hour you are in class. Make a schedule for yourself in which you set aside 2 hours each day to work on algebra. Once you make the schedule, stick to it. Don't just complete your assignments and stop. Use all the time you have set aside. If you complete an assignment and have time left over, read the next section in the book and work more problems.

2 Find Your Mistakes and Correct Them

There is more to studying algebra than just working problems. You must always check your answers with the answers in the back of the book. When you make a mistake, find out what it is and correct it. Making mistakes is part of the process of learning mathematics.

In the Prologue to *The Book of Squares,* Leonardo Fibonacci (ca. 1170–ca. 1250) has this to say about the content of his book:

> I have come to request indulgence if in any place it contains something more or less than right or necessary; for to remember everything and be mistaken in nothing is divine rather than human . . .

Fibonacci knew, as you know, that human beings make mistakes. You cannot learn algebra without making mistakes.

3 Gather Information on Available Resources

You need to anticipate that you will need extra help sometime during the course. There is a form to fill out in Appendix A to help you gather information on resources available to you. One resource is your instructor; you need to know your instructor's office hours and where the office is located. Another resource is the math lab or study center, if they are available at your school. It also helps to have the phone numbers of other students in the class, in case you miss class. You want to anticipate that you will need these resources, so now is the time to gather them together.

1.1 Fundamental Definitions and Notation

OBJECTIVES

A Translate expressions written in English into algebraic expressions.

B Expand and multiply numbers raised to positive integer exponents.

C Simplify expressions using the rule for order of operations.

D Find the value of an expression.

E Find the union and intersection of two sets.

The diagram below is called a *bar chart*. This one shows the net price of a popular intermediate algebra textbook. (The net price is the price the bookstore pays for the book.)

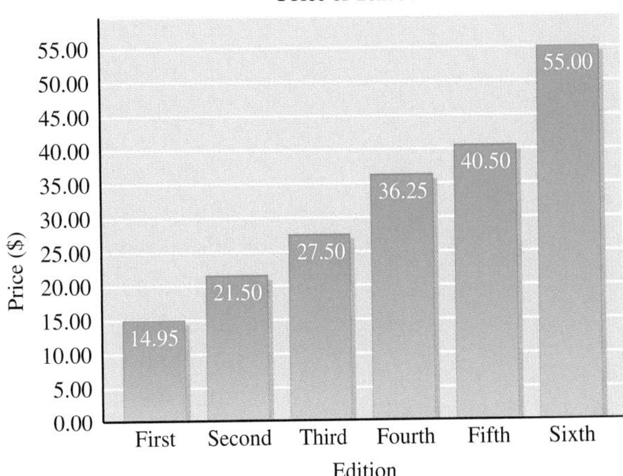

From the chart, we can find many relationships between numbers. We may notice that the price of the third edition was less than the price of the fourth edition. In mathematics we use symbols to represent relationships between quantities. If we let P represent the price of the book, then the relationship just mentioned, between the price of the third edition and the price of the fourth edition, can be written this way:

$$P(3) < P(4)$$

This section is, for the most part, simply a list of many of the basic symbols and definitions we will be using throughout the book.

Comparison Symbols

In Symbols	*In Words*
$a = b$	a is equal to b
$a \neq b$	a is not equal to b
$a < b$	a is less than b
$a \leq b$	a is less than or equal to b
$a \nless b$	a is not less than b
$a > b$	a is greater than b
$a \geq b$	a is greater than or equal to b
$a \ngtr b$	a is not greater than b
$a \Leftrightarrow b$	a is equivalent to b

Operation Symbols

Operation	*In Symbols*	*In Words*
Addition	$a + b$	The sum of a and b
Subtraction	$a - b$	The difference of a and b
Multiplication	ab, $a \cdot b$, $a(b)$, $(a)(b)$, or $(a)b$	The product of a and b
Division	$a \div b$, a/b, or $\frac{a}{b}$	The quotient of a and b

The key words are *sum, difference, product,* and *quotient.* They are used frequently in mathematics. For instance, we may say the product of 3 and 4 is 12. We mean both the expressions $3 \cdot 4$ and 12 are called the products of 3 and 4. The important idea here is that the word *product* implies multiplication, regardless of whether it is written $3 \cdot 4$, 12, 3(4), (3)4, or (3)(4).

When we let a letter, such as x, stand for a number or group of numbers, then we say x is a *variable* because the value it takes on can vary. In the lists of notation and symbols just presented, the letters a and b are variables that represent the numbers we will work with in this book. In the next example we show some translations between sentences written in English and their equivalent expressions in algebra. Note that all the sentences contain at least one variable.

EXAMPLE 1

In English	*In Symbols*
The sum of x and 5 is less than 2.	$x + 5 < 2$
The product of 3 and x is 21.	$3x = 21$
The quotient of y and 6 is 4.	$\frac{y}{6} = 4$
Twice the difference of b and 7 is greater than 5.	$2(b - 7) > 5$
The difference of twice b and 7 is greater than 5.	$2b - 7 > 5$

Exponents

Consider the expression 3^4. The 3 is called the *base* and the 4 is called the *exponent.* The exponent 4 tells us the number of times the base appears in the product. That is:

$$3^4 = 3 \cdot 3 \cdot 3 \cdot 3 = 81$$

The expression 3^4 is said to be in *exponential form*, whereas $3 \cdot 3 \cdot 3 \cdot 3$ is said to be in *expanded form*.

EXAMPLES Expand and multiply.

2. $5^2 = 5 \cdot 5 = 25$ Base 5, exponent 2

3. $2^5 = 2 \cdot 2 \cdot 2 \cdot 2 \cdot 2 = 32$ Base 2, exponent 5

4. $4^3 = 4 \cdot 4 \cdot 4 = 64$ Base 4, exponent 3

Order of Operations

It is important when evaluating arithmetic expressions in mathematics that each expression have only one answer in reduced form. Consider the expression

$$3 \cdot 7 + 2$$

If we find the product of 3 and 7 first, then add 2, the answer is 23. On the other hand, if we first combine the 7 and 2, then multiply by 3, we have 27. The problem seems to have two distinct answers depending on whether we multiply first or add first. To avoid this situation, we will decide that multiplication in a situation like this will always be done before addition. In this case, only the first answer, 23, is correct.

The complete set of rules for evaluating expressions follows.

Note

This rule is very important. We will use it many times throughout the book. It is a simple rule to follow. First we evaluate any numbers with exponents; then we multiply and divide; and finally we add and subtract, always working from left to right when more than one of the same operation symbol occurs in a problem.

Rule (Order of Operations) When evaluating a mathematical expression, we will perform the operations in the following order, beginning with the expression in the innermost parentheses or brackets and working our way out.

1. Simplify all numbers with exponents, working from left to right if more than one of these expressions is present.
2. Then, do all multiplications and divisions left to right.
3. Perform all additions and subtractions left to right.

EXAMPLES Simplify each expression using the rule for order of operations.

5. $5 + 3(2 + 4) = 5 + 3(6)$ **Simplify inside parentheses**
$= 5 + 18$ **Then, multiply**
$= 23$ **Add**

6. $5 \cdot 2^3 - 4 \cdot 3^2 = 5 \cdot 8 - 4 \cdot 9$ **Simplify exponents left to right**
$= 40 - 36$ **Multiply left to right**
$= 4$ **Subtract**

7. $20 - (2 \cdot 5^2 - 30) = 20 - (2 \cdot 25 - 30)$
$= 20 - (50 - 30)$
$= 20 - (20)$
Simplify inside parentheses, evaluating exponents first, then multiplying, and finally subtract
$= 0$

8. $40 - 20 \div 5 + 8 = 40 - 4 + 8$ **Divide first**
$= 36 + 8$
$= 44$
Then, add and subtract left to right

9. $2 + 4[5 + (3 \cdot 2 - 2)] = 2 + 4[5 + (6 - 2)]$
$= 2 + 4(5 + 4)$
$= 2 + 4(9)$
Simplify inside inner-most parentheses
$= 2 + 36$ **Then, multiply**
$= 38$ **Add**

Finding the Value of an Algebraic Expression

An algebraic *expression* is a combination of numbers, variables, and operation symbols. For example, each of the following is an algebraic expression

$$5x \qquad x^2 + 5 \qquad 4a^2b^3 \qquad 5t^2 - 6t + 3$$

An expression such as $x^2 + 5$ will take on different values depending on what number we substitute for x. For example:

When $x = 3$ the expression $x^2 + 5$ becomes $3^2 + 5 = 9 + 5 = 14$

and When $x = 7$ the expression $x^2 + 5$ becomes $7^2 + 5 = 49 + 5 = 54$

As you can see, our expression is 14 when x is 3, and when x is 7, our expression is 54. In each case the value of the expression is found by replacing the variable with a number.

EXAMPLE 10 Evaluate the expression $a^2 + 8a + 16$ when a is 0, 1, 2, 3, and 4.

SOLUTION We can organize our work efficiently by using a table.

When a is	The expression $a^2 + 8a + 16$ becomes
0	$0^2 + 8 \cdot 0 + 16 = 0 + 0 + 16 = 16$
1	$1^2 + 8 \cdot 1 + 16 = 1 + 8 + 16 = 25$
2	$2^2 + 8 \cdot 2 + 16 = 4 + 16 + 16 = 36$
3	$3^2 + 8 \cdot 3 + 16 = 9 + 24 + 16 = 49$
4	$4^2 + 8 \cdot 4 + 16 = 16 + 32 + 16 = 64$

EXAMPLE 11 Find the value of the expression $3x + 4y + 5$ when x is 6 and y is 7.

SOLUTION We substitute the given values of x and y into the expression and simplify the result:

When $x = 6$ and $y = 7$ the expression $3x + 4y + 5$ becomes

$$3 \cdot 6 + 4 \cdot 7 + 5 = 18 + 28 + 5 = 51$$

The next concept we will cover, that of a set, can be considered the starting point for all the branches of mathematics.

Sets

DEFINITION A **set** is a collection of objects or things. The objects in the set are called **elements,** or **members,** of the set.

Sets are usually denoted by capital letters, and elements of sets are denoted by lowercase letters. We use braces, { }, to enclose the elements of a set.

To show that an element is contained in a set we use the symbol $\in$. That is,

$x \in A$ is read "x is an element (member) of set A"

For example, if A is the set $\{1, 2, 3\}$, then $2 \in A$. However, $5 \notin A$ means 5 is not an element of set A.

DEFINITION Set A is a **subset** of set B, written $A \subset B$, if every element in A is also an element of B. That is:

$A \subset B$ if and only if A is contained in B

EXAMPLES

12. The set of numbers used to count things is $\{1, 2, 3, \ldots\}$. The dots mean the set continues indefinitely in the same manner. This is an example of an *infinite set*.

13. The set of all numbers represented by the dots on the faces of a regular die is $\{1, 2, 3, 4, 5, 6\}$. This set is a subset of the set in Example 12. It is an example of a *finite set* because it has a limited number of elements.

DEFINITION The set with no members is called the **empty,** or **null, set.** It is denoted by the symbol $\varnothing$. The empty set is considered a subset of every set.

Operations with Sets

Two basic operations are used to combine sets: union and intersection.

DEFINITION The **union** of two sets A and B, written $A \cup B$, is the set of all elements that are either in A or in B, or in both A and B. The key word here is *or*. For an element to be in $A \cup B$ it must be in A or B. In symbols, the definition looks like this:

$x \in A \cup B$ if and only if $x \in A$ or $x \in B$

DEFINITION The **intersection** of two sets A and B, written $A \cap B$, is the set of elements in both A and B. The key word in this definition is the word *and*. For an element to be in $A \cap B$ it must be in both A and B, or

$x \in A \cap B$ if and only if $x \in A$ and $x \in B$

The diagrams shown here are called *Venn diagrams* after John Venn (1834–1923). They can be used to visualize operations with sets. The region inside the circle labeled A is set A; the region inside the circle labeled B is set B.

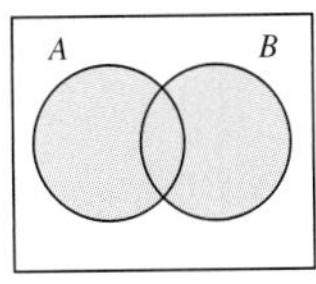

$A \cup B$

FIGURE 1 The union of two sets

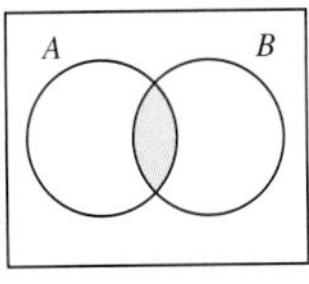

$A \cap B$

FIGURE 2 The intersection of two sets

EXAMPLES

Let $A = \{1, 3, 5\}$, $B = \{0, 2, 4\}$, and $C = \{1, 2, 3, \ldots\}$. Then

14. $A \cup B = \{0, 1, 2, 3, 4, 5\}$

15. $A \cap B = \varnothing$ (A and B have no elements in common.)

16. $A \cap C = \{1, 3, 5\} = A$

17. $B \cup C = \{0, 1, 2, 3, \ldots\}$ 

Another notation we can use to describe sets is called *set-builder* notation. Here is how we write our definition for the union of two sets A and B using set-builder notation:

$$A \cup B = \{x \mid x \in A \text{ or } x \in B\}$$

The right side of this statement is read "the set of all x such that x is a member of A or x is a member of B." As you can see, the vertical line after the first x is read "such that."

EXAMPLE 18

If $A = \{1, 2, 3, 4, 5, 6\}$, find $C = \{x \mid x \in A \text{ and } x \geq 4\}$.

SOLUTION We are looking for all the elements of A that are also greater than or equal to 4. They are 4, 5, and 6. Using set notation, we have

$$C = \{4, 5, 6\}$$

GETTING READY FOR CLASS

Each section of the book will end with some problems and questions like the ones that follow. They are for you to answer after you have read through the section but before you go to class. All of them require that you give written responses in complete sentences. Writing about mathematics is a valuable exercise. If you write with the intention of explaining and communicating what you know to someone else, you will find that you understand the topic you are writing about even better than you did before you started writing. As with all problems in this course, you want to approach these writing exercises with a positive point of view. You will get better at giving written responses to questions as you progress through the course. Even if you are never comfortable writing about mathematics, just the process of attempting to do so will increase your understanding and ability in mathematics.

After reading through the preceding section, respond in your own words and in complete sentences.

1. What is the meaning of the expression 2^3?
2. Why do we have a rule for the order of operations?
3. What is the intersection of two sets?
4. Explain the operations that are associated with the words *sum, difference, product,* and *quotient.*

LINKING OBJECTIVES AND EXAMPLES

Next to each **objective** we have listed the examples that are best described by that objective.

A	1
B	2–4
C	5–9
D	10, 11
E	14–17

Problem Set 1.1

Online support materials can be found at www.thomsonedu.com/login

Translate each of the following statements into symbols.

1. The sum of x and 5 is 2.
2. The sum of y and -3 is 9.
3. The difference of 6 and x is y.
4. The difference of x and 6 is $-y$.
5. The product of t and 2 is less than y.
6. The product of $5x$ and y is equal to z.
7. The sum of x and y is less than the difference of x and y.
8. Twice the sum of a and b is 15.
9. Three times the difference of x and 5 is more than y.
10. The product of x and y is greater than or equal to the quotient of x and y.

Expand and multiply.

11. 6^2
12. 8^2
13. 10^2
14. 10^3
15. 2^3
16. 5^3
17. 2^4
18. 1^4
19. 10^4
20. 4^3
21. 11^2
22. 10^5

The problems that follow are intended to give you practice using the rule for order of operations. Some of them are arranged so they model some of the properties of real numbers.

Simplify each expression.

23. a. $3 \cdot 5 + 4$
    b. $3(5 + 4)$
    c. $3 \cdot 5 + 3 \cdot 4$
24. a. $3 \cdot 7 - 6$
    b. $3(7 - 6)$
    c. $3 \cdot 7 - 3 \cdot 6$
25. a. $6 + 3 \cdot 4 - 2$
    b. $6 + 3(4 - 2)$
    c. $(6 + 3)(4 - 2)$
26. a. $8 + 2 \cdot 7 - 3$
    b. $8 + 2(7 - 3)$
    c. $(8 + 2)(7 - 3)$
27. a. $(7 - 4)(7 + 4)$
    b. $7^2 - 4^2$
28. a. $(8 - 5)(8 + 5)$
    b. $8^2 - 5^2$
29. a. $(5 + 7)^2$
    b. $5^2 + 7^2$
    c. $5^2 + 2 \cdot 5 \cdot 7 + 7^2$
30. a. $(8 - 3)^2$
    b. $8^2 - 3^2$
    c. $8^2 - 2 \cdot 8 \cdot 3 + 3^2$
31. a. $2 + 3 \cdot 2^2 + 3^2$
    b. $2 + 3(2^2 + 3^2)$
    c. $(2 + 3)(2^2 + 3^2)$
32. a. $3 + 4 \cdot 4^2 + 5^2$
    b. $3 + 4(4^2 + 5^2)$
    c. $(3 + 4)(4^2 + 5^2)$
33. a. $40 - 10 \div 5 + 1$
    b. $(40 - 10) \div 5 + 1$
    c. $(40 - 10) \div (5 + 1)$
34. a. $20 - 10 \div 2 + 3$
    b. $(20 - 10) \div 2 + 3$
    c. $(20 - 10) \div (2 + 3)$
35. a. $40 + [10 - (4 - 2)]$
    b. $40 - 10 - 4 - 2$
36. a. $50 - [17 - (8 - 3)]$
    b. $50 - 17 - 8 - 3$
37. a. $3 + 2(2 \cdot 3^2 + 1)$
    b. $(3 + 2)(2 \cdot 3^2 + 1)$
38. a. $4 + 5(3 \cdot 2^2 + 5)$
    b. $(4 + 5)(3 \cdot 2^2 + 5)$

The problems below will make certain you are using the rule for order of operations correctly. Simplify.

39. $5 \cdot 10^3 + 4 \cdot 10^2 + 3 \cdot 10 + 1$
40. $6 \cdot 10^3 + 5 \cdot 10^2 + 4 \cdot 10 + 3$
41. $3[2 + 4(5 + 2 \cdot 3)]$
42. $2[4 + 2(6 + 3 \cdot 5)]$
43. $6[3 + 2(5 \cdot 3 - 10)]$
44. $8[7 + 2(6 \cdot 9 - 14)]$
45. $5(7 \cdot 4 - 3 \cdot 4) + 8(5 \cdot 9 - 4 \cdot 9)$

= Videos available by instructor request

▶ = Online student support materials available at www.thomsonedu.com/login

46. $4(3 \cdot 9 - 2 \cdot 9) + 5(6 \cdot 8 - 5 \cdot 8)$

47. $25 - 17 + 3$

48. $38 - 19 + 1$

49. $109 - 36 + 14$

50. $200 - 150 + 20$

51. $20 - 13 - 3$

52. $57 - 18 - 8$

53. $63 - 37 - 4$

54. $71 - 11 - 1$

55. $36 \div 9 \cdot 4$

56. $28 \div 7 \cdot 2$

57. $75 \div 3 \cdot 25$

58. $48 \div 3 \cdot 2$

59. $64 \div 16 \div 4$

60. $125 \div 25 \div 5$

61. $75 \div 25 \div 5$

62. $36 \div 12 \div 4$

The problems below are problems you will see later in the book. Simplify each expression without using a calculator.

63. $18{,}000 - 9{,}300$

64. $18{,}000 - 4{,}500$

65. $3.45 + 2.6 - 1.004$

66. $24.3 + 6(8.1)$

67. $275 \div 55$

68. $4.8 \div 2.4$

69. $4(2)(4)^2$

70. $230(5) - 20(5)^2$

71. $250(5) - 25(5)^2$

72. $3(3)^2 + 2(3) - 1$

73. $5 \cdot 2^3 - 3 \cdot 2^2 + 4 \cdot 2 - 5$

74. $125 \cdot 2^2$

75. $125 \cdot 2^3$

76. $7.5(10)$

77. $500(1.5)$

78. $39.3(60)$

79. $5(0.10)$

80. $2(0.25)$

81. $0.20(8)$

82. $0.30(12)$

83. $0.08(4{,}000)$

84. $0.09(6{,}000)$

We are assuming that you know how to use a calculator to do simple arithmetic problems. Use a calculator to simplify each expression. If rounding is necessary, round your answers to the nearest ten thousandth (4 places past the decimal point). You will see these problems again later in the book.

85. $0.6931 + 1.0986$

86. $1.6094 + 1.9459$

87. $3(0.6931)$

88. $2(1.9459)$

89. $250(3.14)$

90. $165(3.14)$

91. $4{,}628 \div 25$

92. $7{,}546 \div 35$

93. $65{,}000 \div 5{,}280$

94. $2{,}358 \div 5{,}280$

95. $1 - 0.8413$

96. $1.2052 - 1$

97. $16(3.5)^2$

98. $4(3.14)3^2$

99. $11.5(130) - 0.05(130)^2$

100. $10(130) - 0.04(130)^2$

101. Find the value of each expression when x is 5.

- **a.** $x + 2$
- **b.** $2x$
- **c.** x^2
- **d.** 2^x

102. Find the value of each expression when x is 3.

- **a.** $x + 5$
- **b.** $5x$
- **c.** x^5
- **d.** 5^x

103. Find the value of each expression when x is 10.

- **a.** $x^2 + 2x + 1$
- **b.** $(x + 1)^2$
- **c.** $x^2 + 1$
- **d.** $(x - 1)^2$

104. Find the value of each expression when x is 8.

- **a.** $x^2 - 6x + 9$
- **b.** $(x - 3)^2$
- **c.** $x^2 - 3$
- **d.** $x^2 - 9$

105. Find the value of $b^2 - 4ac$ if

- **a.** $a = 2$, $b = 5$, and $c = 3$
- **b.** $a = 10$, $b = 60$, $c = 30$
- **c.** $a = 0.4$, $b = 1$, and $c = 0.3$

106. Find the value of $6x + 5y + 4$ if

- **a.** $x = 3$ and $y = 2$
- **b.** $x = 2$ and $y = 3$
- **c.** $x = 0$ and $y = 0$

For the following problems, let $A = \{0, 2, 4, 6\}$, $B = \{1, 2, 3, 4, 5\}$, and $C = \{1, 3, 5, 7\}$.

107. $A \cup B$

108. $A \cup C$

109. $A \cap B$

110. $A \cap C$

111. $B \cap C$

112. $B \cup C$

113. $A \cup (B \cap C)$

114. $C \cup (A \cap B)$

115. $\{x \mid x \in A \text{ and } x < 4\}$

116. $\{x \mid x \in B \text{ and } x > 3\}$

117. $\{x \mid x \in A \text{ and } x \notin B\}$

118. $\{x \mid x \in B \text{ and } x \notin C\}$

119. $\{x \mid x \in A \text{ or } x \in C\}$

120. $\{x \mid x \in A \text{ or } x \in B\}$

121. $\{x \mid x \in B \text{ and } x \neq 3\}$

122. $\{x \mid x \in C \text{ and } x \neq 5\}$

Applying the Concepts

123. **Minutes and Seconds** The chart shows the five fastest winning times for the Fifth Avenue Mile run in New York City. The times are given in minutes and seconds, to the nearest hundredth of a second.

a. How much faster was Isaac Viciosa's time than Peter Elliott's time?

b. How much faster was Peter Elliott's time than Stephen Kipkorir's time?

USA TODAY Snapshots®

Fastest on Fifth

The New York Road Runners host the Continental Airlines Fifth Avenue Mile Saturday in New York. The group has offered $10,000 for anyone who can break the record-setting time that has stood for 24 years. Fastest times:

Sydney Maree	1981	3:47.52
Isaac Viciosa	1995	3:47.80
Peter Elliott	1990	3:47.83
Stephen Kipkorir	1995	3:48.20
Steve Cram	1990	3:48.39

Source: New York Road Runners

By Ellen J. Horrow and Sam Ward, USA TODAY

From *USA Today*. Copyright 2005. Reprinted with permission.

124. **Mobile Phone Sales** Use the chart to answer the following questions.

a. For which year are estimated phone sales closest to 900,000,000 phones?

b. Estimate the increase in mobile phone sales from 2005 to 2006.

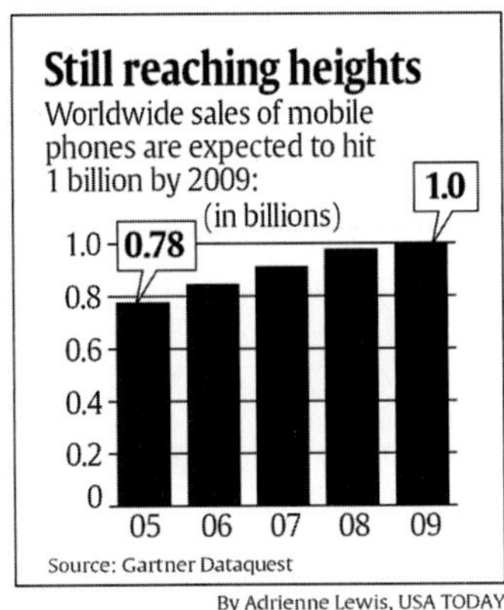

From *USA Today*. Copyright 2005. Reprinted with permission.

Extending the Concepts

Many of the problem sets in this book end with a few problems like the ones that follow. These problems challenge you to extend your knowledge of the material in the problem set. In most cases, there are no examples in the text similar to these problems. You should approach them with a positive point of view because even though you may not complete them correctly, just the process of attempting them will increase your knowledge and ability in algebra.

125. **Fermat's Last Theorem** The postage stamp shown here was issued by France in 2001 to commemorate the 400th anniversary of the birth of the French mathematician Pierre de Fermat. The stamp shows Fermat's last theorem, which states that if n is an integer greater than 2, then there are no positive integers x, y, and z that will make the formula $x^n + y^n = z^n$ true.

However, there are many ways to make the formula $x^n + y^n = z^n$ true when n is 1 or 2. Show that this formula is true for each case below.

a. $n = 1, x = 5, y = 8$, and $z = 13$

b. $n = 1, x = 2, y = 3$, and $z = 5$

c. $n = 2, x = 3, y = 4$, and $z = 5$

d. $n = 2, x = 7, y = 24$, and $z = 25$

1.2 The Real Numbers

OBJECTIVES

A Graph simple and compound inequalities.

B Translate sentences and phrases written in English into inequalities.

C List the elements in subsets of the real numbers.

D Factor positive integers into the product of primes.

E Reduce fractions to lowest terms.

The Real Number Line

The real number line is constructed by drawing a straight line and labeling a convenient point with the number 0. Positive numbers are in increasing order to the right of 0; negative numbers are in decreasing order to the left of 0. The point on the line corresponding to 0 is called the *origin*.

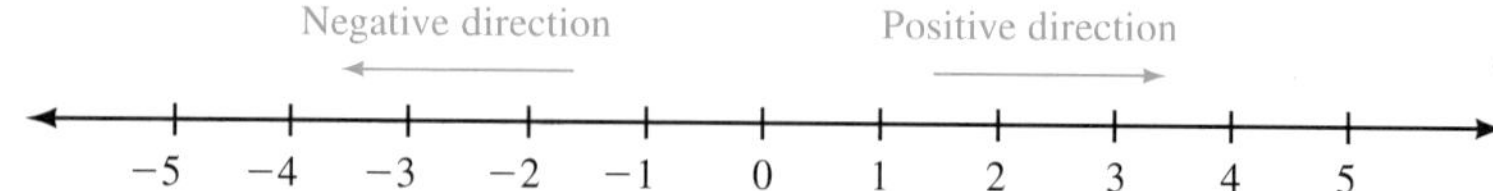

The numbers associated with the points on the line are called *coordinates* of those points. Every point on the line has a number associated with it. The set of all these numbers makes up the set of real numbers.

> **DEFINITION** A **real number** is any number that is the coordinate of a point on the real number line.

EXAMPLE 1 Locate the numbers -4.5, -0.75, $\frac{1}{2}$, $\sqrt{2}$, π, and 4.1 on the real number line.

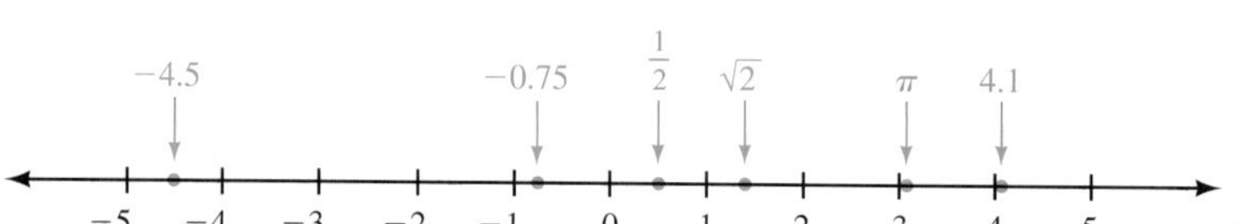

Note The numbers on the number line increase as we move to the right. When we compare the size of two numbers on the number line, the one on the left is always less than the one on the right.

We can use the real number line to give a visual representation to inequality statements.

EXAMPLE 2 Graph $\{x \mid x \leq 3\}$.

SOLUTION We want to graph all the real numbers less than or equal to 3—that is, all the real numbers below 3 and including 3. We label 0 on the number line for reference as well as 3 since the latter is what we call the endpoint. The graph is as follows:

We use a right bracket at 3 to show that 3 is part of the solution set.

Note In this book we will refer to real numbers as being on the real number line. Actually, real numbers are *not* on the line; only the points representing them are on the line. We can save some writing, however, if we simply refer to real numbers as being on the number line.

Note: You may have come from an algebra class in which open and closed circles were used at the endpoints of number line graphs. If so, you would show the graph for Example 2 this way, using a closed circle at 3 to show that 3 is part of the graph.

The two number line graphs above are equivalent; they both show all the real numbers that are less than or equal to 3.

EXAMPLE 3 Graph $\{x \mid x < 3\}$.

SOLUTION The graph is identical to the graph in Example 2 except at the endpoint 3. In this case we use a parenthesis that opens to the left to show that 3 is not part of the graph.

Note: For those of you who are used to open/closed circles at the endpoints of number line graphs, here is the equivalent graph in that format.

The table below further clarifies the relationship between number line graphs that use parentheses and brackets at the endpoints and those that use open and closed circles at the endpoints.

Inequality notation	Graph using parentheses/brackets	Graph using open and closed circles
$x < 2$	0 2	0 2
$x \leq 2$	0 2	0 2
$x \geq -3$	−3 0	−3 0
$x > -3$	−3 0	−3 0

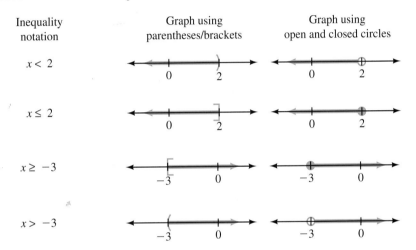

In this book, we will use the parentheses/brackets method of graphing inequalities because that method is better suited to the type of problems we will work in intermediate algebra.

In the previous section we defined the *union* of two sets A and B to be the set of all elements that are in either A or B. The word *or* is the key word in the definition. The *intersection* of two sets A and B is the set of all elements contained in both A and B, the key word here being *and*. We can put the words *and* and *or* together with our methods of graphing inequalities to graph some *compound inequalities*.

EXAMPLE 4 Graph $\{x \mid x \leq -2 \text{ or } x > 3\}$.

SOLUTION The two inequalities connected by the word *or* are referred to as a *compound inequality*. We begin by graphing each inequality separately.

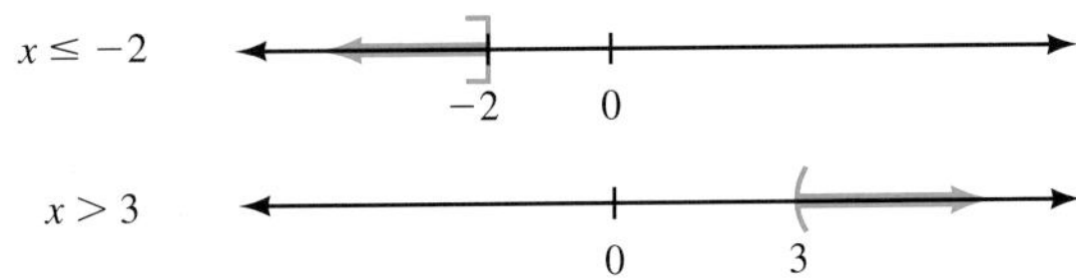

Note
It is not absolutely necessary to show these first two graphs. It is simply helpful to do so. As you get more practice at this type of graphing you can easily omit them.

Because the two inequalities are connected by the word *or,* we graph their union; that is, we graph all points on either graph.

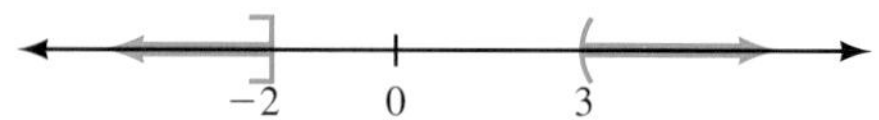

EXAMPLE 5 Graph $\{x \mid x > -1 \text{ and } x < 2\}$.

SOLUTION We first graph each inequality separately.

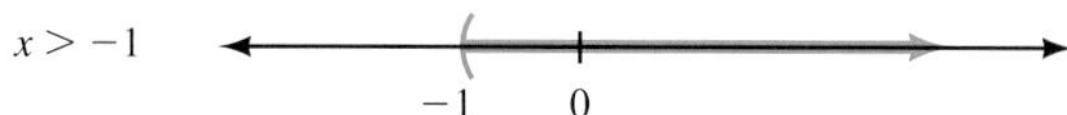

Because the two inequalities are connected by the word *and,* we graph their intersection—the part they have in common.

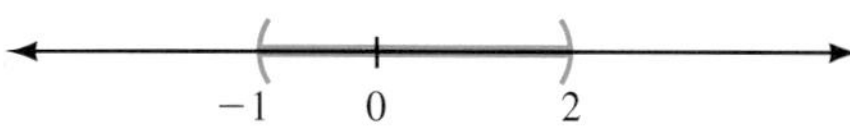

Note: Sometimes compound inequalities that use the word *and* as the connecting word can be written in a shorter form. For example, the compound inequality $-3 \le x$ and $x \le 4$ can be written $-3 \le x \le 4$. The word *and* does not appear when an inequality is written in this form. It is implied. Inequalities of the form $-3 \le x \le 4$ are called *continued inequalities.* This new notation is useful because it takes fewer symbols to write it. The graph of $-3 \le x \le 4$ is

EXAMPLE 6 Graph $\{x \mid 1 \le x < 2\}$.

SOLUTION The word *and* is implied in the continued inequality $1 \le x < 2$; that is, the continued inequality $1 \le x < 2$ is equivalent to $1 \le x$ and $x < 2$. Therefore, we graph all the numbers between 1 and 2 on the number line, including 1 but not including 2.

The table that follows shows the connection between number line graphs for a variety of continued inequalities. Again, we have included the graphs with open and closed circles for those of you who have used this type of graph previously. Remember, however, that in this book we will be using the parentheses/brackets method of graphing.

Inequality notation	Graph using parentheses/brackets	Graph using open and closed circles
$-3 < x < 2$	−3 2	−3 2
$-3 \le x \le 2$	−3 2	−3 2
$-3 \le x < 2$	−3 2	−3 2
$-3 < x \le 2$	−3 2	−3 2

EXAMPLE 7

Graph $\{x \mid x < -2 \text{ or } 2 < x < 6\}$.

SOLUTION Here we have a combination of compound and continued inequalities. We want to graph all real numbers that are either less than -2 or between 2 and 6.

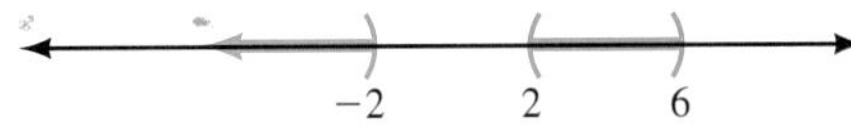

In addition to the phrases that translate directly into inequality statements, we have the following translations:

In Words	In Symbols
x is at least 40	$x \ge 40$
x is at most 30	$x \le 30$
x is no more than 20	$x \le 20$
x is no less than 10	$x \ge 10$
x is between 4 and 5	$4 < x < 5$

In the last case, we can include the endpoints 4 and 5 by using "x is between 4 and 5, inclusive," which translates to $4 \le x \le 5$.

312 Help Wanted

ESPRESSO BAR OPERATOR
Must be dependable, honest, service-oriented. Coffee exp desired. 10–20 hrs per wk. $6.25/hr. Start 5/31. Apply in person: Espresso Yourself, Central Coast Mall. Deadline 5/23.

EXAMPLE 8

Suppose you have a part-time job that requires that you work at least 10 hours, but no more than 20 hours, each week. Use the letter t to write an inequality that shows the number of hours you work per week.

SOLUTION If t is at least 10 but no more than 20, then $10 \le t$ and $t \le 20$, or equivalently, $10 \le t \le 20$. Note that the word *but,* as used here, has the same meaning as the word *and.*

EXAMPLE 9

If the highest temperature on Tuesday was 76°F and the lowest temperature was 55°F, write an inequality using the letter x that gives the range of temperatures on Tuesday.

SOLUTION Since the smallest value of x is 55 and the largest value of x is 76, then $55 \le x \le 76$. We could say that the temperature on Tuesday was between 55°F and 76°F, inclusive.

Subsets of the Real Numbers

> **Note**
> As you can see, the whole numbers are the counting numbers and 0 together, whereas the integers are the whole numbers along with the opposites of all the counting numbers. We can say that the counting numbers are a subset of the whole numbers, and the whole numbers are a subset of the integers.

Next, we consider some of the more important subsets of the real numbers. Each set listed here is a subset of the real numbers:

Counting (or natural) numbers $= \{1, 2, 3, \ldots\}$
Whole numbers $= \{0, 1, 2, 3, \ldots\}$
Integers $= \{\ldots, -3, -2, -1, 0, 1, 2, 3, \ldots\}$

$$\textit{Rational numbers} = \left\{ \frac{a}{b} \,\middle|\, a \text{ and } b \text{ are integers, } b \neq 0 \right\}$$

Remember, the notation used to write the rational numbers is read "the set of numbers a/b, such that a and b are integers and b is not equal to 0." Any number that can be written in the form

$$\frac{\text{Integer}}{\text{Integer}}$$

is a rational number. Rational numbers are numbers that can be written as the ratio of two integers. Each of the following is a rational number:

$\frac{3}{4}$	Because it is the ratio of the integers 3 and 4
-8	Because it can be written as the ratio of -8 to 1
0.75	Because it is the ratio of 75 to 100 (or 3 to 4 if you reduce to lowest terms)
0.333 . . .	Because it can be written as the ratio of 1 to 3

> **Note**
> We can find decimal approximations to some irrational numbers by using a calculator or a table. For example, on an eight-digit calculator
>
> $\sqrt{2} = 1.4142136$
>
> This is not exactly $\sqrt{2}$ but simply an approximation to it. There is no decimal that gives $\sqrt{2}$ exactly.

Still other numbers on the number line are not members of the subsets we have listed so far. They are real numbers, but they cannot be written as the ratio of two integers; that is, they are not rational numbers. For that reason, we call them irrational numbers.

$$\textit{Irrational numbers} = \{x \mid x \text{ is real, but not rational}\}$$

The following are irrational numbers:

$$\sqrt{2}, \quad -\sqrt{3}, \quad 4 + 2\sqrt{3}, \quad \pi, \quad \pi + 5\sqrt{6}$$

EXAMPLE 10

For the set $\{-5, -3.5, 0, \frac{3}{4}, \sqrt{3}, \sqrt{5}, 9\}$, list the numbers that are (a) whole numbers, (b) integers, (c) rational numbers, (d) irrational numbers, and (e) real numbers.

SOLUTION

a. Whole numbers: 0, 9
b. Integers: $-5, 0, 9$
c. Rational numbers: $-5, -3.5, 0, \frac{3}{4}, 9$
d. Irrational numbers: $\sqrt{3}, \sqrt{5}$
e. They are all real numbers.

The following diagram gives a visual representation of the relationships among subsets of the real numbers.

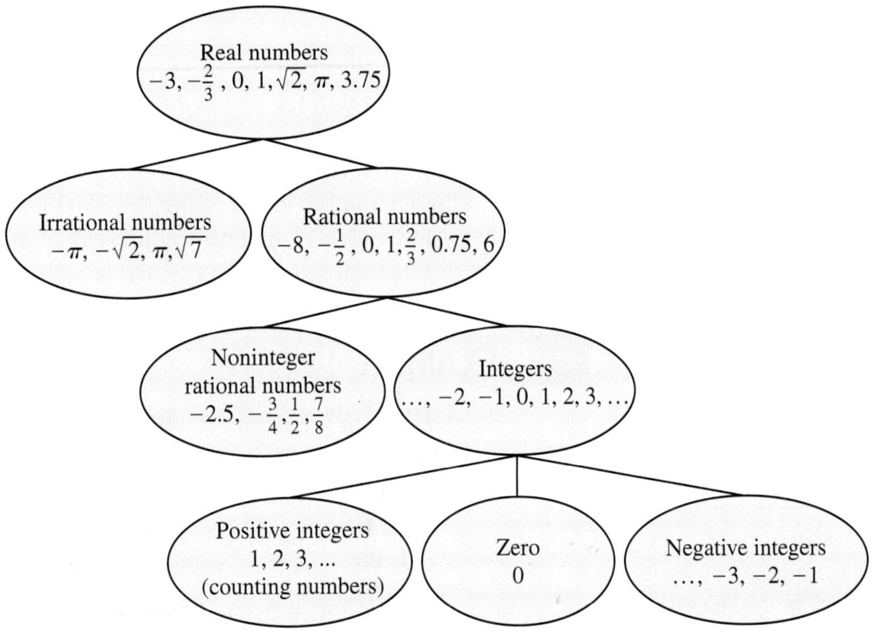

Prime Numbers and Factoring

The following diagram shows the relationship between multiplication and factoring:

Multiplication

Factors $\rightarrow 3 \cdot 4 = 12 \leftarrow$ Product

Factoring

When we read the problem from left to right, we say the *product* of 3 and 4 is 12, or we multiply 3 and 4 to get 12. When we read the problem in the other direction, from right to left, we say we have *factored* 12 into 3 times 4, or 3 and 4 are *factors* of 12.

The number 12 can be factored still further:

$$\begin{aligned} 12 &= 4 \cdot 3 \\ &= 2 \cdot 2 \cdot 3 \\ &= 2^2 \cdot 3 \end{aligned}$$

The numbers 2 and 3 are called *prime* factors of 12 because neither can be factored any further.

DEFINITION If a and b represent integers, then a is said to be a **factor** (or divisor) of b if a divides b evenly—that is, if a divides b with no remainder.

DEFINITION A **prime** number is any positive integer larger than 1 whose only positive factors (divisors) are itself and 1. An integer greater than 1 that is not prime is said to be **composite.**

Note
The number 15 is not a prime number because it has factors of 3 and 5; that is, $15 = 3 \cdot 5$.

Here is a list of the first few prime numbers:

$$\text{Prime numbers} = \{2, 3, 5, 7, 11, 13, 17, 19, 23, 29, 31, 37, 41, \ldots\}$$

When a number is not prime, we can factor it into the product of prime numbers. To factor a number into the product of primes, we simply factor it until it cannot be factored further.

EXAMPLE 11

Factor 525 into the product of primes.

SOLUTION Because 525 ends in 5, it is divisible by 5:

$$\begin{aligned} 525 &= 5 \cdot 105 \\ &= 5 \cdot 5 \cdot 21 \\ &= 5 \cdot 5 \cdot 3 \cdot 7 \\ &= 3 \cdot 5^2 \cdot 7 \end{aligned}$$

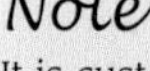
It is customary to write the prime factors in order from smallest to largest.

EXAMPLE 12

Reduce $\frac{210}{231}$ to lowest terms.

SOLUTION First we factor 210 and 231 into the product of prime factors. Then we reduce to lowest terms by dividing the numerator and denominator by any factors they have in common.

$$\begin{aligned} \frac{210}{231} &= \frac{2 \cdot 3 \cdot 5 \cdot 7}{3 \cdot 7 \cdot 11} && \textbf{Factor the numerator and denominator completely} \\ &= \frac{2 \cdot \cancel{3} \cdot 5 \cdot \cancel{7}}{\cancel{3} \cdot \cancel{7} \cdot 11} && \textbf{Divide the numerator and denominator by } \mathbf{3 \cdot 7} \\ &= \frac{2 \cdot 5}{11} \\ &= \frac{10}{11} \end{aligned}$$

The small lines we have drawn through the factors that are common to the numerator and denominator are used to indicate that we have divided the numerator and denominator by those factors.

LINKING OBJECTIVES AND EXAMPLES

Next to each **objective** we have listed the examples that are best described by that objective.

A	2–7
B	8, 9
C	10
D	11
E	12

GETTING READY FOR CLASS

After reading through the preceding section, respond in your own words and in complete sentences.

1. Explain why some, but not all, rational numbers are also integers.
2. Explain the difference between a prime number and a composite number.
3. Give a written description of the set $\{x \mid -2 \le x < 3\}$.
4. What is an irrational number?

Problem Set 1.2

Online support materials can be found at www.thomsonedu.com/login

Graph the following on a real number line.

1. $\{x \mid x < 1\}$
2. $\{x \mid x > -2\}$
3. $\{x \mid x \le 1\}$
4. $\{x \mid x \ge -2\}$
5. $\{x \mid x \ge 4\}$
6. $\{x \mid x \le -3\}$
7. $\{x \mid x > 4\}$
8. $\{x \mid x < -3\}$
9. $\{x \mid x > 0\}$
10. $\{x \mid x < 0\}$
11. $\{x \mid 3 \ge x\}$
12. $\{x \mid 4 \le x\}$
13. $\{x \mid x \le -3 \text{ or } x \ge 1\}$
14. $\{x \mid x < 1 \text{ or } x > 4\}$
15. $\{x \mid -3 \le x \text{ and } x \le 1\}$
16. $\{x \mid 1 < x \text{ and } x < 4\}$
17. $\{x \mid -3 < x \text{ and } x < 1\}$
18. $\{x \mid 1 \le x \text{ and } x \le 4\}$
19. $\{x \mid x < -1 \text{ or } x \ge 3\}$
20. $\{x \mid x < 0 \text{ or } x \ge 3\}$
21. $\{x \mid x \le -1 \text{ and } x \ge 3\}$
22. $\{x \mid x \le 0 \text{ and } x \ge 3\}$
23. $\{x \mid x > -4 \text{ and } x < 2\}$
24. $\{x \mid x > -3 \text{ and } x < 0\}$

Graph the following on a real number line.

25. $\{x \mid -1 \le x \le 2\}$
26. $\{x \mid -2 \le x \le 1\}$
27. $\{x \mid -1 < x < 2\}$
28. $\{x \mid -2 < x < 1\}$
29. $\{x \mid -4 < x \le 1\}$
30. $\{x \mid -1 < x \le 5\}$

Graph each of the following.

31. $\{x \mid x < -3 \text{ or } 2 < x < 4\}$
32. $\{x \mid -4 \le x \le -2 \text{ or } x \ge 3\}$
33. $\{x \mid x \le -5 \text{ or } 0 \le x \le 3\}$
34. $\{x \mid -3 < x < 0 \text{ or } x > 5\}$
35. $\{x \mid -5 < x < -2 \text{ or } 2 < x < 5\}$
36. $\{x \mid -3 \le x \le -1 \text{ or } 1 \le x \le 3\}$

Translate each of the following phrases into an equivalent inequality.

37. x is at least 5
38. x is at least -2
39. x is no more than -3
40. x is no more than 8
41. x is at most 4
42. x is at most -5
43. x is between -4 and 4
44. x is between -3 and 3
45. x is between -4 and 4, inclusive
46. x is between -3 and 3, inclusive

For $\{-6, -5.2, -\sqrt{7}, -\pi, 0, 1, 2, 2.3, \frac{9}{2}, \sqrt{17}\}$, list all the elements of the set that are named in each of the following problems.

47. Counting numbers
48. Whole numbers

= Videos available by instructor request

▶ = Online student support materials available at www.thomsonedu.com/login

49. Rational numbers

50. Integers

51. Irrational numbers

52. Real numbers

53. Nonnegative integers

54. Positive integers

Factor each number into the product of prime factors.

55. 60

56. 154

57. 266

58. 385

59. 111

60. 735

61. 369

62. 1,155

Reduce each fraction to lowest terms.

63. $\frac{165}{385}$

64. $\frac{550}{735}$

65. $\frac{385}{735}$

66. $\frac{266}{285}$

67. $\frac{111}{185}$

68. $\frac{279}{310}$

69. $\frac{525}{630}$

70. $\frac{205}{369}$

71. $\frac{75}{135}$

72. $\frac{38}{30}$

73. $\frac{6}{8}$

74. $\frac{10}{25}$

75. $\frac{200}{5}$

76. $\frac{240}{6}$

77. $\frac{10}{22}$

78. $\frac{39}{13}$

79. Name two numbers that are 5 units from 2 on the number line.

80. Name two numbers that are 6 units from −3 on the number line.

81. Write an inequality that gives all numbers that are less than 5 units from 8 on the number line.

82. Write an inequality that gives all numbers that are less than 3 units from 5 on the number line

83. Write an inequality that gives all numbers that are more than 5 units from 8 on the number line.

84. Write an inequality that gives all numbers that are more than 3 units from 5 on the number line.

Applying the Concepts

85. **Goldbach's Conjecture** The letter shown below was written by Christian Goldbach in 1742. In the letter Goldbach indicates that every even number greater than two can be written as the sum of two prime numbers. For example $4 = 2 + 2$, $6 = 3 + 3$, and $8 = 3 + 5$. His assertion has never been proven, although most mathematicians believe it is true. Because of this, it has come to be known as Goldbach's Conjecture. In March 2000, the publishing firm of Faber and Faber offered a $1 million prize to anyone who could prove Goldbach's Conjecture by March 20, 2002. Show that Goldbach's Conjecture holds for each of the following even numbers.

a. 10

b. 16

c. 24

d. 36

86. **Perfect Numbers** More than 2,200 years ago, the Greek mathematician Euclid (depicted on the stamp shown here) wrote: "A perfect number is a whole number that is equal to the sum of all its divisors, except itself." The first perfect number is 6 because $6 = 1 + 2 + 3$. The next perfect number is between 20 and 30. Find it.

87. **Engine Temperature** The temperature gauge on a car keeps track of the temperature of the water that cools the engine. The temperature gauge shown here registers temperatures from 50°F (read 50 degrees Fahrenheit) to 270° F. If F is the tempera-

ture of the water in the car engine, write an inequality that shows all the temperatures that can register on the temperature gauge below.

88. **Triangle Inequality** The *triangle inequality* is a property that is true for all triangles. It states that in any triangle, the sum of the lengths of any two sides is always greater than the length of the third side. Below is a triangle with sides of length x, y, and z. Use the triangle inequality to write three inequalities using x, y, and z.

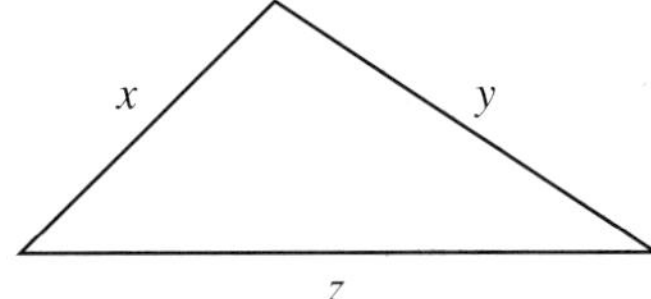

89. **Televisions** Use the chart to answer the questions below.
    a. On average, what percentage of American homes have at least 1 television set?
    b. On average, what percentage of American homes have no more than 3 television sets?
    c. On average, what percentage of American homes have 3 or more television sets?

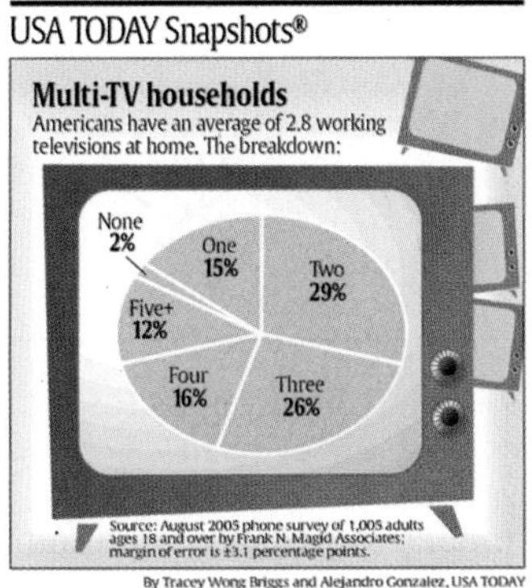

From *USA Today*. Copyright 2005. Reprinted with permission.

90. **Hours of Work** Suppose you have a job that requires that you work at least 20 hours but less than 40 hours per week. Write an inequality, using t, that gives the number of hours you work each week.

91. **Improving Your Quantitative Literacy** Quantitative literacy is a subject discussed by many people involved in teaching mathematics. The person they are concerned with when they discuss it is you. We are going to work at improving your quantitative literacy, but before we do that, we should answer the question: What is quantitative literacy? Lynn Arthur Steen, a noted mathematics educator, has stated that quantitative literacy is "the capacity to deal effectively with the quantitative aspects of life."
    a. Give a definition for the word *quantitative.*
    b. Give a definition for the word *literacy.*
    c. Are there situations that occur in your life that you find distasteful, or that you try to avoid, because they involve numbers and mathematics? If so, list some of them here. (For example, some people find buying a car particularly difficult because they feel the details of financing it are beyond them.)

92. **Improving Your Quantitative Literacy** The chart shows that if the risk of getting in an accident with no alcohol in your system is 1, then the risk of an accident with a blood-alcohol level of 0.24 is 147 times as high.

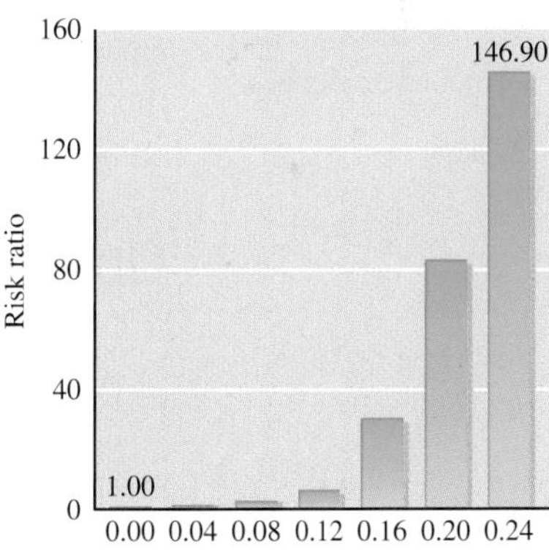

Data from *USA Today.* Copyright 2005.

a. A person driving with a blood-alcohol level of 0.20 is how many times more likely to get in an accident than if she was driving with a blood-alcohol level of 0?
b. If the probability of getting in an accident while driving for an hour on surface streets in a certain city in 0.02%, what is the probability of getting in an accident in the same circumstances with a blood-alcohol level of 0.20?

Extending the Concepts

The expression $n!$ is read "n factorial" and is the product of all the consecutive integers from n down to 1. For example,

$$1! = 1$$
$$2! = 2 \cdot 1 = 2$$
$$3! = 3 \cdot 2 \cdot 1 = 6$$
$$4! = 4 \cdot 3 \cdot 2 \cdot 1 = 24$$

93. Calculate 5!

94. Calculate 6!

95. Show that this statement is true: $6! = 6 \cdot 5!$

96. Show that the following statement is false: $(2 + 3)! = 2! + 3!$

1.3 Properties of Real Numbers

OBJECTIVES

- **A** Find the opposite of a real number.
- **B** Multiply fractions.
- **C** Simplify expressions involving absolute value.
- **D** Recognize and apply the properties of real numbers.
- **E** Recognize and apply the distributive property.
- **F** Add and subtract fractions.
- **G** Simplify algebraic expressions.

The area of the large rectangle shown here can be found in two ways: We can multiply its length a by its width $b + c$, or we can find the areas of the two smaller rectangles and add those areas to find the total area.

Area of large rectangle: $a(b + c)$

Sum of the areas of two smaller rectangles: $ab + ac$

Because the area of the large rectangle is the sum of the areas of the two smaller rectangles, we can write:

$$a(b + c) = ab + ac$$

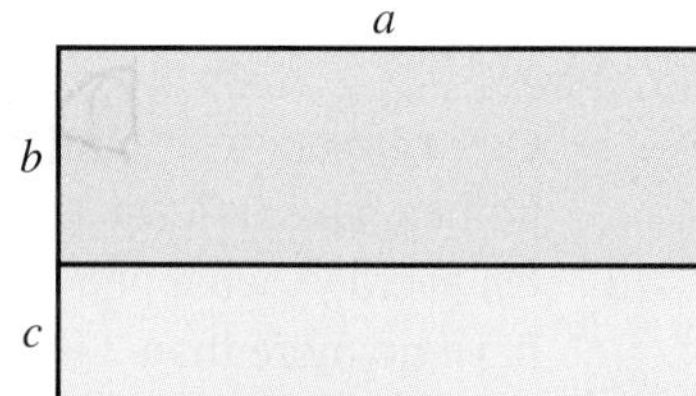

This equation is called the *distributive property.* It is one of the properties we will be discussing in this section. Before we arrive at the distributive property, we need to review some basic definitions and vocabulary.

Opposites and Reciprocals

DEFINITION Any two real numbers the same distance from 0, but in opposite directions from 0 on the number line, are called **opposites,** or **additive inverses.**

EXAMPLE 1 The numbers -3 and 3 are opposites. So are π and $-\pi$, $\frac{3}{4}$ and $-\frac{3}{4}$, and $\sqrt{2}$ and $-\sqrt{2}$.

The negative sign in front of a number can be read in a number of different ways. It can be read as "negative" or "the opposite of." We say -4 is the opposite of 4, or negative 4. The one we use will depend on the situation. For instance, the expression $-(-3)$ is best read "the opposite of negative 3." Because the opposite of -3 is 3, we have $-(-3) = 3$. In general, if a is any positive real number, then

$$-(-a) = a \qquad \textbf{(The opposite of a negative is a positive)}$$

Review of Multiplication with Fractions

Before we go further with our study of the number line, we need to review multiplication with fractions. Recall that for the fraction $\frac{a}{b}$, a is called the numerator and b is called the denominator. To multiply two fractions we simply multiply numerators and multiply denominators.

EXAMPLES Multiply.

2. $\frac{3}{5} \cdot \frac{7}{8} = \frac{3 \cdot 7}{5 \cdot 8} = \frac{21}{40}$

3. $8 \cdot \frac{1}{5} = \frac{8}{1} \cdot \frac{1}{5} = \frac{8 \cdot 1}{1 \cdot 5} = \frac{8}{5}$

4. $\left(\frac{2}{3}\right)^4 = \frac{2}{3} \cdot \frac{2}{3} \cdot \frac{2}{3} \cdot \frac{2}{3} = \frac{16}{81}$

Note

In past math classes you may have written fractions like $\frac{8}{5}$ (improper fractions) as mixed numbers, such as $1\frac{3}{5}$. In algebra, it is usually better to leave them as improper fractions.

The idea of multiplication of fractions is useful in understanding the concept of the reciprocal of a number. Here is the definition.

> **DEFINITION** Any two real numbers whose product is 1 are called **reciprocals,** or **multiplicative inverses.**

EXAMPLES Give the reciprocal of each number.

	Number	*Reciprocal*	
5.	3	$\frac{1}{3}$	Because $3 \cdot \frac{1}{3} = \frac{3}{1} \cdot \frac{1}{3} = \frac{3}{3} = 1$
6.	$\frac{1}{6}$	6	Because $\frac{1}{6} \cdot 6 = \frac{1}{6} \cdot \frac{6}{1} = \frac{6}{6} = 1$
7.	$\frac{4}{5}$	$\frac{5}{4}$	Because $\frac{4}{5} \cdot \frac{5}{4} = \frac{20}{20} = 1$
8.	a	$\frac{1}{a}$	Because $a \cdot \frac{1}{a} = \frac{a}{1} \cdot \frac{1}{a} = \frac{a}{a} = 1$ **($a \neq 0$)**

Although we will not develop multiplication with negative numbers until later in this chapter, you should know that the reciprocal of a negative number is also a negative number. For example, the reciprocal of -5 is $-\frac{1}{5}$.

The Absolute Value of a Real Number

> **DEFINITION** The **absolute value** of a number (also called its **magnitude**) is the distance the number is from 0 on the number line. If x represents a real number, then the absolute value of x is written $|x|$.

This definition of absolute value is geometric in form since it defines absolute value in terms of the number line. Here is an alternative definition of absolute value that is algebraic in form since it involves only symbols.

Note
It is important to recognize that if x is a real number, $-x$ is not necessarily negative. For example, if x is 5, then $-x$ is -5. However, if x were -5, then $-x$ would be $-(-5)$, which is 5.

ALTERNATIVE DEFINITION If x represents a real number, then the **absolute value** of x is written $|x|$, and is given by

$$|x| = \begin{cases} x & \text{if } x \geq 0 \\ -x & \text{if } x < 0 \end{cases}$$

If the original number is positive or 0, then its absolute value is the number itself. If the number is negative, its absolute value is its opposite (which must be positive).

EXAMPLES Write each expression without absolute value symbols.

9. $|5| = 5$

10. $|-2| = 2$

11. $\left|-\frac{1}{2}\right| = \frac{1}{2}$

12. $-|-3| = -3$

13. $-|5| = -5$

14. $-|-\sqrt{2}| = -\sqrt{2}$

Properties of Real Numbers

We know that adding 3 and 7 gives the same answer as adding 7 and 3. The order of two numbers in an addition problem can be changed without changing the result. This fact about numbers and addition is called the *commutative property of addition.*

For all the properties listed in this section, a, b, and c represent real numbers.

Note
The other two basic operations (subtraction and division) are not commutative. If we change the order in which we are subtracting or dividing two numbers, we change the result.

Commutative Property of Addition
In symbols: $a + b = b + a$
In words: The *order* of the numbers in a sum does not affect the result.

Commutative Property of Multiplication
In symbols: $a \cdot b = b \cdot a$
In words: The *order* of the numbers in a product does not affect the result.

EXAMPLES

15. The statement $3 + 7 = 7 + 3$ is an example of the commutative property of addition.

16. The statement $3 \cdot x = x \cdot 3$ is an example of the commutative property of multiplication.

Another property of numbers you have used many times has to do with grouping. When adding $3 + 5 + 7$, we can add the 3 and 5 first and then the 7, or we can add the 5 and 7 first and then the 3. Mathematically, it looks like this: $(3 + 5) + 7 = 3 + (5 + 7)$. Operations that behave in this manner are called *associative* operations.

Associative Property of Addition
In symbols: $a + (b + c) = (a + b) + c$
In words: The *grouping* of the numbers in a sum does not affect the result.

Associative Property of Multiplication

In symbols: $a(bc) = (ab)c$

In words: The *grouping* of the numbers in a product does not affect the result.

The following examples illustrate how the associative properties can be used to simplify expressions that involve both numbers and variables.

EXAMPLES

Simplify by using the associative property.

17. $2 + (3 + y) = (2 + 3) + y$ **Associative property**

$= 5 + y$ **Addition**

18. $5(4x) = (5 \cdot 4)x$ **Associative property**

$= 20x$ **Multiplication**

19. $\frac{1}{4}(4a) = \left(\frac{1}{4} \cdot 4\right)a$ **Associative property**

$= 1a$ **Multiplication**

$= a$

20. $2\left(\frac{1}{2}x\right) = \left(2 \cdot \frac{1}{2}\right)x$ **Associative property**

$= 1x$ **Multiplication**

$= x$

21. $6\left(\frac{1}{3}x\right) = \left(6 \cdot \frac{1}{3}\right)x$ **Associative property**

$= 2x$ **Multiplication**

Our next property involves both addition and multiplication. It is called the *distributive property* and is stated as follows.

Distributive Property

In symbols: $a(b + c) = ab + ac$

In words: Multiplication *distributes* over addition.

Note

Although the properties we are listing are stated for only two or three real numbers, they hold for as many numbers as needed. For example, the distributive property holds for expressions like $3(x + y + z + 2)$. That is:

$3(x + y + z + 2) = 3x + 3y + 3z + 6$

You will see as we progress through the book that the distributive property is used very frequently in algebra. To see that the distributive property works, compare the following:

$3(4 + 5)$	$3(4) + 3(5)$
$3(9)$	$12 + 15$
27	27

In both cases the result is 27. Because the results are the same, the original two expressions must be equal, or $3(4 + 5) = 3(4) + 3(5)$.

EXAMPLES

Apply the distributive property to each expression and then simplify the result.

22. $5(4x + 3) = 5(4x) + 5(3)$ **Distributive property**

$= 20x + 15$ **Multiplication**

23. $6(3x + 2y) = 6(3x) + 6(2y)$ **Distributive property**
$= 18x + 12y$ **Multiplication**

24. $\frac{1}{2}(3x - 6) = \frac{1}{2}(3x) - \frac{1}{2}(6)$ **Distributive property**
$= \frac{3}{2}x - 3$ **Multiplication**

25. $2(3y + 4) + 2 = 2(3y) + 2(4) + 2$ **Distributive property**
$= 6y + 8 + 2$ **Multiplication**
$= 6y + 10$ **Addition**

Note

Example 24 shows that the distributive property holds for subtraction as well as addition. That is, multiplication distributes over subtraction. In symbols,
$a(b - c) = ab - ac$

We can combine our knowledge of the distributive property with multiplication of fractions to manipulate expressions involving fractions. Here are some examples that show how we do this.

EXAMPLES Apply the distributive property, then simplify if possible.

26. $a\left(1 + \frac{1}{a}\right) = a \cdot 1 + a \cdot \frac{1}{a} = a + 1$

27. $3\left(\frac{1}{3}x + 5\right) = 3 \cdot \frac{1}{3}x + 3 \cdot 5 = x + 15$

28. $6\left(\frac{1}{3}x - \frac{1}{2}y\right) = 6 \cdot \frac{1}{3}x - 6 \cdot \frac{1}{2}y = 2x - 3y$

Combining Similar Terms

The distributive property can also be used to combine similar terms. (For now, a *term* is a number, or the product of a number with one or more variables.) Similar terms are terms with the same variable part. The terms $3x$ and $5x$ are similar, as are $2y$, $7y$, and $-3y$, because the variable parts are the same.

EXAMPLES Use the distributive property to combine similar terms.

29. $3x + 5x = (3 + 5)x$ **Distributive property**
$= 8x$ **Addition**

30. $3y + y = (3 + 1)y$ **Distributive property**
$= 4y$ **Addition**

Review of Addition with Fractions

To add fractions, each fraction must have the same denominator.

DEFINITION The **least common denominator** (LCD) for a set of denominators is the smallest number divisible by *all* the denominators.

The first step in adding fractions is to find a common denominator for all the denominators. We then rewrite each fraction (if necessary) as an equivalent fraction with the common denominator. Finally, we add the numerators and reduce to lowest terms if necessary.

EXAMPLE 31 Add $\frac{5}{12} + \frac{7}{18}$.

SOLUTION The least common denominator for the denominators 12 and 18 must be the smallest number divisible by both 12 and 18. We can factor 12 and 18 completely and then build the LCD from these factors.

$$\left.\begin{array}{l}12 = 2 \cdot 2 \cdot 3\\ 18 = 2 \cdot 3 \cdot 3\end{array}\right\} \quad \text{LCD} = \overbrace{2 \cdot 2 \cdot 3}^{\text{12 divides the LCD}} \cdot 3 = 36$$

$$\text{LCD} = 2 \cdot \underbrace{2 \cdot 3 \cdot 3}_{\text{18 divides the LCD}} = 36$$

Next, we rewrite our original fractions as equivalent fractions with denominators of 36. To do so, we multiply each original fraction by an appropriate form of the number 1:

$$\frac{5}{12} + \frac{7}{18} = \frac{5}{12} \cdot \frac{\mathbf{3}}{\mathbf{3}} + \frac{7}{18} \cdot \frac{\mathbf{2}}{\mathbf{2}} = \frac{15}{36} + \frac{14}{36}$$

Finally, we add numerators and place the result over the common denominator, 36.

$$\frac{15}{36} + \frac{14}{36} = \frac{15 + 14}{36} = \frac{29}{36}$$

Simplifying Expressions

We can use the commutative, associative, and distributive properties together to simplify expressions.

EXAMPLE 32 Simplify $7x + 4 + 6x + 3$.

SOLUTION We begin by applying the commutative and associative properties to group similar terms:

$7x + 4 + 6x + 3 = (7x + 6x) + (4 + 3)$	**Commutative and associative properties**
$= (7 + 6)x + (4 + 3)$	**Distributive property**
$= 13x + 7$	**Addition**

EXAMPLE 33 Simplify $4 + 3(2y + 5) + 8y$.

SOLUTION Because our rule for order of operations indicates that we are to multiply before adding, we must distribute the 3 across $2y + 5$ first:

$4 + 3(2y + 5) + 8y = 4 + 6y + 15 + 8y$	**Distributive property**
$= (6y + 8y) + (4 + 15)$	**Commutative and associative properties**
$= (6 + 8)y + (4 + 15)$	**Distributive property**
$= 14y + 19$	**Addition**

Note
0 and 1 are called the *additive identity* and *multiplicative identity*, respectively. Combining 0 with a number, under addition, does not change the identity of the number. Likewise, combining 1 with a number, under multiplication, does not alter the identity of the number. We see that 0 is to addition what 1 is to multiplication.

The remaining properties of real numbers have to do with the numbers 0 and 1.

Additive Identity Property There exists a unique number 0 such that
In symbols: $a + 0 = a$ and $0 + a = a$

Multiplicative Identity Property There exists a unique number 1 such that
In symbols: $a(1) = a$ and $1(a) = a$

Additive Inverse Property For each real number a, there exists a unique real number $-a$ such that
In symbols: $a + (-a) = 0$
In words: Opposites add to 0.

Multiplicative Inverse Property For every real number a, except 0, there exists a unique real number $\frac{1}{a}$ such that
In symbols: $a\left(\frac{1}{a}\right) = 1$
In words: Reciprocals multiply to 1.

EXAMPLES

34. $7(1) = 7$ **Multiplicative identity property**

35. $4 + (-4) = 0$ **Additive inverse property**

36. $6\left(\frac{1}{6}\right) = 1$ **Multiplicative inverse property**

37. $(5 + 0) + 2 = 5 + 2$ **Additive identity property**

LINKING OBJECTIVES AND EXAMPLES

Next to each **objective** we have listed the examples that are best described by that objective.

A	1
B	2–4
C	9–14
D	15–30, 34–37
E	22–30
F	31
G	32, 33

GETTING READY FOR CLASS

After reading through the preceding section, respond in your own words and in complete sentences.

1. Describe the commutative property of multiplication.
2. Give definitions for each of the following:
   a The opposite of a number
   b. The absolute value of a number
3. Explain why subtraction and division are not commutative operations.
4. Explain why zero does not have a multiplicative inverse.

Problem Set 1.3

Online support materials can be found at www.thomsonedu.com/login

Complete the following table.

	Number	Opposite	Reciprocal
1.	4		
2.	-3		
3.	$-\frac{1}{2}$		
4.	$\frac{5}{6}$		
5.		-5	
6.		7	
7.		$-\frac{3}{8}$	
8.		$\frac{1}{2}$	
9.			-6
10.			-3
11.			$\frac{1}{3}$
12.			$-\frac{1}{4}$

13. Name two numbers that are their own reciprocals.

14. Give the number that has no reciprocal.

15. Name the number that is its own opposite.

16. The reciprocal of a negative number is negative—true or false?

Write each of the following without absolute value symbols.

17. $|-2|$
18. $|-7|$
19. $\left|-\frac{3}{4}\right|$
20. $\left|\frac{5}{6}\right|$
21. $|\pi|$
22. $|-\sqrt{2}|$
23. $-|4|$
24. $-|5|$
25. $-|-2|$
26. $-|-10|$
27. $-\left|-\frac{3}{4}\right|$
28. $-\left|\frac{7}{8}\right|$

Multiply the following.

29. $\frac{3}{5} \cdot \frac{7}{8}$
30. $\frac{6}{7} \cdot \frac{9}{5}$
31. $\frac{1}{3} \cdot 6$
32. $\frac{1}{4} \cdot 8$
33. $\left(\frac{2}{3}\right)^3$
34. $\left(\frac{4}{5}\right)^2$
35. $\left(\frac{1}{10}\right)^4$
36. $\left(\frac{1}{2}\right)^5$
37. $\frac{3}{5} \cdot \frac{4}{7} \cdot \frac{6}{11}$
38. $\frac{4}{5} \cdot \frac{6}{7} \cdot \frac{3}{11}$
39. $\frac{4}{3} \cdot \frac{3}{4}$
40. $\frac{5}{8} \cdot \frac{8}{5}$

Use the associative property to rewrite each of the following expressions and then simplify the result.

41. $4 + (2 + x)$
42. $6 + (5 + 3x)$
43. $(a + 3) + 5$
44. $(4a + 5) + 7$
45. $5(3y)$
46. $7(4y)$
47. $\frac{1}{3}(3x)$
48. $\frac{1}{5}(5x)$
49. $4\left(\frac{1}{4}a\right)$
50. $7\left(\frac{1}{7}a\right)$
51. $\frac{2}{3}\left(\frac{3}{2}x\right)$
52. $\frac{4}{3}\left(\frac{3}{4}x\right)$

Apply the distributive property to each expression. Simplify when possible.

53. $3(x + 6)$
54. $5(x + 9)$
55. $2(6x + 4)$
56. $3(7x + 8)$
57. $5(3a + 2b)$
58. $7(2a + 3b)$
59. $\frac{1}{3}(4x + 6)$
60. $\frac{1}{2}(3x + 8)$
61. $\frac{1}{5}(10 + 5y)$
62. $\frac{1}{6}(12 + 6y)$
63. $(5t + 1)8$
64. $(3t + 2)5$
65. $\frac{3}{4}(8x - 4)$
66. $\frac{2}{3}(6x - 9)$
67. $\frac{5}{6}(12x - 18)$
68. $\frac{3}{5}(5x + 10)$

The problems below are problems you will see later in the book. Apply the distributive property, then simplify if possible.

69. $3(3x + y - 2z)$
70. $2(2x - y + z)$
71. $10(0.3x + 0.7y)$
72. $10(0.2x + 0.5y)$

= Videos available by instructor request

▶ = Online student support materials available at www.thomsonedu.com/login

73. $100(0.06x + 0.07y)$

74. $100(0.09x + 0.08y)$

75. $3\left(x + \frac{1}{3}\right)$

76. $5\left(x - \frac{1}{5}\right)$

77. $2\left(x - \frac{1}{2}\right)$

78. $7\left(x + \frac{1}{7}\right)$

79. $x\left(1 + \frac{2}{x}\right)$

80. $x\left(1 - \frac{1}{x}\right)$

81. $a\left(1 - \frac{3}{a}\right)$

82. $a\left(1 + \frac{1}{a}\right)$

83. $8\left(\frac{1}{8}x + 3\right)$

84. $4\left(\frac{1}{4}x - 9\right)$

85. $6\left(\frac{1}{2}x - \frac{1}{3}y\right)$

86. $12\left(\frac{1}{4}x - \frac{1}{6}y\right)$

87. $12\left(\frac{1}{4}x + \frac{2}{3}y\right)$

88. $12\left(\frac{2}{3}x - \frac{1}{4}y\right)$

89. $20\left(\frac{2}{5}x + \frac{1}{4}y\right)$

90. $15\left(\frac{2}{3}x + \frac{2}{5}y\right)$

▶ **91.** $24\left(\frac{2}{3}x + \frac{1}{2}\right)$

▶ **92.** $30\left(\frac{3}{5}x + \frac{1}{3}\right)$

Apply the distributive property to each expression. Simplify when possible.

93. $3(5x + 2) + 4$

94. $4(3x + 2) + 5$

▶ **95.** $4(2y + 6) + 8$

96. $6(2y + 3) + 2$

97. $5(1 + 3t) + 4$

98. $2(1 + 5t) + 6$

99. $3 + (2 + 7x)4$

100. $4 + (1 + 3x)5$

Add the following fractions.

101. $\frac{2}{5} + \frac{1}{15}$

102. $\frac{5}{8} + \frac{1}{4}$

103. $\frac{17}{30} + \frac{11}{42}$

104. $\frac{19}{42} + \frac{13}{70}$

105. $\frac{9}{48} + \frac{3}{54}$

106. $\frac{6}{28} + \frac{5}{42}$

107. $\frac{25}{84} + \frac{41}{90}$

108. $\frac{23}{70} + \frac{29}{84}$

The problems below are problems you will see later in the book. Simplify each expression.

109. $\frac{3}{14} + \frac{7}{30}$

110. $\frac{3}{10} + \frac{11}{42}$

111. $32\left(\frac{3}{4}\right) - 16\left(\frac{3}{4}\right)^2$

112. $32\left(\frac{3}{2}\right) - 16\left(\frac{3}{2}\right)^2$

Use the commutative, associative, and distributive properties to simplify the following.

113. $5a + 7 + 8a + a$

114. $6a + 4 + a + 4a$

▶ **115.** $3y + y + 5 + 2y + 1$

116. $4y + 2y + 3 + y + 7$

117. $2(5x + 1) + 2x$

118. $3(4x + 1) + 9x$

119. $7 + 2(4y + 2)$

120. $6 + 3(5y + 2)$

121. $3 + 4(5a + 3) + 4a$

122. $8 + 2(4a + 2) + 5a$

123. $5x + 2(3x + 8) + 4$

124. $7x + 3(4x + 1) + 7$

Identify the property (or properties) of real numbers that justifies each of the following.

125. $3 + 2 = 2 + 3$

126. $3(ab) = (3a)b$

127. $5 \cdot x = x \cdot 5$

128. $2 + 0 = 2$

129. $4 + (-4) = 0$

130. $1(6) = 6$

131. $x + (y + 2) = (y + 2) + x$

132. $(a + 3) + 4 = a + (3 + 4)$

133. $4(5 \cdot 7) = 5(4 \cdot 7)$

134. $6(xy) = (xy)6$

135. $4 + (x + y) = (4 + y) + x$

136. $(r + 7) + s = (r + s) + 7$

137. $3(4x + 2) = 12x + 6$

138. $5\left(\frac{1}{5}\right) = 1$

Applying the Concepts

139. Rhind Papyrus In approximately 1650 B.C., a mathematical document called the *Rhind Papyrus* (part of which is shown here) was written in ancient Egypt. An "exercise" in this document asked the reader to find "a quantity such that when it is added to one fourth of itself results in 15." Verify this quantity must be 12.

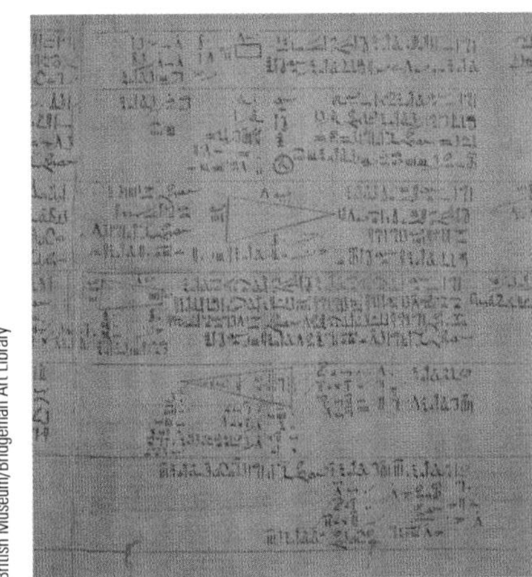

British Museum/Bridgeman Art Library

140. Clock Arithmetic In a normal clock with 12 hours on its face, 12 is the additive identity because adding 12 hours to any time on the clock will not change the hands of the clock. Also, if we think of the hour hand of a clock, the problem $10 + 4$ can be taken to mean: The hour hand is pointing at 10; if we add 4 more hours, it will be pointing at what number? Reasoning this way, we see that in clock arithmetic $10 + 4 = 2$ and $9 + 6 = 3$. Find the following in clock arithmetic:

10(o'clock) + 4(hours) = 2(o'clock)

a. $10 + 5$
b. $10 + 6$
c. $10 + 1$
d. $10 + 12$
e. $x + 12$

141. Improving Your Quantitative Literacy Use the chart to answer the questions below.

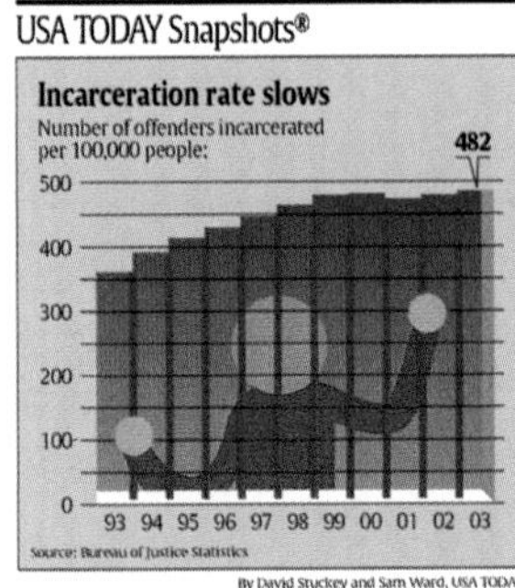

From *USA Today.* Copyright 2005. Reprinted with permission.

a. For the most part, the chart shows that the number of people incarcerated increases over the years. How do you explain the title *Incarceration rate slows*?
b. Which year showed a decline in the number of people incarcerated?
c. Was the rate of increase in incarcerations from 2000 to 2001 a positive number or a negative number?

1.4 Arithmetic with Real Numbers

OBJECTIVES

A Add and subtract real numbers.

B Multiply and divide real numbers.

C Apply the rule for order of operations.

D Simplify algebraic expressions.

E Divide fractions.

F Find the value of an expression.

The temperature at the airport is 70°F. A plane takes off and reaches its cruising altitude of 28,000 feet, where the temperature is −40°F. Find the difference in the temperatures at takeoff and at cruising altitude.

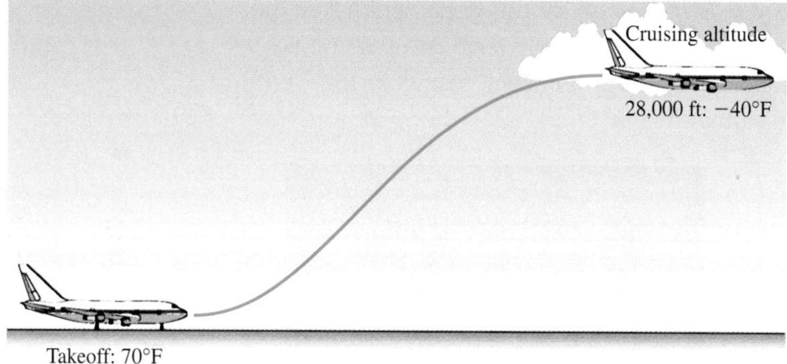

We know intuitively that the difference in temperature is 110°F. If we write this problem using symbols, we have

$$70 - (-40) = 110$$

In this section we review the rules for arithmetic with real numbers, which will include problems such as this one.

Adding Real Numbers

The purpose of this section is to review the rules for arithmetic with real numbers and the justification for those rules. We can justify the rules for addition of real numbers geometrically by use of the real number line. Consider the sum of −5 and 3:

$$-5 + 3$$

We can interpret this expression as meaning "start at the origin and move 5 units in the negative direction and then 3 units in the positive direction." With the aid of a number line we can visualize the process.

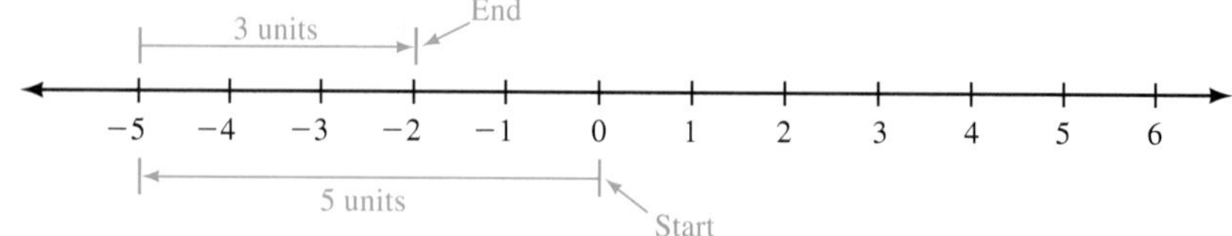

Because the process ends at −2, we say the sum of −5 and 3 is −2:

$$-5 + 3 = -2$$

We can use the real number line in this way to add any combination of positive and negative numbers.

Note

We are showing addition of real numbers on the number line to justify the rule we will write for addition of positive and negative numbers. You may want to skip ahead and read the rule on the next page first, and then come back and read through this discussion again. The discussion here is the "why" behind the rule.

The sum of −4 and −2, $-4 + (-2)$, can be interpreted as starting at the origin, moving 4 units in the negative direction, and then 2 more units in the negative direction:

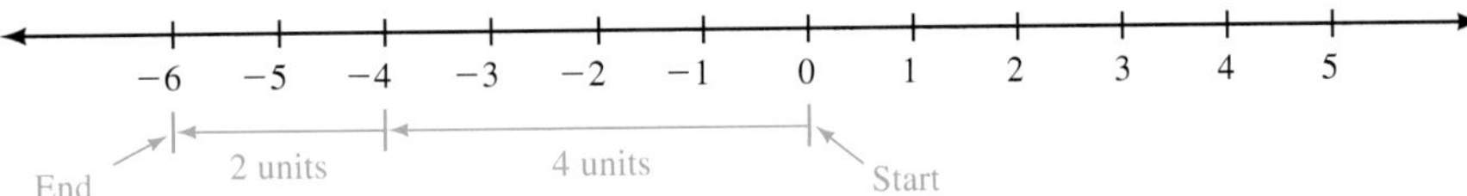

Because the process ends at −6, we say the sum of −4 and −2 is −6:

$$-4 + (-2) = -6$$

We can eliminate actually drawing a number line by simply visualizing it mentally. The following example gives the results of all possible sums of positive and negative 5 and 7.

EXAMPLE 1

Add all combinations of positive and negative 5 and 7.

SOLUTION

$$5 + 7 = 12$$
$$-5 + 7 = 2$$
$$5 + (-7) = -2$$
$$-5 + (-7) = -12$$

Looking closely at the relationships in Example 1 (and trying other similar examples if necessary), we can arrive at the following rule for adding two real numbers.

Note
This rule is the most important rule we have had so far. It is very important that you use it exactly the way it is written. Your goal is to become fast and accurate at adding positive and negative numbers. When you have finished reading this section and working the problems in the problem set, you should have attained that goal.

To Add Two Real Numbers

With the *same* sign:

Step 1: Add their absolute values.

Step 2: Attach their common sign. If both numbers are positive, their sum is positive; if both numbers are negative, their sum is negative.

With *opposite* signs:

Step 1: Subtract the smaller absolute value from the larger.

Step 2: Attach the sign of the number whose absolute value is larger.

Subtracting Real Numbers

In order to have as few rules as possible, we will not attempt to list new rules for the *difference* of two real numbers. We will instead define it in terms of addition and apply the rule for addition.

DEFINITION If a and b are any two real numbers, then the **difference** of a and b, written $a - b$, is given by

$$\underbrace{a - b}_{} = \underbrace{a + (-b)}_{}$$

To subtract b, add the opposite of b.

We define the process of subtracting b from a as being equivalent to adding the opposite of b to a. In short, we say, "subtraction is addition of the opposite."

EXAMPLES Subtract.

2. $5 - 3 = 5 + (-3)$ **Subtracting 3 is equivalent to adding −3**
$= 2$
3. $-7 - 6 = -7 + (-6)$ **Subtracting 6 is equivalent to adding −6**
$= -13$
4. $9 - (-2) = 9 + 2$ **Subtracting −2 is equivalent to adding 2**
$= 11$
5. $-6 - (-5) = -6 + 5$ **Subtracting −5 is equivalent to adding 5**
$= -1$

EXAMPLE 6 Subtract −3 from −9.

SOLUTION Because subtraction is not commutative, we must be sure to write the numbers in the correct order. Because we are subtracting −3, the problem looks like this when translated into symbols:

$-9 - (-3) = -9 + 3$ **Change to addition of the opposite**

$= -6$ **Add**

EXAMPLE 7 Add −4 to the difference of −2 and 5.

SOLUTION The difference of −2 and 5 is written $-2 - 5$. Adding −4 to that difference gives us

$(-2 - 5) + (-4) = -7 + (-4)$ **Simplify inside parentheses**

$= -11$ **Add**

Multiplying Real Numbers

Multiplication with whole numbers is simply a shorthand way of writing repeated addition.

For example, $3(-2)$ can be evaluated as follows:

$$3(-2) = -2 + (-2) + (-2) = -6$$

We can evaluate the product $-3(2)$ in a similar manner if we first apply the commutative property of multiplication:

$$-3(2) = 2(-3) = -3 + (-3) = -6$$

From these results it seems reasonable to say that the product of a positive and a negative is a negative number.

The last case we must consider is the product of two negative numbers, such as $-3(-2)$. To evaluate this product we will look at the expression $-3[2 + (-2)]$ in two different ways. First, since $2 + (-2) = 0$, we know the expression $-3[2 + (-2)]$ is equal to 0. On the other hand, we can apply the distributive property to get

$$-3[2 + (-2)] = -3(2) + (-3)(-2) = -6 + ?$$

Note
This discussion is to show why the rule for multiplication of real numbers is written the way it is. Even if you already know how to multiply positive and negative numbers, it is a good idea to review the "why" that is behind it all.

Because we know the expression is equal to 0, it must be true that our ? is 6 because 6 is the only number we can add to −6 to get 0. Therefore, we have

$$-3(-2) = 6$$

Here is a summary of what we have so far:

Original numbers have		The answer is
The same sign	$3(2) = 6$	Positive
Different signs	$3(-2) = -6$	Negative
Different signs	$-3(2) = -6$	Negative
The same sign	$-3(-2) = 6$	Positive

To Multiply Two Real Numbers

Step 1: Multiply their absolute values.

Step 2: If the two numbers have the *same* sign, the product is positive. If the two numbers have *opposite* signs, the product is negative.

EXAMPLE 8 Multiply all combinations of positive and negative 7 and 3.

SOLUTION

$$7(3) = 21$$
$$7(-3) = -21$$
$$-7(3) = -21$$
$$-7(-3) = 21$$

Dividing Real Numbers

DEFINITION If a and b are any two real numbers, where $b \neq 0$, then the **quotient** of a and b, written $\frac{a}{b}$, is given by

$$\frac{a}{b} = a \cdot \left(\frac{1}{b}\right)$$

Dividing a by b is equivalent to multiplying a by the reciprocal of b. In short, we say, "division is multiplication by the reciprocal."

Because division is defined in terms of multiplication, the same rules hold for assigning the correct sign to a quotient as held for assigning the correct sign to a product; that is, *the quotient of two numbers with like signs is positive, while the quotient of two numbers with unlike signs is negative.*

EXAMPLES Divide.

9. $\frac{6}{3} = 6 \cdot \left(\frac{1}{3}\right) = 2$

10. $\frac{6}{-3} = 6 \cdot \left(-\frac{1}{3}\right) = -2$

Notice these examples indicate that if a and b are positive real numbers then

$$\frac{-a}{b} = \frac{a}{-b} = -\frac{a}{b} \quad \text{and} \quad \frac{-a}{-b} = \frac{a}{b}$$

11. $\frac{-6}{3} = -6 \cdot \left(\frac{1}{3}\right) = -2$

12. $\frac{-6}{-3} = -6 \cdot \left(-\frac{1}{3}\right) = 2$

The second step in the preceding examples is written only to show that each quotient can be written as a product. It is not actually necessary to show this step when working problems.

In the examples that follow, we find a combination of operations. In each case we use the rule for order of operations.

EXAMPLES

Simplify each expression as much as possible.

13. $(-2 - 3)(5 - 9) = (-5)(-4)$ **Simplify inside parentheses**

$= 20$ **Multiply**

14. $2 - 5(7 - 4) - 6 = 2 - 5(3) - 6$ **Simplify inside parentheses**

$= 2 - 15 - 6$ **Then, multiply**

$= -19$ **Finally, subtract, left to right**

15. $2(4 - 7)^3 + 3(-2 - 3)^2 = 2(-3)^3 + 3(-5)^2$ **Simplify inside parentheses**

$= 2(-27) + 3(25)$ **Evaluate numbers with exponents**

$= -54 + 75$ **Multiply**

$= 21$ **Add**

We can combine our knowledge of the properties of multiplication with our definition of division to simplify more expressions involving fractions. Here are two examples:

EXAMPLES

16. $6\left(\frac{t}{3}\right) = 6\left(\frac{1}{3}t\right)$ **Dividing by 3 is the same as multiplying by $\frac{1}{3}$**

$= \left(6 \cdot \frac{1}{3}\right)t$ **Associative property**

$= 2t$ **Multiplication**

17. $3\left(\frac{t}{3} - 2\right) = 3 \cdot \frac{t}{3} - 3 \cdot 2$ **Distributive property**

$= t - 6$ **Multiplication**

Our next examples involve more complicated fractions. The fraction bar works like parentheses to separate the numerator from the denominator. Although we don't write expressions this way, here is one way to think of the fraction bar:

$$\frac{-8 - 8}{-5 - 3} = (-8 - 8) \div (-5 - 3)$$

As you can see, if we apply the rule for order of operations to the expression on the right, we would work inside each set of parentheses first, then divide. Applying this to the expression on the left, we work on the numerator and denominator separately, then we divide or reduce the resulting fraction to lowest terms.

EXAMPLES

Simplify as much as possible.

18. $\dfrac{-8-8}{-5-3} = \dfrac{-16}{-8}$ **Simplify numerator and denominator separately**

$= 2$ **Divide**

19. $\dfrac{-5(-4)+2(-3)}{2(-1)-5} = \dfrac{20-6}{-2-5}$

$= \dfrac{14}{-7}$

$= -2$

20. $\dfrac{2^3+3^3}{2^2-3^2} = \dfrac{8+27}{4-9}$

$= \dfrac{35}{-5}$

$= -7$

Remember, since subtraction is defined in terms of addition, we can restate the distributive property in terms of subtraction; that is, if a, b, and c are real numbers, then $a(b-c) = ab - ac$.

EXAMPLE 21

Simplify $3(2y-1)+y$.

SOLUTION We begin by multiplying the 3 and $2y-1$. Then, we combine similar terms:

$3(2y-1)+y = 6y-3+y$ **Distributive property**

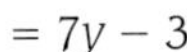

$= 7y-3$ **Combine similar terms**

EXAMPLE 22

Simplify $8-3(4x-2)+5x$.

SOLUTION First we distribute the -3 across the $4x-2$.

$$8-3(4x-2)+5x = 8-12x+6+5x$$

$$= -7x+14$$

EXAMPLE 23

Simplify $5(2a+3)-(6a-4)$.

SOLUTION We begin by applying the distributive property to remove the parentheses. The expression $-(6a-4)$ can be thought of as $-1(6a-4)$. Thinking of it in this way allows us to apply the distributive property.

$$-1(6a-4) = -1(6a)-(-1)(4) = -6a+4$$

Here is the complete problem:

$5(2a+3)-(6a-4) = 10a+15-6a+4$ **Distributive property**

$= 4a+19$ **Combine similar terms**

Dividing Fractions

Next we review division with fractions and division with the number 0.

EXAMPLES

Divide and reduce to lowest terms.

24. $\frac{3}{4} \div \frac{6}{11} = \frac{3}{4} \cdot \frac{11}{6}$ **Definition of division**

$= \frac{33}{24}$ **Multiply numerators, multiply denominators**

$= \frac{11}{8}$ **Divide numerator and denominator by 3**

25. $10 \div \frac{5}{6} = \frac{10}{1} \cdot \frac{6}{5}$ **Definition of division**

$= \frac{60}{5}$ **Multiply numerators, multiply denominators**

$= 12$ **Divide**

26. $-\frac{3}{8} \div 6 = -\frac{3}{8} \cdot \frac{1}{6}$ **Definition of division**

$= -\frac{3}{48}$ **Multiply numerators, multiply denominators**

$= -\frac{1}{16}$ **Divide numerator and denominator by 3**

Division with the Number 0

For every division problem an associated multiplication problem involving the same numbers exists. For example, the following two problems say the same thing about the numbers 2, 3, and 6:

Division	*Multiplication*
$\frac{6}{3} = 2$	$6 = 2(3)$

We can use this relationship between division and multiplication to clarify division involving the number 0.

First of all, dividing 0 by a number other than 0 is allowed and always results in 0. To see this, consider dividing 0 by 5. We know the answer is 0 because of the relationship between multiplication and division. This is how we write it:

$$\frac{0}{5} = 0 \quad \text{because} \quad 0 = 0(5)$$

On the other hand, dividing a nonzero number by 0 is not allowed in the real numbers. Suppose we were attempting to divide 5 by 0. We don't know whether there is an answer to this problem, but if there is, let's say the answer is a number that we can represent with the letter n. If 5 divided by 0 is a number n, then

$$\frac{5}{0} = n \quad \text{and} \quad 5 = n(0)$$

But this is impossible because no matter what number n is, when we multiply it by 0 the answer must be 0. It can never be 5. In algebra, we say expressions like $\frac{5}{0}$ are undefined because there is no answer to them; that is, division by 0 is not allowed in the real numbers.

Finding the Value of an Algebraic Expression

As we mentioned earlier in this chapter, an algebraic expression is a combination of numbers, variables, and operation symbols. Each of the following is an algebraic expression

$$7a \qquad x^2 - y^2 \qquad 2(3t - 4) \qquad \frac{2x - 5}{6}$$

An expression such as $2(3t - 4)$ will take on different values depending on what number we substitute for t. For example, if we substitute -8 for t then the expression $2(3t - 4)$ becomes $2[3(-8) - 4)]$ which simplifies to -56. If we apply the distributive property to $2(3t - 4)$ we have

$$2(3t - 4) = 6t - 8$$

Substituting -8 for t in the simplified expression gives us $6(-8) - 8 = -56$, which is the same result we obtained previously. As you would expect, substituting the same number into an expression, and any simplified form of that expression, will yield the same result.

EXAMPLE 27 Evaluate the expressions $(a + 4)^2$, $a^2 + 16$, and $a^2 + 8a + 16$ when a is -2, 0, and 3.

SOLUTION Organizing our work with a table, we have

a	$(a + 4)^2$	$a^2 + 16$	$a^2 + 8a + 16$
-2	$(-2 + 4)^2 = 4$	$(-2)^2 + 16 = 20$	$(-2)^2 + 8(-2) + 16 = 4$
0	$(0 + 4)^2 = 16$	$0^2 + 16 = 16$	$0^2 + 8(0) + 16 = 16$
3	$(3 + 4)^2 = 49$	$3^2 + 16 = 25$	$3^2 + 8(3) + 16 = 49$

When we study polynomials later in the book, you will see that the expressions $(a + 4)^2$ and $a^2 + 8a + 16$ are equivalent, and that neither one is equivalent to $a^2 + 16$.

LINKING OBJECTIVES AND EXAMPLES

Next to each **objective** we have listed the examples that are best described by that objective.

A	1–7
B	8–12
C	13–20
D	21–23
E	24–26
F	27

GETTING READY FOR CLASS

After reading through the preceding section, respond in your own words and in complete sentences.

1. For each of the following expressions, give an example of an everyday situation that is modeled by the expression.
   a. $35 − $12 = $23
   b. −$35 − $12 = −$47
2. For each of the following expressions, give an example of an everyday situation that is modeled by the expression.
   a. 3(−$25) = −$75
   b. (−$100) ÷ 5 = −$20
3. Why is division by 0 not allowed?
4. Why isn't the statement "two negatives make a positive" true?

Problem Set 1.4

Online support materials can be found at www.thomsonedu.com/login

Find each of the following sums.

1. $6 + (-2)$ **2.** $11 - 5$

3. $-6 + 2$ **4.** $-11 + 5$

Find each of the following differences.

▸ **5.** $-7 - 3$ **6.** $-6 - 9$

▸ **7.** $-7 - (-3)$ **8.** $-6 - (-9)$

9. $\frac{3}{4} - \left(-\frac{5}{6}\right)$ **10.** $\frac{2}{3} - \left(-\frac{7}{5}\right)$

11. $\frac{11}{42} - \frac{17}{30}$ **12.** $\frac{13}{70} - \frac{19}{42}$

13. Subtract 5 from −3.

14. Subtract −3 from 5.

15. Find the difference of −4 and 8.

16. Find the difference of 8 and −4.

17. Subtract $4x$ from $-3x$.

18. Subtract $-5x$ from $7x$.

19. What number do you subtract from 5 to get −8?

20. What number do you subtract from −3 to get 9?

21. Add −7 to the difference of 2 and 9.

22. Add −3 to the difference of 9 and 2.

23. Subtract $3a$ from the sum of $8a$ and a.

24. Subtract $-3a$ from the sum of $3a$ and $5a$.

Find the following products.

25. $3(-5)$ **26.** $-3(5)$

27. $-3(-5)$ **28.** $4(-6)$

29. $2(-3)(4)$ **30.** $-2(3)(-4)$

31. $-2(5x)$ **32.** $-5(4x)$

33. $-\frac{1}{3}(-3x)$ **34.** $-\frac{1}{6}(-6x)$

35. $-\frac{2}{3}\left(-\frac{3}{2}y\right)$ **36.** $-\frac{2}{5}\left(-\frac{5}{2}y\right)$

37. $-2(4x - 3)$ **38.** $-2(-5t + 6)$

39. $-\frac{1}{2}(6a - 8)$ **40.** $-\frac{1}{3}(6a - 9)$

Simplify each expression as much as possible.

▸ **41.** $3(-4) - 2$ **42.** $-3(-4) - 2$

▢ = Videos available by instructor request
▸ = Online student support materials available at www.thomsonedu.com/login

43. $4(-3) - 6(-5)$

44. $-6(-3) - 5(-7)$

45. $2 - 5(-4) - 6$

46. $3 - 8(-1) - 7$

47. $4 - 3(7 - 1) - 5$

48. $8 - 5(6 - 3) - 7$

49. $2(-3)^2 - 4(-2)^3$

50. $5(-2)^2 - 2(-3)^3$

51. $7(3 - 5)^3 - 2(4 - 7)^3$

52. $3(-7 + 9)^3 - 5(-2 + 4)^3$

The problems below are problems you will see later in the book. Simplify.

53. $1(-2) - 2(-16) + 1(9)$

54. $6(1) - 1(-5) + 1(2)$

55. $1(1) - 3(-2) + (-2)(-2)$

56. $-2(-14) + 3(-4) - 1(-10)$

57. $-4(0)(-2) - (-1)(1)(1) - 1(2)(3)$

58. $1(0)(1) + 3(1)(4) + (-2)(2)(-1)$

59. $1[0 - (-1)] - 3(2 - 4) + (-2)(-2 - 0)$

60. $-3(-1 - 1) + 4(-2 + 2) - 5[2 - (-2)]$

61. $3(-2)^2 + 2(-2) - 1$

62. $4(-1)^2 + 3(-1) - 2$

63. $2(-2)^3 - 3(-2)^2 + 4(-2) - 8$

64. $5 \cdot 2^3 - 3 \cdot 2^2 + 4 \cdot 2 - 5$

The problems below are problems you will see later in the book. Multiply, then simplify if possible.

65. $-24\left(\frac{3}{8}\right)$

66. $-30\left(\frac{5}{6}\right)$

67. $24\left(-\frac{3}{8}\right)$

68. $30\left(-\frac{5}{6}\right)$

69. $-15\left(\frac{x}{5}\right)$

70. $63\left(\frac{x}{7}\right)$

71. $-15\left(\frac{y}{-3}\right)$

72. $16\left(\frac{y}{-2}\right)$

73. $-1(5 - x)$

74. $-1(a - b)$

75. $-1(7 - x)$

76. $-1(6 - y)$

77. $-3(2x - 3y)$

78. $-1(x - 2z)$

79. $6\left(\frac{x}{2} - 3\right)$

80. $6\left(\frac{x}{3} + 1\right)$

81. $12\left(\frac{a}{4} + \frac{1}{2}\right)$

82. $15\left(\frac{a}{3} + 2\right)$

83. $15\left(\frac{x}{5} + 4\right)$

84. $10\left(\frac{x}{2} - 9\right)$

85. $8\left(\frac{x}{8} + \frac{y}{2}\right)$

86. $63\left(\frac{x}{7} + \frac{y}{9}\right)$

87. $-15\left(\frac{x}{5} + \frac{y}{-3}\right)$

88. $16\left(\frac{x}{16} + \frac{y}{-2}\right)$

89. $12\left(\frac{y}{2} + \frac{y}{4} + \frac{y}{6}\right)$

90. $12\left(\frac{y}{3} - \frac{y}{6} + \frac{y}{2}\right)$

Simplify each expression.

91. $3(5x + 4) - x$

92. $4(7x + 3) - x$

93. $6 - 7(3 - m)$

94. $3 - 5(5 - m)$

95. $7 - 2(3x - 1) + 4x$

96. $8 - 5(2x - 3) + 4x$

97. $5(3y + 1) - (8y - 5)$

98. $4(6y + 3) - (6y - 6)$

99. $4(2 - 6x) - (3 - 4x)$

100. $7(1 - 2x) - (4 - 10x)$

101. $10 - 4(2x + 1) - (3x - 4)$

102. $7 - 2(3x + 5) - (2x - 3)$

103. $0.06x + 0.05(10{,}000 - x)$

104. $0.08x + 0.10(8{,}000 - x)$

105. $0.12x + 0.10(15{,}000 - x)$

106. $0.09x + 0.11(11{,}000 - x)$

107. $-(a + 1) - 4a$

108. $-(a - 2) - 5a$

Use the definition of division to write each division problem as a multiplication problem, then simplify.

109. $-\frac{3}{4} \div \frac{9}{8}$

110. $-\frac{2}{3} \div \frac{4}{9}$

111. $-8 \div \left(-\frac{1}{4}\right)$

112. $-12 \div \left(-\frac{2}{3}\right)$

113. $\frac{4}{9} \div (-8)$

114. $\frac{3}{7} \div (-6)$

Simplify as much as possible.

115. $\frac{0 - 4}{0 - 2}$

116. $\frac{0 + 6}{0 - 3}$

117. $\frac{-4 - 4}{-4 - 2}$

118. $\frac{6 + 6}{6 - 3}$

119. $\dfrac{-6+6}{-6-3}$

120. $\dfrac{4-4}{4-2}$

121. $\dfrac{2-4}{2-2}$

122. $\dfrac{3+6}{3-3}$

123. $\dfrac{3-(-1)}{-3-3}$

124. $\dfrac{-1-3}{3-(-3)}$

125. $\dfrac{3(-1)-4(-2)}{8-5}$

126. $\dfrac{6(-4)-5(-2)}{7-6}$

127. $8-(-6)\left[\dfrac{2(-3)-5(4)}{-8(6)-4}\right]$

128. $-9-5\left[\dfrac{11(-1)-9}{4(-3)+2(5)}\right]$

129. $6-(-3)\left[\dfrac{2-4(3-8)}{1-5(1-3)}\right]$

130. $8-(-7)\left[\dfrac{6-1(6-10)}{4-3(5-7)}\right]$

Complete each of the following tables.

131.

a	b	Sum $a+b$	Difference $a-b$	Product ab	Quotient $\frac{a}{b}$
3	12				
−3	12				
3	−12				
−3	−12				

132.

a	b	Sum $a+b$	Difference $a-b$	Product ab	Quotient $\frac{a}{b}$
8	2				
−8	2				
8	−2				
−8	−2				

133.

x	$3(5x-2)$	$15x-6$	$15x-2$
−2			
−1			
0			
1			
2			

134.

x	$(x+1)^2$	x^2+1	x^2+2x+1
−2			
−1			
0			
1			
2			

135. Find the value of $-\dfrac{b}{2a}$ when

a. $a=3,\ b=-6$

b. $a=-2,\ b=6$

c. $a=-1,\ b=-2$

d. $a=-0.1,\ b=27$

136. Find the value of b^2-4ac when

a. $a=3,\ b=-2$, and $c=4$

b. $a=1,\ b=-3$, and $c=-28$

c. $a=1,\ b=-6$, and $c=9$

d. $a=0.1,\ b=-27$, and $c=1{,}700$

The next two problems are intended to give you practice reading, and paying attention to, the instructions that accompany the problems you are working. You will see a number of problems like this throughout the book. Working these problems is an excellent way to get ready for a test or quiz.

137. Work each problem according to the instructions given. (Note that each of these instructions could be replaced with the instruction *Simplify*.)

a. Add: $-2.25+7.5$

b. Subtract: $-2.25-7.5$

c. Multiply: $-2.25(7.5)$

d. Divide: $\dfrac{-2.25}{7.5}$

138. Work each problem according to the instructions given.

a. Add: $-\dfrac{5}{6}+\left(-\dfrac{1}{4}\right)$

b. Subtract: $-\dfrac{5}{6}-\left(-\dfrac{1}{4}\right)$

c. Multiply: $-\dfrac{5}{6}\left(-\dfrac{1}{4}\right)$

d. Divide: $-\dfrac{5}{6}\div\left(-\dfrac{1}{4}\right)$

Use a calculator to simplify each expression. If rounding is necessary, round your answers to the nearest ten thousandth (4 places past the decimal point). You will see these problems later in the book.

139. $\dfrac{1.3802}{0.9031}$

140. $\dfrac{1.0792}{0.6990}$

141. $\dfrac{1}{2}(-0.1587)$

142. $\dfrac{1}{2}(-0.7948)$

143. $\dfrac{1}{2}\left(\dfrac{1.2}{1.4} - 1\right)$

144. $\dfrac{1}{2}\left(\dfrac{1.3}{1.1} - 1\right)$

145. $\dfrac{(6.8)(3.9)}{7.8}$

146. $\dfrac{(2.4)(1.8)}{1.2}$

147. $\dfrac{0.0005(200)}{(0.25)^2}$

148. $\dfrac{0.0006(400)}{(0.25)^2}$

149. $-500 + 27(100) - 0.1(100)^2$

150. $-500 + 27(170) - 0.1(170)^2$

151. $-0.05(130)^2 + 9.5(130) - 200$

152. $-0.04(130)^2 + 8.5(130) - 210$

Applying the Concepts

153. Time Zones The continental United States is divided into four time zones. When it is 4:00 in the Pacific zone, it is 5:00 in the Mountain zone, 6:00 in the Central zone, and 7:00 in the Eastern zone.

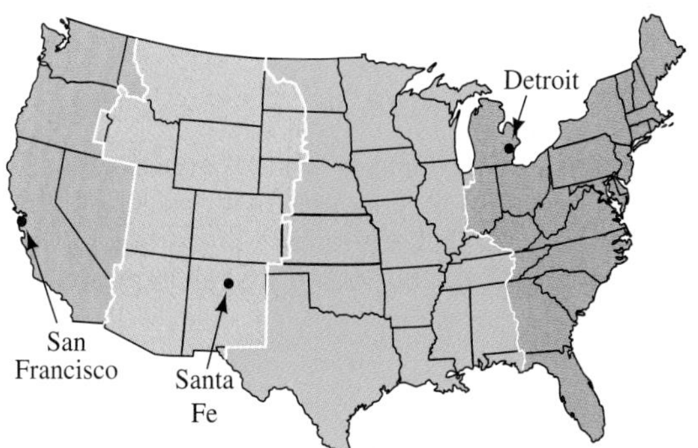

You board a plane at 6:55 P.M. in California (Pacific zone) and take 2 hours and 15 minutes to arrive in Santa Fe, New Mexico (Mountain zone) and another 3 hours and 20 minutes to arrive at your final destination, Detroit, Michigan (Eastern zone). At what local times did you arrive at Santa Fe and Detroit?

154. Oceans and Mountains The deepest ocean depth is 35,840 feet, found in the Pacific Ocean's Mariana Trench. The tallest mountain is Mount Everest, with a height of 29,028 feet. What is the difference between the highest point on earth and the lowest point on earth?

Royalty-Free/Corbis

155. Improving Your Quantitative Literacy In our number system, everything is in terms of powers of 10. With minutes and seconds, we think in terms of 60's. The chart appeared in *USA Today* in July 2005. The format for the times in the chart is:

hours:minutes:seconds

Find the difference between each of the following Triathlon times:

a. Kirill Litovtsenko and Cam Widoff

b. Ryan Bolton and Thomas Hellriegel

c. Cam Widoff and Simon Lessing

USA TODAY Snapshots®
Ironman USA Triathlon time to beat
Lake Placid, N.Y., hosts the Ford Ironman USA Triathlon on Sunday. Fastest winning times in the race's six-year history:
Simon Lessing (Great Britain), 2004 8:23:12
Steve Larsen (Davis, Calif.), 2001 8:33:11
Thomas Hellriegel (Germany), 1999 8:36:59
Ryan Bolton (Boulder, Colo.), 2002 8:39:19
Cam Widoff (Boulder, Colo.), 2000 8:46:05
Kirill Litovtsenko (Estonia), 2003 8:46:15
Source: Ironman USA Triathlon
By Ellen J. Horrow and Adrienne Lewis, USA TODAY

1.5 Recognizing Patterns

OBJECTIVES

- **A** Recognize a pattern in a sequence of numbers.
- **B** Extend an arithmetic sequence.
- **C** Extend a geometric sequence.
- **D** Recognize and extend the Fibonacci sequence.

Much of what we do in mathematics is concerned with recognizing patterns and classifying together groups of numbers that share a common characteristic. For instance, suppose you were asked to give the next number in this sequence:

$$3, 5, 7, \ldots$$

Looking for a pattern, you may observe that each number is 2 more than the number preceding it. That being the case, the next number in the sequence will be 9 because 9 is 2 more than 7. Reasoning in this manner is called *inductive reasoning*. In mathematics, we use inductive reasoning when we notice a pattern to a sequence of numbers and then extend the sequence using the pattern.

EXAMPLE 1 Use inductive reasoning to find the next term in each sequence.

a. $5, 8, 11, 14, \ldots$

b. $\triangle, \triangleright, \triangledown, \triangleleft, \ldots$

c. $1, 4, 9, 16, \ldots$

SOLUTION In each case we use the pattern we observe in the first few terms to write the next term.

a. Each term comes from the previous term by adding 3. Therefore, the next term would be 17.

b. The triangles rotate a quarter turn to the right each time. The next term would be a triangle that points up, $\triangle$.

c. This looks like the sequence of squares, $1^2, 2^2, 3^2, 4^2, \ldots$. The next term is $5^2 = 25$.

Now that we have an intuitive idea of inductive reasoning, here is a formal definition.

> **DEFINITION** **Inductive reasoning** is reasoning in which a conclusion is drawn based on evidence and observations that support that conclusion. In mathematics this usually involves noticing that a few items in a group have a trait or characteristic in common and then concluding that all items in the group have that same trait.

Arithmetic Sequences

We can extend our work with sequences by classifying together sequences that share a common characteristic. Our first classification is for sequences that are constructed by adding the same number each time.

DEFINITION An **arithmetic sequence** is a sequence of numbers in which each number (after the first number) comes from adding the same amount to the number before it.

The sequence

$$4, 7, 10, 13, \ldots$$

is an example of an arithmetic sequence, because each term is obtained from the preceding term by adding 3 each time. The number we add each time—in this case, 3—is the *common difference* because it can be obtained by subtraction.

EXAMPLE 2 Each sequence shown here is an arithmetic sequence. Find the next two numbers in each sequence.

a. $10, 16, 22, \ldots$

b. $\frac{1}{2}, 1, \frac{3}{2}, \ldots$

c. $5, 0, -5, \ldots$

SOLUTION Because we know that each sequence is arithmetic, we know how to look for the number that is added to each term to produce the next consecutive term.

a. $10, 16, 22, \ldots$: Each term is found by adding 6 to the term before it. Therefore, the next two terms will be 28 and 34.

b. $\frac{1}{2}, 1, \frac{3}{2}, \ldots$: Each term comes from the term before it by adding $\frac{1}{2}$. The fourth term will be $\frac{3}{2} + \frac{1}{2} = 2$, while the fifth term will be $2 + \frac{1}{2} = \frac{5}{2}$.

c. $5, 0, -5, \ldots$: Each term comes from adding -5 to the term before it. Therefore, the next two terms will be $-5 + (-5) = -10$, and $-10 + (-5) = -15$.

Geometric Sequences

Our second classification of sequences with a common characteristic involves sequences that are constructed using multiplication.

DEFINITION A **geometric sequence** is a sequence of numbers in which each number (after the first number) comes from the number before it by multiplying by the same amount each time.

The sequence

$$4, 12, 36, 108, \ldots$$

is a geometric sequence. Each term is obtained from the previous term by multiplying by 3. The amount by which we multiply each term to obtain the next term—in this case, 3—is called the *common ratio*.

EXAMPLE 3 Each sequence shown here is a geometric sequence. Find the next number in each sequence.

a. 2, 10, 50, . . .

b. 3, −15, 75, . . .

c. $\frac{1}{8}, \frac{1}{4}, \frac{1}{2}, \ldots$

SOLUTION Because each sequence is a geometric sequence, we know that each term is obtained from the previous term by multiplying by the same number each time.

a. 2, 10, 50, . . . : Starting with 2, each number is obtained from the previous number by multiplying by 5 each time. The next number will be $50 \cdot 5 = 250$.

b. 3, −15, 75, . . . : The sequence starts with 3. After that, each number is obtained by multiplying by −5 each time. The next number will be $75(-5) = -375$.

c. $\frac{1}{8}, \frac{1}{4}, \frac{1}{2}, \ldots$: This sequence starts with $\frac{1}{8}$. Multiplying each number in the sequence by 2 produces the next number in the sequence. To extend the sequence we multiply $\frac{1}{2}$ by 2:

$$\frac{1}{2} \cdot 2 = 1$$

The next number in the sequence is 1. 

The Fibonacci Sequence

In the Study Skills introduction to this chapter we quoted the mathematician Fibonacci. There is a special sequence in mathematics named for Fibonacci.

Fibonacci sequence: 1, 1, 2, 3, 5, 8, . . .

To construct the Fibonacci sequence we start with two 1s. The rest of the numbers in the sequence are found by adding the two previous terms. Adding the first two terms, 1 and 1, we have 2. Then, adding 1 and 2 we have 3. In general, adding any two consecutive terms of the Fibonacci sequence gives us the next term.

A Mathematical Model

One of the reasons we study number sequences is because they can be used to model some of the patterns and events we see in the world around us. The discussion that follows shows how the Fibonacci sequence can be used to predict the number of bees in each generation of the family tree of a male honeybee. It is based on an example from Chapter 2 of the book *Mathematics: A Human Endeavor* by Harold Jacobs. If you find that you enjoy discovering patterns in mathematics, Mr. Jacobs' book has many interesting examples and problems involving patterns in mathematics.

A male honeybee has one parent, its mother, whereas a female honeybee has two parents, a mother and a father. (A male honeybee comes from an unfertilized egg; a female honeybee comes from a fertilized egg.) Using these facts, we construct the family tree of a male honeybee using ♂ to represent a male honeybee and ♀ to represent a female honeybee.

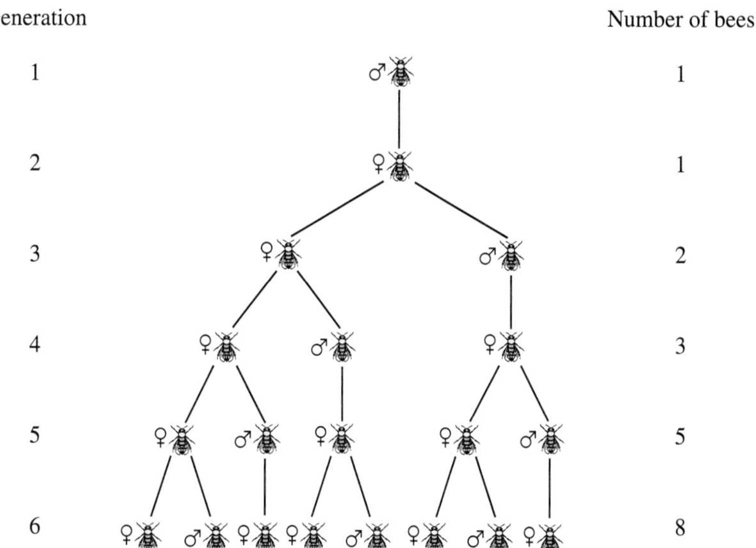

Looking at the numbers in the right column in our diagram, the sequence that gives us the number of bees in each generation of the family tree of a male honeybee is

1 1 2 3 5 8

As you can see, this is the Fibonacci sequence. We have taken our original diagram (the family tree of the male honeybee) and reduced it to a mathematical model (the Fibonacci sequence). The model can be used in place of the diagram to find the number of bees in any generation back from our first bee.

EXAMPLE 4 Find the number of bees in the tenth generation of the family tree of a male honeybee.

SOLUTION We can continue the previous diagram and simply count the number of bees in the tenth generation, or we can use inductive reasoning to conclude that the number of bees in the tenth generation will be the tenth term of the Fibonacci sequence. Let's make it easy on ourselves and find the first ten terms of the Fibonacci sequence.

Generation:	1	2	3	4	5	6	7	8	9	10
Number of bees:	1	1	2	3	5	8	13	21	34	55

As you can see, the number of bees in the tenth generation is 55.

Connecting Two Sequences: Paired Data

Bruce Ayres/Getty Images

We can use the discussion that opened this chapter to illustrate how we connect sequences. Suppose an antidepressant has a half-life of 5 days, and its concentration in a patient's system is 80 ng/mL (nanograms/milliliter) when the patient stops taking the antidepressant. To find the concentration after a half-life passes, we multiply the previous concentration by $\frac{1}{2}$. Here are two sequences that together summarize this information.

Days: 0, 5, 10, 15, . . .

Concentration: 80, 40, 20, 10, . . .

The sequence of days is an arithmetic sequence in which we start with 0 and add 5 each time. The sequence of concentrations is a geometric sequence in which we start with 80 and multiply by $\frac{1}{2}$ each time. Here is the same information displayed in a table.

TABLE 1
Concentration of an antidepressant

Days Since Discontinuing	Concentration (ng/mL)
0	80
5	40
10	20
15	10

The information in Table 1 is shown visually in Figures 1 and 2. The diagram in Figure 1 is called a *scatter diagram.* If we connect the points in the scatter diagram with straight lines, we have the diagram in Figure 2, which is called a *line graph.* In both figures, the horizontal line that shows days is called the *horizontal axis,* whereas the vertical line that shows concentration is called the *vertical axis.*

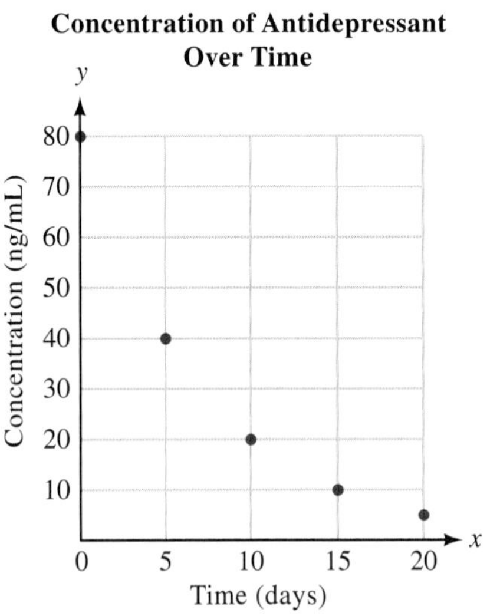

FIGURE 1 **Scatter diagram.**

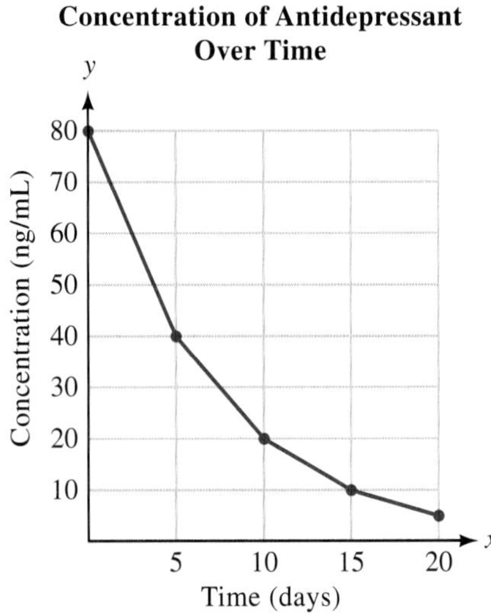

FIGURE 2 **Line graph.**

The data in Table 1 are called *paired data* because each number in the days column is paired with a specific number in the concentration column. For instance, after 15 days, the concentration is 10 ng/mL. Each pair of numbers from Table 1 is associated with one of the dots in the scatter diagram and line graph.

Table 1, Figure 1, and Figure 2 each describe the same relationship. The figures are visual descriptions, whereas the table is a numerical description. The discussion that precedes Table 1 is a written description of that relationship. This gives us three ways to describe the relationship: written, numerical, and visual. As we proceed further into our study of algebra, you will see this same relationship described a fourth way when we give the equation that relates time and concentration: $C = 80 \cdot 2^{-t/5}$.

LINKING OBJECTIVES AND EXAMPLES

Next to each objective we have listed the examples that are best described by that objective.

A	1
B	2
C	3
D	4

GETTING READY FOR CLASS

After reading through the preceding section, respond in your own words and in complete sentences.

1. What is inductive reasoning?
2. If you were to describe the Fibonacci sequence in words, you would start this way: "The first two numbers are ones. After that, each number is found by . . . ." Finish the sentence so that someone reading it will know how to find the numbers in the Fibonacci sequence.
3. Create an arithmetic sequence and explain how it was formed.
4. Create a geometric sequence and explain how it was formed.

Problem Set 1.5

Online support materials can be found at www.thomsonedu.com/login

Here are some sequences that we will be referring to throughout the book. Find the next number in each sequence.

1. 1, 2, 3, 4, . . . (The sequence of counting numbers)

2. 0, 1, 2, 3, . . . (The sequence of whole numbers)

3. 2, 4, 6, 8, . . . (The sequence of even numbers)

4. 1, 3, 5, 7, . . . (The sequence of odd numbers)

5. 1, 4, 9, 16, . . . (The sequence of squares)

▶ **6.** 1, 8, 27, 64, . . .(The sequence of cubes)

Find the next number in each sequence.

▶ **7.** 1, 8, 15, 22, . .

▶ **8.** 1, 8, 64, 512, . . .

9. 1, 8, 14, 19, . . .

▶ **10.** 1, 8, 16, 25, . . .

Give one possibility for the next term in each sequence.

11. △, ◁, ▽, ▷, . . .

12. →, ↓, ←, ↑, . . .

13. △, □, ○, △, ⊡, . . .

14. □, □□, □□ with □ above, ⊞, □□□, . . .

Each sequence shown here is an arithmetic sequence. In each case, find the next two numbers in the sequence.

15. 1, 5, 9, 13, . . .

16. 10, 16, 22, 28, . . .

17. 1, 0, −1, . . .

18. 6, 0, −6, . . .

19. 5, 2, −1, . . .

20. 8, 4, 0, . . .

21. $\frac{1}{4}, 0, -\frac{1}{4}, \ldots$

22. $\frac{2}{5}, 0, -\frac{2}{5}, \ldots$

23. $1, \frac{3}{2}, 2, \ldots$

24. $\frac{1}{3}, 1, \frac{5}{3}, \ldots$

Each sequence shown here is a geometric sequence. In each case, find the next number in the sequence.

25. 1, 3, 9, . . .

26. 1, 7, 49, . . .

27. 10, −30, 90, . . .

28. 10, −20, 40, . . .

29. $1, \frac{1}{2}, \frac{1}{4}, \ldots$

30. $1, \frac{1}{3}, \frac{1}{9}, \ldots$

31. 20, 10, 5, . . .

32. 8, 4, 2, . . .

33. 5, −25, 125, . . .

34. −4, 16, −64, . . .

35. $1, -\frac{1}{5}, \frac{1}{25}, \ldots$

36. $1, -\frac{1}{2}, \frac{1}{4}, \ldots$

37. Find the next number in the sequence 4, 8, . . . if the sequence is
- **a.** An arithmetic sequence
- **b.** A geometric sequence

38. Find the next number in the sequence 1, −4, . . . if the sequence is
- **a.** An arithmetic sequence
- **b.** A geometric sequence

= Videos available by instructor request

▶ = Online student support materials available at www.thomsonedu.com/login

39. Find the 12th term of the Fibonacci sequence.

40. Find the 13th term of the Fibonacci sequence.

Any number in the Fibonacci sequence is a *Fibonacci number.*

41. Name three Fibonacci numbers that are prime numbers.

42. Name three Fibonacci numbers that are composite numbers.

43. In the first ten terms of the Fibonacci sequence, which ones are even numbers?

44. In the first ten terms of the Fibonacci sequence, which ones are odd numbers?

The patterns in the tables below will become important when we do factoring of trinomials later in the book. Complete each table.

45.

Two Numbers a and b	Their Product ab	Their Sum $a + b$
1, −24		
−1, 24		
2, −12		
−2, 12		
3, −8		
−3, 8		
4, −6		
−4, 6		

46.

Two Numbers a and b	Their Product ab	Their Sum $a + b$
1, −54		
−1, 54		
2, −27		
−2, 27		
3, −18		
−3, 18		
6, −9		
−6, 9		

Applying the Concepts

47. Temperature and Altitude A pilot checks the weather conditions before flying and finds that the air temperature drops 3.5°F every 1,000 feet above the surface of the Earth. (The higher he flies, the colder the air.) If the air temperature is 41°F when the plane reaches 10,000 feet, write a sequence of numbers that gives the air temperature every 1,000 feet as the plane climbs from 10,000 feet to 15,000 feet. Is this sequence an arithmetic sequence?

48. Temperature and Altitude For the plane mentioned in Problem 47, at what altitude will the air temperature be 20°F?

Paul S. Bowen

49. Temperature and Altitude The weather conditions on a certain day are such that the air temperature drops 4.5°F every 1,000 feet above the surface of the Earth. If the air temperature is 41°F at 10,000 feet, write a sequence of numbers that gives the air temperature every 1,000 feet starting at 10,000 feet and ending at 5,000 feet. Is this sequence an arithmetic sequence?

50. Value of a Painting Suppose you own a painting that doubles in value every 5 years. If you bought the painting for \$125 in 1990, write a sequence of numbers that gives the value of the painting every 5 years from the time you purchased it until the year 2010. Is this sequence a geometric sequence?

Reread the introduction to this chapter and the discussion of paired data at the end of this section, then work the following problems.

51. **Half-Life** The half-lives of two antidepressants are given below. A patient taking Antidepressant 1 tells his doctor that he begins to feel sick if he misses his morning dose, and a patient taking Antidepressant 2 tells his doctor that he doesn't notice a difference if he misses a day of taking the medication. Explain these situations in terms of the half-life of the medication.

Half-life Antidepressant 1	Antidepressant 2
11 hours	5 days

52. **Half-Life** Two patients are taking the antidepressants mentioned in Problem 51. Both decide to stop their current medications and take another medication. The physician tells the patient on Antidepressant 2 to simply stop taking it but instructs the patient on Antidepressant 1 to take half the normal dose for 3 days, then one-fourth the normal dose for another 3 days before stopping the medication altogether. Use half-life to explain why the doctor uses two different methods of taking the patients off their medications.

53. **Half-Life** The half-life of a drug is 4 hours. A patient has been taking the drug on a regular basis for a few months and then discontinues taking it. The concentration of the drug in a patient's system is 60 ng/mL when the patient stops taking the medication. Complete the table, and then use the results to construct a line graph of those data.

Hours Since Discontinuing	Concentration (ng/mL)
0	60
4	
8	
12	
16	

54. **Half-Life** The half-life of a drug is 8 hours. A patient has been taking the drug on a regular basis for a few months and then discontinues taking it. The concentration of the drug in a patient's system is 120 ng/mL when the patient stops taking the medication. Complete the table, and then use the results to construct a line graph of those data).

Hours Since Discontinuing	Concentration (ng/mL)
0	120
8	
16	
24	
32	

55. **Boiling Point** The boiling point of water at sea level is 212°F. The boiling point of water drops 1.8°F every 1,000 feet above sea level, and it rises 1.8°F every 1,000 feet below sea level. Complete the following table to write the sequence that gives the boiling points of water from 2,000 feet below sea level to 3,000 feet above sea level.

Elevation (ft)	Boiling point (°F)
−2,000	
−1,000	
0	
1,000	
2,000	
3,000	

56. Improving Your Quantitative Literacy Information that comes to us from the media can answer questions and also raise questions. Sometimes the answers are contained in the data itself, and sometimes the answers come from outside the data.

a. How many launches were there in 1982?
b. How many launches were there in 1985?
c. During which years were there 7 launches?
d. Why were there no shuttle launches in 1987 and 2004?

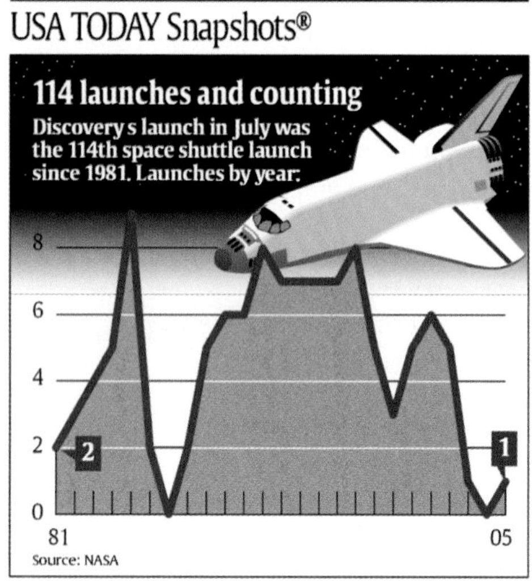

From *USA Today*. Copyright 2005. Reprinted with permission.

Extending the Concepts

57. Reading Graphs A patient is taking a prescribed dose of a medication every 4 hours during the day to relieve the symptoms of a cold. Figure 3 shows how the concentration of that medication in the patient's system changes over time. The 0 on the horizontal axis corresponds to the time the patient takes the first dose of the medication.

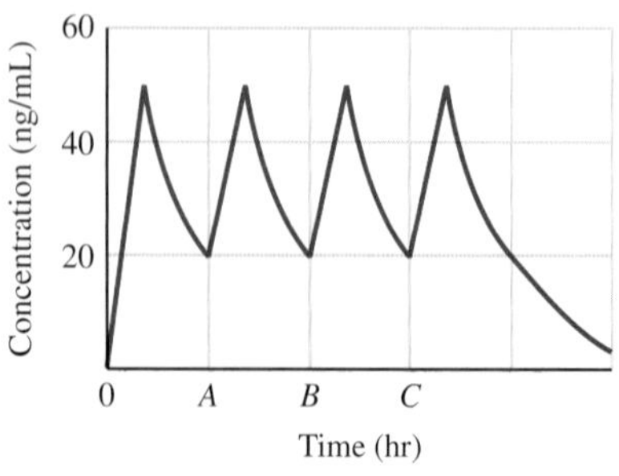

FIGURE 3

a. Explain what the steep straight line segments show with regard to the patient and his medication.
b. What has happened to make the graph fall off on the right?
c. What is the maximum concentration of the medication in the patient's system during the time period shown in Figure 5?
d. Find the values of A, B, and C.

58. Reading Graphs Figure 4 shows the number of people in line at a theater box office to buy tickets for a movie that starts at 7:30. The box office opens at 6:45.

a. How many people are in line at 6:30?
b. How many people are in line when the box office opens?
c. How many people are in line when the show starts?
d. At what times are there 60 people in line?
e. How long after the show starts is there no one left in line?

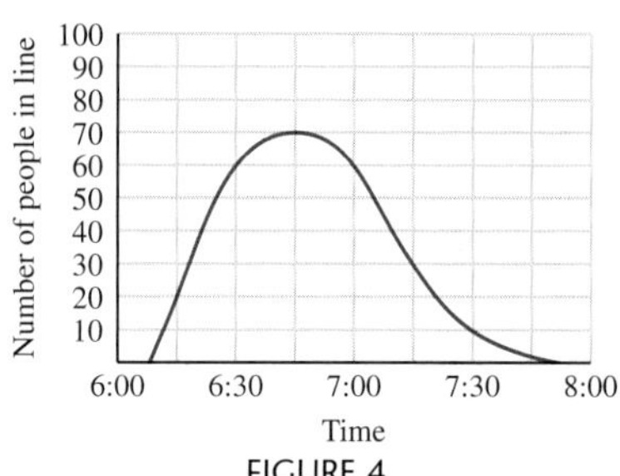

FIGURE 4

59. Shelf Life of Milk The diagram below shows the relationship between the temperature at which the milk is stored and the length of time before it spoils and cannot be consumed. Use the information in the diagram to complete the table. Then use the information in the table to construct a line graph.

Shelf Life of Milk

Temperature (Fahrenheit)	Shelf-Life (days)
32°	24
40°	
50°	
60°	
70°	

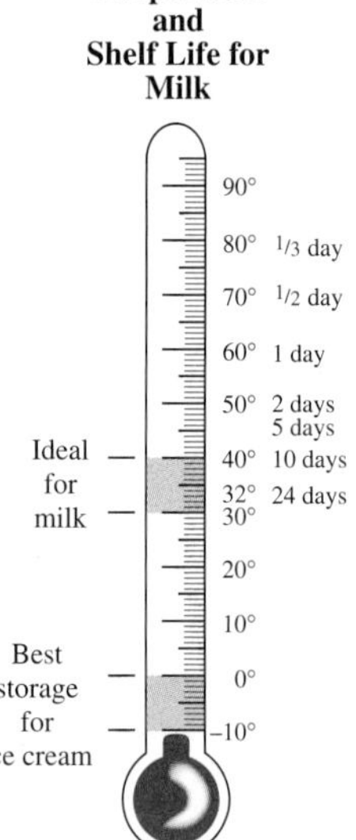

Chapter 1 SUMMARY

The numbers in brackets refer to the section(s) in which the topic can be found.

EXAMPLES

The margins of the chapter summaries will be used for brief examples of the topics being reviewed, whenever it is convenient.

1. $2^5 = 2 \cdot 2 \cdot 2 \cdot 2 \cdot 2 = 32$
$5^2 = 5 \cdot 5 = 25$
$10^3 = 10 \cdot 10 \cdot 10 = 1{,}000$
$1^4 = 1 \cdot 1 \cdot 1 \cdot 1 = 1$

Exponents [1.1]

Exponents represent notation used to indicate repeated multiplication. In the expression 3^4, 3 is the *base* and 4 is the *exponent.*

$$3^4 = 3 \cdot 3 \cdot 3 \cdot 3 = 81$$

The expression 3^4 is said to be in *exponential form,* whereas the expression $3 \cdot 3 \cdot 3 \cdot 3$ is in *expanded form.*

2. $10 + (2 \cdot 3^2 - 4 \cdot 2)$
$= 10 + (2 \cdot 9 - 4 \cdot 2)$
$= 10 + (18 - 8)$
$= 10 + 10$
$= 20$

Order of Operations [1.1]

When evaluating a mathematical expression, we will perform the operations in the following order, beginning with the expression in the innermost parentheses or brackets and working our way out.

1. Simplify all numbers with exponents, working from left to right if more than one of these numbers is present.
2. Then, do all multiplications and divisions left to right.
3. Finally, perform all additions and subtractions left to right.

3. If $A = \{0, 1, 2\}$ and $B = \{2, 3\}$, then $A \cup B = \{0, 1, 2, 3\}$ and $A \cap B = \{2\}$.

Sets [1.1]

A *set* is a collection of objects or things.

The *union* of two sets A and B, written $A \cup B$, is all the elements that are in A *or* are in B.

The *intersection* of two sets A and B, written $A \cap B$, is the set consisting of all elements common to both A *and* B.

Set A is a *subset* of set B, written $A \subset B$, if all elements in set A are also in set B.

4. 5 is a counting number, a whole number, an integer, a rational number, and a real number.
$\frac{3}{4}$ is a rational number and a real number.
$\sqrt{2}$ is an irrational number and a real number.

Special Sets [1.2]

Counting numbers $= \{1, 2, 3, \ldots\}$

Whole numbers $= \{0, 1, 2, 3, \ldots\}$

Integers $= \{\ldots, -3, -2, -1, 0, 1, 2, 3, \ldots\}$

Rational numbers $= \left\{\frac{a}{b} \mid a \text{ and } b \text{ are integers}, b \neq 0\right\}$

Irrational numbers $= \{x \mid x \text{ is real, but not rational}\}$

Real numbers = $\{x \mid x \text{ is rational or } x \text{ is irrational}\}$

Prime numbers = $\{2, 3, 5, 7, 11, \ldots\} = \{x \mid x$ is a positive integer greater than 1 whose only positive divisors are itself and 1$\}$

Opposites [1.3, 1.4]

5. The numbers 5 and -5 are opposites; their sum is 0.

$$5 + (-5) = 0$$

Any two real numbers the same distance from 0 on the number line, but in opposite directions from 0, are called *opposites,* or *additive inverses.* Opposites always add to 0.

Reciprocals [1.3, 1.4]

6. The numbers 3 and $\frac{1}{3}$ are reciprocals; their product is 1.

$$3\left(\frac{1}{3}\right) = 1$$

Any two real numbers whose product is 1 are called *reciprocals.* Every real number has a reciprocal except 0.

Absolute Value [1.3]

7. $|5| = 5$
$|-5| = 5$

The *absolute value* of a real number is its distance from 0 on the number line. If $|x|$ represents the absolute value of x, then

$$|x| = \begin{cases} x & \text{if} \quad x \ge 0 \\ -x & \text{if} \quad x < 0 \end{cases}$$

The absolute value of a real number is never negative.

Inequalities [1.2]

8. Graph each inequality mentioned at the right.

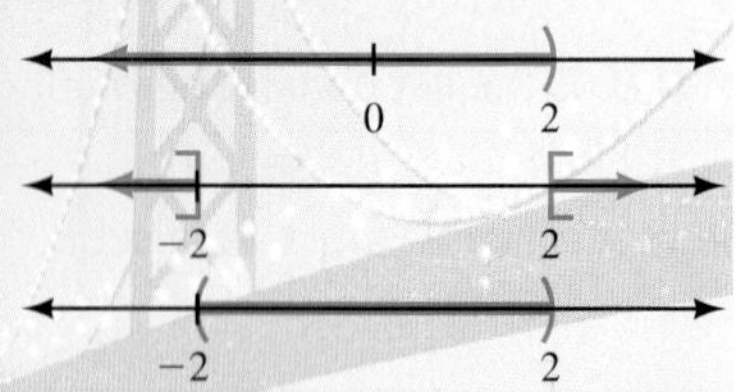

The set $\{x \mid x < 2\}$ is the set of all real numbers that are less than 2. To graph this set we place a right parenthesis at 2 on the real number line and then draw an arrow that starts at 2 and points to the left.

The set $\{x \mid x \le -2 \text{ or } x \ge 2\}$ is the set of all real numbers that are either less than or equal to -2 or greater than or equal to 2.

The set $\{x \mid -2 < x < 2\}$ is the set of all real numbers that are between -2 and 2; that is, the real numbers that are greater than -2 and less than 2.

Properties of Real Numbers [1.3]

	For Addition	*For Multiplication*
Commutative	$a + b = b + a$	$ab = ba$
Associative	$a + (b + c) = (a + b) + c$	$a(bc) = (ab)c$
Identity	$a + 0 = a$	$a \cdot 1 = a$
Inverse	$a + (-a) = 0$	$a\left(\frac{1}{a}\right) = 1$
Distributive	$a(b + c) = ab + ac$	

Addition [1.4]

9. $5 + 3 = 8$
$5 + (-3) = 2$
$-5 + 3 = -2$
$-5 + (-3) = -8$

To add two real numbers with

1. *The same sign:* Simply add absolute values and use the common sign.
2. *Different signs:* Subtract the smaller absolute value from the larger absolute value. The answer has the same sign as the number with the larger absolute value.

Subtraction [1.4]

10. $6 - 2 = 6 + (-2) = 4$
$6 - (-2) = 6 + 2 = 8$

If a and b are real numbers,

$$a - b = a + (-b)$$

To subtract b, add the opposite of b.

Multiplication [1.4]

11. $5(4) = 20$
$5(-4) = -20$
$-5(4) = -20$
$-5(-4) = 20$

To multiply two real numbers, simply multiply their absolute values. Like signs give a positive answer. Unlike signs give a negative answer.

Division [1.4]

12. $\frac{12}{-3} = -4$
$\frac{-12}{-3} = 4$

If a and b are real numbers and $b \neq 0$, then

$$\frac{a}{b} = a \cdot \left(\frac{1}{b}\right)$$

To divide by b, multiply by the reciprocal of b.

Inductive Reasoning [1.5]

13. We use inductive reasoning when we conclude that the next number in the sequence below is 25.

1, 4, 9, 16, . . .

Inductive reasoning is reasoning in which a conclusion is drawn based on evidence and observations that support that conclusion. In mathematics this usually involves noticing that a few items in a group have a trait or characteristic in common and then concluding that all items in the group have that same trait.

Arithmetic Sequence [1.5]

14. The following sequence is an arithmetic sequence because each term is obtained from the preceding term by adding 3 each time.

4, 7, 10, 13, . . .

An *arithmetic sequence* is a sequence of numbers in which each number (after the first number) comes from adding the same amount to the number before it. The number we add to each term to obtain the next term is called the *common difference.*

Geometric Sequence [1.5]

15. The following sequence is a geometric sequence because each term is obtained from the previous term by multiplying by 3 each time.

4, 12, 36, 108, . . .

A *geometric sequence* is a sequence of numbers in which each number (after the first number) comes from the number before it by multiplying by the same amount each time. The amount by which we multiply each term to obtain the next term is called the *common ratio.*

COMMON MISTAKES

1. Interpreting absolute value as changing the sign of the number inside the absolute value symbols; that is, $|-5| = +5$, $|+5| = -5$. To avoid this mistake, remember absolute value is defined as a distance and distance is always measured in positive units.
2. Confusing $-(-5)$ with $-|-5|$. The first answer is $+5$, but the second answer is -5.

Chapter 1 Review Test

The problems below form a comprehensive review of the material in this chapter. They can be used to study for exams. If you would like to take a practice test on this chapter, you can use the odd-numbered problems. Give yourself an hour and work as many of the odd-numbered problems as possible. When you are finished, or when an hour has passed, check your answers with the answers in the back of the book. You can use the even-numbered problems for a second practice test.

The numbers in brackets refer to the section(s) in the text where similar problems can be found.

Translate each expression into symbols. [1.1]

1. The sum of x and 2.

2. The difference of x and 2.

3. The quotient of x and 2.

4. The product of 2 and x.

5. Twice the sum of x and y.

6. The sum of twice x and y.

Expand and multiply. [1.1]

7. 3^3 **8.** 5^3

9. 8^2 **10.** 1^8

11. 2^5 **12.** 3^4

Simplify each expression. [1.1]

13. $2 + 3 \cdot 5$ **14.** $10 - 2 \cdot 3$

15. $20 \div 2 + 3$ **16.** $30 \div 6 + 4 \div 2$

17. $3 + 2(5 - 2)$ **18.** $(10 - 2)(7 - 3)$

19. $3 \cdot 4^2 - 2 \cdot 3^2$ **20.** $3 + 5(2 \cdot 3^2 - 10)$

Let $A = \{1, 3, 5\}$, $B = \{2, 4, 6\}$, and $C = \{0, 1, 2, 3, 4\}$, and find each of the following. [1.1]

21. $A \cup B$ **22.** $A \cap C$

23. $\{x \mid x \in A \text{ and } x \notin C\}$ **24.** $\{x \mid x \in B \text{ and } x > 4\}$

25. Locate the numbers $-4, -2.5, -1, 0, 1.5, 3.1, 4.75$ on the number line. [1.2]

26. Locate the numbers $-4.75, -3, -1, 0, 1.5, 2.3, 3$ on the number line. [1.2]

Give the opposite and reciprocal of each number. [1.3]

27. 2 **28.** $-\frac{2}{5}$

Simplify. [1.3]

29. $|-3|$ **30.** $-|-5|$

31. $|-4|$ **32.** $|10 - 16|$

For the set $\{-7, -4.2, -\sqrt{3}, 0, \frac{3}{4}, \pi, 5\}$ list all the elements that are in the following sets: [1.2]

33. Integers

34. Rational numbers

35. Irrational numbers

36. Factor 4,356 into the product of prime factors. [1.2]

37. Reduce $\frac{4{,}356}{5{,}148}$ to lowest terms. [1.2]

Multiply. [1.3]

38. $\frac{3}{4} \cdot \frac{8}{5} \cdot \frac{5}{6}$ **39.** $\left(\frac{3}{4}\right)^3$

40. $\frac{1}{4} \cdot 8$

Graph each inequality. [1.2]

41. $\{x \mid x < -2 \text{ or } x > 3\}$ **42.** $\{x \mid x > 2 \text{ and } x < 5\}$

43. $\{x \mid -3 \le x \le 4\}$

44. $\{x \mid 0 \le x \le 5 \text{ or } x > 10\}$

Translate each statement into an equivalent inequality. [1.2]

45. x is at least 4

46. x is no more than 5

47. x is between 0 and 8

48. x is between 0 and 8, inclusive

Combine similar terms. [1.4]

49. $-2y + 4y$

50. $-3x - x + 7x$

51. $3x - 2 + 5x + 7$

52. $2y + 4 - y - 2$

Match each expression with the letter of the appropriate property (or properties) below. [1.3]

53. $x + 3 = 3 + x$

54. $(x + 2) + 3 = x + (2 + 3)$

55. $3(x + 4) = 3(4 + x)$

56. $(5x)y = x(5y)$

57. $(x + 2) + y = (x + y) + 2$

58. $3(1) = 3$

59. $5 + 0 = 5$

60. $5 + (-5) = 0$

a. Commutative property of addition

b. Commutative property of multiplication

c. Associative property of addition

d. Associative property of multiplication

e. Additive identity

f. Multiplicative identity

g. Additive inverse

h. Multiplicative inverse

Find the following sums and differences. [1.4]

61. $5 - 3$

62. $-5 - (-3)$

63. $7 + (-2) - 4$

64. $6 - (-3) + 8$

65. $|-4| - |-3| + |-2|$

66. $|7 - 9| - |-3 - 5|$

67. $6 - (-3) - 2 - 5$

68. $2 \cdot 3^2 - 4 \cdot 2^3 + 5 \cdot 4^2$

69. $-\frac{1}{12} - \frac{1}{6} - \frac{1}{4} - \frac{1}{3}$

70. $-\frac{1}{3} - \frac{1}{4} - \frac{1}{6} - \frac{1}{12}$

Find the following products. [1.4]

71. $6(-7)$

72. $-3(5)(-2)$

73. $7(3x)$

74. $-3(2x)$

Apply the distributive property. [1.4]

75. $-2(3x - 5)$

76. $-3(2x - 7)$

77. $-\frac{1}{2}(2x - 6)$

78. $-3(5x - 1)$

Divide. [1.4]

79. $-\frac{5}{8} \div \frac{3}{4}$

80. $-12 \div \frac{1}{3}$

81. $\frac{3}{5} \div 6$

82. $\frac{4}{7} \div (-2)$

Simplify each expression as much as possible. [1.4]

83. $2(-5) - 3$

84. $3(-4) - 5$

85. $6 + 3(-2)$

86. $7 + 2(-4)$

87. $-3(2) - 5(6)$

88. $-4(3)^2 - 2(-1)^3$

89. $8 - 2(6 - 10)$

90. $(8 - 2)(6 - 10)$

91. $\dfrac{3(-4) - 8}{-5 - 5}$

92. $\dfrac{9(-1)^3 - 3(-6)^2}{6 - 9}$

93. $4 - (-2)\left[\dfrac{6 - 3(-4)}{1 + 5(-2)}\right]$

Simplify. [1.3, 1.4]

94. $7 - 2(3y - 1) + 4y$

95. $4(3x - 1) - 5(6x + 2)$

96. $4(2a - 5) - (3a + 2)$

Find the next number in each sequence. Identify any sequences that are arithmetic and any that are geometric. [1.5]

97. $11, 8, 5, 2, \ldots$

98. $1, 1, 2, 3, 5, \ldots$

99. $1, \frac{1}{2}, 0, -\frac{1}{2}, \ldots$

100. $1, -\frac{1}{2}, \frac{1}{4}, -\frac{1}{8}, \ldots$

Chapter 1 Projects

Basic Properties and Definitions

GROUP PROJECT Gauss's Method for Adding Consecutive Integers

Students and Instructors: The end of each chapter in this book will contain a section like this one containing two projects. The group projects are intended to be done in class. The research projects are to be completed outside of class; they can be done in groups or individually. In my classes, I use the research projects for extra credit. I require all research projects to be done on a word processor and to be free of spelling errors.

Bettmann/Corbis

Karl Gauss, 1777–1855

Number of People 3

Time Needed 15–20 minutes

Equipment Paper and pencil

Background There is a popular story about famous mathematician Karl Friedrich Gauss (1777–1855). As the story goes, a 9-year-old Gauss and the rest of his class were given what the teacher had hoped was busy work. They were told to find the sum of all the whole numbers from 1 to 100. While the rest of the class labored under the assignment, Gauss found the answer within a few moments. He may have set up the problem like this:

1	2	3	4	...	98	99	100
100	99	98	97	...	3	2	1

Procedure Copy the preceding illustration to get started.

1. Add the numbers in the columns and write your answers below the line.
2. How many numbers do you have below the line?
3. Suppose you add all the numbers below the line. How will your result be related to the sum of all the whole numbers from 1 to 100?
4. Use multiplication to add all the numbers below the line.
5. What is the sum of all the whole numbers from 1 to 100?
6. Use Gauss's method to add all the whole numbers from 1 to 10. (The answer is 55.)
7. Explain, in your own words, Gauss's method for finding this sum so quickly.

Sofia Kovalevskaya

Sofia Kovalevskaya (1850–1891) was born in Moscow, Russia, around the time that Gauss died. Both Gauss and Kovalevskaya showed talent in mathematics early in their lives, but unlike Gauss, Kovalevskaya had to struggle to receive an education and have her work recognized. The Russian people honored her with the commemorative postage stamp shown here. Research the life of Sofia Kovalevskaya, noting the obstacles she faced in pursuing her dream of studying and producing research in mathematics. Then summarize your results in an essay.

Equations and Inequalities in One Variable

2

Royalty-Free/Corbis

CHAPTER OUTLINE

A recent newspaper article gave the following guideline for college students taking out loans to finance their education: The maximum monthly payment on the amount borrowed should not exceed 8% of their monthly starting salary. In this situation, the maximum monthly payment can be described mathematically with the formula

$$y = 0.08x$$

Using this formula, we can construct a table and a line graph that show the maximum student loan payment that can be made for a variety of starting salaries.

Maximum Student Loan Payments

Monthly Starting Salary	Maximum Loan Payment
$2,000	$160
$2,500	$200
$3,000	$240
$3,500	$280
$4,000	$320
$4,500	$360
$5,000	$400

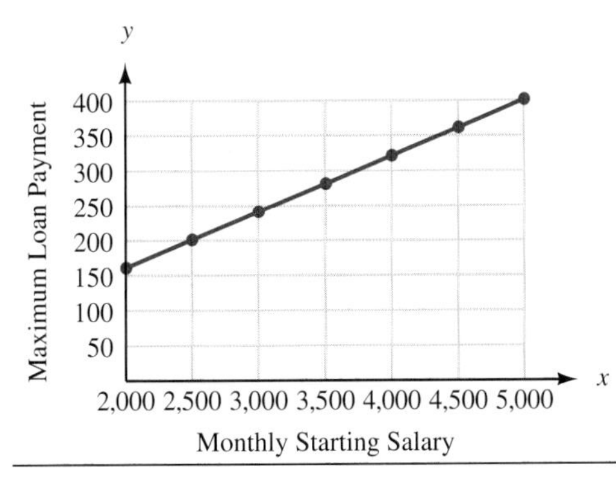

FIGURE 1

In this chapter, we begin our work with connecting tables and line graphs with algebraic formulas.

If you have successfully completed Chapter 1, then you have made a good start at developing the study skills necessary to succeed in all math classes. Here is the list of study skills for this chapter.

1 Imitate Success

Your work should look like the work you see in this book and the work your instructor shows. The person who wrote the steps shown in solving problems in this book has been successful in mathematics. The same is true of your instructor. Your work should imitate the work of people who have been successful in mathematics.

2 List Difficult Problems

Begin to make lists of problems that give you the most difficulty. These are the problems in which you are repeatedly making mistakes.

3 Begin to Develop Confidence with Word Problems

It seems that the main difference between people who are good at working word problems and those who are not is confidence. People with confidence know that no matter how long it takes them, they will eventually be able to solve the problem. Those without confidence begin by saying to themselves, "I'll never be able to work this problem." Are you like that? If you are, what you need to do is put your old ideas about you and word problems aside for a while and make a decision to be successful. Sometimes that's all it takes. Instead of telling yourself that you can't do word problems, that you don't like them, or that they're not good for anything anyway, decide to do whatever it takes to master them.

2.1 Linear Equations in One Variable

OBJECTIVES

A Solve a linear equation in one variable.

A *linear equation in one variable* is any equation that can be put in the form

$$ax + b = c$$

where a, b, and c are constants and $a \neq 0$. For example, each of the equations

$$5x + 3 = 2 \qquad 2x = 7 \qquad 2x + 5 = 0$$

are linear because they can be put in the form $ax + b = c$. In the first equation, $5x$, 3, and 2 are called *terms* of the equation: $5x$ is a variable term; 3 and 2 are constant terms.

> **DEFINITION** The **solution set** for an equation is the set of all numbers that, when used in place of the variable, make the equation a true statement.

EXAMPLE 1 The solution set for $2x - 3 = 9$ is $\{6\}$ since replacing x with 6 makes the equation a true statement.

If	$x = 6$	
then	$2x - 3 = 9$	
becomes	$2(6) - 3 = 9$	
	$12 - 3 = 9$	
	$9 = 9$	**A true statement**

> **DEFINITION** Two or more equations with the same solution set are called **equivalent equations.**

EXAMPLE 2 The equations $2x - 5 = 9$, $x - 1 = 6$, and $x = 7$ are all equivalent equations because the solution set for each is $\{7\}$.

Properties of Equality

The first property states that adding the same quantity to both sides of an equation preserves equality. Or, more importantly, adding the same amount to both sides of an equation *never changes* the solution set. This property is called the *addition property of equality* and is stated in symbols as follows.

> **Addition Property of Equality** For any three algebraic expressions, A, B, and C,
>
> if $\quad A = B$
>
> then $\quad A + C = B + C$
>
> *In words:* Adding the same quantity to both sides of an equation will not change the solution set.

> **Note**
> Because subtraction is defined in terms of addition and division is defined in terms of multiplication, we do not need to introduce separate properties for subtraction and division. The solution set for an equation will never be changed by subtracting the same amount from both sides or by dividing both sides by the same nonzero quantity.

Our second new property is called the *multiplication property of equality* and is stated as follows.

> **Multiplication Property of Equality** For any three algebraic expressions A, B, and C, where $C \neq 0$,
>
> if $\quad A = B$
>
> then $\quad AC = BC$
>
> *In words:* Multiplying both sides of an equation by the same nonzero quantity will not change the solution set.

EXAMPLE 3

Solve $\frac{3}{4}x + 5 = -4$.

SOLUTION We begin by adding -5 to both sides of the equation. Once this has been done, we multiply both sides by the reciprocal of $\frac{3}{4}$, which is $\frac{4}{3}$.

$$\frac{3}{4}x + 5 = -4$$

$$\frac{3}{4}x + 5 + (\mathbf{-5}) = -4 + (\mathbf{-5}) \qquad \textbf{Add } -5 \textbf{ to both sides}$$

$$\frac{3}{4}x = -9$$

$$\mathbf{\frac{4}{3}}\left(\frac{3}{4}x\right) = \mathbf{\frac{4}{3}}(-9) \qquad \textbf{Multiply both sides by } \tfrac{4}{3}$$

$$x = -12 \qquad \tfrac{4}{3}(-9) = \tfrac{4}{3}\left(-\tfrac{9}{1}\right) = -\tfrac{36}{3} = -12$$

EXAMPLE 4

Find the solution set for $3a - 5 = -6a + 1$.

SOLUTION To solve for a we must isolate it on one side of the equation. Let's decide to isolate a on the left side by adding $6a$ to both sides of the equation.

$$3a - 5 = -6a + 1$$

$$3a + \mathbf{6a} - 5 = -6a + \mathbf{6a} + 1 \qquad \textbf{Add } 6a \textbf{ to both sides}$$

$$9a - 5 = 1$$

$$9a - 5 + \mathbf{5} = 1 + \mathbf{5} \qquad \textbf{Add 5 to both sides}$$

$$9a = 6$$

$$\mathbf{\frac{1}{9}}(9a) = \mathbf{\frac{1}{9}}(6) \qquad \textbf{Multiply both sides by } \tfrac{1}{9}$$

$$a = \frac{2}{3} \qquad \tfrac{1}{9}(6) = \tfrac{6}{9} = \tfrac{2}{3}$$

The solution set is $\left\{\frac{2}{3}\right\}$.

> **Note**
> From the previous chapter we know that multiplication by a number and division by its reciprocal always produce the same result. Because of this fact, instead of multiplying each side of our equation by $\frac{1}{9}$, we could just as easily divide each side by 9. If we did so, the last two lines in our solution would look like this:
>
> $$\frac{9a}{9} = \frac{6}{9}$$
>
> $$a = \frac{2}{3}$$

We can check our solution in Example 4 by replacing a in the original equation with $\frac{2}{3}$.

When $\quad a = \frac{2}{3}$

the equation $\quad 3a - 5 = -6a + 1$

becomes $\quad 3\left(\frac{2}{3}\right) - 5 = -6\left(\frac{2}{3}\right) + 1$

$$2 - 5 = -4 + 1$$

$$-3 = -3 \quad \textbf{A true statement}$$

There will be times when we solve equations and end up with a negative sign in front of the variable. The next example shows how to handle this situation.

EXAMPLE 5

Solve each equation.

a. $-x = 4$ **b.** $-y = -8$

SOLUTION Neither equation can be considered solved because of the negative sign in front of the variable. To eliminate the negative signs we simply multiply both sides of each equation by -1.

a. $-x = 4$ **b.** $-y = -8$

$\mathbf{-1}(-x) = \mathbf{-1}(4) \quad \mathbf{-1}(-y) = \mathbf{-1}(-8) \quad$ **Multiply each side by −1**

$x = -4 \quad y = 8$

EXAMPLE 6

Solve $\frac{2}{3}x + \frac{1}{2} = -\frac{3}{8}$.

SOLUTION We can solve this equation by applying our properties and working with fractions, or we can begin by eliminating the fractions. Let's use both methods.

Method 1 Working with the fractions.

$$\frac{2}{3}x + \frac{1}{2} + \left(-\frac{\mathbf{1}}{\mathbf{2}}\right) = -\frac{3}{8} + \left(-\frac{\mathbf{1}}{\mathbf{2}}\right) \quad \textbf{Add } -\tfrac{1}{2} \textbf{ to each side}$$

$$\frac{2}{3}x = -\frac{7}{8} \quad -\tfrac{3}{8} + \left(-\tfrac{1}{2}\right) = -\tfrac{3}{8} + \left(-\tfrac{4}{8}\right)$$

$$\frac{\mathbf{3}}{\mathbf{2}}\left(\frac{2}{3}x\right) = \frac{\mathbf{3}}{\mathbf{2}}\left(-\frac{7}{8}\right) \quad \textbf{Multiply each side by } \tfrac{3}{2}$$

$$x = -\frac{21}{16}$$

Method 2 Eliminating the fractions in the beginning.

Our original equation has denominators of 3, 2, and 8. The least common denominator, abbreviated LCD, for these three denominators is 24, and it has the property that all three denominators will divide it evenly. If we multiply both sides of our equation by 24, each denominator will divide into 24, and we will be left with an equation that does not contain any denominators other than 1.

$$\mathbf{24}\left(\frac{2}{3}x + \frac{1}{2}\right) = \mathbf{24}\left(-\frac{3}{8}\right) \quad \text{Multiply each side by the LCD 24}$$

$$24\left(\frac{2}{3}x\right) + 24\left(\frac{1}{2}\right) = 24\left(-\frac{3}{8}\right) \quad \text{Distributive property on the left side}$$

$$16x + 12 = -9 \quad \text{Multiply}$$

$$16x = -21 \quad \text{Add } -12 \text{ to each side}$$

$$x = -\frac{21}{16} \quad \text{Multiply each side by } \frac{1}{16}$$

Check To check our solution, we substitute $x = -\frac{21}{16}$ back into our original equation to obtain

Note
We are placing a question mark over the equal sign because we don't know yet if the expression on the left will be equal to the expression on the right.

$$\frac{2}{3}\left(-\frac{21}{16}\right) + \frac{1}{2} \stackrel{?}{=} -\frac{3}{8}$$

$$-\frac{7}{8} + \frac{1}{2} \stackrel{?}{=} -\frac{3}{8}$$

$$-\frac{7}{8} + \frac{4}{8} \stackrel{?}{=} -\frac{3}{8}$$

$$-\frac{3}{8} = -\frac{3}{8} \quad \text{A true statement}$$

EXAMPLE 7

Solve the equation $0.06x + 0.05(10{,}000 - x) = 560$.

SOLUTION We can solve the equation in its original form by working with the decimals, or we can eliminate the decimals first by using the multiplication property of equality and solve the resulting equation. Here are both methods.

Method 1 Working with the decimals.

$$0.06x + 0.05(10{,}000 - x) = 560 \quad \text{Original equation}$$

$$0.06x + 0.05(10{,}000) - 0.05x = 560 \quad \text{Distributive property}$$

$$0.01x + 500 = 560 \quad \text{Simplify the left side}$$

$$0.01x + 500 + (\mathbf{-500}) = 560 + (\mathbf{-500}) \quad \text{Add } -500 \text{ to each side}$$

$$0.01x = 60$$

$$\frac{0.01x}{\mathbf{0.01}} = \frac{60}{\mathbf{0.01}} \quad \text{Divide each side by 0.01}$$

$$x = 6{,}000$$

Method 2 Eliminating the decimals in the beginning: To move the decimal point two places to the right in $0.06x$ and 0.05, we multiply each side of the equation by 100.

$$0.06x + 0.05(10{,}000 - x) = 560 \quad \text{Original equation}$$

$$0.06x + 500 - 0.05x = 560 \quad \text{Distributive property}$$

$$\mathbf{100}(0.06x) + \mathbf{100}(500) - \mathbf{100}(0.05x) = \mathbf{100}(560) \quad \text{Multiply each side by 100}$$

$$6x + 50{,}000 - 5x = 56{,}000$$

$$x + 50{,}000 = 56{,}000 \qquad \textbf{Simplify the left side}$$

$$x = 6{,}000 \qquad \textbf{Add } -50{,}000 \textbf{ to each side}$$

Using either method, the solution to our equation is 6,000.

Check We check our work (to be sure we have not made a mistake in applying the properties or in arithmetic) by substituting 6,000 into our original equation and simplifying each side of the result separately, as the following shows.

$$0.06(\mathbf{6{,}000}) + 0.05(10{,}000 - \mathbf{6{,}000}) \stackrel{?}{=} 560$$

$$0.06(6{,}000) + 0.05(4{,}000) \stackrel{?}{=} 560$$

$$360 + 200 \stackrel{?}{=} 560$$

$$560 = 560 \qquad \textbf{A true statement}$$

Here is a list of steps to use as a guideline for solving linear equations in one variable.

Strategy for Solving Linear Equations in One Variable

Step 1: **a.** Use the distributive property to separate terms, if necessary.

b. If fractions are present, consider multiplying both sides by the LCD to eliminate the fractions. If decimals are present, consider multiplying both sides by a power of 10 to clear the equation of decimals.

c. Combine similar terms on each side of the equation.

Step 2: Use the addition property of equality to get all variable terms on one side of the equation and all constant terms on the other side. A **variable term** is a term that contains the variable. A **constant term** is a term that does not contain the variable (the number 3, for example).

Step 3: Use the multiplication property of equality to get the variable by itself on one side of the equation.

Step 4: Check your solution in the original equation to be sure that you have not made a mistake in the solution process.

As you will see as you work through the problems in the problem set, it is not always necessary to use all four steps when solving equations. The number of steps used depends on the equation. In Example 8 there are no fractions or decimals in the original equation, so Step 1b will not be used.

EXAMPLE 8 Solve $3(2y - 1) + y = 5y + 3$.

SOLUTION Applying the steps outlined in the preceding strategy, we have

Step 1: **a.** $3(2y - 1) + y = 5y + 3$ **Original equation**

$6y - 3 + y = 5y + 3$ **Distributive property**

c. $7y - 3 = 5y + 3$ **Simplify**

Step 2: $7y + (\mathbf{-5y}) - 3 = 5y + (\mathbf{-5y}) + 3$ **Add $-5y$ to both sides**

$$2y - 3 = 3$$

$$2y - 3 + \mathbf{3} = 3 + \mathbf{3}$$ **Add +3 to both sides**

$$2y = 6$$

Step 3: $\frac{\mathbf{1}}{\mathbf{2}}(2y) = \frac{\mathbf{1}}{\mathbf{2}}(6)$ **Multiply by $\frac{1}{2}$**

$$y = 3$$

Step 4: *Check* When $y = 3$

the equation $3(2y - 1) + y = 5y + 3$

becomes $3(2 \cdot \mathbf{3} - 1) + \mathbf{3} \stackrel{?}{=} 5 \cdot \mathbf{3} + 3$

$$3(5) + 3 \stackrel{?}{=} 15 + 3$$

$$18 = 18$$ **A true statement**

EXAMPLE 9

Solve the equation $8 - 3(4x - 2) + 5x = 35$.

SOLUTION We must begin by distributing the -3 across the quantity $4x - 2$.

> **Note**
> It would be a mistake to subtract 3 from 8 first because the rule for order of operations indicates we are to do multiplication before subtraction.

Step 1: **a.** $8 - 3(4x - 2) + 5x = 35$ **Original equation**

b. $8 - 12x + 6 + 5x = 35$ **Distributive property**

c. $-7x + 14 = 35$ **Simplify**

Step 2: $-7x = 21$ **Add -14 to each side**

Step 3: $x = -3$ **Multiply by $-\frac{1}{7}$**

Step 4: When x is replaced by -3 in the original equation, a true statement results. Therefore, -3 is the solution to our equation.

Identities and Equations with No Solution

Two special cases are associated with solving linear equations in one variable, each of which is illustrated in the following examples.

EXAMPLE 10

Solve for x: $2(3x - 4) = 3 + 6x$.

SOLUTION Applying the distributive property to the left side gives us

$6x - 8 = 3 + 6x$ **Distributive property**

Now, if we add $-6x$ to each side, we are left with the following

$$-8 = 3$$

which is a false statement. This means that there is no solution to our equation. Any number we substitute for x in the original equation will lead to a similar false statement.

EXAMPLE 11 Solve for x: $-15 + 3x = 3(x - 5)$.

SOLUTION We start by applying the distributive property to the right side.

$$-15 + 3x = 3x - 15 \quad \text{Distributive property}$$

If we add $-3x$ to each side, we are left with the true statement

$$-15 = -15$$

In this case, our result tells us that any number we use in place of x in the original equation will lead to a true statement. Therefore, all real numbers are solutions to our equation. We say the original equation is an *identity* because the left side is always identically equal to the right side.

LINKING OBJECTIVES AND EXAMPLES

Next to each **objective** we have listed the examples that are best described by that objective.

A 3–11

GETTING READY FOR CLASS

After reading through the preceding section, respond in your own words and in complete sentences.

1. What is a solution to an equation?
2. What are equivalent equations?
3. Describe how to eliminate fractions in an equation.
4. Suppose when solving an equation your result is the statement "$3 = -3$." What would you conclude about the solution to the equation?

Problem Set 2.1

Online support materials can be found at www.thomsonedu.com/login

Solve each of the following equations.

1. $x - 5 = 3$
2. $x + 2 = 7$
3. $2x - 4 = 6$
4. $3x - 5 = 4$
5. $7 = 4a - 1$
6. $10 = 3a - 5$
7. $3 - y = 10$
8. $5 - 2y = 11$
9. $-3 - 4x = 15$
10. $-8 - 5x = -6$
11. $-3 = 5 + 2x$
12. $-12 = 6 + 9x$
13. $-300y + 100 = 500$
14. $-20y + 80 = 30$
15. $160 = -50x - 40$
16. $110 = -60x - 50$
17. $-x = 2$
18. $-x = \frac{1}{2}$
19. $-a = -\frac{3}{4}$
20. $-a = -5$
21. $\frac{2}{3}x = 8$
22. $\frac{3}{2}x = 9$
23. $-\frac{3}{5}a + 2 = 8$
24. $-\frac{5}{3}a + 3 = 23$
25. $8 = 6 + \frac{2}{7}y$
26. $1 = 4 + \frac{3}{7}y$
27. $2x - 5 = 3x + 2$
28. $5x - 1 = 4x + 3$
29. $-3a + 2 = -2a - 1$
30. $-4a - 8 = -3a + 7$
31. $5 - 2x = 3x + 1$
32. $7 - 3x = 8x - 4$
33. $11x - 5 + 4x - 2 = 8x$
34. $2x + 7 - 3x + 4 = -2x$
35. $6 - 7(m - 3) = -1$
36. $3 - 5(2m - 5) = -2$
37. $7 + 3(x + 2) = 4(x - 1)$
38. $5 + 2(4x - 4) = 3(2x - 1)$

= Videos available by instructor request
▶ = Online student support materials available at www.thomsonedu.com/login

39. $5 = 7 - 2(3x - 1) + 4x$

40. $20 = 8 - 5(2x - 3) + 4x$

▶ **41.** $\frac{1}{2}x + \frac{1}{4} = \frac{1}{3}x + \frac{5}{4}$

42. $\frac{2}{3}x - \frac{3}{4} = \frac{1}{6}x + \frac{21}{4}$

43. $-\frac{2}{5}x + \frac{2}{15} = \frac{2}{3}$

44. $-\frac{1}{6}x + \frac{2}{3} = \frac{1}{4}$

45. $\frac{3}{4}(8x - 4) = \frac{2}{3}(6x - 9)$

46. $\frac{3}{5}(5x + 10) = \frac{5}{6}(12x - 18)$

47. $\frac{1}{4}(12a + 1) - \frac{1}{4} = 5$

48. $\frac{2}{3}(6x - 1) + \frac{2}{3} = 4$

▶ **49.** $0.35x - 0.2 = 0.15x + 0.1$

50. $0.25x - 0.05 = 0.2x + 0.15$

51. $0.42 - 0.18x = 0.48x - 0.24$

52. $0.3 - 0.12x = 0.18x + 0.06$

53. $3x - 6 = 3(x + 4)$

54. $7x - 14 = 7(x - 2)$

55. $4y + 2 - 3y + 5 = 3 + y + 4$

56. $7y + 5 - 2y - 3 = 6 + 5y - 4$

57. $2(4t - 1) + 3 = 5t + 4 + 3t$

58. $5(2t - 1) + 1 = 2t - 4 + 8t$

Now that you have practiced solving a variety of equations, we can turn our attention to the types of equations you will see as you progress through the book. Each equation appears later in the book exactly as you see it below.

Solve each equation.

59. $3x + 2 = 0$

60. $5x - 4 = 0$

61. $0 = 6{,}400a + 70$

62. $0 = 6{,}400a + 60$

63. $x + 2 = 2x$

64. $x + 2 = 7x$

65. $0.07x = 1.4$

66. $0.02x = 0.3$

67. $5(2x + 1) = 12$

68. $4(3x - 2) = 21$

69. $50 = \frac{K}{48}$

70. $50 = \frac{K}{24}$

71. $100P = 2{,}400$

72. $3.5d = 16(3.5)^2$

73. $x + (3x + 2) = 26$

74. $2(1) + y = 4$

75. $2x - 3(3x - 5) = -6$

76. $2(2y + 6) + 3y = 5$

▶ **77.** $3x + (x - 2) \cdot 2 = 6$

▶ **78.** $2x - (x + 1) = -2$

▶ **79.** $15 - 3(x - 1) = x - 2$

▶ **80.** $4x - 4(x - 3) = x + 3$

▶ **81.** $2(2x - 3) + 2x = 45$

▶ **82.** $2(4x - 10) + 2x = 12.5$

▶ **83.** $2(x + 3) + x = 4(x - 3)$

▶ **84.** $5(y + 2) - 4(y + 1) = 3$

▶ **85.** $6(y - 3) - 5(y + 2) = 8$

▶ **86.** $2(x + 3) + 3(x + 5) = 2x$

▶ **87.** $2(20 + x) = 3(20 - x)$

▶ **88.** $6(7 + x) = 5(9 + x)$

▶ **89.** $2x + 1.5(75 - x) = 127.5$

▶ **90.** $x + 0.06x = 954$

▶ **91.** $0.08x + 0.09(9{,}000 - x) = 750$

▶ **92.** $0.08x + 0.09(9{,}000 - x) = 500$

▶ **93.** $0.12x + 0.10(15{,}000 - x) = 1{,}600$

▶ **94.** $0.09x + 0.11(11{,}000 - x) = 1{,}150$

▶ **95.** $5\left(\frac{19}{15}\right) + 5y = 9$

▶ **96.** $4\left(\frac{19}{15}\right) - 2y = 4$

97. $2\left(\frac{29}{22}\right) - 3y = 4$

98. $2x - 3\left(-\frac{5}{11}\right) = 4$

The next two problems are intended to give you practice reading, and paying attention to, the instructions that accompany the problems you are working. Working these problems is an excellent way to get ready for a test or a quiz.

99. Work each problem according to the instructions given.

a. Solve: $8x - 5 = 0$

b. Solve: $8x - 5 = -5$

c. Add: $(8x - 5) + (2x - 5)$

d. Solve: $8x - 5 = 2x - 5$

e. Multiply: $8(x - 5)$

f. Solve: $8(x - 5) = 2(x - 5)$

100. Work each problem according to the instructions given.

a. Solve: $3x + 6 = 0$

b. Solve: $3x + 6 = 4$

c. Add: $(3x + 6) + (7x + 4)$

d. Solve: $3x + 6 = 7x + 4$

e. Multiply: $3(x + 6)$

f. Solve: $3(x + 6) = 7x + 4$

Applying the Concepts

101. Cost of a Taxi Ride The taximeter was invented in 1891 by Wilhelm Bruhn. The city of Chicago charges \$1.80 plus \$0.40 per mile for a taxi ride.

Jan Halaska/Index Stock Imagery/PictureQuest

a. A woman paid a fare of \$6.60. Write an equation that connects the fare the woman paid, the miles she traveled, n, and the charges the taximeter computes.

b. Solve the equation from part (a) to determine how many miles the woman traveled.

102. Coughs and Earaches In 1992, twice as many people visited their doctor because of a cough than an earache. The total number of doctor's visits for these two ailments was reported to be 45 million.

a. Let x represent the number of earaches reported in 1992, then write an expression using x for the number of coughs reported in 1992.

b. Write an equation that relates 45 million to the variable x.

c. Solve the equation from part (b) to determine the number of people who visited their doctor in 1992 to report an earache.

103. Population Density In July 2001 the population of Puerto Rico was estimated to be 3,937,000 people, with a population density of 1,125 people per square mile.

a. Let A represent the area of Puerto Rico in square miles, and write an equation that shows that the population is equal to the product of the area and the population density.

b. Solve the equation from Part (a), rounding your solution to the nearest square mile.

104. Solving Equations by Trial and Error Sometimes equations can be solved most easily by trial and error. Solve the following equations by trial and error.

a. Find x and y if $x \cdot y + 1 = 36$, and both x and y are prime.

b. Find w, t, and z if $w + t + z + 10 = 52$, and w, t, and z are consecutive terms of a Fibonacci sequence.

c. Find x and y if $x \neq y$ and $x^y = y^x$.

Maintaining Your Skills

From this point on, each problem set will contain a number of problems under the heading *Maintaining Your Skills*. These problems cover the most important skills you have learned in previous sections and chapters. Hopefully, by working these problems on a regular basis, you will keep yourself current on all the topics we have covered, and possibly need less time to study for tests and quizzes.

Identify the property (or properties) that justifies each of the following statements.

105. $ax = xa$

106. $5\left(\frac{1}{5}\right) = 1$

107. $3 + (x + y) = (3 + x) + y$

108. $3 + (x + y) = (x + y) + 3$

109. $3 + (x + y) = (3 + y) + x$

110. $7(3x - 5) = 21x - 35$

111. $5(1) = 5$

112. $5 + 0 = 5$

113. $4(xy) = 4(yx)$

114. $4(xy) = (4y)x$

115. $2 + 0 = 2$

116. $2 + (-2) = 0$

Getting Ready for the Next Section

Problems under the heading *Getting Ready for the Next Section* are problems that you must be able to work in order to understand the material in the next section. In this case, the problems below are variations on the types of problems you have already worked in this problem set. They are exactly the types of problems you will see in explanations and examples in the next section.

Solve each equation.

117. $x \cdot 42 = 21$

118. $x \cdot 84 = 21$

119. $25 = 0.4x$

120. $35 = 0.4x$

121. $12 - 4y = 12$

122. $-6 - 3y = 6$

123. $525 = 900 - 300p$

124. $375 = 900 - 300p$

125. $486.7 = 78.5 + 31.4h$

126. $486.7 = 113.0 + 37.7h$

127. Find the value of $2x - 1$ when x is

a. 2 **b.** 3 **c.** 5

128. Find the value of $\frac{1}{x+1}$ when x is

a. 1 **b.** 2 **c.** 3

Extending the Concepts

Solve for x.

129. $\frac{x+4}{5} - \frac{x+3}{3} = -\frac{7}{15}$

130. $\frac{x+1}{7} - \frac{x-2}{2} = \frac{1}{14}$

131. $\frac{1}{x} - \frac{2}{3} = \frac{2}{x}$

132. $\frac{x+3}{2} - \frac{x-4}{4} = -\frac{1}{8}$

133. $\frac{x-3}{5} - \frac{x+1}{10} = -\frac{1}{10}$

134. $\frac{x-1}{2} - \frac{x+2}{3} = \frac{x+3}{6}$

135. $\frac{x+2}{4} - \frac{x-1}{3} = -\frac{x+2}{6}$

2.2 Formulas

OBJECTIVES

A Solve a formula with numerical replacements for all but one of its variables.

B Solve formulas for the indicated variable.

C Solve basic percent problems by translating them into equations.

A *formula* in mathematics is an equation that contains more than one variable. Some formulas are probably already familiar to you—for example, the formula for the area (A) of a rectangle with length l and width w is $A = lw$.

To begin our work with formulas, we will consider some examples in which we are given numerical replacements for all but one of the variables.

EXAMPLE 1 Find y when x is 4 in the formula $3x - 4y = 2$.

SOLUTION We substitute 4 for x in the formula and then solve for y:

When	$x = 4$	
the formula	$3x - 4y = 2$	
becomes	$3(4) - 4y = 2$	
	$12 - 4y = 2$	**Multiply 3 and 4**
	$-4y = -10$	**Add −12 to each side**
	$y = \frac{5}{2}$	**Divide each side by −4**

Note that in the last line of Example 1 we divided each side of the equation by −4. Remember that this is equivalent to multiplying each side of the equation by $-\frac{1}{4}$.

EXAMPLE 2 A store selling art supplies finds that they can sell x sketch pads each week at a price of p dollars each, according to the formula $x = 900 - 300p$. What price should they charge for each sketch pad if they want to sell 525 pads each week?

SOLUTION Here we are given a formula, $x = 900 - 300p$, and asked to find the value of p if x is 525. To do so, we simply substitute 525 for x and solve for p:

When $x = 525$

the formula $x = 900 - 300p$

becomes $525 = 900 - 300p$

$-375 = -300p$ **Add −900 to each side**

$1.25 = p$ **Divide each side by −300**

To sell 525 sketch pads, the store should charge $1.25 for each pad.

EXAMPLE 3 A boat is traveling upstream against a current. If the speed of the boat in still water is r and the speed of the current is c, then the formula for the distance traveled by the boat is $d = (r - c) \cdot t$, where t is the length of time. Find c if $d = 52$ miles, $r = 16$ miles per hour, and $t = 4$ hours.

SOLUTION Substituting 52 for d, 16 for r, and 4 for t into the formula, we have

$52 = (16 - c) \cdot 4$

$13 = 16 - c$ **Divide each side by 4**

$-3 = -c$ **Add −16 to each side**

$3 = c$ **Divide each side by −1**

The speed of the current is 3 miles per hour.

FACTS FROM GEOMETRY

Formulas for Area and Perimeter

To review, here are the formulas for the area and perimeter of some common geometric objects.

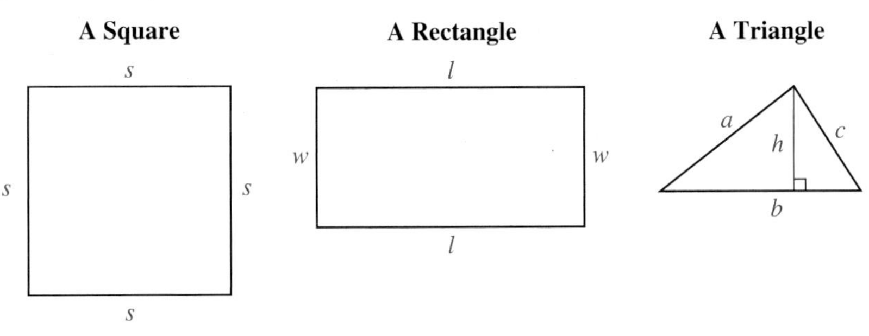

A Square	A Rectangle	A Triangle
Perimeter $= 4s$	Perimeter $= 2l + 2w$	Perimeter $= a + b + c$
Area $= s^2$	Area $= lw$	Area $= \frac{1}{2}bh$

The formula for perimeter gives us the distance around the outside of the object along its sides, whereas the formula for area gives us a measure of the amount of surface the object covers.

Note

The pink line labeled h in the triangle is its height, or altitude. It extends from the top of the triangle down to the base, meeting the base at an angle of 90°. The altitude of a triangle is always perpendicular to the base. The small square shown where the altitude meets the base is used to indicate that the angle formed is 90°.

EXAMPLE 4 Given the formula $P = 2w + 2l$, solve for w.

SOLUTION To solve for w, we must isolate it on one side of the equation. We can accomplish this if we delete the $2l$ term and the coefficient 2 from the right side of the equation.

To begin, we add $-2l$ to both sides:

$$P + (\mathbf{-2l}) = 2w + 2l + (\mathbf{-2l})$$

$$P - 2l = 2w$$

To delete the 2 from the right side, we can multiply both sides by $\frac{1}{2}$:

$$\frac{\mathbf{1}}{\mathbf{2}}(P - 2l) = \frac{\mathbf{1}}{\mathbf{2}}(2w)$$

$$\frac{P - 2l}{2} = w$$

The two formulas

$$P = 2w + 2l \quad \text{and} \quad w = \frac{P - 2l}{2}$$

give the relationship between P, l, and w. They look different, but they both say the same thing about P, l, and w. The first formula gives P in terms of l and w, and the second formula gives w in terms of P and l.

Note

We know we are finished solving a formula for a specified variable when that variable appears alone on one side of the equal sign and not on the other.

Rate Equation and Average Speed

Now we will look at a problem that uses what is called the *rate equation*. You use this equation on an intuitive level when you are estimating how long it will take you to drive long distances. For example, If you drive at 50 miles per hour for 2 hours, you will travel 100 miles. Here is the rate rquation.

$$\text{Distance} = \text{rate} \cdot \text{time, or d} = r \cdot t$$

The rate equation has two equivalent forms, one of which is obtained by solving for r, while the other is obtained by solving for t. Here they are:

$$r = \frac{d}{t} \text{ and } t = \frac{d}{r}$$

The rate in this equation is also referred to as *average speed*.

The *average speed* of a moving object is defined to be the ratio of distance to time. If you drive your car for 5 hours and travel a distance of 200 miles, then your average rate of speed is

$$\text{Average speed} = \frac{200 \text{ miles}}{5 \text{ hours}} = 40 \text{ miles per hour}$$

Our next example involves both the formula for the circumference of a circle and the rate equation. The formula for the circumference of a circle contains the number π. In this book, you can use either 3.14 or $\frac{22}{7}$ as an approximation for π. Generally speaking, if the problem contains decimals, use 3.14 to approximate π.

EXAMPLE 5

The first Ferris wheel was designed and built by George Ferris in 1893. The diameter of the wheel was 250 feet. It had 36 carriages, equally spaced around the wheel, each of which held a maximum of 40 people. One trip around the wheel took 20 minutes. Find the average speed of a rider on the first Ferris wheel. (Use 3.14 as an approximation for π.)

> *Note*
> The circumference of a circle with diameter d is
> $$C = \pi d$$
> Since the diameter is twice the radius ($d = 2r$), we can also write the formula as
> $$C = 2\pi r$$

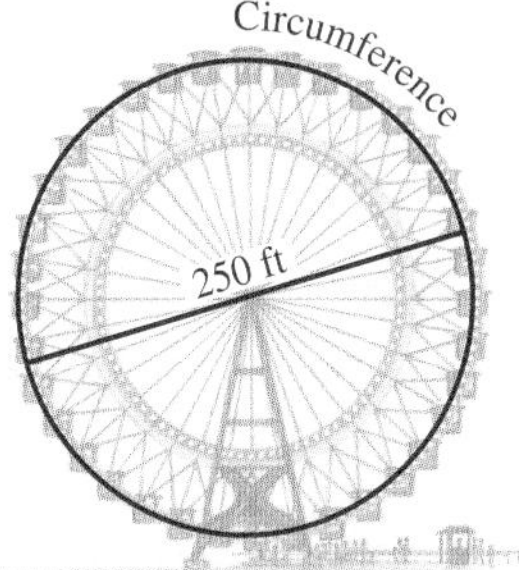

SOLUTION The distance traveled is the circumference of the wheel, which is

$$C = 250\pi = 250(3.14) = 785 \text{ feet}$$

To find the average speed, we divide the distance traveled by the amount of time it took to go once around the wheel.

$$r = \frac{d}{t} = \frac{785 \text{ feet}}{20 \text{ minutes}} = 39.3 \text{ feet per minute (to the nearest tenth)}$$

Later in the book, we will convert this speed into an equivalent speed in miles per hour.

EXAMPLE 6

Solve for x: $ax - 3 = bx + 5$.

SOLUTION In this example, we must begin by collecting all the variable terms on the left side of the equation and all the constant terms on the other side (just like we did when we were solving linear equations):

$$ax - 3 = bx + 5$$

$$ax - bx - 3 = 5 \qquad \textbf{Add } -bx \textbf{ to each side}$$

$$ax - bx = 8 \qquad \textbf{Add 3 to each side}$$

> *Note*
> We are applying the distributive property in the same way we applied it when we first learned how to simplify $7x - 4x$. Recall that $7x - 4x = 3x$ because
> $$7x - 4x = (7 - 4)x = 3x$$
> We are using the same type of reasoning when we write
> $$ax - bx = (a - b)x$$

At this point we need to apply the distributive property to write the left side as $(a - b)x$. After that, we divide each side by $a - b$:

$$(a - b)x = 8 \qquad \textbf{Distributive property}$$

$$x = \frac{8}{a - b} \qquad \textbf{Divide each side by } a - b$$

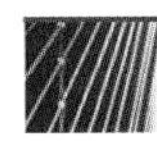

EXAMPLE 7

Solve for y: $\dfrac{y - b}{x - 0} = m$.

SOLUTION Although we will do more extensive work with formulas of this form later in the book, we need to know how to solve this particular formula for y in order to understand some things in the next chapter. We begin by simplifying the denominator on the left side and then multiplying each side of the formula by x. Doing so makes the rest of the solution process simple.

$$\frac{y-b}{x-0} = m \qquad \textbf{Original formula}$$

$$\frac{y-b}{x} = m \qquad x - 0 = x$$

$$x \cdot \frac{y-b}{x} = m \cdot x \qquad \textbf{Multiply each side by } x$$

$$y - b = mx \qquad \textbf{Simplify each side}$$

$$y = mx + b \qquad \textbf{Add } b \textbf{ to each side}$$

This is our solution. If we look back to the first step, we can justify our result on the left side of the equation this way: Dividing by x is equivalent to multiplying by its reciprocal $\frac{1}{x}$. Here is what it looks like when written out completely:

$$x \cdot \frac{y-b}{x} = x \cdot \frac{1}{x} \cdot (y - b) = 1(y - b) = y - b$$

EXAMPLE 8 Solve for y: $\frac{y-4}{x-5} = 3$.

SOLUTION We proceed as we did in the previous example, but this time we clear the formula of fractions by multiplying each side of the formula by $x - 5$.

$$\frac{y-4}{x-5} = 3 \qquad \textbf{Original formula}$$

$$(x-5) \cdot \frac{y-4}{x-5} = 3 \cdot (x - 5) \qquad \textbf{Multiply each side by } (x - 5)$$

$$y - 4 = 3x - 15 \qquad \textbf{Simplify each side}$$

$$y = 3x - 11 \qquad \textbf{Add 4 to each side}$$

We have solved for y. We can justify our result on the left side of the equation this way: Dividing by $x - 5$ is equivalent to multiplying by its reciprocal $\frac{1}{x-5}$. Here are the details:

$$(x-5) \cdot \frac{y-4}{x-5} = (x-5) \cdot \frac{1}{x-5} \cdot (y-4) = 1(y-4) = y - 4$$

Fraction	Decimal	Percent
$\frac{1}{2}$	0.5	50%
$\frac{1}{4}$	0.25	25%
$\frac{3}{4}$	0.75	75%
$\frac{1}{3}$	$0.\overline{3}$	$33\frac{1}{3}\%$
$\frac{2}{3}$	$0.\overline{6}$	$66\frac{2}{3}\%$
$\frac{1}{5}$	0.2	20%
$\frac{2}{5}$	0.4	40%
$\frac{3}{5}$	0.6	60%
$\frac{4}{5}$	0.8	80%

Basic Percent Problems

The next examples in this section show how basic percent problems can be translated directly into equations. To understand these examples, we must recall that percent means "per hundred." That is, 75% is the same as 75/100, 0.75, and, in reduced fraction form, $\frac{3}{4}$. Likewise, the decimal 0.25 is equivalent to 25%. To change a decimal to a percent, we move the decimal point two places to the right and write the % symbol. To change from a percent to a decimal, we drop the % symbol and move the decimal point two places to the left. The table that follows gives some of the most commonly used fractions and decimals and their equivalent percents.

EXAMPLE 9 What number is 15% of 63?

Note
We write 0.15 instead of 15% when we translate the sentence into an equation because we cannot do calculations with the % symbol. Because percent (%) means per hundred, we think of 15% as 15 hundredths, or 0.15.

SOLUTION To solve a problem like this, we let x = the number in question and then translate the sentence directly into an equation. Here is how it is done:

$$\underbrace{\text{What number}}_{x}\ \underset{=}{\text{is}}\ \underset{0.15}{15\%}\ \underset{\cdot}{\text{of}}\ \underset{63}{63?}$$

$$x = 0.15 \cdot 63$$
$$= 9.45$$

The number 9.45 is 15% of 63.

EXAMPLE 10 What percent of 42 is 21?

SOLUTION We translate the sentence as follows:

$$\underbrace{\text{What percent}}_{x}\ \underset{\cdot}{\text{of}}\ \underset{42}{42}\ \underset{=}{\text{is}}\ \underset{21}{21?}$$

$$x \cdot 42 = 21$$

Next, we divide each side by 42:

$$x = \frac{21}{42}$$
$$= 0.50 \text{ or } 50\%$$

EXAMPLE 11 25 is 40% of what number?

SOLUTION Again, we translate the sentence directly:

$$\underset{25}{25}\ \underset{=}{\text{is}}\ \underset{0.40}{40\%}\ \underset{\cdot}{\text{of}}\ \underbrace{\text{what number}}_{x}?$$

$$25 = 0.40 \cdot x$$

We solve the equation by dividing both sides by 0.40:

$$\frac{25}{0.40} = \frac{0.40 \cdot x}{0.40}$$

$$62.5 = x$$

25 is 40% of 62.5.

LINKING OBJECTIVES AND EXAMPLES

Next to each **objective** we have listed the examples that are best described by that objective.

A	1–3, 5
B	4, 6, 7, 8
C	9–11

GETTING READY FOR CLASS

After reading through the preceding section, respond in your own words and in complete sentences.

1. What is a formula in mathematics?
2. Give two equivalent forms of the rate equation $d = rt$.
3. Write a percent problem that can be solved by the equation $30 = 0.25x$.
4. Explain in words the formula for the perimeter of a rectangle.

Problem Set 2.2

Online support materials can be found at www.thomsonedu.com/login

Use the formula $3x - 4y = 12$ to find y if

1. x is 0
2. x is -2
3. x is 4
4. x is -4

Use the formula $y = 2x - 3$ to find x when

5. y is 0
6. y is -3
7. y is 5
8. y is -5

Problems 9 through 26 are all problems that you will see later in the text.

9. If $x - 2y = 4$ and $y = -\frac{6}{5}$, find x.
10. If $x - 2y = 4$ and $x = \frac{8}{5}$, find y.
11. If $2x + 3y = 6$, find y when x is 0.
12. If $2x + 3y = 6$, find x when y is 0.
13. Let $x = 160$ and $y = 0$ in $y = a(x - 80)^2 + 70$ and solve for a.
14. Let $x = 0$ and $y = 0$ in $y = a(x - 80)^2 + 70$ and solve for a.
15. Find R if $p = 1.5$ and $R = (900 - 300p)p$.
16. Find R if $p = 2.5$ and $R = (900 - 300p)p$.
17. Find P if $P = -0.1x^2 + 27x - 1{,}700$ and
    a. $x = 100$ b. $x = 170$
18. Find P if $P = -0.1x^2 + 27x - 1{,}820$ and
    a. $x = 130$ b. $x = 140$
19. Find h if $h = 16 + 32t - 16t^2$ and
    a. $t = \frac{1}{4}$ b. $t = \frac{7}{4}$
20. Find h if $h = 64t - 16t^2$ and
    a. $t = 1$ b. $t = 3$
21. Find y if $x = \frac{3}{2}$ and $y = -2x^2 + 6x - 5$.
22. Find y if $x = \frac{1}{2}$ and $y = -2x^2 + 6x - 5$.
23. If $y = Kx$, find K if $x = 5$ and $y = 15$.
24. If $d = Kt^2$, find K if $t = 2$ and $d = 64$.
25. If $V = \frac{K}{P}$, find K if $P = 48$ and $V = 50$.
26. If $y = Kxz^2$, find K if $x = 5$, $z = 3$, and $y = 180$.

Solve each of the following formulas for the indicated variable.

27. $A = lw$ for l
28. $A = \frac{1}{2}bh$ for b
29. $I = prt$ for t
30. $I = prt$ for r
31. $PV = nRT$ for T
32. $PV = nRT$ for R
33. $y = mx + b$ for x
34. $A = P + Prt$ for t
35. $C = \frac{5}{9}(F - 32)$ for F
36. $F = \frac{9}{5}C + 32$ for C
37. ▶ $h = vt + 16t^2$ for v
38. $h = vt - 16t^2$ for v
39. $A = a + (n - 1)d$ for d
40. $A = a + (n - 1)d$ for n
41. $2x + 3y = 6$ for y
42. $2x - 3y = 6$ for y
43. $-3x + 5y = 15$ for y
44. $-2x - 7y = 14$ for y
45. ▶ $2x - 6y + 12 = 0$ for y
46. $7x - 2y - 6 = 0$ for y
47. ▶ $ax + 4 = bx + 9$ for x
48. $ax - 5 = cx - 2$ for x
49. $A = P + Prt$ for P
50. $ax + b = cx + d$ for x

Solve for y.

51. $\frac{x}{8} + \frac{y}{2} = 1$
52. $\frac{x}{7} + \frac{y}{9} = 1$
53. $\frac{x}{5} + \frac{y}{-3} = 1$
54. $\frac{x}{16} + \frac{y}{-2} = 1$

Problems 55 through 64 are all problems that you will see later in the text. Solve each formula for y.

55. $x = 2y - 3$
56. $x = 4y + 1$
57. $y - 3 = -2(x + 4)$
58. $y - 2 = -3(x + 1)$

■ = Videos available by instructor request

▶ = Online student support materials available at www.thomsonedu.com/login

59. $y - 3 = -\frac{2}{3}(x + 3)$

60. $y + 1 = -\frac{2}{3}(x - 3)$

61. $y - 4 = -\frac{1}{2}(x + 1)$

62. $y - 2 = \frac{1}{3}(x - 1)$

63. Solve for y.

a. $\frac{y+1}{x-0} = 4$

b. $\frac{y+2}{x-4} = -\frac{1}{2}$

c. $\frac{y+3}{x-7} = 0$

64. Solve for y.

a. $\frac{y-1}{x-0} = -3$

b. $\frac{y-2}{x-6} = \frac{2}{3}$

c. $\frac{y-3}{x-1} = 0$

Translate each of the following into a linear equation and then solve the equation.

65. What number is 54% of 38?

66. What number is 11% of 67?

67. What percent of 36 is 9?

68. What percent of 50 is 5?

69. 37 is 4% of what number?

70. 8 is 2% of what number?

The next two problems are intended to give you practice reading, and paying attention to, the instructions that accompany the problems you are working. As we have mentioned previously, working these problems is an excellent way to get ready for a test or a quiz.

71. Work each problem according to the instructions given.

a. Solve: $-4x + 5 = 20$

b. Find the value of $-4x + 5$ when x is 3.

c. Solve for y: $-4x + 5y = 20$

d. Solve for x: $-4x + 5y = 20$

72. Work each problem according to the instructions given.

a. Solve: $2x + 1 = -4$

b. Find the value of $2x + 1$ when x is 8.

c. Solve for y: $2x + y = 20$

d. Solve for x: $2x + y = 20$

Applying the Concepts

73. Devin left a $4 tip for a $25 lunch with his girlfriend. What percent of the cost of lunch was the tip?

74. Janai left a $3 tip for a $15 breakfast with her boyfriend. What percent of the cost of breakfast was the tip?

75. If the sales tax is 6.5% of the purchase price, what is the sales tax on a $50 purchase?

76. If the sales tax is 7.25% of the purchase price, what is the sales tax on a $120 purchase?

77. During the annual sale at the Boot Factory, Fred purchases a pair of $94 boots for only for $56.40. What is the discount rate during the sale?

78. Whole tri-tip is priced at $5.65 per pound at the local supermarket, but costs only $4.52 per pound for members of their frequent shoppers club. What is the discount rate for club members?

Pricing A company that manufactures ink cartridges finds that they can sell x cartridges each week at a price of p dollars each, according to the formula $x = 1{,}300 - 100p$. What price should they charge for each cartridge if they want to sell

79. 800 cartridges each week?

80. 400 cartridges each week?

81. 300 cartridges each week?

82. 900 cartridges each week?

83. **Current** It takes a boat 2 hours to travel 18 miles upstream against the current. If the speed of the boat in still water is 15 miles per hour, what is the speed of the current?

84. **Current** It takes a boat 6.5 hours to travel 117 miles upstream against the current. If the speed of the current is 5 miles per hour, what is the speed of the boat in still water?

85. **Wind** An airplane takes 4 hours to travel 864 miles while flying against the wind. If the speed of the airplane on a windless day is 258 miles per hour, what is the speed of the wind?

86. **Wind** A cyclist takes 3 hours to travel 39 miles while pedaling against the wind. If the speed of the wind is 4 miles per hour, how fast would the cyclist be able to travel on a windless day?

▶ **87. Miles/Hour** A car travels 220 miles in 4 hours. What is the rate of the car in miles per hour?

▶ **88. Miles/Hour** A train travels 360 miles in 5 hours. What is the rate of the train in miles per hour?

▶ **89. Kilometers/Hour** It takes a car 3 hours to travel 252 kilometers. What is the rate in kilometers per hour?

▶ **90. Kilometers/Hour** In 6 hours an airplane travels 4,200 kilometers. What is the rate of the airplane in kilometers per hour?

For problems 91 and 92, use 3.14 as an approximation for π. Round answers to the nearest tenth.

91. Average Speed A person riding a Ferris wheel with a diameter of 65 feet travels once around the wheel in 30 seconds. What is the average speed of the rider in feet per second?

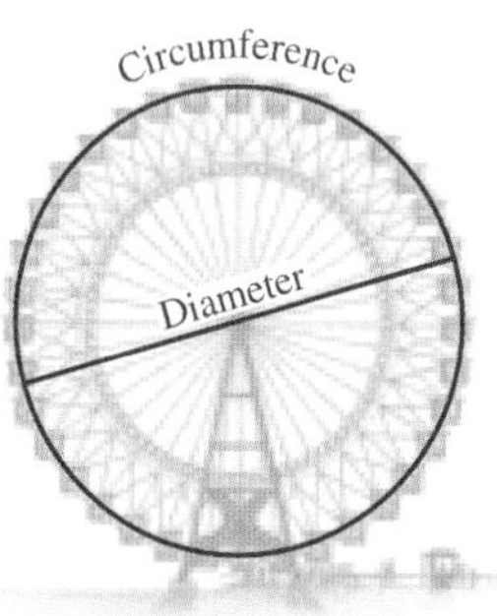

92. Average Speed A person riding a Ferris wheel with a diameter of 102 feet travels once around the wheel in 3.5 minutes. What is the average speed of the rider in feet per minute?

Fermat's Last Theorem As we mentioned in the previous chapter, the postage stamp shows Fermat's last theorem, which states that if n is an integer greater than 2, then there are no positive integers x, y, and z that will make the formula $x^n + y^n = z^n$ true.

Use the formula $x^n + y^n = z^n$ to

93. Find x if $n = 1, y = 7$, and $z = 15$.

94. Find y if $n = 1, x = 23$, and $z = 37$.

Improving Your Quantitative Literacy In exercise physiology, a person's maximum heart rate, in beats per minute, is found by subtracting his age, in years, from 220. So, if A represents your age in years, then your maximum heart rate is

$$M = 220 - A$$

A person's training heart rate, in beats per minute, is her resting heart rate plus 60% of the difference between her maximum heart rate and her resting heart rate. If resting heart rate is R and maximum heart rate is M, then the formula that gives training heart rate is

$$T = R + 0.6(M - R)$$

95. Training Heart Rate Shar is 46 years old. Her daughter, Sara, is 26 years old. If they both have a resting heart rate of 60 beats per minute, find the training heart rate for each.

96. Training Heart Rate Shane is 30 years old and has a resting heart rate of 68 beats per minute. Her mother, Carol, is 52 years old and has the same resting heart rate. Find the training heart rate for Shane and for Carol.

Tom & Dee Ann McCarthy/Corbis

Maintaining Your Skills

Simplify using the rule for order of operations.

97. $38 - 19 + 1$

98. $200 - 150 + 20$

99. $57 - 18 - 8$

100. $71 - 11 - 1$

101. $28 \div 7 \cdot 2$

102. $48 \div 3 \cdot 2$

103. $125 \div 25 \div 5$

104. $36 \div 12 \div 4$

Getting Ready for the Next Section

To understand all of the explanations and examples in the next section, you must be able to work the problems below.

Translate into symbols.

105. Three less than twice a number

106. Ten less than four times a number

107. The sum of x and y is 180.

108. The sum of a and b is 90.

Solve each equation.

109. $x + 2x = 90$

110. $x + 5x = 180$

111. $2(2x - 3) + 2x = 45$

112. $2(4x - 10) + 2x = 12.5$

113. $6x + 5(10{,}000 - x) = 56{,}000$

114. $x + 0.0725x = 17{,}481.75$

Extending the Concepts

115. Solve for x: $\frac{x}{a} + \frac{y}{b} = 1$

116. Solve for y: $\frac{x}{a} + \frac{y}{b} = 1$

117. Solve for a: $\frac{1}{a} + \frac{1}{b} = \frac{1}{c}$

118. Solve for b: $\frac{1}{a} + \frac{1}{b} = \frac{1}{c}$

2.3 Applications

OBJECTIVES

A Apply the Blueprint for Problem Solving to a variety of application problems.

B Use a formula to construct a table of paired data.

In this section we use the skills we have developed for solving equations to solve problems written in words. You may find that some of the examples and problems are more realistic than others. Since we are just beginning our work with application problems, even the ones that seem unrealistic are good practice. What is important in this section is the *method* we use to solve application problems, not the applications themselves. The method, or strategy, that we use to solve application problems is called the *Blueprint for Problem Solving*. It is an outline that will overlay the solution process we use on all application problems.

BLUEPRINT FOR PROBLEM SOLVING

STEP 1: ***Read*** the problem, and then mentally ***list*** the items that are known and the items that are unknown.

STEP 2: ***Assign a variable*** to one of the unknown items. (In most cases this will amount to letting x = the item that is asked for in the problem.) Then ***translate*** the other ***information*** in the problem to expressions involving the variable.

STEP 3: ***Reread*** the problem, and then ***write an equation,*** using the items and variable listed in steps 1 and 2, that describes the situation.

STEP 4: ***Solve the equation*** found in step 3.

STEP 5: ***Write your answer*** using a complete sentence.

STEP 6: ***Reread*** the problem, and ***check*** your solution with the original words in the problem.

A number of substeps occur within each of the steps in our blueprint. For instance, with steps 1 and 2 it is always a good idea to draw a diagram or picture if it helps you visualize the relationship between the items in the problem.

EXAMPLE 1 The length of a rectangle is 3 inches less than twice the width. The perimeter is 45 inches. Find the length and width.

SOLUTION When working problems that involve geometric figures, a sketch of the figure helps organize and visualize the problem.

Step 1: ***Read and list.***

Known items: The figure is a rectangle. The length is 3 inches less than twice the width. The perimeter is 45 inches.

Unknown items: The length and the width

Step 2: ***Assign a variable and translate information.***

Since the length is given in terms of the width (the length is 3 less than twice the width), we let x = the width of the rectangle. The length is 3 less than twice the width, so it must be $2x - 3$. The diagram in Figure 1 is a visual description of the relationships we have listed so far.

FIGURE 1

Step 3: ***Reread and write an equation.***

The equation that describes the situation is

$$\text{Twice the length} + \text{twice the width} = \text{perimeter}$$

$$2(2x - 3) \quad + \quad 2x \quad = \quad 45$$

Step 4: ***Solve the equation.***

$$\begin{aligned} 2(2x - 3) + 2x &= 45 \\ 4x - 6 + 2x &= 45 \\ 6x - 6 &= 45 \\ 6x &= 51 \\ x &= 8.5 \end{aligned}$$

Step 5: ***Write the answer.***

The width is 8.5 inches. The length is $2x - 3 = 2(8.5) - 3 = 14$ inches.

Step 6: ***Reread and check.***

If the length is 14 inches and the width is 8.5 inches, then the perimeter must be $2(14) + 2(8.5) = 28 + 17 = 45$ inches. Also, the length, 14, is 3 less than twice the width.

Remember as you read through the steps in the solutions to the examples in this section that step 1 is done mentally. Read the problem and then *mentally* list the items that you know and the items that you don't know. The purpose of step 1 is to give you direction as you begin to work application problems. Finding the solution to an application problem is a process; it doesn't happen all at once. The first step is to read the problem with a purpose in mind. That purpose is to mentally note the items that are known and the items that are unknown.

EXAMPLE 2 In April 1998, Pat bought a new Ford Mustang with a 5.0-liter engine. The total price, which includes the price of the car plus sales tax, was $17,481.75. If the sales tax rate is 7.25%, what was the price of the car?

SOLUTION

Step 1: ***Read and list.***

Known items: The total price is $17,481.75. The sales tax rate is 7.25%, which is 0.0725 in decimal form.

Unknown item: The price of the car

Step 2: ***Assign a variable and translate information.***

If we let $x =$ the price of the car, then to calculate the sales tax, we multiply the price of the car x by the sales tax rate:

$$\begin{aligned}\text{Sales tax} &= (\text{sales tax rate})(\text{price of the car})\\ &= 0.0725x\end{aligned}$$

Step 3: ***Reread and write an equation.***

$$\begin{aligned}\text{Car price} + \text{sales tax} &= \text{total price}\\ x \quad + 0.0725x &= 17{,}481.75\end{aligned}$$

Step 4: ***Solve the equation.***

$$\begin{aligned}x + 0.0725x &= 17{,}481.75\\ 1.0725x &= 17{,}481.75\\ x &= \frac{17{,}481.75}{1.0725}\\ &= 16{,}300.00\end{aligned}$$

Step 5: ***Write the answer.***

The price of the car is $16,300.00

Step 6: ***Reread and check.***

The price of the car is $16,300.00. The tax is 0.0725(16,300) = $1,181.75. Adding the retail price and the sales tax we have a total bill of $17,481.75.

FACTS FROM GEOMETRY

Angles in General

An angle is formed by two rays with the same endpoint. The common endpoint is called the *vertex* of the angle, and the rays are called the *sides* of the angle.

In Figure 2, angle θ (theta) is formed by the two rays OA and OB. The vertex of θ is O. Angle θ is also denoted as angle AOB, where the letter associated with the vertex is always the middle letter in the three letters used to denote the angle.

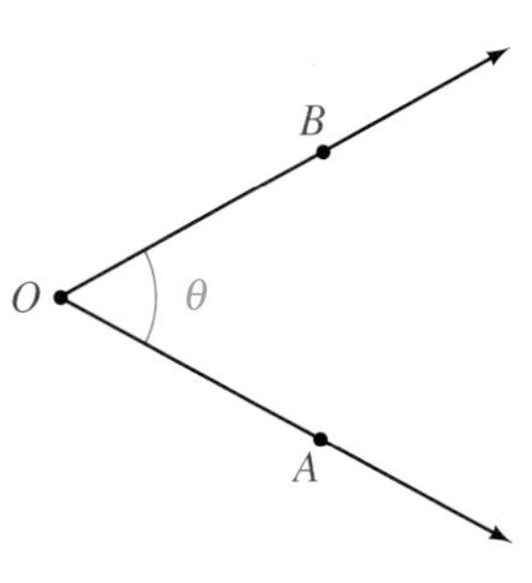

FIGURE 2

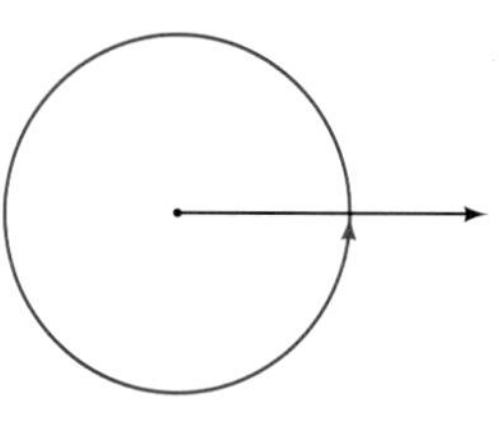

One complete revolution = 360°

FIGURE 3

Degree Measure

The angle formed by rotating a ray through one complete revolution about its endpoint (Figure 3) has a measure of 360 degrees, which we write as 360°.

One degree of angle measure, written 1°, is $\frac{1}{360}$ of a complete rotation of a ray about its endpoint; there are 360° in one full rotation. (The number 360 was decided on by early civilizations because it was believed that the earth was at the center of the universe and the sun would rotate once around the earth every 360 days.) Similarly, 180° is half of a complete rotation, and 90° is a quarter of a full rotation. Angles that measure 90° are called *right angles,* and angles that measure 180° are called *straight angles.* If an angle measures between 0° and 90° it is called an *acute angle,* and an angle that measures between 90° and 180° is an *obtuse angle.* Figure 4 illustrates further.

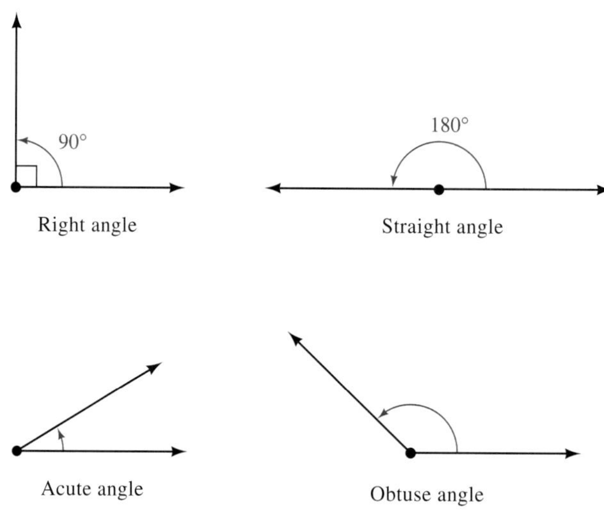

FIGURE 4

Complementary Angles and Supplementary Angles

If two angles add up to 90°, we call them *complementary angles,* and each is called the *complement* of the other. If two angles have a sum of 180°, we call them *supplementary angles,* and each is called the *supplement* of the other. Figure 5 illustrates the relationship between angles that are complementary and angles that are supplementary.

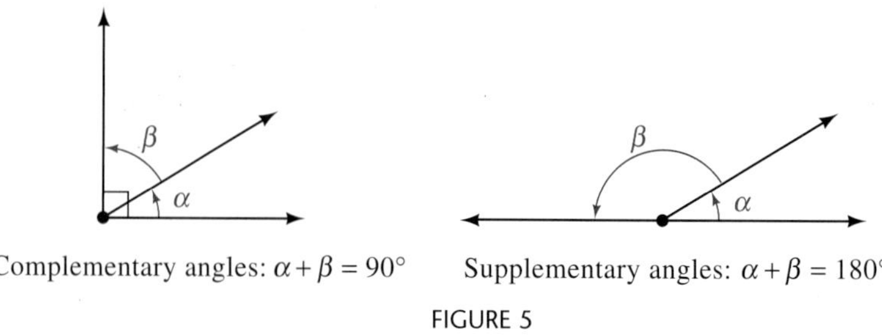

Complementary angles: $\alpha + \beta = 90°$ Supplementary angles: $\alpha + \beta = 180°$

FIGURE 5

Special Triangles

It is not unusual to have the terms we use in mathematics show up in the descriptions of things we find in the world around us. The flag of Puerto Rico

shown here is described on the government website as "Five equal horizontal bands of red (top and bottom) alternating with white; a blue isosceles triangle based on the hoist side bears a large white five-pointed star in the center." An *isosceles triangle* as shown here and in Figure 6, is a triangle with two sides of equal length.

Angles A and B in the isosceles triangle in Figure 6 are called the *base angles:* they are the angles opposite the two equal sides. In every isosceles triangle, the base angles are equal.

Isosceles Triangle

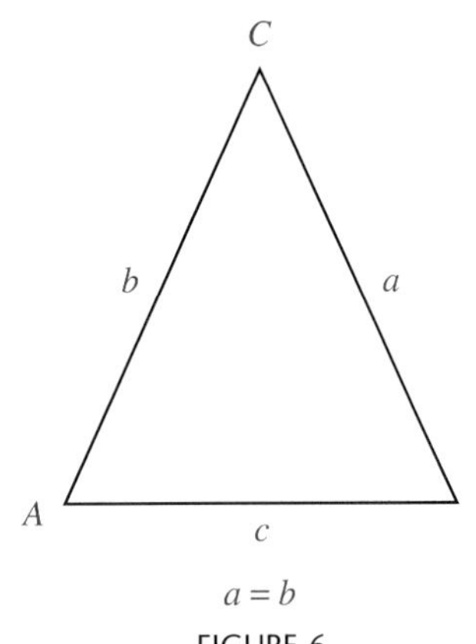

$a = b$

FIGURE 6

Equilateral Triangle

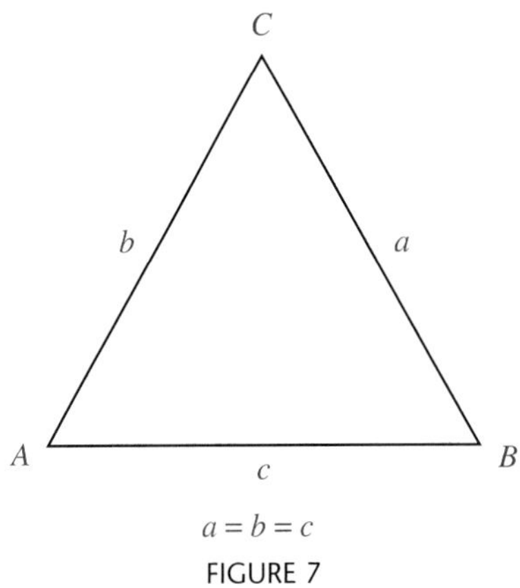

$a = b = c$

FIGURE 7

An *equilateral triangle* (Figure 7) is a triangle with three sides of equal length. If all three sides in a triangle have the same length, then the three interior angles in the triangle also must be equal. Because the sum of the interior angles in a triangle is always 180°, the three interior angles in any equilateral triangle must be 60°.

Note

As you can see from Figures 6 and 7, one way to label the important parts of a triangle is to label the vertices with capital letters and the sides with small letters: side a is opposite vertex A, side b is opposite vertex B, and side c is opposite vertex C.

Also, because each vertex is the vertex of one of the angles of the triangle, we refer to the three interior angles as A, B, and C.

Finally, in any triangle, the sum of the interior angles is 180°. For the triangles shown in Figures 6 and 7, the relationship is written

$$A + B + C = 180°$$

EXAMPLE 3

Two complementary angles are such that one is twice as large as the other. Find the two angles.

SOLUTION Applying the Blueprint for Problem Solving, we have:

Step 1: Read and list.
Known items: Two complementary angles. One is twice as large as the other.
Unknown items: The size of the angles

Step 2: Assign a variable and translate information.
Let x = the smaller angle. The larger angle is twice the smaller so we represent the larger angle with $2x$.

Step 3: Reread and write an equation.
Because the two angles are complementary, their sum is 90. Therefore,

$$x + 2x = 90$$

Step 4: Solve the equation.

$$x + 2x = 90$$
$$3x = 90$$
$$x = 30$$

Step 5: ***Write the answer.***
The smaller angle is 30°, and the larger angle is $2 \cdot 30 = 60°$.

Step 6: ***Reread and check.***
The larger angle is twice the smaller angle, and their sum is 90°. 

Suppose we know that the sum of two numbers is 50. If we let x represent one of the two numbers, how can we represent the other? Let's suppose for a moment that x turns out to be 30. Then the other number will be 20 because their sum is 50; that is, if two numbers add up to 50, and one of them is 30, then the other must be $50 - 30 = 20$. Generalizing this to any number x, we see that if two numbers have a sum of 50 and one of the numbers is x, then the other must be $50 - x$. The following table shows some additional examples:

If two numbers have a sum of	and one of them is	then the other must be
50	x	$50 - x$
10	y	$10 - y$
12	n	$12 - n$

Interest Problem

EXAMPLE 4 Suppose a person invests a total of \$10,000 in two accounts. One account earns 5% annually, and the other earns 6% annually. If the total interest earned from both accounts in a year is \$560, how much is invested in each account?

SOLUTION

Step 1: ***Read and list.***
Known items: Two accounts. One pays interest of 5%, and the other pays 6%. The total invested is \$10,000.
Unknown items: The number of dollars invested in each individual account

Step 2: ***Assign a variable and translate information.***
If we let x equal the amount invested at 6%, then $10{,}000 - x$ is the amount invested at 5%. The total interest earned from both accounts is \$560. The amount of interest earned on x dollars at 6% is $0.06x$, whereas the amount of interest earned on $10{,}000 - x$ dollars at 5% is $0.05(10{,}000 - x)$.

	Dollars at 6%	Dollars at 5%	Total
Number of	x	$10{,}000 - x$	10,000
Interest on	$0.06x$	$0.05(10{,}000 - x)$	560

Step 3: ***Reread and write an equation.***
The last line gives us the equation we are after:

$$0.06x + 0.05(10{,}000 - x) = 560$$

Step 4: ***Solve the equation.***

To make this equation a little easier to solve, we begin by multiplying both sides by 100 to move the decimal point two places to the right:

$$\begin{aligned} 6x + 5(10{,}000 - x) &= 56{,}000 \\ 6x + 50{,}000 - 5x &= 56{,}000 \\ x + 50{,}000 &= 56{,}000 \\ x &= 6{,}000 \end{aligned}$$

Step 5: ***Write the answer.***

The amount of money invested at 6% is \$6,000. The amount of money invested at 5% is \$10,000 − \$6,000 = \$4,000.

Step 6: ***Reread and check.***

To check our results, we find the total interest from the two accounts:

The interest on \$6,000 at 6% is 0.06(6,000)	= 360
The interest on \$4,000 at 5% is 0.05(4,000)	= 200
The total interest	= \$560

Table Building

We can use our knowledge of formulas to build tables of paired data. As you will see, equations or formulas that contain exactly two variables produce pairs of numbers that can be used to construct tables.

EXAMPLE 5 A piece of string 12 inches long is to be formed into a rectangle. Build a table that gives the length of the rectangle if the width is 1, 2, 3, 4, or 5 inches. Then find the area of each of the rectangles formed.

SOLUTION Because the formula for the perimeter of a rectangle is $P = 2l + 2w$ and our piece of string is 12 inches long, then the formula we will use to find the lengths for the given widths is $12 = 2l + 2w$. To solve this formula for l, we divide each side by 2 and then subtract w. The result is $l = 6 - w$. Table 1 organizes our work so that the formula we use to find l for a given value of w is shown, and we have added a last column to give us the areas of the rectangles formed. The units for the first three columns are inches, and the units for the numbers in the last column are square inches.

TABLE 1
Length, Width, and Area

Width (in.) w	Length (in.) $l = 6 - w$	l	Area (in.²) $A = lw$
1	$l = 6 - 1$	5	5
2	$l = 6 - 2$	4	8
3	$l = 6 - 3$	3	9
4	$l = 6 - 4$	2	8
5	$l = 6 - 5$	1	5

Figures 8 and 9 show two bar charts constructed from the information in Table 1.

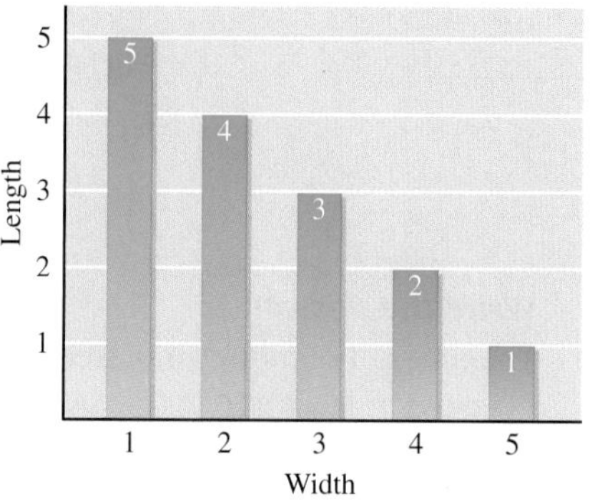

FIGURE 8 **Length and width of rectangles with perimeters fixed at 12 inches.**

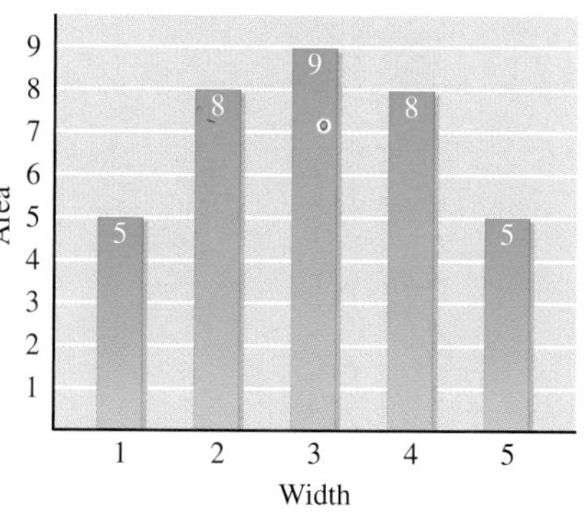

FIGURE 9 **Area and width of rectangles with perimeters fixed at 12 inches.**

USING TECHNOLOGY

A number of graphing calculators have table-building capabilities. We can let the calculator variable X represent the widths of the rectangles in Example 5. To find the lengths, we set variable Y_1 equal to $6 - X$. The area of each rectangle can be found by setting variable Y_2 equal to $X * Y_1$. To have the calculator produce the table automatically, we use a table minimum of 0 and a table increment of 1. Here is a summary of how the graphing calculator is set up:

Table Setup
Table minimum = 0
Table increment = 1
Independent variable: Auto
Dependent variable: Auto

Y Variables Setup
$Y_1 = 6 - X$
$Y_2 = X * Y_1$

The table will look like this:

X	Y_1	Y_2
0	6	0
1	5	5
2	4	8
3	3	9
4	2	8
5	1	5
6	0	0

LINKING OBJECTIVES AND EXAMPLES

Next to each **objective** we have listed the examples that are best described by that objective.

A 1–4

B 5

GETTING READY FOR CLASS

After reading through the preceding section, respond in your own words and in complete sentences.

1. What is the first step in solving an application problem?
2. What is the biggest obstacle between you and success in solving application problems?
3. Write an application problem for which the solution depends on solving the equation $2x + 2 \cdot 3 = 18$.
4. What is the last step in solving an application problem? Why is this step important?

Problem Set 2.3

Online support materials can be found at www.thomsonedu.com/login

Solve each application problem. Be sure to follow the steps in the Blueprint for Problem Solving.

1. **Rectangle** A rectangle is twice as long as it is wide. The perimeter is 60 feet. Find the dimensions.
2. **Rectangle** The length of a rectangle is 5 times the width. The perimeter is 48 inches. Find the dimensions.
3. **Square** A square has a perimeter of 28 feet. Find the length of each side.
4. **Square** A square has a perimeter of 36 centimeters. Find the length of each side.
5. **Triangle** A triangle has a perimeter of 23 inches. The medium side is 3 inches more than the shortest side, and the longest side is twice the shortest side. Find the shortest side.
6. **Triangle** The longest side of a triangle is two times the shortest side, whereas the medium side is 3 meters more than the shortest side. The perimeter is 27 meters. Find the dimensions.
7. **Rectangle** The length of a rectangle is 3 meters less than twice the width. The perimeter is 18 meters. Find the width.
8. **Rectangle** The length of a rectangle is 1 foot more than twice the width. The perimeter is 20 feet. Find the dimensions.
9. **Livestock Pen** A livestock pen is built in the shape of a rectangle that is twice as long as it is wide. The perimeter is 48 feet. If the material used to build the pen is \$1.75 per foot for the longer sides and \$2.25 per foot for the shorter sides (the shorter sides have gates, which increase the cost per foot), find the cost to build the pen.

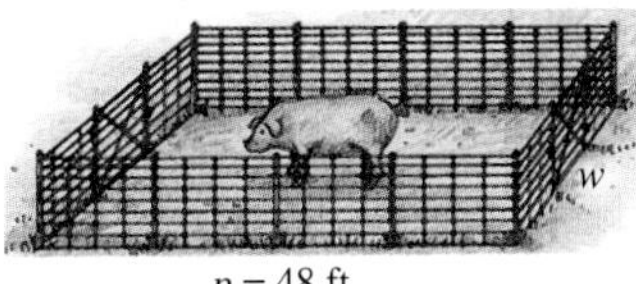

$p = 48$ ft.

10. **Garden** A garden is in the shape of a square with a perimeter of 42 feet. The garden is surrounded by two fences. One fence is around the perimeter of the garden, whereas the second fence is 3 feet from the first fence, as Figure 9 indicates. If the material used to build the two fences is \$1.28 per foot, what was the total cost of the fences?

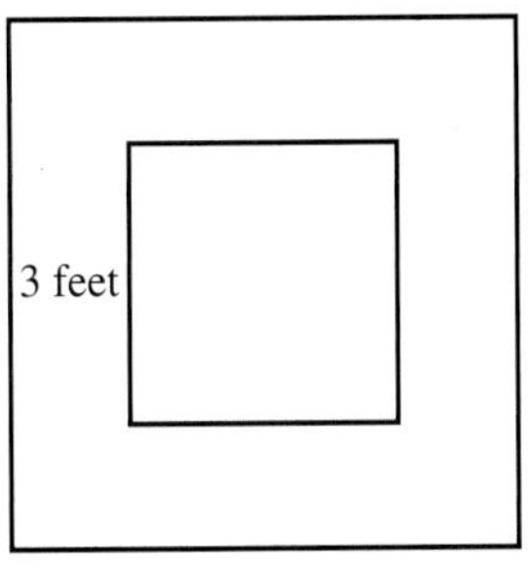

FIGURE 9

= Videos available by instructor request

▶ = Online student support materials available at www.thomsonedu.com/login

Percent Problems

11. **Money** Shane returned from a trip to Las Vegas with \$300.00, which was 50% more money than he had at the beginning of the trip. How much money did Shane have at the beginning of his trip?

12. **Items Sold** Every item in the Just a Dollar store is priced at \$1.00. When Mary Jo opens the store, there is \$125.50 in the cash register. When she counts the money in the cash register at the end of the day, the total is \$1,058.60. If the sales tax rate is 8.5%, how many items were sold that day?

13. **Monthly Salary** An accountant earns \$3,440 per month after receiving a 5.5% raise. What was the accountant's monthly income before the raise? Round your answer to the nearest cent.

14. **Textbook Price** Suppose a college bookstore buys a textbook from a publishing company and then marks up the price they paid for the book 33% and sells it to a student at the marked-up price. If the student pays \$75.00 for the textbook, what did the bookstore pay for it? Round your answer to the nearest cent.

15. **Movies** *Batman Forever* grossed \$52.8 million on its opening weekend and had one of the most successful movie launches in history. If *Batman Forever* accounted for approximately 53% of all box office receipts that weekend, what were the total box office receipts?

16. **Hourly Wage** A sheet metal worker earns \$26.80 per hour after receiving a 4.5% raise. What was the sheet metal worker's hourly pay before the raise? Round your answer to the nearest cent.

More Geometry Problems

17. **Angles** Two supplementary angles are such that one is eight times larger than the other. Find the two angles.

18. **Angles** Two complementary angles are such that one is five times larger than the other. Find the two angles.

19. **Angles** One angle is 12° less than four times another. Find the measure of each angle if
    a. they are complements of each other.
    b. they are supplements of each other.

20. **Angles** One angle is 4° more than three times another. Find the measure of each angle if
    a. they are complements of each other.
    b. they are supplements of each other.

21. **Triangle** A triangle is such that the largest angle is three times the smallest angle. The third angle is 9° less than the largest angle. Find the measure of each angle.

22. **Triangle** The smallest angle in a triangle is half of the largest angle. The third angle is 15° less than the largest angle. Find the measure of all three angles.

23. **Triangle** The smallest angle in a triangle is one-third of the largest angle. The third angle is 10° more than the smallest angle. Find the measure of all three angles.

24. **Triangle** The third angle in an isosceles triangle is half as large as each of the two base angles. Find the measure of each angle.

25. **Isosceles Triangle** The third angle in an isosceles triangle is 8° more than twice as large as each of the two base angles. Find the measure of each angle.

26. **Isosceles Triangle** The third angle in an isosceles triangle is 4° more than one fifth of each of the two base angles. Find the measure of each angle.

Interest Problems

27. **Investing** A woman has a total of \$9,000 to invest. She invests part of the money in an account that pays 8% per year and the rest in an account that pays 9% per year. If the interest earned in the first year is \$750, how much did she invest in each account?

	Dollars at 8%	Dollars at 9%	Total
Number of			
Interest on			

28. **Investing** A man invests \$12,000 in two accounts. If one account pays 10% per year and the other pays 7% per year, how much was invested in each

account if the total interest earned in the first year was \$960?

	Dollars at 10%	Dollars at 7%	Total
Number of			
Interest on			

29. **Investing** A total of \$15,000 is invested in two accounts. One of the accounts earns 12% per year, and the other earns 10% per year. If the total interest earned in the first year is \$1,600, how much was invested in each account?

30. **Investing** A total of \$11,000 is invested in two accounts. One of the two accounts pays 9% per year, and the other account pays 11% per year. If the total interest paid in the first year is \$1,150, how much was invested in each account?

31. **Investing** Stacy has a total of \$6,000 in two accounts. The total amount of interest she earns from both accounts in the first year is \$500. If one of the accounts earns 8% interest per year and the other earns 9% interest per year, how much did she invest in each account?

32. **Investing** Travis has a total of \$6,000 invested in two accounts. The total amount of interest he earns from the accounts in the first year is \$410. If one account pays 6% per year and the other pays 8% per year, how much did he invest in each account?

Miscellaneous Problems

33. **Ticket Prices** Miguel is selling tickets to a barbecue. Adult tickets cost \$6.00 and children's tickets cost \$4.00. He sells six more children's tickets than adult tickets. The total amount of money he collects is \$184. How many adult tickets and how many children's tickets did he sell?

	Adult	Child
Number	x	$x + 6$
Income	$6(x)$	$4(x + 6)$

34. **Working Two Jobs** Maggie has a job working in an office for \$10 an hour and another job driving a tractor for \$12 an hour. One week she works in the office twice as long as she drives the tractor. Her total income for that week is \$416. How many hours did she spend at each job?

Job	Office	Tractor
Hours Worked	$2x$	x
Wages Earned	$10(2x)$	$12x$

35. **Sales Tax** A woman owns a small, cash-only business in a state that requires her to charge 6% sales tax on each item she sells. At the beginning of the day, she has \$250 in the cash register. At the end of the day, she has \$1,204 in the register. How much money should she send to the state government for the sales tax she collected?

36. **Sales Tax** A store is located in a state that requires 6% tax on all items sold. If the store brings in a total of \$3,392 in one day, how much of that total was sales tax?

Table Building

37. Use $h = 32t - 16t^2$ to complete the table.

t	0	$\frac{1}{4}$	1	$\frac{7}{4}$	2
h					

38. Use $s = \dfrac{60}{t}$ to complete the table.

t	4	6	8	10
s				

Coffee Sales The data appeared in *USA Today* in October 2005. Use the information to complete the tables below. Round to the nearest tenth of a billion dollars.

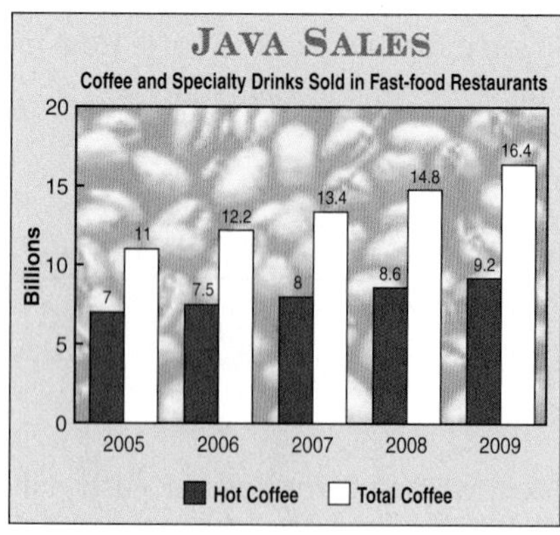

Data from *USA Today*. Copyright 2005.

39.

Hot Coffee Sales

Year	Sales (billions of dollars)
2005	
2006	
2007	
2008	
2009	

40.

Total Coffee Sales

Year	Sales (billions of dollars)
2005	
2006	
2007	
2008	
2009	

41. Distance A search is being conducted for someone guilty of a hit-and-run felony. In order to set up roadblocks at appropriate points, the police must determine how far the guilty party might have traveled during the past half-hour. Use the formula $d = rt$ with $t = 0.5$ hour to complete the following table.

Speed (miles per hour)	Distance (miles)
20	
30	
40	
50	
60	
70	

42. Speed To determine the average speed of a bullet when fired from a rifle, the time is measured from when the gun is fired until the bullet hits a target that is 1,000 feet away. Use the formula $d = rt$ with $d = 1{,}000$ feet to complete the following table.

Time (seconds)	Rate (feet per second)
1.00	
0.80	
0.64	
0.50	
0.40	
0.32	

43. Current A boat that can travel 10 miles per hour in still water is traveling along a stream with a current of 4 miles per hour. The distance the boat will travel upstream is given by the formula $d = (r - c) \cdot t$, and the distance it will travel downstream is given by the formula $d = (r + c) \cdot t$. Use these formulas with $r = 10$ and $c = 4$ to complete the following table.

Time (hours)	Distance Upstream (miles)	Distance Downstream (miles)
1		
2		
3		
4		
5		
6		

44. Wind A plane that can travel 300 miles per hour in still air is traveling in a wind stream with a speed of 20 miles per hour. The distance the plane will travel against the wind is given by the formula $d = (r - w) \cdot t$, and the distance it will travel with the wind is given by the formula $d = (r + w) \cdot t$. Use these formulas with $r = 300$ and $w = 20$ to complete the following table.

Time (hours)	Distance Against the Wind (miles)	Distance With the Wind (miles)
.5		
1		
1.5		
2		
2.5		
3		

Maximum Heart Rate In exercise physiology, a person's maximum heart rate, in beats per minute, is found by subtracting his age in years from 220. So, if A represents your age in years, then your maximum heart rate is

$$M = 220 - A$$

Use this formula to complete the following tables.

45.

Age (years)	Maximum Heart Rate (beats per minute)
18	
19	
20	
21	
22	
23	

46.

Age (years)	Maximum Heart Rate (beats per minute)
15	
20	
25	
30	
35	
40	

Problems 47–48 may be solved using a graphing calculator.

47. Livestock Pen A farmer buys 48 feet of fencing material to build a rectangular livestock pen. Fill in the second column of the table to find the length of the pen if the width is 2, 4, 6, 8, 10, or 12 feet. Then find the area of each of the pens formed.

w	l	A
2		
4		
6		
8		
10		
12		

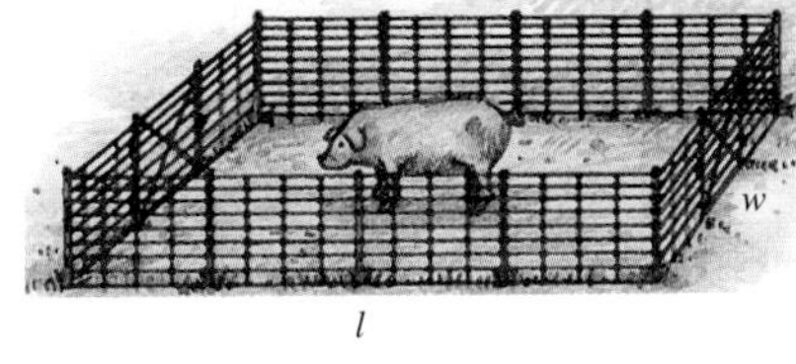

48. Model Rocket A small rocket is projected straight up into the air with a velocity of 128 feet per second. The formula that gives the height h of the rocket t seconds after it is launched is

$$h = -16t^2 + 128t$$

Use this formula to find the height of the rocket after 1, 2, 3, 4, 5, and 6 seconds.

Time (seconds)	Height (feet)
1	
2	
3	
4	
5	
6	

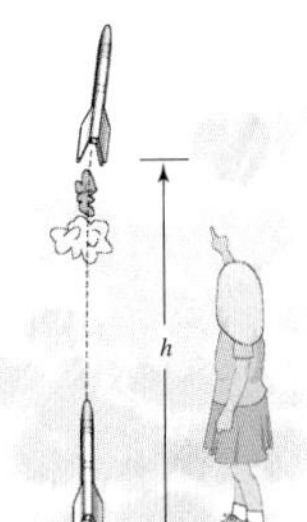

Maintaining Your Skills

Graph each of the following on a number line.

49. $\{x \mid x > -5\}$

50. $\{x \mid x \le 4\}$

51. $\{x \mid x \le -2 \text{ or } x > 5\}$

52. $\{x \mid x < 3 \text{ or } x \ge 5\}$

53. $\{x \mid x > -4 \text{ and } x < 0\}$

54. $\{x \mid x \ge 0 \text{ and } x \le 2\}$

55. $\{x \mid 1 \le x \le 4\}$

56. $\{x \mid -4 < x < -2\}$

Getting Ready for the Next Section

To understand all of the explanations and examples in the next section, you must be able to work the problems below.

Graph each inequality.

57. $x < 2$

58. $x \le 2$

59. $x \ge -3$

60. $x > -3$

Solve each equation.

61. $-2x - 3 = 7$

62. $3x + 3 = 2x - 1$

63. $3(2x - 4) - 7x = -3x$

64. $3(2x + 5) = -3x$

2.4 Linear Inequalities in One Variable

OBJECTIVES

A Solve a linear inequality in one variable and graph the solution set.

B Solve a compound inequality and graph the solution set.

C Write solutions to inequalities using interval notation.

D Solve application problems using inequalities.

A *linear inequality in one variable* is any inequality that can be put in the form

$$ax + b < c \qquad (a, b, \text{and } c \text{ constants}, a \neq 0)$$

where the inequality symbol ($<$) can be replaced with any of the other three inequality symbols ($\leq$, $>$, or $\geq$).

Some examples of linear inequalities are

$$3x - 2 \geq 7 \qquad -5y < 25 \qquad 3(x - 4) > 2x$$

Our first property for inequalities is similar to the addition property we used when solving equations.

> **Addition Property for Inequalities** For any algebraic expressions, A, B, and C,
>
> if $\quad A < B$
>
> then $\quad A + C < B + C$
>
> *In words:* Adding the same quantity to both sides of an inequality will not change the solution set.

EXAMPLE 1

Solve $3x + 3 < 2x - 1$, and graph the solution.

SOLUTION We use the addition property for inequalities to write all the variable terms on one side and all constant terms on the other side:

$$3x + 3 < 2x - 1$$

$$3x + (\mathbf{-2x}) + 3 < 2x + (\mathbf{-2x}) - 1 \qquad \textbf{Add } \mathbf{-2x} \textbf{ to each side}$$

$$x + 3 < -1$$

$$x + 3 + (\mathbf{-3}) < -1 + (\mathbf{-3}) \qquad \textbf{Add } \mathbf{-3} \textbf{ to each side}$$

$$x < -4$$

The solution set is all real numbers that are less than -4. To show this we can use set notation and write

$$\{x \mid x < -4\}$$

Or we can graph the solution set on the number line using a left-opening parenthesis at -4 to show that -4 is not part of the solution set:

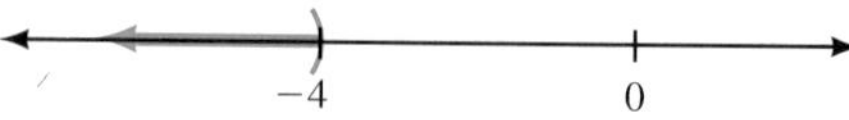

This graph gives rise to the following notation, called *interval notation,* that is an alternative way to write the solution set.

$$(-\infty, -4)$$

The preceding expression indicates that the solution set is all real numbers from negative infinity up to, but not including, -4.

We have three equivalent representations for the solution set to our original inequality. Here are all three together.

Set Notation	Line Graph	Interval Notation
$\{x \mid x < -4\}$	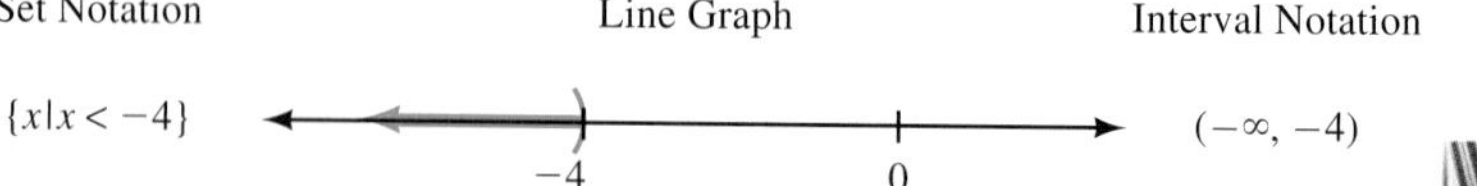	$(-\infty, -4)$

Note Since subtraction is defined as addition of the opposite, our new property holds for subtraction as well as addition; that is, we can subtract the same quantity from each side of an inequality and always be sure that we have not changed the solution.

Note The English mathematician John Wallis (1616–1703) was the first person to use the ∞ symbol to represent infinity. When we encounter the interval $(3, \infty)$, we read it as "the interval from 3 to infinity," and we mean the set of real numbers that are greater than three. Likewise, the interval $(-\infty, -4)$ is read "the interval from negative infinity to -4," which is all real numbers less than -4.

Interval Notation and Graphing

The table below shows the connection between inequalities, interval notation, and number line graphs. We have included the graphs with open and closed circles for those of you who have used this type of graph previously. In this book, we will continue to show our graphs using the parentheses/brackets method.

Inequality notation	Interval notation	Graph using parentheses/brackets	Graph using open and closed circles
$x < 2$	$(-\infty, 2)$		
$x \le 2$	$(-\infty, 2]$		
$x \ge -3$	$[-3, \infty)$		
$x > -3$	$(-3, \infty)$		

> **Note**
> The purpose of this discussion is to justify the multiplication property for inequalities. If you are having trouble understanding it, you may want to read the property itself first and then read through the discussion here. What is important is that you understand and can use the multiplication property for inequalities.

Before we state the multiplication property for inequalities, we will take a look at what happens to an inequality statement when we multiply both sides by a positive number and what happens when we multiply by a negative number.

We begin by writing three true inequality statements:

$$3 < 5 \qquad -3 < 5 \qquad -5 < -3$$

We multiply both sides of each inequality by a positive number—say, 4:

$$4(3) < 4(5) \qquad 4(-3) < 4(5) \qquad 4(-5) < 4(-3)$$
$$12 < 20 \qquad -12 < 20 \qquad -20 < -12$$

Notice in each case that the resulting inequality symbol points in the same direction as the original inequality symbol. Multiplying both sides of an inequality by a positive number preserves the *sense* of the inequality.

Let's take the same three original inequalities and multiply both sides by −4:

$$3 < 5 \qquad -3 < 5 \qquad -5 < -3$$
$$-4(3) > -4(5) \qquad -4(-3) > -4(5) \qquad -4(-5) > -4(-3)$$
$$-12 > -20 \qquad 12 > -20 \qquad 20 > 12$$

Notice in this case that the resulting inequality symbol always points in the opposite direction from the original one. Multiplying both sides of an inequality by a negative number *reverses* the sense of the inequality. Keeping this in mind, we will now state the multiplication property for inequalities.

Note
Because division is defined as multiplication by the reciprocal, we can apply our new property to division as well as to multiplication. We can divide both sides of an inequality by any nonzero number as long as we reverse the direction of the inequality when the number we are dividing by is negative.

Multiplication Property for Inequalities Let A, B, and C represent algebraic expressions.

If	$A < B$		
then	$AC < BC$	if	C is positive $(C > 0)$
or	$AC > BC$	if	C is negative $(C < 0)$

In words: Multiplying both sides of an inequality by a positive number always produces an equivalent inequality. Multiplying both sides of an inequality by a negative number reverses the sense of the inequality.

The multiplication property for inequalities does not limit what we can do with inequalities. We are still free to multiply both sides of an inequality by any nonzero number we choose. If the number we multiply by happens to be *negative,* then we *must also reverse* the direction of the inequality.

EXAMPLE 2
Find the solution set for $-2y - 3 \le 7$.

SOLUTION We begin by adding 3 to each side of the inequality:

$$-2y - 3 \le 7$$

$$-2y \le 10 \quad \textbf{Add 3 to both sides}$$

$$-\frac{1}{2}(-2y) \ge -\frac{1}{2}(10) \quad \textbf{Multiply by } -\tfrac{1}{2} \textbf{ and reverse the direction of the inequality symbol}$$

$$y \ge -5$$

The solution set is all real numbers that are greater than or equal to -5. Below are three equivalent ways to represent this solution set.

Set Notation	Line Graph	Interval Notation
$\{y \mid y \ge -5\}$	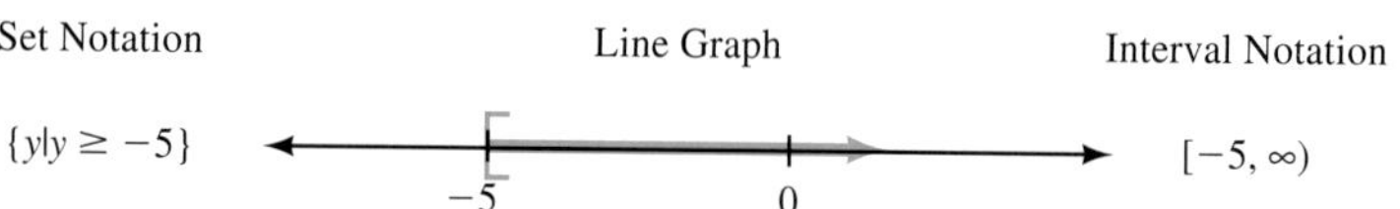	$[-5, \infty)$

Notice how a bracket is used with interval notation to show that -5 is part of the solution set.

When our inequalities become more complicated, we use the same basic steps we used in Section 2.1 when we were solving equations; that is, we simplify each side of the inequality before we apply the addition property or multiplication property. When we have solved the inequality, we graph the solution on a number line.

EXAMPLE 3
Solve $3(2x - 4) - 7x \le -3x$.

SOLUTION We begin by using the distributive property to separate terms. Next, simplify both sides.

$$3(2x - 4) - 7x \le -3x \quad \textbf{Original inequality}$$

$$6x - 12 - 7x \le -3x \quad \textbf{Distributive property}$$

$$-x - 12 \le -3x \quad \mathbf{6x - 7x = (6 - 7)x = -x}$$

$$-12 \le -2x \qquad \textbf{Add } x \textbf{ to both sides}$$

$$-\frac{1}{2}(-12) \ge -\frac{1}{2}(-2x) \qquad \textbf{Multiply both sides by } -\tfrac{1}{2} \textbf{ and reverse the direction of the inequality symbol}$$

$$6 \ge x$$

> **Note**
> In Examples 2 and 3, notice that each time we multiplied both sides of the inequality by a negative number we also reversed the direction of the inequality symbol. Failure to do so would cause our graph to lie on the wrong side of the endpoint.

This last line is equivalent to $x \le 6$. The solution set can be represented with any of the three following items.

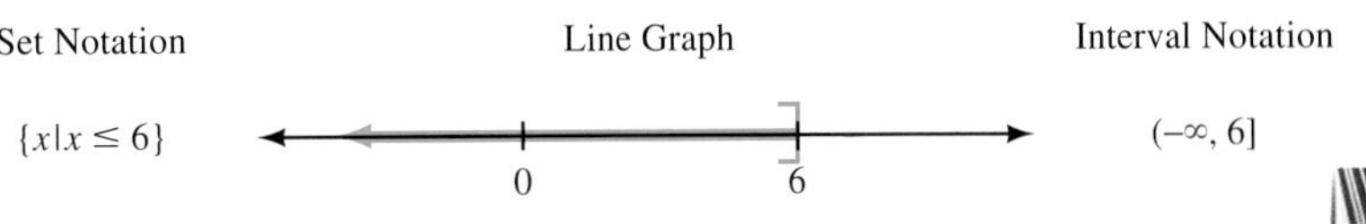

EXAMPLE 4

Solve and graph $-3 \le 2x - 5 \le 3$.

SOLUTION We can extend our properties for addition and multiplication to cover this situation. If we add a number to the middle expression, we must add the same number to the outside expressions. If we multiply the center expression by a number, we must do the same to the outside expressions, remembering to reverse the direction of the inequality symbols if we multiply by a negative number. We begin by adding 5 to all three parts of the inequality:

$$-3 \le 2x - 5 \le 3$$

$$2 \le \quad 2x \quad \le 8 \qquad \textbf{Add 5 to all three members}$$

$$1 \le \quad x \quad \le 4 \qquad \textbf{Multiply through by } \tfrac{1}{2}$$

Here are three ways to write this solution set:

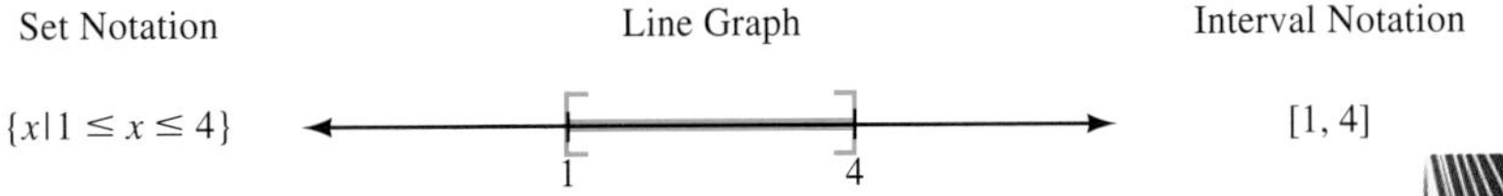

Interval Notation and Graphing

The table below shows the connection between interval notation and number line graphs for a variety of continued inequalities. Again, we have included the graphs with open and closed circles for those of you who have used this type of graph previously. Remember, however, that in this book we will be using the parentheses/brackets method of graphing.

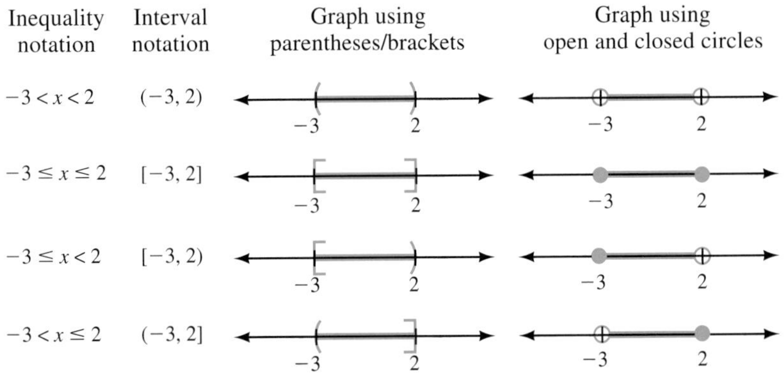

Inequality notation	Interval notation	Graph using parentheses/brackets	Graph using open and closed circles
$-3 < x < 2$	$(-3, 2)$	−3 2	−3 2
$-3 \le x \le 2$	$[-3, 2]$	−3 2	−3 2
$-3 \le x < 2$	$[-3, 2)$	−3 2	−3 2
$-3 < x \le 2$	$(-3, 2]$	−3 2	−3 2

EXAMPLE 5 Solve the compound inequality.

$$3t + 7 \le -4 \quad \text{or} \quad 3t + 7 \ge 4$$

SOLUTION We solve each half of the compound inequality separately, then we graph the solution set:

$3t + 7 \le -4$	or	$3t + 7 \ge 4$	
$3t \le -11$	or	$3t \ge -3$	**Add −7**
$t \le -\frac{11}{3}$	or	$t \ge -1$	**Multiply by $\frac{1}{3}$**

The solution set can be written in any of the following ways:

Set Notation	Line Graph	Interval Notation
$\{t \mid t \le -\frac{11}{3} \text{ or } t \ge -1\}$	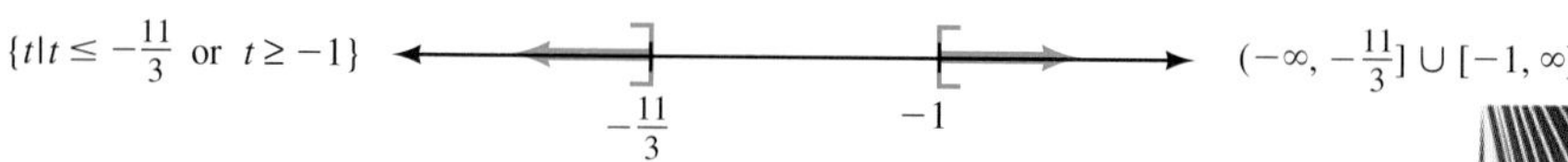	$(-\infty, -\frac{11}{3}] \cup [-1, \infty)$

EXAMPLE 6 A company that manufactures ink cartridges for printers finds that they can sell x cartridges each week at a price of p dollars each, according to the formula $x = 1{,}300 - 100p$. What price should they charge for each cartridge if they want to sell at least 300 cartridges a week?

SOLUTION Because x is the number of cartridges they sell each week, an inequality that corresponds to selling at least 300 cartridges a week is

$$x \ge 300$$

Substituting $1{,}300 - 100p$ for x gives us an inequality in the variable p.

$1{,}300 - 100p \ge 300$	
$-100p \ge -1{,}000$	**Add −1,300 to each side**
$p \le 10$	**Divide each side by −100, and reverse the direction of the inequality symbol**

To sell at least 300 cartridges each week, the price per cartridge should be no more than $10; that is, selling the cartridges for $10 or less will produce weekly sales of 300 or more cartridges.

EXAMPLE 7 The formula $F = \frac{9}{5}C + 32$ gives the relationship between the Celsius and Fahrenheit temperature scales. If the temperature range on a certain day is 86° to 104° Fahrenheit, what is the temperature range in degrees Celsius?

SOLUTION From the given information we can write $86 \le F \le 104$. But, because F is equal to $\frac{9}{5}C + 32$, we can also write

$86 \le \frac{9}{5}C + 32 \le 104$	
$54 \le \frac{9}{5}C \le 72$	**Add −32 to each member**
$\frac{5}{9}(54) \le \frac{5}{9}\left(\frac{9}{5}C\right) \le \frac{5}{9}(72)$	
$30 \le C \le 40$	

A temperature range of 86° to 104° Fahrenheit corresponds to a temperature range of 30° to 40° Celsius.

LINKING OBJECTIVES AND EXAMPLES

Next to each **objective** we have listed the examples that are best described by that objective.

A	1–3
B	4, 5
C	1–5
D	6, 7

GETTING READY FOR CLASS

After reading through the preceding section, respond in your own words and in complete sentences.

1. What is the addition property for inequalities?
2. When we use interval notation to denote a section of the real number line, when do we use parentheses ( ) and when do we use brackets [ ]?
3. Explain the difference between the multiplication property of equality and the multiplication property for inequalities.
4. When solving an inequality, when do we change the direction of the inequality symbol?

Problem Set 2.4

Online support materials can be found at www.thomsonedu.com/login

Solve each of the following inequalities and graph each solution.

1. $2x \le 3$
2. $5x \ge -115$
3. $\frac{1}{2}x > 2$
4. $\frac{1}{3}x > 4$
5. $-5x \le 25$
6. $-7x \ge 35$
7. $-\frac{3}{2}x > -6$
8. $-\frac{2}{3}x < -8$
9. $-12 \le 2x$
10. $-20 \ge 4x$
11. $-1 \ge -\frac{1}{4}x$
12. $-1 \le -\frac{1}{5}x$
13. $-3x + 1 > 10$
14. $-2x - 5 \le 15$
15. $\frac{1}{2} - \frac{m}{12} \le \frac{7}{12}$
16. $\frac{1}{2} - \frac{m}{10} > -\frac{1}{5}$
17. $\frac{1}{2} \ge -\frac{1}{6} - \frac{2}{9}x$
18. $\frac{9}{5} > -\frac{1}{5} - \frac{1}{2}x$
19. $-40 \le 30 - 20y$
20. $-20 > 50 - 30y$
21. $\frac{2}{3}x - 3 < 1$
22. $\frac{3}{4}x - 2 > 7$
23. $10 - \frac{1}{2}y \le 36$
24. $8 - \frac{1}{3}y \ge 20$

Simplify each side first, then solve the following inequalities. Write your answers with interval notation.

25. $2(3y + 1) \le -10$
26. $3(2y - 4) > 0$
27. $-(a + 1) - 4a \le 2a - 8$
28. $-(a - 2) - 5a \le 3a + 7$
29. $\frac{1}{3}t - \frac{1}{2}(5 - t) < 0$
30. $\frac{1}{4}t - \frac{1}{3}(2t - 5) < 0$
31. $-2 \le 5 - 7(2a + 3)$
32. $1 < 3 - 4(3a - 1)$
33. $-\frac{1}{3}(x + 5) \le -\frac{2}{9}(x - 1)$
34. $-\frac{1}{2}(2x + 1) \le -\frac{3}{8}(x + 2)$

Solve each inequality. Write your answer using inequality notation.

35. $20x + 9{,}300 > 18{,}000$
36. $20x + 4{,}800 > 18{,}000$

Solve the following continued inequalities. Use both a line graph and interval notation to write each solution set.

37. $-2 \le m - 5 \le 2$

= Videos available by instructor request

▸ = Online student support materials available at www.thomsonedu.com/login

38. $-3 \le m + 1 \le 3$

39. $-60 < 20a + 20 < 60$

40. $-60 < 50a - 40 < 60$

41. $0.5 \le 0.3a - 0.7 \le 1.1$

42. $0.1 \le 0.4a + 0.1 \le 0.3$

43. $3 < \frac{1}{2}x + 5 < 6$

44. $5 < \frac{1}{4}x + 1 < 9$

45. $4 < 6 + \frac{2}{3}x < 8$

46. $3 < 7 + \frac{4}{5}x < 15$

Graph the solution sets for the following compound inequalities. Then write each solution set using interval notation.

47. $x + 5 \le -2$ or $x + 5 \ge 2$

48. $3x + 2 < -3$ or $3x + 2 > 3$

49. $5y + 1 \le -4$ or $5y + 1 \ge 4$

50. $7y - 5 \le -2$ or $7y - 5 \ge 2$

51. $2x + 5 < 3x - 1$ or $x - 4 > 2x + 6$

52. $3x - 1 > 2x + 4$ or $5x - 2 < 3x + 4$

Translate each of the following phrases into an equivalent inequality statement.

53. x is greater than -2 and at most 4.

54. x is less than 9 and at least -3.

55. x is less than -4 or at least 1.

56. x is at most 1 or more than 6.

57. Write each statement using inequality notation.
   a. x is always positive.
   b. x is never negative.
   c. x is greater than or equal to 0.

58. Match each expression on the left with a phrase on the right.

a. $x^2 \ge 0$	e. never true
b. $x^2 < 0$	f. sometimes true
c. $x^2 \le 0$	g. always true

Solve each inequality by inspection, without showing any work.

59. $x^2 < 0$

60. $x^2 \le 0$

61. $x^2 \ge 0$

62. $\frac{1}{x^2} \ge 0$

63. $\frac{1}{x^2} < 0$

64. $\frac{1}{x^2} = 0$

The next two problems are intended to give you practice reading, and paying attention to, the instructions that accompany the problems you are working.

65. Work each problem according to the instructions given.
   a. Evaluate when $x = 0$: $-\frac{1}{2}x + 1$
   b. Solve: $-\frac{1}{2}x + 1 = -7$
   c. Is 0 a solution to $-\frac{1}{2}x + 1 < -7$?
   d. Solve: $-\frac{1}{2}x + 1 < -7$

66. Work each problem according to the instructions given.
   a. Evaluate when $x = 0$: $-\frac{2}{3}x - 5$
   b. Solve: $-\frac{2}{3}x - 5 = 1$
   c. Is 0 a solution to $-\frac{2}{3}x - 5 > 1$?
   d. Solve: $-\frac{2}{3}x - 5 > 1$

Applying the Concepts

A store selling art supplies finds that they can sell x sketch pads each week at a price of p dollars each, according to the formula $x = 900 - 300p$. What price should they charge if they want to sell

67. at least 300 pads each week?

68. more than 600 pads each week?

69. less than 525 pads each week?

70. at most 375 pads each week?

71. **Amtrak** The average number of passengers carried by Amtrak declined each year for the years

1990–1996 (American Association of Railroads, Washington, DC, *Railroad Facts, Statistics of Railroads of Class 1*.) The linear model for the number of passengers carried each year by Amtrak is given by $P = 22{,}419 - 399x$, where P is the number of passengers, in millions, and x is the number of years after January 1, 1990. In what years did Amtrak have more than 20,500 million passengers?

Courtesy of Amtrak

72. Student Loan When considering how much debt to incur in student loans, you learn that it is wise to keep your student loan payment to 8% or less of your starting monthly income. Suppose you anticipate a starting annual salary of \$24,000. Set up and solve an inequality that represents the amount of monthly debt for student loans that would be considered manageable.

73. Here is what the U.S. Geological Survey has to say about the survival rates of the Apapane, one of the endemic birds of Hawaii.

> *Annual survival rates based on 1,584 recaptures of 429 banded individuals 0.72 ± 0.11 for adults and 0.13 ± 0.07 for juveniles.*

Write the survival rates using inequalities. Then give the survival rates in terms of percent.

Jack Jeffrey Photography

74. Survival Rates for Sea Gulls Here is part of a report concerning the survival rates of Western Gulls that appeared on the website of Cornell University.

> *Survival of eggs to hatching is 70%–80%; of hatched chicks to fledglings 50%–70%; of fledglings to age of first breeding <50%.*

Write the survival rates using inequalities without percent.

Frans Lanting/Minden Pictures

75. Temperature Each of the following temperature ranges is in degrees Fahrenheit. Use the formula $F = \frac{9}{5}C + 32$ to find the corresponding temperature range in degrees Celsius.

a. $95°$ to $113°$

b. $68°$ to $86°$

c. $-13°$ to $14°$

d. $-4°$ to $23°$

76. Improving Your Quantitative Literacy As you can see, the percents shown in the graph are from a survey of 770 families.

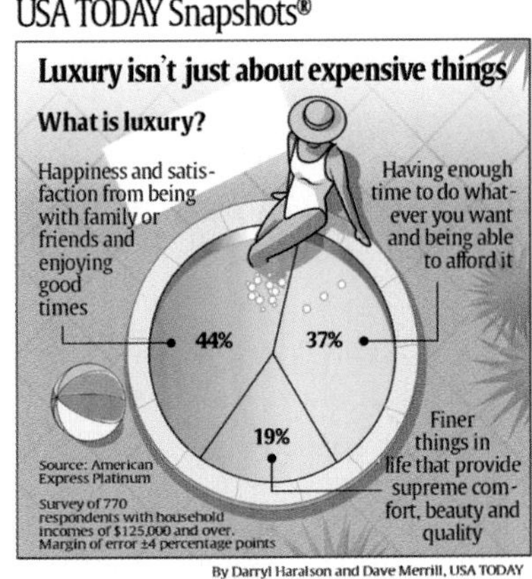

Data from *USA Today*. Copyright 2005. Reprinted with permission.

a. How many families responded that luxury is happiness and satisfaction from being with friends and family and enjoying good times?

b. If x represents the annual income of any of the families surveyed, write an inequality using x that indicates their level of income.

c. The margin of error for the survey is ±4%. What does this mean in relation to the 19% category?

Maintaining Your Skills

The problems that follow review some of the more important skills you have learned in previous sections and chapters. You can consider the time you spend working these problems as time spent studying for exams.

Simplify.

77. $|-3|$

78. $|3|$

79. $-|-3|$

80. $-(-3)$

81. Give a definition for the absolute value of x that involves the number line. (This is the geometric definition.)

82. Give a definition of the absolute value of x that does not involve the number line. (This is the algebraic definition.)

Getting Ready for the Next Section

To understand all of the explanations and examples in the next section, you must be able to work the problems below.

Solve each equation.

83. $2a - 1 = -7$

84. $3x - 6 = 9$

85. $\frac{2}{3}x - 3 = 7$

86. $\frac{2}{3}x - 3 = -7$

87. $x - 5 = x - 7$

88. $x + 3 = x + 8$

89. $x - 5 = -x - 7$

90. $x + 3 = -x + 8$

Extending the Concepts

Assume a, b, and c are positive, and solve each formula for x.

91. $ax + b < c$

92. $\frac{x}{a} + \frac{y}{b} < 1$

93. $-c < ax + b < c$

94. $-1 < \frac{ax + b}{c} < 1$

2.5 Equations with Absolute Value

OBJECTIVES

A Solve equations with absolute value symbols.

Previously we defined the absolute value of x, $|x|$, to be the distance between x and 0 on the number line. The absolute value of a number measures its distance from 0.

EXAMPLE 1 Solve for x: $|x| = 5$.

SOLUTION Using the definition of absolute value, we can read the equation as, "The distance between x and 0 on the number line is 5." If x is 5 units from 0, then x can be 5 or -5:

$$\text{If } |x| = 5 \quad \text{then } x = 5 \quad \text{or} \quad x = -5$$

In general, then, we can see that any equation of the form $|a| = b$ is equivalent to the equations $a = b$ or $a = -b$, as long as $b > 0$.

EXAMPLE 2 Solve $|2a - 1| = 7$.

SOLUTION We can read this question as "$2a - 1$ is 7 units from 0 on the number line." The quantity $2a - 1$ must be equal to 7 or -7:

$$|2a - 1| = 7$$
$$2a - 1 = 7 \quad \text{or} \quad 2a - 1 = -7$$

We have transformed our absolute value equation into two equations that do not involve absolute value. We can solve each equation separately.

$$\begin{aligned} 2a - 1 &= 7 &\text{or}\quad 2a - 1 &= -7 & \\ 2a &= 8 &\text{or}\quad 2a &= -6 & \textbf{Add 1 to both sides} \\ a &= 4 &\text{or}\quad a &= -3 & \textbf{Multiply by } \tfrac{1}{2} \end{aligned}$$

Our solution set is {4, −3}. 

To check our solutions, we put them into the original absolute value equation:

When	$a = 4$	When	$a = -3$
the equation	$\lvert 2a - 1\rvert = 7$	the equation	$\lvert 2a - 1\rvert = 7$
becomes	$\lvert 2(4) - 1\rvert = 7$	becomes	$\lvert 2(-3) - 1\rvert = 7$
	$\lvert 7\rvert = 7$		$\lvert -7\rvert = 7$
	$7 = 7$		$7 = 7$

EXAMPLE 3 Solve $\left|\frac{2}{3}x - 3\right| + 5 = 12$.

SOLUTION To use the definition of absolute value to solve this equation, we must isolate the absolute value on the left side of the equal sign. To do so, we add -5 to both sides of the equation to obtain

$$\left|\frac{2}{3}x - 3\right| = 7$$

Now that the equation is in the correct form, we can write

$$\begin{aligned} \frac{2}{3}x - 3 &= 7 &\text{or}\quad \frac{2}{3}x - 3 &= -7 & \\ \frac{2}{3}x &= 10 &\text{or}\quad \frac{2}{3}x &= -4 & \textbf{Add 3 to both sides} \\ x &= 15 &\text{or}\quad x &= -6 & \textbf{Multiply by } \tfrac{3}{2} \end{aligned}$$

The solution set is {15, −6}. 

EXAMPLE 4 Solve $|3a - 6| = -4$.

SOLUTION The solution set is ∅ because the left side cannot be negative and the right side is negative. No matter what we try to substitute for the variable a, the quantity $|3a - 6|$ will always be positive or zero. It can never be -4.

> **Note**
> Recall that ∅ is the symbol we use to denote the empty set. When we use it to indicate the solutions to an equation, then we are saying the equation has no solution.

Consider the statement $|a| = |b|$. What can we say about a and b? We know they are equal in absolute value. By the definition of absolute value, they are the same distance from 0 on the number line. They must be equal to each other or opposites of each other. In symbols, we write

$$\underset{\text{Equal in absolute value}}{|a| = |b|} \Leftrightarrow \underset{\text{Equals}}{a = b} \quad \text{or} \quad \underset{\text{Opposites}}{a = -b}$$

EXAMPLE 5 Solve $|x - 5| = |x - 7|$.

SOLUTION The quantities $x - 5$ and $x - 7$ must be equal or they must be opposites because their absolute values are equal:

Equals		*Opposites*
$x - 5 = x - 7$	or	$x - 5 = -(x - 7)$
$-5 = -7$		$x - 5 = -x + 7$
No solution here		$2x - 5 = 7$
		$2x = 12$
		$x = 6$

Because the first equation leads to a false statement, it will not give us a solution. (If either of the two equations were to reduce to a true statement, it would mean all real numbers would satisfy the original equation.) In this case, our only solution is $x = 6$.

LINKING OBJECTIVES AND EXAMPLES

Next to each **objective** we have listed the examples that are best described by that objective.

A 1–5

GETTING READY FOR CLASS

After reading through the preceding section, respond in your own words and in complete sentences.

1. Why do some of the equations in this section have two solutions instead of one?
2. Translate $|x| = 6$ into words using the definition of absolute value.
3. Explain in words what the equation $|x - 3| = 4$ means with respect to distance on the number line.
4. When is the statement $|x| = x$ true?

Problem Set 2.5

Online support materials can be found at www.thomsonedu.com/login

Use the definition of absolute value to solve each of the following equations.

1. $|x| = 4$
2. $|x| = 7$
3. $2 = |a|$
4. $5 = |a|$
5. $|x| = -3$
6. $|x| = -4$
7. $|a| + 2 = 3$
8. $|a| - 5 = 2$
9. $|y| + 4 = 3$
10. $|y| + 3 = 1$
11. $4 = |x| - 2$
12. $3 = |x| - 5$
▸ **13.** $|x - 2| = 5$
14. $|x + 1| = 2$
15. $|a - 4| = \frac{5}{3}$
16. $|a + 2| = \frac{7}{5}$
17. $1 = |3 - x|$
18. $2 = |4 - x|$
19. $\left|\frac{3}{5}a + \frac{1}{2}\right| = 1$
20. $\left|\frac{2}{7}a + \frac{3}{4}\right| = 1$
21. $60 = |20x - 40|$
22. $800 = |400x - 200|$
23. $|2x + 1| = -3$
24. $|2x - 5| = -7$
25. $\left|\frac{3}{4}x - 6\right| = 9$
26. $\left|\frac{4}{5}x - 5\right| = 15$
27. $\left|1 - \frac{1}{2}a\right| = 3$
28. $\left|2 - \frac{1}{3}a\right| = 10$

■ = Videos available by instructor request
▸ = Online student support materials available at www.thomsonedu.com/login

Solve each equation.

29. $|3x + 4| + 1 = 7$

30. $|5x - 3| - 4 = 3$

31. $|3 - 2y| + 4 = 3$

32. $|8 - 7y| + 9 = 1$

33. $3 + |4t - 1| = 8$

34. $2 + |2t - 6| = 10$

35. $\left|9 - \frac{3}{5}x\right| + 6 = 12$

36. $\left|4 - \frac{2}{7}x\right| + 2 = 14$

37. $5 = \left|\frac{2x}{7} + \frac{4}{7}\right| - 3$

38. $7 = \left|\frac{3x}{5} + \frac{1}{5}\right| + 2$

39. $2 = -8 + \left|4 - \frac{1}{2}y\right|$

40. $1 = -3 + \left|2 - \frac{1}{4}y\right|$

41. $|3a + 1| = |2a - 4|$

42. $|5a + 2| = |4a + 7|$

43. $\left|x - \frac{1}{3}\right| = \left|\frac{1}{2}x + \frac{1}{6}\right|$

44. $\left|\frac{1}{10}x - \frac{1}{2}\right| = \left|\frac{1}{5}x + \frac{1}{10}\right|$

45. $|y - 2| = |y + 3|$

46. $|y - 5| = |y - 4|$

47. $|3x - 1| = |3x + 1|$

48. $|5x - 8| = |5x + 8|$

49. $|3 - m| = |m + 4|$

50. $|5 - m| = |m + 8|$

51. $|0.03 - 0.01x| = |0.04 + 0.05x|$

52. $|0.07 - 0.01x| = |0.08 - 0.02x|$

53. $|x - 2| = |2 - x|$

54. $|x - 4| = |4 - x|$

55. $\left|\frac{x}{5} - 1\right| = \left|1 - \frac{x}{5}\right|$

56. $\left|\frac{x}{3} - 1\right| = \left|1 - \frac{x}{3}\right|$

57. Work each problem according to the instructions given.

a. Solve: $4x - 5 = 0$

b. Solve: $|4x - 5| = 0$

c. Solve: $4x - 5 = 3$

d. Solve: $|4x - 5| = 3$

e. Solve: $|4x - 5| = |2x + 3|$

58. Work each problem according to the instructions given.

a. Solve: $3x + 6 = 0$

b. Solve: $|3x + 6| = 0$

c. Solve: $3x + 6 = 4$

d. Solve: $|3x + 6| = 4$

e. Solve: $|3x + 6| = |7x + 4|$

Applying the Concepts

59. Amtrak Amtrak's annual passenger revenue for the years 1985–1995 is modeled approximately by the formula

$$R = -60|x - 11| + 962$$

where R is the annual revenue in millions of dollars and x is the number of years since January 1, 1980 (Association of American Railroads, Washington, DC, *Railroad Facts, Statistics of Railroads of Class 1*, annual). In what years was the passenger revenue $722 million?

60. Corporate Profits The corporate profits for various U.S. industries vary from year to year. An approximate model for profits of U.S. "communications companies" during a given year between 1990 and 1997 is given by

$$P = -3{,}400|x - 5.5| + 36{,}000$$

where P is the annual profits (in millions of dollars) and x is the number of years since January 1, 1990 (U.S. Bureau of Economic Analysis, Income and Product Accounts of the U.S. (1929–1994), *Survey of Current Business,* September 1998). Use the model to determine the years in which profits of "communication companies" were $31.5 billion ($31,500 million).

Maintaining Your Skills

The problems that follow review some of the more important skills you have learned in previous sections and chapters. You can consider the time you spend working these problems as time spent studying for exams.

Graph each inequality.

61. $x < -2$ or $x > 8$

62. $1 < x < 4$

63. $-2 \le x \le 1$

64. $x \le -\frac{3}{2}$ or $x \ge 3$

Simplify each expression.

65. $\frac{38}{30}$

66. $\frac{10}{25}$

67. $\frac{240}{6}$

68. $\frac{39}{13}$

69. $\frac{0+6}{0-3}$

70. $\frac{6+6}{6-3}$

71. $\frac{4-4}{4-2}$

72. $\frac{3+6}{3-3}$

Getting Ready for the Next Section

To understand all of the explanations and examples in the next section, you must be able to work the problems below.

Solve each inequality. Do not graph the solution set.

73. $2x - 5 < 3$

74. $-3 < 2x - 5$

75. $-4 \le 3a + 7$

76. $3a + 7 \le 4$

77. $4t - 3 \le -9$

78. $4t - 3 \ge 9$

Extending the Concepts

Solve each formula for x. (Assume a, b, and c are positive.)

79. $|x - a| = b$

80. $|x + a| - b = 0$

81. $|ax + b| = c$

82. $|ax - b| - c = 0$

83. $\left|\frac{x}{a} + \frac{y}{b}\right| = 1$

84. $\left|\frac{x}{a} + \frac{y}{b}\right| = c$

2.6 Inequalities Involving Absolute Value

OBJECTIVES

A Solve inequalities with absolute value and graph the solution set.

In this section we will again apply the definition of absolute value to solve inequalities involving absolute value. Again, the absolute value of x, which is $|x|$, represents the distance that x is from 0 on the number line. We will begin by considering three absolute value equations/inequalities and their English translations:

Expression	*In Words*
$\|x\| = 7$	x is exactly 7 units from 0 on the number line
$\|a\| < 5$	a is less than 5 units from 0 on the number line
$\|y\| \ge 4$	y is greater than or equal to 4 units from 0 on the number line

Once we have translated the expression into words, we can use the translation to graph the original equation or inequality. The graph then is used to write a final equation or inequality that does not involve absolute value.

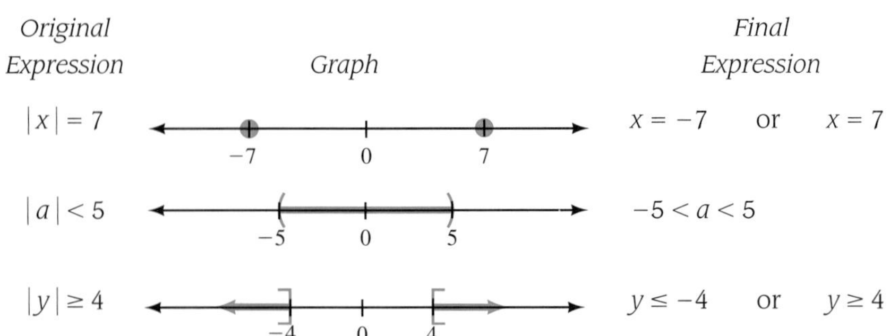

Although we will not always write out the English translation of an absolute value inequality, it is important that we understand the translation. Our second expression, $|a| < 5$, means a is within 5 units of 0 on the number line. The graph of this relationship is

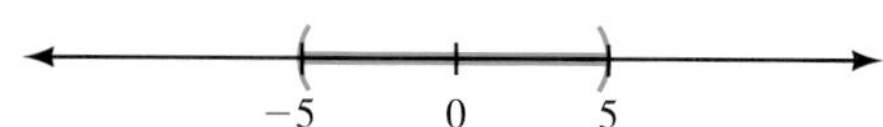

which can be written with the following continued inequality:

$$-5 < a < 5$$

We can follow this same kind of reasoning to solve more complicated absolute value inequalities.

EXAMPLE 1 Graph the solution set: $|2x - 5| < 3$.

SOLUTION The absolute value of $2x - 5$ is the distance that $2x - 5$ is from 0 on the number line. We can translate the inequality as "$2x - 5$ is less than 3 units from 0 on the number line"; that is, $2x - 5$ must appear between -3 and 3 on the number line.

A picture of this relationship is

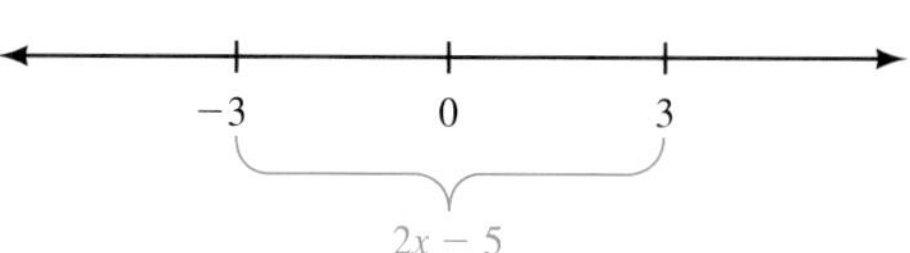

Using the picture, we can write an inequality without absolute value that describes the situation:

$$-3 < 2x - 5 < 3$$

Next, we solve the continued inequality by first adding 5 to all three members and then multiplying all three by $\frac{1}{2}$.

$$-3 < 2x - 5 < 3$$

$2 < 2x < 8$ **Add 5 to all three members**

$1 < x < 4$ **Multiply each member by $\frac{1}{2}$**

The graph of the solution set is

We can see from the solution that for the absolute value of $2x - 5$ to be within 3 units of 0 on the number line, x must be between 1 and 4.

EXAMPLE 2 Solve and graph $|3a + 7| \le 4$.

SOLUTION We can read the inequality as, "The distance between $3a + 7$ and 0 is less than or equal to 4." Or, "$3a + 7$ is within 4 units of 0 on the number line." This relationship can be written without absolute value as

$$-4 \le 3a + 7 \le 4$$

Solving as usual, we have

$$
\begin{aligned}
-4 \le 3a + 7 &\le 4 \\
-11 \le \quad 3a \quad &< -3 \qquad \textbf{Add } -7 \textbf{ to all three members} \\
-\frac{11}{3} \le \quad a \quad &\le -1 \qquad \textbf{Multiply each member by } \tfrac{1}{3}
\end{aligned}
$$

We can see from Examples 1 and 2 that to solve an inequality involving absolute value, we must be able to write an equivalent expression that does not involve absolute value.

EXAMPLE 3

Solve $|x - 3| > 5$ and graph the solution.

SOLUTION We interpret the absolute value inequality to mean that $x - 3$ is more than 5 units from 0 on the number line. The quantity $x - 3$ must be either above $+5$ or below -5. Here is a picture of the relationship:

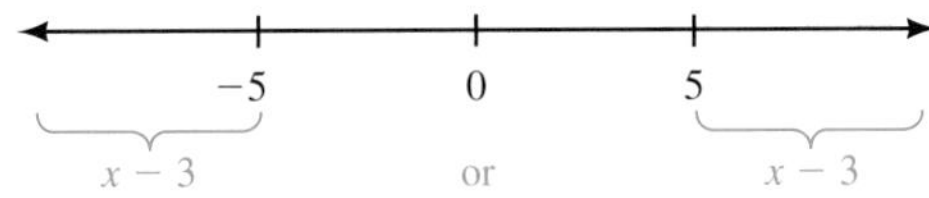

An inequality without absolute value that also describes this situation is

$$x - 3 < -5 \quad \text{or} \quad x - 3 > 5$$

Adding 3 to both sides of each inequality we have

$$x < -2 \quad \text{or} \quad x > 8$$

The graph of which is

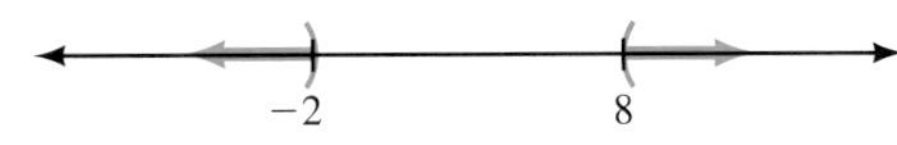

EXAMPLE 4

Graph the solution set: $|4t - 3| \ge 9$.

SOLUTION The quantity $4t - 3$ is greater than or equal to 9 units from 0. It must be either above $+9$ or below -9.

$$
\begin{aligned}
4t - 3 \le -9 \quad &\text{or} \quad 4t - 3 \ge 9 \\
4t \le -6 \quad &\text{or} \quad 4t \ge 12 \qquad \textbf{Add 3} \\
t \le -\frac{6}{4} \quad &\text{or} \quad t \ge \frac{12}{4} \qquad \textbf{Multiply by } \tfrac{1}{4} \\
t \le -\frac{3}{2} \quad &\text{or} \quad t \ge 3
\end{aligned}
$$

We can use the results of our first few examples and the material in the previous section to summarize the information we have related to absolute value equations and inequalities.

Rewriting Absolute Value Equations and Inequalities If c is a positive real number, then each of the following statements on the left is equivalent to the corresponding statement on the right.

With Absolute Value	*Without Absolute Value*
$\lvert x \rvert = c$	$x = -c$ or $x = c$
$\lvert x \rvert < c$	$-c < x < c$
$\lvert x \rvert > c$	$x < -c$ or $x > c$
$\lvert ax + b \rvert = c$	$ax + b = -c$ or $ax + b = c$
$\lvert ax + b \rvert < c$	$-c < ax + b < c$
$\lvert ax + b \rvert > c$	$ax + b < -c$ or $ax + b > c$

EXAMPLE 5

Solve and graph $\lvert 2x + 3 \rvert + 4 < 9$.

SOLUTION Before we can apply the method of solution we used in the previous examples, we must isolate the absolute value on one side of the inequality. To do so, we add $\mathbf{-4}$ to each side.

$$\lvert 2x + 3 \rvert + 4 < 9$$

$$\lvert 2x + 3 \rvert + 4 + (\mathbf{-4}) < 9 + (\mathbf{-4})$$

$$\lvert 2x + 3 \rvert < 5$$

From this last line we know that $2x + 3$ must be between -5 and $+5$.

$$-5 < 2x + 3 < 5$$

$$-8 < 2x < 2 \quad \text{Add } -3 \text{ to each member}$$

$$-4 < x < 1 \quad \text{Multiply each member by } \tfrac{1}{2}$$

The graph is

EXAMPLE 6

Solve and graph $\lvert 4 - 2t \rvert > 2$.

SOLUTION The inequality indicates that $4 - 2t$ is less than -2 or greater than $+2$. Writing this without absolute value symbols, we have

$$4 - 2t < -2 \quad \text{or} \quad 4 - 2t > 2$$

To solve these inequalities we begin by adding $\mathbf{-4}$ to each side.

$$4 + (\mathbf{-4}) - 2t < -2 + (\mathbf{-4}) \quad \text{or} \quad 4 + (\mathbf{-4}) - 2t > 2 + (\mathbf{-4})$$

$$-2t < -6 \quad \text{or} \quad -2t > -2$$

Next we must multiply both sides of each inequality by $-\frac{1}{2}$. When we do so, we must also reverse the direction of each inequality symbol.

$$-2t < -6 \quad \text{or} \quad -2t > -2$$

$$-\tfrac{1}{2}(-2t) > -\tfrac{1}{2}(-6) \quad \text{or} \quad -\tfrac{1}{2}(-2t) < -\tfrac{1}{2}(-2)$$

$$t > 3 \quad \text{or} \quad t < 1$$

Note

Remember, the multiplication property for inequalities requires that we reverse the direction of the inequality symbol *every* time we multiply both sides of an inequality by a negative number.

Although in situations like this we are used to seeing the "less than" symbol written first, the meaning of the solution is clear. We want to graph all real numbers that are either greater than 3 or less than 1. Here is the graph.

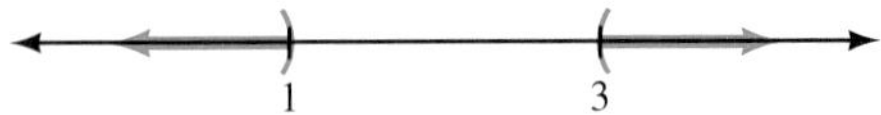

Because absolute value always results in a nonnegative quantity, we sometimes come across special solution sets when a negative number appears on the right side of an absolute value inequality.

EXAMPLE 7

Solve $|7y - 1| < -2$.

SOLUTION The *left* side is never negative because it is an absolute value. The *right* side is negative. We have a positive quantity (or zero) less than a negative quantity, which is impossible. The solution set is the empty set, $\varnothing$. There is no real number to substitute for y to make this inequality a true statement.

EXAMPLE 8

Solve $|6x + 2| > -5$.

SOLUTION This is the opposite case from that in Example 7. No matter what real number we use for x on the *left* side, the result will always be positive, or zero. The *right* side is negative. We have a positive quantity (or zero) greater than a negative quantity. Every real number we choose for x gives us a true statement. The solution set is the set of all real numbers.

LINKING OBJECTIVES AND EXAMPLES

Next to each **objective** we have listed the examples that are best described by that objective.

A 1–8

GETTING READY FOR CLASS

After reading through the preceding section, respond in your own words and in complete sentences.

1. Write an inequality containing absolute value, the solution to which is all the numbers between -5 and 5 on the number line.
2. Translate $|x| \geq 3$ into words using the definition of absolute value.
3. Explain in words what the inequality $|x - 5| < 2$ means with respect to distance on the number line.
4. Why is there no solution to the inequality $|2x - 3| < 0$?

Problem Set 2.6

Online support materials can be found at www.thomsonedu.com/login

Solve each of the following inequalities using the definition of absolute value. Graph the solution set in each case.

1. $|x| < 3$
2. $|x| \leq 7$
3. $|x| \geq 2$
4. $|x| > 4$
5. $|x| + 2 < 5$
6. $|x| - 3 < -1$
7. $|t| - 3 > 4$

= Videos available by instructor request

▸ = Online student support materials available at www.thomsonedu.com/login

8. $|t| + 5 > 8$

9. $|y| < -5$

10. $|y| > -3$

11. $|x| \ge -2$

12. $|x| \le -4$

13. $|x - 3| < 7$

14. $|x + 4| < 2$

15. $|a + 5| \ge 4$

16. $|a - 6| \ge 3$

Solve each inequality and graph the solution set.

17. $|a - 1| < -3$

18. $|a + 2| \ge -5$

19. $|2x - 4| < 6$

20. $|2x + 6| < 2$

▶ 21. $|3y + 9| \ge 6$

22. $|5y - 1| \ge 4$

23. $|2k + 3| \ge 7$

24. $|2k - 5| \ge 3$

25. $|x - 3| + 2 < 6$

26. $|x + 4| - 3 < -1$

27. $|2a + 1| + 4 \ge 7$

28. $|2a - 6| - 1 \ge 2$

▶ 29. $|3x + 5| - 8 < 5$

30. $|6x - 1| - 4 \le 2$

Solve each inequality, and graph the solution set. Keep in mind that if you multiply or divide both sides of an inequality by a negative number, you must reverse the inequality sign.

31. $|5 - x| > 3$

32. $|7 - x| > 2$

33. $\left|3 - \frac{2}{3}x\right| \ge 5$

34. $\left|3 - \frac{3}{4}x\right| \ge 9$

35. $\left|2 - \frac{1}{2}x\right| > 1$

36. $\left|3 - \frac{1}{3}x\right| > 1$

Solve each inequality.

37. $|x - 1| < 0.01$

38. $|x + 1| < 0.01$

39. $|2x + 1| \ge \frac{1}{5}$

40. $|2x - 1| \ge \frac{1}{8}$

41. $\left|\frac{3x - 2}{5}\right| \le \frac{1}{2}$

42. $\left|\frac{4x - 3}{2}\right| \le \frac{1}{3}$

43. $\left|2x - \frac{1}{5}\right| < 0.3$

44. $\left|3x - \frac{3}{5}\right| < 0.2$

45. Write the continued inequality $-4 \le x \le 4$ as a single inequality involving absolute value.

46. Write the continued inequality $-8 \le x \le 8$ as a single inequality involving absolute value.

47. Write $-1 \le x - 5 \le 1$ as a single inequality involving absolute value.

48. Write $-3 \le x + 2 \le 3$ as a single inequality involving absolute value.

49. Work each problem according to the instructions given.
   a. Evaluate when $x = 0$: $|5x + 3|$
   b. Solve: $|5x + 3| = 7$
   c. Is 0 a solution to $|5x + 3| > 7$?
   d. Solve: $|5x + 3| > 7$

50. Work each problem according to the instructions given.
   a. Evaluate when $x = 0$: $|-2x - 5|$
   b. Solve: $|-2x - 5| = 1$
   c. Is 0 a solution to $|-2x - 5| > 1$?
   d. Solve: $|-2x - 5| > 1$

Applying the Concepts

51. **Speed Limits** The interstate speed limit for cars is 75 miles per hour in Nebraska, Nevada, New Mexico, Oklahoma, South Dakota, Utah, and Wyoming and is the highest in the United States. To discourage passing, minimum speeds are also posted, so that the difference between the fastest and slowest

moving traffic is no more than 20 miles per hour. Therefore, the speed x of a car must satisfy the inequality $55 \le x \le 75$. Write this inequality as an absolute value inequality.

52. **Wavelengths of Light** When white light from the sun passes through a prism, it is broken down into bands of light that form colors. The wavelength, v, (in nanometers) of some common colors are:

Blue:	$424 < v < 491$
Green:	$491 < v < 575$
Yellow:	$575 < v < 585$
Orange:	$585 < v < 647$
Red:	$647 < v < 700$

When a fireworks display made of copper is burned, it lets out light with wavelengths, v, that satisfy the relationship $|v - 455| < 23$. Write this inequality without absolute values; find the range of possible values for v; and then, using the preceding list of wavelengths, determine the color of that copper fireworks display.

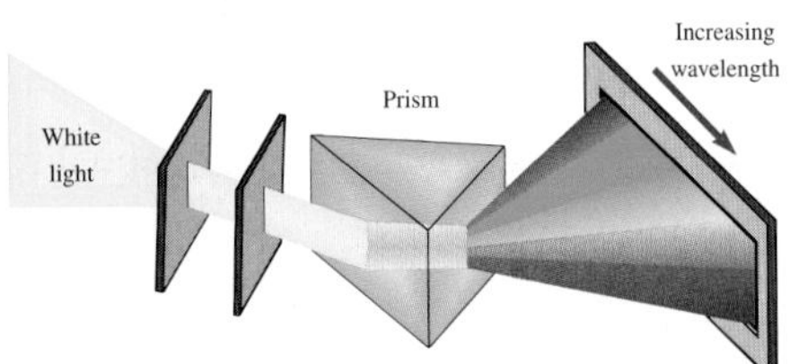

Maintaining Your Skills

The problems that follow review some of the more important skills you have learned in previous sections and chapters. You can consider the time you spend working these problems as time spent studying for exams.

Simplify each expression as much as possible.

53. $-9 \div \frac{3}{2}$

54. $-\frac{4}{5} \div (-4)$

55. $3 - 7(-6 - 3)$

56. $(3 - 7)(-6 - 3)$

57. $-4(-2)^3 - 5(-3)^2$

58. $4(2 - 5)^3 - 3(4 - 5)^5$

59. $-2(-3 + 8) - 7(-9 + 6)$

60. $-3 - 6[5 - 2(-3 - 1)]$

61. $\dfrac{2(-3) - 5(-6)}{-1 - 2 - 3}$

62. $\dfrac{4 - 8(3 - 5)}{2 - 4(3 - 5)}$

63. $6(1) - 1(-5) + 1(2)$

64. $-2(-14) + 3(-4) - 1(-10)$

65. $1(0)(1) + 3(1)(4) + (-2)(2)(-1)$

66. $-3(-1 - 1) + 4(-2 + 2) - 5[2 - (-2)]$

67. $4(-1)^2 + 3(-1) - 2$

68. $5 \cdot 2^3 - 3 \cdot 2^2 + 4 \cdot 2 - 5$

Extending the Concepts

Solve each formula for x. (Assume a, b, and c are positive.)

69. $|x - a| < b$

70. $|x - a| > b$

71. $|ax - b| > c$

72. $|ax - b| < c$

73. $|ax + b| \le c$

74. $|ax + b| \ge c$

Chapter 2 SUMMARY

EXAMPLES

1 We can solve

$$x + 3 = 5$$

by adding $\mathbf{-3}$ to both sides:

$$x + 3 + (\mathbf{-3}) = 5 + (\mathbf{-3})$$
$$x = 2$$

Addition Property of Equality [2.1]

For algebraic expressions A, B, and C,

if $\quad A = B$

then $\quad A + C = B + C$

This property states that we can add the same quantity to both sides of an equation without changing the solution set.

2. We can solve $3x = 12$ by multiplying both sides by $\frac{1}{3}$.

$$3x = 12$$
$$\tfrac{1}{3}(3x) = \tfrac{1}{3}(12)$$
$$x = 4$$

Multiplication Property of Equality [2.1]

For algebraic expressions A, B, and C,

if $\quad A = B$

then $\quad AC = BC \quad C \neq 0$

Multiplying both sides of an equation by the same nonzero quantity never changes the solution set.

3. Solve: $3(2x - 1) = 9$.

$$3(2x - 1) = 9$$
$$6x - 3 = 9$$
$$6x - 3 + \mathbf{3} = 9 + \mathbf{3}$$
$$6x = 12$$
$$\tfrac{1}{6}(6x) = \tfrac{1}{6}(12)$$
$$x = 2$$

Strategy for Solving Linear Equations in One Variation [2.1]

Step 1: **a.** Use the distributive property to separate terms, if necessary.

b. If fractions are present, consider multiplying both sides by the LCD to eliminate the fractions. If decimals are present, consider multiplying both sides by a power of 10 to clear the equation of decimals.

c. Combine similar terms on each side of the equation.

Step 2: Use the addition property of equality to get all variable terms on one side of the equation and all constant terms on the other side. A variable term is a term that contains the variable (for example, $5x$). A constant term is a term that does not contain the variable (the number 3, for example).

Step 3: Use the multiplication property of equality to get x (that is, $1x$) by itself on one side of the equation.

Step 4: Check your solution in the original equation to be sure that you have not made a mistake in the solution process.

Formulas [2.2]

4. Solve for w:

$$P = 2l + 2w$$
$$P - 2l = 2w$$
$$\frac{P - 2l}{2} = w$$

A *formula* in algebra is an equation involving more than one variable. To solve a formula for one of its variables, simply isolate that variable on one side of the equation.

Blueprint for Problem Solving [2.3]

5. The perimeter of a rectangle is 32 inches. If the length is 3 times the width, find the dimensions.

Step 1: This step is done mentally.

Step 2: Let x = the width. Then the length is $3x$.

Step 3: The perimeter is 32; therefore

$$2x + 2(3x) = 32$$

Step 4:

$$8x = 32$$
$$x = 4$$

Step 5: The width is 4 inches. The length is 3(4) = 12 inches.

Step 6: The perimeter is 2(4) + 2(12), which is 32. The length is 3 times the width.

Step 1: ***Read*** the problem, and then mentally ***list*** the items that are known and the items that are unknown.

Step 2: ***Assign a variable*** to one of the unknown items. (In most cases this will amount to letting x = the item that is asked for in the problem.) Then ***translate*** the other ***information*** in the problem to expressions involving the variable.

Step 3: ***Reread*** the problem, and then ***write an equation,*** using the items and variables listed in steps 1 and 2, that describes the situation.

Step 4: ***Solve the equation*** found in step 3.

Step 5: ***Write your answer*** using a complete sentence.

Step 6: ***Reread*** the problem, and ***check*** your solution with the original words in the problem.

Addition Property for Inequalities [2.4]

6. Adding 5 to both sides of the inequality $x - 5 < -2$ gives

$$x - 5 + \mathbf{5} < -2 + \mathbf{5}$$
$$x < 3$$

For expressions A, B, and C,

if $\quad A < B$

then $\quad A + C < B + C$

Adding the same quantity to both sides of an inequality never changes the solution set.

Multiplication Property for Inequalities [2.4]

7. Multiplying both sides of $-2x \geq 6$ by $-\frac{1}{2}$ gives

$$-2x \geq 6$$
$$-\tfrac{1}{2}(-2x) \leq -\tfrac{1}{2}(6)$$
$$x \leq -3$$

For expressions A, B, and C,

if	$A < B$		
then	$AC < BC$	if	$C > 0$ (C is positive)
or	$AC > BC$	if	$C < 0$ (C is negative)

We can multiply both sides of an inequality by the same nonzero number without changing the solution set as long as each time we multiply by a negative number we also reverse the direction of the inequality symbol.

Absolute Value Equations [2.5]

8. To solve

$$|2x - 1| + 2 = 7$$

we first isolate the absolute value on the left side by adding -2 to each side to obtain

$$|2x - 1| = 5$$
$$2x - 1 = 5 \quad \text{or} \quad 2x - 1 = -5$$
$$2x = 6 \quad \text{or} \quad 2x = -4$$
$$x = 3 \quad \text{or} \quad x = -2$$

To solve an equation that involves absolute value, we isolate the absolute value on one side of the equation and then rewrite the absolute value equation as two separate equations that do not involve absolute value. In general, if b is a positive real number, then

$$|a| = b \quad \text{is equivalent to} \quad a = b \quad \text{or} \quad a = -b$$

Absolute Value Inequalities [2.6]

9. To solve

$$|x - 3| + 2 < 6$$

we first add -2 to both sides to obtain

$$|x - 3| < 4$$

which is equivalent to

$$-4 < x - 3 < 4$$
$$-1 < x < 7$$

To solve an inequality that involves absolute value, we first isolate the absolute value on the left side of the inequality symbol. Then we rewrite the absolute value inequality as an equivalent continued or compound inequality that does not contain absolute value symbols. In general, if b is a positive real number, then

$$|a| < b \quad \text{is equivalent to} \quad -b < a < b$$

and

$$|a| > b \quad \text{is equivalent to} \quad a < -b \quad \text{or} \quad a > b$$

! COMMON MISTAKES

A very common mistake in solving inequalities is to forget to reverse the direction of the inequality symbol when multiplying both sides by a negative number. When this mistake occurs, the graph of the solution set is always drawn on the wrong side of the endpoint.

Chapter 2 Review Test

The problems below form a comprehensive review of the material in this chapter. They can be used to study for exams. If you would like to take a practice test on this chapter, you can use the odd-numbered problems. Give yourself an hour and work as many of the odd-numbered problems as possible. When you are finished, or when an hour has passed, check your answers with the answers in the back of the book. You can use the even-numbered problems for a second practice test.

Solve each equation [2.1]

1. $x - 3 = 7$

2. $5x - 2 = 8$

3. $400 - 100a = 200$

4. $5 - \frac{2}{3}a = 7$

5. $4x - 2 = 7x + 7$

6. $\frac{3}{2}x - \frac{1}{6} = -\frac{7}{6}x - \frac{1}{6}$

7. $7y - 5 - 2y = 2y - 3$

8. $\frac{3y}{4} - \frac{1}{2} + \frac{3y}{2} = 2 - y$

9. $3(2x + 1) = 18$

10. $-\frac{1}{2}(4x - 2) = -x$

11. $8 - 3(2t + 1) = 5(t + 2)$

12. $8 + 4(1 - 3t) = -3(t - 4) + 2$

Substitute the given values in each formula and then solve for the variable that does not have a numerical replacement. [2.2]

13. $P = 2b + 2h$: $P = 40$, $b = 3$

14. $A = P + Prt$: $A = 2{,}000$, $P = 1{,}000$, $r = 0.05$

Solve each formula for the indicated variable. [2.2]

15. $I = prt$ for p

16. $y = mx + b$ for x

17. $4x - 3y = 12$ for y

18. $d = vt + 16t^2$ for v

Solve each formula for y. [2.2]

19. $5x + 3y - 6 = 0$

20. $\frac{x}{3} + \frac{y}{2} = 1$

21. $y + 3 = -2(x - 1)$

22. $\frac{y + 2}{x - 1} = -3$

Solve each application. In each case, be sure to show the equation that describes the situation. [2.3]

23. Geometry The length of a rectangle is 3 times the width. The perimeter is 32 feet. Find the length and width.

24. Geometry The three sides of a triangle are given by three consecutive integers. If the perimeter is 12 meters, find the length of each side.

25. Brick Laying The formula $N = 7 \cdot L \cdot H$ gives the number of standard bricks needed in a wall of L feet long and H feet high and is called the *bricklayer's formula.*

a. How many bricks would be required in a wall 45 feet long by 12 feet high?

b. An 8-foot-high wall is to be built from a load of 35,000 bricks. What length wall can be built?

26. Salary A teacher has a salary of $25,920 for her second year on the job. If this is 4.2% more than her first-year salary, how much did she earn her first year?

Solve each inequality. Write your answer using interval notation. [2.4]

27. $-8a > -4$

28. $6 - a \geq -2$

29. $\frac{3}{4}x + 1 \leq 10$

30. $800 - 200x < 1{,}000$

31. $\frac{1}{3} \leq \frac{1}{6}x \leq 1$

32. $-0.01 \leq 0.02x - 0.01 \leq 0.01$

33. $5t + 1 \leq 3t - 2$ or $-7t \leq -21$

34. $3(x + 1) < 2(x + 2)$ or $2(x - 1) \geq x + 2$

Solve each equation. [2.5]

35. $|x| = 2$

36. $|a| - 3 = 1$

37. $|x - 3| = 1$

38. $|2y - 3| = 5$

39. $|4x - 3| + 2 = 11$

40. $\left|\frac{7}{3} - \frac{x}{3}\right| + \frac{4}{3} = 2$

41. $|5t - 3| = |3t - 5|$

42. $\left|\frac{1}{2} - x\right| = \left|x + \frac{1}{2}\right|$

Solve each inequality and graph the solution set. [2.6]

43. $|x| < 5$

44. $|0.01a| \geq 5$

45. $|x| < 0$

46. $|2t + 1| - 3 < 2$

Solve each equation or inequality, if possible. [2.1, 2.4, 2.5, 2.6]

47. $2x - 3 = 2(x - 3)$

48. $3(5x - \frac{1}{2}) = 15x + 2$

49. $|4y + 8| = -1$

50. $|x| > 0$

51. $|5 - 8t| + 4 \leq 1$

52. $|2x + 1| \geq -4$

Chapter 2 Projects

Linear Equations and Inequalities in One Variable

GROUP PROJECT Finding the Maximum Height of a Model Rocket

Number of People 3

Time Needed 20 minutes

Equipment Paper and pencil

Background In this chapter, we used formulas to do some table building. Once we have a table, it is sometimes possible to use just the table information to extend what we know about the situation described by the table. In this project, we take some basic information from a table and then look for patterns among the table entries. Once we have established the patterns, we continue them and, in so doing, solve a realistic application problem.

Procedure A model rocket is launched into the air. Table 1 gives the height of the rocket every second after takeoff for the first 5 seconds. Figure 1 is a graphical representation of the information in Table 1.

TABLE 1
Height of a Model Rocket

Time (seconds)	Height (feet)
0	0
1	176
2	320
3	432
4	512
5	560

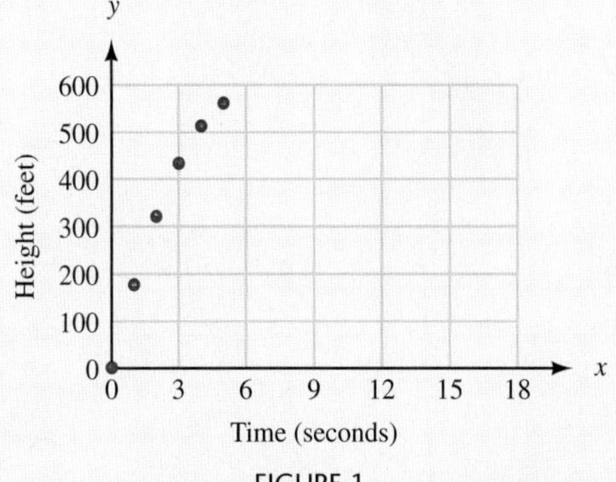

FIGURE 1

1. Table 1 is shown again in the second column with two new columns. Fill in the first five entries in the First Differences column by finding the difference of consecutive heights. For example, the second entry in the First Differences column will be the difference of 320 and 176, which is 144.

TABLE 1
Height of a Model Rocket

Time (seconds)	Height (feet)	First Differences	Second Differences
0	0		
1	176	176	
2	320	144	32
3	432		
4	512		
5	560		

2. Start filling in the Second Differences column by finding the differences of the First Differences.
3. Once you see the pattern in the Second Differences table, fill in the rest of the entries.
4. Now, using the results in the Second Differences table, go back and complete the First Differences table.
5. Now, using the results in the First Differences table, go back and complete the Heights column in the original table.
6. Plot the rest of the points from Table 1 on the graph in Figure 1.
7. What is the maximum height of the rocket?
8. How long was the rocket in the air?

The Equal Sign

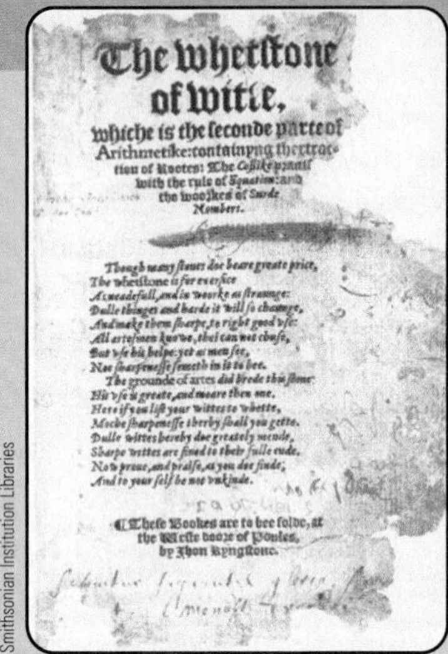
The whetstone of witte,
whiche is the seconde parte of
Arithmetike: containyng thextraction of Rootes: The Cossike practise,
with the rule of Equation: and
the woorkes of Surde
Nombers.

Though many stones doe beare greate price,
The whetstone is for exercise
As neadefull, and in woorke as straunge:
Dulle thinges and harde it will so chaunge,
And make them sharpe, to right good use:
All artesmen knowe, thei can not chuse,
But use his helpe: yet as men see,
Noe sharpenesse semeth in it to bee.
The grounde of artes did brede this stone:
His use is greate, and moare then one.
Here if you list your wittes to whette,
Moche sharpenesse therby shall you gette.
Dulle wittes hereby doe greately mende,
Sharpe wittes are fined to their fulle ende.
Now proue, and praise, as you doe finde,
And to your self be not unkinde.

These Bookes are to bee solde, at
the Weste doore of Poules,
by Jhon Kyngstone.

Smithsonian Institution Libraries

We have been using the equal sign, =, for some time now. The first published use of the symbol was in 1557, with the publication of *The Whetstone of Witte* by the English mathematician and physician Robert Recorde. Research the first use of the symbols we use for addition, subtraction, multiplication, and division and then write an essay on the subject from your results.

Equations and Inequalities in Two Variables

3

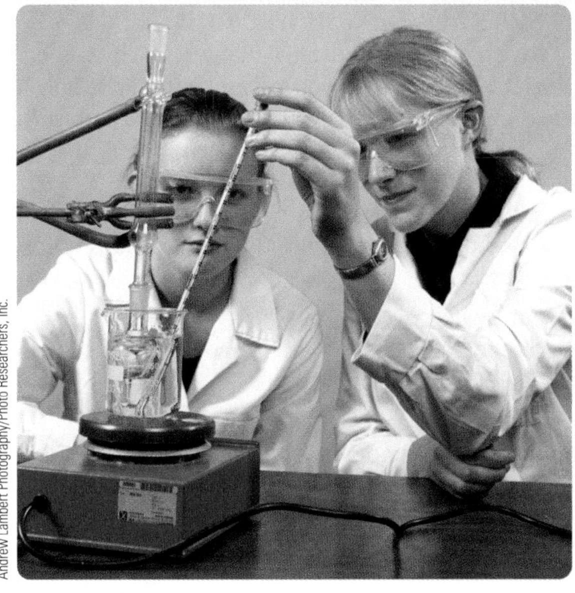

Andrew Lambert Photography/Photo Researchers, Inc.

CHAPTER OUTLINE

A student is heating water in a chemistry lab. As the water heats, she records the temperature readings from two thermometers, one giving temperature in degrees Fahrenheit and the other in degrees Celsius. The table below shows some of the data she collects. Figure 1 is a scatter diagram that gives a visual representation of the data in the table.

Corresponding Temperatures

in Degrees Fahrenheit	in Degrees Celsius
77	25
95	35
167	75
212	100

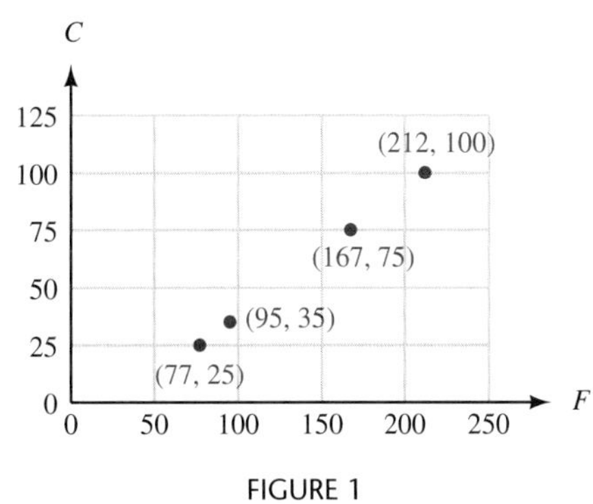

FIGURE 1

The exact relationship between the Fahrenheit and Celsius temperature scales is given by the formula

$$C = \frac{5}{9}(F - 32)$$

We have three ways to describe the relationship between the two temperature scales: a table, a graph, and an equation. But, most important to us, we don't need to accept this formula on faith, because the material we will present in this chapter gives us the ability to derive the formula from the data in the table above.

Try to arrange your daily study habits so that you have very little studying to do the night before your next exam. The next two goals will help you achieve this.

1 Review with the Exam in Mind

Each day you should review material that will be covered on the next exam. Your review should consist of working problems. Preferably, the problems you work should be problems from your list of difficult problems.

2 Pay Attention to Instructions

Each of the following is a valid instruction with respect to the equation $y = 3x - 2$, and the result of applying the instructions will be different in each case:

Find x when y is 10.
Solve for x.
Graph the equation.
Find the intercepts.

There are many things to do with the equation $y = 3x - 2$. If you train yourself to pay attention to the instructions that accompany a problem as you work through the assigned problems, you will not find yourself confused about what to do with a problem when you see it on a test.

3.1 Paired Data and the Rectangular Coordinate System

OBJECTIVES

A Graph ordered pairs on a rectangular coordinate system.

B Graph linear equations by finding intercepts or by making a table.

C Graph horizontal and vertical lines.

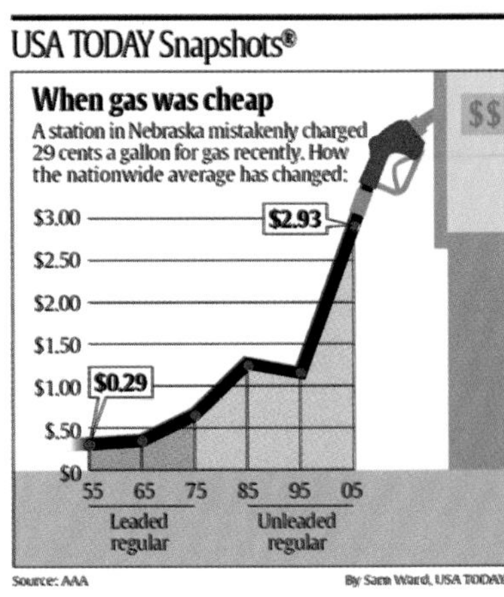

In this section, we place our work with charts and graphs in a more formal setting. Our foundation will be the *rectangular coordinate system*, because it gives us a link between algebra and geometry. With it we notice relationships between certain equations and different lines and curves. As we observe and catalog these relationships, we will, from time to time, go back out to the charts and graphs we see in the media (such as the one shown here) and see how they fit in with the work we are doing in this chapter. The result will be a working knowledge of some very important topics in mathematics and a better understanding of the information that comes to us from the media.

Table 1 gives the net price of a popular intermediate algebra text at the beginning of each year in which a new edition was published. (The net price is the price the bookstore pays for the book, not the price you pay for it.)

TABLE 1
Price of a Textbook

Edition	Year Published	Net Price ($)
First	1991	30.50
Second	1995	39.25
Third	1999	47.50
Fourth	2003	55.00
Fifth	2007	65.75

The information in Table 1 is represented visually in Figures 1 and 2. The diagram in Figure 1 is called a *bar chart.* The diagram in Figure 2 is called a *line graph.* The data in Table 1 are called *paired data* because each number in the year column is paired with a specific number in the price column.

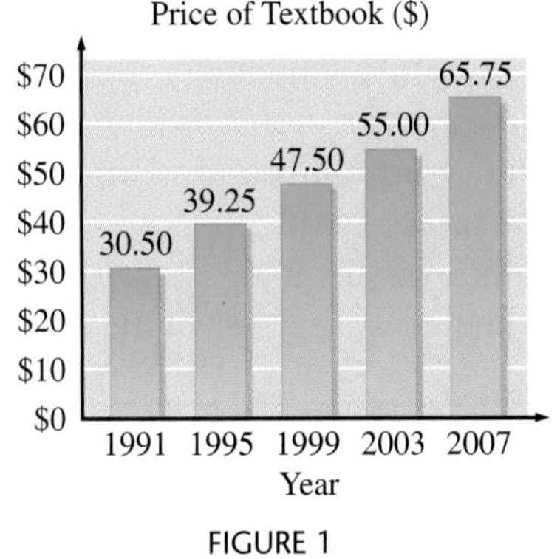

FIGURE 1

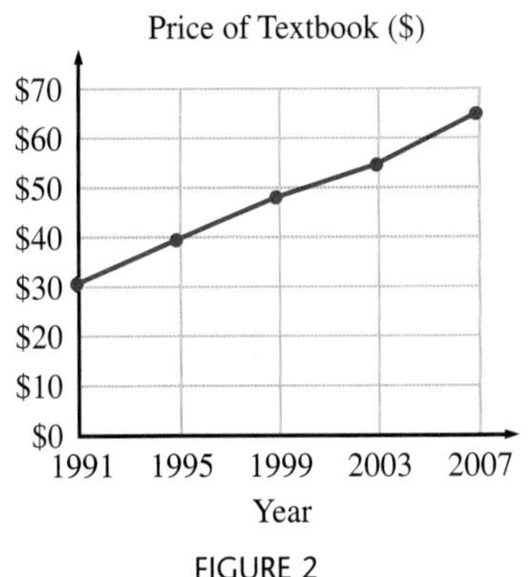

FIGURE 2

Ordered Pairs

Paired data play an important role in equations that contain two variables. Working with these equations is easier if we standardize the terminology and notation associated with paired data. So here is a definition that will do just that.

> **DEFINITION** A pair of numbers enclosed in parentheses and separated by a comma, such as $(-2, 1)$, is called an **ordered pair** of numbers. The first number in the pair is called the ***x*-coordinate** of the ordered pair; the second number is called the ***y*-coordinate.** For the ordered pair $(-2, 1)$, the x-coordinate is -2 and the y-coordinate is 1.

To standardize the way in which we display paired data visually, we use a rectangular coordinate system. A *rectangular coordinate system* is made by drawing two real number lines at right angles to each other. The two number lines, called *axes,* cross each other at 0. This point is called the *origin.* Positive directions are to the right and up. Negative directions are down and to the left. The rectangular coordinate system is shown in Figure 3.

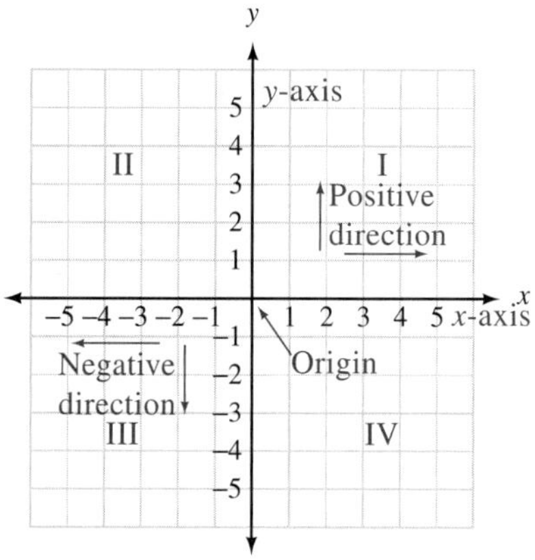

> **Note**
> A rectangular coordinate system allows us to connect algebra and geometry by associating geometric shapes (the curves shown in the diagrams) with algebraic equations. The French philosopher and mathematician René Descartes (1596–1650) usually is credited with the invention of the rectangular coordinate system, which often is referred to as the *Cartesian coordinate system* in his honor. As a philosopher, Descartes is responsible for the statement, "I think, therefore, I am." Until Descartes invented his coordinate system in 1637, algebra and geometry were treated as separate subjects.

FIGURE 3

The horizontal number line is called the *x-axis* and the vertical number line is called the *y-axis.* The two number lines divide the coordinate system into four quadrants, which we number I through IV in a counterclockwise direction. Points on the axes are not considered as being in any quadrant.

To graph the ordered pair (a, b) on a rectangular system, we start at the origin and move a units right or left (right if a is positive, left if a is negative). Then we move b units up or down (up if b is positive and down if b is negative). The point where we end up is the graph of the ordered pair (a, b).

EXAMPLE 1 Plot (graph) the ordered pairs $(2, 5)$, $(-2, 5)$, $(-2, -5)$, and $(2, -5)$.

SOLUTION To graph the ordered pair $(2, 5)$, we start at the origin and move 2 units to the right, then 5 units up. We are now at the point whose coordinates are $(2, 5)$. We graph the other three ordered pairs in a similar manner (see Figure 4).

> *Note*
> From Example 1 we see that any point in quadrant I has both its x- and y-coordinates positive $(+, +)$. Points in quadrant II have negative x-coordinates and positive y-coordinates $(-, +)$. In quadrant III both coordinates are negative $(-, -)$. In quadrant IV the form is $(+, -)$.

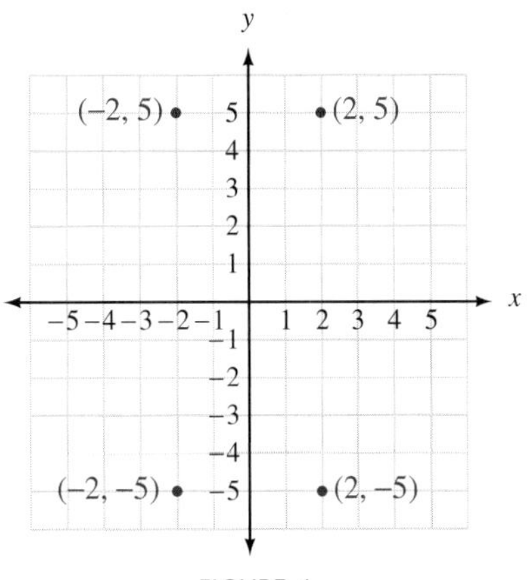

FIGURE 4

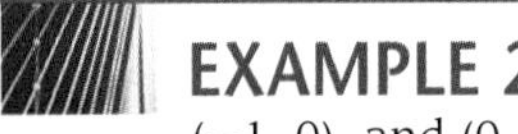

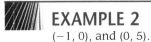

EXAMPLE 2 Graph the ordered pairs $(1, -3)$, $(\frac{1}{2}, 2)$, $(3, 0)$, $(0, -2)$, $(-1, 0)$, and $(0, 5)$.

SOLUTION

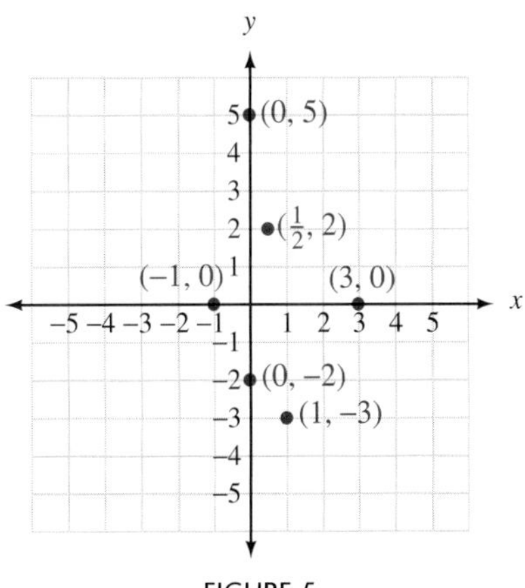

FIGURE 5

From Figure 5 we see that any point on the x-axis has a y-coordinate of 0 (it has no vertical displacement), and any point on the y-axis has an x-coordinate of 0 (no horizontal displacement).

Linear Equations

We can plot a single point from an ordered pair, but to draw a line, we need two points or an equation in two variables.

> **DEFINITION** Any equation that can be put in the form $ax + by = c$, where a, b, and c are real numbers and a and b are not both 0, is called a **linear equation in two variables.** The graph of any equation of this form is a straight line (that is why these equations are called "linear"). The form $ax + by = c$ is called **standard form.**

To graph a linear equation in two variables, we simply graph its solution set; that is, we draw a line through all the points whose coordinates satisfy the equation.

EXAMPLE 3

Graph the equation $y = -\frac{1}{3}x + 2$.

SOLUTION We need to find three ordered pairs that satisfy the equation. To do so, we can let x equal any numbers we choose and find corresponding values of y. But since every value of x we substitute into the equation is going to be multiplied by $-\frac{1}{3}$, let's use numbers for x that are divisible by 3, like -3, 0, and 3. That way, when we multiply them by $-\frac{1}{3}$, the result will be an integer.

$$\text{Let } x = -3; \quad y = -\frac{1}{3}(-3) + 2$$
$$y = 1 + 2$$
$$y = 3$$

The ordered pair $(-3, 3)$ is one solution.

$$\text{Let } x = 0; \quad y = -\frac{1}{3}(0) + 2$$
$$y = 0 + 2$$
$$y = 2$$

The ordered pair $(0, 2)$ is a second solution.

$$\text{Let } x = 3; \quad y = -\frac{1}{3}(3) + 2$$
$$y = -1 + 2$$
$$y = 1$$

The ordered pair $(3, 1)$ is a third solution.

x	y
-3	3
0	2
3	1

Plotting the ordered pairs $(-3, 3)$, $(0, 2)$, and $(3, 1)$ and drawing a straight line through their graphs, we have the graph of the equation $y = -\frac{1}{3}x + 2$, as shown in Figure 6.

Note
It takes only two points to determine a straight line. We have included a third point for "insurance." If all three points do not line up in a straight line, we have made a mistake.

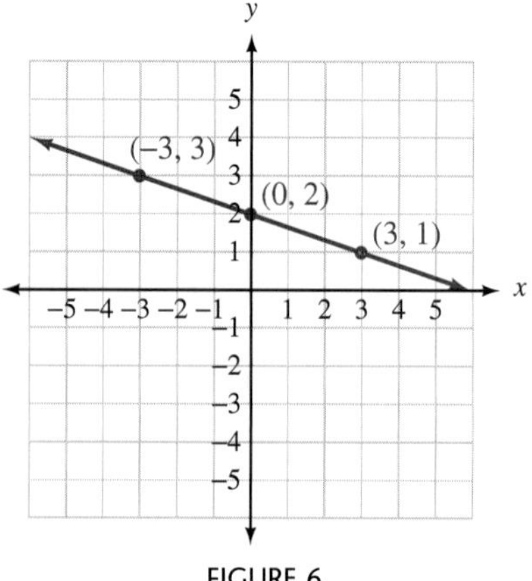

FIGURE 6

Example 3 illustrates again the connection between algebra and geometry that we mentioned earlier in this section. Descartes's rectangular coordinate system allows us to associate the equation $y = -\frac{1}{3}x + 2$ (an algebraic concept) with a specific straight line (a geometric concept). The study of the relationship between equations in algebra and their associated geometric figures is called *analytic geometry*.

Intercepts

Two important points on the graph of a straight line, if they exist, are the points where the graph crosses the axes.

> **DEFINITION** The ***x*-intercept** of the graph of an equation is the *x*-coordinate of the point where the graph crosses the *x*-axis. The ***y*-intercept** is defined similarly.

Because any point on the x-axis has a y-coordinate of 0, we can find the x-intercept by letting $y = 0$ and solving the equation for x. We find the y-intercept by letting $x = 0$ and solving for y.

EXAMPLE 4

Find the x- and y-intercepts for $2x + 3y = 6$; then graph the solution set.

SOLUTION To find the y-intercept we let $x = 0$.

When $x = 0$

we have
$$
\begin{aligned}
2(0) + 3y &= 6 \\
3y &= 6 \\
y &= 2
\end{aligned}
$$

The y-intercept is 2, and the graph crosses the y-axis at the point (0, 2).

When $y = 0$

we have
$$
\begin{aligned}
2x + 3(0) &= 6 \\
2x &= 6 \\
x &= 3
\end{aligned}
$$

The x-intercept is 3, so the graph crosses the x-axis at the point (3, 0). We use these results to graph the solution set for $2x + 3y = 6$. The graph is shown in Figure 7.

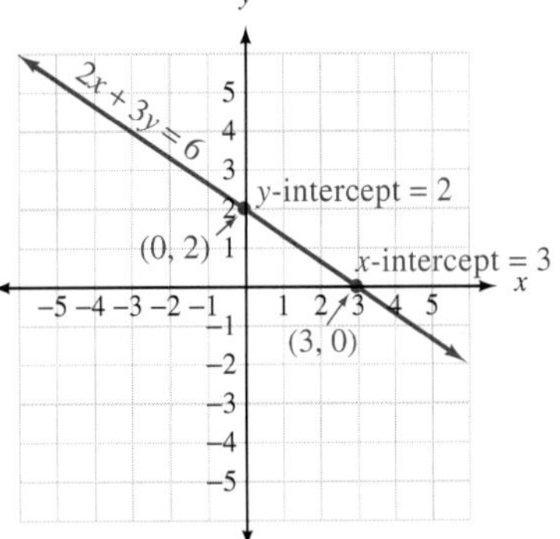

FIGURE 7

Note
Graphing straight lines by finding the intercepts works best when the coefficients of x and y are factors of the constant term.

EXAMPLE 5

Graph each of the following lines.

a. $y = \frac{1}{2}x$ **b.** $x = 3$ **c.** $y = -2$

SOLUTION

a. The line $y = \frac{1}{2}x$ passes through the origin because (0, 0) satisfies the equation. To sketch the graph we need at least one more point on the line. When x is 2, we obtain the point (2, 1), and when x is -4, we obtain the point $(-4, -2)$. The graph of $y = \frac{1}{2}x$ is shown in Figure 8a.

b. The line $x = 3$ is the set of all points whose x-coordinate is 3. The variable y does not appear in the equation, so the y-coordinate can be any number. Note that we can write our equation as a linear equation in two variables by writing it as $x + 0y = 3$. Because the product of 0 and y will always be 0, y can be any number. The graph of $x = 3$ is the vertical line shown in Figure 8b.

c. The line $y = -2$ is the set of all points whose y-coordinate is -2. The variable x does not appear in the equation, so the x-coordinate can be any number. Again, we can write our equation as a linear equation in two variables by writing it as $0x + y = -2$. Because the product of 0 and x will always be 0, x can be any number. The graph of $y = -2$ is the horizontal line shown in Figure 8c.

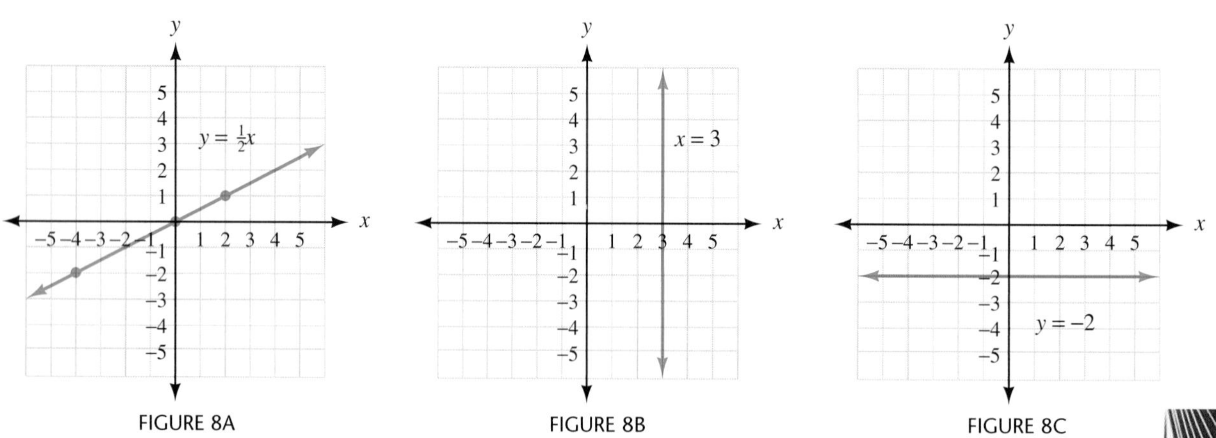

FIGURE 8A FIGURE 8B FIGURE 8C

FACTS FROM GEOMETRY

Special Equations and Their Graphs

For the equations below, m, a, and b are real numbers.

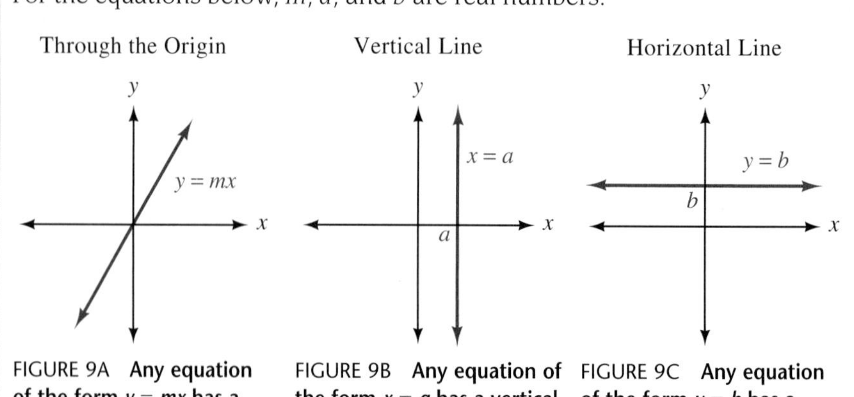

FIGURE 9A **Any equation of the form $y = mx$ has a graph that passes through the origin.**

FIGURE 9B **Any equation of the form $x = a$ has a vertical line for its graph.**

FIGURE 9C **Any equation of the form $y = b$ has a horizontal line for its graph.**

USING TECHNOLOGY

Graphing Calculators and Computer Graphing Programs

A variety of computer programs and graphing calculators are currently available to help us graph equations and then obtain information from those graphs much faster than we could with paper and pencil. We will not give instructions for all the available calculators. Most of the instructions we give are generic in form. You will have to use the manual that came with your calculator to find the specific instructions for your calculator.

Graphing with Trace and Zoom

All graphing calculators have the ability to graph a function and then trace over the points on the graph, giving their coordinates. Furthermore, all graphing calculators can zoom in and out on a graph that has been drawn. To graph a linear equation on a graphing calculator, we first set the graph window. Most calculators call the smallest value of x Xmin and the largest value of x Xmax. The counterpart values of y are Ymin and Ymax. We will use the notation

Window: X from −5 to 4, Y from −3 to 2

to stand for a window in which

Xmin = −5　　Ymin = −3
Xmax = 4　　Ymax = 2

Set your calculator with the following window:

Window: X from −10 to 10, Y from −10 to 10

Graph the equation Y = −X + 8. On the TI-82/83, you use the [Y=] key to enter the equation; you enter a negative sign with the [(−)] key, and a subtraction sign with the [−] key. The graph will be similar to the one shown in Figure 10.

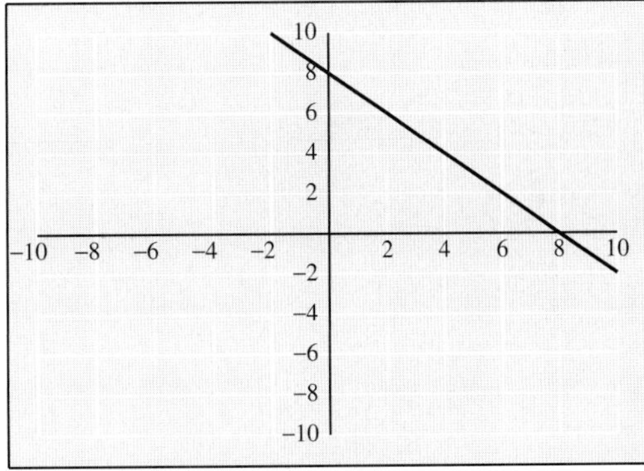

FIGURE 10

Use the Trace feature of your calculator to name three points on the graph. Next, use the Zoom feature of your calculator to zoom out so your window is twice as large.

Solving for *y* First

To graph the equation from Example 4, $2x + 3y = 6$, on a graphing calculator, you must first solve it for y. When you do so, you will get $y = -\frac{2}{3}x + 2$, which you enter into your calculator as Y = −(2/3)X + 2. Graph this equation in the window described here, and compare your results with the graph in Figure 7.

Window: X from −6 to 6, Y from −6 to 6

Hint on Tracing

If you are going to use the Trace feature and you want the x-coordinates to be exact numbers, set your window so that the range of X inputs is a multiple of the number of horizontal pixels on your calculator screen. On the TI-82/83, the screen has 94 pixels across. Here are a few convenient trace windows:

X from −4.7 to 4.7	To trace to the nearest tenth
X from −47 to 47	To trace to the nearest integer
X from 0 to 9.4	To trace to the nearest tenth
X from 0 to 94	To trace to the nearest integer
X from −94 to 94	To trace to the nearest even integer

LINKING OBJECTIVES AND EXAMPLES

Next to each **objective** we have listed the examples that are best described by that objective.

A	1, 2
B	3, 4
C	5

GETTING READY FOR CLASS

After reading through the preceding section, respond in your own words and in complete sentences.

1. Explain how you would construct a rectangular coordinate system from two real number lines.
2. Explain in words how you would graph the ordered pair (2, −3).
3. How can you tell if an ordered pair is a solution to the equation $y = 2x - 5$?
4. If you were looking for solutions to the equation $y = \frac{1}{3}x + 5$, why would it be easier to substitute 6 for x than to substitute 5 for x?

Problem Set 3.1

Online support materials can be found at www.thomsonedu.com/login

Graph each of the following ordered pairs on a rectangular coordinate system.

1. **a.** (1, 2)
b. (−1, −2)
c. (5, 0)
d. (0, 2)
e. (−5, −5)
f. $\left(\frac{1}{2}, 2\right)$

2. **a.** (−1, 2)
b. (1, −2)
c. (0, −3)
d. (4, 0)
e. (−4, −1)
f. $\left(3, \frac{1}{4}\right)$

Give the coordinates of each point.

3.

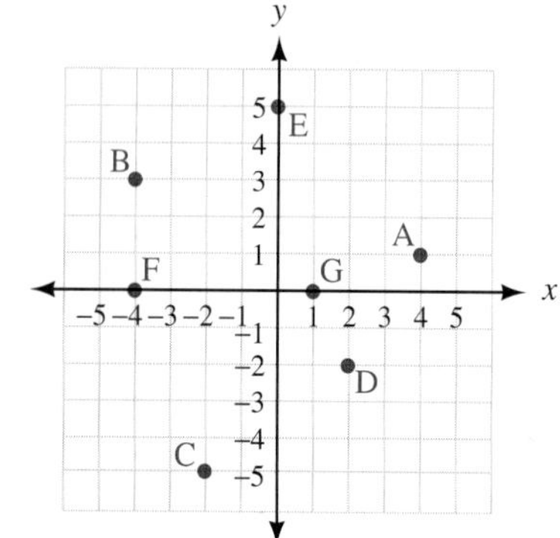

= Videos available by instructor request
▶ = Online student support materials available at www.thomsonedu.com/login

4.

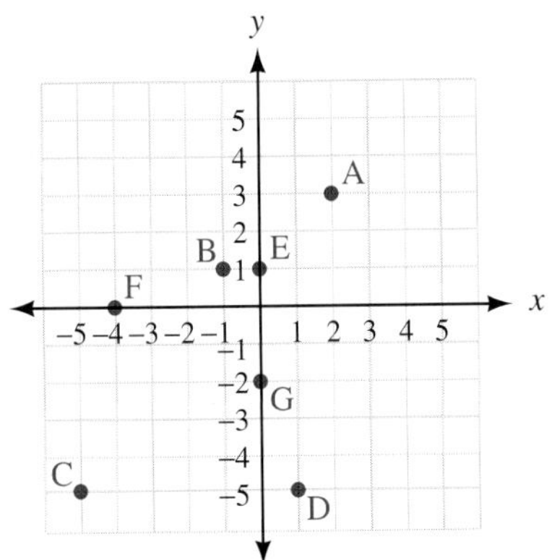

Graph each of the following linear equations by first finding the intercepts.

5. $2x - 3y = 6$
6. $y - 2x = 4$
7. $4x - 5y = 20$
8. $5x - 3y - 15 = 0$
9. $y = 2x + 3$
10. $y = 3x - 2$
11. $-3x + 2y = 12$
12. $5x - 7y = -35$
13. $6x - 5y - 20 = 0$
14. $-4x - 6y - 15 = 0$
15. $y = 3x - 5$
16. $y = -4x + 1$
17. $\frac{x}{2} + \frac{y}{3} = 1$
18. $\frac{x}{4} + \frac{y}{3} = 1$

19. Which of the following tables could be produced from the equation $y = 2x - 6$?

a.

x	y
0	6
1	4
2	2
3	0

b.

x	y
0	−6
1	−4
2	−2
3	0

c.

x	y
0	−6
1	−5
2	−4
3	−3

20. Which of the following tables could be produced from the equation $3x - 5y = 15$?

a.

x	y
0	5
−3	0
10	3

b.

x	y
0	−3
5	0
10	3

c.

x	y
0	−3
−5	0
10	−3

Graph each of the following lines.

21. $y = \frac{1}{3}x$
22. $y = \frac{1}{2}x$
23. $-2x + y = -3$
24. $-3x + y = -2$
25. $y = -\frac{2}{3}x + 1$
26. $y = -\frac{2}{3}x - 1$
27. $\frac{x}{3} + \frac{y}{4} = 1$
28. $\frac{x}{-2} + \frac{y}{3} = 1$

29. The graph shown here is the graph of which of the following equations?
    a. $3x - 2y = 6$
    b. $2x - 3y = 6$
    c. $2x + 3y = 6$

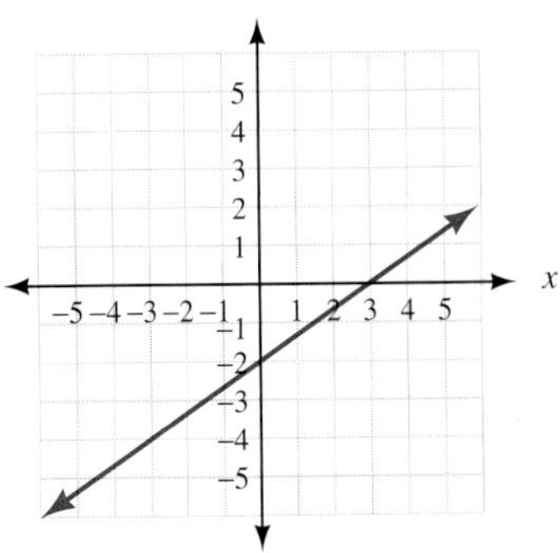

30. The graph shown here is the graph of which of the following equations?
    a. $3x - 2y = 8$
    b. $2x - 3y = 8$
    c. $2x + 3y = 8$

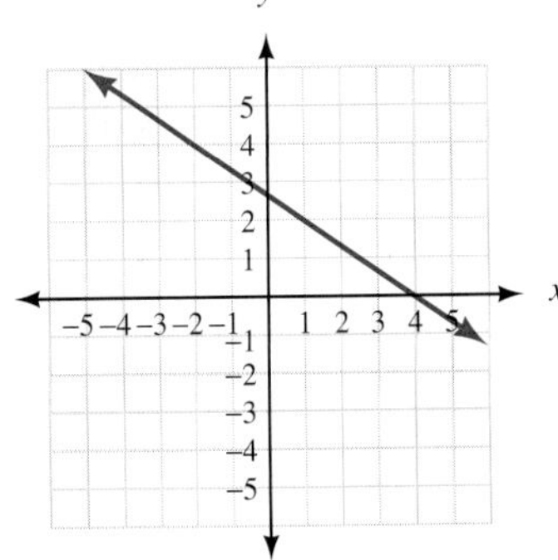

The next two problems are intended to give you practice reading, and paying attention to, the instructions that accompany the problems you are working. Working these problems is an excellent way to get ready for a test or a quiz.

31. Work each problem according to the instructions given.
    a. Solve: $4x + 12 = -16$
    b. Find x when y is 0: $4x + 12y = -16$
    c. Find y when x is 0: $4x + 12y = -16$
    d. Graph: $4x + 12y = -16$
    e. Solve for y: $4x + 12y = -16$

32. Work each problem according to the instructions given.
    a. Solve: $3x - 8 = -12$
    b. Find x when y is 0: $3x - 8y = -12$
    c. Find y when x is 0: $3x - 8y = -12$
    d. Graph: $3x - 8y = -12$
    e. Solve for y: $3x - 8y = -12$

33. Graph each of the following lines.
a. $y = 2x$ **b.** $x = -3$ **c.** $y = 2$

34. Graph each of the following lines.
a. $y = 3x$ **b.** $x = -2$ **c.** $y = 4$

35. Graph each of the following lines.
a. $y = -\frac{1}{2}x$ **b.** $x = 4$ **c.** $y = -3$

36. Graph each of the following lines.
a. $y = -\frac{1}{3}x$ **b.** $x = 1$ **c.** $y = -5$

37. Graph the line $0.02x + 0.03y = 0.06$.

38. Graph the line $0.05x - 0.03y = 0.15$.

39. The ordered pairs that satisfy the equation $y = 3x$ all have the form $(x, 3x)$ because y is always 3 times x. Graph all ordered pairs of the form $(x, 3x)$.

40. Graph all ordered pairs of the form $(x, -3x)$.

Applying the Concepts

41. Hourly Wages Eva takes a job at Brooke's Boutique. Her job pays \$8.00 per hour. The graph shows how much Eva earns for working from 0 to 40 hours in a week.

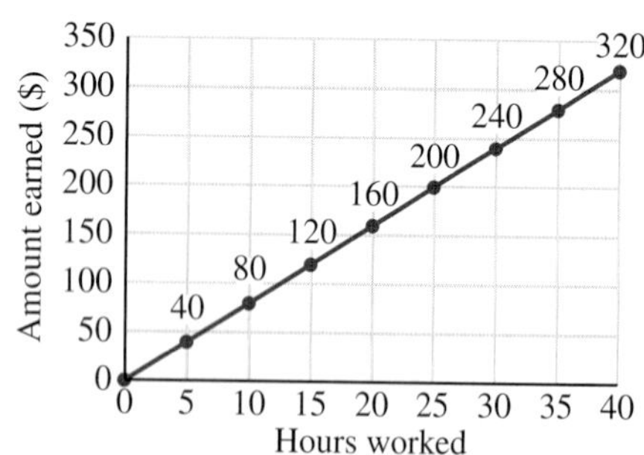

a. List three ordered pairs that lie on the line graph.
b. How much will she earn for working 40 hours?
c. If her check for one week is \$240, how many hours did she work?
d. She works 35 hours one week, but her paycheck before deductions are subtracted out is for \$260. Is this correct? Explain.

42. Hourly Wages Tyler takes a job at Tanner's Toys. His job pays \$6.00 per hour plus \$50 per week in commission. The graph shows how much Tyler earns for working from 0 to 40 hours in a week.

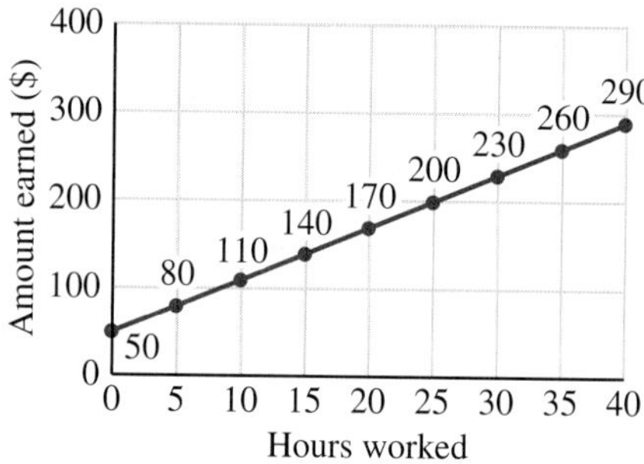

a. List three ordered pairs that lie on the line graph.
b. How much will he earn for working 40 hours?
c. If his check for one week is \$230, how many hours did he work?
d. He works 35 hours one week, but his paycheck before deductions are subtracted out is for \$260. Is this correct? Explain.

43. Non–Camera Phone Sales The table and bar chart shown here show what are the projected sales of non-camera phones for the years 2006–2010. Use the information from the table and chart to construct a line graph.

Year	2006	2007	2008	2009	2010
Sales (in millions)	300	250	175	150	125

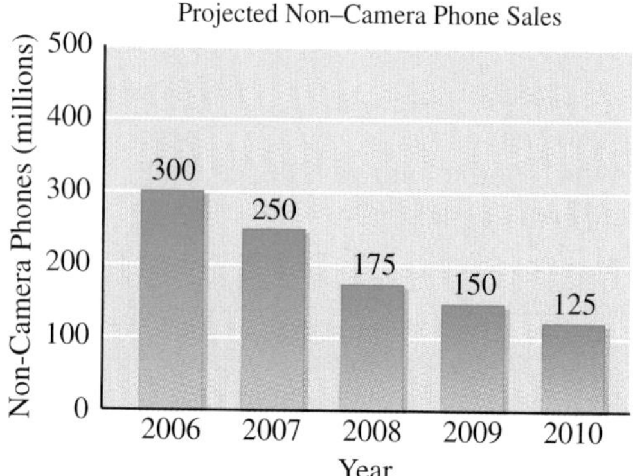

44. Camera Phone Sales The table and bar chart shown here show the projected sales of camera phones from 2006 to 2010. Use the information from the table and chart to construct a line graph.

Year	2006	2007	2008	2009	2010
Sales (in millions)	500	650	750	875	900

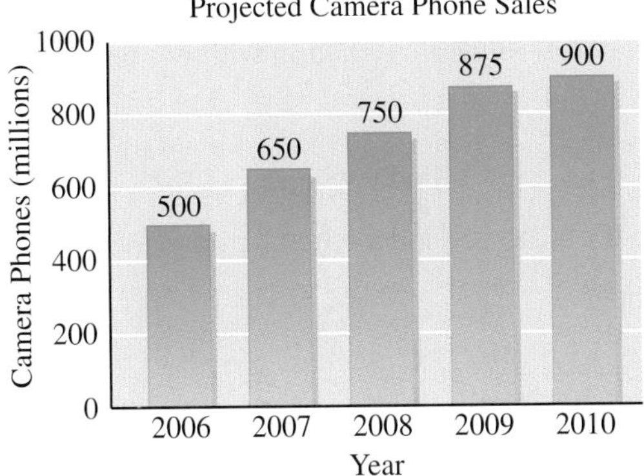

45. Kentucky Derby The graph gives the monetary bets placed at the Kentucky Derby for specific years. If x represents the year in question and y represents the total wagering for that year, write five ordered pairs that describe the information in the graph.

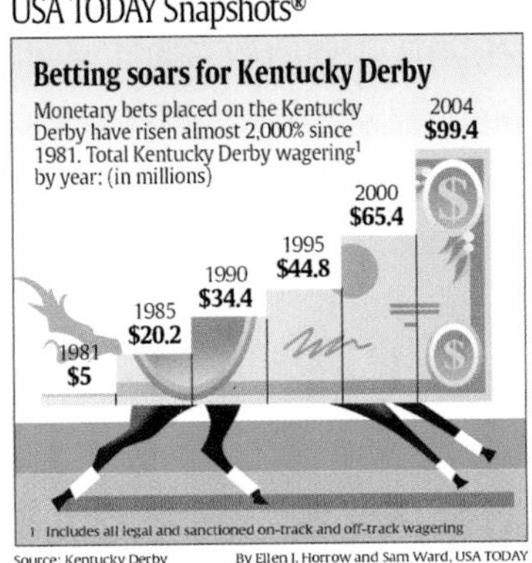

46. Age of New Mothers The graph shows the increase in average age of first-time mothers in the United States since 1970. Estimate from the graph the average age of first-time mothers for the years 1970, 1980, 1990, and 2000 as a list of ordered pairs. (Round ages to the nearest whole number.)

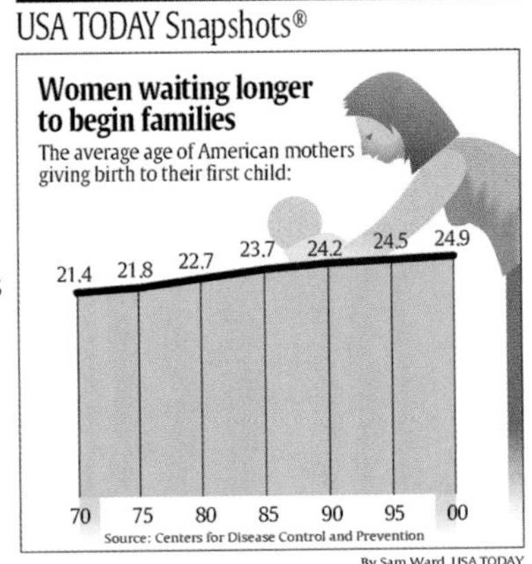

In the figures below, right triangle ABC has legs of length 5. In each case, find the coordinates of A and B.

47. **48.**

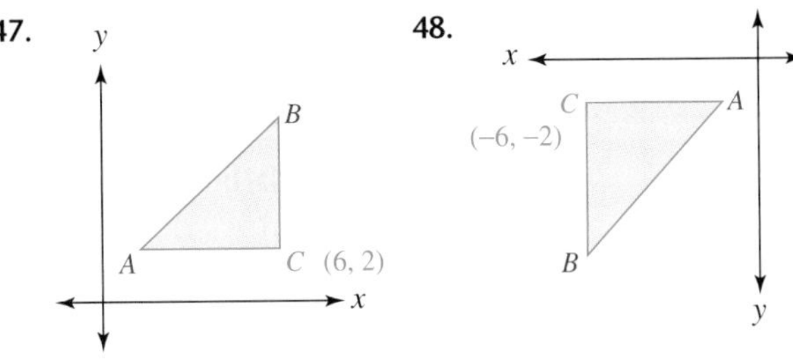

In the figures below, rectangle $ABCD$ has a length of 5 and a width of 3. In each case, find points A, B, and C.

49. **50.**

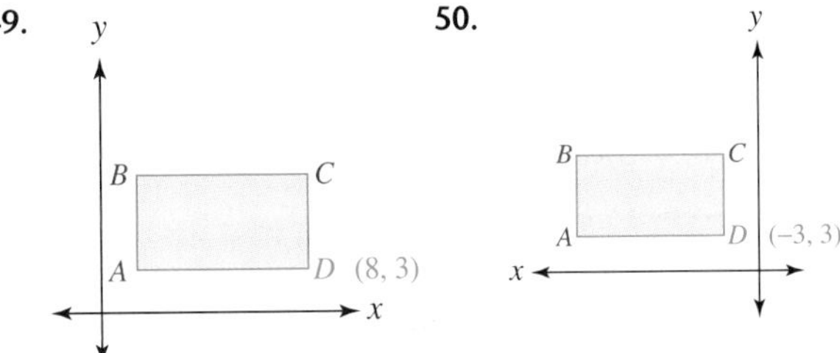

51. Reading Graphs When we see graphs in the media, the axes are not always labeled completely. Use the chart to answer the following questions.

a. Approximately what percentage of people were online in 1998?

b. Approximately what percentage of people were online in 2003?

c. Name the year or years during which 65% of adults were online.

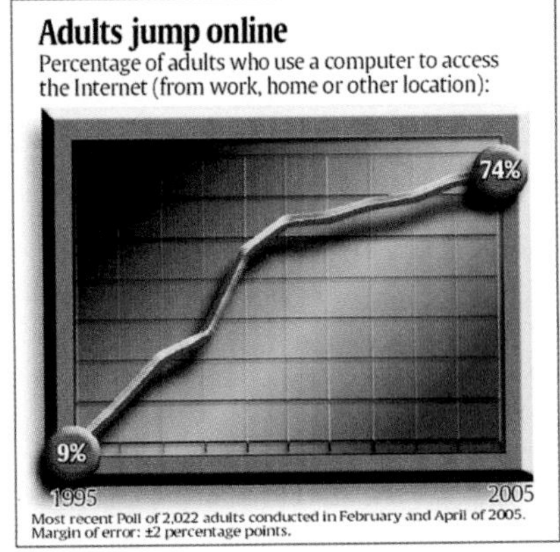

52. **Bridal Books** Use the chart to make a table of values of x and y, where x is years and y is the estimated number of bridal books published.

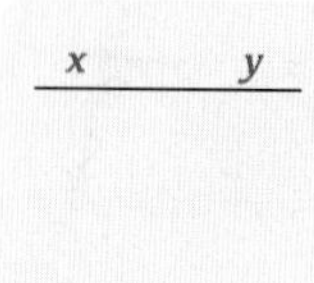

x	y

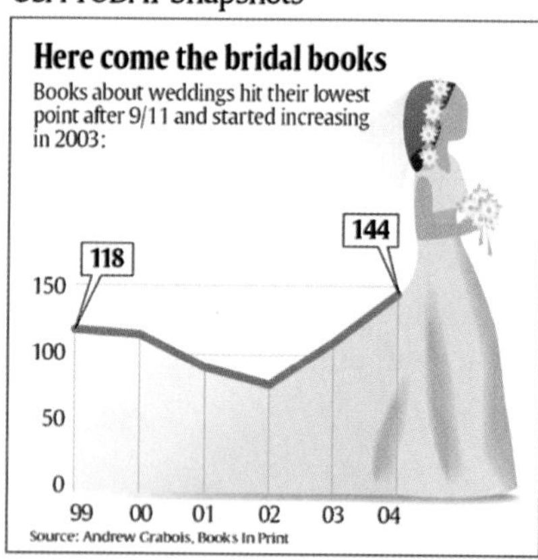

From *USA Today*. Copyright 2005. Reprinted with permission.

Maintaining Your Skills

The problems that follow review some of the more important skills you have learned in previous sections and chapters.

Solve each equation.

53. $5x - 4 = -3x + 12$

54. $\frac{1}{2} - \frac{y}{5} = -\frac{9}{10} + \frac{y}{2}$

55. $\frac{1}{2} - \frac{1}{8}(3t - 4) = -\frac{7}{8}t$

56. $3(5t - 1) - (3 - 2t) = 5t - 8$

57. $50 = \frac{K}{24}$

58. $3.5d = 16(3.5)^2$

59. $2(1) + y = 4$

60. $2(2y + 6) + 3y = 5$

61. $4\left(\frac{19}{15}\right) - 2y = 4$

62. $2x - 3\left(-\frac{5}{11}\right) = 4$

Getting Ready for the Next Section

63. Write -0.06 as a fraction with denominator 100.

64. Write -0.07 as a fraction with denominator 100.

65. If $y = 2x - 3$, find y when $x = 2$.

66. If $y = 2x - 3$, find x when $y = 5$.

Simplify.

67. $\frac{1 - (-3)}{-5 - (-2)}$

68. $\frac{-3 - 1}{-2 - (-5)}$

69. $\frac{-1 - 4}{3 - 3}$

70. $\frac{-3 - (-3)}{2 - (-1)}$

71. The product of $\frac{2}{3}$ and what number will result in

a. 1? **b.** -1?

72. The product of 3 and what number will result in

a. 1? **b.** -1?

Extending the Concepts

Find the x- and y-intercepts for each equation. Your answers will contain the constants a, b, and c.

73. $ax + by = c$

74. $ax - by = c$

75. $\frac{x}{a} + \frac{y}{b} = 1$

76. $y = ax + b$

3.2 The Slope of a Line

OBJECTIVES

A Find the slope of a line from its graph.

B Find the slope of a line given two points on the line.

A highway sign tells us we are approaching a 6% downgrade. As we drive down this hill, each 100 feet we travel horizontally is accompanied by a 6-foot drop in elevation.

Highway sign

Mathematical model

In mathematics we say the slope of the highway is $-0.06 = -\frac{6}{100} = -\frac{3}{50}$. The *slope* is the ratio of the vertical change to the accompanying horizontal change.

In defining the slope of a straight line, we are looking for a number to associate with a straight line that does two things. First, we want the slope of a line to measure the "steepness" of the line; that is, in comparing two lines, the slope of the steeper line should have the larger numerical value. Second, we want a line that *rises* going from left to right to have a *positive* slope. We want a line that *falls* going from left to right to have a *negative* slope. (A line that neither rises nor falls going from left to right must, therefore, have 0 slope.)

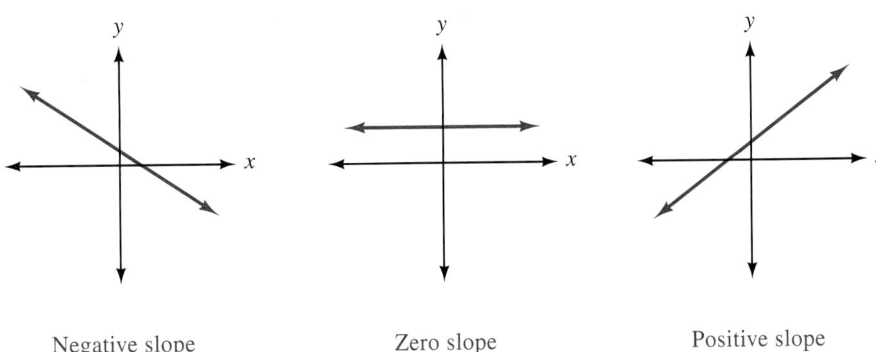

Negative slope

Zero slope

Positive slope

Geometrically, we can define the *slope* of a line as the ratio of the vertical change to the horizontal change encountered when moving from one point to another on the line. The vertical change is sometimes called the *rise*. The horizontal change is called the *run*.

EXAMPLE 1

Find the slope of the line $y = 2x - 3$.

SOLUTION To use our geometric definition, we first graph $y = 2x - 3$ (Figure 1). We then pick any two convenient points and find the ratio of rise to run. By convenient points we mean points with integer coordinates. If we let $x = 2$ in the equation, then $y = 1$. Likewise if we let $x = 4$, then y is 5.

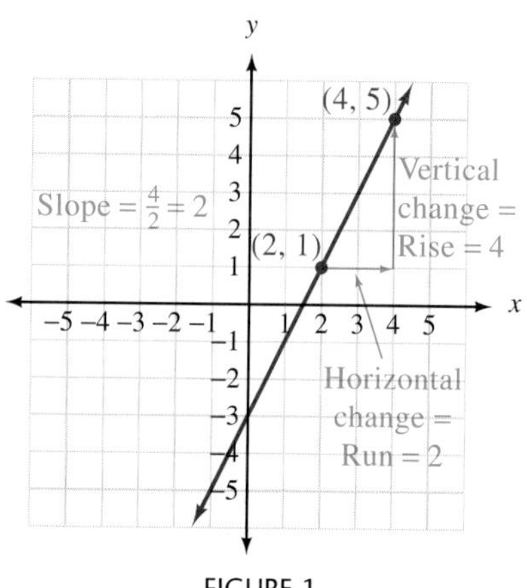

FIGURE 1

Our line has a slope of 2. 

Notice that we can measure the vertical change (rise) by subtracting the y-coordinates of the two points shown in Figure 1: $5 - 1 = 4$. The horizontal change (run) is the difference of the x-coordinates: $4 - 2 = 2$. This gives us a second way of defining the slope of a line.

DEFINITION The *slope* of the line between two points (x_1, y_1) and (x_2, y_2) is given by

$$\text{Slope} = m = \frac{\text{Rise}}{\text{Run}} = \frac{y_2 - y_1}{x_2 - x_1}$$

↗ Geometric Form ↖ Algebraic Form

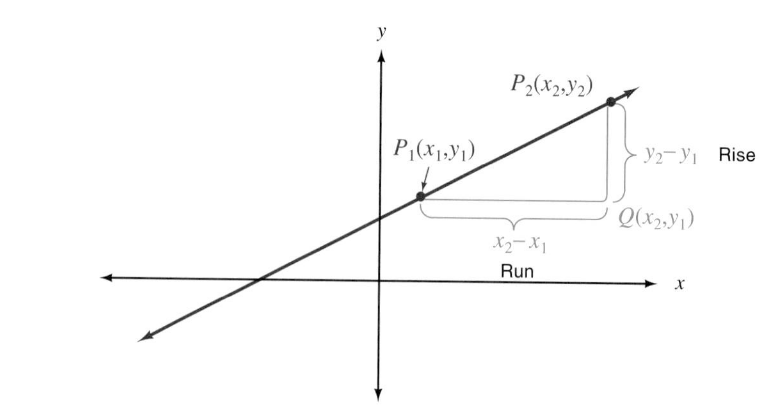

EXAMPLE 2

Find the slope of the line through $(-2, -3)$ and $(-5, 1)$.

SOLUTION

$$m = \frac{y_2 - y_1}{x_2 - x_1} = \frac{1 - (-3)}{-5 - (-2)} = \frac{4}{-3} = -\frac{4}{3}$$

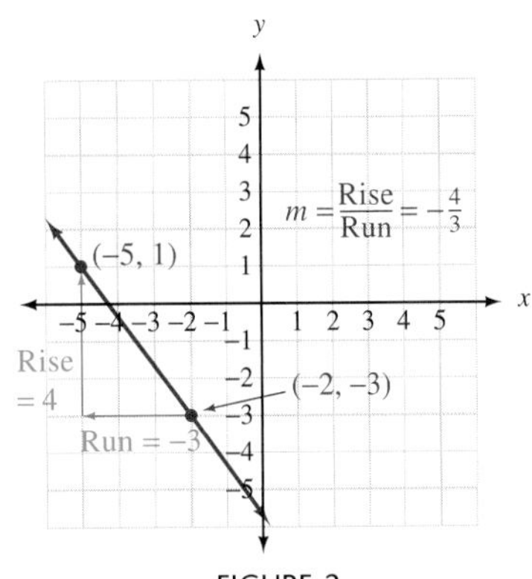

FIGURE 2

Looking at the graph of the line between the two points (Figure 2), we can see our geometric approach does not conflict with our algebraic approach.

We should note here that it does not matter which ordered pair we call (x_1, y_1) and which we call (x_2, y_2). If we were to reverse the order of subtraction of both the *x*- and *y*-coordinates in the preceding example, we would have

$$m = \frac{-3 - 1}{-2 - (-5)} = \frac{-4}{3} = -\frac{4}{3}$$

which is the same as our previous result. 

Note: The two most common mistakes students make when first working with the formula for the slope of a line are

1. Putting the difference of the *x*-coordinates over the difference of the *y*-coordinates.
2. Subtracting in one order in the numerator and then subtracting in the opposite order in the denominator. You would make this mistake in Example 2 if you wrote $1 - (-3)$ in the numerator and then $-2 - (-5)$ in the denominator.

EXAMPLE 3

Find the slope of the line containing $(3, -1)$ and $(3, 4)$.

SOLUTION Using the definition for slope, we have

$$m = \frac{-1 - 4}{3 - 3} = \frac{-5}{0}$$

The expression $\frac{-5}{0}$ is undefined; that is, there is no real number to associate with it. In this case, we say the line *has no slope.*

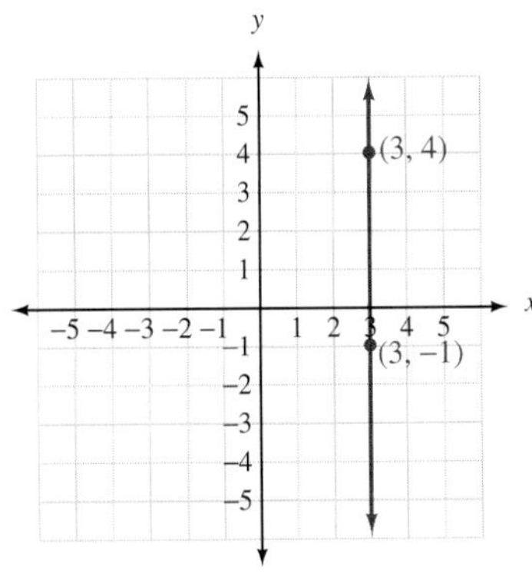

FIGURE 3

The graph of our line is shown in Figure 3. Our line with no slope is a vertical line. All vertical lines have no slope. (And all horizontal lines, as we mentioned earlier, have 0 slope.)

Slopes of Parallel and Perpendicular Lines

In geometry we call lines in the same plane that never intersect parallel. For two lines to be nonintersecting, they must rise or fall at the same rate. In other words, two lines are *parallel* if and only if they have the *same slope.*

Although it is not as obvious, it is also true that two nonvertical lines are *perpendicular* if and only if the *product of their slopes is* -1. This is the same as saying their slopes are negative reciprocals.

We can state these facts with symbols as follows: If line l_1 has slope m_1 and line l_2 has slope m_2, then

$$l_1 \text{ is parallel to } l_2 \Leftrightarrow m_1 = m_2$$

and

$$l_1 \text{ is perpendicular to } l_2 \Leftrightarrow m_1 \cdot m_2 = -1 \left(\text{or } m_1 = \frac{-1}{m_2}\right)$$

For example, if a line has a slope of $\frac{2}{3}$, then any line parallel to it has a slope of $\frac{2}{3}$. Any line perpendicular to it has a slope of $-\frac{3}{2}$ (the negative reciprocal of $\frac{2}{3}$).

Although we cannot give a formal proof of the relationship between the slopes of perpendicular lines at this level of mathematics, we can offer some justification for the relationship. Figure 4 shows the graphs of two lines. One of the lines has a slope of $\frac{2}{3}$; the other has a slope of $-\frac{3}{2}$. As you can see, the lines are perpendicular.

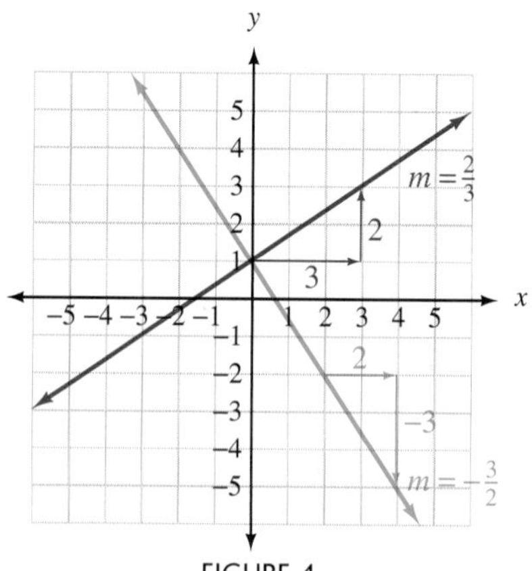

FIGURE 4

Slope and Rate of Change

So far, the slopes we have worked with represent the ratio of the change in y to a corresponding change in x, or, on the graph of the line, the ratio of vertical change to horizontal change in moving from one point on the line to another. However, when our variables represent quantities from the world around us, slope can have additional interpretations.

EXAMPLE 4 On the chart below, find the slope of the line connecting the first point (1955, 0.29) with the last point (2005, 2.93). Explain the significance of the result.

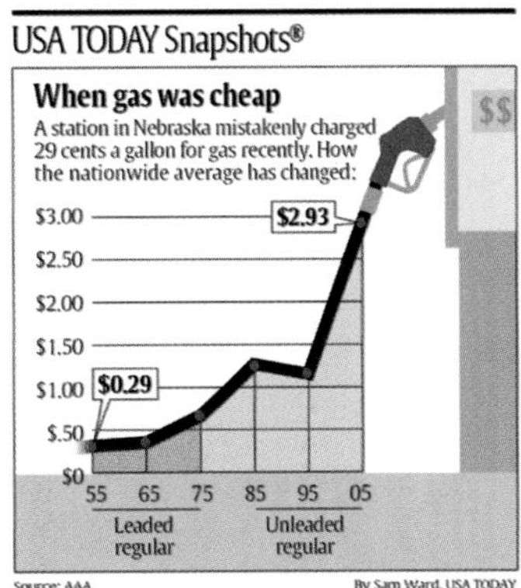

From *USA Today*. Copyright 2005. Reprinted with permission.

SOLUTION The slope of the line connecting the first point (1955, 0.29) with the last point (2005, 2.93), is

$$m = \frac{2.93 - 0.29}{2005 - 1955} = \frac{2.64}{50} = 0.0528$$

The units are dollars/year. If we write this in terms of cents we have

$$m = 5.28 \text{ cents/year}$$

which is the average change in the price of a gallon of gasoline over a 50-year period of time.

For the chart in Example 4, if we connect the points (1995, 1.10) and (2005, 2.93), the line that results has a slope of

$$m = \frac{2.93 - 1.10}{2005 - 1995} = \frac{1.83}{10} = 0.183 \text{ dollars/year} = 18.3 \text{ cents/year}$$

which is the average change in the price of a gallon of gasoline over a 10-year period. As you can imagine by looking at the chart, the line connecting the first and last point is not as steep as the line connecting the points from 1995 and 2005, and this is what we are seeing numerically with our slope calculations. If we were summarizing this information for an article in the newspaper, we could say: Although the price of a gallon of gasoline has increased only 5.28 cents per year over the last 50 years, for the last 10 years the average annual rate of increase is more than triple that at 18.3 cents per year.

Slope and Average Speed

Previously we introduced the rate equation $d = rt$. Suppose that a boat is traveling at a constant speed of 15 miles per hour in still water. The following table shows the distance the boat will have traveled in the specified number of hours.

The graph of these data is shown in Figure 5. Notice that the points all lie along a line.

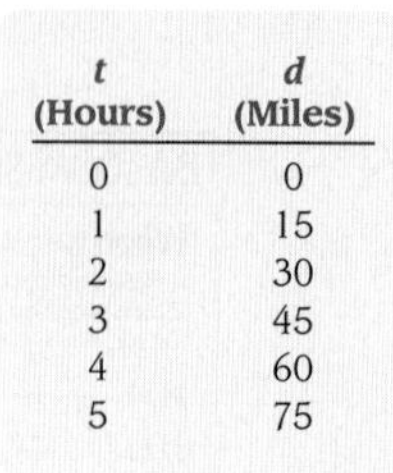

t (Hours)	d (Miles)
0	0
1	15
2	30
3	45
4	60
5	75

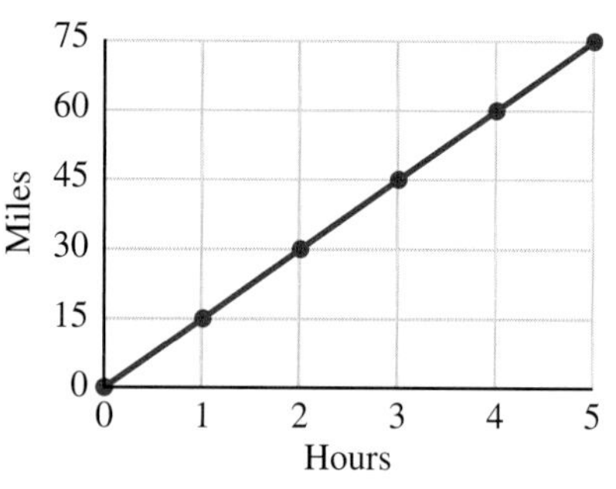

FIGURE 5

We can calculate the slope of this line using any two points from the table. Notice we have graphed the data with t on the horizontal axis and d on the vertical axis. Using the points (2, 30) and (3, 45), the slope will be

$$m = \frac{\text{rise}}{\text{run}} = \frac{45 - 30}{3 - 2} = \frac{15}{1} = 15$$

The units of the rise are miles and the units of the run are hours, so the slope will be in units of miles per hour. We see that the slope is simply the change in distance divided by the change in time, which is how we compute the average speed. Since the speed is constant, the slope of the line represents the speed of 15 miles per hour.

EXAMPLE 5 A car is traveling at a constant speed. A graph of the distance the car has traveled over time is shown in Figure 6. Use the graph to find the speed of the car.

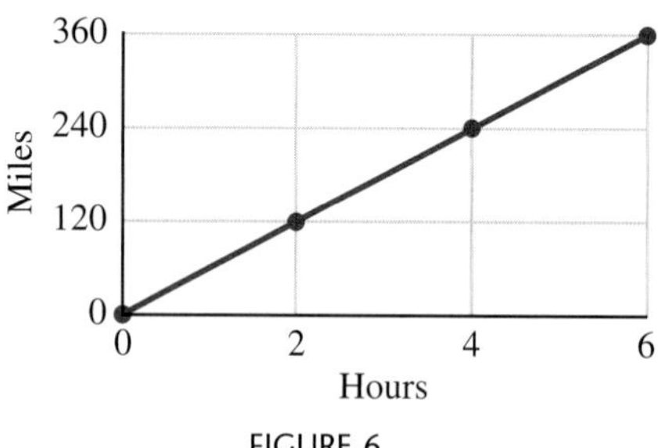

FIGURE 6

SOLUTION Using the second and third points, we see the rise is $240 - 120 = 120$ miles, and the run is $4 - 2 = 2$ hours. The speed is given by the slope, which is

$$\begin{aligned} m &= \frac{\text{rise}}{\text{run}} \\ &= \frac{120 \text{ miles}}{2 \text{ hours}} \\ &= 60 \text{ miles per hour} \end{aligned}$$

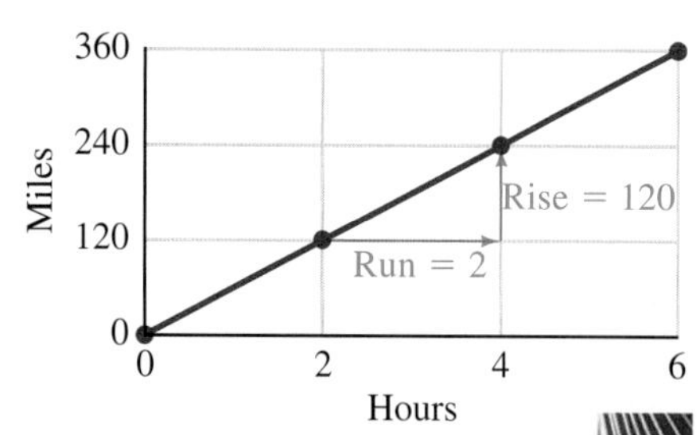

USING TECHNOLOGY

Families of Curves

We can use a graphing calculator to investigate the effects of the numbers a and b on the graph of $y = ax + b$. To see how the number b affects the graph, we can hold a constant and let b vary. Doing so will give us a *family* of curves. Suppose we set $a = 1$ and then let b take on integer values from -3 to 3. The equations we obtain are

$$y = x - 3$$
$$y = x - 2$$
$$y = x - 1$$
$$y = x$$
$$y = x + 1$$
$$y = x + 2$$
$$y = x + 3$$

We will give three methods of graphing this set of equations on a graphing calculator.

Method 1: Y-Variables List

To use the Y-variables list, enter each equation at one of the Y variables, set the graph window, then graph. The calculator will graph the equations in order, starting with Y_1 and ending with Y_7. Following is the Y-variables list, an appropriate window, and a sample of the type of graph obtained (Figure 7).

$Y_1 = X - 3$
$Y_2 = X - 2$
$Y_3 = X - 1$
$Y_4 = X$
$Y_5 = X + 1$
$Y_6 = X + 2$
$Y_7 = X + 3$

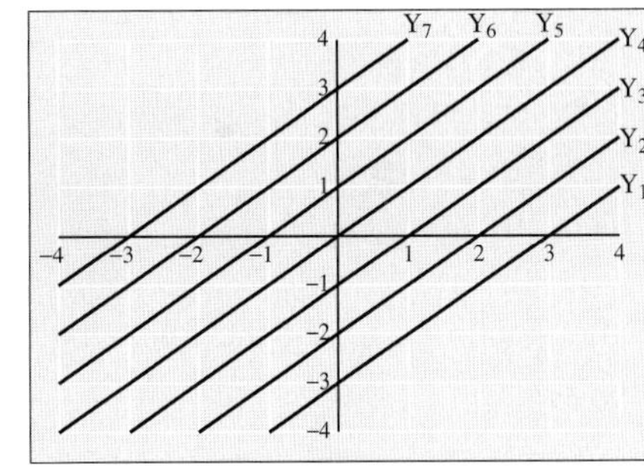

FIGURE 7

Window: X from -4 to 4, Y from -4 to 4

Method 2: Programming

The same result can be obtained by programming your calculator to graph $y = x + b$ for $b = -3, -2, -1, 0, 1, 2,$ and 3. Here is an outline of a program that will do this. Check the manual that came with your calculator to find the commands for your calculator.

Step 1: Clear screen
Step 2: Set window for X from -4 to 4 and Y from -4 to 4
Step 3: $-3 \rightarrow B$
Step 4: Label 1
Step 5: Graph $Y = X + B$
Step 6: $B + 1 \rightarrow B$
Step 7: If $B < 4$, go to 1
Step 8: End

Method 3: Using Lists

On the TI-82/83 you can set Y_1 as follows

$$Y_1 = X + \{-3, -2, -1, 0, 1, 2, 3\}$$

When you press GRAPH, the calculator will graph each line from $y = x + (-3)$ to $y = x + 3$.

Each of the three methods will produce graphs similar to those in Figure 7.

LINKING OBJECTIVES AND EXAMPLES

Next to each **objective** we have listed the examples that are best described by that objective.

A 1, 5

B 2–4

GETTING READY FOR CLASS

After reading through the preceding section, respond in your own words and in complete sentences.

1. If you were looking at a graph that described the performance of a stock you had purchased, why would it be better if the slope of the line were positive, rather than negative?
2. Describe the behavior of a line with a negative slope.
3. Would you rather climb a hill with a slope of $\frac{1}{2}$ or a slope of 3? Explain why.
4. Describe how to obtain the slope of a line if you know the coordinates of two points on the line.

Problem Set 3.2

Online support materials can be found at www.thomsonedu.com/login

Find the slope of each of the following lines from the given graph.

1.

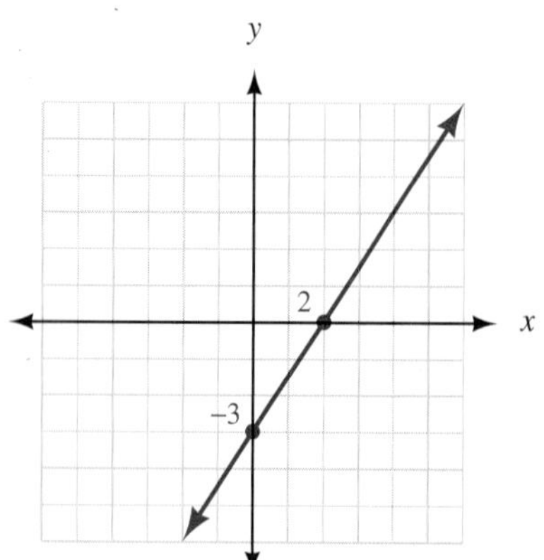

2.

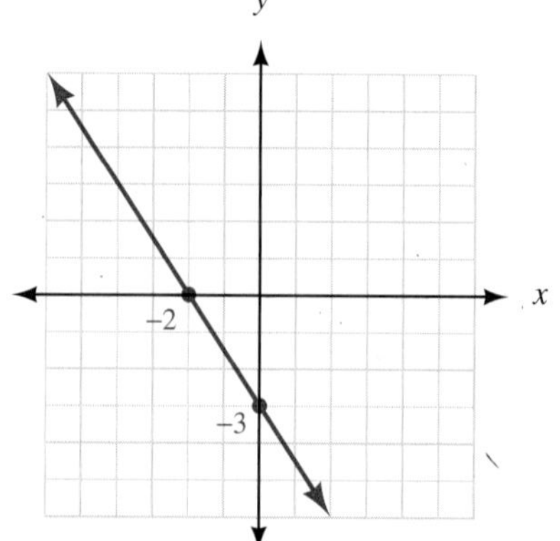

3.

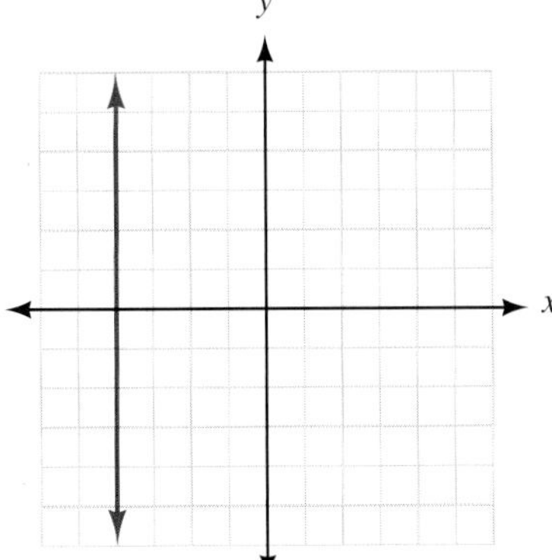

4.

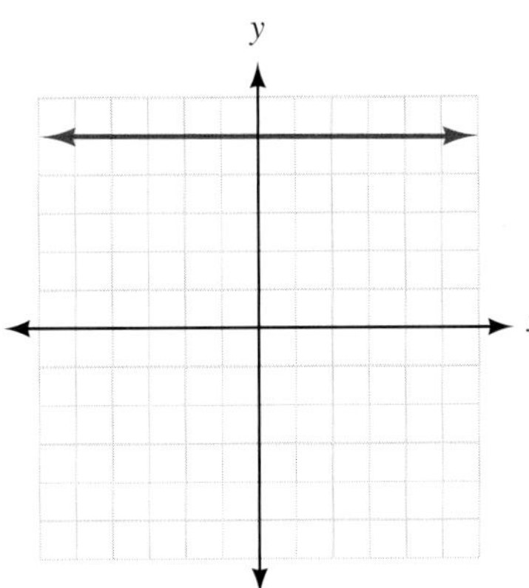

5.

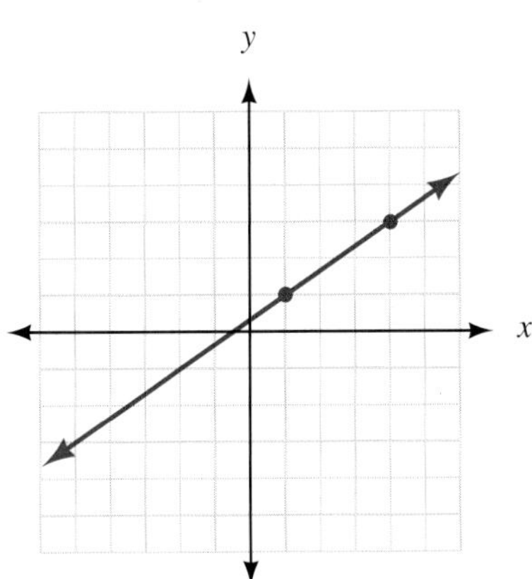

6.

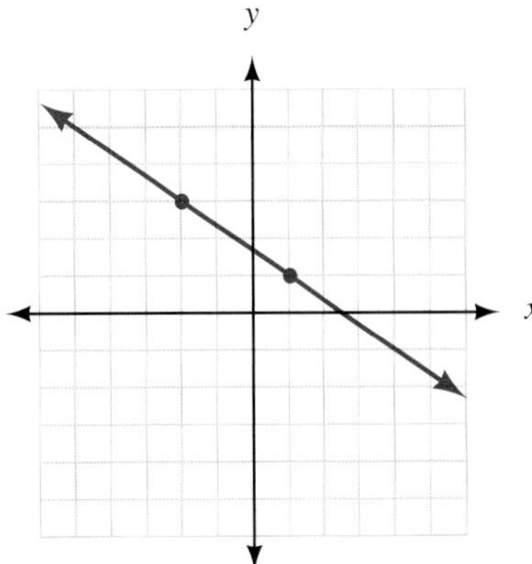

Find the slope of the line through each of the following pairs of points. Then, plot each pair of points, draw a line through them, and indicate the rise and run in the graph in the manner shown in Example 2.

7. (2, 1), (4, 4)

▸ 8. (3, 1), (5, 4)

9. (1, 4), (5, 2)

10. (1, 3), (5, 2)

11. (1, −3), (4, 2)

12. (2, −3), (5, 2)

13. (−3, −2), (1, 3)

14. (−3, −1), (1, 4)

▸ 15. (−3, 2), (3, −2)

16. (−3, 3), (3, −1)

17. (2, −5), (3, −2)

18. (2, −4), (3, −1)

Solve for the indicated variable if the line through the two given points has the given slope.

▸ 19. (5, 2) and $(x, -2)$, $m = 2$.

▸ 20. (−1, −4) and $(x, 5)$, $m = 3$.

▸ 21. $(a, 3)$ and (2, 6), $m = -1$.

▸ 22. $(a, -2)$ and (4, −6), $m = -3$.

▸ 23. $(2, b)$ and $(-1, 4b)$, $m = -2$.

▸ 24. $(-4, y)$ and $(-1, 6y)$, $m = 2$.

For each equation below, complete the table, and then use the results to find the slope of the graph of the equation.

25. $2x + 3y = 6$

x	y
0	
	0

26. $3x - 2y = 6$

x	y
0	
	0

27 $y = \frac{2}{3}x - 5$

x	y
0	
3	

28. $y = -\frac{3}{4}x + 2$

x	y
0	
4	

29. **Finding Slope from Intercepts** Graph the line with x-intercept 4 and y-intercept 2. What is the slope of this line?

30. **Finding Slope from Intercepts** Graph the line with x-intercept -4 and y-intercept -2. What is the slope of this line?

31. **Parallel Lines** Find the slope of any line parallel to the line through (2, 3) and (−8, 1).

32. **Parallel Lines** Find the slope of any line parallel to the line through (2, 5) and (5, −3).

33. **Perpendicular Lines** Line l contains the points (5, −6) and (5, 2). Give the slope of any line perpendicular to l.

34. **Perpendicular Lines** Line l contains the points (3, 4) and (−3, 1). Give the slope of any line perpendicular to l.

35. **Parallel Lines** Line l contains the points (−2, 1) and (4, −5). Find the slope of any line parallel to l.

36. **Parallel Lines** Line l contains the points (3, −4) and (−2, −6). Find the slope of any line parallel to l.

37. **Perpendicular Lines** Line l contains the points (−2, −5) and (1, −3). Find the slope of any line perpendicular to l.

38. **Perpendicular Lines** Line l contains the points (6, −3) and (−2, 7). Find the slope of any line perpendicular to l.

39. Determine if each of the following tables could represent ordered pairs from an equation of a line.

a.

x	y
0	5
1	7
2	9
3	11

b.

x	y
−2	−5
0	−2
2	0
4	1

40. The following lines have slope 2, $\frac{1}{2}$, 0, and −1. Match each line to its slope value.

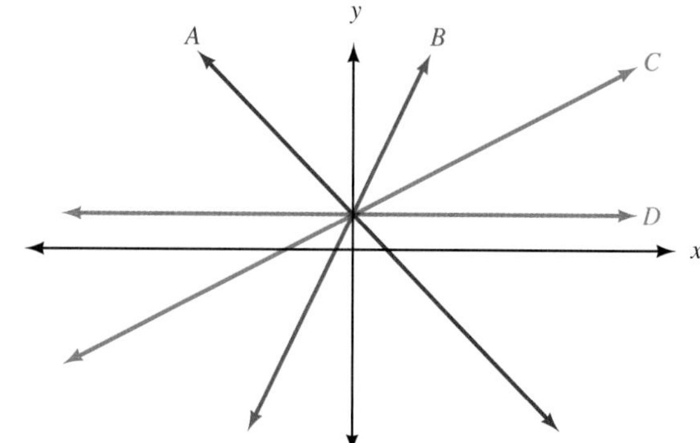

Applying the Concepts

41. An object is traveling at a constant speed. The distance and time data are shown on the given graph. Use the graph to find the speed of the object

a.

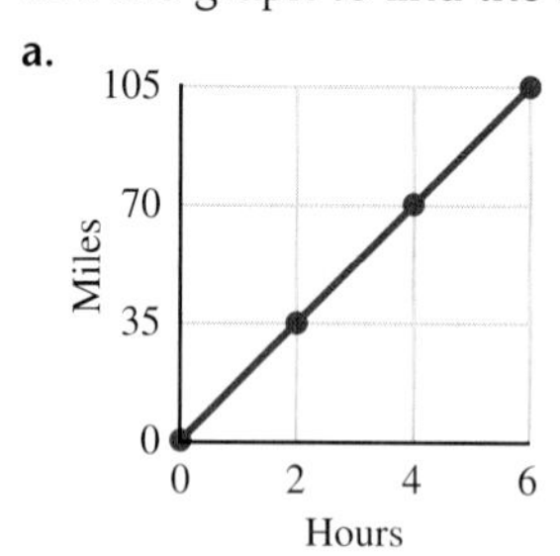

b.

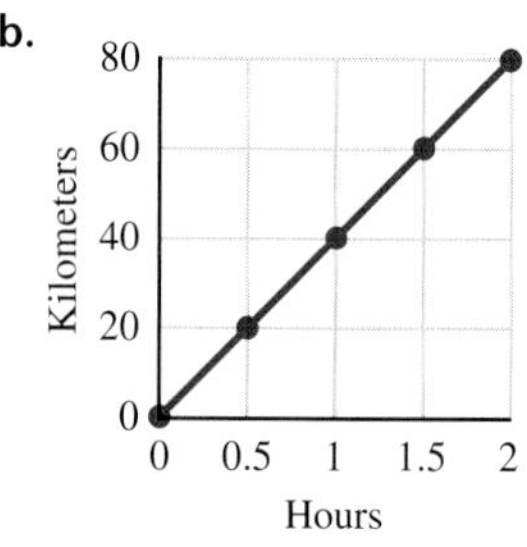

c.

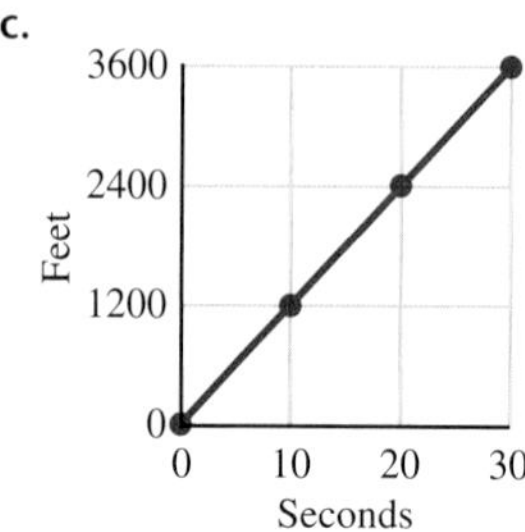

d.

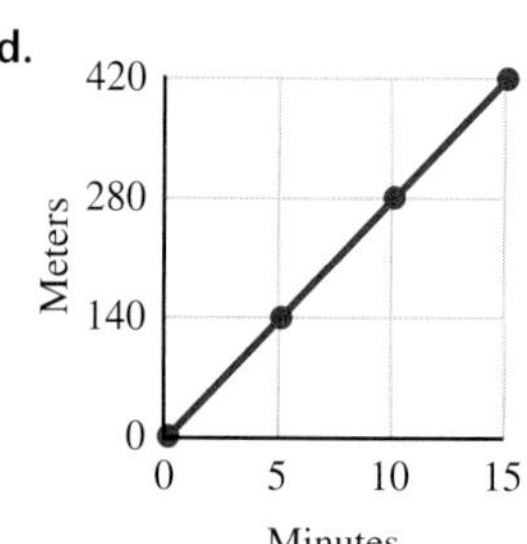

42. Improving Your Quantitative Literacy A cyclist is traveling at a constant speed. The graph shows the distance the cyclist travels over time when there is no wind present, when she travels against the wind, and when she travels with the wind.

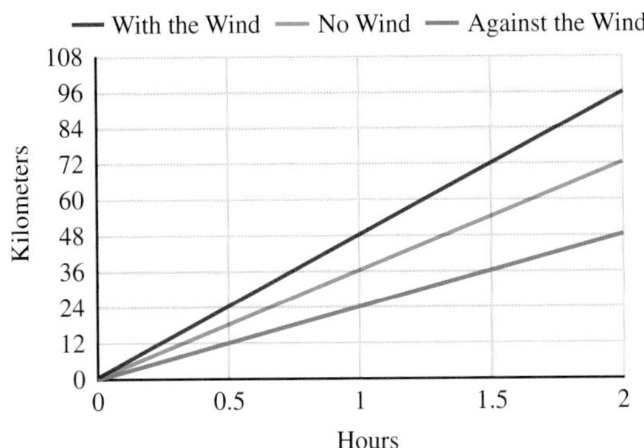

a. Use the concept of slope to find the speed of the cyclist under each of the three conditions.

b. Compare the speed of the cyclist when she is traveling without any wind to when she is riding against the wind. How do the two speeds differ?

c. Compare the speed of the cyclist when she is traveling without any wind to when she is riding with the wind. How do the two speeds differ?

d. What is the speed of the wind?

▶ **43. Non–Camera Phone Sales** The table and line graph here each show the projected non–camera phone sales each year through 2010. Find the slope of each of the three line segments, *A*, *B*, and *C*.

Year	2006	2007	2008	2009	2010
Sales (in millions)	300	250	175	150	125

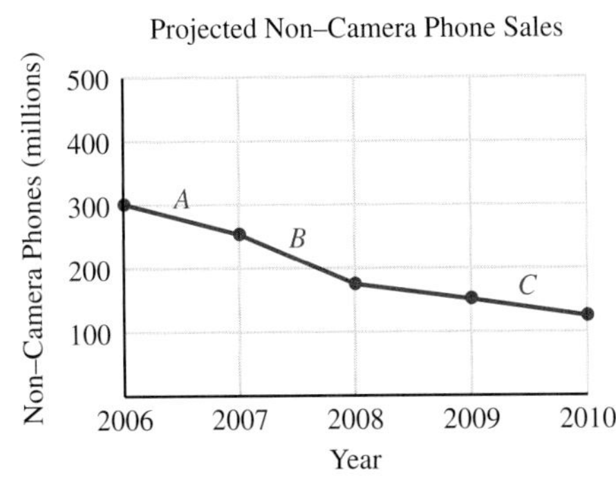

▶ **44. Camera Phone Sales** The table and line graph here each show the projected camera phone sales each year through 2010. Find the slope of each of the three line segments, *A*, *B*, and *C*.

Year	2006	2007	2008	2009	2010
Sales (in millions)	500	650	750	875	900

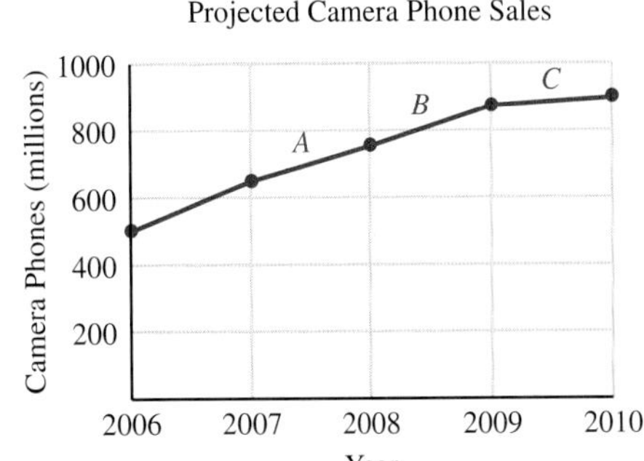

45. Slope of a Highway A sign at the top of the Cuesta Grade, outside of San Luis Obispo, reads "7% downgrade next 3 miles." The diagram shown here is a model of the Cuesta Grade that takes into account the information on that sign.

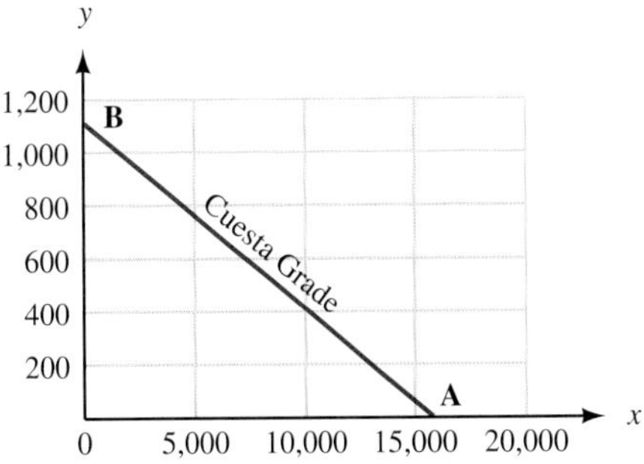

a. At point B, the graph crosses the y-axis at 1,106 feet. How far is it from the origin to point A?

b. What is the slope of the Cuesta Grade?

46. Heating a Block of Ice A block of ice with an initial temperature of $-20°C$ is heated at a steady rate. The graph shows how the temperature changes as the ice melts to become water and the water boils to become steam and water.

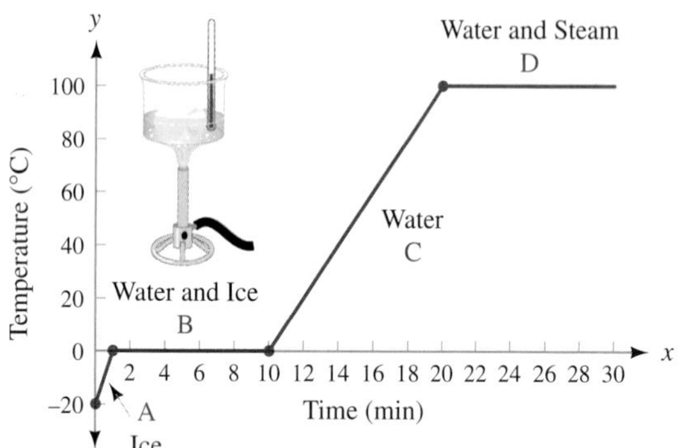
y
Water and Steam
D
100
80
60
40
20
0
-20
Temperature (°C)
Water
C
Water and Ice
B
2 4 6 8 10 12 14 16 18 20 22 24 26 28 30
x
A
Ice
Time (min)

a. How long does it take all the ice to melt?

b. From the time the heat is applied to the block of ice, how long is it before the water boils?

c. Find the slope of the line segment labeled A. What units would you attach to this number?

d. Find the slope of the line segment labeled C. Be sure to attach units to your answer.

e. Is the temperature changing faster during the 1st minute or the 16th minute?

47. Slope and Rate of Change Find the slope of the line connecting the first and last points on the chart. Explain in words what the slope represents.

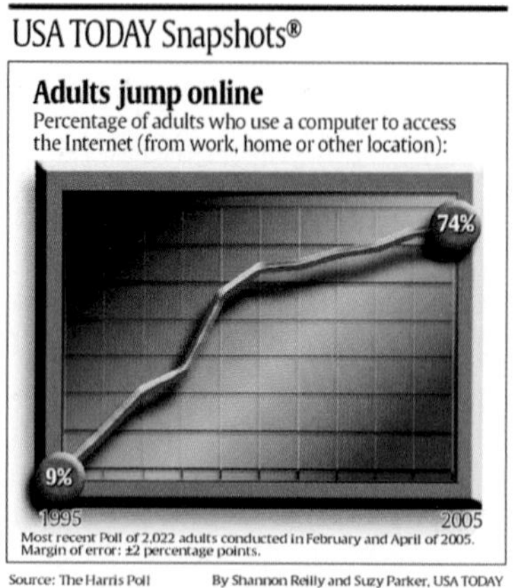
USA TODAY Snapshots®
Adults jump online
Percentage of adults who use a computer to access the Internet (from work, home or other location):
74%
9%
1995
2005
Most recent Poll of 2,022 adults conducted in February and April of 2005. Margin of error: ±2 percentage points.
Source: The Harris Poll
By Shannon Reilly and Suzy Parker, USA TODAY

From *USA Today.* Copyright 2005. Reprinted with permission.

48. Bridal Books Find the slope of the line that connects the points (1999, 118) and (2002, 75). Round to the nearest tenth. Explain in words what the slope represents.

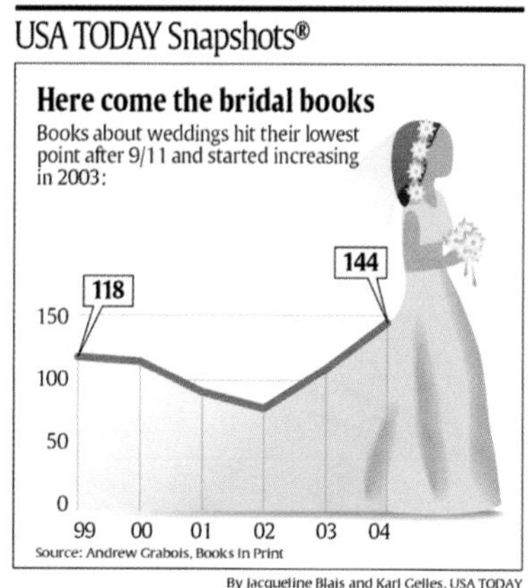
USA TODAY Snapshots®
Here come the bridal books
Books about weddings hit their lowest point after 9/11 and started increasing in 2003:
118
144
150
100
50
0
99 00 01 02 03 04
Source: Andrew Grabois, Books in Print
By Jacqueline Blais and Karl Gelles, USA TODAY

From *USA Today.* Copyright 2005. Reprinted with permission.

49. Health Care Use the chart to work the following problems involving slope.

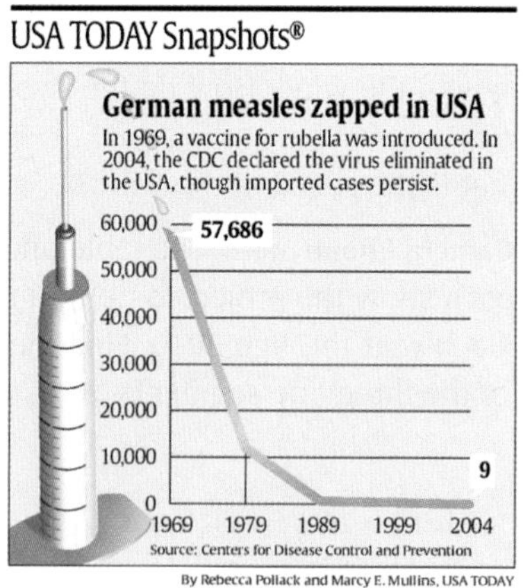
USA TODAY Snapshots®
German measles zapped in USA
In 1969, a vaccine for rubella was introduced. In 2004, the CDC declared the virus eliminated in the USA, though imported cases persist.
60,000
57,686
50,000
40,000
30,000
20,000
10,000
0
9
1969 1979 1989 1999 2004
Source: Centers for Disease Control and Prevention
By Rebecca Pollack and Marcy E. Mullins, USA TODAY

From *USA Today.* Copyright 2005. Reprinted with permission.

a. Find the slope of the line from 1969 to 1979, and then give a written explanation of the significance of that number.

b. Find the slope of the line from 1979 to 1989, and then give a written explanation of the significance of that number.

c. What is the slope of the line connecting (1989, 9) and (2004, 9)? What does this tell us about German measles?

50. Triathlon Entries Use the chart to work the following problems.

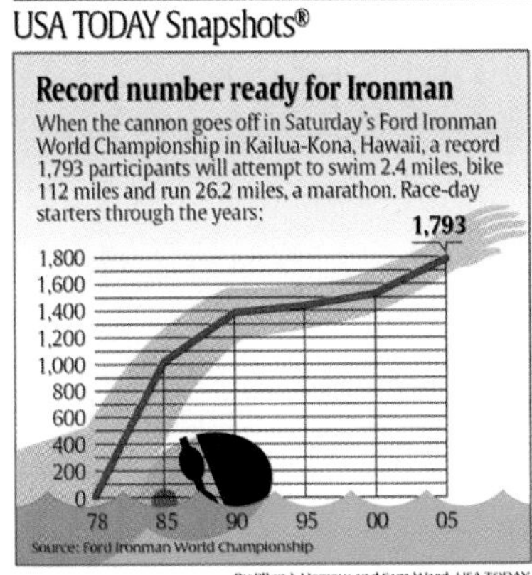
USA TODAY Snapshots®
Record number ready for Ironman
When the cannon goes off in Saturday's Ford Ironman World Championship in Kailua-Kona, Hawaii, a record 1,793 participants will attempt to swim 2.4 miles, bike 112 miles and run 26.2 miles, a marathon. Race-day starters through the years:
1,793
1,800
1,600
1,400
1,200
1,000
800
600
400
200
0
78
85
90
95
00
05
Source: Ford Ironman World Championship
By Ellen J. Horrow and Sam Ward, USA TODAY

From *USA Today*. Copyright 2005. Reprinted with permission.

a. Find the slope of the line that connects the points (1978, 0) and (2005, 1,793). Round your answer to the nearest tenth. Explain in words what the slope represents.

b. Of the line segments shown in the chart, one has a larger slope than the rest. Find that slope and give an interpretation of its meaning.

Maintaining Your Skills

The problems that follow review some of the more important skills you have learned in previous sections and chapters.

51. If $3x + 2y = 12$, find y when x is 4.

52. If $y = 3x - 1$, find x when y is 0.

53. Solve the formula $3x + 2y = 12$ for y.

54. Solve the formula $y = 3x - 1$ for x.

55. Solve the formula $A = P + Prt$ for t.

56. Solve the formula $S = \pi r^2 + 2\pi rh$ for h.

Getting Ready for the Next Section

Simplify.

57. $2\left(-\frac{1}{2}\right)$

58. $\frac{3-(-1)}{-3-3}$

59. $\frac{5-(-3)}{2-6}$

60. $3\left(-\frac{2}{3}x + 1\right)$

Solve for y.

61. $\frac{y-b}{x-0} = m$

62. $2x + 3y = 6$

63. $y - 3 = -2(x + 4)$

64. $y + 1 = -\frac{2}{3}(x - 3)$

65. If $y = -\frac{4}{3}x + 5$, find y when x is 0.

66. If $y = -\frac{4}{3}x + 5$, find y when x is 3.

Extending the Concepts

67. Find a point P on the line $y = 3x - 2$ such that the slope of the line through (1, 2) and P is equal to -1.

68. Find a point P on the line $y = -2x + 3$ such that the slope of the line through (2, 3) and P is equal to 2.

69. Use your Y-variables list or write a program to graph the family of curves $Y = -2X + B$ for $B = -3, -2, -1, 0, 1, 2$, and 3.

70. Use your Y-variables list or write a program to graph the family of curves $Y = -2X - B$ for $B = -3, -2, -1, 0, 1, 2$, and 3.

71. Use your Y-variables list or write a program to graph the family of curves $Y = AX$ for $A = -3, -2, -1, 0, 1, 2$, and 3.

72. Use your Y-variables list or write a program to graph the family of curves $Y = AX$ for $A = \frac{1}{4}, \frac{1}{3}, \frac{1}{2}, 1, 2$, and 3.

73. Use your Y-variables list or write a program to graph the family of curves $Y = AX + 2$ for $A = -3, -2, -1, 0, 1, 2$, and 3.

74. Use your Y-variables list or write a program to graph the family of curves $Y = AX - 2$ for $A = \frac{1}{4}, \frac{1}{3}, \frac{1}{2}, 1, 2$, and 3.

3.3 The Equation of a Line

OBJECTIVES

A Find the equation of a line given its slope and y-intercept.

B Find the slope and y-intercept from the equation of a line.

C Find the equation of a line given the slope and a point on the line.

D Find the equation of a line given two points on the line.

The table and illustrations show some corresponding temperatures on the Fahrenheit and Celsius temperature scales. For example, water freezes at 32°F and 0°C, and boils at 212°F and 100°C.

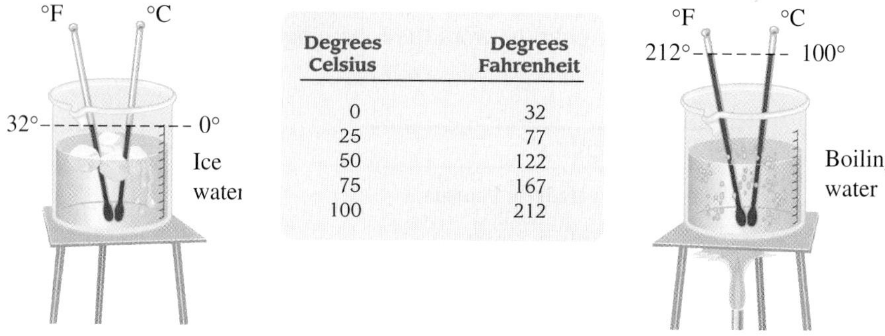

Degrees Celsius	Degrees Fahrenheit
0	32
25	77
50	122
75	167
100	212

If we plot all the points in the table using the x-axis for temperatures on the Celsius scale and the y-axis for temperatures on the Fahrenheit scale, we see that they line up in a straight line (Figure 1). This means that a linear equation in two variables will give a perfect description of the relationship between the two scales. That equation is

$$F = \frac{9}{5}C + 32$$

The techniques we use to find the equation of a line from a set of points is what this section is all about.

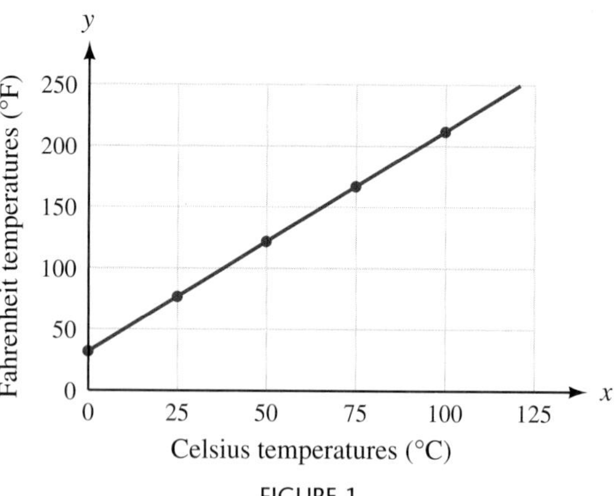

FIGURE 1

Suppose line l has slope m and y-intercept b. What is the equation of l? Because the y-intercept is b, we know the point $(0, b)$ is on the line. If (x, y) is any other point on l, then using the definition for slope, we have

$$\frac{y - b}{x - 0} = m \qquad \textbf{Definition of slope}$$

$$y - b = mx \qquad \textbf{Multiply both sides by } x$$

$$y = mx + b \qquad \textbf{Add } b \textbf{ to both sides}$$

This last equation is known as the *slope-intercept form* of the equation of a straight line.

Slope-Intercept Form of the Equation of a Line The equation of any line with slope m and y-intercept b is given by

$$y = mx + b$$

↗ Slope ↑ y-intercept

When the equation is in this form, the *slope* of the line is always the *coefficent of* x, and the *y-intercept* is always the *constant term.*

EXAMPLE 1

Find the equation of the line with slope $-\frac{4}{3}$ and y-intercept 5. Then graph the line.

SOLUTION Substituting $m = -\frac{4}{3}$ and $b = 5$ into the equation $y = mx + b$, we have

$$y = -\frac{4}{3}x + 5$$

Finding the equation from the slope and y-intercept is just that easy. If the slope is m and the y-intercept is b, then the equation is always $y = mx + b$. Now, let's graph the line.

Because the y-intercept is 5, the graph goes through the point (0, 5). To find a second point on the graph, we start at (0, 5) and move 4 units down (that's a rise of −4) and 3 units to the right (a run of 3). The point we end up at is (3, 1). Drawing a line that passes through (0, 5) and (3, 1), we have the graph of our equation. (Note that we could also let the rise = 4 and the run = −3 and obtain the same graph.) The graph is shown in Figure 2.

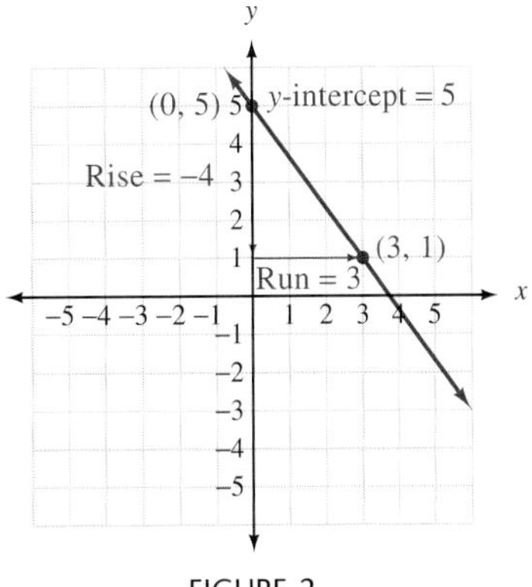

FIGURE 2

EXAMPLE 2

Give the slope and y-intercept for the line $2x - 3y = 5$.

SOLUTION To use the slope-intercept form, we must solve the equation for y in terms of x.

$$2x - 3y = 5$$

$$-3y = -2x + 5 \quad \textbf{Add } -2x \textbf{ to both sides}$$

$$y = \frac{2}{3}x - \frac{5}{3} \quad \textbf{Divide by } -3$$

The last equation has the form $y = mx + b$. The slope must be $m = \frac{2}{3}$, and the y-intercept is $b = -\frac{5}{3}$.

EXAMPLE 3 Graph the equation $2x + 3y = 6$ using the slope and y-intercept.

SOLUTION Although we could graph this equation by finding ordered pairs that are solutions to the equation and drawing a line through their graphs, it is sometimes easier to graph a line using the slope-intercept form of the equation.

Solving the equation for y, we have,

$$2x + 3y = 6$$

$$3y = -2x + 6 \quad \textbf{Add } -2x \textbf{ to both sides}$$

$$y = -\frac{2}{3}x + 2 \quad \textbf{Divide by 3}$$

The slope is $m = -\frac{2}{3}$ and the y-intercept is $b = 2$. Therefore, the point $(0, 2)$ is on the graph and the ratio of rise to run going from $(0, 2)$ to any other point on the line is $-\frac{2}{3}$. If we start at $(0, 2)$ and move 2 units up (that's a rise of 2) and 3 units to the left (a run of -3), we will be at another point on the graph. (We could also go down 2 units and right 3 units and still be assured of ending up at another point on the line because $\frac{2}{-3}$ is the same as $\frac{-2}{3}$.)

Note

As we mentioned earlier, the rectangular coordinate system is the tool we use to connect algebra and geometry. Example 3 illustrates this connection, as do the many other examples in this chapter. In Example 3, Descartes's rectangular coordinate system allows us to associate the equation $2x + 3y = 6$ (an algebraic concept) with the straight line (a geometric concept) shown in Figure 3.

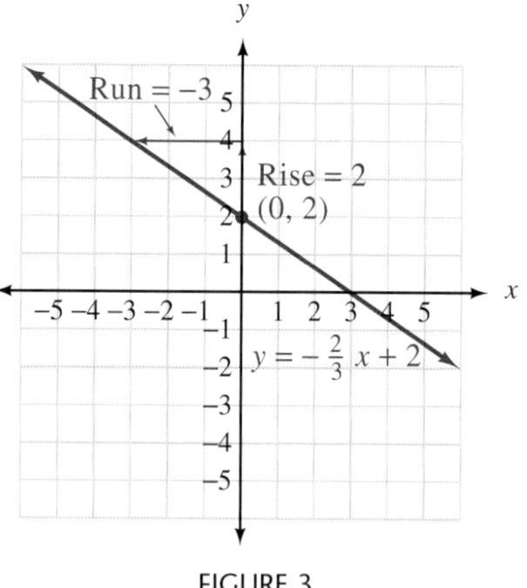

FIGURE 3

A second useful form of the equation of a straight line is the point-slope form.

Let line l contain the point (x_1, y_1) and have slope m. If (x, y) is any other point on l, then by the definition of slope we have

$$\frac{y - y_1}{x - x_1} = m$$

Multiplying both sides by $(x - x_1)$ gives us

$$\mathbf{(x - x_1)} \cdot \frac{y - y_1}{x - x_1} = m\mathbf{(x - x_1)}$$

$$y - y_1 = m(x - x_1)$$

This last equation is known as the *point-slope form* of the equation of a straight line.

> **Point-Slope Form of the Equation of a Line** The equation of the line through (x_1, y_1) with slope m is given by
> $$y - y_1 = m(x - x_1)$$

This form of the equation of a straight line is used to find the equation of a line, either given one point on the line and the slope, or given two points on the line.

EXAMPLE 4

Find the equation of the line with slope -2 that contains the point $(-4, 3)$. Write the answer in slope-intercept form.

SOLUTION

Using $(x_1, y_1) = (-4, 3)$ and $m = -2$

in	$y - y_1 = m(x - x_1)$	**Point-slope form**
gives us	$y - 3 = -2(x + 4)$	***Note:*** **$x - (-4) = x + 4$**
	$y - 3 = -2x - 8$	**Multiply out right side**
	$y = -2x - 5$	**Add 3 to each side**

Figure 4 is the graph of the line that contains $(-4, 3)$ and has a slope of -2. Notice that the y-intercept on the graph matches that of the equation we found.

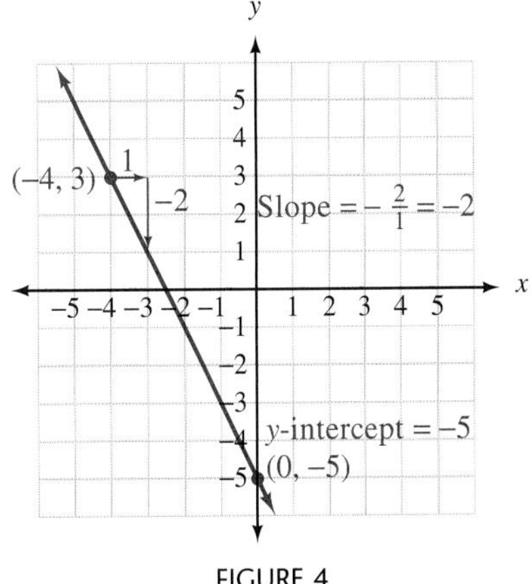

FIGURE 4

EXAMPLE 5

Find the equation of the line that passes through the points $(-3, 3)$ and $(3, -1)$.

SOLUTION We begin by finding the slope of the line:

$$m = \frac{3 - (-1)}{-3 - 3} = \frac{4}{-6} = -\frac{2}{3}$$

Using $(x_1, y_1) = (3, -1)$ and $m = -\frac{2}{3}$ in $y - y_1 = m(x - x_1)$ yields

$$y + 1 = -\frac{2}{3}(x - 3)$$

$$y + 1 = -\frac{2}{3}x + 2 \qquad \textbf{Multiply out right side}$$

$$y = -\frac{2}{3}x + 1 \qquad \textbf{Add −1 to each side}$$

Note

In Example 5 we could have used the point $(-3, 3)$ instead of $(3, -1)$ and obtained the same equation; that is, using $(x_1, y_1) = (-3, 3)$ and $m = -\frac{2}{3}$ in $y - y_1 = m(x - x_1)$ gives us

$$y - 3 = -\frac{2}{3}(x + 3)$$

$$y - 3 = -\frac{2}{3}x - 2$$

$$y = -\frac{2}{3}x + 1$$

which is the same result we obtained using $(3, -1)$.

Figure 5 shows the graph of the line that passes through the points $(-3, 3)$ and $(3, -1)$. As you can see, the slope and y-intercept are $-\frac{2}{3}$ and 1, respectively.

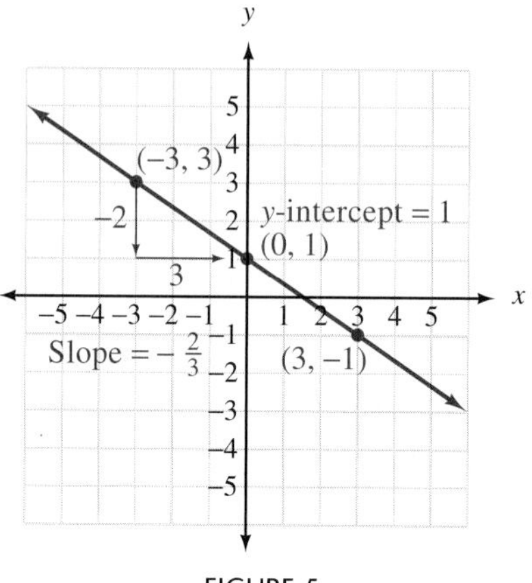

FIGURE 5

The last form of the equation of a line that we will consider in this section is called the standard form. It is used mainly to write equations in a form that is free of fractions and is easy to compare with other equations.

Standard Form for the Equation of a Line If a, b, and c are integers, then the equation of a line is in standard form when it has the form

$$ax + by = c$$

If we were to write the equation

$$y = -\frac{2}{3}x + 1$$

in standard form, we would first multiply both sides by 3 to obtain

$$3y = -2x + 3$$

Then we would add $2x$ to each side, yielding

$$2x + 3y = 3$$

which is a linear equation in standard form.

EXAMPLE 6

Give the equation of the line through $(-1, 4)$ whose graph is perpendicular to the graph of $2x - y = -3$. Write the answer in standard form.

SOLUTION To find the slope of $2x - y = -3$, we solve for y:

$$2x - y = -3$$

$$y = 2x + 3$$

The slope of this line is 2. The line we are interested in is perpendicular to the line with slope 2 and must, therefore, have a slope of $-\frac{1}{2}$.

Using $(x_1, y_1) = (-1, 4)$ and $m = -\frac{1}{2}$, we have

$$y - y_1 = m(x - x_1)$$

$$y - 4 = -\frac{1}{2}(x + 1)$$

Because we want our answer in standard form, we multiply each side by 2.

$$2y - 8 = -1(x + 1)$$

$$2y - 8 = -x - 1$$

$$x + 2y - 8 = -1$$

$$x + 2y = 7$$

The last equation is in standard form.

As a final note, the summary reminds us that all horizontal lines have equations of the form $y = b$ and slopes of 0. Because they cross the y-axis at b, the y-intercept is b; there is no x-intercept. Vertical lines have no slope and equations of the form $x = a$. Each will have an x-intercept at a and no y-intercept. Finally, equations of the form $y = mx$ have graphs that pass through the origin. The slope is always m and both the x-intercept and the y-intercept are 0.

FACTS FROM GEOMETRY

Special Equations and Their Graphs, Slopes, and Intercepts

For the equations below, m, a, and b are real numbers.

Through the Origin	*Vertical Line*	*Horizontal Line*
Equation: $y = mx$	Equation: $x = a$	Equation: $y = b$
Slope $= m$	No slope	Slope $= 0$
x-intercept $= 0$	x-intercept $= a$	No x-intercept
y-intercept $= 0$	No y-intercept	y-intercept $= b$

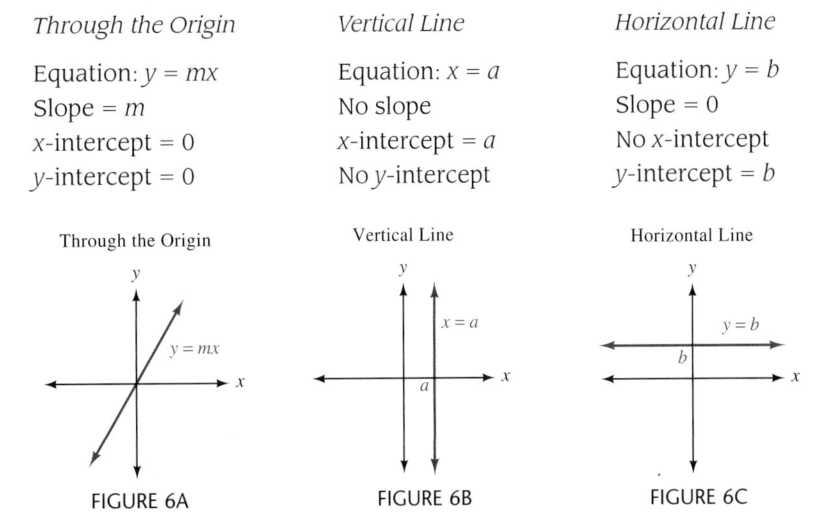

FIGURE 6A FIGURE 6B FIGURE 6C

USING TECHNOLOGY

Graphing Calculators

One advantage of using a graphing calculator to graph lines is that a calculator does not care whether the equation has been simplified or not. To illustrate, in Example 5 we found that the equation of the line with slope $-\frac{2}{3}$ that passes through the point $(3, -1)$ is

$$y + 1 = -\frac{2}{3}(x - 3)$$

Normally, to graph this equation we would simplify it first. With a graphing calculator, we add -1 to each side and enter the equation this way:

$$Y_1 = -(2/3)(X - 3) - 1$$

LINKING OBJECTIVES AND EXAMPLES

Next to each objective we have listed the examples that are best described by that objective.

A	1
B	2, 3
C	4, 6
D	5

GETTING READY FOR CLASS

After reading through the preceding section, respond in your own words and in complete sentences.

1. How would you graph the line $y = \frac{1}{2}x + 3$?
2. What is the slope-intercept form of the equation of a line?
3. Describe how you would find the equation of a line if you knew the slope and the y-intercept of the line.
4. If you had the graph of a line, how would you use it to find the equation of the line?

Problem Set 3.3

Online support materials can be found at www.thomsonedu.com/login

Write the equation of the line with the given slope and y-intercept.

1. $m = 2, b = 3$
2. $m = -4, b = 2$
3. $m = 1, b = -5$
4. $m = -5, b = -3$
5. $m = \frac{1}{2}, b = \frac{3}{2}$
6. $m = \frac{2}{3}, b = \frac{5}{6}$
7. $m = 0, b = 4$
8. $m = 0, b = -2$

Find the slope of a line (a) parallel and (b) perpendicular to the given line.

▸ 9. $y = 3x - 4$
▸ 10. $y = -4x + 1$
▸ 11. $3x + y = -2$
▸ 12. $2x - y = -4$
▸ 13. $2x + 5y = -11$
▸ 14. $3x - 5y = -4$

Give the slope and y-intercept for each of the following equations. Sketch the graph using the slope and y-intercept. Give the slope of any line perpendicular to the given line.

15. $y = 3x - 2$
16. $y = 2x + 3$
17. $2x - 3y = 12$
18. $3x - 2y = 12$
▸ 19. $4x + 5y = 20$
20. $5x - 4y = 20$

For each of the following lines, name the slope and y-intercept. Then write the equation of the line in slope-intercept form.

21.

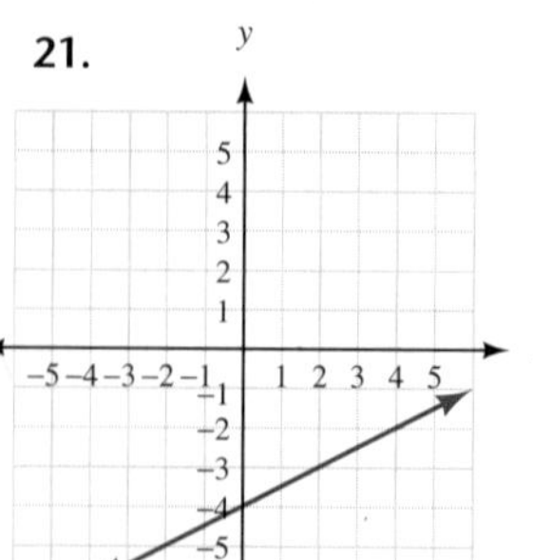

22.

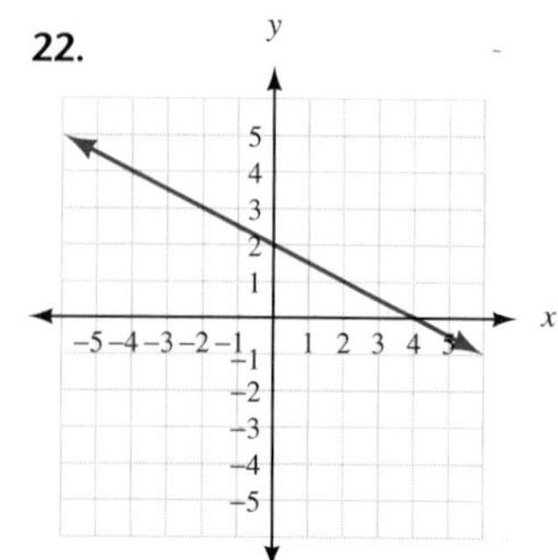

23.

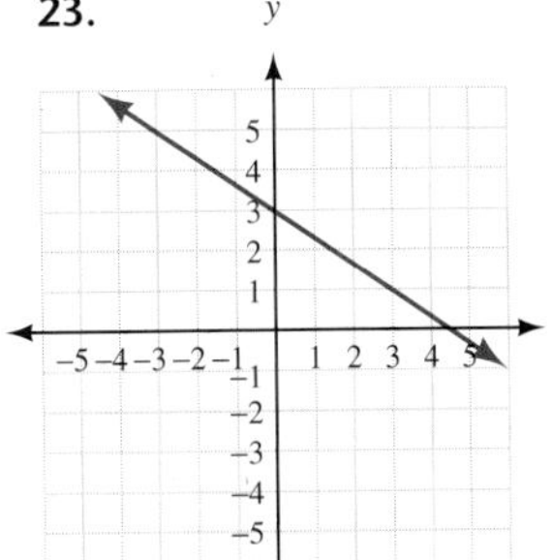

24.

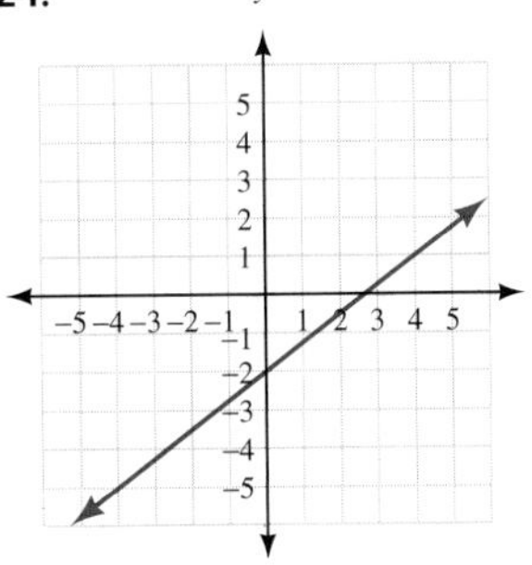

For each of the following problems, the slope and one point on a line are given. In each case, find the equation of that line. (Write the equation for each line in slope-intercept form.)

25. $(-2, -5)$; $m = 2$

26. $(-1, -5)$; $m = 2$

27. $(-4, 1)$; $m = -\frac{1}{2}$

28. $(-2, 1)$; $m = -\frac{1}{2}$

29. $\left(-\frac{1}{3}, 2\right)$; $m = -3$

30. $\left(-\frac{2}{3}, 5\right)$; $m = -3$

31. $(-4, -2)$; $m = \frac{2}{3}$

32. $(3, 4)$; $m = -\frac{1}{3}$

33. $(-5, 2)$; $m = -\frac{1}{4}$

34. $(-4, 3)$; $m = -\frac{1}{6}$

Find the equation of the line that passes through each pair of points. Write your answers in standard form.

35. $(-2, -4)$, $(1, -1)$

36. $(2, 4)$, $(-3, -1)$

37. $(-1, -5)$, $(2, 1)$

38. $(-1, 6)$, $(1, 2)$

39. $\left(\frac{1}{3}, -\frac{1}{5}\right)$, $\left(-\frac{1}{3}, -1\right)$

40. $\left(-\frac{1}{2}, -\frac{1}{2}\right)$, $\left(\frac{1}{2}, \frac{1}{10}\right)$

41. The equation $3x - 2y = 10$ is a linear equation in standard form. From this equation, answer the following:
   a. Find the x- and y-intercepts.
   b. Find a solution to this equation other than the intercepts in part a.
   c. Write this equation in slope-intercept form.
   d. Is the point $(2, 2)$ a solution to the equation?

42. The equation $4x + 3y = 8$ is a linear equation in standard form. From this equation, answer the following:
   a. Find the x- and y-intercepts.
   b. Find a solution to this equation other than the intercepts in part a.
   c. Write this equation in slope-intercept form.
   d. Is the point $(-3, 2)$ a solution to the equation?

43. The equation $\frac{3x}{4} - \frac{y}{2} = 1$ is a linear equation. From this equation, answer the following:
   a. Find the x- and y-intercepts.
   b. Find a solution to this equation other than the intercepts in part a.
   c. Write this equation in slope-intercept form.
   d. Is the point $(1, 2)$ a solution to the equation?

44. The equation $\frac{3x}{5} + \frac{2y}{3} = 1$ is a linear equation. From this equation, answer the following:
   a. Find the x- and y-intercepts.
   b. Find a solution to this equation other than the intercepts in part a.
   c. Write this equation in slope-intercept form.
   d. Is the point $(-5, 3)$ a solution to the equation?

The next two problems are intended to give you practice reading, and paying attention to, the instructions that accompany the problems you are working. Working these problems is an excellent way to get ready for a test or a quiz.

45. Work each problem according to the instructions given.
   a. Solve: $-2x + 1 = -3$
   b. Write in slope-intercept form: $-2x + y = -3$
   c. Find the y-intercept: $-2x + y = -3$
   d. Find the slope: $-2x + y = -3$
   e. Graph: $-2x + y = -3$

46. Work each problem according to the instructions given.
   a. Solve: $\frac{x}{3} + \frac{1}{4} = 1$
   b. Write in slope-intercept form: $\frac{x}{3} + \frac{y}{4} = 1$

c. Find the y-intercept: $\frac{x}{3} + \frac{y}{4} = 1$

d. Find the slope: $\frac{x}{3} + \frac{y}{4} = 1$

e. Graph: $\frac{x}{3} + \frac{y}{4} = 1$

For each of the following lines, name the coordinates of any two points on the line. Then use those two points to find the equation of the line.

47.

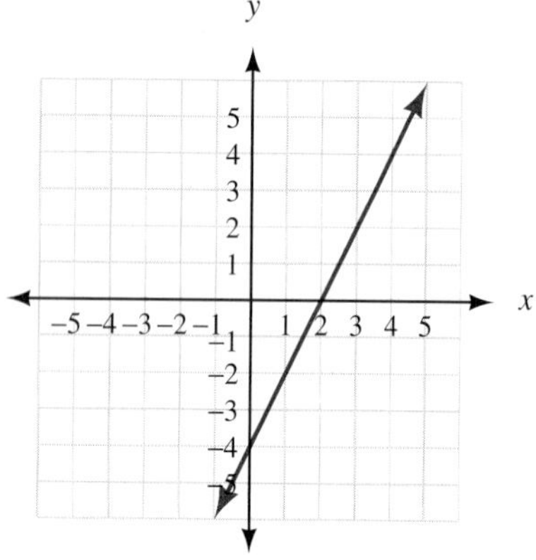

48.

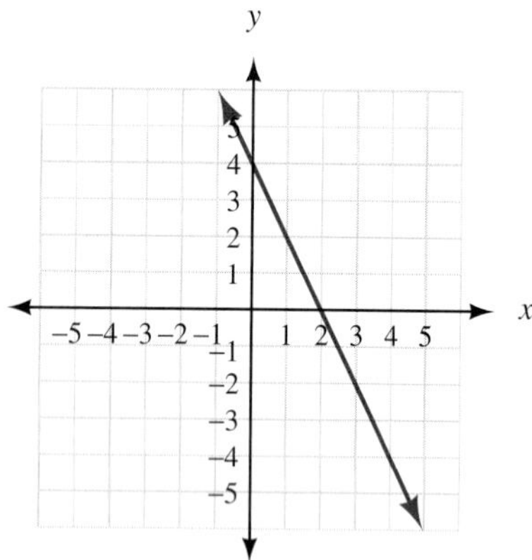

49.

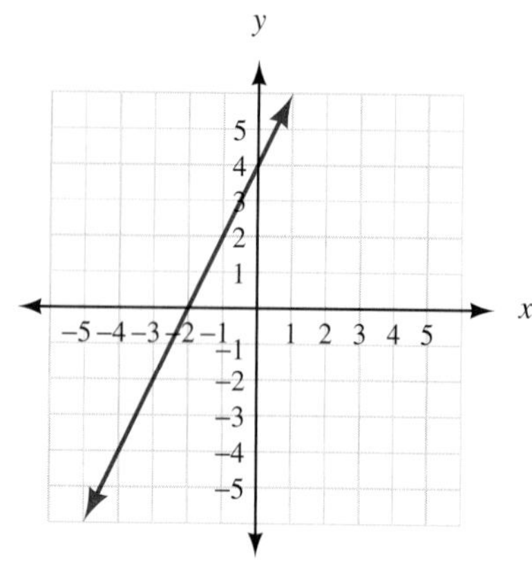

50.

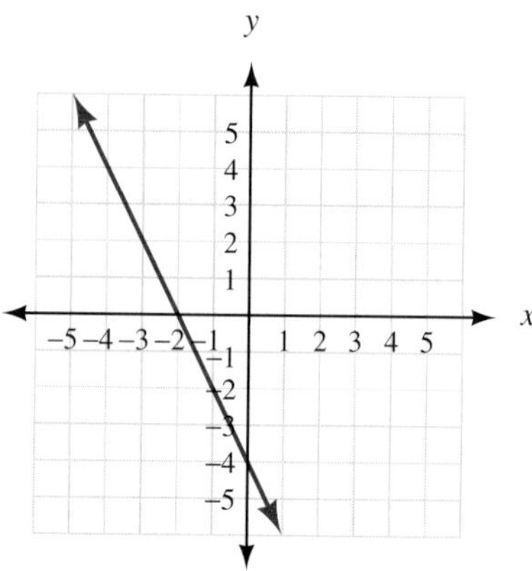

51. Give the slope and y-intercept of $y = -2$. Sketch the graph.

52. For the line $x = -3$, sketch the graph, give the slope, and name any intercepts.

53. Find the equation of the line parallel to the graph of $3x - y = 5$ that contains the point $(-1, 4)$.

54. Find the equation of the line parallel to the graph of $2x - 4y = 5$ that contains the point $(0, 3)$.

55. Line l is perpendicular to the graph of the equation $2x - 5y = 10$ and contains the point $(-4, -3)$. Find the equation for l.

56. Line l is perpendicular to the graph of the equation $-3x - 5y = 2$ and contains the point $(2, -6)$. Find the equation for l.

57. Give the equation of the line perpendicular to the graph of $y = -4x + 2$ that has an x-intercept of -1.

58. Write the equation of the line parallel to the graph of $7x - 2y = 14$ that has an x-intercept of 5.

59. Find the equation of the line with x-intercept 3 and y-intercept 2.

60. Find the equation of the line with x-intercept 2 and y-intercept 3.

Applying the Concepts

61. **Deriving the Temperature Equation** The table from the introduction to this section is repeated here. The rows of the table give us ordered pairs (C, F).

Degrees Celsius C	Degrees Fahrenheit F
0	32
25	77
50	122
75	167
100	212

a. Use any two of the ordered pairs from the table to derive the equation $F = \frac{9}{5}C + 32$.
b. Use the equation from part (a) to find the Fahrenheit temperature that corresponds to a Celsius temperature of 30°.

62. **Maximum Heart Rate** The table gives the maximum heart rate for adults 30, 40, 50, and 60 years old. Each row of the table gives us an ordered pair (A, M).

Age (years) A	Maximum Heart Rate (beats per minute) M
30	190
40	180
50	170
60	160

a. Use any two of the ordered pairs from the table to derive the equation $M = 220 - A$, which gives the maximum heart rate M for an adult whose age is A.
b. Use the equation from part (a) to find the maximum heart rate for a 25-year-old adult.

63. **Textbook Cost** To produce this textbook, suppose the publisher spent $125,000 for typesetting and $6.50 per book for printing and binding. The total cost to produce and print n books can be written as

Total cost to produce n textbooks	=	Fixed cost + Variable cost
C	=	$125,000 + 6.5n$

a. Suppose the number of books printed in the first printing is 10,000. What is the total cost?
b. If the average cost is the total cost divided by the number of books printed, find the average cost of producing 10,000 textbooks.
c. Find the cost to produce one more textbook when you have already produced 10,000 textbooks.

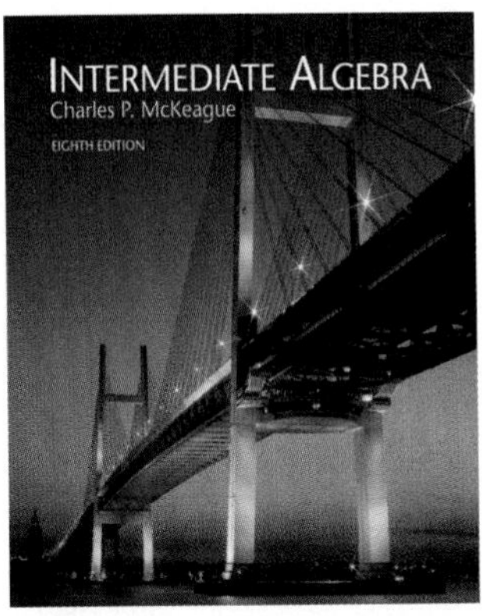

64. **Exercise Heart Rate** In an aerobics class, the instructor indicates that her students' exercise heart rate is 60% of their maximum heart rate, where maximum heart rate is 220 minus their age.
a. Determine the equation that gives exercise heart rate E in terms of age A.
b. Use the equation to find the exercise heart rate of a 22-year-old student.

65. **Bridal Books** Use the chart to work the following problems. Assume 75 books in 2002.
a. Find the equation of the line that runs from 2002 to 2004.
b. Use your answer to part a to estimate the number of bridal books produced in 2006.

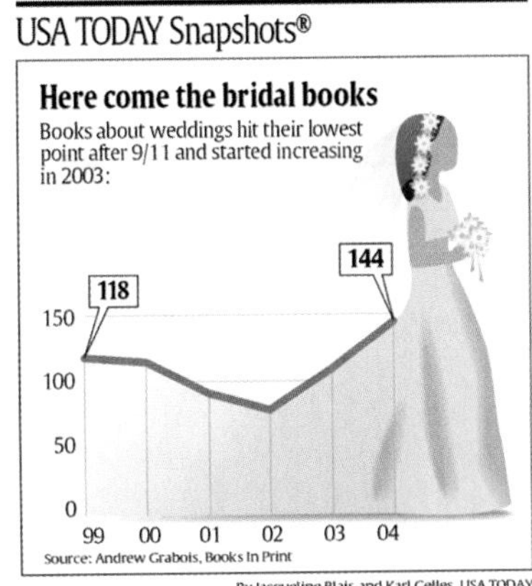

From *USA Today.* Copyright 2005. Reprinted with permission.

66. **Triathlon Entries** Use the chart to work the following problems.

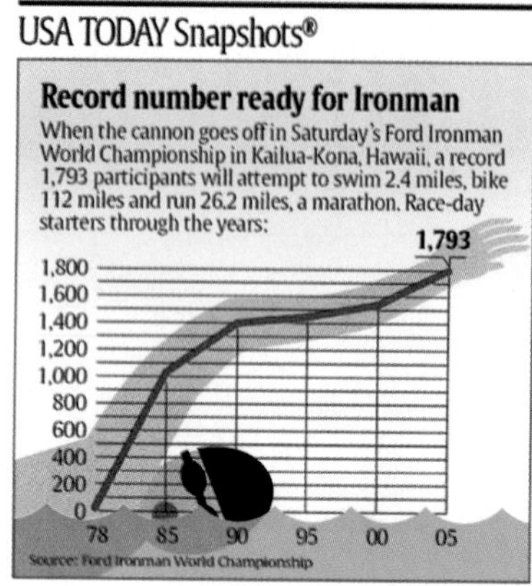
USA TODAY Snapshots®
Record number ready for Ironman
When the cannon goes off in Saturday's Ford Ironman World Championship in Kailua-Kona, Hawaii, a record 1,793 participants will attempt to swim 2.4 miles, bike 112 miles and run 26.2 miles, a marathon. Race-day starters through the years:
1,793
1,800
1,600
1,400
1,200
1,000
800
600
400
200
0
78 85 90 95 00 05
Source: Ford Ironman World Championship
By Ellen J. Horrow and Sam Ward, USA TODAY

From *USA Today*. Copyright 2005. Reprinted with permission.

a. Find the equation of the line that connects the points (1978, 0) and (2005, 1,793).

b. Use your answer to part a to estimate the number of entries in the 2006 Ironman.

c. Do you think the actual number of entries was more or less than your answer to part b? Explain why.

67. **Improving Your Quantitative Literacy** Use the information in the chart to work the following problems.

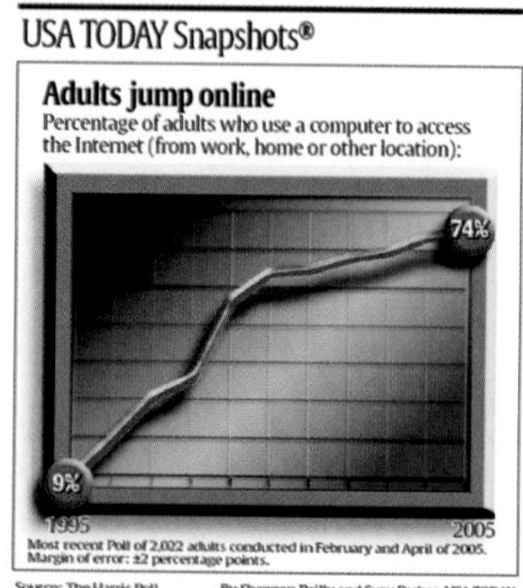
USA TODAY Snapshots®
Adults jump online
Percentage of adults who use a computer to access the Internet (from work, home or other location):
74%
9%
1995
2005
Most recent Poll of 2,022 adults conducted in February and April of 2005. Margin of error: ±2 percentage points.
Source: The Harris Poll
By Shannon Reilly and Suzy Parker, USA TODAY

From *USA Today*. Copyright 2005. Reprinted with permission.

a. Find the equation of the line connecting the first and last points on the chart.

b. Use the equation from part a to estimate the percent of adults accessing the Internet in 2006.

c. What happens when we use our equation to estimate the number of adults online in 2010, and why is this result unacceptable?

68. **Definitions**

a. Use an online dictionary to find definitions for the following words:

Merriam-
Webster
®

interpolation and *extrapolation*

b. Which of the two words describes what you did in problem 67 parts b and c above.

Maintaining Your Skills

The problems that follow review some of the more important skills you have learned in previous sections and chapters.

69. The length of a rectangle is 3 inches more than 4 times the width. The perimeter is 56 inches. Find the length and width.

70. One angle is 10 degrees less than four times another. Find the measure of each angle if:

a. The two angles are complementary.

b. The two angles are supplementary.

71. The cash register in a candy shop contains \$66 at the beginning of the day. At the end of the day, it contains \$732.50. If the sales tax rate is 7.5%, how much of the total is sales tax?

72. The third angle in an isosceles triangle is 20 degrees less than twice as large as each of the two base angles. Find the measure of each angle.

Getting Ready for the Next Section

73. Which of the following are solutions to $x + y \le 4$?

(0, 0) (4, 0) (2, 3)

74. Which of the following are solutions to $y < 2x - 3$?

(0, 0) (3, −2) (−3, 2)

75. Which of the following are solutions to $y \le \frac{1}{2}x$?

(0, 0) (2, 0) (−2, 0)

76. Which of the following are solutions to $y > -2x$?

(0, 0) (2, 0) (−2, 0)

Extending the Concepts

The midpoint M of two points (x_1, y_1) and (x_2, y_2) is defined to be the average of each of their coordinates, so

$$M = \left(\frac{x_1 + x_2}{2}, \frac{y_1 + y_2}{2}\right)$$

For example, the midpoint of (−2, 3) and (6, 8) is given by

$$\left(\frac{-2 + 6}{2}, \frac{3 + 8}{2}\right) = \left(2, \frac{11}{2}\right)$$

For each given pair of points, find the equation of the line that is perpendicular to the line through these points and that passes through their midpoint. Answer using slope-intercept form.

Note: This line is called the perpendicular bisector of the line segment connecting the two points.

77. (1, 4) and (7, 8)

78. (−2, 1) and (6, 7)

79. (−5, 1) and (−1, 4)

80. (−6, −2) and (2, 1)

3.4 Linear Inequalities in Two Variables

OBJECTIVES

A Graph linear inequalities in two variables.

A small movie theater holds 100 people. The owner charges more for adults than for children, so it is important to know the different combinations of adults and children that can be seated at one time. The shaded region in Figure 1 contains all the seating combinations. The line $x + y = 100$ shows the combinations for a full theater: The y-intercept corresponds to a theater full of adults, and the x-intercept corresponds to a theater full of children. In the shaded region below the line $x + y = 100$ are the combinations that occur if the theater is not full.

Shaded regions like the one shown in Figure 1 are produced by linear inequalities in two variables, which is the topic of this section.

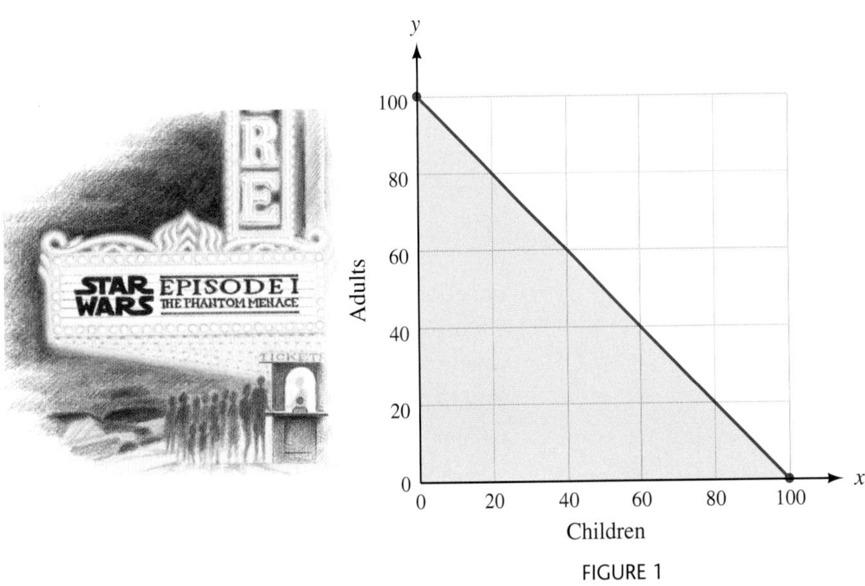

FIGURE 1

A *linear inequality in two variables* is any expression that can be put in the form

$$ax + by < c$$

where a, b, and c are real numbers (a and b not both 0). The inequality symbol can be any one of the following four: $<$, $\leq$, $>$, $\geq$.

Some examples of linear inequalities are

$$2x + 3y < 6 \qquad y \geq 2x + 1 \qquad x - y \leq 0$$

Although not all of these examples have the form $ax + by < c$, each one can be put in that form.

The solution set for a linear inequality is a *section of the coordinate plane.* The *boundary* for the section is found by replacing the inequality symbol with an equal sign and graphing the resulting equation. The boundary is included in the solution set (and represented with a *solid line*) if the inequality symbol used originally is $\leq$ or $\geq$. The boundary is not included (and is represented with a *broken line*) if the original symbol is $<$ or $>$.

EXAMPLE 1

Graph the solution set for $x + y \leq 4$.

SOLUTION The boundary for the graph is the graph of $x + y = 4$. The boundary is included in the solution set because the inequality symbol is $\leq$.

Figure 2 is the graph of the boundary:

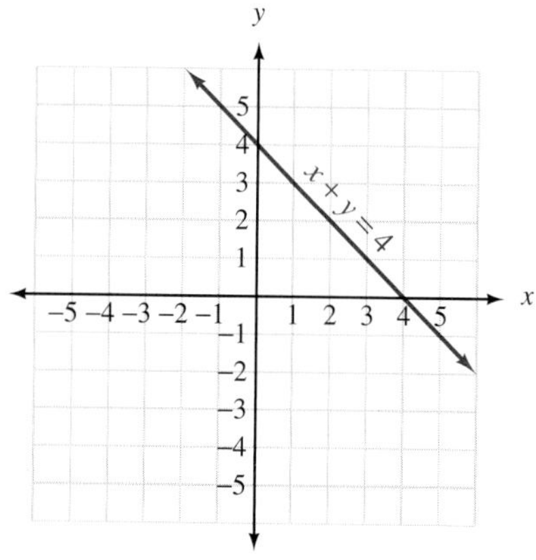

FIGURE 2

The boundary separates the coordinate plane into two regions: the region above the boundary and the region below it. The solution set for $x + y \leq 4$ is one of these two regions along with the boundary. To find the correct region, we simply choose any convenient point that is *not* on the boundary. We then substitute the coordinates of the point into the original inequality $x + y \leq 4$. If the point we choose satisfies the inequality, then it is a member of the solution set, and we can assume that all points on the same side of the boundary as the chosen point are also in the solution set. If the coordinates of our point do not satisfy the original inequality, then the solution set lies on the other side of the boundary.

In this example, a convenient point that is not on the boundary is the origin.

Substituting	$(0, 0)$	
into	$x + y \leq 4$	
gives us	$0 + 0 \leq 4$	
	$0 \leq 4$	**A true statement**

Because the origin is a solution to the inequality $x + y \le 4$ and the origin is below the boundary, all other points below the boundary are also solutions.

Figure 3 is the graph of $x + y \le 4$.

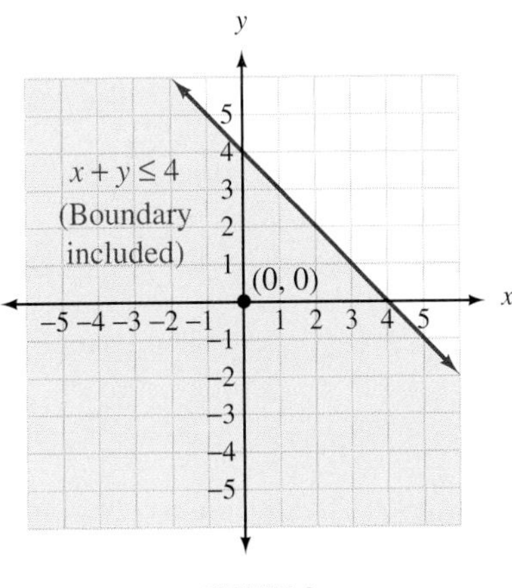

FIGURE 3

The region above the boundary is described by the inequality $x + y > 4$. 

Here is a list of steps to follow when graphing the solution set for a linear inequality in two variables.

To Graph a Linear Inequality in Two Variables

Step 1: Replace the inequality symbol with an equal sign. The resulting equation represents the boundary for the solution set.

Step 2: Graph the boundary found in step 1 using a *solid line* if the boundary is included in the solution set (that is, if the original inequality symbol was either $\le$ or $\ge$). Use a *broken line* to graph the boundary if it is *not* included in the solution set. (It is not included if the original inequality was either $<$ or $>$.)

Step 3: Choose any convenient point not on the boundary and substitute the coordinates into the *original* inequality. If the resulting statement is *true,* the graph lies on the *same* side of the boundary as the chosen point. If the resulting statement is *false,* the solution set lies on the *opposite* side of the boundary.

EXAMPLE 2

Graph the solution set for $y < 2x - 3$.

SOLUTION The boundary is the graph of $y = 2x - 3$, a line with slope 2 and y-intercept -3. The boundary is not included because the original inequality symbol is $<$. We therefore use a broken line to represent the boundary in Figure 4.

A convenient test point is again the origin:

Using	$(0, 0)$	
in	$y < 2x - 3$	
we have	$0 < 2(0) - 3$	
	$0 < -3$	**A false statement**

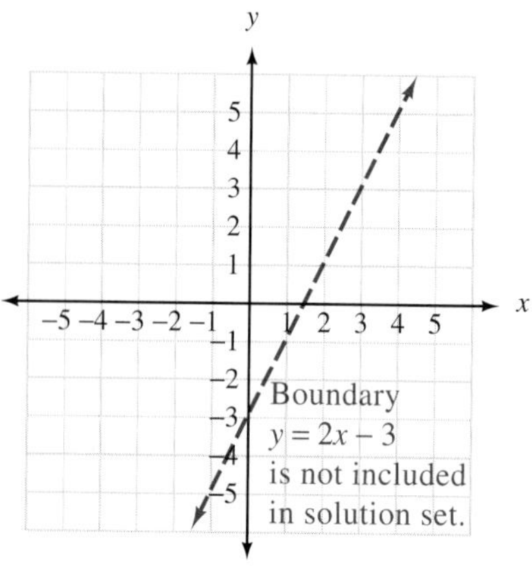

FIGURE 4

Because our test point gives us a false statement and it lies above the boundary, the solution set must lie on the other side of the boundary (Figure 5).

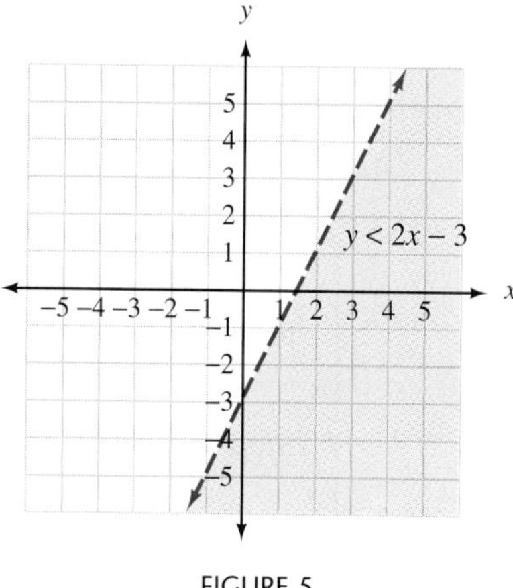

FIGURE 5

USING TECHNOLOGY

Graphing Calculators

Most graphing calculators have a Shade command that allows a portion of a graphing screen to be shaded. With this command we can visualize the solution sets to linear inequalities in two variables. Because most graphing calculators cannot draw a dotted line, however, we are not actually "graphing" the solution set, only visualizing it.

Strategy for Visualizing a Linear Inequality in Two Variables on a Graphing Calculator

Step 1: Solve the inequality for y.

Step 2: Replace the inequality symbol with an equal sign. The resulting equation represents the boundary for the solution set.

Step 3: Graph the equation in an appropriate viewing window.

Step 4: Use the Shade command to indicate the solution set:

For inequalities having the $<$ or $\leq$ sign, use Shade(Ymin, Y_1).

For inequalities having the $>$ or $\geq$ sign, use Shade(Y_1, Ymax).

Note: On the TI-83, step 4 can be done by manipulating the icons in the left column in the list of Y variables.

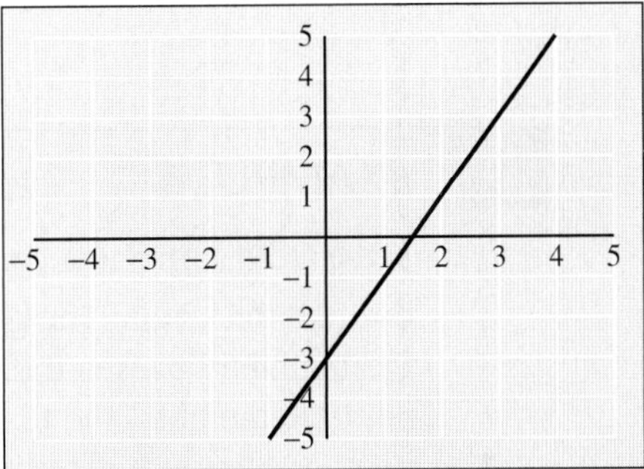

FIGURE 6 $Y_1 = 2X - 3$

Figures 6 and 7 show the graphing calculator screens that help us visualize the solution set to the inequality $y < 2x - 3$ that we graphed in Example 2.

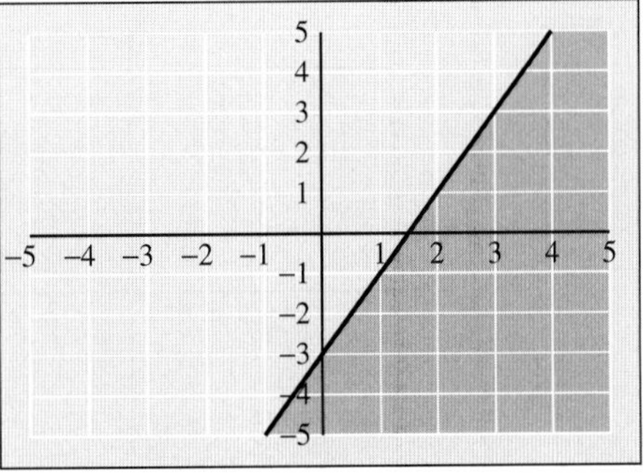

FIGURE 7 **Shade (Ymin, Y_1)**

EXAMPLE 3

Graph the solution set for $x \le 5$.

SOLUTION The boundary is $x = 5$, which is a vertical line. All points in Figure 8 to the left of the boundary have x-coordinates less than 5 and all points to the right have x-coordinates greater than 5.

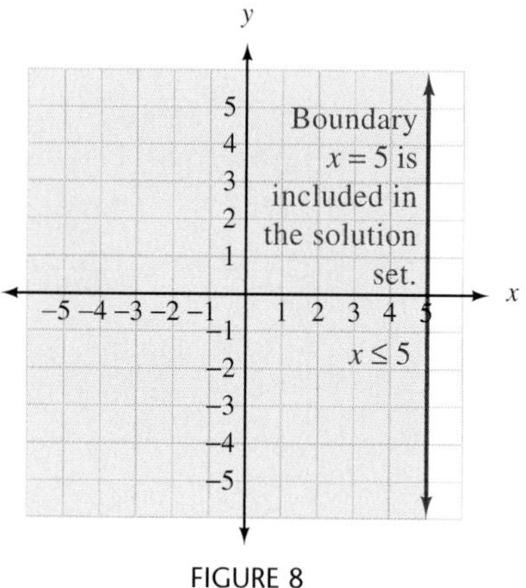

FIGURE 8

EXAMPLE 4

Graph the solution set for $y > \frac{1}{4}x$.

SOLUTION The boundary is the line $y = \frac{1}{4}x$, which has a slope of $\frac{1}{4}$ and passes through the origin. The graph of the boundary line is shown in Figure 9. Since the boundary passes through the origin, we cannot use the origin as our test point. Remember, the test point cannot be on the boundary line. Let's use the point $(0, -4)$ as our test point. It lies below the boundary line. When we substitute the coordinates into our original inequality, the result is the false statement $-4 > 0$. This tells us that the solution set is on the other side of the boundary line. The solution set for our original inequality is shown in Figure 10.

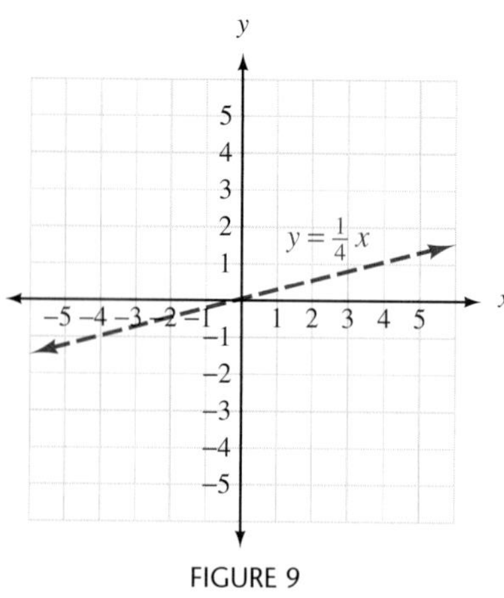

FIGURE 9

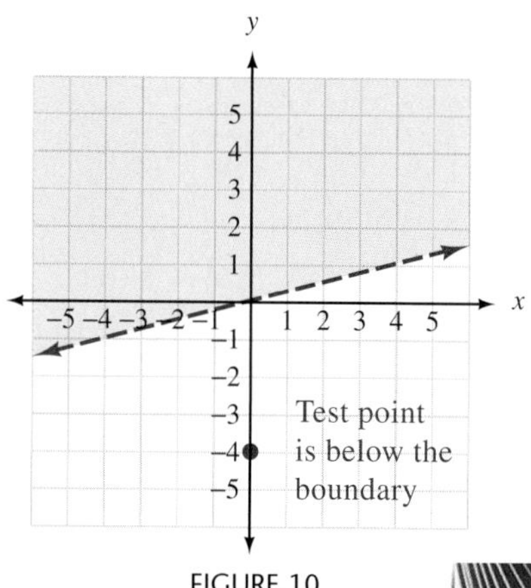

FIGURE 10

LINKING OBJECTIVES AND EXAMPLES

Next to each **objective** we have listed the examples that are best described by that objective.

A 1–4

GETTING READY FOR CLASS

After reading through the preceding section, respond in your own words and in complete sentences.

1. When graphing a linear inequality in two variables, how do you find the equation of the boundary line?
2. What is the significance of a broken line in the graph of an inequality?
3. When graphing a linear inequality in two variables, how do you know which side of the boundary line to shade?
4. Does the graph of $x + y < 4$ include the boundary line? Explain.

Problem Set 3.4

Online support materials can be found at www.thomsonedu.com/login

Use a coordinate system to graph the solution set for each of the following.

1. $x + y < 5$
2. $x + y \le 5$
3. $x - y \ge -3$
4. $x - y > -3$
5. $2x + 3y < 6$
6. $2x - 3y > -6$
7. $-x + 2y > -4$
8. $-x - 2y < 4$
9. $2x + y < 5$
10. $2x + y < -5$
11. $y < 2x - 1$
12. $y \le 2x - 1$
13. $3x - 4y < 12$
14. $-2x + 3y < 6$
15. $-5x + 2y \le 10$
16. $4x - 2y \le 8$
17. $x \ge 3$
18. $x > -2$
19. $y \le 4$
20. $y > -5$
21. $y < 2x$
22. $y > -3x$
23. $y \ge \frac{1}{2}x$
24. $y \le \frac{1}{3}x$
25. $y \ge \frac{3}{4}x - 2$
26. $y > -\dfrac{2}{3}x + 3$
27. $\dfrac{x}{3} + \dfrac{y}{2} > 1$
28. $\dfrac{x}{5} + \dfrac{y}{4} < 1$
29. $\dfrac{x}{3} - \dfrac{y}{2} > 1$
30. $-\dfrac{x}{4} + \dfrac{y}{3} > 1$
31. $y \le -\dfrac{2}{3}x$
32. $y \ge \dfrac{1}{4}x$
33. $5x - 3y < 0$
34. $2x + 3y > 0$
35. $\dfrac{x}{4} + \dfrac{y}{5} \le 1$
36. $\dfrac{x}{2} + \dfrac{y}{3} < 1$

For each graph shown here, name the linear inequality in two variables that is represented by the shaded region.

37.

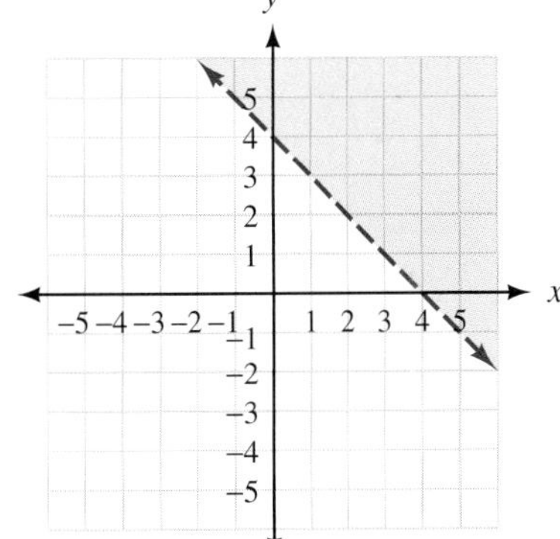

38.

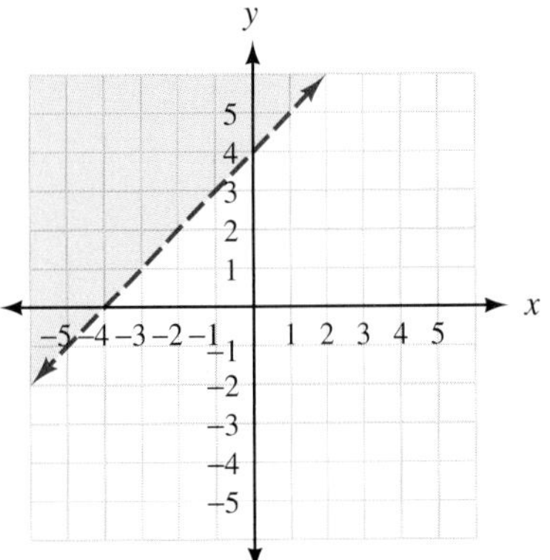

39.

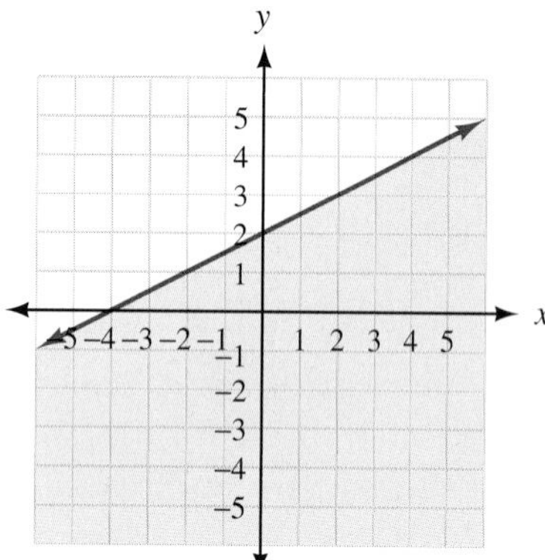

40.

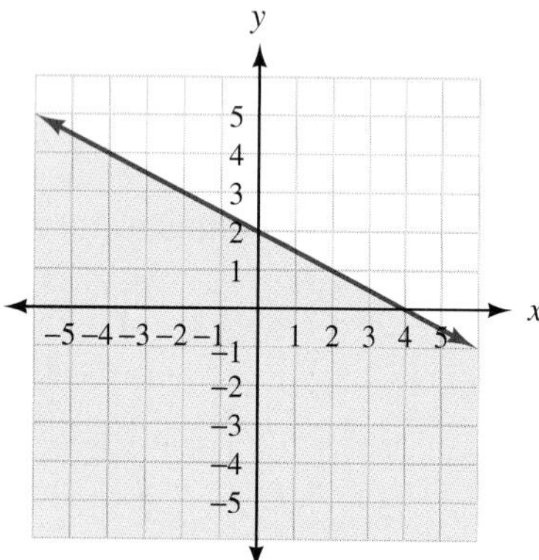

The next two problems are intended to give you practice reading, and paying attention to, the instructions that accompany the problems you are working.

41. Work each problem according to the instructions given.

a. Solve: $\frac{1}{3} + \frac{y}{2} < 1$

b. Solve: $\frac{1}{3} - \frac{y}{2} < 1$

c. Solve for y: $\frac{x}{3} + \frac{y}{2} = 1$

d. Graph: $y < -\frac{4}{3}x + 4$

42. Work each problem according to the instructions given.

a. Solve: $3x + 4 \geq -8$

b. Solve: $-3x + 4 \geq -8$

c. Solve for y: $-3x + 4y = -8$

d. Graph: $y \geq \frac{3}{4}x - 2$

Applying the Concepts

43. Number of People in a Dance Club A dance club holds a maximum of 200 people. The club charges one price for students and a higher price for nonstudents. If the number of students in the club at any time is x and the number of nonstudents is y, shade the region in the first quadrant that contains all combinations of students and nonstudents that are in the club at any time.

44. Many Perimeters Suppose you have 500 feet of fencing that you will use to build a rectangular livestock pen. Let x represent the length of the pen and y represent the width. Shade the region in the first quadrant that contains all possible values of x and y that will give you a rectangle from 500 feet of fencing. (You don't have to use all of the fencing, so the perimeter of the pen could be less than 500 feet.)

45. Gas Mileage You have two cars. The first car travels an average of 12 miles on a gallon of gasoline, and the second averages 22 miles per gallon. Suppose you can afford to buy up to 30 gallons of gasoline this month. If the first car is driven x miles this month, and the second car is driven y miles this month, shade the region in the first quadrant that gives all the possible values of x and y that will keep you from buying more than 30 gallons of gasoline this month.

46. Student Loan Payments When considering how much debt to incur in student loans, it is advisable to keep your student loan payment after graduation to 8% or less of your starting monthly income. Let x represent your starting monthly salary and let y represent your monthly student loan payment, and write an inequality that describes this situation. Shade the region in the first quadrant that is a solution to your inequality.

Maintaining Your Skills

The problems that follow review some of the more important skills you have learned in the previous sections and chapters.

Solve each of the following inequalities.

47. $\frac{1}{3} + \frac{y}{5} \leq \frac{26}{15}$

48. $-\frac{1}{3} \geq \frac{1}{6} - \frac{y}{2}$

49. $5t - 4 > 3t - 8$

50. $-3(t - 2) < 6 - 5(t + 1)$

51. $-9 < -4 + 5t < 6$

52. $-3 < 2t + 1 < 3$

Getting Ready for the Next Section

Complete each table using the given equation.

53. $y = 7.5x$

x	y
0	
10	
20	

54. $h = 32t - 16t^2$

t	h
0	
$\frac{1}{2}$	
1	

55. $x = y^2$

x	y
	0
	1
	-1

56. $y = |x|$

x	y
-3	
0	
3	

Extending the Concepts

Graph each inequality.

57. $y < |x + 2|$

58. $y > |x - 2|$

59. $y > |x - 3|$

60. $y < |x + 3|$

61. $y \le |x - 1|$

62. $y \ge |x + 1|$

63. $y < x^2$

64. $y > x^2 - 2$

65. The Associated Students organization holds a *Night at the Movies* fundraiser. Student tickets are \$1.00 each and nonstudent tickets are \$2.00 each. The theater holds a maximum of 200 people. The club needs to collect at least \$100 to make money. Draw and shade the region in the first quadrant that contains all combinations of students and nonstudents that could attend the movie night so the club makes money.

3.5 Introduction to Functions

OBJECTIVES

A Construct a table or a graph from a function rule.

B Identify the domain and range of a function or a relation.

C Determine whether a relation is also a function.

The ad shown here appeared in the help wanted section of the local newspaper the day I was writing this section of the book. We can use the information in the ad to start an informal discussion of our next topic: functions.

312 Help Wanted

YARD PERSON

Full-time 40 hrs. with weekend work required. Cleaning & loading trucks. \$7.50/hr. Valid CDL with clean record & drug screen required. Submit current MVR to KCI, 225 Suburban Rd., SLO. 805-555-3304.

An Informal Look at Functions

To begin with, suppose you have a job that pays \$7.50 per hour and that you work anywhere from 0 to 40 hours per week. The amount of money you make in one week depends on the number of hours you work that week. In mathematics we say that your weekly earnings are a *function* of the number of hours you work. If we let the variable x represent hours and the variable y represent the money you make, then the relationship between x and y can be written as

$$y = 7.5x \quad \text{for} \quad 0 \le x \le 40$$

EXAMPLE 1

Construct a table and graph for the function

$$y = 7.5x \quad \text{for} \quad 0 \le x \le 40$$

SOLUTION Table 1 gives some of the paired data that satisfy the equation $y = 7.5x$. Figure 1 is the graph of the equation with the restriction $0 \le x \le 40$.

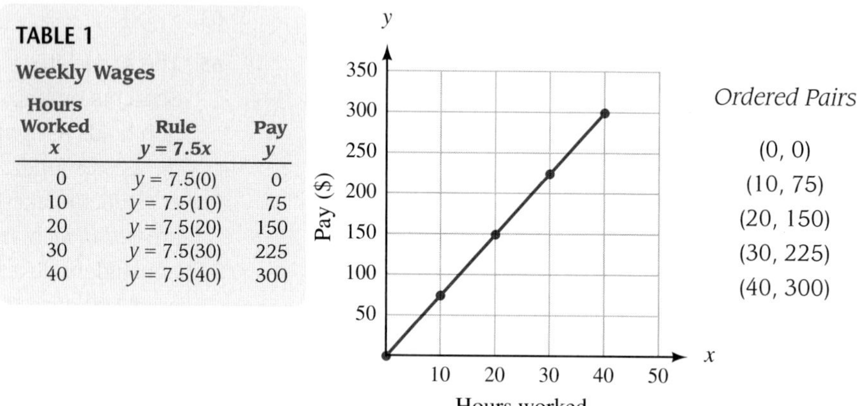

TABLE 1
Weekly Wages

Hours Worked x	Rule $y = 7.5x$	Pay y
0	$y = 7.5(0)$	0
10	$y = 7.5(10)$	75
20	$y = 7.5(20)$	150
30	$y = 7.5(30)$	225
40	$y = 7.5(40)$	300

FIGURE 1 **Weekly wages at $7.50 per hour.**

The equation $y = 7.5x$ with the restriction $0 \le x \le 40$, Table 1, and Figure 1 are three ways to describe the same relationship between the number of hours you work in 1 week and your gross pay for that week. In all three, we *input* values of x, and then use the function rule to *output* values of y.

Domain and Range of a Function

We began this discussion by saying that the number of hours worked during the week was from 0 to 40, so these are the values that x can assume. From the line graph in Figure 1, we see that the values of y range from 0 to 300. We call the complete set of values that x can assume the *domain* of the function. The values that are assigned to y are called the *range* of the function.

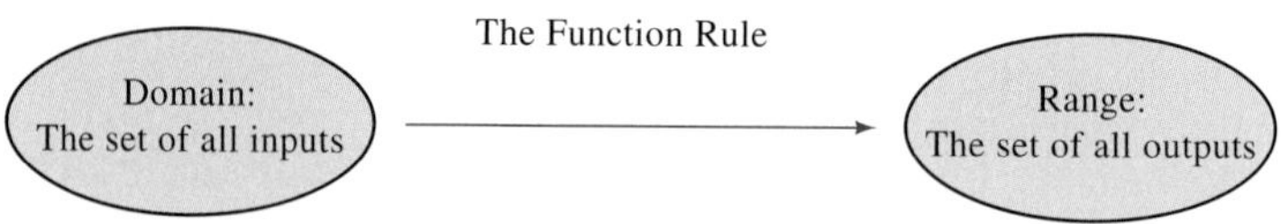

EXAMPLE 2

State the domain and range for the function

$$y = 7.5x, \quad 0 \le x \le 40$$

SOLUTION From the previous discussion we have

$$\text{Domain} = \{x \mid 0 \le x \le 40\}$$

$$\text{Range} = \{y \mid 0 \le y \le 300\}$$

Function Maps

Another way to visualize the relationship between x and y is with the diagram in Figure 2, which we call a *function map:*

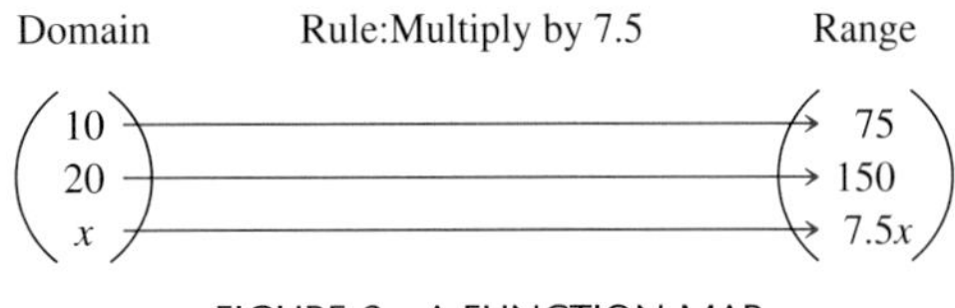

FIGURE 2 A FUNCTION MAP

Although Figure 2 does not show all the values that x and y can assume, it does give us a visual description of how x and y are related. It shows that values of y in the range come from values of x in the domain according to a specific rule (multiply by 7.5 each time).

A Formal Look at Functions

What is apparent from the preceding discussion is that we are working with paired data. The solutions to the equation $y = 7.5x$ are pairs of numbers; the points on the line graph in Figure 1 come from paired data; and the diagram in Figure 2 pairs numbers in the domain with numbers in the range. We are now ready for the formal definition of a function.

DEFINITION A **function** is a rule that pairs each element in one set, called the **domain,** with exactly one element from a second set, called the **range.**

In other words, a function is a rule for which each input is paired with exactly one output.

Functions as Ordered Pairs

The function rule $y = 7.5x$ from Example 1 produces ordered pairs of numbers (x, y). The same thing happens with all functions: The function rule produces ordered pairs of numbers. We use this result to write an alternative definition for a function.

ALTERNATIVE DEFINITION A **function** is a set of ordered pairs in which no two different ordered pairs have the same first coordinate. The set of all first coordinates is called the **domain** of the function. The set of all second coordinates is called the **range** of the function.

The restriction on first coordinates in the alternative definition keeps us from assigning a number in the domain to more than one number in the range.

A Relationship That Is Not a Function

You may be wondering if any sets of paired data fail to qualify as functions. The answer is yes, as the next example reveals.

EXAMPLE 3 Table 2 shows the prices of used Ford Mustangs that were listed in the local newspaper. The diagram in Figure 3 is called a *scatter diagram.* It gives a visual representation of the data in Table 2. Why is this data not a function?

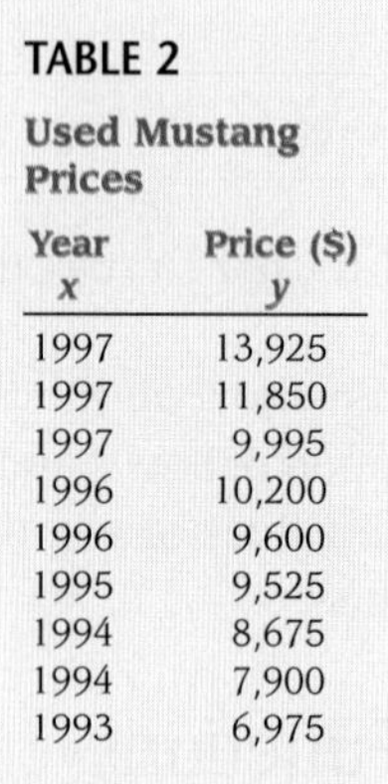

TABLE 2
Used Mustang Prices

Year x	Price ($) y
1997	13,925
1997	11,850
1997	9,995
1996	10,200
1996	9,600
1995	9,525
1994	8,675
1994	7,900
1993	6,975

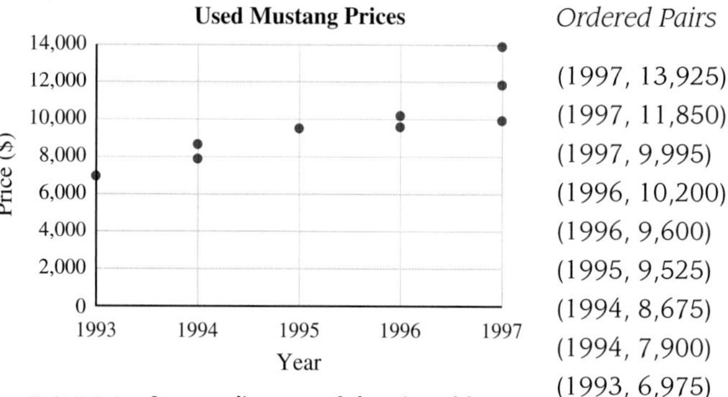

FIGURE 3 **Scatter diagram of data in Table 2**

Ordered Pairs

(1997, 13,925)
(1997, 11,850)
(1997, 9,995)
(1996, 10,200)
(1996, 9,600)
(1995, 9,525)
(1994, 8,675)
(1994, 7,900)
(1993, 6,975)

SOLUTION In Table 2, the year 1997 is paired with three different prices: $13,925, $11,850, and $9,995. That is enough to disqualify the data from belonging to a function. For a set of paired data to be considered a function, each number in the domain must be paired with exactly one number in the range.

Still, there is a relationship between the first coordinates and second coordinates in the used-car data. It is not a function relationship, but it is a relationship. To classify all relationships specified by ordered pairs, whether they are functions or not, we include the following two definitions.

DEFINITION A **relation** is a rule that pairs each element in one set, called the **domain,** with one or more elements from a second set, called the **range.**

ALTERNATIVE DEFINITION A **relation** is a set of ordered pairs. The set of all first coordinates is the **domain** of the relation. The set of all second coordinates is the **range** of the relation.

Here are some facts that will help clarify the distinction between relations and functions.

1. Any rule that assigns numbers from one set to numbers in another set is a relation. If that rule makes the assignment so that no input has more than one output, then it is also a function.
2. Any set of ordered pairs is a relation. If none of the first coordinates of those ordered pairs is repeated, the set of ordered pairs is also a function.
3. Every function is a relation.
4. Not every relation is a function.

Graphing Relations and Functions

To give ourselves a wider perspective on functions and relations, we consider some equations whose graphs are not straight lines.

EXAMPLE 4 Kendra is tossing a softball into the air with an underhand motion. The distance of the ball above her hand at any time is given by the function

$$h = 32t - 16t^2 \quad \text{for} \quad 0 \le t \le 2$$

where h is the height of the ball in feet and t is the time in seconds. Construct a table that gives the height of the ball at quarter-second intervals, starting with $t = 0$ and ending with $t = 2$. Construct a line graph from the table.

SOLUTION We construct Table 3 using the following values of t: 0, $\frac{1}{4}$, $\frac{1}{2}$, $\frac{3}{4}$, 1, $\frac{5}{4}$, $\frac{3}{2}$, $\frac{7}{4}$, 2. The values of h come from substituting these values of t into the equation $h = 32t - 16t^2$. (This equation comes from physics. If you take a physics class, you will learn how to derive this equation.) Then we construct the graph in Figure 4 from the table. The graph appears only in the first quadrant because neither t nor h can be negative.

TABLE 3

Tossing a Softball into the Air

Time (sec)	Function Rule	Distance (ft)
t	$h = 32t - 16t^2$	h
0	$h = 32(0) - 16(0)^2 = 0 - 0 = 0$	0
$\frac{1}{4}$	$h = 32(\frac{1}{4}) - 16(\frac{1}{4})^2 = 8 - 1 = 7$	7
$\frac{1}{2}$	$h = 32(\frac{1}{2}) - 16(\frac{1}{2})^2 = 16 - 4 = 12$	12
$\frac{3}{4}$	$h = 32(\frac{3}{4}) - 16(\frac{3}{4})^2 = 24 - 9 = 15$	15
1	$h = 32(1) - 16(1)^2 = 32 - 16 = 16$	16
$\frac{5}{4}$	$h = 32(\frac{5}{4}) - 16(\frac{5}{4})^2 = 40 - 25 = 15$	15
$\frac{3}{2}$	$h = 32(\frac{3}{2}) - 16(\frac{3}{2})^2 = 48 - 36 = 12$	12
$\frac{7}{4}$	$h = 32(\frac{7}{4}) - 16(\frac{7}{4})^2 = 56 - 49 = 7$	7
2	$h = 32(2) - 16(2)^2 = 64 - 64 = 0$	0

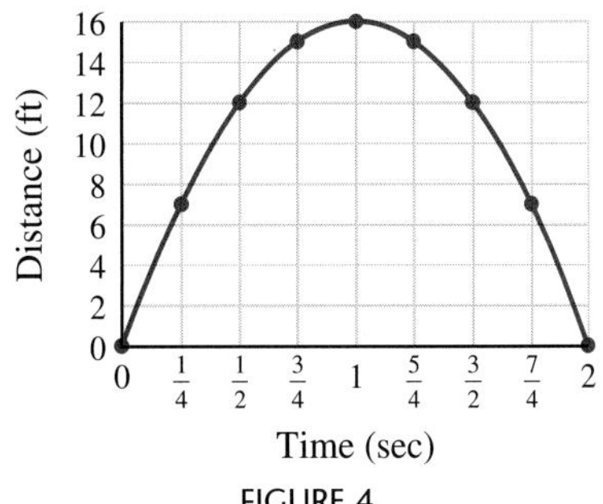

FIGURE 4

Here is a summary of what we know about functions as it applies to this example: We input values of t and output values of h according to the function rule

$$h = 32t - 16t^2 \quad \text{for} \quad 0 \le t \le 2$$

The domain is given by the inequality that follows the equation; it is

$$\text{Domain} = \{t \mid 0 \le t \le 2\}$$

The range is the set of all outputs that are possible by substituting the values of t from the domain into the equation. From our table and graph, it seems that the range is

$$\text{Range} = \{h \mid 0 \le h \le 16\}$$

USING TECHNOLOGY

More About Example 4

Most graphing calculators can easily produce the information in Table 3. Simply set Y_1 equal to $32X - 16X^2$. Then set up the table so it starts at 0 and increases by an increment of 0.25 each time. (On a TI-82/83, use the TBLSET key to set up the table.)

Table Setup

Table minimum = 0
Table increment = .25
Dependent variable: Auto
Independent variable: Auto

Y Variables Setup

$Y_1 = 32X - 16X^2$

The table will look like this:

X	Y_1
0.00	0
0.25	7
0.50	12
0.75	15
1.00	16
1.25	15
1.50	12

EXAMPLE 5 Sketch the graph of $x = y^2$.

SOLUTION Without going into much detail, we graph the equation $x = y^2$ by finding a number of ordered pairs that satisfy the equation, plotting these points, then drawing a smooth curve that connects them. A table of values for x and y that satisfy the equation follows, along with the graph of $x = y^2$ shown in Figure 5.

x	y
0	0
1	1
1	−1
4	2
4	−2
9	3
9	−3

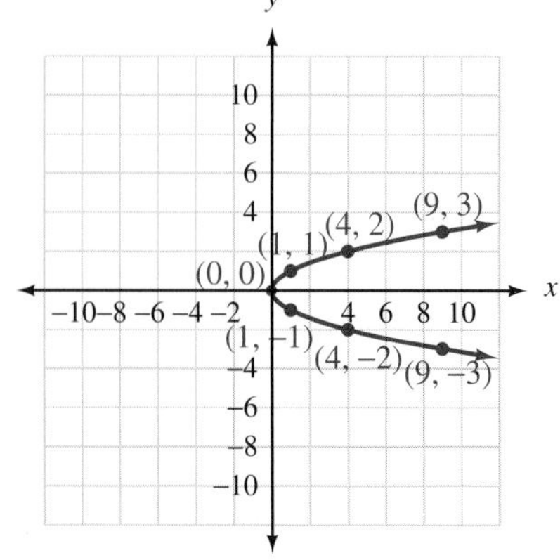

FIGURE 5

As you can see from looking at the table and the graph in Figure 5, several ordered pairs whose graphs lie on the curve have repeated first coordinates. For instance, (1, 1) and (1, −1), (4, 2) and (4, −2), and (9, 3) and (9, −3). The graph is therefore not the graph of a function.

Vertical Line Test

Look back at the scatter diagram for used Mustang prices shown in Figure 3. Notice that some of the points on the diagram lie above and below each other along vertical lines. This is an indication that the data do not constitute a function. Two data points that lie on the same vertical line must have come from two ordered pairs with the same first coordinates.

Now, look at the graph shown in Figure 5. The reason this graph is the graph of a relation, but not of a function, is that some points on the graph have the same first coordinates—for example, the points (4, 2) and (4, −2). Furthermore, any time two points on a graph have the same first coordinates, those points must lie on a vertical line. [To convince yourself, connect the points (4, 2) and (4, −2) with a straight line. You will see that it must be a vertical line.] This allows us to write the following test that uses the graph to determine whether a relation is also a function.

Vertical Line Test If a vertical line crosses the graph of a relation in more than one place, the relation cannot be a function. If no vertical line can be found that crosses a graph in more than one place, then the graph is the graph of a function.

If we look back to the graph of $h = 32t - 16t^2$ as shown in Figure 4, we see that no vertical line can be found that crosses this graph in more than one place. The graph shown in Figure 4 is therefore the graph of a function.

EXAMPLE 6 Graph $y = |x|$. Use the graph to determine whether we have the graph of a function. State the domain and range.

SOLUTION We let x take on values of −4, −3, −2, −1, 0, 1, 2, 3, and 4. The corresponding values of y are shown in the table. The graph is shown in Figure 6.

x	y
−4	4
−3	3
−2	2
−1	1
0	0
1	1
2	2
3	3
4	4

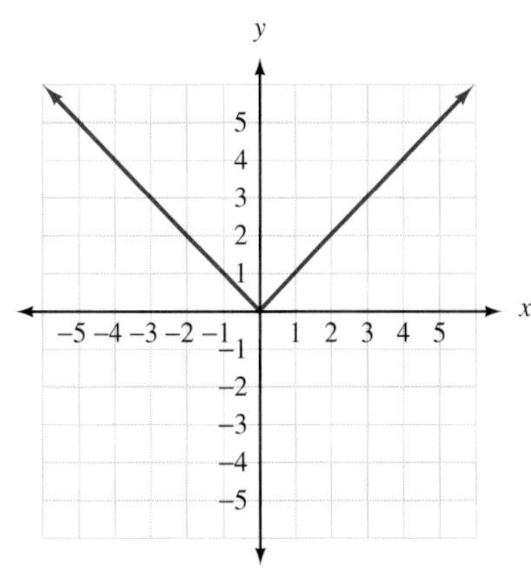

FIGURE 6

Because no vertical line can be found that crosses the graph in more than one place, $y = |x|$ is a function. The domain is all real numbers. The range is $\{y \mid y \geq 0\}$.

LINKING OBJECTIVES AND EXAMPLES

Next to each **objective** we have listed the examples that are best described by that objective.

A	1, 4
B	2, 4
C	3–6

GETTING READY FOR CLASS

After reading through the preceding section, respond in your own words and in complete sentences.

1. What is a function?
2. What is the vertical line test?
3. Is every line the graph of a function? Explain.
4. Which variable is usually associated with the domain of a function?

Problem Set 3.5

Online support materials can be found at www.thomsonedu.com/login

For each of the following relations, give the domain and range, and indicate which are also functions.

1. $\{(1, 3), (2, 5), (4, 1)\}$

2. $\{(3, 1), (5, 7), (2, 3)\}$

3. $\{(-1, 3), (1, 3), (2, -5)\}$

4. $\{(3, -4), (-1, 5), (3, 2)\}$

▸ **5.** $\{(7, -1), (3, -1), (7, 4)\}$

6. $\{(5, -2), (3, -2), (5, -1)\}$

▸ **7.** $\{(a, 3), (b, 4), (c, 3), (d, 5)\}$

▸ **8.** $\{(a, 5), (b, 5), (c, 4), (d, 5)\}$

▸ **9.** $\{(a, 1), (a, 2), (a, 3), (a, 4)\}$

▸ **10.** $\{(a, 1), (b, 1), (c, 1), (d, 1)\}$

State whether each of the following graphs represents a function.

11. 

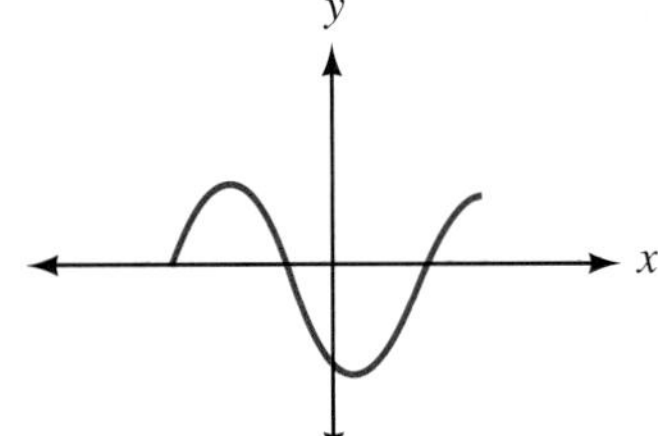

▸ **12.**

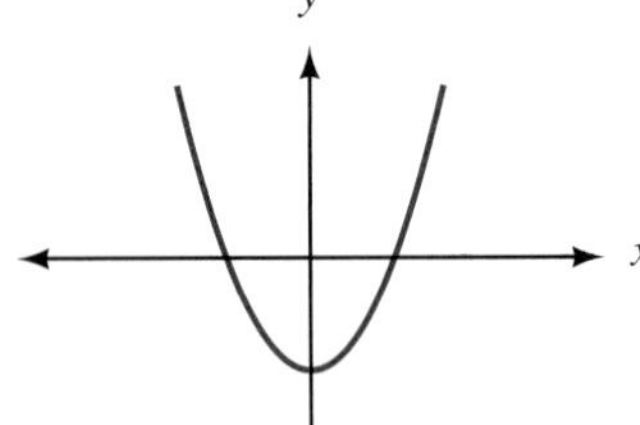

▸ **13.**

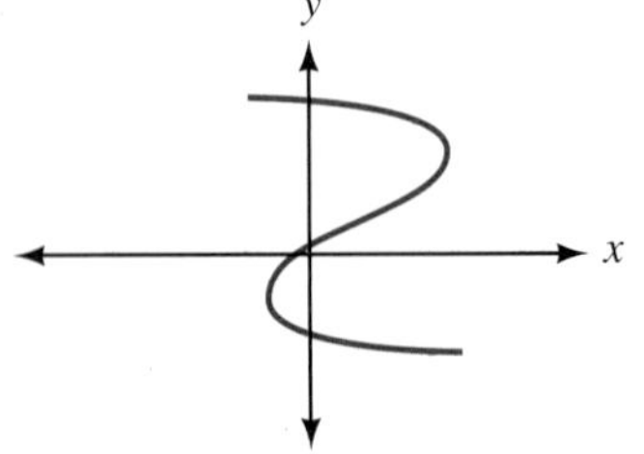

14.

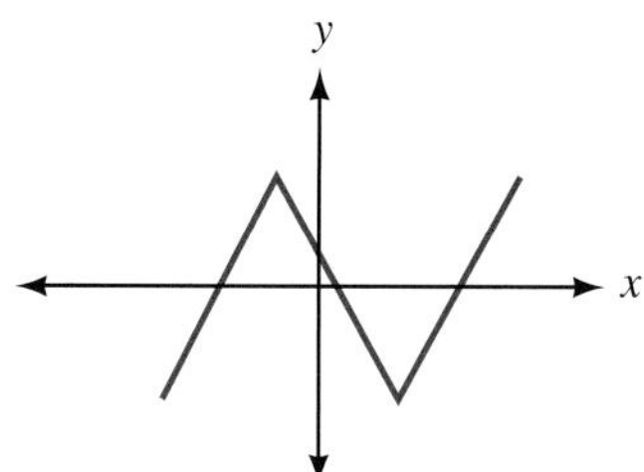

15.

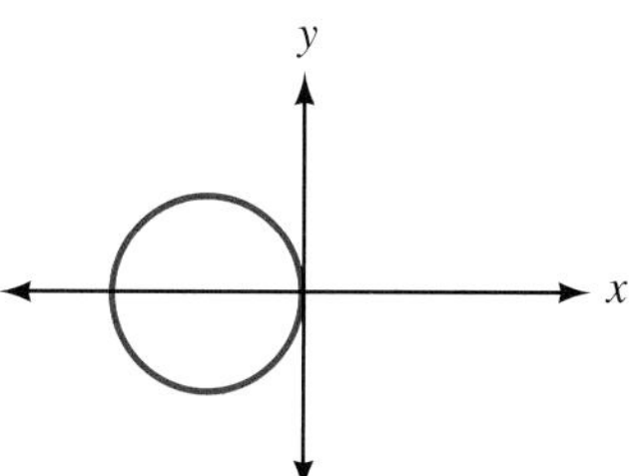

16.

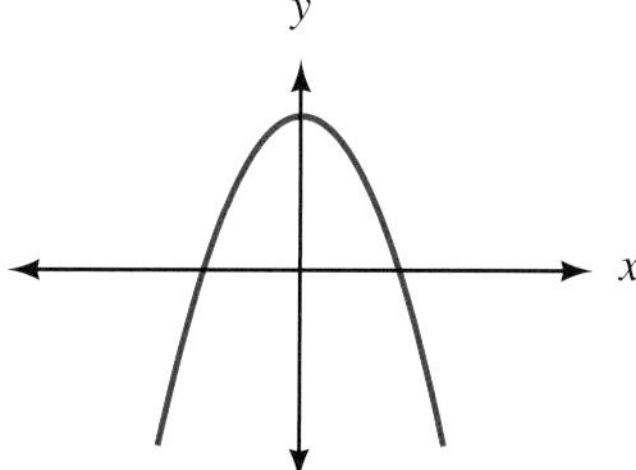

17.

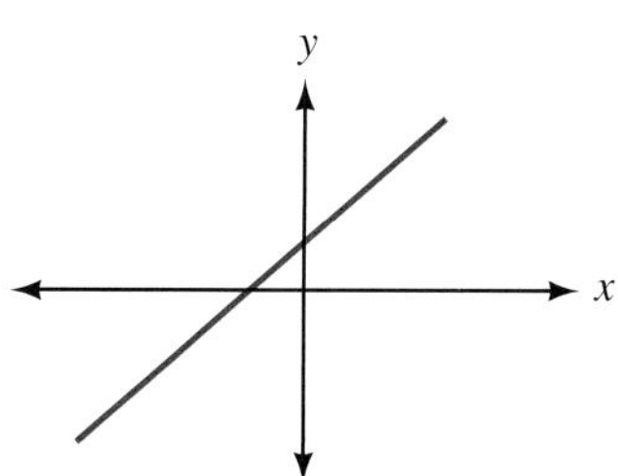

18.

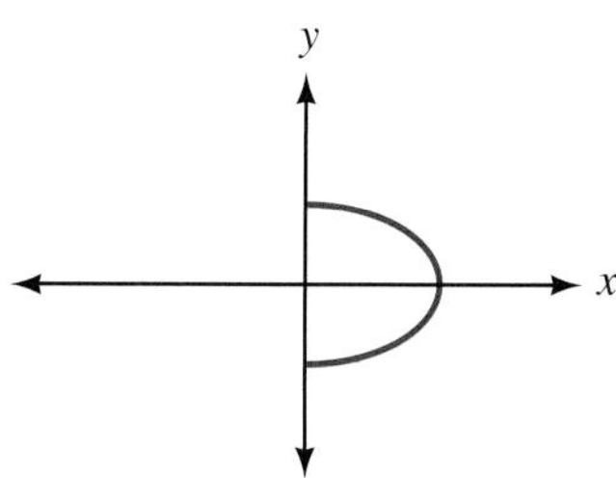

19.

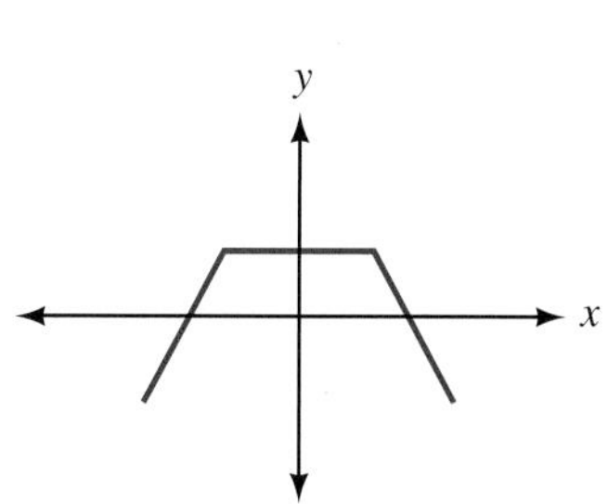

20.

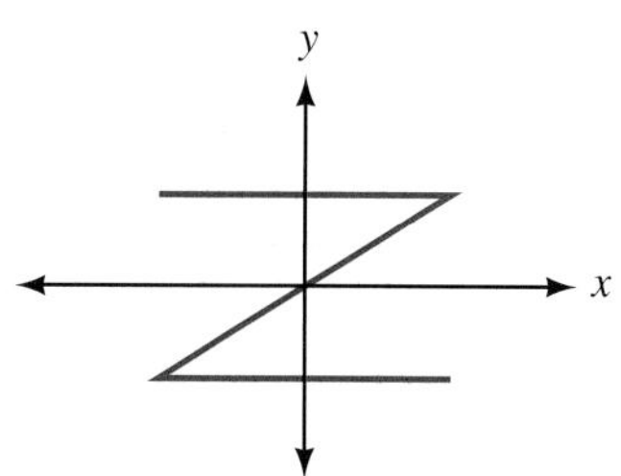

Determine the domain and range of the following functions. Assume the *entire* function is shown.

▸ 21.

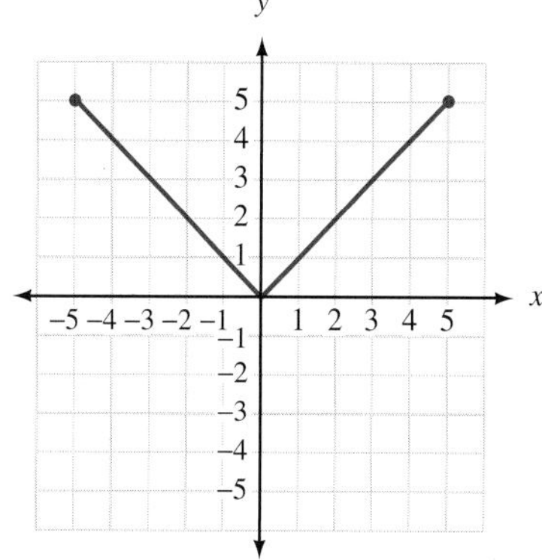

▸ 22.

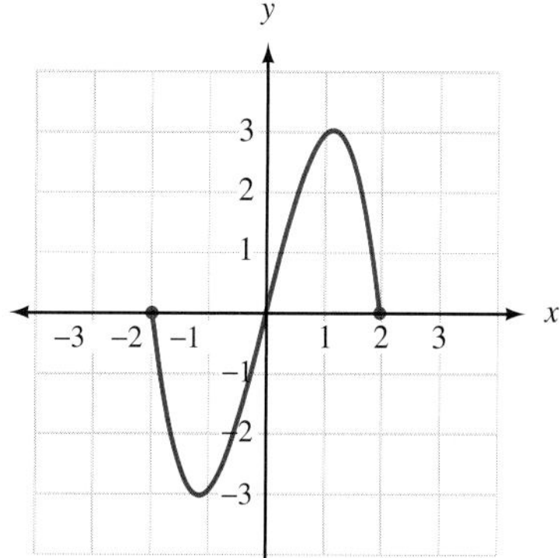

▸ 23.

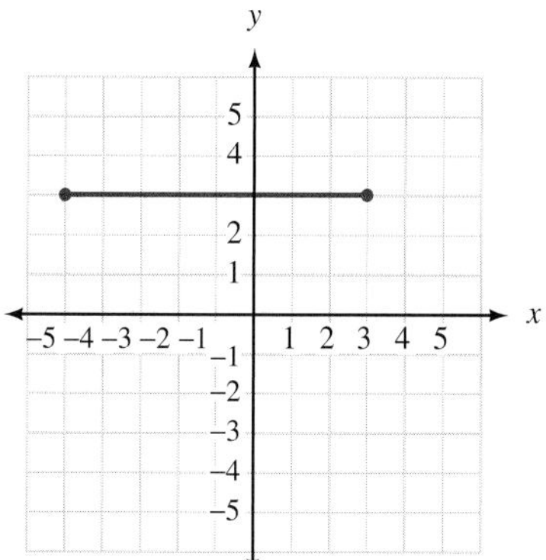

24.

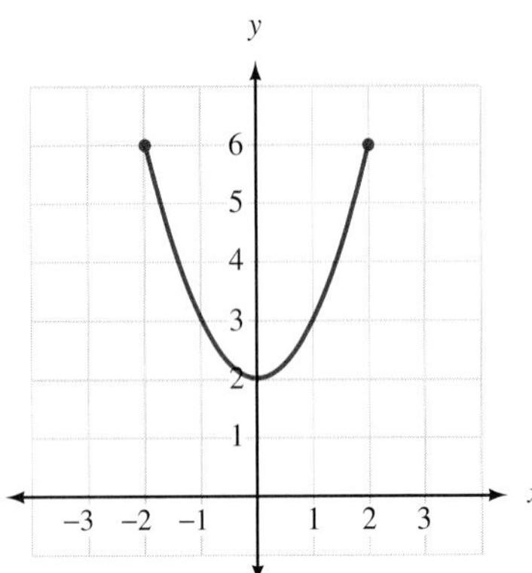

Graph each of the following relations. In each case, use the graph to find the domain and range, and indicate whether the graph is the graph of a function.

25. $y = x^2 - 1$
26. $y = x^2 + 1$
27. $y = x^2 + 4$
28. $y = x^2 - 9$
29. $x = y^2 - 1$
30. $x = y^2 + 1$
31. $x = y^2 + 4$
32. $x = y^2 - 9$
33. $y = |x - 2|$
34. $y = |x + 2|$
35. $y = |x| - 2$
36. $y = |x| + 2$

Applying the Concepts

37. **Weekly Wages** Suppose you have a job that pays \$8.50 per hour and you work anywhere from 10 to 40 hours per week.

a. Write an equation, with a restriction on the variable x, that gives the amount of money, y, you will earn for working x hours in one week.

b. Use the function rule to complete the table.

TABLE 4
Weekly Wages

Hours Worked x	Function Rule $y = 8.5x$	Gross Pay (\$) y
10		
20		
30		
40		

c. Use the data from the table to graph the function.

d. State the domain and range of this function.

e. What is the minimum amount you can earn in a week with this job? What is the maximum amount?

38. **Weekly Wages** The ad shown here was in the local newspaper. Suppose you are hired for the job described in the ad.

312 Help Wanted

ESPRESSO BAR OPERATOR

Must be dependable, honest, service-oriented. Coffee exp desired. 15–30 hrs per wk. \$5.25/hr. Start 5/31. Apply in person: Espresso Yourself, Central Coast Mall. Deadline 5/23.

a. If x is the number of hours you work per week and y is your weekly gross pay, write the equation for y. (Be sure to include any restrictions on the variable x that are given in the ad.)

b. Use the function rule to complete the table.

TABLE 5
Weekly Wages

Hours Worked x	Function Rule $y = 5.25x$	Gross Pay (\$) y
15		
20		
25		
30		

c. Use the data from the table to graph the function.

d. State the domain and range of this function.

e. What is the minimum amount you can earn in a week with this job? What is the maximum amount?

39. **Tossing a Coin** Hali is tossing a quarter into the air with an underhand motion. The distance the quarter is above her hand at any time is given by the function

$$h = 16t - 16t^2 \quad \text{for} \quad 0 \le t \le 1$$

where h is the height of the quarter in feet, and t is the time in seconds.

a. Fill in the table.

Time (sec) t	Function Rule $h = 16t - 16t^2$	Distance (ft) h
0		
0.1		
0.2		
0.3		
0.4		
0.5		
0.6		
0.7		
0.8		
0.9		
1		

b. State the domain and range of this function.

c. Use the data from the table to graph the function.

40. **Intensity of Light** The formula below gives the intensity of light that falls on a surface at various distances from a 100-watt light bulb:

$$I = \frac{120}{d^2} \quad \text{for} \quad d > 0$$

where I is the intensity of light (in lumens per square foot), and d is the distance (in feet) from the light bulb to the surface.

a. Fill in the table.

Distance (ft) D	Function Rule $I = \frac{120}{d^2}$	Intensity I
1		
2		
3		
4		
5		
6		

b. Use the data from the table to graph the function.

41. **Area of a Circle** The formula for the area A of a circle with radius r is given by $A = \pi r^2$. The formula shows that A is a function of r.

a. Graph the function $A = \pi r^2$ for $0 \le r \le 3$. (On the graph, let the horizontal axis be the r-axis, and let the vertical axis be the A-axis.)

b. State the domain and range of the function $A = \pi r^2$, $0 \le r \le 3$.

42. **Area and Perimeter of a Rectangle** A rectangle is 2 inches longer than it is wide. Let x = the width, P = the perimeter, and A = the area of the rectangle.

a. Write an equation that will give the perimeter P in terms of the width x of the rectangle. Are there any restrictions on the values that x can assume?

b. Graph the relationship between P and x.

43. **Tossing a Ball** A ball is thrown straight up into the air from ground level. The relationship between the height h of the ball at any time t is illustrated by the following graph:

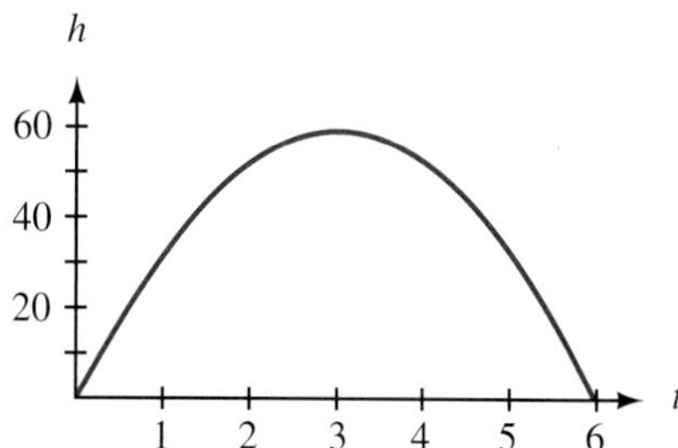

The horizontal axis represents time t, and the vertical axis represents height h.

a. Is this graph the graph of a function?

b. State the domain and range.

c. At what time does the ball reach its maximum height?

d. What is the maximum height of the ball?

e. At what time does the ball hit the ground?

44. **Company Profits** The amount of profit a company earns is based on the number of items it sells. The relationship between the profit P and number of items it sells x, is illustrated by the following graph:

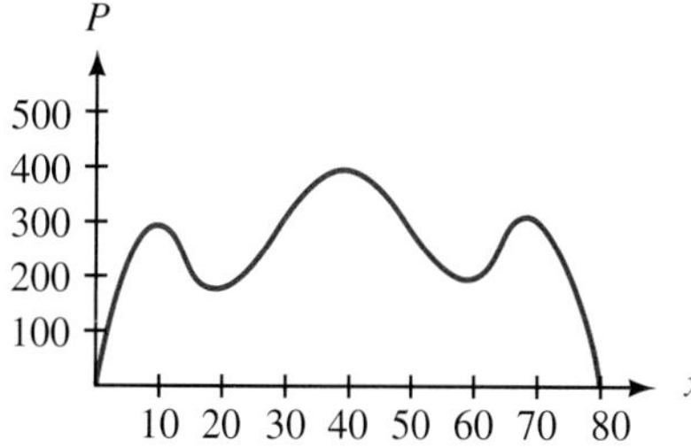

The horizontal axis represents items sold, x, and the vertical axis represents the profit, P.

a. Is this graph the graph of a function?

b. State the domain and range.

c. How many items must the company sell to make their maximum profit?

d. What is their maximum profit?

45. **Profits** Match each of the following statements to the appropriate graph indicated by labels I–IV.

a. Sarah works 25 hours to earn $250.

b. Justin works 35 hours to earn $560.

c. Rosemary works 30 hours to earn \$360.
d. Marcus works 40 hours to earn \$320.

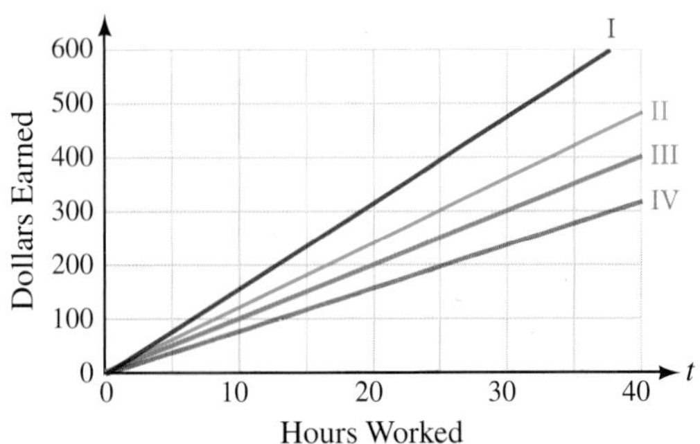

46. Find an equation for each of the functions shown in Problem 45. Show dollars earned, E, as a function of hours worked, t. Then, indicate the domain and range of each function.

a. Graph I: $E =$
Domain = $\{t \mid \quad\}$
Range = $\{E \mid \quad\}$

b. Graph II: $E =$
Domain = $\{t \mid \quad\}$
Range = $\{E \mid \quad\}$

c. Graph III: $E =$
Domain = $\{t \mid \quad\}$
Range = $\{E \mid \quad\}$

d. Graph IV: $E =$
Domain = $\{t \mid \quad\}$
Range = $\{E \mid \quad\}$

Maintaining Your Skills

The problems that follow review some of the more important skills you have learned in previous sections and chapters.

For the equation $y = 3x - 2$:

47. Find y if x is 4. **48.** Find y if x is 0.

49. Find y if x is -4.

50. Find y if x is -2.

For the equation $y = x^2 - 3$:

51. Find y if x is 2. **52.** Find y if x is -2.

53. Find y if x is 0. **54.** Find y if x is -4.

55. If $x - 2y = 4$ and $x = \frac{8}{5}$, find y.

56. If $5x - 10y = 15$, find y when x is 3.

57. Let $x = 0$ and $y = 0$ in $y = a(x - 80)^2 + 70$ and solve for a.

58. Find R if $p = 2.5$ and $R = (900 - 300p)p$.

Getting Ready for the Next Section

Simplify. Round to the nearest whole number if necessary.

59. $7.5(20)$ **60.** $60 \div 7.5$

61. $4(3.14)(9)$ **62.** $\frac{4}{3}(3.14) \cdot 3^3$

63. $4(-2) - 1$ **64.** $3(3)^2 + 2(3) - 1$

65. If $s = \frac{60}{t}$, find s when
a. $t = 10$ **b.** $t = 8$

66. If $y = 3x^2 + 2x - 1$, find y when
a. $x = 0$ **b.** $x = -2$

67. Find the value of $x^2 + 2$ for
a. $x = 5$ **b.** $x = -2$

68. Find the value of $125 \cdot 2^t$ for
a. $t = 0$ **b.** $t = 1$

Extending the Concepts

Graph each of the following relations. In each case, use the graph to find the domain and range, and indicate whether the graph is the graph of a function.

69. $y = 5 - |x|$ **70.** $y = |x| - 3$

71. $x = |y| + 3$ **72.** $x = 2 - |y|$

73. $|x| + |y| = 4$ **74.** $2|x| + |y| = 6$

3.6 Function Notation

OBJECTIVES

A Use function notation to find the value of a function for a given value of the variable.

Let's return to the discussion that introduced us to functions. If a job pays \$7.50 per hour for working from 0 to 40 hours a week, then the amount of money y earned in 1 week is a function of the number of hours worked, x. The exact relationship between x and y is written

$$y = 7.5x \quad \text{for} \quad 0 \le x \le 40$$

Because the amount of money earned y depends on the number of hours worked x, we call y the *dependent variable* and x the *independent variable.* Furthermore, if we let f represent all the ordered pairs produced by the equation, then we can write

$$f = \{(x, y) \mid y = 7.5x \text{ and } 0 \le x \le 40\}$$

312 Help Wanted

YARD PERSON

Full-time 40 hrs. with weekend work required. Cleaning & loading trucks. \$7.50/hr. Valid CDL with clean record & drug screen required. Submit current MVR to KCI, 225 Suburban Rd., SLO. 805-555-3304.

Once we have named a function with a letter, we can use an alternative notation to represent the dependent variable y. The alternative notation for y is $f(x)$. It is read "f of x" and can be used instead of the variable y when working with functions. The notation y and the notation $f(x)$ are equivalent—that is,

$$y = 7.5x \Leftrightarrow f(x) = 7.5x$$

When we use the notation $f(x)$ we are using *function notation.* The benefit of using function notation is that we can write more information with fewer symbols than we can by using just the variable y. For example, asking how much money a person will make for working 20 hours is simply a matter of asking for $f(20)$. Without function notation, we would have to say, "Find the value of y that corresponds to a value of $x = 20$." To illustrate further, using the variable y, we can say, "y is 150 when x is 20." Using the notation $f(x)$, we simply say, "$f(20) = 150$." Each expression indicates that you will earn \$150 for working 20 hours.

EXAMPLE 1 If $f(x) = 7.5x$, find $f(0)$, $f(10)$, and $f(20)$.

SOLUTION To find $f(0)$ we substitute 0 for x in the expression $7.5x$ and simplify. We find $f(10)$ and $f(20)$ in a similar manner—by substitution.

$$\begin{aligned} \text{If} \quad f(x) &= 7.5x \\ \text{then} \quad f(\mathbf{0}) &= 7.5(\mathbf{0}) = 0 \\ f(\mathbf{10}) &= 7.5(\mathbf{10}) = 75 \\ f(\mathbf{20}) &= 7.5(\mathbf{20}) = 150 \end{aligned}$$

Input x

Function machine

Output $f(x)$

Some students like to think of functions as machines. Values of x are put into the machine, which transforms them into values of $f(x)$, which then are output by the machine.

If we changed the example in the discussion that opened this section so that the hourly wage was \$6.50 per hour, we would have a new equation to work with:

$$y = 6.5x \qquad \text{for} \qquad 0 \le x \le 40$$

Suppose we name this new function with the letter g. Then

$$g = \{(x, y) \mid y = 6.5x \text{ and } 0 \le x \le 40\}$$

and

$$g(x) = 6.5x$$

If we want to talk about both functions in the same discussion, having two different letters, f and g, makes it easy to distinguish between them. For example, because $f(x) = 7.5x$ and $g(x) = 6.5x$, asking how much money a person makes for working 20 hours is simply a matter of asking for $f(20)$ or $g(20)$, avoiding any confusion over which hourly wage we are talking about.

The diagrams shown in Figure 1 further illustrate the similarities and differences between the two functions we have been discussing.

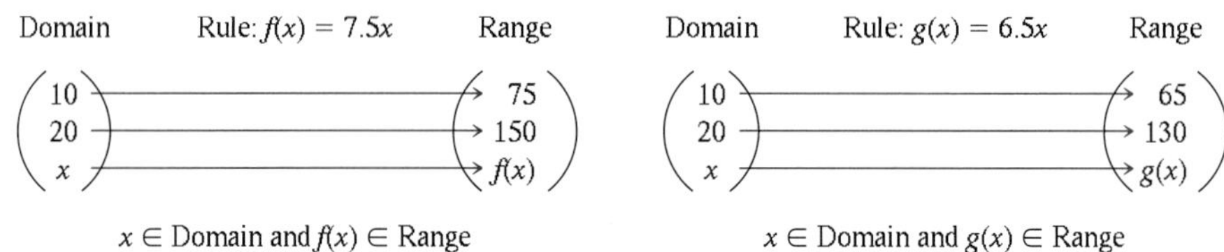

FIGURE 1 **Function maps**

Function Notation and Graphs

We can visualize the relationship between x and $f(x)$ on the graph of the function. Figure 2 shows the graph of $f(x) = 7.5x$ along with two additional line segments. The horizontal line segment corresponds to $x = 20$, and the vertical line segment corresponds to $f(20)$. (Note that the domain is restricted to $0 \le x \le 40$.)

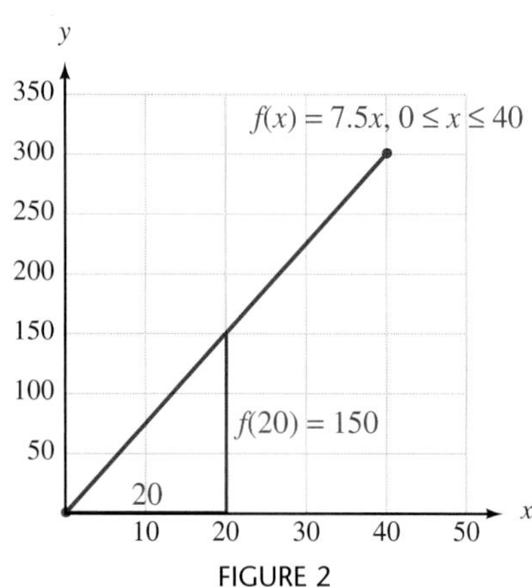

FIGURE 2

We can use functions and function notation to talk about numbers in the chart on gasoline prices. Let's let x represent one of the years in the chart.

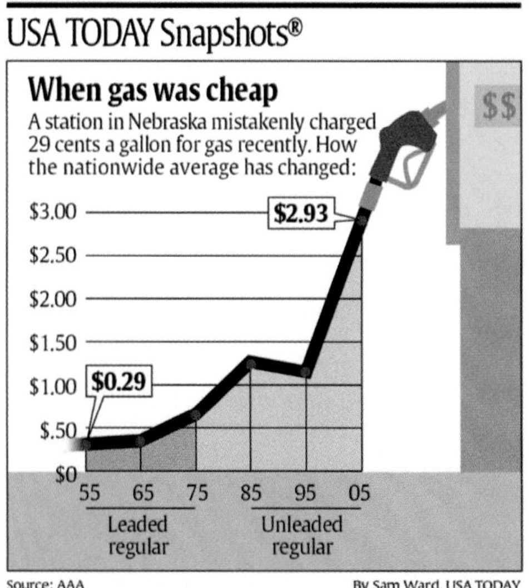

If the function f pairs each year in the chart with the average price of regular gasoline for that year, then each statement below is true:

$f(1955) = \$0.29$

The domain of $f = \{1955, 1965, 1975, 1985, 1995, 2005\}$

$f(1985) > f(1995)$

In general, when we refer to the function f we are referring to the domain, the range, and the rule that takes elements in the domain and outputs elements in the range. When we talk about $f(x)$ we are talking about the rule itself, or an element in the range, or the variable y.

The function f

Domain of f	$y = f(x)$	Range of f
Inputs	*Rule*	*Outputs*

Using Function Notation

The remaining examples in this section show a variety of ways to use and interpret function notation.

EXAMPLE 2 If it takes Lorena t minutes to run a mile, then her average speed $s(t)$ in miles per hour is given by the formula

$$s(t) = \frac{60}{t} \quad \text{for} \quad t > 0$$

Find $s(10)$ and $s(8)$, and then explain what they mean.

SOLUTION To find $s(10)$, we substitute 10 for t in the equation and simplify:

$$s(\mathbf{10}) = \frac{60}{\mathbf{10}} = 6$$

In words: When Lorena runs a mile in 10 minutes, her average speed is 6 miles per hour.

We calculate $s(8)$ by substituting 8 for t in the equation. Doing so gives us

$$s(\mathbf{8}) = \frac{60}{\mathbf{8}} = 7.5$$

In words: Running a mile in 8 minutes is running at a rate of 7.5 miles per hour.

EXAMPLE 3 A painting is purchased as an investment for \$125. If its value increases continuously so that it doubles every 5 years, then its value is given by the function

$$V(t) = 125 \cdot 2^{t/5} \quad \text{for} \quad t \geq 0$$

where t is the number of years since the painting was purchased, and $V(t)$ is its value (in dollars) at time t. Find $V(5)$ and $V(10)$, and explain what they mean.

SOLUTION The expression $V(5)$ is the value of the painting when $t = 5$ (5 years after it is purchased). We calculate $V(5)$ by substituting 5 for t in the equation $V(t) = 125 \cdot 2^{t/5}$. Here is our work:

$$V(\mathbf{5}) = 125 \cdot 2^{\mathbf{5}/5} = 125 \cdot 2^1 = 125 \cdot 2 = 250$$

In words: After 5 years, the painting is worth $250.

The expression $V(10)$ is the value of the painting after 10 years. To find this number, we substitute 10 for t in the equation:

$$V(\mathbf{10}) = 125 \cdot 2^{\mathbf{10}/5} = 125 \cdot 2^2 = 125 \cdot 4 = 500$$

In words: The value of the painting 10 years after it is purchased is $500. 

EXAMPLE 4

A balloon has the shape of a sphere with a radius of 3 inches. Use the following formulas to find the volume and surface area of the balloon.

$$V(r) = \frac{4}{3}\pi r^3 \qquad S(r) = 4\pi r^2$$

SOLUTION As you can see, we have used function notation to write the two formulas for volume and surface area because each quantity is a function of the radius. To find these quantities when the radius is 3 inches, we evaluate $V(3)$ and $S(3)$:

$V(\mathbf{3}) = \frac{4}{3}\pi \mathbf{3}^3 = \frac{4}{3}\pi 27 = 36\pi$ cubic inches, or 113 cubic inches — **To the nearest whole number**

$S(\mathbf{3}) = 4\pi \mathbf{3}^2 = 36\pi$ square inches, or 113 square inches — **To the nearest whole number**

The fact that $V(3) = 36\pi$ means that the ordered pair $(3, 36\pi)$ belongs to the function V. Likewise, the fact that $S(3) = 36\pi$ tells us that the ordered pair $(3, 36\pi)$ is a member of function S.

We can generalize the discussion at the end of Example 4 this way:

$$(a, b) \in f \quad \text{if and only if} \quad f(a) = b$$

USING TECHNOLOGY

More About Example 4

If we look back at Example 4, we see that when the radius of a sphere is 3, the numerical values of the volume and surface area are equal. How unusual is this? Are there other values of r for which $V(r)$ and $S(r)$ are equal? We can answer this question by looking at the graphs of both V and S.

To graph the function $V(r) = \frac{4}{3}\pi r^3$, set $Y_1 = 4\pi X^3/3$. To graph $S(r) = 4\pi r^2$, set $Y_2 = 4\pi X^2$. Graph the two functions in each of the following windows:

Window 1: X from −4 to 4, Y from −2 to 10

Window 2: X from 0 to 4, Y from 0 to 50

Window 3: X from 0 to 4, Y from 0 to 150

Then use the Trace and Zoom features of your calculator to locate the point in the first quadrant where the two graphs intersect. How do the coordinates of this point compare with the results in Example 4?

EXAMPLE 5

If $f(x) = 3x^2 + 2x - 1$, find $f(0)$, $f(3)$, and $f(-2)$.

SOLUTION Because $f(x) = 3x^2 + 2x - 1$, we have

$$f(\mathbf{0}) = 3(\mathbf{0})^2 + 2(\mathbf{0}) - 1 = 0 + 0 - 1 = -1$$

$$f(\mathbf{3}) = 3(\mathbf{3})^2 + 2(\mathbf{3}) - 1 = 27 + 6 - 1 = 32$$

$$f(\mathbf{-2}) = 3(\mathbf{-2})^2 + 2(\mathbf{-2}) - 1 = 12 - 4 - 1 = 7$$

In Example 5, the function f is defined by the equation $f(x) = 3x^2 + 2x - 1$. We could just as easily have said $y = 3x^2 + 2x - 1$; that is, $y = f(x)$. Saying $f(-2) = 7$ is exactly the same as saying y is 7 when x is -2.

EXAMPLE 6

If $f(x) = 4x - 1$ and $g(x) = x^2 + 2$, then

$$f(\mathbf{5}) = 4(\mathbf{5}) - 1 = 19 \quad \text{and} \quad g(\mathbf{5}) = \mathbf{5}^2 + 2 = 27$$

$$f(\mathbf{-2}) = 4(\mathbf{-2}) - 1 = -9 \quad \text{and} \quad g(\mathbf{-2}) = (\mathbf{-2})^2 + 2 = 6$$

$$f(\mathbf{0}) = 4(\mathbf{0}) - 1 = -1 \quad \text{and} \quad g(\mathbf{0}) = \mathbf{0}^2 + 2 = 2$$

$$f(\boldsymbol{z}) = 4\boldsymbol{z} - 1 \quad \text{and} \quad g(\boldsymbol{z}) = \boldsymbol{z}^2 + 2$$

$$f(\boldsymbol{a}) = 4\boldsymbol{a} - 1 \quad \text{and} \quad g(\boldsymbol{a}) = \boldsymbol{a}^2 + 2$$

USING TECHNOLOGY

More About Example 6

Most graphing calculators can use tables to evaluate functions. To work Example 6 using a graphing calculator table, set Y_1 equal to $4X - 1$ and Y_2 equal to $X^2 + 2$. Then set the independent variable in the table to Ask instead of Auto. Go to your table and input 5, -2, and 0. Under Y_1 in the table, you will find $f(5)$, $f(-2)$, and $f(0)$. Under Y_2, you will find $g(5)$, $g(-2)$, and $g(0)$.

Table Setup
Table minimum = 0
Table increment = 1
Independent variable: Ask
Dependent variable: Ask

Y Variables Setup
$Y_1 = 4X - 1$
$Y_2 = X^2 + 2$

The table will look like this:

X	Y_1	Y_2
5	19	27
−2	−9	6
0	−1	2

Although the calculator asks us for a table increment, the increment doesn't matter since we are inputting the X-values ourselves.

EXAMPLE 7 If the function f is given by

$$f = \{(-2, 0), (3, -1), (2, 4), (7, 5)\}$$

then $f(-2) = 0, f(3) = -1, f(2) = 4$, and $f(7) = 5$.

EXAMPLE 8 If $f(x) = 2x^2$ and $g(x) = 3x - 1$, find

a. $f[g(2)]$ **b.** $g[f(2)]$

SOLUTION The expression $f[g(2)]$ is read "f of g of 2."

a. Since $g(2) = 3(2) - 1 = 5$,

$$f[g(2)] = f(5) = 2(5)^2 = 50$$

b. Since $f(2) = 2(2)^2 = 8$,

$$g[f(2)] = g(8) = 3(8) - 1 = 23$$

GETTING READY FOR CLASS

After reading through the preceding section, respond in your own words and in complete sentences.

1. Explain what you are calculating when you find f(2) for a given function f.
2. If $s(t) = \frac{60}{t}$, how do you find s(10)?
3. If f(2) = 3 for a function f, what is the relationship between the numbers 2 and 3 and the graph of f?
4. If f(6) = 0 for a particular function f, then you can immediately graph one of the intercepts. Explain.

LINKING OBJECTIVES AND EXAMPLES

Next to each **objective** we have listed the examples that are best described by that objective.

A 1–8

Problem Set 3.6

Online support materials can be found at www.thomsonedu.com/login

Let $f(x) = 2x - 5$ and $g(x) = x^2 + 3x + 4$. Evaluate the following.

1. $f(2)$ **2.** $f(3)$

3. $f(-3)$ **4.** $g(-2)$

5. $g(-1)$ **6.** $f(-4)$

7. $g(-3)$ **8.** $g(2)$

9. $g(4) + f(4)$ **10.** $f(2) - g(3)$

11. $f(3) - g(2)$ **12.** $g(-1) + f(-1)$

Let $f(x) = 3x^2 - 4x + 1$ and $g(x) = 2x - 1$. Evaluate the following.

13. $f(0)$ **14.** $g(0)$

15. $g(-4)$ **16.** $f(1)$

17. $f(-1)$ **18.** $g(-1)$

19. $g(10)$ **20.** $f(10)$

21. $f(3)$ **22.** $g(3)$

23. $g\left(\frac{1}{2}\right)$ **24.** $g\left(\frac{1}{4}\right)$

25. $f(a)$ **26.** $g(b)$

■ = Videos available by instructor request
▶ = Online student support materials available at www.thomsonedu.com/login

If $f = \{(1, 4), (-2, 0), (3, \frac{1}{2}), (\pi, 0)\}$ and $g = \{(1, 1), (-2, 2), (\frac{1}{2}, 0)\}$, find each of the following values of f and g.

27. $f(1)$
28. $g(1)$
29. $g\left(\frac{1}{2}\right)$
30. $f(3)$
31. $g(-2)$
32. $f(\pi)$

Let $f(x) = 2x^2 - 8$ and $g(x) = \frac{1}{2}x + 1$. Evaluate each of the following.

33. $f(0)$
34. $g(0)$
35. $g(-4)$
36. $f(1)$
37. $f(a)$
38. $g(z)$
39. $f(b)$
40. $g(t)$
41. $f[g(2)]$
42. $g[f(2)]$
43. $g[f(-1)]$
44. $f[g(-2)]$
45. $g[f(0)]$
46. $f[g(0)]$

47. Graph the function $f(x) = \frac{1}{2}x + 2$. Then draw and label the line segments that represent $x = 4$ and $f(4)$.

48. Graph the function $f(x) = -\frac{1}{2}x + 6$. Then draw and label the line segments that represent $x = 4$ and $f(4)$.

49. For the function $f(x) = \frac{1}{2}x + 2$, find the value of x for which $f(x) = x$.

50. For the function $f(x) = -\frac{1}{2}x + 6$, find the value of x for which $f(x) = x$.

51. Graph the function $f(x) = x^2$. Then draw and label the line segments that represent $x = 1$ and $f(1)$, $x = 2$ and $f(2)$, and, finally, $x = 3$ and $f(3)$.

52. Graph the function $f(x) = x^2 - 2$. Then draw and label the line segments that represent $x = 2$ and $f(2)$, and the line segments corresponding to $x = 3$ and $f(3)$.

Applying the Concepts

53. Investing in Art A painting is purchased as an investment for \$150. If its value increases continuously so that it doubles every 3 years, then its value is given by the function

$$V(t) = 150 \cdot 2^{t/3} \quad \text{for} \quad t \geq 0$$

where t is the number of years since the painting was purchased, and $V(t)$ is its value (in dollars) at time t. Find $V(3)$ and $V(6)$, and then explain what they mean.

54. Average Speed If it takes Minke t minutes to run a mile, then her average speed $s(t)$, in miles per hour, is given by the formula

$$s(t) = \frac{60}{t} \quad \text{for} \quad t > 0$$

Find $s(4)$ and $s(5)$, and then explain what they mean.

55. Antidepressant Sales Suppose x represents one of the years in the chart. Suppose further that we have three functions f, g, and h that do the following:

f pairs each year with the total sales of Zoloft in billions of dollars for that year.
g pairs each year with the total sales of Effexor in billions of dollars for that year.
h pairs each year with the total sales of Wellbutrin in billions of dollars for that year.

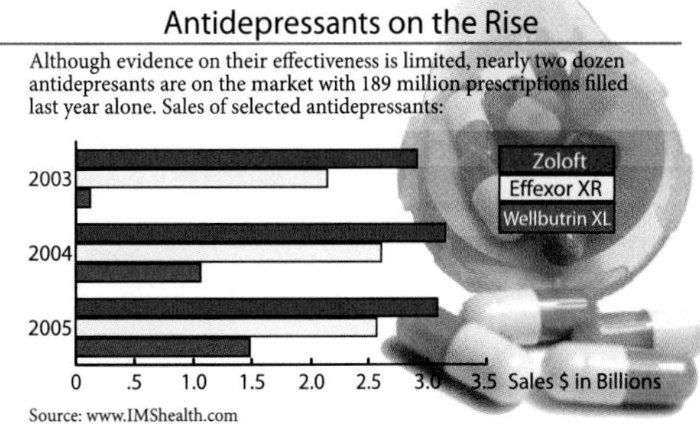

For each statement below, indicate whether the statement is true or false.

a. The domain of g is $\{2003, 2004, 2005\}$.
b. $f(2003) < f(2004)$
c. $f(2004) > g(2004)$
d. $h(2005) > 1.5$
e. $h(2005) > h(2004) > h(2003)$

56. Mobile Phone Sales Suppose x represents one of the years in the chart. Suppose further that we have three functions f, g, and h that do the following:

f pairs each year with the number of camera phones sold that year.

g pairs each year with the number of non-camera phones sold that year.

h is such that $h(x) = f(x) + g(x)$.

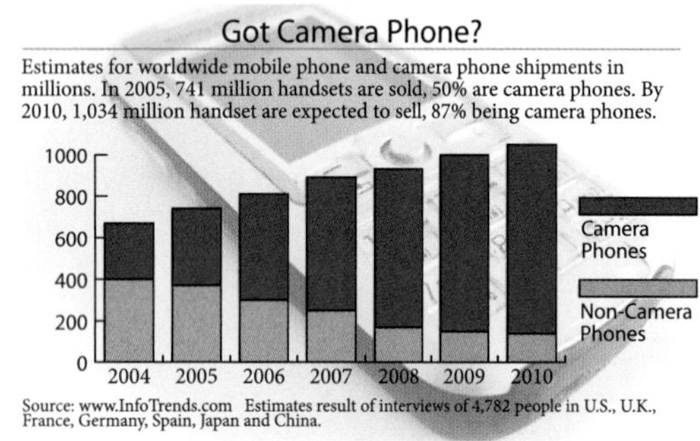

Got Camera Phone?
Estimates for worldwide mobile phone and camera phone shipments in millions. In 2005, 741 million handsets are sold, 50% are camera phones. By 2010, 1,034 million handset are expected to sell, 87% being camera phones.
1000
800
600
400
200
0
2004 2005 2006 2007 2008 2009 2010
Camera Phones
Non-Camera Phones
Source: www.InfoTrends.com Estimates result of interviews of 4,782 people in U.S., U.K., France, Germany, Spain, Japan and China.

For each statement below, indicate whether the statement is true or false.

a. The domain of f is {2004, 2005, 2006, 2007, 2008, 2009, 2010}.

b. $h(2005) = 741{,}000{,}000$

c. $f(2009) > g(2009)$

d. $f(2004) < f(2005)$

e. $h(2010) > h(2007) > h(2004)$

57. Value of a Copy Machine The function $V(t) = -3{,}300t + 18{,}000$, where V is value and t is time in years, can be used to find the value of a large copy machine during the first 5 years of use.

Photo Disc/Getty Images

a. What is the value of the copier after 3 years and 9 months?

b. What is the salvage value of this copier if it is replaced after 5 years?

c. State the domain of this function.

d. Sketch the graph of this function.

e. What is the range of this function?

f. After how many years will the copier be worth only $10,000?

58. Value of a Forklift The function $V = -16{,}500t + 125{,}000$, where V is value and t is time in years, can be used to find the value of an electric forklift during the first 6 years of use.

a. What is the value of the forklift after 2 years and 3 months?

b. What is the salvage value of this forklift if it is replaced after 6 years?

c. State the domain of this function.

d. Sketch the graph of this function.

e. What is the range of this function?

f. After how many years will the forklift be worth only $45,000?

Michael Grecco/The Image Bank/Getty Images

Maintaining Your Skills

Solve each equation.

59. $|3x - 5| = 7$

60. $|0.04 - 0.03x| = 0.02$

61. $|4y + 2| - 8 = -2$

62. $4 = |3 - 2y| - 5$

63. $5 + |6t + 2| = 3$

64. $7 + |3 - \frac{3}{4}t| = 10$

Getting Ready for the Next Section

Simplify.

65. $(35x - 0.1x^2) - (8x + 500)$

66. $70 + 0.6(m - 70)$

67. $(4x^2 + 3x + 2) + (2x^2 - 5x - 6)$

68. $(4x - 3) + (4x^2 - 7x + 3)$

69. $(4x^2 + 3x + 2) - (2x^2 - 5x - 6)$

70. $(x + 5)^2 - 2(x + 5)$

71. $0.6(m - 70)$

72. $x(35 - 0.1x)$

73. $(4x - 3)(x - 1)$

74. $(4x - 3)(4x^2 - 7x + 3)$

Extending the Concepts

The graphs of two functions are shown in Figures 3 and 4. Use the graphs to find the following.

75. a. $f(2)$ b. $f(-4)$

76. a. $g(0)$ b. $g(3)$

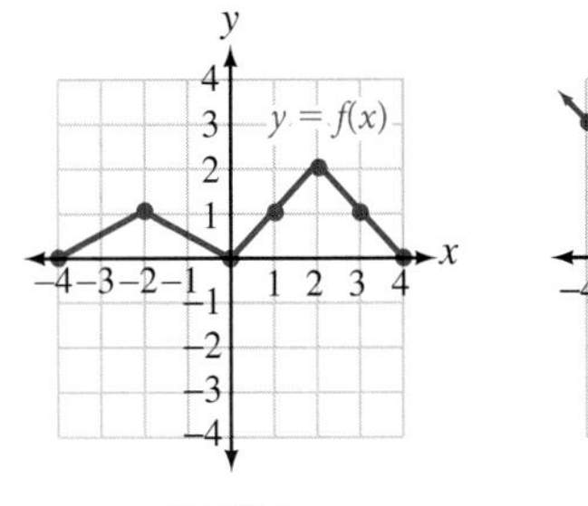

FIGURE 3

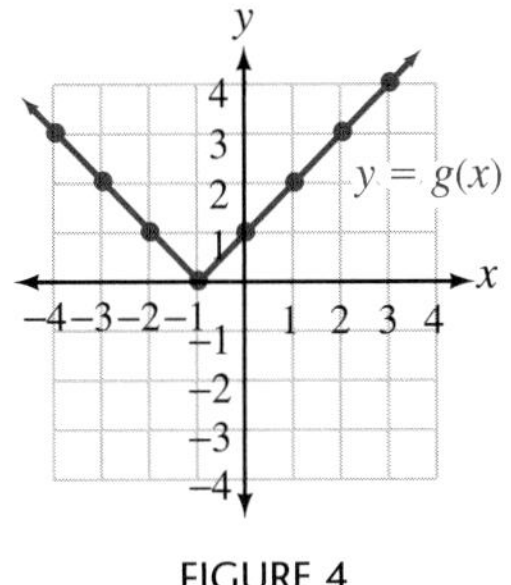

FIGURE 4

77. **Step Function** Figure 5 shows the graph of the step function C that was used to calculate the first-class postage on a letter weighing x ounces in 2006. Use this graph to answer questions (a) through (d).

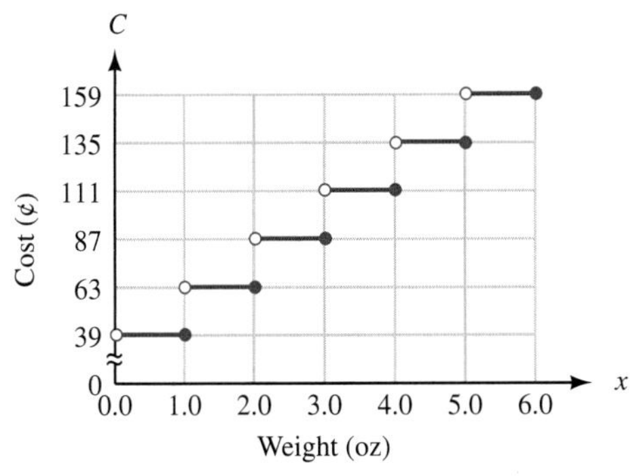

FIGURE 5 **The graph of $C(x)$**

a. Fill in the following table:

Weight (ounces)	0.6	1.0	1.1	2.5	3.0	4.8	5.0	5.3
Cost (cents)								

b. If a letter costs 87 cents to mail, how much does it weigh? State your answer in words. State your answer as an inequality.

c. If the entire function is shown in Figure 5, state the domain.

d. State the range of the function shown in Figure 5.

78. **Step Function** A taxi ride in Boston at the time I am writing this problem is \$1.50 for the first $\frac{1}{4}$ mile, and then \$0.25 for each additional $\frac{1}{8}$ of a mile. The following graph shows how much you will pay for a taxi ride of 1 mile or less.

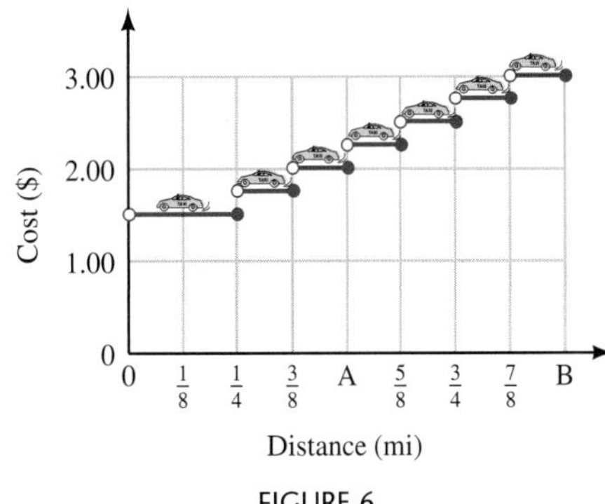

FIGURE 6

a. What is the most you will pay for this taxi ride?

b. How much does it cost to ride the taxi for $\frac{8}{10}$ of a mile?

c. Find the values of A and B on the horizontal axis.

d. If a taxi ride costs \$2.50, what distance was the ride?

e. If the complete function is shown in Figure 6, find the domain and range of the function.

3.7 Algebra and Composition with Functions

OBJECTIVES

A Find the sum, difference, product, and quotient of two functions.

B Find the composition of two functions.

A company produces and sells copies of an accounting program for home computers. The price they charge for the program is related to the number of copies sold by the demand function

$$p(x) = 35 - 0.1x$$

We find the revenue for this business by multiplying the number of items sold by the price per item. When we do so, we are forming a new function by combining two existing functions; that is, if $n(x) = x$ is the number of items sold and $p(x) = 35 - 0.1x$ is the price per item, then revenue is

$$R(x) = n(x) \cdot p(x) = x(35 - 0.1x) = 35x - 0.1x^2$$

In this case, the revenue function is the product of two functions. When we combine functions in this manner, we are applying our rules for algebra to functions.

To carry this situation further, we know the profit function is the difference between two functions. If the cost function for producing x copies of the accounting program is $C(x) = 8x + 500$, then the profit function is

$$P(x) = R(x) - C(x) = (35x - 0.1x^2) - (8x + 500) = -500 + 27x - 0.1x^2$$

The relationship between these last three functions is shown visually in Figure 1.

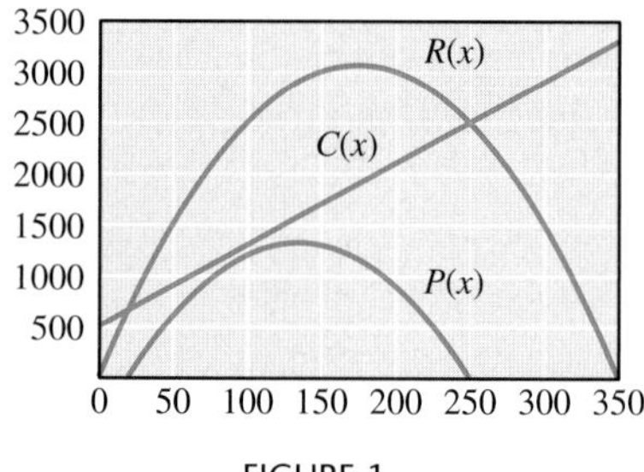

FIGURE 1

Again, when we combine functions in the manner shown, we are applying our rules for algebra to functions. To begin this section, we take a formal look at addition, subtraction, multiplication, and division with functions.

If we are given two functions f and g with a common domain, we can define four other functions as follows.

DEFINITION

$(f + g)(x) = f(x) + g(x)$	The function $f + g$ is the sum of the functions f and g.
$(f - g)(x) = f(x) - g(x)$	The function $f - g$ is the difference of the functions f and g.
$(fg)(x) = f(x)g(x)$	The function fg is the product of the functions f and g.
$\left(\frac{f}{g}\right)(x) = \frac{f(x)}{g(x)}$	The function f/g is the quotient of the functions f and g, where $g(x) \neq 0$.

EXAMPLE 1 If $f(x) = 4x^2 + 3x$ and $g(x) = -5x - 6$, write the formula for the functions $f + g$ and $f - g$.

SOLUTION The function $f + g$ is defined by

$$\begin{aligned}(f+g)(x) &= f(x) + g(x)\\ &= (4x^2 + 3x) + (-5x - 6)\\ &= 4x^2 - 2x - 6\end{aligned}$$

The function $f - g$ is defined by

$$\begin{aligned}(f-g)(x) &= f(x) - g(x)\\ &= (4x^2 + 3x) - (-5x - 6)\\ &= 4x^2 + 3x + 5x + 6\\ &= 4x^2 + 8x + 6\end{aligned}$$

EXAMPLE 2 Let $f(x) = 4x - 3$ and $g(x) = 4x^2$. Find $f + g$, fg, and g/f.

SOLUTION The function $f + g$, the sum of functions f and g, is defined by

$$\begin{aligned}(f+g)(x) &= f(x) + g(x)\\ &= (4x - 3) + 4x^2\\ &= 4x^2 + 4x - 3\end{aligned}$$

The product of the functions f and g, fg, is given by

$$\begin{aligned}(fg)(x) &= f(x)g(x)\\ &= (4x - 3)4x^2\\ &= 16x^3 - 12x^2\end{aligned}$$

The quotient of the functions g and f, g/f, is defined as

$$\begin{aligned}\left(\frac{g}{f}\right)(x) &= \frac{g(x)}{f(x)}\\ &= \frac{4x^2}{4x - 3}\end{aligned}$$

EXAMPLE 3 Let $f(x) = x + 3$, $g(x) = 2x - 1$, and $h(x) = 2x^2 + 5x - 3$. Evaluate $(fg)(2)$, $h(2)$, $\dfrac{h(1)}{g(1)}$, and $f(1)$

SOLUTION

$$\begin{aligned}(fg)(2) &= f(2)g(2)\\ &= 5 \cdot 3\\ &= 15\end{aligned}$$

$$\begin{aligned}h(2) &= 2 \cdot 2^2 + 5 \cdot 2 - 3\\ &= 8 + 10 - 3\\ &= 15\end{aligned}$$

$$\begin{aligned}\frac{h(1)}{g(1)} &= \frac{2 \cdot 1^2 + 5 \cdot 1 - 3}{2 \cdot 1 - 1} \\ &= \frac{2 + 5 - 3}{2 - 1} \\ &= \frac{4}{1} \\ &= 4 \\ f(1) &= 1 + 3 \\ &= 4\end{aligned}$$

Composition of Functions

In addition to the four operations used to combine functions shown so far in this section, there is a fifth way to combine two functions to obtain a new function. It is called *composition of functions.* To illustrate the concept, recall the definition of training heart rate: Training heart rate, in beats per minute, is resting heart rate plus 60% of the difference between maximum heart rate and resting heart rate. If your resting heart rate is 70 beats per minute, then your training heart rate is a function of your maximum heart rate M:

$$T(M) = 70 + 0.6(M - 70) = 70 + 0.6M - 42 = 28 + 0.6M$$

But your maximum heart rate is found by subtracting your age in years from 220. So, if x represents your age in years, then your maximum heart rate is

$$M(x) = 220 - x$$

Therefore, if your resting heart rate is 70 beats per minute and your age in years is x, then your training heart rate can be written as a function of x.

$$T(x) = 28 + 0.6(220 - x)$$

This last line is the composition of functions T and M. We input x into function M, which outputs $M(x)$. Then, we input $M(x)$ into function T, which outputs $T(M(x))$, which is the training heart rate as a function of age x. Here is a diagram of the situation, which is called a function map:

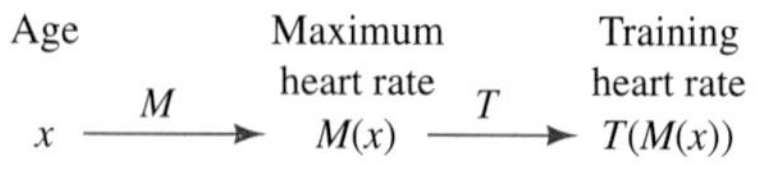

Age		Maximum heart rate		Training heart rate
x	$\xrightarrow{M}$	$M(x)$	$\xrightarrow{T}$	$T(M(x))$

FIGURE 2

Now let's generalize the preceding ideas into a formal development of composition of functions. To find the composition of two functions f and g, we first require that the range of g have numbers in common with the domain of f. Then the composition of f with g, $f \circ g$, is defined this way:

$$(f \circ g)(x) = f(g(x))$$

To understand this new function, we begin with a number x, and we operate on it with g, giving us $g(x)$. Then we take $g(x)$ and operate on it with f, giving us $f(g(x))$. The only numbers we can use for the domain of the composition of f with

g are numbers x in the domain of g, for which $g(x)$ is in the domain of f. The diagrams in Figure 3 illustrate the composition of f with g.

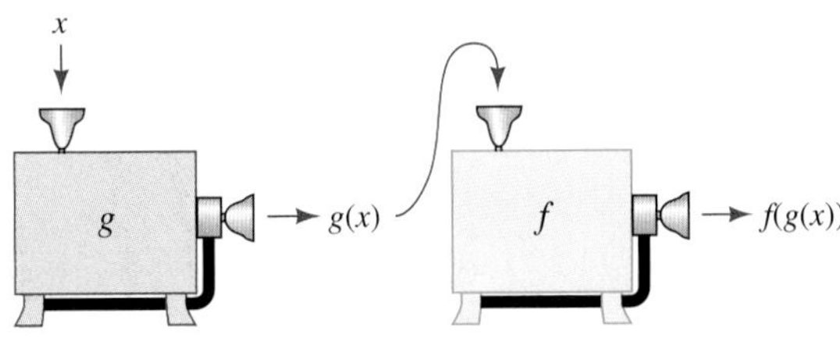

Function machines

FIGURE 3

Composition of functions is not commutative. The composition of f with g, $f \circ g$, may therefore be different from the composition of g with f, $g \circ f$.

$$(g \circ f)(x) = g(f(x))$$

Again, the only numbers we can use for the domain of the composition of g with f are numbers in the domain of f, for which $f(x)$ is in the domain of g. The diagrams in Figure 4 illustrate the composition of g with f.

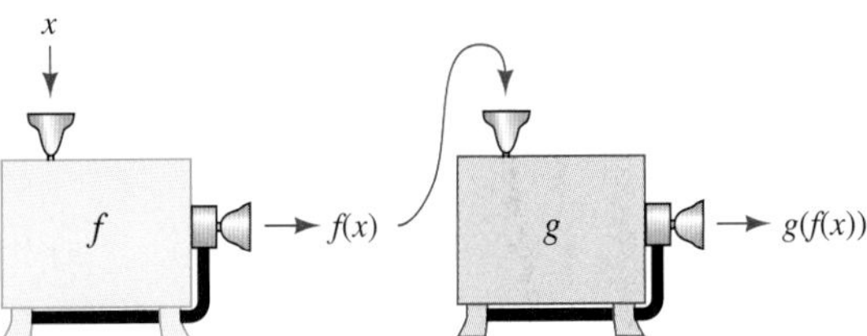

Function machines

FIGURE 4

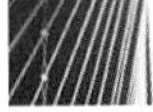

EXAMPLE 4

If $f(x) = x + 5$ and $g(x) = 2x$, find $(f \circ g)(x)$ and $(g \circ f)(x)$.

SOLUTION The composition of f with g is

$$\begin{aligned}(f \circ g)(x) &= f[g(x)] \\ &= f(2x) \\ &= 2x + 5\end{aligned}$$

The composition of g with f is

$$\begin{aligned}(g \circ f)(x) &= g[f(x)] \\ &= g(x + 5) \\ &= 2(x + 5) \\ &= 2x + 10\end{aligned}$$

LINKING OBJECTIVES AND EXAMPLES

Next to each **objective** we have listed the examples that are best described by that objective.

A	1–3
B	4

GETTING READY FOR CLASS

After reading through the preceding section, respond in your own words and in complete sentences.

1. How are profit, revenue, and cost related?
2. How do you find the maximum heart rate?
3. For functions f and g, how do you find the composition of f with g?
4. For functions f and g, how do you find the composition of g with f?

Problem Set 3.7

Online support materials can be found at www.thomsonedu.com/login

Let $f(x) = 4x - 3$ and $g(x) = 2x + 5$. Write a formula for each of the following functions.

1. $f+g$ **2.** $f-g$

3. $g-f$ **4.** $g+f$

If the functions f, g, and h are defined by $f(x) = 3x - 5$, $g(x) = x - 2$, and $h(x) = 3x^2$, write a formula for each of the following functions.

▸ **5.** $g+f$ ▸ **6.** $f+h$

▸ **7.** $g+h$ ▸ **8.** $f-g$

▸ **9.** $g-f$ ▸ **10.** $h-g$

▸ **11.** fh ▸ **12.** gh

▸ **13.** h/f ▸ **14.** h/g

▸ **15.** f/h ▸ **16.** g/h

▸ **17.** $f+g+h$ ▸ **18.** $h-g+f$

Let $f(x) = 2x + 1$, $g(x) = 4x + 2$, and $h(x) = 4x^2 + 4x + 1$, and find the following.

19. $(f+g)(2)$ **20.** $(f-g)(-1)$

21. $(fg)(3)$ **22.** $(f/g)(-3)$

23. $(h/g)(1)$ **24.** $(hg)(1)$

25. $(fh)(0)$ **26.** $(h-g)(-4)$

27. $(f+g+h)(2)$ **28.** $(h-f+g)(0)$

29. $(h+fg)(3)$ **30.** $(h-fg)(5)$

31. Let $f(x) = x^2$ and $g(x) = x + 4$, and find
  a. $(f \circ g)(5)$
  b. $(g \circ f)(5)$
  c. $(f \circ g)(x)$
  d. $(g \circ f)(x)$

32. Let $f(x) = 3 - x$ and $g(x) = x^3 - 1$, and find
  a. $(f \circ g)(0)$
  b. $(g \circ f)(0)$
  c. $(f \circ g)(x)$

33. Let $f(x) = x^2 + 3x$ and $g(x) = 4x - 1$, and find
  a. $(f \circ g)(0)$
  b. $(g \circ f)(0)$
  c. $(g \circ f)(x)$

34. Let $f(x) = (x - 2)^2$ and $g(x) = x + 1$, and find the following
  a. $(f \circ g)(-1)$
  b. $(g \circ f)(-1)$

For each of the following pairs of functions f and g, show that $(f \circ g)(x) = (g \circ f)(x) = x$.

35. $f(x) = 5x - 4$ and $g(x) = \dfrac{x + 4}{5}$

36. $f(x) = \dfrac{x}{6} - 2$ and $g(x) = 6x + 12$

Use the graph to answer problems 37–44.

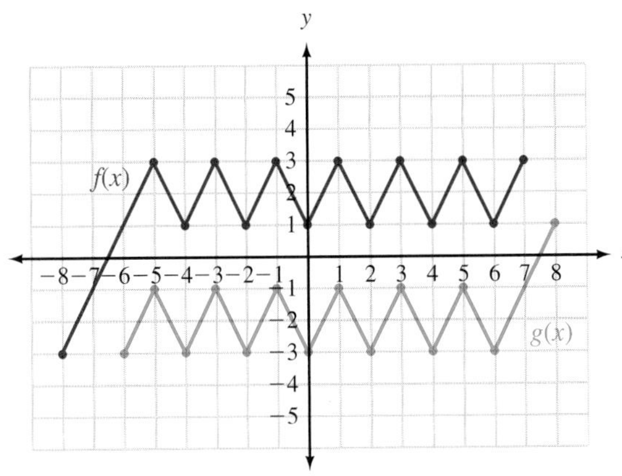

Evaluate.

37. $f(2) + 5$

38. $g(-2) - 5$

39. $f(-3) + g(-3)$

40. $f(5) - g(5)$

41. $(f \circ g)(0)$

42. $(g \circ f)(0)$

43. Find x if $f(x) = -3$.

44. Find x if $g(x) = 1$.

Use the graph to answer problems 45–52.

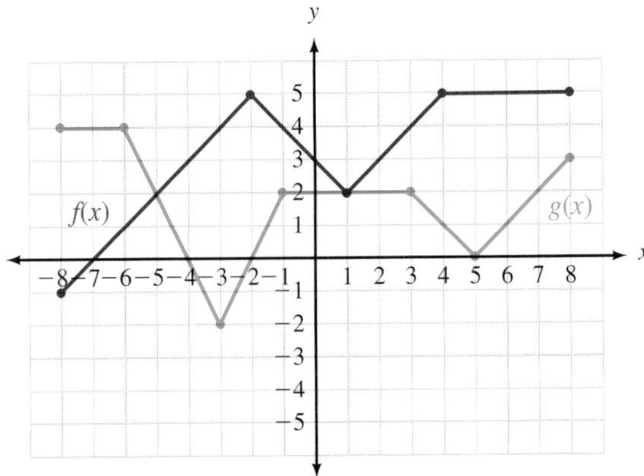

Evaluate.

45. $f(-3) + 2$

46. $g(3) - 3$

47. $f(2) + g(2)$

48. $f(-5) - g(-5)$

49. $(f \circ g)(0)$

50. $(g \circ f)(0)$

51. Find x if $f(x) = 1$.

52. Find x if $g(x) = -2$.

Applying the Concepts

53. Profit, Revenue, and Cost A company manufactures and sells prerecorded videotapes. Here are the equations they use in connection with their business.

Number of tapes sold each day: $n(x) = x$

Selling price for each tape: $p(x) = 11.5 - 0.05x$

Daily fixed costs: $f(x) = 200$

Daily variable costs: $v(x) = 2x$

Find the following functions.

a. Revenue $= R(x) =$ the product of the number of tapes sold each day and the selling price of each tape.

b. Cost $= C(x) =$ the sum of the fixed costs and the variable costs.

c. Profit $= P(x) =$ the difference between revenue and cost.

d. Average cost $= \overline{C}(x) =$ the quotient of cost and the number of tapes sold each day.

54. Profit, Revenue, and Cost A company manufactures and sells diskettes for home computers. Here are the equations they use in connection with their business.

Number of diskettes sold each day: $n(x) = x$

Selling price for each diskette: $p(x) = 3 - \frac{1}{300}x$

Daily fixed costs: $f(x) = 200$

Daily variable costs: $v(x) = 2x$

Find the following functions.

a. Revenue $= R(x) =$ the product of the number of diskettes sold each day and the selling price of each diskette.

b. Cost $= C(x) =$ the sum of the fixed costs and the variable costs.

c. Profit $= P(x) =$ the difference between revenue and cost.

d. Average cost $= \overline{C}(x) =$ the quotient of cost and the number of diskettes sold each day.

55. Training Heart Rate Find the training heart rate function, $T(M)$, for a person with a resting heart rate of 62 beats per minute, then find the following.

a. Find the maximum heart rate function, $M(x)$, for a person x years of age.

b. What is the maximum heart rate for a 24-year-old person?

c. What is the training heart rate for a 24-year-old person with a resting heart rate of 62 beats per minute?

d. What is the training heart rate for a 36-year-old person with a resting heart rate of 62 beats per minute?

e. What is the training heart rate for a 48-year-old person with a resting heart rate of 62 beats per minute?

56. **Training Heart Rate** Find the training heart rate function, $T(M)$, for a person with a resting heart rate of 72 beats per minute, then finding the following to the nearest whole number.
   a. Find the maximum heart rate function, $M(x)$, for a person x years of age.
   b. What is the maximum heart rate for a 20-year-old person?
   c. What is the training heart rate for a 20-year-old person with a resting heart rate of 72 beats per minute?
   d. What is the training heart rate for a 30-year-old person with a resting heart rate of 72 beats per minute?
   e. What is the training heart rate for a 40-year-old person with a resting heart rate of 72 beats per minute?

Maintaining Your Skills

Solve each inequality.

57. $|x - 3| < 1$
58. $|x - 3| > 1$
59. $|6 - x| > 2$
60. $\left|1 - \frac{1}{2}x\right| > 2$
61. $|7x - 1| \le 6$
62. $|7x - 1| \ge 6$

Getting Ready for the Next Section

Simplify.

63. $16(3.5)^2$
64. $\frac{2{,}400}{100}$
65. $\frac{180}{45}$
66. $4(2)(4)^2$
67. $\frac{0.0005(200)}{(0.25)^2}$
68. $\frac{0.2(0.5)^2}{100}$
69. If $y = Kx$, find K if $x = 5$ and $y = 15$.
70. If $d = Kt^2$, find K if $t = 2$ and $d = 64$.
71. If $V = \frac{K}{P}$, find K if $P = 48$ and $V = 50$.
72. If $y = Kxz^2$, find K if $x = 5$, $z = 3$, and $y = 180$.

3.8 Variation

OBJECTIVES

A Set up and solve problems with direct, inverse, or joint variation.

If you are a runner and you average t minutes for every mile you run during one of your workouts, then your speed s in miles per hour is given by the equation and graph shown here. Figure 1 is shown in the first quadrant only because both t and s are positive.

$$s = \frac{60}{t}$$

Input t	Output s
4	15.0
6	10.0
8	7.5
10	6.0
12	5.0
14	4.3

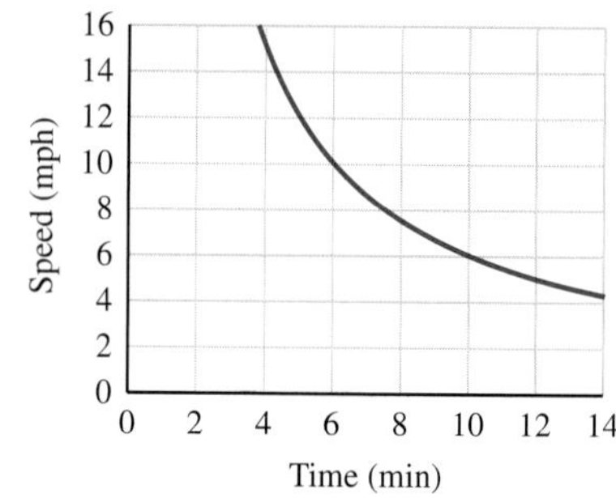

FIGURE 1

You know intuitively that as your average time per mile (t) increases, your speed (s) decreases. Likewise, lowering your time per mile will increase your speed. The equation and Figure 1 also show this to be true: increasing t decreases s, and decreasing t increases s. Quantities that are connected in this way are said to *vary inversely* with each other. Inverse variation is one of the topics we will study in this section.

There are two main types of variation: *direct variation* and *inverse variation.* Variation problems are most common in the sciences, particularly in chemistry and physics.

Direct Variation

When we say the variable y *varies directly* with the variable x, we mean that the relationship can be written in symbols as $y = Kx$, where K is a nonzero constant called the *constant of variation* (or *proportionality constant*). Another way of saying y varies directly with x is to say y is *directly proportional* to x.

Study the following list. It gives the mathematical equivalent of some direct variation statements.

English Phrase	Algebraic Equation
y varies directly with x	$y = Kx$
s varies directly with the square of t	$s = Kt^2$
y is directly proportional to the cube of z	$y = Kz^3$
u is directly proportional to the square root of v	$u = K\sqrt{v}$

EXAMPLE 1 y varies directly with x. If y is 15 when x is 5, find y when x is 7.

SOLUTION The first sentence gives us the general relationship between x and y. The equation equivalent to the statement "y varies directly with x" is

$$y = Kx$$

The first part of the second sentence in our example gives us the information necessary to evaluate the constant K:

When	$y = 15$
and	$x = 5$
the equation	$y = Kx$
becomes	$15 = K \cdot 5$
or	$K = 3$

The equation now can be written specifically as

$$y = 3x$$

Letting $x = 7$, we have

$$y = 3 \cdot 7$$

$$y = 21$$

EXAMPLE 2 A skydiver jumps from a plane. As with any object that falls toward Earth, the distance the skydiver falls is directly proportional to the square of the time he has been falling until he reaches his terminal velocity. If the skydiver falls 64 feet in the first 2 seconds of the jump, then

a. How far will he have fallen after 3.5 seconds?

b. Graph the relationship between distance and time.

c. How long will it take him to fall 256 feet?

SOLUTION We let t represent the time the skydiver has been falling. Then we can let $d(t)$ represent the distance he has fallen.

a. Because $d(t)$ is directly proportional to the square of t, we have the general function that describes this situation:

$$d(t) = Kt^2$$

Next, we use the fact that $d(2) = 64$ to find K.

$$64 = K(2^2)$$

$$K = 16$$

The specific equation that describes this situation is

$$d(t) = 16t^2$$

To find how far a skydiver will have fallen after 3.5 seconds, we find $d(3.5)$:

$$d(3.5) = 16(3.5^2)$$

$$d(3.5) = 196$$

A skydiver will have fallen 196 feet after 3.5 seconds.

b. To graph this equation, we use a table:

Input t	Output $d(t)$
0	0
1	16
2	64
3	144
4	256
5	400

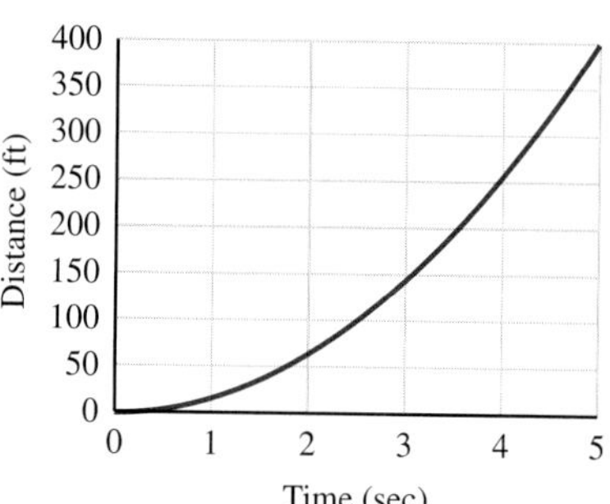

FIGURE 2

c. From the table or the graph (Figure 2), we see that it will take 4 seconds for the skydiver to fall 256 feet.

Inverse Variation

Running From the introduction to this section, we know that the relationship between the number of minutes (t) it takes a person to run a mile and his or her average speed in miles per hour (s) can be described with the following equation, table, and Figure 3.

$$s = \frac{60}{t}$$

Input t	Output s
4	15.0
6	10.0
8	7.5
10	6.0
12	5.0
14	4.3

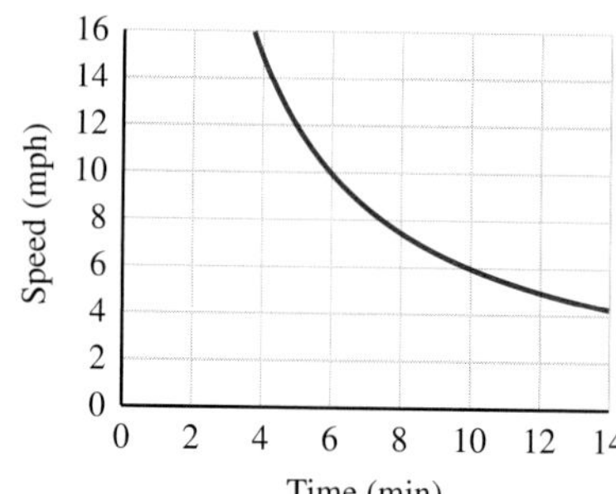

FIGURE 3

If t decreases, then s will increase, and if t increases, then s will decrease. The variable s is *inversely proportional* to the variable t. In this case, the *constant of proportionality* is 60.

Photography If you are familiar with the terminology and mechanics associated with photography, you know that the f-stop for a particular lens will increase as the aperture (the maximum diameter of the opening of the lens) decreases. In mathematics we say that f-stop and aperture vary inversely with each other. The diagram illustrates this relationship.

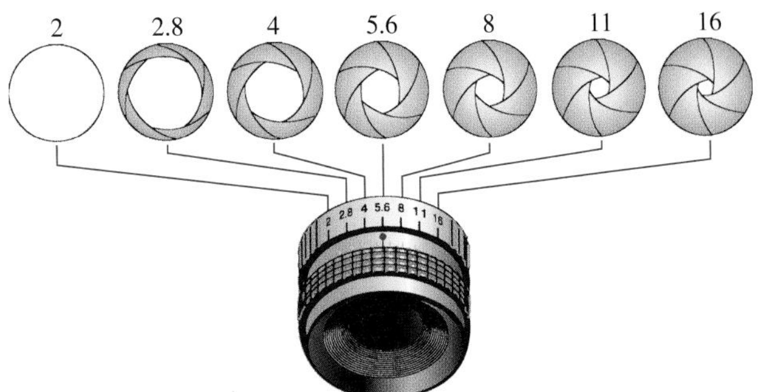

If f is the f-stop and d is the aperture, then their relationship can be written

$$f = \frac{K}{d}$$

In this case, K is the constant of proportionality. (Those of you familiar with photography know that K is also the focal length of the camera lens.)

In General We generalize this discussion of inverse variation as follows: If y varies inversely with x, then

$$y = K\frac{1}{x} \quad \text{or} \quad y = \frac{K}{x}$$

We can also say y is inversely proportional to x. The constant K is again called the constant of variation or proportionality constant.

English Phrase	Algebraic Equation
y is inversely proportional to x	$y = \frac{K}{x}$
s varies inversely with the square of t	$s = \frac{K}{t^2}$
y is inversely proportional to x^4	$y = \frac{K}{x^4}$
z varies inversely with the cube root of t	$z = \frac{K}{\sqrt[3]{t}}$

EXAMPLE 3 The volume of a gas is inversely proportional to the pressure of the gas on its container. If a pressure of 48 pounds per square inch corresponds to a volume of 50 cubic feet, what pressure is needed to produce a volume of 100 cubic feet?

Note

The relationship between pressure and volume as given in this example is known as Boyle's law and applies to situations such as those encountered in a piston-cylinder arrangement. It was Robert Boyle (1627–1691) who, in 1662, published the results of some of his experiments that showed, among other things, that the volume of a gas decreases as the pressure increases. This is an example of inverse variation.

SOLUTION We can represent volume with V and pressure with P:

$$V = \frac{K}{P}$$

Using $P = 48$ and $V = 50$, we have

$$50 = \frac{K}{48}$$

$$K = 50(48)$$

$$K = 2{,}400$$

The equation that describes the relationship between P and V is

$$V = \frac{2{,}400}{P}$$

Here is a graph of this relationship.

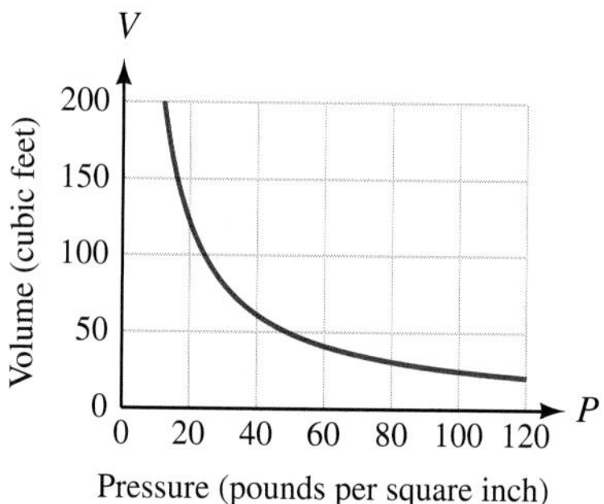

Substituting $V = 100$ into our last equation, we get

$$100 = \frac{2{,}400}{P}$$

$$100P = 2{,}400$$

$$P = \frac{2{,}400}{100}$$

$$P = 24$$

A volume of 100 cubic feet is produced by a pressure of 24 pounds per square inch.

Joint Variation and Other Variation Combinations

Many times relationships among different quantities are described in terms of more than two variables. If the variable y varies directly with *two* other variables, say x and z, then we say y varies *jointly* with x and z. In addition to joint variation, there are many other combinations of direct and inverse variation involving more than two variables. The following table is a list of some variation statements and their equivalent mathematical forms:

English Phrase	Algebraic Equation
y varies jointly with x and z	$y = Kxz$
z varies jointly with r and the square of s	$z = Krs^2$
V is directly proportional to T and inversely proportional to P	$V = \frac{KT}{P}$
F varies jointly with m_1 and m_2 and inversely with the square of r	$F = \frac{Km_1m_2}{r^2}$

EXAMPLE 4 y varies jointly with x and the square of z. When x is 5 and z is 3, y is 180. Find y when x is 2 and z is 4.

SOLUTION The general equation is given by

$$y = Kxz^2$$

Substituting $x = 5$, $z = 3$, and $y = 180$, we have

$$180 = K(5)(3)^2$$

$$180 = 45K$$

$$K = 4$$

The specific equation is

$$y = 4xz^2$$

When $x = 2$ and $z = 4$, the last equation becomes

$$y = 4(2)(4)^2$$

$$y = 128$$

EXAMPLE 5 In electricity, the resistance of a cable is directly proportional to its length and inversely proportional to the square of the diameter. If a 100-foot cable 0.5 inch in diameter has a resistance of 0.2 ohm, what will be the resistance of a cable made from the same material if it is 200 feet long with a diameter of 0.25 inch?

SOLUTION Let R = resistance, l = length, and d = diameter. The equation is

$$R = \frac{Kl}{d^2}$$

When $R = 0.2$, $l = 100$, and $d = 0.5$, the equation becomes

$$0.2 = \frac{K(100)}{(0.5)^2}$$

or

$$K = 0.0005$$

Using this value of K in our original equation, the result is

$$R = \frac{0.0005l}{d^2}$$

When $l = 200$ and $d = 0.25$, the equation becomes

$$R = \frac{0.0005(200)}{(0.25)^2}$$

$$R = 1.6 \text{ ohms}$$

LINKING OBJECTIVES AND EXAMPLES

Next to each **objective** we have listed the examples that are best described by that objective.

A 1–5

GETTING READY FOR CLASS

After reading through the preceding section, respond in your own words and in complete sentences.

1. Give an example of a direct variation statement, and then translate it into symbols.
2. Translate the equation $y = \frac{K}{x}$ into words.
3. For the inverse variation equation $y = \frac{3}{x}$, what happens to the values of y as x gets larger?
4. How are direct variation statements and linear equations in two variables related?

Problem Set 3.8

Online support materials can be found at www.thomsonedu.com/login

Describe each relationship below as direct or inverse.

▶ **1.** The number of homework problems assigned each night and the time it takes to complete the assignment

▶ **2.** The number of years of school and starting salary

▶ **3.** The length of the line for an amusement park ride and the number of people in line

▶ **4.** The age of a home computer and its relative speed

▶ **5.** The age of a bottle of wine and its price

▶ **6.** The speed of an Internet connection and the time to download a song

▶ **7.** The length of a song and the time it takes to download

8. The speed of a rollercoaster and your willingness to ride it

9. The diameter of a Ferris wheel and circumference

10. The measure angle α to the measure of angle β if $\alpha + \beta = 90°$ ($\alpha = 0$)

11. The altitude above sea level and the air temperature

12. The uncommonness of a search term and the number of search results

For the following problems, y varies directly with x.

▶ **13.** If y is 10 when x is 2, find y when x is 6.

14. If y is 20 when x is 5, find y when x is 3.

15. If y is -32 when x is 4, find x when y is -40.

16. If y is -50 when x is 5, find x when y is -70.

For the following problems, r is inversely proportional to s.

17. If r is -3 when s is 4, find r when s is 2.

18. If r is -10 when s is 6, find r when s is -5.

19. If r is 8 when s is 3, find s when r is 48.

20. If r is 12 when s is 5, find s when r is 30.

For the following problems, d varies directly with the square of r.

21. If $d = 10$ when $r = 5$, find d when $r = 10$.

22. If $d = 12$ when $r = 6$, find d when $r = 9$.

23. If $d = 100$ when $r = 2$, find d when $r = 3$.

24. If $d = 50$ when $r = 5$, find d when $r = 7$.

For the following problems, y varies inversely with the absolute value of x.

▶ **25.** If $y = 6$ when $x = 3$, find y when $x = 9$.

▶ **26.** If $y = 6$ when $x = -3$, find y when $x = -9$.

▶ **27.** If $y = 20$ when $x = -5$, find y when $x = 10$.

■ = Videos available by instructor request

▶ = Online student support materials available at www.thomsonedu.com/login

28. If $y = 20$ when $x = 5$, find y when $x = 10$.

For the following problems, y varies inversely with the square of x.

29. If $y = 45$ when $x = 3$, find y when x is 5.

30. If $y = 12$ when $x = 2$, find y when x is 6.

31. If $y = 18$ when $x = 3$, find y when x is 2.

32. If $y = 45$ when $x = 4$, find y when x is 5.

For the following problems, z varies jointly with x and the square of y.

33. If z is 54 when x and y are 3, find z when $x = 2$ and $y = 4$.

34. If z is 80 when x is 5 and y is 2, find z when $x = 2$ and $y = 5$.

35. If z is 64 when $x = 1$ and $y = 4$, find x when $z = 32$ and $y = 1$.

36. If z is 27 when $x = 6$ and $y = 3$, find x when $z = 50$ and $y = 4$.

Applying the Concepts

37. **Length of a Spring** The length a spring stretches is directly proportional to the force applied. If a force of 5 pounds stretches a spring 3 inches, how much force is necessary to stretch the same spring 10 inches?

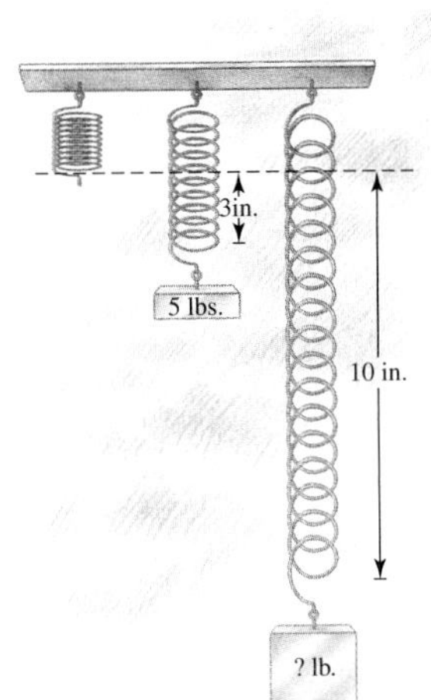

38. **Weight and Surface Area** The weight of a certain material varies directly with the surface area of that material. If 8 square feet weighs half a pound, how much will 10 square feet weigh?

39. **Pressure and Temperature** The temperature of a gas varies directly with its pressure. A temperature of 200 K produces a pressure of 50 pounds per square inch.
   a. Find the equation that relates pressure and temperature.
   b. Graph the equation from part (a) in the first quadrant only.
   c. What pressure will the gas have at 280° K?

40. **Circumference and Diameter** The circumference of a wheel is directly proportional to its diameter. A wheel has a circumference of 8.5 feet and a diameter of 2.7 feet.
   a. Find the equation that relates circumference and diameter.
   b. Graph the equation from part (a) in the first quadrant only.
   c. What is the circumference of a wheel that has a diameter of 11.3 feet?

41. **Volume and Pressure** The volume of a gas is inversely proportional to the pressure. If a pressure of 36 pounds per square inch corresponds to a volume of 25 cubic feet, what pressure is needed to produce a volume of 75 cubic feet?

42. **Wave Frequency** The frequency of an electromagnetic wave varies inversely with the wavelength. If a wavelength of 200 meters has a frequency of 800 kilocycles per second, what frequency will be associated with a wavelength of 500 meters?

43. ***f*-Stop and Aperture Diameter** The relative aperture or *f*-stop for a camera lens is inversely proportional to the diameter of the aperture. An *f*-stop of 2 corresponds to an aperture diameter of 40 millimeters for the lens on an automatic camera.
   a. Find the equation that relates *f*-stop and diameter.
   b. Graph the equation from part (a) in the first quadrant only.
   c. What is the *f*-stop of this camera when the aperture diameter is 10 millimeters?

44. ***f*-Stop and Aperture Diameter** The relative aperture or *f*-stop for a camera lens is inversely proportional to the diameter of the aperture. An *f*-stop of 2.8

corresponds to an aperture diameter of 75 millimeters for a certain telephoto lens.

a. Find the equation that relates *f*-stop and diameter.

b. Graph the equation from part (a) in the first quadrant only.

c. What aperture diameter corresponds to an *f*-stop of 5.6?

45. **Surface Area of a Cylinder** The surface area of a hollow cylinder varies jointly with the height and radius of the cylinder. If a cylinder with radius 3 inches and height 5 inches has a surface area of 94 square inches, what is the surface area of a cylinder with radius 2 inches and height 8 inches?

46. **Capacity of a Cylinder** The capacity of a cylinder varies jointly with its height and the square of its radius. If a cylinder with a radius of 3 centimeters and a height of 6 centimeters has a capacity of 169.56 cubic centimeters, what will be the capacity of a cylinder with radius 4 centimeters and height 9 centimeters?

47. **Electrical Resistance** The resistance of a wire varies directly with its length and inversely with the square of its diameter. If 100 feet of wire with diameter 0.01 inch has a resistance of 10 ohms, what is the resistance of 60 feet of the same type of wire if its diameter is 0.02 inch?

48. **Volume and Temperature** The volume of a gas varies directly with its temperature and inversely with the pressure. If the volume of a certain gas is 30 cubic feet at a temperature of 300 K and a pressure of 20 pounds per square inch, what is the volume of the same gas at 340 K when the pressure is 30 pounds per square inch?

49. **Music** A musical tone's pitch varies inversely with its wavelength. If one tone has a pitch of 420 vibrations each second and a wavelength of 2.2 meters, find the wavelength of a tone that has a pitch of 720 vibrations each second.

50. **Hooke's Law** Hooke's law states that the stress (force per unit area) placed on a solid object varies directly with the strain (deformation) produced.

a. Using the variables S_1 for stress and S_2 for strain, state this law in algebraic form.

b. Find the constant, K, if for one type of material $S_1 = 24$ and $S_2 = 72$.

51. **Gravity** In Book Three of his *Principia*, Isaac Newton (depicted on the postage stamp) states that there is a single force in the universe that holds everything together, called the force of universal gravity. Newton stated that the force of universal gravity, F, is directly proportional with the product of two masses, m_1 and m_2, and inversely proportional with the square of the distance d between them. Write the equation for Newton's force of universal gravity, using the symbol G as the constant of proportionality.

52. **Boyle's Law and Charles's Law** Boyle's law states that for low pressures, the pressure of an ideal gas kept at a constant temperature varies inversely with the volume of the gas. Charles's law states that for low pressures, the density of an ideal gas kept at a constant pressure varies inversely with the absolute temperature of the gas.

a. State Boyle's law as an equation using the symbols P, K, and V.

b. State Charles's law as an equation using the symbols D, K, and T.

Maintaining Your Skills

Solve the following equations.

53. $x - 5 = 7$

54. $3y = -4$

55. $5 - \frac{4}{7}a = -11$

56. $\frac{1}{5}x - \frac{1}{2} - \frac{1}{10}x + \frac{2}{5} = \frac{3}{10}x + \frac{1}{2}$

57. $5(x - 1) - 2(2x + 3) = 5x - 4$

58. $0.07 - 0.02(3x + 1) = -0.04x + 0.01$

Solve for the indicated variable.

59. $P = 2l + 2w$ for w

60. $A = \frac{1}{2}h(b + B)$ for B

Solve the following inequalities. Write the solution set using interval notation, then graph the solution set.

61. $-5t \leq 30$

62. $5 - \frac{3}{2}x > -1$

63. $1.6x - 2 < 0.8x + 2.8$

64. $3(2y + 4) \geq 5(y - 8)$

Solve the following equations.

65. $\left|\frac{1}{4}x - 1\right| = \frac{1}{2}$

66. $\left|\frac{2}{3}a + 4\right| = 6$

67. $|3 - 2x| + 5 = 2$

68. $5 = |3y + 6| - 4$

Solve each inequality and graph the solution set.

69. $\left|\frac{x}{5} + 1\right| \geq \frac{4}{5}$

70. $|2 - 6t| < -5$

71. $|3 - 4t| > -5$

72. $|6y - 1| - 4 \leq 2$

Extending the Concepts

73. **Human Cannonball** A circus company is deciding where to position the net for the human cannonball so that he will land safely during the act. They do this by firing a 100-pound sack of potatoes out of the cannon at different speeds and then measuring how far from the cannon the sack lands. The results are shown in the table.

Speed in Miles/Hour	Distance in Feet
40	108
50	169
60	243
70	331

The Image Bank/Getty Images

a. Does distance vary directly with the speed, or directly with the square of the speed?

b. Write the equation that describes the relationship between speed and distance.

c. If the cannon will fire a human safely at 55 miles/hour, where should they position the net so the cannonball has a safe landing?

d. How much farther will he land if his speed out of the cannon is 56 miles/hour?

Chapter 3 SUMMARY

EXAMPLES

1. The equation $3x + 2y = 6$ is an example of a linear equation in two variables.

Linear Equations in Two Variables [3.1, 3.3]

A *linear equation in two variables* is any equation that can be put in *standard form* $ax + by = c$. The graph of every linear equation is a straight line.

2. To find the x-intercept for $3x + 2y = 6$, we let $y = 0$ and get

$$3x = 6$$
$$x = 2$$

In this case the x-intercept is 2, and the graph crosses the x-axis at (2, 0).

Intercepts [3.1]

The *x-intercept* of an equation is the *x-coordinate* of the point where the graph crosses the x-axis. The *y-intercept* is the y-coordinate of the point where the graph crosses the y-axis. We find the y-intercept by substituting $x = 0$ into the equation and solving for y. The x-intercept is found by letting $y = 0$ and solving for x.

3. The slope of the line through (6, 9) and (1, −1) is

$$m = \frac{9-(-1)}{6-1} = \frac{10}{5} = 2$$

The Slope of a Line [3.2]

The *slope* of the line containing points (x_1, y_1) and (x_2, y_2) is given by

$$\text{Slope} = m = \frac{\text{Rise}}{\text{Run}} = \frac{y_2 - y_1}{x_2 - x_1}$$

Horizontal lines have 0 slope, and vertical lines have no slope.

Parallel lines have equal slopes, and perpendicular lines have slopes that are negative reciprocals.

4. The equation of the line with slope 5 and y-intercept 3 is

$$y = 5x + 3$$

The Slope Intercept Form of a Line [3.3]

The equation of a line with slope m and y-intercept b is given by

$$y = mx + b$$

5. The equation of the line through (3, 2) with slope −4 is

$$y - 2 = -4(x - 3)$$

which can be simplified to

$$y = -4x + 14$$

The Point Slope Form of a Line [3.3]

The equation of the line through (x_1, y_1) that has slope m can be written as

$$y - y_1 = m(x - x_1)$$

Linear Inequalities in Two Variables [3.4]

6. The graph of

$$x - y \le 3$$

is

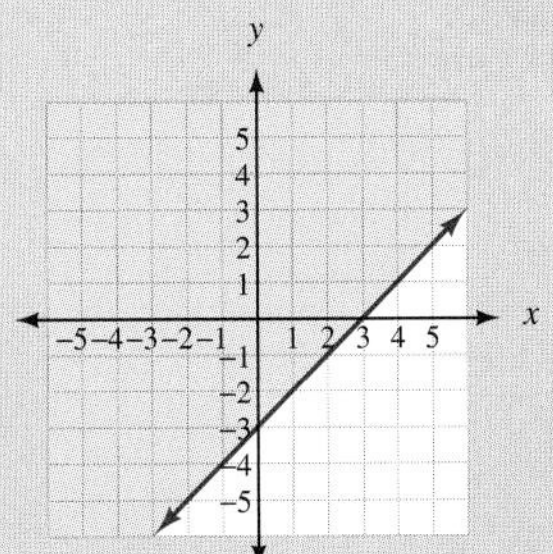

An inequality of the form $ax + by < c$ is a *linear inequality in two variables.* The equation for the boundary of the solution set is given by $ax + by = c$. (This equation is found by simply replacing the inequality symbol with an equal sign.)

To graph a linear inequality, first graph the boundary, using a solid line if the boundary is included in the solution set and a broken line if the boundary is not included in the solution set. Next, choose any point not on the boundary and substitute its coordinates into the original inequality. If the resulting statement is true, the graph lies on the same side of the boundary as the test point. A false statement indicates that the solution set lies on the other side of the boundary.

Relations and Functions [3.5]

7. The relation

$$\{(8, 1), (6, 1), (-3, 0)\}$$

is also a function because no ordered pairs have the same first coordinates. The domain is $\{8, 6, -3\}$ and the range is $\{1, 0\}$.

A *function* is a rule that pairs each element in one set, called the *domain,* with exactly one element from a second set, called the *range.*

A *relation* is any set of ordered pairs. The set of all first coordinates is called the *domain* of the relation, and the set of all second coordinates is the *range* of the relation. A function is a relation in which no two different ordered pairs have the same first coordinates.

Vertical Line Test [3.6]

8. The graph of $x = y^2$ shown in Figure 5 in Section 3.5 fails the vertical line test. It is not the graph of a function.

If a vertical line crosses the graph of a relation in more than one place, the relation cannot be a function. If no vertical line can be found that crosses the graph in more than one place, the relation must be a function.

Function Notation [3.6]

9. If $f(x) = 5x - 3$ then

$$\begin{aligned} f(0) &= 5(0) - 3 \\ &= -3 \\ f(1) &= 5(1) - 3 \\ &= 2 \\ f(-2) &= 5(-2) - 3 \\ &= -13 \\ f(a) &= 5a - 3 \end{aligned}$$

The alternative notation for y is $f(x)$. It is read "f of x" and can be used instead of the variable y when working with functions. The notation y and the notation $f(x)$ are equivalent; that is, $y = f(x)$.

Algebra with Functions [3.7]

10. If $f(x) = 4x$ and $g(x) = x^2 - 3$, then

$(f + g)(x) = x^2 + 4x - 3$

$(f - g)(x) = -x^2 + 4x + 3$

$(fg)(x) = 4x^3 - 12x$

$\frac{f}{g}(x) = \frac{4x}{x^2 - 3}$

If f and g are any two functions with a common domain, then:

$(f + g)(x) = f(x) + g(x)$ The function $f + g$ is the sum of the functions f and g.

$(f - g)(x) = f(x) - g(x)$ The function $f - g$ is the difference of the functions f and g.

$(fg)(x) = f(x)g(x)$ The function fg is the product of the functions f and g.

$\frac{f}{g}(x) = \frac{f(x)}{g(x)}$ The function $\frac{f}{g}$ is the quotient of the functions f and g, where $g(x) \neq 0$

Composition of Functions [3.7]

11. If $f(x) = 4x$ and $g(x) = x^2 - 3$, then

$(f \circ g)(x) = f[g(x)] = f(x^2 - 3)$
$= 4x^2 - 12$

$(g \circ f)(x) = g[f(x)] = g(4x)$
$= 16x^2 - 3$

If f and g are two functions for which the range of each has numbers in common with the domain of the other, then we have the following definitions:

The composition of f with g: $(f \circ g)(x) = f[g(x)]$

The composition of g with f: $(g \circ f)(x) = g[f(x)]$

Variation [3.8]

12. If y varies directly with x, then

$$y = Kx$$

Then if y is 18 when x is 6,

$$18 = K \cdot 6$$

or

$$K = 3$$

So the equation can be written more specifically as

$$y = 3x$$

If we want to know what y is when x is 4, we simply substitute:

$$y = 3 \cdot 4$$
$$y = 12$$

If y *varies directly* with x (y is directly proportional to x), then

$$y = Kx$$

If y *varies inversely* with x (y is inversely proportional to x), then

$$y = \frac{K}{x}$$

If z *varies jointly* with x and y (z is directly proportional to both x and y), then

$$z = Kxy$$

In each case, K is called the *constant of variation.*

COMMON MISTAKES

1. When graphing ordered pairs, the most common mistake is to associate the first coordinate with the y-axis and the second with the x-axis. If you make this mistake you would graph (3, 1) by going up 3 and to the right 1, which is just the reverse of what you should do. Remember, the first coordinate is always associated with the horizontal axis, and the second coordinate is always associated with the vertical axis.
2. The two most common mistakes students make when first working with the formula for the slope of a line are the following:
   a. Putting the difference of the x-coordinates over the difference of the y-coordinates.
   b. Subtracting in one order in the numerator and then subtracting in the opposite order in the denominator.
3. When graphing linear inequalities in two variables, remember to graph the boundary with a broken line when the inequality symbol is $<$ or $>$. The only time you use a solid line for the boundary is when the inequality symbol is $\leq$ or $\geq$.

Chapter 3 Review Test

The problems below form a comprehensive review of the material in this chapter. They can be used to study for exams. If you would like to take a practice test on this chapter, you can use the odd-numbered problems. Give yourself an hour and work as many of the odd-numbered problems as possible. When you are finished, or when an hour has passed, check your answers with the answers in the back of the book. You can use the even-numbered problems for a second practice test.

Graph each line. [3.1]

1. $3x + 2y = 6$ **2.** $y = -\frac{3}{2}x + 1$

3. $x = 3$

Find the slope of the line through the following pairs of points. [3.2]

4. (5, 2), (3, 6) **5.** (−4, 2), (3, 2)

Find x if the line through the two given points has the given slope. [3.2]

6. $(4, x)$, $(1, -3)$; $m = 2$ **7.** $(-4, 7)$, $(2, x)$; $m = -\frac{1}{3}$

8. Find the slope of any line parallel to the line through (3, 8) and (5, −2). [3.2]

9. The line through $(5, 3y)$ and $(2, y)$ is parallel to a line with slope 4. What is the value of y? [3.2]

Give the equation of the line with the following slope and y-intercept. [3.3]

10. $m = 3$, $b = 5$ **11.** $m = -2$, $b = 0$

Give the slope and y-intercept of each equation. [3.3]

12. $3x - y = 6$ **13.** $2x - 3y = 9$

Find the equation of the line that contains the given point and has the given slope. [3.3]

14. (2, 4), $m = 2$ **15.** (−3, 1), $m = -\frac{1}{3}$

Find the equation of the line that contains the given pair of points. [3.3]

16. (2, 5), (−3, −5) **17.** (−3, 7), (4, 7)

18. (−5, −1), (−3, −4)

19. Find the equation of the line that is parallel to $2x - y = 4$ and contains the point (2, −3). [3.3]

20. Find the equation of the line perpendicular to $y = -3x + 1$ that has an x-intercept of 2. [3.3]

Graph each linear inequality. [3.4]

21. $y \leq 2x - 3$ **22.** $x \geq -1$

State the domain and range of each relation, and then indicate which relations are also functions. [3.5]

23. {(2, 4), (3, 3), (4, 2)} **24.** {(6, 3), (−4, 3), (−2, 0)}

If $f = \{(2, -1), (-3, 0), (4, \frac{1}{2}), (\pi, 2)\}$ and $g = \{(2, 2), (-1, 4), (0, 0)\}$, find the following. [3.6]

25. $f(-3)$ **26.** $f(2) + g(2)$

Let $f(x) = 2x^2 - 4x + 1$ and $g(x) = 3x + 2$, and evaluate each of the following. [3.6]

27. $f(0)$ **28.** $g(a)$

29. $f[g(0)]$ **30.** $f[g(1)]$

For the following problems, y varies directly with x. [3.8]

31. If y is 6 when x is 2, find y when x is 8.

32. If y is −3 when x is 5, find y when x is −10.

For the following problems, y varies inversely with the square of x. [3.8]

33. If y is 9 when x is 2, find y when x is 3.

34. If y is 4 when x is 5, find y when x is 2.

Solve each application problem. [3.8]

35. Tension in a Spring The tension t in a spring varies directly with the distance d the spring is stretched. If the tension is 42 pounds when the spring is stretched 2 inches, find the tension when the spring is stretched twice as far.

36. Light Intensity The intensity of a light source varies inversely with the square of the distance from the source. Four feet from the source the intensity is 9 foot-candles. What is the intensity 3 feet from the source?

Chapter 3 Projects

Equations and Inequalities in Two Variables

GROUP PROJECT Light Intensity

Number of People 2–3

Time Needed 15 minutes

Equipment Paper and pencil

Background I found the following diagram while shopping for some track lighting for my home. I was impressed by the diagram because it displays a lot of useful information in a very efficient manner. As the diagram indicates, the amount of light that falls on a surface depends on how far above the surface the light is placed and how much the light spreads out on the surface. Assume that this light illuminates a circle on a flat surface, and work the following problems.

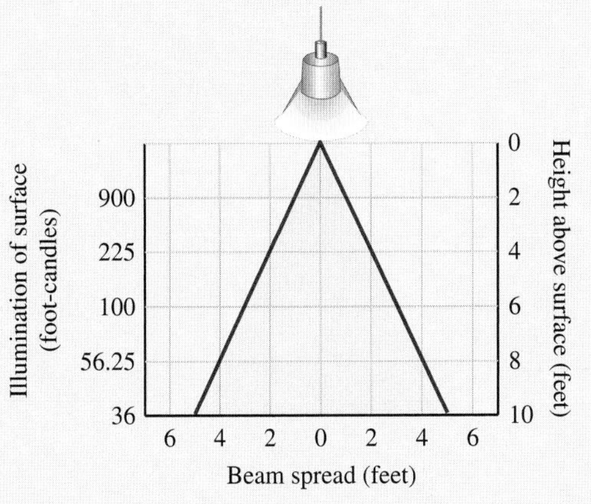

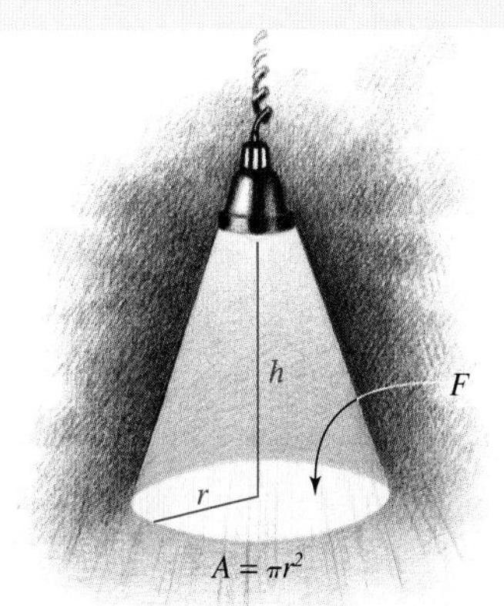

Procedure

a. Fill in each table.

Height Above Surface (ft)	Illumination (foot-candles)
2	
4	
6	
8	
10	

Distance Above Surface (ft)	Area of Illuminated Region (ft²)
2	
4	
6	
8	
10	

b. Construct line graphs from the data in the tables.

c. Which of the relationships is direct variation, and which is inverse variation?

d. Let F represent the number of foot-candles that fall on the surface, h the distance the light source is above the surface, and A the area of the illuminated region. Write an equation that shows the relationship between A and h, then write another equation that gives the relationship between F and h.

RESEARCH PROJECT

Descartes and Pascal

David Eugene Smith Collection/Rare Book and Manuscript Library/Columbia University

René Descartes, 1596–1650

David Eugene Smith Collection/Rare Book and Manuscript Library/Columbia University

Blaise Pascal, 1623–1662

In this chapter, we mentioned that René Descartes, the inventor of the rectangular coordinate system, is the person who made the statement, "I think, therefore, I am." Blaise Pascal, another French philosopher, is responsible for the statement, "The heart has its reasons which reason does not know." Although Pascal and Descartes were contemporaries, the philosophies of the two men differed greatly. Research the philosophy of both Descartes and Pascal, and then write an essay that gives the main points of each man's philosophy. In the essay, show how the quotations given here fit in with the philosophy of the man responsible for the quotation.

Systems of Linear Equations and Inequalities

4

CHAPTER OUTLINE

Suppose you decide to buy a cellular phone and are trying to decide between two rate plans. Plan A is \$18.95 per month plus \$0.48 for each minute, or fraction of a minute, that you use the phone. Plan B is \$34.95 per month plus \$0.36 for each minute, or fraction of a minute. The monthly cost $C(x)$ for each plan can be represented with a linear equation in two variables:

Plan A: $C(x) = 0.48x + 18.95$

Plan B: $C(x) = 0.36x + 34.95$

To compare the two plans, we use the table and graph shown below.

Monthly Cellular Phone Charges

Number of Minutes x	Monthly Cost Plan A (\$)	Monthly Cost Plan B (\$)
0	18.95	34.95
40	38.15	49.35
80	57.35	63.75
120	76.55	78.15
160	95.75	92.55
200	114.95	106.95
240	134.15	121.35

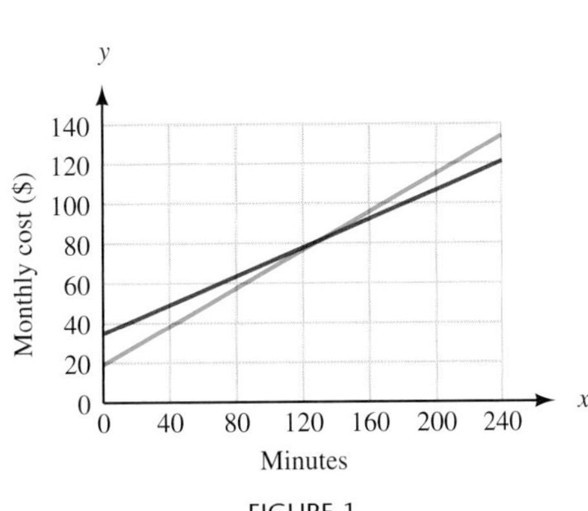

FIGURE 1

The point of intersection of the two lines in Figure 1 is the point at which the monthly costs of the two plans are equal. In this chapter, we will develop methods of finding that point of intersection.

STUDY SKILLS

The study skills for this chapter concern the way you approach new situations in mathematics. The first study skill applies to your natural instincts for what does and doesn't work in mathematics. The second study skill gives you a way of testing your instincts.

1 Don't Let Your Intuition Fool You

As you become more experienced and more successful in mathematics, you will be able to trust your mathematical intuition. For now, though, it can get in the way of success. For example, if you ask a beginning algebra student to "subtract 3 from -5" many will answer -2 or 2. Both answers are incorrect, even though they may seem intuitively true.

2 Test Properties About Which You Are Unsure

From time to time you will be in a situation in which you would like to apply a property or rule, but you are not sure if it is true. You can always test a property or statement by substituting numbers for variables. For instance, I always have students who rewrite $(x + 3)^2$ as $x^2 + 9$, thinking that the two expressions are equivalent. The fact that the two expressions are not equivalent becomes obvious when we substitute 10 for x in each one.

When $x = 10$, the expression $(x + 3)^2$ is $(10 + 3)^2 = 13^2 = 169$

When $x = 10$, the expression $x^2 + 9 = 10^2 + 9 = 100 + 9 = 109$

It is not unusual, nor is it wrong, to try occasionally to apply a property that doesn't exist. If you have any doubt about generalizations you are making, test them by replacing variables with numbers and simplifying.

4.1 Systems of Linear Equations in Two Variables

OBJECTIVES

A Solve systems of linear equations in two variables by graphing.

B Solve systems of linear equations in two variables by the addition method.

C Solve systems of linear equation in two variables by the substitution method.

Previously we found the graph of an equation of the form $ax + by = c$ to be a straight line. Since the graph is a straight line, the equation is said to be a linear equation. Two linear equations considered together form a *linear system* of equations. For example,

$$3x - 2y = 6$$
$$2x + 4y = 20$$

is a linear system. The solution set to the system is the set of all ordered pairs that satisfy both equations. If we graph each equation on the same set of axes, we can see the solution set (Figure 1).

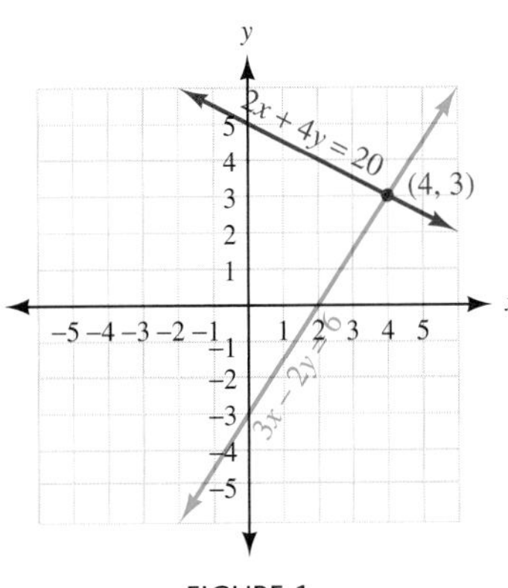

FIGURE 1

The point (4, 3) lies on both lines and therefore must satisfy both equations. It is obvious from the graph that it is the only point that does so. The solution set for the system is {(4, 3)}.

More generally, if $a_1x + b_1y = c_1$ and $a_2x + b_2y = c_2$ are linear equations, then the solution set for the system

$$a_1x + b_1y = c_1$$
$$a_2x + b_2y = c_2$$

can be illustrated through one of the graphs in Figure 2.

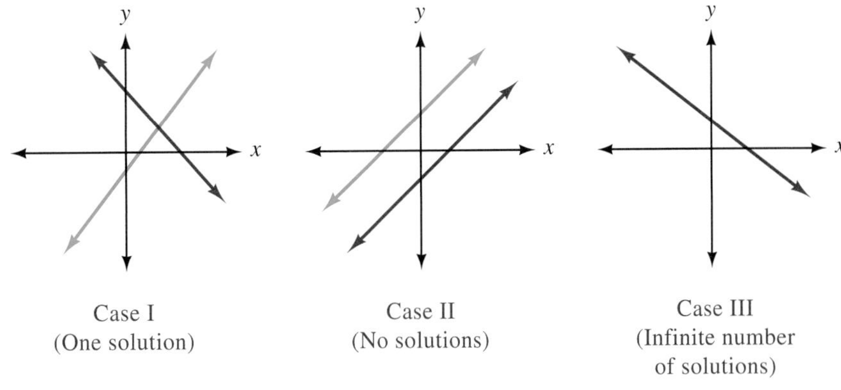

FIGURE 2

Note
It is important that you write solutions to systems of equations in two variables as ordered pairs, like (4, 3); that is, you should always enclose the ordered pairs in parentheses.

Case I The two lines intersect at one and only one point. The coordinates of the point give the solution to the system. This is what usually happens.
Case II The lines are parallel and therefore have no points in common. The solution set to the system is the empty set, ∅. In this case, we say the system is *inconsistent.*
Case III The lines coincide; that is, their graphs represent the same line. The solution set consists of all ordered pairs that satisfy either equation. In this case, the equations are said to be *dependent.*

In the beginning of this section we found the solution set for the system

$$3x - 2y = 6$$
$$2x + 4y = 20$$

by graphing each equation and then reading the solution set from the graph. Solving a system of linear equations by graphing is the least accurate method. If the coordinates of the point of intersection are not integers, it can be very difficult to read the solution set from the graph. There is another method of solving a linear system that does not depend on the graph. It is called the *addition method.*

The Addition Method

EXAMPLE 1

Solve the system.

$$4x + 3y = 10$$
$$2x + y = 4$$

SOLUTION If we multiply the bottom equation by -3, the coefficients of y in the resulting equation and the top equation will be opposites:

$$4x + 3y = 10 \xrightarrow{\text{No change}} 4x + 3y = 10$$
$$2x + y = 4 \xrightarrow[\text{Multiply by } -3]{} -6x - 3y = -12$$

Adding the left and right sides of the resulting equations, we have

$$\begin{aligned} 4x + 3y &= 10 \\ -6x - 3y &= -12 \\ \hline -2x \quad &= -2 \end{aligned}$$

The result is a linear equation in one variable. We have eliminated the variable y from the equations by addition. (It is for this reason we call this method of solving a linear system the *addition method.*) Solving $-2x = -2$ for x, we have

$$x = 1$$

This is the x-coordinate of the solution to our system. To find the y-coordinate, we substitute $x = 1$ into any of the equations containing both the variables x and y. Let's try the second equation in our original system:

$$2(1) + y = 4$$
$$2 + y = 4$$
$$y = 2$$

This is the y-coordinate of the solution to our system. The ordered pair $(1, 2)$ is the solution to the system.

Checking Solutions We can check our solution by substituting it into both of our equations.

Substituting $x = 1$ and $y = 2$ into $4x + 3y = 10$, we have

$$4(1) + 3(2) \stackrel{?}{=} 10$$
$$4 + 6 \stackrel{?}{=} 10$$
$$10 = 10 \quad \textbf{A true statement}$$

Substituting $x = 1$ and $y = 2$ into $2x + y = 4$, we have

$$2(1) + 2 \stackrel{?}{=} 4$$
$$2 + 2 \stackrel{?}{=} 4$$
$$4 = 4 \quad \textbf{A true statement}$$

Our solution satisfies both equations; therefore, it is a solution to our system of equations.

EXAMPLE 2 Solve the system.

$$3x - 5y = -2$$
$$2x - 3y = 1$$

SOLUTION We can eliminate either variable. Let's decide to eliminate the variable x. We can do so by multiplying the top equation by 2 and the bottom equation by -3, and then adding the left and right sides of the resulting equations:

$$\begin{array}{lcr} 3x - 5y = -2 & \xrightarrow{\text{Multiply by 2}} & 6x - 10y = -4 \\ 2x - 3y = 1 & \xrightarrow[\text{Multiply by } -3]{} & -6x + 9y = -3 \\ \hline & & -y = -7 \\ & & y = 7 \end{array}$$

The y-coordinate of the solution to the system is 7. Substituting this value of y into any of the equations with both x- and y-variables gives $x = 11$. The solution to the system is $(11, 7)$. It is the only ordered pair that satisfies both equations.

Checking Solutions Checking $(11, 7)$ in each equation looks like this

Substituting $x = 11$ and $y = 7$ into $3x - 5y = -2$, we have

$$3(11) - 5(7) \stackrel{?}{=} -2$$
$$33 - 35 \stackrel{?}{=} -2$$
$$-2 = -2 \quad \textbf{A true statement}$$

Substituting $x = 11$ and $y = 7$ into $2x - 3y = 1$, we have

$$2(11) - 3(7) \stackrel{?}{=} 1$$
$$22 - 21 \stackrel{?}{=} 1$$
$$1 = 1 \quad \textbf{A true statement}$$

Our solution satisfies both equations; therefore, $(11, 7)$ is a solution to our system.

EXAMPLE 3 Solve the system.

$$2x - 3y = 4$$
$$4x + 5y = 3$$

SOLUTION We can eliminate x by multiplying the top equation by -2 and adding it to the bottom equation:

$$\begin{array}{lcr} 2x - 3y = 4 & \xrightarrow{\text{Multiply by } -2} & -4x + 6y = -8 \\ 4x + 5y = 3 & \xrightarrow[\text{No change}]{} & 4x + 5y = 3 \\ \hline & & 11y = -5 \\ & & y = -\dfrac{5}{11} \end{array}$$

The y-coordinate of our solution is $-\frac{5}{11}$. If we were to substitute this value of y back into either of our original equations, we would find the arithmetic necessary to solve for x cumbersome. For this reason, it is probably best to go back to the original system and solve it a second time—for x instead of y. Here is how we do that:

$$\begin{aligned} 2x - 3y = 4 &\xrightarrow{\text{Multiply by 5}} 10x - 15y = 20 \\ 4x + 5y = 3 &\xrightarrow[\text{Multiply by 3}]{} \underline{12x + 15y = 9} \\ &\qquad\qquad 22x \qquad\; = 29 \\ &\qquad\qquad\qquad\quad x = \frac{29}{22} \end{aligned}$$

The solution to our system is $\left(\frac{29}{22}, -\frac{5}{11}\right)$.

The main idea in solving a system of linear equations by the addition method is to use the multiplication property of equality on one or both of the original equations, if necessary, to make the coefficients of either variable opposites. The following box shows some steps to follow when solving a system of linear equations by the addition method.

Strategy for Solving a System of Linear Equations by the Addition Method

Step 1: Decide which variable to eliminate. (In some cases, one variable will be easier to eliminate than the other. With some practice, you will notice which one it is.)

Step 2: Use the multiplication property of equality on each equation separately to make the coefficients of the variable that is to be eliminated opposites.

Step 3: Add the respective left and right sides of the system together.

Step 4: Solve for the remaining variable.

Step 5: Substitute the value of the variable from step 4 into an equation containing both variables and solve for the other variable. (Or repeat steps 2–4 to eliminate the other variable.)

Step 6: Check your solution in both equations, if necessary.

EXAMPLE 4

Solve the system.

$$\begin{aligned} 5x - 2y &= 5 \\ -10x + 4y &= 15 \end{aligned}$$

SOLUTION We can eliminate y by multiplying the first equation by 2 and adding the result to the second equation:

$$\begin{aligned} 5x - 2y = 5 &\xrightarrow{\text{Multiply by 2}} 10x - 4y = 10 \\ -10x + 4y = 15 &\xrightarrow[\text{No change}]{} \underline{-10x + 4y = 15} \\ &\qquad\qquad\qquad 0 = 25 \end{aligned}$$

The result is the false statement $0 = 25$, which indicates there is no solution to the system. If we were to graph the two lines, we would find that they are parallel. In a case like this, we say the system is *inconsistent.* Whenever both variables have been eliminated and the resulting statement is false, the solution set for the system will be the empty set, $\varnothing$.

EXAMPLE 5 Solve the system.

$$4x + 3y = 2$$
$$8x + 6y = 4$$

SOLUTION Multiplying the top equation by -2 and adding, we can eliminate the variable x:

$$\begin{array}{rcr} 4x + 3y = 2 & \xrightarrow{\text{Multiply by } -2} & -8x - 6y = -4 \\ 8x + 6y = 4 & \xrightarrow[\text{No change}]{} & 8x + 6y = 4 \\ \hline & & 0 = 0 \end{array}$$

Both variables have been eliminated and the resulting statement $0 = 0$ is true. In this case, the lines coincide and the equations are said to be *dependent*. The solution set consists of all ordered pairs that satisfy either equation. We can write the solution set as $\{(x, y) \mid 4x + 3y = 2\}$ or $\{(x, y) \mid 8x + 6y = 4\}$.

Special Cases

The previous two examples illustrate the two special cases in which the graphs of the equations in the system either coincide or are parallel. In both cases the left-hand sides of the equations were multiples of each other. In the case of the dependent equations the right-hand sides were also multiples. We can generalize these observations for the system

$$a_1x + b_1y = c_1$$
$$a_2x + b_2y = c_2$$

Inconsistent System

What Happens	*Geometric Interpretation*	*Algebraic Interpretation*
Both variables are eliminated, and the resulting statement is false.	The lines are parallel, and there is no solution to the system.	$\frac{a_1}{a_2} = \frac{b_1}{b_2} \neq \frac{c_1}{c_2}$

Dependent Equations

What Happens	*Geometric Interpretation*	*Algebraic Interpretation*
Both variables are eliminated, and the resulting statement is true.	The lines coincide, and there are an infinite number of solutions to the system.	$\frac{a_1}{a_2} = \frac{b_1}{b_2} = \frac{c_1}{c_2}$

EXAMPLE 6 Solve the system.

$$\frac{1}{2}x - \frac{1}{3}y = 2$$

$$\frac{1}{4}x + \frac{2}{3}y = 6$$

SOLUTION Although we could solve this system without clearing the equations of fractions, there is probably less chance for error if we have only integer coefficients to work with. So let's begin by multiplying both sides of the top equation by 6 and both sides of the bottom equation by 12 to clear each equation of fractions:

$$\frac{1}{2}x - \frac{1}{3}y = 2 \xrightarrow{\text{Multiply by 6}} 3x - 2y = 12$$

$$\frac{1}{4}x + \frac{2}{3}y = 6 \xrightarrow[\text{Multiply by 12}]{} 3x + 8y = 72$$

Now we can eliminate x by multiplying the top equation by -1 and leaving the bottom equation unchanged:

$$\begin{aligned} 3x - 2y = 12 &\xrightarrow{\text{Multiply by } -1} & -3x + 2y &= -12 \\ 3x + 8y = 72 &\xrightarrow[\text{No change}]{} & 3x + 8y &= 72 \\ & & \hline 10y &= 60 \\ & & y &= 6 \end{aligned}$$

We can substitute $y = 6$ into any equation that contains both x and y. Let's use $3x - 2y = 12$.

$$3x - 2(6) = 12$$

$$3x - 12 = 12$$

$$3x = 24$$

$$x = 8$$

The solution to the system is $(8, 6)$. 

The Substitution Method

EXAMPLE 7

Solve the system.

$$2x - 3y = -6$$

$$y = 3x - 5$$

SOLUTION The second equation tells us y is $3x - 5$. Substituting the expression $3x - 5$ for y in the first equation, we have

$$2x - 3(3x - 5) = -6$$

The result of the substitution is the elimination of the variable y. Solving the resulting linear equation in x as usual, we have

$$2x - 9x + 15 = -6$$

$$-7x + 15 = -6$$

$$-7x = -21$$

$$x = 3$$

Putting $x = 3$ into the second equation in the original system, we have

$$\begin{aligned} y &= 3(3) - 5 \\ &= 9 - 5 \\ &= 4 \end{aligned}$$

The solution to the system is (3, 4).

Checking Solutions Checking (3, 4) in each equation looks like this

Substituting $x = 3$ and $y = 4$ into $2x - 3y = -6$, we have

$$2(3) - 3(4) \stackrel{?}{=} -6$$
$$6 - 12 \stackrel{?}{=} -6$$
$$-6 = -6 \quad \textbf{A true statement}$$

Substituting $x = 3$ and $y = 4$ into $y = 3x - 5$, we have

$$4 \stackrel{?}{=} 3(3) - 5$$
$$4 \stackrel{?}{=} 9 - 5$$
$$4 = 4 \quad \textbf{A true statement}$$

Our solution satisfies both equations; therefore, (3, 4) is a solution to our system.

Strategy for Solving a System of Equations by the Substitution Method

Step 1: Solve either one of the equations for x or y. (This step is not necessary if one of the equations is already in the correct form, as in Example 7.)

Step 2: Substitute the expression for the variable obtained in step 1 into the other equation and solve it.

Step 3: Substitute the solution for step 2 into any equation in the system that contains both variables and solve it.

Step 4: Check your results, if necessary.

EXAMPLE 8

Solve by substitution.

$$2x + 3y = 5$$
$$x - 2y = 6$$

SOLUTION To use the substitution method, we must solve one of the two equations for x or y. We can solve for x in the second equation by adding $2y$ to both sides:

$$x - 2y = 6$$
$$x = 2y + 6 \quad \textbf{Add } 2y \textbf{ to both sides.}$$

Substituting the expression $2y + 6$ for x in the first equation of our system, we have

$$\begin{aligned} 2(2y + 6) + 3y &= 5 \\ 4y + 12 + 3y &= 5 \\ 7y + 12 &= 5 \\ 7y &= -7 \\ y &= -1 \end{aligned}$$

Using $y = -1$ in either equation in the original system, we get $x = 4$. The solution is $(4, -1)$.

Note

Both the substitution method and the addition method can be used to solve any system of linear equations in two variables. Systems like the one in Example 7, however, are easier to solve using the substitution method because one of the variables is already written in terms of the other. A system like the one in Example 2 is easier to solve using the addition method because solving for one of the variables would lead to an expression involving fractions. The system in Example 8 could be solved easily by either method because solving the second equation for x is a one-step process.

USING TECHNOLOGY

Graphing Calculators

Solving Systems That Intersect in Exactly One Point

A graphing calculator can be used to solve a system of equations in two variables if the equations intersect in exactly one point. To solve the system shown in Example 3, we first solve each equation for y. Here is the result:

$$2x - 3y = 4 \quad \text{becomes} \quad y = \frac{4 - 2x}{-3}$$

$$4x + 5y = 3 \quad \text{becomes} \quad y = \frac{3 - 4x}{5}$$

Graphing these two functions on the calculator gives a diagram similar to the one in Figure 6.

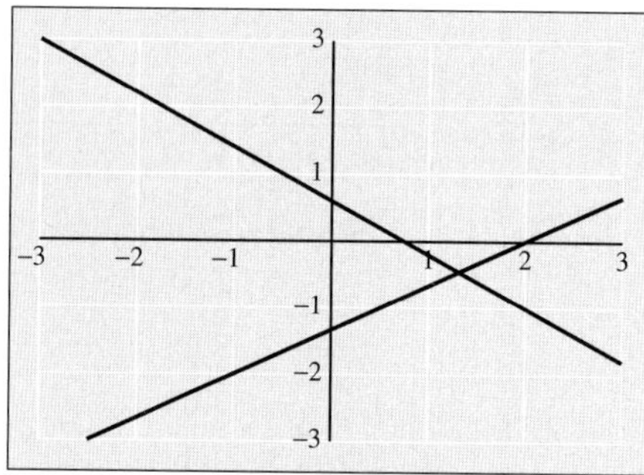

FIGURE 6

Using the Trace and Zoom features, we find that the two lines intersect at $x = 1.32$ and $y = -0.45$, which are the decimal equivalents (accurate to the nearest hundredth) of the fractions found in Example 3.

GETTING READY FOR CLASS

After reading through the preceding section, respond in your own words and in complete sentences.

1. Two linear equations, each with the same two variables, form a system of equations. How do we define a solution to this system? That is, what form will a solution have, and what properties does a solution possess?
2. When would substitution be more efficient than the addition method in solving two linear equations?
3. Explain what an inconsistent system of linear equations looks like graphically and what would result algebraically when attempting to solve the system.
4. When might the graphing method of solving a system of equations be more desirable than the other techniques, and when might it be less desirable?

LINKING OBJECTIVES AND EXAMPLES

Next to each **objective** we have listed the examples that are best described by that objective.

A	None
B	1–6
C	7, 8

Problem Set 4.1

Online support materials can be found at www.thomsonedu.com/login

Solve each system by graphing both equations on the same set of axes and then reading the solution from the graph.

1. $3x - 2y = 6$, $x - y = 1$
2. $5x - 2y = 10$, $x - y = -1$
3. $y = \frac{3}{5}x - 3$, $2x - y = -4$
4. $y = \frac{1}{2}x - 2$, $2x - y = -1$
5. $y = \frac{1}{2}x$, $y = -\frac{3}{4}x + 5$
6. $y = \frac{2}{3}x$, $y = -\frac{1}{3}x + 6$
7. $3x + 3y = -2$, $y = -x + 4$
8. $2x - 2y = 6$, $y = x - 3$
9. $2x - y = 5$, $y = 2x - 5$
10. $x + 2y = 5$, $y = -\frac{1}{2}x + 3$

Solve each of the following systems by the addition method.

11. $x + y = 5$, $3x - y = 3$
12. $x - y = 4$, $-x + 2y = -3$
13. $3x + y = 4$, $4x + y = 5$
14. $6x - 2y = -10$, $6x + 3y = -15$
15. $3x - 2y = 6$, $6x - 4y = 12$
16. $4x + 5y = -3$, $-8x - 10y = 3$
17. $x + 2y = 0$, $2x - 6y = 5$
18. $x + 3y = 3$, $2x - 9y = 1$
19. $2x - 5y = 16$, $4x - 3y = 11$
20. $5x - 3y = -11$, $7x + 6y = -12$
21. $6x + 3y = -1$, $9x + 5y = 1$
22. $5x + 4y = -1$, $7x + 6y = -2$
23. $4x + 3y = 14$, $9x - 2y = 14$
24. $7x - 6y = 13$, $6x - 5y = 11$
25. $2x - 5y = 3$, $-4x + 10y = 3$
26. $3x - 2y = 1$, $-6x + 4y = -2$
27. $\frac{1}{4}x - \frac{1}{6}y = -2$, $-\frac{1}{6}x + \frac{1}{5}y = 4$
28. $-\frac{1}{3}x + \frac{1}{4}y = 0$, $\frac{1}{5}x - \frac{1}{10}y = 1$
29. $\frac{1}{2}x + \frac{1}{3}y = 13$, $\frac{2}{5}x + \frac{1}{4}y = 10$
30. $\frac{1}{2}x + \frac{1}{3}y = \frac{2}{3}$, $\frac{2}{3}x + \frac{2}{5}y = \frac{14}{15}$

Solve each of the following systems by the substitution method.

31. $7x - y = 24$, $x = 2y + 9$
32. $3x - y = -8$, $y = 6x + 3$
33. $6x - y = 10$, $y = -\frac{3}{4}x - 1$
34. $2x - y = 6$, $y = -\frac{4}{3}x + 1$
35. $3y + 4z = 23$, $6y + z = 32$
36. $2x - y = 650$, $3.5x - y = 1{,}400$
37. $y = 3x - 2$, $y = 4x - 4$
38. $y = 5x - 2$, $y = -2x + 5$
39. $2x - y = 5$, $4x - 2y = 10$
40. $-10x + 8y = -6$, $y = \frac{5}{4}x$
41. $\frac{1}{3}x - \frac{1}{2}y = 0$, $x = \frac{3}{2}y$
42. $\frac{2}{5}x - \frac{2}{3}y = 0$, $y = \frac{3}{5}x$

You may want to read Example 3 again before solving the systems that follow.

43. $4x - 7y = 3$, $5x + 2y = -3$
44. $3x - 4y = 7$, $6x - 3y = 5$
45. $9x - 8y = 4$, $2x + 3y = 6$
46. $4x - 7y = 10$, $-3x + 2y = -9$
47. $3x - 5y = 2$, $7x + 2y = 1$
48. $4x - 3y = -1$, $5x + 8y = 2$

Solve each of the following systems by using either the addition or substitution method. Choose the method that is most appropriate for the problem.

▸ 49. $x - 3y = 7$, $2x + y = -6$
▸ 50. $2x - y = 9$, $x + 2y = -11$
▸ 51. $y = \frac{1}{2}x + \frac{1}{3}$, $y = -\frac{1}{3}x + 2$
▸ 52. $y = \frac{3}{4}x - \frac{4}{5}$, $y = \frac{1}{2}x - \frac{1}{2}$
▸ 53. $3x - 4y = 12$, $x = \frac{2}{3}y - 4$
▸ 54. $-5x + 3y = -15$, $x = \frac{4}{5}y - 2$
55. $4x - 3y = -7$, $-8x + 6y = -11$
56. $3x - 4y = 8$, $y = \frac{3}{4}x - 2$

▸ **57.** $3y + z = 17$
$5y + 20z = 65$

▸ **58.** $x + y = 850$
$1.5x + y = 1{,}100$

▸ **59.** $\frac{3}{4}x - \frac{1}{3}y = 1$
$y = \frac{1}{4}x$

▸ **60.** $-\frac{2}{3}x + \frac{1}{2}y = -1$
$y = -\frac{1}{3}x$

▸ **61.** $\frac{1}{4}x - \frac{1}{2}y = \frac{1}{3}$
$\frac{1}{3}x - \frac{1}{4}y = -\frac{2}{3}$

▸ **62.** $\frac{1}{5}x - \frac{1}{10}y = -\frac{1}{5}$
$\frac{2}{3}x - \frac{1}{2}y = -\frac{1}{6}$

The next two problems are intended to give you practice reading, and paying attention to, the instructions that accompany the problems you are working.

63. Work each problem according to the instructions given.
a. Simplify: $(3x - 4y) - 3(x - y)$
b. Find y when x is 0 in $3x - 4y = 8$.
c. Find the y-intercept: $3x - 4y = 8$
d. Graph: $3x - 4y = 8$
e. Find the point where the graphs of $3x - 4y = 8$ and $x - y = 2$ cross.

64. Work each problem according to the instructions given.
a. Solve: $4x - 5 = 20$
b. Solve for y: $4x - 5y = 20$
c. Solve for x: $x - y = 5$
d. Solve the system: $4x - 5y = 20$
$x - y = 5$

65. Multiply both sides of the second equation in the following system by 100, and then solve as usual.
$$x + y = 10{,}000$$
$$0.06x + 0.05y = 560$$

66. Multiply both sides of the second equation in the following system by 10, and then solve as usual.
$$x + y = 12$$
$$0.20x + 0.50y = 0.30(12)$$

67. What value of c will make the following system a dependent system (one in which the lines coincide)?
$$6x - 9y = 3$$
$$4x - 6y = c$$

68. What value of c will make the following system a dependent system?
$$5x - 7y = c$$
$$-15x + 21y = 9$$

69. Where do the graphs of the lines $x + y = 4$ and $x - 2y = 4$ intersect?

70. Where do the graphs of the line $x = -1$ and $x - 2y = 4$ intersect?

Maintaining Your Skills

71. Find the slope of the line that contains $(-4, -1)$ and $(-2, 5)$.

72. A line has a slope of $\frac{2}{3}$. Find the slope of any line
a. Parallel to it.
b. Perpendicular to it.

73. Give the slope and y-intercept of the line $2x - 3y = 6$.

74. Give the equation of the line with slope -3 and y-intercept 5.

75. Find the equation of the line with slope $\frac{2}{3}$ that contains the point $(-6, 2)$.

76. Find the equation of the line through $(1, 3)$ and $(-1, -5)$.

77. Find the equation of the line with x-intercept 3 and y-intercept -2.

78. Find the equation of the line through $(-1, 4)$ whose graph is perpendicular to the graph of $y = 2x + 3$.

Getting Ready for the Next Section

Simplify.

79. $2 - 2(6)$

80. $2(1) - 2 + 3$

81. $(x + 3y) - 1(x - 2z)$

82. $(x + y + z) + (2x - y + z)$

Solve.

83. $-9y = -9$

84. $30x = 38$

85. $3(1) + 2z = 9$

86. $4\left(\frac{19}{15}\right) - 2y = 4$

Apply the distributive property, then simplify if possible.

87. $2(5x - z)$

88. $-1(x - 2z)$

89. $3(3x + y - 2z)$

90. $2(2x - y + z)$

Extending the Concepts

91. Find a and b so that the line $ax + by = 7$ passes through the points $(1, -2)$ and $(3, 1)$.

92. Find a and b so that the line $ax + by = 2$ passes through the points $(2, 2)$ and $(6, 7)$.

93. The height of an object thrown in the air is given by the function $h(t) = at^2 + bt$, where t is measured in seconds and $h(t)$ is measured in feet. After 3 seconds, the object reaches a height of 24 feet; after 6 seconds, it hits the ground. Find the values of a and b.

94. Improving Your Quantitative Literacy The graph here shows the percentage of alcohol- and nonalcohol-related car crash fatalities for the years 1982 through 2004. Although these graphs are not linear, they do intersect, and so we can apply some of the same reasoning we have used in this section to interpret what the intersections mean.

a. How many points of intersection are there on the chart?

b. During what period of years do the graphs intersect?

c. Explain the significance of the points of intersection in terms of car crash fatalities.

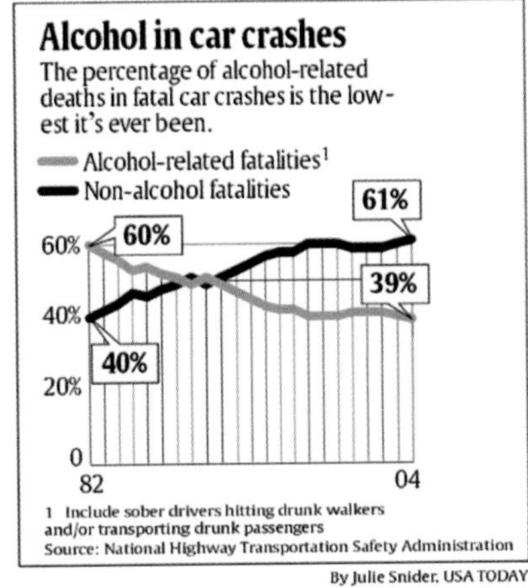

From *USA Today*. Copyright 2005. Reprinted with permission.

4.2 Systems of Linear Equations in Three Variables

OBJECTIVES

A Solve systems of linear equations in three variables.

A solution to an equation in three variables such as

$$2x + y - 3z = 6$$

Is an ordered triple of numbers (x, y, z). For example, the ordered triples $(0, 0, -2)$, $(2, 2, 0)$, and $(0, 9, 1)$ are solutions to the equation $2x + y - 3z = 6$ since they produce a true statement when their coordinates are substituted for x, y, and z in the equation.

DEFINITION The **solution set** for a system of three linear equations in three variables is the set of ordered triples that satisfy all three equations.

EXAMPLE 1 Solve the system.

$$x + y + z = 6 \quad (1)$$
$$2x - y + z = 3 \quad (2)$$
$$x + 2y - 3z = -4 \quad (3)$$

SOLUTION We want to find the ordered triple (x, y, z) that satisfies all three equations. We have numbered the equations so it will be easier to keep track of where they are and what we are doing.

There are many ways to proceed. The main idea is to take two different pairs of equations and eliminate the same variable from each pair. We begin by adding equations (1) and (2) to eliminate the y-variable. The resulting equation is numbered (4):

$$x + y + z = 6 \quad \textbf{(1)}$$
$$2x - y + z = 3 \quad \textbf{(2)}$$
$$3x + 2z = 9 \quad \textbf{(4)}$$

Adding twice equation (2) to equation (3) will also eliminate the variable y. The resulting equation is numbered (5):

$$4x - 2y + 2z = 6 \quad \textbf{Twice (2)}$$
$$x + 2y - 3z = -4 \quad \textbf{(3)}$$
$$5x - z = 2 \quad \textbf{(5)}$$

Equations (4) and (5) form a linear system in two variables. By multiplying equation (5) by 2 and adding the result to equation (4), we succeed in eliminating the variable z from the new pair of equations:

$$3x + 2z = 9 \quad \textbf{(4)}$$
$$10x - 2z = 4 \quad \textbf{Twice (5)}$$
$$13x = 13$$
$$x = 1$$

Substituting $x = 1$ into equation (4), we have

$$3(1) + 2z = 9$$
$$2z = 6$$
$$z = 3$$

Using $x = 1$ and $z = 3$ in equation (1) gives us

$$1 + y + 3 = 6$$
$$y + 4 = 6$$
$$y = 2$$

The solution is the ordered triple $(1, 2, 3)$. 

EXAMPLE 2

Solve the system.

$$2x + y - z = 3 \quad \textbf{(1)}$$
$$3x + 4y + z = 6 \quad \textbf{(2)}$$
$$2x - 3y + z = 1 \quad \textbf{(3)}$$

SOLUTION It is easiest to eliminate z from the equations. The equation produced by adding (1) and (2) is

$$5x + 5y = 9 \quad \textbf{(4)}$$

The equation that results from adding (1) and (3) is

$$4x - 2y = 4 \qquad \textbf{(5)}$$

Equations (4) and (5) form a linear system in two variables. We can eliminate the variable y from this system as follows:

$$5x + 5y = 9 \xrightarrow{\text{Multiply by 2}} 10x + 10y = 18$$

$$4x - 2y = 4 \xrightarrow[\text{Multiply by 5}]{} \underline{20x - 10y = 20}$$

$$30x \qquad = 38$$

$$x = \frac{38}{30}$$

$$= \frac{19}{15}$$

Substituting $x = \frac{19}{15}$ into equation (5) or equation (4) and solving for y gives

$$y = \frac{8}{15}$$

Using $x = \frac{19}{15}$ and $y = \frac{8}{15}$ in equation (1), (2), or (3) and solving for z results in

$$z = \frac{1}{15}$$

The ordered triple that satisfies all three equations is $(\frac{19}{15}, \frac{8}{15}, \frac{1}{15})$.

EXAMPLE 3

Solve the system.

$$2x + 3y - z = 5 \qquad \textbf{(1)}$$
$$4x + 6y - 2z = 10 \qquad \textbf{(2)}$$
$$x - 4y + 3z = 5 \qquad \textbf{(3)}$$

SOLUTION Multiplying equation (1) by -2 and adding the result to equation (2) looks like this:

$$-4x - 6y + 2z = -10 \qquad \textbf{−2 times (1)}$$
$$\underline{4x + 6y - 2z = 10} \qquad \textbf{(2)}$$
$$0 = 0$$

> **Note**
> On the next page, you will find a discussion of the geometric interpretations associated with systems of equations in three variables.

All three variables have been eliminated, and we are left with a true statement. As was the case in the previous section, this implies that the two equations are dependent. With a system of three equations in three variables, however, a system such as this one can have no solution or an infinite number of solutions. In either case, we have no unique solution, meaning there is no single ordered triple that is the only solution to the system.

EXAMPLE 4

Solve the system.

$$x - 5y + 4z = 8 \qquad \textbf{(1)}$$
$$3x + y - 2z = 7 \qquad \textbf{(2)}$$
$$-9x - 3y + 6z = 5 \qquad \textbf{(3)}$$

SOLUTION Multiplying equation (2) by 3 and adding the result to equation (3) produces

$$\begin{array}{rl} 9x + 3y - 6z = 21 & \textbf{3 times (2)} \\ -9x - 3y + 6z = 5 & \textbf{(3)} \\ \hline 0 = 26 & \end{array}$$

In this case all three variables have been eliminated, and we are left with a false statement. The system is inconsistent: there are no ordered triples that satisfy both equations. The solution set for the system is the empty set, ∅. If equations (2) and (3) have no ordered triples in common, then certainly (1), (2), and (3) do not either.

EXAMPLE 5

Solve the system.

$$\begin{array}{rl} x + 3y = 5 & \textbf{(1)} \\ 6y + z = 12 & \textbf{(2)} \\ x - 2z = -10 & \textbf{(3)} \end{array}$$

SOLUTION It may be helpful to rewrite the system as

$$\begin{array}{rl} x + 3y \qquad = 5 & \textbf{(1)} \\ 6y + z = 12 & \textbf{(2)} \\ x - \qquad 2z = -10 & \textbf{(3)} \end{array}$$

Equation (2) does not contain the variable x. If we multiply equation (3) by -1 and add the result to equation (1), we will be left with another equation that does not contain the variable x:

$$\begin{array}{rl} x + 3y \qquad = 5 & \textbf{(1)} \\ -x \qquad + 2z = 10 & \textbf{−1 times (3)} \\ \hline 3y + 2z = 15 & \textbf{(4)} \end{array}$$

Equations (2) and (4) form a linear system in two variables. Multiplying equation (2) by -2 and adding the result to equation (4) eliminates the variable z:

$$\begin{array}{rcr} 6y + z = 12 & \xrightarrow{\text{Multiply by } -2} & -12y - 2z = -24 \\ 3y + 2z = 15 & \xrightarrow[\text{No change}]{} & 3y + 2z = 15 \\ \cline{3-3} & & -9y \qquad = -9 \\ & & y = 1 \end{array}$$

Using $y = 1$ in equation (4) and solving for z, we have

$$z = 6$$

Substituting $y = 1$ into equation (1) gives

$$x = 2$$

The ordered triple that satisfies all three equations is (2, 1, 6).

The Geometry Behind Equations in Three Variables

We can graph an ordered triple on a coordinate system with three axes. The graph will be a point in space. The coordinate system is drawn in perspective; you have to imagine that the x-axis comes out of the paper and is perpendicular to both the y-axis and the z-axis. To graph the point (3, 4, 5), we move 3 units in the x-direction, 4 units in the y-direction, and then 5 units in the z-direction, as shown in Figure 1.

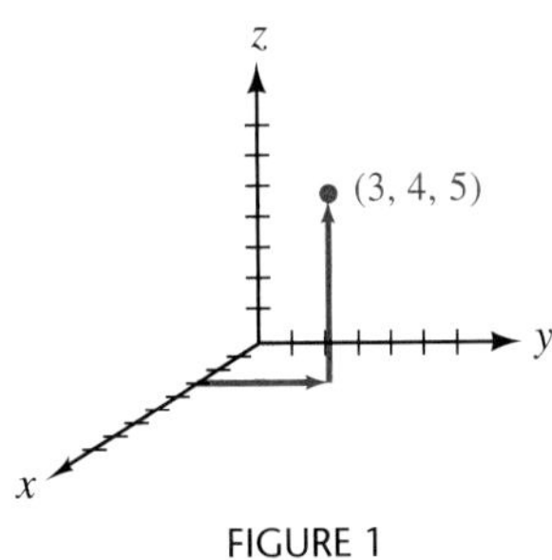

FIGURE 1

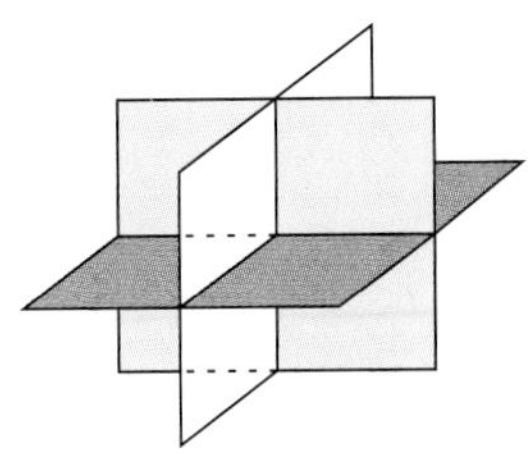

Case 1 The three planes have exactly one point in common. In this case we get one solution to our system, as in Examples 1, 2, and 5.

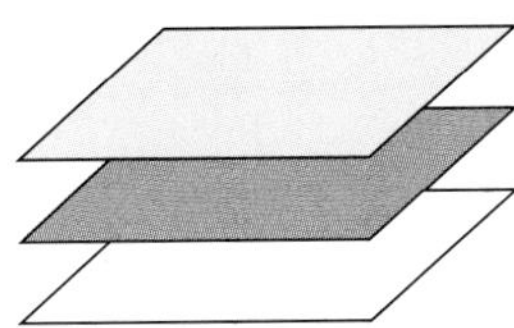

Case 2 The three planes have no points in common because they are all parallel to each other. The system they represent is an inconsistent system.

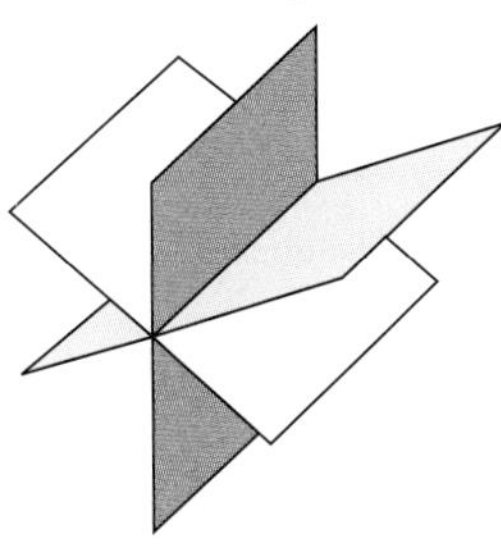

Case 3 The three planes intersect in a line. Any point on the line is a solution to the system.

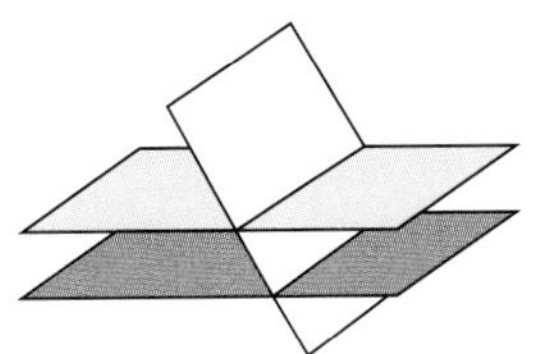

Case 4 In this case the three planes have no points in common. There is no solution to the system; it is an inconsistent system.

Although in actual practice it is sometimes difficult to graph equations in three variables, if we have to graph a linear equation in three variables, we would find that the graph was a plane in space. A system of three equations in three variables is represented by three planes in space.

There are a number of possible ways in which these three planes can intersect, some of which are shown in the margin on this page. There are still other possibilities that are not among those shown in the margin.

In Example 3 we found that equations 1 and 2 were dependent equations. They represent the same plane; that is, they have all their points in common. But the system of equations that they came from has either no solution or an infinite number of solutions. It all depends on the third plane. If the third plane coincides with the first two, then the solution to the system is a plane. If the third plane is distinct from but parallel to the first two, then there is no solution to the system. And, finally, if the third plane intersects the first two, but does not coincide with them, then the solution to the system is that line of intersection.

In Example 4 we found that trying to eliminate a variable from the second and third equations resulted in a false statement. This means that the two planes represented by these equations are parallel. It makes no difference where the third plane is; there is no solution to the system in Example 4. (If we were to graph the three planes from Example 4, we would obtain a diagram similar to Case 4 in the margin.)

If, in the process of solving a system of linear equations in three variables, we eliminate all the variables from a pair of equations and are left with a false statement, we will say the system is inconsistent. If we eliminate all the variables and are left with a true statement, then we will say the system has no unique solution.

LINKING OBJECTIVES AND EXAMPLES

Next to each **objective** we have listed the examples that are best described by that objective.

A 1–5

GETTING READY FOR CLASS

After reading through the preceding section, respond in your own words and in complete sentences.

1. What is an ordered triple of numbers?
2. Explain what it means for (1, 2, 3) to be a solution to a system of linear equations in three variables.
3. Explain in a general way the procedure you would use to solve a system of three linear equations in three variables.
4. How do you know when a system of linear equations in three variables has no solution?

Problem Set 4.2

Online support materials can be found at www.thomsonedu.com/login

Solve the following systems.

1. $x + y + z = 4$
   $x - y + 2z = 1$
   $x - y - 3z = -4$

2. $x - y - 2z = -1$
   $x + y + z = 6$
   $x + y - z = 4$

3. $x + y + z = 6$
   $x - y + 2z = 7$
   $2x - y - 4z = -9$

4. $x + y + z = 0$
   $x + y - z = 6$
   $x - y + 2z = -7$

5. $x + 2y + z = 3$
   $2x - y + 2z = 6$
   $3x + y - z = 5$

6. $2x + y - 3z = -14$
   $x - 3y + 4z = 22$
   $3x + 2y + z = 0$

7. $2x + 3y - 2z = 4$
   $x + 3y - 3z = 4$
   $3x - 6y + z = -3$

8. $4x + y - 2z = 0$
   $2x - 3y + 3z = 9$
   $-6x - 2y + z = 0$

9. $-x + 4y - 3z = 2$
   $2x - 8y + 6z = 1$
   $3x - y + z = 0$

10. $4x + 6y - 8z = 1$
    $-6x - 9y + 12z = 0$
    $x - 2y - 2z = 3$

11. $\frac{1}{2}x - y + z = 0$
    $2x + \frac{1}{3}y + z = 2$
    $x + y + z = -4$

12. $\frac{1}{3}x + \frac{1}{2}y + z = -1$
    $x - y + \frac{1}{5}z = 1$
    $x + y + z = 5$

13. $2x - y - 3z = 1$
    $x + 2y + 4z = 3$
    $4x - 2y - 6z = 2$

14. $3x + 2y + z = 3$
    $x - 3y + z = 4$
    $-6x - 4y - 2z = 1$

15. $2x - y + 3z = 4$
    $x + 2y - z = -3$
    $4x + 3y + 2z = -5$

16. $6x - 2y + z = 5$
    $3x + y + 3z = 7$
    $x + 4y - z = 4$

17. $x + y = 9$
    $y + z = 7$
    $x - z = 2$

18. $x - y = -3$
    $x + z = 2$
    $y - z = 7$

19. $2x + y = 2$
    $y + z = 3$
    $4x - z = 0$

20. $2x + y = 6$
    $3y - 2z = -8$
    $x + z = 5$

21. $2x - 3y = 0$
    $6y - 4z = 1$
    $x + 2z = 1$

22. $3x + 2y = 3$
    $y + 2z = 2$
    $6x - 4z = 1$

23. $x + y - z = 2$
    $2x + y + 3z = 4$
    $x - 2y + 2z = 6$

24. $x + 2y - 2z = 4$
    $3x + 4y - z = -2$
    $2x + 3y - 3z = -5$

▸ 25. $2x + 3y = -\frac{1}{2}$
    $4x + 8z = 2$
    $3y + 2z = -\frac{3}{4}$

▸ 26. $3x - 5y = 2$
    $4x + 6z = \frac{1}{3}$
    $5y - 7z = \frac{1}{6}$

▸ 27. $\frac{1}{3}x + \frac{1}{2}y - \frac{1}{6}z = 4$
    $\frac{1}{4}x - \frac{3}{4}y + \frac{1}{2}z = \frac{3}{2}$
    $\frac{1}{2}x - \frac{2}{3}y - \frac{1}{4}z = -\frac{16}{3}$

▸ 28. $-\frac{1}{4}x + \frac{3}{8}y + \frac{1}{2}z = -1$
    $\frac{2}{3}x - \frac{1}{6}y - \frac{1}{2}z = 2$
    $\frac{3}{4}x - \frac{1}{2}y - \frac{1}{8}z = 1$

▸ 29. $x - \frac{1}{2}y - \frac{1}{3}z = -\frac{4}{3}$
    $\frac{1}{3}x + y - \frac{1}{2}z = 5$
    $-\frac{1}{4}x + \frac{2}{3}y - z = -\frac{3}{4}$

▸ 30. $x + \frac{1}{3}y - \frac{1}{2}z = -\frac{3}{2}$
    $\frac{1}{2}x - y + \frac{1}{3}z = 8$
    $\frac{1}{3}x - \frac{1}{4}y - z = -\frac{5}{6}$

31. $\frac{1}{2}x + \frac{2}{3}y = \frac{5}{2}$
    $\frac{1}{5}x - \frac{1}{2}z = -\frac{3}{10}$
    $\frac{1}{3}y - \frac{1}{4}z = \frac{3}{4}$

32. $\frac{1}{2}x - \frac{1}{3}y = \frac{1}{6}$
    $\frac{1}{3}y - \frac{1}{3}z = 1$
    $\frac{1}{5}x - \frac{1}{2}z = -\frac{4}{5}$

33. $\frac{1}{2}x - \frac{1}{4}y + \frac{1}{2}z = -2$
    $\frac{1}{4}x - \frac{1}{12}y - \frac{1}{3}z = \frac{1}{4}$
    $\frac{1}{6}x + \frac{1}{3}y - \frac{1}{2}z = \frac{3}{2}$

34. $\frac{1}{2}x + \frac{1}{2}y + z = \frac{1}{2}$
    $\frac{1}{2}x - \frac{1}{4}y - \frac{1}{4}z = 0$
    $\frac{1}{4}x + \frac{1}{12}y + \frac{1}{6}z = \frac{1}{6}$

▢ = Videos available by instructor request
▸ = Online student support materials available at www.thomsonedu.com/login

Applying the Concepts

35. Electric Current In the following diagram of an electrical circuit, x, y, and z represent the amount of current (in amperes) flowing across the 5-ohm, 20-ohm, and 10-ohm resistors, respectively. (In circuit diagrams resistors are represented by -\/\/\/- and potential differences by —| |—.)

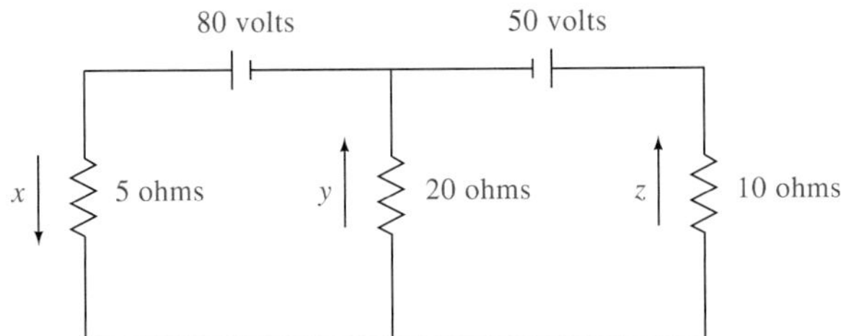

The system of equations used to find the three currents x, y, and z is

$$x - y - z = 0$$
$$5x + 20y = 80$$
$$20y - 10z = 50$$

Solve the system for all variables.

36. Cost of a Rental Car If a car rental company charges \$10 a day and 8¢ a mile to rent one of its cars, then the cost z, in dollars, to rent a car for x days and drive y miles can be found from the equation

$$z = 10x + 0.08y$$

a. How much does it cost to rent a car for 2 days and drive it 200 miles under these conditions?

b. A second company charges \$12 a day and 6¢ a mile for the same car. Write an equation that gives the cost z, in dollars, to rent a car from this company for x days and drive it y miles.

c. A car is rented from each of the companies mentioned in (a) and (b) for 2 days. To find the mileage at which the cost of renting the cars from each of the two companies will be equal, solve the following system for y:

$$z = 10x + 0.08y$$
$$z = 12x + 0.06y$$
$$x = 2$$

Maintaining Your Skills

37. If y varies directly with square of x, and y is 75 when x is 5, find y when x is 7.

38. Suppose y varies directly with the cube of x. If y is 16 when x is 2, find y when x is 3.

39. Suppose y varies inversely with x. If y is 10 when x is 25, find x when y is 5.

40. If y varies inversely with the cube of x, and y is 2 when x is 2, find y when x is 4.

41. Suppose z varies jointly with x and the square of y. If z is 40 when x is 5 and y is 2, find z when x is 2 and y is 5.

42. Suppose z varies jointly with x and the cube of y. If z is 48 when x is 3 and y is 2, find z when x is 4 and y is $\frac{1}{2}$.

Getting Ready for the Next Section

Simplify.

43. $1(4) - 3(2)$

44. $3(7) - (-2)(5)$

45. $1(1) - 3(-2) + (-2)(-2)$

46. $-4(0)(-2) - (-1)(1)(1) - 1(2)(3)$

47. $-3(-1 - 1) + 4(-2 + 2) - 5[2 - (-2)]$

48. $12 + 4 - (-1) - 6$

Solve.

49. $-5x = 20$

50. $4x - 2x = 8$

Extending the Concepts

Solve each system for the solution (x, y, z, w).

51.
$$\begin{aligned} x + y + z + w &= 10 \\ x + 2y - z + w &= 6 \\ x - y - z + 2w &= 4 \\ x - 2y + z - 3w &= -12 \end{aligned}$$

52.
$$\begin{aligned} x + y + z + w &= 16 \\ x - y + 2z - w &= 1 \\ x + 3y - z - w &= -2 \\ x - 3y - 2z + 2w &= -4 \end{aligned}$$

4.3 Introduction to Determinants

OBJECTIVES

A Find the value of a 2×2 determinant.

B Find the value of a 3×3 determinant.

In this section we will expand and evaluate *determinants.* The purpose of this section is simply to be able to find the value of a given determinant. As we will see in the next section, determinants are very useful in solving systems of linear equations. Before we apply determinants to systems of linear equations, however, we must practice calculating the value of some determinants.

DEFINITION The value of the **2×2 (2 by 2) determinant**

$$\begin{vmatrix} a & c \\ b & d \end{vmatrix}$$

is given by

$$\begin{vmatrix} a & c \\ b & d \end{vmatrix} = ad - bc$$

From the preceding definition we see that a determinant is simply a square array of numbers with two vertical lines enclosing it. The value of a 2×2 determinant is found by cross-multiplying on the diagonals and then subtracting, a diagram of which looks like

$$\begin{vmatrix} a & c \\ b & d \end{vmatrix} = ad - bc$$

EXAMPLE 1 Find the value of the following 2×2 determinants:

a. $\begin{vmatrix} 1 & 2 \\ 3 & 4 \end{vmatrix} = 1(4) - 3(2) = 4 - 6 = -2$

b. $\begin{vmatrix} 3 & 5 \\ -2 & 7 \end{vmatrix} = 3(7) - (-2)5 = 21 + 10 = 31$

EXAMPLE 2 Solve for x if

$$\begin{vmatrix} x & 2 \\ x & 4 \end{vmatrix} = 8$$

SOLUTION We expand the determinant on the left side to get

$$x(4) - x(2) = 8$$

$$4x - 2x = 8$$

$$2x = 8$$

$$x = 4$$

We now turn our attention to 3×3 determinants. A 3×3 determinant is also a square array of numbers enclosed by vertical lines, the value of which is given by the following definition.

DEFINITION The value of the 3×3 determinant

$$\begin{vmatrix} a_1 & b_1 & c_1 \\ a_2 & b_2 & c_2 \\ a_3 & b_3 & c_3 \end{vmatrix}$$

is given by

$$\begin{vmatrix} a_1 & b_1 & c_1 \\ a_2 & b_2 & c_2 \\ a_3 & b_3 & c_3 \end{vmatrix} = a_1b_2c_3 + a_3b_1c_2 + a_2b_3c_1 - a_3b_2c_1 - a_1b_3c_2 - a_2b_1c_3$$

At first glance, the expansion of a 3×3 determinant looks a little complicated. There are actually two different methods used to find the six products in the preceding definition that simplify matters somewhat.

Method I We begin by writing the determinant with the first two columns repeated on the right:

$$\left|\begin{matrix} a_1 & b_1 & c_1 \\ a_2 & b_2 & c_2 \\ a_3 & b_3 & c_3 \end{matrix}\right|\begin{matrix} a_1 & b_1 \\ a_2 & b_2 \\ a_3 & b_3 \end{matrix}$$

The positive products in the definition come from multiplying down the three full diagonals:

$$\left|\begin{matrix} a_1 & b_1 & c_1 \\ a_2 & b_2 & c_2 \\ a_3 & b_3 & c_3 \end{matrix}\right|\begin{matrix} a_1 & b_1 \\ a_2 & b_2 \\ a_3 & b_3 \end{matrix}$$
$$+ \quad + \quad +$$

The negative products come from multiplying up the three full diagonals:

$$- \quad - \quad -$$
$$\left|\begin{matrix} a_1 & b_1 & c_1 \\ a_2 & b_2 & c_2 \\ a_3 & b_3 & c_3 \end{matrix}\right|\begin{matrix} a_1 & b_1 \\ a_2 & b_2 \\ a_3 & b_3 \end{matrix}$$

Note
Check the products found by multiplying up and down the diagonals given here with the products given in the definition of a 3×3 determinant to see that they match.

EXAMPLE 3

Find the value of

$$\begin{vmatrix} 1 & 3 & -2 \\ 2 & 0 & 1 \\ 4 & -1 & 1 \end{vmatrix}$$

SOLUTION Repeating the first two columns and then finding the products down the diagonals and the products up the diagonals as given in Method 1, we have

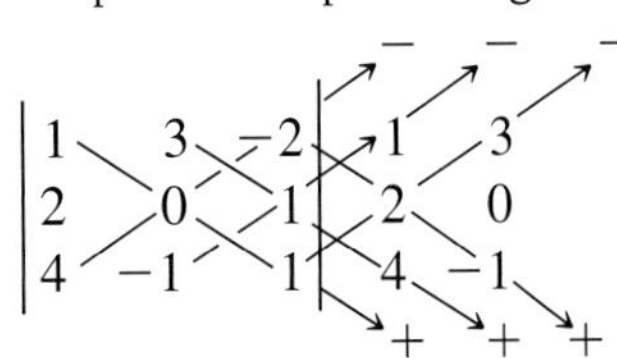

$$= 1(0)(1) + 3(1)(4) + (-2)(2)(-1)$$
$$-4(0)(-2) - (-1)(1)(1) - 1(2)(3)$$
$$= 0 + 12 + 4 - 0 - (-1) - 6$$
$$= 11$$

Method 2 The second method of evaluating a 3×3 determinant is called *expansion by minors.*

DEFINITION The **minor** for an element in a 3×3 determinant is the determinant consisting of the elements remaining when the row and column to which the element belongs are deleted. For example, in the determinant

$$\begin{vmatrix} a_1 & b_1 & c_1 \\ a_2 & b_2 & c_2 \\ a_3 & b_3 & c_3 \end{vmatrix}$$

Minor for element $a_1 = \begin{vmatrix} b_2 & c_2 \\ b_3 & c_3 \end{vmatrix}$

Minor for element $b_2 = \begin{vmatrix} a_1 & c_1 \\ a_3 & c_3 \end{vmatrix}$

Minor for element $c_3 = \begin{vmatrix} a_1 & b_1 \\ a_2 & b_2 \end{vmatrix}$

Note
If you have read this far and are confused, hang on. After you have done a couple of examples you will find expansion by minors to be a fairly simple process. It just takes a lot of writing to explain it.

Before we can evaluate a 3×3 determinant by Method 2, we must first define what is known as the sign array for a 3×3 determinant.

DEFINITION The **sign array** for a 3×3 determinant is a 3×3 array of signs in the following pattern:

$$\begin{vmatrix} + & - & + \\ - & + & - \\ + & - & + \end{vmatrix}$$

The sign array begins with a plus sign in the upper left-hand corner. The signs then alternate between plus and minus across every row and down every column.

To Evaluate a 3×3 Determinant by Expansion of Minors We can evaluate a 3×3 determinant by expanding across any row or down any column as follows:

Step 1: Choose a row or column to expand.

Step 2: Write the product of each element in the row or column chosen in step 1 with its minor.

Step 3: Connect the three products in step 2 with the signs in the corresponding row or column in the sign array.

To illustrate the procedure, we will use the same determinant we used in Example 3.

EXAMPLE 4

Expand across the first row:

$$\begin{vmatrix} 1 & 3 & -2 \\ 2 & 0 & 1 \\ 4 & -1 & 1 \end{vmatrix}$$

SOLUTION The products of the three elements in row 1 with their minors are

$$1\begin{vmatrix} 0 & 1 \\ -1 & 1 \end{vmatrix} \qquad 3\begin{vmatrix} 2 & 1 \\ 4 & 1 \end{vmatrix} \qquad (-2)\begin{vmatrix} 2 & 0 \\ 4 & -1 \end{vmatrix}$$

Connecting these three products with the signs from the first row of the sign array, we have

$$+1\begin{vmatrix} 0 & 1 \\ -1 & 1 \end{vmatrix} - 3\begin{vmatrix} 2 & 1 \\ 4 & 1 \end{vmatrix} + (-2)\begin{vmatrix} 2 & 0 \\ 4 & -1 \end{vmatrix}$$

> **Note**
> This method of evaluating a determinant is actually more valuable than our first method because it works with any size determinant from 3×3 to 4×4 to any higher order determinant. Method 1 works only on 3×3 determinants. It cannot be used on a 4×4 determinant.

We complete the problem by evaluating each of the three 2×2 determinants and then simplifying the resulting expression:

$$\begin{aligned} &+1[0 - (-1)] - 3(2 - 4) + (-2)(-2 - 0) \\ &= 1(1) - 3(-2) + (-2)(-2) \\ &= 1 + 6 + 4 \\ &= 11 \end{aligned}$$

The results of Examples 3 and 4 match. It makes no difference which method we use—the value of a 3×3 determinant is unique.

EXAMPLE 5

Expand down column 2:

$$\begin{vmatrix} 2 & 3 & -2 \\ 1 & 4 & 1 \\ 1 & 5 & -1 \end{vmatrix}$$

SOLUTION We connect the products of elements in column 2 and their minors with the signs from the second column in the sign array:

$$\begin{aligned} \begin{vmatrix} 2 & 3 & -2 \\ 1 & 4 & 1 \\ 1 & 5 & -1 \end{vmatrix} &= -3\begin{vmatrix} 1 & 1 \\ 1 & -1 \end{vmatrix} + 4\begin{vmatrix} 2 & -2 \\ 1 & -1 \end{vmatrix} - 5\begin{vmatrix} 2 & -2 \\ 1 & 1 \end{vmatrix} \\ &= -3(-1 - 1) + 4[-2 - (-2)] - 5[2 - (-2)] \\ &= -3(-2) + 4(0) - 5(4) \\ &= 6 + 0 - 20 \\ &= -14 \end{aligned}$$

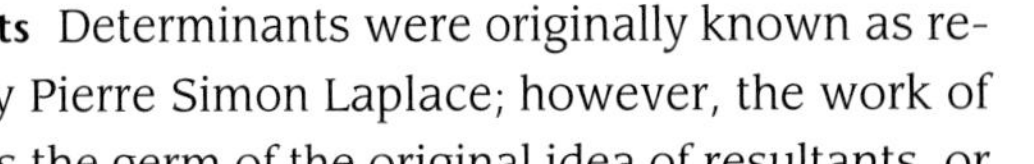

A Note on the History of Determinants Determinants were originally known as resultants, a name given to them by Pierre Simon Laplace; however, the work of Gottfried Wilhelm Leibniz contains the germ of the original idea of resultants, or determinants.

LINKING OBJECTIVES AND EXAMPLES

Next to each **objective** we have listed the examples that are best described by that objective.

A	1, 2
B	3–5

GETTING READY FOR CLASS

After reading through the preceding section, respond in your own words and in complete sentences.

1. Describe how you evaluate a 2×2 determinant.
2. Name the row and column that the number 3 is in: $\begin{vmatrix} 1 & 3 \\ 5 & 7 \end{vmatrix}$
3. What is the sign array for a 3×3 determinant?
4. If the value of a determinant is 0, does one of the elements have to be 0? Explain.

Problem Set 4.3

Online support materials can be found at www.thomsonedu.com/login

Find the value of the following 2×2 determinants.

1. $\begin{vmatrix} 1 & 0 \\ 2 & 3 \end{vmatrix}$
2. $\begin{vmatrix} 5 & 4 \\ 3 & 2 \end{vmatrix}$
3. $\begin{vmatrix} 2 & 1 \\ 3 & 4 \end{vmatrix}$
4. $\begin{vmatrix} 4 & 1 \\ 5 & 2 \end{vmatrix}$
5. $\begin{vmatrix} 0 & 1 \\ 1 & 0 \end{vmatrix}$
6. $\begin{vmatrix} 1 & 0 \\ 0 & 1 \end{vmatrix}$
7. $\begin{vmatrix} -3 & 2 \\ 6 & -4 \end{vmatrix}$
8. $\begin{vmatrix} 8 & -3 \\ -2 & 5 \end{vmatrix}$

Solve each of the following for x.

9. $\begin{vmatrix} 2x & 1 \\ x & 3 \end{vmatrix} = 10$
10. $\begin{vmatrix} 3x & -2 \\ 2x & 3 \end{vmatrix} = 26$
11. $\begin{vmatrix} 1 & 2x \\ 2 & -3x \end{vmatrix} = 21$
12. $\begin{vmatrix} -5 & 4x \\ 1 & -x \end{vmatrix} = 27$
13. $\begin{vmatrix} 2x & -4 \\ x & 2 \end{vmatrix} = -16$
14. $\begin{vmatrix} 3x & -2 \\ x & 4 \end{vmatrix} = -28$
15. $\begin{vmatrix} 11x & -7x \\ 3 & -2 \end{vmatrix} = 3$
16. $\begin{vmatrix} -3x & -5x \\ 4 & 6 \end{vmatrix} = -14$

Find the value of each of the following 3×3 determinants by using Method 1 of this section.

17. $\begin{vmatrix} 1 & 2 & 0 \\ 0 & 2 & 1 \\ 1 & 1 & 1 \end{vmatrix}$
18. $\begin{vmatrix} -1 & 0 & 2 \\ 3 & 0 & 1 \\ 0 & 1 & 3 \end{vmatrix}$
19. $\begin{vmatrix} 1 & 2 & 3 \\ 3 & 2 & 1 \\ 1 & 1 & 1 \end{vmatrix}$
20. $\begin{vmatrix} -1 & 2 & 0 \\ 3 & -2 & 1 \\ 0 & 5 & 4 \end{vmatrix}$

Find the value of each determinant by using Method 2 and expanding across the first row.

21. $\begin{vmatrix} 0 & 1 & 2 \\ 1 & 0 & 1 \\ -1 & 2 & 0 \end{vmatrix}$
22. $\begin{vmatrix} 3 & -2 & 1 \\ 0 & -1 & 0 \\ 2 & 0 & 1 \end{vmatrix}$
23. $\begin{vmatrix} 3 & 0 & 2 \\ 0 & -1 & -1 \\ 4 & 0 & 0 \end{vmatrix}$
24. $\begin{vmatrix} 1 & 1 & 1 \\ 1 & -1 & 1 \\ 1 & 1 & -1 \end{vmatrix}$

Find the value of each of the following determinants.

25. $\begin{vmatrix} 2 & -1 & 0 \\ 1 & 0 & -2 \\ 0 & 1 & 2 \end{vmatrix}$
26. $\begin{vmatrix} 5 & 0 & -4 \\ 0 & 1 & 3 \\ -1 & 2 & -1 \end{vmatrix}$
27. $\begin{vmatrix} 1 & 3 & 7 \\ -2 & 6 & 4 \\ 3 & 7 & -1 \end{vmatrix}$
28. $\begin{vmatrix} 2 & 1 & 5 \\ 6 & -3 & 4 \\ 8 & 9 & -2 \end{vmatrix}$

Applying the Concepts

29. **Slope-Intercept Form** Show that the following determinant equation is another way to write the slope-intercept form of the equation of a line.

$$\begin{vmatrix} y & x \\ m & 1 \end{vmatrix} = b$$

= Videos available by instructor request

▸ = Online student support materials available at www.thomsonedu.com/login

30. **Temperature Conversion** Show that the following determinant equation is another way to write the equation $F = \frac{9}{5}C + 32$.

$$\begin{vmatrix} C & F & 1 \\ 5 & 41 & 1 \\ -10 & 14 & 1 \end{vmatrix} = 0$$

31. **Amusement Park Income** From 1986 to 1990, the annual income of amusement parks was linearly increasing, after which time it remained fairly constant. The annual income y, in billions of dollars, may be found for one of these years by evaluating the following determinant equation, in which x represents the number of years past January 1, 1986.

$$\begin{vmatrix} x & -1.7 \\ 2 & 0.3 \end{vmatrix} = y$$

a. Write the determinant equation in slope-intercept form.

b. Use the equation from part (a) to find the approximate income for amusement parks in the year 1988.

32. **Women in Armed Forces** From 1981, the enrollment of women in the United States armed forces was linearly increasing until 1990, after which it declined. The approximate number of women, w, enrolled in the armed forces from 1981 to 1990 may be found by evaluating the following determinant equation, in which x represents the number of years past January 1, 1981.

$$\begin{vmatrix} 6{,}509 & -2 \\ 85{,}709 & x \end{vmatrix} = w$$

Use this equation to determine the number of women enrolled in the armed forces on January 1, 1985.

Maintaining Your Skills

For each relation that follows, state the domain and the range, and indicate which are also functions.

33. $\{(1, 2), (3, 4), (4, 2)\}$

34. $\{(0, 0), (1, 1), (0, 1)\}$

35. $\{(3, 1), (2, 3), (1, 2)\}$

36. $\{(-1, 1), (2, -2), (-3, -3)\}$

State whether each of the following graphs is the graph of a function

37.

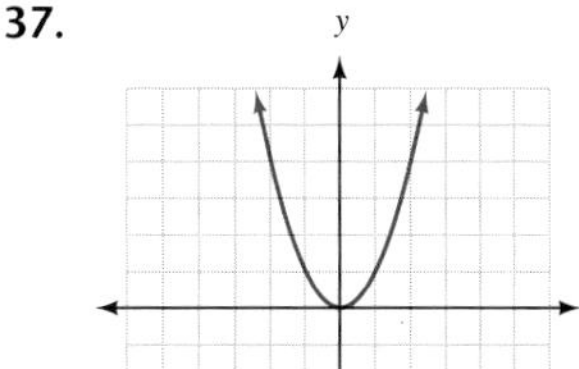

38.

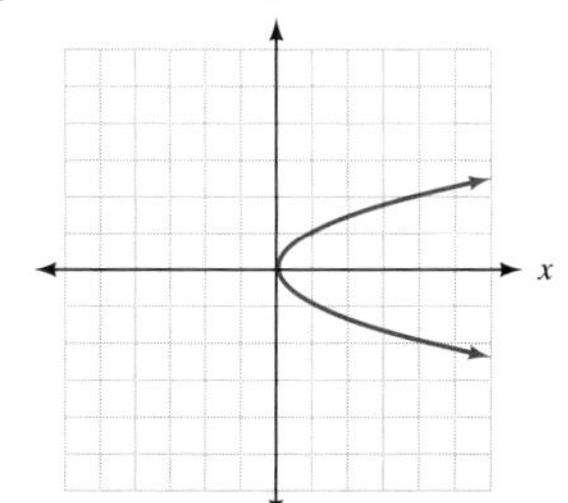

39.

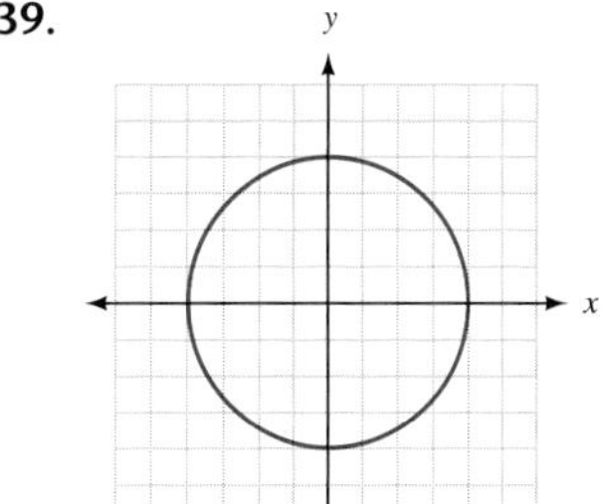

40.

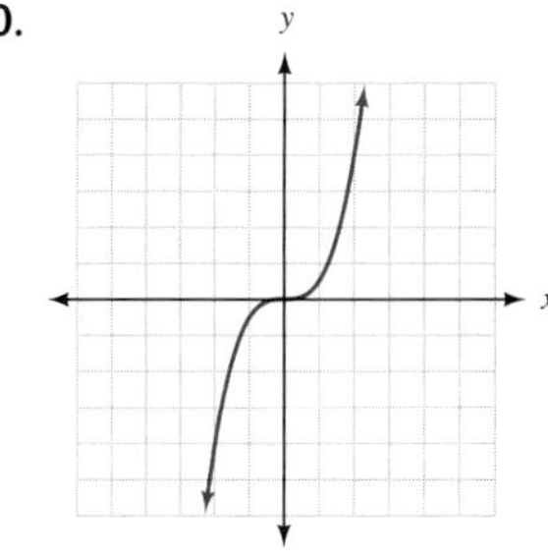

Getting Ready for the Next Section

Simplify.

41. $2(3) - 4(4)$

42. $2(5) - 4(-3)$

43. $1(-1)(-3)$

44. $-(-3)(2)(1)$

45. $-\frac{10}{22}$

46. $\frac{39}{13}$

47. $6(1) - 1(-5) + 1(2)$

48. $-2(-14) + 3(-4) - 1(-10)$

Find the value of each determinant.

49. $\begin{vmatrix} 3 & -5 \\ 2 & 4 \end{vmatrix}$

50. $\begin{vmatrix} 3 & 2 \\ 2 & 1 \end{vmatrix}$

51. $\begin{vmatrix} 6 & 1 & 1 \\ 3 & -1 & 1 \\ -4 & 2 & -3 \end{vmatrix}$

52. $\begin{vmatrix} 1 & 1 & 0 \\ 2 & 0 & 1 \\ 0 & 1 & 2 \end{vmatrix}$

Extending the Concepts

A 4×4 determinant can be evaluated only by using Method 2, expansion by minors; Method 1 will not work. Below is a 4×4 determinant and its associated sign array.

$$\begin{vmatrix} 2 & 0 & 1 & -3 \\ -1 & 2 & 0 & 1 \\ -3 & 0 & 1 & 0 \\ 1 & 1 & 0 & 0 \end{vmatrix} \qquad \begin{vmatrix} + & - & + & - \\ - & + & - & + \\ + & - & + & - \\ - & + & - & + \end{vmatrix}$$

4×4 determinant $\qquad$ 4×4 sign array

53. Use expansion by minors to evaluate the preceding 4×4 determinant by expanding it across row 1.

54. Evaluate the preceding determinant by expanding it down column 4.

55. Use expansion by minors down column 3 to evaluate the preceding determinant.

56. Evaluate the preceding determinant by expanding it across row 4.

4.4 Cramer's Rule

OBJECTIVES

A Solve a system of linear equations in two variables using Cramer's rule.

B Solve a system of linear equations in three variables using Cramer's rule.

We begin this section with a look at how determinants can be used to solve a system of linear equations in two variables. The method we use is called *Cramer's rule.* We state it here as a theorem without proof.

> **Cramer's Rule I** The solution to the system
>
> $$a_1x + b_1y = c_1$$
> $$a_2x + b_2y = c_2$$
>
> is given by
>
> $$x = \frac{D_x}{D}, \qquad y = \frac{D_y}{D}$$
>
> where
>
> $$D = \begin{vmatrix} a_1 & b_1 \\ a_2 & b_2 \end{vmatrix} \qquad D_x = \begin{vmatrix} c_1 & b_1 \\ c_2 & b_2 \end{vmatrix} \qquad D_y = \begin{vmatrix} a_1 & c_1 \\ a_2 & c_2 \end{vmatrix} \qquad (D \neq 0)$$

The determinant D is made up of the coefficients of x and y in the original system. The determinants D_x and D_y are found by replacing the coefficients of x or y by the constant terms in the original system. Notice also that Cramer's rule does not apply if $D = 0$. In this case the equations are dependent, or the system is inconsistent.

EXAMPLE 1

Use Cramer's rule to solve

$$2x - 3y = 4$$
$$4x + 5y = 3$$

SOLUTION We begin by calculating the determinants D, D_x, and D_y.

$$D = \begin{vmatrix} 2 & -3 \\ 4 & 5 \end{vmatrix} = 2(5) - 4(-3) = 22$$

$$D_x = \begin{vmatrix} 4 & -3 \\ 3 & 5 \end{vmatrix} = 4(5) - 3(-3) = 29$$

$$D_y = \begin{vmatrix} 2 & 4 \\ 4 & 3 \end{vmatrix} = 2(3) - 4(4) = -10$$

$$x = \frac{D_x}{D} = \frac{29}{22} \quad \text{and} \quad y = \frac{D_y}{D} = \frac{-10}{22} = -\frac{5}{11}$$

The solution set for the system is $\left\{\left(\frac{29}{22}, -\frac{5}{11}\right)\right\}$.

Cramer's rule can also be applied to systems of linear equations in three variables.

Cramer's Rule II The solution set to the system

$$a_1x + b_1y + c_1z = d_1$$
$$a_2x + b_2y + c_2z = d_2$$
$$a_3x + b_3y + c_3z = d_3$$

is given by

$$x = \frac{D_x}{D}, \quad y = \frac{D_y}{D}, \quad \text{and} \quad z = \frac{D_z}{D}$$

where

$$D = \begin{vmatrix} a_1 & b_1 & c_1 \\ a_2 & b_2 & c_2 \\ a_3 & b_3 & c_3 \end{vmatrix} \quad D_x = \begin{vmatrix} d_1 & b_1 & c_1 \\ d_2 & b_2 & c_2 \\ d_3 & b_3 & c_3 \end{vmatrix} \quad (D \neq 0)$$

$$D_y = \begin{vmatrix} a_1 & d_1 & c_1 \\ a_2 & d_2 & c_2 \\ a_3 & d_3 & c_3 \end{vmatrix} \quad D_z = \begin{vmatrix} a_1 & b_1 & d_1 \\ a_2 & b_2 & d_2 \\ a_3 & b_3 & d_3 \end{vmatrix}$$

Again, the determinant D consists of the coefficients of x, y, and z in the original system. The determinants D_x, D_y, and D_z are found by replacing the coefficients of x, y, and z, respectively, with the constant terms from the original system. If $D = 0$, there is no unique solution to the system.

EXAMPLE 2

Use Cramer's rule to solve

$$x + y + z = 6$$
$$2x - y + z = 3$$
$$x + 2y - 3z = -4$$

SOLUTION This is the same system used in Example 1 in the second section of this chapter, so we can compare Cramer's rule with our previous methods of solving a system in three variables. We begin by setting up and evaluating D, D_x, D_y, and D_z. (Recall that there are a number of ways to evaluate a 3×3 determinant. Since we have four of these determinants, we can use both Methods 1 and 2 from the previous section.) We evaluate D using Method 1 from the previous section.

> **Note**
> When we are solving a system of linear equations by Cramer's rule, it is best to find the determinant D first. If $D = 0$, then there is no unique solution to the system and we may not want to go further.

$$D = \begin{vmatrix} 1 & 1 & 1 \\ 2 & -1 & 1 \\ 1 & 2 & -3 \end{vmatrix} \begin{matrix} 1 & 1 \\ 2 & -1 \\ 1 & 2 \end{matrix}$$

$$= 3 + 1 + 4 - (-1) - (2) - (-6) = 13$$

We evaluate D_x using Method 2 from the previous section and expanding across row 1:

$$D_x = \begin{vmatrix} 6 & 1 & 1 \\ 3 & -1 & 1 \\ -4 & 2 & -3 \end{vmatrix} = 6\begin{vmatrix} -1 & 1 \\ 2 & -3 \end{vmatrix} - 1\begin{vmatrix} 3 & 1 \\ -4 & -3 \end{vmatrix} + 1\begin{vmatrix} 3 & -1 \\ -4 & 2 \end{vmatrix}$$

$$= 6(1) - 1(-5) + 1(2)$$

$$= 13$$

> **Note**
> We are solving each of these determinants by expanding about different rows or columns just to show the different ways these determinants can be evaluated.

Find D_y by expanding across row 2:

$$D_y = \begin{vmatrix} 1 & 6 & 1 \\ 2 & 3 & 1 \\ 1 & -4 & -3 \end{vmatrix} = -2\begin{vmatrix} 6 & 1 \\ -4 & -3 \end{vmatrix} + 3\begin{vmatrix} 1 & 1 \\ 1 & -3 \end{vmatrix} - 1\begin{vmatrix} 1 & 6 \\ 1 & -4 \end{vmatrix}$$

$$= -2(-14) + 3(-4) - 1(-10)$$

$$= 26$$

Find D_z by expanding down column 1:

$$D_z = \begin{vmatrix} 1 & 1 & 6 \\ 2 & -1 & 3 \\ 1 & 2 & -4 \end{vmatrix} = 1\begin{vmatrix} -1 & 3 \\ 2 & -4 \end{vmatrix} - 2\begin{vmatrix} 1 & 6 \\ 2 & -4 \end{vmatrix} + 1\begin{vmatrix} 1 & 6 \\ -1 & 3 \end{vmatrix}$$

$$= 1(-2) - 2(-16) + 1(9)$$

$$= 39$$

$$x = \frac{D_x}{D} = \frac{13}{13} = 1 \qquad y = \frac{D_y}{D} = \frac{26}{13} = 2 \qquad z = \frac{D_z}{D} = \frac{39}{13} = 3$$

The solution set is $\{(1, 2, 3)\}$.

EXAMPLE 3

Use Cramer's rule to solve

$$\begin{aligned} x + y &= -1 \\ 2x - z &= 3 \\ y + 2z &= -1 \end{aligned}$$

SOLUTION It is helpful to rewrite the system using zeros for the coefficients of those variables not shown:

$$\begin{aligned} x + \ y + 0z &= -1 \\ 2x + 0y - \ z &= \ \ 3 \\ 0x + \ y + 2z &= -1 \end{aligned}$$

The four determinants used in Cramer's rule are

$$D = \begin{vmatrix} 1 & 1 & 0 \\ 2 & 0 & -1 \\ 0 & 1 & 2 \end{vmatrix} = -3$$

$$D_x = \begin{vmatrix} -1 & 1 & 0 \\ 3 & 0 & -1 \\ -1 & 1 & 2 \end{vmatrix} = -6$$

$$D_y = \begin{vmatrix} 1 & -1 & 0 \\ 2 & 3 & -1 \\ 0 & -1 & 2 \end{vmatrix} = 9$$

$$D_z = \begin{vmatrix} 1 & 1 & -1 \\ 2 & 0 & 3 \\ 0 & 1 & -1 \end{vmatrix} = -3$$

$$x = \frac{D_x}{D} = \frac{-6}{-3} = 2 \qquad y = \frac{D_y}{D} = \frac{9}{-3} = -3 \qquad z = \frac{D_z}{D} = \frac{-3}{-3} = 1$$

The solution set is $\{(2, -3, 1)\}$.

Finally, we should mention the possible situations that can occur when the determinant D is 0 when we are using Cramer's rule.

If $D = 0$ and at least one of the other determinants, D_x or D_y (or D_z), is not 0, then the system is inconsistent. In this case there is no solution to the system.

However, if $D = 0$ and both D_x and D_y (and D_z in a system of three equations in three variables) are 0, then the equations are dependent.

A Note on the History of Cramer's Rule Cramer's rule is named after the Swiss mathematician Gabriel Cramer (1704–1752). Cramer's rule appeared in the appendix of an algebraic work of his classifying curves, but the basic idea behind his now-famous rule was formulated earlier by Leibniz and Chinese mathematicians. It was actually Cramer's superior notation that helped to popularize the technique.

Cramer has a respectable reputation as a mathematician, but he does not rank with the great mathematicians of his time, although through his extensive travels he met many of them, such as the Bernoullis, Euler, and D'Alembert. Cramer had very broad interests. He wrote on philosophy, law, and government, as well as mathematics; served in public office; and was an expert on cathedrals, often instructing workers about their repair and coordinating excavations to recover cathedral archives. Cramer never married, and a fall from a carriage eventually led to his death.

LINKING OBJECTIVES AND EXAMPLES

Next to each **objective** we have listed the examples that are best described by that objective.

A	1
B	2, 3

GETTING READY FOR CLASS

After reading through the preceding section, respond in your own words and in complete sentences.

1. When applying Cramer's rule, when will you see 2×2 determinants?
2. Why would it be impossible to use Cramer's rule if the determinant $D = 0$?
3. When applying Cramer's rule to solve a system of two linear equations in two variables, how many numbers should you obtain? How do these numbers relate to the system?
4. What will happen when you apply Cramer's rule to a system of equations made up of two parallel lines?

Problem Set 4.4

Online support materials can be found at www.thomsonedu.com/login

Solve each of the following systems using Cramer's rule.

1. $2x - 3y = 3$
   $4x - 2y = 10$
2. $3x + y = -2$
   $-3x + 2y = -4$
3. $5x - 2y = 4$
   $-10x + 4y = 1$
4. $-4x + 3y = -11$
   $5x + 4y = 6$
5. $4x - 7y = 3$
   $5x + 2y = -3$
6. $3x - 4y = 7$
   $6x - 2y = 5$
7. $9x - 8y = 4$
   $2x + 3y = 6$
8. $4x - 7y = 10$
   $-3x + 2y = -9$
9. $x + y + z = 4$
   $x - y - z = 2$
   $2x + 2y - z = 2$
10. $-x + y + 3z = 6$
    $x + y + 2z = 7$
    $2x + 3y + z = 4$
11. $x + y - z = 2$
    $-x + y + z = 3$
    $x + y + z = 4$
12. $-x - y + z = 1$
    $x - y + z = 3$
    $x + y - z = 4$
13. $3x - y + 2z = 4$
    $6x - 2y + 4z = 8$
    $x - 5y + 2z = 1$
14. $2x - 3y + z = 1$
    $3x - y - z = 4$
    $4x - 6y + 2z = 3$
15. $2x - y + 3z = 4$
    $x - 5y - 2z = 1$
    $-4x - 2y + z = 3$
16. $4x - y + 5z = 1$
    $2x + 3y + 4z = 5$
    $x + y + 3z = 2$
17. $-x - 7y = 1$
    $x + 3z = 11$
    $2y + z = 0$
18. $x + y = 2$
    $-x + 3z = 0$
    $2y + z = 3$
19. $x - y = 2$
    $3x + z = 11$
    $y - 2z = -3$
20. $4x + 5y = -1$
    $2y + 3z = -5$
    $x + 2z = -1$

Applying the Concepts

21. **Break-Even Point** If a company has fixed costs of \$100 per week and each item it produces costs \$10 to manufacture, then the total cost y per week to produce x items is

$$y = 10x + 100$$

If the company sells each item it manufactures for \$12, then the total amount of money y the company brings in for selling x items is

$$y = 12x$$

Use Cramer's rule to solve the system

$$y = 10x + 100$$
$$y = 12x$$

for x to find the number of items the company must sell per week to break even.

■ = Videos available by instructor request

▸ = Online student support materials available at www.thomsonedu.com/login

22. Break-Even Point Suppose a company has fixed costs of \$200 per week and each item it produces costs \$20 to manufacture.

a. Write an equation that gives the total cost per week y to manufacture x items.

b. If each item sells for \$25, write an equation that gives the total amount of money y the company brings in for selling x items.

c. Use Cramer's rule to find the number of items the company must sell each week to break even.

Maintaining Your Skills

Let $f(x) = \frac{1}{2}x + 3$ and $g(x) = x^2 - 4$, and find

23. $f(0)$

24. $g(0)$

25. $g(2)$

26. $f(2)$

27. $f(-4)$

28. $g(-6)$

29. $f[g(2)]$

30. $g[f(2)]$

Getting Ready for the Next Section

Translate into symbols.

31. Two more than 3 times a number

32. One less than twice a number

Simplify.

33. $25 - \frac{385}{9}$

34. $0.30(12)$

35. $0.08(4{,}000)$

36. $500(1.5)$

Apply the distributive property, then simplify.

37. $10(0.2x + 0.5y)$

38. $100(0.09x + 0.08y)$

Solve.

39. $x + (3x + 2) = 26$

40. $5x = 2{,}500$

Solve each system.

41. $\begin{aligned} -2y - 4z &= -18 \\ -7y + 4z &= 27 \end{aligned}$

42. $\begin{aligned} -x + 2y &= 200 \\ 4x - 2y &= 1{,}300 \end{aligned}$

Extending the Concepts

43. Name the system of equations for which Cramer's rule yields the following determinants.

$$D = \begin{vmatrix} 1 & 2 \\ 3 & 4 \end{vmatrix} \qquad D_x = \begin{vmatrix} 1 & 2 \\ 0 & 4 \end{vmatrix}$$

44. Name the system of equations for which Cramer's rule yields the following determinants.

$$D = \begin{vmatrix} 1 & 3 & 2 \\ -1 & 0 & 4 \\ 2 & 5 & -1 \end{vmatrix} \qquad D_y = \begin{vmatrix} 1 & 1 & 2 \\ -1 & 3 & 4 \\ 2 & 5 & -1 \end{vmatrix}$$

4.5 Applications

OBJECTIVES

A Solve application problems whose solutions are found through systems of linear equations.

Many times word problems involve more than one unknown quantity. If a problem is stated in terms of two unknowns and we represent each unknown quantity with a different variable, then we must write the relationships between the variables with two equations. The two equations written in terms of the two variables form a system of linear equations that we solve using the methods developed in this chapter. If we find a problem that relates three unknown quantities, then we need three equations to form a linear system we can solve.

Here is our Blueprint for Problem Solving, modified to fit the application problems that you will find in this section.

BLUEPRINT FOR PROBLEM SOLVING USING A SYSTEM OF EQUATIONS

STEP 1: ***Read*** the problem, and then mentally ***list*** the items that are known and the items that are unknown.

STEP 2: ***Assign variables*** to each of the unknown items; that is, let x = one of the unknown items and y = the other unknown item (and z = the third unknown item, if there is a third one). Then ***translate*** the other ***information*** in the problem to expressions involving the two (or three) variables.

STEP 3: ***Reread*** the problem, and then ***write a system of equations,*** using the items and variables listed in steps 1 and 2, that describes the situation.

STEP 4: ***Solve the system*** found in step 3.

STEP 5: ***Write your answers*** using complete sentences.

STEP 6: ***Reread*** the problem, and ***check*** your solution with the original words in the problem.

EXAMPLE 1

One number is 2 more than 3 times another. Their sum is 26. Find the two numbers.

SOLUTION Applying the steps from our Blueprint, we have:

Step 1: ***Read and list.***
We know that we have two numbers, whose sum is 26. One of them is 2 more than 3 times the other. The unknown quantities are the two numbers.

Step 2: ***Assign variables and translate information.***
Let x = one of the numbers and y = the other number.

Step 3: ***Write a system of equations.***
The first sentence in the problem translates into $y = 3x + 2$. The second sentence gives us a second equation: $x + y = 26$. Together, these two equations give us the following system of equations:

$$x + y = 26$$

$$y = 3x + 2$$

Step 4: ***Solve the system.***

Substituting the expression for y from the second equation into the first and solving for x yields

$$x + (3x + 2) = 26$$

$$4x + 2 = 26$$

$$4x = 24$$

$$x = 6$$

Using $x = 6$ in $y = 3x + 2$ gives the second number:

$$y = 3(6) + 2$$

$$y = 20$$

Step 5: ***Write answers.***

The two numbers are 6 and 20.

Step 6: ***Reread and check.***

The sum of 6 and 20 is 26, and 20 is 2 more than 3 times 6.

EXAMPLE 2

Suppose 850 tickets were sold for a game for a total of \$1,100. If adult tickets cost \$1.50 and children's tickets cost \$1.00, how many of each kind of ticket were sold?

SOLUTION

Step 1: ***Read and list.***

The total number of tickets sold is 850. The total income from tickets is \$1,100. Adult tickets are \$1.50 each. Children's tickets are \$1.00 each. We don't know how many of each type of ticket have been sold.

Step 2: ***Assign variables and translate information.***

We let x = the number of adult tickets and y = the number of children's tickets.

Step 3: ***Write a system of equations.***

The total number of tickets sold is 850, giving us our first equation.

$$x + y = 850$$

Because each adult ticket costs \$1.50 and each children's ticket costs \$1.00 and the total amount of money paid for tickets was \$1,100, a second equation is

$$1.50x + 1.00y = 1,100$$

	Adult Tickets	Children's Tickets	Total
Number	x	y	850
Value	$1.50x$	$1.00y$	1,100

The same information can also be obtained by summarizing the problem with a table. One such table appears at left. Notice that the two equations we obtained previously are given by the two rows of the table.

Whether we use a table to summarize the information in the problem or just talk our way through the problem, the system of equations that describes the situation is

$$x + y = 850$$
$$1.50x + 1.00y = 1{,}100$$

Step 4: Solve the system.

If we multiply the second equation by 10 to clear it of decimals, we have the system

$$x + y = 850$$
$$15x + 10y = 11{,}000$$

Multiplying the first equation by -10 and adding the result to the second equation eliminates the variable y from the system:

$$\begin{aligned} -10x - 10y &= -8{,}500 \\ 15x + 10y &= 11{,}000 \\ \hline 5x \qquad &= 2{,}500 \\ x &= 500 \end{aligned}$$

The number of adult tickets sold was 500. To find the number of children's tickets, we substitute $x = 500$ into $x + y = 850$ to get

$$500 + y = 850$$
$$y = 350$$

Step 5: Write answers.

The number of children's tickets is 350, and the number of adult tickets is 500.

Step 6: Reread and check.

The total number of tickets is $350 + 500 = 850$. The amount of money from selling the two types of tickets is

350 children's tickets at \$1.00 each is $350(1.00) = \$350$
500 adult tickets at \$1.50 each is $500(1.50) = \$750$
The total income from ticket sales is \$1,100

EXAMPLE 3 Suppose a person invests a total of \$10,000 in two accounts. One account earns 8% annually, and the other earns 9% annually. If the total interest earned from both accounts in a year is \$860, how much was invested in each account?

SOLUTION

Step 1: Read and list.

The total investment is \$10,000 split between two accounts. One account earns 8% annually, and the other earns 9% annually. The interest from both accounts is

$860 in 1 year. We don't know how much is in each account.

Step 2: ***Assign variables and translate information.***
We let x equal the amount invested at 9% and y be the amount invested at 8%.

Step 3: ***Write a system of equations.***
Because the total investment is $10,000, one relationship between x and y can be written as

$$x + y = 10{,}000$$

The total interest earned from both accounts is $860. The amount of interest earned on x dollars at 9% is $0.09x$, whereas the amount of interest earned on y dollars at 8% is $0.08y$. This relationship is represented by the equation

$$0.09x + 0.08y = 860$$

The two equations we have just written can also be found by first summarizing the information from the problem in a table. Again, the two rows of the table yield the two equations just written. Here is the table.

	Dollars at 9%	**Dollars at 8%**	**Total**
Number	x	y	10,000
Interest	$0.09x$	$0.08y$	860

The system of equations that describes this situation is given by

$$\begin{aligned} x + \quad y &= 10{,}000 \\ 0.09x + 0.08y &= 860 \end{aligned}$$

Step 4: ***Solve the system.***
Multiplying the second equation by 100 will clear it of decimals. The system that results after doing so is

$$\begin{aligned} x + y &= 10{,}000 \\ 9x + 8y &= 86{,}000 \end{aligned}$$

We can eliminate y from this system by multiplying the first equation by -8 and adding the result to the second equation.

$$\begin{aligned} -8x - 8y &= -80{,}000 \\ 9x + 8y &= 86{,}000 \\ \hline x \qquad &= 6{,}000 \end{aligned}$$

The amount of money invested at 9% is $6,000. Because the total investment was $10,000, the amount invested at 8% must be $4,000.

Step 5: ***Write answers.***

The amount invested at 8% is \$4,000, and the amount invested at 9% is \$6,000.

Step 6: ***Reread and check.***

The total investment is \$4,000 + \$6,000 = \$10,000. The amount of interest earned from the two accounts is

In 1 year, \$4,000 invested at 8% earns 0.08(4,000) = \$320
In 1 year, \$6,000 invested at 9% earns 0.09(6,000) = \$540
The total interest from the two accounts is \$860

EXAMPLE 4 How much 20% alcohol solution and 50% alcohol solution must be mixed to get 12 gallons of 30% alcohol solution?

SOLUTION To solve this problem, we must first understand that a 20% alcohol solution is 20% alcohol and 80% water.

Step 1: ***Read and list.***

We will mix two solutions to obtain 12 gallons of solution that is 30% alcohol. One of the solutions is 20% alcohol and the other is 50% alcohol. We don't know how much of each solution we need.

Step 2: ***Assign variables and translate information.***

Let x = the number of gallons of 20% alcohol solution needed and y = the number of gallons of 50% alcohol solution needed.

Step 3: ***Write a system of equations.***

Because we must end up with a total of 12 gallons of solution, one equation for the system is

$$x + y = 12$$

The amount of alcohol in the x gallons of 20% solution is $0.20x$, whereas the amount of alcohol in the y gallons of 50% solution is $0.50y$. Because the total amount of alcohol in the 20% and 50% solutions must add up to the amount of alcohol in the 12 gallons of 30% solution, the second equation in our system can be written as

$$0.20x + 0.50y = 0.30(12)$$

Again, let's make a table that summarizes the information we have to this point in the problem.

	20% Solution	**50% Solution**	**Final Solution**
Total number of gallons	x	y	12
Gallons of alcohol	$0.20x$	$0.50y$	0.30(12)

Our system of equations is

$$x + \quad y = 12$$

$$0.20x + 0.50y = 0.30(12) = 3.6$$

Step 4: ***Solve the system.***

Multiplying the second equation by 10 gives us an equivalent system:

$$x + y = 12$$

$$2x + 5y = 36$$

Multiplying the top equation by -2 to eliminate the x-variable, we have

$$\begin{aligned} -2x - 2y &= -24 \\ 2x + 5y &= 36 \\ \hline 3y &= 12 \\ y &= 4 \end{aligned}$$

Substituting $y = 4$ into $x + y = 12$, we solve for x:

$$x + 4 = 12$$

$$x = 8$$

Step 5: ***Write answers.***

It takes 8 gallons of 20% alcohol solution and 4 gallons of 50% alcohol solution to produce 12 gallons of 30% alcohol solution.

Step 6: ***Reread and check.***

If we mix 8 gallons of 20% solution and 4 gallons of 50% solution, we end up with a total of 12 gallons of solution. To check the percentages, we look for the total amount of alcohol in the two initial solutions and in the final solution.

In the initial solutions

The amount of alcohol in 8 gallons of 20% solution is 0.20(8) = 1.6 gallons
The amount of alcohol in 4 gallons of 50% solution is 0.50(4) = 2.0 gallons
The total amount of alcohol in the initial solutions is 3.6 gallons

In the final solution

The amount of alcohol in 12 gallons of 30% solution is 0.30(12) = 3.6 gallons.

EXAMPLE 5 It takes 2 hours for a boat to travel 28 miles downstream (with the current). The same boat can travel 18 miles upstream (against the current) in 3 hours. What is the speed of the boat in still water, and what is the speed of the current of the river?

SOLUTION

Step 1: ***Read and list.***
A boat travels 18 miles upstream and 28 miles downstream. The trip upstream takes 3 hours. The trip downstream takes 2 hours. We don't know the speed of the boat or the speed of the current.

Step 2: ***Assign variables and translate information.***
Let x = the speed of the boat in still water and let y = the speed of the current. The average speed (rate) of the boat upstream is $x - y$ because it is traveling against the current. The rate of the boat downstream is $x + y$ because the boat is traveling with the current.

Step 3: ***Write a system of equations.***
Putting the information into a table, we have

	d (Distance, miles)	r (Rate, mph)	t (Time, h)
Upstream	18	$x - y$	3
Downstream	28	$x + y$	2

The formula for the relationship between distance d, rate r, and time t is $d = rt$ (the rate equation). Because $d = r \cdot t$, the system we need to solve the problem is

$$18 = (x - y) \cdot 3$$
$$28 = (x + y) \cdot 2$$

which is equivalent to

$$6 = x - y$$
$$14 = x + y$$

Step 4: ***Solve the system.***
Adding the two equations, we have

$$20 = 2x$$
$$x = 10$$

Substituting $x = 10$ into $14 = x + y$, we see that

$$y = 4$$

Step 5: ***Write answers.***
The speed of the boat in still water is 10 miles per hour; the speed of the current is 4 miles per hour.

Step 6: ***Reread and check.***
The boat travels at $10 + 4 = 14$ miles per hour downstream, so in 2 hours it will travel $14 \cdot 2 = 28$ miles. The boat travels at $10 - 4 = 6$ miles per hour upstream, so in 3 hours it will travel $6 \cdot 3 = 18$ miles.

EXAMPLE 6 A coin collection consists of 14 coins with a total value of \$1.35. If the coins are nickels, dimes, and quarters, and the number of nickels is 3 less than twice the number of dimes, how many of each coin is there in the collection?

SOLUTION This problem will require three variables and three equations.

Step 1: ***Read and list.***

We have 14 coins with a total value of \$1.35. The coins are nickels, dimes, and quarters. The number of nickels is 3 less than twice the number of dimes. We do not know how many of each coin we have.

Step 2: ***Assign variables and translate information.***

Because we have three types of coins, we will have to use three variables. Let's let $x =$ the number of nickels, $y =$ the number of dimes, and $z =$ the number of quarters.

Step 3: ***Write a system of equations.***

Because the total number of coins is 14, our first equation is

$$x + y + z = 14$$

Because the number of nickels is 3 less than twice the number of dimes, a second equation is

$$x = 2y - 3 \quad \text{which is equivalent to} \quad x - 2y = -3$$

Our last equation is obtained by considering the value of each coin and the total value of the collection. Let's write the equation in terms of cents, so we won't have to clear it of decimals later.

$$5x + 10y + 25z = 135$$

Here is our system, with the equations numbered for reference:

$$\begin{aligned} x + y + z &= 14 \quad &\textbf{(1)} \\ x - 2y &= -3 \quad &\textbf{(2)} \\ 5x + 10y + 25z &= 135 \quad &\textbf{(3)} \end{aligned}$$

Step 4: ***Solve the system.***

Let's begin by eliminating x from the first and second equations and the first and third equations. Adding -1 times the second equation to the first equation gives us an equation in only y and z. We call this equation (4).

$$3y + z = 17 \quad \textbf{(4)}$$

Adding -5 times equation (1) to equation (3) gives us

$$5y + 20z = 65 \quad \textbf{(5)}$$

We can eliminate z from equations (4) and (5) by adding -20 times (4) to (5). Here is the result:

$$-55y = -275$$

$$y = 5$$

Substituting $y = 5$ into equation (4) gives us $z = 2$. Substituting $y = 5$ and $z = 2$ into equation (1) gives us $x = 7$.

Step 5: ***Write answers.***
The collection consists of 7 nickels, 5 dimes, and 2 quarters.

Step 6: ***Reread and check.***
The total number of coins is $7 + 5 + 2 = 14$. The number of nickels, 7, is 3 less than twice the number of dimes, 5. To find the total value of the collection, we have

The value of the 7 nickels is $7(0.05) = \$0.35$
The value of the 5 dimes is $5(0.10) = \$0.50$
The value of the 2 quarters is $2(0.25) = \$0.50$

The total value of the collection is \$1.35

If you go on to take a chemistry class, you may see the next example (or one much like it).

EXAMPLE 7

In a chemistry lab, students record the temperature of water at room temperature and find that it is 77° on the Fahrenheit temperature scale and 25° on the Celsius temperature scale. The water is then heated until it boils. The temperature of the boiling water is 212°F and 100°C. Assume that the relationship between the two temperature scales is a linear one, then use the preceding data to find the formula that gives the Celsius temperature C in terms of the Fahrenheit temperature F.

SOLUTION The data are summarized in Table 1.

TABLE 1
Corresponding Temperatures

In Degrees Fahrenheit	In Degrees Celsius
77	25
212	100

If we assume the relationship is linear, then the formula that relates the two temperature scales can be written in slope-intercept form as

$$C = mF + b$$

Substituting $C = 25$ and $F = 77$ into this formula gives us

$$25 = 77m + b$$

Substituting $C = 100$ and $F = 212$ into the formula yields

$$100 = 212m + b$$

Together, the two equations form a system of equations, which we can solve using the addition method.

$$25 = 77m + b \xrightarrow{\text{Multiply by } -1} -25 = -77m - b$$

$$100 = 212m + b \xrightarrow[\text{No change}]{} 100 = 212m + b$$

$$75 = 135m$$

$$m = \frac{75}{135} = \frac{5}{9}$$

To find the value of b, we substitute $m = \frac{5}{9}$ into $25 = 77m + b$ and solve for b.

$$25 = 77\left(\frac{5}{9}\right) + b$$

$$25 = \frac{385}{9} + b$$

$$b = 25 - \frac{385}{9} = \frac{225}{9} - \frac{385}{9} = -\frac{160}{9}$$

The equation that gives C in terms of F is

$$C = \frac{5}{9}F - \frac{160}{9}$$

GETTING READY FOR CLASS

After reading through the preceding section, respond in your own words and in complete sentences.

1. To apply the Blueprint for Problem Solving to the examples in this section, what is the first step?
2. To apply the Blueprint for Problem Solving to the examples in this section, what is the last step?
3. When working application problems involving boats moving in rivers, how does the current of the river affect the speed of the boat?
4. Write an application problem for which the solution depends on solving the system

$$x + y = 1{,}000$$
$$0.05x + 0.06y = 55$$

LINKING OBJECTIVES AND EXAMPLES

Next to each **objective** we have listed the examples that are best described by that objective.

A 1–7

Problem Set 4.5

Online support materials can be found at www.thomsonedu.com/login

Number Problems

▸ **1.** One number is 3 more than twice another. The sum of the numbers is 18. Find the two numbers.

2. The sum of two numbers is 32. One of the numbers is 4 less than 5 times the other. Find the two numbers.

3. The difference of two numbers is 6. Twice the smaller is 4 more than the larger. Find the two numbers.

4. The larger of two numbers is 5 more than twice the smaller. If the smaller is subtracted from the larger, the result is 12. Find the two numbers.

5. The sum of three numbers is 8. Twice the smallest is 2 less than the largest, and the sum of the largest and smallest is 5. Use a linear system in three variables to find the three numbers.

6. The sum of three numbers is 14. The largest is 4 times the smallest, the sum of the smallest and twice the largest is 18. Use a linear system in three variables to find the three numbers.

Ticket and Interest Problems

7. A total of 925 tickets were sold for a game for a total of $1,150. If adult tickets sold for $2.00 and children's tickets sold for $1.00, how many of each kind of ticket were sold?

8. If tickets for a show cost $2.00 for adults and $1.50 for children, how many of each kind of ticket were sold if a total of 300 tickets were sold for $525?

9. Mr. Jones has $20,000 to invest. He invests part at 6% and the rest at 7%. If he earns $1,280 in interest after 1 year, how much did he invest at each rate?

10. A man invests $17,000 in two accounts. One account earns 5% interest per year and the other earns 6.5%. If his total interest after one year is $970, how much did he invest at each rate?

▸ **11.** Susan invests twice as much money at 7.5% as she does at 6%. If her total interest after a year is $840, how much does she have invested at each rate?

12. A woman earns $1,350 in interest from two accounts in a year. If she has three times as much invested at 7% as she does at 6%, how much does she have in each account?

13. A man invests $2,200 in three accounts that pay 6%, 8%, and 9% in annual interest, respectively. He has three times as much invested at 9% as he does at 6%. If his total interest for the year is $178, how much is invested at each rate?

14. A student has money in three accounts that pay 5%, 7%, and 8% in annual interest. She has three times as much invested at 8% as she does at 5%. If the total amount she has invested is $1,600 and her interest for the year comes to $115, how much money does she have in each account?

Mixture Problems

▸ **15.** How many gallons of 20% alcohol solution and 50% alcohol solution must be mixed to get 9 gallons of 30% alcohol solution?

16. How many ounces of 30% hydrochloric acid solution and 80% hydrochloric acid solution must be mixed to get 10 ounces of 50% hydrochloric acid solution?

17. A mixture of 16% disinfectant solution is to be made from 20% and 14% disinfectant solutions. How much of each solution should be used if 15 gallons of the 16% solution are needed?

18. Paul mixes nuts worth $1.55 per pound with oats worth $1.35 per pound to get 25 pounds of trail mix worth $1.45 per pound. How many pounds of nuts and how many pounds of oats did he use?

19. **Metal Alloys** Metal workers solve systems of equations when forming metal alloys. If a certain metal alloy is 40% copper and another alloy is 60% copper, then a system of equations may be written to determine the amount of each alloy necessary to make 50 pounds of a metal alloy that is 55% copper. Write the system and determine this amount.

▢ = Videos available by instructor request
▸ = Online student support materials available at www.thomsonedu.com/login

20. A chemist has three different acid solutions. The first acid solution contains 20% acid, the second contains 40%, and the third contains 60%. He wants to use all three solutions to obtain a mixture of 60 liters containing 50% acid, using twice as much of the 60% solution as the 40% solution. How many liters of each solution should be used?

Rate Problems

21. It takes about 2 hours to travel 24 miles downstream and 3 hours to travel 18 miles upstream. What is the speed of the boat in still water? What is the speed of the current of the river?

22. A boat on a river travels 20 miles downstream in only 2 hours. It takes the same boat 6 hours to travel 12 miles upstream. What are the speed of the boat and the speed of the current?

23. An airplane flying with the wind can cover a certain distance in 2 hours. The return trip against the wind takes $2\frac{1}{2}$ hours. How fast is the plane and what is the speed of the wind, if the one-way distance is 600 miles?

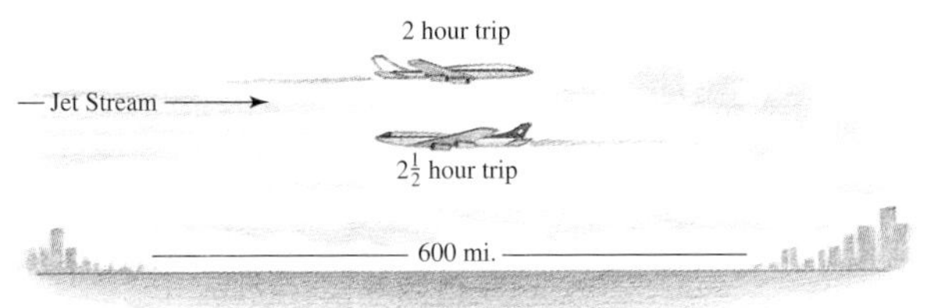

24. An airplane covers a distance of 1,500 miles in 3 hours when it flies with the wind and $3\frac{1}{3}$ hours when it flies against the wind. What is the speed of the plane in still air?

Coin Problems

25. Bob has 20 coins totaling $1.40. If he has only dimes and nickels, how many of each coin does he have?

26. If Amy has 15 coins totaling $2.70, and the coins are quarters and dimes, how many of each coin does she have?

27. A collection of nickels, dimes, and quarters consists of 9 coins with a total value of $1.20. If the number of dimes is equal to the number of nickels, find the number of each type of coin.

28. A coin collection consists of 12 coins with a total value of $1.20. If the collection consists only of nickels, dimes, and quarters, and the number of dimes is two more than twice the number of nickels, how many of each type of coin are in the collection?

29. A collection of nickels, dimes, and quarters amounts to $10.00. If there are 140 coins in all and there are twice as many dimes as there are quarters, find the number of nickels.

30. A cash register contains a total of 95 coins consisting of pennies, nickels, dimes, and quarters. There are only 5 pennies and the total value of the coins is $12.05. Also, there are 5 more quarters than dimes. How many of each coin is in the cash register?

Additional Problems

31. **Price and Demand** A manufacturing company finds that they can sell 300 items if the price per item is $2.00, and 400 items if the price is $1.50 per item. If the relationship between the number of items sold x and the price per item p is a linear one, find a formula that gives x in terms of p. Then use the formula to find the number of items they will sell if the price per item is $3.00.

32. **Price and Demand** A company manufactures and sells bracelets. They have found from experience that they can sell 300 bracelets each week if the price per bracelet is $2.00, but only 150 bracelets are sold if the price is $2.50 per bracelet. If the relationship between the number of bracelets sold x and the price per bracelet p is a linear one, find a

formula that gives x in terms of p. Then use the formula to find the number of bracelets they will sell at \$3.00 each.

33. Height of a Ball A ball is tossed into the air so that the height after 1, 3, and 5 seconds is as given in the following table. If the relationship between the height of the ball h and the time t is quadratic, then the relationship can be written as

$$h = at^2 + bt + c$$

Use the information in the table to write a system of three equations in three variables a, b, and c. Solve the system to find the exact relationship between h and t.

t (sec)	h (ft)
1	128
3	128
5	0

34. Height of a Ball A ball is tossed into the air and its height above the ground after 1, 3, and 4 seconds is recorded as shown in the following table. The relationship between the height of the ball h and the time t is quadratic and can be written as

$$h = at^2 + bt + c$$

Use the information in the table to write a system of three equations in three variables a, b, and c. Solve the system to find the exact relationship between the variables h and t.

t (sec)	h (ft)
1	96
3	64
4	0

Maintaining Your Skills

Graph each inequality.

35. $2x + 3y < 6$

36. $2x + y < -5$

37. $y \geq -3x - 4$

38. $y \geq 2x - 1$

39. $x \geq 3$

40. $y > -5$

Getting Ready for the Next Section

41. Does the graph of $x + y < 4$ include the boundary line?

42. Does the graph of $-x + y \leq 3$ include the boundary line?

43. Where do the graphs of the lines $x + y = 4$ and $x - 2y = 4$ intersect?

44. Where do the graphs of the line $x = -1$ and $x - 2y = 4$ intersect?

Solve.

45. $20x + 9{,}300 > 18{,}000$

46. $20x + 4{,}800 > 18{,}000$

Extending the Concepts

47. High School Dropout Rate The high school dropout rates for males and females over the years 1965 to 1990 are shown in the following table.

Year	Female Dropout Rate (%)	Male Dropout Rate (%)
1965	18	16
1970	16	14
1975	15	13
1980	13	16
1985	12	15
1990	13	14

Plotting the years along the horizontal axis and the dropout rates along the vertical axis, draw a line graph for these data. Draw the female line dashed and the male line solid for easier reading and comparison.

48. High School Dropout Rate Refer to the data in the preceding exercise to answer the following questions.

a. Using the slope, determine the time interval when the decline in the female dropout rate was the steepest.

b. Using the slope, determine the time interval when the increase in the male dropout rate was the steepest.

c. What appears unusual about the time period from 1975 to 1980?

d. Are the dropout rates for males and females generally increasing or generally decreasing?

49. **High School Dropout Rate** Refer to the data about high school dropout rates for males and females.
   a. Write a linear equation for the dropout rate of males, M, based on the year, x, for the years 1975 and 1980.
   b. Write a linear equation for the dropout rate of females, F, based on the year, x, for the years 1975 and 1980.
   c. Using your results from parts (a) and (b), determine when the dropout rates for males and females were the same.

4.6 Systems of Linear Inequalities

OBJECTIVES

A Graph the solution to a system of linear inequalities in two variables.

In the previous chapter, we graphed linear inequalities in two variables. To review, we graph the boundary line, using a solid line if the boundary is part of the solution set and a broken line if the boundary is not part of the solution set. Then we test any point that is not on the boundary line in the original inequality. A true statement tells us that the point lies in the solution set; a false statement tells us the solution set is the other region.

Figure 1 shows the graph of the inequality $x + y < 4$. Note that the boundary is not included in the solution set and is therefore drawn with a broken line. Figure 2 shows the graph of $-x + y \leq 3$. Note that the boundary is drawn with a solid line because it is part of the solution set.

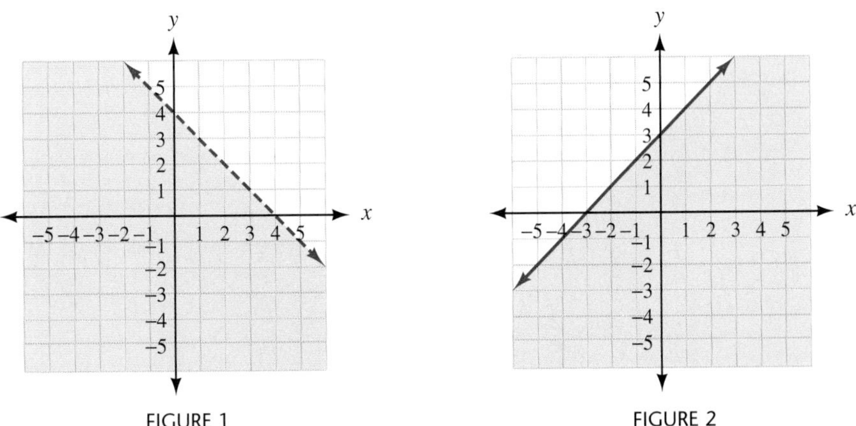

FIGURE 1 FIGURE 2

If we form a system of inequalities with the two inequalities, the solution set will be all the points common to both solution sets shown in the two figures above: It is the intersection of the two solution sets. Therefore, the solution set for the system of inequalities.

$$x + y < 4$$
$$-x + y \leq 3$$

is all the ordered pairs that satisfy both inequalities. It is the set of points that are below the line $x + y = 4$ and also below (and including) the line $-x + y = 3$.

The graph of the solution set to this system is shown in Figure 3. We have written the system in Figure 3 with the word *and* just to remind you that the solution set to a system of equations or inequalities is all the points that satisfy both equations or inequalities.

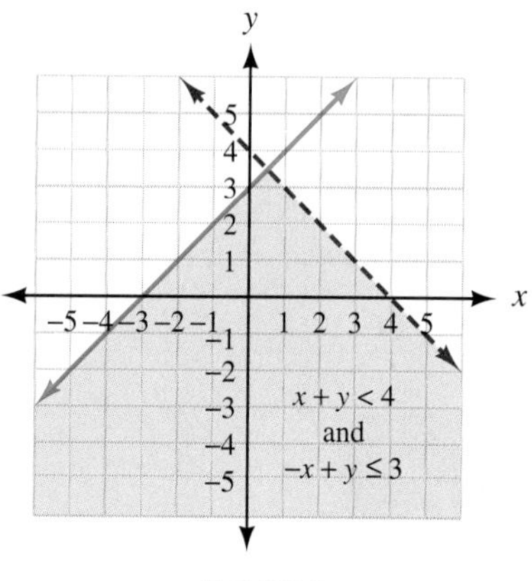

FIGURE 3

EXAMPLE 1

Graph the solution to the system of linear inequalities.

$$y < \frac{1}{2}x + 3$$

$$y \ge \frac{1}{2}x - 2$$

SOLUTION Figures 4 and 5 show the solution set for each of the inequalities separately.

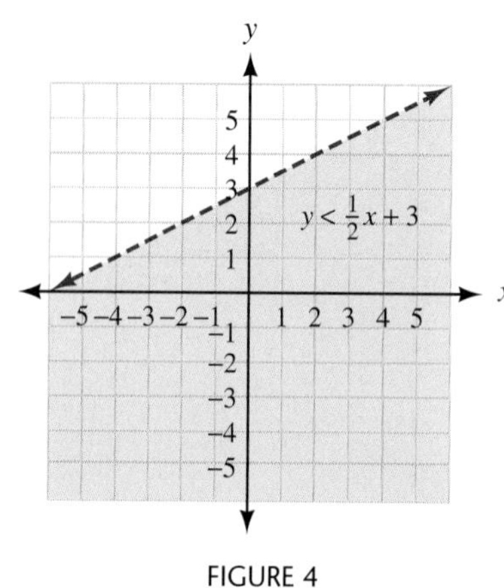

FIGURE 4

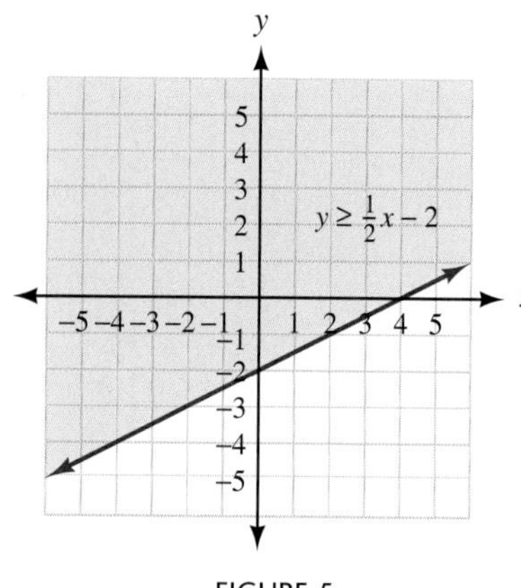

FIGURE 5

Figure 6 is the solution set to the system of inequalities. It is the region consisting of points whose coordinates satisfy both inequalities.

FIGURE 6

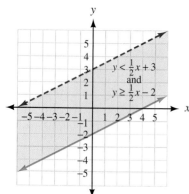

EXAMPLE 2 Graph the solution to the system of linear inequalities.

$$x + y < 4$$
$$x \geq 0$$
$$y \geq 0$$

SOLUTION We graphed the first inequality, $x + y < 4$, in Figure 1 at the beginning of this section. The solution set to the inequality $x \geq 0$, shown in Figure 7, is all the points to the right of the y-axis; that is, all the points with x-coordinates that are greater than or equal to 0. Figure 8 shows the graph of $y \geq 0$. It consists of all points with y-coordinates greater than or equal to 0; that is, all points from the x-axis up.

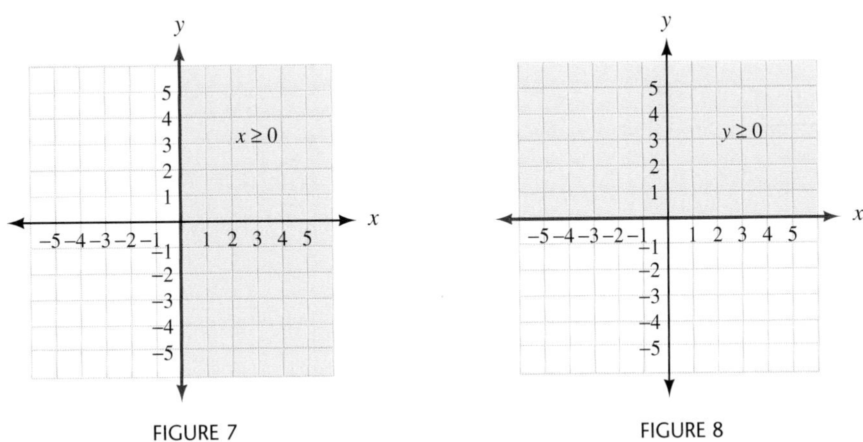

FIGURE 7

FIGURE 8

The regions shown in Figures 7 and 8 overlap in the first quadrant. Therefore, putting all three regions together we have the points in the first quadrant that are below the line $x + y = 4$. This region is shown in Figure 9, and it is the solution to our system of inequalities.

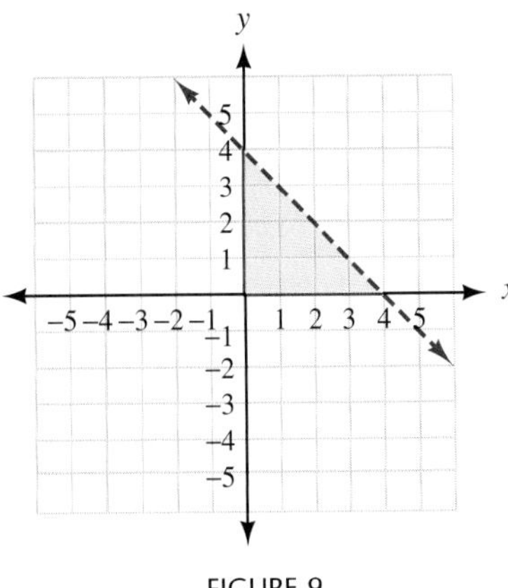

FIGURE 9

Extending the discussion in Example 2 we can name the points in each of the four quadrants using systems of inequalities.

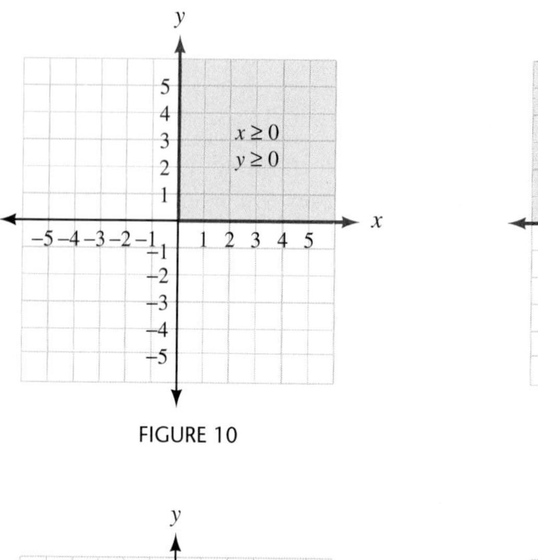

FIGURE 10

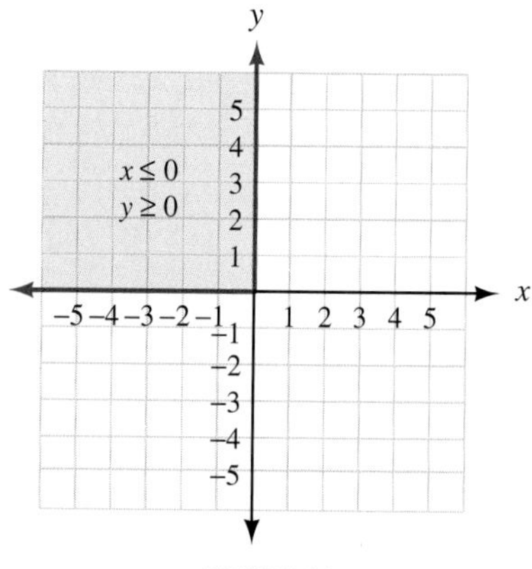

FIGURE 11

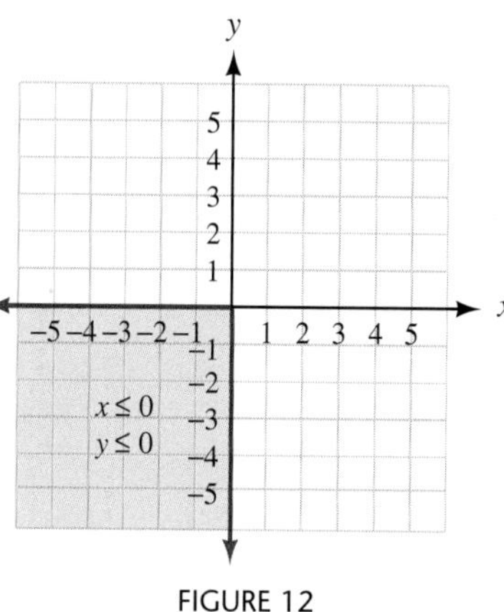

FIGURE 12

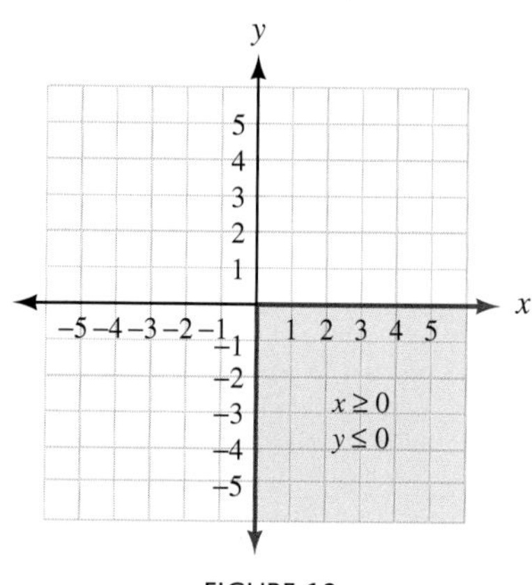

FIGURE 13

EXAMPLE 3

Graph the solution to the system of linear inequalities.

$$x \leq 4$$

$$y \geq -3$$

SOLUTION The solution to this system will consist of all points to the left of and including the vertical line $x = 4$ that intersect with all points above and including the horizontal line $y = -3$. The solution set is shown in Figure 14.

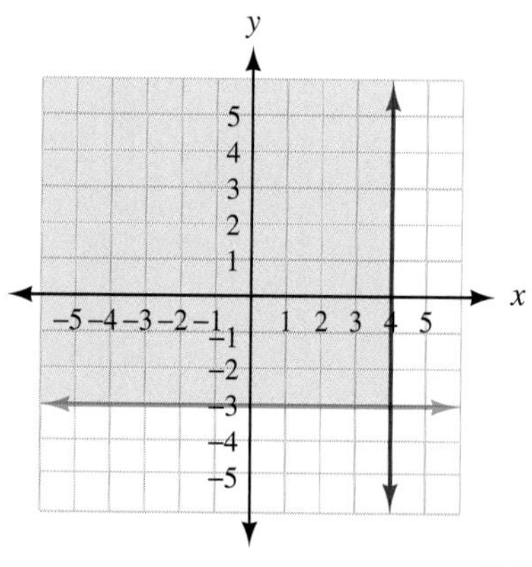

FIGURE 14

EXAMPLE 4 Graph the solution set for the following system.

$$x - 2y \le 4$$
$$x + y \le 4$$
$$x \ge -1$$

SOLUTION We have three linear inequalities, representing three sections of the coordinate plane. The graph of the solution set for this system will be the intersection of these three sections. The graph of $x - 2y \le 4$ is the section above and including the boundary $x - 2y = 4$. The graph of $x + y \le 4$ is the section below and including the boundary line $x + y = 4$. The graph of $x \ge -1$ is all the points to the right of, and including, the vertical line $x = -1$. The intersection of these three graphs is shown in Figure 15.

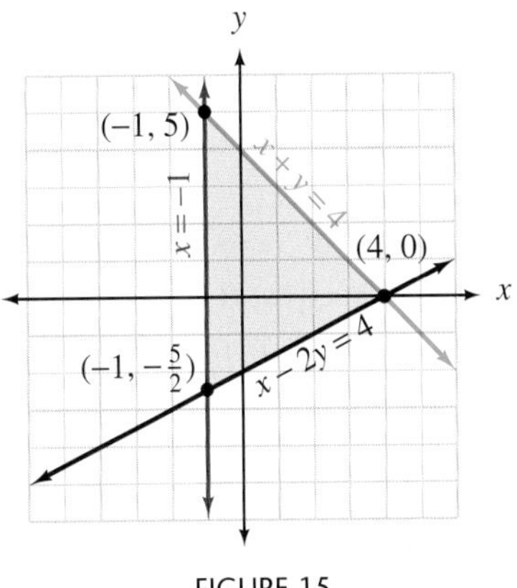

FIGURE 15

EXAMPLE 5 A college basketball arena plans on charging \$20 for certain seats and \$15 for others. They want to bring in more than \$18,000 from all ticket sales and have reserved at least 500 tickets at the \$15 rate. Find a system of inequalities describing all possibilities and sketch the graph. If 620 tickets are sold for \$15, at least how many tickets are sold for \$20?

© Peter Jones/Reuters New Media/Corbis

SOLUTION Let x = the number of \$20 tickets and y = the number of \$15 tickets. We need to write a list of inequalities that describe this situation. That list will form our system of inequalities. First of all, we note that we cannot use negative numbers for either x or y. So, we have our first inequalities:

$$x \geq 0$$
$$y \geq 0$$

Next, we note that they are selling at least 500 tickets for \$15, so we can replace our second inequality with $y \geq 500$. Now our system is

$$x \geq 0$$
$$y \geq 500$$

Now, the amount of money brought in by selling \$20 tickets is $20x$, and the amount of money brought in by selling \$15 tickets is $15y$. If the total income from ticket sales is to be more than \$18,000, then $20x + 15y$ must be greater than 18,000. This gives us our last inequality and completes our system.

$$20x + 15y > 18{,}000$$
$$x \geq 0$$
$$y \geq 500$$

We have used all the information in the problem to arrive at this system of inequalities. The solution set contains all the values of x and y that satisfy all the conditions given in the problem. Here is the graph of the solution set.

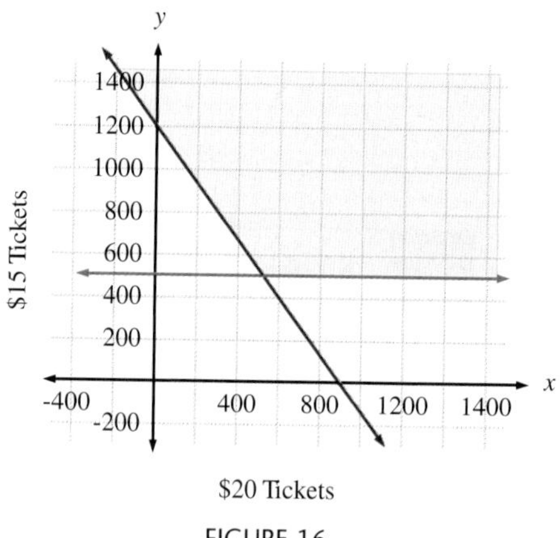

FIGURE 16

If 620 tickets are sold for \$15, then we substitute 620 for y in our first inequality to obtain

$20x + 15(620) > 18{,}000$	**Substitute 620 for y**
$20x + 9{,}300 > 18{,}000$	**Multiply**
$20x > 8{,}700$	**Add −9,300 to each side**
$x > 435$	**Divide each side by 20**

If they sell 620 tickets for \$15 each, then they need to sell more than 435 tickets at \$20 each to bring in more than \$18,000.

GETTING READY FOR CLASS

After reading through the preceding section, respond in your own words and in complete sentences.

1. What is the solution set to a system of inequalities?
2. When graphing a system of linear inequalities, how do you find the equations of the boundary lines?
3. For the boundary lines of a system of linear inequalities, when do you use a dotted line rather than a solid line?
4. Once you have graphed the solution set for each inequality in a system, how do you determine the region to shade for the solution to the system of inequalities?

LINKING OBJECTIVES AND EXAMPLES

Next to each **objective** we have listed the examples that are best described by that objective.

A 1–5

Problem Set 4.6

Online support materials can be found at www.thomsonedu.com/login

Graph the solution set for each system of linear inequalities.

1. $x + y < 5$
   $2x - y > 4$
2. $x + y < 5$
   $2x - y < 4$
3. $y < \frac{1}{3}x + 4$
   $y \geq \frac{1}{3}x - 3$
4. $y < 2x + 4$
   $y \geq 2x - 3$
5. $x \geq -3$
   $y < 2$
6. $x \leq 4$
   $y > -2$
7. $1 \leq x \leq 3$
   $2 \leq y \leq 4$
8. $-4 \leq x \leq -2$
   $1 \leq y \leq 3$
9. $x + y \leq 4$
   $x \geq 0$
   $y \geq 0$
10. $x - y \leq 2$
    $x \geq 0$
    $y \leq 0$
11. $x + y \leq 3$
    $x - 3y \leq 3$
    $x \geq -2$
12. $x - y \leq 4$
    $x + 2y \leq 4$
    $x \geq -1$
13. $x + y \leq 2$
    $-x + y \leq 2$
    $y \geq -2$
14. $x - y \leq 3$
    $-x - y \leq 3$
    $y \leq -1$
15. $x + y < 5$
    $y > x$
    $y \geq 0$
16. $x + y < 5$
    $y > x$
    $x \geq 0$
17. $2x + 3y \leq 6$
    $x \geq 0$
    $y \geq 0$
18. $x + 2y \leq 10$
    $3x + y \leq 12$
    $x \geq 0$

For each figure below, find a system of inequalities that describes the shaded region.

19.

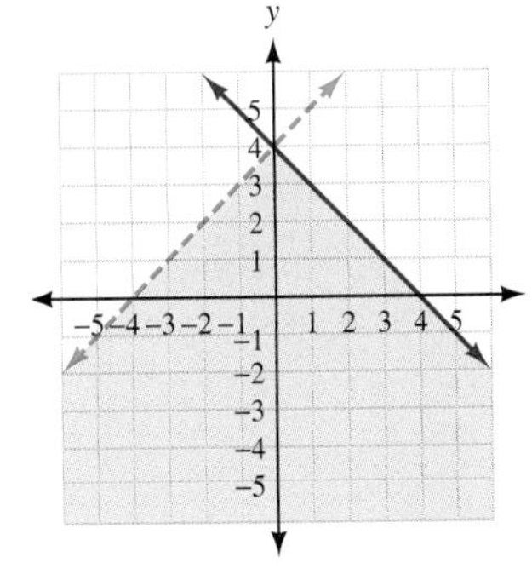

FIGURE 17

20.

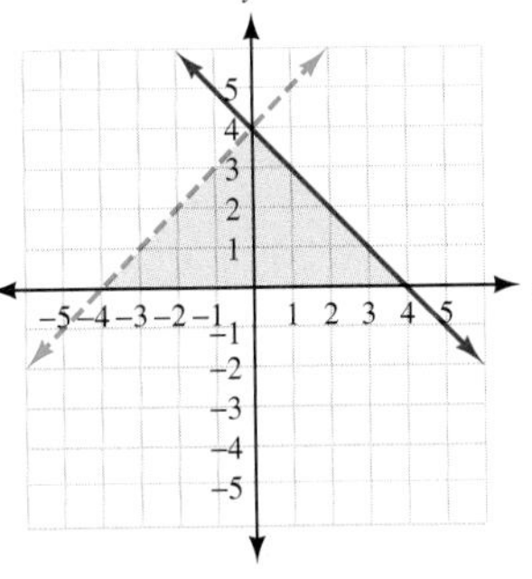

FIGURE 18

21.

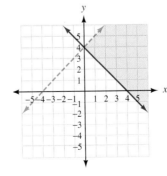

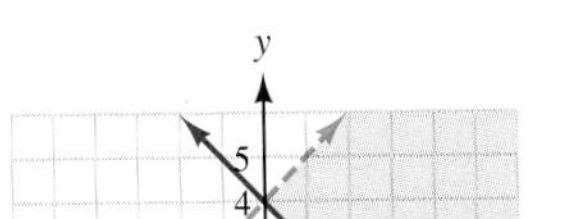

FIGURE 19

22.

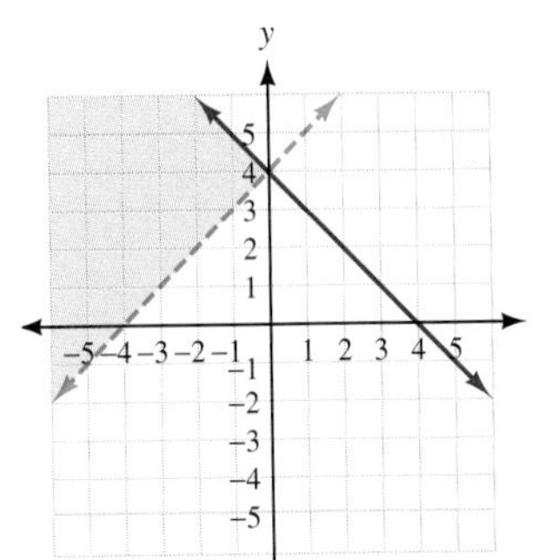

FIGURE 20

▢ = Videos available by instructor request

▶ = Online student support materials available at www.thomsonedu.com/login

Applying the Concepts

23. **Office Supplies** An office worker wants to purchase some \$0.55 postage stamps and also some \$0.65 postage stamps totaling no more than \$40. It also is desired to have at least twice as many \$0.55 stamps and more than 15 \$0.55 stamps.
    a. Find a system of inequalities describing all the possibilities and sketch the graph.
    b. If he purchases 20 \$0.55 stamps, what is the maximum number of \$0.65 stamps he can purchase?

24. **Inventory** A store sells two brands of VCRs. Customer demand indicates that it is necessary to stock at least twice as many VCRs of brand A as of brand B. At least 30 of brand A and 15 of brand B must be on hand. There is room for not more than 100 VCRs in the store.
    a. Find a system of inequalities describing all possibilities, then sketch the graph.
    b. If there are 35 VCRs of brand A, what is the maximum number of brand B VCRs on hand?

Maintaining Your Skills

For each of the following straight lines, identify the x-intercept, y-intercept, and slope, and sketch the graph.

25. $2x + y = 6$

26. $y = \frac{3}{2}x + 4$

27. $x = -2$

Find the equation for each line.

28. Give the equation of the line through $(-1, 3)$ that has slope $m = 2$.

29. Give the equation of the line through $(-3, 2)$ and $(4, -1)$.

30. Line l contains the point $(5, -3)$ and has a graph parallel to the graph of $2x - 5y = 10$. Find the equation for l.

31. Give the equation of the vertical line through $(4, -7)$.

State the domain and range for the following relations, and indicate which relations are also functions.

32. $\{(-2, 0), (-3, 0), (-2, 1)\}$

33. $y = x^2 - 9$

Let $f(x) = x - 2$, $g(x) = 3x + 4$ and $h(x) = 3x^2 - 2x - 8$, and find the following.

34. $f(3) + g(2)$

35. $h(0) + g(0)$

36. $f[g(2)]$

37. $g[f(2)]$

Solve the following variation problems.

38. **Direct Variation** Quantity y varies directly with the square of x. If y is 50 when x is 5, find y when x is 3.

39. **Joint Variation** Quantity z varies jointly with x and the cube of y. If z is 15 when x is 5 and y is 2, find z when x is 2 and y is 3.

Chapter 4 SUMMARY

EXAMPLES

1. The solution to the system
$$x + 2y = 4$$
$$x - y = 1$$
is the ordered pair (2, 1). It is the only ordered pair that satisfies both equations.

Systems of Linear Equations [4.1, 4.2]

A system of linear equations consists of two or more linear equations considered simultaneously. The solution set to a linear system in two variables is the set of ordered pairs that satisfy both equations. The solution set to a linear system in three variables consists of all the ordered triples that satisfy each equation in the system.

2. We can eliminate the y-variable from the system in Example 1 by multiplying both sides of the second equation by 2 and adding the result to the first equation:

$$\begin{aligned} x + 2y = 4 &\xrightarrow{\text{No change}} & x + 2y &= 4 \\ x - y = 1 &\xrightarrow[\text{Multiply by 2}]{} & 2x - 2y &= 2 \\ & & \hline 3x &= 6 \\ & & x &= 2 \end{aligned}$$

Substituting $x = 2$ into either of the original two equations gives $y = 1$. The solution is (2, 1).

To Solve Systems by the Addition Method [4.1]

Step 1: Look the system over to decide which variable will be easier to eliminate.

Step 2: Use the multiplication property of equality on each equation separately, if necessary, to ensure that the coefficients of the variable to be eliminated are opposites.

Step 3: Add the left and right sides of the system produced in step 2, and solve the resulting equation.

Step 4: Substitute the solution from step 3 back into any equation with both x- and y-variables, and solve.

Step 5: Check your solution in both equations if necessary.

3. We can apply the substitution method to the system in Example 1 by first solving the second equation for x to get
$$x = y + 1$$
Substituting this expression for x into the first equation we have
$$y + 1 + 2y = 4$$
$$3y + 1 = 4$$
$$3y = 3$$
$$y = 1$$
Using $y = 1$ in either of the original equations gives $x = 2$.

To Solve Systems by the Substitution Method [4.1]

Step 1: Solve either of the equations for one of the variables (this step is not necessary if one of the equations has the correct form already).

Step 2: Substitute the results of step 1 into the other equation, and solve.

Step 3: Substitute the results of step 2 into an equation with both x- and y-variables, and solve. (The equation produced in step 1 is usually a good one to use.)

Step 4: Check your solution if necessary.

Inconsistent and Dependent Equations [4.1, 4.2]

4. If the two lines are parallel, then the system will be inconsistent and the solution is $\varnothing$. If the two lines coincide, then the equations are dependent.

A system of two linear equations that have no solutions in common is said to be an *inconsistent* system, whereas two linear equations that have all their solutions in common are said to be *dependent* equations.

2 × 2 Determinants [4.3]

5. $\begin{vmatrix} 3 & 4 \\ -2 & 5 \end{vmatrix} = 15 - (-8) = 23$

The value of a 2 × 2 determinant is as follows:

$$\begin{vmatrix} a & c \\ b & d \end{vmatrix} = ad - bc$$

3 × 3 Determinants [4.3]

6. Expanding $\begin{vmatrix} 1 & 3 & -2 \\ 2 & 0 & 1 \\ 4 & -1 & 1 \end{vmatrix}$ across the first row gives us

$$1\begin{vmatrix} 0 & 1 \\ -1 & 1 \end{vmatrix} - 3\begin{vmatrix} 2 & 1 \\ 4 & 1 \end{vmatrix} - 2\begin{vmatrix} 2 & 0 \\ 4 & -1 \end{vmatrix}$$

$= 1(1) - 3(-2) - 2(-2)$
$= 11$

The value of a 3 × 3 determinant is given by

$$\begin{vmatrix} a_1 & b_1 & c_1 \\ a_2 & b_2 & c_2 \\ a_3 & b_3 & c_3 \end{vmatrix} = a_1b_2c_3 + a_3b_1c_2 + a_2b_3c_1 - a_3b_2c_1 - a_1b_3c_2 - a_2b_1c_3$$

There are two methods of finding the six products in the expansion of a 3 × 3 determinant. One method involves a cross-multiplication scheme. The other method involves expanding the determinant by minors.

Cramer's Rule for a Linear System in Two Variables [4.4]

7. For the system

$$x + y = 6$$
$$3x - 2y = -2$$

we have

$$D = \begin{vmatrix} 1 & 1 \\ 3 & -2 \end{vmatrix} = -5$$

$$D_x = \begin{vmatrix} 6 & 1 \\ -2 & -2 \end{vmatrix} = -10$$

$$x = \frac{-10}{-5} = 2$$

$$D_y = \begin{vmatrix} 1 & 6 \\ 3 & -2 \end{vmatrix} = -20$$

$$y = \frac{-20}{-5} = 4$$

The solution to the system

$$a_1x + b_1y = c_1$$
$$a_2x + b_2y = c_2$$

is given by

$$x = \frac{D_x}{D} \quad \text{and} \quad y = \frac{D_y}{D} \quad (D \neq 0)$$

where

$$D = \begin{vmatrix} a_1 & b_1 \\ a_2 & b_2 \end{vmatrix}, \quad D_x = \begin{vmatrix} c_1 & b_1 \\ c_2 & b_2 \end{vmatrix}, \quad D_y = \begin{vmatrix} a_1 & c_1 \\ a_2 & c_2 \end{vmatrix}$$

Cramer's Rule for a Linear System in Three Variables [4.4]

The solution to the system

$$a_1x + b_1y + c_1z = d_1$$
$$a_2x + b_2y + c_2z = d_2$$
$$a_3x + b_3y + c_3z = d_3$$

is given by

$$x = \frac{D_x}{D}, \quad y = \frac{D_y}{D}, \quad \text{and} \quad z = \frac{D_z}{D} \qquad (D \neq 0)$$

where

$$D = \begin{vmatrix} a_1 & b_1 & c_1 \\ a_2 & b_2 & c_2 \\ a_3 & b_3 & c_3 \end{vmatrix} \qquad D_x = \begin{vmatrix} d_1 & b_1 & c_1 \\ d_2 & b_2 & c_2 \\ d_3 & b_3 & c_3 \end{vmatrix}$$

$$D_y = \begin{vmatrix} a_1 & d_1 & c_1 \\ a_2 & d_2 & c_2 \\ a_3 & d_3 & c_3 \end{vmatrix} \qquad D_z = \begin{vmatrix} a_1 & b_1 & d_1 \\ a_2 & b_2 & d_2 \\ a_3 & b_3 & d_3 \end{vmatrix}$$

8. For the system

$$x + y = -1$$
$$2x - z = 3$$
$$y + 2z = -1$$

$$D = \begin{vmatrix} 1 & 1 & 0 \\ 2 & 0 & -1 \\ 0 & 1 & 2 \end{vmatrix} = -3$$

$$D_x = \begin{vmatrix} -1 & 1 & 0 \\ 3 & 0 & -1 \\ -1 & 1 & 2 \end{vmatrix} = -6$$

$$x = \frac{-6}{-3} = 2$$

$$D_y = \begin{vmatrix} 1 & -1 & 0 \\ 2 & 3 & -1 \\ 0 & -1 & 2 \end{vmatrix} = 9$$

$$y = \frac{9}{-3} = -3$$

$$D_z = \begin{vmatrix} 1 & 1 & -1 \\ 2 & 0 & 3 \\ 0 & 1 & -1 \end{vmatrix} = -3$$

$$z = \frac{-3}{-3} = 1$$

Using a System of Equations [4.5]

9. One number is 2 more than 3 times another. Their sum is 26. Find the numbers.

Step 1: ***Read*** the problem, and then mentally ***list*** the items that are known and the items that are unknown.

Step 1: Read and List.
Known items: two numbers, whose sum is 26. One is 2 more than 3 times the other. Unknown items are the two numbers.

Step 2: ***Assign variables*** to each of the unknown items; that is, let x = one of the unknown items and y = the other unknown item (and z = the third unknown item, if there is a third one). Then ***translate*** the other ***information*** in the problem to expressions involving the two (or three) variables.

Step 2: Assign variables and translate information.
Let x = one of the numbers.
Then y = the other number.

Step 3: ***Reread*** the problem, and then ***write a system of equations,*** using the items and variables listed in steps 1 and 2, that describes the situation.

Step 3: Write a system of equations.

$$x + y = 26$$
$$y = 3x + 2$$

Step 4: Solve the system. Substituting the expression for y from the second equation into the first and solving for x yields

$$x + (3x + 2) = 26$$
$$4x + 2 = 26$$
$$4x = 24$$
$$x = 6$$

Using $x = 6$ in $y = 3x + 2$ gives the second number:

$$y = 20$$

Step 5: Write answers. The two numbers are 6 and 20.

Step 6: Reread and check. The sum of 6 and 20 is 26, and 20 is 2 more than 3 times 6.

Step 4: Solve the system found in step 3.

Step 5: Write your answers using complete sentences.

Step 6: Reread the problem, and ***check*** your solution with the original words in the problem.

Systems of Linear Inequalities [4.6]

10. The solution set for the system

$$x + y < 4$$
$$-x + y \le 3$$

is shown below.

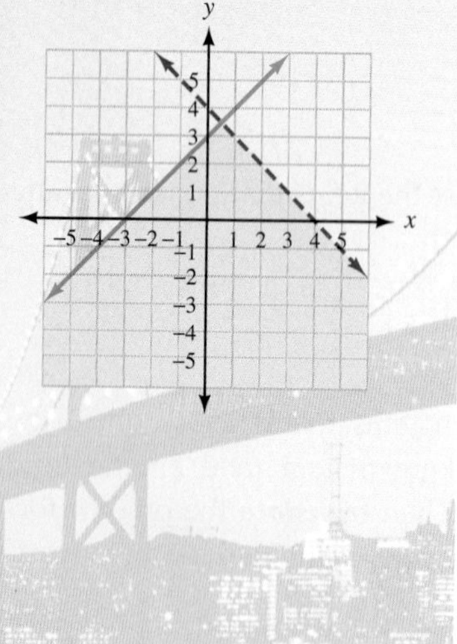

A system of linear inequalities is two or more linear inequalities considered at the same time. To find the solution set to the system, we graph each of the inequalities on the same coordinate system. The solution set is the region that is common to all the regions graphed.

Chapter 4 Review Test

The problems below form a comprehensive review of the material in this chapter. They can be used to study for exams. If you would like to take a practice test on this chapter, you can use the odd-numbered problems. Give yourself an hour and work as many of the odd-numbered problems as possible. When you are finished, or when an hour has passed, check your answers with the answers in the back of the book. You can use the even-numbered problems for a second practice test.

Solve each system using the addition method. [4.1]

1. $x + y = 4$
$2x - y = 14$

2. $3x + y = 2$
$2x + y = 0$

3. $2x - 4y = 5$
$-x + 2y = 3$

4. $5x - 2y = 7$
$3x + y = 2$

5. $6x - 5y = -5$
$3x + y = 1$

6. $6x + 4y = 8$
$9x + 6y = 12$

7. $3x - 7y = 2$
$-4x + 6y = -6$

8. $6x + 5y = 9$
$4x + 3y = 6$

9. $-7x + 4y = -1$
$5x - 3y = 0$

10. $\frac{1}{2}x - \frac{3}{4}y = -4$
$\frac{1}{4}x + \frac{3}{2}y = 13$

11. $\frac{2}{3}x - \frac{1}{6}y = 0$
$\frac{4}{3}x + \frac{5}{6}y = 14$

12. $-\frac{1}{2}x + \frac{1}{3}y = -\frac{13}{6}$
$\frac{4}{5}x + \frac{3}{4}y = \frac{9}{10}$

Solve each system by the substitution method. [4.1]

13. $x + y = 2$
$y = x - 1$

14. $2x - 3y = 5$
$y = 2x - 7$

15. $x + y = 4$
$2x + 5y = 2$

16. $x + y = 3$
$2x + 5y = -6$

17. $3x + 7y = 6$
$x = -3y + 4$

18. $5x - y = 4$
$y = 5x - 3$

Solve each system. [4.2]

19. $x + y + z = 6$
$x - y - 3z = -8$
$x + y - 2z = -6$

20. $3x + 2y + z = 4$
$2x - 4y + z = -1$
$x + 6y + 3z = -4$

21. $5x + 8y - 4z = -7$
$7x + 4y + 2z = -2$
$3x - 2y + 8z = 8$

22. $5x - 3y - 6z = 5$
$4x - 6y - 3z = 4$
$-x + 9y + 9z = 7$

23. $5x - 2y + z = 6$
$-3x + 4y - z = 2$
$6x - 8y + 2z = -4$

26. $x - y = 2$
$y - z = -3$
$x - z = -1$

25. $2x - y = 5$
$3x - 2z = -2$
$5y + z = -1$

24. $4x - 6y + 8z = 4$
$5x + y - 2z = 4$
$6x - 9y + 12z = 6$

Evaluate each determinant. [4.3]

27. $\begin{vmatrix} 2 & 3 \\ -5 & 4 \end{vmatrix}$

28. $\begin{vmatrix} 3 & 0 \\ 5 & -1 \end{vmatrix}$

29. $\begin{vmatrix} 1 & 0 \\ -7 & -3 \end{vmatrix}$

30. $\begin{vmatrix} 3 & 0 & 2 \\ -1 & 4 & 0 \\ 2 & 0 & 0 \end{vmatrix}$

31. $\begin{vmatrix} 3 & -1 & 0 \\ 0 & 2 & -4 \\ 6 & 0 & 2 \end{vmatrix}$

32. $\begin{vmatrix} -3 & -2 & 0 \\ 0 & -4 & 2 \\ 5 & 1 & 1 \end{vmatrix}$

Solve for x. [4.3]

33. $\begin{vmatrix} 2 & 3x \\ -1 & 2x \end{vmatrix} = 4$

34. $\begin{vmatrix} 4x & x \\ 3 & 1 \end{vmatrix} = -4$

Use Cramer's rule to solve each system. [4.4]

35. $3x - 5y = 4$
$7x - 2y = 3$

36. $7x - 5y = 8$
$4x + 3y = 2$

37. $3x - 6y = 9$
$2x - 4y = 6$

38. $6x - 9y = 5$
$7x + 3y = 4$

39. $-6x + 3y = 7$
$5x - 8y = -2$

40. $2x - y + 3z = 4$
$5x + 2y - z = 3$
$-x - 3y + 2z = 1$

41. $4x - 5y = -3$
$2x + 3z = 4$
$3y - z = 8$

42. $2x - 4y = 2$
$4x - 2z = 3$
$4y - z = 2$

Use systems of equations to solve each application problem. In each case, be sure to show the system used. [4.5]

43. Ticket Prices Tickets for the show cost \$2.00 for adults and \$1.50 for children. How many adult tickets and how many children's tickets were sold if a total of 127 tickets were sold for \$214?

44. Coin Collection John has 20 coins totaling \$3.20. If he has only dimes and quarters, how many of each coin does he have?

45. Investments Ms. Jones invests money in two accounts, one of which pays 12% per year, and the other pays 15% per year. If her total investment is \$12,000 and the interest after 1 year is \$1,650, how much is invested in each account?

46. Speed It takes a boat on a river 2 hours to travel 28 miles downstream and 3 hours to travel 30 miles upstream. What is the speed of the boat and the current of the river?

Graph the solution set for each system of linear inequalities. [4.6]

47. $3x + 4y < 12$
$-3x + 2y \le 6$

48. $3x + 4y < 12$
$-3x + 2y \le 6$
$y \ge 0$

49. $3x + 4y < 12$
$x \ge 0$
$y \ge 0$

50. $x > -3$
$y > -2$

Chapter 4 Projects

Systems of Linear Equations and Inequalities

GROUP PROJECT Break-Even Point

Number of People 2 or 3

Time Needed 10–15 minutes

Equipment Pencil and paper

Background The break-even point for a company occurs when the revenue from sales of a product equals the cost of producing the product. This group project is designed to give you more insight into revenue, cost, and break-even point.

Procedure A company is planning to open a factory to manufacture calculators.

1. It costs them \$120,000 to open the factory, and it will cost \$10 for each calculator they make. What is the expression for $C(x)$, the cost of making x calculators?
2. They can sell the calculators for \$50 each. What is the expression for $R(x)$, their revenue from selling x calculators? Remember that $R = px$, where p is the price per calculator.
3. Graph both the cost equation $C(x)$ and the revenue equation $R(x)$ on a coordinate system like the one below.

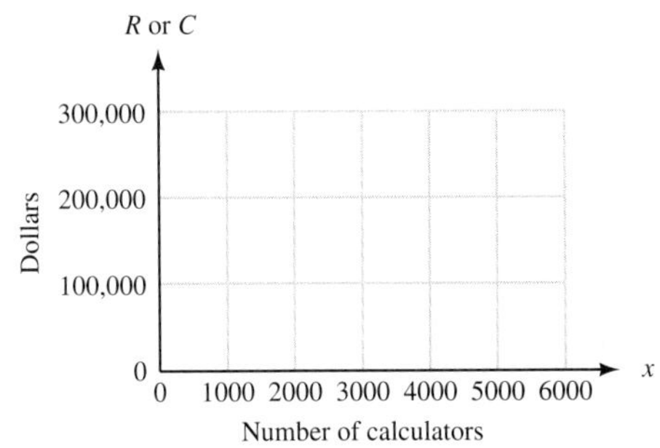

4. The break-even point is the value of x (the number of calculators) for which the revenue is equal to the cost. Where is the break-even point on the graph you produced in Part 3? Estimate the break-even point from the graph.
5. Set the cost equal to the revenue and solve to find x to find the exact value of the break-even point. How many calculators do they need to make and sell to exactly break even? What will be their revenue and their cost for that many calculators?
6. Write an inequality that gives the values of x that will produce a profit for the company. (A profit occurs when the revenue is larger than the cost.)
7. Write an inequality that gives the values of x that will produce a loss for the company. (A loss occurs when the cost is larger than the revenue.)
8. Profit is the difference between revenue and cost, or $P(x) = R(x) - C(x)$. Write the equation for profit and then graph it on a coordinate system like the one below.

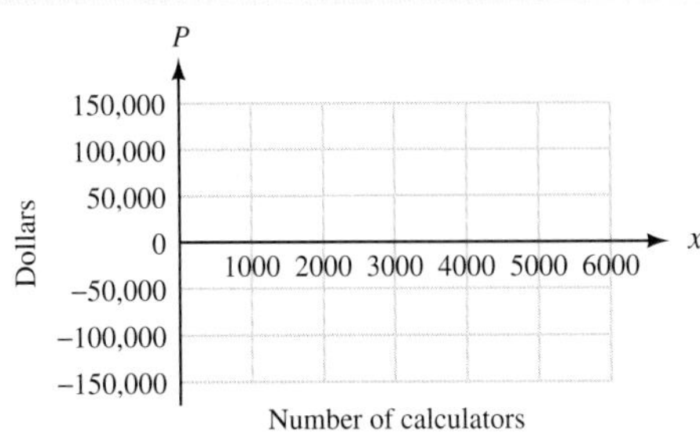

9. How do you recognize the break-even point and the regions of loss and profit on the graph you produced above?

RESEARCH PROJECT

Zeno's Paradoxes

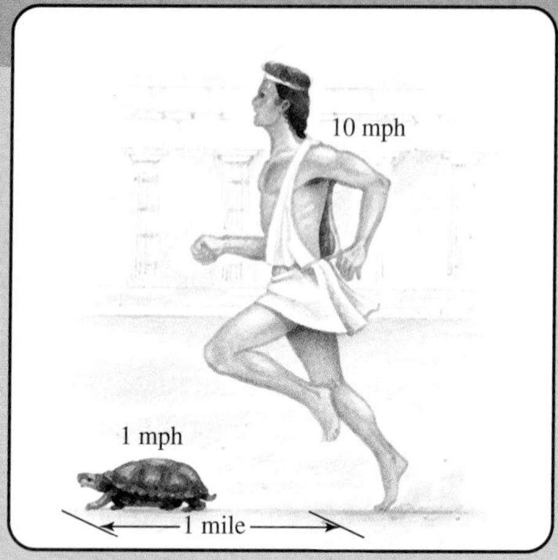

Zeno of Elea was born at about the same time that Pythagoras died. He is responsible for three paradoxes that have come to be known as Zeno's paradoxes. One of the three has to do with a race between Achilles and a tortoise. Achilles is much faster than the tortoise, but the tortoise has a head start. According to Zeno's method of reasoning, Achilles can never pass the tortoise because each time he reaches the place where the tortoise was, the tortoise is gone. Research Zeno's paradox concerning Achilles and the tortoise. Put your findings into essay form that begins with a definition for the word "paradox." Then use Zeno's method of reasoning to describe a race between Achilles and the tortoise—if Achilles runs at 10 miles per hour, the tortoise runs at 1 mile per hour, and the tortoise has a 1-mile head start. Next, use the methods shown in this chapter to find the distance at which Achilles reaches the tortoise and the time at which Achilles reaches the tortoise. Conclude your essay by summarizing what you have done and showing how the two results you have obtained form a paradox.

Exponents and Polynomials

5

Digital Vision/Getty Images

CHAPTER OUTLINE

If you go on to take a business course or an economics course, you will find yourself spending lots of time with the three expressions that form the mathematical foundation of business: profit, revenue, and cost. Many times these expressions are given as polynomials, the topic of this chapter. The relationship between the three expressions is known as the profit equation:

$$\text{Profit} = \text{Revenue} - \text{Cost}$$

$$P(x) = R(x) - C(x)$$

The table and graphs shown here were produced on a graphing calculator. They give numerical and graphical descriptions of revenue, profit, and cost for a company that manufactures and sells prerecorded videotapes according to the equations

$$R(x) = 11.5x - 0.05x^2 \quad \text{and} \quad C(x) = 200 + 2x$$

Revenue, Cost, and Profit on a Graphing Calculator

Number of Videotapes X	Revenue Y_1	Cost Y_2	Profit Y_3
0	0	200	−200
50	450	300	150
100	650	400	250
150	600	500	100
200	300	600	−300

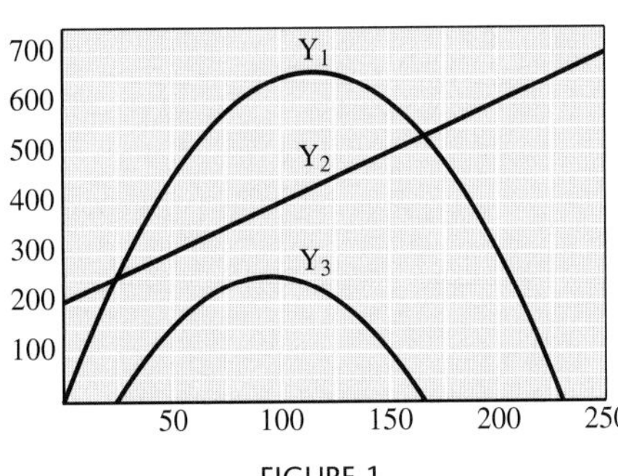

FIGURE 1

By studying the material in this chapter, you will get a head start on learning the equations and relationships that are emphasized in business and economics.

The study skills for this chapter are about attitude. They are points of view that point toward success.

1 Be Focused, Not Distracted

I have students who begin their assignments by asking themselves, "Why am I taking this class?" Or, "When am I ever going to use this stuff?" If you are asking yourself similar questions, you may be distracting yourself away from doing the things that will produce the results you want in this course. Don't dwell on questions and evaluations of the class that can be used as excuses for not doing well. If you want to succeed in this course, focus your energy and efforts toward success, rather than distracting yourself away from your goals.

2 Be Resilient

Don't let setbacks keep you from your goals. You want to put yourself on the road to becoming a person who can succeed in this class or any class in college. Failing a test or quiz, or having a difficult time on some topics, is normal. No one goes through college without some setbacks. Don't let a temporary disappointment keep you from succeeding in this course. A low grade on a test or quiz is simply a signal that some reevaluation of your study habits needs to take place.

3 Intend to Succeed

I always have a few students who simply go through the motions of studying without intending to master the material. It is more important to them to look like they are studying than to actually study. You need to study with the intention of being successful in the course. Intend to master the material, no matter what it takes.

5.1 Properties and Exponents

OBJECTIVES

A Simplify expressions using the properties of exponents.

B Convert back and forth between scientific notation and expanded form.

C Multiply and divide expressions written in scientific notation.

The figure shows a square and a cube, each with a side of length 1.5 centimeters. To find the area of the square, we raise 1.5 to the second power: 1.5^2. To find the volume of the cube, we raise 1.5 to the third power: 1.5^3.

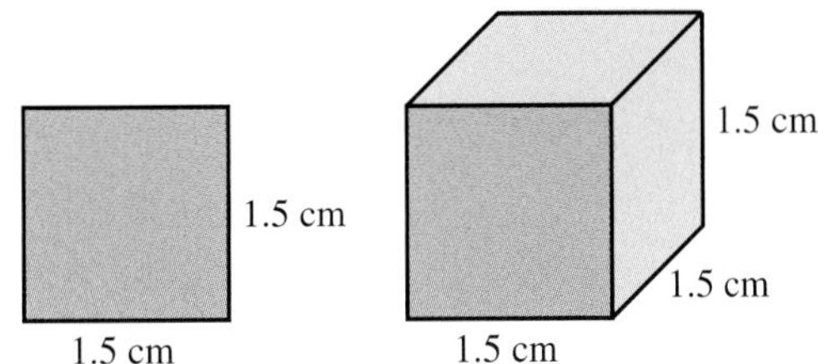

Because the area of the square is 1.5^2, we say second powers are *squares;* that is, x^2 is read "x squared." Likewise, since the volume of the cube is 1.5^3, we say third powers are *cubes,* that is, x^3 is read "x cubed." Exponents and the vocabulary associated with them are topics we will study in this section.

Properties of Exponents

In this section, we will be concerned with the simplification of expressions that involve exponents. We begin by making some generalizations about exponents, based on specific examples.

EXAMPLE 1 Write the product $x^3 \cdot x^4$ with a single exponent.

SOLUTION

$$x^3 \cdot x^4 = (x \cdot x \cdot x)(x \cdot x \cdot x \cdot x)$$
$$= (x \cdot x \cdot x \cdot x \cdot x \cdot x \cdot x)$$
$$= x^7 \quad \textit{Notice: } 3 + 4 = 7$$

We can generalize this result into the first property of exponents.

Property 1 for Exponents If a is a real number and r and s are integers, then

$$a^r \cdot a^s = a^{r+s}$$

EXAMPLE 2 Write $(5^3)^2$ with a single exponent.

SOLUTION

$$(5^3)^2 = 5^3 \cdot 5^3$$
$$= 5^6 \quad \textit{Notice: } 3 \cdot 2 = 6$$

Generalizing this result, we have a second property of exponents.

Property 2 for Exponents If a is a real number and r and s are integers, then

$$(a^r)^s = a^{r \cdot s}$$

A third property of exponents arises when we have the product of two or more numbers raised to an integer power.

EXAMPLE 3 Expand $(3x)^4$ and then multiply.

SOLUTION
$$\begin{aligned}(3x)^4 &= (3x)(3x)(3x)(3x)\\ &= (3 \cdot 3 \cdot 3 \cdot 3)(x \cdot x \cdot x \cdot x)\\ &= 3^4 \cdot x^4 \quad \textit{Notice:} \text{ The exponent 4 distributes over the product } 3x\\ &= 81x^4\end{aligned}$$

Generalizing Example 3 we have a third property for exponents.

> **Property 3 for Exponents** If a and b are any two real numbers and r is an integer, then
>
> $$(ab)^r = a^r \cdot b^r$$

Here are some examples that use combinations of the first three properties of exponents to simplify expressions involving exponents.

EXAMPLES Simplify each expression using the properties of exponents.

4. $(-3x^2)(5x^4) = -3(5)(x^2 \cdot x^4)$ — **Commutative and associative properties**
$= -15x^6$ — **Property 1 for exponents**

5. $(-2x^2)^3(4x^5) = (-2)^3(x^2)^3(4x^5)$ — **Property 3**
$= -8x^6 \cdot (4x^5)$ — **Property 2**
$= (-8 \cdot 4)(x^6 \cdot x^5)$ — **Commutative and associative properties**
$= -32x^{11}$ — **Property 1**

6. $(x^2)^4(x^2y^3)^2(y^4)^3 = x^8 \cdot x^4 \cdot y^6 \cdot y^{12}$ — **Properties 2 and 3**
$= x^{12}y^{18}$ — **Property 1**

The next property of exponents deals with negative integer exponents.

> **Property 4 for Exponents** If a is any nonzero real number and r is a positive integer, then
>
> $$a^{-r} = \frac{1}{a^r}$$

Note
This property is actually a definition; that is, we are defining negative-integer exponents as indicating reciprocals. Doing so gives us a way to write an expression with a negative exponent as an equivalent expression with a positive exponent.

EXAMPLES Write with positive exponents, then simplify.

7. $5^{-2} = \dfrac{1}{5^2} = \dfrac{1}{25}$

8. $(-2)^{-3} = \dfrac{1}{(-2)^3} = \dfrac{1}{-8} = -\dfrac{1}{8}$

9. $\left(\dfrac{3}{4}\right)^{-2} = \dfrac{1}{\left(\frac{3}{4}\right)^2} = \dfrac{1}{\frac{9}{16}} = \dfrac{16}{9}$

If we generalize the result in Example 9, we have the following extension of Property 4,

$$\left(\frac{a}{b}\right)^{-r} = \left(\frac{b}{a}\right)^{r}$$

which indicates that raising a fraction to a negative power is equivalent to raising the reciprocal of the fraction to the positive power.

Property 3 indicated that exponents distribute over products. Since division is defined in terms of multiplication, we can expect that exponents will distribute over quotients as well. Property 5 is the formal statement of this fact.

Property 5 for Exponents If a and b are any two real numbers with $b \neq 0$, and r is an integer, then

$$\left(\frac{a}{b}\right)^{r} = \frac{a^r}{b^r}$$

Proof of Property 5

$$\left(\frac{a}{b}\right)^{r} = \underbrace{\left(\frac{a}{b}\right)\left(\frac{a}{b}\right)\left(\frac{a}{b}\right)\cdots\left(\frac{a}{b}\right)}_{r \text{ factors}}$$

$$= \frac{a \cdot a \cdot a \cdots a \leftarrow r \text{ factors}}{b \cdot b \cdot b \cdots b \leftarrow r \text{ factors}}$$

$$= \frac{a^r}{b^r}$$

Since multiplication with the same base resulted in addition of exponents, it seems reasonable to expect division with the same base to result in subtraction of exponents.

Property 6 for Exponents If a is any nonzero real number, and r and s are any two integers, then

$$\frac{a^r}{a^s} = a^{r-s}$$

Notice again that we have specified r and s to be any integers. Our definition of negative exponents is such that the properties of exponents hold for all integer exponents, whether positive or negative integers. Here is proof of Property 6.

Proof of Property 6

Our proof is centered on the fact that division by a number is equivalent to multiplication by the reciprocal of the number.

$$\frac{a^r}{a^s} = a^r \cdot \frac{1}{a^s} \quad \textbf{Dividing by } a^s \textbf{ is equivalent to multiplying by } \frac{1}{a^s}$$

$$= a^r a^{-s} \quad \textbf{Property 4}$$

$$= a^{r+(-s)} \quad \textbf{Property 1}$$

$$= a^{r-s} \quad \textbf{Definition of subtraction}$$

EXAMPLES Apply Property 6 to each expression, and then simplify the result. All answers that contain exponents should contain positive exponents only.

10. $\dfrac{2^8}{2^3} = 2^{8-3} = 2^5 = 32$

11. $\dfrac{x^2}{x^{18}} = x^{2-18} = x^{-16} = \dfrac{1}{x^{16}}$

12. $\dfrac{a^6}{a^{-8}} = a^{6-(-8)} = a^{14}$

13. $\dfrac{m^{-5}}{m^{-7}} = m^{-5-(-7)} = m^2$

Let's complete our list of properties by looking at how the numbers 0 and 1 behave when used as exponents.

We can use the original definition for exponents when the number 1 is used as an exponent.

$$a^1 = \underbrace{a}_{\text{1 factor}}$$

For 0 as an exponent, consider the expression $\frac{3^4}{3^4}$. Since $3^4 = 81$, we have

$$\frac{3^4}{3^4} = \frac{81}{81} = 1$$

However, because we have the quotient of two expressions with the same base, we can subtract exponents.

$$\frac{3^4}{3^4} = 3^{4-4} = 3^0$$

Hence, 3^0 must be the same as 1.

Summarizing these results, we have our last property for exponents.

Property 7 for Exponents If a is any real number, then

$$a^1 = a$$

and

$$a^0 = 1 \qquad \textbf{(as long as } a \neq 0\textbf{)}$$

EXAMPLES Simplify (assume all variables represent non-zero real numbers).

14. $(2x^2y^4)^0 = 1$

15. $(2x^2y^4)^1 = 2x^2y^4$

Here are some examples that use many of the properties of exponents. There are a number of ways to proceed on problems like these. You should use the method that works best for you.

EXAMPLES Simplify.

16. $\dfrac{(x^3)^{-2}(x^4)^5}{(x^{-2})^7} = \dfrac{x^{-6}x^{20}}{x^{-14}}$ **Property 2**

$= \dfrac{x^{14}}{x^{-14}}$ **Property 1**

$= x^{28}$ **Property 6:** $x^{14-(-14)} = x^{28}$

17. $\dfrac{6a^5b^{-6}}{12a^3b^{-9}} = \dfrac{6}{12} \cdot \dfrac{a^5}{a^3} \cdot \dfrac{b^{-6}}{b^{-9}}$ **Write as separate fractions**

$= \dfrac{1}{2}a^2b^3$ **Property 6**

Note: This last answer also can be written as $\frac{a^2b^3}{2}$. Either answer is correct.

18. $\dfrac{(4x^{-5}y^3)^2}{(x^4y^{-6})^{-3}} = \dfrac{16x^{-10}y^6}{x^{-12}y^{18}}$ **Properties 2 and 3**

$= 16x^2y^{-12}$ **Property 6**

$= 16x^2 \cdot \dfrac{1}{y^{12}}$ **Property 4**

$= \dfrac{16x^2}{y^{12}}$ **Multiplication**

Scientific Notation

Scientific notation is a way in which to write very large or very small numbers in a more manageable form. Here is the definition.

> **DEFINITION** A number is written in **scientific notation** if it is written as the product of a number between 1 and 10 and an integer power of 10. A number written in scientific notation has the form
>
> $$n \times 10^r$$
>
> where $1 \leq n < 10$ and $r =$ an integer.

EXAMPLE 19 Write 376,000 in scientific notation.

SOLUTION We must rewrite 376,000 as the product of a number between 1 and 10 and a power of 10. To do so, we move the decimal point five places to the left so that it appears between the 3 and the 7. Then we multiply this number by 10^5. The number that results has the same value as our original number and is written in scientific notation.

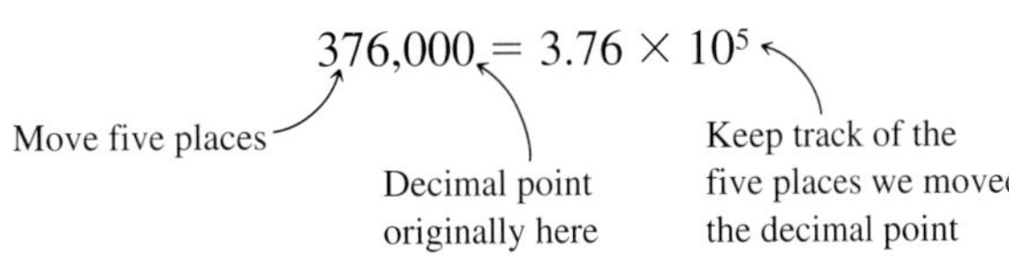
376,000 = 3.76 × 10^5
Move five places
Decimal point originally here
Keep track of the five places we moved the decimal point

If a number written in expanded form is greater than or equal to 10, then when the number is written in scientific notation the exponent on 10 will be positive. A number that is less than 1 will have a negative exponent when written in scientific notation.

EXAMPLE 20

Write 4.52×10^3 in expanded form.

SOLUTION Since 10^3 is 1,000, we can think of this as simply a multiplication problem; that is,

$$4.52 \times 10^3 = 4.52 \times 1{,}000 = 4{,}520$$

However, we can think of the exponent 3 as indicating the number of places we need to move the decimal point to write our number in expanded form. Since our exponent is positive 3, we move the decimal point three places to the right.

$$4.52 \times 10^3 = 4{,}520$$

The following table lists some additional examples of numbers written in expanded form and in scientific notation. In each case, note the relationship between the number of places the decimal point is moved and the exponent on 10.

Number Written in Expanded Notation		Number Written Again in Scientific Notation
376,000	=	3.76×10^5
49,500	=	4.95×10^4
3,200	=	3.2×10^3
591	=	5.91×10^2
46	=	4.6×10^1
8	=	8×10^0
0.47	=	4.7×10^{-1}
0.093	=	9.3×10^{-2}
0.00688	=	6.88×10^{-3}
0.0002	=	2×10^{-4}
0.000098	=	9.8×10^{-5}

Calculator Note Many calculators have a key that allows you to enter numbers in scientific notation. The key is labeled

[EXP] or [EE] or [SCI]

To enter the number 3.45×10^6, you first enter the decimal number, then press the scientific notation key, and finally enter the exponent.

3.45 [EXP] 6

We can use our properties of exponents to do arithmetic with numbers written in scientific notation. Here are some examples.

EXAMPLES

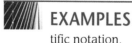

Simplify each expression, and write all answers in scientific notation.

21. $(2 \times 10^8)(3 \times 10^{-3}) = (2)(3) \times (10^8)(10^{-3})$

$$= 6 \times 10^5$$

22. $\dfrac{4.8 \times 10^9}{2.4 \times 10^{-3}} = \dfrac{4.8}{2.4} \times \dfrac{10^9}{10^{-3}}$

$= 2 \times 10^{9-(-3)}$

$= 2 \times 10^{12}$

23. $\dfrac{(6.8 \times 10^5)(3.9 \times 10^{-7})}{7.8 \times 10^{-4}} = \dfrac{(6.8)(3.9)}{7.8} \times \dfrac{(10^5)(10^{-7})}{10^{-4}}$

$= 3.4 \times 10^2$

Note Remember, on some calculators the scientific notation key may be labeled [EE] or [SCI].

Calculator Note On a scientific calculator with a scientific notation key, you would use the following sequence of keys to do Example 22:

4.8 [EXP] 9 [÷] 2.4 [EXP] 3 [+/−] [=]

LINKING OBJECTIVES AND EXAMPLES

Next to each **objective** we have listed the examples that are best described by that objective.

A	1–18
B	19, 20
C	21–23

GETTING READY FOR CLASS

After reading through the preceding section, respond in your own words and in complete sentences.

1. Explain the difference between -2^4 and $(-2)^4$.
2. Explain the difference between 2^5 and 2^{-5}.
3. If a positive base is raised to a negative exponent, can the result be a negative number?
4. State Property 1 for exponents in your own words.

Problem Set 5.1

Online support materials can be found at www.thomsonedu.com/login

1. 4^2
2. $(-4)^2$
3. -4^2
4. $-(-4)^2$
5. -0.3^3
6. $(-0.3)^3$
7. 2^5
8. 2^4
9. $\left(\frac{1}{2}\right)^3$
10. $\left(\frac{3}{4}\right)^2$
11. $\left(-\frac{5}{6}\right)^2$
12. $\left(-\frac{7}{8}\right)^2$

Use the properties of exponents to simplify each of the following as much as possible.

13. $x^5 \cdot x^4$
14. $x^6 \cdot x^3$
15. $(2^3)^2$
16. $(3^2)^2$
17. $\left(-\frac{2}{3}x^2\right)^3$
18. $\left(-\frac{3}{5}x^4\right)^3$
19. $-3a^2(2a^4)$
20. $5a^7(-4a^6)$

Write each of the following with positive exponents. Then simplify as much as possible.

21. 3^{-2}
22. $(-5)^{-2}$
23. $(-2)^{-5}$
24. 2^{-5}
25. $\left(\frac{3}{4}\right)^{-2}$
26. $\left(\frac{3}{5}\right)^{-2}$
27. $\left(\frac{1}{3}\right)^{-2} + \left(\frac{1}{2}\right)^{-3}$
28. $\left(\frac{1}{2}\right)^{-2} + \left(\frac{1}{3}\right)^{-3}$

■ = Videos available by instructor request
▶ = Online student support materials available at www.thomsonedu.com/login

Simplify each expression. Write all answers with positive exponents only. (Assume all variables are nonzero.)

29. $x^{-4}x^7$
30. $x^{-3}x^8$
31. $(a^2b^{-5})^3$
32. $(a^4b^{-3})^3$
33. $(5y^4)^{-3}(2y^{-2})^3$
34. $(3y^5)^{-2}(2y^{-4})^3$
35. $\left(\frac{1}{2}x^3\right)\left(\frac{2}{3}x^4\right)\left(\frac{3}{5}x^{-7}\right)$
36. $\left(\frac{1}{7}x^{-3}\right)\left(\frac{7}{8}x^{-5}\right)\left(\frac{8}{9}x^8\right)$
37. $(4a^5b^2)(2b^{-5}c^2)(3a^7c^4)$
38. $(3a^{-2}c^3)(5b^{-6}c^5)(4a^6b^{-2})$
39. $(2x^2y^{-5})^3(3x^{-4}y^2)^{-4}$
40. $(4x^{-4}y^9)^{-2}(5x^4y^{-3})^2$

Use the properties of exponents to simplify each expression. Write all answers with positive exponents only. (Assume all variables are nonzero.)

41. $\dfrac{x^{-1}}{x^9}$
42. $\dfrac{x^{-3}}{x^5}$
43. $\dfrac{a^4}{a^{-6}}$
44. $\dfrac{a^5}{a^{-2}}$
45. $\dfrac{t^{-10}}{t^{-4}}$
46. $\dfrac{t^{-8}}{t^{-5}}$
47. $\left(\dfrac{x^5}{x^3}\right)^6$
48. $\left(\dfrac{x^7}{x^4}\right)^5$
49. $\dfrac{(x^5)^6}{(x^3)^4}$
50. $\dfrac{(x^7)^3}{(x^4)^5}$
51. $\dfrac{(x^{-2})^3(x^3)^{-2}}{x^{10}}$
52. $\dfrac{(x^{-4})^3(x^3)^{-4}}{x^{10}}$
53. $\dfrac{5a^8b^3}{20a^5b^{-4}}$
54. $\dfrac{7a^6b^{-2}}{21a^2b^{-5}}$
55. $\dfrac{(3x^{-2}y^8)^4}{(9x^4y^{-3})^2}$
56. $\dfrac{(6x^{-3}y^{-5})^2}{(3x^{-4}y^{-3})^4}$
57. $\left(\dfrac{8x^2y}{4x^4y^{-3}}\right)^4$
58. $\left(\dfrac{5x^4y^5}{10xy^{-2}}\right)^3$
59. $\left(\dfrac{x^{-5}y^2}{x^{-3}y^5}\right)^{-2}$
60. $\left(\dfrac{x^{-8}y^{-3}}{x^{-5}y^6}\right)^{-1}$

Write each expression as a perfect square.

61. $x^4y^2 = (\quad)^2$
62. $x^8y^6 = (\quad)^2$
63. $9a^2b^4 = (\quad)^2$
64. $225x^6y^{12} = (\quad)^2$

Write each expression as a perfect cube.

65. $8a^3 = (\quad)^3$
66. $27b^3 = (\quad)^3$
67. $64x^3y^{12} = (\quad)^3$
68. $216x^{15}y^{21} = (\quad)^3$

69. Let $x = 2$ in each of the following expressions and simplify.
   a. x^3x^2
   b. $(x^3)^2$
   c. x^5
   d. x^6

70. Let $x = -1$ in each of the following expressions and simplify.
   a. x^3x^4
   b. $(x^3)^4$
   c. x^7
   d. x^{12}

71. Let $x = 2$ in each of the following expressions and simplify.
   a. $\dfrac{x^5}{x^2}$
   b. x^3
   c. $\dfrac{x^2}{x^6}$
   d. x^{-4}

72. Let $x = -1$ in each of the following expressions and simplify.
   a. $\dfrac{x^{14}}{x^9}$
   b. x^5
   c. $\dfrac{x^{13}}{x^9}$
   d. x^4

73. Write each expression as a perfect square.
   a. $\dfrac{1}{49} = (\quad)^2$
   b. $\dfrac{1}{121} = (\quad)^2$
   c. $\dfrac{1}{4x^2} = (\quad)^2$
   d. $\dfrac{1}{64x^4} = (\quad)^2$

74. Write each expression as a perfect cube.
   a. $\dfrac{1}{125x^3} = (\quad)^3$
   b. $\dfrac{1}{64y^{12}} = (\quad)^3$
   c. $\dfrac{x^6}{216y^9} = (\quad)^3$
   d. $\dfrac{8a^9}{27b^{15}} = (\quad)^3$

Simplify.

75. $2 \cdot 2^{n-1}$

76. $3 \cdot 3^{n-1}$

77. $\frac{ar^6}{ar^3}$

78. $\frac{ar^7}{ar^4}$

Write each number in scientific notation.

79. 378,000

80. 3,780,000

81. 4,900

82. 490

83. 0.00037

84. 0.000037

85. 0.00495

86. 0.0495

Write each number in expanded form.

87. 5.34×10^3

88. 5.34×10^2

89. 7.8×10^6

90. 7.8×10^4

91. 3.44×10^{-3}

92. 3.44×10^{-5}

93. 4.9×10^{-1}

94. 4.9×10^{-2}

Use the properties of exponents to simplify each of the following expressions. Write all answers in scientific notation.

95. $(4 \times 10^{10})(2 \times 10^{-6})$

96. $(3 \times 10^{-12})(3 \times 10^4)$

97. $\frac{8 \times 10^{14}}{4 \times 10^5}$

98. $\frac{6 \times 10^8}{2 \times 10^3}$

99. $\frac{(5 \times 10^6)(4 \times 10^{-8})}{8 \times 10^4}$

100. $\frac{(6 \times 10^{-7})(3 \times 10^9)}{5 \times 10^6}$

Problems 101–110 are problems you will see later in the book.

Multiply.

101. $8x^3 \cdot 10y^6$

102. $5y^2 \cdot 4x^2$

103. $8x^3 \cdot 9y^3$

104. $4y^3 \cdot 3x^2$

105. $3x \cdot 5y$

106. $3xy \cdot 5z$

107. $4x^6y^6 \cdot 3x$

108. $16x^4y^4 \cdot 3y$

109. $27a^6c^3 \cdot 2b^2c$

110. $8a^3b^3 \cdot 5a^2b$

Divide.

111. $\frac{10x^5}{5x^2}$

112. $\frac{-15x^4}{5x^2}$

113. $\frac{20x^3}{5x^2}$

114. $\frac{25x^7}{-5x^2}$

115. $\frac{8x^3y^5}{-2x^2y}$

116. $\frac{-16x^2y^2}{-2x^2y}$

117. $\frac{4x^4y^3}{-2x^2y}$

118. $\frac{10a^4b^2}{4a^2b^2}$

Simplify. Write answers without using scientific notation, rounding to the nearest whole number.

119. $\frac{2.00 \times 10^8}{3.98 \times 10^6}$

120. $\frac{2.00 \times 10^8}{3.16 \times 10^5}$

Use a calculator to find each of the following. Write your answer in scientific notation with the first number in each answer rounded to the nearest tenth.

121. $10^{-4.1}$

122. $10^{-5.6}$

Applying the Concepts

123. **Large Numbers** If you are 20 years old, you have been alive for more than 630,000,000 seconds. Write this last number in scientific notation.

124. **Fingerprints** The FBI has been collecting fingerprint cards since 1924. Their collection has grown to over 200 million cards. They are digitizing the fingerprints. Each fingerprint card turns into about 10 MB of data. (A megabyte [MB] is $2^{20} \approx$ one million bytes.)

   a. How many bytes of storage will they need?

   b. A compression routine called the WSQ method will compress the bytes by ratio of 12.9 to 1. Approximately how many bytes of storage will the FBI need for the compressed data? (Hint: Divide by 12.9.)

125. Search Engines The chart shows the number of people using different search engines in January of 2005. For each of the following search engines, write the number of visitors in scientific notation.

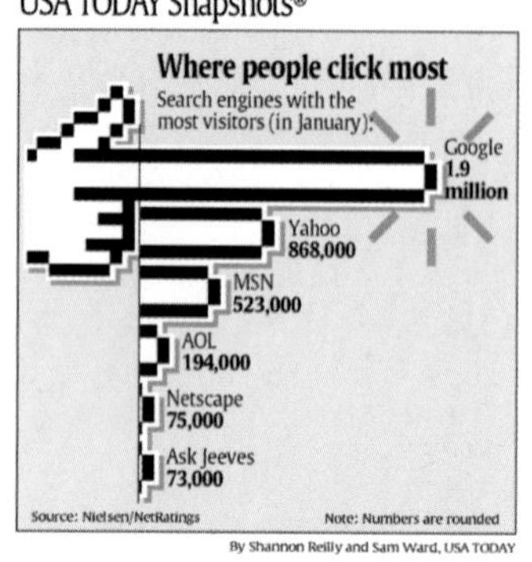

From *USA Today.* Copyright 2005. Reprinted with permission.

a. Ask Jeeves
b. AOL
c. Google

126. Our Galaxy The galaxy the Earth resides in is called the Milky Way galaxy. It is a spiral galaxy that contains approximately 200,000,000,000 stars (our Sun is one of them). Write this number in words and in scientific notation.

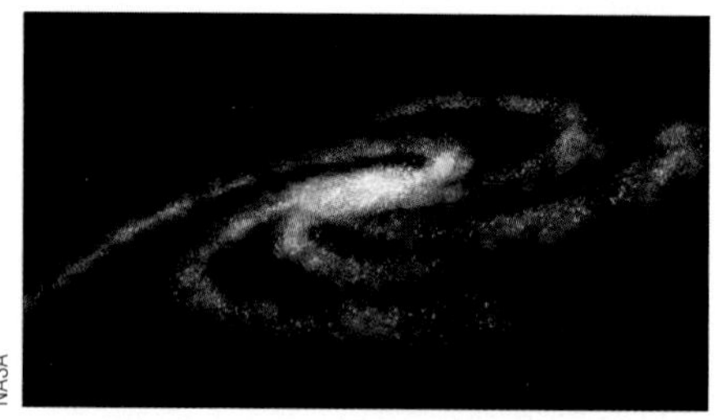

NASA

127. Light Year A light year, the distance light travels in 1 year, is approximately 5.9×10^{12} miles. The Andromeda galaxy is approximately 1.7×10^{6} light years from our galaxy. Find the distance in miles between our galaxy and the Andromeda galaxy.

128. Distance to the Sun The distance from the Earth to the sun is approximately 9.3×10^{7} miles. If light travels 1.2×10^{7} miles in 1 minute, how many minutes does it take the light from the sun to reach the Earth?

Credit Card Debt Outstanding credit-card debt in the United States is over $422 billion.

Stockbyte/SuperStock

129. Write the number 422 billion in scientific notation.

130. If there are approximately 60 million households with at least one credit card, find the average credit-card debt per household, to the nearest dollar.

131. Cone Nebula The photograph was taken by the Hubble telescope in April 2002. The object in the photograph is called the *Cone Nebula.* The distance across the photograph is about 2.5 light-years, which is 14,664,240,000,000 miles. Round this number to the nearest trillion and then write the result in scientific notation.

NASA

132. Computer Science We all use the language of computers to indicate how much memory our computers hold or how much information we can put on a storage device such as a keychain drive. Scientific notation gives us a way to compare the actual numbers associated with the words we use to describe data storage in computers. The smallest amount of data that a computer can hold is measured in bits. A byte is the next largest unit and is equal to 8, or 2^3, bits. Fill in the table below.

Number of Bytes

Unit	Exponential Form	Scientific Notation
Kilobyte	$2^{10} = 1{,}024$	
Megabyte	$2^{20} \approx 1{,}048{,}000$	
Gigabyte	$2^{30} \approx 1{,}074{,}000{,}000$	
Terabyte	$2^{40} \approx 1{,}099{,}500{,}000{,}000$	

Maintaining Your Skills

Solve each system by the addition method.

133. $4x + 3y = 10$
$2x + y = 4$

134. $3x - 5y = -2$
$2x - 3y = 1$

135. $4x + 5y = 5$
$\frac{6}{5}x + y = 2$

136. $4x + 2y = -2$
$\frac{1}{2}x + y = 0$

Solve each system by the substitution method.

137. $x + y = 3$
$y = x + 3$

138. $x + y = 6$
$y = x - 4$

139. $2x - 3y = -6$
$y = 3x - 5$

140. $7x - y = 24$
$x = 2y + 9$

Getting Ready for the Next Section

Simplify.

141. $-4x + 9x$
142. $-6x - 2x$
143. $5x^2 + 3x^2$
144. $7x^2 + 3x^2$
145. $-8x^3 + 10x^3$
146. $4x^3 - 7x^3$
147. $2x + 3 - 2x - 8$
148. $9x - 4 - 9x - 10$
149. $-1(2x - 3)$
150. $-1(-3x + 1)$
151. $-3(-3x - 2)$
152. $-4(-5x + 3)$
153. $-500 + 27(100) - 0.1(100)^2$
154. $-500 + 27(170) - 0.1(170)^2$

Extending the Concepts

Assume all variable exponents represent positive integers and simplify each expression.

155. $x^{m+2} \cdot x^{-2m} \cdot x^{m-5}$
156. $x^{m-4}x^{m+9}x^{-2m}$
157. $(y^m)^2(y^{-3})^m(y^{m+3})$
158. $(y^m)^{-4}(y^3)^m(y^{m+6})$
159. $\dfrac{x^{n+2}}{x^{n-3}}$
160. $\dfrac{x^{n-3}}{x^{n-7}}$

5.2 Polynomials, Sums, and Differences

OBJECTIVES

A Give the degree of a polynomial.

B Add and subtract polynomials.

C Evaluate a polynomial for a given value of its variable.

The chart is from a company that duplicates videotapes. It shows the revenue and cost to duplicate a 30-minute video. From the chart you can see that 300 copies will bring in \$900 in revenue, with a cost of \$600. The profit is the difference between revenue and cost, or \$300.

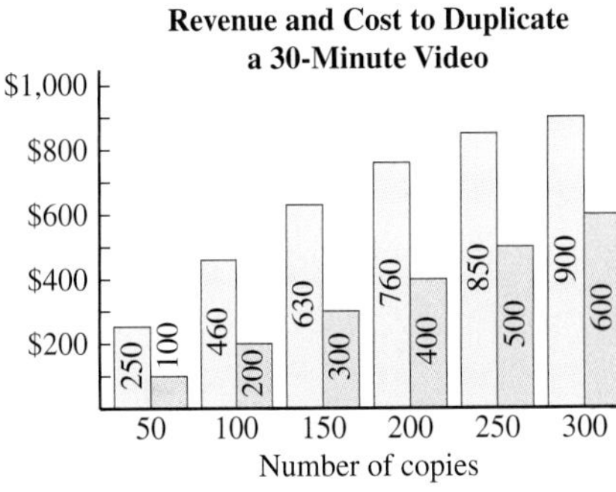

The relationship between profit, revenue, and cost is one application of the polynomials we will study in this section. Let's begin with a definition that we will use to build polynomials.

Polynomials in General

DEFINITION A **term,** or **monomial,** is a constant or the product of a constant and one or more variables raised to whole-number exponents.

The following are monomials, or terms:

$$-16 \qquad 3x^2y \qquad -\frac{2}{5}a^3b^2c \qquad xy^2z$$

The numerical part of each monomial is called the *numerical coefficient,* or just *coefficient* for short. For the preceding terms, the coefficients are -16, 3, $-\frac{2}{5}$, and 1. Notice that the coefficient for xy^2z is understood to be 1.

DEFINITION A **polynomial** is any finite sum of terms. Because subtraction can be written in terms of addition, finite differences are also included in this definition.

The following are polynomials:

$$2x^2 - 6x + 3 \qquad -5x^2y + 2xy^2 \qquad 4a - 5b + 6c + 7d$$

Polynomials can be classified further according to the number of terms present. If a polynomial consists of two terms, it is said to be a *binomial.* If it has three terms, it is called a *trinomial.* And, as stated, a polynomial with only one term is said to be a *monomial.*

DEFINITION The **degree** of a polynomial with one variable is the highest power to which the variable is raised in any one term.

EXAMPLES

1. $6x^2 + 2x - 1$ — **A trinomial of degree 2**
2. $5x - 3$ — **A binomial of degree 1**
3. $7x^6 - 5x^3 + 2x - 4$ — **A polynomial of degree 6**
4. $-7x^4$ — **A monomial of degree 4**
5. 15 — **A monomial of degree 0**

Polynomials in one variable are usually written in decreasing powers of the variable. When this is the case, the coefficient of the first term is called the *leading coefficient.* In Example 1, the leading coefficient is 6. In Example 2, it is 5. The leading coefficient in Example 3 is 7.

DEFINITION Two or more terms that differ only in their numerical coefficients are called **similar,** or **like,** terms. Since similar terms differ only in their coefficients, they have identical variable parts.

Addition and Subtraction of Polynomials

To add two polynomials, we simply apply the commutative and associative properties to group similar terms together and then use the distributive property as we have in the following example.

EXAMPLE 6 Add $5x^2 - 4x + 2$ and $3x^2 + 9x - 6$.

SOLUTION

$$(5x^2 - 4x + 2) + (3x^2 + 9x - 6)$$

$= (5x^2 + 3x^2) + (-4x + 9x) + (2 - 6)$ — **Commutative and associative properties**

$= (5 + 3)x^2 + (-4 + 9)x + (2 - 6)$ — **Distributive property**

$= 8x^2 + 5x + (-4)$

$= 8x^2 + 5x - 4$

Note

In practice it is not necessary to show all the steps shown in Example 6. It is important to understand that addition of polynomials is equivalent to combining similar terms.

EXAMPLE 7

Find the sum of $-8x^3 + 7x^2 - 6x + 5$ and $10x^3 + 3x^2 - 2x - 6$.

SOLUTION We can add the two polynomials using the method of Example 6, or we can arrange similar terms in columns and add vertically. Using the column method, we have

$$\begin{array}{r} -8x^3 + 7x^2 - 6x + 5 \\ \underline{10x^3 + 3x^2 - 2x - 6} \\ 2x^3 + 10x^2 - 8x - 1 \end{array}$$

To find the difference of two polynomials, we need to use the fact that the opposite of a sum is the sum of the opposites; that is,

$$-(a + b) = -a + (-b)$$

One way to remember this is to observe that $-(a + b)$ is equivalent to $-1(a + b) = (-1)a + (-1)b = -a + (-b)$.

If a negative sign directly precedes the parentheses surrounding a polynomial, we may remove the parentheses and the preceding negative sign by changing the sign of each term within the parentheses. For example:

$$-(3x + 4) = -3x + (-4) = -3x - 4$$
$$-(5x^2 - 6x + 9) = -5x^2 + 6x - 9$$
$$-(-x^2 + 7x - 3) = x^2 - 7x + 3$$

EXAMPLE 8

Subtract $(9x^2 - 3x + 5) - (4x^2 + 2x - 3)$.

SOLUTION We subtract by adding the opposite of each term in the polynomial that follows the subtraction sign:

$(9x^2 - 3x + 5) - (4x^2 + 2x - 3)$

$= 9x^2 - 3x + 5 + (-4x^2) + (-2x) + 3$ — **The opposite of a sum is the sum of the opposites**

$= (9x^2 - 4x^2) + (-3x - 2x) + (5 + 3)$ — **Commutative and associative properties**

$= 5x^2 - 5x + 8$ — **Combine similar terms**

EXAMPLE 9

Subtract $4x^2 - 9x + 1$ from $-3x^2 + 5x - 2$.

SOLUTION Again, to subtract, we add the opposite:

$$\begin{aligned} (-3x^2 + 5x - 2) - (4x^2 - 9x + 1) &= -3x^2 + 5x - 2 - 4x^2 + 9x - 1 \\ &= (-3x^2 - 4x^2) + (5x + 9x) + (-2 - 1) \\ &= -7x^2 + 14x - 3 \end{aligned}$$

EXAMPLE 10

Simplify $4x - 3[2 - (3x + 4)]$.

SOLUTION Removing the innermost parentheses first, we have

$$\begin{aligned} 4x - 3[2 - (3x + 4)] &= 4x - 3(2 - 3x - 4) \\ &= 4x - 3(-3x - 2) \\ &= 4x + 9x + 6 \\ &= 13x + 6 \end{aligned}$$

EXAMPLE 11 Simplify $(2x + 3) - [(3x + 1) - (x - 7)]$.

SOLUTION

$$
\begin{aligned}
(2x + 3) - [(3x + 1) - (x - 7)] &= (2x + 3) - (3x + 1 - x + 7) \\
&= (2x + 3) - (2x + 8) \\
&= 2x + 3 - 2x - 8 \\
&= -5
\end{aligned}
$$

In the example that follows we will find the value of a polynomial for a given value of the variable.

EXAMPLE 12 Find the value of $5x^3 - 3x^2 + 4x - 5$ when x is 2.

SOLUTION We begin by substituting 2 for x in the original polynomial:

When $x = 2$

the polynomial $5x^3 - 3x^2 + 4x - 5$

becomes

$$
\begin{aligned}
5 \cdot 2^3 - 3 \cdot 2^2 + 4 \cdot 2 - 5 &= 5 \cdot 8 - 3 \cdot 4 + 4 \cdot 2 - 5 \\
&= 40 - 12 + 8 - 5 \\
&= 31
\end{aligned}
$$

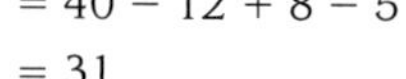

Polynomials and Function Notation

Example 12 can be restated using function notation by calling the polynomial $P(x)$ and asking for $P(2)$. The solution would look like this:

If $P(x) = 5x^3 - 3x^2 + 4x - 5$

then

$$
\begin{aligned}
P(2) &= 5 \cdot 2^3 - 3 \cdot 2^2 + 4 \cdot 2 - 5 \\
&= 31
\end{aligned}
$$

Our next example is stated in terms of function notation.

As we mentioned in the introduction to this chapter, three functions that occur very frequently in business and economics classes are profit, revenue, and cost functions. If a company manufactures and sells x items, then the revenue $R(x)$ is the total amount of money obtained by selling all x items. The cost $C(x)$ is the total amount of money it costs the company to manufacture the x items. The profit $P(x)$ obtained by selling all x items is the difference between the revenue and the cost and is given by the equation

$$P(x) = R(x) - C(x)$$

EXAMPLE 13 A company produces and sells copies of an accounting program for home computers. The total weekly cost (in dollars) to produce x copies of the program is $C(x) = 8x + 500$. Find its weekly profit if the total revenue obtained from selling all x programs is $R(x) = 35x - 0.1x^2$. How much profit will the company make if it produces and sells 100 programs a week? That is, find $P(100)$.

SOLUTION Using the equation $P(x) = R(x) - C(x)$ and the information given in the problem, we have

$$
\begin{aligned}
P(x) &= R(x) - C(x) \\
&= 35x - 0.1x^2 - (8x + 500) \\
&= 35x - 0.1x^2 - 8x - 500 \\
&= -500 + 27x - 0.1x^2
\end{aligned}
$$

If the company produces and sells 100 copies of the program, its weekly profit will be

$$\begin{aligned} P(100) &= -500 + 27(100) - 0.1(100)^2 \\ &= -500 + 27(100) - 0.1(10{,}000) \\ &= -500 + 2{,}700 - 1{,}000 \\ &= 1{,}200 \end{aligned}$$

The weekly profit is $1,200.

USING TECHNOLOGY

Graphing Calculators

Your graphing calculator allows you access to a function list in which you can input formulas for a number of functions. There are many ways to use the function list. For the problems in this section, we can use the function list to evaluate a polynomial. Let's use the profit function from Example 13 to illustrate. To input $P(x) = -500 + 27x - 0.1x^2$, we assign $P(x)$ to the first function in our function list, which is Y_1. Call up the function list for your calculator and enter the profit function this way:

$$Y_1 = -500 + 27X - 0.1X^2$$

Now that you have an expression for Y_1, you can evaluate it by substituting a value for X. To do so, quit the function list and store the number 100 into the variable X:

100 → [X] [ENTER] or 100 [STO] [X] [ENTER]

The display shows 100. Next, display the variable Y_1 (on many calculators, individual variables are on the Y-VAR menu), then press ENTER:

Y_1 [ENTER]

The display will show 1,200.

Now that you have the profit function stored in function Y_1, you can evaluate it for any value of X. Find the weekly profit if 50, 100, 150, 200, 250, and 300 programs are sold.

LINKING OBJECTIVES AND EXAMPLES

Next to each **objective** we have listed the examples that are best described by that objective.

A	1–5
B	6–11
C	12, 13

GETTING READY FOR CLASS

After reading through the preceding section, respond in your own words and in complete sentences.

1. Is $3x^2 + 2x - \frac{1}{x}$ a polynomial? Explain.
2. What are similar terms?
3. Explain in words how you subtract one polynomial from another.
4. What is revenue?

Problem Set 5.2

Online support materials can be found at www.thomsonedu.com/login

Identify those of the following that are monomials, binomials, or trinomials. Give the degree of each, and name the leading coefficient.

1. $5x^2 - 3x + 2$
2. $2x^2 + 4x - 1$
3. $3x - 5$
4. $5y + 3$
5. $8a^2 + 3a - 5$
6. $9a^2 - 8a - 4$
7. $4x^3 - 6x^2 + 5x - 3$
8. $9x^4 + 4x^3 - 2x^2 + x$
9. $-\frac{3}{4}$
10. -16
11. $4x - 5 + 6x^3$
12. $9x + 2 + 3x^3$

Simplify each of the following by combining similar terms.

13. $(4x + 2) + (3x - 1)$
14. $(8x - 5) + (-5x + 4)$
15. $2x^2 - 3x + 10x - 15$
16. $6x^2 - 4x - 15x + 10$
17. $12a^2 + 8ab - 15ab - 10b^2$
18. $28a^2 - 8ab + 7ab - 2b^2$
19. $(5x^2 - 6x + 1) - (4x^2 + 7x - 2)$
20. $(11x^2 - 8x) - (4x^2 - 2x - 7)$
21. $(\frac{1}{2}x^2 - \frac{1}{3}x - \frac{1}{6}) - (\frac{1}{4}x^2 + \frac{7}{12}x) + (\frac{1}{3}x - \frac{1}{12})$
22. $(\frac{2}{3}x^2 - \frac{1}{2}x) - (\frac{1}{4}x^2 + \frac{1}{6}x + \frac{1}{12}) - (\frac{1}{2}x^2 + \frac{1}{4})$
23. $(y^3 - 2y^2 - 3y + 4) - (2y^3 - y^2 + y - 3)$
24. $(8y^3 - 3y^2 + 7y + 2) - (-4y^3 + 6y^2 - 5y - 8)$
25. $(5x^3 - 4x^2) - (3x + 4) + (5x^2 - 7) - (3x^3 + 6)$
26. $(x^3 - x) - (x^2 + x) + (x^3 - 1) - (-3x + 2)$
27. $(\frac{4}{7}x^2 - \frac{1}{7}xy + \frac{1}{14}y^2) - (\frac{1}{2}x^2 - \frac{2}{7}xy - \frac{9}{14}y^2)$
28. $(\frac{1}{5}x^2 - \frac{1}{2}xy + \frac{1}{10}y^2) - (-\frac{3}{10}x^2 + \frac{2}{5}xy - \frac{1}{2}y^2)$
29. $(3a^3 + 2a^2b + ab^2 - b^3) - (6a^3 - 4a^2b + 6ab^2 - b^3)$
30. $(a^3 - 3a^2b + 3ab^2 - b^3) - (a^3 + 3a^2b + 3ab^2 + b^3)$
31. Subtract $2x^2 - 4x$ from $2x^2 - 7x$.
32. Subtract $-3x + 6$ from $-3x + 9$.
33. Find the sum of $x^2 - 6xy + y^2$ and $2x^2 - 6xy - y^2$.
34. Find the sum of $9x^3 - 6x^2 + 2$ and $3x^2 - 5x + 4$.
35. Subtract $-8x^5 - 4x^3 + 6$ from $9x^5 - 4x^3 - 6$.
36. Subtract $4x^4 - 3x^3 - 2x^2$ from $2x^4 + 3x^3 + 4x^2$.
37. Find the sum of $11a^2 + 3ab + 2b^2$, $9a^2 - 2ab + b^2$, and $-6a^2 - 3ab + 5b^2$.
38. Find the sum of $a^2 - ab - b^2$, $a^2 + ab - b^2$, and $a^2 + 2ab + b^2$.

Simplify each of the following. Begin by working on the innermost parentheses.

39. $-[2 - (4 - x)]$
40. $-[-3 - (x - 6)]$
41. $-5[-(x - 3) - (x + 2)]$
42. $-6[(2x - 5) - 3(8x - 2)]$
43. $4x - 5[3 - (x - 4)]$
44. $x - 7[3x - (2 - x)]$
45. $-(3x - 4y) - [(4x + 2y) - (3x + 7y)]$
46. $(8x - y) - [-(2x + y) - (-3x - 6y)]$
47. $4a - \{3a + 2[a - 5(a + 1) + 4]\}$
48. $6a - \{-2a - 6[2a + 3(a - 1) - 6]\}$
49. Find the value of $2x^2 - 3x - 4$ when x is 2.
50. Find the value of $4x^2 + 3x - 2$ when x is -1.
51. If $P(x) = \frac{3}{2}x^2 - \frac{3}{4}x + 1$, find
    a. $P(12)$
    b. $P(-8)$
52. If $P(x) = \frac{2}{5}x^2 - \frac{1}{10}x + 2$, find
    a. $P(10)$
    b. $P(-10)$
53. If $Q(x) = x^3 - x^2 + x - 1$, find
    a. $Q(5)$
    b. $Q(-2)$

= Videos available by instructor request
▶ = Online student support materials available at www.thomsonedu.com/login

▸ **54.** If $Q(x) = x^3 + x^2 + x + 1$, find
  a. $Q(5)$
  b. $Q(-2)$

▸ **55.** If $R(x) = 11.5x - 0.05x^2$, find
  a. $R(10)$
  b. $R(-10)$

▸ **56.** If $R(x) = 11.5x - 0.01x^2$, find
  a. $R(10)$
  b. $R(-10)$

▸ **57.** If $P(x) = 600 + 1{,}000x - 100x^2$, find
  a. $P(-4)$
  b. $P(4)$

▸ **58.** If $P(x) = 500 + 800x - 100x^2$, find
  a. $P(-6)$
  b. $P(8)$

Simplify.

59. a. $(3x - 5) - (3a - 5)$
  b. $(2x + 3) - (2a + 3)$

60. a. $(x^2 - 4) - (a^2 - 4)$
  b. $(x^2 - 1) - (a^2 - 1)$

Applying the Concepts

Problems 61–66 may be solved using a graphing calculator.

61. Height of an Object If an object is thrown straight up into the air with a velocity of 128 feet/second, then its height $h(t)$ above the ground t seconds later is given by the formula

$$h(t) = -16t^2 + 128t$$

Find the height after 3 seconds and after 5 seconds. [Find $h(3)$ and $h(5)$.]

62. Height of an Object The formula for the height of an object that has been thrown straight up with a velocity of 64 feet/second is

$$h(t) = -16t^2 + 64t$$

Find the height after 1 second and after 3 seconds. [Find $h(1)$ and $h(3)$.]

63. Profits The total cost (in dollars) for a company to manufacture and sell x items per week is $C(x) = 60x + 300$. If the revenue brought in by selling all x items is $R(x) = 100x - 0.5x^2$, find the weekly profit. How much profit will be made by producing and selling 60 items each week?

64. Profits The total cost (in dollars) for a company to produce and sell x items per week is $C(x) = 200x + 1{,}600$. If the revenue brought in by selling all x items is $R(x) = 300x - 0.6x^2$, find the weekly profit. How much profit will be made by producing and selling 50 items each week?

65. Profits Suppose it costs a company selling patterns $C(x) = 800 + 6.5x$ dollars to produce and sell x patterns a month. If the revenue obtained by selling x patterns is $R(x) = 10x - 0.002x^2$, what is the profit equation? How much profit will be made if 1,000 patterns are produced and sold in May?

66. Profits Suppose a company manufactures and sells x picture frames each month with a total cost of $C(x) = 1{,}200 + 3.5x$ dollars. If the revenue obtained by selling x frames is $R(x) = 9x - 0.003x^2$, find the profit equation. How much profit will be made if 1,000 frames are manufactured and sold in June?

Maintaining Your Skills

Simplify each expression.

67. $-1(5 - x)$

68. $-1(a - b)$

69. $-1(7 - x)$

70. $-1(6 - y)$

71. $5\left(x - \frac{1}{5}\right)$

72. $7\left(x + \frac{1}{7}\right)$

73. $x\left(1 - \frac{1}{x}\right)$

74. $a\left(1 + \frac{1}{a}\right)$

75. $12\left(\frac{1}{4}x + \frac{2}{3}y\right)$

76. $20\left(\frac{2}{5}x + \frac{1}{4}y\right)$

Getting Ready for the Next Section

Simplify.

77. $2x^2 - 3x + 10x - 15$

78. $12a^2 + 8ab - 15ab - 10b^2$

79. $(6x^3 - 2x^2y + 8xy^2) + (-9x^2y + 3xy^2 - 12y^3)$

80. $(3x^3 - 15x^2 + 18x) + (2x^2 - 10x + 12)$

81. $4x^3(-3x)$
82. $5x^2(-4x)$
83. $4x^3(5x^2)$
84. $5x^2(3x^2)$
85. $(a^3)^2$
86. $(a^4)^2$
87. $11.5(130) - 0.05(130)^2$
88. $-0.05(130)^2 + 9.5(130) - 200$

Extending the Concepts

89. The graphs of two polynomial functions are given in Figures 1 and 2. Use the graphs to find the following.

a. $f(-3)$ b. $f(0)$ c. $f(1)$
d. $g(-1)$ e. $g(0)$ f. $g(2)$
g. $f[g(2)]$ h. $g[f(2)]$

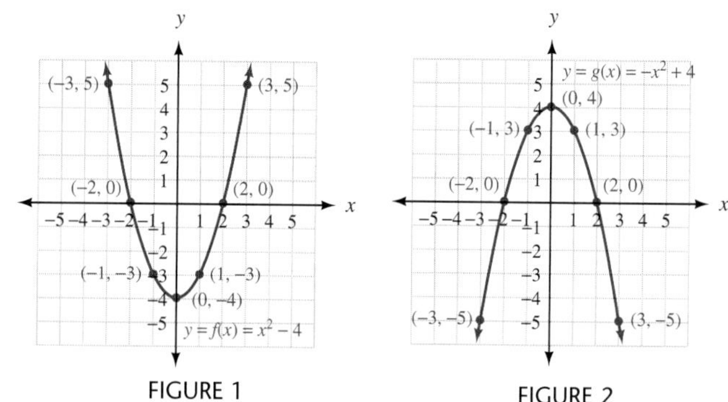

FIGURE 1 FIGURE 2

5.3 Multiplication of Polynomials

OBJECTIVES

A Multiply polynomials.

In the previous section we found the relationship between profit, revenue, and cost to be

$$P(x) = R(x) - C(x)$$

Revenue itself can be broken down further by another formula common in the business world. The revenue obtained from selling all x items is the product of the number of items sold and the price per item; that is,

$$\text{Revenue} = (\text{number of items sold})(\text{price of each item})$$

For example, if 100 items are sold for \$9 each, the revenue is $100(9) = \$900$. Likewise, if 500 items are sold for \$11 each, then the revenue is $500(11) = \$5{,}500$. In general, if x is the number of items sold and p is the selling price of each item, then we can write

$$R = xp$$

Many times, x and p are polynomials, which means that the expression xp is the product of two polynomials. In this section we learn how to multiply polynomials, and in so doing, increase our understanding of the equations and formulas that describe business applications.

EXAMPLE 1 Find the product of $4x^3$ and $5x^2 - 3x + 1$.

SOLUTION To multiply, we apply the distributive property:

$$
\begin{aligned}
&4x^3(5x^2 - 3x + 1)\\
&\quad = 4x^3(5x^2) + 4x^3(-3x) + 4x^3(1) && \textbf{Distributive property}\\
&\quad = 20x^5 - 12x^4 + 4x^3
\end{aligned}
$$

Notice that we multiply coefficients and add exponents.

EXAMPLE 2 Multiply $2x - 3$ and $x + 5$.

SOLUTION Distributing the $2x - 3$ across the sum $x + 5$ gives us

$$
\begin{aligned}
&(2x - 3)(x + 5)\\
&\quad = (2x - 3)x + (2x - 3)5 && \textbf{Distributive property}\\
&\quad = 2x(x) + (-3)x + 2x(5) + (-3)5 && \textbf{Distributive property}\\
&\quad = 2x^2 - 3x + 10x - 15\\
&\quad = 2x^2 + 7x - 15 && \textbf{Combine like terms}
\end{aligned}
$$

Notice the third line in this example. It consists of all possible products of terms in the first binomial and those of the second binomial. We can generalize this into a rule for multiplying two polynomials.

> **Rule** To multiply two polynomials, multiply each term in the first polynomial by each term in the second polynomial.

Multiplying polynomials can be accomplished by a method that looks very similar to long multiplication with whole numbers.

EXAMPLE 3 Multiply $(2x - 3y)$ and $(3x^2 - xy + 4y^2)$ vertically.

SOLUTION

$$
\begin{array}{rrrrl}
3x^2 - & xy + & 4y^2 & & \\
 & 2x - & 3y & & \\
\hline
6x^3 - & 2x^2y + & 8xy^2 & & \textbf{Multiply } (3x^2 - xy + 4y^2) \textbf{ by } 2x\\
 & -\ 9x^2y + & 3xy^2 - & 12y^3 & \textbf{Multiply } (3x^2 - xy + 4y^2) \textbf{ by } -3y\\
\hline
6x^3 - & 11x^2y + & 11xy^2 - & 12y^3 & \textbf{Add similar terms}
\end{array}
$$

Note
The vertical method of multiplying polynomials does not directly show the use of the distributive property. It is, however, very useful since it always gives the correct result and is easy to remember.

Multiplying Binomials—The FOIL Method

Consider the product of $(2x - 5)$ and $(3x - 2)$. Distributing $(3x - 2)$ over $2x$ and -5, we have

$$
\begin{aligned}
(2x - 5)(3x - 2) &= (2x)(3x - 2) + (-5)(3x - 2)\\
&= (2x)(3x) + (2x)(-2) + (-5)(3x) + (-5)(-2)\\
&= 6x^2 - 4x - 15x + 10\\
&= 6x^2 - 19x + 10
\end{aligned}
$$

Looking closely at the second and third lines, we notice the following:

Note
The FOIL method does not show the properties used in multiplying two binomials. It is simply a way of finding products of binomials quickly. Remember, the FOIL method applies only to products of two binomials. The vertical method applies to all products of polynomials with two or more terms.

1. $6x^2$ comes from multiplying the *first* terms in each binomial:

$$(2x - 5)(3x - 2) \qquad 2x(3x) = 6x^2 \qquad \textit{First} \text{ terms}$$

2. $-4x$ comes from multiplying the *outside* terms in the product:

$$(2x - 5)(3x - 2) \qquad 2x(-2) = -4x \qquad \textit{Outside} \text{ terms}$$

3. $-15x$ comes from multiplying the *inside* terms in the product:

$$(2x - 5)(3x - 2) \qquad -5(3x) = -15x \qquad \textit{Inside} \text{ terms}$$

4. 10 comes from multiplying the *last* two terms in the product:

$$(2x - 5)(3x - 2) \qquad -5(-2) = 10 \qquad \textit{Last} \text{ terms}$$

Once we know where the terms in the answer come from, we can reduce the number of steps used in finding the product:

$$(2x - 5)(3x - 2) = \underset{\text{First}}{6x^2} \underset{\text{Outside}}{- 4x} \underset{\text{Inside}}{- 15x} \underset{\text{Last}}{+ 10} = 6x^2 - 19x + 10$$

EXAMPLES

Multiply using the FOIL method.

4. $(4a - 5b)(3a + 2b) = \underset{\text{F}}{12a^2} + \underset{\text{O}}{8ab} - \underset{\text{I}}{15ab} - \underset{\text{L}}{10b^2}$
$= 12a^2 - 7ab - 10b^2$

5. $(3 - 2t)(4 + 7t) = \underset{\text{F}}{12} + \underset{\text{O}}{21t} - \underset{\text{I}}{8t} - \underset{\text{L}}{14t^2}$
$= 12 + 13t - 14t^2$

6. $(2x + \frac{1}{2})(4x - \frac{1}{2}) = \underset{\text{F}}{8x^2} - \underset{\text{O}}{x} + \underset{\text{I}}{2x} - \underset{\text{L}}{\frac{1}{4}}$
$= 8x^2 + x - \frac{1}{4}$

7. $(a^5 + 3)(a^5 - 7) = \underset{\text{F}}{a^{10}} - \underset{\text{O}}{7a^5} + \underset{\text{I}}{3a^5} - \underset{\text{L}}{21}$
$= a^{10} - 4a^5 - 21$

8. $(2x + 3)(5y - 4) = \underset{\text{F}}{10xy} - \underset{\text{O}}{8x} + \underset{\text{I}}{15y} - \underset{\text{L}}{12}$

The Square of a Binomial

EXAMPLE 9

Find $(4x - 6)^2$.

SOLUTION Applying the definition of exponents and then the FOIL method, we have

$$(4x - 6)^2 = (4x - 6)(4x - 6)$$
$$= \underset{\text{F}}{16x^2} - \underset{\text{O}}{24x} - \underset{\text{I}}{24x} + \underset{\text{L}}{36}$$
$$= 16x^2 - 48x + 36$$

The type of product shown in Example 9 occurs frequently enough in algebra that we have special formulas for it. Here are the formulas for binomial squares:

$$(a + b)^2 = (a + b)(a + b) = a^2 + ab + ab + b^2 = a^2 + 2ab + b^2$$
$$(a - b)^2 = (a - b)(a - b) = a^2 - ab - ab + b^2 = a^2 - 2ab + b^2$$

Observing the results in both cases, we have the following rule.

Note
From the rule and examples that follow, it should be obvious that $(a + b)^2$ is not equal to $a^2 + b^2$; that is, the square of a sum is not the same as the sum of the squares.

Rule The square of a binomial is the sum of the square of the first term, twice the product of the two terms, and the square of the last term. Or:

$(a + b)^2 =$	a^2	$+$	$2ab$	$+$	b^2
	Square of first term		Twice the product of the two terms		Square of last term
$(a - b)^2 =$	a^2	$-$	$2ab$	$+$	b^2

EXAMPLES

Use the preceding formulas to expand each binomial square.

10. $(x + 7)^2 = x^2 + 2(x)(7) + 7^2 = x^2 + 14x + 49$

11. $(3t - 5)^2 = (3t)^2 - 2(3t)(5) + 5^2 = 9t^2 - 30t + 25$

12. $(4x + 2y)^2 = (4x)^2 + 2(4x)(2y) + (2y)^2 = 16x^2 + 16xy + 4y^2$

13. $(5 - a^3)^2 = 5^2 - 2(5)(a^3) + (a^3)^2 = 25 - 10a^3 + a^6$ 

Products Resulting in the Difference of Two Squares

Another frequently occurring kind of product is found when multiplying two binomials that differ only in the sign between their terms.

EXAMPLE 14

Multiply $(3x - 5)$ and $(3x + 5)$.

SOLUTION Applying the FOIL method, we have

$$(3x - 5)(3x + 5) = \underset{F}{9x^2} + \underset{O}{15x} - \underset{I}{15x} - \underset{L}{25} \qquad \textbf{Two middle terms add to 0}$$
$$= 9x^2 - 25$$

The outside and inside products in Example 14 are opposites and therefore add to 0. Here it is in general:

$$(a - b)(a + b) = a^2 + ab - ab - b^2 \qquad \textbf{Two middle terms add to 0}$$
$$= a^2 - b^2$$

Rule To multiply two binomials that differ only in the sign between their two terms, simply subtract the square of the second term from the square of the first term:

$$(a + b)(a - b) = a^2 - b^2$$

The expression $a^2 - b^2$ is called the *difference of two squares*.

EXAMPLES Find the following products.

15. $(x - 5)(x + 5) = x^2 - 25$

16. $(2a - 3)(2a + 3) = 4a^2 - 9$

17. $(x^2 + 4)(x^2 - 4) = x^4 - 16$

18. $(x^3 - 2a)(x^3 + 2a) = x^6 - 4a^2$

More About Function Notation

From the introduction to this chapter, we know that the revenue obtained from selling x items at p dollars per item is

$$R = \text{Revenue} = xp \qquad \textbf{(The number of items} \times \textbf{price per item)}$$

For example, if a store sells 100 items at \$4.50 per item, the revenue is $100(4.50) = \$450$. If we have an equation that gives the relationship between x and p, then we can write the revenue in terms of x or in terms of p. With function notation, we would write the revenue as either $R(x)$ or $R(p)$, where

> $R(x)$ is the revenue function that gives the revenue R in terms of the number of items x.
>
> $R(p)$ is the revenue function that gives the revenue R in terms of the price per item p.

With function notation we can see exactly which variables we want our formulas written in terms of.

In the next two examples, we will use function notation to combine a number of problems we have worked previously.

EXAMPLE 19 A company manufactures and sells prerecorded videotapes. They find that they can sell x videotapes each day at p dollars per tape, according to the equation $x = 230 - 20p$. Find $R(x)$ and $R(p)$.

SOLUTION The notation $R(p)$ tells us we are to write the revenue equation in terms of the variable p. To do so, we use the formula $R(p) = xp$ and substitute $230 - 20p$ for x to obtain

$$R(p) = xp = (230 - 20p)p = 230p - 20p^2$$

The notation $R(x)$ indicates that we are to write the revenue in terms of the variable x. We need to solve the equation $x = 230 - 20p$ for p. Let's begin by interchanging the two sides of the equation:

$$230 - 20p = x$$

$$-20p = -230 + x \qquad \textbf{Add } -230 \textbf{ to each side}$$

$$p = \frac{-230 + x}{-20} \qquad \textbf{Divide each side by } -20$$

$$p = 11.5 - 0.05x \qquad \frac{\mathbf{230}}{\mathbf{20}} \mathbf{= 11.5 \text{ and } } \frac{\mathbf{1}}{\mathbf{20}} \mathbf{= 0.05}$$

Now we can find $R(x)$ by substituting $11.5 - 0.05x$ for p in the formula $R(x) = xp$:

$$R(x) = xp = x(11.5 - 0.05x) = 11.5x - 0.05x^2$$

Our two revenue functions are actually equivalent. To offer some justification for this, suppose that the company decides to sell each tape for \$5. The equation $x = 230 - 20p$ indicates that, at \$5 per tape, they will sell $x = 230 - 20(5) = 230 - 100 = 130$ tapes per day. To find the revenue from selling the tapes for \$5 each, we use $R(p)$ with $p = 5$:

$$\begin{aligned} \text{If} \qquad p &= 5 \\ \text{then} \qquad R(p) &= R(5) \\ &= 230(5) - 20(5)^2 \\ &= 1{,}150 - 500 \\ &= \$650 \end{aligned}$$

However, to find the revenue from selling 130 tapes, we use $R(x)$ with $x = 130$:

$$\begin{aligned} \text{If} \qquad x &= 130 \\ \text{then} \qquad R(x) &= R(130) \\ &= 11.5(130) - 0.05(130)^2 \\ &= 1{,}495 - 845 \\ &= \$650 \end{aligned}$$

EXAMPLE 20

Suppose the daily cost function for the videotapes in Example 19 is $C(x) = 200 + 2x$. Find the profit function $P(x)$ and then find $P(130)$.

SOLUTION Since profit is equal to the difference of the revenue and the cost, we have

$$\begin{aligned} P(x) &= R(x) - C(x) \\ &= 11.5x - 0.05x^2 - (200 + 2x) \\ &= -0.05x^2 + 9.5x - 200 \end{aligned}$$

Notice that we used the formula for $R(x)$ from Example 19 instead of the formula for $R(p)$. We did so because we were asked to find $P(x)$, meaning we want the profit P only in terms of the variable x.

Next, we use the formula we just obtained to find $P(130)$:

$$\begin{aligned} P(130) &= -0.05(130)^2 + 9.5(130) - 200 \\ &= -0.05(16{,}900) + 9.5(130) - 200 \\ &= -845 + 1{,}235 - 200 \\ &= \$190 \end{aligned}$$

Because $P(130) = \$190$, the company will make a profit of \$190 per day by selling 130 tapes per day.

USING TECHNOLOGY

Graphing Calculators

More About Example 20

We can visualize the three functions given in Example 20 if we set up the functions list and graphing window on our calculator this way:

$Y_1 = 11.5X - 0.05X^2$ **This gives the graph of $R(x)$**

$Y_2 = 200 + 2X$ **This gives the graph of $C(x)$**

$Y_3 = Y_1 - Y_2$ **This gives the graph of $P(x)$**

Window: X from 0 to 250, Y from 0 to 750

The graphs in Figure 1 are similar to what you will obtain using the functions list and window shown here. Next, find the value of $P(x)$ when $R(x)$ and $C(x)$ intersect.

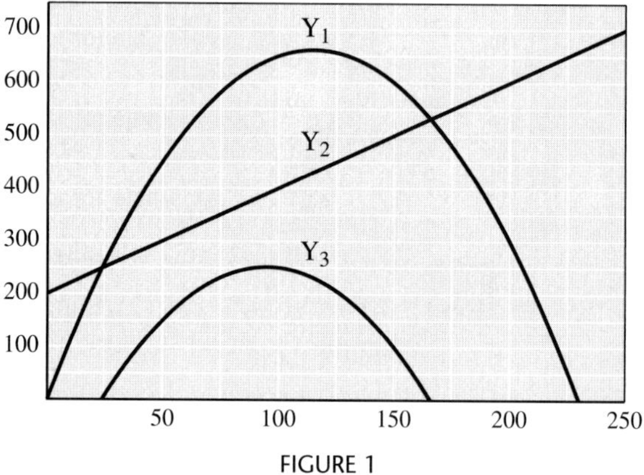

FIGURE 1

LINKING OBJECTIVES AND EXAMPLES

Next to each objective we have listed the examples that are best described by that objective.

A 1–18

GETTING READY FOR CLASS

After reading through the preceding section, respond in your own words and in complete sentences.

1. Describe how the distributive property is used to multiply a monomial and a polynomial.
2. Describe how you would use the FOIL method to multiply two binomials.
3. Explain why $(x + 3)^2$ is not equal to $x^2 + 9$.
4. When will the product of two binomials result in a binomial?

Problem Set 5.3

Online support materials can be found at www.thomsonedu.com/login

Multiply the following by applying the distributive property.

1. $2x(6x^2 - 5x + 4)$
2. $-3x(5x^2 - 6x - 4)$
3. $-3a^2(a^3 - 6a^2 + 7)$
4. $4a^3(3a^2 - a + 1)$
5. $2a^2b(a^3 - ab + b^3)$
6. $5a^2b^2(8a^2 - 2ab + b^2)$

Multiply the following vertically.

7. $(x - 5)(x + 3)$
8. $(x + 4)(x + 6)$
9. $(2x^2 - 3)(3x^2 - 5)$
10. $(3x^2 + 4)(2x^2 - 5)$
11. $(x + 3)(x^2 + 6x + 5)$
12. $(x - 2)(x^2 - 5x + 7)$
13. $(a - b)(a^2 + ab + b^2)$
14. $(a + b)(a^2 - ab + b^2)$
15. $(2x + y)(4x^2 - 2xy + y^2)$
16. $(x - 3y)(x^2 + 3xy + 9y^2)$
17. $(2a - 3b)(a^2 + ab + b^2)$
18. $(5a - 2b)(a^2 - ab - b^2)$

Multiply the following using the FOIL method.

19. $(x - 2)(x + 3)$
20. $(x + 2)(x - 3)$
21. $(2a + 3)(3a + 2)$
22. $(5a - 4)(2a + 1)$
23. $(5 - 3t)(4 + 2t)$
24. $(7 - t)(6 - 3t)$
25. $(x^3 + 3)(x^3 - 5)$
26. $(x^3 + 4)(x^3 - 7)$
27. $(5x - 6y)(4x + 3y)$
28. $(6x - 5y)(2x - 3y)$
29. $(3t + \frac{1}{3})(6t - \frac{2}{3})$
30. $(5t - \frac{1}{5})(10t + \frac{3}{5})$

Find the following special products.

31. $(5x + 2y)^2$
32. $(3x - 4y)^2$
33. $(5 - 3t^3)^2$
34. $(7 - 2t^4)^2$
35. $(2a + 3b)(2a - 3b)$
36. $(6a - 1)(6a + 1)$
37. $(3r^2 + 7s)(3r^2 - 7s)$
38. $(5r^2 - 2s)(5r^2 + 2s)$
39. $\left(y + \frac{3}{2}\right)^2$
40. $\left(y - \frac{7}{2}\right)^2$
41. $\left(a - \frac{1}{2}\right)^2$
42. $\left(a - \frac{5}{2}\right)^2$
43. $\left(x + \frac{1}{4}\right)^2$
44. $\left(x - \frac{3}{8}\right)^2$
45. $\left(t + \frac{1}{3}\right)^2$
46. $\left(t - \frac{2}{5}\right)^2$
47. $(\frac{1}{3}x - \frac{2}{5})(\frac{1}{3}x + \frac{2}{5})$
48. $(\frac{3}{4}x - \frac{1}{7})(\frac{3}{4}x + \frac{1}{7})$

Find the following products.

49. $(x - 2)^3$
50. $(4x + 1)^3$
51. $(x - \frac{1}{2})^3$
52. $(x + \frac{1}{4})^3$
53. $3(x - 1)(x - 2)(x - 3)$
54. $2(x + 1)(x + 2)(x + 3)$
55. $(b^2 + 8)(a^2 + 1)$
56. $(b^2 + 1)(a^4 - 5)$
57. $(x + 1)^2 + (x + 2)^2 + (x + 3)^2$
58. $(x - 1)^2 + (x - 2)^2 + (x - 3)^2$
59. $(2x + 3)^2 - (2x - 3)^2$
60. $(x - 3)^3 - (x + 3)^3$

Here are some problems you will see later in the book.

Simplify.

61. $(x + 3)^2 - 2(x + 3) - 8$
62. $(x - 2)^2 - 3(x - 2) - 10$

= Videos available by instructor request

▸ = Online student support materials available at www.thomsonedu.com/login

63. $(2a-3)^2-9(2a-3)+20$

64. $(3a-2)^2+2(3a-2)-3$

65. $2(4a+2)^2-3(4a+2)-20$

66. $6(2a+4)^2-(2a+4)-2$

67. Multiply $(x+y-4)(x+y+5)$ by first writing it like this:

$$[(x+y)-4][(x+y)+5]$$

and then applying the FOIL method.

68. Multiply $(x-5-y)(x-5+y)$ by first writing it like this:

$$[(x-5)-y][(x-5)+y]$$

and then applying the FOIL method.

69. Let $a=2$ and $b=3$, and evaluate each of the following expressions.

$$a^4-b^4 \quad (a-b)^4 \quad (a^2+b^2)(a+b)(a-b)$$

70. Let $a=2$ and $b=3$, and evaluate each of the following expressions.

$$a^3+b^3 \quad (a+b)^3 \quad a^3+3a^2b+3ab^2+b^3$$

Applying the Concepts

71. **Revenue** A store selling art supplies finds that it can sell x sketch pads per week at p dollars each, according to the formula $x=900-300p$. Write formulas for $R(p)$ and $R(x)$. Then find the revenue obtained by selling the pads for \$1.60 each.

72. **Revenue** A company selling diskettes for home computers finds that it can sell x diskettes per day at p dollars per diskette, according to the formula $x=800-100p$. Write formulas for $R(p)$ and $R(x)$. Then find the revenue obtained by selling the diskettes for \$3.80 each.

73. **Revenue** A company sells an inexpensive accounting program for home computers. If it can sell x programs per week at p dollars per program, according to the formula $x=350-10p$, find formulas for $R(p)$ and $R(x)$. How much will the weekly revenue be if it sells 65 programs?

74. **Revenue** A company sells boxes of greeting cards through the mail. It finds that it can sell x boxes of cards each week at p dollars per box, according to the formula $x=1{,}475-250p$. Write formulas for $R(p)$ and $R(x)$. What revenue will it bring in each week if it sells 200 boxes of cards?

75. **Profit** If the cost to produce the x programs in Problem 73 is $C(x)=5x+500$, find $P(x)$ and $P(60)$.

76. **Profit** If the cost to produce the x diskettes in Problem 72 is $C(x)=2x+200$, find $P(x)$ and $P(40)$.

77. **Interest** If you deposit \$100 in an account with an interest rate r that is compounded annually, then the amount of money in that account at the end of 4 years is given by the formula $A=100(1+r)^4$. Expand the right side of this formula.

78. **Interest** If you deposit P dollars in an account with an annual interest rate r that is compounded twice a year, then at the end of a year the amount of money in that account is given by the formula

$$A=P\left(1+\frac{r}{2}\right)^2$$

Expand the right side of this formula.

Maintaining Your Skills

Solve each system.

79. $\begin{aligned} x+y+z&=6\\ 2x-y+z&=3\\ x+2y-3z&=-4 \end{aligned}$

80. $\begin{aligned} x+y+z&=6\\ x-y+2z&=7\\ 2x-y-z&=0 \end{aligned}$

81. $\begin{aligned} 3x+4y&=15\\ 2x-5z&=-3\\ 4y-3z&=9 \end{aligned}$

82. $\begin{aligned} x+3y&=5\\ 6y+z&=12\\ x-2z&=-10 \end{aligned}$

Getting Ready for the Next Section

Simplify.

83. $\frac{8a^3}{a}$

84. $\frac{-8a^2}{a}$

85. $\frac{-48a}{a}$

86. $\frac{-32a}{a}$

87. $\frac{16a^5b^4}{8a^2b^3}$

88. $\frac{12x^4y^5}{3x^3y^3}$

89. $\frac{-24a^5b^5}{8a^5b^3}$

90. $\frac{-15x^5y^3}{3x^3y^3}$

91. $\frac{x^3y^4}{-x^3}$

92. $\frac{x^2y^2}{-x^2}$

Extending the Concepts

Assume n is a positive integer and multiply.

93. $(x^n - 2)(x^n - 3)$

94. $(x^{2n} + 3)(x^{2n} - 3)$

95. $(2x^n + 3)(5x^n - 1)$

96. $(4x^n - 3)(7x^n + 2)$

97. $(x^n + 5)^2$

98. $(x^n - 2)^2$

99. $(x^n + 1)(x^{2n} - x^n + 1)$

100. $(x^{3n} - 3)(x^{6n} + 3x^{3n} + 9)$

5.4 The Greatest Common Factor and Factoring by Grouping

OBJECTIVES

A Factor by factoring out the greatest common factor.

B Factor by grouping.

Multiplication

Factors → $3 \cdot 7 = 21$ ← Product

Factoring

In general, factoring is the reverse of multiplication. The diagram in the margin illustrates the relationship between factoring and multiplication. Reading from left to right, we say the product of 3 and 7 is 21. Reading in the other direction, from right to left, we say 21 factors into 3 times 7. Or, 3 and 7 are factors of 21.

> **DEFINITION** The **greatest common factor** for a polynomial is the largest monomial that divides (is a factor of) each term of the polynomial.

The greatest common factor for the polynomial $25x^5 + 20x^4 - 30x^3$ is $5x^3$ since it is the largest monomial that is a factor of each term. We can apply the distributive property and write

$$\begin{aligned} 25x^5 + 20x^4 - 30x^3 &= 5x^3(5x^2) + 5x^3(4x) - 5x^3(6) \\ &= 5x^3(5x^2 + 4x - 6) \end{aligned}$$

The last line is written in factored form.

EXAMPLE 1 Factor the greatest common factor from

$$8a^3 - 8a^2 - 48a$$

SOLUTION The greatest common factor is $8a$. It is the largest monomial that divides each term of our polynomial. We can write each term in our polynomial as the product of $8a$ and another monomial. Then, we apply the distributive property to factor $8a$ from each term:

$$\begin{aligned} 8a^3 - 8a^2 - 48a &= 8a(a^2) - 8a(a) - 8a(6) \\ &= 8a(a^2 - a - 6) \end{aligned}$$

> *Note*
> The term *largest monomial,* as used here, refers to the monomial with the largest integer exponents whose coefficient has the greatest absolute value.

EXAMPLE 2 Factor the greatest common factor from

$$16a^5b^4 - 24a^2b^5 - 8a^3b^3$$

SOLUTION The largest monomial that divides each term is $8a^2b^3$. We write each term of the original polynomial in terms of $8a^2b^3$ and apply the distributive property to write the polynomial in factored form:

$$\begin{aligned} 16a^5b^4 - 24a^2b^5 - 8a^3b^3 &= 8a^2b^3(2a^3b) - 8a^2b^3(3b^2) - 8a^2b^3(a) \\ &= 8a^2b^3(2a^3b - 3b^2 - a) \end{aligned}$$

EXAMPLE 3 A company manufacturing prerecorded videotapes finds that the total daily revenue for selling x tapes is given by

$$R(x) = 11.5x - 0.05x^2$$

Factor x from each term on the right side of the equation to find the formula that gives the price p in terms of x.

SOLUTION We begin by factoring x from the right side of the equation:

$$\begin{aligned} &\text{If} \quad & R(x) &= 11.5x - 0.05x^2 \\ &\text{then} \quad & R(x) &= x(11.5 - 0.05x) \end{aligned}$$

Because R is always xp, the quantity in parentheses must be p. The price it should charge if it wants to sell x items per day is therefore

$$p = 11.5 - 0.05x$$

Factoring by Grouping

The polynomial $5x + 5y + x^2 + xy$ can be factored by noticing that the first two terms have a 5 in common, whereas the last two have an x in common. Applying the distributive property, we have

$$5x + 5y + x^2 + xy = 5(x + y) + x(x + y)$$

This last expression can be thought of as having two terms, $5(x + y)$ and $x(x + y)$, each of which has a common factor $(x + y)$. We apply the distributive property again to factor $(x + y)$ from each term:

$$\begin{aligned} &5(x + y) + x(x + y) \\ &\quad = (x + y)(5 + x) \end{aligned}$$

EXAMPLE 4 Factor $a^2b^2 + b^2 + 8a^2 + 8$.

SOLUTION The first two terms have b^2 in common; the last two have 8 in common:

$$\begin{aligned} a^2b^2 + b^2 + 8a^2 + 8 &= b^2(a^2 + 1) + 8(a^2 + 1) \\ &= (a^2 + 1)(b^2 + 8) \end{aligned}$$

EXAMPLE 5

Factor $15 - 5y^4 - 3x^3 + x^3y^4$.

SOLUTION Let's try factoring a 5 from the first two terms and a $-x^3$ from the last two terms:

$$15 - 5y^4 - 3x^3 + x^3y^4 = 5(3 - y^4) - x^3(3 - y^4)$$
$$= (3 - y^4)(5 - x^3)$$

EXAMPLE 6

Factor by grouping $x^3 + 2x^2 + 9x + 18$.

SOLUTION We begin by factoring x^2 from the first two terms and 9 from the second two terms:

$$x^3 + 2x^2 + 9x + 18 = x^2(x + 2) + 9(x + 2)$$
$$= (x + 2)(x^2 + 9)$$

LINKING OBJECTIVES AND EXAMPLES

Next to each **objective** we have listed the examples that are best described by that objective.

A	1–3
B	4–6

GETTING READY FOR CLASS

After reading through the preceding section, respond in your own words and in complete sentences.

1. What is the greatest common factor for a polynomial?
2. After factoring a polynomial, how can you check your result?
3. When would you try to factor by grouping?
4. What is the relationship between multiplication and factoring?

Problem Set 5.4

Online support materials can be found at www.thomsonedu.com/login

Factor the greatest common factor from each of the following. (The answers in the back of the book all show greatest common factors whose coefficients are positive.)

1. $10x^3 - 15x^2$
2. $12x^5 + 18x^7$
3. $9y^6 + 18y^3$
4. $24y^4 - 8y^2$
5. $9a^2b - 6ab^2$
6. $30a^3b^4 + 20a^4b^3$
7. $21xy^4 + 7x^2y^2$
8. $14x^6y^3 - 6x^2y^4$
9. $3a^2 - 21a + 30$
10. $3a^2 - 3a - 6$
11. $4x^3 - 16x^2 - 20x$
12. $2x^3 - 14x^2 + 20x$
13. ▶ $10x^4y^2 + 20x^3y^3 - 30x^2y^4$
14. $6x^4y^2 + 18x^3y^3 - 24x^2y^4$
15. $-x^2y + xy^2 - x^2y^2$
16. $-x^3y^2 - x^2y^3 - x^2y^2$
17. $4x^3y^2z - 8x^2y^2z^2 + 6xy^2z^3$
18. $7x^4y^3z^2 - 21x^2y^2z^2 - 14x^2y^3z^4$
19. $20a^2b^2c^2 - 30ab^2c + 25a^2bc^2$
20. $8a^3bc^5 - 48a^2b^4c + 16ab^3c^5$
21. $5x(a - 2b) - 3y(a - 2b)$
22. $3a(x - y) - 7b(x - y)$
23. $3x^2(x + y)^2 - 6y^2(x + y)^2$

■ = Videos available by instructor request
▶ = Online student support materials available at www.thomsonedu.com/login

24. $10x^3(2x - 3y) - 15x^2(2x - 3y)$
25. $2x^2(x + 5) + 7x(x + 5) + 6(x + 5)$
26. $2x^2(x + 2) + 13x(x + 2) + 15(x + 2)$

Factor each of the following by grouping.

27. $3xy + 3y + 2ax + 2a$
28. $5xy^2 + 5y^2 + 3ax + 3a$
29. $x^2y + x + 3xy + 3$
30. $x^3y^3 + 2x^3 + 5x^2y^3 + 10x^2$
31. $3xy^2 - 6y^2 + 4x - 8$
32. $8x^2y - 4x^2 + 6y - 3$
33. $x^2 - ax - bx + ab$
34. $ax - x^2 - bx + ab$
35. $ab + 5a - b - 5$
36. $x^2 - xy - ax + ay$
37. $a^4b^2 + a^4 - 5b^2 - 5$
38. $2a^2 - a^2b - bc^2 + 2c^2$
39. $x^3 + 3x^2 - 4x - 12$
40. $x^3 + 5x^2 - 4x - 20$
41. $x^3 + 2x^2 - 25x - 50$
42. $x^3 + 4x^2 - 9x - 36$
43. $2x^3 + 3x^2 - 8x - 12$
44. $3x^3 + 2x^2 - 27x - 18$
45. $4x^3 + 12x^2 - 9x - 27$
46. $9x^3 + 18x^2 - 4x - 8$
47. The greatest common factor of the binomial $3x - 9$ is 3. The greatest common factor of the binomial $6x - 2$ is 2. What is the greatest common factor of their product, $(3x - 9)(6x - 2)$, when it has been multiplied out?
48. The greatest common factors of the binomials $5x - 10$ and $2x + 4$ are 5 and 2, respectively. What is the greatest common factor of their product, $(5x - 10)(2x + 4)$, when it has been multiplied out?

Applying the Concepts

49. **Investing** If P dollars are placed in a savings account in which the rate of interest r is compounded yearly, then at the end of 1 year the amount of money in the account can be written as $P + Pr$. At the end of 2 years the amount of money in the account is

$$P + Pr + (P + Pr)r$$

Use factoring by grouping to show that this last expression can be written as $P(1 + r)^2$.

50. **Investing** At the end of 3 years, the amount of money in the savings account in Problem 49 will be

$$P(1 + r)^2 + P(1 + r)^2r$$

Use factoring to show that this last expression can be written as $P(1 + r)^3$.

Use Example 3 as a guide in solving the next four problems.

51. **Price** A company manufacturing prerecorded videotapes finds that the total daily revenue R for selling x tapes at p dollars per tape is given by

$$R(x) = 11.5x - 0.05x^2$$

Factor x from each term on the right side of the equation to find the formula that gives the price p in terms of x. Then, use it to find the price they should charge if they want to sell 125 videotapes per day.

52. **Price** A company producing diskettes for home computers finds that the total daily revenue for selling x diskettes at p dollars per diskette is given by

$$R(x) = 8x - 0.01x^2$$

Use the fact that $R = xp$ and your knowledge of factoring to find a formula that gives the price p in terms of x. Then, use it to find the price they should charge if they want to sell 420 diskettes per day.

53. **Price** The weekly revenue equation for a company selling an inexpensive accounting program for home computers is given by the equation

$$R(x) = 35x - 0.1x^2$$

where x is the number of programs they sell per week. What price p should they charge if they want to sell 65 programs per week?

54. **Price** The weekly revenue equation for a small mail-order company selling boxes of greeting cards is

$$R(x) = 5.9x - 0.004x^2$$

where x is the number of boxes they sell per week. What price p should they charge if they want to sell 200 boxes each week?

Maintaining Your Skills

Evaluate each determinant.

55. $\begin{vmatrix} 3 & 5 \\ -6 & 2 \end{vmatrix}$

56. $\begin{vmatrix} -2 & 0 \\ 0 & -1 \end{vmatrix}$

57. $\begin{vmatrix} 1 & -2 & 3 \\ 0 & 4 & -1 \\ 2 & -4 & 6 \end{vmatrix}$

58. $\begin{vmatrix} 2 & 0 & 0 \\ 0 & -3 & 0 \\ 0 & 0 & 4 \end{vmatrix}$

Getting Ready for the Next Section

Factor out the greatest common factor.

59. $3x^4 - 9x^3y - 18x^2y^2$

60. $5x^2 + 10x + 30$

61. $2x^2(x - 3) - 4x(x - 3) - 3(x - 3)$

62. $3x^2(x - 2) - 8x(x - 2) + 2(x - 2)$

Multiply.

63. $(x + 2)(3x - 1)$

64. $(x - 2)(3x + 1)$

65. $(x - 1)(3x - 2)$

66. $(x + 1)(3x + 2)$

67. $(x + 2)(x + 3)$

68. $(x - 2)(x - 3)$

69. $(2y + 5)(3y - 7)$

70. $(2y - 5)(3y + 7)$

71. $(4 - 3a)(5 - a)$

72. $(4 - 3a)(5 + a)$

Complete each table.

73.

Two Numbers *a* and *b*	Their Product *ab*	Their Sum *a* + *b*
1, −24		
−1, 24		
2, −12		
−2, 12		
3, −8		
−3, 8		
4, −6		
−4, 6		

74.

Two Numbers *a* and *b*	Their Product *ab*	Their Sum *a* + *b*
1, −54		
−1, 54		
2, −27		
−2, 27		
3, −18		
−3, 18		
6, −9		
−6, 9		

5.5 Factoring Trinomials

OBJECTIVES

A Factor trinomials in which the leading coefficient is 1.

B Factor trinomials in which the leading coefficient is a number other than 1.

Factoring Trinomials with a Leading Coefficient of 1

Earlier in this chapter we multiplied binomials:

$$(x - 2)(x + 3) = x^2 + x - 6$$

$$(x + 5)(x + 2) = x^2 + 7x + 10$$

In each case the product of two binomials is a trinomial. The first term in the resulting trinomial is obtained by multiplying the first term in each binomial. The middle term comes from adding the product of the two inside terms with the product of the two outside terms. The last term is the product of the last term in each binomial.

In general,

$$(x + a)(x + b) = x^2 + ax + bx + ab$$
$$= x^2 + (a + b)x + ab$$

Writing this as a factoring problem, we have

$$x^2 + (a + b)x + ab = (x + a)(x + b)$$

To factor a trinomial with a leading coefficient of 1, we simply find the two numbers a and b whose sum is the coefficient of the middle term and whose product is the constant term.

EXAMPLE 1 Factor $x^2 + 2x - 15$.

SOLUTION Again the leading coefficient is 1. We need two integers whose product is -15 and whose sum is $+2$. The integers are $+5$ and -3.

$$x^2 + 2x - 15 = (x + 5)(x - 3)$$

In the preceding example we found factors of $x + 5$ and $x - 3$. These are the only two such factors for $x^2 + 2x - 15$. There is no other pair of binomials $x + a$ and $x + b$ whose product is $x^2 + 2x - 15$.

EXAMPLE 2 Factor $x^2 - xy - 12y^2$.

SOLUTION We need two expressions whose product is $-12y^2$ and whose sum is $-y$. The expressions are $-4y$ and $3y$:

$$x^2 - xy - 12y^2 = (x - 4y)(x + 3y)$$

Checking this result gives

$$(x - 4y)(x + 3y) = x^2 + 3xy - 4xy - 12y^2$$
$$= x^2 - xy - 12y^2$$

EXAMPLE 3 Factor $x^2 - 8x + 6$.

SOLUTION Since there is no pair of integers whose product is 6 and whose sum is -8, the trinomial $x^2 - 8x + 6$ is not factorable. We say it is a *prime polynomial.*

EXAMPLE 4 Factor $3x^4 - 15x^3y - 18x^2y^2$.

SOLUTION The leading coefficient is not 1. Each term is divisible by $3x^2$, however. Factoring this out to begin with we have

$$3x^4 - 15x^3y - 18x^2y^2 = 3x^2(x^2 - 5xy - 6y^2)$$

Factoring the resulting trinomial as in the previous examples gives

$$3x^2(x^2 - 5xy - 6y^2) = 3x^2(x - 6y)(x + y)$$

Note
As a general rule, it is best to factor out the greatest common factor first.

Factoring Other Trinomials by Trial and Error

We want to turn our attention now to trinomials with leading coefficients other than 1 and with no greatest common factor other than 1.

Suppose we want to factor $3x^2 - x - 2$. The factors will be a pair of binomials. The product of the first terms will be $3x^2$, and the product of the last terms will be -2. We can list all the possible factors along with their products as follows.

Possible Factors	First Term	Middle Term	Last Term
$(x + 2)(3x - 1)$	$3x^2$	$+5x$	-2
$(x - 2)(3x + 1)$	$3x^2$	$-5x$	-2
$(x + 1)(3x - 2)$	$3x^2$	$+x$	-2
$(x - 1)(3x + 2)$	$3x^2$	$-x$	-2

From the last line we see that the factors of $3x^2 - x - 2$ are $(x - 1)(3x + 2)$. That is,

$$3x^2 - x - 2 = (x - 1)(3x + 2)$$

To factor trinomials with leading coefficients other than 1, when the greatest common factor is 1, we must use trial and error or list all the possible factors. In either case the idea is this: Look only at pairs of binomials whose products give the correct first and last terms, then look for the combination that will give the correct middle term.

EXAMPLE 5 Factor $2x^2 + 13xy + 15y^2$.

SOLUTION Listing all possible factors the product of whose first terms is $2x^2$ and the product of whose last terms is $+15y^2$ yields

Possible Factors	Middle Term of Product
$(2x - 5y)(x - 3y)$	$-11xy$
$(2x - 3y)(x - 5y)$	$-13xy$
$(2x + 5y)(x + 3y)$	$+11xy$
$(2x + 3y)(x + 5y)$	$+13xy$
$(2x + 15y)(x + y)$	$+17xy$
$(2x - 15y)(x - y)$	$-17xy$
$(2x + y)(x + 15y)$	$+31xy$
$(2x - y)(x - 15y)$	$-31xy$

The fourth line has the correct middle term:

$$2x^2 + 13xy + 15y^2 = (2x + 3y)(x + 5y)$$

Actually, we did not need to check the first two pairs of possible factors in the preceding list. Because all the signs in the trinomial $2x^2 + 13xy + 15y^2$ are positive, the binomial factors must be of the form $(ax + b)(cx + d)$, where a, b, c, and d are all positive.

There are other ways to reduce the number of possible factors to consider. For example, if we were to factor the trinomial $2x^2 - 11x + 12$, we would not have to consider the pair of possible factors $(2x - 4)(x - 3)$. If the original trinomial has no greatest common factor other than 1, then neither of its binomial factors will either. The trinomial $2x^2 - 11x + 12$ has a greatest common factor of 1, but the possible factor $2x - 4$ has a greatest common factor of 2: $2x - 4 = 2(x - 2)$. Therefore, we do not need to consider $2x - 4$ as a possible factor.

EXAMPLE 6 Factor $12x^4 + 17x^2 + 6$.

SOLUTION This is a trinomial in x^2:

$$12x^4 + 17x^2 + 6 = (4x^2 + 3)(3x^2 + 2)$$

EXAMPLE 7 Factor $2x^2(x - 3) - 5x(x - 3) - 3(x - 3)$.

SOLUTION We begin by factoring out the greatest common factor $(x - 3)$. Then we factor the trinomial that remains.

$$\begin{aligned} 2x^2(x - 3) - 5x(x - 3) - 3(x - 3) &= (x - 3)(2x^2 - 5x - 3) \\ &= (x - 3)(2x + 1)(x - 3) \\ &= (x - 3)^2(2x + 1) \end{aligned}$$

Another Method of Factoring Trinomials

As an alternative to the trial-and-error method of factoring trinomials, we present the following method. The new method does not require as much trial and error. To use this new method, we must rewrite our original trinomial in such a way that the factoring by grouping method can be applied.

Here are the steps we use to factor $ax^2 + bx + c$.

Step 1: Form the product ac.
Step 2: Find a pair of numbers whose product is ac and whose sum is b.
Step 3: Rewrite the polynomial to be factored so that the middle term bx is written as the sum of two terms whose coefficients are the two numbers found in step 2.
Step 4: Factor by grouping.

EXAMPLE 8 Factor $3x^2 - 10x - 8$ using these steps.

SOLUTION The trinomial $3x^2 - 10x - 8$ has the form $ax^2 + bx + c$, where $a = 3$, $b = -10$, and $c = -8$.

Step 1: The product ac is $3(-8) = -24$.
Step 2: We need to find two numbers whose product is -24 and whose sum is -10. Let's list all the pairs of numbers whose product is -24 to find the pair whose sum is -10.

Product	*Sum*
$1(-24) = -24$	$1 + (-24) = -23$
$-1(24) = -24$	$-1 + 24 = 23$
$2(-12) = -24$	$2 + (-12) = -10$
$-2(12) = -24$	$-2 + 12 = 10$
$3(-8) = -24$	$3 + (-8) = -5$
$-3(8) = -24$	$-3 + 8 = 5$
$4(-6) = -24$	$4 + (-6) = -2$
$-4(6) = -24$	$-4 + 6 = 2$

As you can see, of all the pairs of numbers whose product is -24, only 2 and -12 have a sum of -10.

Step 3: We now rewrite our original trinomial so the middle term $-10x$ is written as the sum of $-12x$ and $2x$:

$$3x^2 - 10x - 8 = 3x^2 - 12x + 2x - 8$$

Step 4: Factoring by grouping, we have

$$\begin{aligned} 3x^2 - 12x + 2x - 8 &= 3x(x - 4) + 2(x - 4) \\ &= (x - 4)(3x + 2) \end{aligned}$$

You can check that this method works by multiplying $x - 4$ and $3x + 2$ to get

$$3x^2 - 10x - 8$$

EXAMPLE 9

Factor $9x^2 + 15x + 4$.

SOLUTION In this case $a = 9$, $b = 15$, and $c = 4$. The product ac is $9 \cdot 4 = 36$. Listing pairs of numbers whose product is 36 with their corresponding sums, we have

Product	*Sum*
$1(36) = 36$	$1 + 36 = 37$
$2(18) = 36$	$2 + 18 = 20$
$3(12) = 36$	$3 + 12 = 15$
$4(9) = 36$	$4 + 9 = 13$
$6(6) = 36$	$6 + 6 = 12$

Notice we list only positive numbers since both the product and sum we are looking for are positive. The numbers 3 and 12 are the numbers we are looking for. Their product is 36, and their sum is 15. We now rewrite the original polynomial $9x^2 + 15x + 4$ with the middle term written as $3x + 12x$. We then factor by grouping:

$$\begin{aligned} 9x^2 + 15x + 4 &= 9x^2 + 3x + 12x + 4 \\ &= 3x(3x + 1) + 4(3x + 1) \\ &= (3x + 1)(3x + 4) \end{aligned}$$

The polynomial $9x^2 + 15x + 4$ factors into the product

$$(3x + 1)(3x + 4)$$

EXAMPLE 10

Factor $8x^2 - 2x - 15$.

SOLUTION The product ac is $8(-15) = -120$. There are many pairs of numbers whose product is -120. We are looking for the pair whose sum is also -2. The numbers are -12 and 10. Writing $-2x$ as $-12x + 10x$ and then factoring by grouping, we have

$$\begin{aligned} 8x^2 - 2x - 15 &= 8x^2 - 12x + 10x - 15 \\ &= 4x(2x - 3) + 5(2x - 3) \\ &= (2x - 3)(4x + 5) \end{aligned}$$

LINKING OBJECTIVES AND EXAMPLES

Next to each **objective** we have listed the examples that are best described by that objective.

A	1–3
B	4–10

GETTING READY FOR CLASS

After reading through the preceding section, respond in your own words and in complete sentences.

1. What is a prime polynomial?
2. When factoring polynomials, what should you look for first?
3. How can you check to see that you have factored a trinomial correctly?
4. Describe how to determine the binomial factors of $6x^2 + 5x - 25$.

Problem Set 5.5

Online support materials can be found at www.thomsonedu.com/login

Factor each of the following trinomials.

1. $x^2 + 7x + 12$
2. $x^2 - 7x + 12$
3. $x^2 - x - 12$
4. $x^2 + x - 12$
5. $y^2 + y - 6$
6. $y^2 - y - 6$
7. $16 - 6x - x^2$
8. $3 + 2x - x^2$
9. $12 + 8x + x^2$
10. $15 - 2x - x^2$

Factor completely by first factoring out the greatest common factor and then factoring the trinomial that remains.

11. $3a^2 - 21a + 30$
12. $3a^2 - 3a - 6$
13. $4x^3 - 16x^2 - 20x$
14. $2x^3 - 14x^2 + 20x$

Factor.

15. $x^2 + 3xy + 2y^2$
16. $x^2 - 5xy - 24y^2$
17. $a^2 + 3ab - 18b^2$
18. $a^2 - 8ab - 9b^2$
19. $x^2 - 2xa - 48a^2$
20. $x^2 + 14xa + 48a^2$
21. $x^2 - 12xb + 36b^2$
22. $x^2 + 10xb + 25b^2$

Factor completely. Be sure to factor out the greatest common factor first if it is other than 1.

23. $3x^2 - 6xy - 9y^2$
24. $5x^2 + 25xy + 20y^2$
25. $2a^5 + 4a^4b + 4a^3b^2$
26. $3a^4 - 18a^3b + 27a^2b^2$
27. $10x^4y^2 + 20x^3y^3 - 30x^2y^4$
28. $6x^4y^2 + 18x^3y^3 - 24x^2y^4$
29. $2x^2 + 7x - 15$
30. $2x^2 - 7x - 15$
31. $2x^2 + x - 15$
32. $2x^2 - x - 15$
33. $2x^2 - 13x + 15$
34. $2x^2 + 13x + 15$
35. $2x^2 - 11x + 15$
36. $2x^2 + 11x + 15$
37. $2x^2 + 7x + 15$
38. $2x^2 + x + 15$
39. $2 + 7a + 6a^2$
40. $2 - 7a + 6a^2$
41. $60y^2 - 15y - 45$
42. $72y^2 + 60y - 72$

= Videotapes available by instructor request

▸ = Online student support materials available at www.thomsonedu.com/login

43. $6x^4 - x^3 - 2x^2$

44. $3x^4 + 2x^3 - 5x^2$

45. $40r^3 - 120r^2 + 90r$

46. $40r^3 + 200r^2 + 250r$

47. $4x^2 - 11xy - 3y^2$

48. $3x^2 + 19xy - 14y^2$

49. $10x^2 - 3xa - 18a^2$

50. $9x^2 + 9xa - 10a^2$

51. $18a^2 + 3ab - 28b^2$

52. $6a^2 - 7ab - 5b^2$

53. $600 + 800t - 800t^2$

54. $200 - 600t - 350t^2$

55. $9y^4 + 9y^3 - 10y^2$

56. $4y^5 + 7y^4 - 2y^3$

57. $24a^2 - 2a^3 - 12a^4$

58. $60a^2 + 65a^3 - 20a^4$

59. $8x^4y^2 - 2x^3y^3 - 6x^2y^4$

60. $8x^4y^2 - 47x^3y^3 - 6x^2y^4$

61. $300x^4 + 1{,}000x^2 + 300$

62. $600x^4 - 100x^2 - 200$

63. $20a^4 + 37a^2 + 15$

64. $20a^4 + 13a^2 - 15$

65. $9 + 3r^2 - 12r^4$

66. $2 - 4r^2 - 30r^4$

Factor each of the following by first factoring out the greatest common factor and then factoring the trinomial that remains.

67. $2x^2(x + 5) + 7x(x + 5) + 6(x + 5)$

68. $2x^2(x + 2) + 13x(x + 2) + 15(x + 2)$

69. $x^2(2x + 3) + 7x(2x + 3) + 10(2x + 3)$

70. $2x^2(x + 1) + 7x(x + 1) + 6(x + 1)$

71. $3x^2(x - 3) + 7x(x - 3) - 20(x - 3)$

72. $4x^2(x + 6) + 23x(x + 6) + 15(x + 6)$

73. $6x^2(x - 2) - 17x(x - 2) + 12(x - 2)$

74. $10x^2(x + 4) - 33x(x + 4) - 7(x + 4)$

75. $12x^2(x + 3) + 7x(x + 3) - 45(x + 3)$

76. $24x^2(x - 6) + 38x(x - 6) + 15(x - 6)$

77. $6x^2(5x - 2) - 11x(5x - 2) - 10(5x - 2)$

78. $14x^2(3x + 4) - 39x(3x + 4) + 10(3x + 4)$

79. $20x^2(2x + 3) + 47x(2x + 3) + 21(2x + 3)$

80. $15x^2(4x - 5) - 2x(4x - 5) - 24(4x - 5)$

81. What polynomial, when factored, gives $(3x + 5y)(3x - 5y)$?

82. What polynomial, when factored, gives $(7x + 2y)(7x - 2y)$?

83. One factor of the trinomial $a^2 + 260a + 2{,}500$ is $a + 10$. What is the other factor?

84. One factor of the trinomial $a^2 - 75a - 2{,}500$ is $a + 25$. What is the other factor?

85. One factor of the trinomial $12x^2 - 107x + 210$ is $x - 6$. What is the other factor?

86. One factor of the trinomial $36x^2 + 134x - 40$ is $2x + 8$. What is the other factor?

87. One factor of the trinomial $54x^2 + 111x + 56$ is $6x + 7$. What is the other factor?

88. One factor of the trinomial $63x^2 + 110x + 48$ is $7x + 6$. What is the other factor?

89. One factor of the trinomial $35x^2 + 19x - 24$ is $5x - 3$. What is the other factor?

90. One factor of the trinomial $36x^2 + 43x - 35$ is $4x + 7$. What is the other factor?

91. Factor the right side of the equation $y = 4x^2 + 18x - 10$, and then use the result to find y when x is $\frac{1}{2}$, when x is -5, and when x is 2.

92. Factor the right side of the equation $y = 9x^2 + 33x - 12$, and use the result to find y when x is $\frac{1}{3}$, when x is -4, and when x is 3.

Maintaining Your Skills

Multiply.

93. $(2x - 3)(2x + 3)$

94. $(4 - 5x)(4 + 5x)$

95. $(2x - 3)^2$

96. $(4 - 5x)^2$

97. $(2x - 3)(4x^2 + 6x + 9)$

98. $(2x + 3)(4x^2 - 6x + 9)$

Getting Ready for the Next Section

For each problem below, place a number or expression inside the parentheses so that the resulting statement is true.

99. $\frac{25}{64} = (\quad)^2$

100. $\frac{4}{9} = (\quad)^2$

101. $x^6 = (\quad)^2$

102. $x^8 = (\quad)^2$

103. $16x^4 = (\quad)^2$

104. $81y^4 = (\quad)^2$

Write as a perfect cube.

105. $\frac{1}{8} = (\quad)^3$

106. $\frac{1}{27} = (\quad)^3$

107. $x^6 = (\quad)^3$

108. $x^{12} = (\quad)^3$

109. $27x^3 = (\quad)^3$

110. $125y^3 = (\quad)^3$

111. $8y^3 = (\quad)^3$

112. $1{,}000x^3 = (\quad)^3$

Extending the Concepts

Factor completely.

113. $8x^6 + 26x^3y^2 + 15y^4$

114. $24x^4 + 6x^2y^3 - 45y^6$

115. $3x^2 + 295x - 500$

116. $3x^2 + 594x - 1{,}200$

117. $\frac{1}{8}x^2 + x + 2$

118. $\frac{1}{9}x^2 + x + 2$

119. $2x^2 + 1.5x + 0.25$

120. $6x^2 + 2x + 0.16$

5.6 Special Factoring

OBJECTIVES

A Factor perfect square trinomials.

B Factor the difference of two squares.

C Factor the sum or difference of two cubes.

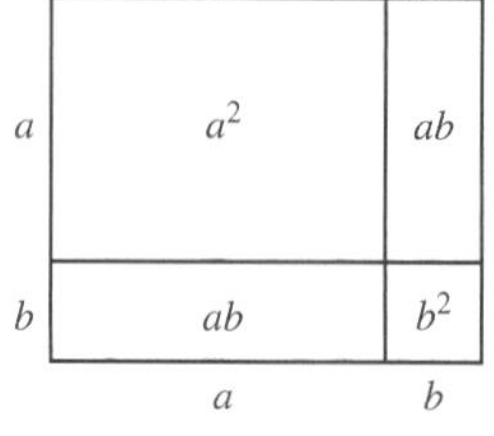

To find the area of the large square in the margin, we can square the length of its side, giving us $(a + b)^2$. However, we can add the areas of the four smaller figures to arrive at the same result.

Since the area of the large square is the same whether we find it by squaring a side or by adding the four smaller areas, we can write the following relationship:

$$(a + b)^2 = a^2 + 2ab + b^2$$

This is the formula for the square of a binomial. The figure gives us a geometric interpretation for this special multiplication formula. We begin this section by looking at the special multiplication formulas from a factoring perspective.

Perfect Square Trinomials

We previously listed some special products found in multiplying polynomials. Two of the formulas looked like this:

$$(a + b)^2 = a^2 + 2ab + b^2$$

$$(a - b)^2 = a^2 - 2ab + b^2$$

If we exchange the left and right sides of each formula, we have two special formulas for factoring:

$$a^2 + 2ab + b^2 = (a + b)^2$$

$$a^2 - 2ab + b^2 = (a - b)^2$$

The left side of each formula is called a *perfect square trinomial.* The right sides are binomial squares. Perfect square trinomials can always be factored using the usual methods for factoring trinomials. However, if we notice that the first and last terms of a trinomial are perfect squares, it is wise to see whether the trinomial factors as a binomial square before attempting to factor by the usual method.

EXAMPLE 1 Factor $x^2 - 6x + 9$.

SOLUTION Since the first and last terms are perfect squares, we attempt to factor according to the preceding formulas:

$$x^2 - 6x + 9 = (x - 3)^2$$

If we expand $(x - 3)^2$, we have $x^2 - 6x + 9$, indicating we have factored correctly.

EXAMPLES Factor each of the following perfect square trinomials.

2. $16a^2 + 40ab + 25b^2 = (4a + 5b)^2$
3. $49 - 14t + t^2 = (7 - t)^2$
4. $9x^4 - 12x^2 + 4 = (3x^2 - 2)^2$
5. $(y + 3)^2 + 10(y + 3) + 25 = [(y + 3) + 5]^2 = (y + 8)^2$

EXAMPLE 6 Factor $8x^2 - 24xy + 18y^2$.

SOLUTION We begin by factoring the greatest common factor 2 from each term:

$$\begin{aligned} 8x^2 - 24xy + 18y^2 &= 2(4x^2 - 12xy + 9y^2) \\ &= 2(2x - 3y)^2 \end{aligned}$$

The Difference of Two Squares

Recall the formula that results in the difference of two squares: $(a + b)(a - b) = a^2 - b^2$. Writing this as a factoring formula, we have

$$a^2 - b^2 = (a + b)(a - b)$$

EXAMPLES Each of the following is the difference of two squares. Use the formula $a^2 - b^2 = (a + b)(a - b)$ to factor each one.

7. $x^2 - 25 = x^2 - 5^2 = (x + 5)(x - 5)$
8. $49 - t^2 = 7^2 - t^2 = (7 + t)(7 - t)$
9. $81a^2 - 25b^2 = (9a)^2 - (5b)^2 = (9a + 5b)(9a - 5b)$
10. $4x^6 - 1 = (2x^3)^2 - 1^2 = (2x^3 + 1)(2x^3 - 1)$
11. $x^2 - \frac{4}{9} = x^2 - (\frac{2}{3})^2 = (x + \frac{2}{3})(x - \frac{2}{3})$

As our next example shows, the difference of two fourth powers can be factored as the difference of two squares.

EXAMPLE 12 Factor $16x^4 - 81y^4$.

SOLUTION The first and last terms are perfect squares. We factor according to the preceding formula:

$$16x^4 - 81y^4 = (4x^2)^2 - (9y^2)^2$$
$$= (4x^2 + 9y^2)(4x^2 - 9y^2)$$

Notice that the second factor is also the difference of two squares. Factoring completely, we have

$$16x^4 - 81y^4 = (4x^2 + 9y^2)(2x + 3y)(2x - 3y)$$

Note
The sum of two squares never factors into the product of two binomials; that is, if we were to attempt to factor $(4x^2 + 9y^2)$ in the last example, we would be unable to find two binomials (or any other polynomials) whose product is $4x^2 + 9y^2$. The factors do not exist as polynomials.

Here is another example of the difference of two squares.

EXAMPLE 13 Factor $(x - 3)^2 - 25$.

SOLUTION This example has the form $a^2 - b^2$, where a is $x - 3$ and b is 5. We factor it according to the formula for the difference of two squares:

$(x - 3)^2 - 25 = (x - 3)^2 - 5^2$	**Write 25 as 5^2**
$= [(x - 3) + 5][(x - 3) - 5]$	**Factor**
$= (x + 2)(x - 8)$	**Simplify**

Notice in this example we could have expanded $(x - 3)^2$, subtracted 25, and then factored to obtain the same result:

$(x - 3)^2 - 25 = x^2 - 6x + 9 - 25$	**Expand $(x - 3)^2$**
$= x^2 - 6x - 16$	**Simplify**
$= (x - 8)(x + 2)$	**Factor**

EXAMPLE 14 Factor $x^2 - 10x + 25 - y^2$.

SOLUTION Notice the first three items form a perfect square trinomial; that is, $x^2 - 10x + 25 = (x - 5)^2$. If we replace the first three terms by $(x - 5)^2$, the expression that results has the form $a^2 - b^2$. We can factor as we did in Example 13:

$x^2 - 10x + 25 - y^2 = (x^2 - 10x + 25) - y^2$	**Group first three terms together**
$= (x - 5)^2 - y^2$	**This has the form $a^2 - b^2$**
$= [(x - 5) + y][(x - 5) - y]$	**Factor according to the formula $a^2 - b^2 = (a + b)(a - b)$**
$= (x - 5 + y)(x - 5 - y)$	**Simplify**

We could check this result by multiplying the two factors together. (You may want to do that to convince yourself that we have the correct result.)

EXAMPLE 15 Factor completely $x^3 + 2x^2 - 9x - 18$.

SOLUTION We use factoring by grouping to begin and then factor the difference of two squares:

$$\begin{aligned} x^3 + 2x^2 - 9x - 18 &= x^2(x + 2) - 9(x + 2) \\ &= (x + 2)(x^2 - 9) \\ &= (x + 2)(x + 3)(x - 3) \end{aligned}$$

The Sum and Difference of Two Cubes

Here are the formulas for factoring the sum and difference of two cubes:

$$a^3 + b^3 = (a + b)(a^2 - ab + b^2)$$

$$a^3 - b^3 = (a - b)(a^2 + ab + b^2)$$

Since these formulas are unfamiliar, it is important that we verify them.

EXAMPLE 16 Verify the two formulas.

SOLUTION We verify the formulas by multiplying the right sides and comparing the results with the left sides:

$$\begin{array}{llll} a^2 & - ab & + b^2 & \\ & a & + b & \\ \hline a^3 & - a^2b & + ab^2 & \\ & a^2b & - ab^2 & + b^3 \\ \hline a^3 & & & + b^3 \end{array}$$

The first formula is correct.

$$\begin{array}{llll} a^2 & + ab & + b^2 & \\ & a & - b & \\ \hline a^3 & + a^2b & + ab^2 & \\ & - a^2b & - ab^2 & - b^3 \\ \hline a^3 & & & - b^3 \end{array}$$

The second formula is correct.

Here are some examples using the formulas for factoring the sum and difference of two cubes.

EXAMPLE 17 Factor $64 + t^3$.

SOLUTION The first term is the cube of 4 and the second term is the cube of t. Therefore,

$$\begin{aligned} 64 + t^3 &= 4^3 + t^3 \\ &= (4 + t)(16 - 4t + t^2) \end{aligned}$$

EXAMPLE 18 Factor $27x^3 + 125y^3$.

SOLUTION Writing both terms as perfect cubes, we have

$$
\begin{aligned}
27x^3 + 125y^3 &= (3x)^3 + (5y)^3 \\
&= (3x + 5y)(9x^2 - 15xy + 25y^2)
\end{aligned}
$$

EXAMPLE 19 Factor $a^3 - \frac{1}{8}$.

SOLUTION The first term is the cube of a, whereas the second term is the cube of $\frac{1}{2}$:

$$
\begin{aligned}
a^3 - \tfrac{1}{8} &= a^3 - (\tfrac{1}{2})^3 \\
&= (a - \tfrac{1}{2})(a^2 + \tfrac{1}{2}a + \tfrac{1}{4})
\end{aligned}
$$

EXAMPLE 20 Factor $x^6 - y^6$.

SOLUTION We have a choice of how we want to write the two terms to begin. We can write the expression as the difference of two squares, $(x^3)^2 - (y^3)^2$, or as the difference of two cubes, $(x^2)^3 - (y^2)^3$. It is better to use the difference of two squares if we have a choice:

$$
\begin{aligned}
x^6 - y^6 &= (x^3)^2 - (y^3)^2 \\
&= (x^3 - y^3)(x^3 + y^3) \\
&= (x - y)(x^2 + xy + y^2)(x + y)(x^2 - xy + y^2)
\end{aligned}
$$

Try this example again writing the first line as the difference of two cubes instead of the difference of two squares. It will become apparent why it is better to use the difference of two squares.

LINKING OBJECTIVES AND EXAMPLES

Next to each **objective** we have listed the examples that are best described by that objective.

A	1–6
B	7–15
C	16–20

GETTING READY FOR CLASS

After reading through the preceding section, respond in your own words and in complete sentences.

1. In what cases can you factor a binomial?
2. What is a perfect square trinomial?
3. Is it possible to factor the sum of two squares?
4. Write the formula you use to factor the sum of two cubes.

Problem Set 5.6

Online support materials can be found at www.thomsonedu.com/login

Factor each perfect square trinomial.

1. $x^2 - 6x + 9$
2. $x^2 + 10x + 25$
3. $a^2 - 12a + 36$
4. $36 - 12a + a^2$
5. $25 - 10t + t^2$
6. $64 + 16t + t^2$
7. $\frac{1}{9}x^2 + 2x + 9$
8. $\frac{1}{4}x^2 - 2x + 4$
9. $4y^4 - 12y^2 + 9$
10. $9y^4 + 12y^2 + 4$
11. $16a^2 + 40ab + 25b^2$
12. $25a^2 - 40ab + 16b^2$
13. $\frac{1}{25} + \frac{1}{10}t^2 + \frac{1}{16}t^4$
14. $\frac{1}{9} - \frac{1}{3}t^3 + \frac{1}{4}t^6$
15. $y^2 + 3y + \frac{9}{4}$
16. $y^2 - 7y + \frac{49}{4}$
17. $a^2 - a + \frac{1}{4}$
18. $a^2 - 5a + \frac{25}{4}$
19. $x^2 - \frac{1}{2}x + \frac{1}{16}$
20. $x^2 - \frac{3}{4}x + \frac{9}{64}$
21. $t^2 + \frac{2}{3}t + \frac{1}{9}$
22. $t^2 - \frac{4}{5}t + \frac{4}{25}$
23. $16x^2 - 48x + 36$
24. $36x^2 + 48x + 16$
25. $75a^3 + 30a^2 + 3a$
26. $45a^4 - 30a^3 + 5a^2$
27. $(x + 2)^2 + 6(x + 2) + 9$
28. $(x + 5)^2 + 4(x + 5) + 4$

Factor each as the difference of two squares. Be sure to factor completely.

29. $x^2 - 9$
30. $x^2 - 16$
31. $49x^2 - 64y^2$
32. $81x^2 - 49y^2$
33. $4a^2 - \frac{1}{4}$
34. $25a^2 - \frac{1}{25}$
35. $x^2 - \frac{9}{25}$
36. $x^2 - \frac{25}{36}$
37. $9x^2 - 16y^2$
38. $25x^2 - 49y^2$
39. $250 - 10t^2$
40. $640 - 10t^2$

Factor each as the difference of two squares. Be sure to factor completely.

41. $x^4 - 81$
42. $x^4 - 16$
43. $9x^6 - 1$
44. $25x^6 - 1$
45. $16a^4 - 81$
46. $81a^4 - 16b^4$
47. $\frac{1}{81} - \frac{y^4}{16}$
48. $\frac{1}{25} - \frac{y^4}{64}$

Factor completely.

49. $x^6 - y^6$
50. $x^6 - 1$
51. $2a^7 - 128a$
52. $128a^8 - 2a^2$
53. $(x - 2)^2 - 9$
54. $(x + 2)^2 - 9$
55. $(y + 4)^2 - 16$
56. $(y - 4)^2 - 16$
57. $x^2 - 10x + 25 - y^2$
58. $x^2 - 6x + 9 - y^2$
59. $a^2 + 8a + 16 - b^2$
60. $a^2 + 12a + 36 - b^2$
61. $x^2 + 2xy + y^2 - a^2$
62. $a^2 + 2ab + b^2 - y^2$
63. $x^3 + 3x^2 - 4x - 12$
64. $x^3 + 5x^2 - 4x - 20$
65. $x^3 + 2x^2 - 25x - 50$
66. $x^3 + 4x^2 - 9x - 36$
67. $2x^3 + 3x^2 - 8x - 12$
68. $3x^3 + 2x^2 - 27x - 18$
69. $4x^3 + 12x^2 - 9x - 27$
70. $9x^3 + 18x^2 - 4x - 8$
71. $(2x - 5)^2 - 100$

= Videotapes available by instructor request

▶ = Online student support materials available at www.thomsonedu.com/login

72. $(7a + 5)^2 - 64$

73. $(a - 3)^2 - (4b)^2$

74. $(2x - 5)^2 - (6y)^2$

75. $a^2 - 6a + 9 - 16b^2$

76. $x^2 - 10x + 25 - 9y^2$

77. $x^2(x + 4) - 6x(x + 4) + 9(x + 4)$

78. $x^2(x - 6) + 8x(x - 6) + 16(x - 6)$

Factor each of the following as the sum or difference of two cubes.

79. $x^3 - y^3$

80. $x^3 + y^3$

81. $a^3 + 8$

82. $a^3 - 8$

83. $27 + x^3$

84. $27 - x^3$

85. $y^3 - 1$

86. $y^3 + 1$

87. $10r^3 - 1{,}250$

88. $10r^3 + 1{,}250$

89. $64 + 27a^3$

90. $27 - 64a^3$

91. $8x^3 - 27y^3$

92. $27x^3 - 8y^3$

93. $t^3 + \frac{1}{27}$

94. $t^3 - \frac{1}{27}$

95. $27x^3 - \frac{1}{27}$

96. $8x^3 + \frac{1}{8}$

97. $64a^3 + 125b^3$

98. $125a^3 - 27b^3$

99. Find two values of b that will make $9x^2 + bx + 25$ a perfect square trinomial.

100. Find a value of c that will make $49x^2 - 42x + c$ a perfect square trinomial.

Maintaining Your Skills

Solve each system by using Cramer's rule.

101. $4x - 7y = 3$
$5x + 2y = -3$

102. $9x - 8y = 4$
$2x + 3y = 6$

103. $3x + 4y = 15$
$2x - 5z = -3$
$4y - 3z = 9$

104. $x + 3y = 5$
$6y + z = 12$
$x - 2z = -10$

Getting Ready for the Next Section

Factor out the greatest common factor.

105. $y^3 + 25y$

106. $y^4 + 36y^2$

107. $2ab^5 + 8ab^4 + 2ab^3$

108. $3a^2b^3 + 6a^2b^2 - 3a^2b$

Factor by grouping.

109. $4x^2 - 6x + 2ax - 3a$

110. $6x^2 - 4x + 3ax - 2a$

Factor the difference of squares.

111. $x^2 - 4$

112. $x^2 - 9$

Factor the perfect square trinomial.

113. $x^2 - 6x + 9$

114. $x^2 - 10x + 25$

Factor.

115. $6a^2 - 11a + 4$

116. $6x^2 - x - 15$

Factor the sum or difference of cubes.

117. $x^3 + 8$

118. $x^3 - 27$

Extending the Concepts

Factor completely.

119. $a^2 - b^2 + 6b - 9$

120. $a^2 - b^2 - 18b - 81$

121. $(x - 3)^2 - (y + 5)^2$

122. $(a + 7)^2 - (b - 9)^2$

Find k such that each trinomial becomes a perfect square trinomial.

123. $kx^2 - 168xy + 49y^2$

124. $kx^2 + 110xy + 121y^2$

125. $49x^2 + kx + 81$

126. $64x^2 + kx + 169$

5.7 Factoring: A General Review

OBJECTIVES

A Factor a variety of polynomials.

In this section we will review the different methods of factoring that we have presented in the previous sections of this chapter. This section is important because it will give you an opportunity to factor a variety of polynomials.

We begin this section by listing the steps that can be used to factor polynomials of any type.

To Factor a Polynomial

Step 1: If the polynomial has a greatest common factor other than 1, then factor out the greatest common factor.

Step 2: If the polynomial has two terms (it is a binomial), then see if it is the difference of two squares or the sum or difference of two cubes, and then factor accordingly. Remember, if it is the sum of two squares it will not factor.

Step 3: If the polynomial has three terms (a trinomial), then it is either a perfect square trinomial, which will factor into the square of a binomial, or it is not a perfect square trinomial, in which case we try to write it as the product of two binomials using the methods developed in this chapter.

Step 4: If the polynomial has more than three terms, then try to factor it by grouping.

Step 5: As a final check, see if any of the factors you have written can be factored further. If you have overlooked a common factor, you can catch it here.

Here are some examples illustrating how we use the steps in our list. There are no new factoring problems in this section. The problems here are all similar to the problems you have seen before. What is different is that they are not all of the same type.

EXAMPLE 1

Factor $2x^5 - 8x^3$.

SOLUTION First we check to see if the greatest common factor is other than 1. Since the greatest common factor is $2x^3$, we begin by factoring it out. Once we have done so, we notice that the binomial that remains is the difference of two squares, which we factor according to the formula $a^2 - b^2 = (a + b)(a - b)$.

$2x^5 - 8x^3 = 2x^3(x^2 - 4)$ **Factor out the greatest common factor, $2x^3$**

$= 2x^3(x + 2)(x - 2)$ **Factor the difference of two squares**

EXAMPLE 2 Factor $3x^4 - 18x^3 + 27x^2$.

SOLUTION Step 1 is to factor out the greatest common factor $3x^2$. After we have done so, we notice that the trinomial that remains is a perfect square trinomial, which will factor as the square of a binomial.

$$3x^4 - 18x^3 + 27x^2 = 3x^2(x^2 - 6x + 9) \quad \textbf{Factor out } 3x^2$$

$$= 3x^2(x - 3)^2 \quad x^2 - 6x + 9 \textbf{ is the square of } x - 3$$

EXAMPLE 3 Factor $y^3 + 25y$.

SOLUTION We begin by factoring out the y that is common to both terms. The binomial that remains after we have done so is the sum of two squares, which does not factor, so after the first step, we are finished.

$$y^3 + 25y = y(y^2 + 25)$$

EXAMPLE 4 Factor $6a^2 - 11a + 4$.

SOLUTION Here we have a trinomial that does not have a greatest common factor other than 1. Since it is not a perfect square trinomial, we factor it by trial and error. Without showing all the different possibilities, here is the answer.

$$6a^2 - 11a + 4 = (3a - 4)(2a - 1)$$

EXAMPLE 5 Factor $2x^4 + 16x$.

SOLUTION This binomial has a greatest common factor of $2x$. The binomial that remains after the $2x$ has been factored from each term is the sum of two cubes, which we factor according to the formula $a^3 + b^3 = (a + b)(a^2 - ab + b^2)$.

$$2x^4 + 16x = 2x(x^3 + 8) \quad \textbf{Factor } 2x \textbf{ from each term}$$

$$= 2x(x + 2)(x^2 - 2x + 4) \quad \textbf{The sum of two cubes}$$

EXAMPLE 6 Factor $2ab^5 + 8ab^4 + 2ab^3$.

SOLUTION The greatest common factor is $2ab^3$. We begin by factoring it from each term. After that we find that the trinomial that remains cannot be factored further.

$$2ab^5 + 8ab^4 + 2ab^3 = 2ab^3(b^2 + 4b + 1)$$

EXAMPLE 7 Factor $4x^2 - 6x + 2ax - 3a$.

SOLUTION Our polynomial has four terms, so we factor by grouping.

$$4x^2 - 6x + 2ax - 3a = 2x(2x - 3) + a(2x - 3)$$

$$= (2x - 3)(2x + a)$$

LINKING OBJECTIVES AND EXAMPLES

Next to each **objective** we have listed the examples that are best described by that objective.

A 1–7

GETTING READY FOR CLASS

After reading through the preceding section, respond in your own words and in complete sentences.

1. How do you know when you've factored completely?
2. If a polynomial has four terms, what method of factoring should you try?
3. What is the first step in factoring a polynomial?
4. What do we call a polynomial that does not factor?

Problem Set 5.7

Online support materials can be found at www.thomsonedu.com/login

Factor each of the following polynomials completely. Once you are finished factoring, none of the factors you obtain should be factorable. Also, note that the even-numbered problems are not necessarily similar to the odd-numbered problems that precede them in this problem set.

▶ **1.** $x^2 - 81$ **2.** $x^2 - 18x + 81$

3. $x^2 + 2x - 15$

4. $15x^2 + 13x - 6$

5. $x^2(x + 2) + 6x(x + 2) + 9(x + 2)$

▶ **6.** $12x^2 - 11x + 2$

7. $x^2y^2 + 2y^2 + x^2 + 2$

8. $21y^2 - 25y - 4$

9. $2a^3b + 6a^2b + 2ab$

10. $6a^2 - ab - 15b^2$

11. $x^2 + x + 1$

12. $x^2y + 3y + 2x^2 + 6$

13. $12a^2 - 75$

14. $18a^2 - 50$

15. $9x^2 - 12xy + 4y^2$

▶ **16.** $x^3 - x^2$

17. $25 - 10t + t^2$

18. $t^2 + 4t + 4 - y^2$

19. $4x^3 + 16xy^2$

20. $16x^2 + 49y^2$

21. $2y^3 + 20y^2 + 50y$

22. $x^2 + 5bx - 2ax - 10ab$

23. $a^7 + 8a^4b^3$

24. $5a^2 - 45b^2$

25. $t^2 + 6t + 9 - x^2$

26. $36 + 12t + t^2$

Factor completely.

▶ **27.** $x^3 + 5x^2 - 9x - 45$ **28.** $x^3 + 5x^2 - 16x - 80$

29. $5a^2 + 10ab + 5b^2$ **30.** $3a^3b^2 + 15a^2b^2 + 3ab^2$

▶ **31.** $x^2 + 49$ **32.** $16 - x^4$

33. $3x^2 + 15xy + 18y^2$

34. $3x^2 + 27xy + 54y^2$

35. $9a^2 + 2a + \frac{1}{9}$

36. $18 - 2a^2$

37. $x^2(x - 3) - 14x(x - 3) + 49(x - 3)$

38. $x^2 + 3ax - 2bx - 6ab$

39. $x^2 - 64$

40. $9x^2 - 4$

41. $8 - 14x - 15x^2$

42. $5x^4 + 14x^2 - 3$

43. $49a^7 - 9a^5$

44. $a^6 - b^6$

45. $r^2 - \frac{1}{25}$

▶ **46.** $27 - r^3$

▢ = Videotapes available by instructor request

▶ = Online student support materials available at www.thomsonedu.com/login

47. $49x^2 + 9y^2$

48. $12x^4 - 62x^3 + 70x^2$

49. $100x^2 - 100x - 600$

50. $100x^2 - 100x - 1{,}200$

51. $25a^3 + 20a^2 + 3a$

52. $16a^5 - 54a^2$

53. $3x^4 - 14x^2 - 5$

54. $8 - 2x - 15x^2$

55. $24a^5b - 3a^2b$

56. $18a^4b^2 - 24a^3b^3 + 8a^2b^4$

57. $64 - r^3$

58. $r^2 - \frac{1}{9}$

59. $20x^4 - 45x^2$

60. $16x^3 + 16x^2 + 3x$

61. $400t^2 - 900$

62. $900 - 400t^2$

63. $16x^5 - 44x^4 + 30x^3$

64. $16x^2 + 16x - 1$

65. $y^6 - 1$

66. $25y^7 - 16y^5$

67. $50 - 2a^2$

68. $4a^2 + 2a + \frac{1}{4}$

69. $12x^4y^2 + 36x^3y^3 + 27x^2y^4$

70. $16x^3y^2 - 4xy^2$

71. $x^2 - 4x + 4 - y^2$

72. $x^2 - 12x + 36 - b^2$

73. $a^2 - \frac{4}{3}ab + \frac{4}{9}b^2$

74. $a^2 + \frac{3}{2}ab + \frac{9}{16}b^2$

75. $x^2 - \frac{4}{5}xy + \frac{4}{25}y^2$

76. $x^2 + \frac{8}{7}xy + \frac{16}{49}y^2$

77. $a^2 - \frac{5}{3}ab + \frac{25}{36}b^2$

78. $a^2 + \frac{5}{4}ab + \frac{25}{64}b^2$

79. $x^2 - \frac{8}{5}xy + \frac{16}{25}y^2$

80. $a^2 + \frac{3}{5}ab + \frac{9}{100}b^2$

81. $2x^2(x + 2) - 13x(x + 2) + 15(x + 2)$

82. $5x^2(x - 4) - 14x(x - 4) - 3(x - 4)$

83. $(x - 4)^3 + (x - 4)^4$

84. $(2x - 7)^5 + (2x - 7)^6$

85. $2y^3 - 54$

86. $81 + 3y^3$

87. $2a^3 - 128b^3$

88. $128a^3 + 2b^3$

89. $2x^3 + 432y^3$

90. $432x^3 - 2y^3$

Maintaining Your Skills

The following problems are taken from the book *Algebra for the Practical Man,* written by J. E. Thompson and published by D. Van Nostrand Company in 1931.

91. A man spent $112.80 for 108 geese and ducks, each goose costing 14 dimes and each duck 6 dimes. How many of each did he buy?

92. If 15 pounds of tea and 10 pounds of coffee together cost $15.50, while 25 pounds of tea and 13 pounds of coffee at the same prices cost $24.55, find the price per pound of each.

93. A number of oranges at the rate of three for $0.10 and apples at $0.15 a dozen cost, together, $6.80. Five times as many oranges and one-fourth as many apples at the same rates would have cost $25.45. How many of each were bought?

94. An estate is divided among three persons: *A*, *B*, and *C*. *A*'s share is three times that of *B* and *B*'s share is twice that of *C*. If *A* receives $9,000 more than *C*, how much does each receive?

Getting Ready for the Next Section

Simplify.

95. $x^2 + (x + 1)^2$

96. $x^2 + (x + 3)^2$

97. $\dfrac{16t^2 - 64t + 48}{16}$

98. $\dfrac{100p^2 - 1{,}300p + 4{,}000}{100}$

Factor each of the following.

99. $x^2 - 2x - 24$

100. $x^2 - x - 6$

101. $2x^3 - 5x^2 - 3x$

102. $3x^3 - 5x^2 - 2x$

103. $x^3 + 2x^2 - 9x - 18$

104. $x^3 + 5x^2 - 4x - 20$

Solve.

105. $x - 6 = 0$

106. $x + 4 = 0$

107. $2x + 1 = 0$

108. $3x + 1 = 0$

5.8 Solving Equations by Factoring

OBJECTIVES

A Solve equations by factoring.

B Apply the Blueprint for Problem Solving to solve application problems whose solutions involve quadratic equations.

C Solve problems that contain formulas that are quadratic.

In this section we will use our knowledge of factoring to solve equations. Most of the equations we will solve in this section are *quadratic equations.* Here is the definition of a quadratic equation.

DEFINITION Any equation that can be written in the form

$$ax^2 + bx + c = 0$$

where a, b, and c are constants and a is not 0 ($a \neq 0$) is called a **quadratic equation.** The form $ax^2 + bx + c = 0$ is called **standard form** for quadratic equations.

Each of the following is a quadratic equation:

$$2x^2 = 5x + 3 \qquad 5x^2 = 75 \qquad 4x^2 - 3x + 2 = 0$$

Note
The third equation is clearly a quadratic equation since it is in standard form. (Notice that a is 4, b is -3, and c is 2.) The first two equations are also quadratic because they could be put in the form $ax^2 + bx + c = 0$ by using the addition property of equality.

Notation For a quadratic equation written in standard form, the first term ax^2 is called the *quadratic term;* the second term bx is the *linear term;* and the last term c is called the *constant term.*

In the past we have noticed that the number 0 is a special number. There is another property of 0 that is the key to solving quadratic equations. It is called the *zero-factor property.*

Zero-Factor Property For all real numbers r and s,

$$r \cdot s = 0 \quad \text{if and only if} \quad r = 0 \quad \text{or} \quad s = 0 \quad \text{(or both)}$$

EXAMPLE 1 Solve $x^2 - 2x - 24 = 0$.

SOLUTION We begin by factoring the left side as $(x - 6)(x + 4)$ and get

$$(x - 6)(x + 4) = 0$$

Now both $(x - 6)$ and $(x + 4)$ represent real numbers. We notice that their product is 0. By the zero-factor property, one or both of them must be 0:

$$x - 6 = 0 \quad \text{or} \quad x + 4 = 0$$

> **Note**
> What the zero-factor property says in words is that we can't multiply and get 0 without multiplying by 0; that is, if we multiply two numbers and get 0, then one or both of the original two numbers we multiplied must have been 0.

We have used factoring and the zero-factor property to rewrite our original second-degree equation as two first-degree equations connected by the word *or*. Completing the solution, we solve the two first-degree equations:

$$x - 6 = 0 \quad \text{or} \quad x + 4 = 0$$
$$x = 6 \quad \text{or} \quad x = -4$$

We check our solutions in the original equation as follows:

Check $x = 6$	Check $x = -4$
$6^2 - 2(6) - 24 \stackrel{?}{=} 0$	$(-4)^2 - 2(-4) - 24 \stackrel{?}{=} 0$
$36 - 12 - 24 \stackrel{?}{=} 0$	$16 + 8 - 24 \stackrel{?}{=} 0$
$0 = 0$	$0 = 0$

> **Note**
> We are placing a question mark over the equal sign because we don't know yet if the expression on the left will be equal to the expression on the right.

In both cases the result is a true statement, which means that both 6 and -4 are solutions to the original equation.

Although the next equation is not quadratic, the method we use is similar.

EXAMPLE 2 Solve $\frac{1}{3}x^3 = \frac{5}{6}x^2 + \frac{1}{2}x$.

SOLUTION We can simplify our work if we clear the equation of fractions. Multiplying both sides by the LCD, 6, we have

$$\mathbf{6} \cdot \tfrac{1}{3}x^3 = \mathbf{6} \cdot \tfrac{5}{6}x^2 + \mathbf{6} \cdot \tfrac{1}{2}x$$
$$2x^3 = 5x^2 + 3x$$

Next we add $-5x^2$ and $-3x$ to each side so that the right side will become 0.

$$2x^3 - 5x^2 - 3x = 0 \qquad \textbf{Standard form}$$

We factor the left side and then use the zero-factor property to set each factor to 0.

$$x(2x^2 - 5x - 3) = 0 \qquad \textbf{Factor out the greatest common factor}$$
$$x(2x + 1)(x - 3) = 0 \qquad \textbf{Continue factoring}$$
$$x = 0 \quad \text{or} \quad 2x + 1 = 0 \quad \text{or} \quad x - 3 = 0 \qquad \textbf{Zero-factor property}$$

Solving each of the resulting equations, we have

$$x = 0 \quad \text{or} \quad x = -\tfrac{1}{2} \quad \text{or} \quad x = 3$$

To generalize the preceding example, here are the steps used in solving a quadratic equation by factoring.

To Solve an Equation by Factoring

Step 1: Write the equation in standard form.

Step 2: Factor the left side.

Step 3: Use the zero-factor property to set each factor equal to 0.

Step 4: Solve the resulting linear equations.

EXAMPLE 3

Solve $100x^2 = 300x$.

SOLUTION We begin by writing the equation in standard form and factoring:

$$100x^2 = 300x$$

$$100x^2 - 300x = 0 \quad \textbf{Standard form}$$

$$100x(x - 3) = 0 \quad \textbf{Factor}$$

Using the zero-factor property to set each factor to 0, we have

$$100x = 0 \quad \text{or} \quad x - 3 = 0$$

$$x = 0 \quad \text{or} \quad x = 3$$

The two solutions are 0 and 3.

EXAMPLE 4

Solve $(x - 2)(x + 1) = 4$.

SOLUTION We begin by multiplying the two factors on the left side. (Notice that it would be incorrect to set each of the factors on the left side equal to 4. The fact that the product is 4 does not imply that either of the factors must be 4.)

$$(x - 2)(x + 1) = 4$$

$$x^2 - x - 2 = 4 \quad \textbf{Multiply the left side}$$

$$x^2 - x - 6 = 0 \quad \textbf{Standard form}$$

$$(x - 3)(x + 2) = 0 \quad \textbf{Factor}$$

$$x - 3 = 0 \quad \text{or} \quad x + 2 = 0 \quad \textbf{Zero-factor property}$$

$$x = 3 \quad \text{or} \quad x = -2$$

EXAMPLE 5

Solve for x: $x^3 + 2x^2 - 9x - 18 = 0$.

SOLUTION We start with factoring by grouping.

$$x^3 + 2x^2 - 9x - 18 = 0$$

$$x^2(x + 2) - 9(x + 2) = 0$$

$$(x + 2)(x^2 - 9) = 0$$

$$(x + 2)(x - 3)(x + 3) = 0 \quad \textbf{The difference of two squares}$$

$$x + 2 = 0 \quad \text{or} \quad x - 3 = 0 \quad \text{or} \quad x + 3 = 0 \quad \textbf{Set factors to 0}$$

$$x = -2 \quad \text{or} \quad x = 3 \quad \text{or} \quad x = -3$$

We have three solutions: -2, 3, and -3.

EXAMPLE 6 The sum of the squares of two consecutive integers is 25. Find the two integers.

SOLUTION We apply the Blueprint for Problem Solving to solve this application problem. Remember, step 1 in the blueprint is done mentally.

Step 1: ***Read and list.***
Known items: Two consecutive integers. If we add their squares, the result is 25.
Unknown items: The two integers

Step 2: ***Assign a variable and translate information.***
Let x = the first integer; then $x + 1$ = the next consecutive integer.

Step 3: ***Reread and write an equation.***
Since the sum of the squares of the two integers is 25, the equation that describes the situation is

$$x^2 + (x + 1)^2 = 25$$

Step 4: ***Solve the equation.***

$$x^2 + (x + 1)^2 = 25$$

$$x^2 + (x^2 + 2x + 1) = 25$$

$$2x^2 + 2x - 24 = 0$$

$$x^2 + x - 12 = 0 \quad \text{Divide each side by 2}$$

$$(x + 4)(x - 3) = 0$$

$$x = -4 \quad \text{or} \quad x = 3$$

Step 5: ***Write the answer.***
If $x = -4$, then $x + 1 = -3$. If $x = 3$, then $x + 1 = 4$. The two integers are -4 and -3, or the two integers are 3 and 4.

Step 6: ***Reread and check.***
The two integers in each pair are consecutive integers, and the sum of the squares of either pair is 25.

Another application of quadratic equations involves the Pythagorean theorem, an important theorem from geometry. The theorem gives the relationship between the sides of any right triangle (a triangle with a 90-degree angle). We state it here without proof.

Pythagorean Theorem In any right triangle, the square of the longest side (hypotenuse) is equal to the sum of the squares of the other two sides (legs).

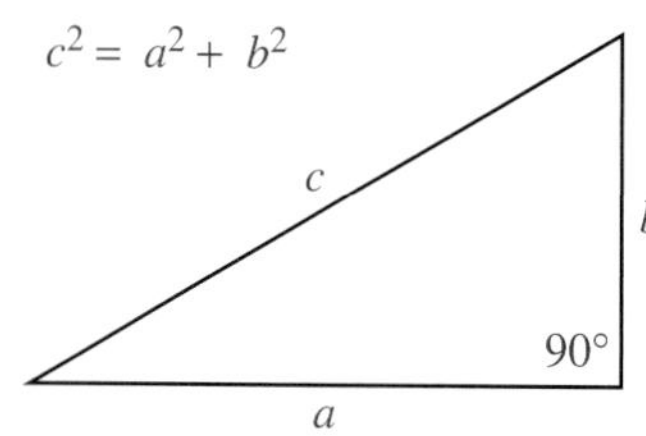

EXAMPLE 7

The lengths of the three sides of a right triangle are given by three consecutive integers. Find the lengths of the three sides.

SOLUTION

Step 1: ***Read and list.***

Known items: A right triangle. The three sides are three consecutive integers.

Unknown items: The three sides

Step 2: ***Assign a variable and translate information.***

Let x = first integer (shortest side).

Then $x + 1$ = next consecutive integer

$x + 2$ = last consecutive integer (longest side)

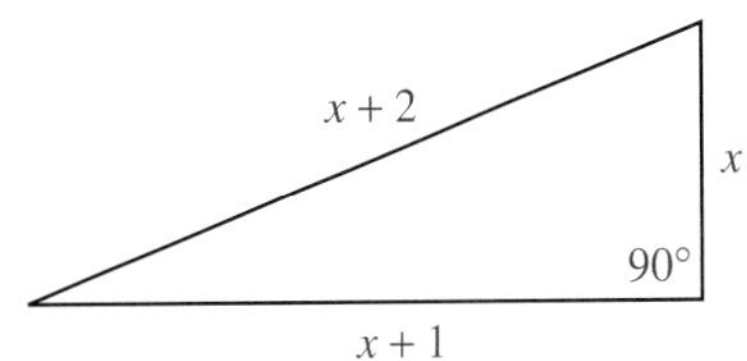

Step 3: ***Reread and write an equation.***

By the Pythagorean theorem, we have

$$(x + 2)^2 = (x + 1)^2 + x^2$$

Step 4: ***Solve the equation.***

$$x^2 + 4x + 4 = x^2 + 2x + 1 + x^2$$

$$x^2 - 2x - 3 = 0$$

$$(x - 3)(x + 1) = 0$$

$$x = 3 \quad \text{or} \quad x = -1$$

Step 5: ***Write the answer.***

Since x is the length of a side in a triangle, it must be a positive number. Therefore, $x = -1$ cannot be used.

The shortest side is 3. The other two sides are 4 and 5.

Step 6: ***Reread and check.***

The three sides are given by consecutive integers. The square of the longest side is equal to the sum of the squares of the two shorter sides.

EXAMPLE 8

Two boats leave from an island port at the same time. One travels due north at a speed of 12 miles per hour, and the other travels due west at a speed of 16 miles per hour. How long until the distance between the boats is 60 miles?

SOLUTION

Step 1: ***Read and list.***

Known items: The speed and direction of both boats. The distance between the boats.

Unknown items: The distance traveled by each boat, and the time

Step 2: ***Assign a variable and translate information.***

Let t = the time.

Then $12t$ = the distance traveled by boat going north

$16t$ = the distance traveled by boat going west

If we draw a diagram for the problem, we see that the distances traveled by the two boats form the legs of a right triangle. The hypotenuse of the triangle will be the distance between the boats, which is 60 miles.

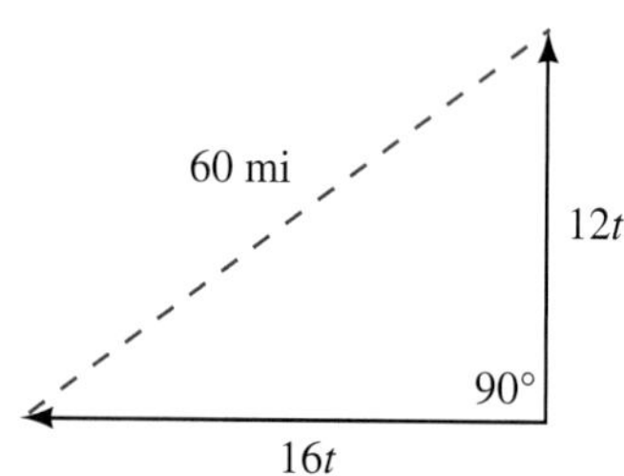

Step 3: ***Reread and write an equation.***

By the Pythagorean theorem, we have

$$(16t)^2 + (12t)^2 = 60^2$$

Step 4: ***Solve the equation.***

$$256t^2 + 144t^2 = 3600$$

$$400t^2 = 3600$$

$$400t^2 - 3600 = 0$$

$$t^2 - 9 = 0 \quad \text{Divide each side by 400}$$

$$(t + 3)(t - 3) = 0$$

$$t = -3 \quad \text{or} \quad t = 3$$

Step 5: ***Write the answer.***

Because t is measuring time, it must be a positive number. Therefore, $t = -3$ cannot be used.

The two boats will be 60 miles apart after 3 hours.

Step 6: ***Reread and check.***

The boat going north will travel $12 \cdot 3 = 36$ miles in 3 hours, and the boat going west wil travel $16 \cdot 3 = 48$ miles. The distance between them after 3 hours will be 60 miles ($48^2 + 36^2 = 60^2$).

Our next two examples involve formulas that are quadratic.

EXAMPLE 9 If an object is projected into the air with an initial vertical velocity of v feet/second, its height h, in feet, above the ground after t seconds will be given by

$$h = vt - 16t^2$$

Find t if $v = 64$ feet/second and $h = 48$ feet.

SOLUTION Substituting $v = 64$ and $h = 48$ into the preceding formula, we have

$$48 = 64t - 16t^2$$

which is a quadratic equation. We write it in standard form and solve by factoring:

$$16t^2 - 64t + 48 = 0$$

$$t^2 - 4t + 3 = 0 \quad \text{Divide each side by 16}$$

$$(t - 1)(t - 3) = 0$$

$$t - 1 = 0 \quad \text{or} \quad t - 3 = 0$$

$$t = 1 \quad \text{or} \quad t = 3$$

Here is how we interpret our results: If an object is projected upward with an initial vertical velocity of 64 feet/second, it will be 48 feet above the ground after 1 second and after 3 seconds; that is, it passes 48 feet going up and also coming down.

EXAMPLE 10 A manufacturer of small portable radios knows that the number of radios she can sell each week is related to the price of the radios by the equation $x = 1{,}300 - 100p$, where x is the number of radios and p is the price per radio. What price should she charge for each radio if she wants the weekly revenue to be $4,000?

SOLUTION The formula for total revenue is $R = xp$. Since we want R in terms of p, we substitute $1{,}300 - 100p$ for x in the equation $R = xp$:

If $\quad R = xp$

and $\quad x = 1{,}300 - 100p$

then $\quad R = (1{,}300 - 100p)p$

We want to find p when R is 4,000. Substituting 4,000 for R in the formula gives us

$$4{,}000 = (1{,}300 - 100p)p$$

$$4{,}000 = 1{,}300p - 100p^2$$

which is a quadratic equation. To write it in standard form, we add $100p^2$ and $-1{,}300p$ to each side, giving us

$$100p^2 - 1{,}300p + 4{,}000 = 0$$

$$p^2 - 13p + 40 = 0 \quad \text{Divide each side by 100}$$

$$(p - 5)(p - 8) = 0$$

$$p - 5 = 0 \quad \text{or} \quad p - 8 = 0$$

$$p = 5 \quad \text{or} \quad p = 8$$

If she sells the radios for $5 each or for $8 each she will have a weekly revenue of $4,000.

LINKING OBJECTIVES AND EXAMPLES

Next to each objective we have listed the examples that are best described by that objective.

A	1–5
B	6–8
C	9, 10

GETTING READY FOR CLASS

After reading through the preceding section, respond in your own words and in complete sentences.

1. Explain the Pythagorean theorem in words.
2. What is the first step in solving an equation by factoring?
3. Describe the zero-factor property in your own words.
4. Write an application problem for which the solution depends on solving the equation $x^2 + (x + 1)^2 = 313$.

Problem Set 5.8

Online support materials can be found at www.thomsonedu.com/login

Solve each equation.

1. $x^2 - 5x - 6 = 0$
2. $x^2 + 5x - 6 = 0$
3. $x^3 - 5x^2 + 6x = 0$
4. $x^3 + 5x^2 + 6x = 0$
5. $3y^2 + 11y - 4 = 0$
6. $3y^2 - y - 4 = 0$
7. $60x^2 - 130x + 60 = 0$
8. $90x^2 + 60x - 80 = 0$
9. $\frac{1}{10}t^2 - \frac{5}{2} = 0$
10. $\frac{2}{7}t^2 - \frac{7}{2} = 0$
11. $100x^4 = 400x^3 + 2{,}100x^2$
12. $100x^4 = -400x^3 + 2{,}100x^2$
13. $\frac{1}{5}y^2 - 2 = -\frac{3}{10}y$
14. $\frac{1}{2}y^2 + \frac{5}{3} = \frac{17}{6}y$
15. $9x^2 - 12x = 0$
16. $4x^2 + 4x = 0$
17. $0.02r + 0.01 = 0.15r^2$
18. $0.02r - 0.01 = -0.08r^2$
19. $9a^3 = 16a$
20. $16a^3 = 25a$
21. $-100x = 10x^2$
22. $800x = 100x^2$
23. $(x + 6)(x - 2) = -7$
24. $(x - 7)(x + 5) = -20$
25. $(y - 4)(y + 1) = -6$
26. $(y - 6)(y + 1) = -12$
27. $(x + 1)^2 = 3x + 7$
28. $(x + 2)^2 = 9x$
29. $(2r + 3)(2r - 1) = -(3r + 1)$
30. $(3r + 2)(r - 1) = -(7r - 7)$
31. $x^3 + 3x^2 - 4x - 12 = 0$
32. $x^3 + 5x^2 - 4x - 20 = 0$
33. $x^3 + 2x^2 - 25x - 50 = 0$
34. $x^3 + 4x^2 - 9x - 36 = 0$
35. $2x^3 + 3x^2 - 8x - 12 = 0$
36. $3x^3 + 2x^2 - 27x - 18 = 0$
37. $4x^3 + 12x^2 - 9x - 27 = 0$
38. $9x^3 + 18x^2 - 4x - 8 = 0$

Problems 39–48 are problems you will see later in the book. Solve each equation.

39. $3x^2 + x = 10$
40. $y^2 + y - 20 = 2y$
41. $12(x + 3) + 12(x - 3) = 3(x^2 - 9)$
42. $8(x + 2) + 8(x - 2) = 3(x^2 - 4)$
43. $(y + 3)^2 + y^2 = 9$
44. $(2y + 4)^2 + y^2 = 4$
45. $(x + 3)^2 + 1^2 = 2$
46. $(x - 3)^2 + (-1)^2 = 10$
47. $(x + 2)x = 2^3$
48. $(x + 3)x = 2^2$
49. Let $f(x) = \left(x + \frac{3}{2}\right)^2$. Find all values for the variable x for which $f(x) = 0$.

= Videotapes available by instructor request

▶ = Online student support materials available at www.thomsonedu.com/login

▸ **50.** Let $f(x) = \left(x - \frac{5}{2}\right)^2$. Find all values for the variable x for which $f(x) = 0$.

▸ **51.** Let $f(x) = (x - 3)^2 - 25$. Find all values for the variable x for which $f(x) = 0$.

▸ **52.** Let $f(x) = 9x^3 + 18x^2 - 4x - 8$. Find all values for the variable x for which $f(x) = 0$.

Let $f(x) = x^2 + 6x + 3$. Find all values for the variable x, for which $f(x) = g(x)$.

▸ **53.** $g(x) = -6$

▸ **54.** $g(x) = 19$

▸ **55.** $g(x) = 10$

▸ **56.** $g(x) = -2$

Let $h(x) = x^2 - 5x$. Find all values for the variable x, for which $h(x) = f(x)$.

▸ **57.** $f(x) = 0$

▸ **58.** $f(x) = -6$

▸ **59.** $f(x) = 2x + 8$

▸ **60.** $f(x) = -2x + 10$

▸ **61.** Solve each equation

a. $9x - 25 = 0$
b. $9x^2 - 25 = 0$
c. $9x^2 - 25 = 56$
d. $9x^2 - 25 = 30x - 50$

▸ **62.** Solve each equation

a. $5x - 6 = 0$
b. $(5x - 6)^2 = 0$
c. $25x^2 - 36 = 0$
d. $25x^2 - 36 = 28$

Applying the Concepts

▸ **63. Distance** Two cyclists leave from an intersection at the same time. One travels due north at a speed of 15 miles per hour, and the other travels due east at a speed of 20 miles per hour. How long until the distance between the two cyclists is 75 miles?

▸ **64. Distance** Two airplanes leave from an airport at the same time. One travels due south at a speed of 480 miles per hour, and the other travels due west at a speed of 360 miles per hour. How long until the distance between the two airplanes is 2,400 miles?

65. Consecutive Integers The square of the sum of two consecutive integers is 81. Find the two integers.

66. Consecutive Integers Find two consecutive even integers whose sum squared is 100.

67. Right Triangle A 25-foot ladder is leaning against a building. The base of the ladder is 7 feet from the side of the building. How high does the ladder reach along the side of the building?

68. Right Triangle Noreen wants to place her 13-foot ramp against the side of her house so that the top of the ramp rests on a ledge that is 5 feet above the ground. How far will the base of the ramp be from the house?

69. Right Triangle The lengths of the three sides of a right triangle are given by three consecutive even integers. Find the lengths of the three sides.

70. Right Triangle The longest side of a right triangle is 3 less than twice the shortest side. The third side measures 12 inches. Find the length of the shortest side.

71. Geometry The length of a rectangle is 2 feet more than 3 times the width. If the area is 16 square feet, find the width and the length.

72. Geometry The length of a rectangle is 4 yards more than twice the width. If the area is 70 square yards, find the width and the length.

73. Geometry The base of a triangle is 2 inches more than 4 times the height. If the area is 36 square inches, find the base and the height.

74. Geometry The height of a triangle is 4 feet less than twice the base. If the area is 48 square feet, find the base and the height.

75. Projectile Motion If an object is thrown straight up into the air with an initial velocity of 32 feet per second, then its height above the ground at any time t is given by the formula $h = 32t - 16t^2$. Find the times at which the object is on the ground by letting $h = 0$ in the equation and solving for t.

76. Projectile Motion An object is projected into the air with an initial velocity of 64 feet per second. Its height at any time t is given by the formula $h = 64t - 16t^2$. Find the times at which the object is on the ground.

The formula $h = vt - 16t^2$ gives the height h, in feet, of an object projected into the air with an initial vertical velocity v, in feet per second, after t seconds.

77. Projectile Motion If an object is projected upward with an initial velocity of 48 feet per second, at what times will it reach a height of 32 feet above the ground?

78. Projectile Motion If an object is projected upward into the air with an initial velocity of 80 feet per second, at what times will it reach a height of 64 feet above the ground?

79. Projectile Motion An object is projected into the air with a vertical velocity of 24 feet per second. At what times will the object be on the ground? (It is on the ground when h is 0.)

80. Projectile Motion An object is projected into the air with a vertical velocity of 20 feet per second. At what times will the object be on the ground?

81. Height of a Bullet A bullet is fired into the air with an initial upward velocity of 80 feet per second from the top of a building 96 feet high. The equation that gives the height of the bullet at any time t is $h = 96 + 80t - 16t^2$. At what times will the bullet be 192 feet in the air?

82. Height of an Arrow An arrow is shot into the air with an upward velocity of 48 feet per second from a hill 32 feet high. The equation that gives the height of the arrow at any time t is $h = 32 + 48t - 16t^2$. Find the times at which the arrow will be 64 feet above the ground.

83. Price and Revenue A company that manufactures typewriter ribbons knows that the number of ribbons x it can sell each week is related to the price per ribbon p by the equation $x = 1{,}200 - 100p$. At what price should it sell the ribbons if it wants the weekly revenue to be \$3,200? (*Remember:* The equation for revenue is $R = xp$.)

84. Price and Revenue A company manufactures diskettes for home computers. It knows from past experience that the number of diskettes x it can sell each day is related to the price per diskette p by the equation $x = 800 - 100p$. At what price should it sell its diskettes if it wants the daily revenue to be \$1,200?

85. Price and Revenue The relationship between the number of calculators x a company sells per day and the price of each calculator p is given by the equation $x = 1{,}700 - 100p$. At what price should the calculators be sold if the daily revenue is to be \$7,000?

86. Price and Revenue The relationship between the number of pencil sharpeners x a company can sell each week and the price of each sharpener p is given by the equation $x = 1{,}800 - 100p$. At what price should the sharpeners be sold if the weekly revenue is to be \$7,200?

Maintaining Your Skills

Solve each system.

87. $2x - 5y = -8$
$3x + y = 5$

88. $4x - 7y = -2$
$-5x + 6y = -3$

89. $\frac{1}{3}x - \frac{1}{6}y = 3$
$-\frac{1}{5}x + \frac{1}{4}y = 0$

90. $2x - 5y = 14$
$y = 3x + 8$

91. $2x - y + z = 9$
$x + y - 3z = -2$
$3x + y - z = 6$

92. Number Problem A number is 1 less than twice another. Their sum is 14. Find the two numbers.

93. Investing John invests twice as much money at 6% as he does at 5%. If his investments earn a total of

\$680 in 1 year, how much does he have invested at each rate?

94. Speed of a Boat A boat can travel 20 miles downstream in 2 hours. The same boat can travel 18 miles upstream in 3 hours. What is the speed of the boat in still water, and what is the speed of the current?

Graph the solution set for each system.

95. $3x + 2y < 6$
$-2x + 3y < 6$

96. $y \le x + 3$
$y > x - 4$

97. $x \le 4$
$y < 2$

98. $2x + y < 4$
$x \ge 0$
$y \ge 0$

Chapter 5 SUMMARY

Properties of Exponents [5.1]

EXAMPLES

1. These expressions illustrate the properties of exponents.
   a. $x^2 \cdot x^3 = x^{2+3} = x^5$
   b. $(x^2)^3 = x^{2 \cdot 3} = x^6$
   c. $(3x)^2 = 3^2 \cdot x^2 = 9x^2$
   d. $2^{-3} = \frac{1}{2^3} = \frac{1}{8}$
   e. $\left(\frac{x}{5}\right)^2 = \frac{x^2}{5^2} = \frac{x^2}{25}$
   f. $\frac{x^7}{x^5} = x^{7-5} = x^2$
   g. $3^1 = 3$
   $3^0 = 1$

If a and b represent real numbers and r and s represent integers, then

1. $a^r \cdot a^s = a^{r+s}$
2. $(a^r)^s = a^{r \cdot s}$
3. $(ab)^r = a^r \cdot b^r$
4. $a^{-r} = \dfrac{1}{a^r} \qquad (a \neq 0)$
5. $\left(\dfrac{a}{b}\right)^r = \dfrac{a^r}{b^r} \qquad (b \neq 0)$
6. $\dfrac{a^r}{a^s} = a^{r-s} \qquad (a \neq 0)$
7. $a^1 = a$
   $a^0 = 1 \qquad (a \neq 0)$

Scientific Notation [5.1]

2. $49{,}800{,}000 = 4.98 \times 10^7$
   $0.00462 = 4.62 \times 10^{-3}$

A number is written in scientific notation when it is written as the product of a number between 1 and 10 and an integer power of 10; that is, when it has the form

$$n \times 10^r$$

where $1 \le n < 10$ and $r =$ an integer.

Addition of Polynomials [5.2]

3. $(3x^2 + 2x - 5) + (4x^2 - 7x + 2)$
   $= 7x^2 - 5x - 3$

To add two polynomials, simply combine the coefficients of similar terms.

Negative Signs Preceding Parentheses [5.2]

4. $-(2x^2 - 8x - 9)$
   $= -2x^2 + 8x + 9$

If there is a negative sign directly preceding the parentheses surrounding a polynomial, we may remove the parentheses and preceding negative sign by changing the sign of each term within the parentheses. (This procedure is actually just another application of the distributive property.)

Multiplication of Polynomials [5.3]

5. $(3x - 5)(x + 2)$
   $= 3x^2 + 6x - 5x - 10$
   $= 3x^2 + x - 10$

To multiply two polynomials, multiply each term in the first by each term in the second.

Special Products [5.3]

6. The following are examples of the three special products:

$(x + 3)^2 = x^2 + 6x + 9$
$(5 - x)^2 = 25 - 10x + x^2$
$(x + 7)(x - 7) = x^2 - 49$

$$(a + b)^2 = a^2 + 2ab + b^2$$
$$(a - b)^2 = a^2 - 2ab + b^2$$
$$(a + b)(a - b) = a^2 - b^2$$

Business Applications [5.2, 5.3, 5.4]

7. A company makes x items each week and sells them for p dollars each, according to the equation $p = 35 - 0.1x$. Then, the revenue is

$R = x(35 - 0.1x)$
$= 35x - 0.1x^2$

If the total cost to make all x items is $C = 8x + 500$, then the profit gained by selling the x items is

$P = 35x - 0.1x^2 - (8x + 500)$
$= -500 + 27x - 0.1x^2$

If a company manufactures and sells x items at p dollars per item, then the revenue R is given by the formula

$$R = xp$$

If the total cost to manufacture all x items is C, then the profit obtained from selling all x items is

$$P = R - C$$

Greatest Common Factor [5.4]

8. The greatest common factor of $10x^5 - 15x^4 + 30x^3$ is $5x^3$. Factoring it out of each term, we have

$5x^3(2x^2 - 3x + 6)$

The greatest common factor of a polynomial is the largest monomial (the monomial with the largest coefficient and highest exponent) that divides each term of the polynomial. The first step in factoring a polynomial is to factor the greatest common factor (if it is other than 1) out of each term.

Factoring Trinomials [5.5]

9. $x^2 + 5x + 6 = (x + 2)(x + 3)$
$x^2 - 5x + 6 = (x - 2)(x - 3)$
$x^2 + x - 6 = (x - 2)(x + 3)$
$x^2 - x - 6 = (x + 2)(x - 3)$

We factor a trinomial by writing it as the product of two binomials. (This refers to trinomials whose greatest common factor is 1.) Each factorable trinomial has a unique set of factors. Finding the factors is sometimes a matter of trial and error.

Factoring Trinomials [5.5]

10. Here are some binomials that have been factored this way.

$x^2 + 6x + 9 = (x + 3)^2$
$x^2 - 6x + 9 = (x - 3)^2$
$x^2 - 9 = (x + 3)(x - 3)$
$x^3 - 27 = (x - 3)(x^2 + 3x + 9)$
$x^3 + 27 = (x + 3)(x^2 - 3x + 9)$

$a^2 + 2ab + b^2 = (a + b)^2$ **Perfect square trinomials**

$a^2 - 2ab + b^2 = (a - b)^2$

$a^2 - b^2 = (a - b)(a + b)$ **Difference of two squares**

$a^3 - b^3 = (a - b)(a^2 + ab + b^2)$ **Difference of two cubes**

$a^3 + b^3 = (a + b)(a^2 - ab + b^2)$ **Sum of two cubes**

To Factor Polynomials in General [5.7]

11. Factor completely.

a. $3x^3 - 6x^2 = 3x^2(x - 2)$

b. $x^2 - 9 = (x + 3)(x - 3)$
$x^3 - 8 = (x - 2)(x^2 + 2x + 4)$
$x^3 + 27 = (x + 3)(x^2 - 3x + 9)$

c. $x^2 - 6x + 9 = (x - 3)^2$
$6x^2 - 7x - 5 = (2x + 1)(3x - 5)$

d. $x^2 + ax + bx + ab$
$= x(x + a) + b(x + a)$
$= (x + a)(x + b)$

Step 1: If the polynomial has a greatest common factor other than 1, then factor out the greatest common factor.

Step 2: If the polynomial has two terms (it is a binomial), then see if it is the difference of two squares, or the sum or difference of two cubes, and then factor accordingly. Remember, if it is the sum of two squares it will not factor.

Step 3: If the polynomial has three terms (a trinomial), then it is either a perfect square trinomial, which will factor into the square of a binomial, or it is not a perfect square trinomial, in which case you use one of the methods developed in Section 5.5.

Step 4: If the polynomial has more than three terms, then try to factor it by grouping.

Step 5: As a final check, see if any of the factors you have written can be factored further. If you have overlooked a common factor, you can catch it here.

To Solve an Equation by Factoring [5.8]

12. Solve $x^2 - 5x = -6$.

$$x^2 - 5x + 6 = 0$$
$$(x - 3)(x - 2) = 0$$
$$x - 3 = 0 \quad \text{or} \quad x - 2 = 0$$
$$x = 3 \quad \text{or} \quad x = 2$$

Step 1: Write the equation in standard form.

Step 2: Factor the left side.

Step 3: Use the zero-factor property to set each factor equal to zero.

Step 4: Solve the resulting linear equations.

! COMMON MISTAKES

When we subtract one polynomial from another, it is common to forget to add the opposite of each term in the second polynomial. For example

$$(6x - 5) - (3x + 4) = 6x - 5 - 3x + 4 \qquad \textbf{Mistake}$$

This mistake occurs if the negative sign outside the second set of parentheses is not distributed over all terms inside the parentheses. To avoid this mistake, remember: The opposite of a sum is the sum of the opposites, or,

$$-(3x + 4) = -3x + (-4)$$

Chapter 5 Review Test

The problems below form a comprehensive review of the material in this chapter. They can be used to study for exams. If you would like to take a practice test on this chapter, you can use the odd-numbered problems. Give yourself an hour and work as many of the odd-numbered problems as possible. When you are finished, or when an hour has passed, check your answers with the answers in the back of the book. You can use the even-numbered problems for a second practice test.

Simplify each of the following. [5.1]

1. $x^3 \cdot x^7$ **2.** $(5x^3)^2$

3. $(2x^3y)^2(-2x^4y^2)^3$

Write with positive exponents, and then simplify. [5.1]

4. 2^{-3} **5.** $\left(\frac{2}{3}\right)^{-2}$

6. $2^{-2} + 4^{-1}$

Write in scientific notation. [5.1]

7. 34,500,000 **8.** 0.00357

Write in expanded form. [5.1]

9. 4.45×10^4 **10.** 4.45×10^{-4}

Simplify each expression. All answers should contain positive exponents only. (Assume all variables are nonnegative.) [5.1]

11. $\dfrac{a^{-4}}{a^5}$ **12.** $\dfrac{(4x^2)(-3x^3)^2}{(12x^{-2})^2}$

13. $\dfrac{x^n x^{3n}}{x^{4n-2}}$

Simplify each expression as much as possible. Write all answers in scientific notation. [5.1]

14. $(2 \times 10^3)(4 \times 10^{-5})$ **15.** $\dfrac{(600{,}000)(0.000008)}{(4{,}000)(3{,}000{,}000)}$

Simplify by combining similar terms. [5.2]

16. $(6x^2 - 3x + 2) - (4x^2 + 2x - 5)$

17. $(x^3 - x) - (x^2 + x) + (x^3 - 3) - (x^2 + 1)$

18. Subtract $2x^2 - 3x + 1$ from $3x^2 - 5x - 2$.

19. Simplify $-3[2x - 4(3x + 1)]$.

20. Find the value of $2x^2 - 3x + 1$ when x is -2.

Multiply. [5.3]

21. $3x(4x^2 - 2x + 1)$ **22.** $2a^2b^3(a^2 + 2ab + b^2)$

23. $(6 - y)(3 - y)$ **24.** $(2x^2 - 1)(3x^2 + 4)$

25. $2t(t + 1)(t - 3)$ **26.** $(x + 3)(x^2 - 3x + 9)$

27. $(2x - 3)(4x^2 + 6x + 9)$ **28.** $(a^2 - 2)^2$

29. $(3x + 5)^2$ **30.** $(4x - 3y)^2$

31. $\left(x - \frac{1}{3}\right)\left(x + \frac{1}{3}\right)$ **32.** $(2a + b)(2a - b)$

33. $(x - 1)^3$ **34.** $(x^m + 2)(x^m - 2)$

Factor out the greatest common factor. [5.4]

35. $6x^4y - 9xy^4 + 18x^3y^3$ **36.** $4x^2(x + y)^2 - 8y^2(x + y)^2$

Factor by grouping. [5.4, 5.6]

37. $8x^2 + 10 - 4x^2y - 5y$ **38.** $x^3 + 8b^2 - x^3y^2 - 8y^2b^2$

Factor completely. [5.4, 5.5]

39. $x^2 - 5x + 6$ **40.** $2x^3 + 4x^2 - 30x$

41. $20a^2 - 41ab + 20b^2$ **42.** $6x^4 - 11x^3 - 10x^2$

43. $24x^2y - 6xy - 45y$

Factor completely. [5.6]

44. $x^4 - 16$ **45.** $3a^4 + 18a^2 + 27$

46. $a^3 - 8$ **47.** $5x^3 + 30x^2y + 45xy^2$

48. $3a^3b - 27ab^3$ **49.** $x^2 - 10x + 25 - y^2$

50. $36 - 25a^2$ **51.** $x^3 + 4x^2 - 9x - 36$

Solve each equation. [5.8]

52. $x^2 + 5x + 6 = 0$ **53.** $\frac{5}{6}y^2 = \frac{1}{4}y + \frac{1}{3}$

54. $9x^2 - 25 = 0$ **55.** $5x^2 = -10x$

56. $(x + 2)(x - 5) = 8$ **57.** $x^3 + 4x^2 - 9x - 36 = 0$

Solve each application. In each case be sure to show the equation used. [5.8]

58. **Consecutive Numbers** The product of two consecutive even integers is 80. Find the two integers.

59. **Consecutive Numbers** The sum of the squares of two consecutive integers is 41. Find the two integers.

60. **Geometry** The lengths of the three sides of a right triangle are given by three consecutive integers. Find the three sides.

61. **Geometry** The lengths of three sides of a right triangle are given by three consecutive even integers. Find the three sides.

GROUP PROJECT Discovering Pascal's Triangle

Number of People 3

Time Needed 20 minutes

Equipment Paper and pencils

Background The triangular array of numbers shown here is known as Pascal's triangle, after the French philosopher Blaise Pascal (1623–1662).

$$\begin{array}{ccccccccccc} & & & & & 1 & & & & & \\ & & & & 1 & & 1 & & & & \\ & & & 1 & & 2 & & 1 & & & \\ & & 1 & & 3 & & 3 & & 1 & & \\ & 1 & & 4 & & 6 & & 4 & & 1 & \\ 1 & & 5 & & 10 & & 10 & & 5 & & 1 \end{array}$$

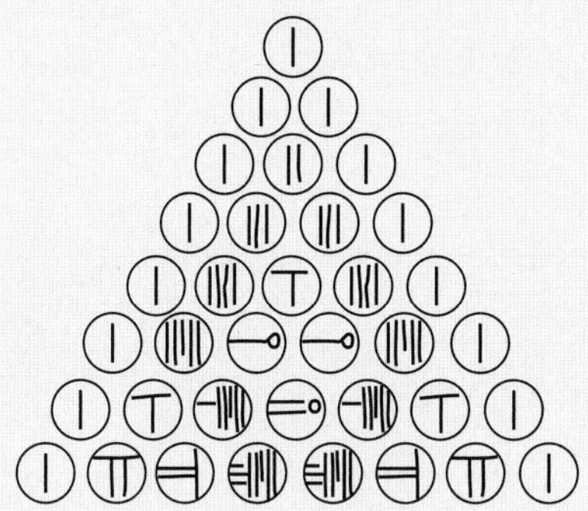

Pascal's triangle in Japanese (1781)

Procedure Look at Pascal's triangle and discover how the numbers in each row of the triangle are obtained from the numbers in the row above it.

1. Once you have discovered how to extend the triangle, write the next two rows.
2. Pascal's triangle can be linked to the Fibonacci sequence by rewriting Pascal's triangle so that the 1's on the left side of the triangle line up under one another and the other columns are equally spaced to the right of the first column. Rewrite Pascal's triangle as indicated and then look along the diagonals of the new array until you discover how the Fibonacci sequence can be obtained from it.
3. The diagram above shows Pascal's triangle as written in Japanese in 1781. Use your knowledge of Pascal's triangle to translate the numbers written in Japanese into our number system. Then write down the Japanese numbers from 1 to 20.

Binomial Expansions

The title on the following diagram is *Binomial Expansions* because each line gives the expansion of the binomial $x + y$ raised to a whole-number power.

Binomial Expansions

$$(x + y)^0 = 1$$

$$(x + y)^1 = x + y$$

$$(x + y)^2 = x^2 + 2xy + y^2$$

$$(x + y)^3 = x^3 + 3x^2y + 3xy^2 + y^3$$

$$(x + y)^4 =$$

$$(x + y)^5 =$$

The fourth row in the diagram was completed by expanding $(x + y)^3$ using the methods developed in this chapter. Next, complete the diagram by expanding the binomials $(x + y)^4$ and $(x + y)^5$ using the multiplication procedures you have learned in this chapter. Finally, study the completed diagram until you see patterns that will allow you to continue the diagram one more row without using multiplication. (One pattern that you will see is Pascal's triangle, which we mentioned in the preceding group project.) When you are finished, write an essay in which you describe what you have done and the results you have obtained.

Rational Expressions and Rational Functions

6

Michael A. Keller/Corbis

CHAPTER OUTLINE

If you have ever put yourself on a weight loss diet, you know that you lose more weight at the beginning of the diet than you do later. If we let $W(x)$ represent a person's weight after x weeks on the diet, then the rational function

$$W(x) = \frac{80(2x + 15)}{x + 6}$$

is a mathematical model of the person's weekly progress on a diet intended to take them from 200 pounds to about 160 pounds. Rational functions are good models for quantities that fall off rapidly to begin with and then level off over time. The table shows some values for this function, while Figure 1 shows the graph of this function.

Weekly Weight Loss

Weeks Since Starting Diet	Weight (nearest pound)
0	200
4	184
8	177
12	173
16	171
20	169
24	168

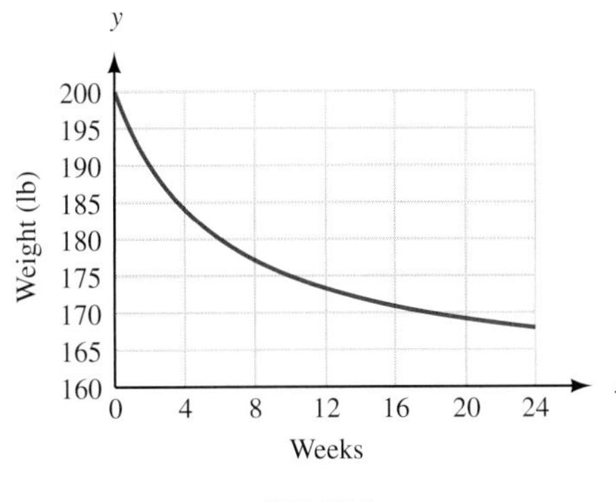

FIGURE 1

As you progress through this chapter, you will acquire an intuitive feel for these types of functions, and as a result, you will see why they are good models for situations such as dieting.

The study skills for this chapter are a continuation of skills from previous chapters.

1 Continue to Set and Keep a Schedule

Sometimes I find students do well on a test and then become overconfident. They begin to put in less time with their homework. Don't do it. Keep to the same schedule.

2 Increase Effectiveness

You want to become more and more effective with the time you spend on your homework. You want to increase the amount of learning you obtain in the time you have set aside. Increase those activities that you feel are the most beneficial, and decrease those that have not given you the results you want.

3 Continue to List Difficult Problems

This study skill was started in a previous chapter. You should continue to list and rework the problems that give you the most difficulty. It is this list that you will use to study for the next exam. Your goal is to go into the next exam knowing that you successfully can work any problem from your list of hard problems.

6.1 Basic Properties and Reducing to Lowest Terms

OBJECTIVES

A Reduce rational expressions to lowest terms.

B Find function values for rational functions.

C Work with ratios.

We will begin this section with the definition of a rational expression. We then will state the two basic properties associated with rational expressions, and go on to apply one of the properties to reduce rational expressions to lowest terms.

Recall from Chapter 1 that a *rational number* is any number that can be expressed as the ratio of two integers:

$$\text{Rational numbers} = \left\{ \frac{a}{b} \,\middle|\, a \text{ and } b \text{ are integers, } b \neq 0 \right\}$$

A rational expression is defined similarly as any expression that can be written as the ratio of two polynomials:

$$\text{Rational expressions} = \left\{ \frac{P}{Q} \,\middle|\, P \text{ and } Q \text{ are polynomials, } Q \neq 0 \right\}$$

Some examples of rational expressions are

$$\frac{2x-3}{x+5} \qquad \frac{x^2-5x-6}{x^2-1} \qquad \frac{a-b}{b-a}$$

Basic Properties

For rational expressions, multiplying the numerator and denominator by the same nonzero expression may change the form of the rational expression, but it will always produce an expression equivalent to the original one. The same is true when dividing the numerator and denominator by the same nonzero quantity.

Note

These two statements are equivalent since division is defined as multiplication by the reciprocal. We choose to state them separately for clarity.

Properties of Rational Expressions

If P, Q, and K are polynomials with $Q \neq 0$ and $K \neq 0$, then

$$\frac{P}{Q} = \frac{PK}{QK} \quad \text{and} \quad \frac{P}{Q} = \frac{P/K}{Q/K}$$

Reducing to Lowest Terms

The fraction $\frac{6}{8}$ can be written in lowest terms as $\frac{3}{4}$. The process is shown here:

$$\frac{6}{8} = \frac{3 \cdot \overset{1}{\cancel{2}}}{4 \cdot \underset{1}{\cancel{2}}} = \frac{3}{4}$$

Reducing $\frac{6}{8}$ to $\frac{3}{4}$ involves dividing the numerator and denominator by 2, the factor they have in common. Before dividing out the common factor 2, we must notice that the common factor *is* 2! (This may not be obvious since we are very familiar with the numbers 6 and 8 and therefore do not have to put much thought into finding what number divides both of them.)

We reduce rational expressions to lowest terms by first factoring the numerator and denominator and then dividing both numerator and denominator by any factors they have in common.

EXAMPLE 1 Reduce $\frac{x^2 - 9}{x - 3}$ to lowest terms.

SOLUTION Factoring, we have

$$\frac{x^2 - 9}{x - 3} = \frac{(x + 3)(x - 3)}{x - 3}$$

The numerator and denominator have the factor $x - 3$ in common. Dividing the numerator and denominator by $x - 3$, we have

$$\frac{(x + 3)\overset{1}{\cancel{(x - 3)}}}{\underset{1}{\cancel{x - 3}}} = \frac{x + 3}{1} = x + 3$$

Note

The lines drawn through the $(x - 3)$ in the numerator and denominator indicate that we have divided through by $(x - 3)$. As the problems become more involved, these lines will help keep track of which factors have been divided out and which have not.

Note For the problem in Example 1, there is an implied restriction on the variable x: it cannot be 3. If x were 3, the expression $(x^2 - 9)/(x - 3)$ would become $0/0$, an expression that we cannot associate with a real number. For all problems involving rational expressions, we restrict the variable to only those values that result in a nonzero denominator. When we state the relationship

$$\frac{x^2 - 9}{x - 3} = x + 3$$

we are assuming that it is true for all values of x except $x = 3$.

Here are some other examples of reducing rational expressions to lowest terms.

EXAMPLES Reduce to lowest terms.

Note

Beginning with Example 2 we are no longer showing the 1's that result when we divide out our common factors. If you need to show the 1's to remind yourself that you are *dividing*, then be sure to do so.

2. $\frac{y^2 - 5y - 6}{y^2 - 1} = \frac{(y - 6)\cancel{(y + 1)}}{(y - 1)\cancel{(y + 1)}}$ **Factor numerator and denominator**

$= \frac{y - 6}{y - 1}$ **Divide out common factor $y + 1$**

3. $\frac{2a^3 - 16}{4a^2 - 12a + 8} = \frac{2(a^3 - 8)}{4(a^2 - 3a + 2)}$

$= \frac{2\cancel{(a - 2)}(a^2 + 2a + 4)}{4\cancel{(a - 2)}(a - 1)}$ **Factor numerator and denominator**

$= \frac{a^2 + 2a + 4}{2(a - 1)}$ **Divide out common factor $2(a - 2)$**

4. $\frac{x^2 - 3x + ax - 3a}{x^2 - ax - 3x + 3a} = \frac{x(x - 3) + a(x - 3)}{x(x - a) - 3(x - a)}$

$= \frac{\cancel{(x - 3)}(x + a)}{(x - a)\cancel{(x - 3)}}$ **Factor numerator and denominator**

$= \frac{x + a}{x - a}$ **Divide out common factor $x - 3$**

The answer to Example 4 is $(x + a)/(x - a)$. The problem cannot be reduced further. It is a fairly common mistake to attempt to divide out an x or an a in this last expression. Remember, we can divide out only the factors common to the numerator and denominator of a rational expression. For the last expression in

Example 4, neither the numerator nor the denominator can be factored further; x is not a factor of the numerator or the denominator, and neither is a. The expression is in lowest terms.

The next example involves what we call a trick. The trick is to reverse the order of the terms in a difference by factoring -1 from each term. The next examples illustrate how this is done.

EXAMPLE 5 Reduce to lowest terms: $\frac{a-b}{b-a}$.

SOLUTION The relationship between $a - b$ and $b - a$ is that they are opposites. We can show this fact by factoring -1 from each term in the numerator:

$$\frac{a-b}{b-a} = \frac{-1(-a+b)}{b-a} \quad \text{Factor } -1 \text{ from each term in the numerator}$$

$$= \frac{-1(b-a)}{b-a} \quad \text{Reverse the order of the terms in the numerator}$$

$$= -1 \quad \text{Divide out common factor } b-a$$

EXAMPLE 6 Reduce to lowest terms: $\frac{x^2-25}{5-x}$.

SOLUTION Begin by factoring the numerator:

$$\frac{x^2-25}{5-x} = \frac{(x-5)(x+5)}{5-x}$$

The factors $x - 5$ and $5 - x$ are similar but are not exactly the same. We can reverse the order of either by factoring -1 from it; that is: $5 - x = -1(-5 + x) = -1(x - 5)$.

$$\frac{(x-5)(x+5)}{5-x} = \frac{\cancel{(x-5)}(x+5)}{-1\cancel{(x-5)}}$$

$$= \frac{x+5}{-1}$$

$$= -(x+5)$$

Applications

Ratios You may recall from previous math classes that the ratio of a to b is the same as the fraction $\frac{a}{b}$. Here are two ratios that are used frequently in mathematics:

1. The number π is defined as the ratio of the circumference of a circle to the diameter of a circle; that is

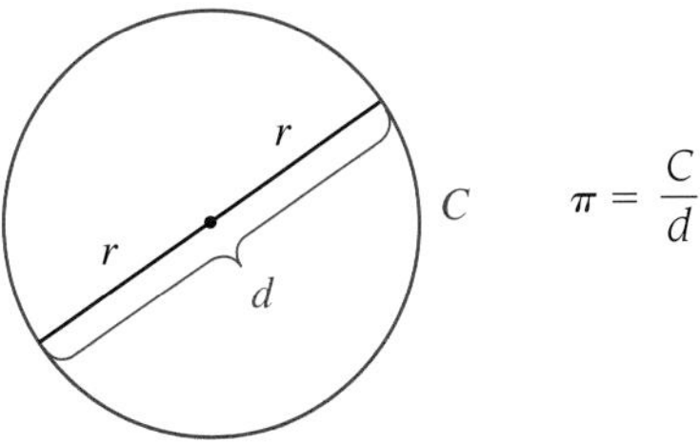

$$\pi = \frac{C}{d}$$

Multiplying both sides of this formula by d, we have the more common form $C = \pi d$.

Note
Since the diameter of a circle is twice the radius, the formula for circumference is sometimes written $C = 2\pi r$. Neither formula should be confused with the formula for the area of a circle, which is $A = \pi r^2$.

2. The *average speed* of a moving object is defined to be the ratio of distance to time. If you drive your car for 5 hours and travel a distance of 200 miles, then your average rate of speed is

$$\text{Average speed} = \frac{200 \text{ miles}}{5 \text{ hours}} = 40 \text{ miles per hour}$$

The formula we use for the relationship between average speed r, distance d, and time t is

$$r = \frac{d}{t}$$

The formula is sometimes called the *rate equation.* Multiplying both sides by t, we have an equivalent form of the rate equation, $d = rt$.

In the next example we graph the relationship between the average speed of a person riding a Ferris wheel and the amount of time it takes the wheel to complete one revolution.

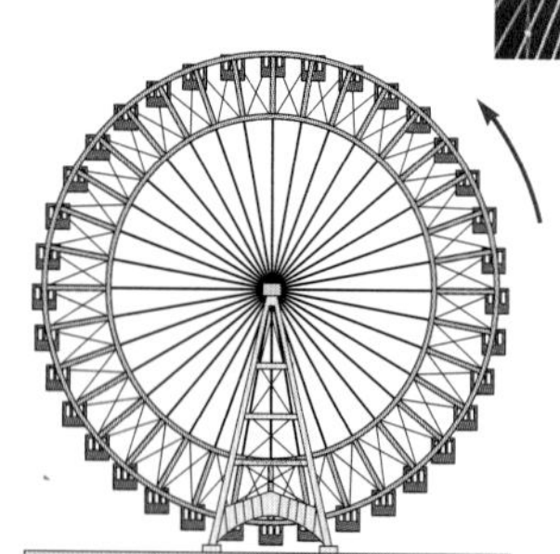

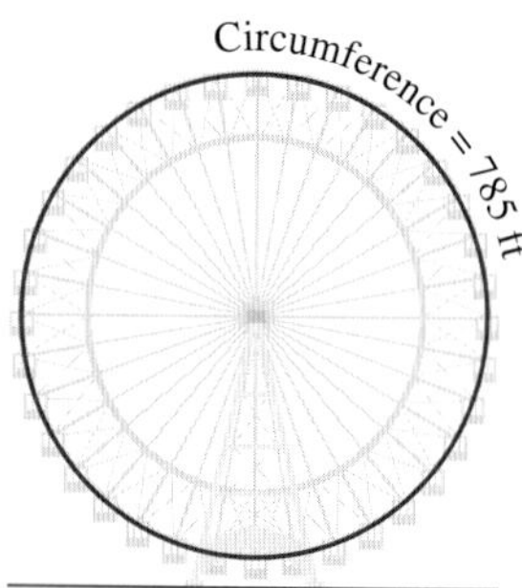

EXAMPLE 7

A Ferris wheel has a circumference of 785 feet. If one complete revolution of the wheel takes 10 to 30 minutes, then the relationship between the average speed of a rider on the wheel and the amount of time it takes the wheel to complete one revolution is given by the function

$$r(t) = \frac{785}{t} \qquad 10 \le t \le 30$$

where $r(t)$ is the average speed (in feet per minute) and t is the amount of time (in minutes) it takes the wheel to complete one revolution. Graph the function.

SOLUTION Since the variables r and t represent speed and time, both must be positive quantities. Therefore, the graph of this function will lie in the first quadrant only. The following table displays the values of t and $r(t)$ found from the function, along with the graph of the function (Figure 1). (Some of the numbers in the table have been rounded to the nearest tenth.)

Time to Complete One Revolution t	Speed (ft/min) $r(t)$
10	78.5
15	52.3
20	39.3
25	31.4
30	26.2

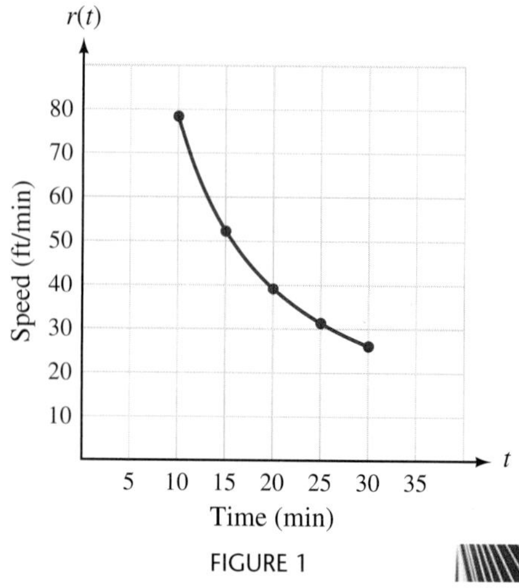

FIGURE 1

USING TECHNOLOGY

More About Example 7

If we use a graphing calculator to graph the equation in Example 8, it is not necessary to construct the table first. In fact, if we graph

$Y_1 = 785/X$ Window: X from 0 to 40, Y from 0 to 90

we can use the Trace and Zoom features together to produce the numbers in the table next to Figure 1. Graph the preceding equation, and zoom in on the point with x-coordinate 20 until you are convinced that the table values for x and y are correct.

Rational Functions

The function shown in Example 7 is called a *rational function* because the right side, $785/t$, is a rational expression (the numerator, 785, is a polynomial of degree 0). We can extend our knowledge of rational expressions to functions with the following definition:

DEFINITION A **rational function** is any function that can be written in the form

$$f(x) = \frac{P(x)}{Q(x)}$$

where $P(x)$ and $Q(x)$ are polynomials and $Q(x) \neq 0$.

EXAMPLE 8 For the rational function $f(x) = \dfrac{x-4}{x-2}$, find $f(0)$, $f(-4)$, $f(4)$, $f(-2)$, and $f(2)$.

SOLUTION To find these function values, we substitute the given value of x into the rational expression, and then simplify if possible.

$$f(0) = \frac{0-4}{0-2} = \frac{-4}{-2} = 2 \qquad f(-2) = \frac{-2-4}{-2-2} = \frac{-6}{-4} = \frac{3}{2}$$

$$f(-4) = \frac{-4-4}{-4-2} = \frac{-8}{-6} = \frac{4}{3} \qquad f(2) = \frac{2-4}{2-2} = \frac{-2}{0} \quad \text{Undefined}$$

$$f(4) = \frac{4-4}{4-2} = \frac{0}{2} = 0$$

Because the rational function in Example 8 is not defined when x is 2, the domain of that function does not include 2. We have more to say about the domain of a rational function next.

The Domain of a Rational Function

In Example 7 the domain of the rational function is specified as $10 \le t \le 30$, and the function is defined for all values of t in that domain. If the domain of a rational function is not specified, it is assumed to be all real numbers for which the function is defined; that is, the domain of the rational function

$$f(x) = \frac{P(x)}{Q(x)}$$

is all x for which $Q(x)$ is nonzero. For example:

The domain for $r(t) = \frac{785}{t}$, $10 \le t \le 30$, is $\{t \mid 10 \le t \le 30\}$.

The domain for $f(x) = \frac{x-4}{x-2}$ is $\{x \mid x \ne 2\}$.

The domain for $g(x) = \frac{x^2+5}{x+1}$ is $\{x \mid x \ne -1\}$.

The domain for $h(x) = \frac{x}{x^2-9}$ is $\{x \mid x \ne -3, x \ne 3\}$.

Notice that, for these functions, $f(2)$, $g(-1)$, $h(-3)$, and $h(3)$ are all undefined, and that is why the domains are written as shown.

EXAMPLE 9

Graph the equation $y = \frac{x^2-9}{x-3}$. How is this graph different from the graph of $y = x + 3$?

SOLUTION We know from the discussion in Example 1 that

$$y = \frac{x^2-9}{x-3} = \frac{(x+3)(x-3)}{x-3} = x + 3$$

This relationship is true for all x except $x = 3$ because the rational expression with $x - 3$ in the denominator is undefined when x is 3. However, for all other values of x, the expressions

$$\frac{x^2-9}{x-3} \quad \text{and} \quad x + 3$$

are equal. Therefore, the graphs of

$$y = \frac{x^2-9}{x-3} \quad \text{and} \quad y = x + 3$$

will be the same except when x is 3. In the first equation, there is no value of y to correspond to $x = 3$. In the second equation, $y = x + 3$, so y is 6 when x is 3.

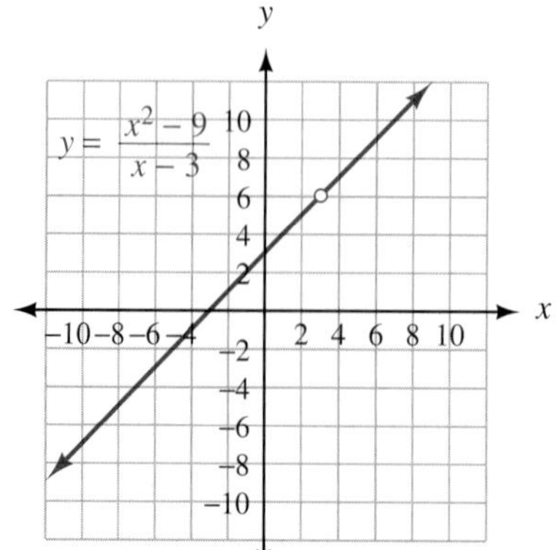

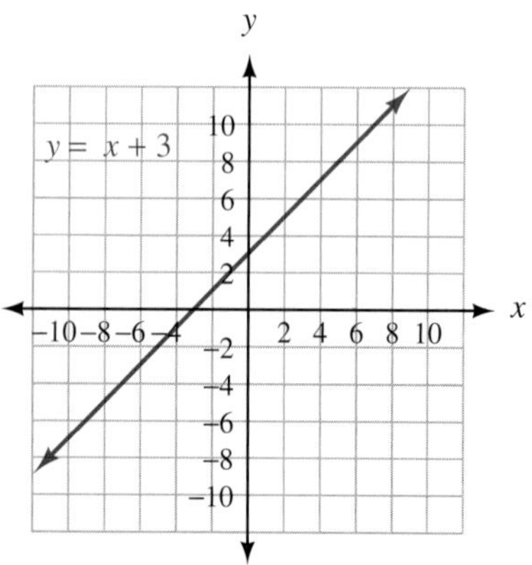

FIGURE 2 **Graphs of $y = \frac{x^2-9}{x-3}$ (left) and $y = x + 3$ (right)**

Now you can see the difference in the graphs of the two equations. To show that there is no y value for $x = 3$ in the graph on the left in Figure 2, we draw an open circle at that point on the line.

Notice that the two graphs shown in Figure 2 are both graphs of functions. Suppose we use function notation to designate them as follows:

$$f(x) = \frac{x^2 - 9}{x - 3} \quad \text{and} \quad g(x) = x + 3$$

The two functions, f and g, are equivalent except when $x = 3$ because $f(3)$ is undefined, while $g(3) = 6$. The domain of the function f is all real numbers except $x = 3$, while the domain for g is all real numbers, with no restrictions.

LINKING OBJECTIVES AND EXAMPLES

Next to each objective we have listed the examples that are best described by that objective.

A	1–6
B	8
C	7

GETTING READY FOR CLASS

After reading through the preceding section, respond in your own words and in complete sentences.

1. What is a rational expression?
2. Explain how to determine if a rational expression is in "lowest terms."
3. When is a rational expression undefined?
4. Explain the process we use to reduce a rational expression or a fraction to lowest terms.

Problem Set 6.1

Online support materials can be found at www.thomsonedu.com/login

1. Simplify each expression. State any restrictions on the variables.

a. $\dfrac{6 + 1}{36 - 1}$

b. $\dfrac{x + 3}{x^2 - 9}$

c. $\dfrac{x^2 - 3x}{x^2 - 9}$

d. $\dfrac{x^3 - 27}{x^2 - 9}$

e. $\dfrac{x^3 - 27}{x^3 - 3x^2}$

2. Simplify each expression. State any restrictions on the variables.

a. $\dfrac{64 - 80 + 25}{64 - 25}$

b. $\dfrac{x^2 - 10x + 25}{x^2 - 25}$

c. $\dfrac{x^2 - 26x + 25}{x^2 - 25x}$

d. $\dfrac{x^2 + 5x + ax + 5a}{x^2 - 25}$

e. $\dfrac{x^3 + 125}{x^3 - 25x}$

3. If $h(t) = \dfrac{t - 3}{t + 1}$, find $h(0)$, $h(-3)$, $h(3)$, $h(-1)$, and $h(1)$, if possible.

4. If $h(t) = \dfrac{t - 2}{t + 1}$, find $h(0)$, $h(-2)$, $h(2)$, $h(-1)$, and $h(1)$, if possible.

State the domain for each rational function.

5. $f(x) = \dfrac{x - 3}{x - 1}$

6. $g(x) = \dfrac{x^2 - 4}{x - 2}$

= Videos available by instructor request

▶ = Online student support materials available at www.thomsonedu.com/login

7. $h(t) = \dfrac{t-4}{t^2-16}$

8. $h(t) = \dfrac{t-5}{t^2-25}$

▶ 9. $f(x) = \dfrac{3(x^2-25)}{3x-15}$

▶ 10. $g(x) = -\dfrac{4x-36}{4x+28}$

▶ 11. $f(x) = -\dfrac{x^2+25}{10}$

▶ 12. $g(x) = -\dfrac{x-9}{45}$

▶ 13. $f(x) = \dfrac{2(x^2+49)}{7x}$

▶ 14. $g(x) = -\dfrac{x^3-27}{4x}$

▶ 15. $h(x) = \dfrac{x^3-8}{x^2-x-20}$

▶ 16. $f(x) = \dfrac{x+\pi-5}{x^2+x-12}$

Reduce each rational expression to lowest terms.

17. $\dfrac{x^2-16}{6x+24}$

▶ 18. $\dfrac{5x+25}{x^2-25}$

19. $\dfrac{12x-9y}{3x^2+3xy}$

20. $\dfrac{x^3-xy^2}{4x+4y}$

21. $\dfrac{a^4-81}{a-3}$

22. $\dfrac{a+4}{a^2-16}$

23. $\dfrac{a^2-4a-12}{a^2+8a+12}$

24. $\dfrac{a^2-7a+12}{a^2-9a+20}$

25. $\dfrac{20y^2-45}{10y^2-5y-15}$

26. $\dfrac{54y^2-6}{18y^2-60y+18}$

▶ 27. $\dfrac{a^3+b^3}{a^2-b^2}$

28. $\dfrac{a^2-b^2}{a^3-b^3}$

Reduce to lowest terms.

29. $\dfrac{8x^4-8x}{4x^4+4x^3+4x^2}$

30. $\dfrac{6x^5-48x^2}{12x^3+24x^2+48x}$

31. $\dfrac{6x^2+7xy-3y^2}{6x^2+xy-y^2}$

32. $\dfrac{4x^2-y^2}{4x^2-8xy-5y^2}$

33. $\dfrac{ax+2x+3a+6}{ay+2y-4a-8}$

34. $\dfrac{ax-x-5a+5}{ax+x-5a-5}$

▶ 35. $\dfrac{x^2+bx-3x-3b}{x^2-2bx-3x+6b}$

36. $\dfrac{x^2-3ax-2x+6a}{x^2-3ax+2x-6a}$

37. $\dfrac{x^3+3x^2-4x-12}{x^2+x-6}$

38. $\dfrac{x^3+5x^2-4x-20}{x^2+7x+10}$

39. $\dfrac{4x^4-25}{6x^3-4x^2+15x-10}$

40. $\dfrac{16x^4-49}{8x^3-12x^2+14x-21}$

▶ 41. $\dfrac{x^3-8}{x^2-4}$

▶ 42. $\dfrac{y^2-9}{y^3+27}$

▶ 43. $\dfrac{64+t^3}{16-4t+t^2}$

▶ 44. $\dfrac{25+5a+a^2}{125-a^3}$

▶ 45. $\dfrac{8x^3-27}{4x^2-9}$

▶ 46. $\dfrac{25y^2-4}{125y^3+8}$

Refer to Examples 5 and 6 in this section, and reduce the following to lowest terms.

▶ 47. $\dfrac{x-4}{4-x}$

48. $\dfrac{6-x}{x-6}$

49. $\dfrac{y^2-36}{6-y}$

50. $\dfrac{1-y}{y^2-1}$

51. $\dfrac{1-9a^2}{9a^2-6a+1}$

52. $\dfrac{1-a^2}{a^2-2a+1}$

Simplify each expression.

53. $\dfrac{(3x-5)-(3a-5)}{x-a}$

54. $\dfrac{(2x+3)-(2a+3)}{x-a}$

55. $\dfrac{(x^2-4)-(a^2-4)}{x-a}$

56. $\dfrac{(x^2-1)-(a^2-1)}{x-a}$

Let $f(x) = \dfrac{x^2-4}{x-2}$ and $g(x) = x+2$, and evaluate the following expressions, if possible.

57. $f(0)$ and $g(0)$

58. $f(1)$ and $g(1)$

59. $f(2)$ and $g(2)$

60. $f(3)$ and $g(3)$

Let $f(x) = \dfrac{x^2-1}{x-1}$ and $g(x) = x+1$, and evaluate the following expressions, if possible.

61. $f(0)$ and $g(0)$

62. $f(1)$ and $g(1)$

63. $f(2)$ and $g(2)$

64. $f(-1)$ and $g(-1)$

65. Graph the equation $y = \dfrac{x^2-4}{x-2}$. Then explain how this graph is different from the graph of $y = x+2$.

66. Graph the equation $y = \frac{x^2 - 1}{x - 1}$. Then explain how this graph is different from the graph of $y = x + 1$.

The next two problems are intended to give you practice reading, and paying attention to, the instructions that accompany the problems you are working. Working these problems is an excellent way to get ready for a test or a quiz.

67. Work each problem according to the instructions given.
  a. Add: $(x^2 - 7x) + (7x - 49)$
  b. Subtract: $(x^2 - 7x) - (7x - 49)$
  c. Multiply: $(x^2 - 7x)(7x - 49)$
  d. Reduce: $\frac{x^2 - 7x}{7x - 49}$

68. Work each problem according to the instructions given.
  a. Add: $(16x^2 + 4x) + (12x + 3)$
  b. Subtract: $(16x^2 + 4x) - (12x + 3)$
  c. Multiply: $(16x^2 + 4x)(12x + 3)$
  d. Reduce: $\frac{16x^2 + 4x}{12x + 3}$

Applying the Concepts

69. **Diet** The following rational function is the one we mentioned in the introduction to this chapter. The quantity $W(x)$ is the weight (in pounds) of the person after x weeks of dieting. Use the function to fill in the table, rounding to the nearest tenth. Then compare your results with the graph in the chapter introduction.

$$W(x) = \frac{80(2x + 15)}{x + 6}$$

Weeks x	Weight (pounds) $W(x)$
0	
1	
4	
12	
24	

70. **Drag Racing** The following rational function gives the speed $V(x)$, in miles per hour, of a dragster at each second x during a quarter-mile race. Use the function to fill in the table, rounding to the nearest tenth.

$$V(x) = \frac{340x}{x + 3}$$

Time (sec) x	Speed (mi/hr) $V(x)$
0	
1	
2	
3	
4	
5	
6	

Average Speed For Problems 71 and 72, use 3.14 as an approximation for π. Round answers to the nearest tenth.

71. A person riding a Ferris wheel with a diameter of 50 feet travels once around the wheel in 20 seconds. What is the average speed of the rider in feet per second?

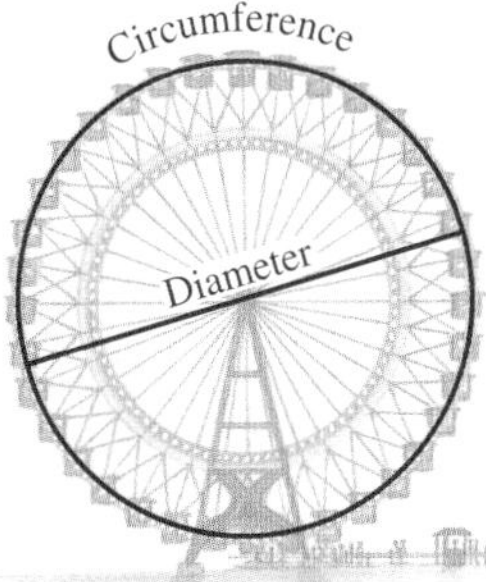

72. A person riding a Ferris wheel with a diameter of 150 feet travels once around the wheel in 3 minutes. What is the average speed of the rider in feet per minute?

73. **Average Speed** A Ferris wheel has a circumference of 204 feet (to the nearest foot). If a ride on the wheel takes from 20 to 50 seconds, then the relationship between the average speed of a rider and the amount of time it takes to complete one revolution is given by the function

$$r(t) = \frac{204}{t} \qquad 20 \le t \le 50$$

where $r(t)$ is in feet per second and t is in seconds.
  a. State the domain for this function.
  b. Graph this function.

74. **Average Speed** A Ferris wheel has a circumference of 320 feet (to the nearest foot). If a ride on the wheel takes from 3 to 5 minutes, then the relationship between the average speed of a rider and the

amount of time it takes to complete one revolution is given by the function

$$r(t) = \frac{320}{t} \qquad 3 \le t \le 5$$

where $r(t)$ is in feet per minute and t is in minutes.

a. State the domain for this function.

b. Graph this function.

75. **Intensity of Light** The relationship between the intensity of light that falls on a surface from a 100-watt light bulb and the distance from that surface is given by the rational function

$$I(d) = \frac{120}{d^2} \qquad \text{for } 1 \le d \le 6$$

where $I(d)$ is the intensity of light (in lumens per square foot) and d is the distance (in feet) from the light bulb to the surface.

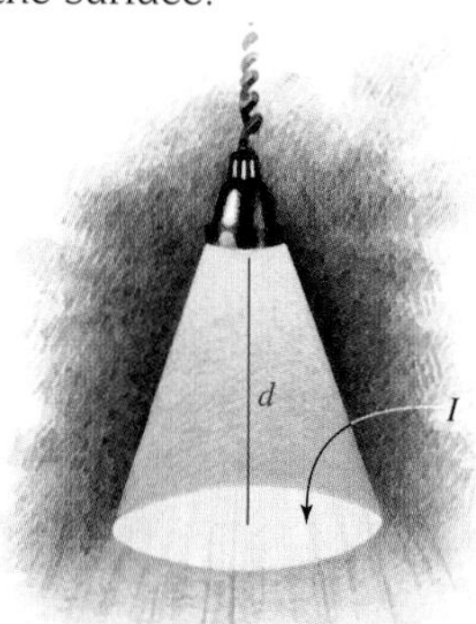

a. State the domain for this function.

b. Graph this function.

76. **Average Speed** If it takes Maria t minutes to run a mile, then her average speed $s(t)$ is given by the rational function

$$s(t) = \frac{60}{t} \qquad \text{for } 6 \le t \le 12$$

where $s(t)$ is in miles per hour and t is in minutes.

a. State the domain for this function.

b. Graph this function.

Maintaining Your Skills

Subtract as indicated.

77. Subtract $x^2 + 2x + 1$ from $4x^2 - 5x + 5$.

78. Subtract $3x^2 - 5x + 2$ from $7x^2 + 6x + 4$.

79. Subtract $10x - 20$ from $10x - 11$.

80. Subtract $-6x - 18$ from $-6x + 5$.

81. Subtract $4x^3 - 8x^2$ from $4x^3$.

82. Subtract $2x^2 + 6x$ from $2x^2$.

Getting Ready for the Next Section

Divide.

83. $\frac{10x^5}{5x^2}$

84. $\frac{-15x^4}{5x^2}$

85. $\frac{4x^4y^3}{-2x^2y}$

86. $\frac{10a^4b^2}{4a^2b^2}$

87. $4{,}628 \div 25$

88. $7{,}546 \div 35$

Multiply.

89. $2x^2(2x - 4)$

90. $3x^2(x - 2)$

91. $(2x - 4)(2x^2 + 4x + 5)$

92. $(x - 2)(3x^2 + 6x + 15)$

Subtract.

93. $(2x^2 - 7x + 9) - (2x^2 - 4x)$

94. $(x^2 - 6xy - 7y^2) - (x^2 + xy)$

Factor.

95. $x^2 - a^2$

96. $x^2 - 1$

97. $x^2 - 6xy - 7y^2$

98. $2x^2 - 5xy + 3y^2$

Extending the Concepts

99. The graphs of two rational functions are given in Figures 3 and 4. Use the graphs to find the following.

a. $f(2)$

b. $f(-1)$

c. $f(0)$

d. $g(3)$

e. $g(6)$

f. $g(-1)$

g. $f(g(6))$

h. $g(f(-2))$

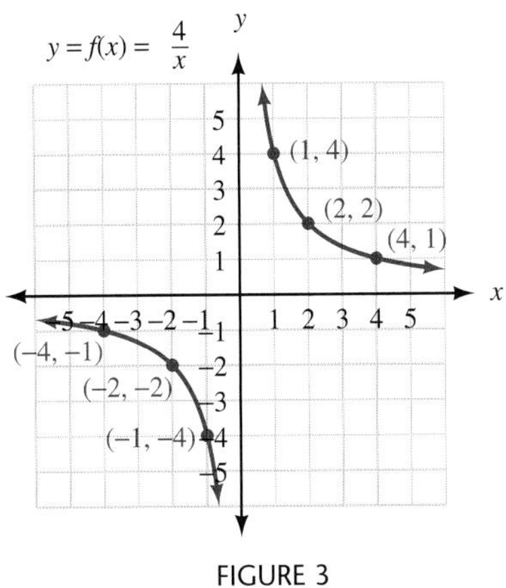

FIGURE 3

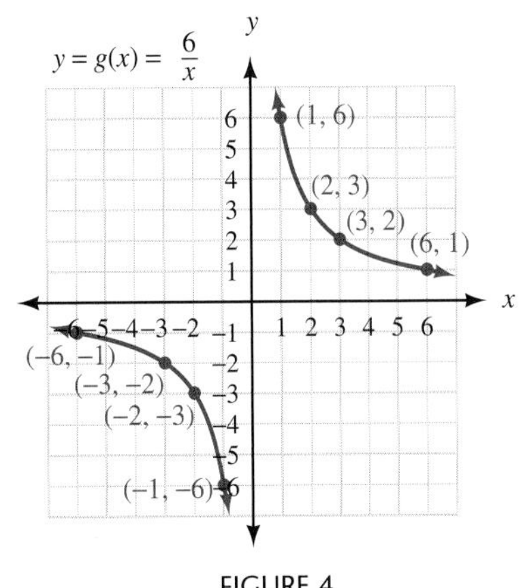

FIGURE 4

6.2 Division of Polynomials and Difference Quotients

OBJECTIVES

A Divide a polynomial by a monomial.

B Divide a polynomial by a polynomial.

First Bank of San Luis Obispo charges \$2.00 per month and \$0.15 per check for a regular checking account. So, if you write x checks in one month, the total monthly cost of the checking account will be $C(x) = 2.00 + 0.15x$. From this formula, we see that the more checks we write in a month, the more we pay for the account. But it is also true that the more checks we write in a month, the lower the cost per check. To find the cost per check, we use the average cost function. To find the average cost function, we divide the total cost by the number of checks written.

$$\text{Average Cost} = \overline{C}(x) = \frac{C(x)}{x} = \frac{2.00 + 0.15x}{x}$$

This last expression gives us the average cost per check for each of the x checks written. To work with this last expression, we need to know something about division with polynomials, and that is what we will cover in this section.

We begin this section by considering division of a polynomial by a monomial. This is the simplest kind of polynomial division. The rest of the section is devoted to division of a polynomial by a polynomial. This kind of division is similar to long division with whole numbers.

Dividing a Polynomial by a Monomial

To divide a polynomial by a monomial, we use the definition of division and apply the distributive property. The following example illustrates the procedure.

EXAMPLE 1 Divide $\dfrac{10x^5 - 15x^4 + 20x^3}{5x^2}$.

SOLUTION

$$= (10x^5 - 15x^4 + 20x^3) \cdot \frac{1}{5x^2}$$

Dividing by $5x^2$ is the same as multiplying by $\frac{1}{5x^2}$

$$= 10x^5 \cdot \frac{1}{5x^2} - 15x^4 \cdot \frac{1}{5x^2} + 20x^3 \cdot \frac{1}{5x^2}$$ **Distributive property**

$$= \frac{10x^5}{5x^2} - \frac{15x^4}{5x^2} + \frac{20x^3}{5x^2}$$ **Multiplying by $\frac{1}{5x^2}$ is the same as dividing by $5x^2$**

$$= 2x^3 - 3x^2 + 4x$$ **Divide coefficients, subtract exponents**

Notice that division of a polynomial by a monomial is accomplished by dividing each term of the polynomial by the monomial. The first two steps are usually not shown in a problem like this. They are part of Example 1 to justify distributing $5x^2$ under all three terms of the polynomial $10x^5 - 15x^4 + 20x^3$.

Here are some more examples of this kind of division.

EXAMPLES

Divide. Write all results with positive exponents.

2. $$\frac{8x^3y^5 - 16x^2y^2 + 4x^4y^3}{-2x^2y} = \frac{8x^3y^5}{-2x^2y} + \frac{-16x^2y^2}{-2x^2y} + \frac{4x^4y^3}{-2x^2y}$$
$$= -4xy^4 + 8y - 2x^2y^2$$

3. $$\frac{10a^4b^2 + 8ab^3 - 12a^3b + 6ab}{4a^2b^2} = \frac{10a^4b^2}{4a^2b^2} + \frac{8ab^3}{4a^2b^2} - \frac{12a^3b}{4a^2b^2} + \frac{6ab}{4a^2b^2}$$
$$= \frac{5a^2}{2} + \frac{2b}{a} - \frac{3a}{b} + \frac{3}{2ab}$$

Notice in Example 3 that the result is not a polynomial because of the last three terms. If we were to write each term as a product, some of the variables would have negative exponents. For example, the second term would be

$$\frac{2b}{a} = 2a^{-1}b$$

The divisor in each of the previous examples was a monomial. We now want to turn our attention to division of polynomials in which the divisor has two or more terms.

Dividing a Polynomial by a Polynomial

EXAMPLE 4

Divide $\frac{x^2 - 6xy - 7y^2}{x + y}$.

SOLUTION In this case, we can factor the numerator and perform our division by simply dividing out common factors, just like we did in the previous section:

$$\frac{x^2 - 6xy - 7y^2}{x + y} = \frac{\cancel{(x + y)}(x - 7y)}{\cancel{x + y}}$$
$$= x - 7y$$

Difference Quotients The diagram in Figure 1 is an important diagram from calculus. Although it may look complicated, the point of it is simple: The slope of the line passing through the points P and Q is given by the formula

$$\text{Slope of line through } PQ = m = \frac{f(x) - f(a)}{x - a}$$

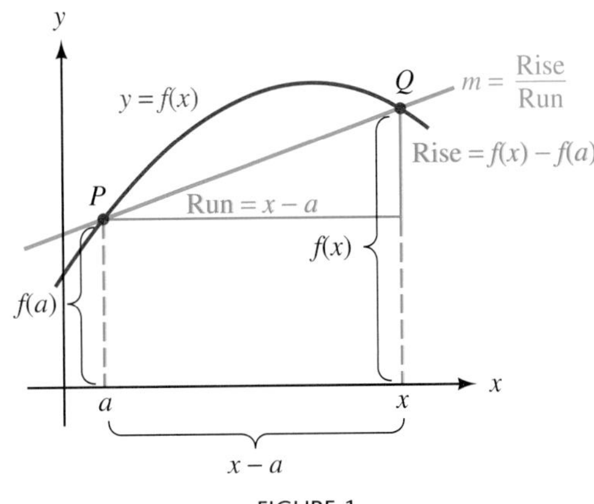

FIGURE 1

The expression $\frac{f(x) - f(a)}{x - a}$ is called a difference quotient. When $f(x)$ is a polynomial, the difference quotient will be a rational expression.

EXAMPLE 5

If $f(x) = 3x - 5$, find $\frac{f(x) - f(a)}{x - a}$.

SOLUTION

$$\begin{aligned} \frac{f(x) - f(a)}{x - a} &= \frac{(3x - 5) - (3a - 5)}{x - a} \\ &= \frac{3x - 5 - 3a + 5}{x - a} \\ &= \frac{3x - 3a}{x - a} \\ &= \frac{3(x - a)}{x - a} \\ &= 3 \end{aligned}$$

EXAMPLE 6

If $f(x) = x^2 - 4$, find $\frac{f(x) - f(a)}{x - a}$ and simplify.

SOLUTION Because $f(x) = x^2 - 4$ and $f(a) = a^2 - 4$, we have

$$\begin{aligned} \frac{f(x) - f(a)}{x - a} &= \frac{(x^2 - 4) - (a^2 - 4)}{x - a} \\ &= \frac{x^2 - 4 - a^2 + 4}{x - a} \\ &= \frac{x^2 - a^2}{x - a} \\ &= \frac{(x + a)\cancel{(x - a)}}{\cancel{x - a}} \qquad \text{Factor and divide out common factor} \\ &= x + a \end{aligned}$$

Long Division

For the type of division shown in Examples 4 through 6, the denominator must be a factor of the numerator. When the denominator is not a factor of the numerator, or in the case where we can't factor the numerator, the method used in Examples 4 through 6 won't work. We need to develop a new method for these cases. Since this new method is very similar to long division with whole numbers, we will review it here.

EXAMPLE 7

Divide $25\overline{)4{,}628}$.

SOLUTION

$$\begin{array}{rl} 1 & \leftarrow \textbf{Estimate 25 into 46} \\ 25\overline{)4{,}628} & \\ \underline{2\,5\;\;} & \leftarrow \textbf{Multiply } 1 \times 25 = 25 \\ 2\,1\;\; & \leftarrow \textbf{Subtract } 46 - 25 = 21 \end{array}$$

$$\begin{array}{rl} 1\;\; & \\ 25\overline{)4{,}628} & \\ \underline{2\,5\downarrow} & \\ 2\,12 & \leftarrow \textbf{Bring down the 2} \end{array}$$

Note You may realize when looking over this example that you don't have a very good idea why you proceed as you do with the steps in long division. What you do know is the process always works. We are going to approach the explanation for division of two polynomials with this in mind; that is, we won't always be sure why the steps we use are important, only that they always produce the correct result.

These are the four basic steps in long division: estimate, multiply, subtract, and bring down the next term. To complete the problem, we simply perform the same four steps:

$$\begin{array}{rl} 18\;\; & \leftarrow \textbf{8 is the estimate} \\ 25\overline{)4{,}628} & \\ \underline{2\,5\;\;\;} & \\ 2\,12\; & \\ \underline{2\,00\;} & \leftarrow \textbf{Multiply to get 200} \\ 128 & \leftarrow \textbf{Subtract to get 12, then bring down the 8} \end{array}$$

One more time:

$$\begin{array}{rl} 185 & \leftarrow \textbf{5 is the estimate} \\ 25\overline{)4{,}628} & \\ \underline{2\,5\;\;\;} & \\ 2\,12\; & \\ \underline{2\,00\;} & \\ 128 & \\ \underline{125} & \leftarrow \textbf{Multiply to get 125} \\ 3 & \leftarrow \textbf{Subtract to get 3} \end{array}$$

Since 3 is less than 25 and we have no more terms to bring down, we have our answer:

$$\frac{4{,}628}{25} = 185 + \frac{3}{25}$$

To check our answer, we multiply 185 by 25 and then add 3 to the result:

$$25(185) + 3 = 4{,}625 + 3 = 4{,}628$$

Long division with polynomials is very similar to long division with whole numbers. Both use the same four basic steps: estimate, multiply, subtract, and bring down the next term. We use long division with polynomials when the denominator has two or more terms and is not a factor of the numerator. Here is an example.

EXAMPLE 8

Divide $\dfrac{2x^2 - 7x + 9}{x - 2}$.

SOLUTION

$$\begin{array}{r|l} & 2x \qquad\qquad \leftarrow \textbf{Estimate } 2x^2 \div x = 2x \\ x - 2 & \overline{2x^2 - 7x + 9} \\ & \underline{\not+ 2x^2 \not- 4x} \qquad \leftarrow \textbf{Multiply } 2x(x - 2) = 2x^2 - 4x \\ & \quad -3x \qquad \leftarrow \textbf{Subtract } (2x^2 - 7x) - (2x^2 - 4x) = -3x \end{array}$$

$$\begin{array}{r|l} & 2x \\ x - 2 & \overline{2x^2 - 7x + 9} \\ & \underline{\not+ 2x^2 \not- 4x} \\ & \quad -3x + 9 \leftarrow \textbf{Bring down the 9} \end{array}$$

Notice we change the signs on $2x^2 - 4x$ and add in the subtraction step. Subtracting a polynomial is equivalent to adding its opposite.

We repeat the four steps.

$$\begin{array}{r|l} & 2x - 3 \qquad \leftarrow \textbf{-3 is the estimate: } -3x \div x = -3 \\ x - 2 & \overline{2x^2 - 7x + 9} \\ & \underline{\not+ 2x^2 \not- 4x} \\ & \quad -3x + 9 \\ & \quad \underline{\not- 3x \not+ 6} \leftarrow \textbf{Multiply } -3(x - 2) = -3x + 6 \\ & \qquad\qquad 3 \leftarrow \textbf{Subtract } (-3x + 9) - (-3x + 6) = 3 \end{array}$$

Since we have no other term to bring down, we have our answer:

$$\frac{2x^2 - 7x + 9}{x - 2} = 2x - 3 + \frac{3}{x - 2}$$

To check, we multiply $(2x - 3)(x - 2)$ to get $2x^2 - 7x + 6$; then, adding the remainder 3 to this result, we have $2x^2 - 7x + 9$.

In setting up a division problem involving two polynomials, we must remember two things: (1) both polynomials should be in decreasing powers of the variable, and (2) neither should skip any powers from the highest power down to the constant term. If there are any missing terms, they can be filled in using a coefficient of 0.

EXAMPLE 9

Divide $2x - 4\overline{)4x^3 - 6x - 11}$.

SOLUTION Since the trinomial is missing a term in x^2, we can fill it in with $0x^2$:

$$4x^3 - 6x - 11 = 4x^3 + 0x^2 - 6x - 11$$

Adding $0x^2$ does not change our original problem.

$$
\begin{array}{r|l}
 & 2x^2 + 4x + 5 \\
2x - 4 & 4x^3 + 0x^2 - 6x - 11 \\
 & \underline{\mp 4x^3 \pm 8x^2} \\
 & \quad + 8x^2 - 6x \\
 & \quad \underline{\mp 8x^2 \pm 16x} \\
 & \qquad + 10x - 11 \\
 & \qquad \underline{\mp 10x \pm 20} \\
 & \qquad\qquad + 9
\end{array}
$$

Notice: **Adding the $0x^2$ term gives us a column in which to write $-8x^2$**

$$\frac{4x^3 - 6x - 11}{2x - 4} = 2x^2 + 4x + 5 + \frac{9}{2x - 4}$$

To check this result, we multiply $2x - 4$ and $2x^2 + 4x + 5$:

$$
\begin{array}{r}
2x^2 + 4x + 5 \\
\underline{2x - 4} \\
4x^3 + 8x^2 + 10x \qquad \\
\underline{- 8x^2 - 16x - 20} \\
4x^3 \qquad - 6x - 20
\end{array}
$$

Adding 9 (the remainder) to this result gives us the polynomial $4x^3 - 6x - 11$. Our answer checks.

For our last example in this section, let's do Example 4 again, but this time use long division.

EXAMPLE 10 Divide $\dfrac{x^2 - 6xy - 7y^2}{x + y}$.

SOLUTION

$$
\begin{array}{r|l}
 & x - 7y \\
x + y & x^2 - 6xy - 7y^2 \\
 & \underline{\mp x^2 \mp xy} \\
 & \quad - 7xy - 7y^2 \\
 & \quad \underline{\pm 7xy \pm 7y^2} \\
 & \qquad\qquad 0
\end{array}
$$

In this case, the remainder is 0 and we have

$$\frac{x^2 - 6xy - 7y^2}{x + y} = x - 7y$$

which is easy to check since

$$(x + y)(x - 7y) = x^2 - 6xy - 7y^2$$

LINKING OBJECTIVES AND EXAMPLES

Next to each **objective** we have listed the examples that are best described by that objective.

A	1–3
B	4–6, 8–10

GETTING READY FOR CLASS

After reading through the preceding section, respond in your own words and in complete sentences.

1. What are the four steps used in long division with polynomials?
2. What does it mean to have a remainder of 0?
3. When must long division be performed, and when can factoring be used to divide polynomials?
4. What property of real numbers is the key to dividing a polynomial by a monomial?

Problem Set 6.2

Online support materials can be found at www.thomsonedu.com/login

Find the following quotients.

1. $\dfrac{4x^3 - 8x^2 + 6x}{2x}$
2. $\dfrac{6x^3 + 12x^2 - 9x}{3x}$
3. $\dfrac{10x^4 + 15x^3 - 20x^2}{-5x^2}$
4. $\dfrac{12x^5 - 18x^4 - 6x^3}{6x^3}$
5. $\dfrac{8y^5 + 10y^3 - 6y}{4y^3}$
6. $\dfrac{6y^4 - 3y^3 + 18y^2}{9y^2}$
7. $\dfrac{5x^3 - 8x^2 - 6x}{-2x^2}$
8. $\dfrac{-9x^5 + 10x^3 - 12x}{-6x^4}$
9. $\dfrac{28a^3b^5 + 42a^4b^3}{7a^2b^2}$
10. $\dfrac{a^2b + ab^2}{ab}$
11. $\dfrac{10x^3y^2 - 20x^2y^3 - 30x^3y^3}{-10x^2y}$
12. $\dfrac{9x^4y^4 + 18x^3y^4 - 27x^2y^4}{-9xy^3}$

Divide by factoring numerators and then dividing out common factors.

13. $\dfrac{x^2 - x - 6}{x - 3}$
14. $\dfrac{x^2 - x - 6}{x + 2}$
15. $\dfrac{2a^2 - 3a - 9}{2a + 3}$
16. $\dfrac{2a^2 + 3a - 9}{2a - 3}$
17. $\dfrac{5x^2 - 14xy - 24y^2}{x - 4y}$
18. $\dfrac{5x^2 - 26xy - 24y^2}{5x + 4y}$
19. $\dfrac{x^3 - y^3}{x - y}$
20. $\dfrac{x^3 + 8}{x + 2}$
21. $\dfrac{y^4 - 16}{y - 2}$
22. $\dfrac{y^4 - 81}{y - 3}$
23. $\dfrac{x^3 + 2x^2 - 25x - 50}{x - 5}$
24. $\dfrac{x^3 + 2x^2 - 25x - 50}{x + 5}$

Divide using the long division method.

25. $\dfrac{x^2 - 5x - 7}{x + 2}$
26. $\dfrac{x^2 + 4x - 8}{x - 3}$
27. $\dfrac{2x^3 - 3x^2 - 4x + 5}{x + 1}$
28. $\dfrac{3x^3 - 5x^2 + 2x - 1}{x - 2}$
29. $\dfrac{2y^3 - 9y^2 - 17y + 39}{2y - 3}$
30. $\dfrac{3y^3 - 19y^2 + 17y + 4}{3y - 4}$
31. $\dfrac{6y^3 - 8y + 5}{2y - 4}$

■ = Videos available by instructor request
▶ = Online student support materials available at www.thomsonedu.com/login

32. $\dfrac{9y^3 - 6y^2 + 8}{3y - 3}$

33. $\dfrac{a^4 - 2a + 5}{a - 2}$

34. $\dfrac{a^4 + a^3 - 1}{a + 2}$

35. $\dfrac{y^4 - 16}{y - 2}$

36. $\dfrac{y^4 - 81}{y - 3}$

37. Let $f(x) = x^2 - 36$ and $g(x) = 4x - 24$. If $h(x) = \dfrac{f(x)}{g(x)}$, find $h(x)$, then state the domain.

38. Let $f(x) = x^2 - 49$ and $g(x) = 2x + 14$. If $h(x) = \dfrac{f(x)}{g(x)}$, find $h(x)$, then state the domain.

39. Let $f(x) = x^2 - 16x + 64$ and $g(x) = x^2 - 4x - 32$. If $h(x) = \dfrac{f(x)}{g(x)}$, find $h(x)$, then state the domain.

40. Let $f(x) = x^2 + 20x + 100$ and $g(x) = x^2 + 5x - 50$. If $h(x) = \dfrac{f(x)}{g(x)}$, find $h(x)$, then state the domain.

41. Let $f(x) = x^3 - 27$ and $g(x) = x - 3$. If $h(x) = \dfrac{f(x)}{g(x)}$, find $h(x)$, then state the domain.

42. Let $f(x) = x^3 + 125$ and $g(x) = x + 5$. If $h(x) = \dfrac{f(x)}{g(x)}$, find $h(x)$, then state the domain.

For the functions below, evaluate

a. $\dfrac{f(x+h) - f(x)}{h}$ b. $\dfrac{f(x) - f(a)}{x - a}$

43. $f(x) = 4x$

44. $f(x) = -3x$

45. $f(x) = 5x + 3$

46. $f(x) = 6x - 5$

47. $f(x) = x^2$

48. $f(x) = 3x^2$

49. $f(x) = x^2 + 1$

50. $f(x) = x^2 - 3$

51. $f(x) = x^2 - 3x + 4$

52. $f(x) = x^2 + 4x - 7$

53. $f(x) = 2x^2 + 3x - 4$

54. $f(x) = 5x^2 + 3x - 7$

55. **The Factor Theorem** The factor theorem of algebra states that if $x - a$ is a factor of a polynomial, $P(x)$, then $P(a) = 0$. Verify the following.
   a. That $x - 2$ is a factor of $P(x) = x^3 - 3x^2 + 5x - 6$, and that $P(2) = 0$
   b. That $x - 5$ is a factor of $P(x) = x^4 - 5x^3 - x^2 + 6x - 5$, and that $P(5) = 0$

56. One factor of $x^3 + 6x^2 + 11x + 6$ is $x + 3$.
   a. Factor $x^3 + 6x^2 + 11x + 6$ completely.
   b. Reduce $\dfrac{x^3 + 6x^2 + 11x + 6}{x + 3}$.
   c. Why are the problems in parts a and b similar?

57. One factor of $x^3 + 10x^2 + 29x + 20$ is $x + 4$.
   a. Factor $x^3 + 10x^2 + 29x + 20$ completely.
   b. Reduce $\dfrac{x^3 + 10x^2 + 29x + 20}{x + 4}$.
   c. Why are the problems in parts a and b similar?

58. One factor of $x^3 + 5x^2 - 2x - 24$ is $x + 3$.
   a. Factor $x^3 + 5x^2 - 2x - 24$ completely.
   b. Reduce $\dfrac{x^3 + 5x^2 - 2x - 24}{x + 3}$.
   c. Why are the problems in parts a and b similar?

59. One factor of $x^3 + 3x^2 - 10x - 24$ is $x + 2$.
   a. Factor $x^3 + 3x^2 - 10x - 24$ completely.
   b. Reduce $\dfrac{x^3 + 3x^2 - 10x - 24}{x + 2}$.
   c. Why are the problems in parts a and b similar?

60. Find $P(-2)$ if $P(x) = x^2 - 5x - 7$. Compare it with the remainder in Problem 25.

61. Find $P(3)$ if $P(x) = x^2 + 4x - 8$. Compare it with the remainder in Problem 26.

62. **The Remainder Theorem** The remainder theorem of algebra states that if a polynomial, $P(x)$, is divided by $x - a$, then the remainder is $P(a)$. Verify the remainder theorem by showing that when $P(x) = x^2 - x + 3$ is divided by $x - 2$ the remainder is 5 and that $P(2) = 5$.

63. **Checking Account** First Bank of San Luis Obispo charges \$2.00 per month and \$0.15 per check for a regular checking account. As we mentioned in the introduction to this section, the total monthly cost

of this account is $C(x) = 2.00 + 0.15x$. To find the average cost of each of the x checks, we divide the total cost by the number of checks written; that is,

$$\overline{C}(x) = \frac{C(x)}{x}$$

a. Use the total cost function to fill in the following table.

x	1	5	10	15	20
$C(x)$					

b. Find the formula for the average cost function, $\overline{C}(x)$.

c. Use the average cost function to fill in the following table. Round to the nearest cent.

x	1	5	10	15	20
$\overline{C}(x)$					

d. What happens to the average cost as more checks are written?

64. Average Cost A company that manufactures computer diskettes uses the function $C(x) = 200 + 2x$ to represent the daily cost of producing x diskettes.

a. Find the average cost function, $\overline{C}(x)$.

b. Use the average cost function to fill in the following table.

x	1	5	10	20	50
$\overline{C}(x)$					

c. What happens to the average cost as more items are produced?

Maintaining Your Skills

Divide.

65. $\frac{3}{5} \div \frac{2}{7}$

66. $\frac{2}{7} \div \frac{3}{5}$

67. $\frac{3}{4} \div \frac{6}{11}$

68. $\frac{6}{8} \div \frac{2}{5}$

69. $\frac{4}{9} \div 8$

70. $\frac{3}{7} \div 6$

71. $8 \div \frac{1}{4}$

72. $12 \div \frac{2}{3}$

Write each expression with positive exponents and simplify as much as possible.

73. $\left(\frac{1}{3}\right)^{-2} + \left(\frac{1}{2}\right)^{-3}$

74. $\left(\frac{1}{2}\right)^{-3} - \left(\frac{1}{3}\right)^{-3}$

Simplify, and write your answers with positive exponents only.

75. $(9x^{-4}y^{9})^{-2}(3x^{2}y^{-1})^{4}$

76. $(4x^{4}y^{-3})^{2}(2x^{-6}y^{4})^{-3}$

Getting Ready for the Next Section

Multiply or divide as indicated.

77. $\frac{6}{7} \cdot \frac{14}{18}$

78. $\frac{6}{8} \div \frac{3}{5}$

79. $5y^2 \cdot 4x^2$

80. $4y^3 \cdot 3x^2$

81. $9x^4 \cdot 8y^5$

82. $6x^4 \cdot 12y^5$

Factor.

83. $x^2 - 4$

84. $x^2 - 6x + 9$

85. $x^3 - x^2y$

86. $a^2 - 5a + 6$

87. $2y^2 - 2$

88. $xa + xb + ya + yb$

Extending the Concepts

Divide.

89. $\dfrac{4x^5 - x^4 - 20x^3 + 8x^2 - 15}{x^2 - 5}$

90. $\dfrac{4x^5 + 2x^4 + x^3 - 20x^2 - 10x - 5}{x^3 - 5}$

91. $\dfrac{0.5x^3 - 0.3x^2 + 0.22x + 0.06}{x + 0.2}$

92. $\dfrac{0.6x^3 - 1.1x^2 - 0.1x + 0.6}{0.3x + 0.2}$

93. $\dfrac{3x^2 + x - 9}{2x + 4}$

94. $\dfrac{2x^2 - x - 9}{3x + 6}$

95. $\dfrac{2x^2 + \frac{1}{3}x + \frac{5}{3}}{3x - 1}$

96. $\dfrac{x^2 + \frac{3}{5}x + \frac{8}{5}}{5x - 2}$

6.3 Multiplication and Division of Rational Expressions

OBJECTIVES

A Multiply and divide rational expressions.

If you have ever taken a home videotape to be duplicated, you know the amount you pay for the duplication service depends on the number of copies you have made: The more copies you have made, the lower the charge per copy. The following demand function gives the price (in dollars) per tape $p(x)$ a company charges for making x copies of a 30-minute videotape. As you can see, it is a rational function.

$$p(x) = \frac{2(x + 60)}{x + 5}$$

The graph in Figure 1 shows this function from $x = 0$ to $x = 100$. As you can see, the more copies that are made, the lower the price per copy.

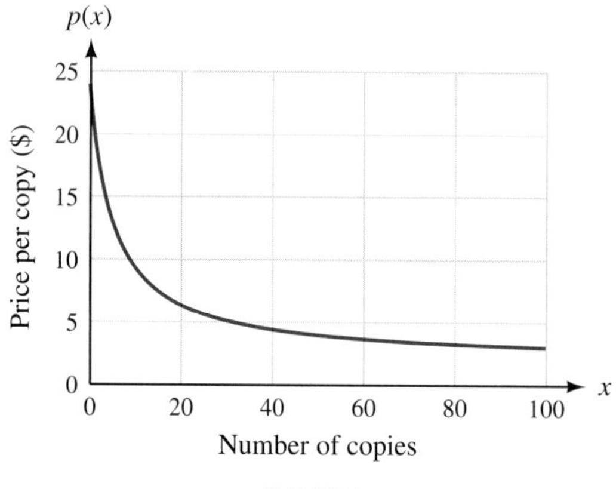

FIGURE 1

If we were interested in finding the revenue function for this situation, we would multiply the number of copies made x by the price per copy $p(x)$. This involves multiplication with a rational expression, which is one of the topics we cover in this section.

In the previous section we found the process of reducing rational expressions to lowest terms to be the same process used in reducing fractions to lowest terms. The similarity also holds for the process of multiplication or division of rational expressions.

Multiplication with fractions is the simplest of the four basic operations. To multiply two fractions we simply multiply numerators and multiply denominators; that is, if a, b, c, and d are real numbers, with $b \neq 0$ and $d \neq 0$, then

$$\frac{a}{b} \cdot \frac{c}{d} = \frac{ac}{bd}$$

EXAMPLE 1

Multiply $\frac{6}{7} \cdot \frac{14}{18}$.

SOLUTION

$$\frac{6}{7} \cdot \frac{14}{18} = \frac{6(14)}{7(18)} \qquad \textbf{Multiply numerators and denominators}$$

$$= \frac{\cancel{2} \cdot \cancel{3}(2 \cdot \cancel{7})}{\cancel{7}(\cancel{2} \cdot \cancel{3} \cdot 3)} \qquad \textbf{Factor}$$

$$= \frac{2}{3} \qquad \textbf{Divide out common factors}$$

Our next example is similar to some of the problems we worked in an earlier chapter. We multiply fractions whose numerators and denominators are monomials by multiplying numerators and multiplying denominators and then reducing to lowest terms. Here is how it looks.

EXAMPLE 2

Multiply $\frac{8x^3}{27y^8} \cdot \frac{9y^3}{12x^2}$.

SOLUTION We multiply numerators and denominators without actually carrying out the multiplication:

$$\frac{8x^3}{27y^8} \cdot \frac{9y^3}{12x^2} = \frac{8 \cdot 9x^3y^3}{27 \cdot 12x^2y^8} \qquad \textbf{Multiply numerators, Multiply denominators}$$

$$= \frac{\cancel{4} \cdot 2 \cdot \cancel{9}x^3y^3}{\cancel{9} \cdot 3 \cdot \cancel{4} \cdot 3x^2y^8} \qquad \textbf{Factor coefficients}$$

$$= \frac{2x}{9y^5} \qquad \textbf{Divide out common factors}$$

> ***Note***
> Notice how we factor the coefficients just enough so that we can see the factors they have in common. If you want to show this step without showing the factoring, it would look like this:
> $$\frac{\overset{2}{\cancel{8}} \cdot \overset{1}{\cancel{9}}x^3y^3}{\underset{3}{\cancel{27}} \cdot \underset{3}{\cancel{12}}x^2y^8}$$

The product of two rational expressions is the product of their numerators over the product of their denominators.

Once again, we should mention that the little slashes we have drawn through the factors are used to denote the factors we have divided out of the numerator and denominator.

EXAMPLE 3

Multiply $\frac{x-3}{x^2-4} \cdot \frac{x+2}{x^2-6x+9}$.

SOLUTION We begin by multiplying numerators and denominators. We then factor all polynomials and divide out factors common to the numerator and denominator:

$$\frac{x-3}{x^2-4} \cdot \frac{x+2}{x^2-6x+9} = \frac{(x-3)(x+2)}{(x^2-4)(x^2-6x+9)} \qquad \textbf{Multiply}$$

$$= \frac{\cancel{(x-3)}\cancel{(x+2)}}{\cancel{(x+2)}(x-2)\cancel{(x-3)}(x-3)} \qquad \textbf{Factor}$$

$$= \frac{1}{(x-2)(x-3)} \qquad \textbf{Divide out common factors}$$

The first two steps can be combined to save time. We can perform the multiplication and factoring steps together.

EXAMPLE 4

Multiply $\dfrac{2y^2 - 4y}{2y^2 - 2} \cdot \dfrac{y^2 - 2y - 3}{y^2 - 5y + 6}$.

SOLUTION

$$\frac{2y^2 - 4y}{2y^2 - 2} \cdot \frac{y^2 - 2y - 3}{y^2 - 5y + 6} = \frac{\cancel{2}y\cancel{(y-2)}\cancel{(y-3)}\cancel{(y+1)}}{\cancel{2}\cancel{(y+1)}(y-1)\cancel{(y-3)}\cancel{(y-2)}}$$

$$= \frac{y}{y-1}$$

Notice in both of the preceding examples that we did not actually multiply the polynomials as we did in the chapter on exponents and polynomials. It would be senseless to do that since we would then have to factor each of the resulting products to reduce them to lowest terms.

The quotient of two rational expressions is the product of the first and the reciprocal of the second; that is, we find the quotient of two rational expressions the same way we find the quotient of two fractions. Here is an example that reviews division with fractions.

EXAMPLE 5

Divide $\dfrac{6}{8} \div \dfrac{3}{5}$.

SOLUTION

$$\frac{6}{8} \div \frac{3}{5} = \frac{6}{8} \cdot \frac{5}{3} \quad \textbf{Write division in terms of multiplication}$$

$$= \frac{6(5)}{8(3)} \quad \textbf{Multiply numerators and denominators}$$

$$= \frac{\cancel{2} \cdot \cancel{3}(5)}{\cancel{2} \cdot 2 \cdot 2(\cancel{3})} \quad \textbf{Factor}$$

$$= \frac{5}{4} \quad \textbf{Divide out common factors}$$

To divide one rational expression by another, we use the definition of division to multiply by the reciprocal of the expression that follows the division symbol.

EXAMPLE 6

Divide $\dfrac{8x^3}{5y^2} \div \dfrac{4x^2}{10y^6}$.

SOLUTION First we rewrite the problem in terms of multiplication. Then we multiply.

$$\frac{8x^3}{5y^2} \div \frac{4x^2}{10y^6} = \frac{8x^3}{5y^2} \cdot \frac{10y^6}{4x^2}$$

$$= \frac{\overset{2}{\cancel{8}} \cdot \overset{2}{\cancel{10}}x^3y^6}{\cancel{4} \cdot \cancel{5}x^2y^2}$$

$$= 4xy^4$$

EXAMPLE 7

Divide $\dfrac{x^2 - y^2}{x^2 - 2xy + y^2} \div \dfrac{x^3 + y^3}{x^3 - x^2y}$.

SOLUTION We begin by writing the problem as the product of the first and the reciprocal of the second and then proceed as in the previous two examples:

$$\frac{x^2 - y^2}{x^2 - 2xy + y^2} \div \frac{x^3 + y^3}{x^3 - x^2y}$$

Multiply by the reciprocal of the divisor

$$= \frac{x^2 - y^2}{x^2 - 2xy + y^2} \cdot \frac{x^3 - x^2y}{x^3 + y^3}$$

$$= \frac{\cancel{(x - y)}\cancel{(x + y)}(x^2)\cancel{(x - y)}}{\cancel{(x - y)}\cancel{(x - y)}\cancel{(x + y)}(x^2 - xy + y^2)}$$

Factor and multiply

$$= \frac{x^2}{x^2 - xy + y^2}$$

Divide out common factors

Here are some more examples of multiplication and division with rational expressions.

EXAMPLE 8

Perform the indicated operations.

$$\frac{a^2 - 8a + 15}{a + 4} \cdot \frac{a + 2}{a^2 - 5a + 6} \div \frac{a^2 - 3a - 10}{a^2 + 2a - 8}$$

SOLUTION First we rewrite the division as multiplication by the reciprocal. Then we proceed as usual.

$$\frac{a^2 - 8a + 15}{a + 4} \cdot \frac{a + 2}{a^2 - 5a + 6} \div \frac{a^2 - 3a - 10}{a^2 + 2a - 8}$$

$$= \frac{(a^2 - 8a + 15)(a + 2)(a^2 + 2a - 8)}{(a + 4)(a^2 - 5a + 6)(a^2 - 3a - 10)}$$

Change division to multiplication by the reciprocal

$$= \frac{\cancel{(a - 5)}\cancel{(a - 3)}\cancel{(a + 2)}\cancel{(a + 4)}\cancel{(a - 2)}}{\cancel{(a + 4)}\cancel{(a - 3)}\cancel{(a - 2)}\cancel{(a - 5)}\cancel{(a + 2)}}$$

Factor

$$= 1$$

Divide out common factors

Our next example involves factoring by grouping. As you may have noticed, working the problems in this chapter gives you a very detailed review of factoring.

EXAMPLE 9

Multiply $\dfrac{xa + xb + ya + yb}{xa - xb - ya + yb} \cdot \dfrac{xa + xb - ya - yb}{xa - xb + ya - yb}$

SOLUTION We will factor each polynomial by grouping, which takes two steps.

$$\frac{xa + xb + ya + yb}{xa - xb - ya + yb} \cdot \frac{xa + xb - ya - yb}{xa - xb + ya - yb}$$

$$= \frac{x(a + b) + y(a + b)}{x(a - b) - y(a - b)} \cdot \frac{x(a + b) - y(a + b)}{x(a - b) + y(a - b)}$$

$$= \frac{(a + b)\cancel{(x + y)}(a + b)\cancel{(x - y)}}{(a - b)\cancel{(x - y)}(a - b)\cancel{(x + y)}}$$

Factor by grouping

$$= \frac{(a + b)^2}{(a - b)^2}$$

EXAMPLE 10 Multiply $(4x^2 - 36) \cdot \frac{12}{4x + 12}$.

SOLUTION We can think of $4x^2 - 36$ as having a denominator of 1. Thinking of it in this way allows us to proceed as we did in the previous examples.

$$(4x^2 - 36) \cdot \frac{12}{4x + 12}$$

$$= \frac{4x^2 - 36}{1} \cdot \frac{12}{4x + 12} \quad \text{Write } 4x^2 - 36 \text{ with denominator 1}$$

$$= \frac{\cancel{4}(x - 3)\cancel{(x + 3)}12}{\cancel{4}\cancel{(x + 3)}} \quad \text{Factor}$$

$$= 12(x - 3) \quad \text{Divide out common factors}$$

LINKING OBJECTIVES AND EXAMPLES

Next to each **objective** we have listed the examples that are best described by that objective.

A 1–10

GETTING READY FOR CLASS

After reading through the preceding section, respond in your own words and in complete sentences.

1. Summarize the steps used to multiply fractions.
2. What is the first step in multiplying two rational expressions?
3. Why is factoring important when multiplying and dividing rational expressions?
4. How is division with rational expressions different than multiplication of rational expressions?

Problem Set 6.3

Online support materials can be found at www.thomsonedu.com/login

Perform the indicated operations involving fractions.

1. $\frac{2}{9} \cdot \frac{3}{4}$
2. $\frac{5}{6} \cdot \frac{7}{8}$
3. $\frac{3}{4} \div \frac{1}{3}$
4. $\frac{3}{8} \div \frac{5}{4}$
5. $\frac{3}{7} \cdot \frac{14}{24} \div \frac{1}{2}$
6. $\frac{6}{5} \cdot \frac{10}{36} \div \frac{3}{4}$
7. $\frac{10x^2}{5y^2} \cdot \frac{15y^3}{2x^4}$
8. $\frac{8x^3}{7y^4} \cdot \frac{14y^6}{16x^2}$
9. $\frac{11a^2b}{5ab^2} \div \frac{22a^3b^2}{10ab^4}$
10. $\frac{8ab^3}{9a^2b} \div \frac{16a^2b^2}{18ab^3}$
11. $\frac{6x^2}{5y^3} \cdot \frac{11z^2}{2x^2} \div \frac{33z^5}{10y^8}$
12. $\frac{4x^3}{7y^2} \cdot \frac{6z^5}{5x^6} \div \frac{24z^2}{35x^6}$

Perform the indicated operations. Be sure to write all answers in lowest terms.

13. $\frac{x^2 - 9}{x^2 - 4} \cdot \frac{x - 2}{x - 3}$
14. $\frac{x^2 - 16}{x^2 - 25} \cdot \frac{x - 5}{x - 4}$
15. $\frac{y^2 - 1}{y + 2} \cdot \frac{y^2 + 5y + 6}{y^2 + 2y - 3}$
16. $\frac{y - 1}{y^2 - y - 6} \cdot \frac{y^2 + 5y + 6}{y^2 - 1}$
17. $\frac{3x - 12}{x^2 - 4} \cdot \frac{x^2 + 6x + 8}{x - 4}$
18. $\frac{x^2 + 5x + 1}{4x - 4} \cdot \frac{x - 1}{x^2 + 5x + 1}$

19. $\dfrac{5x + 2y}{25x^2 - 5xy - 6y^2} \cdot \dfrac{20x^2 - 7xy - 3y^2}{4x + y}$

20. $\dfrac{7x + 3y}{42x^2 - 17xy - 15y^2} \cdot \dfrac{12x^2 - 4xy - 5y^2}{2x + y}$

21. $\dfrac{a^2 - 5a + 6}{a^2 - 2a - 3} \div \dfrac{a - 5}{a^2 + 3a + 2}$

22. $\dfrac{a^2 + 7a + 12}{a - 5} \div \dfrac{a^2 + 9a + 18}{a^2 - 7a + 10}$

23. $\dfrac{4t^2 - 1}{6t^2 + t - 2} \div \dfrac{8t^3 + 1}{27t^3 + 8}$

24. $\dfrac{9t^2 - 1}{6t^2 + 7t - 3} \div \dfrac{27t^3 + 1}{8t^3 + 27}$

25. $\dfrac{2x^2 - 5x - 12}{4x^2 + 8x + 3} \div \dfrac{x^2 - 16}{2x^2 + 7x + 3}$

26. $\dfrac{x^2 - 2x + 1}{3x^2 + 7x - 20} \div \dfrac{x^2 + 3x - 4}{3x^2 - 2x - 5}$

27. $\dfrac{6a^2b + 2ab^2 - 20b^3}{4a^2b - 16b^3} \cdot \dfrac{10a^2 - 22ab + 4b^2}{27a^3 - 125b^3}$

28. $\dfrac{12a^2b - 3ab^2 - 42b^3}{9a^2 - 36b^2} \cdot \dfrac{6a^2 - 15ab + 6b^2}{8a^3b - b^4}$

29. $\dfrac{360x^3 - 490x}{36x^2 + 84x + 49} \cdot \dfrac{30x^2 + 83x + 56}{150x^3 + 65x^2 - 280x}$

30. $\dfrac{490x^2 - 640}{49x^2 - 112x + 64} \cdot \dfrac{28x^2 - 95x + 72}{56x^3 - 62x^2 - 144x}$

31. $\dfrac{x^5 - x^2}{5x^5 - 5x} \cdot \dfrac{10x^4 - 10x^2}{2x^4 + 2x^3 + 2x^2}$

32. $\dfrac{2x^4 - 16x}{3x^6 - 48x^2} \cdot \dfrac{6x^5 + 24x^3}{4x^4 + 8x^3 + 16x^2}$

33. $\dfrac{a^2 - 16b^2}{a^2 - 8ab + 16b^2} \cdot \dfrac{a^2 - 9ab + 20b^2}{a^2 - 7ab + 12b^2} \div \dfrac{a^2 - 25b^2}{a^2 - 6ab + 9b^2}$

34. $\dfrac{a^2 - 6ab + 9b^2}{a^2 - 4b^2} \cdot \dfrac{a^2 - 5ab + 6b^2}{(a - 3b)^2} \div \dfrac{a^2 - 9b^2}{a^2 - ab - 6b^2}$

35. $\dfrac{2y^2 - 7y - 15}{42y^2 - 29y - 5} \cdot \dfrac{12y^2 - 16y + 5}{7y^2 - 36y + 5} \div \dfrac{4y^2 - 9}{49y^2 - 1}$

36. $\dfrac{8y^2 + 18y - 5}{21y^2 - 16y + 3} \cdot \dfrac{35y^2 - 22y + 3}{6y^2 + 17y + 5} \div \dfrac{16y^2 - 1}{9y^2 - 1}$

37. $\dfrac{xy - 2x + 3y - 6}{xy + 2x - 4y - 8} \cdot \dfrac{xy + x - 4y - 4}{xy - x + 3y - 3}$

38. $\dfrac{ax + bx + 2a + 2b}{ax - 3a + bx - 3b} \cdot \dfrac{ax - bx - 3a + 3b}{ax - bx - 2a + 2b}$

39. $\dfrac{xy^2 - y^2 + 4xy - 4y}{xy - 3y + 4x - 12} \div \dfrac{xy^3 + 2xy^2 + y^3 + 2y^2}{xy^2 - 3y^2 + 2xy - 6y}$

40. $\dfrac{4xb - 8b + 12x - 24}{xb^2 + 3b^2 + 3xb + 9b} \div \dfrac{4xb - 8b - 8x + 16}{xb^2 + 3b^2 - 2xb - 6b}$

41. $\dfrac{2x^3 + 10x^2 - 8x - 40}{x^3 + 4x^2 - 9x - 36} \cdot \dfrac{x^2 + x - 12}{2x^2 + 14x + 20}$

42. $\dfrac{x^3 + 2x^2 - 9x - 18}{x^4 + 3x^3 - 4x^2 - 12x} \cdot \dfrac{x^3 + 5x^2 + 6x}{x^2 - x - 6}$

The next two problems are intended to give you practice reading, and paying attention to, the instructions that accompany the problems you are working. Working these problems is an excellent way to get ready for a test or a quiz.

43. Work each problem according to the instructions given.

   a. Simplify: $\dfrac{16 - 1}{64 - 1}$

   b. Reduce: $\dfrac{25x^2 - 9}{125x^3 - 27}$

   c. Multiply: $\dfrac{25x^2 - 9}{125x^3 - 27} \cdot \dfrac{5x - 3}{5x + 3}$

   d. Divide: $\dfrac{25x^2 - 9}{125x^3 - 27} \div \dfrac{5x - 3}{25x^2 + 15x + 9}$

44. Work each problem according to the instructions given.

   a. Simplify: $\dfrac{64 - 49}{64 + 112 + 49}$

   b. Reduce: $\dfrac{9x^2 - 49}{9x^2 + 42x + 49}$

   c. Multiply: $\dfrac{9x^2 - 49}{9x^2 + 42x + 49} \cdot \dfrac{3x + 7}{3x - 7}$

   d. Divide: $\dfrac{9x^2 - 49}{9x^2 + 42x + 49} \div \dfrac{3x + 7}{3x - 7}$

The work you did with the algebra of functions will help with these:

45. Let $f(x) = \dfrac{x^2 - x - 6}{x - 1}$ and $g(x) = \dfrac{x + 2}{x^2 - 4x + 3}$, find

   a. $f(x) \cdot g(x)$

   b. $f(x) \div g(x)$

46. Let $f(x) = \dfrac{x^2 - x - 12}{x^2 - 4x + 3}$ and $g(x) = \dfrac{x^2 - x - 12}{x^2 - 5x + 4}$; find

a. $f(x) \cdot g(x)$

b. $f(x) \div g(x)$

47. Let $f(x) = \dfrac{x^3 - 9x^2 - 3x + 27}{4x^2 - 12}$ and $g(x) = \dfrac{x^2 - 2x - 8}{x^2 - 81}$; find $f(x) \cdot g(x)$

48. Let $f(x) = \dfrac{x^2 - 7x + 12}{x^2 - 16}$ and $g(x) = \dfrac{x^2 - 4x + 3}{x^2 - 6x + 9}$; find $f(x) \cdot g(x)$

49. Let $f(x) = \dfrac{x^3 - 3x^2 - 4x + 12}{x + 2}$ and $g(x) = \dfrac{x^2 + 7x + 12}{x^2 - 5x + 6}$; find $f(x) \cdot g(x)$

50. Let $f(x) = 2x^2 - 9x + 9$ and $g(x) = \dfrac{2x + 8}{x^2 + x - 12}$; find $f(x) \cdot g(x)$

Use the method shown in Example 10 to find the following products.

▸ **51.** $(3x - 6) \cdot \dfrac{x}{x - 2}$

52. $(4x + 8) \cdot \dfrac{x}{x + 2}$

53. $(x^2 - 25) \cdot \dfrac{2}{x - 5}$

54. $(x^2 - 49) \cdot \dfrac{5}{x + 7}$

55. $(x^2 - 3x + 2) \cdot \dfrac{3}{3x - 3}$

56. $(x^2 - 3x + 2) \cdot \dfrac{-1}{x - 2}$

57. $(y - 3)(y - 4)(y + 3) \cdot \dfrac{-1}{y^2 - 9}$

58. $(y + 1)(y + 4)(y - 1) \cdot \dfrac{3}{y^2 - 1}$

59. $a(a + 5)(a - 5) \cdot \dfrac{a + 1}{a^2 + 5a}$

60. $a(a + 3)(a - 3) \cdot \dfrac{a - 1}{a^2 - 3a}$

Divide.

▸ **61.** $(x^2 - 2x - 8) \div \dfrac{x^2 - x - 6}{x - 4}$

▸ **62.** $(2x^2 + 7x - 15) \div \dfrac{6x^2 + 21x - 45}{2x - 3}$

▸ **63.** $(3 - x) \div \dfrac{x^2 - 9}{x - 1}$

▸ **64.** $(-x^2 + 3x - 2) \div \dfrac{x^2 + 2x - 8}{x^2 + 3x - 4}$

▸ **65.** $(xy - 2x - 3y + 6) \div \dfrac{x^2 - 2x - 3}{x^2 - 6x - 7}$

▸ **66.** $(x^3 - x^2 - 9x + 9) \div \dfrac{x^3 - 3x^2 - 4x + 12}{6x - 12}$

Simplify.

▸ **67.** $\dfrac{x^2(x + 2) + 6x(x + 2) + 9(x + 2)}{x^2 - 2x - 8}$

▸ **68.** $\dfrac{x^2(x - 3) - 14x(x - 3) + 49(x - 3)}{x^2 - 3x - 28}$

▸ **69.** $\dfrac{2x^2(x + 3) - 5x(x + 3) + 3(x + 3)}{x^2 + x - 6}$

▸ **70.** $\dfrac{3x^2(x + 1) - 4x(x + 1) + (x + 1)}{x^3 + 5x^2 - x - 5}$

Applying the Concepts

At the beginning of this section we introduced the demand equation shown here. Use it to work Problems 71–74.

$$p(x) = \frac{2(x + 60)}{x + 5}$$

71. Demand Equation Use the demand equation to fill in the table to the nearest cent. Then compare your results with the graph shown in Figure 1 of this section.

Number of Copies	Price per Copy ($)
1	
10	
20	
50	
100	

72. Demand Equation To find the revenue for selling 50 copies of a tape, we multiply the price per tape by 50. Find the revenue for selling 50 tapes.

73. Revenue Find the revenue for selling 100 tapes.

74. Revenue Find the revenue equation $R(x)$.

75. Area The following box has a square top. The front face of the box has an area of $A = x^3 - 2x^2 - 2x - 3$.

The height of the box is $h = x^2 + x + 1$. Find a formula for the area of the top square in terms of x.

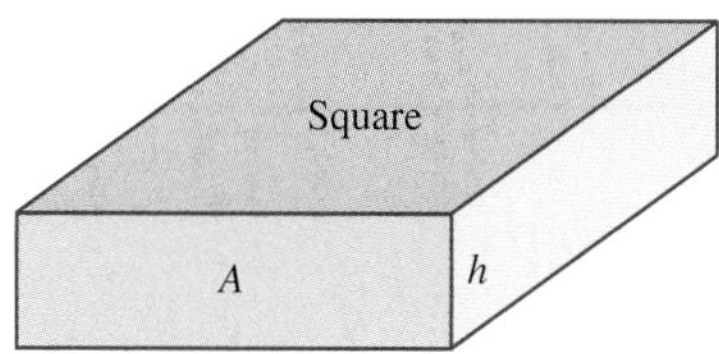

76. **Surface Area of a Cylinder** The surface area of the cylinder in the figure is defined as the area of its two circular bases and the lateral, or side, area. The surface area may be found by the formula

$$A = 2\pi r^2 + 2\pi rh$$

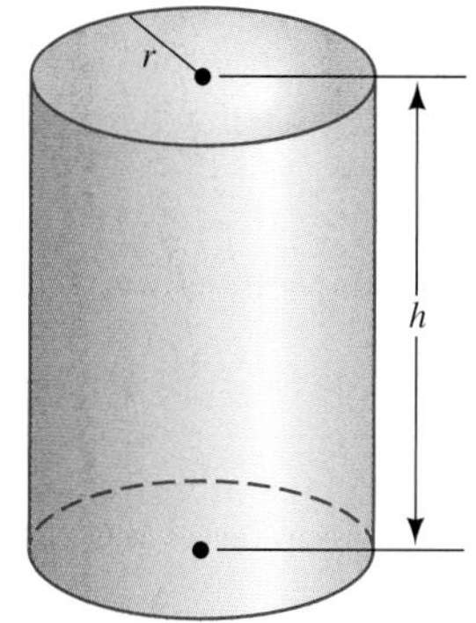

If the surface area is 6π, and $h = 2$, find r.

Maintaining Your Skills

Multiply.

77. $2x^2(5x^3 + 4x - 3)$

78. $3x^3(7x^2 - 4x - 8)$

79. $(3a - 1)(4a + 5)$

80. $(6a - 3)(2a + 1)$

81. $(3x + 7)(4y - 2)$

82. $(x + 2a)(2 - 3b)$

83. $(3 - t^2)^2$

84. $(2 - t^3)^2$

85. $3(x + 1)(x + 2)(x + 3)$

86. $4(x - 1)(x - 2)(x - 3)$

Getting Ready for the Next Section

Combine.

87. $\frac{4}{9} + \frac{2}{9}$

88. $\frac{3}{8} + \frac{1}{8}$

89. $\frac{3}{14} + \frac{7}{30}$

90. $\frac{3}{10} + \frac{11}{42}$

Multiply.

91. $-1(7 - x)$

92. $-1(3 - x)$

Factor.

93. $x^2 - 1$

94. $x^2 - 2x - 3$

95. $2x + 10$

96. $x^2 + 4x + 3$

97. $a^3 - b^3$

98. $8y^3 - 27$

Extending the Concepts

Divide.

99. $\frac{x^6 + y^6}{x^4 + 4x^2y^2 + 3y^4} \div \frac{x^4 + 3x^2y^2 + 2y^4}{x^4 + 5x^2y^2 + 6y^4}$

100. $\frac{x^2 + 9xy + 8y^2}{x^2 + 7xy - 8y^2} \div \frac{x^2 - y^2}{x^2 + 5xy - 6y^2}$

101. $\frac{a^2(2a + b) + 6a(2a + b) + 5(2a + b)}{3a^2(2a + b) - 2a(2a + b) + (2a + b)} \div \frac{a + 1}{a - 1}$

102. $\frac{2x^2(x - 3z) - 5x(x - 3z) + 2(x - 3z)}{4x^2(x - 3z) - 11x(x - 3z) + 6(x - 3z)} \div \frac{4x - 3}{4x + 1}$

103. $\frac{a^3 - a^2b}{ac - a} \div \left(\frac{a - b}{c - 1}\right)^2$

104. $\frac{p^3 + q^3}{q - p} \div \frac{(p + q)^2}{p^2 - q^2}$

6.4 Addition and Subtraction of Rational Expressions

OBJECTIVES

A Add and subtract rational expressions with the same denominator.

B Add and subtract rational expressions with different denominators.

This section is concerned with addition and subtraction of rational expressions. In the first part of this section we will look at addition of expressions that have the same denominator. In the second part of this section we will look at addition of expressions that have different denominators.

Addition and Subtraction with the Same Denominator

To add two expressions that have the same denominator, we simply add numerators and put the sum over the common denominator. Since the process we use to add and subtract rational expressions is the same process used to add and subtract fractions, we will begin with an example involving fractions.

EXAMPLE 1 Add $\frac{4}{9} + \frac{2}{9}$.

SOLUTION We add fractions with the same denominator by using the distributive property. Here is a detailed look at the steps involved.

$$\begin{aligned} \frac{4}{9} + \frac{2}{9} &= 4\left(\frac{1}{9}\right) + 2\left(\frac{1}{9}\right) \\ &= (4 + 2)\left(\frac{1}{9}\right) && \textbf{Distributive property} \\ &= 6\left(\frac{1}{9}\right) \\ &= \frac{6}{9} \\ &= \frac{2}{3} && \textbf{Divide numerator and denominator by common factor 3} \end{aligned}$$

Note that the important thing about the fractions in this example is that they each have a denominator of 9. If they did not have the same denominator, we could not have written them as two terms with a factor of $\frac{1}{9}$ in common. Without the $\frac{1}{9}$ common to each term, we couldn't apply the distributive property. And without the distributive property, we would not have been able to add the two fractions.

In the examples that follow, we will not show all the steps we showed in Example 1. The steps are shown in Example 1 so that you will see why both fractions must have the same denominator before we can add them. In practice we simply add numerators and place the result over the common denominator.

We add and subtract rational expressions with the same denominator by combining numerators and writing the result over the common denominator. Then we reduce the result to lowest terms, if possible. Example 2 shows this process in detail. If you see the similarities between operations on rational numbers and operations on rational expressions, this chapter will look like an extension of rational numbers rather than a completely new set of topics.

EXAMPLE 2 Add $\frac{x}{x^2-1} + \frac{1}{x^2-1}$.

SOLUTION Since the denominators are the same, we simply add numerators:

$$\frac{x}{x^2-1} + \frac{1}{x^2-1} = \frac{x+1}{x^2-1} \qquad \textbf{Add numerators}$$

$$= \frac{\cancel{x+1}}{(x-1)\cancel{(x+1)}} \qquad \textbf{Factor denominator}$$

$$= \frac{1}{x-1} \qquad \textbf{Divide out common factor } x+1$$

Our next example involves subtraction of rational expressions. Pay careful attention to what happens to the signs of the terms in the numerator of the second expression when we subtract it from the first expression.

EXAMPLE 3 Subtract $\frac{2x-5}{x-2} - \frac{x-3}{x-2}$.

SOLUTION Since each expression has the same denominator, we simply subtract the numerator in the second expression from the numerator in the first expression and write the difference over the common denominator $x - 2$. We must be careful, however, that we subtract both terms in the second numerator. To ensure that we do, we will enclose that numerator in parentheses.

$$\frac{2x-5}{x-2} - \frac{x-3}{x-2} = \frac{2x-5-(x-3)}{x-2} \qquad \textbf{Subtract numerators}$$

$$= \frac{2x-5-x+3}{x-2} \qquad \textbf{Remove parentheses}$$

$$= \frac{x-2}{x-2} \qquad \textbf{Combine similar terms in the numerator}$$

$$= 1 \qquad \textbf{Reduce (or divide)}$$

Note the $+3$ in the numerator of the second step. It is a very common mistake to write that as -3 by forgetting to subtract both terms in the numerator of the second expression. Whenever the expression we are subtracting has two or more terms in its numerator, we have to watch for this mistake.

Next we consider addition and subtraction of fractions and rational expressions that have different denominators.

Addition and Subtraction with Different Denominators

Before we look at an example of addition of fractions with different denominators, we need to review the definition for the least common denominator.

> **DEFINITION** The **least common denominator,** abbreviated LCD, for a set of denominators is the smallest expression that is divisible by each of the denominators.

The first step in combining two fractions is to find the LCD. Once we have the common denominator, we rewrite each fraction as an equivalent fraction with the common denominator. After that, we simply add or subtract as we did in our first three examples.

Example 4 is a review of the step-by-step procedure used to add two fractions with different denominators.

EXAMPLE 4

Add $\frac{3}{14} + \frac{7}{30}$.

SOLUTION

Step 1: Find the LCD.

To do this, we first factor both denominators into prime factors.

Factor 14: $14 = 2 \cdot 7$

Factor 30: $30 = 2 \cdot 3 \cdot 5$

Since the LCD must be divisible by 14, it must have factors of $2 \cdot 7$. It must also be divisible by 30 and, therefore, have factors of $2 \cdot 3 \cdot 5$. We do not need to repeat the 2 that appears in both the factors of 14 and those of 30. Therefore,

$$\text{LCD} = 2 \cdot 3 \cdot 5 \cdot 7 = 210$$

Step 2: Change to equivalent fractions.

Since we want each fraction to have a denominator of 210 and at the same time keep its original value, we multiply each by 1 in the appropriate form.

Change $\frac{3}{14}$ to a fraction with denominator 210:

$$\frac{3}{14} \cdot \frac{\mathbf{15}}{\mathbf{15}} = \frac{45}{210}$$

Change $\frac{7}{30}$ to a fraction with denominator 210:

$$\frac{7}{30} \cdot \frac{\mathbf{7}}{\mathbf{7}} = \frac{49}{210}$$

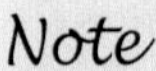

Note

When we multiply $\frac{3}{14}$ by $\frac{15}{15}$ we obtain a fraction with the same value as $\frac{3}{14}$ (because we multiplied by 1) but with the common denominator 210.

Step 3: Add numerators of equivalent fractions found in step 2:

$$\frac{45}{210} + \frac{49}{210} = \frac{94}{210}$$

Step 4: Reduce to lowest terms if necessary:

$$\frac{94}{210} = \frac{47}{105}$$

The main idea in adding fractions is to write each fraction again with the LCD for a denominator. In doing so, we must be sure not to change the value of either of the original fractions.

EXAMPLE 5

Add $\dfrac{-2}{x^2 - 2x - 3} + \dfrac{3}{x^2 - 9}$.

SOLUTION

Step 1: Factor each denominator and build the LCD from the factors:

$$\left.\begin{aligned} x^2 - 2x - 3 &= (x - 3)(x + 1) \\ x^2 - 9 &= (x - 3)(x + 3) \end{aligned}\right\} \quad \text{LCD} = (x - 3)(x + 3)(x + 1)$$

Step 2: Change each rational expression to an equivalent expression that has the LCD for a denominator:

$$\frac{-2}{x^2 - 2x - 3} = \frac{-2}{(x - 3)(x + 1)} \cdot \frac{\mathbf{(x + 3)}}{\mathbf{(x + 3)}} = \frac{-2x - 6}{(x - 3)(x + 3)(x + 1)}$$

$$\frac{3}{x^2 - 9} = \frac{3}{(x - 3)(x + 3)} \cdot \frac{\mathbf{(x + 1)}}{\mathbf{(x + 1)}} = \frac{3x + 3}{(x - 3)(x + 3)(x + 1)}$$

Step 3: Add numerators of the rational expressions found in step 2:

$$\frac{-2x - 6}{(x - 3)(x + 3)(x + 1)} + \frac{3x + 3}{(x - 3)(x + 3)(x + 1)} = \frac{x - 3}{(x - 3)(x + 3)(x + 1)}$$

Step 4: Reduce to lowest terms by dividing out the common factor $x - 3$:

$$= \frac{1}{(x + 3)(x + 1)}$$

EXAMPLE 6

Subtract $\dfrac{x + 4}{2x + 10} - \dfrac{5}{x^2 - 25}$.

SOLUTION We begin by factoring each denominator:

$$\frac{x + 4}{2x + 10} - \frac{5}{x^2 - 25} = \frac{x + 4}{2(x + 5)} - \frac{5}{(x + 5)(x - 5)}$$

The LCD is $2(x + 5)(x - 5)$. Completing the solution we have

$$= \frac{x + 4}{2(x + 5)} \cdot \frac{\mathbf{(x - 5)}}{\mathbf{(x - 5)}} - \frac{5}{(x + 5)(x - 5)} \cdot \frac{\mathbf{2}}{\mathbf{2}}$$

$$= \frac{x^2 - x - 20}{2(x + 5)(x - 5)} - \frac{10}{2(x + 5)(x - 5)}$$

$$= \frac{x^2 - x - 30}{2(x + 5)(x - 5)}$$

To see if this expression will reduce, we factor the numerator into $(x - 6)(x + 5)$.

$$= \frac{(x - 6)\cancel{(x + 5)}}{2\cancel{(x + 5)}(x - 5)}$$

$$= \frac{x - 6}{2(x - 5)}$$

EXAMPLE 7 Subtract $\dfrac{2x - 2}{x^2 + 4x + 3} - \dfrac{x - 1}{x^2 + 5x + 6}$.

SOLUTION We factor each denominator and build the LCD from those factors:

$$\frac{2x - 2}{x^2 + 4x + 3} - \frac{x - 1}{x^2 + 5x + 6}$$

$$= \frac{2x - 2}{(x + 3)(x + 1)} - \frac{x - 1}{(x + 3)(x + 2)}$$

$$= \frac{2x - 2}{(x + 3)(x + 1)} \cdot \frac{\mathbf{(x + 2)}}{\mathbf{(x + 2)}} - \frac{x - 1}{(x + 3)(x + 2)} \cdot \frac{\mathbf{(x + 1)}}{\mathbf{(x + 1)}} \qquad \text{Build the LCD}$$

$$= \frac{2x^2 + 2x - 4}{(x + 1)(x + 2)(x + 3)} - \frac{x^2 - 1}{(x + 1)(x + 2)(x + 3)} \qquad \text{Multiply out each numerator}$$

$$= \frac{(2x^2 + 2x - 4) - (x^2 - 1)}{(x + 1)(x + 2)(x + 3)}$$

$$= \frac{x^2 + 2x - 3}{(x + 1)(x + 2)(x + 3)} \qquad \text{Subtract numerators}$$

$$= \frac{\cancel{(x + 3)}(x - 1)}{(x + 1)(x + 2)\cancel{(x + 3)}} \qquad \text{Factor numerator to see if we can reduce}$$

$$= \frac{x - 1}{(x + 1)(x + 2)} \qquad \text{Reduce}$$

EXAMPLE 8 Add $\dfrac{x^2}{x - 7} + \dfrac{6x + 7}{7 - x}$.

SOLUTION In the first section of this chapter we were able to reverse the terms in a factor such as $7 - x$ by factoring -1 from each term. In a problem like this, the same result can be obtained by multiplying the numerator and denominator by -1:

$$\frac{x^2}{x - 7} + \frac{6x + 7}{7 - x} \cdot \frac{\mathbf{-1}}{\mathbf{-1}} = \frac{x^2}{x - 7} + \frac{-6x - 7}{x - 7}$$

$$= \frac{x^2 - 6x - 7}{x - 7} \qquad \text{Add numerators}$$

$$= \frac{\cancel{(x - 7)}(x + 1)}{\cancel{(x - 7)}} \qquad \text{Factor numerator}$$

$$= x + 1 \qquad \text{Divide out } x - 7$$

For our next example we will look at a problem in which we combine a whole number and a rational expression.

EXAMPLE 9 Subtract $2 - \dfrac{9}{3x + 1}$.

SOLUTION To subtract these two expressions, we think of 2 as a rational expression with a denominator of 1.

$$2 - \frac{9}{3x + 1} = \frac{2}{1} - \frac{9}{3x + 1}$$

The LCD is $3x + 1$. Multiplying the numerator and denominator of the first expression by $3x + 1$ gives us a rational expression equivalent to 2 but with a denominator of $3x + 1$.

$$\frac{2}{1} \cdot \frac{\mathbf{(3x + 1)}}{\mathbf{(3x + 1)}} - \frac{9}{3x + 1} = \frac{6x + 2 - 9}{3x + 1}$$

$$= \frac{6x - 7}{3x + 1}$$

The numerator and denominator of this last expression do not have any factors in common other than 1, so the expression is in lowest terms.

EXAMPLE 10 Write an expression for the sum of a number and twice its reciprocal. Then, simplify that expression.

SOLUTION If x is the number, then its reciprocal is $\frac{1}{x}$. Twice its reciprocal is $\frac{2}{x}$. The sum of the number and twice its reciprocal is

$$x + \frac{2}{x}$$

To combine these two expressions, we think of the first term x as a rational expression with a denominator of 1. The least common denominator is x:

$$x + \frac{2}{x} = \frac{x}{1} + \frac{2}{x}$$

$$= \frac{x}{1} \cdot \frac{\mathbf{x}}{\mathbf{x}} + \frac{2}{x}$$

$$= \frac{x^2 + 2}{x}$$

LINKING OBJECTIVES AND EXAMPLES

Next to each **objective** we have listed the examples that are best described by that objective.

A	1–3
B	4–10

GETTING READY FOR CLASS

After reading through the preceding section, respond in your own words and in complete sentences.

1. Briefly describe how you would add two rational expressions that have the same denominator.
2. Why is factoring important in finding a least common denominator?
3 What is the last step in adding or subtracting two rational expressions?
4 Explain how you would change the fraction $\frac{5}{x - 3}$ to an equivalent fraction with denominator $x^2 - 9$.

Problem Set 6.4

Online support materials can be found at www.thomsonedu.com/login

Combine the following fractions.

1. $\frac{3}{4} + \frac{1}{2}$
2. $\frac{5}{6} + \frac{1}{3}$
3. $\frac{2}{5} - \frac{1}{15}$
4. $\frac{5}{8} - \frac{1}{4}$
5. $\frac{5}{6} + \frac{7}{8}$
6. $\frac{3}{4} + \frac{2}{3}$
7. $\frac{9}{48} - \frac{3}{54}$
8. $\frac{6}{28} - \frac{5}{42}$
9. $\frac{3}{4} - \frac{1}{8} + \frac{2}{3}$
10. $\frac{1}{3} - \frac{5}{6} + \frac{5}{12}$

Combine the following rational expressions. Reduce all answers to lowest terms.

11. $\frac{x}{x+3} + \frac{3}{x+3}$
12. $\frac{5x}{5x+2} + \frac{2}{5x+2}$
13. $\frac{4}{y-4} - \frac{y}{y-4}$
14. $\frac{8}{y+8} + \frac{y}{y+8}$
15. $\frac{x}{x^2-y^2} - \frac{y}{x^2-y^2}$
16. $\frac{x}{x^2-y^2} + \frac{y}{x^2-y^2}$
17. $\frac{2x-3}{x-2} - \frac{x-1}{x-2}$
18. $\frac{2x-4}{x+2} - \frac{x-6}{x+2}$
19. $\frac{1}{a} + \frac{2}{a^2} - \frac{3}{a^3}$
20. $\frac{3}{a} + \frac{2}{a^2} - \frac{1}{a^3}$
21. $\frac{7x-2}{2x+1} - \frac{5x-3}{2x+1}$
22. $\frac{7x-1}{3x+2} - \frac{4x-3}{3x+2}$
23. Work each problem according to the instructions given.
    a. Multiply: $\frac{3}{8} \cdot \frac{1}{6}$
    b. Divide: $\frac{3}{8} \div \frac{1}{6}$
    c. Add: $\frac{3}{8} + \frac{1}{6}$
    d. Multiply: $\frac{x+3}{x-3} \cdot \frac{5x+15}{x^2-9}$
    e. Divide: $\frac{x+3}{x-3} \div \frac{5x+15}{x^2-9}$
    f. Subtract: $\frac{x+3}{x-3} - \frac{5x+15}{x^2-9}$
24. Work each problem according to the instructions given.
    a. Multiply: $\frac{16}{49} \cdot \frac{1}{28}$
    b. Divide: $\frac{16}{49} \div \frac{1}{28}$
    c. Subtract: $\frac{16}{49} - \frac{1}{28}$
    d. Multiply: $\frac{3x-2}{3x+2} \cdot \frac{15x+6}{9x^2-4}$
    e. Divide: $\frac{3x-2}{3x+2} \div \frac{15x+6}{9x^2-4}$
    f. Subtract: $\frac{3x+2}{3x-2} - \frac{15x+6}{9x^2-4}$

Combine the following rational expressions. Reduce all answers to lowest terms.

25. $\frac{2}{t^2} - \frac{3}{2t}$
26. $\frac{5}{3t} - \frac{4}{t^2}$
27. $\frac{3x+1}{2x-6} - \frac{x+2}{x-3}$
28. $\frac{x+1}{x-2} - \frac{4x+7}{5x-10}$
29. $\frac{x+1}{2x-2} - \frac{2}{x^2-1}$
30. $\frac{x+7}{2x+12} + \frac{6}{x^2-36}$
31. $\frac{1}{a-b} - \frac{3ab}{a^3-b^3}$
32. $\frac{1}{a+b} + \frac{3ab}{a^3+b^3}$
33. $\frac{1}{2y-3} - \frac{18y}{8y^3-27}$
34. $\frac{1}{3y-2} - \frac{18y}{27y^3-8}$
35. $\frac{x}{x^2-5x+6} - \frac{3}{3-x}$
36. $\frac{x}{x^2+4x+4} - \frac{2}{2+x}$
37. $\frac{2}{4t-5} + \frac{9}{8t^2-38t+35}$
38. $\frac{3}{2t-5} + \frac{21}{8t^2-14t-15}$
39. $\frac{1}{a^2-5a+6} + \frac{3}{a^2-a-2}$
40. $\frac{-3}{a^2+a-2} + \frac{5}{a^2-a-6}$

41. $\frac{1}{8x^3 - 1} - \frac{1}{4x^2 - 1}$

42. $\frac{1}{27x^3 - 1} - \frac{1}{9x^2 - 1}$

43. $\frac{4}{4x^2 - 9} - \frac{6}{8x^2 - 6x - 9}$

44. $\frac{9}{9x^2 + 6x - 8} - \frac{6}{9x^2 - 4}$

45. $\frac{4a}{a^2 + 6a + 5} - \frac{3a}{a^2 + 5a + 4}$

46. $\frac{3a}{a^2 + 7a + 10} - \frac{2a}{a^2 + 6a + 8}$

47. $\frac{2x - 1}{x^2 + x - 6} - \frac{x + 2}{x^2 + 5x + 6}$

48. $\frac{4x + 1}{x^2 + 5x + 4} - \frac{x + 3}{x^2 + 4x + 3}$

49. $\frac{2x - 8}{3x^2 + 8x + 4} + \frac{x + 3}{3x^2 + 5x + 2}$

50. $\frac{5x + 3}{2x^2 + 5x + 3} - \frac{3x + 9}{2x^2 + 7x + 6}$

51. $\frac{2}{x^2 + 5x + 6} - \frac{4}{x^2 + 4x + 3} + \frac{3}{x^2 + 3x + 2}$

52. $\frac{-5}{x^2 + 3x - 4} + \frac{5}{x^2 + 2x - 3} + \frac{1}{x^2 + 7x + 12}$

53. $\frac{2x + 8}{x^2 + 5x + 6} - \frac{x + 5}{x^2 + 4x + 3} - \frac{x - 1}{x^2 + 3x + 2}$

54. $\frac{2x + 11}{x^2 + 9x + 20} - \frac{x + 1}{x^2 + 7x + 12} - \frac{x + 6}{x^2 + 8x + 15}$

55. $2 + \frac{3}{2x + 1}$

56. $3 - \frac{2}{2x + 3}$

57. $5 + \frac{2}{4 - t}$

58. $7 + \frac{3}{5 - t}$

59. $x - \frac{4}{2x + 3}$

60. $x - \frac{5}{3x + 4} + 1$

61. $\frac{x}{x + 2} + \frac{1}{2x + 4} - \frac{3}{x^2 + 2x}$

62. $\frac{x}{x + 3} + \frac{7}{3x + 9} - \frac{2}{x^2 + 3x}$

63. $\frac{1}{x} + \frac{x}{2x + 4} - \frac{2}{x^2 + 2x}$

64. $\frac{1}{x} + \frac{x}{3x + 9} - \frac{3}{x^2 + 3x}$

▶ 65. Let $f(x) = \frac{2x - 1}{4x - 16}$ and $g(x) = \frac{x - 3}{x - 4}$; find $f(x) - g(x)$.

▶ 66. Let $f(x) = \frac{2}{2x + 4}$ and $g(x) = \frac{x}{x - 4}$; find $f(x) + g(x)$.

▶ 67. Let $f(x) = \frac{2}{x + 4}$ and $g(x) = \frac{x - 1}{x^2 + 3x - 4}$; find $f(x) + g(x)$.

▶ 68. Let $f(t) = \frac{5}{3t - 2}$ and $g(t) = \frac{t - 3}{3t^2 + 7t - 6}$; find $f(t) - g(t)$.

▶ 69. Let $f(x) = \frac{2x}{x^2 - x - 2}$ and $g(x) = \frac{5}{x^2 + x - 6}$; find $f(x) + g(x)$.

▶ 70. Let $f(x) = \frac{7}{x^2 - x - 12}$ and $g(x) = \frac{5}{x^2 + x - 6}$; find $f(x) - g(x)$.

▶ 71. Let $f(x) = \frac{x}{9x^2 - 4}$ and $g(x) = \frac{1}{3x^2 - 4x - 4}$; find $f(x) + g(x)$.

▶ 72. Let $f(x) = \frac{1}{16x^2 - 1}$ and $g(x) = \frac{1}{64x^3 - 1}$; find $f(x) - g(x)$.

▶ 73. Let $f(x) = \frac{3x}{2x^2 - x - 1}$ and $g(x) = \frac{6}{2x^2 - 5x - 3}$; find $f(x) + g(x)$.

▶ 74. Let $f(x) = \frac{5x}{x^2 - 8x - 9}$ and $g(x) = \frac{4x}{x^2 - 10x + 9}$; find $f(x) - g(x)$.

▶ 75. Let $f(x) = \frac{5x}{x^2 - 13x + 36}$ and $g(x) = \frac{3x}{x^2 - 11x + 28}$; find $f(x) - g(x)$.

76. Let $f(x) = \dfrac{x + 4}{x^2 - 6x - 16}$ and $g(x) = \dfrac{x - 2}{x^2 - 11x + 24}$; find $f(x) - g(x)$.

Applying the Concepts

77. Number Problem Write an expression for the sum of a number and 4 times its reciprocal. Then, simplify that expression.

78. Number Problem Write an expression for the sum of the reciprocals of two consecutive integers. Then, simplify that expression.

79. Optometry The formula

$$P = \frac{1}{a} + \frac{1}{b}$$

is used by optometrists to help determine how strong to make the lenses for a pair of eyeglasses. If a is 10 and b is 0.2, find the corresponding value of P.

80. Quadratic Formula Later in the book we will work with the quadratic formula. The derivation of the formula requires that you can add the fractions below. Add the fractions.

$$-\frac{c}{a} + \left(\frac{b}{2a}\right)^2$$

81. Elliptical Orbits Consider two objects, A and B, that move in the same direction along an elliptical path at constant but different velocities.

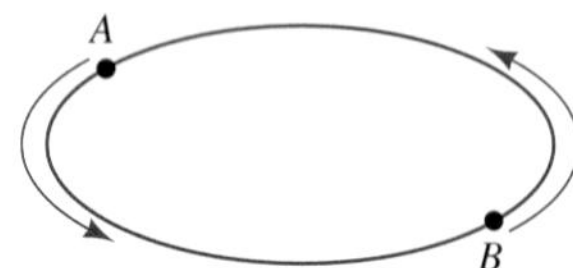

It can be shown that the time, T, it takes for the two objects to meet can be found from the formula

$$\frac{1}{T} = \frac{1}{t_A} - \frac{1}{t_B}$$

where t_A = time required for object A to orbit, and t_B = time required for object B to orbit.

a. If $t_A = 24$ months and $t_B = 30$ months, when will these two objects meet?

b. If $t_A = t_B$ what can one conclude?

82. Average Velocity If a car travels at a constant velocity, v_1, for 10 miles and then at a constant, but different velocity, v_2, for the next 10 miles, it can be shown that the cars average velocity, v_{avg}, over these 20 miles satisfies the equation

$$\frac{2}{v_{avg}} = \frac{1}{v_1} + \frac{1}{v_2}$$

Find the average velocity of a car that travels a constant 45 miles per hour for 10 miles and then increases to a constant 60 miles per hour for the next 10 miles. Round to the nearest tenth.

Maintaining Your Skills

Write each number in scientific notation.

83. 54,000

84. 768,000

85. 0.00034

86. 0.0359

Write each number in expanded form.

87. 6.44×10^3

88. 2.5×10^2

89. 6.44×10^{-3}

90. 2.5×10^{-2}

Simplify each expression as much as possible. Write all answers in scientific notation.

91. $(3 \times 10^8)(4 \times 10^{-5})$

92. $\dfrac{8 \times 10^{-3}}{4 \times 10^{-6}}$

Getting Ready for the Next Section

Divide.

93. $\dfrac{3}{4} \div \dfrac{5}{8}$

94. $\dfrac{2}{3} \div \dfrac{5}{6}$

Multiply.

95. $x\left(1 + \dfrac{2}{x}\right)$

96. $3\left(x + \dfrac{1}{3}\right)$

97. $3x\left(\dfrac{1}{x} - \dfrac{1}{3}\right)$

98. $3x\left(\dfrac{1}{x} + \dfrac{1}{3}\right)$

Factor.

99. $x^2 - 4$

100. $x^2 - x - 6$

Extending the Concepts

Simplify.

101. $\left(1 - \frac{1}{x}\right)\left(1 - \frac{1}{x+1}\right)\left(1 - \frac{1}{x+2}\right)\left(1 - \frac{1}{x+3}\right)$

102. $\left(1 + \frac{1}{x}\right)\left(1 + \frac{1}{x+1}\right)\left(1 + \frac{1}{x+2}\right)\left(1 + \frac{1}{x+3}\right)$

103. $\left(\frac{a^2 - b^2}{u^2 - v^2}\right)\left(\frac{av - au}{b - a}\right) + \left(\frac{a^2 - av}{u + v}\right)\left(\frac{1}{a}\right)$

104. $\left(\frac{6r^2}{r^2 - 1}\right)\left(\frac{r + 1}{3}\right) - \frac{2r^2}{r - 1}$

105. $\frac{18x - 19}{4x^2 + 27x - 7} - \frac{12x - 41}{3x^2 + 17x - 28}$

106. $\frac{42 - 22y}{3y^2 - 13y - 10} - \frac{21 - 13y}{2y^2 - 9y - 5}$

107. $\left(\frac{1}{y^2 - 1} \div \frac{1}{y^2 + 1}\right)\left(\frac{y^3 + 1}{y^4 - 1}\right) + \frac{1}{(y + 1)^2(y - 1)}$

108. $\left(\frac{a^3 - 64}{a^2 - 16} \div \frac{a^2 - 4a + 16}{a^2 - 4} \div \frac{a^2 + 4a + 16}{a^3 + 64}\right) + 4 - a^2$

6.5 Complex Fractions

OBJECTIVES

A Simplify complex fractions.

The quotient of two fractions or two rational expressions is called a *complex fraction.* This section is concerned with the simplification of complex fractions.

EXAMPLE 1

Simplify $\dfrac{\frac{3}{4}}{\frac{5}{8}}$.

SOLUTION There are generally two methods that can be used to simplify complex fractions.

Method 1 We can multiply the numerator and denominator of the complex fraction by the LCD for both of the fractions, which in this case is 8.

$$\frac{\frac{3}{4}}{\frac{5}{8}} = \frac{\frac{3}{4} \cdot \mathbf{8}}{\frac{5}{8} \cdot \mathbf{8}} = \frac{6}{5}$$

Method 2 To divide by $\frac{5}{8}$, we multiply by $\frac{8}{5}$.

$$\frac{\frac{3}{4}}{\frac{5}{8}} = \frac{3}{4} \cdot \frac{8}{5} = \frac{24}{20} = \frac{6}{5}$$

Note

You should become proficient at both methods. Each method can be useful in simplifying complex fractions.

Here are some examples of complex fractions involving rational expressions. Most can be solved using either of the two methods shown in Example 1.

EXAMPLE 2

Simplify $\dfrac{\frac{1}{x} + \frac{1}{y}}{\frac{1}{x} - \frac{1}{y}}$.

SOLUTION This problem is most easily solved using Method 1. We begin by multiplying both the numerator and denominator by the quantity xy, which is the LCD for all the fractions:

$$\frac{\frac{1}{x}+\frac{1}{y}}{\frac{1}{x}-\frac{1}{y}} = \frac{\left(\frac{1}{x}+\frac{1}{y}\right)\cdot \mathbf{xy}}{\left(\frac{1}{x}-\frac{1}{y}\right)\cdot \mathbf{xy}}$$

$$= \frac{\frac{1}{x}(xy)+\frac{1}{y}(xy)}{\frac{1}{x}(xy)-\frac{1}{y}(xy)}$$ **Apply the distributive property to distribute xy over both terms in the numerator and denominator**

$$= \frac{y+x}{y-x}$$

EXAMPLE 3

Simplify $\dfrac{\frac{x-2}{x^2-9}}{\frac{x^2-4}{x+3}}$.

> **Note**
> We could have used method 1 just as easily: we would have multiplied the numerator and denominator of the complex fraction by $(x+3)(x-3)$.

SOLUTION Applying method 2, we have

$$\frac{\frac{x-2}{x^2-9}}{\frac{x^2-4}{x+3}} = \frac{x-2}{x^2-9}\cdot\frac{x+3}{x^2-4}$$

$$= \frac{\cancel{(x-2)}\cancel{(x+3)}}{\cancel{(x+3)}(x-3)(x+2)\cancel{(x-2)}}$$

$$= \frac{1}{(x-3)(x+2)}$$

EXAMPLE 4

Simplify $\dfrac{1-\frac{4}{x^2}}{1-\frac{1}{x}-\frac{6}{x^2}}$.

SOLUTION The simplest way to simplify this complex fraction is to multiply the numerator and denominator by the LCD, x^2:

$$\frac{1-\frac{4}{x^2}}{1-\frac{1}{x}-\frac{6}{x^2}} = \frac{\mathbf{x^2}\left(1-\frac{4}{x^2}\right)}{\mathbf{x^2}\left(1-\frac{1}{x}-\frac{6}{x^2}\right)}$$ **Multiply numerator and denominator by x^2**

$$= \frac{x^2\cdot 1 - x^2\cdot\frac{4}{x^2}}{x^2\cdot 1 - x^2\cdot\frac{1}{x} - x^2\cdot\frac{6}{x^2}}$$ **Distributive property**

$$= \frac{x^2-4}{x^2-x-6}$$ **Simplify**

$$= \frac{(x-2)\cancel{(x+2)}}{(x-3)\cancel{(x+2)}}$$ **Factor**

$$= \frac{x-2}{x-3}$$ **Reduce**

EXAMPLE 5

Simplify $2 - \dfrac{3}{x + \frac{1}{3}}$.

SOLUTION First we simplify the expression that follows the subtraction sign.

$$2 - \frac{3}{x + \frac{1}{3}} = 2 - \frac{\mathbf{3} \cdot 3}{\mathbf{3}\left(x + \frac{1}{3}\right)} = 2 - \frac{9}{3x + 1}$$

Now we subtract by rewriting the first term, 2, with the LCD, $3x + 1$.

$$2 - \frac{9}{3x + 1} = \frac{2}{1} \cdot \frac{\mathbf{3x + 1}}{\mathbf{3x + 1}} - \frac{9}{3x + 1}$$

$$= \frac{6x + 2 - 9}{3x + 1} = \frac{6x - 7}{3x + 1}$$

GETTING READY FOR CLASS

After reading through the preceding section, respond in your own words and in complete sentences.

1. What is a complex fraction?
2. Explain how a least common denominator can be used to simplify a complex fraction.
3. Explain how some complex fractions can be converted to division problems. When is it more efficient to convert a complex fraction to a division problem of rational expressions?
4. Which method of simplifying complex fractions do you prefer? Why?

LINKING OBJECTIVES AND EXAMPLES

Next to each **objective** we have listed the examples that are best described by that objective.

A 1–5

Problem Set 6.5

Online support materials can be found at www.thomsonedu.com/login

Simplify each of the following as much as possible.

1. $\dfrac{\frac{3}{4}}{\frac{2}{3}}$

2. $\dfrac{\frac{5}{9}}{\frac{7}{12}}$

3. $\dfrac{\frac{1}{3} - \frac{1}{4}}{\frac{1}{2} + \frac{1}{8}}$

4. $\dfrac{\frac{1}{6} - \frac{1}{3}}{\frac{1}{4} - \frac{1}{8}}$

5. $\dfrac{3 + \frac{2}{5}}{1 - \frac{3}{7}}$

6. $\dfrac{2 + \frac{5}{6}}{1 - \frac{7}{8}}$

7. $\dfrac{\frac{1}{x}}{1 + \frac{1}{x}}$

8. $\dfrac{1 - \frac{1}{x}}{\frac{1}{x}}$

9. $\dfrac{1 + \frac{1}{a}}{1 - \frac{1}{a}}$

10. $\dfrac{1 - \frac{2}{a}}{1 - \frac{3}{a}}$

11. $\dfrac{\frac{1}{x} - \frac{1}{y}}{\frac{1}{x} + \frac{1}{y}}$

12. $\dfrac{\frac{1}{x} + \frac{2}{y}}{\frac{2}{x} + \frac{1}{y}}$

= Videos available by instructor request

▶ = Online student support materials available at www.thomsonedu.com/login

13. $\dfrac{\dfrac{x-5}{x^2-4}}{\dfrac{x^2-25}{x+2}}$

14. $\dfrac{\dfrac{3x+1}{x^2-49}}{\dfrac{9x^2-1}{x-7}}$

15. $\dfrac{\dfrac{4a}{2a^3+2}}{\dfrac{8a}{4a+4}}$

16. $\dfrac{\dfrac{2a}{3a^3-3}}{\dfrac{4a}{6a-6}}$

17. $\dfrac{1-\dfrac{9}{x^2}}{1-\dfrac{1}{x}-\dfrac{6}{x^2}}$

18. $\dfrac{4-\dfrac{1}{x^2}}{4+\dfrac{4}{x}+\dfrac{1}{x^2}}$

19. $\dfrac{2+\dfrac{5}{a}-\dfrac{3}{a^2}}{2-\dfrac{5}{a}+\dfrac{2}{a^2}}$

20. $\dfrac{3+\dfrac{5}{a}-\dfrac{2}{a^2}}{3-\dfrac{10}{a}+\dfrac{3}{a^2}}$

21. $\dfrac{27-\dfrac{8}{x^3}}{3+\dfrac{1}{x}-\dfrac{2}{x^2}}$

22. $\dfrac{64+\dfrac{1}{x^3}}{4-\dfrac{11}{x}-\dfrac{3}{x^2}}$

23. $\dfrac{1+\dfrac{2}{x}+\dfrac{4}{x^2}+\dfrac{8}{x^3}}{1-\dfrac{16}{x^4}}$

24. $\dfrac{27+\dfrac{9}{x}+\dfrac{3}{x^2}+\dfrac{1}{x^3}}{81-\dfrac{1}{x^4}}$

25. $\dfrac{2+\dfrac{3}{x}-\dfrac{18}{x^2}-\dfrac{27}{x^3}}{2+\dfrac{9}{x}+\dfrac{9}{x^2}}$

26. $\dfrac{3+\dfrac{5}{x}-\dfrac{12}{x^2}-\dfrac{20}{x^3}}{3+\dfrac{11}{x}+\dfrac{10}{x^2}}$

27. $\dfrac{1+\dfrac{1}{x+3}}{1-\dfrac{1}{x+3}}$

28. $\dfrac{1+\dfrac{1}{x-2}}{1-\dfrac{1}{x-2}}$

29. $\dfrac{1-\dfrac{1}{a+1}}{1+\dfrac{1}{a-1}}$

30. $\dfrac{\dfrac{1}{a-1}+1}{\dfrac{1}{a+1}-1}$

31. $\dfrac{\dfrac{1}{x+3}+\dfrac{1}{x-3}}{\dfrac{1}{x+3}-\dfrac{1}{x-3}}$

32. $\dfrac{\dfrac{1}{x+a}+\dfrac{1}{x-a}}{\dfrac{1}{x+a}-\dfrac{1}{x-a}}$

33. $\dfrac{\dfrac{y+1}{y-1}+\dfrac{y-1}{y+1}}{\dfrac{y+1}{y-1}-\dfrac{y-1}{y+1}}$

34. $\dfrac{\dfrac{y-1}{y+1}-\dfrac{y+1}{y-1}}{\dfrac{y-1}{y+1}+\dfrac{y+1}{y-1}}$

35. $1-\dfrac{x}{1-\dfrac{1}{x}}$

36 $x-\dfrac{1}{x-\dfrac{1}{2}}$

37. $1+\dfrac{1}{1+\dfrac{1}{1+1}}$

38. $1-\dfrac{1}{1-\dfrac{1}{1-\frac{1}{2}}}$

39. $\dfrac{1-\dfrac{1}{x+\dfrac{1}{2}}}{1+\dfrac{1}{x+\dfrac{1}{2}}}$

40. $\dfrac{2+\dfrac{1}{x-\dfrac{1}{3}}}{2-\dfrac{1}{x-\dfrac{1}{3}}}$

41. $\dfrac{\dfrac{1}{x+h}-\dfrac{1}{x}}{h}$

42. $\dfrac{\dfrac{1}{(x+h)^2}-\dfrac{1}{x^2}}{h}$

43. $\dfrac{\dfrac{3}{ab}+\dfrac{4}{bc}-\dfrac{2}{ac}}{\dfrac{5}{abc}}$

44. $\dfrac{\dfrac{x}{yz}-\dfrac{y}{xz}+\dfrac{z}{xy}}{\dfrac{1}{x^2y^2}-\dfrac{1}{x^2z^2}+\dfrac{1}{y^2z^2}}$

45. $\dfrac{\dfrac{t^2-2t-8}{t^2+7t+6}}{\dfrac{t^2-t-6}{t^2+2t+1}}$

▶ 46. $\dfrac{\dfrac{y^2 - 5y - 14}{y^2 + 3y - 10}}{\dfrac{y^2 - 8y + 7}{y^2 + 6y + 5}}$

▶ 47. $\dfrac{5 + \dfrac{4}{b - 1}}{\dfrac{7}{b + 5} - \dfrac{3}{b - 1}}$

▶ 48. $\dfrac{\dfrac{6}{x + 5} - 7}{\dfrac{8}{x + 5} - \dfrac{9}{x + 3}}$

▶ 49. $\dfrac{\dfrac{3}{x^2 - x - 6}}{\dfrac{2}{x + 2} - \dfrac{4}{x - 3}}$

▶ 50. $\dfrac{\dfrac{9}{a - 7} + \dfrac{8}{2a + 3}}{\dfrac{10}{2a^2 - 11a - 21}}$

▶ 51. $\dfrac{\dfrac{1}{m - 4} + \dfrac{1}{m - 5}}{\dfrac{1}{m^2 - 9m + 20}}$

▶ 52. $\dfrac{\dfrac{1}{k^2 - 7k + 12}}{\dfrac{1}{k - 3} + \dfrac{1}{k - 4}}$

Applying the Concepts

53. **Difference Quotient** For each rational function below, find the difference quotient

$$\frac{f(x) - f(a)}{x - a}$$

a. $f(x) = \dfrac{4}{x}$

b. $f(x) = \dfrac{1}{x + 1}$

c. $f(x) = \dfrac{1}{x^2}$

54. **Difference Quotient** For each rational function below, find the difference quotient

$$\frac{f(x) - f(2)}{x - 2}$$

a. $f(x) = \dfrac{4}{x}$

b. $f(x) = \dfrac{1}{x + 1}$

c. $f(x) = \dfrac{1}{x^2}$

55. **Optics** The formula $f = \dfrac{ab}{a + b}$ is used in optics to find the focal length of a lens. Show that the formula $f = (a^{-1} + b^{-1})^{-1}$ is equivalent to the preceding formula by rewriting it without the negative exponents and then simplifying the results.

56. **Optics** Show that the expression $(a^{-1} - b^{-1})^{-1}$ can be simplified to $\dfrac{ab}{b - a}$ by first writing it without the negative exponents and then simplifying the result.

57. **Doppler Effect** The change in the pitch of a sound (such as a train whistle) as an object passes is called the Doppler effect, named after C. J. Doppler (1803–1853). A person will *hear* a sound with a frequency, h, according to the formula

$$h = \frac{f}{1 + \dfrac{v}{s}}$$

where f is the actual frequency of the sound being produced, s is the speed of sound (about 740 miles per hour), and v is the velocity of the moving object.

a. Examine this fraction, and then explain why h and f approach the same value as v becomes smaller and smaller.

b. Solve this formula for v.

58. **Work Problem** A water storage tank has two drains. It can be shown that the time it takes to empty the tank if both drains are open is given by the formula

$$\frac{1}{\dfrac{1}{a} + \dfrac{1}{b}}$$

where a = time it takes for the first drain to empty the tank, and b = time for the second drain to empty the tank.

a. Simplify this complex fraction.

b. Find the amount of time needed to empty the tank using both drains if, used alone, the first drain empties the tank in 4 hours and the second drain can empty the tank in 3 hours.

Maintaining Your Skills

Solve each equation.

59. $3x + 60 = 15$
60. $3x - 18 = 4$
61. $3(y - 3) = 2(y - 2)$
62. $5(y + 2) = 4(y + 1)$
63. $10 - 2(x + 3) = x + 1$
64. $15 - 3(x - 1) = x - 2$
65. $x^2 - x - 12 = 0$
66. $3x^2 + x - 10 = 0$
67. $(x + 1)(x - 6) = -12$
68. $(x + 1)(x - 4) = -6$

Getting Ready for the Next Section

Multiply.

69. $x(y - 2)$
70. $x(y - 1)$
71. $6\left(\frac{x}{2} - 3\right)$
72. $6\left(\frac{x}{3} + 1\right)$
73. $xab \cdot \frac{1}{x}$
74. $xab\left(\frac{1}{b} + \frac{1}{a}\right)$

Factor.

75. $y^2 - 25$
76. $x^2 - 3x + 2$
77. $xa + xb$
78. $xy - y$

Solve.

79. $5x - 4 = 6$
80. $y^2 + y - 20 = 2y$

Extending the Concepts

Simplify each expression.

81. $\dfrac{\left(\frac{1}{3}\right) - \left(\frac{1}{3}\right)^2}{1 - \frac{1}{3}}$
82. $\dfrac{\left(\frac{1}{2}\right) - \left(\frac{1}{2}\right)^2}{1 - \frac{1}{2}}$
83. $\dfrac{\left(\frac{1}{9}\right) - \frac{1}{9}\left(\frac{1}{3}\right)^4}{1 - \frac{1}{3}}$
84. $\dfrac{\left(\frac{1}{6}\right) - \frac{1}{6}\left(\frac{1}{2}\right)^4}{1 - \frac{1}{2}}$
85. $\dfrac{1 + \dfrac{1}{1 - \frac{a}{b}}}{1 - \dfrac{3}{1 - \frac{a}{b}}}$
86. $\dfrac{1 - \dfrac{1}{\frac{a}{b} + 2}}{1 + \dfrac{3}{\frac{a}{2b} + 1}}$
87. $\dfrac{a^{-1} + b^{-1}}{(ab)^{-1}}$
88. $\dfrac{(r^{-1} - s^{-1})^{-1}}{(rs)^{-2}}$
89. $\dfrac{(q^{-2} - t^{-2})^{-1}}{(t^{-1} - q^{-1})^{-1}}$
90. $\dfrac{(q^{-2} + t^{-2})^{-1}}{(t^{-1} + q^{-1})^{-1}}$

6.6 Equations Involving Rational Expressions

OBJECTIVES

A Solve equations containing rational expressions.

B Solve formulas containing rational expressions.

The first step in solving an equation that contains one or more rational expressions is to find the LCD for all denominators in the equation. We then multiply both sides of the equation by the LCD to clear the equation of all fractions—that is, after we have multiplied through by the LCD, each term in the resulting equation will have a denominator of 1.

EXAMPLE 1 Solve $\frac{x}{2} - 3 = \frac{2}{3}$.

SOLUTION The LCD for 2 and 3 is 6. Multiplying both sides by 6, we have

$$\mathbf{6}\left(\frac{x}{2} - 3\right) = \mathbf{6}\left(\frac{2}{3}\right)$$
$$6\left(\frac{x}{2}\right) - 6(3) = 6\left(\frac{2}{3}\right)$$
$$3x - 18 = 4$$
$$3x = 22$$
$$x = \frac{22}{3}$$

Multiplying both sides of an equation by the LCD clears the equation of fractions because the LCD has the property that all the denominators divide it evenly.

EXAMPLE 2 Solve $\frac{6}{a-4} = \frac{3}{8}$.

SOLUTION The LCD for $a - 4$ and 8 is $8(a - 4)$. Multiplying both sides by this quantity yields

$$\mathbf{8(a-4)} \cdot \frac{6}{a-4} = \mathbf{8(a-4)} \cdot \frac{3}{8}$$

$$48 = (a-4) \cdot 3$$

$$48 = 3a - 12$$

$$60 = 3a$$

$$20 = a$$

The solution set is {20}, which checks in the original equation. 

When we multiply both sides of an equation by an expression containing the variable, we must be sure to check our solutions. The multiplication property of equality does not allow multiplication by 0. If the expression we multiply by contains the variable, then it has the possibility of being 0. In the last example we multiplied both sides by $8(a - 4)$. This gives a restriction $a \neq 4$ for any solution we come up with.

EXAMPLE 3 Solve $\frac{x}{x-2} + \frac{2}{3} = \frac{2}{x-2}$.

SOLUTION The LCD is $3(x - 2)$. We are assuming $x \neq 2$ when we multiply both sides of the equation by $3(x - 2)$:

$$\mathbf{3(x-2)} \cdot \left[\frac{x}{x-2} + \frac{2}{3}\right] = \mathbf{3(x-2)} \cdot \frac{2}{x-2}$$

$$3x + (x-2) \cdot 2 = 3 \cdot 2$$

$$3x + 2x - 4 = 6$$

$$5x - 4 = 6$$

$$5x = 10$$

$$x = 2$$

The only possible solution is $x = 2$. Checking this value back in the original equation gives

$$\frac{2}{2-2} + \frac{2}{3} \stackrel{?}{=} \frac{2}{2-2}$$

$$\frac{2}{0} + \frac{2}{3} \stackrel{?}{=} \frac{2}{0}$$

The first and last terms are undefined. The proposed solution, $x = 2$, does not check in the original equation. The solution set is the empty set. There is no solution to the original equation.

Note
In the process of solving the equation, we multiplied both sides by $3(x - 2)$, solved for x, and got $x = 2$ for our solution. But when x is 2, the quantity $3(x - 2) = 3(2 - 2) = 3(0) = 0$, which means we multiplied both sides of our equation by 0, which is not allowed under the multiplication property of equality.

When the proposed solution to an equation is not actually a solution, it is called an *extraneous* solution. In the last example, $x = 2$ is an extraneous solution.

EXAMPLE 4 Solve $\frac{5}{x^2 - 3x + 2} - \frac{1}{x - 2} = \frac{1}{3x - 3}$.

SOLUTION Writing the equation again with the denominators in factored form, we have

$$\frac{5}{(x-2)(x-1)} - \frac{1}{x-2} = \frac{1}{3(x-1)}$$

The LCD is $3(x-2)(x-1)$. Multiplying through by the LCD, we have

$$\mathbf{3(x-2)(x-1)} \cdot \frac{5}{(x-2)(x-1)} - \mathbf{3(x-2)(x-1)} \cdot \frac{1}{(x-2)} = \mathbf{3(x-2)(x-1)} \cdot \frac{1}{3(x-1)}$$

$$\begin{aligned} 3 \cdot 5 - 3(x-1) \cdot 1 &= (x-2) \cdot 1 \\ 15 - 3x + 3 &= x - 2 \\ -3x + 18 &= x - 2 \\ -4x + 18 &= -2 \\ -4x &= -20 \\ x &= 5 \end{aligned}$$

Checking the proposed solution $x = 5$ in the original equation yields a true statement. Try it and see.

Note
We can check the proposed solution in any of the equations obtained before multiplying through by the LCD. We cannot check the proposed solution in an equation obtained *after* multiplying through by the LCD since, if we have multiplied by 0, the resulting equations will not be equivalent to the original one.

EXAMPLE 5 Solve $3 + \frac{1}{x} = \frac{10}{x^2}$.

SOLUTION To clear the equation of denominators, we multiply both sides by x^2:

$$\begin{aligned} \mathbf{x^2}\left(3 + \frac{1}{x}\right) &= \mathbf{x^2}\left(\frac{10}{x^2}\right) \\ 3(x^2) + \left(\frac{1}{x}\right)(x^2) &= \left(\frac{10}{x^2}\right)(x^2) \\ 3x^2 + x &= 10 \end{aligned}$$

Rewrite in standard form, and solve:

$$3x^2 + x - 10 = 0$$

$$(3x - 5)(x + 2) = 0$$

$$3x - 5 = 0 \quad \text{or} \quad x + 2 = 0$$

$$x = \frac{5}{3} \quad \text{or} \quad x = -2$$

The solution set is $\{-2, \frac{5}{3}\}$. Both solutions check in the original equation. Remember: We have to check *all solutions* any time we multiply both sides of the equation by an expression that contains the variable, just to be sure we haven't multiplied by 0.

EXAMPLE 6

Solve $\dfrac{y-4}{y^2-5y} = \dfrac{2}{y^2-25}$.

SOLUTION Factoring each denominator, we find the LCD is $y(y-5)(y+5)$. Multiplying each side of the equation by the LCD clears the equation of denominators and leads us to our possible solutions:

$$\mathbf{y(y-5)(y+5)} \cdot \frac{y-4}{y(y-5)} = \frac{2}{(y-5)(y+5)} \cdot \mathbf{y(y-5)(y+5)}$$

$$(y+5)(y-4) = 2y$$

$$y^2 + y - 20 = 2y \quad \textbf{Multiply out the left side}$$

$$y^2 - y - 20 = 0 \quad \textbf{Add } -2y \textbf{ to each side}$$

$$(y-5)(y+4) = 0$$

$$y - 5 = 0 \quad \text{or} \quad y + 4 = 0$$

$$y = 5 \quad \text{or} \quad y = -4$$

The two possible solutions are 5 and -4. If we substitute -4 for y in the original equation, we find that it leads to a true statement. It is, therefore, a solution. However, if we substitute 5 for y in the original equation, we find that both sides of the equation are undefined. The only solution to our original equation is $y = -4$. The other possible solution $y = 5$ is extraneous.

EXAMPLE 7

Solve for y: $x = \dfrac{y-4}{y-2}$.

SOLUTION To solve for y, we first multiply each side by $y - 2$ to obtain

$$x(y-2) = y - 4$$

$$xy - 2x = y - 4 \quad \textbf{Distributive property}$$

$$xy - y = 2x - 4 \quad \textbf{Collect all terms containing } y \textbf{ on the left side}$$

$$y(x-1) = 2x - 4 \quad \textbf{Factor } y \textbf{ from each term on the left side}$$

$$y = \frac{2x-4}{x-1} \quad \textbf{Divide each side by } x - 1$$

EXAMPLE 8

Solve the formula $\frac{1}{x} = \frac{1}{b} + \frac{1}{a}$ for x.

SOLUTION We begin by multiplying both sides by the least common denominator xab. As you can see from our previous examples, multiplying both sides of an equation by the LCD is equivalent to multiplying each term of both sides by the LCD:

$$\mathbf{xab} \cdot \frac{1}{x} = \frac{1}{b} \cdot \mathbf{xab} + \frac{1}{a} \cdot \mathbf{xab}$$

$$ab = xa + xb$$

$$ab = (a+b)x \quad \textbf{Factor } x \textbf{ from the right side}$$

$$\frac{ab}{a+b} = x$$

We know we are finished because the variable we were solving for is alone on one side of the equation and does not appear on the other side.

LINKING OBJECTIVES AND EXAMPLES

Next to each **objective** we have listed the examples that are best described by that objective.

A	1–6
B	7, 8

GETTING READY FOR CLASS

After reading through the preceding section, respond in your own words and in complete sentences.

1. Explain how a least common denominator can be used to simplify an equation.
2. What is an extraneous solution?
3. Is it possible for an equation containing rational expressions to have no solutions?
4. What is the last step in solving an equation that contains rational expressions?

Problem Set 6.6

Online support materials can be found at www.thomsonedu.com/login

Solve each of the following equations.

1. $\frac{x}{5} + 4 = \frac{5}{3}$

2. $\frac{x}{5} = \frac{x}{2} - 9$

3. $\frac{a}{3} + 2 = \frac{4}{5}$

4. $\frac{a}{4} + \frac{1}{2} = \frac{2}{3}$

5. $\frac{y}{2} + \frac{y}{4} + \frac{y}{6} = 3$

6. $\frac{y}{3} - \frac{y}{6} + \frac{y}{2} = 1$

7. $\frac{5}{2x} = \frac{1}{x} + \frac{3}{4}$

8. $\frac{1}{2a} = \frac{2}{a} - \frac{3}{8}$

9. $\frac{1}{x} = \frac{1}{3} - \frac{2}{3x}$

10. $\frac{5}{2x} = \frac{2}{x} - \frac{1}{12}$

11. $\frac{2x}{x-3} + 2 = \frac{2}{x-3}$

12. $\frac{2}{x+5} = \frac{2}{5} - \frac{x}{x+5}$

13. $1 - \frac{1}{x} = \frac{12}{x^2}$

14. $2 + \frac{5}{x} = \frac{3}{x^2}$

15. $y - \frac{4}{3y} = -\frac{1}{3}$

16. $\frac{y}{2} - \frac{4}{y} = -\frac{7}{2}$

Let $f(x) = \frac{1}{x-3}$ and $g(x) = \frac{1}{x+3}$, and find x if

▸ **17.** $f(x) + g(x) = \frac{5}{8}$

▸ **18.** $f(x) - g(x) = \frac{2}{9}$

▸ **19.** $\frac{f(x)}{g(x)} = 5$

▸ **20.** $\frac{g(x)}{f(x)} = 5$

▸ **21.** $f(x) = g(x)$

▸ **22.** $f(x) = -g(x)$

Let $f(x) = \frac{4}{x+2}$ and $g(x) = \frac{4}{x-2}$, and find x if

▸ **23.** $f(x) + g(x) = \frac{24}{5}$

▸ **24.** $f(x) - g(x) = -\frac{4}{3}$

▸ **25.** $\frac{f(x)}{g(x)} = -5$

▸ **26.** $\frac{g(x)}{f(x)} = -7$

▸ **27.** $f(x) = g(x)$

▸ **28.** $f(x) = -g(x)$

29. Solve each equation.

a. $6x - 2 = 0$

b. $\frac{6}{x} - 2 = 0$

c. $\frac{x}{6} - 2 = -\frac{1}{2}$

d. $\frac{6}{x} - 2 = -\frac{1}{2}$

e. $\frac{6}{x^2} + 6 = \frac{20}{x}$

30. Solve each equation.

a. $5x - 2 = 0$

b. $5 - \frac{2}{x} = 0$

c. $\frac{x}{2} - 5 = -\frac{3}{4}$

d. $\frac{2}{x} - 5 = -\frac{3}{4}$

e. $-\frac{3}{x} + \frac{2}{x^2} = 5$

= Videos available by instructor request

▸ = Online student support materials available at www.thomsonedu.com/login

31. Work each problem according to the instructions given.

a. Divide: $\frac{6}{x^2 - 2x - 8} \div \frac{x+3}{x+2}$

b. Add: $\frac{6}{x^2 - 2x - 8} + \frac{x+3}{x+2}$

c. Solve: $\frac{6}{x^2 - 2x - 8} + \frac{x+3}{x+2} = 2$

32. Work each problem according to the instructions given.

a. Divide: $\frac{-10}{x^2 - 25} \div \frac{x-4}{x-5}$

b. Add: $\frac{-10}{x^2 - 25} + \frac{x-4}{x-5}$

c. Solve: $\frac{-10}{x^2 - 25} + \frac{x-4}{x-5} = \frac{4}{5}$

Solve each equation.

33. $\frac{x+2}{x+1} = \frac{1}{x+1} + 2$

34. $\frac{x+6}{x+3} = \frac{3}{x+3} + 2$

35. $\frac{3}{a-2} = \frac{2}{a-3}$

36. $\frac{5}{a+1} = \frac{4}{a+2}$

37. $6 - \frac{5}{x^2} = \frac{7}{x}$

38. $10 - \frac{3}{x^2} = -\frac{1}{x}$

39. $\frac{1}{x-1} - \frac{1}{x+1} = \frac{3x}{x^2 - 1}$

40. $\frac{5}{x-1} + \frac{2}{x-1} = \frac{4}{x+1}$

41. $\frac{2}{x-3} + \frac{x}{x^2 - 9} = \frac{4}{x+3}$

42. $\frac{2}{x+5} + \frac{3}{x+4} = \frac{2x}{x^2 + 9x + 20}$

43. $\frac{3}{2} - \frac{1}{x-4} = \frac{-2}{2x-8}$

44. $\frac{2}{x} - \frac{1}{x+1} = \frac{-2}{5x+5}$

45. $\frac{t-4}{t^2 - 3t} = \frac{-2}{t^2 - 9}$

46. $\frac{t+3}{t^2 - 2t} = \frac{10}{t^2 - 4}$

47. $\frac{3}{y-4} - \frac{2}{y+1} = \frac{5}{y^2 - 3y - 4}$

48. $\frac{1}{y+2} - \frac{2}{y-3} = \frac{-2y}{y^2 - y - 6}$

49. $\frac{2}{1+a} = \frac{3}{1-a} + \frac{5}{a}$

50. $\frac{1}{a+3} - \frac{a}{a^2 - 9} = \frac{2}{3-a}$

51. $\frac{3}{2x-6} - \frac{x+1}{4x-12} = 4$

52. $\frac{2x-3}{5x+10} + \frac{3x-2}{4x+8} = 1$

53. $\frac{y+2}{y^2 - y} - \frac{6}{y^2 - 1} = 0$

54. $\frac{y+3}{y^2 - y} - \frac{8}{y^2 - 1} = 0$

55. $\frac{4}{2x-6} - \frac{12}{4x+12} = \frac{12}{x^2 - 9}$

56. $\frac{1}{x+2} + \frac{1}{x-2} = \frac{4}{x^2 - 4}$

57. $\frac{2}{y^2 - 7y + 12} - \frac{1}{y^2 - 9} = \frac{4}{y^2 - y - 12}$

58. $\frac{1}{y^2 + 5y + 4} + \frac{3}{y^2 - 1} = \frac{-1}{y^2 + 3y - 4}$

59. Solve the equation $6x^{-1} + 4 = 7$ by multiplying both sides by x. (Remember, $x^{-1} \cdot x = x^{-1} \cdot x^1 = x^0 = 1$.)

60. Solve the equation $3x^{-1} - 5 = 2x^{-1} - 3$ by multiplying both sides by x.

61. Solve the equation $1 + 5x^{-2} = 6x^{-1}$ by multiplying both sides by x^2.

62. Solve the equation $1 + 3x^{-2} = 4x^{-1}$ by multiplying both sides by x^2.

63. Solve the formula $\frac{1}{x} = \frac{1}{b} - \frac{1}{a}$ for x.

64. Solve $\frac{1}{x} = \frac{1}{a} - \frac{1}{b}$ for x.

65. Solve for R in the formula $\frac{1}{R} = \frac{1}{R_1} + \frac{1}{R_2}$.

66. Solve for R in the formula $\frac{1}{R} = \frac{1}{R_1} + \frac{1}{R_2} + \frac{1}{R_3}$.

Solve for y.

67. $x = \dfrac{y-3}{y-1}$

68. $x = \dfrac{y-2}{y-3}$

69. $x = \dfrac{2y+1}{3y+1}$

70. $x = \dfrac{3y+2}{5y+1}$

Applying the Concepts

71. **An Identity** An identity is an equation that is true for any value of the variable for which the expression is defined. Verify the following expression is an identity by simplifying the left side of the expression.

$$\frac{2}{x-y} - \frac{1}{y-x} = \frac{3}{x-y}$$

72. **Harmonic Mean** A number, h, is the harmonic mean of two numbers, n_1 and n_2, if $1/h$ is the mean (average) of $1/n_1$ and $1/n_2$.
    a. Write an equation relating the harmonic mean, h, to two numbers, n_1 and n_2 then solve the equation for h.
    b. Find the harmonic mean of 3 and 5.

73. **Kayak Race** In a kayak race, the participants must paddle a kayak 450 meters down a river and then return 450 meters up the river to the starting point (Figure 1). Susan has correctly deduced that the total time t (in seconds) depends on the speed c (in meters per second) of the water according to the following expression:

$$t = \frac{450}{v+c} + \frac{450}{v-c}$$

where v is the speed of the kayak relative to the water (the speed of the kayak in still water).

Fill in the following table. Round to the nearest whole number, if necessary.

Time t (sec)	Speed of Kayak Relative to the Water v (m/sec)	Current of the River c (m/sec)
240		1
300		2
	4	3
	3	1
540	3	
	3	3

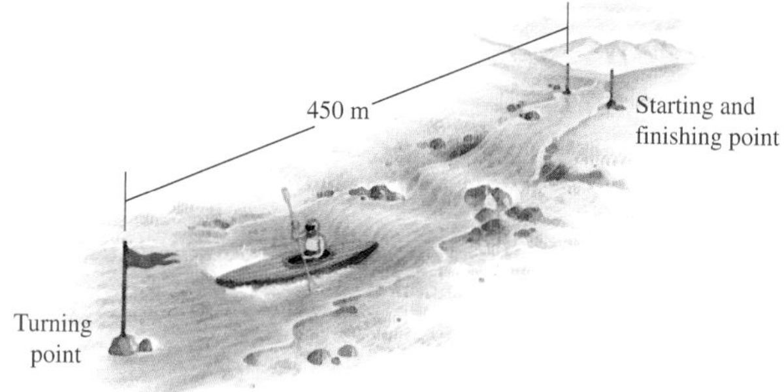

FIGURE 1

74. **Geometry** From plane geometry and the principle of similar triangles, the relationship between y_1, y_2, and h shown in Figure 2 can be expressed as

$$\frac{1}{h} = \frac{1}{y_1} + \frac{1}{y_2}$$

Two poles are 12 feet high and 8 feet high. If a wire is attached to the top of each one and stretched to the bottom of the other, what is the height above the ground at which the two wires will meet?

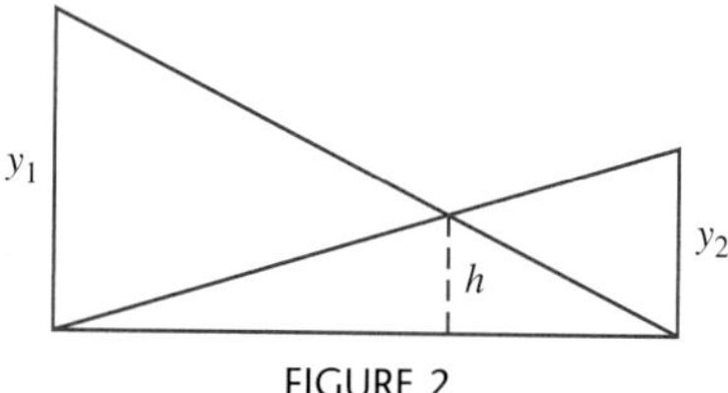

FIGURE 2

Maintaining Your Skills

75. **Number Problem** Twice the sum of a number and 3 is 16. Find the number.

76. **Number Problem** The sum of two consecutive odd integers is 48. Find the two integers.

77. **Geometry** The length of a rectangle is 3 less than twice the width. The perimeter is 42 meters. Find the length and width.

78. **Geometry** The smaller angle in a triangle is one fourth as large as the largest angle. The third angle is 9 degrees more than the smallest angle. Find the measure of all three angles.

79. **Consecutive Integers** The sum of the squares of two consecutive integers is 61. Find the integers.

80. **Consecutive Integers** The square of the sum of two consecutive integers is 121. Find the two integers.

81. **Geometry** The lengths of the sides of a right triangle are given by three consecutive integers. Find the lengths of the three sides.

82. **Geometry** The longest side of a right triangle is 8 inches more than the shortest side. The other side is 7 inches more than the shortest side. Find the lengths of the three sides.

Getting Ready for the Next Section

Multiply.

83. $39.3 \cdot 60$

84. $1{,}100 \cdot 60 \cdot 60$

Divide. Round to the nearest tenth, if necessary.

85. $65{,}000 \div 5{,}280$

86. $3{,}960{,}000 \div 5{,}280$

Multiply.

87. $2x\left(\frac{1}{x} + \frac{1}{2x}\right)$

88. $3x\left(\frac{1}{x} + \frac{1}{3x}\right)$

Solve.

89. $12(x + 3) + 12(x - 3) = 3(x^2 - 9)$

90. $40 + 2x = 60 - 3x$

91. $\frac{1}{10} - \frac{1}{12} = \frac{1}{x}$

92. $\frac{1}{x} + \frac{1}{2x} = 2$

Extending the Concepts

Solve each equation.

93. $\frac{12}{x} + \frac{8}{x^2} - \frac{75}{x^3} - \frac{50}{x^4} = 0$

94. $\frac{45}{x} + \frac{18}{x^2} - \frac{80}{x^3} - \frac{32}{x^4} = 0$

95. $\frac{1}{x^3} - \frac{1}{3x^2} - \frac{1}{4x} + \frac{1}{12} = 0$

96. $\frac{1}{x^3} - \frac{1}{2x^2} - \frac{1}{9x} + \frac{1}{18} = 0$

97. Solve for x. $\frac{2}{x} + \frac{4}{x + a} = \frac{-6}{a - x}$

98. Solve for x. $\frac{1}{b - x} - \frac{1}{x} = \frac{-2}{b + x}$

99. Solve for v. $\frac{s - vt}{t^2} = -16$

100. Solve for r. $A = P\left(1 + \frac{r}{n}\right)$

101. Solve for f. $\frac{1}{p} = \frac{1}{f} + \frac{1}{g}$

102. Solve for p. $h = \frac{v^2}{2g} + \frac{p}{c}$

6.7 Applications

OBJECTIVES

A Solve application problems using equations containing rational expressions.

B Solve conversion problems using unit analysis.

We begin this section with some application problems, the solutions to which involve equations that contain rational expressions. As you will see, the solutions to the examples show only the essential steps from our Blueprint for Problem Solving. Recall that step 1 was done mentally; we read the problem and mentally list the items that are known and the items that are unknown. This is an essential part of problem solving. Now that you have had experience with application problems, however, you are doing step 1 automatically.

Also in this section we will look at a method of solving conversion problems that is called *unit analysis.* With unit analysis, we can convert expressions with units of feet per minute to equivalent expressions in miles per hour. This method of converting between different units of measure is used often in chemistry, physics, and engineering classes.

EXAMPLE 1 One number is twice another. The sum of their reciprocals is 2. Find the numbers.

SOLUTION Let x = the smaller number. The larger number is $2x$. Their reciprocals are $\frac{1}{x}$ and $\frac{1}{2x}$. The equation that describes the situation is

$$\frac{1}{x} + \frac{1}{2x} = 2$$

Multiplying both sides by the LCD $2x$, we have

$$\mathbf{2x} \cdot \frac{1}{x} + \mathbf{2x} \cdot \frac{1}{2x} = \mathbf{2x}(2)$$

$$2 + 1 = 4x$$

$$3 = 4x$$

$$x = \frac{3}{4}$$

The smaller number is $\frac{3}{4}$. The larger is $2\left(\frac{3}{4}\right) = \frac{6}{4} = \frac{3}{2}$. Adding their reciprocals, we have

$$\frac{4}{3} + \frac{2}{3} = \frac{6}{3} = 2$$

The sum of the reciprocals of $\frac{3}{4}$ and $\frac{3}{2}$ is 2.

EXAMPLE 2 Two families from the same neighborhood plan a ski trip together. The first family makes the 455-mile trip at a speed 5 miles per hour faster than the second family. The second family takes a half-hour longer to make the trip. What are the speeds of the two families?

SOLUTION The following table will be helpful in finding the equation necessary to solve this problem.

	d(distance)	r(rate)	t(time)
First Family			
Second Family			

If we let x be the speed of the second family, then the speed of the first family will be $x + 5$. Both families travel the same distance of 455 miles. Putting this information into the table we have

	d(distance)	r(rate)	t(time)
First Family	455	$x + 5$	
Second Family	455	x	

To fill in the last two spaces in the table, we use the relationship $d = r \cdot t$. Since the last column of the table is the time, we solve the equation $d = r \cdot t$ for t and get

$$t = \frac{d}{r}$$

Taking the distance and dividing by the rate (speed) for each family, we complete the table.

	d(distance)	*r*(rate)	*t*(time)
First Family	455	$x + 5$	$\frac{455}{x+5}$
Second Family	455	x	$\frac{455}{x}$

Reading the problem again, we find that the time for the second family is longer than the time for the first family by one-half hour. In other words, the time for the second family can be found by adding one-half hour to the time for the first family, or

$$\frac{455}{x+5} + \frac{1}{2} = \frac{455}{x}$$

Multiplying both sides by the LCD of $2x(x + 5)$ gives

$$2x \cdot (455) + x(x + 5) \cdot 1 = 455 \cdot 2(x + 5)$$
$$910x + x^2 + 5x = 910x + 4{,}550$$
$$x^2 + 5x - 4{,}550 = 0$$
$$(x + 70)(x - 65) = 0$$
$$x = -70 \quad \text{or} \quad x = 65$$

Since we cannot have a negative speed, the only solution is $x = 65$. Then

$$x + 5 = 65 + 5 = 70$$

The speed of the first family is 70 miles per hour, and the speed of the second family is 65 miles per hour.

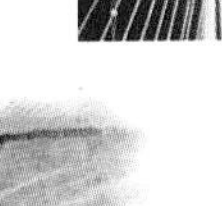

EXAMPLE 3 The speed of a boat in still water is 20 miles per hour. It takes the same amount of time for the boat to travel 3 miles downstream (with the current) as it does to travel 2 miles upstream (against the current). Find the speed of the current.

SOLUTION The following table will be helpful in finding the equation necessary to solve this problem.

	d (distance)	*r* (rate)	*t* (time)
Upstream			
Downstream			

If we let x = the speed of the current, the speed (rate) of the boat upstream is $(20 - x)$ since it is traveling against the current. The rate downstream is $(20 + x)$ since the boat is then traveling with the current. The distance traveled upstream is 2 miles, and the distance traveled downstream is 3 miles. Putting the information given here into the table, we have

	d	*r*	*t*
Upstream	2	$20 - x$	
Downstream	3	$20 + x$	

To fill in the last two spaces in the table we must use the relationship $d = r \cdot t$. Since we know the spaces to be filled in are in the time column, we solve the equation $d = r \cdot t$ for t and get

$$t = \frac{d}{r}$$

The completed table then is

	d	r	t
Upstream	2	$20 - x$	$\frac{2}{20 - x}$
Downstream	3	$20 + x$	$\frac{3}{20 + x}$

Reading the problem again, we find that the time moving upstream is equal to the time moving downstream, or

$$\frac{2}{20 - x} = \frac{3}{20 + x}$$

Multiplying both sides by the LCD $(20 - x)(20 + x)$ gives

$$(20 + x) \cdot 2 = 3(20 - x)$$

$$40 + 2x = 60 - 3x$$

$$5x = 20$$

$$x = 4$$

The speed of the current is 4 miles per hour. 

EXAMPLE 4 The current of a river is 3 miles per hour. It takes a motorboat a total of 3 hours to travel 12 miles upstream and return 12 miles downstream. What is the speed of the boat in still water?

SOLUTION This time we let x = the speed of the boat in still water. Then, we fill in as much of the table as possible using the information given in the problem. For instance, since we let x = the speed of the boat in still water, the rate upstream (against the current) must be $x - 3$. The rate downstream (with the current) is $x + 3$.

	d	r	t
Upstream	12	$x - 3$	
Downstream	12	$x + 3$	

The last two boxes can be filled in using the relationship

$t = \frac{d}{r}$

	d	r	t
Upstream	12	$x - 3$	$\frac{12}{x-3}$
Downstream	12	$x + 3$	$\frac{12}{x+3}$

The total time for the trip up and back is 3 hours:

$$\text{Time upstream} + \text{Time downstream} = \text{Total time}$$

$$\frac{12}{x-3} + \frac{12}{x+3} = 3$$

Multiplying both sides by $(x - 3)(x + 3)$, we have

$$12(x + 3) + 12(x - 3) = 3(x^2 - 9)$$

$$12x + 36 + 12x - 36 = 3x^2 - 27$$

$$3x^2 - 24x - 27 = 0$$

$$x^2 - 8x - 9 = 0 \qquad \textbf{Divide both sides by 3}$$

$$(x - 9)(x + 1) = 0$$

$$x = 9 \quad \text{or} \quad x = -1$$

The speed of the motorboat in still water is 9 miles per hour. (We don't use $x = -1$ because the speed of the motorboat cannot be a negative number.)

EXAMPLE 5 An inlet pipe can fill a pool in 10 hours, and the drain can empty it in 12 hours. If the pool is empty and both the inlet pipe and drain are open, how long will it take to fill the pool?

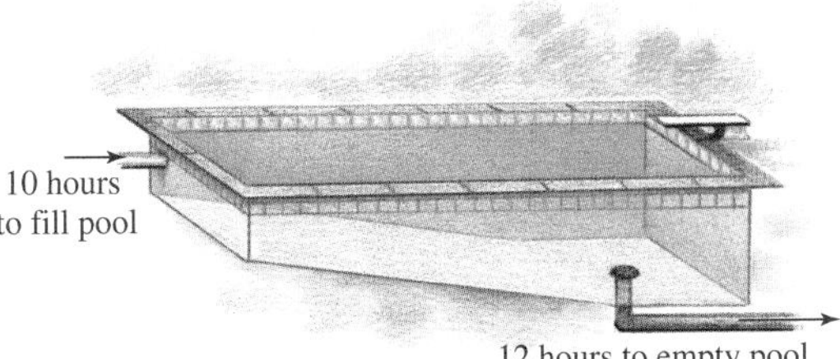

SOLUTION It is helpful to think in terms of how much work is done by each pipe in 1 hour.

Let x = the time it takes to fill the pool with both pipes open.

If the inlet pipe can fill the pool in 10 hours, then in 1 hour it is $\frac{1}{10}$ full. If the outlet pipe empties the pool in 12 hours, then in 1 hour it is $\frac{1}{12}$ empty. If the pool can be filled in x hours with both the inlet pipe and the drain open, then in 1 hour it is $\frac{1}{x}$ full when both pipes are open.

Here is the equation:

In 1 hour

$$\left[\begin{array}{c}\text{Amount filled by}\\ \text{inlet pipe}\end{array}\right] - \left[\begin{array}{c}\text{Amount emptied by}\\ \text{the drain}\end{array}\right] = \left[\begin{array}{c}\text{Fraction of pool filled}\\ \text{with both pipes open}\end{array}\right]$$

$$\frac{1}{10} - \frac{1}{12} = \frac{1}{x}$$

Multiplying through by $60x$, we have

$$\mathbf{60x} \cdot \frac{1}{10} - \mathbf{60x} \cdot \frac{1}{12} = \mathbf{60x} \cdot \frac{1}{x}$$

$$6x - 5x = 60$$

$$x = 60$$

It takes 60 hours to fill the pool if both the inlet pipe and the drain are open.

Unit Analysis

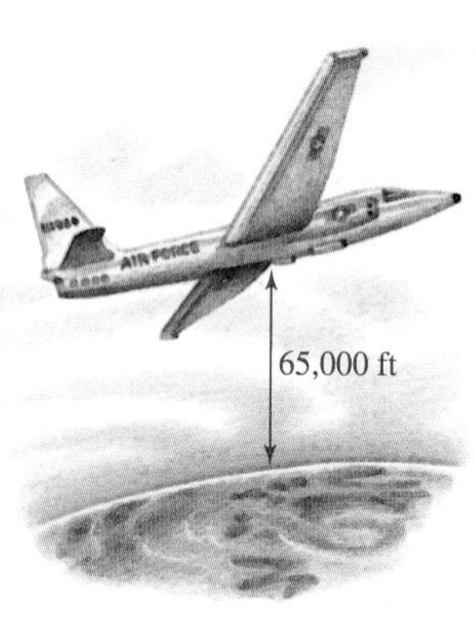

In the 1950s the United States had a spy plane, the U-2, that could fly at an altitude of 65,000 feet. Do you know how many miles are in 65,000 feet?

We can solve problems like this by using a method called *unit analysis.* With unit analysis, we analyze the units we are given and the units for which we are asked, and then multiply by the appropriate *conversion factor.* Since 1 mile is 5,280 feet, the conversion factor we use is

$$\frac{1 \text{ mile}}{5{,}280 \text{ feet}}$$

which is the number 1. Multiplying 65,000 feet by this conversion factor we have the following:

$$65{,}000 \text{ feet} = \frac{65{,}000 \text{ feet}}{1} \cdot \frac{1 \text{ mile}}{5{,}280 \text{ feet}}$$

We treat the units common to the numerator and denominator in the same way we treat factors common to the numerator and denominator: We divide out common units, just as we divide out common factors. In the preceding expression we have feet common to the numerator and denominator. Dividing them out leaves us with miles only. Here is the complete problem.

$$65{,}000 \text{ feet} = \frac{65{,}000 \text{ \sout{feet}}}{1} \cdot \frac{1 \text{ mile}}{5{,}280 \text{ \sout{feet}}}$$

$$= \frac{65{,}000}{5{,}280} \text{ mile}$$

$$= 12.3 \text{ miles, to the nearest tenth of a mile}$$

The key to solving a problem like this one lies in choosing the appropriate conversion factor. The fact that 1 mile = 5,280 feet yields two conversion factors, each of which is equal to the number 1. They are

$$\frac{1 \text{ mile}}{5{,}280 \text{ feet}} \quad \text{and} \quad \frac{5{,}280 \text{ feet}}{1 \text{ mile}}$$

The conversion factor we choose depends on the units we are given and the units with which we want to end up. Multiplying any expression by either of the two conversion factors leaves the value of the original expression unchanged because each of the conversion factors is simply the number 1.

EXAMPLE 6

Previously, we found a rider on the first Ferris wheel was traveling at approximately 39.3 feet per minute. Convert 39.3 feet per minute to miles per hour.

SOLUTION We know that 5,280 feet = 1 mile and 60 minutes = 1 hour. Therefore, we have the following conversion factors, each of which is equal to 1.

$$\frac{5{,}280 \text{ feet}}{1 \text{ mile}} \qquad \frac{1 \text{ mile}}{5{,}280 \text{ feet}} \qquad \frac{60 \text{ minutes}}{1 \text{ hour}} \qquad \frac{1 \text{ hour}}{60 \text{ minutes}}$$

The conversion factors we choose to multiply by are the ones that will allow us to divide out the units we are converting from and leave us with the units we are converting to. Specifically, we want to get rid of feet and be left with miles. Likewise, we want to get rid of minutes and be left with hours. Here is the conversion process that will accomplish these goals:

$$\begin{aligned} 39.3 \text{ feet per minute} &= \frac{39.3 \cancel{\text{feet}}}{1 \cancel{\text{minute}}} \cdot \frac{1 \text{ mile}}{5{,}280 \cancel{\text{feet}}} \cdot \frac{60 \cancel{\text{minutes}}}{1 \text{ hour}} \\ &= \frac{39.3 \cdot 60 \text{ miles}}{5{,}280 \text{ hours}} \\ &= 0.45 \text{ miles per hour, to the nearest hundredth} \end{aligned}$$

EXAMPLE 7

In 1993, a ski resort in Vermont advertised their new high-speed chair lift as "the world's fastest chair lift, with a speed of 1,100 feet per second." Show why the speed cannot be correct.

SOLUTION To solve this problem, we can convert feet per second into miles per hour, a unit of measure we are more familiar with on an intuitive level.

$$\begin{aligned} 1{,}100 \text{ feet per second} &= \frac{1{,}100 \cancel{\text{feet}}}{1 \cancel{\text{second}}} \cdot \frac{1 \text{ mile}}{5{,}280 \cancel{\text{feet}}} \cdot \frac{60 \cancel{\text{seconds}}}{1 \cancel{\text{minute}}} \cdot \frac{60 \cancel{\text{minutes}}}{1 \text{ hour}} \\ &= \frac{1{,}100 \cdot 60 \cdot 60 \text{ miles}}{5{,}280 \text{ hours}} \\ &= 750 \text{ miles per hour} \end{aligned}$$

Obviously, there is a mistake in the advertisement.

EXAMPLE 8

This Snapshot appeared in *USA Today* in November of 2005. It gives three rates that together account for the average change in the population of the United States over any given period of time. Use the information in the table to find the average number of births in one week, and the average number of deaths in one week.

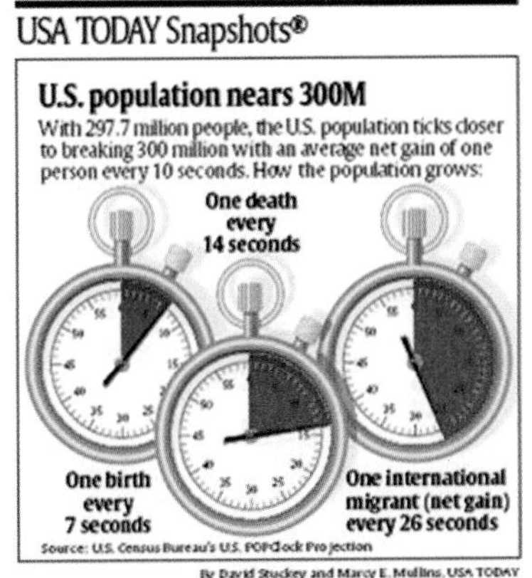

USA TODAY Snapshots®
U.S. population nears 300M
With 297.7 million people, the U.S. population ticks closer to breaking 300 million with an average net gain of one person every 10 seconds. How the population grows:
One death every 14 seconds
One birth every 7 seconds
One international migrant (net gain) every 26 seconds
Source: U.S. Census Bureau's U.S. POPClock Projection
By David Stuckey and Marcy E. Mullins, USA TODAY

From *USA Today*. Copyright 2005. Reprinted with Permission

SOLUTION Along with the rates given in the Snapshot, we need the following relationships

1 minute = 60 seconds 1 hour = 60 minutes
1 day = 24 hours 1 week = 7 days

Writing each of these relationships as a conversion factor equal to 1, we have

$$\frac{60 \text{ sec}}{1 \text{ min}} \quad \frac{60 \text{ min}}{1 \text{ hr}} \quad \frac{24 \text{ hr}}{1 \text{ day}} \quad \frac{7 \text{ days}}{1 \text{ week}}$$

To find the average number of births in one week, we convert from births per second to births per week:

$$\frac{1 \text{ birth}}{7 \text{ seconds}} = \frac{1 \text{ birth}}{7 \cancel{\text{sec}}} \cdot \frac{60 \cancel{\text{sec}}}{1 \cancel{\text{min}}} \cdot \frac{60 \cancel{\text{min}}}{1 \cancel{\text{hr}}} \cdot \frac{24 \cancel{\text{hr}}}{1 \cancel{\text{day}}} \cdot \frac{7 \cancel{\text{days}}}{1 \text{ week}}$$

$$= 86{,}400 \text{ births/week}$$

To find the number of deaths per week, we could apply the same procedure as shown above, or we could simply note that there are half as many deaths every 7 seconds as births and simply divide our last result by 2 to obtain

43,200 deaths/week

More About Graphing Rational Functions

We continue our investigation of the graphs of rational functions by considering the graph of a rational function with binomials in the numerator and denominator.

EXAMPLE 9 Graph the rational function $y = \frac{x-4}{x-2}$.

SOLUTION In addition to making a table to find some points on the graph, we can analyze the graph as follows:

1. The graph will have a y-intercept of 2, because when $x = 0$, $y = \frac{-4}{-2} = 2$.
2. To find the x-intercept, we let $y = 0$ to get

   $$0 = \frac{x-4}{x-2}$$

   The only way this expression can be 0 is if the numerator is 0, which happens when $x = 4$. (If you want to solve this equation, multiply both sides by $x - 2$. You will get the same solution, $x = 4$.)
3. The graph will have a *vertical asymptote* at $x = 2$, because $x = 2$ will make the denominator of the function 0, meaning y is undefined when x is 2.
4. The graph will have a *horizontal asymptote* at $y = 1$ because for very large values of x, the expression $\frac{x-4}{x-2}$ is very close to 1. The larger x is, the closer $\frac{x-4}{x-2}$ is to 1. Likewise, for values of x such as $-1{,}000$ and $-10{,}000$, the expression gets closer to 1.

Putting this information together with the ordered pairs in the table next to the figure, we have the graph shown in Figure 1.

x	y
-1	$\frac{5}{3}$
0	2
1	3
2	Undefined
3	-1
4	0
5	$\frac{1}{3}$

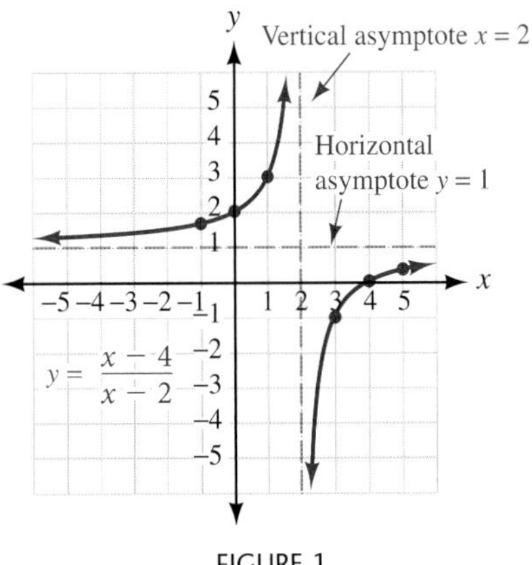

FIGURE 1

USING TECHNOLOGY

More About Example 9

Let's use technology to explore the graph near the horizontal asymtote. In Figure 1, the horizontal asymptote is at $y = 1$. To show that the graph approaches this line as x becomes very large, we use the table function on our graphing calculator, with X taking values of 100, 1,000, and 10,000. To show that the graph approaches the line $y = 1$ on the left side of the coordinate system, we let X become -100, $-1{,}000$, and $-10{,}000$.

Table Setup
Table minimum = 0
Table increment = 1
Independent variable: Ask
Dependent variable: Auto

Y Variables Setup
$Y_1 = (X - 4)/(X - 2)$

The table will look like this:

X	Y_1
100	.97959
1000	.998
10000	.9998
−100	1.0196
−1000	1.002
−100000	1.0002

As you can see, as x becomes very large in the positive direction, the graph approaches the line $y = 1$ from below. As x moves farther from zero in the negative direction, the graph approaches the line $y = 1$ from above.

LINKING OBJECTIVES AND EXAMPLES

Next to each **objective** we have listed the examples that are best described by that objective.

A	1–5
B	6–8

GETTING READY FOR CLASS

After reading through the preceding section, respond in your own words and in complete sentences.

1. Briefly list the steps in the Blueprint for Problem Solving that you have used previously to solve application problems.
2. Write an application problem for which the solution depends on solving the equation $\frac{1}{2} + \frac{1}{3} = \frac{1}{x}$.
3. What is a conversion factor in unit analysis?
4. What conversion factors would you use to convert feet per second to miles per hour?

Problem Set 6.7

Online support materials can be found at www.thomsonedu.com/login

Solve each of the following word problems. Be sure to show the equation in each case.

Number Problems

1. One number is 3 times another. The sum of their reciprocals is $\frac{20}{3}$. Find the numbers.
2. One number is 3 times another. The sum of their reciprocals is $\frac{4}{9}$. Find the numbers.
3. The sum of a number and its reciprocal is $\frac{10}{3}$. Find the number.
4. The sum of a number and twice its reciprocal is $\frac{27}{5}$. Find the number.
5. The sum of the reciprocals of two consecutive integers is $\frac{7}{12}$. Find the two integers.
6. Find two consecutive even integers, the sum of whose reciprocals is $\frac{3}{4}$.
7. If a certain number is added to the numerator and denominator of $\frac{7}{9}$, the result is $\frac{5}{6}$. Find the number.
8. Find the number you would add to both the numerator and denominator of $\frac{8}{11}$ so the result would be $\frac{6}{7}$.

Rate Problems

9. The speed of a boat in still water is 5 miles per hour. If the boat travels 3 miles downstream in the same amount of time it takes to travel 1.5 miles upstream, what is the speed of the current?
10. A boat, which moves at 18 miles per hour in still water, travels 14 miles downstream in the same amount of time it takes to travel 10 miles upstream. Find the speed of the current.
11. The current of a river is 2 miles per hour. A boat travels to a point 8 miles upstream and back again in 3 hours. What is the speed of the boat in still water?
12. A motorboat travels at 4 miles per hour in still water. It goes 12 miles upstream and 12 miles back again in a total of 8 hours. Find the speed of the current of the river.
13. Train A has a speed 15 miles per hour greater than that of train B. If train A travels 150 miles in the same time train B travels 120 miles, what are the speeds of the two trains?

■ = Videos available by instructor request

▶ = Online student support materials available at www.thomsonedu.com/login

14. A train travels 30 miles per hour faster than a car. If the train covers 120 miles in the same time the car covers 80 miles, what is the speed of each of them?

15. A small airplane flies 810 miles from Los Angeles to Portland, Oregon, with an average speed of 270 miles per hour. An hour and a half after the plane leaves, a Boeing 747 leaves Los Angeles for Portland. Both planes arrive in Portland at the same time. What was the average speed of the 747?

16. Lou leaves for a cross-country excursion traveling on a bicycle at 20 miles per hour. His friends are driving the trip and will meet him at several rest stops along the way. The first stop is scheduled 30 miles from the original starting point. If the people driving leave 15 minutes after Lou from the same place, how fast will they have to drive to reach the first rest stop at the same time as Lou?

17. A tour bus leaves Sacramento every Friday evening at 5:00 P.M. for a 270-mile trip to Las Vegas. This week, however, the bus leaves at 5:30 P.M. To arrive in Las Vegas on time, the driver drives 6 miles per hour faster than usual. What is the bus's usual speed?

18. A bakery delivery truck leaves the bakery at 5:00 A.M. each morning on its 140-mile route. One day the driver gets a late start and does not leave the bakery until 5:30 A.M. To finish her route on time the driver drives 5 miles per hour faster than usual. At what speed does she usually drive?

Work Problems

19. A water tank can be filled by an inlet pipe in 8 hours. It takes twice that long for the outlet pipe to empty the tank. How long will it take to fill the tank if both pipes are open?

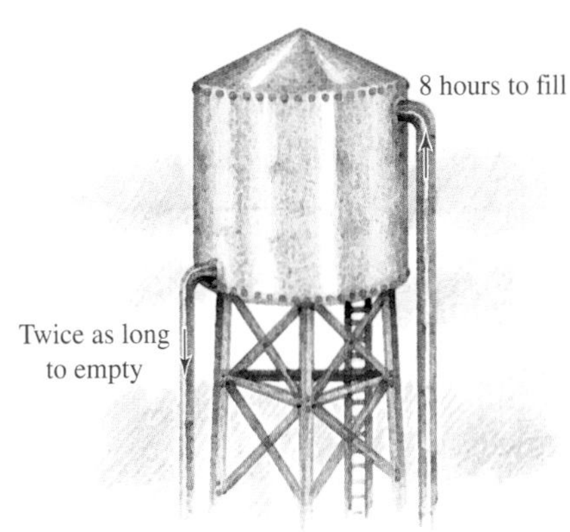

20. It takes 10 hours to fill a pool with the inlet pipe. It can be emptied in 15 hours with the outlet pipe. If the pool is half full to begin with, how long will it take to fill it from there if both pipes are open?

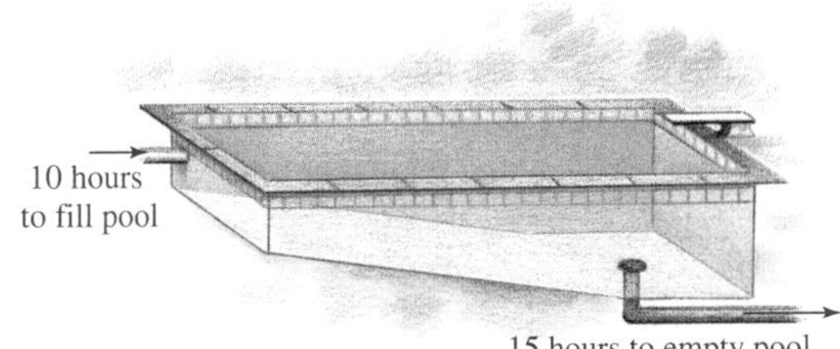

21. A sink can be filled from the faucet in 5 minutes. It takes only 3 minutes to empty the sink when the drain is open. If the sink is full and both the faucet and the drain are open, how long will it take to empty the sink?

22. A sink is one quarter full when both the faucet and the drain are opened. The faucet alone can fill the sink in 6 minutes, whereas it takes 8 minutes to empty it with the drain. How long will it take to fill the remaining three quarters of the sink?

23. A sink has two faucets: one for hot water and one for cold water. The sink can be filled by a cold-water faucet in 3.5 minutes. If both faucets are open, the sink is filled in 2.1 minutes. How long does it take to fill the sink with just the hot-water faucet open?

24. A water tank is being filled by two inlet pipes. Pipe A can fill the tank in $4\frac{1}{2}$ hours, but both pipes together can fill the tank in 2 hours. How long does it take to fill the tank using only pipe B?

Unit Analysis Problems

Give your answers to the following problems to the nearest tenth.

25. The South Coast Shopping Mall in Costa Mesa, California, covers an area of 2,224,750 square feet. If 1 acre = 43,560 square feet, how many acres does the South Coast Shopping Mall cover?

26. The relationship between liters and cubic inches, both of which are measures of volume, is 0.0164 liters = 1 cubic inch. If a Ford Mustang has a motor with a displacement of 4.9 liters, what is the displacement in cubic inches?

27. The Forest chair lift at the Northstar ski resort in Lake Tahoe is 5,750 feet long. If a ride on this chair lift takes 11 minutes, what is the average speed of the lift in miles per hour?

28. The Bear Paw chair lift at the Northstar ski resort in Lake Tahoe is 790 feet long. If a ride on this chair lift takes 2.2 minutes, what is the average speed of the lift in miles per hour?

▸ 29. A sprinter runs 100 meters in 10.8 seconds. What is the sprinter's average speed in miles per hour? (1 meter = 3.28 feet)

30. A runner covers 400 meters in 49.8 seconds. What is the average speed of the runner in miles per hour?

31. A person riding a Ferris wheel with a diameter of 65 feet travels once around the wheel in 30 seconds. What is the average speed of the rider in miles per hour?

32. A person riding a Ferris wheel with a diameter of 102 feet travels once around the wheel in 3.5 minutes. What is the average speed of the rider in miles per hour?

33. A $3\frac{1}{2}$-inch diskette, when placed in the disk drive of a computer, rotates at 300 rpm (meaning one revolution takes 1/300 minute). Find the average speed of a point 2 inches from the center of the diskette in miles per hour.

34. **Golf** The length of a golf course is the sum of the lengths of all 18 holes on the course. It is measured in yards. Round each of the lengths in the chart to the nearest hundred yards, then convert the result to miles, and round to the nearest tenth of a mile.

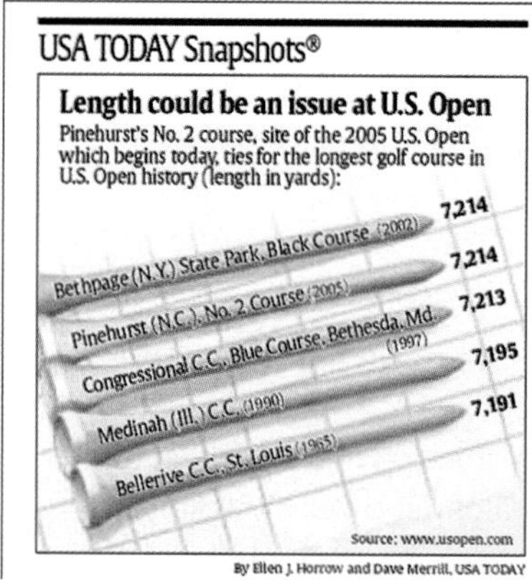

From *USA Today*. Copyright 2005. Reprinted with Permission

Miscellaneous Problems

35. **Rhind Papyrus** Nearly 4,000 years ago, Egyptians worked mathematical exercises involving reciprocals. The *Rhind Papyrus* contains a wealth of such problems, and one of them is as follows:

British Museum/Bridgeman Art Library

> A quantity and its two thirds are added together, one third of this is added, then one third of the sum is taken, and the result is 10.

Write an equation and solve this exercise.

36. **Photography** For clear photographs, a camera must be focused properly. Professional photographers use a mathematical relationship relating the distance from the camera lens to the object being photographed, a; the distance from the lens to the film, b; and the focal length of the lens, f. These quantities, a, b, and f, are related by the equation

$$\frac{1}{a} + \frac{1}{b} = \frac{1}{f}$$

A camera has a focal length of 3 inches. If the lens is 5 inches from the film, how far should the lens be placed from the object being photographed for the camera to be perfectly focused?

Maintaining Your Skills

Perform the indicated operations.

37. $\dfrac{2a + 10}{a^3} \cdot \dfrac{a^2}{3a + 15}$

38. $\dfrac{4a + 8}{a^2 - a - 6} \div \dfrac{a^2 + 7a + 12}{a^2 - 9}$

39. $(x^2 - 9)\left(\dfrac{x + 2}{x + 3}\right)$

40. $\dfrac{1}{x + 4} + \dfrac{8}{x^2 - 16}$

41. $\dfrac{2x - 7}{x - 2} - \dfrac{x - 5}{x - 2}$

42. $2 + \dfrac{25}{5x - 1}$

Simplify each expression.

43. $\dfrac{\frac{1}{x} - \frac{1}{3}}{\frac{1}{x} + \frac{1}{3}}$

44. $\dfrac{1 - \frac{9}{x^2}}{1 - \frac{1}{x} - \frac{6}{x^2}}$

Solve each equation.

45. $\dfrac{x}{x - 3} + \dfrac{3}{2} = \dfrac{3}{x - 3}$

46. $1 - \dfrac{3}{x} = \dfrac{-2}{x^2}$

Chapter 6 SUMMARY

EXAMPLES

1. $\frac{3}{4}$ is a rational number. $\dfrac{x-3}{x^2-9}$ is a rational expression.

Rational Numbers and Expressions [6.1]

A *rational number* is any number that can be expressed as the ratio of two integers:

$$\text{Rational numbers} = \left\{ \frac{a}{b} \,\middle|\, a \text{ and } b \text{ are integers, } b \neq 0 \right\}$$

A *rational expression* is any quantity that can be expressed as the ratio of two polynomials:

$$\text{Rational expressions} = \left\{ \frac{P}{Q} \,\middle|\, P \text{ and } Q \text{ are polynomials, } Q \neq 0 \right\}$$

Properties of Rational Expressions [6.1]

If P, Q, and K are polynomials with $Q \neq 0$ and $K \neq 0$, then

$$\frac{P}{Q} = \frac{PK}{QK} \quad \text{and} \quad \frac{P}{Q} = \frac{P/K}{Q/K}$$

which is to say that multiplying or dividing the numerator and denominator of a rational expression by the same nonzero quantity always produces an equivalent rational expression.

2. $\dfrac{x-3}{x^2-9} = \dfrac{\cancel{x-3}}{(\cancel{x-3})(x+3)}$

$= \dfrac{1}{x+3}$

Reducing to Lowest Terms [6.1]

To reduce a rational expression to lowest terms, we first factor the numerator and denominator and then divide the numerator and denominator by any factors they have in common.

3. $\dfrac{15x^3 - 20x^2 + 10x}{5x}$

$= 3x^2 - 4x + 2$

Dividing a Polynomial by a Monomial [6.2]

To divide a polynomial by a monomial, divide each term of the polynomial by the monomial.

4.
$$\begin{array}{r} x - 2 \\ x-3 \overline{) \; x^2 - 5x + 8} \\ \underline{\not{+}x^2 \not{-} 3x} \quad \\ -2x + 8 \\ \underline{\not{-}2x \not{+} 6} \\ 2 \end{array}$$

Long Division with Polynomials [6.2]

If division with polynomials cannot be accomplished by dividing out factors common to the numerator and denominator, then we use a process similar to long division with whole numbers. The steps in the process are estimate, multiply, subtract, and bring down the next term.

Multiplication [6.3]

5. $\frac{x+1}{x^2-4} \cdot \frac{x+2}{3x+3}$

$= \frac{(x+1)(x+2)}{(x-2)(x+2)(3)(x+1)}$

$= \frac{1}{3(x-2)}$

To multiply two rational numbers or rational expressions, multiply numerators and multiply denominators. In symbols,

$$\frac{P}{Q} \cdot \frac{R}{S} = \frac{PR}{QS} \qquad (Q \neq 0 \text{ and } S \neq 0)$$

In practice, we don't really multiply, but rather, we factor and then divide out common factors.

Division [6.3]

6. $\frac{x^2-y^2}{x^3+y^3} \div \frac{x-y}{x^2-xy+y^2}$

$= \frac{x^2-y^2}{x^3+y^3} \cdot \frac{x^2-xy+y^2}{x-y}$

$= \frac{(x+y)(x-y)(x^2-xy+y^2)}{(x+y)(x^2-xy+y^2)(x-y)}$

$= 1$

To divide one rational expression by another, we use the definition of division to rewrite our division problem as an equivalent multiplication problem. To divide by a rational expression we multiply by its reciprocal. In symbols,

$$\frac{P}{Q} \div \frac{R}{S} = \frac{P}{Q} \cdot \frac{S}{R} = \frac{PS}{QR} \qquad (Q \neq 0, S \neq 0, R \neq 0)$$

Least Common Denominator [6.4]

7. The LCD for $\frac{2}{x-3}$ and $\frac{3}{5}$ is $5(x-3)$.

The *least common denominator,* LCD, for a set of denominators is the smallest quantity divisible by each of the denominators.

Addition and Subtraction [6.4]

8. $\frac{2}{x-3} + \frac{3}{5}$

$= \frac{2}{x-3} \cdot \frac{\mathbf{5}}{\mathbf{5}} + \frac{3}{5} \cdot \frac{\mathbf{x-3}}{\mathbf{x-3}}$

$= \frac{3x+1}{5(x-3)}$

If P, Q, and R represent polynomials, $R \neq 0$, then

$$\frac{P}{R} + \frac{Q}{R} = \frac{P+Q}{R} \quad \text{and} \quad \frac{P}{R} - \frac{Q}{R} = \frac{P-Q}{R}$$

When adding or subtracting rational expressions with different denominators, we must find the LCD for all denominators and change each rational expression to an equivalent expression that has the LCD.

Complex Fractions [6.5]

9. $\frac{\frac{1}{x}+\frac{1}{y}}{\frac{1}{x}-\frac{1}{y}} = \frac{\mathbf{xy}\left(\frac{1}{x}+\frac{1}{y}\right)}{\mathbf{xy}\left(\frac{1}{x}-\frac{1}{y}\right)}$

$= \frac{y+x}{y-x}$

A rational expression that contains, in its numerator or denominator, other rational expressions is called a *complex fraction.* One method of simplifying a complex fraction is to multiply the numerator and denominator by the LCD for all denominators.

Equations Involving Rational Expressions [6.6]

10. Solve $\frac{x}{2} + 3 = \frac{1}{3}$.

$$\mathbf{6}\left(\frac{x}{2}\right) + \mathbf{6} \cdot 3 = \mathbf{6} \cdot \frac{1}{3}$$
$$3x + 18 = 2$$
$$x = -\frac{16}{3}$$

To solve an equation involving rational expressions, we first find the LCD for all denominators appearing on either side of the equation. We then multiply both sides by the LCD to clear the equation of all fractions and solve as usual.

! COMMON MISTAKES

1. Attempting to divide the numerator and denominator of a rational expression by a quantity that is not a factor of both. Like this:

$$\frac{\cancel{x^2} - \overset{3}{\cancel{9x}} + \overset{2}{\cancel{20}}}{\cancel{x^2} - \underset{1}{\cancel{3x}} - \underset{1}{\cancel{10}}} \qquad \textbf{Mistake}$$

   This makes no sense at all. The numerator and denominator must be factored completely before any factors they have in common can be recognized:

$$\frac{x^2 - 9x + 20}{x^2 - 3x - 10} = \frac{\cancel{(x-5)}(x-4)}{\cancel{(x-5)}(x+2)}$$
$$= \frac{x-4}{x+2}$$

2. Forgetting to check solutions to equations involving rational expressions. When we multiply both sides of an equation by a quantity containing the variable, we must be sure to check for extraneous solutions.

Chapter 6 Review Test

The problems below form a comprehensive review of the material in this chapter. They can be used to study for exams. If you would like to take a practice test on this chapter, you can use the odd-numbered problems. Give yourself an hour and work as many of the odd-numbered problems as possible. When you are finished, or when an hour has passed, check your answers with the answers in the back of the book. You can use the even-numbered problems for a second practice test.

Reduce to lowest terms. [6.1]

1. $\dfrac{125x^4yz^3}{35x^2y^4z^3}$ **2.** $\dfrac{a^3 - ab^2}{4a + 4b}$

3. $\dfrac{x^2 - 25}{x^2 + 10x + 25}$ **4.** $\dfrac{ax + x - 5a - 5}{ax - x - 5a + 5}$

Divide. If the denominator is a factor of the numerator, you may want to factor the numerator and divide out the common factor. [6.2]

5. $\dfrac{12x^3 + 8x^2 + 16x}{4x^2}$

6. $\dfrac{27a^2b^3 - 15a^3b^2 + 21a^4b^4}{-3a^2b^2}$

7. $\dfrac{x^{6n} - x^{5n}}{x^{3n}}$ **8.** $\dfrac{x^2 - x - 6}{x - 3}$

9. $\dfrac{5x^2 - 14xy - 24y^2}{x - 4y}$ **10.** $\dfrac{y^4 - 16}{y - 2}$

11. $\dfrac{8x^2 - 26x - 9}{2x - 7}$ **12.** $\dfrac{2y^3 - 9y^2 - 17y + 39}{2y - 3}$

Multiply and divide as indicated. [6.3]

13. $\dfrac{3}{4} \cdot \dfrac{12}{15} \div \dfrac{1}{3}$ **14.** $\dfrac{15x^2y}{8xy^2} \div \dfrac{10xy}{4x}$

15. $\dfrac{x^3 - 1}{x^4 - 1} \cdot \dfrac{x^2 - 1}{x^2 + x + 1}$

16. $\dfrac{a^2 + 5a + 6}{a + 1} \cdot \dfrac{a + 5}{a^2 + 2a - 3} \div \dfrac{a^2 + 7a + 10}{a^2 - 1}$

17. $\dfrac{ax + bx + 2a + 2b}{ax - 3a + bx - 3b} \div \dfrac{ax - bx - 2a + 2b}{ax - bx - 3a + 3b}$

18. $(4x^2 - 9) \cdot \dfrac{x + 3}{2x + 3}$

Add and subtract as indicated. [6.4]

19. $\dfrac{3}{5} - \dfrac{1}{10} + \dfrac{8}{15}$ **20.** $\dfrac{5}{x - 5} - \dfrac{x}{x - 5}$

21. $\dfrac{1}{x} + \dfrac{1}{x^2} + \dfrac{1}{x^3}$ **22.** $\dfrac{8}{y^2 - 16} - \dfrac{7}{y^2 - y - 12}$

23. $\dfrac{x - 2}{x^2 + 5x + 4} - \dfrac{x - 4}{2x^2 + 12x + 16}$

24. $3 + \dfrac{4}{5x - 2}$

Simplify each complex fraction. [6.5]

25. $\dfrac{1 + \dfrac{2}{3}}{1 - \dfrac{2}{3}}$ **26.** $\dfrac{\dfrac{4a}{2a^3 + 2}}{\dfrac{8a}{4a + 4}}$

27. $1 + \dfrac{1}{x + \dfrac{1}{x}}$ **28.** $\dfrac{1 - \dfrac{9}{x^2}}{1 - \dfrac{1}{x} - \dfrac{6}{x^2}}$

Solve each equation. [6.6]

29. $\dfrac{3}{x - 1} = \dfrac{3}{5}$ **30.** $\dfrac{x + 1}{3} + \dfrac{x - 3}{4} = \dfrac{1}{6}$

31. $\dfrac{5}{y + 1} = \dfrac{4}{y + 2}$ **32.** $\dfrac{x + 6}{x + 3} - 2 = \dfrac{3}{x + 3}$

33. $\dfrac{4}{x^2 - x - 12} + \dfrac{1}{x^2 - 9} = \dfrac{2}{x^2 - 7x + 12}$

34. $\dfrac{a + 4}{a^2 + 5a} = \dfrac{-2}{a^2 - 25}$

35. Distance, Rate, and Time A car makes a 120-mile trip 10 miles per hour faster than a truck. The truck takes 2 hours longer to make the trip. What are the speeds of the car and the truck? [6.7]

36. Average Speed A jogger covers 3.5 miles in 28 minutes. Find the average speed of the jogger in miles per hour. [6.7]

37. Unit Analysis The speed of sound is 1,088 feet per second. Convert the speed of sound to miles per hour. Round your answer to the nearest whole number. [6.7]

GROUP PROJECT Rational Expressions

Number of People 3

Time Needed 10–15 minutes

Equipment Pencil and paper

Procedure The four problems shown here all involve the same rational expressions. Often, students who have worked problems successfully on their homework have trouble when they take a test on rational expressions because the problems are mixed up and do not have similar instructions. Noticing similarities and differences between the types of problems involving rational expressions can help with this situation.

Problem 1: Add: $\dfrac{-2}{x^2 - 2x - 3} + \dfrac{3}{x^2 - 9}$

Problem 2: Divide: $\dfrac{-2}{x^2 - 2x - 3} \div \dfrac{3}{x^2 - 9}$

Problem 3: Solve: $\dfrac{-2}{x^2 - 2x - 3} + \dfrac{3}{x^2 - 9} = -1$

Problem 4: Simplify: $\dfrac{\dfrac{-2}{x^2 - 2x - 3}}{\dfrac{3}{x^2 - 9}}$

1. Which problems here do not require the use of a least common denominator?

2. Which two problems involve multiplying by the least common denominator?

3. Which of the problems will have an answer that is one or two numbers but no variables?

4. Work each of the four problems.

Ferris Wheel and *The Third Man*

© Hulton Getty/Stone/Getty Images

Among the large Ferris wheels built around the turn of the twentieth century was one built in Vienna in 1897. It is the only one of those large wheels that is still in operation today. Known as the Riesenrad, it has a diameter of 197 feet and can carry a total of 800 people. A brochure that gives some statistics associated with the Riesenrad indicates that passengers riding it travel at 2 feet 6 inches per second. You can check the accuracy of this number by watching the movie *The Third Man*. In the movie, Orson Welles rides the Riesenrad through one complete revolution. Play *The Third Man* on a VCR so you can view the Riesenrad in operation. Use the pause button and the timer on the VCR to time how long it takes Orson Welles to ride once around the wheel. Then calculate his average speed during the ride. Use your results to either prove or disprove the claim that passengers travel at 2 feet 6 inches per second on the Riesenrad. When you have finished, write your procedures and results in essay form.

© Selznick Films/The Kobal Collection

Rational Exponents and Roots

7

CHAPTER OUTLINE

Ecology and conservation are topics that interest most college students. If our rivers and oceans are to be preserved for future generations, we need to work to eliminate pollution from our waters. If a river is flowing at 1 meter per second and a pollutant is entering the river at a constant rate, the shape of the pollution plume can often be modeled by the simple equation

$$y = \sqrt{x}$$

The following table and graph were produced from the equation.

Width of a Pollutant Plume

Distance from Source (meters) x	Width of Plume (meters) y
0	0
1	1
4	2
9	3
16	4

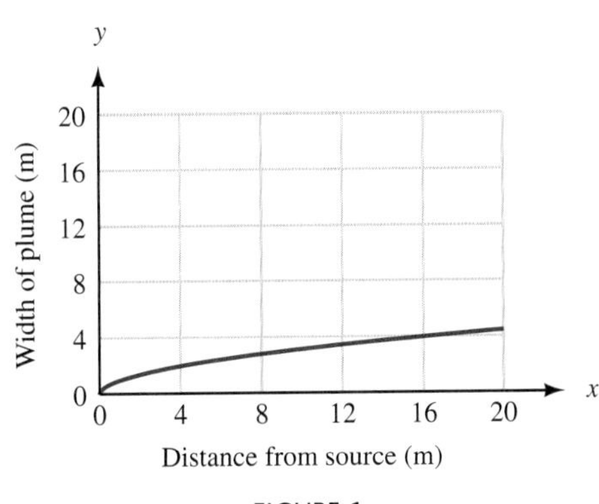

FIGURE 1

To visualize how Figure 1 models the pollutant plume, imagine that the river is flowing from left to right, parallel to the x-axis, with the x-axis as one of its banks. The pollutant is entering the river from the bank at (0, 0).

By modeling pollution with mathematics, we can use our knowledge of mathematics to help control and eliminate pollution.

This is the last chapter in which we will mention study skills. You know by now what works best for you and what you have to do to achieve your goals for this course. From now on, it is simply a matter of sticking with the things that work for you and avoiding the things that do not work. It seems simple, but as with anything that takes effort, it is up to you to see that you maintain the skills that get you where you want to be in the course.

If you intend to take more classes in mathematics, and you want to ensure your success in those classes, then you can work toward this goal: *Become a student who can learn mathematics on your own.* Most people who have degrees in mathematics were students who could learn mathematics on their own. This doesn't mean that you have to learn it all on your own; it simply means that if you have to, you can learn it on your own. Attaining this goal gives you independence and puts you in control of your success in any math class you take.

7.1 Rational Exponents

OBJECTIVES

A Simplify radical expressions using the definition for roots.

B Simplify expressions with rational exponents.

Figure 1 shows a square in which each of the four sides is 1 inch long. To find the square of the length of the diagonal c, we apply the Pythagorean theorem:

$$c^2 = 1^2 + 1^2$$

$$c^2 = 2$$

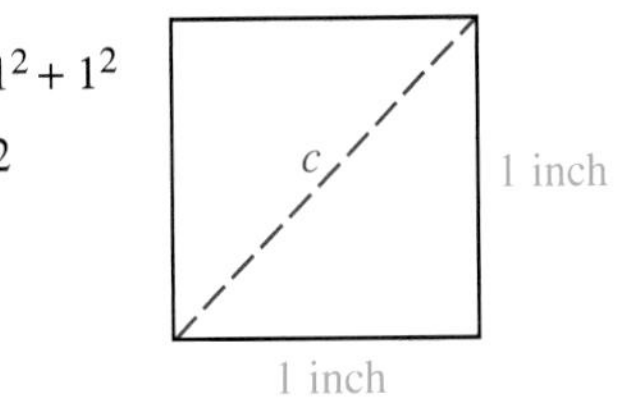

FIGURE 1

Because we know that c is positive and that its square is 2, we call c the *positive square root* of 2, and we write $c = \sqrt{2}$. Associating numbers, such as $\sqrt{2}$, with the diagonal of a square or rectangle allows us to analyze some interesting items from geometry. One particularly interesting geometric object that we will study in this section is shown in Figure 2. It is constructed from a right triangle, and the length of the diagonal is found from the Pythagorean theorem. We will come back to this figure at the end of this section.

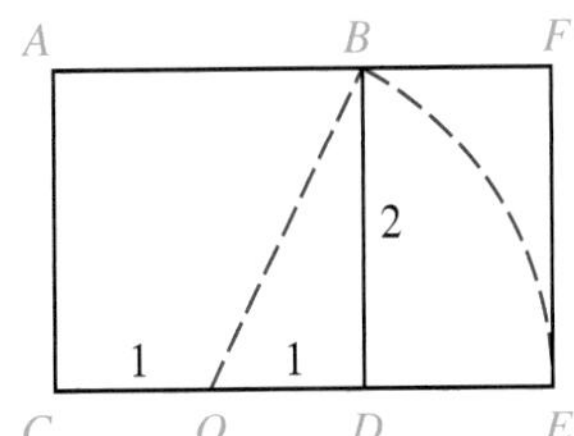

The Golden Rectangle

FIGURE 2

Previously, we developed notation (exponents) to give us the square, cube, or any other power of a number. For instance, if we wanted the square of 3, we wrote $3^2 = 9$. If we wanted the cube of 3, we wrote $3^3 = 27$. In this section, we will develop notation that will take us in the reverse direction—that is, from the square of a number, say 25, back to the original number, 5.

Note

It is a common mistake to assume that an expression like $\sqrt{25}$ indicates both square roots, 5 and -5. The expression $\sqrt{25}$ indicates only the positive square root of 25, which is 5. If we want the negative square root, we must use a negative sign: $-\sqrt{25} = -5$.

DEFINITION If x is a nonnegative real number, then the expression $\sqrt{x}$ is called the **positive square root** of x and is the nonnegative number such that

$$(\sqrt{x})^2 = x$$

In words: $\sqrt{x}$ is the nonnegative number we square to get x.

The negative square root of x, $-\sqrt{x}$, is defined in a similar manner.

EXAMPLE 1 The positive square root of 64 is 8 because 8 is the positive number with the property $8^2 = 64$. The negative square root of 64 is -8 since -8 is the negative number whose square is 64. We can summarize both of these facts by saying

$$\sqrt{64} = 8 \quad \text{and} \quad -\sqrt{64} = -8$$

The higher roots, cube roots, fourth roots, and so on are defined by definitions similar to that of square roots.

> **Note**
> We have restricted the even roots in this definition to nonnegative numbers. Even roots of negative numbers exist but are not represented by real numbers; that is, $\sqrt{-4}$ is not a real number since there is no real number whose square is -4.

DEFINITION If x is a real number and n is a positive integer, then

Positive square root of x, $\sqrt{x}$, is such that $(\sqrt{x})^2 = x \quad x \ge 0$

Cube root of x, $\sqrt[3]{x}$, is such that $(\sqrt[3]{x})^3 = x$

Positive fourth root of x, $\sqrt[4]{x}$, is such that $(\sqrt[4]{x})^4 = x \quad x \ge 0$

Fifth root of x, $\sqrt[5]{x}$, is such that $(\sqrt[5]{x})^5 = x$

$\vdots \qquad \vdots$

The ***n*th root of *x***, $\sqrt[n]{x}$, is such that $(\sqrt[n]{x})^n = x \quad x \ge 0$ if n is even

The following is a table of the most common roots used in this book. Any of the roots that are unfamiliar should be memorized.

Square Roots		Cube Roots	Fourth Roots
$\sqrt{0} = 0$	$\sqrt{49} = 7$	$\sqrt[3]{0} = 0$	$\sqrt[4]{0} = 0$
$\sqrt{1} = 1$	$\sqrt{64} = 8$	$\sqrt[3]{1} = 1$	$\sqrt[4]{1} = 1$
$\sqrt{4} = 2$	$\sqrt{81} = 9$	$\sqrt[3]{8} = 2$	$\sqrt[4]{16} = 2$
$\sqrt{9} = 3$	$\sqrt{100} = 10$	$\sqrt[3]{27} = 3$	$\sqrt[4]{81} = 3$
$\sqrt{16} = 4$	$\sqrt{121} = 11$	$\sqrt[3]{64} = 4$	
$\sqrt{25} = 5$	$\sqrt{144} = 12$	$\sqrt[3]{125} = 5$	
$\sqrt{36} = 6$	$\sqrt{169} = 13$		

Notation An expression like $\sqrt[3]{8}$ that involves a root is called a *radical expression.* In the expression $\sqrt[3]{8}$, the ^{3} is called the *index,* the $\sqrt{\ }$ is the *radical sign,* and 8 is called the *radicand.* The index of a radical must be a positive integer greater than 1. If no index is written, it is assumed to be 2.

Roots and Negative Numbers

When dealing with negative numbers and radicals, the only restriction concerns negative numbers under even roots. We can have negative signs in front of radicals and negative numbers under odd roots and still obtain real numbers. Here are some examples to help clarify this. In the last section of this chapter we will see how to deal with even roots of negative numbers.

EXAMPLES Simplify each expression, if possible.

2. $\sqrt[3]{-8} = -2$ because $(-2)^3 = -8$.

3. $\sqrt{-4}$ is not a real number since there is no real number whose square is -4.

4. $-\sqrt{25} = -5$ is the negative square root of 25.

5. $\sqrt[5]{-32} = -2$ because $(-2)^5 = -32$.

6. $\sqrt[4]{-81}$ is not a real number since there is no real number we can raise to the fourth power and obtain -81.

Variables Under a Radical

From the preceding examples it is clear that we must be careful that we do not try to take an even root of a negative number. For this reason, we will assume that all variables appearing under a radical sign represent nonnegative numbers.

EXAMPLES Assume all variables represent nonnegative numbers and simplify each expression as much as possible.

7. $\sqrt{25a^4b^6} = 5a^2b^3$ because $(5a^2b^3)^2 = 25a^4b^6$.

8. $\sqrt[3]{x^6y^{12}} = x^2y^4$ because $(x^2y^4)^3 = x^6y^{12}$.

9. $\sqrt[4]{81r^8s^{20}} = 3r^2s^5$ because $(3r^2s^5)^4 = 81r^8s^{20}$.

Rational Numbers as Exponents

Next we develop a second kind of notation involving exponents that will allow us to designate square roots, cube roots, and so on in another way.

Consider the equation $x = 8^{1/3}$. Although we have not encountered fractional exponents before, let's assume that all the properties of exponents hold in this case. Cubing both sides of the equation, we have

$$x^3 = (8^{1/3})^3$$

$$x^3 = 8^{(1/3)(3)}$$

$$x^3 = 8^1$$

$$x^3 = 8$$

The last line tells us that x is the number whose cube is 8. It must be true, then, that x is the cube root of 8, $x = \sqrt[3]{8}$. Since we started with $x = 8^{1/3}$, it follows that

$$8^{1/3} = \sqrt[3]{8}$$

It seems reasonable, then, to define fractional exponents as indicating roots. Here is the formal definition.

DEFINITION If x is a real number and n is a positive integer greater than 1, then

$$x^{1/n} = \sqrt[n]{x} \qquad (x \geq 0 \text{ when } n \text{ is even})$$

In words: The quantity $x^{1/n}$ is the nth root of x.

With this definition we have a way of representing roots with exponents. Here are some examples.

EXAMPLES Write each expression as a root and then simplify, if possible.

10. $8^{1/3} = \sqrt[3]{8} = 2$

11. $36^{1/2} = \sqrt{36} = 6$

12. $-25^{1/2} = -\sqrt{25} = -5$

13. $(-25)^{1/2} = \sqrt{-25}$, which is not a real number

14. $\left(\frac{4}{9}\right)^{1/2} = \sqrt{\frac{4}{9}} = \frac{2}{3}$

The properties of exponents developed in a previous chapter were applied to integer exponents only. We will now extend these properties to include rational exponents also. We do so without proof.

Properties of Exponents If a and b are real numbers and r and s are rational numbers, and a and b are nonnegative whenever r and s indicate even roots, then

1. $a^r \cdot a^s = a^{r+s}$
2. $(a^r)^s = a^{rs}$
3. $(ab)^r = a^r b^r$
4. $a^{-r} = \frac{1}{a^r}$ $(a \neq 0)$
5. $\left(\frac{a}{b}\right)^r = \frac{a^r}{b^r}$ $(b \neq 0)$
6. $\frac{a^r}{a^s} = a^{r-s}$ $(a \neq 0)$

There are times when rational exponents can simplify our work with radicals. Here are Examples 8 and 9 again, but this time we will work them using rational exponents.

EXAMPLES Write each radical with a rational exponent and then simplify.

> **Note**
> Unless we state otherwise, assume all variables throughout this section and problem set represent nonnegative numbers.

15. $\sqrt[3]{x^6y^{12}} = (x^6y^{12})^{1/3}$

$= (x^6)^{1/3}(y^{12})^{1/3}$

$= x^2y^4$

16. $\sqrt[4]{81r^8s^{20}} = (81r^8s^{20})^{1/4}$

$= 81^{1/4}(r^8)^{1/4}(s^{20})^{1/4}$

$= 3r^2s^5$

So far, the numerators of all the rational exponents we have encountered have been 1. The next theorem extends the work we can do with rational exponents to rational exponents with numerators other than 1.

We can extend our properties of exponents with the following theorem.

The Rational Exponent Theorem If a is a nonnegative real number, m is an integer, and n is a positive integer, then

$$a^{m/n} = (a^{1/n})^m = (a^m)^{1/n}$$

Proof We can prove this theorem using the properties of exponents. Since $\frac{m}{n} = m\left(\frac{1}{n}\right)$ we have

$a^{m/n} = a^{m(1/n)}$	$a^{m/n} = a^{(1/n)(m)}$
$= (a^m)^{1/n}$	$= (a^{1/n})^m$

Here are some examples that illustrate how we use this theorem.

EXAMPLES

Simplify as much as possible.

> **Note**
> On a scientific calculator, Example 17 would look like this:
> 8 [y^x] [( ] [2] [÷] [3] [ )] [=]

17. $8^{2/3} = (8^{1/3})^2$ — **Rational exponent theorem**

$= 2^2$ — **Definition of fractional exponents**

$= 4$ — **The square of 2 is 4**

18. $25^{3/2} = (25^{1/2})^3$ — **Rational exponent theorem**

$= 5^3$ — **Definition of fractional exponents**

$= 125$ — **The cube of 5 is 125**

19. $9^{-3/2} = (9^{1/2})^{-3}$ — **Rational exponent theorem**

$= 3^{-3}$ — **Definition of fractional exponents**

$= \dfrac{1}{3^3}$ — **Property 4 for exponents**

$= \dfrac{1}{27}$ — **The cube of 3 is 27**

20. $\left(\dfrac{27}{8}\right)^{-4/3} = \left[\left(\dfrac{27}{8}\right)^{1/3}\right]^{-4}$ — **Rational exponent theorem**

$= \left(\dfrac{3}{2}\right)^{-4}$ — **Definition of fractional exponents**

$= \left(\dfrac{2}{3}\right)^{4}$ — **Property 4 for exponents**

$= \dfrac{16}{81}$ — **The fourth power of $\frac{2}{3}$ is $\frac{16}{81}$**

EXAMPLES

Assume all variables represent positive quantities and simplify as much as possible.

21. $x^{1/3} \cdot x^{5/6} = x^{1/3+5/6}$ — **Property 1**

$= x^{2/6+5/6}$ — **LCD is 6**

$= x^{7/6}$ — **Add fractions**

22. $(y^{2/3})^{3/4} = y^{(2/3)(3/4)}$ — **Property 2**

$= y^{1/2}$ — **Multiply fractions: $\frac{2}{3} \cdot \frac{3}{4} = \frac{6}{12} = \frac{1}{2}$**

23. $\dfrac{z^{1/3}}{z^{1/4}} = z^{1/3-1/4}$ — **Property 6**

$= z^{4/12-3/12}$ — **LCD is 12**

$= z^{1/12}$ — **Subtract fractions**

24. $\dfrac{(x^{-3}y^{1/2})^4}{x^{10}y^{3/2}} = \dfrac{(x^{-3})^4(y^{1/2})^4}{x^{10}y^{3/2}}$ — **Property 3**

$= \dfrac{x^{-12}y^2}{x^{10}y^{3/2}}$ — **Property 2**

$= x^{-22}y^{1/2}$ — **Property 6**

$= \dfrac{y^{1/2}}{x^{22}}$ — **Property 4**

FACTS FROM GEOMETRY

The Pythagorean Theorem (Again) and the Golden Rectangle

Now that we have had some experience working with square roots, we can rewrite the Pythagorean theorem using a square root. If triangle ABC is a right triangle with $C = 90°$, then the length of the longest side is the *positive square root* of the sum of the squares of the other two sides (see Figure 3).

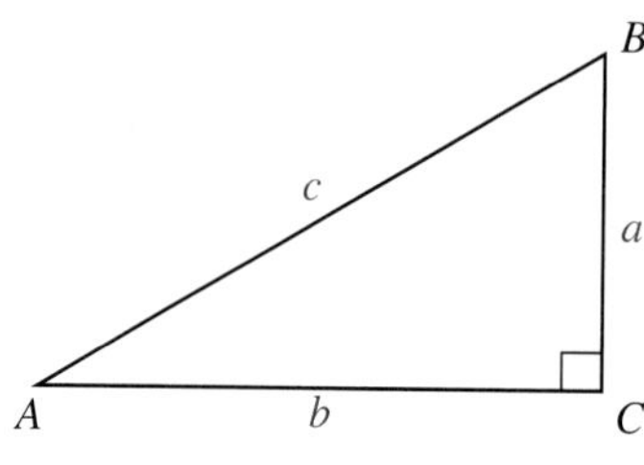

$$c = \sqrt{a^2 + b^2}$$

FIGURE 3

Constructing a Golden Rectangle from a Square of Side 2

Step 1: Draw a square with a side of length two. Connect the midpoint of side CD to corner B. (Note that we have labeled the midpoint of segment CD with the letter O.)

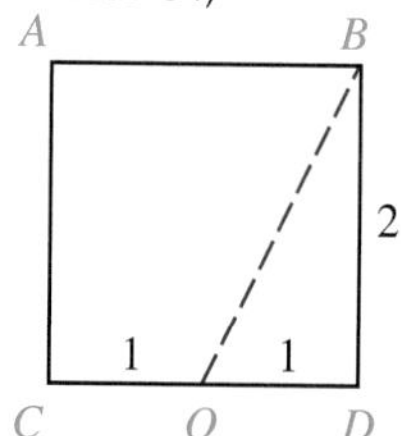

Step 2: Drop the diagonal from step 1 down so it aligns with side CD.

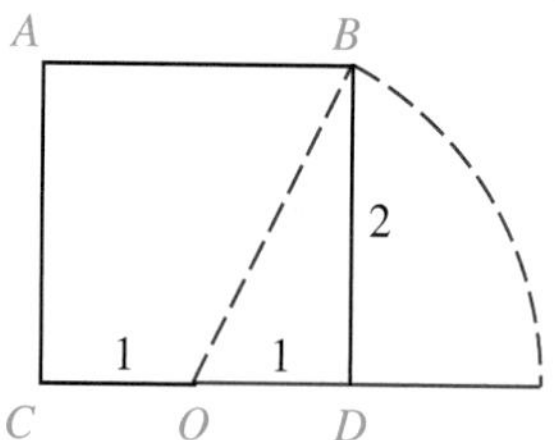

Step 3: Form rectangle $ACEF$. This is a golden rectangle.

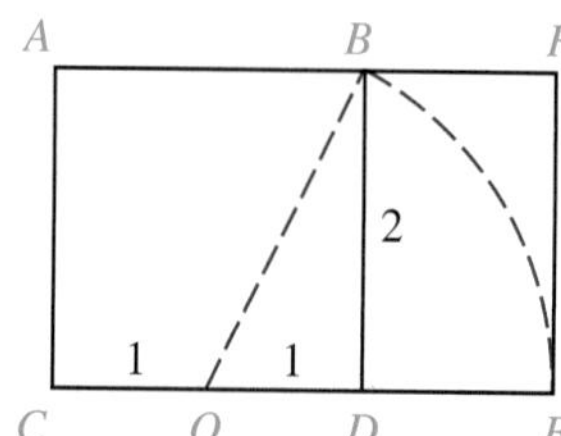

All golden rectangles are constructed from squares. Every golden rectangle, no matter how large or small it is, will have the same shape. To associate a number with the shape of the golden rectangle, we use the ratio of its length to its width. This ratio is called the *golden ratio.* To calculate the golden ratio, we must first find the length of the diagonal we used to construct the golden rec-

Note In the introduction to this section we mentioned the golden rectangle. Its origins can be traced back more than 2,000 years to the Greek civilization that produced Pythagoras, Socrates, Plato, Aristotle, and Euclid. The most important mathematical work to come from that Greek civilization was Euclid's *Elements,* an elegantly written summary of all that was known about geometry at that time in history. Euclid's *Elements,* according to Howard Eves, an authority on the history of mathematics, exercised a greater influence on scientific thinking than any other work. Here is how we construct a golden rectangle from a square of side 2, using the same method that Euclid used in his *Elements.*

tangle. Figure 4 shows the golden rectangle that we constructed from a square of side 2. The length of the diagonal OB is found by applying the Pythagorean theorem to triangle OBD.

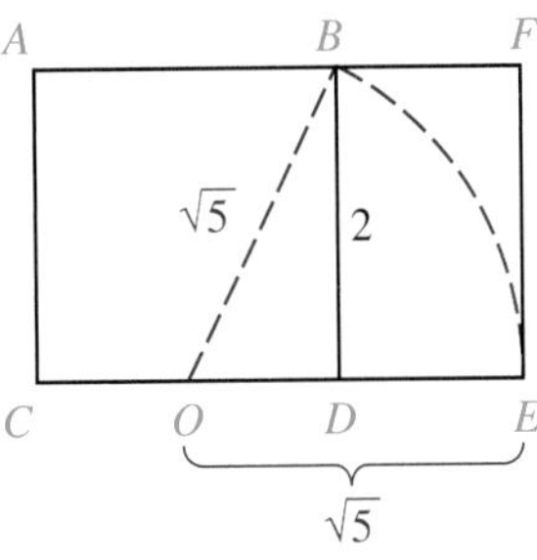

FIGURE 4

The length of segment OE is equal to the length of diagonal OB; both are $\sqrt{5}$. Since the distance from C to O is 1, the length CE of the golden rectangle is $1 + \sqrt{5}$. Now we can find the golden ratio:

$$\text{Golden ratio} = \frac{\text{length}}{\text{width}} = \frac{CE}{EF} = \frac{1+\sqrt{5}}{2}$$

USING TECHNOLOGY

Graphing Calculators—A Word of Caution

Some graphing calculators give surprising results when evaluating expressions such as $(-8)^{2/3}$. As you know from reading this section, the expression $(-8)^{2/3}$ simplifies to 4, either by taking the cube root first and then squaring the result, or by squaring the base first and then taking the cube root of the result. Here are three different ways to evaluate this expression on your calculator:

1. (−8)^(2/3) **To evaluate $(-8)^{2/3}$**
2. ((−8)^2)^(1/3) **To evaluate $((-8)^2)^{1/3}$**
3. ((−8)^(1/3))^2 **To evaluate $((-8)^{1/3})^2$**

Note any difference in the results.

Next, graph each of the following functions, one at a time.

1. $Y_1 = X^{2/3}$ 2. $Y_2 = (X^2)^{1/3}$ 3. $Y_3 = (X^{1/3})^2$

The correct graph is shown in Figure 5. Note which of your graphs match the correct graph.

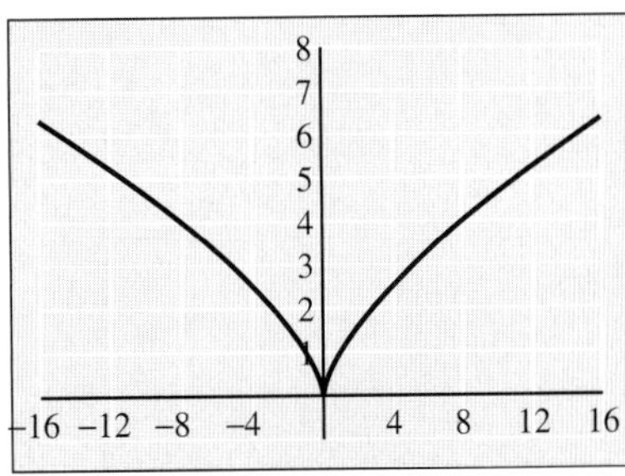

FIGURE 5

Different calculators evaluate exponential expressions in different ways. You should use the method (or methods) that gave you the correct graph.

LINKING OBJECTIVES AND EXAMPLES

Next to each **objective** we have listed the examples that are best described by that objective.

A	1–9
B	10–24

GETTING READY FOR CLASS

After reading through the preceding section, respond in your own words and in complete sentences.

1. Every real number has two square roots. Explain the notation we use to tell them apart. Use the square roots of 3 for examples.
2. Explain why a square root of -4 is not a real number.
3. We use the notation $\sqrt{2}$ to represent the positive square root of 2. Explain why there isn't a simpler way to express the positive square root of 2.
4. For the expression $a^{m/n}$, explain the significance of the numerator m and the significance of the denominator n in the exponent.

Problem Set 7.1

Online support materials can be found at www.thomsonedu.com/login

Find each of the following roots, if possible.

1. $\sqrt{144}$
2. $-\sqrt{144}$
3. $\sqrt{-144}$
4. $\sqrt{-49}$
5. $-\sqrt{49}$
6. $\sqrt{49}$
7. $\sqrt[3]{-27}$
8. $-\sqrt[3]{27}$
9. $\sqrt[4]{16}$
10. $-\sqrt[4]{16}$
11. $\sqrt[4]{-16}$
12. $-\sqrt[4]{-16}$
13. $\sqrt{0.04}$
14. $\sqrt{0.81}$
15. $\sqrt[3]{0.008}$
16. $\sqrt[3]{0.125}$
17. $\sqrt[3]{125}$
18. $\sqrt[3]{-125}$
19. ▸ $-\sqrt[3]{216}$
20. ▸ $-\sqrt[3]{-216}$
21. ▸ $\sqrt{\frac{1}{36}}$
22. ▸ $\sqrt{\frac{9}{25}}$
23. ▸ $\sqrt[3]{\frac{8}{125}}$
24. ▸ $\sqrt[3]{-\frac{27}{216}}$

Simplify each expression. Assume all variables represent nonnegative numbers.

25. $\sqrt{36a^8}$
26. $\sqrt{49a^{10}}$
27. ▸ $\sqrt[3]{27a^{12}}$
28. $\sqrt[3]{8a^{15}}$
29. $\sqrt[5]{32x^{10}y^5}$
30. $\sqrt[5]{32x^5y^{10}}$
31. ▸ $\sqrt[4]{16a^{12}b^{20}}$
32. $\sqrt[4]{81a^{24}b^8}$

Use the definition of rational exponents to write each of the following with the appropriate root. Then simplify.

33. $36^{1/2}$
34. $49^{1/2}$
35. $-9^{1/2}$
36. $-16^{1/2}$
37. $8^{1/3}$
38. $-8^{1/3}$
39. ▸ $(-8)^{1/3}$
40. $-27^{1/3}$
41. $32^{1/5}$
42. $81^{1/4}$
43. $\left(\frac{81}{25}\right)^{1/2}$
44. $\left(\frac{64}{125}\right)^{1/3}$

Use the rational exponent theorem to simplify each of the following as much as possible.

45. $27^{2/3}$
46. $8^{4/3}$
47. ▸ $25^{3/2}$
48. $81^{3/4}$

Simplify each expression. Remember, negative exponents give reciprocals.

49. $27^{-1/3}$
50. $9^{-1/2}$
51. ▸ $81^{-3/4}$
52. $4^{-3/2}$
53. $\left(\frac{25}{36}\right)^{-1/2}$
54. $\left(\frac{16}{49}\right)^{-1/2}$
55. $\left(\frac{81}{16}\right)^{-3/4}$
56. $\left(\frac{27}{8}\right)^{-2/3}$
57. $16^{1/2} + 27^{1/3}$
58. $25^{1/2} + 100^{1/2}$
59. $8^{-2/3} + 4^{-1/2}$
60. $49^{-1/2} + 25^{-1/2}$

■ = Videos available by instructor request

▸ = Online student support materials available at www.thomsonedu.com/login

Use the properties of exponents to simplify each of the following as much as possible. Assume all bases are positive.

61. $x^{3/5} \cdot x^{1/5}$
62. $x^{3/4} \cdot x^{5/4}$
63. $(a^{3/4})^{4/3}$
64. $(a^{2/3})^{3/4}$
65. $\dfrac{x^{1/5}}{x^{3/5}}$
66. $\dfrac{x^{2/7}}{x^{5/7}}$
67. $\dfrac{x^{5/6}}{x^{2/3}}$
68. $\dfrac{x^{7/8}}{x^{8/7}}$
69. $(x^{3/5}y^{5/6}z^{1/3})^{3/5}$
70. $(x^{3/4}y^{1/8}z^{5/6})^{4/5}$
71. $\dfrac{a^{3/4}b^2}{a^{7/8}b^{1/4}}$
72. $\dfrac{a^{1/3}b^4}{a^{3/5}b^{1/3}}$
73. $\dfrac{(y^{2/3})^{3/4}}{(y^{1/3})^{3/5}}$
74. $\dfrac{(y^{5/4})^{2/5}}{(y^{1/4})^{4/3}}$
75. $\left(\dfrac{a^{-1/4}}{b^{1/2}}\right)^8$
76. $\left(\dfrac{a^{-1/5}}{b^{1/3}}\right)^{15}$
77. $\dfrac{(r^{-2}s^{1/3})^6}{r^8s^{3/2}}$
78. $\dfrac{(r^{-5}s^{1/2})^4}{r^{12}s^{5/2}}$
79. $\dfrac{(25a^6b^4)^{1/2}}{(8a^{-9}b^3)^{-1/3}}$
80. $\dfrac{(27a^3b^6)^{1/3}}{(81a^8b^{-4})^{1/4}}$

Applying the Concepts

81. **Maximum Speed** The maximum speed (v) that an automobile can travel around a curve of radius r without skidding is given by the equation

$$v = \left(\frac{5r}{2}\right)^{1/2}$$

where v is in miles per hour and r is measured in feet. What is the maximum speed a car can travel around a curve with a radius of 250 feet without skidding?

82. **Golden Ratio** The golden ratio is the ratio of the length to the width in any golden rectangle. The exact value of this number is $\dfrac{1+\sqrt{5}}{2}$. Use a calculator to find a decimal approximation to this number and round it to the nearest thousandth.

83. **Chemistry** Figure 6 shows part of a model of a magnesium oxide (MgO) crystal. Each corner of the square is at the center of one oxygen ion (O^{2-}), and the center of the middle ion is at the center of the square. The radius for each oxygen ion is 126 picometers (pm), and the radius for each magnesium ion (Mg^{2+}) is 86 picometers.
   a. Find the length of the side of the square. Write your answer in picometers.
   b. Find the length of the diagonal of the square to the nearest picometer.
   c. If 1 meter is 10^{12} picometers, give the length of the diagonal of the square in meters.

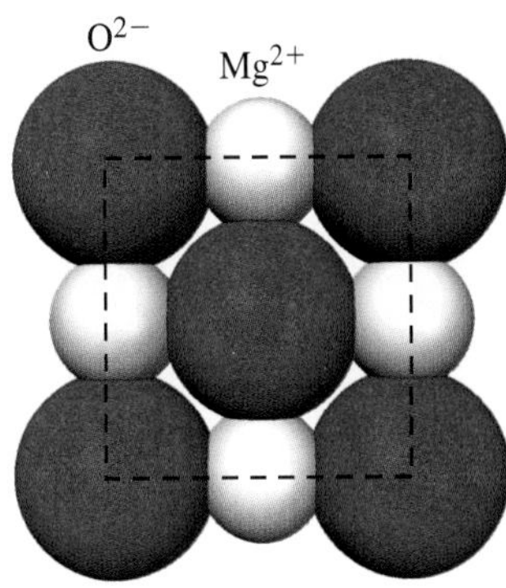

FIGURE 6 **(Susan M. Young)**

84. **Geometry** The length of each side of the cube shown in Figure 7 is 1 inch.
   a. Find the length of the diagonal CH.
   b. Find the length of the diagonal CF.

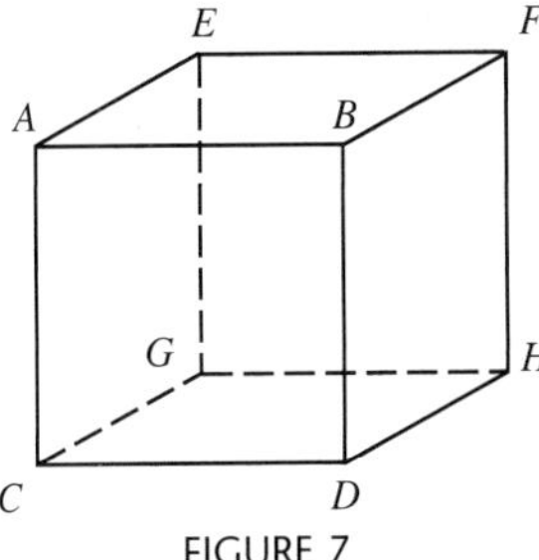

FIGURE 7

85. **Comparing Graphs** Identify the graph with the correct equation.
   a. $y = x$
   b. $y = x^2$
   c. $y = x^{2/3}$
   d. What are the two points of intersection of all three graphs?

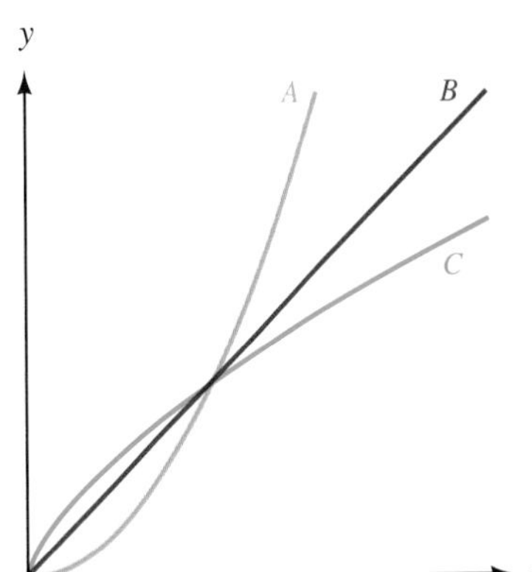

86. **Falling Objects** The time t in seconds it takes an object to fall d feet is given by the equation

$$t = \frac{1}{4}\sqrt{d}$$

a. The Sears Tower in Chicago is 1,450 feet tall. How long would it take a penny to fall to the ground from the top of the Sears Tower? Round to the nearest hundredth.

b. An object took 30 seconds to fall to the ground. From what distance must it have been dropped?

Maintaining Your Skills

Multiply.

87. $x^2(x^4 - x)$

88. $5x^2(2x^3 - x)$

89. $(x - 3)(x + 5)$

90. $(x - 2)(x + 2)$

91. $(x^2 - 5)^2$

92. $(x^2 + 5)^2$

93. $(x - 3)(x^2 + 3x + 9)$

94. $(x + 3)(x^2 - 3x + 9)$

Getting Ready for the Next Section

Simplify.

95. $x^2(x^4 - x^3)$

96. $(x^2 - 3)(x^2 + 5)$

97. $(3a - 2b)(4a - b)$

98. $(x^2 + 3)^2$

99. $(x^3 - 2)(x^3 + 2)$

100. $(a - b)(a^2 + ab + b^2)$

101. $\dfrac{15x^2y - 20x^4y^2}{5xy}$

102. $\dfrac{12a^3b^2 - 24a^2b^4}{3ab}$

Factor.

103. $x^2 - 3x - 10$

104. $x^2 + x - 12$

105. $6x^2 + 11x - 10$

106. $10x^2 - x - 3$

Use rules of exponents to simplify.

107. $x^{2/3} \cdot x^{4/3}$

108. $x^{1/4} \cdot x^{3/4}$

109. $(t^{1/2})^2$

110. $(x^{3/2})^2$

111. $\dfrac{x^{2/3}}{x^{1/3}}$

112. $\dfrac{x^{1/2}}{x^{1/2}}$

Extending the Concepts

113. Show that the expression $(a^{1/2} + b^{1/2})^2$ is not equal to $a + b$ by replacing a with 9 and b with 4 in both expressions and then simplifying each.

114. Show that the statement $(a^2 + b^2)^{1/2} = a + b$ is not, in general, true by replacing a with 3 and b with 4 and then simplifying both sides.

115. You may have noticed, if you have been using a calculator to find roots, that you can find the fourth root of a number by pressing the square root button twice. Written in symbols, this fact looks like this:

$$\sqrt{\sqrt{a}} = \sqrt[4]{a} \qquad (a \ge 0)$$

Show that this statement is true by rewriting each side with exponents instead of radical notation and then simplifying the left side.

116. Show that the following statement is true by rewriting each side with exponents instead of radical notation and then simplifying the left side.

$$\sqrt[3]{\sqrt{a}} = \sqrt[6]{a} \qquad (a \ge 0)$$

7.2 More Expressions Involving Rational Exponents

OBJECTIVES

A Multiply expressions with rational exponents.

B Divide expressions with rational exponents.

C Factor expressions with rational exponents.

D Add and subtract expressions with rational exponents.

Suppose you purchased 10 silver proof coin sets in 1997 for \$21 each, for a total investment of \$210. Three years later, in 2000, you find that each set is worth \$30, which means that your 10 sets have a total value of \$300.

You can calculate the annual rate of return on this investment using a formula that involves rational exponents. The annual rate of return will tell you at what interest rate you would have to invest your original \$210 for it to be worth \$300 three years later. As you will see at the end of this section, the annual rate of return on this investment is 12.6%, which is a good return on your money.

In this section we will look at multiplication, division, factoring, and simplification of some expressions that resemble polynomials but contain rational exponents. The problems in this section will be of particular interest to you if you are planning to take either an engineering calculus class or a business calculus class. As was the case in the previous section, we will assume all variables represent nonnegative real numbers. That way, we will not have to worry about the possibility of introducing undefined terms—even roots of negative numbers—into any of our examples. Let's begin this section with a look at multiplication of expressions containing rational exponents.

EXAMPLE 1 Multiply $x^{2/3}(x^{4/3} - x^{1/3})$.

SOLUTION Applying the distributive property and then simplifying the resulting terms, we have:

$$
\begin{aligned}
x^{2/3}(x^{4/3} - x^{1/3}) &= x^{2/3}x^{4/3} - x^{2/3}x^{1/3} && \textbf{Distributive property} \\
&= x^{6/3} - x^{3/3} && \textbf{Add exponents} \\
&= x^2 - x && \textbf{Simplify}
\end{aligned}
$$

EXAMPLE 2 Multiply $(x^{2/3} - 3)(x^{2/3} + 5)$.

SOLUTION Applying the FOIL method, we multiply as if we were multiplying two binomials:

$$
\begin{aligned}
(x^{2/3} - 3)(x^{2/3} + 5) &= x^{2/3}x^{2/3} + 5x^{2/3} - 3x^{2/3} - 15 \\
&= x^{4/3} + 2x^{2/3} - 15
\end{aligned}
$$

EXAMPLE 3 Multiply $(3a^{1/3} - 2b^{1/3})(4a^{1/3} - b^{1/3})$.

SOLUTION Again, we use the FOIL method to multiply:

$$\begin{aligned}(3a^{1/3} - 2b^{1/3})(4a^{1/3} - b^{1/3}) \\ &= 3a^{1/3}4a^{1/3} - 3a^{1/3}b^{1/3} - 2b^{1/3}4a^{1/3} + 2b^{1/3}b^{1/3} \\ &= 12a^{2/3} - 11a^{1/3}b^{1/3} + 2b^{2/3}\end{aligned}$$

EXAMPLE 4 Expand $(t^{1/2} - 5)^2$.

SOLUTION We can use the definition of exponents and the FOIL method:

$$\begin{aligned}(t^{1/2} - 5)^2 &= (t^{1/2} - 5)(t^{1/2} - 5) \\ &= t^{1/2}t^{1/2} - 5t^{1/2} - 5t^{1/2} + 25 \\ &= t - 10t^{1/2} + 25\end{aligned}$$

We can obtain the same result by using the formula for the square of a binomial, $(a - b)^2 = a^2 - 2ab + b^2$.

$$\begin{aligned}(t^{1/2} - 5)^2 &= (t^{1/2})^2 - 2t^{1/2} \cdot 5 + 5^2 \\ &= t - 10t^{1/2} + 25\end{aligned}$$

EXAMPLE 5 Multiply $(x^{3/2} - 2^{3/2})(x^{3/2} + 2^{3/2})$.

SOLUTION This product has the form $(a - b)(a + b)$, which will result in the difference of two squares, $a^2 - b^2$:

$$\begin{aligned}(x^{3/2} - 2^{3/2})(x^{3/2} + 2^{3/2}) &= (x^{3/2})^2 - (2^{3/2})^2 \\ &= x^3 - 2^3 \\ &= x^3 - 8\end{aligned}$$

EXAMPLE 6 Multiply $(a^{1/3} - b^{1/3})(a^{2/3} + a^{1/3}b^{1/3} + b^{2/3})$.

SOLUTION We can find this product by multiplying in columns:

$$\begin{array}{llll} a^{2/3} + a^{1/3}b^{1/3} + b^{2/3} \\ \quad\quad a^{1/3} - b^{1/3} \\ \hline a \quad + a^{2/3}b^{1/3} + a^{1/3}b^{2/3} \\ \quad\quad - a^{2/3}b^{1/3} - a^{1/3}b^{2/3} - b \\ \hline a \quad\quad\quad\quad\quad\quad\quad - b \end{array}$$

The product is $a - b$.

Our next example involves division with expressions that contain rational exponents. As you will see, this kind of division is very similar to division of a polynomial by a monomial.

EXAMPLE 7

Divide $\dfrac{15x^{2/3}y^{1/3} - 20x^{4/3}y^{2/3}}{5x^{1/3}y^{1/3}}$.

Note
For Example 7, assume x and y are positive numbers.

SOLUTION We can approach this problem in the same way we approached division by a monomial. We simply divide each term in the numerator by the term in the denominator:

$$\frac{15x^{2/3}y^{1/3} - 20x^{4/3}y^{2/3}}{5x^{1/3}y^{1/3}} = \frac{15x^{2/3}y^{1/3}}{5x^{1/3}y^{1/3}} - \frac{20x^{4/3}y^{2/3}}{5x^{1/3}y^{1/3}}$$

$$= 3x^{1/3} - 4xy^{1/3}$$

The next three examples involve factoring. In the first example, we are told what to factor from each term of an expression.

EXAMPLE 8

Factor $3(x-2)^{1/3}$ from $12(x-2)^{4/3} - 9(x-2)^{1/3}$, and then simplify, if possible.

SOLUTION This solution is similar to factoring out the greatest common factor:

$$12(x-2)^{4/3} - 9(x-2)^{1/3} = 3(x-2)^{1/3}[4(x-2) - 3]$$

$$= 3(x-2)^{1/3}(4x - 11)$$

Although an expression containing rational exponents is not a polynomial—remember, a polynomial must have exponents that are whole numbers—we are going to treat the expressions that follow as if they were polynomials.

EXAMPLE 9

Factor $x^{2/3} - 3x^{1/3} - 10$ as if it were a trinomial.

SOLUTION We can think of $x^{2/3} - 3x^{1/3} - 10$ as if it is a trinomial in which the variable is $x^{1/3}$. To see this, replace $x^{1/3}$ with y to get

$$y^2 - 3y - 10$$

Since this trinomial in y factors as $(y-5)(y+2)$, we can factor our original expression similarly:

$$x^{2/3} - 3x^{1/3} - 10 = (x^{1/3} - 5)(x^{1/3} + 2)$$

Remember, with factoring, we can always multiply our factors to check that we have factored correctly.

EXAMPLE 10

Factor $6x^{2/5} + 11x^{1/5} - 10$ as if it were a trinomial.

SOLUTION We can think of the expression in question as a trinomial in $x^{1/5}$.

$$6x^{2/5} + 11x^{1/5} - 10 = (3x^{1/5} - 2)(2x^{1/5} + 5)$$

In our next example, we combine two expressions by applying the methods we used to add and subtract fractions or rational expressions.

EXAMPLE 11 Subtract $(x^2+4)^{1/2} - \dfrac{x^2}{(x^2+4)^{1/2}}$.

SOLUTION To combine these two expressions, we need to find a least common denominator, change to equivalent fractions, and subtract numerators. The least common denominator is $(x^2+4)^{1/2}$.

$$(x^2+4)^{1/2} - \frac{x^2}{(x^2+4)^{1/2}} = \frac{(x^2+4)^{1/2}}{1} \cdot \frac{\mathbf{(x^2+4)^{1/2}}}{\mathbf{(x^2+4)^{1/2}}} - \frac{x^2}{(x^2+4)^{1/2}}$$

$$= \frac{x^2+4-x^2}{(x^2+4)^{1/2}}$$

$$= \frac{4}{(x^2+4)^{1/2}}$$

EXAMPLE 12 If you purchase an investment for P dollars and t years later it is worth A dollars, then the annual rate of return r on that investment is given by the formula

$$r = \left(\frac{A}{P}\right)^{1/t} - 1$$

Find the annual rate of return on a coin collection that was purchased for $210 and sold 3 years later for $300.

SOLUTION Using $A = 300$, $P = 210$, and $t = 3$ in the formula, we have

$$r = \left(\frac{300}{210}\right)^{1/3} - 1$$

The easiest way to simplify this expression is with a calculator.

[(] 300 [÷] 210 [)] [^] [(] 1 [÷] 3 [)] [−] 1 [=]

Allowing three decimal places, the result is 0.126. The annual return on the coin collection is approximately 12.6%. To do as well with a savings account, we would have to invest the original $210 in an account that paid 12.6%, compounded annually.

LINKING OBJECTIVES AND EXAMPLES

Next to each objective we have listed the examples that are best described by that objective.

A	1–6
B	7
C	8–10
D	11

GETTING READY FOR CLASS

After reading through the preceding section, respond in your own words and in complete sentences.

1. When multiplying expressions with fractional exponents, when do we add the fractional exponents?
2. Is it possible to multiply two expressions with fractional exponents and end up with an expression containing only integer exponents? Support your answer with examples.
3. Write an application modeled by the equation $r = \left(\dfrac{1{,}000}{600}\right)^{1/8} - 1$.
4. When can you use the FOIL method with expressions that contain rational exponents?

Problem Set 7.2

Online support materials can be found at www.thomsonedu.com/login

Multiply. (Assume all variables in this problem set represent nonnegative real numbers.)

1. $x^{2/3}(x^{1/3} + x^{4/3})$
2. $x^{2/5}(x^{3/5} - x^{8/5})$
3. $a^{1/2}(a^{3/2} - a^{1/2})$
4. $a^{1/4}(a^{3/4} + a^{7/4})$
5. $2x^{1/3}(3x^{8/3} - 4x^{5/3} + 5x^{2/3})$
6. $5x^{1/2}(4x^{5/2} + 3x^{3/2} + 2x^{1/2})$
7. $4x^{1/2}y^{3/5}(3x^{3/2}y^{-3/5} - 9x^{-1/2}y^{7/5})$
8. $3x^{4/5}y^{1/3}(4x^{6/5}y^{-1/3} - 12x^{-4/5}y^{5/3})$
9. $(x^{2/3} - 4)(x^{2/3} + 2)$
10. $(x^{2/3} - 5)(x^{2/3} + 2)$
11. $(a^{1/2} - 3)(a^{1/2} - 7)$
12. $(a^{1/2} - 6)(a^{1/2} - 2)$
13. $(4y^{1/3} - 3)(5y^{1/3} + 2)$
14. $(5y^{1/3} - 2)(4y^{1/3} + 3)$
15. $(5x^{2/3} + 3y^{1/2})(2x^{2/3} + 3y^{1/2})$
16. $(4x^{2/3} - 2y^{1/2})(5x^{2/3} - 3y^{1/2})$
17. $(t^{1/2} + 5)^2$
18. $(t^{1/2} - 3)^2$
19. $(x^{3/2} + 4)^2$
20. $(x^{3/2} - 6)^2$
21. $(a^{1/2} - b^{1/2})^2$
22. $(a^{1/2} + b^{1/2})^2$
23. $(2x^{1/2} - 3y^{1/2})^2$
24. $(5x^{1/2} + 4y^{1/2})^2$
25. $(a^{1/2} - 3^{1/2})(a^{1/2} + 3^{1/2})$
26. $(a^{1/2} - 5^{1/2})(a^{1/2} + 5^{1/2})$
27. $(x^{3/2} + y^{3/2})(x^{3/2} - y^{3/2})$
28. $(x^{5/2} + y^{5/2})(x^{5/2} - y^{5/2})$
29. $(t^{1/2} - 2^{3/2})(t^{1/2} + 2^{3/2})$
30. $(t^{1/2} - 5^{3/2})(t^{1/2} + 5^{3/2})$
31. $(2x^{3/2} + 3^{1/2})(2x^{3/2} - 3^{1/2})$
32. $(3x^{1/2} + 2^{3/2})(3x^{1/2} - 2^{3/2})$
33. $(x^{1/3} + y^{1/3})(x^{2/3} - x^{1/3}y^{1/3} + y^{2/3})$
34. $(x^{1/3} - y^{1/3})(x^{2/3} + x^{1/3}y^{1/3} + y^{2/3})$
35. $(a^{1/3} - 2)(a^{2/3} + 2a^{1/3} + 4)$
36. $(a^{1/3} + 3)(a^{2/3} - 3a^{1/3} + 9)$
37. $(2x^{1/3} + 1)(4x^{2/3} - 2x^{1/3} + 1)$
38. $(3x^{1/3} - 1)(9x^{2/3} + 3x^{1/3} + 1)$
39. $(t^{1/4} - 1)(t^{1/4} + 1)(t^{1/2} + 1)$
40. $(t^{1/4} - 2)(t^{1/4} + 2)(t^{1/2} + 4)$

Divide. (Assume all variables represent positive numbers.)

41. $\dfrac{18x^{3/4} + 27x^{1/4}}{9x^{1/4}}$
42. $\dfrac{25x^{1/4} + 30x^{3/4}}{5x^{1/4}}$
43. $\dfrac{12x^{2/3}y^{1/3} - 16x^{1/3}y^{2/3}}{4x^{1/3}y^{1/3}}$
44. $\dfrac{12x^{4/3}y^{1/3} - 18x^{1/3}y^{4/3}}{6x^{1/3}y^{1/3}}$
45. $\dfrac{21a^{7/5}b^{3/5} - 14a^{2/5}b^{8/5}}{7a^{2/5}b^{3/5}}$
46. $\dfrac{24a^{9/5}b^{3/5} - 16a^{4/5}b^{8/5}}{8a^{4/5}b^{3/5}}$
47. Factor $3(x - 2)^{1/2}$ from $12(x - 2)^{3/2} - 9(x - 2)^{1/2}$.
48. Factor $4(x + 1)^{1/3}$ from $4(x + 1)^{4/3} + 8(x + 1)^{1/3}$.
49. Factor $5(x - 3)^{7/5}$ from $5(x - 3)^{12/5} - 15(x - 3)^{7/5}$.
50. Factor $6(x + 3)^{8/7}$ from $6(x + 3)^{15/7} - 12(x + 3)^{8/7}$.
51. Factor $3(x + 1)^{1/2}$ from $9x(x + 1)^{3/2} + 6(x + 1)^{1/2}$.
52. Factor $4x(x + 1)^{1/2}$ from $4x^2(x + 1)^{1/2} + 8x(x + 1)^{3/2}$.

Factor each of the following as if it were a trinomial.

53. $x^{2/3} - 5x^{1/3} + 6$
54. $x^{2/3} - x^{1/3} - 6$
55. $a^{2/5} - 2a^{1/5} - 8$
56. $a^{2/5} + 2a^{1/5} - 8$
57. $2y^{2/3} - 5y^{1/3} - 3$
58. $3y^{2/3} + 5y^{1/3} - 2$
59. $9t^{2/5} - 25$

= Videos available by instructor request

▶ = Online student support materials available at www.thomsonedu.com/login

60. $16t^{2/5} - 49$

61. $4x^{2/7} + 20x^{1/7} + 25$

62. $25x^{2/7} - 20x^{1/7} + 4$

Evaluate the following functions for the given value.

63. If $f(x) = x - 2\sqrt{x} - 8$, find $f(4)$.

64. If $g(x) = x - 2\sqrt{x} - 3$, find $g(9)$.

65. If $f(x) = 2x + 9\sqrt{x} - 5$, find $f(25)$.

66. If $f(t) = t - 2\sqrt{t} - 15$, find $f(9)$.

67. If $g(x) = 2x - \sqrt{x} - 6$, find $g\left(\frac{9}{4}\right)$.

68. If $f(x) = 2x + \sqrt{x} - 15$, find $f(9)$.

69. If $f(x) = x^{2/3} - 2x^{1/3} - 8$, find $f(-8)$.

70. If $g(x) = x^{2/3} + 4x^{1/3} - 12$, find $g(8)$.

Simplify each of the following to a single fraction. (Assume all variables represent positive numbers.)

71. $\frac{3}{x^{1/2}} + x^{1/2}$

72. $\frac{2}{x^{1/2}} - x^{1/2}$

73. $x^{2/3} + \frac{5}{x^{1/3}}$

74. $x^{3/4} - \frac{7}{x^{1/4}}$

75. $\frac{3x^2}{(x^3 + 1)^{1/2}} + (x^3 + 1)^{1/2}$

76. $\frac{x^3}{(x^2 - 1)^{1/2}} + 2x(x^2 - 1)^{1/2}$

77. $\frac{x^2}{(x^2 + 4)^{1/2}} - (x^2 + 4)^{1/2}$

78. $\frac{x^5}{(x^2 - 2)^{1/2}} + 4x^3(x^2 - 2)^{1/2}$

Applying the Concepts

79. **Investing** A coin collection is purchased as an investment for \$500 and sold 4 years later for \$900. Find the annual rate of return on the investment.

80. **Investing** An investor buys stock in a company for \$800. Five years later, the same stock is worth \$1,600. Find the annual rate of return on the stocks.

Maintaining Your Skills

Reduce to lowest terms.

81. $\frac{x^2 - 9}{x^4 - 81}$

82. $\frac{6 - a - a^2}{3 - 2a - a^2}$

Divide.

83. $\frac{15x^2y - 20x^4y^2}{5xy}$

84. $\frac{12x^3y^2 - 24x^2y^3}{6xy}$

Divide using long division.

85. $\frac{10x^2 + 7x - 12}{2x + 3}$

86. $\frac{6x^2 - x - 35}{2x - 5}$

87. $\frac{x^3 - 125}{x - 5}$

88. $\frac{x^3 + 64}{x + 4}$

Getting Ready for the Next Section

Simplify. Assume all variable are positive real numbers.

89. $\sqrt{25}$

90. $\sqrt{4}$

91. $\sqrt{6^2}$

92. $\sqrt{3^2}$

93. $\sqrt{16x^4y^2}$

94. $\sqrt{4x^6y^8}$

95. $\sqrt{(5y)^2}$

96. $\sqrt{(8x^3)^2}$

97. $\sqrt[3]{27}$

98. $\sqrt[3]{-8}$

99. $\sqrt[3]{2^3}$

100. $\sqrt[3]{(-5)^3}$

101. $\sqrt[3]{8a^3b^3}$

102. $\sqrt[3]{64a^6b^3}$

Fill in the blank.

103. $50 = \underline{\qquad} \cdot 2$

104. $12 = \underline{\qquad} \cdot 3$

105. $48x^4y^3 = \underline{\qquad} \cdot y$

106. $40a^5b^4 = \underline{\qquad} \cdot 5a^2b$

107. $12x^7y^6 = \underline{\qquad} \cdot 3x$

108. $54a^6b^2c^4 = \underline{\qquad} \cdot 2b^2c$

7.3 Simplified Form for Radicals

OBJECTIVES

A Write radical expressions in simplified form.

B Rationalize a denominator that contains only one term.

Earlier in this chapter, we showed how the Pythagorean theorem can be used to construct a golden rectangle. In a similar manner, the Pythagorean theorem can be used to construct the attractive spiral shown here.

The Spiral of Roots

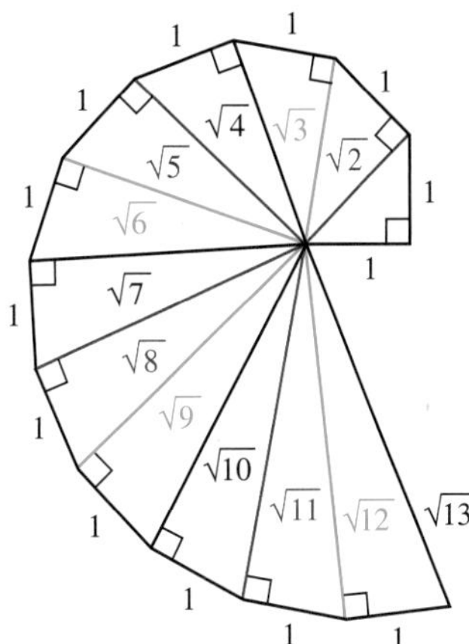

This spiral is called the Spiral of Roots because each of the diagonals is the positive square root of one of the positive integers. At the end of this section, we will use the Pythagorean theorem and some of the material in this section to construct this spiral.

In this section we will use radical notation instead of rational exponents. We will begin by stating two properties of radicals. Following this, we will give a definition for simplified form for radical expressions. The examples in this section show how we use the properties of radicals to write radical expressions in simplified form.

There are two properties of radicals. For these two properties, we will assume a and b are nonnegative real numbers whenever n is an even number.

Property 1 for Radicals

$$\sqrt[n]{ab} = \sqrt[n]{a}\,\sqrt[n]{b}$$

In words: The nth root of a product is the product of the nth roots.

Proof of Property 1

$$\sqrt[n]{ab} = (ab)^{1/n} \quad \text{Definition of fractional exponents}$$

$$= a^{1/n}b^{1/n} \quad \text{Exponents distribute over products}$$

$$= \sqrt[n]{a}\,\sqrt[n]{b} \quad \text{Definition of fractional exponents}$$

Note

There is no property for radicals that says the nth root of a sum is the sum of the nth roots; that is, in general

$$\sqrt[n]{a+b} \neq \sqrt[n]{a} + \sqrt[n]{b}$$

Property 2 for Radicals

$$\sqrt[n]{\frac{a}{b}} = \frac{\sqrt[n]{a}}{\sqrt[n]{b}} \qquad (b \neq 0)$$

In words: The nth root of a quotient is the quotient of the nth roots.

The proof of property 2 is similar to the proof of property 1.

The two properties of radicals allow us to change the form of and simplify radical expressions without changing their value.

Note
Writing a radical expression in simplified form does not always result in a simpler-looking expression. Simplified form for radicals is a way of writing radicals so they are easiest to work with.

Simplified Form for Radical Expressions A radical expression is in *simplified form* if

1. None of the factors of the radicand (the quantity under the radical sign) can be written as powers greater than or equal to the index—that is, no perfect squares can be factors of the quantity under a square root sign, no perfect cubes can be factors of what is under a cube root sign, and so forth;
2. There are no fractions under the radical sign; and
3. There are no radicals in the denominator.

Satisfying the first condition for simplified form actually amounts to taking as much out from under the radical sign as possible. The following examples illustrate the first condition for simplified form.

EXAMPLE 1 Write $\sqrt{50}$ in simplified form.

SOLUTION The largest perfect square that divides 50 is 25. We write 50 as $25 \cdot 2$ and apply property 1 for radicals:

$$\begin{aligned} \sqrt{50} &= \sqrt{25 \cdot 2} && \mathbf{50 = 25 \cdot 2} \\ &= \sqrt{25}\,\sqrt{2} && \textbf{Property 1} \\ &= 5\sqrt{2} && \mathbf{\sqrt{25} = 5} \end{aligned}$$

We have taken as much as possible out from under the radical sign—in this case, factoring 25 from 50 and then writing $\sqrt{25}$ as 5.

Note
Unless we state otherwise, assume all variables throughout this section represent nonnegative numbers. Further, if a variable appears in a denominator, assume the variable cannot be 0.

EXAMPLE 2 Write in simplified form: $\sqrt{48x^4y^3}$, where $x, y \geq 0$.

SOLUTION The largest perfect square that is a factor of the radicand is $16x^4y^2$. Applying property 1 again, we have

$$\begin{aligned} \sqrt{48x^4y^3} &= \sqrt{16x^4y^2 \cdot 3y} \\ &= \sqrt{16x^4y^2}\,\sqrt{3y} \\ &= 4x^2y\sqrt{3y} \end{aligned}$$

EXAMPLE 3 Write $\sqrt[3]{40a^5b^4}$ in simplified form.

SOLUTION We now want to factor the largest perfect cube from the radicand. We write $40a^5b^4$ as $8a^3b^3 \cdot 5a^2b$ and proceed as we did in Examples 1 and 2.

$$\begin{aligned} \sqrt[3]{40a^5b^4} &= \sqrt[3]{8a^3b^3 \cdot 5a^2b} \\ &= \sqrt[3]{8a^3b^3}\,\sqrt[3]{5a^2b} \\ &= 2ab\sqrt[3]{5a^2b} \end{aligned}$$

Here are some further examples concerning the first condition for simplified form.

EXAMPLES Write each expression in simplified form.

4. $\sqrt{12x^7y^6} = \sqrt{4x^6y^6 \cdot 3x}$

$= \sqrt{4x^6y^6}\sqrt{3x}$

$= 2x^3y^3\sqrt{3x}$

5. $\sqrt[3]{54a^6b^2c^4} = \sqrt[3]{27a^6c^3 \cdot 2b^2c}$

$= \sqrt[3]{27a^6c^3}\sqrt[3]{2b^2c}$

$= 3a^2c\sqrt[3]{2b^2c}$

The second property of radicals is used to simplify a radical that contains a fraction.

EXAMPLE 6 Simplify $\sqrt{\frac{3}{4}}$.

SOLUTION Applying property 2 for radicals, we have

$$\sqrt{\frac{3}{4}} = \frac{\sqrt{3}}{\sqrt{4}} \quad \textbf{Property 2}$$

$$= \frac{\sqrt{3}}{2} \quad \sqrt{4} = 2$$

The last expression is in simplified form because it satisfies all three conditions for simplified form.

EXAMPLE 7 Write $\sqrt{\frac{5}{6}}$ in simplified form.

SOLUTION Proceeding as in Example 6, we have

$$\sqrt{\frac{5}{6}} = \frac{\sqrt{5}}{\sqrt{6}}$$

> *Note*
> The idea behind rationalizing the denominator is to produce a perfect square under the square root sign in the denominator. This is accomplished by multiplying both the numerator and denominator by the appropriate radical.

The resulting expression satisfies the second condition for simplified form since neither radical contains a fraction. It does, however, violate condition 3 since it has a radical in the denominator. Getting rid of the radical in the denominator is called *rationalizing the denominator* and is accomplished, in this case, by multiplying the numerator and denominator by $\sqrt{6}$:

$$\frac{\sqrt{5}}{\sqrt{6}} = \frac{\sqrt{5}}{\sqrt{6}} \cdot \frac{\sqrt{6}}{\sqrt{6}}$$

$$= \frac{\sqrt{30}}{\sqrt{6^2}}$$

$$= \frac{\sqrt{30}}{6}$$

EXAMPLES

Rationalize the denominator.

8. $\dfrac{4}{\sqrt{3}} = \dfrac{4}{\sqrt{3}} \cdot \dfrac{\sqrt{\mathbf{3}}}{\sqrt{\mathbf{3}}}$

$$= \frac{4\sqrt{3}}{\sqrt{3^2}}$$

$$= \frac{4\sqrt{3}}{3}$$

9. $\dfrac{2\sqrt{3x}}{\sqrt{5y}} = \dfrac{2\sqrt{3x}}{\sqrt{5y}} \cdot \dfrac{\sqrt{\mathbf{5y}}}{\sqrt{\mathbf{5y}}}$

$$= \frac{2\sqrt{15xy}}{\sqrt{(5y)^2}}$$

$$= \frac{2\sqrt{15xy}}{5y}$$

When the denominator involves a cube root, we must multiply by a radical that will produce a perfect cube under the cube root sign in the denominator, as our next example illustrates.

EXAMPLE 10

Rationalize the denominator in $\dfrac{7}{\sqrt[3]{4}}$.

SOLUTION Since $4 = 2^2$, we can multiply both numerator and denominator by $\sqrt[3]{2}$ and obtain $\sqrt[3]{2^3}$ in the denominator.

$$\frac{7}{\sqrt[3]{4}} = \frac{7}{\sqrt[3]{2^2}}$$

$$= \frac{7}{\sqrt[3]{2^2}} \cdot \frac{\sqrt[3]{\mathbf{2}}}{\sqrt[3]{\mathbf{2}}}$$

$$= \frac{7\sqrt[3]{2}}{\sqrt[3]{2^3}}$$

$$= \frac{7\sqrt[3]{2}}{2}$$

EXAMPLE 11

Simplify $\sqrt{\dfrac{12x^5y^3}{5z}}$.

SOLUTION We use property 2 to write the numerator and denominator as two separate radicals:

$$\sqrt{\frac{12x^5y^3}{5z}} = \frac{\sqrt{12x^5y^3}}{\sqrt{5z}}$$

Simplifying the numerator, we have

$$\frac{\sqrt{12x^5y^3}}{\sqrt{5z}} = \frac{\sqrt{4x^4y^2}\,\sqrt{3xy}}{\sqrt{5z}}$$

$$= \frac{2x^2y\sqrt{3xy}}{\sqrt{5z}}$$

To rationalize the denominator, we multiply the numerator and denominator by $\sqrt{5z}$:

$$\frac{2x^2y\sqrt{3xy}}{\sqrt{5z}} \cdot \frac{\sqrt{\mathbf{5z}}}{\sqrt{\mathbf{5z}}} = \frac{2x^2y\sqrt{15xyz}}{\sqrt{(5z)^2}}$$

$$= \frac{2x^2y\sqrt{15xyz}}{5z}$$

The Square Root of a Perfect Square

So far in this chapter we have assumed that all our variables are nonnegative when they appear under a square root symbol. There are times, however, when this is not the case.

Consider the following two statements:

$$\sqrt{3^2} = \sqrt{9} = 3 \qquad \text{and} \qquad \sqrt{(-3)^2} = \sqrt{9} = 3$$

Whether we operate on 3 or −3, the result is the same: Both expressions simplify to 3. The other operation we have worked with in the past that produces the same result is absolute value; that is,

$$|3| = 3 \qquad \text{and} \qquad |-3| = 3$$

This leads us to the next property of radicals.

Property 3 for Radicals If a is a real number, then $\sqrt{a^2} = |a|$.

The result of this discussion and property 3 is simply this:

If we know a is positive, then $\sqrt{a^2} = a$.

If we know a is negative, then $\sqrt{a^2} = |a|$.

If we don't know if a is positive or negative, then $\sqrt{a^2} = |a|$.

EXAMPLES Simplify each expression. Do *not* assume the variables represent positive numbers.

12. $\sqrt{9x^2} = 3|x|$

13. $\sqrt{x^3} = |x|\sqrt{x}$

14. $\sqrt{x^2 - 6x + 9} = \sqrt{(x-3)^2} = |x - 3|$

15. $\sqrt{x^3 - 5x^2} = \sqrt{x^2(x-5)} = |x|\sqrt{x - 5}$

As you can see, we must use absolute value symbols when we take a square root of a perfect square, unless we know the base of the perfect square is a positive number. The same idea holds for higher even roots, but not for odd roots. With odd roots, no absolute value symbols are necessary.

EXAMPLES Simplify each expression.

16. $\sqrt[3]{(-2)^3} = \sqrt[3]{-8} = -2$

17. $\sqrt[3]{(-5)^3} = \sqrt[3]{-125} = -5$

We can extend this discussion to all roots as follows:

Extending Property 3 for Radicals If a is a real number, then

$$\sqrt[n]{a^n} = |a| \quad \text{if} \quad n \text{ is even}$$

$$\sqrt[n]{a^n} = a \quad \text{if} \quad n \text{ is odd}$$

LINKING OBJECTIVES AND EXAMPLES

Next to each **objective** we have listed the examples that are best described by that objective.

A	1–6, 12–17
B	7–11

GETTING READY FOR CLASS

After reading through the preceding section, respond in your own words and in complete sentences.

1. Explain why this statement is false: "The square root of a sum is the sum of the square roots."
2. What is simplified form for an expression that contains a square root?
3. Why is it not necessarily true that $\sqrt{a^2} = a$?
4. What does it mean to rationalize the denominator in an expression?

Problem Set 7.3

Online support materials can be found at www.thomsonedu.com/login

Use property 1 for radicals to write each of the following expressions in simplified form. (Assume all variables are nonnegative through Problem 84.)

1. $\sqrt{8}$
2. $\sqrt{32}$
3. $\sqrt{98}$
4. $\sqrt{75}$
5. $\sqrt{288}$
6. $\sqrt{128}$
7. $\sqrt{80}$
8. $\sqrt{200}$
9. $\sqrt{48}$
10. $\sqrt{27}$
11. $\sqrt{675}$
12. $\sqrt{972}$
13. $\sqrt[3]{54}$
14. $\sqrt[3]{24}$
15. $\sqrt[3]{128}$
16. $\sqrt[3]{162}$
17. $\sqrt[3]{432}$
18. $\sqrt[3]{1{,}536}$
19. $\sqrt[5]{64}$
20. $\sqrt[4]{48}$
21. $\sqrt{18x^3}$
22. $\sqrt{27x^5}$
23. $\sqrt[4]{32y^7}$
24. $\sqrt[5]{32y^7}$
25. $\sqrt[3]{40x^4y^7}$
26. $\sqrt[3]{128x^6y^2}$
27. $\sqrt{48a^2b^3c^4}$
28. $\sqrt{72a^4b^3c^2}$
29. $\sqrt[3]{48a^2b^3c^4}$
30. $\sqrt[3]{72a^4b^3c^2}$
31. $\sqrt[5]{64x^8y^{12}}$
32. $\sqrt[4]{32x^9y^{10}}$
33. $\sqrt[5]{243x^7y^{10}z^5}$
34. $\sqrt[5]{64x^8y^4z^{11}}$

Substitute the given number into the expression $\sqrt{b^2 - 4ac}$, and then simplify.

35. $a = 2, b = -6, c = 3$
36. $a = 6, b = 7, c = -5$
37. $a = 1, b = 2, c = 6$
38. $a = 2, b = 5, c = 3$
39. $a = \frac{1}{2}, b = -\frac{1}{2}, c = -\frac{5}{4}$
40. $a = \frac{7}{4}, b = -\frac{3}{4}, c = -2$

Rationalize the denominator in each of the following expressions.

41. $\frac{2}{\sqrt{3}}$
42. $\frac{3}{\sqrt{2}}$
43. $\frac{5}{\sqrt{6}}$
44. $\frac{7}{\sqrt{5}}$

■ = Videos available by instructor request

▶ = Online student support materials available at www.thomsonedu.com/login

45. $\sqrt{\frac{1}{2}}$

46. $\sqrt{\frac{1}{3}}$

47. $\sqrt{\frac{1}{5}}$

48. $\sqrt{\frac{1}{6}}$

49. $\frac{4}{\sqrt[3]{2}}$

50. $\frac{5}{\sqrt[3]{3}}$

51. $\frac{2}{\sqrt[3]{9}}$

52. $\frac{3}{\sqrt[3]{4}}$

53. $\sqrt[4]{\frac{3}{2x^2}}$

54. $\sqrt[4]{\frac{5}{3x^2}}$

55. $\sqrt[4]{\frac{8}{y}}$

56. $\sqrt[4]{\frac{27}{y}}$

57. $\sqrt[3]{\frac{4x}{3y}}$

58. $\sqrt[3]{\frac{7x}{6y}}$

59. $\sqrt[3]{\frac{2x}{9y}}$

60. $\sqrt[3]{\frac{5x}{4y}}$

61. $\sqrt[4]{\frac{1}{8x^3}}$

62. $\sqrt[4]{\frac{8}{9x^3}}$

Write each of the following in simplified form.

63. $\sqrt{\frac{27x^3}{5y}}$

64. $\sqrt{\frac{12x^5}{7y}}$

65. $\sqrt{\frac{75x^3y^2}{2z}}$

66. $\sqrt{\frac{50x^2y^3}{3z}}$

67. $\sqrt[3]{\frac{16a^4b^3}{9c}}$

68. $\sqrt[3]{\frac{54a^5b^4}{25c^2}}$

69. $\sqrt[3]{\frac{8x^3y^6}{9z}}$

70. $\sqrt[3]{\frac{27x^6y^3}{2z^2}}$

71. $\sqrt{\sqrt{x^2}}$

72. $\sqrt{\sqrt{2x^3}}$

73. $\sqrt[3]{\sqrt{xy}}$

74. $\sqrt{\sqrt{4x}}$

75. $\sqrt[3]{\sqrt[4]{a}}$

76. $\sqrt[6]{\sqrt[4]{x}}$

77. $\sqrt[3]{\sqrt[3]{6x^{10}}}$

78. $\sqrt[5]{\sqrt{x^{14}y^{11}z}}$

79. $\sqrt[4]{\sqrt[3]{a^{12}b^{24}c^{14}}}$

80. $\sqrt{\sqrt{4a^{17}}}$

81. $\sqrt[3]{\sqrt[5]{3a^{17}b^{16}c^{30}}}$

82. $\left(\sqrt{\sqrt[4]{x^4y^8z^9}}\right)^2$

83. $\left(\sqrt{\sqrt[3]{8ab^6}}\right)^2$

84. $\left(\sqrt[4]{\sqrt[3]{16x^8y^{12}z^3}}\right)^3$

Simplify each expression. Do *not* assume the variables represent positive numbers.

85. $\sqrt{25x^2}$

86. $\sqrt{49x^2}$

87. $\sqrt{27x^3y^2}$

88. $\sqrt{40x^3y^2}$

89. $\sqrt{x^2 - 10x + 25}$

90. $\sqrt{x^2 - 16x + 64}$

91. $\sqrt{4x^2 + 12x + 9}$

92. $\sqrt{16x^2 + 40x + 25}$

93. $\sqrt{4a^4 + 16a^3 + 16a^2}$

94. $\sqrt{9a^4 + 18a^3 + 9a^2}$

95. $\sqrt{4x^3 - 8x^2}$

96. $\sqrt{18x^3 - 9x^2}$

97. Show that the statement $\sqrt{a+b} = \sqrt{a} + \sqrt{b}$ is not true by replacing a with 9 and b with 16 and simplifying both sides.

98. Find a pair of values for a and b that will make the statement $\sqrt{a+b} = \sqrt{a} + \sqrt{b}$ true.

Applying the Concepts

99. **Diagonal Distance** The distance d between opposite corners of a rectangular room with length l and width w is given by

$$d = \sqrt{l^2 + w^2}$$

How far is it between opposite corners of a living room that measures 10 by 15 feet?

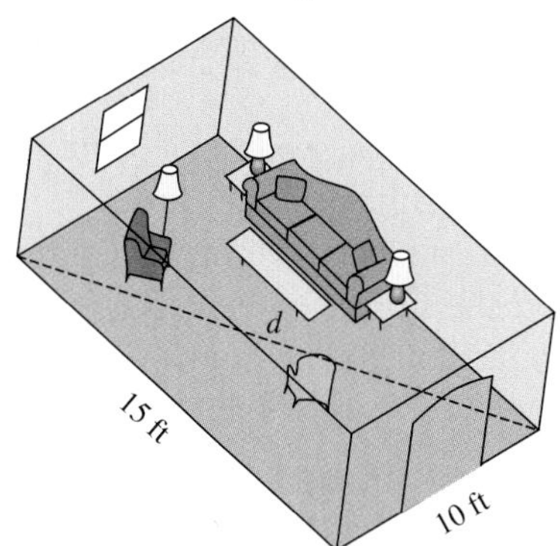

100. **Radius of a Sphere** The radius r of a sphere with volume V can be found by using the formula

$$r = \sqrt[3]{\frac{3V}{4\pi}}$$

Find the radius of a sphere with volume 9 cubic feet. Write your answer in simplified form. (Use $\frac{22}{7}$ for π.)

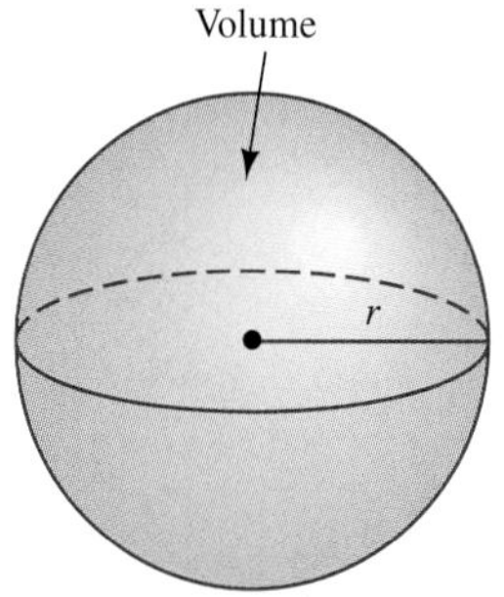

101. **Diagonal of a Box** The length of the diagonal of a rectangular box with length l, width w, and height h is given by $d = \sqrt{l^2 + w^2 + h^2}$.

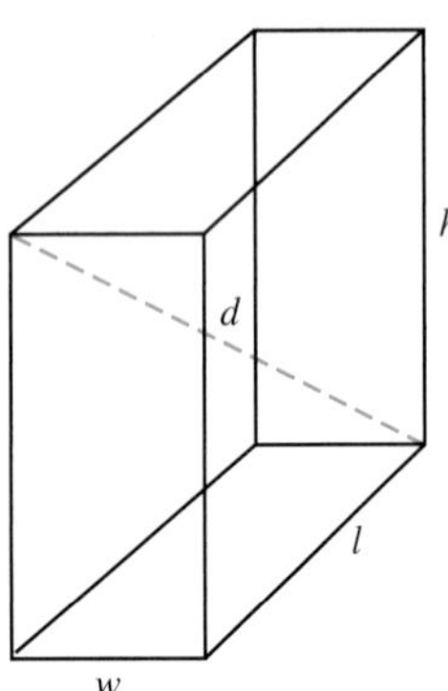

a. Find the length of the diagonal of a rectangular box that is 3 feet wide, 4 feet long, and 12 feet high.
b. Find the length of the diagonal of a rectangular box that is 2 feet wide, 4 feet high, and 6 feet long.

102. **Distance to the Horizon** If you are at a point k miles above the surface of the Earth, the distance you can see, in miles, is approximated by the equation $d = \sqrt{8000k + k^2}$.
a. How far can you see from a point that is 1 mile above the surface of the Earth?
b. How far can you see from a point that is 2 miles above the surface of the Earth?
c. How far can you see from a point that is 3 miles above the surface of the Earth?

103. **Spiral of Roots** Construct your own spiral of roots by using a ruler. Draw the first triangle by using two 1-inch lines. The first diagonal will have a length of $\sqrt{2}$ inches. Each new triangle will be formed by drawing a 1-inch line segment at the end of the previous diagonal so that the angle formed is 90°.

104. **Spiral of Roots** Construct a spiral of roots by using line segments of length 2 inches. The length of the first diagonal will be $2\sqrt{2}$ inches. The length of the second diagonal will be $2\sqrt{3}$ inches.

Maintaining Your Skills

Perform the indicated operations.

105. $\dfrac{8xy^3}{9x^2y} \div \dfrac{16x^2y^2}{18xy^3}$

106. $\dfrac{25x^2}{5y^4} \cdot \dfrac{30y^3}{2x^5}$

107. $\dfrac{12a^2 - 4a - 5}{2a + 1} \cdot \dfrac{7a + 3}{42a^2 - 17a - 15}$

108. $\dfrac{20a^2 - 7a - 3}{4a + 1} \cdot \dfrac{25a^2 - 5a - 6}{5a + 2}$

109. $\dfrac{8x^3 + 27}{27x^3 + 1} \div \dfrac{6x^2 + 7x - 3}{9x^2 - 1}$

110. $\dfrac{27x^3 + 8}{8x^3 + 1} \div \dfrac{6x^2 + x - 2}{4x^2 - 1}$

Getting Ready for the Next Section

Simplify the following.

111. $5x - 4x + 6x$

112. $12x + 8x - 7x$

113. $35xy^2 - 8xy^2$

114. $20a^2b + 33a^2b$

115. $\frac{1}{2}x + \frac{1}{3}x$

116. $\frac{2}{3}x + \frac{5}{8}x$

Write in simplified form for radicals.

117. $\sqrt{18}$

118. $\sqrt{8}$

119. $\sqrt{75xy^3}$

120. $\sqrt{12xy}$

121. $\sqrt[3]{8a^4b^2}$

122. $\sqrt[3]{27ab^2}$

Extending the Concepts

Factor each radicand into the product of prime factors. Then simplify each radical.

123. $\sqrt[3]{8,640}$

124. $\sqrt{8,640}$

125. $\sqrt[3]{10,584}$

126. $\sqrt{10,584}$

Assume a is a positive number, and rationalize each denominator.

127. $\dfrac{1}{\sqrt[10]{a^3}}$

128. $\dfrac{1}{\sqrt[12]{a^7}}$

129. $\dfrac{1}{\sqrt[20]{a^{11}}}$

130. $\dfrac{1}{\sqrt[15]{a^{13}}}$

131. Show that the two expressions $\sqrt{x^2 + 1}$ and $x + 1$ are not, in general, equal to each other by graphing $y = \sqrt{x^2 + 1}$ and $y = x + 1$ in the same viewing window.

132. Show that the two expressions $\sqrt{x^2+9}$ and $x+3$ are not, in general, equal to each other by graphing $y=\sqrt{x^2+9}$ and $y=x+3$ in the same viewing window.

133. Approximately how far apart are the graphs in Problem 131 when $x=2$?

134. Approximately how far apart are the graphs in Problem 132 when $x=2$?

135. For what value of x are the expressions $\sqrt{x^2+1}$ and $x+1$ equal in Problem 131?

136. For what value of x are the expressions $\sqrt{x^2+9}$ and $x+3$ equal in Problem 132?

7.4 Addition and Subtraction of Radical Expressions

OBJECTIVES

A Add and subtract radicals.

B Construct golden rectangles from squares.

We have been able to add and subtract polynomials by combining similar terms. The same idea applies to addition and subtraction of radical expressions.

> **DEFINITION** Two radicals are said to be **similar radicals** if they have the same index and the same radicand.

The expressions $5\sqrt[3]{7}$ and $-8\sqrt[3]{7}$ are similar since the index is ^{3} in both cases and the radicands are 7. The expressions $3\sqrt[4]{5}$ and $7\sqrt[3]{5}$ are not similar since they have different indices, and the expressions $2\sqrt[5]{8}$ and $3\sqrt[5]{9}$ are not similar because the radicands are not the same.

EXAMPLE 1 Combine $5\sqrt{3}-4\sqrt{3}+6\sqrt{3}$.

SOLUTION All three radicals are similar. We apply the distributive property to get

$$\begin{aligned} 5\sqrt{3}-4\sqrt{3}+6\sqrt{3} &= (5-4+6)\sqrt{3} \\ &= 7\sqrt{3} \end{aligned}$$

EXAMPLE 2 Combine $3\sqrt{8}+5\sqrt{18}$.

SOLUTION The two radicals do not seem to be similar. We must write each in simplified form before applying the distributive property.

$$\begin{aligned} 3\sqrt{8}+5\sqrt{18} &= 3\sqrt{4\cdot 2}+5\sqrt{9\cdot 2} \\ &= 3\sqrt{4}\sqrt{2}+5\sqrt{9}\sqrt{2} \\ &= 3\cdot 2\sqrt{2}+5\cdot 3\sqrt{2} \\ &= 6\sqrt{2}+15\sqrt{2} \\ &= (6+15)\sqrt{2} \\ &= 21\sqrt{2} \end{aligned}$$

The result of Example 2 can be generalized to the following rule for sums and differences of radical expressions.

> **Rule** To add or subtract radical expressions, put each in simplified form and apply the distributive property if possible. We can add only similar radicals. We must write each expression in simplified form for radicals before we can tell if the radicals are similar.

EXAMPLE 3 Combine $7\sqrt{75xy^3} - 4y\sqrt{12xy}$, where $x, y \geq 0$.

SOLUTION We write each expression in simplified form and combine similar radicals:

$$\begin{aligned} 7\sqrt{75xy^3} - 4y\sqrt{12xy} &= 7\sqrt{25y^2}\,\sqrt{3xy} - 4y\sqrt{4}\,\sqrt{3xy} \\ &= 35y\sqrt{3xy} - 8y\sqrt{3xy} \\ &= (35y - 8y)\sqrt{3xy} \\ &= 27y\sqrt{3xy} \end{aligned}$$

EXAMPLE 4 Combine $10\sqrt[3]{8a^4b^2} + 11a\sqrt[3]{27ab^2}$.

SOLUTION Writing each radical in simplified form and combining similar terms, we have

$$\begin{aligned} 10\sqrt[3]{8a^4b^2} + 11a\sqrt[3]{27ab^2} &= 10\sqrt[3]{8a^3}\,\sqrt[3]{ab^2} + 11a\sqrt[3]{27}\,\sqrt[3]{ab^2} \\ &= 20a\sqrt[3]{ab^2} + 33a\sqrt[3]{ab^2} \\ &= 53a\sqrt[3]{ab^2} \end{aligned}$$

EXAMPLE 5 Combine $\dfrac{\sqrt{3}}{2} + \dfrac{1}{\sqrt{3}}$.

SOLUTION We begin by writing the second term in simplified form.

$$\begin{aligned} \frac{\sqrt{3}}{2} + \frac{1}{\sqrt{3}} &= \frac{\sqrt{3}}{2} + \frac{1}{\sqrt{3}} \cdot \frac{\mathbf{\sqrt{3}}}{\mathbf{\sqrt{3}}} \\ &= \frac{\sqrt{3}}{2} + \frac{\sqrt{3}}{3} \\ &= \frac{1}{2}\sqrt{3} + \frac{1}{3}\sqrt{3} \\ &= \left(\frac{1}{2} + \frac{1}{3}\right)\sqrt{3} \\ &= \frac{5}{6}\sqrt{3} = \frac{5\sqrt{3}}{6} \end{aligned}$$

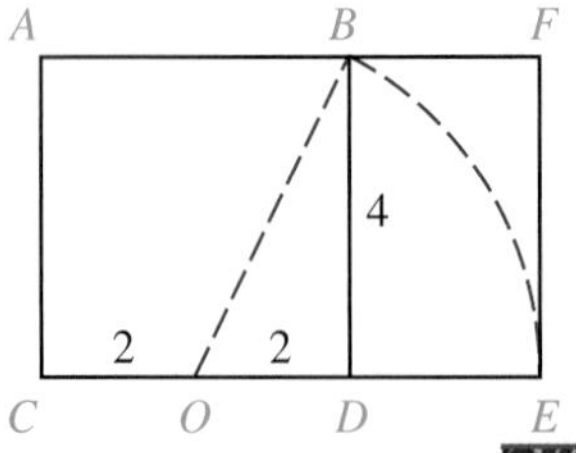

FIGURE 1

EXAMPLE 6 Construct a golden rectangle from a square of side 4. Then show that the ratio of the length to the width is the golden ratio $\dfrac{1 + \sqrt{5}}{2}$.

SOLUTION Figure 1 shows the golden rectangle constructed from a square of side 4.

The length of the diagonal OB is found from the Pythagorean theorem.

$$OB = \sqrt{2^2 + 4^2} = \sqrt{4 + 16} = \sqrt{20} = 2\sqrt{5}$$

The ratio of the length to the width for the rectangle is the golden ratio.

$$\text{Golden ratio} = \frac{CE}{EF} = \frac{2 + 2\sqrt{5}}{4} = \frac{2(1 + \sqrt{5})}{2 \cdot 2} = \frac{1 + \sqrt{5}}{2}$$

LINKING OBJECTIVES AND EXAMPLES

Next to each **objective** we have listed the examples that are best described by that objective.

A	1–5
B	6

GETTING READY FOR CLASS

After reading through the preceding section, respond in your own words and in complete sentences.

1. What are similar radicals?
2. When can we add two radical expressions?
3. What is the first step when adding or subtracting expressions containing radicals?
4. What is the golden ratio, and where does it come from?

Problem Set 7.4

Online support materials can be found at www.thomsonedu.com/login

Combine the following expressions. (Assume any variables under an even root are nonnegative.)

1. $3\sqrt{5} + 4\sqrt{5}$
2. $6\sqrt{3} - 5\sqrt{3}$
3. $3x\sqrt{7} - 4x\sqrt{7}$
4. $6y\sqrt{a} + 7y\sqrt{a}$
5. $5\sqrt[3]{10} - 4\sqrt[3]{10}$
6. $6\sqrt[4]{2} + 9\sqrt[4]{2}$
7. $8\sqrt[5]{6} - 2\sqrt[5]{6} + 3\sqrt[5]{6}$
8. $7\sqrt[6]{7} - \sqrt[6]{7} + 4\sqrt[6]{7}$
9. $3x\sqrt{2} - 4x\sqrt{2} + x\sqrt{2}$
10. $5x\sqrt{6} - 3x\sqrt{6} - 2x\sqrt{6}$
11. $\sqrt{20} - \sqrt{80} + \sqrt{45}$
12. $\sqrt{8} - \sqrt{32} - \sqrt{18}$
13. $4\sqrt{8} - 2\sqrt{50} - 5\sqrt{72}$
14. $\sqrt{48} - 3\sqrt{27} + 2\sqrt{75}$
15. $5x\sqrt{8} + 3\sqrt{32x^2} - 5\sqrt{50x^2}$
16. $2\sqrt{50x^2} - 8x\sqrt{18} - 3\sqrt{72x^2}$
17. $5\sqrt[3]{16} - 4\sqrt[3]{54}$
18. $\sqrt[3]{81} + 3\sqrt[3]{24}$
19. $\sqrt[3]{x^4y^2} + 7x\sqrt[3]{xy^2}$
20. $2\sqrt[3]{x^8y^6} - 3y^2\sqrt[3]{8x^8}$
21. $5a^2\sqrt{27ab^3} - 6b\sqrt{12a^5b}$
22. $9a\sqrt{20a^3b^2} + 7b\sqrt{45a^5}$
23. $b\sqrt[3]{24a^5b} + 3a\sqrt[3]{81a^2b^4}$
24. $7\sqrt[3]{a^4b^3c^2} - 6ab\sqrt[3]{ac^2}$
25. $5x\sqrt[4]{3y^5} + y\sqrt[4]{243x^4y} + \sqrt[4]{48x^4y^5}$
26. $x\sqrt[4]{5xy^8} + y\sqrt[4]{405x^5y^4} + y^2\sqrt[4]{80x^5}$
27. $\dfrac{\sqrt{2}}{2} + \dfrac{1}{\sqrt{2}}$
28. $\dfrac{\sqrt{3}}{3} + \dfrac{1}{\sqrt{3}}$
29. $\dfrac{\sqrt{5}}{3} + \dfrac{1}{\sqrt{5}}$
30. $\dfrac{\sqrt{6}}{2} + \dfrac{1}{\sqrt{6}}$
31. $\sqrt{x} - \dfrac{1}{\sqrt{x}}$
32. $\sqrt{x} + \dfrac{1}{\sqrt{x}}$
33. $\dfrac{\sqrt{18}}{6} + \sqrt{\dfrac{1}{2}} + \dfrac{\sqrt{2}}{2}$
34. $\dfrac{\sqrt{12}}{6} + \sqrt{\dfrac{1}{3}} + \dfrac{\sqrt{3}}{3}$
35. $\sqrt{6} - \sqrt{\dfrac{2}{3}} + \sqrt{\dfrac{1}{6}}$
36. $\sqrt{15} - \sqrt{\dfrac{3}{5}} + \sqrt{\dfrac{5}{3}}$

■ = Videos available by instructor request

▶ = Online student support materials available at www.thomsonedu.com/login

37. $\sqrt[3]{25} + \dfrac{3}{\sqrt[3]{5}}$

38. $\sqrt[4]{8} + \dfrac{1}{\sqrt[4]{2}}$

The following problems apply to what you have learned with the algebra of functions.

39. Let $f(x) = \sqrt{8x}$ and $g(x) = \sqrt{72x}$, then find
   a. $f(x) + g(x)$
   b. $f(x) - g(x)$

40. Let $f(x) = x + \sqrt{3}$ and $g(x) = x + 2\sqrt{3}$, then find
   a. $f(x) + g(x)$
   b. $f(x) - g(x)$

41. Let $f(x) = 3\sqrt{2x}$ and $g(x) = \sqrt{2x}$, then find
   a. $f(x) + g(x)$
   b. $f(x) - g(x)$

42. Let $f(x) = x\sqrt[3]{64}$ and $g(x) = x\sqrt{81}$, then find
   a. $f(x) + g(x)$
   b. $f(x) - g(x)$

43. Let $f(x) = x\sqrt{2}$ and $g(x) = 2x\sqrt{2}$, then find
   a. $f(x) + g(x)$
   b. $f(x) - g(x)$

44. Let $f(x) = 5 + 2\sqrt{5x}$ and $g(x) = 3\sqrt{5x}$, then find
   a. $f(x) + g(x)$
   b. $f(x) - g(x)$

45. Let $f(x) = \sqrt{2x} - 2$ and $g(x) = 2\sqrt{2x} + 5$, then find
   a. $f(x) + g(x)$
   b. $f(x) - g(x)$

46. Let $f(x) = 1 - 2\sqrt[3]{3x}$ and $g(x) = 2 + 3\sqrt[3]{3x}$, then find
   a. $f(x) + g(x)$
   b. $f(x) - g(x)$

47. Use a calculator to find a decimal approximation for $\sqrt{12}$ and for $2\sqrt{3}$.

48. Use a calculator to find decimal approximations for $\sqrt{50}$ and $5\sqrt{2}$.

49. Use a calculator to find a decimal approximation for $\sqrt{8} + \sqrt{18}$. Is it equal to the decimal approximation for $\sqrt{26}$ or $\sqrt{50}$?

50. Use a calculator to find a decimal approximation for $\sqrt{3} + \sqrt{12}$. Is it equal to the decimal approximation for $\sqrt{15}$ or $\sqrt{27}$?

Applying the Concepts

51. **Golden Rectangle** Construct a golden rectangle from a square of side 8. Then show that the ratio of the length to the width is the golden ratio $\dfrac{1 + \sqrt{5}}{2}$.

52. **Golden Rectangle** Construct a golden rectangle from a square of side 10. Then show that the ratio of the length to the width is the golden ratio $\dfrac{1 + \sqrt{5}}{2}$.

53. **Golden Rectangle** To show that all golden rectangles have the same ratio of length to width, construct a golden rectangle from a square of side $2x$. Then show that the ratio of the length to the width is the golden ratio.

54. **Golden Rectangle** To show that all golden rectangles have the same ratio of length to width, construct a golden rectangle from a square of side x. Then show that the ratio of the length to the width is the golden ratio.

55. **Equilateral Triangles** A triangle is equilateral if it has three equal sides. The triangle in the figure is equilateral with each side of length $2x$. Find the ratio of the height to a side.

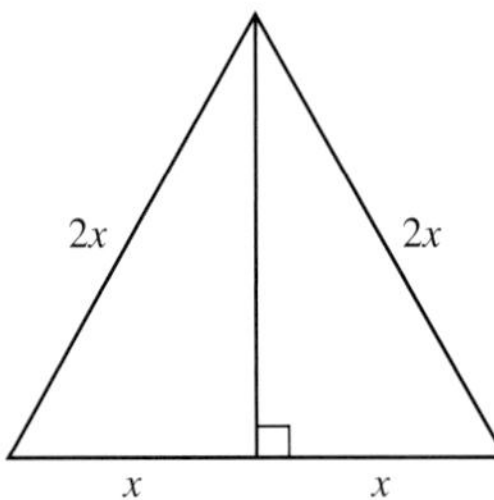

56. **Pyramids** Refer to the diagram of a square pyramid below. Find the ratio of the height h of the pyramid to the altitude a.

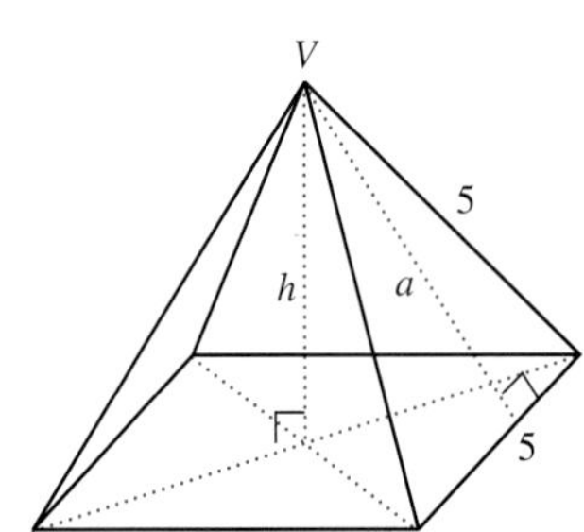

Maintaining Your Skills

Add or subtract as indicated.

57. $\frac{2a-4}{a+2} - \frac{a-6}{a+2}$

58. $\frac{2a-3}{a-2} - \frac{a-1}{a-2}$

59. $3 + \frac{4}{3-t}$

60. $6 + \frac{2}{5-t}$

61. $\frac{3}{2x-5} - \frac{39}{8x^2-14x-15}$

62. $\frac{2}{4x-5} + \frac{9}{8x^2-38x+35}$

63. $\frac{1}{x-y} - \frac{3xy}{x^3-y^3}$

64. $\frac{1}{x+y} + \frac{3xy}{x^3+y^3}$

Getting Ready for the Next Section

Simplify the following.

65. $3 \cdot 2$

66. $5 \cdot 7$

67. $(x+y)(4x-y)$

68. $(2x+y)(x-y)$

69. $(x+3)^2$

70. $(3x-2y)^2$

71. $(x-2)(x+2)$

72. $(2x+5)(2x-5)$

Simplify the following expressions.

73. $2\sqrt{18}$

74. $5\sqrt{36}$

75. $(\sqrt{6})^2$

76. $(\sqrt{2})^2$

77. $(3\sqrt{x})^2$

78. $(2\sqrt{y})^2$

Rationalize the denominator.

79. $\frac{\sqrt{3}}{\sqrt{2}}$

80. $\frac{\sqrt{5}}{\sqrt{6}}$

Extending the Concepts

Assume all variables represent positive numbers. Simplify.

81. $\sqrt[5]{32x^5y^5} - y\sqrt[3]{27x^3}$

82. $\sqrt[6]{x^4} + 4\sqrt[3]{8x^2}$

83. $3\sqrt[9]{x^9y^{18}z^{27}} - 4\sqrt[6]{x^6y^{12}z^{18}}$

84. $4a\sqrt{b^4c^6} + 3b\sqrt[3]{a^3b^3c^9}$

85. $3c\sqrt[8]{4a^6b^{18}} + b\sqrt[4]{32a^3b^5c^4}$

86. $4x\sqrt[6]{16y^6z^8} - y\sqrt[3]{32x^3z^4}$

87. $3\sqrt[9]{8a^{12}b^9} + b\sqrt[3]{16a^4} - 8\sqrt[6]{4a^8b^6}$

88. $-ac\sqrt{108bc^3} - 4\sqrt[6]{27a^6b^3c^{15}} + 3\sqrt[4]{9a^4b^2c^{10}}$

7.5 Multiplication and Division of Radical Expressions

OBJECTIVES

A Multiply expressions containing radicals.

B Rationalize a denominator containing two terms.

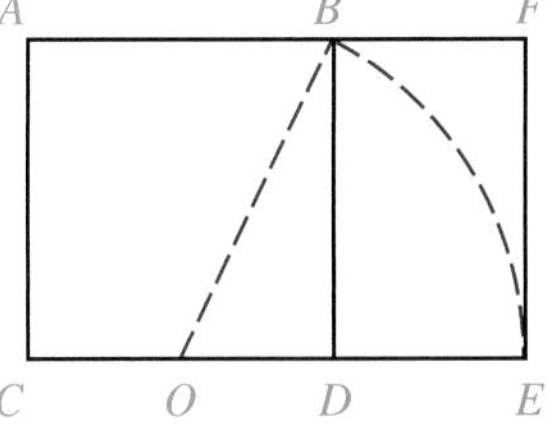

We have worked with the golden rectangle more than once in this chapter. The following is one such golden rectangle.

By now you know that in any golden rectangle constructed from a square (of any size) the ratio of the length to the width will be

$$\frac{1+\sqrt{5}}{2}$$

which we call the golden ratio. What is interesting is that the smaller rectangle on the right, *BFED*, is also a golden rectangle. We will use the mathematics developed in this section to confirm this fact.

In this section we will look at multiplication and division of expressions that contain radicals. As you will see, multiplication of expressions that contain radicals is very similar to multiplication of polynomials. The division problems in this section are just an extension of the work we did previously when we rationalized denominators.

EXAMPLE 1 Multiply $(3\sqrt{5})(2\sqrt{7})$.

SOLUTION We can rearrange the order and grouping of the numbers in this product by applying the commutative and associative properties. Following this, we apply property 1 for radicals and multiply:

$$\begin{aligned}(3\sqrt{5})(2\sqrt{7}) &= (3 \cdot 2)(\sqrt{5}\,\sqrt{7}) && \textbf{Commutative and associative properties}\\ &= (3 \cdot 2)(\sqrt{5 \cdot 7}) && \textbf{Property 1 for radicals}\\ &= 6\sqrt{35} && \textbf{Multiplication}\end{aligned}$$

In practice, it is not necessary to show the first two steps.

EXAMPLE 2 Multiply $\sqrt{3}(2\sqrt{6} - 5\sqrt{12})$.

SOLUTION Applying the distributive property, we have

$$\begin{aligned}\sqrt{3}(2\sqrt{6} - 5\sqrt{12}) &= \sqrt{3} \cdot 2\sqrt{6} - \sqrt{3} \cdot 5\sqrt{12}\\ &= 2\sqrt{18} - 5\sqrt{36}\end{aligned}$$

Writing each radical in simplified form gives

$$\begin{aligned}2\sqrt{18} - 5\sqrt{36} &= 2\sqrt{9}\,\sqrt{2} - 5\sqrt{36}\\ &= 6\sqrt{2} - 30\end{aligned}$$

EXAMPLE 3 Multiply $(\sqrt{3} + \sqrt{5})(4\sqrt{3} - \sqrt{5})$.

SOLUTION The same principle that applies when multiplying two binomials applies to this product. We must multiply each term in the first expression by each term in the second one. Any convenient method can be used. Let's use the FOIL method.

$$\begin{aligned}(\sqrt{3} + \sqrt{5})(4\sqrt{3} - \sqrt{5}) &= \overset{\text{F}}{\sqrt{3} \cdot 4\sqrt{3}} \overset{\text{O}}{- \sqrt{3}\,\sqrt{5}} \overset{\text{I}}{+ \sqrt{5} \cdot 4\sqrt{3}} \overset{\text{L}}{- \sqrt{5}\,\sqrt{5}}\\ &= 4 \cdot 3 - \sqrt{15} + 4\sqrt{15} - 5\\ &= 12 + 3\sqrt{15} - 5\\ &= 7 + 3\sqrt{15}\end{aligned}$$

EXAMPLE 4 Expand and simplify $(\sqrt{x} + 3)^2$.

SOLUTION 1 We can write this problem as a multiplication problem and proceed as we did in Example 3:

$$\begin{aligned}(\sqrt{x} + 3)^2 &= (\sqrt{x} + 3)(\sqrt{x} + 3)\\ &= \overset{\text{F}}{\sqrt{x} \cdot \sqrt{x}} \overset{\text{O}}{+ 3\sqrt{x}} \overset{\text{I}}{+ 3\sqrt{x}} \overset{\text{L}}{+ 3 \cdot 3}\\ &= x + 3\sqrt{x} + 3\sqrt{x} + 9\\ &= x + 6\sqrt{x} + 9\end{aligned}$$

SOLUTION 2 We can obtain the same result by applying the formula for the square of a sum: $(a + b)^2 = a^2 + 2ab + b^2$.

$$(\sqrt{x} + 3)^2 = (\sqrt{x})^2 + 2(\sqrt{x})(3) + 3^2$$
$$= x + 6\sqrt{x} + 9$$

EXAMPLE 5

Expand $(3\sqrt{x} - 2\sqrt{y})^2$ and simplify the result.

SOLUTION Let's apply the formula for the square of a difference, $(a - b)^2 = a^2 - 2ab + b^2$.

$$(3\sqrt{x} - 2\sqrt{y})^2 = (3\sqrt{x})^2 - 2(3\sqrt{x})(2\sqrt{y}) + (2\sqrt{y})^2$$
$$= 9x - 12\sqrt{xy} + 4y$$

EXAMPLE 6

Expand and simplify $(\sqrt{x + 2} - 1)^2$.

SOLUTION Applying the formula $(a - b)^2 = a^2 - 2ab + b^2$, we have

$$(\sqrt{x + 2} - 1)^2 = (\sqrt{x + 2})^2 - 2\sqrt{x + 2}(1) + 1^2$$
$$= x + 2 - 2\sqrt{x + 2} + 1$$
$$= x + 3 - 2\sqrt{x + 2}$$

EXAMPLE 7

Multiply $(\sqrt{6} + \sqrt{2})(\sqrt{6} - \sqrt{2})$.

SOLUTION We notice the product is of the form $(a + b)(a - b)$, which always gives the difference of two squares, $a^2 - b^2$:

$$(\sqrt{6} + \sqrt{2})(\sqrt{6} - \sqrt{2}) = (\sqrt{6})^2 - (\sqrt{2})^2$$
$$= 6 - 2$$
$$= 4$$

> **Note**
>
> We can prove that conjugates always multiply to yield a rational number as follows: If a and b are positive integers, then
> $(\sqrt{a} + \sqrt{b})(\sqrt{a} - \sqrt{b})$
> $= \sqrt{a}\sqrt{a} - \sqrt{a}\sqrt{b} + \sqrt{a}\sqrt{b} - \sqrt{b}\sqrt{b}$
> $= a - \sqrt{ab} + \sqrt{ab} - b$
> $= a - b$
> which is rational if a and b are rational.

The two expressions $(\sqrt{6} + \sqrt{2})$ and $(\sqrt{6} - \sqrt{2})$ are called *conjugates.* In general, the conjugate of $\sqrt{a} + \sqrt{b}$ is $\sqrt{a} - \sqrt{b}$. If a and b are integers, multiplying conjugates of this form always produces a rational number.

Division with radical expressions is the same as rationalizing the denominator. In a previous section we were able to divide $\sqrt{3}$ by $\sqrt{2}$ by rationalizing the denominator:

$$\frac{\sqrt{3}}{\sqrt{2}} = \frac{\sqrt{3}}{\sqrt{2}} \cdot \frac{\sqrt{\mathbf{2}}}{\sqrt{\mathbf{2}}} = \frac{\sqrt{6}}{2}$$

We can accomplish the same result with expressions such as

$$\frac{6}{\sqrt{5} - \sqrt{3}}$$

by multiplying the numerator and denominator by the conjugate of the denominator.

EXAMPLE 8 Divide $\frac{6}{\sqrt{5}-\sqrt{3}}$. (Rationalize the denominator.)

SOLUTION Since the product of two conjugates is a rational number, we multiply the numerator and denominator by the conjugate of the denominator.

$$\frac{6}{\sqrt{5}-\sqrt{3}} = \frac{6}{\sqrt{5}-\sqrt{3}} \cdot \frac{(\sqrt{5}+\sqrt{3})}{(\sqrt{5}+\sqrt{3})}$$

$$= \frac{6\sqrt{5}+6\sqrt{3}}{(\sqrt{5})^2-(\sqrt{3})^2}$$

$$= \frac{6\sqrt{5}+6\sqrt{3}}{5-3}$$

$$= \frac{6\sqrt{5}+6\sqrt{3}}{2}$$

The numerator and denominator of this last expression have a factor of 2 in common. We can reduce to lowest terms by factoring 2 from the numerator and then dividing both the numerator and denominator by 2:

$$= \frac{\cancel{2}(3\sqrt{5}+3\sqrt{3})}{\cancel{2}}$$

$$= 3\sqrt{5}+3\sqrt{3}$$

EXAMPLE 9 Rationalize the denominator $\frac{\sqrt{5}-2}{\sqrt{5}+2}$.

SOLUTION To rationalize the denominator, we multiply the numerator and denominator by the conjugate of the denominator:

$$\frac{\sqrt{5}-2}{\sqrt{5}+2} = \frac{\sqrt{5}-2}{\sqrt{5}+2} \cdot \frac{(\sqrt{5}-2)}{(\sqrt{5}-2)}$$

$$= \frac{5-2\sqrt{5}-2\sqrt{5}+4}{(\sqrt{5})^2-2^2}$$

$$= \frac{9-4\sqrt{5}}{5-4}$$

$$= \frac{9-4\sqrt{5}}{1}$$

$$= 9-4\sqrt{5}$$

EXAMPLE 10 A golden rectangle constructed from a square of side 2 is shown in Figure 1. Show that the smaller rectangle *BDEF* is also a golden rectangle by finding the ratio of its length to its width.

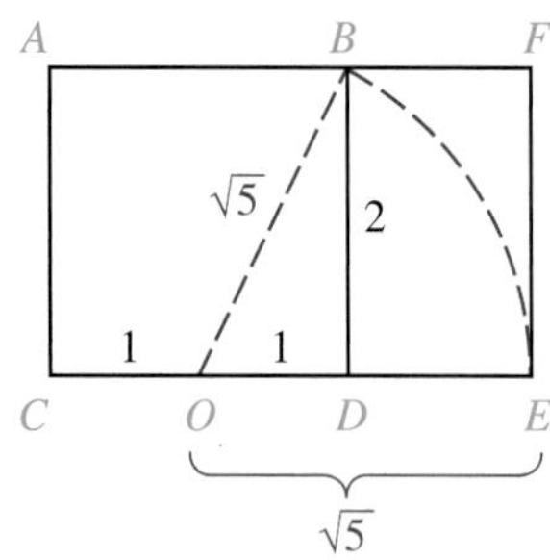

FIGURE 1

SOLUTION First we find expressions for the length and width of the smaller rectangle.

$$\text{Length} = EF = 2$$

$$\text{Width} = DE = \sqrt{5} - 1$$

Next, we find the ratio of length to width.

$$\text{Ratio of length to width} = \frac{EF}{DE} = \frac{2}{\sqrt{5} - 1}$$

To show that the small rectangle is a golden rectangle, we must show that the ratio of length to width is the golden ratio. We do so by rationalizing the denominator.

$$\frac{2}{\sqrt{5} - 1} = \frac{2}{\sqrt{5} - 1} \cdot \frac{\mathbf{\sqrt{5} + 1}}{\mathbf{\sqrt{5} + 1}}$$

$$= \frac{2(\sqrt{5} + 1)}{5 - 1}$$

$$= \frac{2(\sqrt{5} + 1)}{4}$$

$$= \frac{\sqrt{5} + 1}{2} \quad \text{Divide out common factor 2}$$

Since addition is commutative, this last expression is the golden ratio. Therefore, the small rectangle in Figure 1 is a golden rectangle.

GETTING READY FOR CLASS

After reading through the preceding section, respond in your own words and in complete sentences.

1. Explain why $(\sqrt{5} + \sqrt{2})^2 \neq 5 + 2$.
2. Explain in words how you would rationalize the denominator in the expression $\frac{\sqrt{3}}{\sqrt{5} - \sqrt{2}}$
3. What are conjugates?
4. What result is guaranteed when multiplying radical expressions that are conjugates?

LINKING OBJECTIVES AND EXAMPLES

Next to each **objective** we have listed the examples that are best described by that objective.

A	1–7
B	8–10

Problem Set 7.5

Online support materials can be found at www.thomsonedu.com/login

Multiply. (Assume all expressions appearing under a square root symbol represent nonnegative numbers throughout this problem set.)

1. $\sqrt{6}\sqrt{3}$
2. $\sqrt{6}\sqrt{2}$
3. $(2\sqrt{3})(5\sqrt{7})$
4. $(3\sqrt{5})(2\sqrt{7})$
5. $(4\sqrt{6})(2\sqrt{15})(3\sqrt{10})$
6. $(4\sqrt{35})(2\sqrt{21})(5\sqrt{15})$
7. $(3\sqrt[3]{3})(6\sqrt[3]{9})$
8. $(2\sqrt[3]{2})(6\sqrt[3]{4})$
9. $\sqrt{3}(\sqrt{2} - 3\sqrt{3})$
10. $\sqrt{2}(5\sqrt{3} + 4\sqrt{2})$
11. $6\sqrt[3]{4}(2\sqrt[3]{2} + 1)$
12. $7\sqrt[3]{5}(3\sqrt[3]{25} - 2)$
13. $\sqrt[3]{4}(\sqrt[3]{2} + \sqrt[3]{6})$
14. $\sqrt[3]{5}(\sqrt[3]{8} - \sqrt[3]{25})$
15. $\sqrt[3]{x}(\sqrt[3]{x^2y^4} + \sqrt[3]{x^5y})$
16. $\sqrt[3]{x^2y}(2\sqrt[3]{x^4y^2} - \sqrt[3]{x^2y^4})$
17. $\sqrt[4]{2x^3}(\sqrt[4]{8x^6} + \sqrt[4]{16x^9})$
18. $\sqrt[4]{8y^2}(\sqrt[4]{6y^6} - \sqrt[4]{8y^9})$
19. $(\sqrt{3} + \sqrt{2})(3\sqrt{3} - \sqrt{2})$
20. $(\sqrt{5} - \sqrt{2})(3\sqrt{5} + 2\sqrt{2})$
21. $(\sqrt{x} + 5)(\sqrt{x} - 3)$
22. $(\sqrt{x} + 4)(\sqrt{x} + 2)$
23. $(3\sqrt{6} + 4\sqrt{2})(\sqrt{6} + 2\sqrt{2})$
24. $(\sqrt{7} - 3\sqrt{3})(2\sqrt{7} - 4\sqrt{3})$
25. $(\sqrt{3} + 4)^2$
26. $(\sqrt{5} - 2)^2$
27. $(\sqrt{x} - 3)^2$
28. $(\sqrt{x} + 4)^2$
29. $(2\sqrt{a} - 3\sqrt{b})^2$
30. $(5\sqrt{a} - 2\sqrt{b})^2$
31. $(\sqrt{x-4} + 2)^2$
32. $(\sqrt{x-3} + 2)^2$
33. $(\sqrt{x-5} - 3)^2$
34. $(\sqrt{x-3} - 4)^2$
35. $(\sqrt{3} - \sqrt{2})(\sqrt{3} + \sqrt{2})$
36. $(\sqrt{5} - \sqrt{2})(\sqrt{5} + \sqrt{2})$
37. $(\sqrt{a} + 7)(\sqrt{a} - 7)$
38. $(\sqrt{a} + 5)(\sqrt{a} - 5)$
39. $(5 - \sqrt{x})(5 + \sqrt{x})$
40. $(3 - \sqrt{x})(3 + \sqrt{x})$
41. $(\sqrt{x-4} + 2)(\sqrt{x-4} - 2)$
42. $(\sqrt{x+3} + 5)(\sqrt{x+3} - 5)$
43. $(\sqrt{3} + 1)^3$
44. $(\sqrt{5} - 2)^3$
45. $(\sqrt[3]{3} + \sqrt[3]{2})(\sqrt[3]{9} + \sqrt[3]{4})$
46. $(\sqrt[3]{5} + \sqrt[3]{3})(\sqrt[3]{9} + \sqrt[3]{25})$
47. $(\sqrt[3]{x^5} + \sqrt[3]{y})(\sqrt[3]{x} + \sqrt[3]{y^2})$
48. $(\sqrt[3]{x^7} - \sqrt[3]{y^4})(\sqrt[3]{x^2} - \sqrt[3]{y^2})$

Rationalize the denominator in each of the following.

49. $\dfrac{1}{\sqrt{2}}$
50. $\dfrac{x}{\sqrt{2}}$
51. $\dfrac{1}{\sqrt{x}}$
52. $\dfrac{3}{\sqrt{x}}$
53. $\dfrac{4}{\sqrt{3}}$
54. $\dfrac{4}{\sqrt{2x}}$
55. $\dfrac{2x}{\sqrt{6}}$
56. $\dfrac{6x}{\sqrt{3}}$
57. $\dfrac{4}{\sqrt{10x}}$
58. $\sqrt{\dfrac{4x^2}{3x}}$
59. $\dfrac{2}{\sqrt{8}}$
60. $\sqrt{\dfrac{16xy}{4x^2y}}$
61. $\sqrt{\dfrac{32x^3y}{4xy^2}}$
62. $\dfrac{4}{\sqrt{6x^2y^5}}$
63. $\sqrt{\dfrac{12a^3b^3c}{8ab^5c^2}}$
64. $\sqrt{\dfrac{18x^3y^2z^4}{16xy^3z^2}}$
65. $\sqrt[3]{\dfrac{16a^4b^2c^4}{2ab^3c}}$
66. $\sqrt[3]{\dfrac{9x^5y^5z}{3x^3y^7z^2}}$
67. $\dfrac{\sqrt{2}}{\sqrt{6} - \sqrt{2}}$
68. $\dfrac{\sqrt{5}}{\sqrt{5} + \sqrt{3}}$
69. $\dfrac{\sqrt{5}}{\sqrt{5} + 1}$
70. $\dfrac{\sqrt{7}}{\sqrt{7} - 1}$
71. $\dfrac{\sqrt{x}}{\sqrt{x} - 3}$
72. $\dfrac{\sqrt{x}}{\sqrt{x} + 2}$
73. $\dfrac{\sqrt{5}}{2\sqrt{5} - 3}$
74. $\dfrac{\sqrt{7}}{3\sqrt{7} - 2}$
75. $\dfrac{3}{\sqrt{x} - \sqrt{y}}$
76. $\dfrac{2}{\sqrt{x} + \sqrt{y}}$
77. $\dfrac{\sqrt{6} + \sqrt{2}}{\sqrt{6} - \sqrt{2}}$
78. $\dfrac{\sqrt{5} - \sqrt{3}}{\sqrt{5} + \sqrt{3}}$
79. $\dfrac{\sqrt{7} - 2}{\sqrt{7} + 2}$
80. $\dfrac{\sqrt{11} + 3}{\sqrt{11} - 3}$
81. $\dfrac{\sqrt{a} + \sqrt{b}}{\sqrt{a} - \sqrt{b}}$

■ = Videos available by instructor request

▶ = Online student support materials available at www.thomsonedu.com/login

82. $\dfrac{\sqrt{a}-\sqrt{b}}{\sqrt{a}+\sqrt{b}}$

83. $\dfrac{\sqrt{x}+2}{\sqrt{x}-2}$

84. $\dfrac{\sqrt{x}-3}{\sqrt{x}+3}$

85. $\dfrac{2\sqrt{3}-\sqrt{7}}{3\sqrt{3}+\sqrt{7}}$

86. $\dfrac{5\sqrt{6}+2\sqrt{2}}{\sqrt{6}-\sqrt{2}}$

87. $\dfrac{3\sqrt{x}+2}{1+\sqrt{x}}$

88. $\dfrac{5\sqrt{x}-1}{2+\sqrt{x}}$

▸ **89.** $\dfrac{2}{\sqrt{3}}+\sqrt{12}$

▸ **90.** $\sqrt{2}+\dfrac{5}{\sqrt{8}}$

91. $\dfrac{1}{\sqrt{5}}+\sqrt{20}$

92. $\dfrac{4}{\sqrt{2}}+\sqrt{72}$

93. $\dfrac{6}{\sqrt{12}}-\sqrt{75}$

94. $\dfrac{5}{\sqrt{7}}-\sqrt{63}+\sqrt{28}$

95. $\dfrac{1}{\sqrt{3}}+\sqrt{48}+\dfrac{4}{\sqrt{12}}$

96. $\dfrac{4}{\sqrt{2}}+\sqrt{8}+\dfrac{10}{\sqrt{50}}$

97. Show that the product $(\sqrt[3]{2}+\sqrt[3]{3})(\sqrt[3]{4}-\sqrt[3]{6}+\sqrt[3]{9})$ is 5.

98. Show that the product $(\sqrt[3]{x}+2)(\sqrt[3]{x^2}-2\sqrt[3]{x}+4)$ is $x+8$.

Each statement below is false. Correct the right side of each one.

99. $5(2\sqrt{3})=10\sqrt{15}$

100. $3(2\sqrt{x})=6\sqrt{3x}$

101. $(\sqrt{x}+3)^2=x+9$

102. $(\sqrt{x}-7)^2=x-49$

103. $(5\sqrt{3})^2=15$

104. $(3\sqrt{5})^2=15$

Applying the Concepts

105. Gravity If an object is dropped from the top of a 100-foot building, the amount of time t (in seconds) that it takes for the object to be h feet from the ground is given by the formula

$$t=\frac{\sqrt{100-h}}{4}$$

How long does it take before the object is 50 feet from the ground? How long does it take to reach the ground? (When it is on the ground, h is 0.)

106. Gravity Use the formula given in Problem 105 to determine h if t is 1.25 seconds.

107. Golden Rectangle Rectangle $ACEF$ in Figure 2 is a golden rectangle. If side AC is 6 inches, show that the smaller rectangle $BDEF$ is also a golden rectangle.

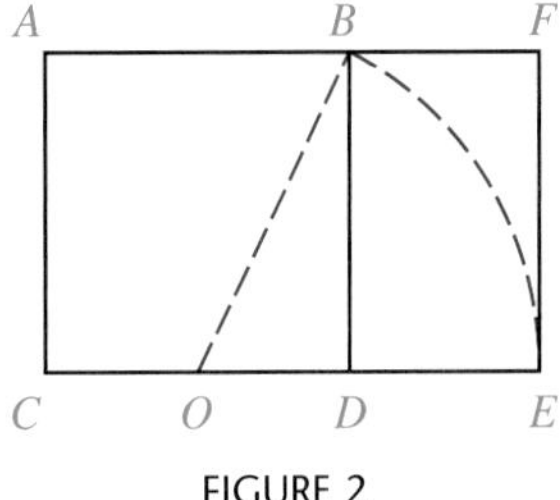

FIGURE 2

108. Golden Rectangle Rectangle $ACEF$ in Figure 2 is a golden rectangle. If side AC is 1 inch, show that the smaller rectangle $BDEF$ is also a golden rectangle.

109. Golden Rectangle If side AC in Figure 2 is $2x$, show the rectangle $BDEF$ is a golden rectangle.

110. Golden Rectangle If side AC in Figure 2 is x, show that rectangle $BDEF$ is a golden rectangle.

Maintaining Your Skills

Simplify each complex fraction.

111. $\dfrac{\frac{1}{4}-\frac{1}{3}}{\frac{1}{2}+\frac{1}{6}}$

112. $\dfrac{\frac{1}{8}-\frac{1}{3}}{\frac{1}{4}-\frac{1}{3}}$

113. $\dfrac{1-\frac{2}{y}}{1+\frac{2}{y}}$

114. $\dfrac{1+\frac{3}{y}}{1-\frac{3}{y}}$

115. $\dfrac{4+\frac{4}{x}+\frac{1}{x^2}}{4-\frac{1}{x^2}}$

116. $\dfrac{1-\frac{1}{x}-\frac{6}{x^2}}{1-\frac{9}{x^2}}$

Getting Ready for the Next Section

Simplify.

117. $(t + 5)^2$

118. $(x - 4)^2$

119. $\sqrt{x} \cdot \sqrt{x}$

120. $\sqrt{3x} \cdot \sqrt{3x}$

Solve.

121. $3x + 4 = 5^2$

122. $4x - 7 = 3^2$

123. $t^2 + 7t + 12 = 0$

124. $x^2 - 3x - 10 = 0$

125. $t^2 + 10t + 25 = t + 7$

126. $x^2 - 4x + 4 = x - 2$

127. $(x + 4)^2 = x + 6$

128. $(x - 6)^2 = x - 4$

Extending the Concepts

Rationalize the denominator in each of the following.

129. $\dfrac{x}{\sqrt{x-2} + 4}$

130. $\dfrac{3}{\sqrt{x-4} - 7}$

131. $\dfrac{x}{\sqrt{x+5} - 5}$

132. $\dfrac{2}{\sqrt{2x-1} + 3}$

133. $\dfrac{3x}{\sqrt{5x} + x}$

134. $\dfrac{4x}{\sqrt{5x^3} + 2x}$

7.6 Equations with Radicals

OBJECTIVES

A Solve equations containing radicals.

B Graph simple square root and cube root equations in two variables.

This section is concerned with solving equations that involve one or more radicals. The first step in solving an equation that contains a radical is to eliminate the radical from the equation. To do so, we need an additional property.

> **Squaring Property of Equality** If both sides of an equation are squared, the solutions to the original equation are solutions to the resulting equation.

We will never lose solutions to our equations by squaring both sides. We may, however, introduce *extraneous solutions.* Extraneous solutions satisfy the equation obtained by squaring both sides of the original equation, but they do not satisfy the original equation.

We know that if two real numbers a and b are equal, then so are their squares:

If $\quad a = b$

then $\quad a^2 = b^2$

However, extraneous solutions are introduced when we square opposites; that is, even though opposites are not equal, their squares are. For example,

$5 = -5$ **A false statement**

$(5)^2 = (-5)^2$ **Square both sides**

$25 = 25$ **A true statement**

We are free to square both sides of an equation any time it is convenient. We must be aware, however, that doing so may introduce extraneous solutions. We must, therefore, check all our solutions in the original equation if at any time we square both sides of the original equation.

EXAMPLE 1

Solve for x: $\sqrt{3x + 4} = 5$.

SOLUTION We square both sides and proceed as usual:

$$\sqrt{3x + 4} = 5$$
$$(\sqrt{3x + 4})^2 = 5^2$$
$$3x + 4 = 25$$
$$3x = 21$$
$$x = 7$$

Checking $x = 7$ in the original equation, we have

$$\sqrt{3(7) + 4} \stackrel{?}{=} 5$$
$$\sqrt{21 + 4} \stackrel{?}{=} 5$$
$$\sqrt{25} \stackrel{?}{=} 5$$
$$5 = 5$$

The solution $x = 7$ satisfies the original equation.

EXAMPLE 2

Solve $\sqrt{4x - 7} = -3$.

SOLUTION Squaring both sides, we have

$$\sqrt{4x - 7} = -3$$
$$(\sqrt{4x - 7})^2 = (-3)^2$$
$$4x - 7 = 9$$
$$4x = 16$$
$$x = 4$$

Checking $x = 4$ in the original equation gives

$$\sqrt{4(4) - 7} \stackrel{?}{=} -3$$
$$\sqrt{16 - 7} \stackrel{?}{=} -3$$
$$\sqrt{9} \stackrel{?}{=} -3$$
$$3 = -3$$

The solution $x = 4$ produces a false statement when checked in the original equation. Since $x = 4$ was the only possible solution, there is no solution to the original equation. The possible solution $x = 4$ is an extraneous solution. It satisfies the equation obtained by squaring both sides of the original equation, but it does not satisfy the original equation.

Note

The fact that there is no solution to the equation in Example 2 was obvious to begin with. Notice that the left side of the equation is the *positive* square root of $4x - 7$, which must be a nonnegative number. The right side of the equation is -3. Since we cannot have a number that is nonnegative equal to a negative number, there is no solution to the equation.

EXAMPLE 3

Solve $\sqrt{5x-1}+3=7$.

SOLUTION We must isolate the radical on the left side of the equation. If we attempt to square both sides without doing so, the resulting equation will also contain a radical. Adding -3 to both sides, we have

$$\sqrt{5x-1}+3=7$$
$$\sqrt{5x-1}=4$$

We can now square both sides and proceed as usual:

$$(\sqrt{5x-1})^2=4^2$$
$$5x-1=16$$
$$5x=17$$
$$x=\frac{17}{5}$$

Checking $x=\frac{17}{5}$, we have

$$\sqrt{5\left(\frac{17}{5}\right)-1}+3\stackrel{?}{=}7$$
$$\sqrt{17-1}+3\stackrel{?}{=}7$$
$$\sqrt{16}+3\stackrel{?}{=}7$$
$$4+3\stackrel{?}{=}7$$
$$7=7$$

EXAMPLE 4

Solve $t+5=\sqrt{t+7}$.

SOLUTION This time, squaring both sides of the equation results in a quadratic equation:

$$(t+5)^2=(\sqrt{t+7})^2 \qquad \textbf{Square both sides}$$
$$t^2+10t+25=t+7$$
$$t^2+9t+18=0 \qquad \textbf{Standard form}$$
$$(t+3)(t+6)=0 \qquad \textbf{Factor the left side}$$
$$t+3=0 \quad \text{or} \quad t+6=0 \qquad \textbf{Set factors equal to 0}$$
$$t=-3 \quad \text{or} \quad t=-6$$

We must check each solution in the original equation:

Check $t=-3$	Check $t=-6$
$-3+5\stackrel{?}{=}\sqrt{-3+7}$	$-6+5\stackrel{?}{=}\sqrt{-6+7}$
$2\stackrel{?}{=}\sqrt{4}$	$-1\stackrel{?}{=}\sqrt{1}$
$2=2$	$-1=1$
A true statement	A false statement

Since $t=-6$ does not check, our only solution is $t=-3$.

EXAMPLE 5

Solve $\sqrt{x-3} = \sqrt{x} - 3$.

SOLUTION We begin by squaring both sides. Note what happens when we square the right side of the equation, and compare the square of the right side with the square of the left side. You must convince yourself that these results are correct. (The note in the margin will help if you are having trouble convincing yourself that what is written below is true.)

$$(\sqrt{x-3})^2 = (\sqrt{x} - 3)^2$$

$$x - 3 = x - 6\sqrt{x} + 9$$

> **Note**
> It is very important that you realize that the square of $(\sqrt{x} - 3)$ is not $x + 9$. Remember, when we square a difference with two terms, we use the formula
> $$(a - b)^2 = a^2 - 2ab + b^2$$
> Applying this formula to $(\sqrt{x} - 3)^2$, we have
> $$(\sqrt{x} - 3)^2 = (\sqrt{x})^2 - 2(\sqrt{x})(3) + 3^2$$
> $$= x - 6\sqrt{x} + 9$$

Now we still have a radical in our equation, so we will have to square both sides again. Before we do, though, let's isolate the remaining radical.

$$x - 3 = x - 6\sqrt{x} + 9$$

$$-3 = -6\sqrt{x} + 9 \qquad \textbf{Add } -x \textbf{ to each side}$$

$$-12 = -6\sqrt{x} \qquad \textbf{Add } -9 \textbf{ to each side}$$

$$2 = \sqrt{x} \qquad \textbf{Divide each side by } -6$$

$$4 = x \qquad \textbf{Square each side}$$

Our only possible solution is $x = 4$, which we check in our original equation as follows:

$$\sqrt{4-3} \stackrel{?}{=} \sqrt{4} - 3$$

$$\sqrt{1} \stackrel{?}{=} 2 - 3$$

$$1 = -1 \qquad \textbf{A false statement}$$

Substituting 4 for x in the original equation yields a false statement. Since 4 was our only possible solution, there is no solution to our equation.

Here is another example of an equation for which we must apply our squaring property twice before all radicals are eliminated.

EXAMPLE 6

Solve $\sqrt{x+1} = 1 - \sqrt{2x}$.

SOLUTION This equation has two separate terms involving radical signs. Squaring both sides gives

$$x + 1 = 1 - 2\sqrt{2x} + 2x$$

$$-x = -2\sqrt{2x} \qquad \textbf{Add } -2x \textbf{ and } -1 \textbf{ to both sides}$$

$$x^2 = 4(2x) \qquad \textbf{Square both sides}$$

$$x^2 - 8x = 0 \qquad \textbf{Standard form}$$

Our equation is a quadratic equation in standard form. To solve for x, we factor the left side and set each factor equal to 0.

$$x(x - 8) = 0 \qquad \textbf{Factor left side}$$

$$x = 0 \quad \text{or} \quad x - 8 = 0 \qquad \textbf{Set factors equal to 0}$$

$$x = 8$$

Since we squared both sides of our equation, we have the possibility that one or both of the solutions are extraneous. We must check each one in the original equation:

Check $x = 8$	Check $x = 0$
$\sqrt{8+1} \stackrel{?}{=} 1 - \sqrt{2 \cdot 8}$	$\sqrt{0+1} \stackrel{?}{=} 1 - \sqrt{2 \cdot 0}$
$\sqrt{9} \stackrel{?}{=} 1 - \sqrt{16}$	$\sqrt{1} \stackrel{?}{=} 1 - \sqrt{0}$
$3 \stackrel{?}{=} 1 - 4$	$1 \stackrel{?}{=} 1 - 0$
$3 = -3$	$1 = 1$
A false statement	A true statement

Since $x = 8$ does not check, it is an extraneous solution. Our only solution is $x = 0$.

EXAMPLE 7 Solve $\sqrt{x+1} = \sqrt{x+2} - 1$.

SOLUTION Squaring both sides we have

$$(\sqrt{x+1})^2 = (\sqrt{x+2} - 1)^2$$

$$x + 1 = x + 2 - 2\sqrt{x+2} + 1$$

Once again we are left with a radical in our equation. Before we square each side again, we must isolate the radical on the right side of the equation.

$x + 1 = x + 3 - 2\sqrt{x+2}$	**Simplify the right side**
$1 = 3 - 2\sqrt{x+2}$	**Add $-x$ to each side**
$-2 = -2\sqrt{x+2}$	**Add -3 to each side**
$1 = \sqrt{x+2}$	**Divide each side by -2**
$1 = x + 2$	**Square both sides**
$-1 = x$	**Add -2 to each side**

Checking our only possible solution, $x = -1$, in our original equation, we have

$$\sqrt{-1+1} \stackrel{?}{=} \sqrt{-1+2} - 1$$

$$\sqrt{0} \stackrel{?}{=} \sqrt{1} - 1$$

$$0 \stackrel{?}{=} 1 - 1$$

$$0 = 0 \quad \textbf{A true statement}$$

Our solution checks.

It is also possible to raise both sides of an equation to powers greater than 2. We only need to check for extraneous solutions when we raise both sides of an equation to an even power. Raising both sides of an equation to an odd power will not produce extraneous solutions.

EXAMPLE 8

Solve $\sqrt[3]{4x+5} = 3$.

SOLUTION Cubing both sides we have

$$(\sqrt[3]{4x+5})^3 = 3^3$$

$$4x + 5 = 27$$

$$4x = 22$$

$$x = \frac{22}{4}$$

$$x = \frac{11}{2}$$

We do not need to check $x = \frac{11}{2}$ since we raised both sides to an odd power.

We end this section by looking at graphs of some equations that contain radicals.

EXAMPLE 9

Graph $y = \sqrt{x}$ and $y = \sqrt[3]{x}$.

SOLUTION The graphs are shown in Figures 1 and 2. Notice that the graph of $y = \sqrt{x}$ appears in the first quadrant only because in the equation $y = \sqrt{x}$, x and y cannot be negative.

The graph of $y = \sqrt[3]{x}$ appears in quadrants 1 and 3 since the cube root of a positive number is also a positive number and the cube root of a negative number is a negative number; that is, when x is positive, y will be positive and when x is negative, y will be negative.

The graphs of both equations will contain the origin since $y = 0$ when $x = 0$ in both equations.

x	y
−4	undefined
−1	undefined
0	0
1	1
4	2
9	3
16	4

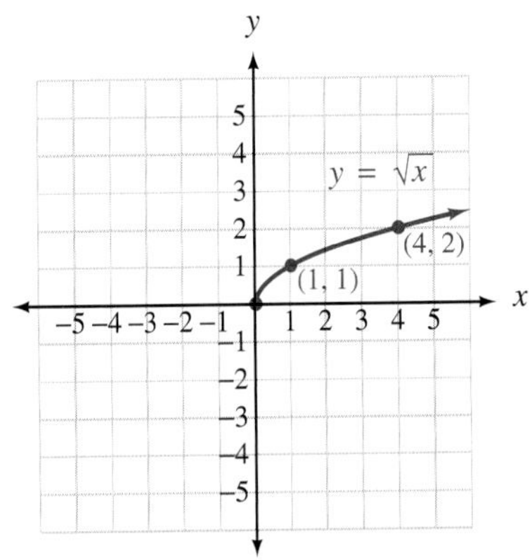

FIGURE 1

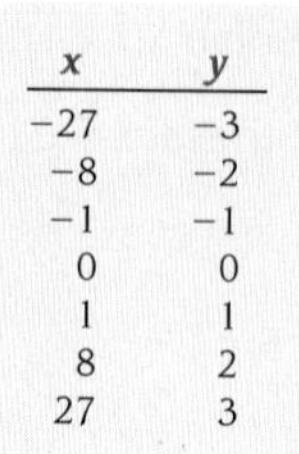

x	y
−27	−3
−8	−2
−1	−1
0	0
1	1
8	2
27	3

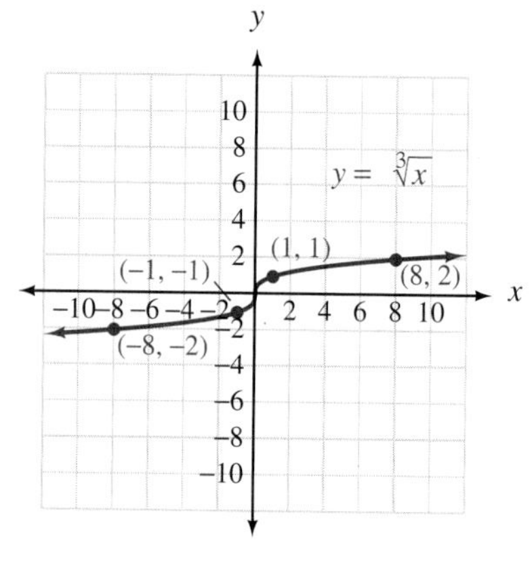

FIGURE 2

LINKING OBJECTIVES AND EXAMPLES

Next to each **objective** we have listed the examples that are best described by that objective.

A	1–8
B	9

GETTING READY FOR CLASS

After reading through the preceding section, respond in your own words and in complete sentences.

1. What is the squaring property of equality?
2. Under what conditions do we obtain extraneous solutions to equations that contain radical expressions?
3. If we have raised both sides of an equation to a power, when is it not necessary to check for extraneous solutions?
4. When will you need to apply the squaring property of equality twice in the process of solving an equation containing radicals?

Problem Set 7.6

Online support materials can be found at www.thomsonedu.com/login

Solve each of the following equations.

1. $\sqrt{2x+1} = 3$
2. $\sqrt{3x+1} = 4$
3. $\sqrt{4x+1} = -5$
4. $\sqrt{6x+1} = -5$
5. $\sqrt{2y-1} = 3$
6. $\sqrt{3y-1} = 2$
7. $\sqrt{5x-7} = -1$
8. $\sqrt{8x+3} = -6$
9. $\sqrt{2x-3} - 2 = 4$
10. $\sqrt{3x+1} - 4 = 1$
11. $\sqrt{4x+1} + 3 = 2$
12. $\sqrt{5a-3} + 6 = 2$
13. $\sqrt[4]{3x+1} = 2$
14. $\sqrt[4]{4x+1} = 3$
15. $\sqrt[3]{2x-5} = 1$
16. $\sqrt[3]{5x+7} = 2$
17. $\sqrt[3]{3a+5} = -3$
18. $\sqrt[3]{2a+7} = -2$
19. $\sqrt{y-3} = y - 3$
20. $\sqrt{y+3} = y - 3$
21. $\sqrt{a+2} = a + 2$
22. $\sqrt{a+10} = a - 2$
23. ▸ $\sqrt{2x+3} = \dfrac{2x-7}{3}$
24. ▸ $\sqrt{3x-2} = \dfrac{2x-3}{3}$
25. ▸ $\sqrt{4x-3} = \dfrac{x+3}{2}$
26. ▸ $\sqrt{4x+5} = \dfrac{x+3}{2}$
27. ▸ $\sqrt{7x+2} = \dfrac{2x+2}{3}$
28. ▸ $\sqrt{7x-3} = \dfrac{4x-1}{3}$
29. $\sqrt{2x+4} = \sqrt{1-x}$
30. $\sqrt{3x+4} = -\sqrt{2x+3}$

■ = Videos available by instructor request

▸ = Online student support materials available at www.thomsonedu.com/login

31. $\sqrt{4a+7} = -\sqrt{a+2}$
32. $\sqrt{7a-1} = \sqrt{2a+4}$
33. $\sqrt[4]{5x-8} = \sqrt[4]{4x-1}$
34. $\sqrt[4]{6x+7} = \sqrt[4]{x+2}$
35. $x+1 = \sqrt{5x+1}$
36. $x-1 = \sqrt{6x+1}$
37. $t+5 = \sqrt{2t+9}$
38. $t+7 = \sqrt{2t+13}$
39. $\sqrt{y-8} = \sqrt{8-y}$
40. $\sqrt{2y+5} = \sqrt{5y+2}$
41. $\sqrt[3]{3x+5} = \sqrt[3]{5-2x}$
42. $\sqrt[3]{4x+9} = \sqrt[3]{3-2x}$

The following equations will require that you square both sides twice before all the radicals are eliminated. Solve each equation using the methods shown in Examples 5, 6, and 7.

43. $\sqrt{x-8} = \sqrt{x}-2$
44. $\sqrt{x+3} = \sqrt{x}-3$
45. $\sqrt{x+1} = \sqrt{x}+1$
46. $\sqrt{x-1} = \sqrt{x}-1$
47. $\sqrt{x+8} = \sqrt{x-4}+2$
48. $\sqrt{x+5} = \sqrt{x-3}+2$
49. $\sqrt{x-5}-3 = \sqrt{x-8}$
50. $\sqrt{x-3}-4 = \sqrt{x-3}$
51. $\sqrt{x+4} = 2-\sqrt{2x}$
52. $\sqrt{5x+1} = 1+\sqrt{5x}$
53. $\sqrt{2x+4} = \sqrt{x+3}+1$
54. $\sqrt{2x-1} = \sqrt{x-4}+2$

Let $f(x) = \sqrt{2x-1}$. Find all values for the variable x that produce the following values of $f(x)$.

55. $f(x) = 0$
56. $f(x) = -10$
57. $f(x) = 2x-1$
58. $f(x) = \sqrt{5x-10}$
59. $f(x) = \sqrt{x-4}+2$
60. $f(x) = 9$

Let $g(x) = \sqrt{2x+3}$. Find all values for the variable x that produce the following values of $g(x)$.

61. $g(x) = 0$
62. $g(x) = x$
63. $g(x) = -\sqrt{5x}$
64. $g(x) = \sqrt{x^2-5}$

Let $f(x) = \sqrt{2x}-1$. Find all values for the variable x for which $f(x) = g(x)$.

65. $g(x) = 0$
66. $g(x) = 5$
67. $g(x) = \sqrt{2x+5}$
68. $g(x) = x-1$

Let $h(x) = \sqrt[3]{3x+5}$. Find all values for the variable x for which $h(x) = f(x)$.

69. $f(x) = 2$
70. $f(x) = -1$

Let $h(x) = \sqrt[3]{5-2x}$. Find all values for the variable x for which $h(x) = f(x)$.

71. $f(x) = 3$
72. $f(x) = -1$

Graph each equation.

73. $y = 2\sqrt{x}$
74. $y = -2\sqrt{x}$
75. $y = \sqrt{x}-2$
76. $y = \sqrt{x}+2$
77. $y = \sqrt{x-2}$
78. $y = \sqrt{x+2}$
79. $y = 3\sqrt[3]{x}$
80. $y = -3\sqrt[3]{x}$
81. $y = \sqrt[3]{x}+3$
82. $y = \sqrt[3]{x}-3$
83. $y = \sqrt[3]{x+3}$
84. $y = \sqrt[3]{x-3}$

Applying the Concepts

85. **Solving a Formula** Solve the following formula for h:
$$t = \frac{\sqrt{100-h}}{4}$$

86. **Solving a Formula** Solve the following formula for h:
$$t = \sqrt{\frac{2h-40t}{g}}$$

87. **Pendulum Clock** The length of time (T) in seconds it takes the pendulum of a grandfather clock to swing through one complete cycle is given by the formula
$$T = 2\pi\sqrt{\frac{L}{32}}$$
where L is the length, in feet, of the pendulum, and π is approximately $\frac{22}{7}$. How long must the pendulum be if one complete cycle takes 2 seconds?

88. **Pendulum Clock** Solve the formula in Problem 87 for L.

Pollution A long straight river, 100 meters wide, is flowing at 1 meter per second. A pollutant is entering the river at a constant rate from one of its banks. As the pollutant disperses in the water, it forms a plume that is modeled by the equation $y = \sqrt{x}$. Use this information to answer the following questions.

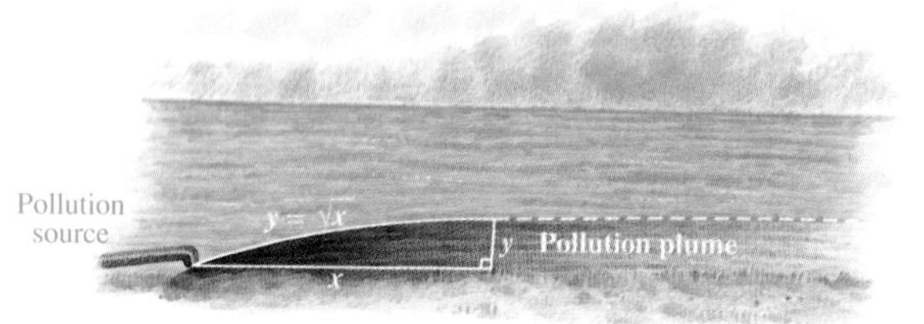

89. How wide is the plume 25 meters down river from the source of the pollution?

90. How wide is the plume 100 meters down river from the source of the pollution?

91. How far down river from the source of the pollution does the plume reach halfway across the river?

92. How far down river from the source of the pollution does the plume reach the other side of the river?

Maintaining Your Skills

Multiply.

93. $\sqrt{2}(\sqrt{3} - \sqrt{2})$

94. $(\sqrt{x} - 4)(\sqrt{x} + 5)$

95. $(\sqrt{x} + 5)^2$

Rationalize the denominator.

97. $\dfrac{\sqrt{x}}{\sqrt{x} + 3}$

98. $\dfrac{\sqrt{5} - \sqrt{3}}{\sqrt{5} + \sqrt{3}}$

Getting Ready for the Next Section

Simplify.

99. $\sqrt{25}$

100. $\sqrt{49}$

101. $\sqrt{12}$

102. $\sqrt{50}$

103. $(-1)^{15}$

104. $(-1)^{20}$

105. $(-1)^{50}$

106. $(-1)^5$

Solve.

107. $3x = 12$

108. $4 = 8y$

109. $4x - 3 = 5$

110. $7 = 2y - 1$

Perform the indicated operation.

111. $(3 + 4x) + (7 - 6x)$

112. $(2 - 5x) + (-1 + 7x)$

113. $(7 + 3x) - (5 + 6x)$

114. $(5 - 2x) - (9 - 4x)$

115. $(3 - 4x)(2 + 5x)$

116. $(8 + x)(7 - 3x)$

117. $2x(4 - 6x)$

118. $3x(7 + 2x)$

119. $(2 + 3x)^2$

120. $(3 + 5x)^2$

121. $(2 - 3x)(2 + 3x)$

122. $(4 - 5x)(4 + 5x)$

7.7 Complex Numbers

OBJECTIVES

A Simplify square roots of negative numbers.

B Simplify powers of i.

C Solve for unknown variables by equating real parts and equating imaginary parts of two complex numbers.

D Add and subtract complex numbers.

E Multiply complex numbers.

F Divide complex numbers.

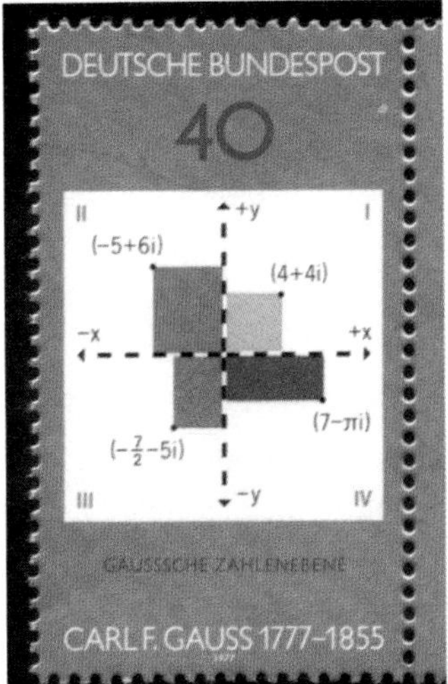

The stamp shown in the margin was issued by Germany in 1977 to commemorate the 200th anniversary of the birth of mathematician Carl Gauss. The number $-5+6i$ shown on the stamp is a complex number, as are the other numbers on the stamp. Working with complex numbers gives us a way to solve a wider variety of equations. For example, the equation $x^2 = -9$ has no real number solutions since the square of a real number is always positive. We have been unable to work with square roots of negative numbers like $\sqrt{-25}$ and $\sqrt{-16}$ for the same reason. Complex numbers allow us to expand our work with radicals to include square roots of negative numbers and solve equations like $x^2 = -9$ and $x^2 = -64$. Our work with complex numbers is based on the following definition.

> **DEFINITION The number i** is such that $i = \sqrt{-1}$ (which is the same as saying $i^2 = -1$).

The number i, as we have defined it here, is not a real number. Because of the way we have defined i, we can use it to simplify square roots of negative numbers.

> **Square Roots of Negative Numbers** If a is a positive number, then $\sqrt{-a}$ can always be written as $i\sqrt{a}$; that is,
>
> $$\sqrt{-a} = i\sqrt{a} \quad \text{if } a \text{ is a positive number}$$

To justify our rule, we simply square the quantity $i\sqrt{a}$ to obtain $-a$. Here is what it looks like when we do so:

$$(i\sqrt{a})^2 = i^2 \cdot (\sqrt{a})^2$$
$$= -1 \cdot a$$
$$= -a$$

Here is some examples that illustrate the use of our new rule.

EXAMPLES

Write each square root in terms of the number i.

1. $\sqrt{-25} = i\sqrt{25} = i \cdot 5 = 5i$
2. $\sqrt{-49} = i\sqrt{49} = i \cdot 7 = 7i$
3. $\sqrt{-12} = i\sqrt{12} = i \cdot 2\sqrt{3} = 2i\sqrt{3}$
4. $\sqrt{-17} = i\sqrt{17}$

Note

In Examples 3 and 4 we wrote i before the radical simply to avoid confusion. If we were to write the answer to 3 as $2\sqrt{3}i$, some people would think the i was under the radical sign and it is not.

If we assume all the properties of exponents hold when the base is i, we can write any power of i as i, -1, $-i$, or 1. Using the fact that $i^2 = -1$, we have

$$i^1 = i$$
$$i^2 = -1$$
$$i^3 = i^2 \cdot i = -1(i) = -i$$
$$i^4 = i^2 \cdot i^2 = -1(-1) = 1$$

Since $i^4 = 1$, i^5 will simplify to i, and we will begin repeating the sequence i, -1, $-i$, 1 as we simplify higher powers of i: Any power of i simplifies to i, -1, $-i$, or 1. The easiest way to simplify higher powers of i is to write them in terms of i^2. For instance, to simplify i^{21}, we would write it as

$$(i^2)^{10} \cdot i \qquad \text{because } 2 \cdot 10 + 1 = 21$$

Then, since $i^2 = -1$, we have

$$(-1)^{10} \cdot i = 1 \cdot i = i$$

EXAMPLES Simplify as much as possible.

5. $i^{30} = (i^2)^{15} = (-1)^{15} = -1$

6. $i^{11} = (i^2)^5 \cdot i = (-1)^5 \cdot i = (-1)i = -i$

7. $i^{40} = (i^2)^{20} = (-1)^{20} = 1$

DEFINITION A **complex number** is any number that can be put in the form

$$a + bi$$

where a and b are real numbers and $i = \sqrt{-1}$. The form $a + bi$ is called **standard form** for complex numbers. The number a is called the **real part** of the complex number. The number b is called the **imaginary part** of the complex number.

Every real number is a complex number. For example, 8 can be written as $8 + 0i$. Likewise, $-\frac{1}{2}$, π, $\sqrt{3}$, and -9 are complex numbers because they can all be written in the form $a + bi$:

$$-\frac{1}{2} = -\frac{1}{2} + 0i \qquad \pi = \pi + 0i$$

$$\sqrt{3} = \sqrt{3} + 0i \qquad -9 = -9 + 0i$$

The real numbers occur when $b = 0$. When $b \neq 0$, we have complex numbers that contain i, such as $2 + 5i$, $6 - i$, $4i$, and $\frac{1}{2}i$. These numbers are called *imaginary numbers.* The diagram explains this further.

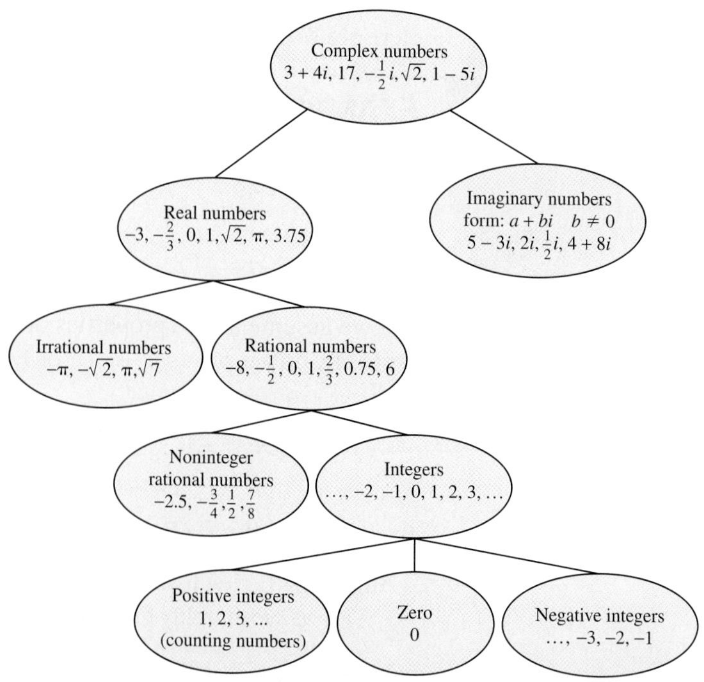

Equality for Complex Numbers

Two complex numbers are equal if and only if their real parts are equal and their imaginary parts are equal; that is, for real numbers a, b, c, and d,

$$a + bi = c + di \quad \text{if and only if} \quad a = c \quad \text{and} \quad b = d$$

EXAMPLE 8 Find x and y if $3x + 4i = 12 - 8yi$.

SOLUTION Since the two complex numbers are equal, their real parts are equal and their imaginary parts are equal:

$$3x = 12 \qquad \text{and} \qquad 4 = -8y$$

$$x = 4 \qquad\qquad y = -\frac{1}{2}$$

EXAMPLE 9 Find x and y if $(4x - 3) + 7i = 5 + (2y - 1)i$.

SOLUTION The real parts are $4x - 3$ and 5. The imaginary parts are 7 and $2y - 1$:

$$4x - 3 = 5 \qquad \text{and} \qquad 7 = 2y - 1$$

$$4x = 8 \qquad\qquad 8 = 2y$$

$$x = 2 \qquad\qquad y = 4$$

Addition and Subtraction of Complex Numbers

To add two complex numbers, add their real parts and add their imaginary parts; that is, if a, b, c, and d are real numbers, then

$$(a + bi) + (c + di) = (a + c) + (b + d)i$$

If we assume that the commutative, associative, and distributive properties hold for the number i, then the definition of addition is simply an extension of these properties.

We define subtraction in a similar manner. If a, b, c, and d are real numbers, then

$$(a + bi) - (c + di) = (a - c) + (b - d)i$$

EXAMPLES Add or subtract as indicated.

10. $(3 + 4i) + (7 - 6i) = (3 + 7) + (4 - 6)i = 10 - 2i$

11. $(7 + 3i) - (5 + 6i) = (7 - 5) + (3 - 6)i = 2 - 3i$

12. $(5 - 2i) - (9 - 4i) = (5 - 9) + (-2 + 4)i = -4 + 2i$

Multiplication of Complex Numbers

Since complex numbers have the same form as binomials, we find the product of two complex numbers the same way we find the product of two binomials.

EXAMPLE 13 Multiply $(3 - 4i)(2 + 5i)$.

SOLUTION Multiplying each term in the second complex number by each term in the first, we have

$$(3 - 4i)(2 + 5i) = 3 \cdot 2 + 3 \cdot 5i - 2 \cdot 4i - 5i(4i)$$
$$= 6 + 15i - 8i - 20i^2$$

Combining similar terms and using the fact that $i^2 = -1$, we can simplify as follows:

$$6 + 15i - 8i - 20i^2 = 6 + 7i - 20(-1)$$
$$= 6 + 7i + 20$$
$$= 26 + 7i$$

The product of the complex numbers $3 - 4i$ and $2 + 5i$ is the complex number $26 + 7i$.

EXAMPLE 14 Multiply $2i(4 - 6i)$.

SOLUTION Applying the distributive property gives us

$$2i(4 - 6i) = 2i \cdot 4 - 2i \cdot 6i$$
$$= 8i - 12i^2$$
$$= 12 + 8i$$

EXAMPLE 15 Expand $(3 + 5i)^2$.

SOLUTION We treat this like the square of a binomial. Remember, $(a + b)^2 = a^2 + 2ab + b^2$.

$$(3 + 5i)^2 = 3^2 + 2(3)(5i) + (5i)^2$$
$$= 9 + 30i + 25i^2$$
$$= 9 + 30i - 25$$
$$= -16 + 30i$$

Note
We can obtain the same result by writing $(3 + 5i)^2$ as $(3 + 5i)$ times $(3 + 5i)$ and applying the FOIL method as we did with the problem in Example 13.

EXAMPLE 16 Multiply $(2 - 3i)(2 + 3i)$.

SOLUTION This product has the form $(a - b)(a + b)$, which we know results in the difference of two squares, $a^2 - b^2$:

$$(2 - 3i)(2 + 3i) = 2^2 - (3i)^2$$
$$= 4 - 9i^2$$
$$= 4 + 9$$
$$= 13$$

The product of the two complex numbers $2 - 3i$ and $2 + 3i$ is the real number 13. The two complex numbers $2 - 3i$ and $2 + 3i$ are called complex conjugates. The fact that their product is a real number is very useful.

DEFINITION The complex numbers $a + bi$ and $a - bi$ are called **complex conjugates.** One important property they have is that their product is the real number $a^2 + b^2$. Here's why:

$$\begin{aligned}(a + bi)(a - bi) &= a^2 - (bi)^2 \\ &= a^2 - b^2i^2 \\ &= a^2 - b^2(-1) \\ &= a^2 + b^2\end{aligned}$$

Division with Complex Numbers

The fact that the product of two complex conjugates is a real number is the key to division with complex numbers.

EXAMPLE 17 Divide $\dfrac{2 + i}{3 - 2i}$.

SOLUTION We want a complex number in standard form that is equivalent to the quotient $\dfrac{2 + i}{3 - 2i}$. We need to eliminate i from the denominator. Multiplying the numerator and denominator by $3 + 2i$ will give us what we want:

$$\begin{aligned}\frac{2 + i}{3 - 2i} &= \frac{2 + i}{3 - 2i} \cdot \frac{\mathbf{(3 + 2i)}}{\mathbf{(3 + 2i)}} \\ &= \frac{6 + 4i + 3i + 2i^2}{9 - 4i^2} \\ &= \frac{6 + 7i - 2}{9 + 4} \\ &= \frac{4 + 7i}{13} \\ &= \frac{4}{13} + \frac{7}{13}i\end{aligned}$$

Dividing the complex number $2 + i$ by $3 - 2i$ gives the complex number $\frac{4}{13} + \frac{7}{13}i$.

EXAMPLE 18 Divide $\dfrac{7 - 4i}{i}$.

SOLUTION The conjugate of the denominator is $-i$. Multiplying numerator and denominator by this number, we have

$$\begin{aligned}\frac{7 - 4i}{i} &= \frac{7 - 4i}{i} \cdot \frac{\mathbf{-i}}{\mathbf{-i}} \\ &= \frac{-7i + 4i^2}{-i^2} \\ &= \frac{-7i + 4(-1)}{-(-1)} \\ &= -4 - 7i\end{aligned}$$

LINKING OBJECTIVES AND EXAMPLES

Next to each **objective** we have listed the examples that are best described by that objective.

A	1–4
B	5–7
C	8, 9
D	10–12
E	13–16
F	17, 18

GETTING READY FOR CLASS

After reading through the preceding section, respond in your own words and in complete sentences.

1. What is the number i?
2. What is a complex number?
3. What kind of number results when we multiply complex conjugates?
4. Explain how to divide complex numbers.

Problem Set 7.7

Online support materials can be found at www.thomsonedu.com/login

Write the following in terms of i, and simplify as much as possible.

1. $\sqrt{-36}$
2. $\sqrt{-49}$
3. $-\sqrt{-25}$
4. $-\sqrt{-81}$
▸ **5.** $\sqrt{-72}$
6. $\sqrt{-48}$
7. $-\sqrt{-12}$
8. $-\sqrt{-75}$

Write each of the following as i, -1, $-i$, or 1.

9. i^{28}
10. i^{31}
11. i^{26}
12. i^{37}
13. i^{75}
14. i^{42}

Find x and y so each of the following equations is true.

15. $2x + 3yi = 6 - 3i$
16. $4x - 2yi = 4 + 8i$
17. $2 - 5i = -x + 10yi$
18. $4 + 7i = 6x - 14yi$
19. $2x + 10i = -16 - 2yi$
20. $4x - 5i = -2 + 3yi$
21. $(2x - 4) - 3i = 10 - 6yi$
22. $(4x - 3) - 2i = 8 + yi$
23. $(7x - 1) + 4i = 2 + (5y + 2)i$
24. $(5x + 2) - 7i = 4 + (2y + 1)i$

Combine the following complex numbers.

25. $(2 + 3i) + (3 + 6i)$
26. $(4 + i) + (3 + 2i)$
27. $(3 - 5i) + (2 + 4i)$
28. $(7 + 2i) + (3 - 4i)$
29. $(5 + 2i) - (3 + 6i)$
▸ **30.** $(6 + 7i) - (4 + i)$
31. $(3 - 5i) - (2 + i)$
32. $(7 - 3i) - (4 + 10i)$
33. $[(3 + 2i) - (6 + i)] + (5 + i)$
34. $[(4 - 5i) - (2 + i)] + (2 + 5i)$
35. $[(7 - i) - (2 + 4i)] - (6 + 2i)$
36. $[(3 - i) - (4 + 7i)] - (3 - 4i)$

▢ = Videos available by instructor request
▸ = Online student support materials available at www.thomsonedu.com/login

37. $(3 + 2i) - [(3 - 4i) - (6 + 2i)]$

38. $(7 - 4i) - [(-2 + i) - (3 + 7i)]$

39. $(4 - 9i) + [(2 - 7i) - (4 + 8i)]$

40. $(10 - 2i) - [(2 + i) - (3 - i)]$

Find the following products.

41. $3i(4 + 5i)$

42. $2i(3 + 4i)$

43. $6i(4 - 3i)$

44. $11i(2 - i)$

45. $(3 + 2i)(4 + i)$

46. $(2 - 4i)(3 + i)$

47. $(4 + 9i)(3 - i)$

48. $(5 - 2i)(1 + i)$

49. $(1 + i)^3$

50. $(1 - i)^3$

51. $(2 - i)^3$

52. $(2 + i)^3$

53. $(2 + 5i)^2$

54. $(3 + 2i)^2$

55. $(1 - i)^2$

56. $(1 + i)^2$

57. $(3 - 4i)^2$

58. $(6 - 5i)^2$

59. $(2 + i)(2 - i)$

60. $(3 + i)(3 - i)$

61. $(6 - 2i)(6 + 2i)$

62. $(5 + 4i)(5 - 4i)$

63. $(2 + 3i)(2 - 3i)$

64. $(2 - 7i)(2 + 7i)$

65. $(10 + 8i)(10 - 8i)$

66. $(11 - 7i)(11 + 7i)$

Find the following quotients. Write all answers in standard form for complex numbers.

67. $\dfrac{2 - 3i}{i}$

68. $\dfrac{3 + 4i}{i}$

69. $\dfrac{5 + 2i}{-i}$

70. $\dfrac{4 - 3i}{-i}$

71. $\dfrac{4}{2 - 3i}$

72. $\dfrac{3}{4 - 5i}$

73. $\dfrac{6}{-3 + 2i}$

74. $\dfrac{-1}{-2 - 5i}$

75. $\dfrac{2 + 3i}{2 - 3i}$

76. $\dfrac{4 - 7i}{4 + 7i}$

77. $\dfrac{5 + 4i}{3 + 6i}$

78. $\dfrac{2 + i}{5 - 6i}$

Applying the Concepts

79. **Electric Circuits** Complex numbers may be applied to electrical circuits. Electrical engineers use the fact that resistance R to electrical flow of the electrical current I and the voltage V are related by the formula $V = RI$. (Voltage is measured in volts, resistance in ohms, and current in amperes.) Find the resistance to electrical flow in a circuit that has a voltage $V = (80 + 20i)$ volts and current $I = (-6 + 2i)$ amps.

80. **Electric Circuits** Refer to the information about electrical circuits in Problem 79, and find the current in a circuit that has a resistance of $(4 + 10i)$ ohms and a voltage of $(5 - 7i)$ volts.

Maintaining Your Skills

Solve each equation.

81. $\dfrac{t}{3} - \dfrac{1}{2} = -1$

82. $\dfrac{x}{x - 2} + \dfrac{2}{3} = \dfrac{2}{x - 2}$

83. $2 + \dfrac{5}{y} = \dfrac{3}{y^2}$

84. $1 - \dfrac{1}{y} = \dfrac{12}{y^2}$

Solve each application problem.

85. The sum of a number and its reciprocal is $\frac{41}{20}$. Find the number.

86. It takes an inlet pipe 8 hours to fill a tank. The drain can empty the tank in 6 hours. If the tank is full and both the inlet pipe and drain are open, how long will it take to drain the tank?

Extending the Concepts

87. Show that $-i$ and $\dfrac{1}{i}$ (the opposite and the reciprocal of i) are the same number.

88. Show that i^{2n+1} is the same as i for all positive even integers n.

89. Show that $x = 1 + i$ is a solution to the equation $x^2 - 2x + 2 = 0$.

90. Show that $x = 1 - i$ is a solution to the equation $x^2 - 2x + 2 = 0$.

91. Show that $x = 2 + i$ is a solution to the equation $x^3 - 11x + 20 = 0$.

92. Show that $x = 2 - i$ is a solution to the equation $x^3 - 11x + 20 = 0$.

Chapter 7 SUMMARY

EXAMPLES

1. The number 49 has two square roots, 7 and −7. They are written like this: $\sqrt{49} = 7 \qquad -\sqrt{49} = -7$

Square Roots [7.1]

Every positive real number x has two square roots. The *positive square root* of x is written $\sqrt{x}$, and the *negative square root* of x is written $-\sqrt{x}$. Both the positive and the negative square roots of x are numbers we square to get x; that is,

$$\left.\begin{aligned} (\sqrt{x})^2 &= x \\ \text{and} \quad (-\sqrt{x})^2 &= x \end{aligned}\right\} \quad \text{for } x \ge 0$$

2. $\sqrt[3]{8} = 2$
   $\sqrt[3]{-27} = -3$

Higher Roots [7.1]

In the expression $\sqrt[n]{a}$, n is the *index*, a is the *radicand*, and $\sqrt{}$ is the *radical sign*. The expression $\sqrt[n]{a}$ is such that

$$(\sqrt[n]{a})^n = a \qquad a \ge 0 \text{ when } n \text{ is even}$$

3. $25^{1/2} = \sqrt{25} = 5$
   $8^{2/3} = (\sqrt[3]{8})^2 = 2^2 = 4$
   $9^{3/2} = (\sqrt{9})^3 = 3^3 = 27$

Rational Exponents [7.1, 7.2]

Rational exponents are used to indicate roots. The relationship between rational exponents and roots is as follows:

$$a^{1/n} = \sqrt[n]{a} \qquad \text{and} \qquad a^{m/n} = (a^{1/n})^m = (a^m)^{1/n}$$

$$a \ge 0 \text{ when } n \text{ is even}$$

4. $\sqrt{4 \cdot 5} = \sqrt{4}\,\sqrt{5} = 2\sqrt{5}$
   $\sqrt{\dfrac{7}{9}} = \dfrac{\sqrt{7}}{\sqrt{9}} = \dfrac{\sqrt{7}}{3}$

Properties of Radicals [7.3]

If a and b are nonnegative real numbers whenever n is even, then

1. $\sqrt[n]{ab} = \sqrt[n]{a}\,\sqrt[n]{b}$
2. $\sqrt[n]{\dfrac{a}{b}} = \dfrac{\sqrt[n]{a}}{\sqrt[n]{b}} \qquad (b \ne 0)$

5. $\sqrt{\dfrac{4}{5}} = \dfrac{\sqrt{4}}{\sqrt{5}}$
   $= \dfrac{2}{\sqrt{5}} \cdot \dfrac{\sqrt{5}}{\sqrt{5}}$
   $= \dfrac{2\sqrt{5}}{5}$

Simplified Form for Radicals [7.3]

A radical expression is said to be in *simplified form*

1. If there is no factor of the radicand that can be written as a power greater than or equal to the index;
2. If there are no fractions under the radical sign; and
3. If there are no radicals in the denominator.

Addition and Subtraction of Radical Expressions [7.4]

6. $5\sqrt{3} - 7\sqrt{3} = (5 - 7)\sqrt{3}$
$= -2\sqrt{3}$
$\sqrt{20} + \sqrt{45} = 2\sqrt{5} + 3\sqrt{5}$
$= (2 + 3)\sqrt{5}$
$= 5\sqrt{5}$

We add and subtract radical expressions by using the distributive property to combine similar radicals. Similar radicals are radicals with the same index and the same radicand.

Multiplication of Radical Expressions [7.5]

7. $(\sqrt{x} + 2)(\sqrt{x} + 3)$
$= \sqrt{x}\sqrt{x} + 3\sqrt{x} + 2\sqrt{x} + 2 \cdot 3$
$= x + 5\sqrt{x} + 6$

We multiply radical expressions in the same way that we multiply polynomials. We can use the distributive property and the FOIL method.

Rationalizing the Denominator [7.3, 7.5]

8. $\frac{3}{\sqrt{2}} = \frac{3}{\sqrt{2}} \cdot \frac{\sqrt{2}}{\sqrt{2}} = \frac{3\sqrt{2}}{2}$

$\frac{3}{\sqrt{5} - \sqrt{3}} = \frac{3}{\sqrt{5} - \sqrt{3}} \cdot \frac{\sqrt{5} + \sqrt{3}}{\sqrt{5} + \sqrt{3}}$

$= \frac{3\sqrt{5} + 3\sqrt{3}}{5 - 3}$

$= \frac{3\sqrt{5} + 3\sqrt{3}}{2}$

When a fraction contains a square root in the denominator, we rationalize the denominator by multiplying numerator and denominator by

1. The square root itself if there is only one term in the denominator, or
2. The conjugate of the denominator if there are two terms in the denominator.

Rationalizing the denominator is also called division of radical expressions.

Squaring Property of Equality [7.6]

9. $\sqrt{2x + 1} = 3$
$(\sqrt{2x + 1})^2 = 3^2$
$2x + 1 = 9$
$x = 4$

We may square both sides of an equation any time it is convenient to do so, as long as we check all resulting solutions in the original equation.

Complex Numbers [7.7]

10. $3 + 4i$ is a complex number.
Addition
$(3 + 4i) + (2 - 5i) = 5 - i$
Multiplication
$(3 + 4i)(2 - 5i)$
$= 6 - 15i + 8i - 20i^2$
$= 6 - 7i + 20$
$= 26 - 7i$

A *complex number* is any number that can be put in the form

$$a + bi$$

where a and b are real numbers and $i = \sqrt{-1}$. The *real part* of the complex number is a, and b is the *imaginary part.*

If a, b, c, and d are real numbers, then we have the following definitions associated with complex numbers:

1. Equality

$$a + bi = c + di \quad \text{if and only if} \quad a = c \text{ and } b = d$$

Division

$$\frac{2}{3+4i} = \frac{2}{3+4i} \cdot \frac{\mathbf{3-4i}}{\mathbf{3-4i}}$$

$$= \frac{6-8i}{9+16}$$

$$= \frac{6}{25} - \frac{8}{25}i$$

2. Addition and subtraction

$$(a + bi) + (c + di) = (a + c) + (b + d)i$$

$$(a + bi) - (c + di) = (a - c) + (b - d)i$$

3. Multiplication

$$(a + bi)(c + di) = (ac - bd) + (ad + bc)i$$

4. Division is similar to rationalizing the denominator.

Chapter 7 Review Test

The problems below form a comprehensive review of the material in this chapter. They can be used to study for exams. If you would like to take a practice test on this chapter, you can use the odd-numbered problems. Give yourself an hour and work as many of the odd-numbered problems as possible. When you are finished, or when an hour has passed, check your answers with the answers in the back of the book. You can use the even-numbered problems for a second practice test.

Simplify each expression as much as possible. [7.1]

1. $\sqrt{49}$
2. $(-27)^{1/3}$
3. $16^{1/4}$
4. $9^{3/2}$
5. $\sqrt[5]{32x^{15}y^{10}}$
6. $8^{-4/3}$

Use the properties of exponents to simplify each expression. Assume all bases represent positive numbers. [7.1]

7. $x^{2/3} \cdot x^{4/3}$
8. $(a^{2/3}b^{4/3})^3$
9. $\dfrac{a^{3/5}}{a^{1/4}}$
10. $\dfrac{a^{2/3}b^3}{a^{1/4}b^{1/3}}$

Multiply. [7.2]

11. $(3x^{1/2} + 5y^{1/2})(4x^{1/2} - 3y^{1/2})$
12. $(a^{1/3} - 5)^2$
13. Divide: $\dfrac{28x^{5/6} + 14x^{7/6}}{7x^{1/3}}$. (Assume $x > 0$.) [7.2]
14. Factor $2(x-3)^{1/4}$ from $8(x-3)^{5/4} - 2(x-3)^{1/4}$. [7.2]
15. Simplify $x^{3/4} + \dfrac{5}{x^{1/4}}$ into a single fraction. (Assume $x > 0$.) [7.2]

Write each expression in simplified form for radicals. (Assume all variables represent nonnegative numbers.) [7.3]

16. $\sqrt{12}$
17. $\sqrt{50}$
18. $\sqrt[3]{16}$
19. $\sqrt{18x^2}$
20. $\sqrt{80a^3b^4c^2}$
21. $\sqrt[4]{32a^4b^5c^6}$

Rationalize the denominator in each expression. [7.3]

22. $\dfrac{3}{\sqrt{2}}$
23. $\dfrac{6}{\sqrt[3]{2}}$

Write each expression in simplified form. (Assume all variables represent positive numbers.) [7.3]

24. $\sqrt{\dfrac{48x^3}{7y}}$
25. $\sqrt[3]{\dfrac{40x^2y^3}{3z}}$

Combine the following expressions. (Assume all variables represent positive numbers.) [7.4]

26. $5x\sqrt{6} + 2x\sqrt{6} - 9x\sqrt{6}$
27. $\sqrt{12} + \sqrt{3}$
28. $\dfrac{3}{\sqrt{5}} + \sqrt{5}$
29. $3\sqrt{8} - 4\sqrt{72} + 5\sqrt{50}$
30. $3b\sqrt{27a^5b} + 2a\sqrt{3a^3b^3}$
31. $2x\sqrt[3]{xy^3z^2} - 6y\sqrt[3]{x^4z^2}$

Multiply. [7.5]

32. $\sqrt{2}(\sqrt{3} - 2\sqrt{2})$
33. $(\sqrt{x} - 2)(\sqrt{x} - 3)$

Rationalize the denominator. (Assume $x, y > 0$.) [7.5]

34. $\dfrac{3}{\sqrt{5} - 2}$
35. $\dfrac{\sqrt{7} + \sqrt{5}}{\sqrt{7} - \sqrt{5}}$
36. $\dfrac{3\sqrt{7}}{3\sqrt{7} - 4}$

Solve each equation. [7.6]

37. $\sqrt{4a + 1} = 1$
38. $\sqrt[3]{3x - 8} = 1$
39. $\sqrt{3x + 1} - 3 = 1$
40. $\sqrt{x + 4} = \sqrt{x} - 2$

Graph each equation. [7.6]

41. $y = 3\sqrt{x}$
42. $y = \sqrt[3]{x} + 2$

Write each of the following as i, -1, $-i$, or 1. [7.7]

43. i^{24}
44. i^{27}

Find x and y so that each of the following equations is true. [7.7]

45. $3 - 4i = -2x + 8yi$
46. $(3x + 2) - 8i = -4 + 2yi$

Combine the following complex numbers. [7.7]

47. $(3 + 5i) + (6 - 2i)$
48. $(2 + 5i) - [(3 + 2i) + (6 - i)]$

Multiply. [7.7]

49. $3i(4 + 2i)$

50. $(2 + 3i)(4 + i)$

51. $(4 + 2i)^2$

52. $(4 + 3i)(4 - 3i)$

Divide. Write all answers in standard form for complex numbers. [7.7]

53. $\dfrac{3 + i}{i}$

54. $\dfrac{-3}{2 + i}$

55. Construction The roof of the house shown in Figure 1 is to extend up 13.5 feet above the ceiling, which is 36 feet across. Find the length of one side of the roof.

FIGURE 1

56. Surveying A surveyor is attempting to find the distance across a pond. From a point on one side of the pond he walks 25 yards to the end of the pond and then makes a 90-degree turn and walks another 60 yards before coming to a point directly across the pond from the point at which he started. What is the distance across the pond? (See Figure 2.)

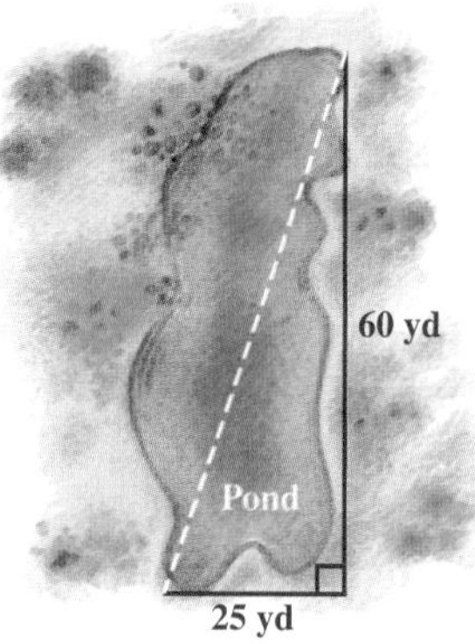

FIGURE 2

Chapter 7 Projects

Rational Exponents and Roots

GROUP PROJECT Constructing the Spiral of Roots

Number of People 3

Time Needed 20 minutes

Equipment Two sheets of graph paper (4 or 5 squares per inch) and pencils.

Background The spiral of roots gives us a way to visualize the positive square roots of the counting numbers, and in so doing, we see many line segments whose lengths are irrational numbers.

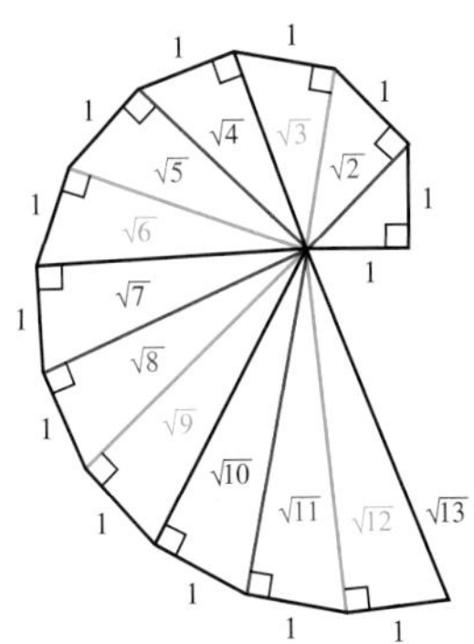

Procedure You are to construct a spiral of roots from a line segment 1 inch long. The graph paper you have contains either 4 or 5 squares per inch, allowing you to accurately draw 1-inch line segments. Because the lines on the graph paper are perpendicular to one another, if you are careful, you can use the graph paper to connect one line segment to another so that they form a right angle.

1. Fold one of the pieces of graph paper so it can be used as a ruler.
2. Use the folded paper to draw a line segment 1-inch long, just to the right of the middle of the unfolded paper. On the end of this segment, attach another segment of 1-inch length at a right angle to the first one. Connect the end points of the segments to form a right triangle. Label each side of this triangle. When you are finished, your work should resemble Figure 1.

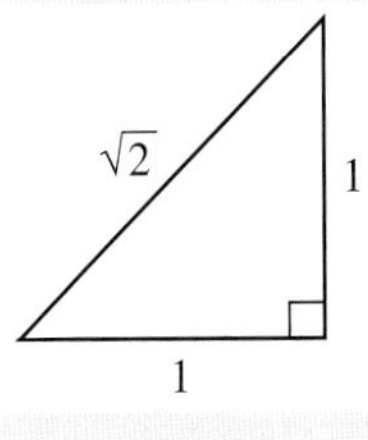

FIGURE 1

3. On the end of the hypotenuse of the triangle, attach a 1-inch line segment so that the two segments form a right angle. (Use the folded paper to do this.) Draw the hypotenuse of this triangle. Label all the sides of this second triangle. Your work should resemble Figure 2.

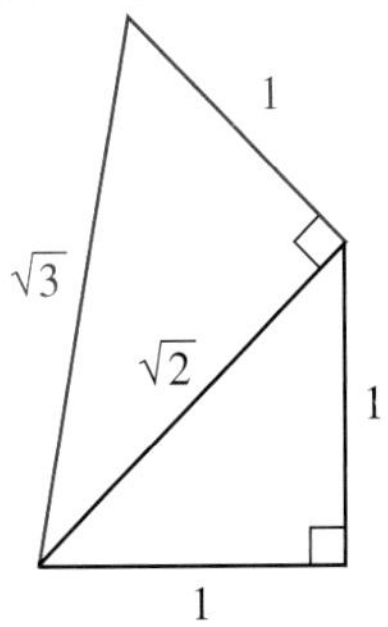

FIGURE 2

4. Continue to draw a new right triangle by attaching 1-inch line segments at right angles to the previous hypotenuse. Label all the sides of each triangle.
5. Stop when you have drawn a hypotenuse $\sqrt{8}$ inches long.

RESEARCH PROJECT

Maria Gaetana Agnesi (1718–1799)

January 9, 1999, was the 200th anniversary of the death of Maria Agnesi, the author of *Instituzioni analitiche ad uso della gioventu italiana* (1748), a calculus textbook considered to be the best book of its time and the first surviving mathematical work written by a woman. Maria Agnesi is also famous for a curve that is named for her. The curve is called the Witch of Agnesi in English, but its actual translation is the Locus of Agnesi. The foundation of the curve is shown in the figure. Research the Witch of Agnesi and then explain, in essay form, how the diagram in the figure is used to produce the Witch of Agnesi. Include a rough sketch of the curve, starting with a circle of diameter a as shown in the figure. Then, for comparison, sketch the curve again, starting with a circle of diameter $2a$.

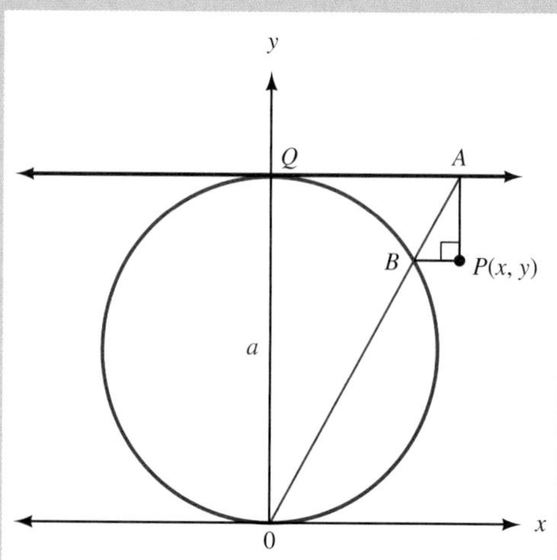

Quadratic Functions

8

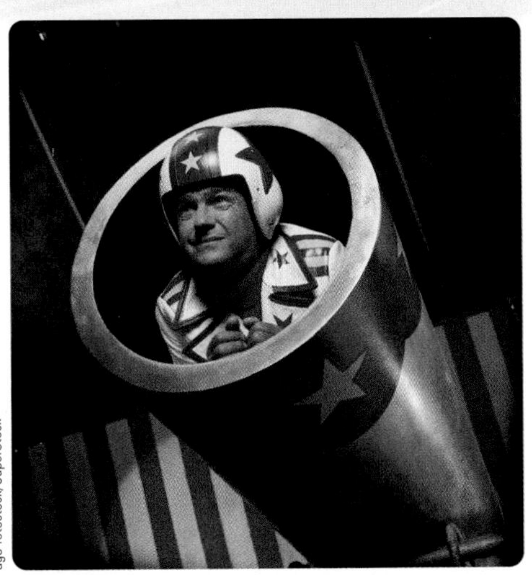

age fotostock/SuperStock

CHAPTER OUTLINE

If you have been to the circus or the county fair, you may have witnessed one of the more spectacular acts, the human cannonball. The human cannonball shown in the photograph will reach a height of 70 feet and travel a distance of 160 feet before landing in a safety net. In this chapter, we use this information to derive the equation

$$f(x) = -\frac{7}{640}(x - 80)^2 + 70 \quad \text{for } 0 \le x \le 160$$

which describes the path flown by this particular cannonball. The table and graph below were constructed from this equation.

Path of a Human Cannonball

x (feet)	$f(x)$ (nearest foot)
0	0
40	53
80	70
120	53
160	0

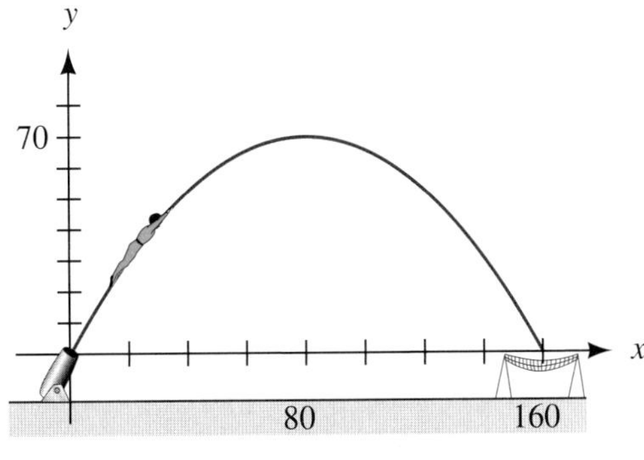

FIGURE 1

All objects that are projected into the air, whether they are basketballs, bullets, arrows, or coins, follow parabolic paths like the one shown in Figure 1. Studying the material in this chapter will give you a more mathematical hold on the world around you.

8.1 Completing the Square

OBJECTIVES

A Solve quadratic equations by taking the square root of both sides.

B Solve quadratic equations by completing the square.

C Use quadratic equations to solve for missing parts of right triangles.

Table 1 is taken from the trail map given to skiers at the Northstar at Tahoe Ski Resort in Lake Tahoe, California. The table gives the length of each chair lift at Northstar, along with the change in elevation from the beginning of the lift to the end of the lift.

Right triangles are good mathematical models for chair lifts. In this section, we will use our knowledge of right triangles, along with the new material developed in the section, to solve problems involving chair lifts and a variety of other examples.

TABLE 1
From the Trail Map for the Northstar at Tahoe Ski Resort

Lift Information		
Lift	**Vertical Rise (feet)**	**Length (feet)**
Big Springs Gondola	480	4,100
Bear Paw Double	120	790
Echo Triple	710	4,890
Aspen Express Quad	900	5,100
Forest Double	1,170	5,750
Lookout Double	960	4,330
Comstock Express Quad	1,250	5,900
Rendezvous Triple	650	2,900
Schaffer Camp Triple	1,860	6,150
Chipmunk Tow Lift	28	280
Bear Cub Tow Lift	120	750

In this section, we will develop the first of our new methods of solving quadratic equations. The new method is called *completing the square.* Completing the square on a quadratic equation allows us to obtain solutions, regardless of whether the equation can be factored. Before we solve equations by completing the square, we need to learn how to solve equations by taking square roots of both sides.

Consider the equation

$$x^2 = 16$$

We could solve it by writing it in standard form, factoring the left side, and proceeding as we have done previously. We can shorten our work considerably, however, if we simply notice that x must be either the positive square root of 16 or the negative square root of 16; that is,

$$\text{If} \quad x^2 = 16$$

$$\text{then} \quad x = \sqrt{16} \quad \text{or} \quad x = -\sqrt{16}$$

$$x = 4 \quad \text{or} \quad x = -4$$

We can generalize this result into a theorem as follows.

> **Theorem 1** If $a^2 = b$ where b is a real number, then $a = \sqrt{b}$ or $a = -\sqrt{b}$.

Notation The expression $a = \sqrt{b}$ or $a = -\sqrt{b}$ can be written in shorthand form as $a = \pm\sqrt{b}$. The symbol $\pm$ is read "plus or minus."

We can apply Theorem 1 to some fairly complicated quadratic equations.

EXAMPLE 1

Solve $(2x-3)^2 = 25$.

SOLUTION

$$(2x-3)^2 = 25$$

$$2x - 3 = \pm\sqrt{25} \quad \textbf{Theorem 1}$$

$$2x - 3 = \pm 5 \quad \sqrt{25} = 5$$

$$2x = 3 \pm 5 \quad \textbf{Add 3 to both sides}$$

$$x = \frac{3 \pm 5}{2} \quad \textbf{Divide both sides by 2}$$

The last equation can be written as two separate statements:

$$x = \frac{3+5}{2} \quad \text{or} \quad x = \frac{3-5}{2}$$

$$= \frac{8}{2} \qquad\qquad = \frac{-2}{2}$$

$$= 4 \quad \text{or} \quad = -1$$

The solution set is $\{4, -1\}$.

Notice that we could have solved the equation in Example 1 by expanding the left side, writing the resulting equation in standard form, and then factoring. The problem would look like this:

$$(2x-3)^2 = 25 \quad \textbf{Original equation}$$

$$4x^2 - 12x + 9 = 25 \quad \textbf{Expand the left side}$$

$$4x^2 - 12x - 16 = 0 \quad \textbf{Add } -25 \textbf{ to each side}$$

$$4(x^2 - 3x - 4) = 0 \quad \textbf{Begin factoring}$$

$$4(x-4)(x+1) = 0 \quad \textbf{Factor completely}$$

$$x - 4 = 0 \quad \text{or} \quad x + 1 = 0 \quad \textbf{Set variable factors equal to 0}$$

$$x = 4 \quad \text{or} \quad x = -1$$

As you can see, solving the equation by factoring leads to the same two solutions.

> **Note**
> We cannot solve the equation in Example 2 by factoring. If we expand the left side and write the resulting equation in standard form, we are left with a quadratic equation that does not factor:
> $(3x-1)^2 = -12$ Equation from Example 2
> $9x^2 - 6x + 1 = -12$ Expand the left side
> $9x^2 - 6x + 13 = 0$ Standard form, but not factorable

EXAMPLE 2

Solve for x: $(3x-1)^2 = -12$.

SOLUTION

$$(3x-1)^2 = -12$$

$$3x - 1 = \pm\sqrt{-12} \quad \textbf{Theorem 1}$$

$$3x - 1 = \pm 2i\sqrt{3} \quad \sqrt{-12} = 2i\sqrt{3}$$

$$3x = 1 \pm 2i\sqrt{3} \quad \textbf{Add 1 to both sides}$$

$$x = \frac{1 \pm 2i\sqrt{3}}{3} \quad \textbf{Divide both sides by 3}$$

The solution set is $\left\{\frac{1+2i\sqrt{3}}{3}, \frac{1-2i\sqrt{3}}{3}\right\}$

Both solutions are complex. Here is a check of the first solution:

When $\qquad x = \dfrac{1 + 2i\sqrt{3}}{3}$

the equation $\qquad (3x - 1)^2 = -12$

becomes $\qquad \left(3 \cdot \dfrac{1 + 2i\sqrt{3}}{3} - 1\right)^2 \overset{?}{=} -12$

or $\qquad (1 + 2i\sqrt{3} - 1)^2 \overset{?}{=} -12$

$$(2i\sqrt{3})^2 \overset{?}{=} -12$$

$$4 \cdot i^2 \cdot 3 \overset{?}{=} -12$$

$$12(-1) \overset{?}{=} -12$$

$$-12 = -12$$

EXAMPLE 3

Solve $x^2 + 6x + 9 = 12$.

SOLUTION We can solve this equation as we have the equations in Examples 1 and 2 if we first write the left side as $(x + 3)^2$.

$x^2 + 6x + 9 = 12$	**Original equation**
$(x + 3)^2 = 12$	**Write $x^2 + 6x + 9$ as $(x + 3)^2$**
$x + 3 = \pm 2\sqrt{3}$	**Theorem 1**
$x = -3 \pm 2\sqrt{3}$	**Add -3 to each side**

We have two irrational solutions: $-3 + 2\sqrt{3}$ and $-3 - 2\sqrt{3}$. What is important about this problem, however, is the fact that the equation was easy to solve because the left side was a perfect square trinomial.

Completing the Square

The method of completing the square is simply a way of transforming any quadratic equation into an equation of the form found in the preceding three examples.

The key to understanding the method of completing the square lies in recognizing the relationship between the last two terms of any perfect square trinomial whose leading coefficient is 1.

Consider the following list of perfect square trinomials and their corresponding binomial squares:

$$x^2 - 6x + 9 = (x - 3)^3$$

$$x^2 + 8x + 16 = (x + 4)^2$$

$$x^2 - 10x + 25 = (x - 5)^2$$

$$x^2 + 12x + 36 = (x + 6)^2$$

In each case the leading coefficient is 1. A more important observation comes from noticing the relationship between the linear and constant terms (middle and last terms) in each trinomial. Observe that the constant term in each case is the square of half the coefficient of x in the middle term. For example, in the last expression, the constant term 36 is the square of half of 12, where 12 is the coefficient of x in the middle term. (Notice also that the second terms in all the binomials on the right side are half the coefficients of the middle terms of the trinomials on the left side.) We can use these observations to build our own perfect square trinomials and, in doing so, solve some quadratic equations.

Consider the following equation:

$$x^2 + 6x = 3$$

We can think of the left side as having the first two terms of a perfect square trinomial. We need only add the correct constant term. If we take half the coefficient of x, we get 3. If we then square this quantity, we have 9. Adding the 9 to both sides, the equation becomes

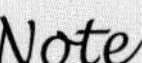

Note
This is the step in which we actually complete the square.

$$x^2 + 6x + \mathbf{9} = 3 + \mathbf{9}$$

The left side is the perfect square $(x + 3)^2$; the right side is 12:

$$(x + 3)^2 = 12$$

The equation is now in the correct form. We can apply Theorem 1 and finish the solution:

$$\begin{aligned} (x + 3)^2 &= 12 \\ x + 3 &= \pm\sqrt{12} && \textbf{Theorem 1} \\ x + 3 &= \pm 2\sqrt{3} \\ x &= -3 \pm 2\sqrt{3} \end{aligned}$$

The solution set is $\{-3 + 2\sqrt{3}, -3 - 2\sqrt{3}\}$. The method just used is called *completing the square* since we complete the square on the left side of the original equation by adding the appropriate constant term.

EXAMPLE 4

Solve by completing the square: $x^2 + 5x - 2 = 0$.

SOLUTION We must begin by adding 2 to both sides. (The left side of the equation, as it is, is not a perfect square because it does not have the correct constant term. We will simply "move" that term to the other side and use our own constant term.)

$$x^2 + 5x = 2 \qquad \textbf{Add 2 to each side}$$

We complete the square by adding the square of half the coefficient of the linear term to both sides:

$$\begin{aligned} x^2 + 5x + \mathbf{\frac{25}{4}} &= 2 + \mathbf{\frac{25}{4}} && \textbf{Half of 5 is } \tfrac{5}{2}\textbf{, the square of which is } \tfrac{25}{4} \\ \left(x + \frac{5}{2}\right)^2 &= \frac{33}{4} && 2 + \tfrac{25}{4} = \tfrac{8}{4} + \tfrac{25}{4} = \tfrac{33}{4} \\ x + \frac{5}{2} &= \pm\sqrt{\frac{33}{4}} && \textbf{Theorem 1} \\ x + \frac{5}{2} &= \pm\frac{\sqrt{33}}{2} && \textbf{Simplify the radical} \\ x &= -\frac{5}{2} \pm \frac{\sqrt{33}}{2} && \textbf{Add } -\tfrac{5}{2} \textbf{ to both sides} \\ &= \frac{-5 \pm \sqrt{33}}{2} \end{aligned}$$

Note
We can use a calculator to get decimal approximations to these solutions. If $\sqrt{33} \approx 5.74$, then

$$\frac{-5 + 5.74}{2} = 0.37$$

$$\frac{-5 - 5.74}{2} = -5.37$$

The solution set is $\left\{\dfrac{-5 + \sqrt{33}}{2}, \dfrac{-5 - \sqrt{33}}{2}\right\}$.

EXAMPLE 5

Solve for x: $3x^2 - 8x + 7 = 0$.

SOLUTION

$$3x^2 - 8x + 7 = 0$$

$$3x^2 - 8x = -7 \qquad \textbf{Add } -7 \textbf{ to both sides}$$

We cannot complete the square on the left side because the leading coefficient is not 1. We take an extra step and divide both sides by 3:

$$\frac{3x^2}{3} - \frac{8x}{3} = -\frac{7}{3}$$

$$x^2 - \frac{8}{3}x = -\frac{7}{3}$$

Half of $\frac{8}{3}$ is $\frac{4}{3}$, the square of which is $\frac{16}{9}$.

$$x^2 - \frac{8}{3}x + \mathbf{\frac{16}{9}} = -\frac{7}{3} + \mathbf{\frac{16}{9}} \qquad \textbf{Add } \frac{16}{9} \textbf{ to both sides}$$

$$\left(x - \frac{4}{3}\right)^2 = -\frac{5}{9} \qquad \textbf{Simplify right side}$$

$$x - \frac{4}{3} = \pm\sqrt{-\frac{5}{9}} \qquad \textbf{Theorem 1}$$

$$x - \frac{4}{3} = \pm\frac{i\sqrt{5}}{3} \qquad \sqrt{-\frac{5}{9}} = \frac{\sqrt{-5}}{3} = \frac{i\sqrt{5}}{3}$$

$$x = \frac{4}{3} \pm \frac{i\sqrt{5}}{3} \qquad \textbf{Add } \frac{4}{3} \textbf{ to both sides}$$

$$x = \frac{4 \pm i\sqrt{5}}{3}$$

The solution set is $\left\{\frac{4 + i\sqrt{5}}{3}, \frac{4 - i\sqrt{5}}{3}\right\}$.

To Solve a Quadratic Equation by Completing the Square To summarize the method used in the preceding two examples, we list the following steps:

Step 1: Write the equation in the form $ax^2 + bx = c$.

Step 2: If the leading coefficient is not 1, divide both sides by the coefficient so that the resulting equation has a leading coefficient of 1; that is, if $a \neq 1$, then divide both sides by a.

Step 3: Add the square of half the coefficient of the linear term to both sides of the equation.

Step 4: Write the left side of the equation as the square of a binomial, and simplify the right side if possible.

Step 5: Apply Theorem 1, and solve as usual.

FACTS FROM GEOMETRY

More Special Triangles

The triangles shown in Figures 1 and 2 occur frequently in mathematics.

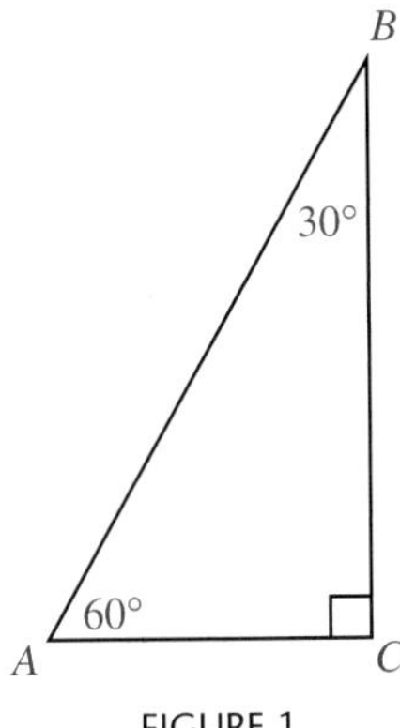

FIGURE 1

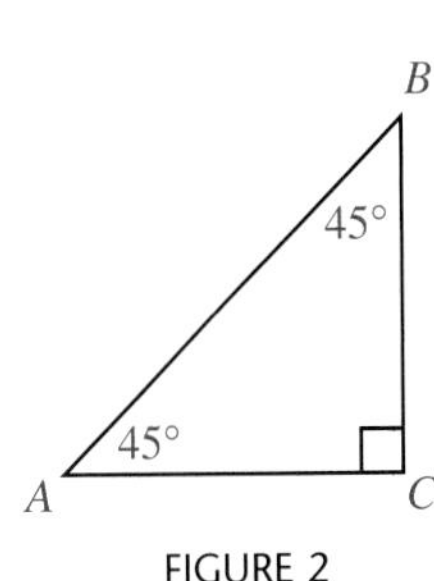

FIGURE 2

Note that both of the triangles are right triangles. We refer to the triangle in Figure 1 as a 30°–60°–90° triangle and the triangle in Figure 2 as a 45°–45°–90° triangle.

EXAMPLE 6

If the shortest side in a 30°–60°–90° triangle is 1 inch, find the lengths of the other two sides.

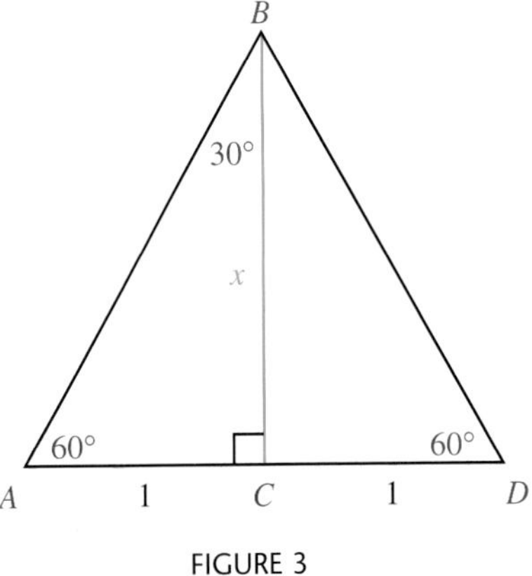

FIGURE 3

SOLUTION In Figure 3 triangle *ABC* is a 30°–60°–90° triangle in which the shortest side *AC* is 1 inch long. Triangle *DBC* is also a 30°–60°–90° triangle in which the shortest side *DC* is 1 inch long.

Notice that the large triangle *ABD* is an equilateral triangle because each of its interior angles is 60°. Each side of triangle *ABD* is 2 inches long. Side *AB* in triangle *ABC* is therefore 2 inches. To find the length of side *BC*, we use the Pythagorean theorem.

$$BC^2 + AC^2 = AB^2$$

$$x^2 + 1^2 = 2^2$$

$$x^2 + 1 = 4$$

$$x^2 = 3$$

$$x = \sqrt{3} \text{ inches}$$

Note that we write only the positive square root because x is the length of a side in a triangle and is therefore a positive number.

EXAMPLE 7 Table 1 in the introduction to this section gives the vertical rise of the Forest Double chair lift as 1,170 feet and the length of the chair lift as 5,750 feet. To the nearest foot, find the horizontal distance covered by a person riding this lift.

SOLUTION Figure 4 is a model of the Forest Double chair lift. A rider gets on the lift at point A and exits at point B. The length of the lift is AB.

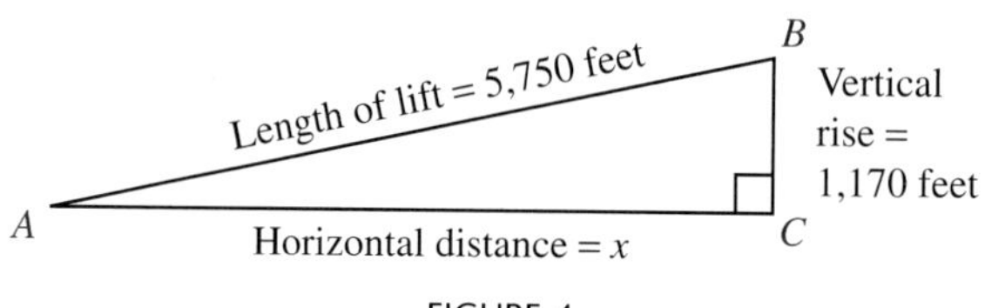

FIGURE 4

To find the horizontal distance covered by a person riding the chair lift, we use the Pythagorean theorem.

$5{,}750^2 = x^2 + 1{,}170^2$	**Pythagorean theorem**
$33{,}062{,}500 = x^2 + 1{,}368{,}900$	**Simplify squares**
$x^2 = 33{,}062{,}500 - 1{,}368{,}900$	**Solve for x^2**
$x^2 = 31{,}693{,}600$	**Simplify the right side**
$x = \sqrt{31{,}693{,}600}$	**Theorem 1**
$= 5{,}630$ feet (to the nearest foot)	

A rider getting on the lift at point A and riding to point B will cover a horizontal distance of approximately 5,630 feet.

LINKING OBJECTIVES AND EXAMPLES

Next to each **objective** we have listed the examples that are best described by that objective.

A	1–3
B	4, 5
C	6, 7

GETTING READY FOR CLASS

After reading through the preceding section, respond in your own words and in complete sentences.

1. What kind of equation do we solve using the method of completing the square?
2. Explain in words how you would complete the square on $x^2 - 16x = 4$.
3. What is the relationship between the shortest side and the longest side in a 30°–60°–90° triangle?
4. What two expressions together are equivalent to $x = \pm 4$?

Problem Set 8.1

Online support materials can be found at www.thomsonedu.com/login

Solve the following equations.

1. $x^2 = 25$
2. $x^2 = 16$
3. $y^2 = \frac{3}{4}$
4. $y^2 = \frac{5}{9}$
5. $x^2 + 12 = 0$
6. $x^2 + 8 = 0$
7. $4a^2 - 45 = 0$
8. $9a^2 - 20 = 0$
9. $(2y - 1)^2 = 25$
10. $(3y + 7)^2 = 1$
11. $(2a + 3)^2 = -9$
12. $(3a - 5)^2 = -49$
13. $x^2 + 8x + 16 = -27$
14. $x^2 - 12x + 36 = -8$
15. $4a^2 - 12a + 9 = -4$
16. $9a^2 - 12a + 4 = -9$

Copy each of the following, and fill in the blanks so that the left side of each is a perfect square trinomial; that is, complete the square.

17. $x^2 + 12x +$ _____ $= (x +$ _____$)^2$
18. $x^2 + 6x +$ _____ $= (x +$ _____$)^2$
19. $x^2 - 4x +$ _____ $= (x -$ _____$)^2$
20. $x^2 - 2x +$ _____ $= (x -$ _____$)^2$
21. $a^2 - 10a +$ _____ $= (a -$ _____$)^2$
22. $a^2 - 8a +$ _____ $= (a -$ _____$)^2$
23. $x^2 + 5x +$ _____ $= (x +$ _____$)^2$
24. $x^2 + 3x +$ _____ $= (x +$ _____$)^2$
25. $y^2 - 7y +$ _____ $= (y -$ _____$)^2$
26. $y^2 - y +$ _____ $= (y -$ _____$)^2$
27. $x^2 + \frac{1}{2}x +$ _____ $= (x +$ _____$)^2$
28. $x^2 - \frac{3}{4}x +$ _____ $= (x -$ _____$)^2$
29. $x^2 + \frac{2}{3}x +$ _____ $= (x +$ _____$)^2$
30. $x^2 - \frac{4}{5}x +$ _____ $= (x -$ _____$)^2$

Solve each of the following quadratic equations by completing the square.

31. $x^2 + 12x = -27$
32. $x^2 - 6x = 16$
33. $a^2 - 2a + 5 = 0$
34. $a^2 + 10a + 22 = 0$
35. $y^2 - 8y + 1 = 0$
36. $y^2 + 6y - 1 = 0$
37. $x^2 - 5x - 3 = 0$
38. $x^2 - 5x - 2 = 0$
39. $2x^2 - 4x - 8 = 0$
40. $3x^2 - 9x - 12 = 0$
41. $3t^2 - 8t + 1 = 0$
42. $5t^2 + 12t - 1 = 0$
43. $4x^2 - 3x + 5 = 0$
44. $7x^2 - 5x + 2 = 0$
45. For the equation $x^2 = -9$
    a. Can it be solved by factoring?
    b. Solve it.
46. For the equation $x^2 - 10x + 18 = 0$
    a. Can it be solved by factoring?
    b. Solve it.
47. Solve each equation below by the indicated method.
    a. $x^2 - 6x = 0$ Factoring
    b. $x^2 - 6x = 0$ Completing the square
48. Solve each equation below by the indicated method.
    a. $x^2 + ax = 0$ Factoring
    b. $x^2 + ax = 0$ Completing the square
49. Solve the equation $x^2 + 2x = 35$
    a. By factoring.
    b. By completing the square.
50. Solve the equation $8x^2 - 10x - 25 = 0$
    a. By factoring.
    b. By completing the square.
51. Is $x = -3 + \sqrt{2}$ a solution to $x^2 - 6x = 7$?
52. Is $x = 2 - \sqrt{5}$ a solution to $x^2 - 4x = 1$?
53. Solve each equation
    a. $5x - 7 = 0$
    b. $5x - 7 = 8$
    c. $(5x - 7)^2 = 8$
    d. $\sqrt{5x - 7} = 8$
    e. $\frac{5}{2} - \frac{7}{2x} = \frac{4}{x}$

▇ = Videos available by instructor request

▶ = Online student support materials available at www.thomsonedu.com/login

54. Solve each equation

a. $5x + 11 = 0$

b. $5x + 11 = 9$

c. $(5x + 11)^2 = 9$

d. $\sqrt{5x + 11} = 9$

e. $\frac{5}{3} - \frac{11}{3x} = \frac{3}{x}$

Simplify the left side of each equation, and then solve for x.

55. $(x + 5)^2 + (x - 5)^2 = 52$

56. $(2x + 1)^2 + (2x - 1)^2 = 10$

57. $(2x + 3)^2 + (2x - 3)^2 = 26$

58. $(3x + 2)^2 + (3x - 2)^2 = 26$

59. $(3x + 4)(3x - 4) - (x + 2)(x - 2) = -4$

60. $(5x + 2)(5x - 2) - (x + 3)(x - 3) = 29$

61. Fill in the table below given the following functions. $f(x) = (2x - 3)^2$, $g(x) = 4x^2 - 12x + 9$, $h(x) = 4x^2 + 9$

x	$f(x)$	$g(x)$	$h(x)$
−2			
−1			
0			
1			
2			

62. Fill in the table below given the following functions. $f(x) = \left(x + \frac{1}{2}\right)^2$, $g(x) = x^2 + x + \frac{1}{4}$, $h(x) = x^2 + \frac{1}{4}$

x	$f(x)$	$g(x)$	$h(x)$
−2			
−1			
0			
1			
2			

▸ **63.** If $f(x) = (x - 3)^2$, find x if $f(x) = 0$.

▸ **64.** If $f(x) = (3x - 5)^2$, find x if $f(x) = 0$.

▸ **65.** If $f(x) = x^2 - 5x - 6$ find

a. The x-intercepts

b. The value of x for which $f(x) = 0$

c. $f(0)$

d. $f(1)$

▸ **66.** If $f(x) = 9x^2 - 12x + 4$ find

a. The x-intercepts

b. The value of x for which $f(x) = 0$

c. $f(0)$

d. $f(\frac{2}{3})$

Applying the Concepts

67. Geometry If the shortest side in a 30°–60°–90° triangle is $\frac{1}{2}$ inch long, find the lengths of the other two sides.

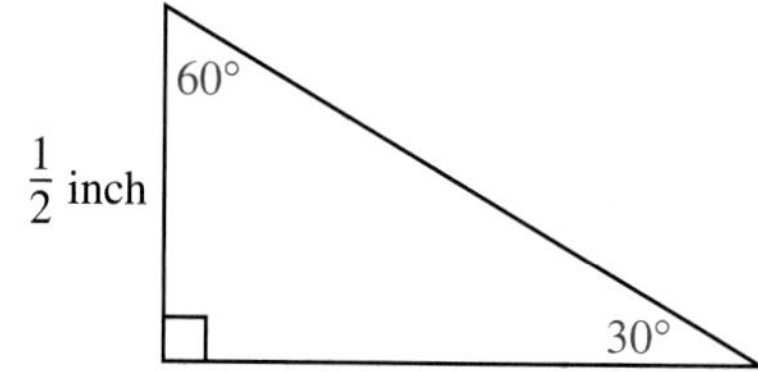

68. Geometry If the length of the longest side of a 30°–60°–90° triangle is x, find the length of the other two sides in terms of x.

69. Geometry If the length of the shorter sides of a 45°–45°–90° triangle is 1 inch, find the length of the hypotenuse.

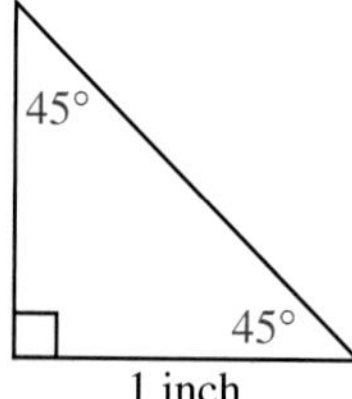

70. Geometry If the length of the shorter sides of a 45°–45°–90° triangle is x, find the length of the hypotenuse, in terms of x.

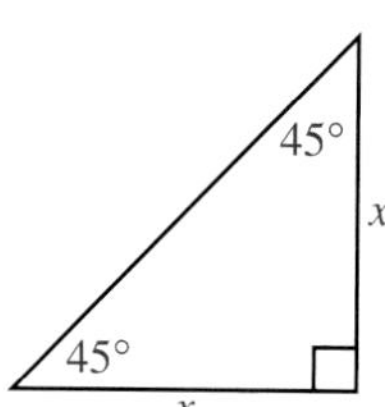

71. **Chair Lift** Use Table 1 from the introduction to this chapter to find the horizontal distance covered by a person riding the Bear Paw Double chair lift. Round your answer to the nearest foot.

72. **Fermat's Last Theorem** As we mentioned in a previous chapter, the postage stamp shows Fermat's last theorem, which states that if n is an integer greater than 2, then there are no positive integers x, y, and z that will make the formula $x^n + y^n = z^n$ true.

Use the formula $x^n + y^n = z^n$ to

a. find z if $n = 2$, $x = 6$, and $y = 8$.

b. find y if $n = 2$, $x = 5$, and $z = 13$.

73. **Interest Rate** Suppose a deposit of \$3,000 in a savings account that paid an annual interest rate r (compounded yearly) is worth \$3,456 after 2 years. Using the formula $A = P(1 + r)^t$, we have

$$3{,}456 = 3{,}000(1 + r)^2$$

Solve for r to find the annual interest rate.

74. **Special Triangles** In Figure 5, triangle ABC has angles 45° and 30° and height x. Find the lengths of sides AB, BC, and AC, in terms of x.

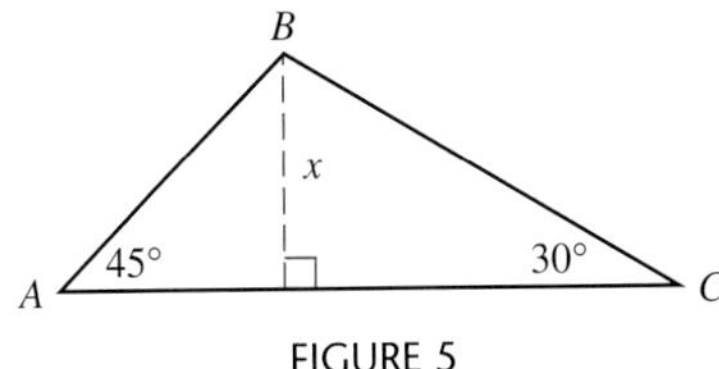

FIGURE 5

75. **Length of an Escalator** An escalator in a department store is to carry people a vertical distance of 20 feet between floors. How long is the escalator if it makes an angle of 45° with the ground? (See Figure 6.)

FIGURE 6

76. **Dimensions of a Tent** A two-person tent is to be made so the height at the center is 4 feet. If the sides of the tent are to meet the ground at an angle of 60° and the tent is to be 6 feet in length, how many square feet of material will be needed to make the tent? (Figure 7, assume that the tent has a floor and is closed at both ends.) Give your answer to the nearest tenth of a square foot.

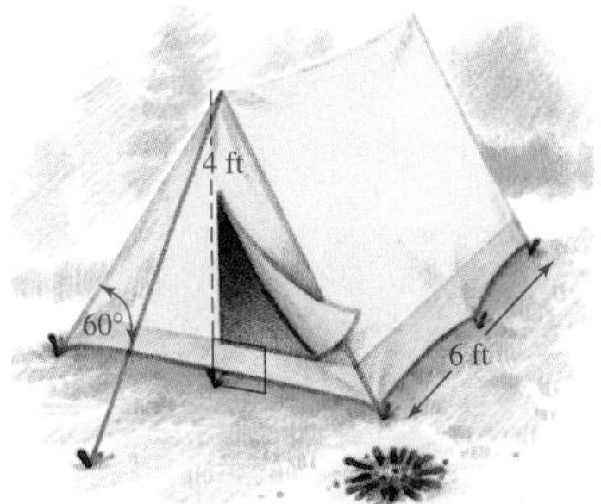

FIGURE 7

Maintaining Your Skills

Write each of the following in simplified form for radicals.

77. $\sqrt{45}$

78. $\sqrt{24}$

79. $\sqrt{27y^5}$

80. $\sqrt{8y^3}$

81. $\sqrt[3]{54x^6y^5}$

82. $\sqrt[3]{16x^9y^7}$

Rationalize the denominator.

83. $\dfrac{3}{\sqrt{2}}$

84. $\dfrac{5}{\sqrt{3}}$

85. $\dfrac{2}{\sqrt[3]{4}}$

86. $\dfrac{3}{\sqrt[3]{2}}$

Getting Ready for the Next Section

Simplify.

87. $\sqrt{49 - 4(6)(-5)}$

88. $\sqrt{49 - 4(6)2}$

89. $\sqrt{(-27)^2 - 4(0.1)(1{,}700)}$

90. $\sqrt{25 - 4(4)(-10)}$

91. $\dfrac{-7 + \sqrt{169}}{12}$

92. $\dfrac{-7 - \sqrt{169}}{12}$

93. Simplify $\sqrt{b^2 - 4ac}$ when $a = 6$, $b = 7$, and $c = -5$.

94. Simplify $\sqrt{b^2 - 4ac}$ when $a = 2$, $b = -6$, and $c = 3$.

Factor.

95. $27t^3 - 8$

96. $125t^3 + 1$

Extending the Concepts

Solve for x.

97. $(x + a)^2 + (x - a)^2 = 10a^2$

98. $(ax + 1)^2 + (ax - 1)^2 = 10$

Assume p and q are positive numbers and solve for x by completing the square on x.

99. $x^2 + px + q = 0$

100. $x^2 - px + q = 0$

101. $3x^2 + px + q = 0$

102. $3x^2 + 2px + q = 0$

Complete the square on x and also on y so that each equation below is written in the form

$$(x - a)^2 + (y - b)^2 = r^2$$

which you will see later in the book as the equation of a circle with center (a,b) and radius r.

103. $x^2 - 10x + y^2 - 6y = -30$

104. $x^2 - 2x + y^2 - 4y = 20$

8.2 The Quadratic Formula

OBJECTIVES

A Solve quadratic equations by the quadratic formula.

B Solve application problems using quadratic equations.

In this section we will use the method of completing the square from the preceding section to derive the quadratic formula. The *quadratic formula* is a very useful tool in mathematics. It allows us to solve all types of quadratic equations.

The Quadratic Theorem For any quadratic equation in the form $ax^2 + bx + c = 0$, where $a \neq 0$, the two solutions are

$$x = \frac{-b + \sqrt{b^2 - 4ac}}{2a} \quad \text{and} \quad x = \frac{-b - \sqrt{b^2 - 4ac}}{2a}$$

Proof We will prove the quadratic theorem by completing the square on $ax^2 + bx + c = 0$;

$$ax^2 + bx + c = 0$$

$$ax^2 + bx = -c \qquad \textbf{Add } -c \textbf{ to both sides}$$

$$x^2 + \frac{b}{a}x = -\frac{c}{a} \qquad \textbf{Divide both sides by } a$$

To complete the square on the left side, we add the square of $\frac{1}{2}$ of $\frac{b}{a}$ to both sides. $\left(\frac{1}{2} \text{ of } \frac{b}{a} \text{ is } \frac{b}{2a}.\right)$

$$x^2 + \frac{b}{a}x + \left(\frac{b}{2a}\right)^2 = -\frac{c}{a} + \left(\frac{b}{2a}\right)^2$$

We now simplify the right side as a separate step. We square the second term and combine the two terms by writing each with the least common denominator $4a^2$.

$$-\frac{c}{a} + \left(\frac{b}{2a}\right)^2 = -\frac{c}{a} + \frac{b^2}{4a^2} = \frac{\mathbf{4a}}{\mathbf{4a}}\left(\frac{-c}{a}\right) + \frac{b^2}{4a^2} = \frac{-4ac + b^2}{4a^2}$$

It is convenient to write this last expression as

$$\frac{b^2 - 4ac}{4a^2}$$

Continuing with the proof, we have

$$x^2 + \frac{b}{a}x + \left(\frac{b}{2a}\right)^2 = \frac{b^2 - 4ac}{4a^2}$$

$$\left(x + \frac{b}{2a}\right)^2 = \frac{b^2 - 4ac}{4a^2} \quad \text{Write left side as a binomial square}$$

$$x + \frac{b}{2a} = \pm\frac{\sqrt{b^2 - 4ac}}{2a} \quad \text{Theorem 1}$$

$$x = -\frac{b}{2a} \pm \frac{\sqrt{b^2 - 4ac}}{2a} \quad \text{Add } -\frac{b}{2a} \text{ to both sides}$$

$$= \frac{-b \pm \sqrt{b^2 - 4ac}}{2a}$$

Our proof is now complete. What we have is this: If our equation is in the form $ax^2 + bx + c = 0$ (standard form), where $a \neq 0$, the two solutions are always given by the formula

$$x = \frac{-b \pm \sqrt{b^2 - 4ac}}{2a}$$

This formula is known as the *quadratic formula.* If we substitute the coefficients a, b, and c of any quadratic equation in standard form into the formula, we need only perform some basic arithmetic to arrive at the solution set.

EXAMPLE 1

Use the quadratic formula to solve $6x^2 + 7x - 5 = 0$.

SOLUTION Using $a = 6$, $b = 7$, and $c = -5$ in the formula

$$x = \frac{-b \pm \sqrt{b^2 - 4ac}}{2a}$$

we have

$$x = \frac{-7 \pm \sqrt{49 - 4(6)(-5)}}{2(6)}$$

or

$$x = \frac{-7 \pm \sqrt{49 + 120}}{12}$$

$$= \frac{-7 \pm \sqrt{169}}{12}$$

$$= \frac{-7 \pm 13}{12}$$

We separate the last equation into the two statements

$$x = \frac{-7 + 13}{12} \quad \text{or} \quad x = \frac{-7 - 13}{12}$$

$$x = \frac{1}{2} \quad \text{or} \quad x = -\frac{5}{3}$$

The solution set is $\{\frac{1}{2}, -\frac{5}{3}\}$.

Whenever the solutions to a quadratic equation are rational numbers, as they are in Example 1, it means that the original equation was solvable by factoring. To illustrate, let's solve the equation from Example 1 again but this time by factoring:

$$6x^2 + 7x - 5 = 0 \qquad \textbf{Equation in standard form}$$

$$(3x + 5)(2x - 1) = 0 \qquad \textbf{Factor the left side}$$

$$3x + 5 = 0 \quad \text{or} \quad 2x - 1 = 0 \qquad \textbf{Set factors equal to 0}$$

$$x = -\frac{5}{3} \quad \text{or} \quad x = \frac{1}{2}$$

When an equation can be solved by factoring, then factoring is usually the faster method of solution. It is best to try to factor first, and then if you have trouble factoring, go to the quadratic formula. It always works.

EXAMPLE 2 Solve $\frac{x^2}{3} - x = -\frac{1}{2}$.

SOLUTION Multiplying through by 6 and writing the result in standard form, we have

$$2x^2 - 6x + 3 = 0$$

the left side of which is not factorable. Therefore, we use the quadratic formula with $a = 2$, $b = -6$, and $c = 3$. The two solutions are given by

$$x = \frac{-(-6) \pm \sqrt{36 - 4(2)(3)}}{2(2)}$$

$$= \frac{6 \pm \sqrt{12}}{4}$$

$$= \frac{6 \pm 2\sqrt{3}}{4} \qquad \boldsymbol{\sqrt{12} = \sqrt{4 \cdot 3} = \sqrt{4}\sqrt{3} = 2\sqrt{3}}$$

We can reduce this last expression to lowest terms by factoring 2 from the numerator and denominator and then dividing the numerator and denominator by 2:

$$x = \frac{\cancel{2}(3 \pm \sqrt{3})}{\cancel{2} \cdot 2} = \frac{3 \pm \sqrt{3}}{2}$$

EXAMPLE 3

Solve $\dfrac{1}{x+2} - \dfrac{1}{x} = \dfrac{1}{3}$.

SOLUTION To solve this equation, we must first put it in standard form. To do so, we must clear the equation of fractions by multiplying each side by the LCD for all the denominators, which is $3x(x + 2)$. Multiplying both sides by the LCD, we have

$$\mathbf{3x(x+2)}\left(\frac{1}{x+2} - \frac{1}{x}\right) = \frac{1}{3}\cdot \mathbf{3x(x+2)} \qquad \textbf{Multiply each by the LCD}$$

$$3x\cancel{(x+2)}\cdot\frac{1}{\cancel{x+2}} - 3\cancel{x}(x+2)\cdot\frac{1}{\cancel{x}} = \frac{1}{\cancel{3}}\cdot\cancel{3}x(x+2)$$

$$3x - 3(x+2) = x(x+2)$$

$$3x - 3x - 6 = x^2 + 2x \qquad \textbf{Multiplication}$$

$$-6 = x^2 + 2x \qquad \textbf{Simplify left side}$$

$$0 = x^2 + 2x + 6 \qquad \textbf{Add 6 to each side}$$

Since the right side of our last equation is not factorable, we use the quadratic formula. From our last equation, we have $a = 1$, $b = 2$, and $c = 6$. Using these numbers for a, b, and c in the quadratic formula gives us

$$x = \frac{-2 \pm \sqrt{4 - 4(1)(6)}}{2(1)}$$

$$= \frac{-2 \pm \sqrt{4 - 24}}{2} \qquad \textbf{Simplify inside the radical}$$

$$= \frac{-2 \pm \sqrt{-20}}{2} \qquad \mathbf{4 - 24 = -20}$$

$$= \frac{-2 \pm 2i\sqrt{5}}{2} \qquad \mathbf{-20 = i\sqrt{20} = i\sqrt{4}\sqrt{5} = 2i\sqrt{5}}$$

$$= \frac{\cancel{2}(-1 \pm i\sqrt{5})}{\cancel{2})} \qquad \textbf{Factor 2 from the numerator}$$

$$= -1 \pm i\sqrt{5} \qquad \textbf{Divide numerator and denominator by 2}$$

Since neither of the two solutions, $-1 + i\sqrt{5}$ nor $-1 - i\sqrt{5}$, will make any of the denominators in our original equation 0, they are both solutions.

Although the equation in our next example is not a quadratic equation, we solve it by using both factoring and the quadratic formula.

EXAMPLE 4

Solve $27t^3 - 8 = 0$.

SOLUTION It would be a mistake to add 8 to each side of this equation and then take the cube root of each side because we would lose two of our solutions. Instead, we factor the left side, and then set the factors equal to 0:

$$27t^3 - 8 = 0 \qquad \textbf{Equation in standard form}$$

$$(3t - 2)(9t^2 + 6t + 4) = 0 \qquad \textbf{Factor as the difference of two cubes}$$

$$3t - 2 = 0 \quad \text{or} \quad 9t^2 + 6t + 4 = 0 \qquad \textbf{Set each factor equal to 0}$$

The first equation leads to a solution of $t = \frac{2}{3}$. The second equation does not factor, so we use the quadratic formula with $a = 9$, $b = 6$, and $c = 4$:

$$t = \frac{-6 \pm \sqrt{36 - 4(9)(4)}}{2(9)}$$

$$= \frac{-6 \pm \sqrt{36 - 144}}{18}$$

$$= \frac{-6 \pm \sqrt{-108}}{18}$$

$$= \frac{-6 \pm 6i\sqrt{3}}{18} \qquad \sqrt{-108} = i\sqrt{36 \cdot 3} = 6i\sqrt{3}$$

$$= \frac{\not{6}(-1 \pm i\sqrt{3})}{\not{6} \cdot 3} \qquad \text{Factor 6 from the numerator and denominator}$$

$$= \frac{-1 \pm i\sqrt{3}}{3} \qquad \text{Divide out common factor 6}$$

The three solutions to our original equation are

$$\frac{2}{3}, \quad \frac{-1 + i\sqrt{3}}{3}, \quad \text{and} \quad \frac{-1 - i\sqrt{3}}{3}$$

EXAMPLE 5

If an object is thrown downward with an initial velocity of 20 feet per second, the distance $s(t)$, in feet, it travels in t seconds is given by the function $s(t) = 20t + 16t^2$. How long does it take the object to fall 40 feet?

SOLUTION We let $s(t) = 40$, and solve for t:

When $s(t) = 40$

the function $s(t) = 20t + 16t^2$

becomes $40 = 20t + 16t^2$

or $16t^2 + 20t - 40 = 0$

$4t^2 + 5t - 10 = 0$ **Divide by 4**

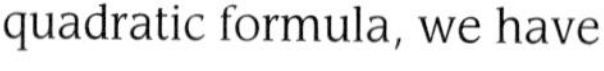

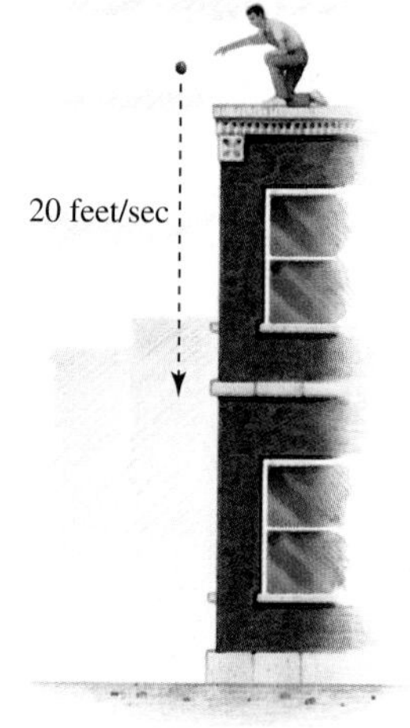

Using the quadratic formula, we have

$$t = \frac{-5 \pm \sqrt{25 - 4(4)(-10)}}{2(4)}$$

$$= \frac{-5 \pm \sqrt{185}}{8}$$

$$= \frac{-5 + \sqrt{185}}{8} \quad \text{or} \quad \frac{-5 - \sqrt{185}}{8}$$

The second solution is impossible since it is a negative number and time t must be positive. It takes

$$t = \frac{-5 + \sqrt{185}}{8} \quad \text{or approximately} \quad \frac{-5 + 13.60}{8} \approx 1.08 \text{ seconds}$$

for the object to fall 40 feet.

Recall that the relationship between profit, revenue, and cost is given by the formula

$$P(x) = R(x) - C(x)$$

where $P(x)$ is the profit, $R(x)$ is the total revenue, and $C(x)$ is the total cost of producing and selling x items.

EXAMPLE 6 A company produces and sells copies of an accounting program for home computers. The total weekly cost (in dollars) to produce x copies of the program is $C(x) = 8x + 500$, and the weekly revenue for selling all x copies of the program is $R(x) = 35x - 0.1x^2$. How many programs must be sold each week for the weekly profit to be $1,200?

SOLUTION Substituting the given expressions for $R(x)$ and $C(x)$ in the equation $P(x) = R(x) - C(x)$, we have a polynomial in x that represents the weekly profit $P(x)$:

$$\begin{aligned} P(x) &= R(x) - C(x) \\ &= 35x - 0.1x^2 - (8x + 500) \\ &= 35x - 0.1x^2 - 8x - 500 \\ &= -500 + 27x - 0.1x^2 \end{aligned}$$

Setting this expression equal to 1,200, we have a quadratic equation to solve that gives us the number of programs x that need to be sold each week to bring in a profit of $1,200:

$$1{,}200 = -500 + 27x - 0.1x^2$$

We can write this equation in standard form by adding the opposite of each term on the right side of the equation to both sides of the equation. Doing so produces the following equation:

$$0.1x^2 - 27x + 1{,}700 = 0$$

Applying the quadratic formula to this equation with $a = 0.1$, $b = -27$, and $c = 1{,}700$, we have

$$\begin{aligned} x &= \frac{27 \pm \sqrt{(-27)^2 - 4(0.1)(1{,}700)}}{2(0.1)} \\ &= \frac{27 \pm \sqrt{729 - 680}}{0.2} \\ &= \frac{27 \pm \sqrt{49}}{0.2} \\ &= \frac{27 \pm 7}{0.2} \end{aligned}$$

Writing this last expression as two separate expressions, we have our two solutions:

$$x = \frac{27 + 7}{0.2} = \frac{34}{0.2} = 170 \quad \text{or} \quad x = \frac{27 - 7}{0.2} = \frac{20}{0.2} = 100$$

The weekly profit will be $1,200 if the company produces and sells 100 programs or 170 programs.

What is interesting about the equation we solved in Example 6 is that it has rational solutions, meaning it could have been solved by factoring. But looking back at the equation, factoring does not seem like a reasonable method of solution because the coefficients are either very large or very small. So, there are times when using the quadratic formula is a faster method of solution, even though the equation you are solving is factorable.

USING TECHNOLOGY

Graphing Calculators

More About Example 5

We can solve the problem discussed in Example 5 by graphing the function $Y_1 = 20X + 16X^2$ in a window with X from 0 to 2 (because X is taking the place of t and we know t is a positive quantity) and Y from 0 to 50 (because we are looking for X when Y_1 is 40). Graphing Y_1 gives a graph similar to the graph in Figure 1. Using the Zoom and Trace features at $Y_1 = 40$ gives us $X = 1.08$ to the nearest hundredth, matching the results we obtained by solving the original equation algebraically.

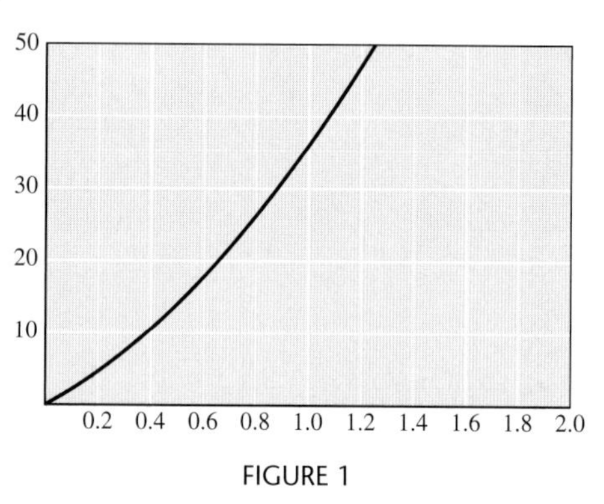

FIGURE 1

More About Example 6

To visualize the functions in Example 6, we set up our calculator this way:

$Y_1 = 35X - .1X^2$ **Revenue function**

$Y_2 = 8X + 500$ **Cost function**

$Y_3 = Y_1 - Y_2$ **Profit function**

Window: X from 0 to 350, Y from 0 to 3500

Graphing these functions produces graphs similar to the ones shown in Figure 2. The lower graph is the graph of the profit function. Using the Zoom and Trace features on the lower graph at $Y_3 =$ 1,200 produces two corresponding values of X, 170 and 100, which match the results in Example 6.

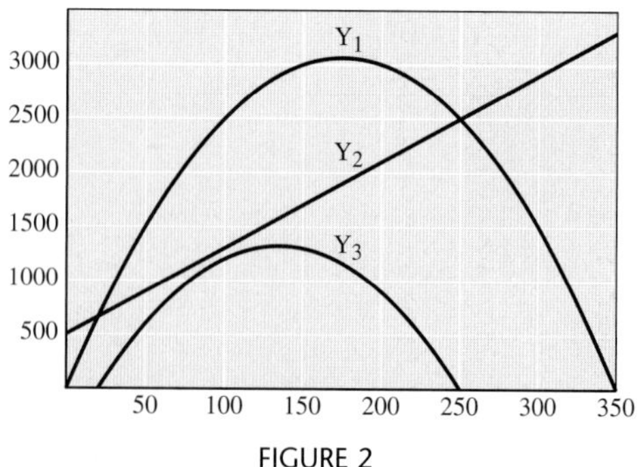

FIGURE 2

We will continue this discussion of the relationship between graphs of functions and solutions to equations in the Using Technology material in the next section.

LINKING OBJECTIVES AND EXAMPLES

Next to each **objective** we have listed the examples that are best described by that objective.

A	1–4
B	5, 6

GETTING READY FOR CLASS

After reading through the preceding section, respond in your own words and in complete sentences.

1. What is the quadratic formula?
2. Under what circumstances should the quadratic formula be applied?
3. When would the quadratic formula result in complex solutions?
4. When will the quadratic formula result in only one solution?

Problem Set 8.2

Online support materials can be found at www.thomsonedu.com/login

Solve each equation in each problem using the quadratic formula.

1. a. $3x^2 + 4x - 2 = 0$
   b. $3x^2 - 4x - 2 = 0$
   c. $3x^2 + 4x + 2 = 0$
   d. $2x^2 + 4x - 3 = 0$
   e. $2x^2 - 4x + 3 = 0$
2. a. $3x^2 + 6x - 2 = 0$
   b. $3x^2 - 6x - 2 = 0$
   c. $3x^2 + 6x + 2 = 0$
   d. $2x^2 + 6x + 3 = 0$
   e. $2x^2 + 6x - 3 = 0$
3. a. $x^2 - 2x + 2 = 0$
   b. $x^2 - 2x + 5 = 0$
   c. $x^2 + 2x + 2 = 0$
4. a. $x^2 - 4x + 5 = 0$
   b. $x^2 + 4x + 5 = 0$
   c. $a^2 + 4a + 1 = 0$

Solve each equation. Use factoring or the quadratic formula, whichever is appropriate. (Try factoring first. If you have any difficulty factoring, then go right to the quadratic formula.)

5. $\frac{1}{6}x^2 - \frac{1}{2}x + \frac{1}{3} = 0$
6. $\frac{1}{4}x^2 + \frac{1}{4}x - \frac{1}{2} = 0$
7. $\frac{x^2}{2} + 1 = \frac{2x}{3}$
8. $\frac{x^2}{2} + \frac{2}{3} = -\frac{2x}{3}$
9. $y^2 - 5y = 0$
10. $2y^2 + 10y = 0$
11. $30x^2 + 40x = 0$
12. $50x^2 - 20x = 0$
13. $\frac{2t^2}{3} - t = -\frac{1}{6}$
14. $\frac{t^2}{3} - \frac{t}{2} = -\frac{3}{2}$
15. $0.01x^2 + 0.06x - 0.08 = 0$
16. $0.02x^2 - 0.03x + 0.05 = 0$
17. $2x + 3 = -2x^2$
18. $2x - 3 = 3x^2$

■ = Videos available by instructor request

▶ = Online student support materials available at www.thomsonedu.com/login

19. $100x^2 - 200x + 100 = 0$

20. $100x^2 - 600x + 900 = 0$

21. $\frac{1}{2}r^2 = \frac{1}{6}r - \frac{2}{3}$

22. $\frac{1}{4}r^2 = \frac{2}{5}r + \frac{1}{10}$

23. $(x - 3)(x - 5) = 1$

24. $(x - 3)(x + 1) = -6$

25. $(x + 3)^2 + (x - 8)(x - 1) = 16$

26. $(x - 4)^2 + (x + 2)(x + 1) = 9$

27. $\frac{x^2}{3} - \frac{5x}{6} = \frac{1}{2}$

28. $\frac{x^2}{6} + \frac{5}{6} = -\frac{x}{3}$

Multiply both sides of each equation by its LCD. Then solve the resulting equation.

29. $\frac{1}{x+1} - \frac{1}{x} = \frac{1}{2}$

30. $\frac{1}{x+1} + \frac{1}{x} = \frac{1}{3}$

31. $\frac{1}{y-1} + \frac{1}{y+1} = 1$

32. $\frac{2}{y+2} + \frac{3}{y-2} = 1$

33. $\frac{1}{x+2} + \frac{1}{x+3} = 1$

34. $\frac{1}{x+3} + \frac{1}{x+4} = 1$

35. $\frac{6}{r^2-1} - \frac{1}{2} = \frac{1}{r+1}$

36. $2 + \frac{5}{r-1} = \frac{12}{(r-1)^2}$

Solve each equation. In each case you will have three solutions.

37. $x^3 - 8 = 0$

38. $x^3 - 27 = 0$

39. $8a^3 + 27 = 0$

40. $27a^3 + 8 = 0$

41. $125t^3 - 1 = 0$

42. $64t^3 + 1 = 0$

Each of the following equations has three solutions. Look for the greatest common factor, then use the quadratic formula to find all solutions.

43. $2x^3 + 2x^2 + 3x = 0$

44. $6x^3 - 4x^2 + 6x = 0$

45. $3y^4 = 6y^3 - 6y^2$

46. $4y^4 = 16y^3 - 20y^2$

47. $6t^5 + 4t^4 = -2t^3$

48. $8t^5 + 2t^4 = -10t^3$

49. Which two of the expressions below are equivalent?

a. $\frac{6 + 2\sqrt{3}}{4}$

b. $\frac{3 + \sqrt{3}}{2}$

c. $6 + \frac{\sqrt{3}}{2}$

50. Which two of the expressions below are equivalent?

a. $\frac{8 - 4\sqrt{2}}{4}$

b. $2 - 4\sqrt{3}$

c. $2 - \sqrt{2}$

51. Solve $3x^2 - 5x = 0$
   a. By factoring
   b. By the quadratic formula

52. Solve $3x^2 + 23x - 70 = 0$
   a. By factoring
   b. By the quadratic formula

53. Can the equation $x^2 - 4x + 7 = 0$ be solved by factoring? Solve the equation.

54. Can the equation $x^2 = 5$ be solved by factoring? Solve the equation.

55. Is $x = -1 + i$ a solution to $x^2 + 2x = -2$?

56. Is $x = 2 + 2i$ a solution to $(x - 2)^2 = -4$?

57. Let $f(x) = x^2 - 2x - 3$. Find all values for the variable x, which produce the following values of $f(x)$.
    a. $f(x) = 0$
    b. $f(x) = -11$
    c. $f(x) = -2x + 1$
    d. $f(x) = 2x + 1$

58. Let $g(x) = x^2 + 16$. Find all values for the variable x, which produce the following values of $g(x)$.
    a. $g(x) = 0$
    b. $g(x) = 20$
    c. $g(x) = 8x$
    d. $g(x) = -8x$

59. Let $f(x) = \dfrac{10}{x^2}$. Find all values for the variable x, for which $f(x) = g(x)$.
    a. $g(x) = 3 + \dfrac{1}{x}$
    b. $g(x) = 8x - \dfrac{17}{x^2}$
    c. $g(x) = 0$
    d. $g(x) = 10$

60. Let $h(x) = \dfrac{x + 2}{x}$. Find all values for the variable x, for which $h(x) = f(x)$.
    a. $f(x) = 2$
    b. $f(x) = x + 2$
    c. $f(x) = x - 2$
    d. $f(x) = \dfrac{x - 2}{4}$

Applying the Concepts

61. **Falling Object** An object is thrown downward with an initial velocity of 5 feet per second. The relationship between the distance s it travels and time t is given by $s = 5t + 16t^2$. How long does it take the object to fall 74 feet?

62. **Coin Toss** A coin is tossed upward with an initial velocity of 32 feet per second from a height of 16 feet above the ground. The equation giving the object's height h at any time t is $h = 16 + 32t - 16t^2$. Does the object ever reach a height of 32 feet?

63. **Profit** The total cost (in dollars) for a company to manufacture and sell x items per week is $C = 60x + 300$, whereas the revenue brought in by selling all x items is $R = 100x - 0.5x^2$. How many items must be sold to obtain a weekly profit of $300?

Kathleen Olson

64. **Profit** Suppose a company manufactures and sells x picture frames each month with a total cost of $C = 1{,}200 + 3.5x$ dollars. If the revenue obtained by selling x frames is $R = 9x - 0.002x^2$, find the number of frames it must sell each month if its monthly profit is to be $2,300.

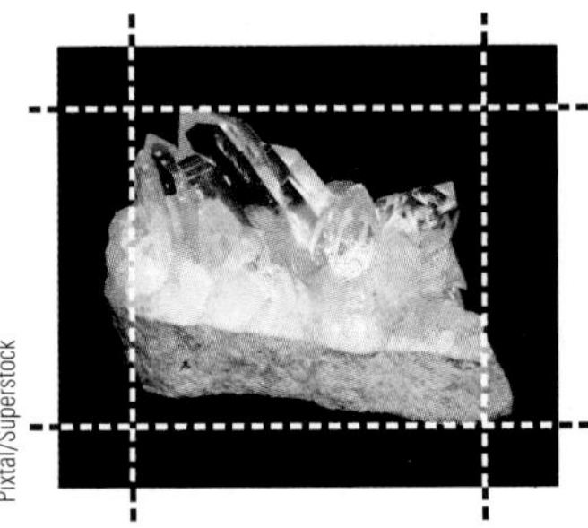
Pixtal/Superstock

65. **Photograph Cropping** The following figure shows a photographic image on a 10.5-centimeter by 8.2-centimeter background. The overall area of the background is to be reduced to 80% of its original area by cutting off (cropping) equal strips on all four sides. What is the width of the strip that is cut from each side?

66. **Area of a Garden** A garden measures 20.3 meters by 16.4 meters. To double the area of the garden, strips of equal width are added to all four sides.
    a. Draw a diagram that illustrates these conditions.
    b. What are the new overall dimensions of the garden?

67. **Area and Perimeter** A rectangle has a perimeter of 20 yards and an area of 15 square yards.
    a. Write two equations that state these facts in terms of the rectangle's length, l, and its width, w.
    b. Solve the two equations from part (a) to determine the actual length and width of the rectangle.
    c. Explain why two answers are possible to part (b).

68. **Population Size** Writing in 1829, former President James Madison made some predictions about the

growth of the population of the United States. The populations he predicted fit the equation

$$y = 0.029x^2 - 1.39x + 42$$

where y is the population in millions of people x years from 1829.

Library of Congress

a. Use the equation to determine the approximate year President Madison would have predicted that the U.S. population would reach 100,000,000.

b. If the U.S. population in 2006 was approximately 300 million, were President Madison's predictions accurate in the long term? Explain why or why not.

Maintaining Your Skills

Divide, using long division.

69. $\dfrac{8y^2 - 26y - 9}{2y - 7}$

70. $\dfrac{6y^2 + 7y - 18}{3y - 4}$

71. $\dfrac{x^3 + 9x^2 + 26x + 24}{x + 2}$

72. $\dfrac{x^3 + 6x^2 + 11x + 6}{x + 3}$

Simplify each expression. (Assume $x, y > 0$.)

73. $25^{1/2}$

74. $8^{1/3}$

75. $\left(\dfrac{9}{25}\right)^{3/2}$

76. $\left(\dfrac{16}{81}\right)^{3/4}$

77. $8^{-2/3}$

78. $4^{-3/2}$

79. $\dfrac{(49x^8y^{-4})^{1/2}}{(27x^{-3}y^9)^{-1/3}}$

80. $\dfrac{(x^{-2}y^{1/3})^6}{x^{-10}y^{3/2}}$

Getting Ready for the Next Section

Find the value of $b^2 - 4ac$ when

81. $a = 1, b = -3, c = -40$

82. $a = 2, b = 3, c = 4$

83. $a = 4, b = 12, c = 9$

84. $a = -3, b = 8, c = -1$

Solve.

85. $k^2 - 144 = 0$

86. $36 - 20k = 0$

Multiply.

87. $(x - 3)(x + 2)$

88. $(t - 5)(t + 5)$

89. $(x - 3)(x - 3)(x + 2)$

90. $(t - 5)(t + 5)(t - 3)$

Extending the Concepts

So far, all the equations we have solved have had coefficients that were rational numbers. Here are some equations that have irrational coefficients and some that have complex coefficients. Solve each equation. (Remember, $i^2 = -1$.)

91. $x^2 + \sqrt{3}x - 6 = 0$

92. $x^2 - \sqrt{5}x - 5 = 0$

93. $\sqrt{2}x^2 + 2x - \sqrt{2} = 0$

94. $\sqrt{7}x^2 + 2\sqrt{2}x - \sqrt{7} = 0$

95. $x^2 + ix + 2 = 0$

96. $x^2 + 3ix - 2 = 0$

8.3 Additional Items Involving Solutions to Equations

OBJECTIVES

A Find the number and kind of solutions to a quadratic equation by using the discriminant.

B Find an unknown constant in quadratic equation so that there is exactly one solution.

C Find an equation from its solutions.

In this section we will do two things. First, we will define the discriminant and use it to find the kind of solutions a quadratic equation has without solving the equation. Second, we will use the zero-factor property to build equations from their solutions.

The Discriminant

The quadratic formula

$$x = \frac{-b \pm \sqrt{b^2 - 4ac}}{2a}$$

gives the solutions to any quadratic equation in standard form. When working with quadratic equations, there are times when it is important only to know what kind of solutions the equation has.

> **DEFINITION** The expression under the radical in the quadratic formula is called the **discriminant:**
>
> $$\text{Discriminant} = D = b^2 - 4ac$$

The discriminant indicates the number and type of solutions to a quadratic equation, when the original equation has integer coefficients. For example, if we were to use the quadratic formula to solve the equation $2x^2 + 2x + 3 = 0$, we would find the discriminant to be

$$b^2 - 4ac = 2^2 - 4(2)(3) = -20$$

Since the discriminant appears under a square root symbol, we have the square root of a negative number in the quadratic formula. Our solutions therefore would be complex numbers. Similarly, if the discriminant were 0, the quadratic formula would yield

$$x = \frac{-b \pm \sqrt{0}}{2a} = \frac{-b \pm 0}{2a} = \frac{-b}{2a}$$

and the equation would have one rational solution, the number $\frac{-b}{2a}$.

The following table gives the relationship between the discriminant and the type of solutions to the equation.

For the equation $ax^2 + bx + c = 0$ where a, b, and c are integers and $a \neq 0$:

If the Discriminant $b^2 - 4ac$ is	Then the Equation Will Have
Negative	Two complex solutions containing i
Zero	One rational solution
A positive number that is also a perfect square	Two rational solutions
A positive number that is not a perfect square	Two irrational solutions

In the second and third cases, when the discriminant is 0 or a positive perfect square, the solutions are rational numbers. The quadratic equations in these two cases are the ones that can be factored.

EXAMPLES For each equation, give the number and kind of solutions.

1. $x^2 - 3x - 40 = 0$

SOLUTION Using $a = 1$, $b = -3$, and $c = -40$ in $b^2 - 4ac$, we have $(-3)^2 - 4(1)(-40) = 9 + 160 = 169$.

The discriminant is a perfect square. The equation therefore has two rational solutions.

2. $2x^2 - 3x + 4 = 0$

SOLUTION Using $a = 2$, $b = -3$, and $c = 4$, we have

$$b^2 - 4ac = (-3)^2 - 4(2)(4) = 9 - 32 = -23$$

The discriminant is negative, implying the equation has two complex solutions that contain i.

3. $4x^2 - 12x + 9 = 0$

SOLUTION Using $a = 4$, $b = -12$, and $c = 9$, the discriminant is

$$b^2 - 4ac = (-12)^2 - 4(4)(9) = 144 - 144 = 0$$

Since the discriminant is 0, the equation will have one rational solution.

4. $x^2 + 6x = 8$

SOLUTION We first must put the equation in standard form by adding -8 to each side. If we do so, the resulting equation is

$$x^2 + 6x - 8 = 0$$

Now we identify a, b, and c as 1, 6, and -8, respectively:

$$b^2 - 4ac = 6^2 - 4(1)(-8) = 36 + 32 = 68$$

The discriminant is a positive number but not a perfect square. Therefore, the equation will have two irrational solutions.

EXAMPLE 5 Find an appropriate k so that the equation $4x^2 - kx = -9$ has exactly one rational solution.

SOLUTION We begin by writing the equation in standard form:

$$4x^2 - kx + 9 = 0$$

Using $a = 4$, $b = -k$, and $c = 9$, we have

$$\begin{aligned} b^2 - 4ac &= (-k)^2 - 4(4)(9) \\ &= k^2 - 144 \end{aligned}$$

An equation has exactly one rational solution when the discriminant is 0. We set the discriminant equal to 0 and solve:

$$\begin{aligned} k^2 - 144 &= 0 \\ k^2 &= 144 \\ k &= \pm 12 \end{aligned}$$

Choosing k to be 12 or -12 will result in an equation with one rational solution.

Building Equations from Their Solutions

Suppose we know that the solutions to an equation are $x = 3$ and $x = -2$. We can find equations with these solutions by using the zero-factor property. First, let's write our solutions as equations with 0 on the right side:

If	$x = 3$	**First solution**
then	$x - 3 = 0$	**Add −3 to each side**
and if	$x = -2$	**Second solution**
then	$x + 2 = 0$	**Add 2 to each side**

Now, since both $x - 3$ and $x + 2$ are 0, their product must be 0 also. Therefore, we can write

$$(x - 3)(x + 2) = 0 \quad \textbf{Zero-factor property}$$

$$x^2 - x - 6 = 0 \quad \textbf{Multiply out the left side}$$

Many other equations have 3 and −2 as solutions. For example, any constant multiple of $x^2 - x - 6 = 0$, such as $5x^2 - 5x - 30 = 0$, also has 3 and −2 as solutions. Similarly, any equation built from positive integer powers of the factors $x - 3$ and $x + 2$ will also have 3 and −2 as solutions. One such equation is

$$(x - 3)^2(x + 2) = 0$$

$$(x^2 - 6x + 9)(x + 2) = 0$$

$$x^3 - 4x^2 - 3x + 18 = 0$$

In mathematics we distinguish between the solutions to this last equation and those to the equation $x^2 - x - 6 = 0$ by saying $x = 3$ is a solution of *multiplicity 2* in the equation $x^3 - 4x^2 - 3x + 18 = 0$ and a solution of *multiplicity 1* in the equation $x^2 - x - 6 = 0$.

EXAMPLE 6 Find an equation that has solutions $t = 5$, $t = -5$, and $t = 3$.

SOLUTION First, we use the given solutions to write equations that have 0 on their right sides:

If	$t = 5$	$t = -5$	$t = 3$
then	$t - 5 = 0$	$t + 5 = 0$	$t - 3 = 0$

Since $t - 5$, $t + 5$, and $t - 3$ are all 0, their product is also 0 by the zero-factor property. An equation with solutions of 5, −5, and 3 is

$$(t - 5)(t + 5)(t - 3) = 0 \quad \textbf{Zero-factor property}$$

$$(t^2 - 25)(t - 3) = 0 \quad \textbf{Multiply first two binomials}$$

$$t^3 - 3t^2 - 25t + 75 = 0 \quad \textbf{Complete the multiplication}$$

The last line gives us an equation with solutions of 5, −5, and 3. Remember, many other equations have these same solutions.

EXAMPLE 7 Find an equation with solutions $x = -\frac{2}{3}$ and $x = \frac{4}{5}$.

SOLUTION The solution $x = -\frac{2}{3}$ can be rewritten as $3x + 2 = 0$ as follows:

$x = -\frac{2}{3}$ **The first solution**

$3x = -2$ **Multiply each side by 3**

$3x + 2 = 0$ **Add 2 to each side**

Similarly, the solution $x = \frac{4}{5}$ can be rewritten as $5x - 4 = 0$:

$x = \frac{4}{5}$ **The second solution**

$5x = 4$ **Multiply each side by 5**

$5x - 4 = 0$ **Add −4 to each side**

Since both $3x + 2$ and $5x - 4$ are 0, their product is 0 also, giving us the equation we are looking for:

$(3x + 2)(5x - 4) = 0$ **Zero-factor property**

$15x^2 - 2x - 8 = 0$ **Multiplication**

USING TECHNOLOGY

Graphing Calculators

Solving Equations

Now that we have explored the relationship between equations and their solutions, we can look at how a graphing calculator can be used in the solution process. To begin, let's solve the equation $x^2 = x + 2$ using techniques from algebra: writing it in standard form, factoring, and then setting each factor equal to 0.

$x^2 - x - 2 = 0$ **Standard form**

$(x - 2)(x + 1) = 0$ **Factor**

$x - 2 = 0$ or $x + 1 = 0$ **Set each factor equal to 0**

$x = 2$ or $x = -1$ **Solve**

Our original equation, $x^2 = x + 2$, has two solutions: $x = 2$ and $x = -1$. To solve the equation using a graphing calculator, we need to associate it with an equation (or equations) in two variables. One way to do this is to associate the left side with the equation $y = x^2$ and the right side of the equation with $y = x + 2$. To do so, we set up the functions list in our calculator this way:

$Y_1 = X^2$

$Y_2 = X + 2$

Window: X from −5 to 5, Y from −5 to 5

Graphing these functions in this window will produce a graph similar to the one shown in Figure 1.

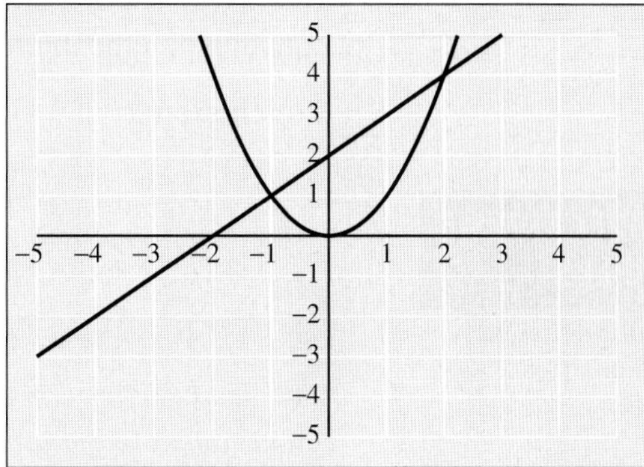

Figure 1

If we use the Trace feature to find the coordinates of the points of intersection, we find that the two curves intersect at $(-1, 1)$ and $(2, 4)$. We note that the x-coordinates of these two points match the solutions to the equation $x^2 = x + 2$, which we found using algebraic techniques. This makes sense because if two graphs intersect at a point (x, y), then the coordinates of that point satisfy both equations. If a point (x, y) satisfies both $y = x^2$ and $y = x + 2$, then, for that particular point, $x^2 = x + 2$. From this we conclude that the x-coordinates of the points of intersection are solutions to our original equation. Here is a summary of what we have discovered:

Conclusion 1 If the graph of two functions $y = f(x)$ and $y = g(x)$ intersect in the coordinate plane, then the x-coordinates of the points of intersection are solutions to the equation $f(x) = g(x)$.

A second method of solving our original equation $x^2 = x + 2$ graphically requires the use of one function instead of two. To begin, we write the equation in standard form as $x^2 - x - 2 = 0$. Next, we graph the function $y = x^2 - x - 2$. The x-intercepts of the graph are the points with y-coordinates of 0. They therefore satisfy the equation $0 = x^2 - x - 2$, which is equivalent to our original equation. The graph in Figure 2 shows $Y_1 = X^2 - X - 2$ in a window with X from -5 to 5 and Y from -5 to 5.

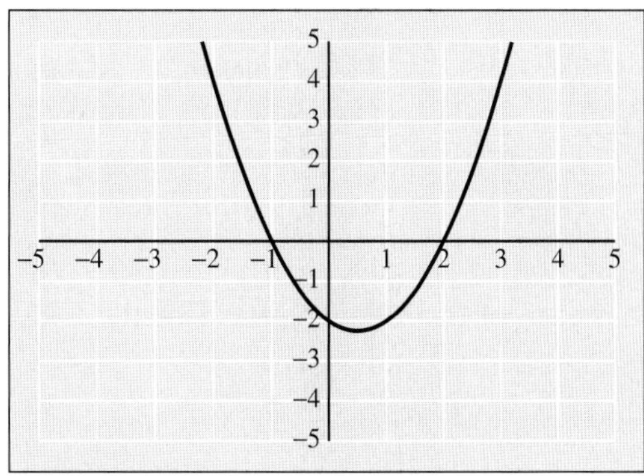

Figure 2

Using the Trace feature, we find that the x-intercepts of the graph are $x = -1$ and $x = 2$, which match the solutions to our original equation $x^2 = x + 2$. We can summarize the relationship between solutions to an equation and the intercepts of its associated graph this way:

Conclusion 2 If $y = f(x)$ is a function, then any x-intercept on the graph of $y = f(x)$ is a solution to the equation $f(x) = 0$.

LINKING OBJECTIVES AND EXAMPLES

Next to each **objective** we have listed the examples that are best described by that objective.

A	1–4
B	5
C	6, 7

GETTING READY FOR CLASS

After reading through the preceding section, respond in your own words and in complete sentences.

1. What is the discriminant?
2. What kind of solutions do we get to a quadratic equation when the discriminant is negative?
3. What does it mean for a solution to have multiplicity 3?
4. When will a quadratic equation have two rational solutions?

Problem Set 8.3

Online support materials can be found at www.thomsonedu.com/login

Use the discriminant to find the number and kind of solutions for each of the following equations.

1. $x^2 - 6x + 5 = 0$
2. $x^2 - x - 12 = 0$
3. $4x^2 - 4x = -1$
4. $9x^2 + 12x = -4$
5. $x^2 + x - 1 = 0$
6. $x^2 - 2x + 3 = 0$
7. $2y^2 = 3y + 1$
8. $3y^2 = 4y - 2$
9. $x^2 - 9 = 0$
10. $4x^2 - 81 = 0$
11. $5a^2 - 4a = 5$
12. $3a = 4a^2 - 5$

Determine k so that each of the following has exactly one real solution.

13. $x^2 - kx + 25 = 0$
14. $x^2 + kx + 25 = 0$
15. $x^2 = kx - 36$
16. $x^2 = kx - 49$
17. $4x^2 - 12x + k = 0$
18. $9x^2 + 30x + k = 0$
19. $kx^2 - 40x = 25$
20. $kx^2 - 2x = -1$
21. $3x^2 - kx + 2 = 0$
22. $5x^2 + kx + 1 = 0$

For each of the following problems, find an equation that has the given solutions.

23. $x = 5, x = 2$
24. $x = -5, x = -2$
25. $t = -3, t = 6$
26. $t = -4, t = 2$
27. $y = 2, y = -2, y = 4$
28. $y = 1, y = -1, y = 3$
29. $x = \frac{1}{2}, x = 3$
30. $x = \frac{1}{3}, x = 5$

■ = Videos available by instructor request
▸ = Online student support materials available at www.thomsonedu.com/login

31. $t = -\frac{3}{4}, t = 3$

32. $t = -\frac{4}{5}, t = 2$

33. $x = 3, x = -3, x = \frac{5}{6}$

34. $x = 5, x = -5, x = \frac{2}{3}$

35. $a = -\frac{1}{2}, a = \frac{3}{5}$

36. $a = -\frac{1}{3}, a = \frac{4}{7}$

37. $x = -\frac{2}{3}, x = \frac{2}{3}, x = 1$

38. $x = -\frac{4}{5}, x = \frac{4}{5}, x = -1$

39. $x = 2, x = -2, x = 3, x = -3$

40. $x = 1, x = -1, x = 5, x = -5$

41. Find $f(x)$ if $f(1) = 0$ and $f(-2) = 0$.

42. Find $f(x)$ if $f(4) = 0$ and $f(-3) = 0$.

43. Find $f(x)$ if $f(3 + \sqrt{2}) = 0$ and $f(3 - \sqrt{2}) = 0$.

44. Find $f(x)$ if $f(-3 + \sqrt{15}) = 0$ and $f(-3 - \sqrt{15}) = 0$.

45. Find $f(x)$ if $f(2 + i) = 0$ and $f(2 - i) = 0$.

46. Find $f(x)$ if $f(3 + i\sqrt{5}) = 0$ and $f(3 - i\sqrt{5}) = 0$.

47. Find $f(x)$ if $f\left(\frac{5 + \sqrt{7}}{2}\right) = 0$ and $f\left(\frac{5 - \sqrt{7}}{2}\right) = 0$.

48. Find $f(x)$ if $f\left(\frac{-4 + \sqrt{11}}{3}\right) = 0$ and $f\left(\frac{-4 - \sqrt{11}}{3}\right) = 0$.

49. Find $f(x)$ if $f\left(\frac{3 + i\sqrt{5}}{2}\right) = 0$ and $f\left(\frac{3 - i\sqrt{5}}{2}\right) = 0$.

50. Find $f(x)$ if $f\left(\frac{-7 + i\sqrt{14}}{3}\right) = 0$ and $f\left(\frac{-7 - i\sqrt{14}}{3}\right) = 0$.

51. Find $f(x)$ if $f(2 + i\sqrt{3}) = 0$ and $f(2 - i\sqrt{3}) = 0$, and $f(0) = 0$.

52. Find $f(x)$ if $f(5 + i\sqrt{7}) = 0$ and $f(5 - i\sqrt{7}) = 0$, and $f(0) = 0$.

53. Indicate which of the graphs represent the following types of solutions.
   a. One real solution
   b. No real solutions
   c. Two real solutions

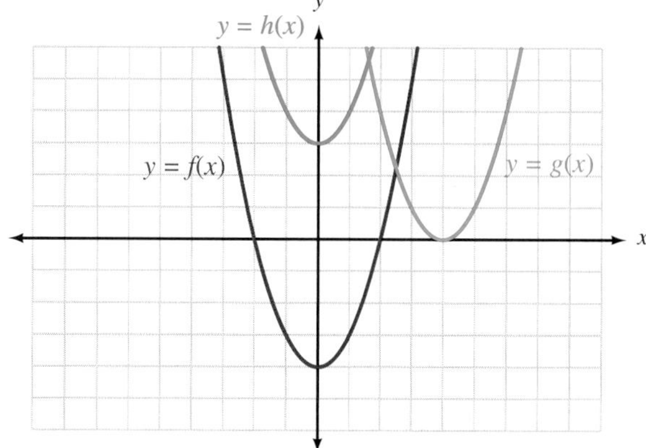

Maintaining Your Skills

Multiply. (Assume all variables represent positive numbers for the rest of the problems in this section.)

54. $a^4(a^{3/2} - a^{1/2})$

55. $(a^{1/2} - 5)(a^{1/2} + 3)$

56. $(x^{3/2} - 3)^2$

57. $(x^{1/2} - 8)(x^{1/2} + 8)$

Divide.

58. $\dfrac{30x^{3/4} - 25x^{5/4}}{5x^{1/4}}$

59. $\dfrac{45x^{5/3}y^{7/3} - 36x^{8/3}y^{4/3}}{9x^{2/3}y^{1/3}}$

60. Factor $5(x - 3)^{1/2}$ from $10(x - 3)^{3/2} - 15(x - 3)^{1/2}$.

61. Factor $2(x + 1)^{1/3}$ from $8(x + 1)^{4/3} - 2(x + 1)^{1/3}$.

Factor each of the following.

62. $2x^{2/3} - 11x^{1/3} + 12$

63. $9x^{2/3} + 12x^{1/3} + 4$

Getting Ready for the Next Section

Simplify.

64. $(x + 3)^2 - 2(x + 3) - 8$

65. $(x - 2)^2 - 3(x - 2) - 10$

66. $(2a - 3)^2 - 9(2a - 3) + 20$

67. $(3a-2)^2+2(3a-2)-3$

68. $2(4a+2)^2-3(4a+2)-20$

69. $6(2a+4)^2-(2a+4)-2$

Solve.

70. $x^2=\frac{1}{4}$

71. $x^2=-2$

72. $\sqrt{x}=-3$

73. $\sqrt{x}=2$

74. $x+3=4$

75. $x+3=-2$

76. $y^2-2y-8=0$

77. $y^2+y-6=0$

78. $4y^2+7y-2=0$

79. $6x^2-13x-5=0$

Extending the Concepts

Find all solutions to the following equations. Solve using algebra and by graphing. If rounding is necessary, round to the nearest hundredth. A calculator can be used in these problems.

80. $x^2=4x+5$

81. $4x^2=8x+5$

82. $x^2-1=2x$

83. $4x^2-1=4x$

Find all solutions to each equation. If rounding is necessary, round to the nearest hundredth.

84. $2x^3-x^2-2x+1=0$

85. $3x^3-2x^2-3x+2=0$

86. $2x^3+2=x^2+4x$

87. $3x^3-9x=2x^2-6$

8.4 Equations Quadratic in Form

OBJECTIVES

A Solve equations that are reducible to a quadratic equation.

B Solve application problems using equations quadratic in form.

We are now in a position to put our knowledge of quadratic equations to work to solve a variety of equations.

EXAMPLE 1

Solve $(x+3)^2-2(x+3)-8=0$.

SOLUTION We can see that this equation is quadratic in form by replacing $x+3$ with another variable, say y. Replacing $x+3$ with y we have

$$y^2-2y-8=0$$

We can solve this equation by factoring the left side and then setting each factor equal to 0.

$$y^2-2y-8=0$$

$$(y-4)(y+2)=0 \quad \textbf{Factor}$$

$$y-4=0 \quad \text{or} \quad y+2=0 \quad \textbf{Set factors to 0}$$

$$y=4 \quad \text{or} \quad y=-2$$

Since our original equation was written in terms of the variable x, we would like our solutions in terms of x also. Replacing y with $x+3$ and then solving for x we have

$$x+3=4 \quad \text{or} \quad x+3=-2$$

$$x=1 \quad \text{or} \quad x=-5$$

The solutions to our original equation are 1 and -5.

The method we have just shown lends itself well to other types of equations that are quadratic in form, as we will see. In this example, however, there is another method that works just as well. Let's solve our original equation again, but this time, let's begin by expanding $(x + 3)^2$ and $2(x + 3)$.

$$(x + 3)^2 - 2(x + 3) - 8 = 0$$
$$x^2 + 6x + 9 - 2x - 6 - 8 = 0 \quad \textbf{Multiply}$$
$$x^2 + 4x - 5 = 0 \quad \textbf{Combine similar terms}$$
$$(x - 1)(x + 5) = 0 \quad \textbf{Factor}$$
$$x - 1 = 0 \quad \text{or} \quad x + 5 = 0 \quad \textbf{Set factors to 0}$$
$$x = 1 \quad \text{or} \quad x = -5$$

As you can see, either method produces the same result.

EXAMPLE 2

Solve $4x^4 + 7x^2 = 2$.

SOLUTION This equation is quadratic in x^2. We can make it easier to look at by using the substitution $y = x^2$. (The choice of the letter y is arbitrary. We could just as easily use the substitution $m = x^2$.) Making the substitution $y = x^2$ and then solving the resulting equation we have

$$4y^2 + 7y = 2$$
$$4y^2 + 7y - 2 = 0 \quad \textbf{Standard form}$$
$$(4y - 1)(y + 2) = 0 \quad \textbf{Factor}$$
$$4y - 1 = 0 \quad \text{or} \quad y + 2 = 0 \quad \textbf{Set factors to 0}$$
$$y = \frac{1}{4} \quad \text{or} \quad y = -2$$

Now we replace y with x^2 to solve for x:

$$x^2 = \frac{1}{4} \quad \text{or} \quad x^2 = -2$$
$$x = \pm\sqrt{\frac{1}{4}} \quad \text{or} \quad x = \pm\sqrt{-2} \quad \textbf{Theorem 1}$$
$$x = \pm\frac{1}{2} \quad \text{or} \quad x = \pm i\sqrt{2}$$

The solution set is $\{\frac{1}{2}, -\frac{1}{2}, i\sqrt{2}, -i\sqrt{2}\}$.

EXAMPLE 3

Solve for x: $x + \sqrt{x} - 6 = 0$.

SOLUTION To see that this equation is quadratic in form, we have to notice that $(\sqrt{x})^2 = x$; that is, the equation can be rewritten as

$$(\sqrt{x})^2 + \sqrt{x} - 6 = 0$$

Replacing $\sqrt{x}$ with y and solving as usual, we have

$$y^2 + y - 6 = 0$$
$$(y + 3)(y - 2) = 0$$
$$y + 3 = 0 \quad \text{or} \quad y - 2 = 0$$
$$y = -3 \quad \text{or} \quad y = 2$$

Again, to find x, we replace with $\sqrt{x}$ and solve:

$$\sqrt{x} = -3 \quad \text{or} \quad \sqrt{x} = 2$$
$$x = 9 \qquad x = 4 \quad \textbf{Square both sides of each equation}$$

Since we squared both sides of each equation, we have the possibility of obtaining extraneous solutions. We have to check both solutions in our original equation.

	When $x = 9$		When $x = 4$
the equation	$x + \sqrt{x} - 6 = 0$	the equation	$x + \sqrt{x} - 6 = 0$
becomes	$9 + \sqrt{9} - 6 \stackrel{?}{=} 0$	becomes	$4 + \sqrt{4} - 6 \stackrel{?}{=} 0$
	$9 + 3 - 6 \stackrel{?}{=} 0$		$4 + 2 - 6 \stackrel{?}{=} 0$
	$6 \neq 0$		$0 = 0$
	This means 9 is extraneous.		This means 4 is a solution.

The only solution to the equation $x + \sqrt{x} - 6 = 0$ is $x = 4$. 

We should note here that the two possible solutions, 9 and 4, to the equation in Example 3 can be obtained by another method. Instead of substituting for $\sqrt{x}$, we can isolate it on one side of the equation and then square both sides to clear the equation of radicals.

$$x + \sqrt{x} - 6 = 0$$

$$\sqrt{x} = -x + 6 \quad \textbf{Isolate } x$$

$$x = x^2 - 12x + 36 \quad \textbf{Square both sides}$$

$$0 = x^2 - 13x + 36 \quad \textbf{Add } -x \textbf{ to both sides}$$

$$0 = (x - 4)(x - 9) \quad \textbf{Factor}$$

$$x - 4 = 0 \quad \text{or} \quad x - 9 = 0$$

$$x = 4 \qquad\qquad x = 9$$

We obtain the same two possible solutions. Since we squared both sides of the equation to find them, we would have to check each one in the original equation. As was the case in Example 3, only $x = 4$ is a solution; $x = 9$ is extraneous.

EXAMPLE 4

If an object is tossed into the air with an upward velocity of 12 feet per second from the top of a building h feet high, the time it takes for the object to hit the ground below is given by the formula

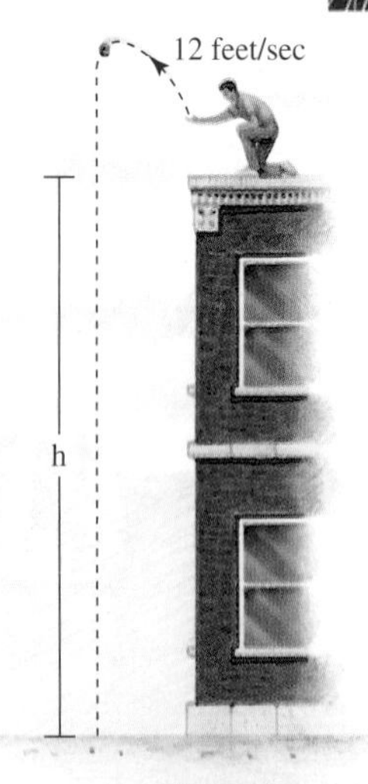

$$16t^2 - 12t - h = 0$$

Solve this formula for t.

SOLUTION The formula is in standard form and is quadratic in t. The coefficients a, b, and c that we need to apply to the quadratic formula are $a = 16$, $b = -12$, and $c = -h$. Substituting these quantities into the quadratic formula, we have

$$t = \frac{12 \pm \sqrt{144 - 4(16)(-h)}}{2(16)}$$

$$= \frac{12 \pm \sqrt{144 + 64h}}{32}$$

We can factor the perfect square 16 from the two terms under the radical and simplify our radical somewhat:

$$t = \frac{12 \pm \sqrt{16(9 + 4h)}}{32}$$

$$= \frac{12 \pm 4\sqrt{9 + 4h}}{32}$$

Now we can reduce to lowest terms by factoring a 4 from the numerator and denominator.

$$t = \frac{\not{4}(3 \pm \sqrt{9 + 4h})}{\not{4} \cdot 8}$$

$$= \frac{3 \pm \sqrt{9 + 4h}}{8}$$

If we were given a value of h, we would find that one of the solutions to this last formula would be a negative number. Since time is always measured in positive units, we wouldn't use that solution.

More About the Golden Ratio

Previously, we derived the golden ratio $\frac{1 + \sqrt{5}}{2}$ by finding the ratio of length to width for a golden rectangle. The golden ratio was actually discovered before the golden rectangle by the Greeks who lived before Euclid. The early Greeks found the golden ratio by dividing a line segment into two parts so that the ratio of the shorter part to the longer part was the same as the ratio of the longer part to the whole segment. When they divided a line segment in this manner, they said it was divided in "extreme and mean ratio." Figure 1 illustrates a line segment divided this way.

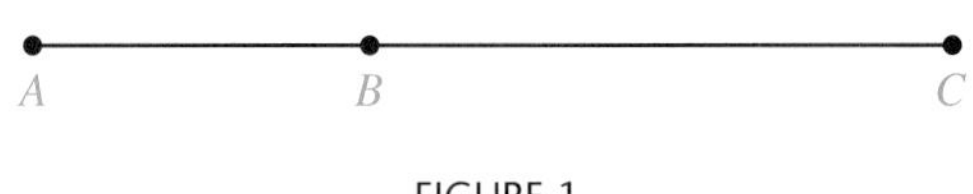

FIGURE 1

If point B divides segment AC in "extreme and mean ratio," then

$$\frac{\text{Length of shorter segment}}{\text{Length of longer segment}} = \frac{\text{Length of longer segment}}{\text{Length of whole segment}}$$

$$\frac{AB}{BC} = \frac{BC}{AC}$$

EXAMPLE 5 If the length of segment AB in Figure 1 is 1 inch, find the length of BC so that the whole segment AC is divided in "extreme and mean ratio."

SOLUTION Using Figure 1 as a guide, if we let x = the length of segment BC, then the length of AC is $x + 1$. If B divides AC into "extreme and mean ratio," then the ratio of AB to BC must equal the ratio of BC to AC. Writing this relationship using the variable x, we have

$$\frac{1}{x} = \frac{x}{x+1}$$

If we multiply both sides of this equation by the LCD $x(x + 1)$ we have

$$x + 1 = x^2$$

$$0 = x^2 - x - 1 \quad \textbf{Write equation in standard form}$$

Since this last equation is not factorable, we apply the quadratic formula.

$$x = \frac{1 \pm \sqrt{(-1)^2 - 4(1)(-1)}}{2}$$

$$= \frac{1 \pm \sqrt{5}}{2}$$

Our equation has two solutions, which we approximate using decimals:

$$\frac{1 + \sqrt{5}}{2} \approx 1.618 \qquad \frac{1 - \sqrt{5}}{2} \approx -0.618$$

Since we originally let x equal the length of segment BC, we use only the positive solution to our equation. As you can see, the positive solution is the golden ratio.

USING TECHNOLOGY

Graphing Calculators

More About Example 1

As we mentioned earlier, algebraic expressions entered into a graphing calculator do not have to be simplified to be evaluated. This fact applies to equations as well. We can graph the equation $y = (x + 3)^2 - 2(x + 3) - 8$ to assist us in solving the equation in Example 1. The graph is shown in Figure 2. Using the Zoom and Trace features at the x-intercepts gives us $x = 1$ and $x = -5$ as the solutions to the equation $0 = (x + 3)^2 - 2(x + 3) - 8$.

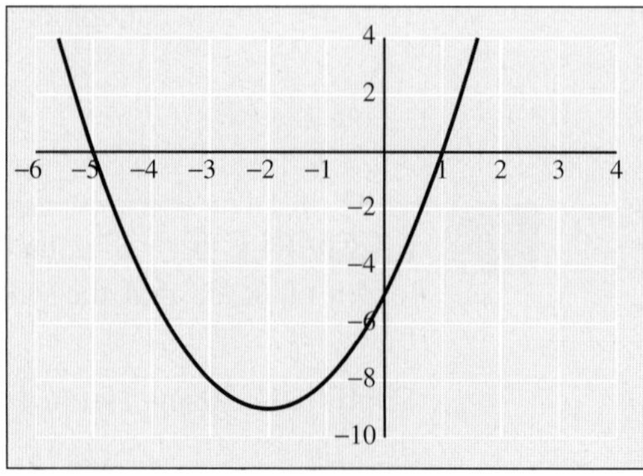

FIGURE 2

More About Example 2

Figure 3 shows the graph of $y = 4x^4 + 7x^2 - 2$. As we expect, the x-intercepts give the real number solutions to the equation $0 = 4x^4 + 7x^2 - 2$. The complex solutions do not appear on the graph.

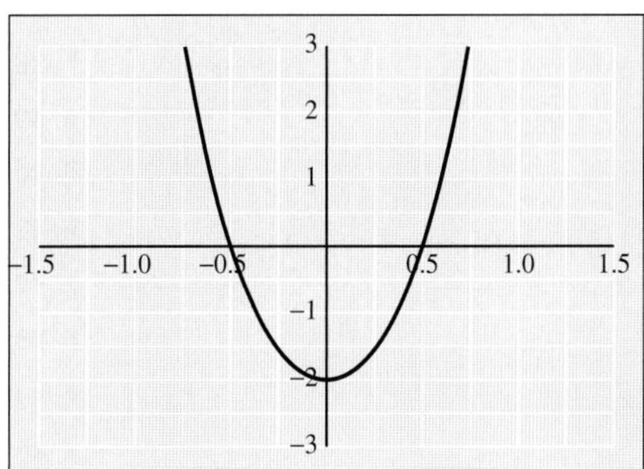

FIGURE 3

More About Example 3

In solving the equation in Example 3, we found that one of the possible solutions was an extraneous solution. If we solve the equation $x + \sqrt{x} - 6 = 0$ by graphing the function $y = x + \sqrt{x} - 6$, we find that the extraneous solution, 9, is not an x-intercept. Figure 4 shows that the only solution to the equation occurs at the x-intercept 4.

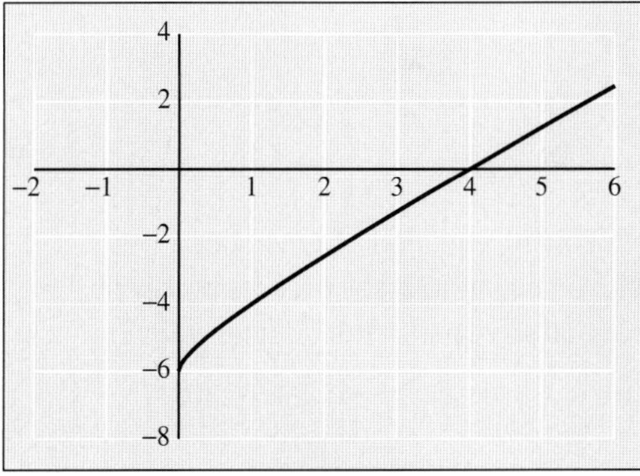

FIGURE 4

LINKING OBJECTIVES AND EXAMPLES

Next to each **objective** we have listed the examples that are best described by that objective.

A	1–3
B	4, 5

GETTING READY FOR CLASS

After reading through the preceding section, respond in your own words and in complete sentences.

1. What does it mean for an equation to be quadratic in form?
2. What are all the circumstances in solving equations (that we have studied) in which it is necessary to check for extraneous solutions?
3. How would you start to solve the equation $x + \sqrt{x} - 6 = 0$?
4. What does it mean for a line segment to be divided in "extreme and mean ratio"?

Problem Set 8.4

Online support materials can be found at www.thomsonedu.com/login

Solve each equation.

1. $(x-3)^2+3(x-3)+2=0$
2. $(x+4)^2-(x+4)-6=0$
3. $2(x+4)^2+5(x+4)-12=0$
4. $3(x-5)^2+14(x-5)-5=0$
5. $x^4-10x^2+9=0$
6. $x^4-29x^2+100=0$
7. $x^4-7x^2+12=0$
8. $x^4-14x^2+45=0$
9. $x^4-6x^2-27=0$
10. $x^4+2x^2-8=0$
11. $x^4+9x^2=-20$
12. $x^4-11x^2=-30$
13. $(2a-3)^2-9(2a-3)=-20$
14. $(3a-2)^2+2(3a-2)=3$
15. $2(4a+2)^2=3(4a+2)+20$
16. $6(2a+4)^2=(2a+4)+2$
17. $6t^4=-t^2+5$
18. $3t^4=-2t^2+8$
19. $9x^4-49=0$
20. $25x^4-9=0$

Solve each of the following equations. Remember, if you square both sides of an equation in the process of solving it, you have to check all solutions in the original equation.

21. $x-7\sqrt{x}+10=0$
22. $x-6\sqrt{x}+8=0$
23. $t-2\sqrt{t}-15=0$
24. $t-3\sqrt{t}-10=0$
25. $6x+11\sqrt{x}=35$
26. $2x+\sqrt{x}=15$
27. $x-2\sqrt{x}-8=0$
28. $x+2\sqrt{x}-3=0$
29. $x+3\sqrt{x}-18=0$
30. $x-5\sqrt{x}-14=0$
31. $2x+9\sqrt{x}-5=0$
32. $2x-\sqrt{x}-6=0$
33. $(a-2)-11\sqrt{a-2}+30=0$
34. $(a-3)-9\sqrt{a-3}+20=0$
35. $(2x+1)-8\sqrt{2x+1}+15=0$
36. $(2x-3)-7\sqrt{2x-3}+12=0$
37. $(x^2+1)^2-2(x^2+1)-15=0$
38. $(x^2-3)^2-4(x^2-3)-12=0$
39. $(x^2+5)^2-6(x^2+5)-27=0$
40. $(x^2-2)^2-6(x^2-2)-7=0$
41. $x^{-2}-3x^{-1}+2=0$
42. $x^{-2}+x^{-1}-2=0$
43. $y^{-4}-5y^{-2}=0$
44. $2y^{-4}+10y^{-2}=0$
45. $x^{2/3}+4x^{1/3}-12=0$
46. $x^{2/3}-2x^{1/3}-8=0$
47. $x^{4/3}+6x^{2/3}+9=0$
48. $x^{4/3}-10x^{2/3}+25=0$
49. Solve the formula $16t^2-vt-h=0$ for t.
50. Solve the formula $16t^2+vt+h=0$ for t.
51. Solve the formula $kx^2+8x+4=0$ for x.
52. Solve the formula $k^2x^2+kx+4=0$ for x.
53. Solve $x^2+2xy+y^2=0$ for x by using the quadratic formula with $a=1$, $b=2y$, and $c=y^2$.
54. Solve $x^2-2xy+y^2=0$ for x by using the quadratic formula, with $a=1$, $b=-2y$, and $c=y^2$.

Applying the Concepts

For Problems 55–56, t is in seconds.

55. **Falling Object** An object is tossed into the air with an upward velocity of 8 feet per second from the top of a building h feet high. The time it takes for the object to hit the ground below is given by the formula $16t^2-8t-h=0$. Solve this formula for t.

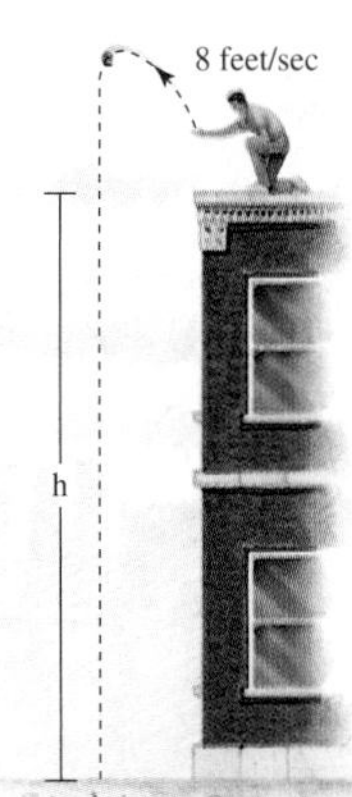

■ = Videos available by instructor request
▶ = Online student support materials available

56. **Falling Object** An object is tossed into the air with an upward velocity of 6 feet per second from the top of a building h feet high. The time it takes for the object to hit the ground below is given by the formula $16t^2 - 6t - h = 0$. Solve this formula for t.

57. **Saint Louis Arch** The shape of the famous "Gateway to the West" arch in Saint Louis can be modeled by a parabola. The equation for one such parabola is:

$$y = -\frac{1}{150}x^2 + \frac{21}{5}x$$

where x and y are in feet.

Royalty-Free/Corbis

a. Sketch the graph of the arch's equation on a coordinate axis.
b. Approximately how far do you have to walk to get from one side of the arch to the other?

58. **Area and Perimeter** A total of 160 yards of fencing is to be used to enclose part of a lot that borders on a river. This situation is shown in the following diagram.

a. Write an equation that gives the relationship between the length and width and the 160 yards of fencing.
b. The formula for the area that is enclosed by the fencing and the river is $A = lw$. Solve the equation in part (a) for l, and then use the result to write the area in terms of w only.
c. Make a table that gives at least five possible values of w and associated area A.
d. From the pattern in your table shown in part (c), what is the largest area that can be enclosed by the 160 yards of fencing? (Try some other table values if necessary.)

Golden Ratio Use Figure 1 from this section as a guide to working Problems 59–60.

59. If AB in Figure 1 is 4 inches, and B divides AC in "extreme and mean ratio," find BC, and then show that BC is 4 times the golden ratio.

60. If AB in Figure 1 is $\frac{1}{2}$ inch, and B divides AC in "extreme and mean ratio," find BC, and then show that the ratio of BC to AB is the golden ratio.

Maintaining Your Skills

Combine, if possible.

61. $5\sqrt{7} - 2\sqrt{7}$
62. $6\sqrt{2} - 9\sqrt{2}$
63. $\sqrt{18} - \sqrt{8} + \sqrt{32}$
64. $\sqrt{50} + \sqrt{72} - \sqrt{8}$
65. $9x\sqrt{20x^3y^2} + 7y\sqrt{45x^5}$
66. $5x^2\sqrt{27xy^3} - 6y\sqrt{12x^5y}$

Multiply.

67. $(\sqrt{5} - 2)(\sqrt{5} + 8)$
68. $(2\sqrt{3} - 7)(2\sqrt{3} + 7)$
69. $(\sqrt{x} + 2)^2$
70. $(3 - \sqrt{x})(3 + \sqrt{x})$

Rationalize the denominator.

71. $\dfrac{\sqrt{7}}{\sqrt{7} - 2}$
72. $\dfrac{\sqrt{5} - \sqrt{2}}{\sqrt{5} + \sqrt{2}}$

Getting Ready for the Next Section

73. Evaluate $y = 3x^2 - 6x + 1$ for $x = 1$.
74. Evaluate $y = -2x^2 + 6x - 5$ for $x = \frac{3}{2}$.
75. Let $P(x) = -0.1x^2 + 27x - 500$ and find $P(135)$.
76. Let $P(x) = -0.1x^2 + 12x - 400$ and find $P(600)$.

Solve.

77. $0 = a(80)^2 + 70$
78. $0 = a(80)^2 + 90$
79. $x^2 - 6x + 5 = 0$
80. $x^2 - 3x - 4 = 0$
81. $-x^2 - 2x + 3 = 0$
82. $-x^2 + 4x + 12 = 0$
83. $2x^2 - 6x + 5 = 0$
84. $x^2 - 4x + 5 = 0$

Fill in the blanks to complete the square.

85. $x^2 - 6x + \square = (x - \square)^2$

86. $x^2 + 10x + \square = (x + \square)^2$

87. $y^2 + 2y + \square = (y + \square)^2$

88. $y^2 - 12y + \square = (y - \square)^2$

Extending the Concepts

Find the x- and y-intercepts.

89. $y = x^3 - 4x$

90. $y = x^4 - 10x^2 + 9$

91. $y = 3x^3 + x^2 - 27x - 9$

92. $y = 2x^3 + x^2 - 8x - 4$

93. The graph of $y = 2x^3 - 7x^2 - 5x + 4$ crosses the x-axis at $x = 4$. Where else does it cross the x-axis?

94. The graph of $y = 6x^3 + x^2 - 12x + 5$ crosses the x-axis at $x = 1$. Where else does it cross the x-axis?

8.5 Graphing Parabolas

OBJECTIVES

A Graph a parabola.

B Solve application problems using information from a graph.

C Find an equation from its graph.

The solution set to the equation

$$y = x^2 - 3$$

consists of ordered pairs. One method of graphing the solution set is to find a number of ordered pairs that satisfy the equation and to graph them. We can obtain some ordered pairs that are solutions to $y = x^2 - 3$ by use of a table as follows:

x	$y = x^2 - 3$	y	**Solutions**
−3	$y = (-3)^2 - 3 = 9 - 3 = 6$	6	(−3, 6)
−2	$y = (-2)^2 - 3 = 4 - 3 = 1$	1	(−2, 1)
−1	$y = (-1)^2 - 3 = 1 - 3 = -2$	−2	(−1, −2)
0	$y = 0^2 - 3 = 0 - 3 = -3$	−3	(0, −3)
1	$y = 1^2 - 3 = 1 - 3 = -2$	−2	(1, −2)
2	$y = 2^2 - 3 = 4 - 3 = 1$	1	(2, 1)
3	$y = 3^2 - 3 = 9 - 3 = 6$	6	(3, 6)

Graphing these solutions and then connecting them with a smooth curve, we have the graph of $y = x^2 - 3$ (Figure 1).

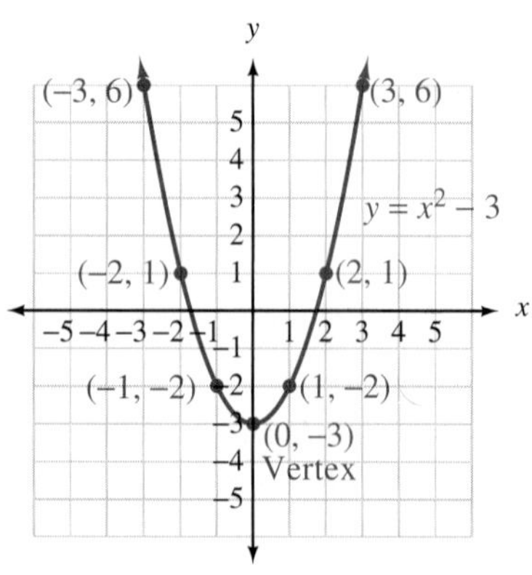

FIGURE 1

This graph is an example of a *parabola*. All equations of the form $y = ax^2 + bx + c$, $a \neq 0$ have parabolas for graphs.

Although it is always possible to graph parabolas by making a table of values of x and y that satisfy the equation, there are other methods that are faster and, in some cases, more accurate.

The important points associated with the graph of a parabola are the highest (or lowest) point on the graph and the x-intercepts. The y-intercepts can also be useful.

Intercepts for Parabolas

The graph of the equation $y = ax^2 + bx + c$ crosses the y-axis at $y = c$ since substituting $x = 0$ into $y = ax^2 + bx + c$ yields $y = c$.

Since the graph crosses the x-axis when $y = 0$, the x-intercepts are those values of x that are solutions to the quadratic equation $0 = ax^2 + bx + c$.

The Vertex of a Parabola

The highest or lowest point on a parabola is called the *vertex.* The vertex for the graph of $y = ax^2 + bx + c$ will occur when

$$x = \frac{-b}{2a}$$

To see this, we must transform the right side of $y = ax^2 + bx + c$ into an expression that contains x in just one of its terms. This is accomplished by completing the square on the first two terms. Here is what it looks like:

$$y = ax^2 + bx + c$$

$$y = a\left(x^2 + \frac{b}{a}x\right) + c$$

$$y = a\left[x^2 + \frac{b}{a}x + \left(\frac{b}{2a}\right)^2\right] + c - a\left(\frac{b}{2a}\right)^2$$

$$y = a\left(x + \frac{b}{2a}\right)^2 + \frac{4ac - b^2}{4a}$$

It may not look like it, but this last line indicates that the vertex of the graph of $y = ax^2 + bx + c$ has an x-coordinate of $\frac{-b}{2a}$. Since a, b, and c are constants, the only quantity that is varying in the last expression is the x in $\left(x + \frac{b}{2a}\right)^2$. Since the quantity $\left(x + \frac{b}{2a}\right)^2$ is the square of $x + \frac{b}{2a}$, the smallest it will ever be is 0, and that will happen when $x = \frac{-b}{2a}$.

Note
What we are doing here is attempting to explain why the vertex of a parabola always has an x-coordinate of $\frac{-b}{2a}$. But the explanation may not be easy to understand the first time you see it. It may be helpful to look over the examples in this section and the notes that accompany these examples and then come back and read over this discussion again.

We can use the vertex point along with the x- and y-intercepts to sketch the graph of any equation of the form $y = ax^2 + bx + c$. Here is a summary of the preceding information.

Graphing Parabolas I The graph of $y = ax^2 + bx + c$, $a \neq 0$, will have

1. A y-intercept at $y = c$
2. x-intercepts (if they exist) at
$$x = \frac{-b \pm \sqrt{b^2 - 4ac}}{2a}$$
3. A vertex when $x = \frac{-b}{2a}$

EXAMPLE 1 Sketch the graph of $y = x^2 - 6x + 5$.

SOLUTION To find the x-intercepts, we let $y = 0$ and solve for x:

$$0 = x^2 - 6x + 5$$

$$0 = (x - 5)(x - 1)$$

$$x = 5 \quad \text{or} \quad x = 1$$

To find the coordinates of the vertex, we first find

$$x = \frac{-b}{2a} = \frac{-(-6)}{2(1)} = 3$$

The x-coordinate of the vertex is 3. To find the y-coordinate, we substitute 3 for x in our original equation:

$$y = 3^2 - 6(3) + 5 = 9 - 18 + 5 = -4$$

The graph crosses the x-axis at 1 and 5 and has its vertex at $(3, -4)$. Plotting these points and connecting them with a smooth curve, we have the graph shown in Figure 2. The graph is a parabola that opens up, so we say the graph is *concave up*. The vertex is the lowest point on the graph. (Note that the graph crosses the y-axis at 5, which is the value of y we obtain when we let $x = 0$.)

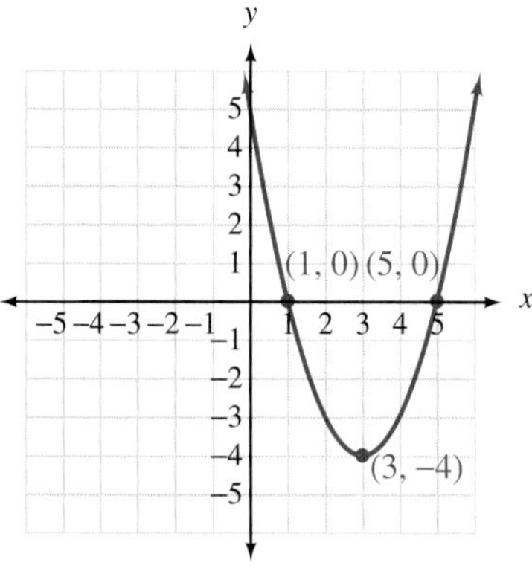

FIGURE 2

Finding the Vertex by Completing the Square

Another way to locate the vertex of the parabola in Example 1 is by completing the square on the first two terms on the right side of the equation $y = x^2 - 6x + 5$. In this case, we would do so by adding 9 to and subtracting 9 from the right side of the equation. This amounts to adding 0 to the equation, so we know we haven't changed its solutions. This is what it looks like:

$$y = (x^2 - 6x \qquad) + 5$$

$$y = (x^2 - 6x + \mathbf{9}) + 5 - \mathbf{9}$$

$$y = (x - 3)^2 - 4$$

You may have to look at this last equation a while to see this, but when $x = 3$, then $y = (x - 3)^2 - 4 = 0^2 - 4 = -4$ is the smallest y will ever be. And that is why the vertex is at (3, −4). As a matter of fact, this is the same kind of reasoning we used when we derived the formula $x = \frac{-b}{2a}$ for the x-coordinate of the vertex.

EXAMPLE 2

Graph $y = -x^2 - 2x + 3$.

SOLUTION To find the x-intercepts, we let $y = 0$:

$$0 = -x^2 - 2x + 3$$

$$0 = x^2 + 2x - 3 \qquad \textbf{Multiply each side by } -1$$

$$0 = (x + 3)(x - 1)$$

$$x = -3 \quad \text{or} \quad x = 1$$

The x-coordinate of the vertex is given by

$$x = \frac{-b}{2a} = \frac{-(-2)}{2(-1)} = \frac{2}{-2} = -1$$

To find the y-coordinate of the vertex, we substitute −1 for x in our original equation to get

$$y = -(-1)^2 - 2(-1) + 3 = -1 + 2 + 3 = 4$$

Our parabola has x-intercepts at −3 and 1 and a vertex at (−1, 4). Figure 3 shows the graph. We say the graph is *concave down* since it opens downward.

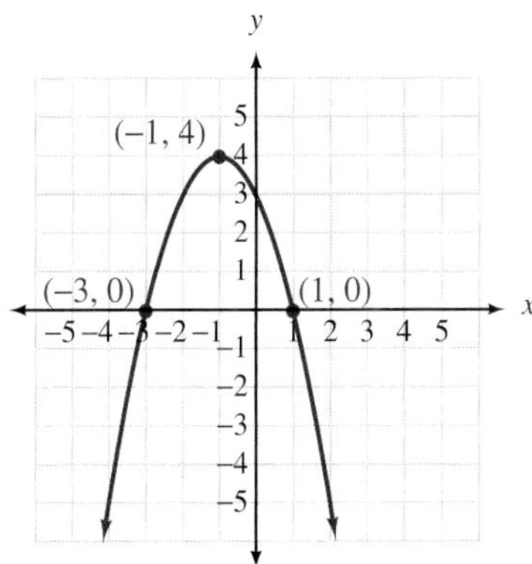

FIGURE 3

Again, we could have obtained the coordinates of the vertex by completing the square on the first two terms on the right side of our equation. To do so, we must first factor −1 from the first two terms. (Remember, the leading coefficients must be 1 to complete the square.) When we complete the square, we add 1 inside the parentheses, which actually decreases the right side of the equation by −1 since everything in the parentheses is multiplied by −1. To make up for it, we add 1 outside the parentheses.

$$y = -1(x^2 + 2x) + 3$$

$$y = -1(x^2 + 2x + \mathbf{1}) + 3 + \mathbf{1}$$

$$y = -1(x + 1)^2 + 4$$

The last line tells us that the *largest* value of y will be 4, and that will occur when $x = -1$.

EXAMPLE 3

Graph $y = 3x^2 - 6x + 1$.

SOLUTION To find the x-intercepts, we let $y = 0$ and solve for x:

$$0 = 3x^2 - 6x + 1$$

Since the right side of this equation does not factor, we can look at the discriminant to see what kind of solutions are possible. The discriminant for this equation is

$$b^2 - 4ac = 36 - 4(3)(1) = 24$$

Since the discriminant is a positive number but not a perfect square, the equation will have irrational solutions. This means that the x-intercepts are irrational numbers and will have to be approximated with decimals using the quadratic formula. Rather than use the quadratic formula, we will find some other points on the graph, but first let's find the vertex.

Here are both methods of finding the vertex:

Using the formula that gives us the x-coordinate of the vertex, we have:

$$x = \frac{-b}{2a} = \frac{-(-6)}{2(3)} = 1$$

Substituting 1 for x in the equation gives us the y-coordinate of the vertex:

$$y = 3 \cdot 1^2 - 6 \cdot 1 + 1 = -2$$

To complete the square on the right side of the equation, we factor 3 from the first two terms, add 1 inside the parentheses, and add -3 outside the parentheses (this amounts to adding 0 to the right side):

$$y = 3(x^2 - 2x \qquad) + 1$$

$$y = 3(x^2 - 2x + \mathbf{1}) + 1 - \mathbf{3}$$

$$y = 3(x - 1)^2 - 2$$

In either case, the vertex is $(1, -2)$.

If we can find two points, one on each side of the vertex, we can sketch the graph. Let's let $x = 0$ and $x = 2$ since each of these numbers is the same distance from $x = 1$ and $x = 0$ will give us the y-intercept.

When $x = 0$

$$\begin{aligned} y &= 3(0)^2 - 6(0) + 1 \\ &= 0 - 0 + 1 \\ &= 1 \end{aligned}$$

When $x = 2$

$$\begin{aligned} y &= 3(2)^2 - 6(2) + 1 \\ &= 12 - 12 + 1 \\ &= 1 \end{aligned}$$

The two points just found are $(0, 1)$ and $(2, 1)$. Plotting these two points along with the vertex $(1, -2)$, we have the graph shown in Figure 4.

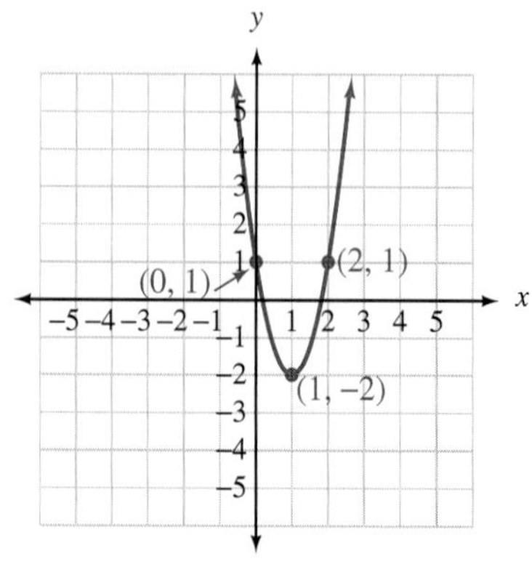

FIGURE 4

EXAMPLE 4

Graph $y = -2x^2 + 6x - 5$.

SOLUTION Letting $y = 0$, we have

$$0 = -2x^2 + 6x - 5$$

Again, the right side of this equation does not factor. The discriminant is $b^2 - 4ac = 36 - 4(-2)(-5) = -4$, which indicates that the solutions are complex numbers. This means that our original equation does not have x-intercepts. The graph does not cross the x-axis.

Let's find the vertex.

Using our formula for the x-coordinate of the vertex, we have

$$x = \frac{-b}{2a} = \frac{-6}{2(-2)} = \frac{6}{4} = \frac{3}{2}$$

To find the y-coordinate, we let $x = \frac{3}{2}$:

$$y = -2\left(\frac{3}{2}\right)^2 + 6\left(\frac{3}{2}\right) - 5$$

$$= \frac{-18}{4} + \frac{18}{2} - 5$$

$$= \frac{-18 + 36 - 20}{4}$$

$$= -\frac{1}{2}$$

Finding the vertex by completing the square is a more complicated matter. To make the coefficient of x^2 a 1, we must factor -2 from the first two terms. To complete the square inside the parentheses, we add $\frac{9}{4}$. Since each term inside the parentheses is multiplied by -2, we add $\frac{9}{2}$ outside the parentheses so that the net result is the same as adding 0 to the right side:

$$y = -2(x^2 - 3x \qquad) - 5$$

$$y = -2\left(x^2 - 3x + \mathbf{\frac{9}{4}}\right) - 5 + \mathbf{\frac{9}{2}}$$

$$y = -2\left(x - \frac{3}{2}\right)^2 - \frac{1}{2}$$

The vertex is $(\frac{3}{2}, -\frac{1}{2})$. Since this is the only point we have so far, we must find two others, Let's let $x = 3$ and $x = 0$ since each point is the same distance from $x = \frac{3}{2}$ and on either side:

When $x = 3$

$$y = -2(3)^2 + 6(3) - 5$$
$$= -18 + 18 - 5$$
$$= -5$$

When $x = 0$

$$y = -2(0)^2 + 6(0) - 5$$
$$= 0 + 0 - 5$$
$$= -5$$

The two additional points on the graph are $(3, -5)$ and $(0, -5)$. Figure 5 shows the graph.

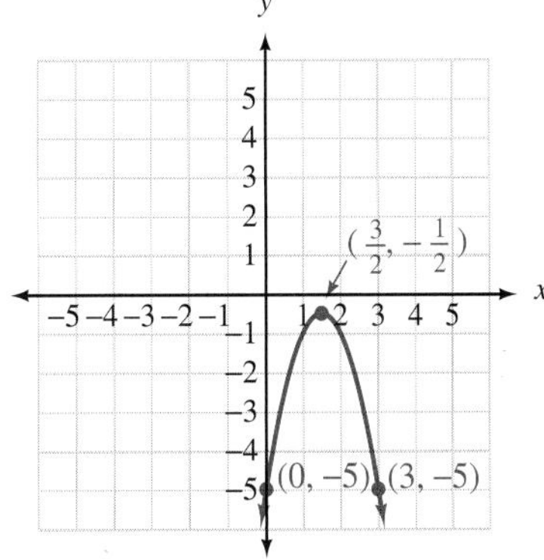

FIGURE 5

The graph is concave down. The vertex is the highest point on the graph. 

By looking at the equations and graphs in Examples 1 through 4, we can conclude that the graph of $y = ax^2 + bx + c$ will be concave up when a is positive and concave down when a is negative. Taking this even further, if $a > 0$, then the vertex is the lowest point on the graph, and if $a < 0$, the vertex is the highest point on the graph. Finally, if we complete the square on x in the equation $y = ax^2 + bx + c$, $a \neq 0$, we can rewrite the equation of our parabola as $y = a(x - h)^2 + k$. When the equation is in this form, the vertex is at the point (h, k). Here is a summary:

Graphing Parabolas II The graph of

$$y = a(x - h)^2 + k, a \neq 0$$

will be a parabola with a vertex at (h, k). The vertex will be the highest point on the graph when $a < 0$, and it will be the lowest point on the graph when $a > 0$.

EXAMPLE 5

A company selling copies of an accounting program for home computers finds that it will make a weekly profit of P dollars from selling x copies of the program, according to the equation

$$P(x) = -0.1x^2 + 27x - 500$$

How many copies of the program should it sell to make the largest possible profit, and what is the largest possible profit?

SOLUTION Since the coefficient of x^2 is negative, we know the graph of this parabola will be concave down, meaning that the vertex is the highest point of the curve. We find the vertex by first finding its x-coordinate:

$$x = \frac{-b}{2a} = \frac{-27}{2(-0.1)} = \frac{27}{0.2} = 135$$

This represents the number of programs the company needs to sell each week to make a maximum profit. To find the maximum profit, we substitute 135 for x in the original equation. (A calculator is helpful for these kinds of calculations.)

$$\begin{aligned} P(135) &= -0.1(135)^2 + 27(135) - 500 \\ &= -0.1(18{,}225) + 3{,}645 - 500 \\ &= -1{,}822.5 + 3{,}645 - 500 \\ &= 1{,}322.5 \end{aligned}$$

The maximum weekly profit is \$1,322.50 and is obtained by selling 135 programs a week.

EXAMPLE 6

An art supply store finds that they can sell x sketch pads each week at p dollars each, according to the equation $x = 900 - 300p$. Graph the revenue equation $R = xp$. Then use the graph to find the price p that will bring in the maximum revenue. Finally, find the maximum revenue.

SOLUTION As it stands, the revenue equation contains three variables. Since we are asked to find the value of p that gives us the maximum value of R, we rewrite the equation using just the variables R and p. Since $x = 900 - 300p$, we have

$$R = xp = (900 - 300p)p$$

The graph of this equation is shown in Figure 6. The graph appears in the first quadrant only since R and p are both positive quantities.

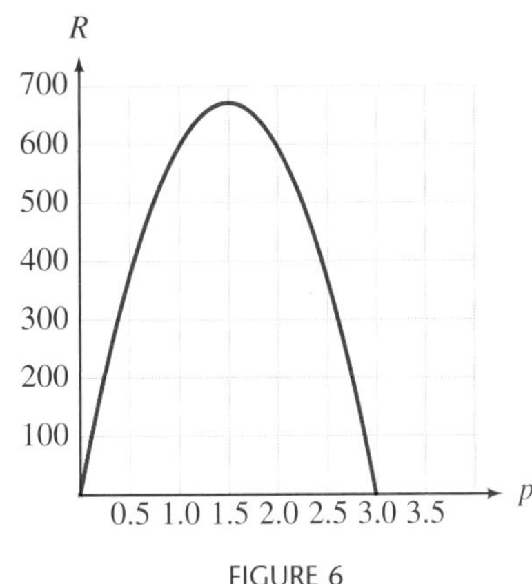

FIGURE 6

From the graph we see that the maximum value of R occurs when $p = \$1.50$. We can calculate the maximum value of R from the equation:

When	$p = 1.5$
The equation	$R = (900 - 300p)p$
becomes	$R = (900 - 300 \cdot 1.5)1.5$
	$= (900 - 450)1.5$
	$= 450 \cdot 1.5$
	$= 675$

The maximum revenue is \$675. It is obtained by setting the price of each sketch pad at $p = \$1.50$.

USING TECHNOLOGY

Graphing Calculators

If you have been using a graphing calculator for some of the material in this course, you are well aware that your calculator can draw all the graphs in this section very easily. It is important, however, that you be able to recognize and sketch the graph of any parabola by hand. It is a skill that all successful intermediate algebra students should possess, even if they are proficient in the use of a graphing calculator. My suggestion is that you work the problems in this section and problem set without your calculator. Then use your calculator to check your results.

Finding the Equation from the Graph

EXAMPLE 7 At the 1997 Washington County Fair in Oregon, David Smith, Jr., The Bullet, was shot from a cannon. As a human cannonball, he reached a height of 70 feet before landing in a net 160 feet from the cannon. Sketch the graph of his path, and then find the equation of the graph.

SOLUTION We assume that the path taken by the human cannonball is a parabola. If the origin of the coordinate system is at the opening of the cannon, then the net that catches him will be at 160 on the x-axis. Figure 7 shows the graph:

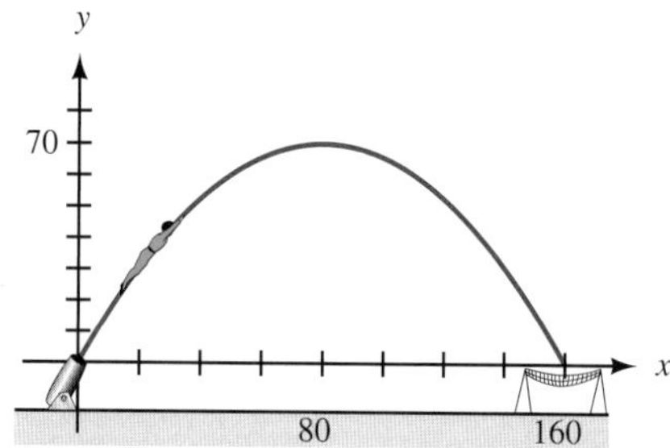

FIGURE 7

Since the curve is a parabola, we know the equation will have the form

$$y = a(x - h)^2 + k$$

Since the vertex of the parabola is at (80, 70), we can fill in two of the three constants in our equation, giving us

$$y = a(x - 80)^2 + 70$$

To find a, we note that the landing point will be (160, 0). Substituting the coordinates of this point into the equation, we solve for a:

$$0 = a(160 - 80)^2 + 70$$

$$0 = a(80)^2 + 70$$

$$0 = 6400a + 70$$

$$a = -\frac{70}{6400} = -\frac{7}{640}$$

The equation that describes the path of the human cannonball is

$$y = -\frac{7}{640}(x - 80)^2 + 70 \text{ for } 0 \le x \le 160$$

USING TECHNOLOGY

Graphing Calculators

Graph the equation found in Example 7 by graphing it on a graphing calculator using the window shown below. (We will use this graph later in the book to find the angle between the cannon and the horizontal.)

Window: X from 0 to 180, increment 20

Y from 0 to 80, increment 10

On the TI-83, an increment of 20 for X means Xscl=20.

LINKING OBJECTIVES AND EXAMPLES

Next to each **objective** we have listed the examples that are best described by that objective.

A	1–4
B	5
C	6

GETTING READY FOR CLASS

After reading through the preceding section, respond in your own words and in complete sentences.

1. What is a parabola?
2. What part of the equation of a parabola determines whether the graph is concave up or concave down?
3. Suppose $f(x) = ax^2 + bx + c$ is the equation of a parabola. Explain how $f(4) = 1$ relates to the graph of the parabola.
4. A line can be graphed with two points. How many points are necessary to get a reasonable sketch of a parabola? Explain.

Problem Set 8.5

Online support materials can be found at www.thomsonedu.com/login

For each of the following equations, give the x-intercepts and the coordinates of the vertex, and sketch the graph.

1. $y = x^2 + 2x - 3$
2. $y = x^2 - 2x - 3$
3. $y = -x^2 - 4x + 5$
4. $y = x^2 + 4x - 5$
5. $y = x^2 - 1$
6. $y = x^2 - 4$
7. $y = -x^2 + 9$
8. $y = -x^2 + 1$
9. $y = 2x^2 - 4x - 6$
10. $y = 2x^2 + 4x - 6$
11. $y = x^2 - 2x - 4$
12. $y = x^2 - 2x - 2$

Find the vertex and any two convenient points to sketch the graphs of the following.

13. $y = x^2 - 4x - 4$
14. $y = x^2 - 2x + 3$
15. $y = -x^2 + 2x - 5$
16. $y = -x^2 + 4x - 2$
17. $y = x^2 + 1$
18. $y = x^2 + 4$
19. $y = -x^2 - 3$
20. $y = -x^2 - 2$

For each of the following equations, find the coordinates of the vertex, and indicate whether the vertex is the highest point on the graph or the lowest point on the graph. (Do not graph.)

21. $y = x^2 - 6x + 5$
22. $y = -x^2 + 6x - 5$
23. $y = -x^2 + 2x + 8$
24. $y = x^2 - 2x - 8$
25. $y = 12 + 4x - x^2$
26. $y = -12 - 4x + x^2$
27. $y = -x^2 - 8x$
28. $y = x^2 + 8x$

Applying the Concepts

29. **Maximum Profit** A company finds that it can make a profit of P dollars each month by selling x patterns, according to the formula $P(x) = -0.002x^2 + 3.5x - 800$. How many patterns must it sell each month to have a maximum profit? What is the maximum profit?

■ = Videos available by instructor request
▸ = Online student support materials available at www.thomsonedu.com/login

30. **Maximum Profit** A company selling picture frames finds that it can make a profit of P dollars each month by selling x frames, according to the formula $P(x) = -0.002x^2 + 5.5x - 1{,}200$. How many frames must it sell each month to have a maximum profit? What is the maximum profit?

Kathleen Olson

31. **Maximum Height** Chaudra is tossing a softball into the air with an underhand motion. The distance of the ball above her hand at any time is given by the function

$$h(t) = 32t - 16t^2 \quad \text{for } 0 \le t \le 2$$

where $h(t)$ is the height of the ball (in feet) and t is the time (in seconds). Find the times at which the ball is in her hand and the maximum height of the ball.

32. **Maximum Area** Justin wants to fence 3 sides of a rectangular exercise yard for his dog. The fourth side of the exercise yard will be a side of the house. He has 80 feet of fencing available. Find the dimensions of the exercise yard that will enclose the maximum area.

33. **Maximum Revenue** A company that manufactures typewriter ribbons knows that the number of ribbons x it can sell each week is related to the price p of each ribbon by the equation $x = 1{,}200 - 100p$. Graph the revenue equation $R = xp$. Then use the graph to find the price p that will bring in the maximum revenue. Finally, find the maximum revenue.

34. **Maximum Revenue** A company that manufactures diskettes for home computers finds that it can sell x diskettes each day at p dollars per diskette, according to the equation $x = 800 - 100p$. Graph the revenue equation $R = xp$. Then use the graph to find the price p that will bring in the maximum revenue. Finally, find the maximum revenue.

35. **Maximum Revenue** The relationship between the number of calculators x a company sells each day and the price p of each calculator is given by the equation $x = 1{,}700 - 100p$. Graph the revenue equation $R = xp$, and use the graph to find the price p that will bring in the maximum revenue. Then find the maximum revenue.

36. **Maximum Revenue** The relationship between the number x of pencil sharpeners a company sells each week and the price p of each sharpener is given by the equation $x = 1{,}800 - 100p$. Graph the revenue equation $R = xp$, and use the graph to find the price p that will bring in the maximum revenue. Then find the maximum revenue.

37. **Human Cannonball** A human cannonball is shot from a cannon at the county fair. He reaches a height of 60 feet before landing in a net 180 feet from the cannon. Sketch the graph of his path, and then find the equation of the graph.

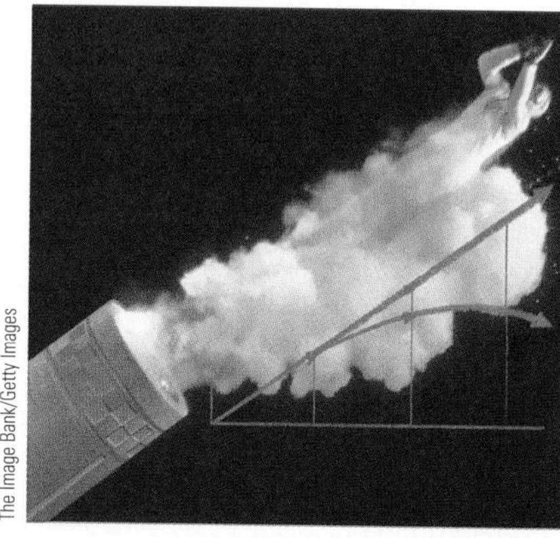
The Image Bank/Getty Images

38. Improving Your Quantitative Literacy The graph below shows the different paths taken by the human cannonball when his velocity out of the cannon is 50 miles/hour and his cannon is inclined at varying angles.

a. If his landing net is placed 108 feet from the cannon, at what angle should the cannon be inclined so that he lands in the net?

b. Approximately where do you think he would land if the cannon was inclined at 45°?

c. If the cannon was inclined at 45°, approximately what height do you think he would attain?

d. Do you think there is another angle for which he would travel the same distance he travels at 80°? Give an estimate of that angle.

e. The fact that every landing point can come from two different paths makes us think that the equations that give us the landing points must be what type of equations?

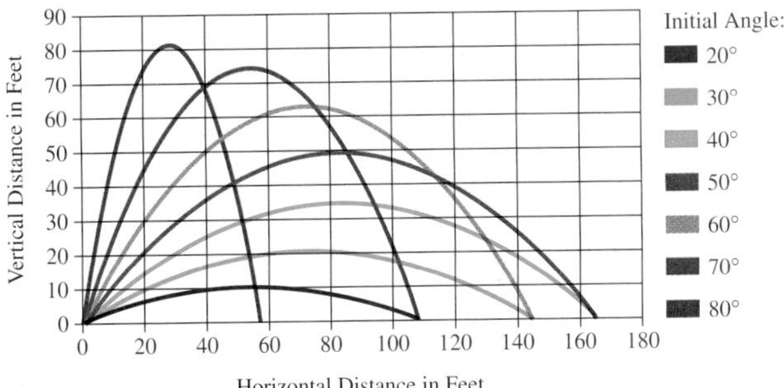

Maintaining Your Skills

Perform the indicated operations.

39. $(3 - 5i) - (2 - 4i)$

40. $2i(5 - 6i)$

41. $(3 + 2i)(7 - 3i)$

42. $(4 + 5i)^2$

43. $\dfrac{i}{3 + i}$

44. $\dfrac{2 + 3i}{2 - 3i}$

Getting Ready for the Next Section

Solve.

45. $x^2 - 2x - 8 = 0$

46. $x^2 - x - 12 = 0$

47. $6x^2 - x = 2$

48. $3x^2 - 5x = 2$

49. $x^2 - 6x + 9 = 0$

50. $x^2 + 8x + 16 = 0$

Extending the Concepts

Finding the Equation from the Graph For each problem below, the graph is a parabola. In each case, find an equation in the form $y = a(x - h)^2 + k$ that describes the graph.

51.

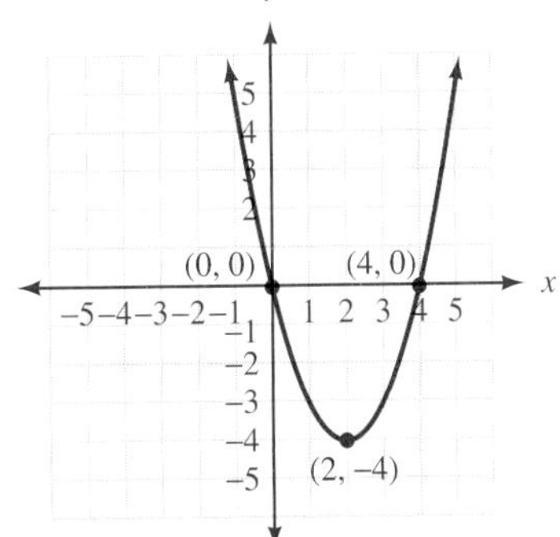

52.

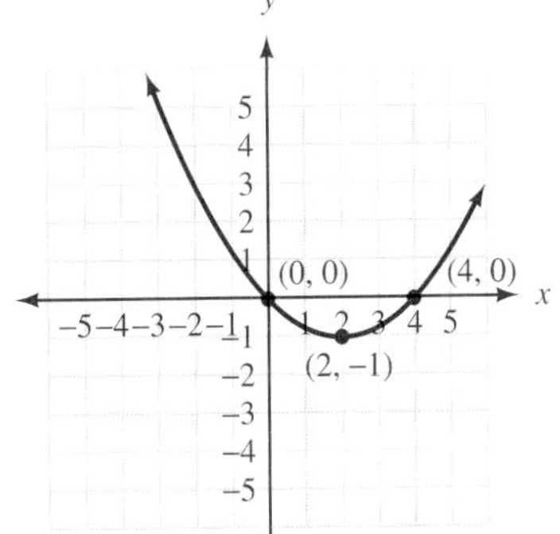

8.6 Quadratic Inequalities

OBJECTIVES

A Solve quadratic inequalities and graph the solution set.

Quadratic inequalities in one variable are inequalities of the form

$$ax^2 + bx + c < 0 \qquad ax^2 + bx + c > 0$$

$$ax^2 + bx + c \le 0 \qquad ax^2 + bx + c \ge 0$$

where a, b, and c are constants, with $a \neq 0$. The technique we will use to solve inequalities of this type involves graphing. Suppose, for example, we wish to find the solution set for the inequality $x^2 - x - 6 > 0$. We begin by factoring the left side to obtain

$$(x - 3)(x + 2) > 0$$

We have two real numbers $x - 3$ and $x + 2$ whose product $(x - 3)(x + 2)$ is greater than zero; that is, their product is positive. The only way the product can be positive is either if both factors, $(x - 3)$ and $(x + 2)$, are positive or if they are both negative. To help visualize where $x - 3$ is positive and where it is negative, we draw a real number line and label it accordingly:

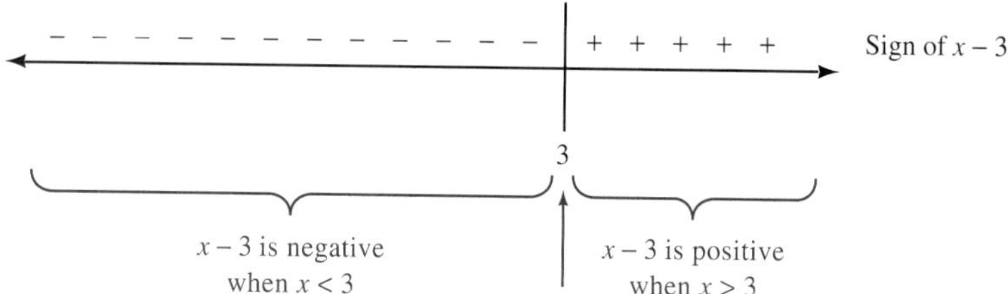

Here is a similar diagram showing where the factor $x + 2$ is positive and where it is negative:

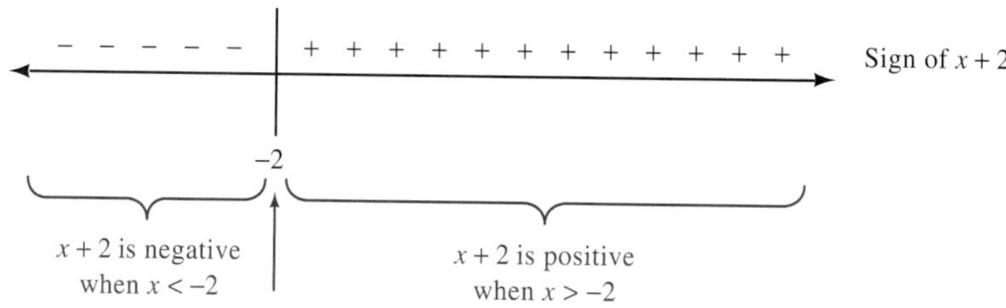

Drawing the two number lines together and eliminating the unnecessary numbers, we have

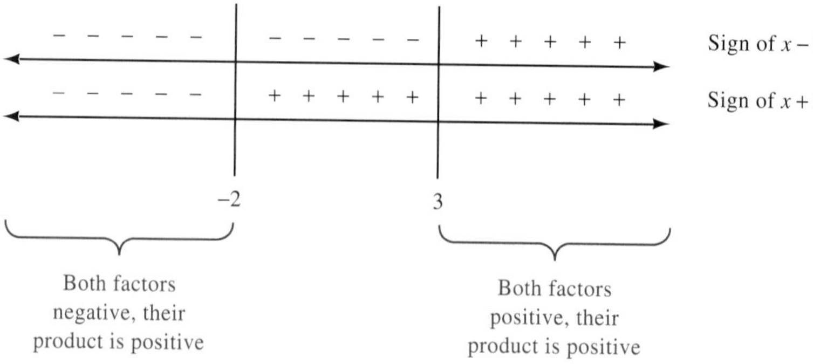

We can see from the preceding diagram that the graph of the solution to $x^2 - x - 6 > 0$ is

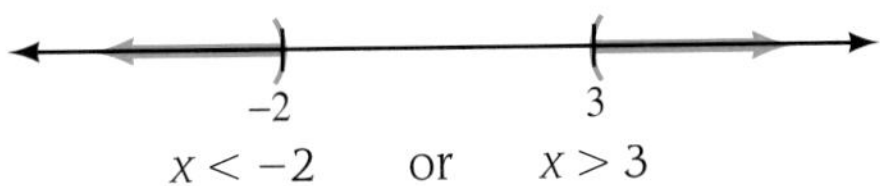
–2
3

$x < -2$ or $x > 3$

USING TECHNOLOGY

Graphical Solutions to Quadratic Inequalities

We can solve the preceding problem by using a graphing calculator to visualize where the product $(x - 3)(x + 2)$ is positive. First, we graph the function $y = (x - 3)(x + 2)$ as shown in Figure 1

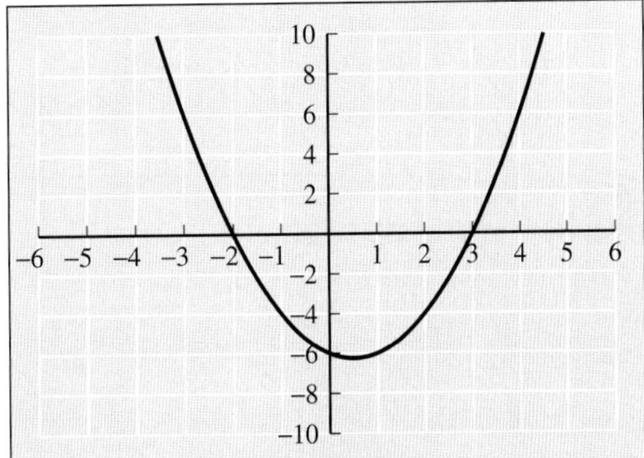

FIGURE 1

Next, we observe where the graph is above the x-axis. As you can see, the graph is above the x-axis to the right of 3 and to the left of –2, as shown in Figure 2.

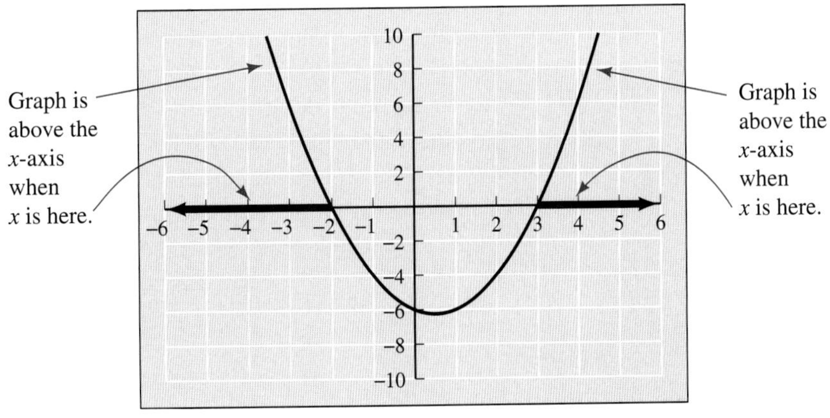
Graph is above the x-axis when x is here.
Graph is above the x-axis when x is here.

FIGURE 2

When the graph is above the x-axis, we have points whose y-coordinates are positive. Since these y-coordinates are the same as the expression $(x - 3)(x + 2)$, the values of x for which the graph of $y = (x - 3)(x + 2)$ is above the x-axis are the values of x for which the inequality $(x - 3)(x + 2) > 0$ is true. Our solution set is therefore

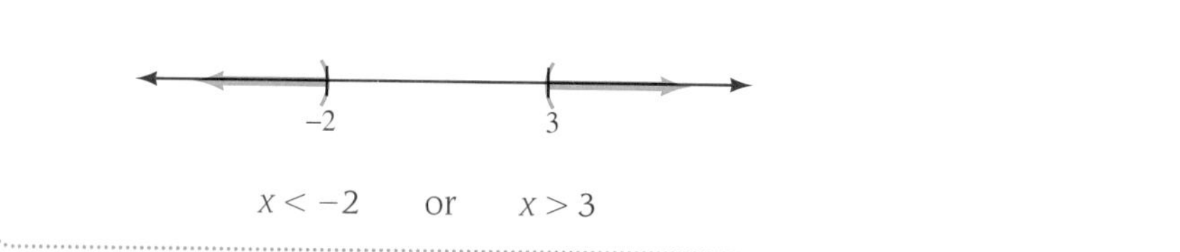

$x < -2$ or $x > 3$

EXAMPLE 1 Solve for x: $x^2 - 2x - 8 \le 0$.

ALGEBRAIC SOLUTION We begin by factoring:

$$x^2 - 2x - 8 \le 0$$
$$(x - 4)(x + 2) \le 0$$

The product $(x - 4)(x + 2)$ is negative or zero. The factors must have opposite signs. We draw a diagram showing where each factor is positive and where each factor is negative:

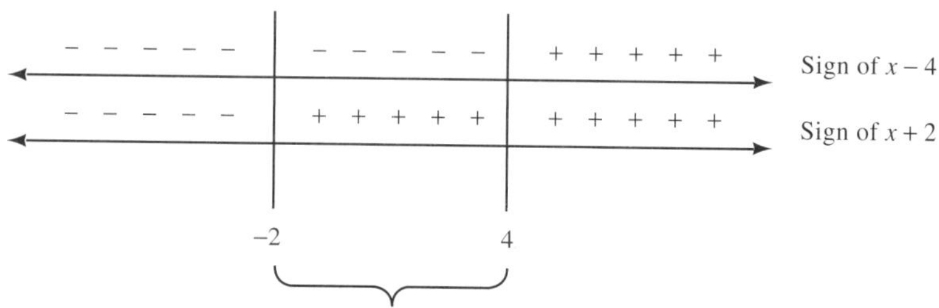

From the diagram we have the graph of the solution set:

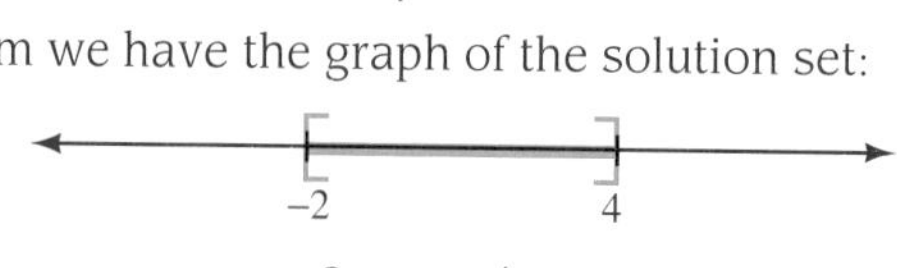
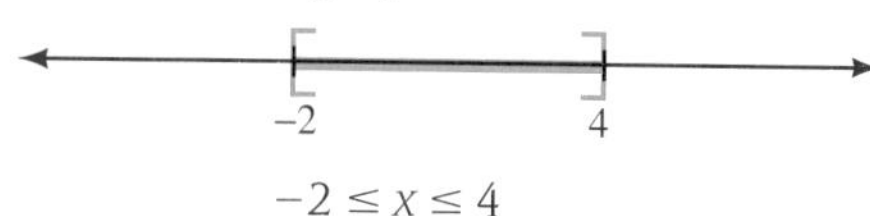

$$-2 \le x \le 4$$

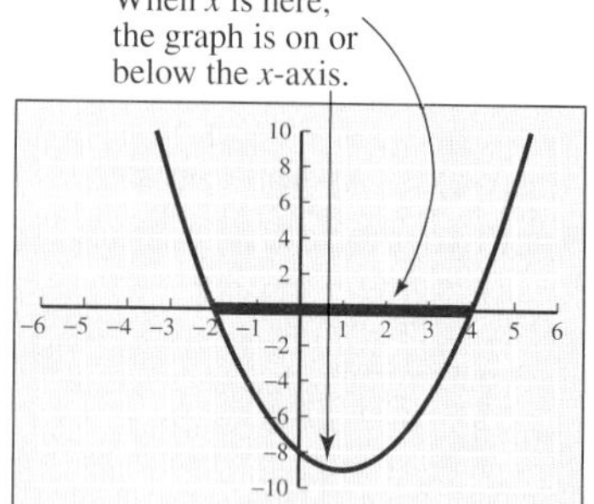

When x is here, the graph is on or below the x-axis.

FIGURE 3

GRAPHIC SOLUTION To solve this inequality with a graphing calculator, we graph the function $y = (x - 4)(x + 2)$ and observe where the graph is below the x-axis. These points have negative y-coordinates, which means that the product $(x - 4)(x + 2)$ is negative for these points. Figure 3 shows the graph of $y = (x - 4)(x + 2)$, along with the region on the x-axis where the graph contains points with negative y-coordinates.

As you can see, the graph is below the x-axis when x is between -2 and 4. Since our original inequality includes the possibility that $(x - 4)(x + 2)$ is 0, we include the endpoints, -2 and 4, with our solution set.

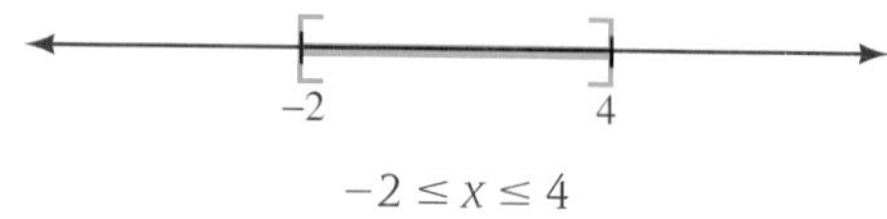

$$-2 \le x \le 4$$

EXAMPLE 2 Solve for x: $6x^2 - x \ge 2$.

ALGEBRAIC SOLUTION

$$6x^2 - x \ge 2$$
$$6x^2 - x - 2 \ge 0 \leftarrow \textbf{Standard form}$$
$$(3x - 2)(2x + 1) \ge 0$$

The product is positive, so the factors must agree in sign. Here is the diagram showing where that occurs:

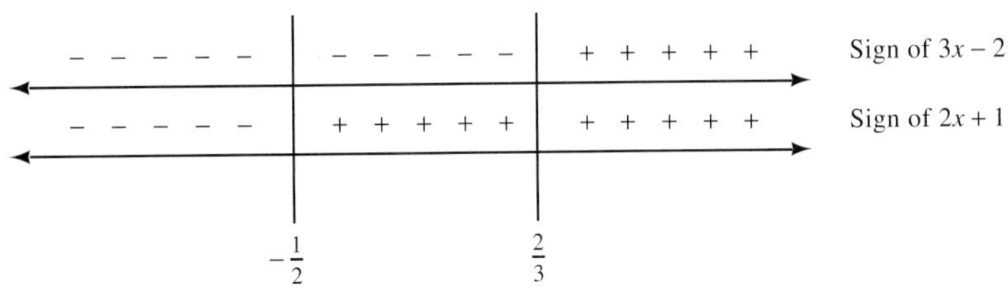

Sign of 3x − 2
Sign of 2x + 1
$-\frac{1}{2}$
$\frac{2}{3}$

Since the factors agree in sign below $-\frac{1}{2}$ and above $\frac{2}{3}$, the graph of the solution set is

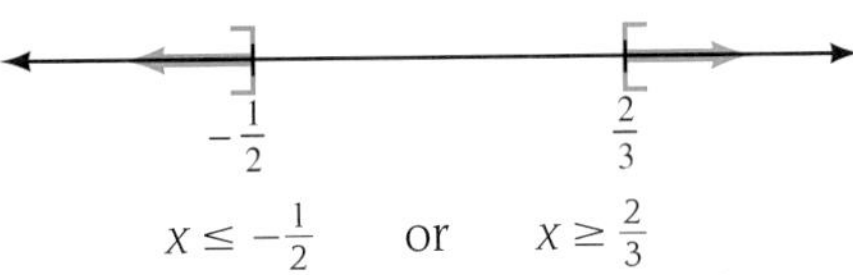

$-\frac{1}{2}$
$\frac{2}{3}$

$$x \le -\frac{1}{2} \quad \text{or} \quad x \ge \frac{2}{3}$$

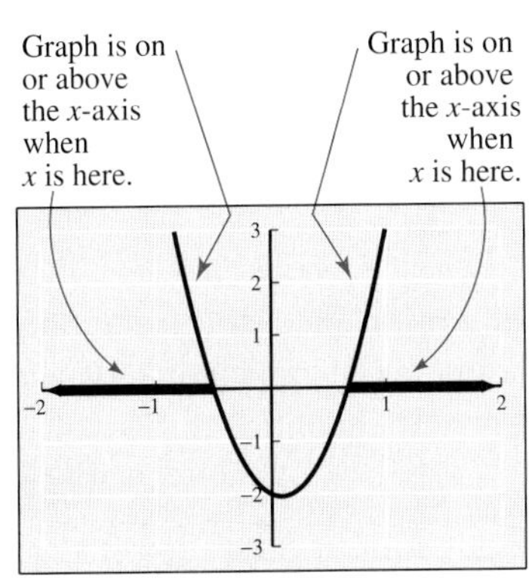

Graph is on or above the x-axis when x is here.
Graph is on or above the x-axis when x is here.

FIGURE 4

GRAPHICAL SOLUTION To solve this inequality with a graphing calculator, we graph the function $y = (3x - 2)(2x + 1)$ and observe where the graph is above the x-axis. These are the points that have positive y-coordinates, which means that the product $(3x - 2)(2x + 1)$ is positive for these points. Figure 4 shows the graph of $y = (3x - 2)(2x + 1)$, along with the regions on the x-axis where the graph is on or above the x-axis.

To find the points where the graph crosses the x-axis, we need to use either the Trace and Zoom features to zoom in on each point, or the calculator function that finds the intercepts automatically (on the TI-82/83 this is the root/zero function under the CALC key). Whichever method we use, we will obtain the following result:

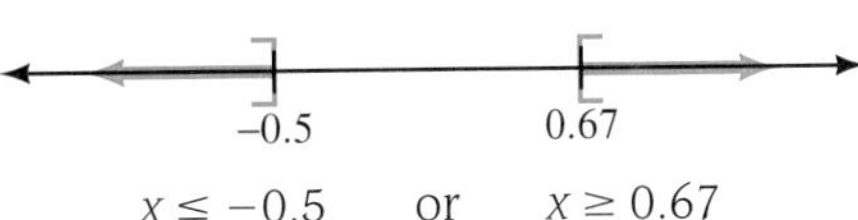

−0.5
0.67

$$x \le -0.5 \quad \text{or} \quad x \ge 0.67$$

EXAMPLE 3

Solve $x^2 - 6x + 9 \ge 0$.

ALGEBRAIC SOLUTION

$$x^2 - 6x + 9 \ge 0$$

$$(x - 3)^2 \ge 0$$

This is a special case in which both factors are the same. Since $(x - 3)^2$ is always positive or zero, the solution set is all real numbers; that is, any real number that is used in place of x in the original inequality will produce a true statement.

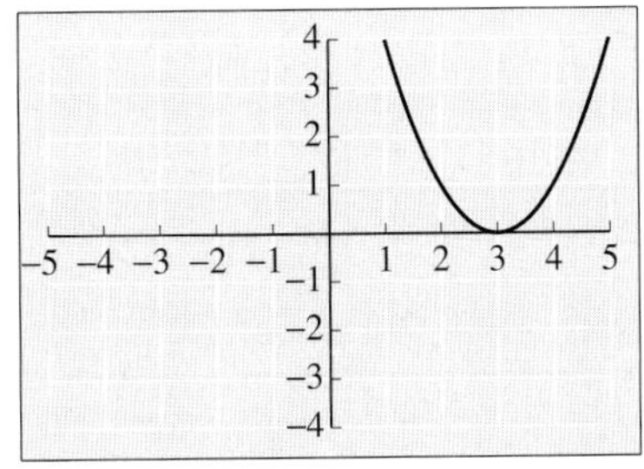

FIGURE 5

GRAPHICAL SOLUTION The graph of $y = (x - 3)^2$ is shown in Figure 5. Notice that it touches the x-axis at 3 and is above the x-axis everywhere else. This means that every point on the graph has a y-coordinate greater than or equal to 0, no matter what the value of x. The conclusion that we draw from the graph is that the inequality $(x - 3)^2 \ge 0$ is true for all values of x.

EXAMPLE 4

Solve: $\dfrac{x - 4}{x + 1} \le 0$

SOLUTION The inequality indicates that the quotient of $(x - 4)$ and $(x + 1)$ is negative or 0 (less than or equal to 0). We can use the same reasoning we used to solve the first three examples, because quotients are positive or negative under the same conditions that products are positive or negative. Here is the diagram that shows where each factor is positive and where each factor is negative:

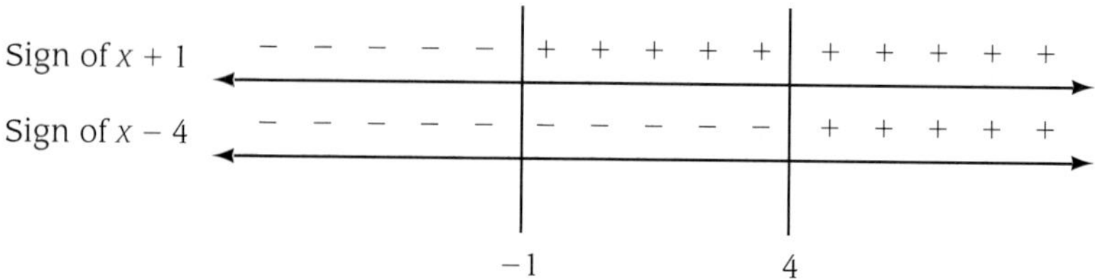

Between −1 and 4 the factors have opposite signs, making the quotient negative. Thus, the region between −1 and 4 is where the solutions lie, because the original inequality indicates the quotient $\frac{x-4}{x+1}$ is negative. The solution set and its graph are shown here:

$$-1 < x \le 4$$

Notice that the left endpoint is open—that is, it is not included in the solution set—because $x = -1$ would make the denominator in the original inequality 0. It is important to check all endpoints of solution sets to inequalities that involve rational expressions.

EXAMPLE 5

Solve: $\frac{3}{x-2} - \frac{2}{x-3} > 0$

SOLUTION We begin by adding the two rational expressions on the left side. The common denominator is $(x-2)(x-3)$:

$$\frac{3}{x-2}\cdot\frac{\mathbf{(x-3)}}{\mathbf{(x-3)}} - \frac{2}{x-3}\cdot\frac{\mathbf{(x-2)}}{\mathbf{(x-2)}} > 0$$

$$\frac{3x-9-2x+4}{(x-2)(x-3)} > 0$$

$$\frac{x-5}{(x-2)(x-3)} > 0$$

This time the quotient involves three factors. Here is the diagram that shows the signs of the three factors:

	below 2	2 to 3	3 to 5	above 5
Sign of $(x-5)$	− − − −	− − − −	− − − − −	+ + + +
Sign of $(x-3)$	− − − −	− − − −	+ + + + +	+ + + +
Sign of $(x-2)$	− − − −	+ + + +	+ + + + +	+ + + +

The original inequality indicates that the quotient is positive. For this to happen, either all three factors must be positive, or exactly two factors must be negative.

Looking back at the diagram, we see the regions that satisfy these conditions are between 2 and 3 or above 5. Here is our solution set:

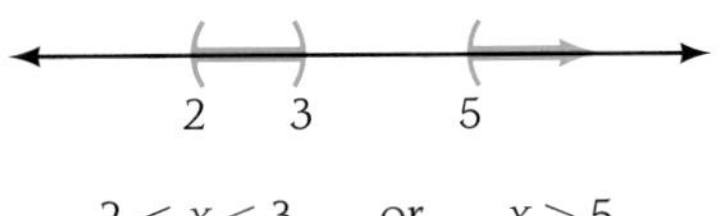

$2 < x < 3 \quad \text{or} \quad x > 5$

LINKING OBJECTIVES AND EXAMPLES

Next to each **objective** we have listed the examples that are best described by that objective.

A 1–3

GETTING READY FOR CLASS

After reading through the preceding section, respond in your own words and in complete sentences.

1. What is the first step in solving a quadratic inequality?
2. How do you show that the endpoint of a line segment is not part of the graph of a quadratic inequality?
3. How would you use the graph of $y = ax^2 + bx + c$ to help you find the graph of $ax^2 + bx + c < 0$?
4. Can a quadratic inequality have exactly one solution? Give an example.

Problem Set 8.6

Online support materials can be found at www.thomsonedu.com/login

Solve each of the following inequalities and graph the solution set.

▶ **1.** $x^2 + x - 6 > 0$ **2.** $x^2 + x - 6 < 0$

3. $x^2 - x - 12 \le 0$ **4.** $x^2 - x - 12 \ge 0$

5. $x^2 + 5x \ge -6$ **6.** $x^2 - 5x > 6$

▶ **7.** $6x^2 < 5x - 1$ **8.** $4x^2 \ge -5x + 6$

▶ **9.** $x^2 - 9 < 0$ **10.** $x^2 - 16 \ge 0$

11. $4x^2 - 9 \ge 0$ **12.** $9x^2 - 4 < 0$

13. $2x^2 - x - 3 < 0$ **14.** $3x^2 + x - 10 \ge 0$

15. $x^2 - 4x + 4 \ge 0$ **16.** $x^2 - 4x + 4 < 0$

17. $x^2 - 10x + 25 < 0$ **18.** $x^2 - 10x + 25 > 0$

▶ **19.** $(x - 2)(x - 3)(x - 4) > 0$ **20.** $(x - 2)(x - 3)(x - 4) < 0$

21. $(x + 1)(x + 2)(x + 3) \le 0$ **22.** $(x + 1)(x + 2)(x + 3) \ge 0$

▶ **23.** $\frac{x - 1}{x + 4} \le 0$ ▶ **24.** $\frac{x + 4}{x - 1} \le 0$

▶ **25.** $\frac{3x - 8}{x + 6} < 0$ ▶ **26.** $\frac{5x - 3}{x + 1} < 0$

▶ **27.** $\frac{x - 2}{x - 6} > 0$ ▶ **28.** $\frac{x - 1}{x - 3} \ge 0$

▶ **29.** $\frac{x - 2}{(x + 3)(x - 4)} < 0$ ▶ **30.** $\frac{x - 1}{(x + 2)(x - 5)} < 0$

▶ **31.** $\frac{2}{x - 4} - \frac{1}{x - 3} > 0$ **32.** $\frac{4}{x + 3} - \frac{3}{x + 2} > 0$

33. Write each statement using inequality notation.
 a. $x - 1$ is always positive.
 b. $x - 1$ is never negative.
 c. $x - 1$ is greater than or equal to 0.

34. Match each expression on the left with a phrase on the right.
 a. $(x - 1)^2 \ge 0$ **i.** Never true
 b. $(x - 1)^2 < 0$ **ii.** Sometimes true
 c. $(x - 1)^2 \le 0$ **iii.** Always true

Solve each inequality by inspection without showing any work.

▶ **35.** $(x - 1)^2 < 0$ ▶ **36.** $(x + 2)^2 < 0$

▶ **37.** $(x - 1)^2 \le 0$ ▶ **38.** $(x + 2)^2 \le 0$

39. $(x - 1)^2 \ge 0$ **40.** $(x + 2)^2 \ge 0$

41. $\frac{1}{(x - 1)^2} \ge 0$ **42.** $\frac{1}{(x + 2)^2} > 0$

▢ = Videos available by instructor request
▶ = Online student support materials available at www.thomsonedu.com/login

43. $x^2 - 6x + 9 < 0$

44. $x^2 - 6x + 9 \le 0$

45. $x^2 - 6x + 9 > 0$

46. $\dfrac{1}{x^2 - 6x + 9} > 0$

47. The graph of $y = x^2 - 4$ is shown in Figure 6. Use the graph to write the solution set for each of the following:

a. $x^2 - 4 < 0$

b. $x^2 - 4 > 0$

c. $x^2 - 4 = 0$

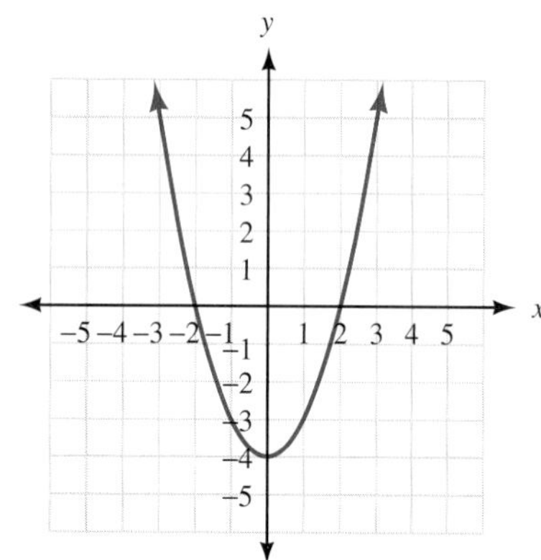

FIGURE 6

48. The graph of $y = 4 - x^2$ is shown in Figure 7. Use the graph to write the solution set for each of the following:

a. $4 - x^2 < 0$

b. $4 - x^2 > 0$

c. $4 - x^2 = 0$

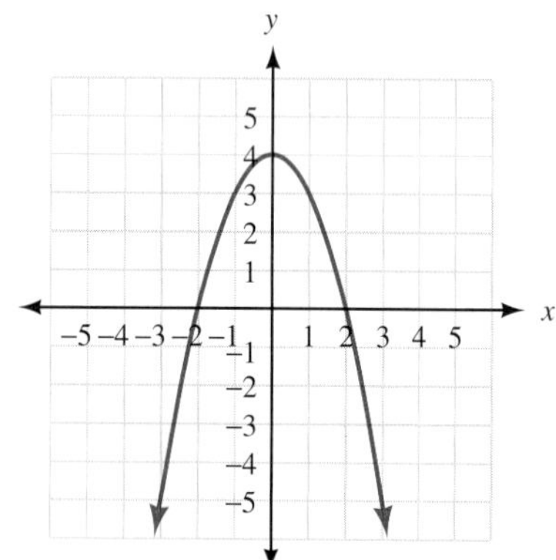

FIGURE 7

49. The graph of $y = x^2 - 3x - 10$ is shown in Figure 8. Use the graph to write the solution set for each of the following:

a. $x^2 - 3x - 10 < 0$

b. $x^2 - 3x - 10 > 0$

c. $x^2 - 3x - 10 = 0$

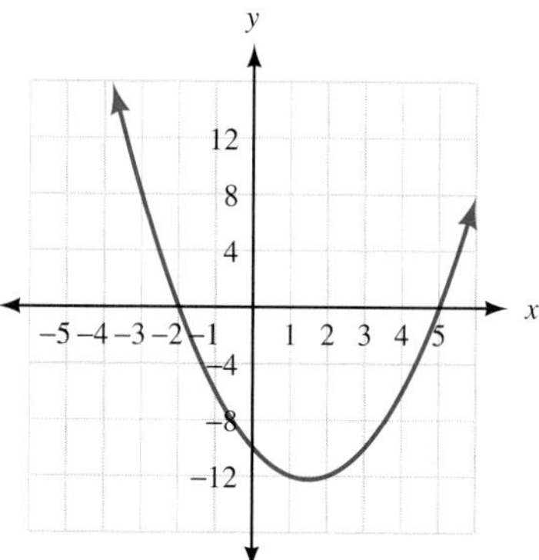

FIGURE 8

50. The graph of $y = x^2 + x - 12$ is shown in Figure 9. Use the graph to write the solution set for each of the following:

a. $x^2 + x - 12 < 0$

b. $x^2 - x - 12 > 0$

c. $x^2 + x - 12 = 0$

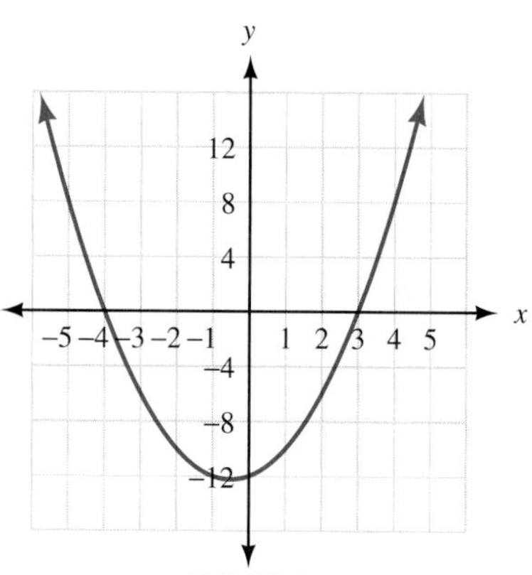

FIGURE 9

51. The graph of $y = x^3 - 3x^2 - x + 3$ is shown in Figure 10. Use the graph to write the solution set for each of the following:

a. $x^3 - 3x^2 - x + 3 < 0$

b. $x^3 - 3x^2 - x + 3 > 0$

c. $x^3 - 3x^2 - x + 3 = 0$

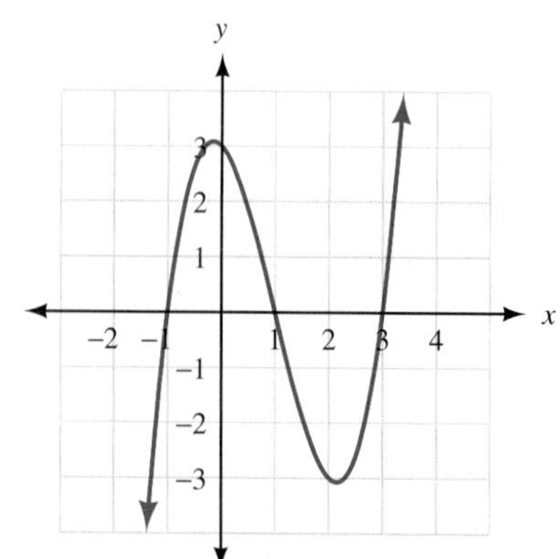

FIGURE 10

52. The graph of $y = x^3 + 4x^2 - 4x - 16$ is shown in Figure 11. Use the graph to write the solution set for each of the following:

a. $x^3 + 4x^2 - 4x - 16 < 0$

b. $x^3 + 4x^2 - 4x - 16 > 0$

c. $x^3 + 4x^2 - 4x - 16 = 0$

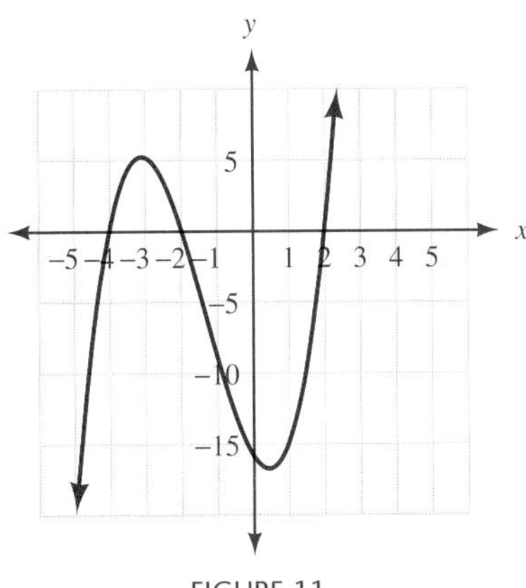

FIGURE 11

Applying the Concepts

53. **Dimensions of a Rectangle** The length of a rectangle is 3 inches more than twice the width. If the area is to be at least 44 square inches, what are the possibilities for the width?

54. **Dimensions of a Rectangle** The length of a rectangle is 5 inches less than 3 times the width. If the area is to be less than 12 square inches, what are the possibilities for the width?

55. **Revenue** A manufacturer of portable radios knows that the weekly revenue produced by selling x radios is given by the equation $R = 1{,}300p - 100p^2$, where p is the price of each radio. What price should she charge for each radio if she wants her weekly revenue to be at least \$4,000?

56. **Revenue** A manufacturer of small calculators knows that the weekly revenue produced by selling x calculators is given by the equation $R = 1{,}700p - 100p^2$, where p is the price of each calculator. What price should be charged for each calculator if the revenue is to be at least \$7,000 each week?

Maintaining Your Skills

Use a calculator to evaluate. Give answers to 4 decimal places.

57. $\dfrac{50{,}000}{32{,}000}$

58. $\dfrac{2.4362}{1.9758} - 1$

59. $\dfrac{1}{2}\left(\dfrac{4.5926}{1.3876} - 2\right)$

60. $1 + \dfrac{0.06}{12}$

Solve each equation.

61. $\sqrt{3t - 1} = 2$

62. $\sqrt{4t + 5} + 7 = 3$

63. $\sqrt{x + 3} = x - 3$

64. $\sqrt{x + 3} = \sqrt{x} - 3$

Graph each equation.

65. $y = \sqrt[3]{x - 1}$

66. $y = \sqrt[3]{x} - 1$

Extending the Concepts

Graph the solution set for each inequality.

67. $x^2 - 2x - 1 < 0$

68. $x^2 - 6x + 7 < 0$

69. $x^2 - 8x + 13 > 0$

70. $x^2 - 10x + 18 > 0$

Chapter 8 SUMMARY

EXAMPLES

1. If $(x-3)^2 = 25$
   then $x - 3 = \pm 5$
   $x = 3 \pm 5$
   $x = 8$ or $x = -2$

Theorem 1 [8.1]

If $a^2 = b$, where b is a real number, then

$$a = \sqrt{b} \qquad \text{or} \qquad a = -\sqrt{b}$$

which can be written as $a = \pm\sqrt{b}$.

2. Solve $x^2 - 6x - 6 = 0$
   $x^2 - 6x = 6$
   $x^2 - 6x + \mathbf{9} = 6 + \mathbf{9}$
   $(x-3)^2 = 15$
   $x - 3 = \pm\sqrt{15}$
   $x = 3 \pm \sqrt{15}$

To Solve a Quadratic Equation by Completing the Square [8.1]

Step 1: Write the equation in the form $ax^2 + bx = c$.

Step 2: If $a \neq 1$, divide through by the constant a so the coefficient of x^2 is 1.

Step 3: Complete the square on the left side by adding the square of $\frac{1}{2}$ the coefficient of x to both sides.

Step 4: Write the left side of the equation as the square of a binomial. Simplify the right side if possible.

Step 5: Apply Theorem 1, and solve as usual.

3. If $2x^2 + 3x - 4 = 0$, then
   $$x = \frac{-3 \pm \sqrt{9 - 4(2)(-4)}}{2(2)}$$
   $$= \frac{-3 \pm \sqrt{41}}{4}$$

The Quadratic Theorem [8.2]

For any quadratic equation in the form $ax^2 + bx + c = 0$, $a \neq 0$, the two solutions are

$$x = \frac{-b \pm \sqrt{b^2 - 4ac}}{2a}$$

This last expression is known as the *quadratic formula.*

4. The discriminant for
   $x^2 + 6x + 9 = 0$
   is $D = 36 - 4(1)(9) = 0$, which means the equation has one rational solution.

The Discriminant [8.3]

The expression $b^2 - 4ac$ that appears under the radical sign in the quadratic formula is known as the *discriminant.*

We can classify the solutions to $ax^2 + bx + c = 0$:

The Solutions Are	When the Discriminant Is
Two complex numbers containing i	Negative
One rational number	Zero
Two rational numbers	A positive perfect square
Two irrational numbers	A positive number, but not a perfect square

Equations Quadratic in Form [8.4]

5. The equation $x^4 - x^2 - 12 = 0$ is quadratic in x^2. Letting $y = x^2$ we have
$$y^2 - y - 12 = 0$$
$$(y - 4)(y + 3) = 0$$
$$y = 4 \quad \text{or} \quad y = -3$$
Resubstituting x^2 for y, we have
$$x^2 = 4 \quad \text{or} \quad x^2 = -3$$
$$x = \pm 2 \quad \text{or} \quad x = \pm i\sqrt{3}$$

There are a variety of equations whose form is quadratic. We solve most of them by making a substitution so the equation becomes quadratic, and then solving the equation by factoring or the quadratic formula. For example,

The equation	*is quadratic in*
$(2x - 3)^2 + 5(2x - 3) - 6 = 0$	$2x - 3$
$4x^4 - 7x^2 - 2 = 0$	x^2
$2x - 7\sqrt{x} + 3 = 0$	$\sqrt{x}$

Graphing Parabolas [8.5]

6. The graph of $y = x^2 - 4$ will be a parabola. It will cross the x-axis at 2 and -2, and the vertex will be $(0, -4)$.

The graph of any equation of the form
$$y = ax^2 + bx + c \qquad a \neq 0$$
is a *parabola*. The graph is *concave up* if $a > 0$ and *concave down* if $a < 0$. The highest or lowest point on the graph is called the *vertex* and always has an x-coordinate of $x = \frac{-b}{2a}$.

Quadratic Inequalities [8.6]

7. Solve $x^2 - 2x - 8 > 0$. We factor and draw the sign diagram:
$$(x - 4)(x + 2) > 0$$

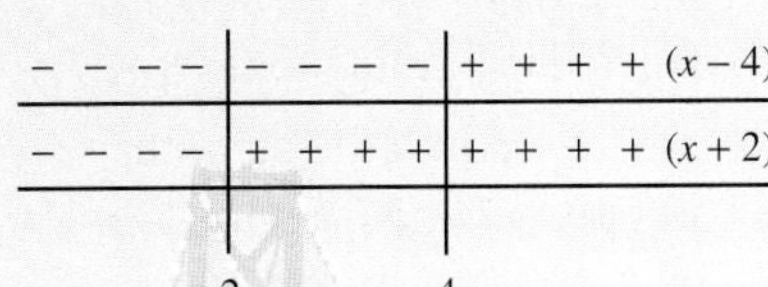

The solution is $x < -2$ or $x > 4$.

We solve quadratic inequalities by manipulating the inequality to get 0 on the right side and then factoring the left side. We then make a diagram that indicates where the factors are positive and where they are negative. From this sign diagram and the original inequality we graph the appropriate solution set.

Chapter 8 Review Test

The problems below form a comprehensive review of the material in this chapter. They can be used to study for exams. If you would like to take a practice test on this chapter, you can use the odd-numbered problems. Give yourself an hour and work as many of the odd-numbered problems as possible. When you are finished, or when an hour has passed, check your answers with the answers in the back of the book. You can use the even-numbered problems for a second practice test.

Solve each equation. [8.1]

1. $(2t-5)^2 = 25$
2. $(3t-2)^2 = 4$
3. $(3y-4)^2 = -49$
4. $(2x+6)^2 = 12$

Solve by completing the square. [8.1]

5. $2x^2 + 6x - 20 = 0$
6. $3x^2 + 15x = -18$
7. $a^2 + 9 = 6a$
8. $a^2 + 4 = 4a$
9. $2y^2 + 6y = -3$
10. $3y^2 + 3 = 9y$

Solve each equation. [8.2]

11. $\frac{1}{6}x^2 + \frac{1}{2}x - \frac{5}{3} = 0$
12. $8x^2 - 18x = 0$
13. $4t^2 - 8t + 19 = 0$
14. $100x^2 - 200x = 100$
15. $0.06a^2 + 0.05a = 0.04$
16. $9 - 6x = -x^2$
17. $(2x+1)(x-5) - (x+3)(x-2) = -17$
18. $2y^3 + 2y = 10y^2$
19. $5x^2 = -2x + 3$
20. $x^3 - 27 = 0$
21. $3 - \frac{2}{x} + \frac{1}{x^2} = 0$
22. $\frac{1}{x-3} + \frac{1}{x+2} = 1$
23. **Profit** The total cost (in dollars) for a company to produce x items per week is $C = 7x + 400$. The revenue for selling all x items is $R = 34x - 0.1x^2$. How many items must it produce and sell each week for its weekly profit to be $1,300? [8.2]
24. **Profit** The total cost (in dollars) for a company to produce x items per week is $C = 70x + 300$. The revenue for selling all x items is $R = 110x - 0.5x^2$. How many items must it produce and sell each week for its weekly profit to be $300? [8.2]

Use the discriminant to find the number and kind of solutions for each equation. [8.3]

25. $2x^2 - 8x = -8$
26. $4x^2 - 8x = -4$
27. $2x^2 + x - 3 = 0$
28. $5x^2 + 11x = 12$
29. $x^2 - x = 1$
30. $x^2 - 5x = -5$
31. $3x^2 + 5x = -4$
32. $4x^2 - 3x = -6$

Determine k so that each equation has exactly one real solution. [8.3]

33. $25x^2 - kx + 4 = 0$
34. $4x^2 + kx + 25 = 0$
35. $kx^2 + 12x + 9 = 0$
36. $kx^2 - 16x + 16 = 0$
37. $9x^2 + 30x + k = 0$
38. $4x^2 + 28x + k = 0$

For each of the following problems, find an equation that has the given solutions. [8.3]

39. $x = 3, x = 5$
40. $x = -2, x = 4$
41. $y = \frac{1}{2}, y = -4$
42. $t = 3, t = -3, t = 5$

Find all solutions. [8.4]

43. $(x-2)^2 - 4(x-2) - 60 = 0$
44. $6(2y+1)^2 - (2y+1) - 2 = 0$
45. $x^4 - x^2 = 12$
46. $x - \sqrt{x} - 2 = 0$
47. $2x - 11\sqrt{x} = -12$
48. $\sqrt{x+5} = \sqrt{x} + 1$
49. $\sqrt{y+21} + \sqrt{y} = 7$
50. $\sqrt{y+9} - \sqrt{y-6} = 3$
51. **Projectile Motion** An object is tossed into the air with an upward velocity of 10 feet per second from the top of a building h feet high. The time it takes for the object to hit the ground below is given by the formula $16t^2 - 10t - h = 0$. Solve this formula for t. [8.4]
52. **Projectile Motion** An object is tossed into the air with an upward velocity of v feet per second from the top of a 10-foot wall. The time it takes for the object to hit the ground below is given by the formula $16t^2 - vt - 10 = 0$. Solve this formula for t. [8.4]

Solve each inequality and graph the solution set. [8.6]

53. $x^2 - x - 2 < 0$
54. $3x^2 - 14x + 8 \le 0$
55. $2x^2 + 5x - 12 \ge 0$
56. $(x+2)(x-3)(x+4) > 0$

Find the x-intercepts, if they exist, and the vertex for each parabola. Then use them to sketch the graph. [8.5]

57. $y = x^2 - 6x + 8$
58. $y = x^2 - 4$

Chapter 8 Projects

Quadratic Functions

GROUP PROJECT Maximum Volume of a Box

Number of People 5

Time Needed 30 minutes

Equipment Graphing calculator and five pieces of graph paper

Background For many people, having a concrete model to work with allows them to visualize situations that they would have difficulty with if they had only a written description to work with. The purpose of this project is to rework a problem we have worked previously but this time with a concrete model.

Procedure You are going to make boxes of varying dimensions from rectangles that are 11 centimeters wide and 17 centimeters long.

1. Cut a rectangle from your graph paper that is 11 squares for the width and 17 squares for the length. Pretend that each small square is 1 centimeter by 1 centimeter. Do this with five pieces of paper.
2. One person tears off one square from each corner of their paper, then folds up the sides to form a box. Write down the length, width, and height of this box. Then calculate its volume.
3. The next person tears a square that is two units on a side from each corner of their paper, then folds up the sides to form a box. Write down the length, width, and height of this box. Then calculate its volume.

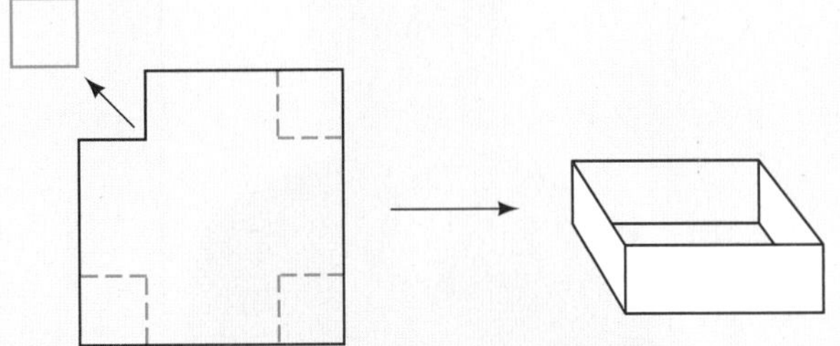

4. The next person follows the same procedure, tearing a still larger square from each corner of their paper. This continues until the squares that are to be torn off are larger than the original piece of paper.
5. Enter the data from each box you have created into the table. Then graph all the points (x, V) from the table.
6. Using what you have learned from filling in the table, write a formula for the volume of the box that is created when a square of side x is cut from each corner of the original piece of paper.
7. Graph the equation you found in Problem 6 on a graphing calculator. Use the result to connect the points you plotted in the graph.
8. Use the graphing calculator to find the value of x that will give the maximum volume of the box. Your answer should be accurate to the nearest hundredth.

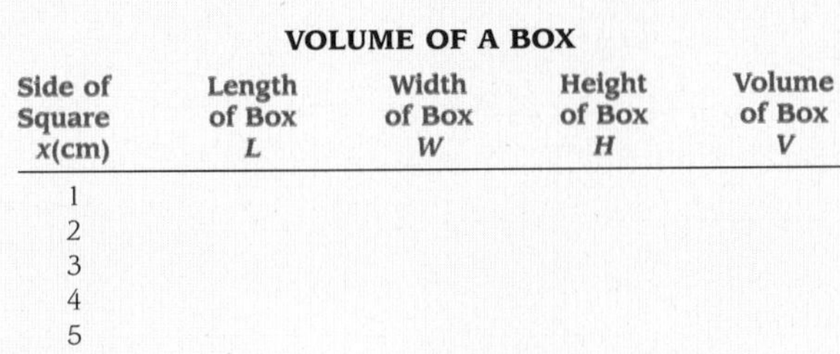

VOLUME OF A BOX

Side of Square x(cm)	Length of Box L	Width of Box W	Height of Box H	Volume of Box V
1				
2				
3				
4				
5				

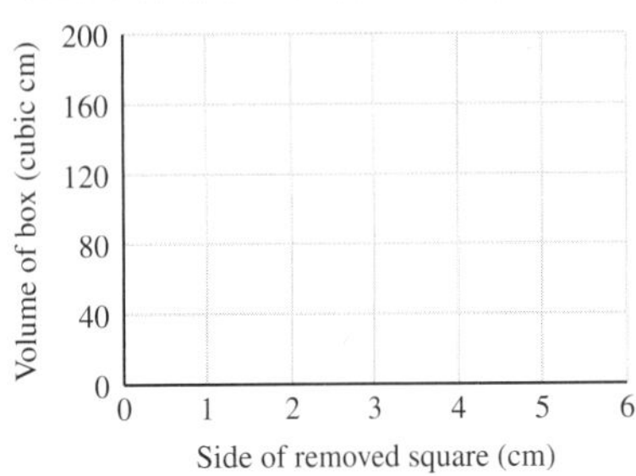

RESEARCH PROJECT

Arlie O. Petters

Arlie O. Petters, Massachusetts Institute of Technology

It can seem at times as if all the mathematicians of note lived 100 or more years ago. However, that is not the case. There are mathematicians doing research today, who are discovering new mathematical ideas and extending what is known about mathematics. One of the current group of research mathematicians is Arlie O. Petters. Dr. Petters earned his Ph.D. from the Massachusetts Institute of Technology in 1991 and currently is working in the mathematics department at Duke University. Use the Internet to find out more about Dr. Petters' education, awards, and research interests. Then use your results to give a profile, in essay form, of a present-day working mathematician.

Exponential and Logarithmic Functions

9

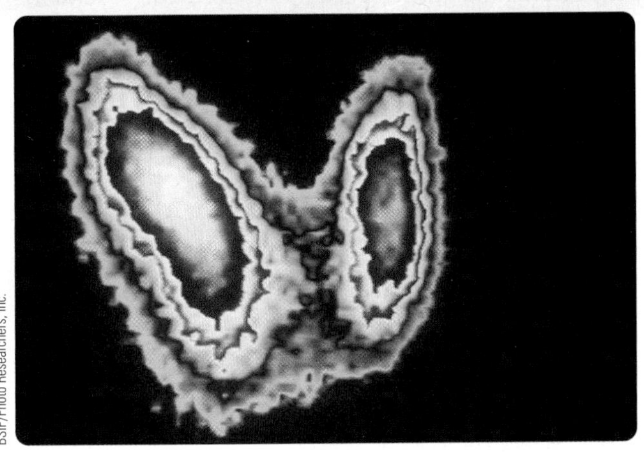

BSIP/Photo Researchers, Inc.

CHAPTER OUTLINE

If you have had any problems with or had testing done on your thyroid gland, then you may have come in contact with radioactive iodine-131. Like all radioactive elements, iodine-131 decays naturally. The half-life of iodine-131 is 8 days, which means that every 8 days a sample of iodine-131 will decrease to half of its original amount. The following table and graph show what happens to a 1,600-microgram sample of iodine-131 over time.

Iodine-131 as a Function of Time

t (days)	A (micrograms)
0	1,600
8	800
16	400
24	200
32	100

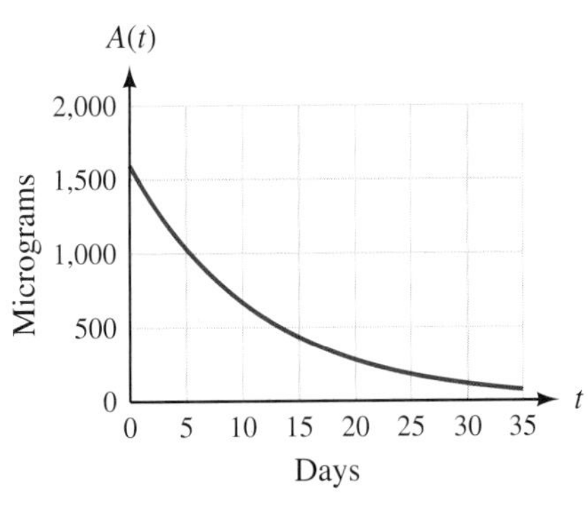

FIGURE 1

The function represented by the information in the table and Figure 1 is

$$A(t) = 1{,}600 \cdot 2^{-t/8}$$

It is one of the types of functions we will study in this chapter.

▶ Improve your grade and save time!
Go online to **www.thomsonedu.com/login** where you can

- Watch videos of instructors working through the in-text examples
- Follow step-by-step online tutorials of in-text examples and review questions
- Work practice problems
- Check your readiness for an exam by taking a pre-test and exploring the modules recommended in your Personalized Study plan
- Receive help from a live tutor online through vMentor™

Try it out! Log in with an access code or purchase access at **www.ichapters.com.**

9.1 Exponential Functions

OBJECTIVES

A Find function values for exponential functions.

B Graph exponential functions.

C Work problems involving exponential growth and decay.

To obtain an intuitive idea of how exponential functions behave, we can consider the heights attained by a bouncing ball. When a ball used in the game of racquetball is dropped from any height, the first bounce will reach a height that is $\frac{2}{3}$ of the original height. The second bounce will reach $\frac{2}{3}$ of the height of the first bounce, and so on, as shown in Figure 1.

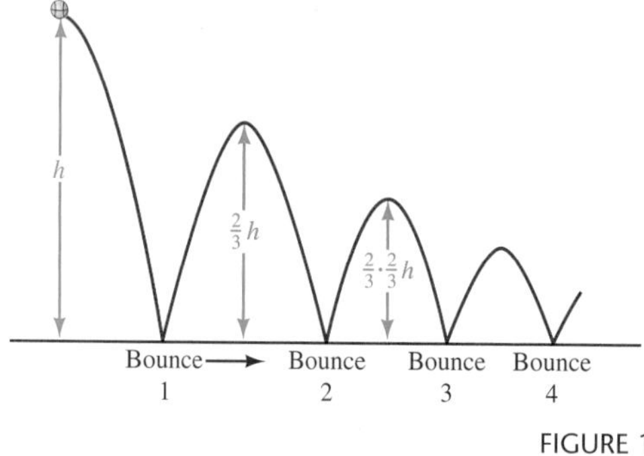

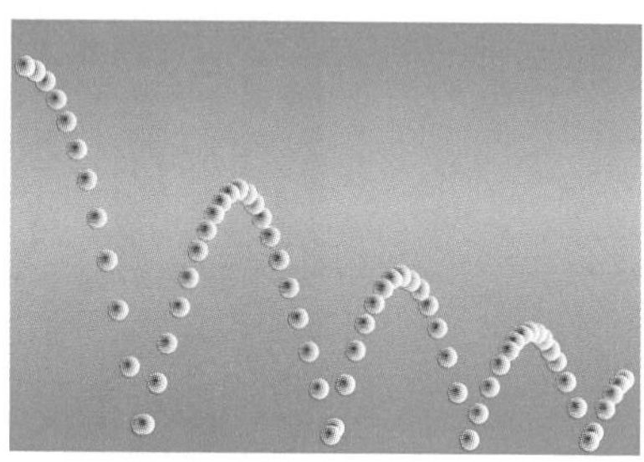

FIGURE 1

If the ball is dropped initially from a height of 1 meter, then during the first bounce it will reach a height of $\frac{2}{3}$ meter. The height of the second bounce will reach $\frac{2}{3}$ of the height reached on the first bounce. The maximum height of any bounce is $\frac{2}{3}$ of the height of the previous bounce.

Initial height: $h = 1$

Bounce 1: $h = \frac{2}{3}(1) = \frac{2}{3}$

Bounce 2: $h = \frac{2}{3}\left(\frac{2}{3}\right) = \left(\frac{2}{3}\right)^2$

Bounce 3: $h = \frac{2}{3}\left(\frac{2}{3}\right)^2 = \left(\frac{2}{3}\right)^3$

Bounce 4: $h = \frac{2}{3}\left(\frac{2}{3}\right)^3 = \left(\frac{2}{3}\right)^4$

⋮

Bounce n: $h = \frac{2}{3}\left(\frac{2}{3}\right)^{n-1} = \left(\frac{2}{3}\right)^n$

This last equation is exponential in form. We classify all exponential functions together with the following definition.

> **DEFINITION** An **exponential function** is any function that can be written in the form
>
> $$f(x) = b^x$$
>
> where b is a positive real number other than 1.

Each of the following is an exponential function:

$$f(x) = 2^x \qquad y = 3^x \qquad f(x) = \left(\frac{1}{4}\right)^x$$

The first step in becoming familiar with exponential functions is to find some values for specific exponential functions.

EXAMPLE 1 If the exponential functions f and g are defined by

$$f(x) = 2^x \quad \text{and} \quad g(x) = 3^x$$

then

$$f(0) = 2^0 = 1 \qquad g(0) = 3^0 = 1$$
$$f(1) = 2^1 = 2 \qquad g(1) = 3^1 = 3$$
$$f(2) = 2^2 = 4 \qquad g(2) = 3^2 = 9$$
$$f(3) = 2^3 = 8 \qquad g(3) = 3^3 = 27$$
$$f(-2) = 2^{-2} = \frac{1}{2^2} = \frac{1}{4} \qquad g(-2) = 3^{-2} = \frac{1}{3^2} = \frac{1}{9}$$
$$f(-3) = 2^{-3} = \frac{1}{2^3} = \frac{1}{8} \qquad g(-3) = 3^{-3} = \frac{1}{3^3} = \frac{1}{27}$$

In the introduction to this chapter we indicated that the half-life of iodine-131 is 8 days, which means that every 8 days a sample of iodine-131 will decrease to half of its original amount. If we start with A_0 micrograms of iodine-131, then after t days the sample will contain

$$A(t) = A_0 \cdot 2^{-t/8}$$

micrograms of iodine-131.

EXAMPLE 2 A patient is administered a 1,200-microgram dose of iodine-131. How much iodine-131 will be in the patient's system after 10 days and after 16 days?

SOLUTION The initial amount of iodine-131 is $A_0 = 1{,}200$, so the function that gives the amount left in the patient's system after t days is

Note
Recall that the symbol $\approx$ is read "is approximately equal to."

$$A(t) = 1{,}200 \cdot 2^{-t/8}$$

After 10 days, the amount left in the patient's system is

$$A(10) = 1{,}200 \cdot 2^{-10/8}$$
$$= 1{,}200 \cdot 2^{-1.25}$$
$$\approx 504.5 \text{ micrograms}$$

After 16 days, the amount left in the patient's system is

$$A(16) = 1{,}200 \cdot 2^{-16/8}$$
$$= 1{,}200 \cdot 2^{-2}$$
$$= 300 \text{ micrograms}$$

We will now turn our attention to the graphs of exponential functions. Since the notation y is easier to use when graphing, and $y = f(x)$, for convenience we will write the exponential functions as

$$y = b^x$$

EXAMPLE 3

Sketch the graph of the exponential function $y = 2^x$.

SOLUTION Using the results of Example 1, we have the following table. Graphing the ordered pairs given in the table and connecting them with a smooth curve, we have the graph of $y = 2^x$ shown in Figure 2.

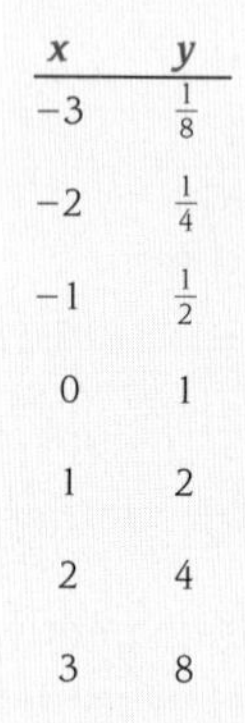

x	y
-3	$\frac{1}{8}$
-2	$\frac{1}{4}$
-1	$\frac{1}{2}$
0	1
1	2
2	4
3	8

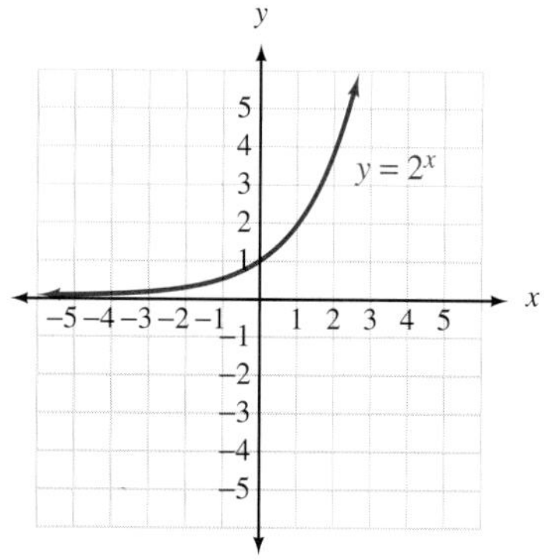

FIGURE 2

Notice that the graph does not cross the x-axis. It *approaches* the x-axis—in fact, we can get it as close to the x-axis as we want without it actually intersecting the x-axis. For the graph of $y = 2^x$ to intersect the x-axis, we would have to find a value of x that would make $2^x = 0$. Because no such value of x exists, the graph of $y = 2^x$ cannot intersect the x-axis.

EXAMPLE 4

Sketch the graph of $y = \left(\frac{1}{3}\right)^x$.

SOLUTION The table shown here gives some ordered pairs that satisfy the equation. Using the ordered pairs from the table, we have the graph shown in Figure 3.

x	y
-3	27
-2	9
-1	3
0	1
1	$\frac{1}{3}$
2	$\frac{1}{9}$
3	$\frac{1}{27}$

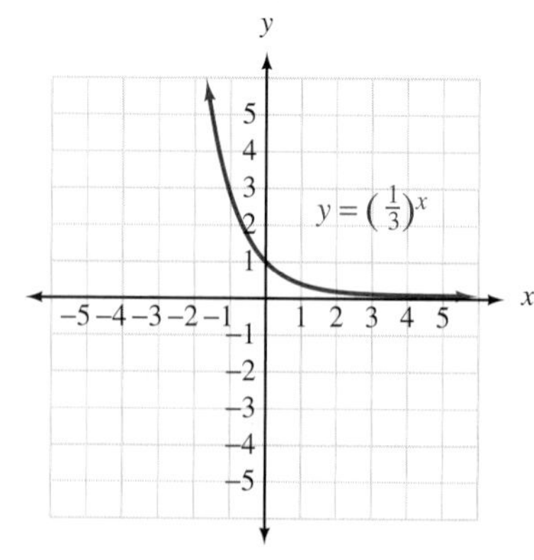

FIGURE 3

The graphs of all exponential functions have two things in common: (1) each crosses the y-axis at (0, 1) since $b^0 = 1$; and (2) none can cross the x-axis since $b^x = 0$ is impossible because of the restrictions on b.

Figures 4 and 5 show some families of exponential curves to help you become more familiar with them on an intuitive level.

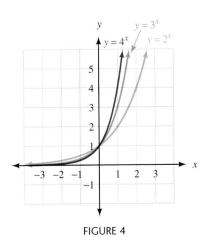

FIGURE 4

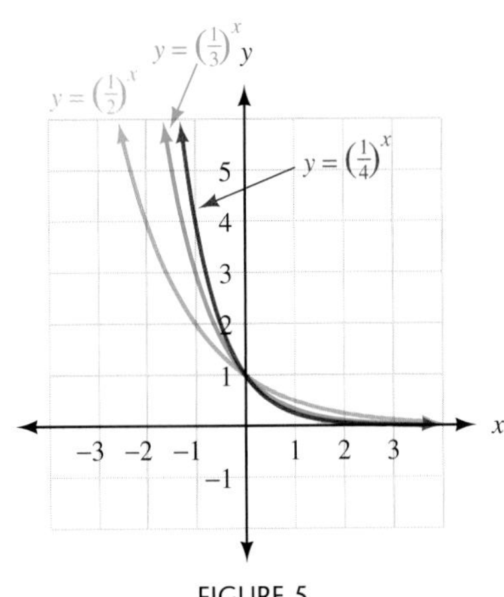

FIGURE 5

Among the many applications of exponential functions are the applications having to do with interest-bearing accounts. Here are the details.

Compound Interest If P dollars are deposited in an account with annual interest rate r, compounded n times per year, then the amount of money in the account after t years is given by the formula

$$A(t) = P\left(1 + \frac{r}{n}\right)^{nt}$$

EXAMPLE 5

Suppose you deposit \$500 in an account with an annual interest rate of 8% compounded quarterly. Find an equation that gives the amount of money in the account after t years. Then find

a. The amount of money in the account after 5 years.

b. The number of years it will take for the account to contain \$1,000.

SOLUTION First we note that $P = 500$ and $r = 0.08$. Interest that is compounded quarterly is compounded 4 times a year, giving us $n = 4$. Substituting these numbers into the preceding formula, we have our function

$$A(t) = 500\left(1 + \frac{0.08}{4}\right)^{4t} = 500(1.02)^{4t}$$

a. To find the amount after 5 years, we let $t = 5$:

$$\begin{aligned} A(5) &= 500(1.02)^{4 \cdot 5} \\ &= 500(1.02)^{20} \\ &\approx \$742.97 \end{aligned}$$

Our answer is found on a calculator and then rounded to the nearest cent.

b. To see how long it will take for this account to total \$1,000, we graph the equation $Y_1 = 500(1.02)^{4X}$ on a graphing calculator and then look to see where it intersects the line $Y_2 = 1{,}000$. The two graphs are shown in Figure 6.

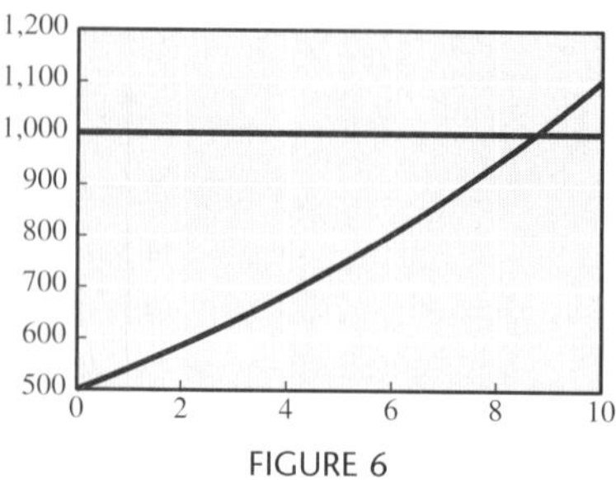

FIGURE 6

Using Zoom and Trace, or the Intersect function on the graphing calculator, we find that the two curves intersect at X ≈ 8.75 and Y = 1,000. This means that our account will contain $1,000 after the money has been on deposit for 8.75 years.

The Natural Exponential Function

A very commonly occurring exponential function is based on a special number we denote with the letter e. The number e is a number like π. It is irrational and occurs in many formulas that describe the world around us. Like π, it can be approximated with a decimal number. Whereas π is approximately 3.1416, e is approximately 2.7183. (If you have a calculator with a key labeled $\boxed{e^x}$, you can use it to find e^1 to find a more accurate approximation to e.) We cannot give a more precise definition of the number e without using some of the topics taught in calculus. For the work we are going to do with the number e, we only need to know that it is an irrational number that is approximately 2.7183.

Here are a table and graph for the natural exponential function.

$$y = f(x) = e^x$$

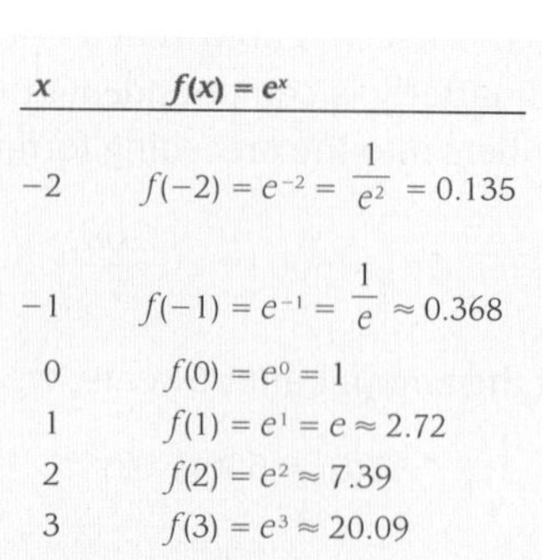

x	$f(x) = e^x$
−2	$f(-2) = e^{-2} = \frac{1}{e^2} = 0.135$
−1	$f(-1) = e^{-1} = \frac{1}{e} \approx 0.368$
0	$f(0) = e^0 = 1$
1	$f(1) = e^1 = e \approx 2.72$
2	$f(2) = e^2 \approx 7.39$
3	$f(3) = e^3 \approx 20.09$

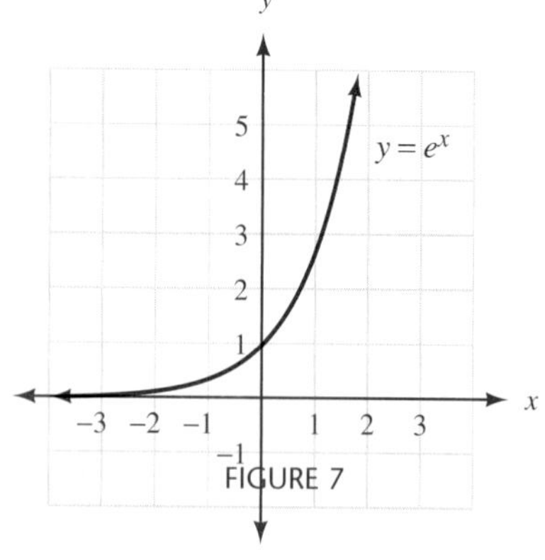

FIGURE 7

One common application of natural exponential functions is with interest-bearing accounts. In Example 5 we worked with the formula

$$A = P\left(1 + \frac{r}{n}\right)^{nt}$$

which gives the amount of money in an account if P dollars are deposited for t years at annual interest rate r, compounded n times per year. In Example 5 the number of compounding periods was 4. What would happen if we let the number of compounding periods become larger and larger, so that we compounded the interest every day, then every hour, then every second, and so on? If we take this as far as it can go, we end up compounding the interest every moment. When this happens, we have an account with interest that is compounded continuously, and the amount of money in such an account depends on the number e.

Here are the details.

Continuously Compounded Interest If P dollars are deposited in an account with annual interest rate r, compounded continuously, then the amount of money in the account after t years is given by the formula

$$A(t) = Pe^{rt}$$

EXAMPLE 6

Suppose you deposit \$500 in an account with an annual interest rate of 8% compounded continously. Find an equation that gives the amount of money in the account after t years. Then find the amount of money in the account after 5 years.

SOLUTION Since the interest is compounded continuously, we use the formula $A(t) = Pe^{rt}$. Substituting $P = 500$ and $r = 0.08$ into this formula we have

$$A(t) = 500e^{0.08t}$$

After 5 years, this account will contain

$$\begin{aligned} A(5) &= 500e^{0.08 \cdot 5} \\ &= 500e^{0.4} \\ &\approx \$745.91 \end{aligned}$$

to the nearest cent. Compare this result with the answer to Example 5a. 

LINKING OBJECTIVES AND EXAMPLES

Next to each objective we have listed the examples that are best described by that objective.

A	1, 2
B	3–5
C	5, 6

GETTING READY FOR CLASS

After reading through the preceding section, respond in your own words and in complete sentences.

1. What is an exponential function?
2. In an exponential function, explain why the base b cannot equal 1. (What kind of function would you get if the base was equal to 1?)
3. Explain continuously compounded interest.
4. What characteristics do the graphs of $y = 2^x$ and $y = (\frac{1}{2})^x$ have in common?

Problem Set 9.1

Online support materials can be found at www.thomsonedu.com/login

Let $f(x) = 3^x$ and $g(x) = \left(\frac{1}{2}\right)^x$, and evaluate each of the following.

▸ 1. $g(0)$ 2. $f(0)$

3. $g(-1)$ 4. $g(-4)$

▸ 5. $f(-3)$ 6. $f(-1)$

▸ 7. $f(2) + g(-2)$ 8. $f(2) - g(-2)$

Graph each of the following functions.

9. $y = 4^x$ 10. $y = 2^{-x}$

11. $y = 3^{-x}$ 12. $y = \left(\frac{1}{3}\right)^{-x}$

13. $y = 2^{x+1}$ 14. $y = 2^{x-3}$

15. $y = e^x$ 16. $y = e^{-x}$

Graph each of the following functions on the same coordinate system for positive values of x only.

17. $y = 2x, y = x^2, y = 2^x$ 18. $y = 3x, y = x^3, y = 3^x$

19. On a graphing calculator, graph the family of curves $y = b^x$, $b = 2, 4, 6, 8$.

20. On a graphing calculator, graph the family of curves $y = b^x$, $b = \frac{1}{2}, \frac{1}{4}, \frac{1}{6}, \frac{1}{8}$.

Applying the Concepts

21. **Bouncing Ball** Suppose the ball mentioned in the introduction to this section is dropped from a height of 6 feet above the ground. Find an exponential equation that gives the height h the ball will attain during the nth bounce. How high will it bounce on the fifth bounce?

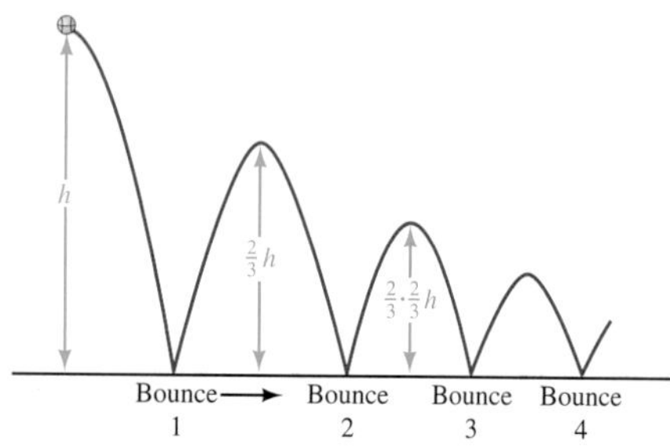

22. **Bouncing Ball** A golf ball is manufactured so that if it is dropped from A feet above the ground onto a hard surface, the maximum height of each bounce will be $\frac{1}{2}$ of the height of the previous bounce. Find an exponential equation that gives the height h the ball will attain during the nth bounce. If the ball is dropped from 10 feet above the ground onto a hard surface, how high will it bounce on the 8th bounce?

23. **Exponential Decay** The half-life of iodine-131 is 8 days. If a patient is administered a 1,400-microgram dose of iodine-131, how much iodine-131 will be in the patient's system after 8 days and after 11 days? (See Example 2.)

24. **Exponential Growth** Automobiles built before 1993 use Freon in their air conditioners. The federal government now prohibits the manufacture of Freon. Because the supply of Freon is decreasing, the price per pound is increasing exponentially. Current estimates put the formula for the price per pound of Freon at $p(t) = 1.89(1.25)^t$, where t is the number of years since 1990. Find the price of Freon in 1995 and 1990. How much will Freon cost in the year 2005?

25. **Compound Interest** Suppose you deposit \$1,200 in an account with an annual interest rate of 6% compounded quarterly. (See Example 5.)
   a. Find an equation that gives the amount of money in the account after t years.
   b. Find the amount of money in the account after 8 years.
   c. How many years will it take for the account to contain \$2,400?
   d. If the interest were compounded continuously, how much money would the account contain after 8 years?

26. **Compound Interest** Suppose you deposit \$500 in an account with an annual interest rate of 8% compounded monthly.
   a. Find an equation that gives the amount of money in the account after t years.
   b. Find the amount of money in the account after 5 years.
   c. How many years will it take for the account to contain \$1,000?
   d. If the interest were compounded continuously, how much money would the account contain after 5 years?

27. **Health Care** In 1990, \$699 billion were spent on health care expenditures. The amount of money, E, in billions spent on health care expenditures can be esti-

■ = Videos available by instructor request

▸ = Online student support materials available at www.thomsonedu.com/login

mated using the function $E(t) = 78.16(1.11)^t$, where t is time in years since 1970 (U.S. Census Bureau).

a. How close was the estimate determined by the function in estimating the actual amount of money spent on health care expenditures in 1990?

b. What are the expected health care expenditures in 2008, 2009, and 2010? Round to the nearest billion.

28. **Exponential Growth** The cost of a can of Coca-Cola on January 1, 1960, was 10 cents. The function below gives the cost of a can of Coca-Cola t years after that.

$$C(t) = 0.10e^{0.0576t}$$

a. Use the function to fill in the table below. (Round to the nearest cent.)

Years Since 1960 t	Cost $C(t)$
0	$0.10
15	
40	
50	
90	

"Coca-Cola" is a registered trademark of the Coca-Cola Company. Used with the express permission of the Coca-Cola Company.

b. Use the table to find the cost of a can of Coca-Cola at the beginning of the year 2000.

c. In what year will a can of Coca-Cola cost $17.84?

29. **Value of a Painting** A painting is purchased as an investment for $150. If the painting's value doubles every 3 years, then its value is given by the function

$$V(t) = 150 \cdot 2^{t/3} \quad \text{for } t \geq 0$$

where t is the number of years since it was purchased, and $V(t)$ is its value (in dollars) at that time. Graph this function.

30. **Value of a Painting** A painting is purchased as an investment for $125. If the painting's value doubles every 5 years, then its value is given by the function

$$V(t) = 125 \cdot 2^{t/5} \quad \text{for} \quad t \geq 0$$

where t is the number of years since it was purchased, and $V(t)$ is its value (in dollars) at that time. Graph this function.

Christian Pierre/SuperStock

31. **Value of a Painting** When will the painting mentioned in Problem 29 be worth $600?

32. **Value of a Painting** When will the painting mentioned in Problem 30 be worth $250?

33. **Value of a Crane** The function

$$V(t) = 450{,}000(1 - 0.30)^t$$

where V is value and t is time in years, can be used to find the value of a crane for the first 6 years of use.

Sandy Felsenthal/Corbis

a. What is the value of the crane after 3 years and 6 months?

b. State the domain of this function.

c. Sketch the graph of this function.

d. State the range of this function.

e. After how many years will the crane be worth only $85,000?

34. **Value of a Printing Press** The function $V(t) = 375{,}000(1 - 0.25)^t$, where V is value and t is time in years, can be used to find the value of a printing press during the first 7 years of use.

Lester Lefkowitz/Taxi/Getty Images

a. What is the value of the printing press after 4 years and 9 months?

b. State the domain of this function.

c. Sketch the graph of this function.

d. State the range of this function.

e. After how many years will the printing press be worth only $65,000?

Maintaining Your Skills

For each of the following relations, specify the domain and range, then indicate which are also functions.

35. $\{(1, 2), (3, 4), (4, 1)\}$

36. $\{(-2, 6), (-2, 8), (2, 3)\}$

State the domain for each of the following functions.

37. $y = \sqrt{3x + 1}$

38. $y = \dfrac{-4}{x^2 + 2x - 35}$

If $f(x) = 2x^2 - 18$ and $g(x) = 2x - 6$, find

39. $f(0)$

40. $g[f(0)]$

41. $\dfrac{g(x + h) - g(x)}{h}$

42. $\dfrac{g}{f}(x)$

Getting Ready for the Next Section

Solve each equation for y.

43. $x = 2y - 3$

44. $x = \dfrac{y + 7}{5}$

45. $x = y^2 - 2$

46. $x = (y + 4)^3$

47. $x = \dfrac{y - 4}{y - 2}$

48. $x = \dfrac{y + 5}{y - 3}$

49. $x = \sqrt{y - 3}$

50. $x = \sqrt{y} + 5$

Extending the Concepts

51. Drag Racing Previously we mentioned the dragster equipped with a computer. The table gives the speed of the dragster every second during one race at the 1993 Winternationals. Figure 8 is a line graph constructed from the data in the table.

Speed of a Dragster

Elapsed Time (sec)	Speed (mph)
0	0.0
1	72.7
2	129.9
3	162.8
4	192.2
5	212.4
6	228.1

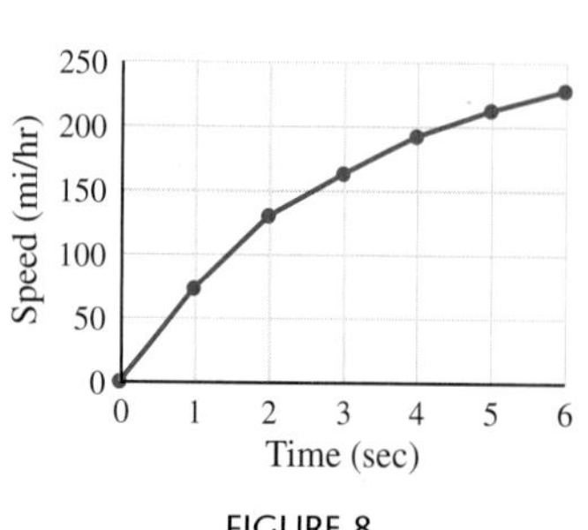

FIGURE 8

The graph of the following function contains the first point and the last point shown in Figure 8; that is, both (0, 0) and (6, 228.1) satisfy the function. Graph the function to see how close it comes to the other points in Figure 8.

$$s(t) = 250(1 - 1.5^{-t})$$

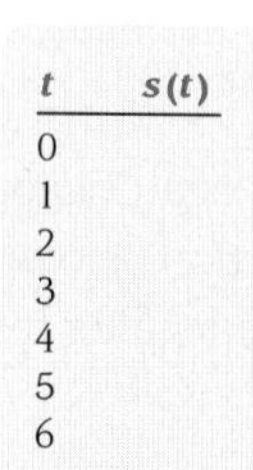

t	$s(t)$
0	
1	
2	
3	
4	
5	
6	

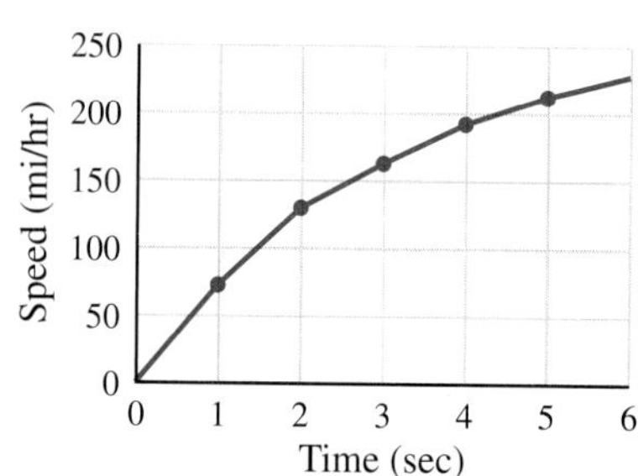

52. The graphs of two exponential functions are given in Figures 9 and 10. Use the graph to find the following:

a. $f(0)$

b. $f(-1)$

c. $f(1)$

d. $g(0)$

e. $g(1)$

f. $g(-1)$

g. $f[g(0)]$

h. $g[f(0)]$

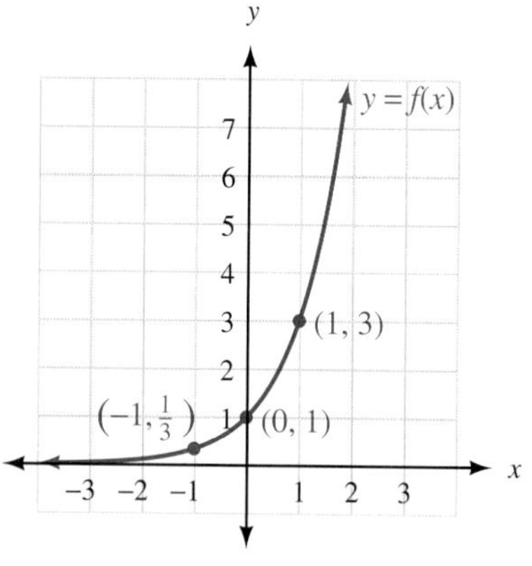

FIGURE 9

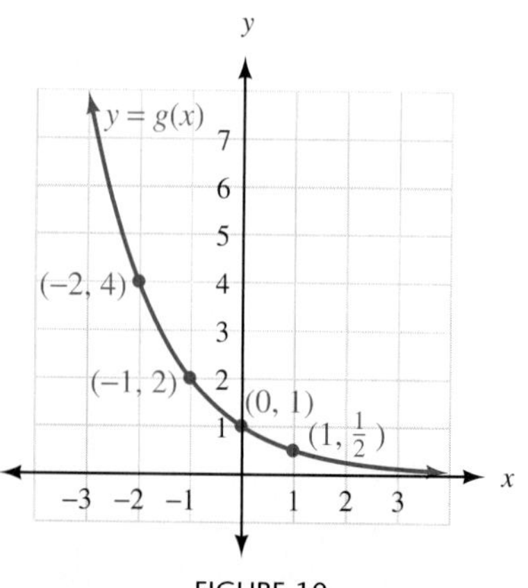

FIGURE 10

9.2 The Inverse of a Function

OBJECTIVES

A Find the equation of the inverse of a function.

B Sketch a graph of a function and its inverse.

The following diagram (Figure 1) shows the route Justin takes to school. He leaves his home and drives 3 miles east and then turns left and drives 2 miles north. When he leaves school to drive home, he drives the same two segments but in the reverse order and the opposite direction; that is, he drives 2 miles south, turns right, and drives 3 miles west. When he arrives home from school, he is right where he started. His route home "undoes" his route to school, leaving him where he began.

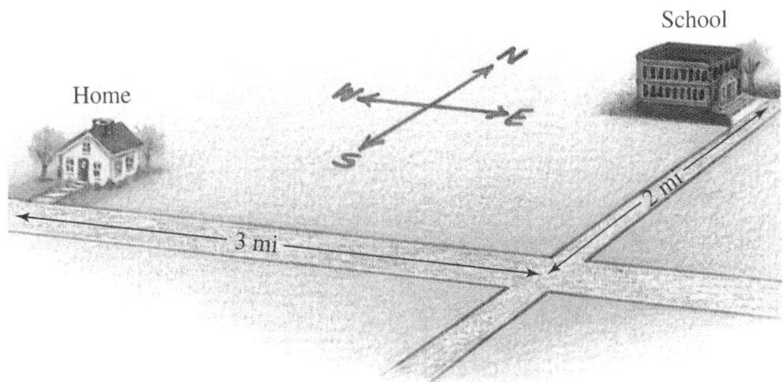

FIGURE 1

As you will see, the relationship between a function and its inverse function is similar to the relationship between Justin's route from home to school and his route from school to home.

Suppose the function f is given by

$$f = \{(1, 4), (2, 5), (3, 6), (4, 7)\}$$

The inverse of f is obtained by reversing the order of the coordinates in each ordered pair in f. The inverse of f is the relation given by

$$g = \{(4, 1), (5, 2), (6, 3), (7, 4)\}$$

It is obvious that the domain of f is now the range of g, and the range of f is now the domain of g. Every function (or relation) has an inverse that is obtained from the original function by interchanging the components of each ordered pair.

Suppose a function f is defined with an equation instead of a list of ordered pairs. We can obtain the equation of the inverse of f by interchanging the role of x and y in the equation for f.

EXAMPLE 1 If the function f is defined by $f(x) = 2x - 3$, find the equation that represents the inverse of f.

SOLUTION Since the inverse of f is obtained by interchanging the components of all the ordered pairs belonging to f, and each ordered pair in f satisfies the equation $y = 2x - 3$, we simply exchange x and y in the equation $y = 2x - 3$ to get the formula for the inverse of f:

$$x = 2y - 3$$

We now solve this equation for y in terms of x:

$$x + 3 = 2y$$

$$\frac{x + 3}{2} = y$$

$$y = \frac{x + 3}{2}$$

The last line gives the equation that defines the inverse of f. Let's compare the graphs of f and its inverse as given above. (See Figure 2.)

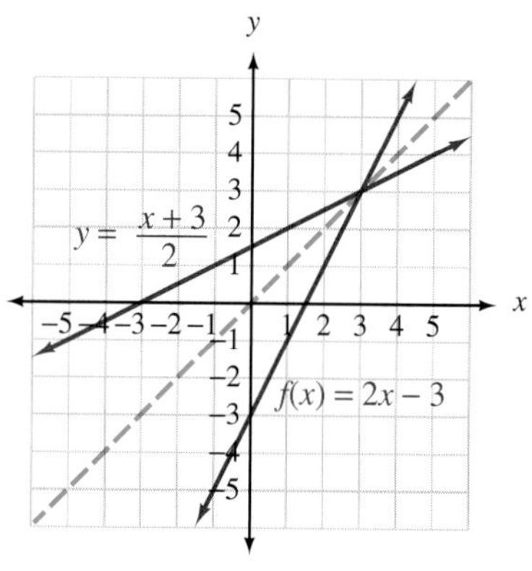

FIGURE 2

The graphs of f and its inverse have symmetry about the line $y = x$. This is a reasonable result since the one function was obtained from the other by interchanging x and y in the equation. The ordered pairs (a, b) and (b, a) always have symmetry about the line $y = x$.

EXAMPLE 2 Graph the function $y = x^2 - 2$ and its inverse. Give the equation for the inverse.

SOLUTION We can obtain the graph of the inverse of $y = x^2 - 2$ by graphing $y = x^2 - 2$ by the usual methods and then reflecting the graph about the line $y = x$.

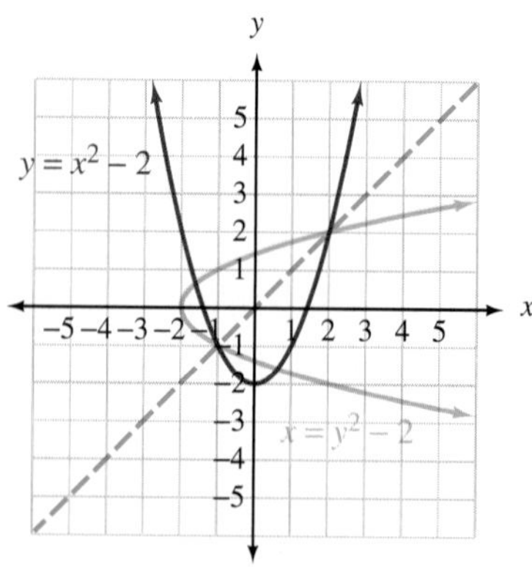

FIGURE 3

The equation that corresponds to the inverse of $y = x^2 - 2$ is obtained by interchanging x and y to get $x = y^2 - 2$.

We can solve the equation $x = y^2 - 2$ for y in terms of x as follows:

$$x = y^2 - 2$$

$$x + 2 = y^2$$

$$y = \pm\sqrt{x + 2}$$

USING TECHNOLOGY

Graphing an Inverse

One way to graph a function and its inverse is to use parametric equations. To graph the function $y = x^2 - 2$ and its inverse from Example 2, first set your graphing calculator to parametric mode. Then define the following set of parametric equations (Figure 4).

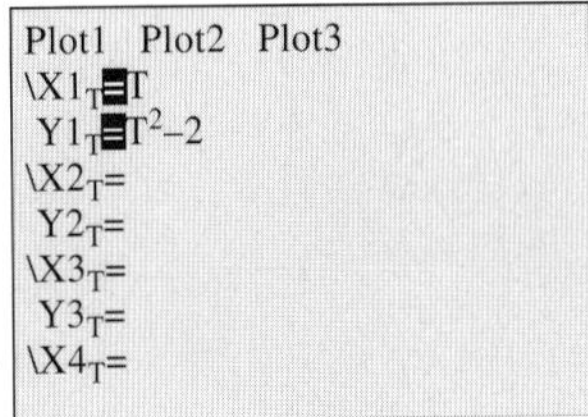

FIGURE 4

$$X_1 = t;\ Y_1 = t^2 - 2$$

Set the window variables so that

$$-3 \le t \le 3,\ \text{step} = 0.05;\ -4 \le x \le 4;\ -4 \le y \le 4$$

Graph the function using the zoom-square command. Your graph should look similar to Figure 5.

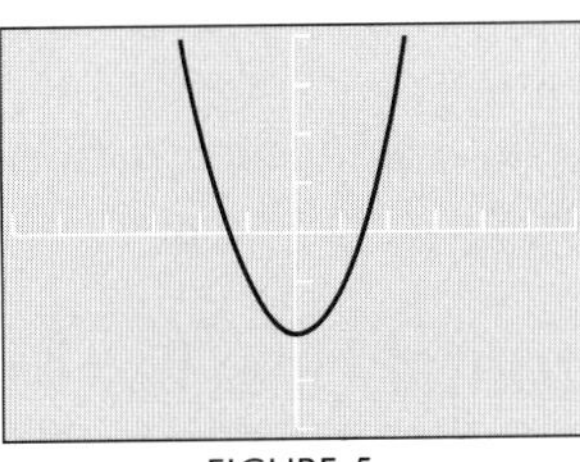

FIGURE 5

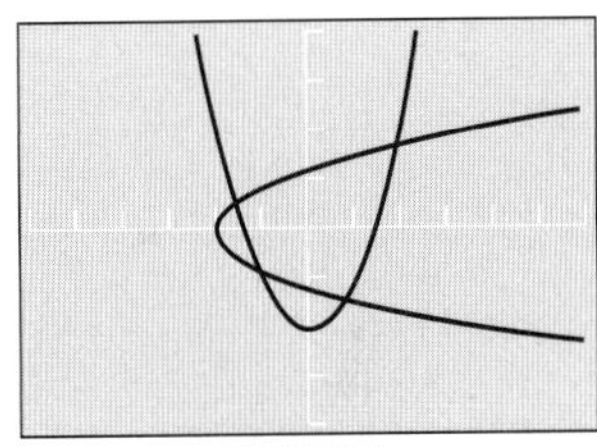

FIGURE 6

To graph the inverse, we need to interchange the roles of x and y for the original function. This is easily done by defining a new set of parametric equations that is just the reverse of the pair given above:

$$X_2 = t^2 - 2,\ Y_2 = t$$

Press GRAPH again, and you should now see the graphs of the original function and its inverse (Figure 6). If you trace to any point on either graph, you can alternate between the two curves to see how the coordinates of the corresponding ordered pairs compare. As Figure 7 illustrates, the coordinates of a point on one graph are reversed for the other graph.

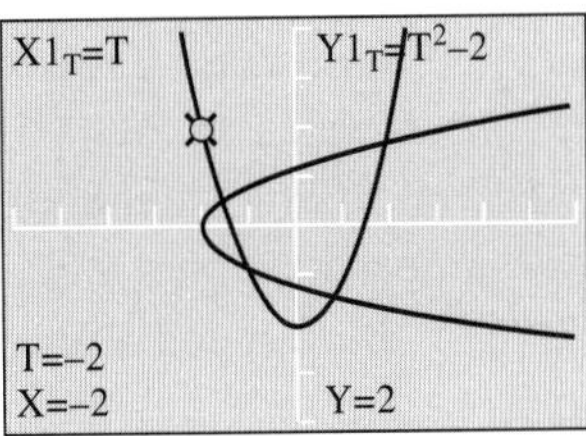

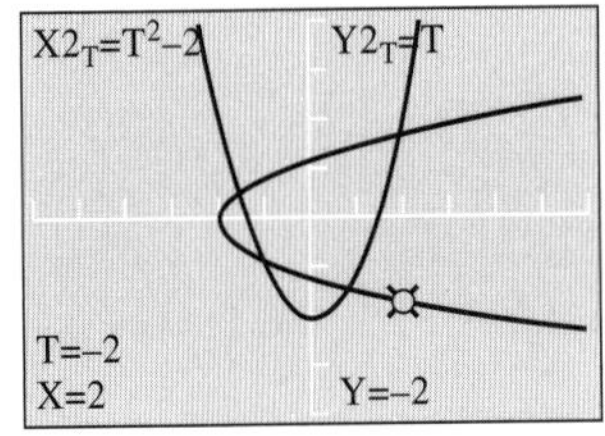

FIGURE 7

Comparing the graphs from Examples 1 and 2, we observe that the inverse of a function is not always a function. In Example 1, both f and its inverse have graphs that are nonvertical straight lines and therefore both represent functions. In Example 2, the inverse of function f is not a function since a vertical line crosses it in more than one place.

One-to-One Functions

We can distinguish between those functions with inverses that are also functions and those functions with inverses that are not functions with the following definition.

DEFINITION A function is a **one-to-one function** if every element in the range comes from exactly one element in the domain.

This definition indicates that a one-to-one function will yield a set of ordered pairs in which no two different ordered pairs have the same second coordinates. For example, the function

$$f = \{(2, 3), (-1, 3), (5, 8)\}$$

is not one-to-one because the element 3 in the range comes from both 2 and -1 in the domain. On the other hand, the function

$$g = \{(5, 7), (3, -1), (4, 2)\}$$

is a one-to-one function because every element in the range comes from only one element in the domain.

Horizontal Line Test

If we have the graph of a function, we can determine if the function is one-to-one with the following test. If a horizontal line crosses the graph of a function in more than one place, then the function is not a one-to-one function because the points at which the horizontal line crosses the graph will be points with the same y-coordinates but different x-coordinates. Therefore, the function will have an element in the range (the y-coordinate) that comes from more than one element in the domain (the x-coordinates).

Of the functions we have covered previously, all the linear functions and exponential functions are one-to-one functions because no horizontal lines can be found that will cross their graphs in more than one place.

Functions Whose Inverses Are Also Functions

Because one-to-one functions do not repeat second coordinates, when we reverse the order of the ordered pairs in a one-to-one function, we obtain a relation in which no two ordered pairs have the same first coordinate—by definition, this relation must be a function. In other words, every one-to-one function has an inverse that is itself a function. Because of this, we can use function notation to represent that inverse.

Inverse Function Notation If $y = f(x)$ is a one-to-one function, then the inverse of f is also a function and can be denoted by $y = f^{-1}(x)$.

Note
The notation f^{-1} does not represent the reciprocal of f; that is, the -1 in this notation is not an exponent. The notation f^{-1} is defined as representing the inverse function for a one-to-one function.

To illustrate, in Example 1 we found the inverse of $f(x) = 2x - 3$ was the function $y = \dfrac{x+3}{2}$. We can write this inverse function with inverse function notation as

$$f^{-1}(x) = \frac{x+3}{2}$$

However, the inverse of the function in Example 2 is not itself a function, so we do not use the notation $f^{-1}(x)$ to represent it.

EXAMPLE 3 Find the inverse of $g(x) = \dfrac{x-4}{x-2}$.

SOLUTION To find the inverse for g, we begin by replacing $g(x)$ with y to obtain

$$y = \frac{x-4}{x-2} \qquad \textbf{The original function}$$

To find an equation for the inverse, we exchange x and y.

$$x = \frac{y-4}{y-2} \qquad \textbf{The inverse of the original function}$$

To solve for y, we first multiply each side by $y - 2$ to obtain

$$x(y-2) = y - 4$$

$$xy - 2x = y - 4 \qquad \textbf{Distributive property}$$

$$xy - y = 2x - 4 \qquad \textbf{Collect all terms containing } y \textbf{ on the left side}$$

$$y(x-1) = 2x - 4 \qquad \textbf{Factor } y \textbf{ from each term on the left side}$$

$$y = \frac{2x-4}{x-1} \qquad \textbf{Divide each side by } x - 1$$

Because our original function is one-to-one, as verified by the graph in Figure 8, its inverse is also a function. Therefore, we can use inverse function notation to write

$$g^{-1}(x) = \frac{2x-4}{x-1}$$

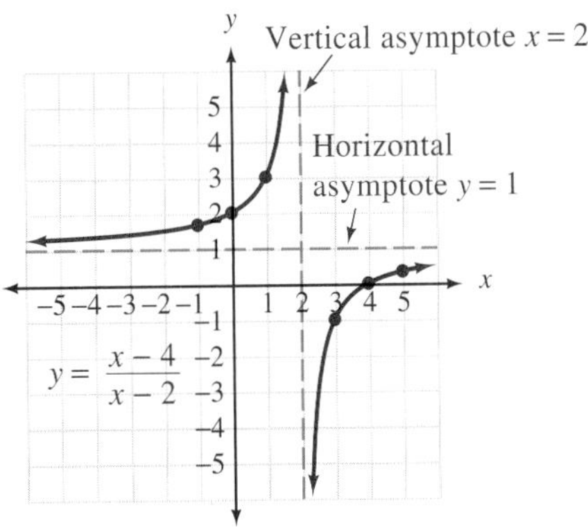

FIGURE 8

EXAMPLE 4 Graph the function $y = 2^x$ and its inverse $x = 2^y$.

SOLUTION We graphed $y = 2^x$ in the preceding section. We simply reflect its graph about the line $y = x$ to obtain the graph of its inverse $x = 2^y$. (See Figure 9.)

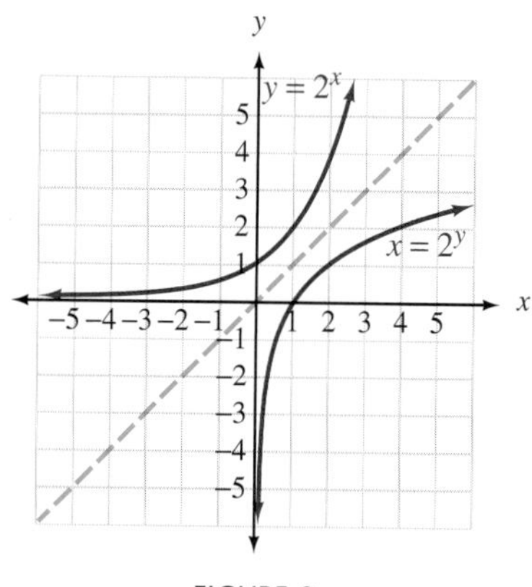

FIGURE 9

As you can see from the graph, $x = 2^y$ is a function. We do not have the mathematical tools to solve this equation for y, however. Therefore, we are unable to use the inverse function notation to represent this function. In the next section, we will give a definition that solves this problem. For now, we simply leave the equation as $x = 2^y$.

Functions, Relations, and Inverses—A Summary

Here is a summary of some of the things we know about functions, relations, and their inverses:

1. Every function is a relation, but not every relation is a function.
2. Every function has an inverse, but only one-to-one functions have inverses that are also functions.
3. The domain of a function is the range of its inverse, and the range of a function is the domain of its inverse.
4. If $y = f(x)$ is a one-to-one function, then we can use the notation $y = f^{-1}(x)$ to represent its inverse function.
5. The graph of a function and its inverse have symmetry about the line $y = x$.
6. If (a, b) belongs to the function f, then the point (b, a) belongs to its inverse.

LINKING OBJECTIVES AND EXAMPLES

Next to each **objective** we have listed the examples that are best described by that objective.

A	1–3
B	1–4

GETTING READY FOR CLASS

After reading through the preceding section, respond in your own words and in complete sentences.

1. What is the inverse of a function?
2. What is the relationship between the graph of a function and the graph of its inverse?
3. Explain why only one-to-one functions have inverses that are also functions.
4. Describe the vertical line test, and explain the difference between the vertical line test and the horizontal line test.

Problem Set 9.2

Online support materials can be found at www.thomsonedu.com/login

For each of the following one-to-one functions, find the equation of the inverse. Write the inverse using the notation $f^{-1}(x)$.

1. $f(x) = 3x - 1$
2. $f(x) = 2x - 5$
3. $f(x) = x^3$
4. $f(x) = x^3 - 2$
5. $f(x) = \dfrac{x-3}{x-1}$
6. $f(x) = \dfrac{x-2}{x-3}$
7. $f(x) = \dfrac{x-3}{4}$
8. $f(x) = \dfrac{x+7}{2}$
9. $f(x) = \dfrac{1}{2}x - 3$
10. $f(x) = \dfrac{1}{3}x + 1$
11. $f(x) = \dfrac{2x+1}{3x+1}$
12. $f(x) = \dfrac{3x+2}{5x+1}$

For each of the following relations, sketch the graph of the relation and its inverse, and write an equation for the inverse.

13. $y = 2x - 1$
14. $y = 3x + 1$
15. $y = x^2 - 3$
16. $y = x^2 + 1$
17. $y = x^2 - 2x - 3$
18. $y = x^2 + 2x - 3$
19. $y = 3^x$
20. $y = \left(\dfrac{1}{2}\right)^x$
21. $y = 4$
22. $y = -2$
23. $y = \dfrac{1}{2}x^3$
24. $y = x^3 - 2$
25. $y = \dfrac{1}{2}x + 2$
26. $y = \dfrac{1}{3}x - 1$
27. $y = \sqrt{x+2}$
28. $y = \sqrt{x} + 2$
29. Determine if the following functions are one-to-one.

a.

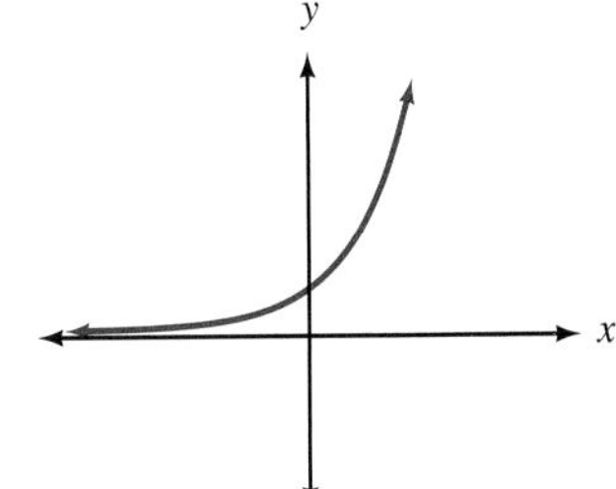

b.

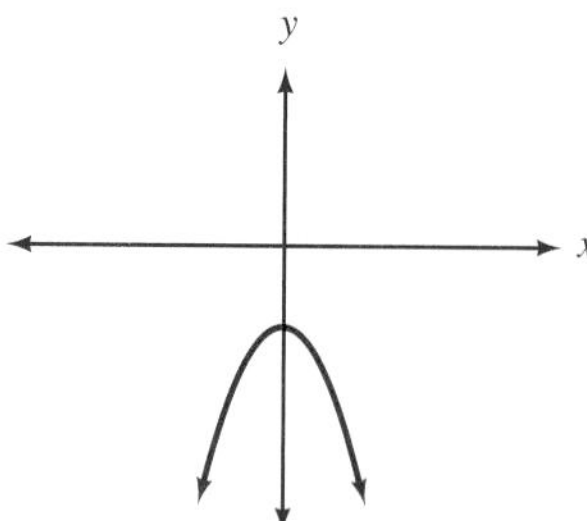

c.

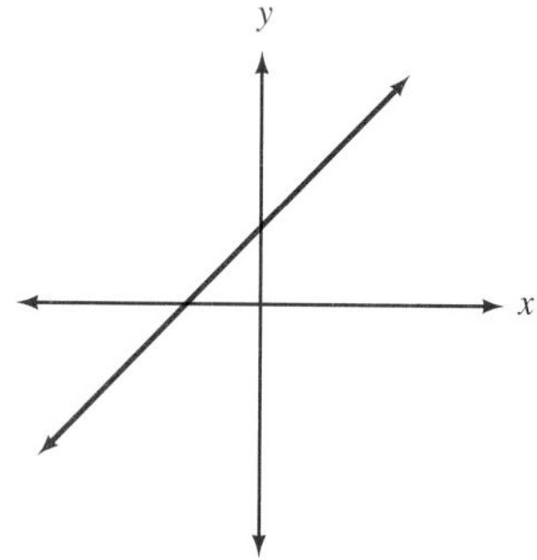

30. Could the following tables of values represent ordered pairs from one-to-one functions? Explain your answer.

a.

x	y
−2	5
−1	4
0	3
1	4
2	5

b.

x	y
1.5	0.1
2.0	0.2
2.5	0.3
3.0	0.4
3.5	0.5

31. If $f(x) = 3x - 2$, then $f^{-1}(x) = \dfrac{x+2}{3}$. Use these two functions to find

a. $f(2)$ **b.** $f^{-1}(2)$
c. $f[f^{-1}(2)]$ **d.** $f^{-1}[f(2)]$

32. If $f(x) = \frac{1}{2}x + 5$, then $f^{-1}(x) = 2x - 10$. Use these two functions to find

a. $f(-4)$ **b.** $f^{-1}(-4)$
c. $f[f^{-1}(-4)]$ **d.** $f^{-1}[f(-4)]$

33. Let $f(x) = \dfrac{1}{x}$, and find $f^{-1}(x)$.

34. Let $f(x) = \dfrac{a}{x}$, and find $f^{-1}(x)$. (a is a real number constant.)

Applying the Concepts

35. Reading Tables Evaluate each of the following functions using the functions defined by Tables 1 and 2.

a. $f[g(-3)]$ **b.** $g[f(-6)]$
c. $g[f(2)]$ **d.** $f[g(3)]$
e. $f[g(-2)]$ **f.** $g[f(3)]$
g. What can you conclude about the relationship between functions f and g?

TABLE 1

x	$f(x)$
−6	3
2	−3
3	−2
6	4

TABLE 2

x	$g(x)$
−3	2
−2	3
3	−6
4	6

36. Reading Tables Use the functions defined in Tables 1 and 2 in Problem 35 to answer the following questions.

a. What are the domain and range of f?

b. What are the domain and range of g?

c. How are the domain and range of f related to the domain and range of g?

d. Is f a one-to-one function?

e. Is g a one-to-one function?

Maintaining Your Skills

Solve each equation.

37. $(2x - 1)^2 = 25$

38. $(3x + 5)^2 = -12$

39. What number would you add to $x^2 - 10x$ to make it a perfect square trinomial?

40. What number would you add to $x^2 - 5x$ to make it a perfect square trinomial?

Solve by completing the square.

41. $x^2 - 10x + 8 = 0$

42. $x^2 - 5x + 4 = 0$

43. $3x^2 - 6x + 6 = 0$

44. $4x^2 - 16x - 8 = 0$

Getting Ready for the Next Section

Simplify.

45. 3^{-2}

46. 2^3

Solve.

47. $2 = 3x$

48. $3 = 5x$

49. $4 = x^3$

50. $12 = x^2$

Fill in the blanks to make each statement true.

51. $8 = 2^{\square}$

52. $27 = 3^{\square}$

53. $10{,}000 = 10^{\square}$

54. $1{,}000 = 10^{\square}$

55. $81 = 3^{\square}$

56. $81 = 9^{\square}$

57. $6 = 6^{\square}$

58. $1 = 5^{\square}$

Extending the Concepts

For each of the following functions, find $f^{-1}(x)$. Then show that $f[f^{-1}(x)] = x$.

59. $f(x) = 3x + 5$

60. $f(x) = 6 - 8x$

61. $f(x) = x^3 + 1$

62. $f(x) = x^3 - 8$

63. $f(x) = \dfrac{x - 4}{x - 2}$

64. $f(x) = \dfrac{x - 3}{x - 1}$

65. The graphs of a function and its inverse are shown in the figure. Use the graphs to find the following:

a. $f(0)$

b. $f(1)$

c. $f(2)$

d. $f^{-1}(1)$

e. $f^{-1}(2)$

f. $f^{-1}(5)$

g. $f^{-1}[f(2)]$

h. $f[f^{-1}(5)]$

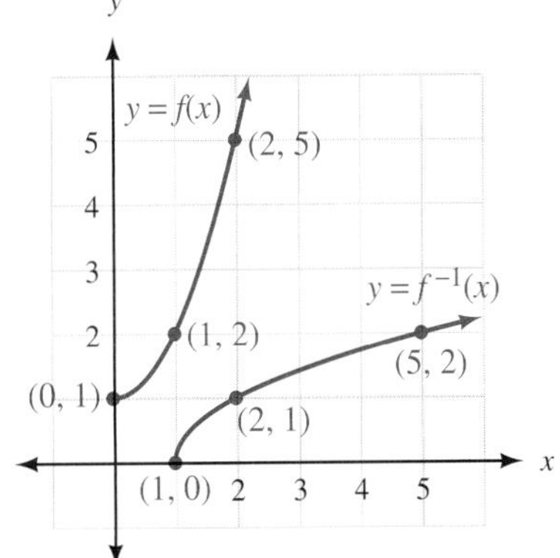

9.3 Logarithms Are Exponents

OBJECTIVES

A Convert between logarithmic form and exponential form.

B Use the definition of logarithms to solve simple logarithmic equations.

C Sketch the graph of a logarithmic function.

D Simplify expressions involving logarithms.

In January 1999, ABC News reported that an earthquake had occurred in Colombia, causing massive destruction. They reported the strength of the quake by indicating that it measured 6.0 on the Richter scale. For comparison, Table 1 gives the Richter magnitude of a number of other earthquakes.

TABLE 1
Earthquakes

Year	Earthquake	Richter Magnitude
1971	Los Angeles	6.6
1985	Mexico City	8.1
1989	San Francisco	7.1
1992	Kobe, Japan	7.2
1994	Northridge	6.6
1999	Armenia, Colombia	6.0

Reuters/Jose Miguel Gomez/Corbis

Although the sizes of the numbers in the table do not seem to be very different, the intensity of the earthquakes they measure can be very different. For example, the 1989 San Francisco earthquake was more than 10 times stronger than the 1999 earthquake in Colombia. The reason behind this is that the Richter scale is a *logarithmic scale.* In this section, we start our work with logarithms, which will give you an understanding of the Richter scale. Let's begin.

As you know from your work in the previous sections, equations of the form

$$y = b^x \qquad b > 0, b \neq 1$$

are called exponential functions. Because the equation of the inverse of a function can be obtained by exchanging x and y in the equation of the original function, the inverse of an exponential function must have the form

$$x = b^y \qquad b > 0, b \neq 1$$

Now, this last equation is actually the equation of a logarithmic function, as the following definition indicates:

DEFINITION The equation $y = \log_b x$ is read "y is the logarithm to the base b of x" and is equivalent to the equation

$$x = b^y \qquad b > 0, b \neq 1$$

In words, we say "y is the number we raise b to in order to get x."

Notation When an equation is in the form $x = b^y$, it is said to be in exponential form. On the other hand, if an equation is in the form $y = \log_b x$, it is said to be in logarithmic form.

Here are some equivalent statements written in both forms.

Note
How to change from logarithmic form to exponential form is the first important thing to know about logarithms.

Exponential Form		Logarithmic Form
$8 = 2^3$	$\Leftrightarrow$	$\log_2 8 = 3$
$25 = 5^2$	$\Leftrightarrow$	$\log_5 25 = 2$
$0.1 = 10^{-1}$	$\Leftrightarrow$	$\log_{10} 0.1 = -1$
$\frac{1}{8} = 2^{-3}$	$\Leftrightarrow$	$\log_2 \frac{1}{8} = -3$
$r = z^s$	$\Leftrightarrow$	$\log_z r = s$

EXAMPLE 1 Solve for x: $\log_3 x = -2$

SOLUTION In exponential form the equation looks like this:

$$x = 3^{-2}$$

or $$x = \frac{1}{9}$$

The solution is $\frac{1}{9}$.

EXAMPLE 2 Solve $\log_x 4 = 3$.

SOLUTION Again, we use the definition of logarithms to write the equation in exponential form:

$$4 = x^3$$

Taking the cube root of both sides, we have

$$\sqrt[3]{4} = \sqrt[3]{x^3}$$

$$x = \sqrt[3]{4}$$

The solution set is $\{\sqrt[3]{4}\}$.

EXAMPLE 3 Solve $\log_8 4 = x$.

SOLUTION We write the equation again in exponential form:

$$4 = 8^x$$

Since both 4 and 8 can be written as powers of 2, we write them in terms of powers of 2:

$$2^2 = (2^3)^x$$

$$2^2 = 2^{3x}$$

The only way the left and right sides of this last line can be equal is if the exponents are equal—that is, if

$$2 = 3x$$

$$\text{or} \quad x = \frac{2}{3}$$

The solution is $\frac{2}{3}$. We check as follows:

$$\log_8 4 = \frac{2}{3} \Leftrightarrow 4 = 8^{2/3}$$

$$4 = (\sqrt[3]{8})^2$$

$$4 = 2^2$$

$$4 = 4$$

The solution checks when used in the original equation.

> **Note**
> The first step in each of these first three examples is the same. In each case the first step in solving the equation is to put the equation in exponential form.

Graphing Logarithmic Functions

Graphing logarithmic functions can be done using the graphs of exponential functions and the fact that the graphs of inverse functions have symmetry about the line $y = x$. Here's an example to illustrate.

EXAMPLE 4 Graph the equation $y = \log_2 x$.

SOLUTION The equation $y = \log_2 x$ is, by definition, equivalent to the exponential equation

$$x = 2^y$$

which is the equation of the inverse of the function

$$y = 2^x$$

We simply reflect the graph of $y = 2^x$ about the line $y = x$ to get the graph of $x = 2^y$, which is also the graph of $y = \log_2 x$. (See Figure 1.)

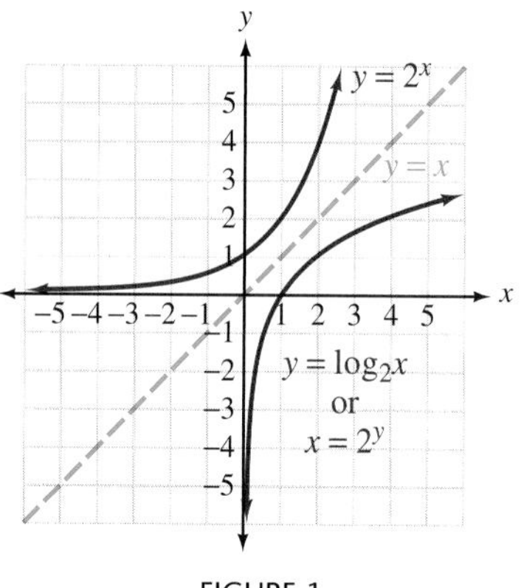

FIGURE 1

Note

From the graph of $y = \log_2 x$ it is apparent that $x > 0$ since the graph does not appear to the left of the y-axis. In general, the only variable in the expression $y = \log_b x$ that can be negative is y.

It is apparent from the graph that $y = \log_2 x$ is a function since no vertical line will cross its graph in more than one place. The same is true for all logarithmic equations of the form $y = \log_b x$, where b is a positive number other than 1. Note also that the graph of $y = \log_b x$ will always appear to the right of the y-axis, meaning that x will always be positive in the equation $y = \log_b x$.

USING TECHNOLOGY

Graphing Logarithmic Functions

As demonstrated in Example 4, we can graph the logarithmic function $y = \log_2 x$ as the inverse of the exponential function $y = 2x$. Your graphing calculator most likely has a command to do this. First, define the exponential function as $Y_1 = 2x$. To see the line of symmetry, define a second function $Y_2 = x$. Set the window variables so that

$$-6 \le x \le 6; \ -6 \le y \le 6$$

and use your zoom-square command to graph both functions. Your graph should look similar to the one shown in Figure 2. Now use the appropriate command to graph the inverse of the exponential function defined as Y_1 (Figure 3).

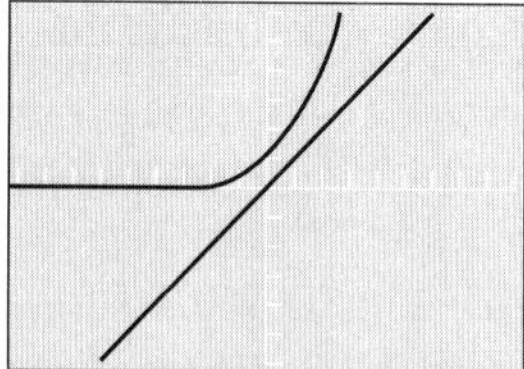

FIGURE 2

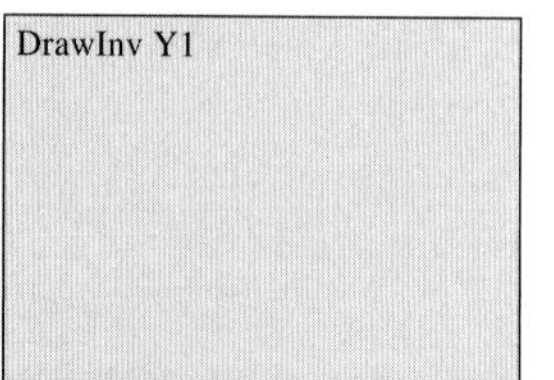

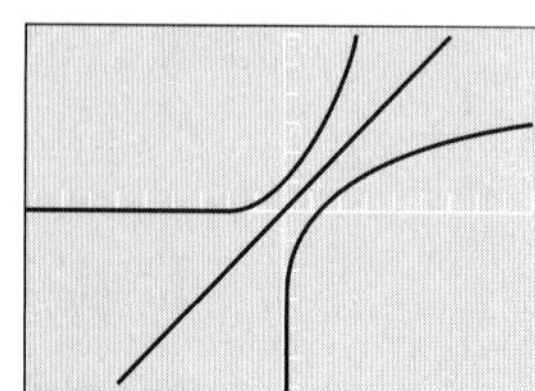

FIGURE 3

Two Special Identities

If b is a positive real number other than 1, then each of the following is a consequence of the definition of a logarithm:

$$(1)\ b^{\log_b x} = x \quad \text{and} \quad (2)\ \log_b b^x = x$$

The justifications for these identities are similar. Let's consider only the first one. Consider the equation

$$y = \log_b x$$

By definition, it is equivalent to

$$x = b^y$$

Substituting $\log_b x$ for y in the last line gives us

$$x = b^{\log_b x}$$

The next examples in this section show how these two special properties can be used to simplify equations involving logarithms.

EXAMPLE 5 Simplify $\log_2 8$.

SOLUTION Substitute 2^3 for 8:

$$\log_2 8 = \log_2 2^3$$
$$= 3$$

EXAMPLE 6 Simplify $\log_{10} 10{,}000$.

SOLUTION 10,000 can be written as 10^4:

$$\log_{10} 10{,}000 = \log_{10} 10^4$$
$$= 4$$

EXAMPLE 7 Simplify $\log_b b$ $(b > 0, b \neq 1)$.

SOLUTION Since $b^1 = b$, we have

$$\log_b b = \log_b b^1$$
$$= 1$$

EXAMPLE 8 Simplify $\log_b 1$ $(b > 0, b \neq 1)$.

SOLUTION Since $1 = b^0$, we have

$$\log_b 1 = \log_b b^0$$
$$= 0$$

EXAMPLE 9 Simplify $\log_4(\log_5 5)$.

SOLUTION Since $\log_5 5 = 1$,

$$\log_4(\log_5 5) = \log_4 1$$
$$= 0$$

Application

As we mentioned in the introduction to this section, one application of logarithms is in measuring the magnitude of an earthquake. If an earthquake has a shock wave T times greater than the smallest shock wave that can be measured on a seismograph, then the magnitude M of the earthquake, as measured on the Richter scale, is given by the formula

$$M = \log_{10} T$$

(When we talk about the size of a shock wave, we are talking about its amplitude. The amplitude of a wave is half the difference between its highest point and its lowest point.)

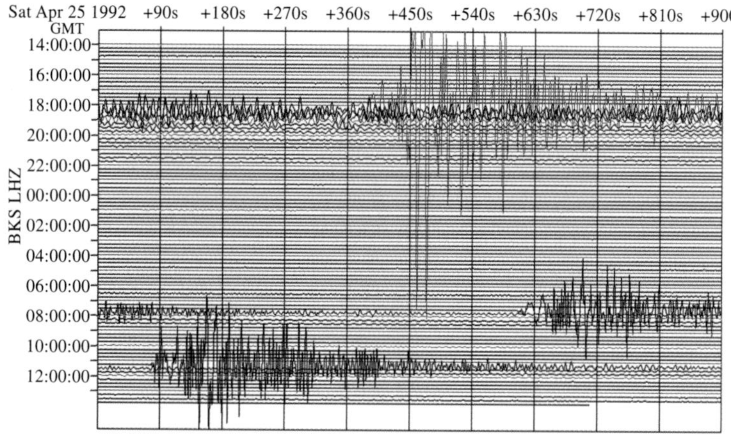

FIGURE 4

To illustrate the discussion, an earthquake that produces a shock wave that is 10,000 times greater than the smallest shock wave measurable on a seismograph will have a magnitude M on the Richter scale of

$$M = \log_{10} 10{,}000 = 4$$

EXAMPLE 10 If an earthquake has a magnitude of $M = 5$ on the Richter scale, what can you say about the size of its shock wave?

SOLUTION To answer this question, we put $M = 5$ into the formula $M = \log_{10} T$ to obtain

$$5 = \log_{10} T$$

Writing this equation in exponential form, we have

$$T = 10^5 = 100{,}000$$

We can say that an earthquake that measures 5 on the Richter scale has a shock wave 100,000 times greater than the smallest shock wave measurable on a seismograph.

From Example 10 and the discussion that preceded it, we find that an earthquake of magnitude 5 has a shock wave that is 10 times greater than an earthquake of magnitude 4 because 100,000 is 10 times 10,000.

LINKING OBJECTIVES AND EXAMPLES

Next to each objective we have listed the examples that are best described by that objective.

A	1–4
B	1–3
C	4
D	5–10

GETTING READY FOR CLASS

After reading through the preceding section, respond in your own words and in complete sentences.

1. What is a logarithm?
2. What is the relationship between $y = 2^x$ and $y = \log_2 x$? How are their graphs related?
3. Will the graph of $y = \log_b x$ ever appear in the second or third quadrants? Explain why or why not.
4. Explain why $\log_2 0 = x$ has no solution for x.

Problem Set 9.3

Online support materials can be found at www.thomsonedu.com/login

Write each of the following equations in logarithmic form.

1. $2^4 = 16$
2. $3^2 = 9$
3. $125 = 5^3$
4. $16 = 4^2$
5. $0.01 = 10^{-2}$
6. $0.001 = 10^{-3}$
7. $2^{-5} = \frac{1}{32}$
8. $4^{-2} = \frac{1}{16}$
9. $\left(\frac{1}{2}\right)^{-3} = 8$
10. $\left(\frac{1}{3}\right)^{-2} = 9$
11. $27 = 3^3$
12. $81 = 3^4$

Write each of the following equations in exponential form.

13. $\log_{10} 100 = 2$
14. $\log_2 8 = 3$
15. $\log_2 64 = 6$
16. $\log_2 32 = 5$
17. $\log_8 1 = 0$
18. $\log_9 9 = 1$
19. $\log_{10} 0.001 = -3$
20. $\log_{10} 0.0001 = -4$
21. $\log_6 36 = 2$
22. $\log_7 49 = 2$
23. $\log_5 \frac{1}{25} = -2$
24. $\log_3 \frac{1}{81} = -4$

Solve each of the following equations for x.

25. $\log_3 x = 2$
26. $\log_4 x = 3$
27. $\log_5 x = -3$
28. $\log_2 x = -4$
29. $\log_2 16 = x$
30. $\log_3 27 = x$
31. $\log_8 2 = x$
32. $\log_{25} 5 = x$
33. $\log_x 4 = 2$
34. $\log_x 16 = 4$
35. $\log_x 5 = 3$
36. $\log_x 8 = 2$

= Videos available by instructor request

▸ = Online student support materials available at www.thomsonedu.com/login

Sketch the graph of each of the following logarithmic equations.

37. $y = \log_3 x$

38. $y = \log_{1/2} x$

39. $y = \log_{1/3} x$

40. $y = \log_4 x$

41. $y = \log_5 x$

42. $y = \log_{1/5} x$

43. $y = \log_{10} x$

44. $y = \log_{1/4} x$

Simplify each of the following.

45. $\log_2 16$

46. $\log_3 9$

47. $\log_{25} 125$

48. $\log_9 27$

49. $\log_{10} 1{,}000$

50. $\log_{10} 10{,}000$

51. $\log_3 3$

52. $\log_4 4$

53. $\log_5 1$

54. $\log_{10} 1$

55. $\log_3(\log_6 6)$

56. $\log_5(\log_3 3)$

57. $\log_4[\log_2(\log_2 16)]$

58. $\log_4[\log_3(\log_2 8)]$

Applying the Concepts

$\text{pH} = -\log_{10}[H^+]$, where $[H^+]$ is the concentration of the hydrogen ions in solution. An acid solution has a pH below 7, and a basic solution has a pH higher than 7.

59. In distilled water, the concentration of hydrogen ions is $[H^+] = 10^{-7}$. What is the pH?

60. Find the pH of a bottle of vinegar if the concentration of hydrogen ions is $[H^+] = 10^{-3}$.

61. A hair conditioner has a pH of 6. Find the concentration of hydrogen ions, $[H^+]$, in the conditioner.

62. If a glass of orange juice has a pH of 4, what is the concentration of hydrogen ions, $[H^+]$, in the juice?

63. **Magnitude of an Earthquake** Find the magnitude M of an earthquake with a shock wave that measures $T = 100$ on a seismograph.

64. **Magnitude of an Earthquake** Find the magnitude M of an earthquake with a shock wave that measures $T = 100{,}000$ on a seismograph.

65. **Shock Wave** If an earthquake has a magnitude of 8 on the Richter scale, how many times greater is its shock wave than the smallest shock wave measurable on a seismograph?

66. **Shock Wave** If an earthquake has a magnitude of 6 on the Richter scale, how many times greater is its shock wave than the smallest shock wave measurable on a seismograph?

Maintaining Your Skills

Fill in the blanks to complete the square.

67. $x^2 + 10x + \square = (x + \square)^2$

68. $x^2 + 4x + \square = (x + \square)^2$

69. $y^2 - 2y + \square = (y - \square)^2$

70. $y^2 + 3y + \square = \left(y + \square\right)^2$

Solve.

71. $-y^2 = 9$

72. $7 + y^2 = 11$

73. $-x^2 - 8 = -4$

74. $10x^2 = 100$

75. $2x^2 + 4x - 3 = 0$

76. $3x^2 + 4x - 2 = 0$

77. $(2y - 3)(2y - 1) = -4$

78. $(y - 1)(3y - 3) = 10$

79. $t^3 - 125 = 0$

80. $8t^3 + 1 = 0$

81. $4x^5 - 16x^4 = 20x^3$

82. $3x^4 + 6x^2 = 6x^3$

83. $\dfrac{1}{x-3} + \dfrac{1}{x+2} = 1$

84. $\dfrac{1}{x+3} + \dfrac{1}{x-2} = 1$

Getting Ready for the Next Section

Simplify.

85. $8^{2/3}$

86. $27^{2/3}$

Solve.

87. $(x + 2)(x) = 2^3$

88. $(x + 3)(x) = 2^2$

89. $\dfrac{x-2}{x+1} = 9$

90. $\dfrac{x+1}{x-4} = 25$

Write in exponential form.

91. $\log_2 [(x + 2)(x)] = 3$

92. $\log_4 [x(x - 6)] = 2$

93. $\log_3\left(\dfrac{x-2}{x+1}\right) = 4$

94. $\log_5\left(\dfrac{x-1}{x-4}\right) = 2$

Extending the Concepts

95. The graph of the exponential function $y = f(x) = b^x$ is shown here. Use the graph to complete parts a through d.

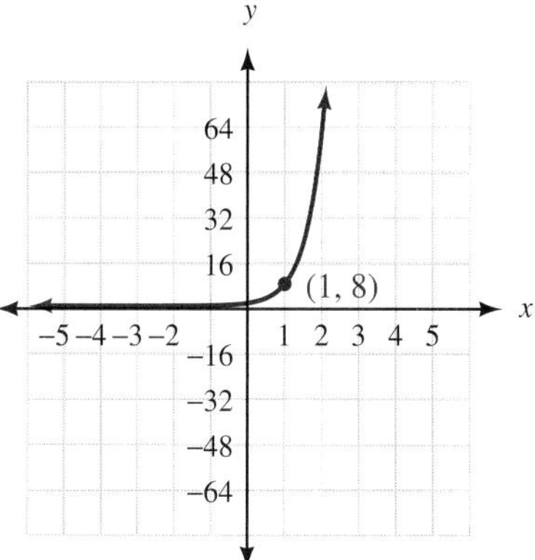

a. Fill in the table.

x	$f(x)$
−1	
0	
1	
2	

b. Fill in the table.

x	$f^{-1}(x)$
	−1
	0
	1
	2

c. Find the equation for $f(x)$.

d. Find the equation for $f^{-1}(x)$.

96. The graph of the exponential function $y = f(x) = b^x$ is shown here. Use the graph to complete parts a through d.

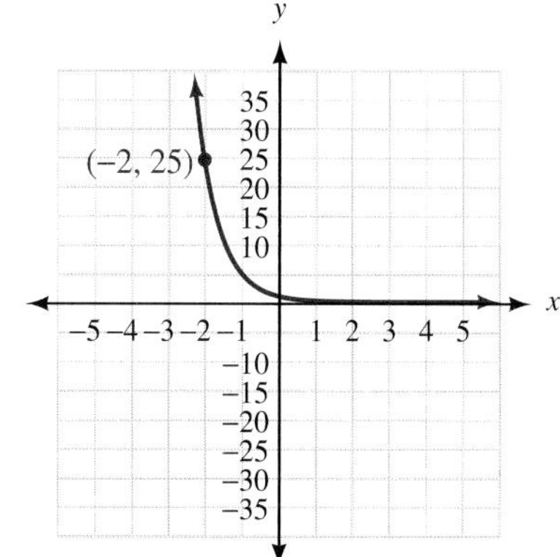

a. Fill in the table.

x	$f(x)$
−1	
0	
1	
2	

b. Fill in the table.

x	$f^{-1}(x)$
	−1
	0
	1
	2

c. Find the equation for $f(x)$.

d. Find the equation for $f^{-1}(x)$.

9.4 Properties of Logarithms

OBJECTIVES

A Use the properties of logarithms to convert between expanded form and single logarithms.

B Use the properties of logarithms to solve equations that contain logarithms.

If we search for the word *decibel* in *Microsoft Bookshelf,* we find the following definition:

> A unit used to express relative difference in power or intensity, usually between two acoustic or electric signals, equal to ten times the common logarithm of the ratio of the two levels.

Decibels	Comparable to
10	A light whisper
20	Quiet conversation
30	Normal conversation
40	Light traffic
50	Typewriter, loud conversation
60	Noisy office
70	Normal traffic, quiet train
80	Rock music, subway
90	Heavy traffic, thunder
100	Jet plane at takeoff

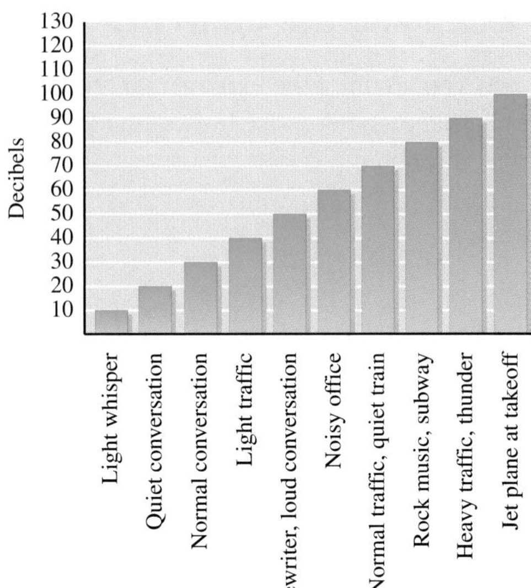

The precise definition for a *decibel* is

$$D = 10 \log_{10}\left(\frac{I}{I_0}\right)$$

where I is the intensity of the sound being measured, and I_0 is the intensity of the least audible sound. (Sound intensity is related to the amplitude of the sound wave that models the sound and is given in units of watts per meter^2.) In this section, we will see that the preceding formula can also be written as

$$D = 10(\log_{10} I - \log_{10} I_0)$$

The rules we use to rewrite expressions containing logarithms are called the *properties of logarithms.* There are three of them.

For the following three properties, x, y, and b are all positive real numbers, $b \neq 1$, and r is any real number.

Property 1

$$\log_b (xy) = \log_b x + \log_b y$$

In words: The logarithm of a *product* is the *sum* of the logarithms.

Property 2

$$\log_b \left(\frac{x}{y}\right) = \log_b x - \log_b y$$

In words: The logarithm of a *quotient* is the *difference* of the logarithms.

Property 3

$$\log_b x^r = r \log_b x$$

In words: The logarithm of a number raised to a *power* is the *product* of the power and the logarithm of the number.

Proof of Property 1 To prove property 1, we simply apply the first identity for logarithms given in the preceding section:

$$b^{\log_b xy} = xy = (b^{\log_b x})(b^{\log_b y}) = b^{\log_b x + \log_b y}$$

Since the first and last expressions are equal and the bases are the same, the exponents $\log_b xy$ and $\log_b x + \log_b y$ must be equal. Therefore,

$$\log_b xy = \log_b x + \log_b y$$

The proofs of properties 2 and 3 proceed in much the same manner, so we will omit them here. The examples that follow show how the three properties can be used.

EXAMPLE 1

Expand, using the properties of logarithms: $\log_5 \dfrac{3xy}{z}$

SOLUTION Applying property 2, we can write the quotient of $3xy$ and z in terms of a difference:

$$\log_5 \frac{3xy}{z} = \log_5 3xy - \log_5 z$$

Applying property 1 to the product $3xy$, we write it in terms of addition:

$$\log_5 \frac{3xy}{z} = \log_5 3 + \log_5 x + \log_5 y - \log_5 z$$

EXAMPLE 2

Expand, using the properties of logarithms:

$$\log_2 \frac{x^4}{\sqrt{y} \cdot z^3}$$

SOLUTION We write $\sqrt{y}$ as $y^{1/2}$ and apply the properties:

$$\log_2 \frac{x^4}{\sqrt{y} \cdot z^3} = \log_2 \frac{x^4}{y^{1/2}z^3} \qquad \sqrt{y} = y^{1/2}$$

$$= \log_2 x^4 - \log_2(y^{1/2} \cdot z^3) \qquad \textbf{Property 2}$$

$$= \log_2 x^4 - (\log_2 y^{1/2} + \log_2 z^3) \qquad \textbf{Property 1}$$

$$= \log_2 x^4 - \log_2 y^{1/2} - \log_2 z^3 \qquad \textbf{Remove parentheses}$$

$$= 4 \log_2 x - \frac{1}{2} \log_2 y - 3 \log_2 z \qquad \textbf{Property 3}$$

We can also use the three properties to write an expression in expanded form as just one logarithm.

EXAMPLE 3

Write as a single logarithm:

$$2 \log_{10} a + 3 \log_{10} b - \frac{1}{3} \log_{10} c$$

SOLUTION We begin by applying property 3:

$$2 \log_{10} a + 3 \log_{10} b - \frac{1}{3} \log_{10} c$$

$$= \log_{10} a^2 + \log_{10} b^3 - \log_{10} c^{1/3} \quad \textbf{Property 3}$$

$$= \log_{10} (a^2 \cdot b^3) - \log_{10} c^{1/3} \quad \textbf{Property 1}$$

$$= \log_{10} \frac{a^2b^3}{c^{1/3}} \quad \textbf{Property 2}$$

$$= \log_{10} \frac{a^2b^3}{\sqrt[3]{c}} \quad c^{1/3} = \sqrt[3]{c}$$

The properties of logarithms along with the definition of logarithms are useful in solving equations that involve logarithms.

EXAMPLE 4

Solve for x: $\log_2(x + 2) + \log_2 x = 3$

SOLUTION Applying property 1 to the left side of the equation allows us to write it as a single logarithm:

$$\log_2(x + 2) + \log_2 x = 3$$

$$\log_2[(x + 2)(x)] = 3$$

The last line can be written in exponential form using the definition of logarithms:

$$(x + 2)(x) = 2^3$$

Solve as usual:

$$x^2 + 2x = 8$$

$$x^2 + 2x - 8 = 0$$

$$(x + 4)(x - 2) = 0$$

$$x + 4 = 0 \quad \text{or} \quad x - 2 = 0$$

$$x = -4 \quad \text{or} \quad x = 2$$

In the previous section we noted the fact that x in the expression $y = \log_b x$ cannot be a negative number. Since substitution of $x = -4$ into the original equation gives

$$\log_2(-2) + \log_2(-4) = 3$$

which contains logarithms of negative numbers, we cannot use -4 as a solution. The solution set is $\{2\}$.

LINKING OBJECTIVES AND EXAMPLES

Next to each **objective** we have listed the examples that are best described by that objective.

A	1–3
B	4

GETTING READY FOR CLASS

After reading through the preceding section, respond in your own words and in complete sentences.

1. Explain why the following statement is false: "The logarithm of a product is the product of the logarithms."
2. Explain why the following statement is false: "The logarithm of a quotient is the quotient of the logarithms."
3. Explain the difference between $\log_b m + \log_b n$ and $\log_b(m + n)$. Are they equivalent?
4. Explain the difference between $\log_b(mn)$ and $(\log_b m)(\log_b n)$. Are they equivalent?

Problem Set 9.4

Online support materials can be found at www.thomsonedu.com/login

Use the three properties of logarithms given in this section to expand each expression as much as possible.

1. $\log_3 4x$
2. $\log_2 5x$
3. $\log_6 \frac{5}{x}$
4. $\log_3 \frac{x}{5}$
5. $\log_2 y^5$
6. $\log_7 y^3$
7. $\log_9 \sqrt[3]{z}$
8. $\log_8 \sqrt{z}$
9. $\log_6 x^2y^4$
10. $\log_{10} x^2y^4$
11. $\log_5 \sqrt{x} \cdot y^4$
12. $\log_8 \sqrt[3]{xy^6}$
13. $\log_b \frac{xy}{z}$
14. $\log_b \frac{3x}{y}$
15. $\log_{10} \frac{4}{xy}$
16. $\log_{10} \frac{5}{4y}$
17. $\log_{10} \frac{x^2y}{\sqrt{z}}$
18. $\log_{10} \frac{\sqrt{x} \cdot y}{z^3}$
19. $\log_{10} \frac{x^3\sqrt{y}}{z^4}$
20. $\log_{10} \frac{x^4\sqrt[3]{y}}{\sqrt{z}}$
21. $\log_b \sqrt[3]{\frac{x^2y}{z^4}}$
22. $\log_b \sqrt[4]{\frac{x^4y^3}{z^5}}$

Write each expression as a single logarithm.

23. $\log_b x + \log_b z$
24. $\log_b x - \log_b z$
25. $2 \log_3 x - 3 \log_3 y$
26. $4 \log_2 x + 5 \log_2 y$
27. $\frac{1}{2} \log_{10} x + \frac{1}{3} \log_{10} y$
28. $\frac{1}{3} \log_{10} x - \frac{1}{4} \log_{10} y$
29. $3 \log_2 x + \frac{1}{2} \log_2 y - \log_2 z$
30. $2 \log_3 x + 3 \log_3 y - \log_3 z$
31. $\frac{1}{2} \log_2 x - 3 \log_2 y - 4 \log_2 z$

= Videos available by instructor request

▸ = Online student support materials available at www.thomsonedu.com/login

32. $3 \log_{10} x - \log_{10} y - \log_{10} z$

33. $\frac{3}{2} \log_{10} x - \frac{3}{4} \log_{10} y - \frac{4}{5} \log_{10} z$

34. $3 \log_{10} x - \frac{4}{3} \log_{10} y - 5 \log_{10} z$

Solve each of the following equations.

35. $\log_2 x + \log_2 3 = 1$

36. $\log_3 x + \log_3 3 = 1$

37. $\log_3 x - \log_3 2 = 2$

38. $\log_3 x + \log_3 2 = 2$

39. $\log_3 x + \log_3(x - 2) = 1$

40. $\log_6 x + \log_6(x - 1) = 1$

41. $\log_3(x + 3) - \log_3(x - 1) = 1$

42. $\log_4(x - 2) - \log_4(x + 1) = 1$

43. $\log_2 x + \log_2(x - 2) = 3$

44. $\log_4 x + \log_4(x + 6) = 2$

45. $\log_8 x + \log_8(x - 3) = \frac{2}{3}$

46. $\log_{27} x + \log_{27}(x + 8) = \frac{2}{3}$

47. $\log_5 \sqrt{x} + \log_5 \sqrt{6x + 5} = 1$

48. $\log_2 \sqrt{x} + \log_2 \sqrt{6x + 5} = 1$

Applying the Concepts

49. **Decibel Formula** Use the properties of logarithms to rewrite the decibel formula $D = 10 \log_{10}(\frac{I}{I_0})$ as

$$D = 10(\log_{10} I - \log_{10} I_0).$$

50. **Decibel Formula** In the decibel formula $D = 10 \log_{10}(\frac{I}{I_0})$, the threshold of hearing, I_0, is

$$I_0 = 10^{-12} \text{ watts/meter}^2$$

Substitute 10^{-12} for I_0 in the decibel formula, and then show that it simplifies to

$$D = 10(\log_{10} I + 12)$$

51. **Acoustic Powers** The formula $N = \log_{10}\frac{P_1}{P_2}$ is used in radio electronics to find the ratio of the acoustic powers of two electric circuits in terms of their electric powers. Find N if P_1 is 100 and P_2 is 1. Then use the same two values of P_1 and P_2 to find N in the formula $N = \log_{10} P_1 - \log_{10} P_2$.

52. **Henderson–Hasselbalch Formula** Doctors use the Henderson–Hasselbalch formula to calculate the pH of a person's blood. pH is a measure of the acidity and/or the alkalinity of a solution. This formula is represented as

$$\text{pH} = 6.1 + \log_{10}\left(\frac{x}{y}\right)$$

where x is the base concentration and y is the acidic concentration. Rewrite the Henderson–Hasselbalch formula so that the logarithm of a quotient is not involved.

Maintaining Your Skills

Divide.

53. $\frac{12x^2 + y^2}{36}$

54. $\frac{x^2 + 4y^2}{16}$

55. Divide $25x^2 + 4y^2$ by 100

56. Divide $4x^2 + 9y^2$ by 36

Use the discriminant to find the number and kind of solutions to the following equations.

57. $2x^2 - 5x + 4 = 0$

58. $4x^2 - 12x = -9$

For each of the following problems, find an equation with the given solutions.

59. $x = -3, x = 5$

60. $x = 2, x = -2, x = 1$

61. $y = \frac{2}{3}, y = 3$

62. $y = -\frac{3}{5}, y = 2$

Getting Ready for the Next Section

Simplify.

63. 5^0

64. 4^1

65. $\log_3 3$

66. $\log_5 5$

67. $\log_b b^4$

68. $\log_a a^k$

Use a calculator to find each of the following. Write your answer in scientific notation with the first number in each answer rounded to the nearest tenth.

69. $10^{-5.6}$

70. $10^{-4.1}$

Divide and round to the nearest whole number.

71. $\dfrac{2.00 \times 10^8}{3.96 \times 10^6}$

72. $\dfrac{3.25 \times 10^{12}}{1.72 \times 10^{10}}$

9.5 Common Logarithms and Natural Logarithms

OBJECTIVES

A Use a calculator to find common logarithms.

B Use a calculator to find a number given its common logarithm.

C Simplify expressions containing natural logarithms.

Acid rain was first discovered in the 1960s by Gene Likens and his research team, who studied the damage caused by acid rain to Hubbard Brook in New Hampshire. Acid rain is rain with a pH of 5.6 and below. As you will see as you work your way through this section, pH is defined in terms of common logarithms—one of the topics we present in this section. So, when you are finished with this section, you will have a more detailed knowledge of pH and acid rain.

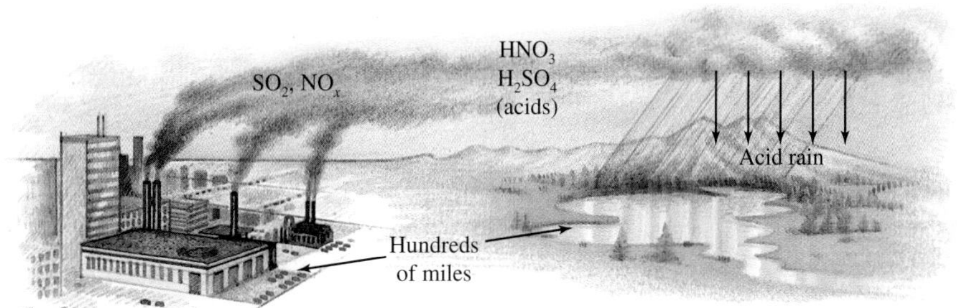

There are two kinds of logarithms that occur more frequently than other logarithms. They are logarithms with base 10 and natural logarithms, or logarithms with base e. Logarithms with a base of 10 are very common because our number system is a base-10 number system. For this reason, we call base-10 logarithms *common logarithms.*

DEFINITION A **common logarithm** is a logarithm with a base of 10. Because common logarithms are used so frequently, it is customary, in order to save time, to omit notating the base; that is,

$$\log_{10} x = \log x$$

When the base is not shown, it is assumed to be 10.

Common Logarithms

Common logarithms of powers of 10 are simple to evaluate. We need only recognize that $\log 10 = \log_{10} 10 = 1$ and apply the third property of logarithms: $\log_b x^r = r \log_b x$.

$$\begin{aligned}
\log 1{,}000 &= \log 10^3 = 3 \log 10 = 3(1) = 3 \\
\log 100 &= \log 10^2 = 2 \log 10 = 2(1) = 2 \\
\log 10 &= \log 10^1 = 1 \log 10 = 1(1) = 1 \\
\log 1 &= \log 10^0 = 0 \log 10 = 0(1) = 0 \\
\log 0.1 &= \log 10^{-1} = -1 \log 10 = -1(1) = -1 \\
\log 0.01 &= \log 10^{-2} = -2 \log 10 = -2(1) = -2 \\
\log 0.001 &= \log 10^{-3} = -3 \log 10 = -3(1) = -3
\end{aligned}$$

Note Remember, when the base is not written it is assumed to be 10.

To find common logarithms of numbers that are not powers of 10, we use a calculator with a [log] key.

Check the following logarithms to be sure you know how to use your calculator. (These answers have been rounded to the nearest ten-thousandth.)

$$\begin{aligned}
\log 7.02 &\approx 0.8463 \\
\log 1.39 &\approx 0.1430 \\
\log 6.00 &\approx 0.7782 \\
\log 9.99 &\approx 0.9996
\end{aligned}$$

EXAMPLE 1 Use a calculator to find log 2,760.

SOLUTION $\log 2{,}760 \approx 3.4409$

To work this problem on a scientific calculator, we simply enter the number 2,760 and press the key labeled [log]. On a graphing calculator we press the [log] key first, then 2,760.

The 3 in the answer is called the *characteristic,* and the decimal part of the logarithm is called the *mantissa.*

EXAMPLE 2 Find log 0.0391.

SOLUTION $\log 0.0391 \approx -1.4078$

EXAMPLE 3 Find log 0.00523.

SOLUTION $\log 0.00523 \approx -2.2815$

EXAMPLE 4 Find x if $\log x = 3.8774$.

SOLUTION We are looking for the number whose logarithm is 3.8774. On a scientific calculator, we enter 3.8774 and press the key labeled [10^x]. On a graphing calculator we press $\boxed{10^x}$ first, then 3.8774. The result is 7,540 to four significant digits. Here's why:

$$\begin{aligned} \text{If} \qquad \log x &= 3.8774 \\ \text{then} \qquad x &= 10^{3.8774} \\ &\approx 7{,}540 \end{aligned}$$

The number 7,540 is called the *antilogarithm* or just *antilog* of 3.8774, that is, 7,540 is the number whose logarithm is 3.8774.

EXAMPLE 5 Find x if $\log x = -2.4179$.

SOLUTION Using the $\boxed{10^x}$ key, the result is 0.00382.

$$\begin{aligned} \text{If} \qquad \log x &= -2.4179 \\ \text{then} \qquad x &= 10^{-2.4179} \\ &\approx 0.00382 \end{aligned}$$

The antilog of -2.4179 is 0.00382; that is, the logarithm of 0.00382 is -2.4179.

Applications

Previously, we found that the magnitude M of an earthquake that produces a shock wave T times larger than the smallest shock wave that can be measured on a seismograph is given by the formula

$$M = \log_{10} T$$

We can rewrite this formula using our shorthand notation for common logarithms as

$$M = \log T$$

EXAMPLE 6 The San Francisco earthquake of 1906 is estimated to have measured 8.3 on the Richter scale. The San Fernando earthquake of 1971 measured 6.6 on the Richter scale. Find T for each earthquake, and then give some indication of how much stronger the 1906 earthquake was than the 1971 earthquake.

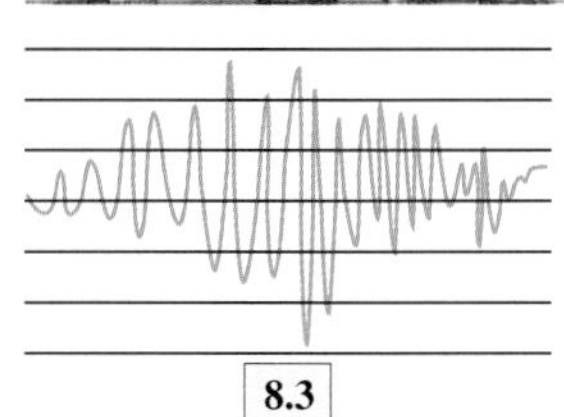

SOLUTION For the 1906 earthquake:

$$\text{If } \log T = 8.3, \text{ then } T = 2.00 \times 10^8.$$

For the 1971 earthquake:

$$\text{If } \log T = 6.6, \text{ then } T = 3.98 \times 10^6.$$

Dividing the two values of T and rounding our answer to the nearest whole number, we have

$$\frac{2.00 \times 10^8}{3.98 \times 10^6} \approx 50$$

The shock wave for the 1906 earthquake was approximately 50 times stronger than the shock wave for the 1971 earthquake.

In chemistry, the pH of a solution is the measure of the acidity of the solution. The definition for pH involves common logarithms. Here it is:

$$\text{pH} = -\log[\text{H}^+]$$

where $[\text{H}^+]$ is the concentration of the hydrogen ion in moles per liter. The range for pH is from 0 to 14. Pure water, a neutral solution, has a pH of 7. An acidic solution, such as vinegar, will have a pH less than 7, and an alkaline solution, such as ammonia, has a pH above 7.

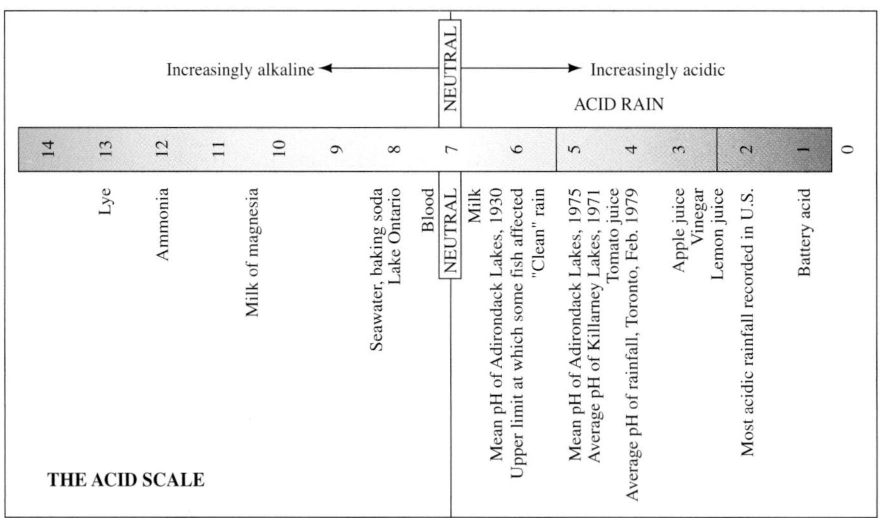

EXAMPLE 7 Normal rainwater has a pH of 5.6. What is the concentration of the hydrogen ion in normal rainwater?

SOLUTION Substituting 5.6 for pH in the formula $\text{pH} = -\log[\text{H}^+]$, we have

$5.6 = -\log[\text{H}^+]$	**Substitution**
$\log[\text{H}^+] = -5.6$	**Isolate the logarithm**
$[\text{H}^+] = 10^{-5.6}$	**Write in exponential form**
$\approx 2.5 \times 10^{-6}$ mole per liter	**Answer in scientific notation**

EXAMPLE 8 The concentration of the hydrogen ion in a sample of acid rain known to kill fish is 3.2×10^{-5} mole per liter. Find the pH of this acid rain to the nearest tenth.

SOLUTION Substituting 3.2×10^{-5} for $[\text{H}^+]$ in the formula $\text{pH} = -\log[\text{H}^+]$, we have

$\text{pH} = -\log[3.2 \times 10^{-5}]$	**Substitution**
$\approx -(-4.5)$	**Evaluate the logarithm**
$= 4.5$	**Simplify**

Natural Logarithms

DEFINITION A **natural logarithm** is a logarithm with a base of e. The natural logarithm of x is denoted by $\ln x$; that is,

$$\ln x = \log_e x$$

The postage stamp shown here contains one of the two special identities we mentioned previously in this chapter, but stated in terms of natural logarithms.

We can assume that all our properties of exponents and logarithms hold for expressions with a base of e since e is a real number. Here are some examples intended to make you more familiar with the number e and natural logarithms.

EXAMPLE 9

Simplify each of the following expressions.

a. $e^0 = 1$

b. $e^1 = e$

c. $\ln e = 1$ **In exponential form, $e^1 = e$**

d. $\ln 1 = 0$ **In exponential form, $e^0 = 1$**

e. $\ln e^3 = 3$

f. $\ln e^{-4} = -4$

g. $\ln e^t = t$

EXAMPLE 10

Use the properties of logarithms to expand the expression $\ln Ae^{5t}$.

SOLUTION Since the properties of logarithms hold for natural logarithms, we have

$$\begin{aligned}\ln Ae^{5t} &= \ln A + \ln e^{5t} \\ &= \ln A + 5t \ln e \\ &= \ln A + 5t\end{aligned}$$

Because $\ln e = 1$

EXAMPLE 11 If $\ln 2 = 0.6931$ and $\ln 3 = 1.0986$, find

a. $\ln 6$ **b.** $\ln 0.5$ **c.** $\ln 8$

SOLUTION

a. Since $6 = 2 \cdot 3$, we have

$$\begin{aligned} \ln 6 &= \ln (2 \cdot 3) \\ &= \ln 2 + \ln 3 \\ &= 0.6931 + 1.0986 \\ &= 1.7917 \end{aligned}$$

b. Writing 0.5 as $\frac{1}{2}$ and applying property 2 for logarithms gives us

$$\begin{aligned} \ln 0.5 &= \ln \frac{1}{2} \\ &= \ln 1 - \ln 2 \\ &= 0 - 0.6931 \\ &= -0.6931 \end{aligned}$$

c. Writing 8 as 2^3 and applying property 3 for logarithms, we have

$$\begin{aligned} \ln 8 &= \ln 2^3 \\ &= 3 \ln 2 \\ &= 3(0.6931) \\ &= 2.0793 \end{aligned}$$

LINKING OBJECTIVES AND EXAMPLES

Next to each **objective** we have listed the examples that are best described by that objective.

A	1–3
B	4–8
C	9–11

GETTING READY FOR CLASS

After reading through the preceding section, respond in your own words and in complete sentences.

1. What is a common logarithm?
2. What is a natural logarithm?
3. Is e a rational number? Explain.
4. Find ln e, and explain how you arrived at your answer.

Problem Set 9.5

Online support materials can be found at www.thomsonedu.com/login

Find the following logarithms. Round to four decimal places.

1. log 378
2. log 426
3. log 37.8
4. log 42,600
5. log 3,780
6. log 0.4260
7. log 0.0378
8. log 0.0426
9. log 37,800
10. log 4,900
11. log 600
12. log 900
13. log 2,010
14. log 10,200
15. log 0.00971
16. log 0.0312
17. log 0.0314
18. log 0.00052
19. log 0.399
20. log 0.111

Find x in the following equations.

21. $\log x = 2.8802$
22. $\log x = 4.8802$
23. $\log x = -2.1198$
24. $\log x = -3.1198$
25. $\log x = 3.1553$
26. $\log x = 5.5911$
27. $\log x = -5.3497$
28. $\log x = -1.5670$

Find x.

29. $\log x = -7.0372$
30. $\log x = -4.2000$
31. $\log x = 10$
32. $\log x = -1$
33. $\log x = -10$
34. $\log x = 1$
35. $\log x = 20$
36. $\log x = -20$
37. $\log x = -2$
38. $\log x = 4$
39. $\log x = \log_2 8$
40. $\log x = \log_3 9$

Simplify each of the following expressions.

41. $\ln e$
42. $\ln 1$
43. $\ln e^5$
44. $\ln e^{-3}$
45. $\ln e^x$
46. $\ln e^y$

Use the properties of logarithms to expand each of the following expressions.

47. $\ln 10e^{3t}$
48. $\ln 10e^{4t}$
49. $\ln Ae^{-2t}$
50. $\ln Ae^{-3t}$

If $\ln 2 = 0.6931$, $\ln 3 = 1.0986$, and $\ln 5 = 1.6094$, find each of the following.

51. $\ln 15$
52. $\ln 10$
53. $\ln \frac{1}{3}$
54. $\ln \frac{1}{5}$
55. $\ln 9$
56. $\ln 25$
57. $\ln 16$
58. $\ln 81$

Use a calculator to evaluate each expression. Round your answers 4 decimal places, if necessary.

59. **a.** $\frac{\log 25}{\log 15}$ **b.** $\log \frac{25}{15}$
60. **a.** $\frac{\log 12}{\log 7}$ **b.** $\log \frac{12}{7}$
61. **a.** $\frac{\log 4}{\log 8}$ **b.** $\log \frac{4}{8}$
62. **a.** $\frac{\log 3}{\log 9}$ **b.** $\log \frac{3}{9}$

Applying the Concepts

Measuring Acidity Previously we indicated that the pH of a solution is defined in terms of logarithms as

$$\text{pH} = -\log[\text{H}^+]$$

where $[\text{H}^+]$ is the concentration of the hydrogen ion in that solution.

Round to the nearest hundredth.

63. Find the pH of orange juice if the concentration of the hydrogen ion in the juice is $[\text{H}^+] = 6.50 \times 10^{-4}$.
64. Find the pH of milk if the concentration of the hydrogen ion in milk is $[\text{H}^+] = 1.88 \times 10^{-6}$.

65. Find the concentration of hydrogen ions in a glass of wine if the pH is 4.75.

66. Find the concentration of hydrogen ions in a bottle of vinegar if the pH is 5.75.

The Richter Scale Find the relative size T of the shock wave of earthquakes with the following magnitudes, as measured on the Richter scale. Round to the nearest hundredth.

67. 5.5 **68.** 6.6

69. 8.3 **70.** 8.7

71. Shock Wave How much larger is the shock wave of an earthquake that measures 6.5 on the Richter scale than one that measures 5.5 on the same scale?

72. Shock Wave How much larger is the shock wave of an earthquake that measures 8.5 on the Richter scale than one that measures 5.5 on the same scale?

73. Earthquake The chart below is a partial listing of earthquakes that were recorded in Canada in 2000. Complete the chart by computing the magnitude on the Richter scale, M, or the number of times the associated shock wave is larger than the smallest measurable shock wave, T.

Location	Date	Magnitude, M	Shock Wave, T
Moresby Island	January 23	4.0	
Vancouver Island	April 30		1.99×10^5
Quebec City	June 29	3.2	
Mould Bay	November 13	5.2	
St. Lawrence	December 14		5.01×10^3

Source: National Resources Canada, National Earthquake Hazards Program.

74. Earthquake On January 6, 2001, an earthquake with a magnitude of 7.7 on the Richter scale hit southern India (*National Earthquake Information Center*). By what factor was this earthquake's shock wave greater than the smallest measurable shock wave?

Depreciation The annual rate of depreciation r on a car that is purchased for P dollars and is worth W dollars t years later can be found from the formula

$$\log(1 - r) = \frac{1}{t} \log \frac{W}{P}$$

75. Find the annual rate of depreciation on a car that is purchased for \$9,000 and sold 5 years later for \$4,500.

76. Find the annual rate of depreciation on a car that is purchased for \$9,000 and sold 4 years later for \$3,000.

Two cars depreciate in value according to the following depreciation tables. In each case, find the annual rate of depreciation.

77.

Age in Years	Value in Dollars
new	7,550
5	5,750

78.

Age in Years	Value in Dollars
new	7,550
3	5,750

79. Getting Close to *e* Use a calculator to complete the following table.

x	$(1 + x)^{1/x}$
1	
0.5	
0.1	
0.01	
0.001	
0.0001	
0.00001	

What number does the expression $(1 + x)^{1/x}$ seem to approach as x gets closer and closer to zero?

80. Getting Close to *e* Use a calculator to complete the following table.

x	$\left(1 + \frac{1}{x}\right)^x$
1	
10	
50	
100	
500	
1,000	
10,000	
1,000,000	

What number does the expression $\left(1 + \frac{1}{x}\right)^x$ seem to approach as x gets larger and larger?

Maintaining Your Skills

Solve each equation.

81. $(y + 3)^2 + y^2 = 9$

82. $(2y + 4)^2 + y^2 = 4$

83. $(x + 3)^2 + 1^2 = 2$

84. $(x - 3)^2 + (-1)^2 = 10$

85. $x^4 - 2x^2 - 8 = 0$

86. $x^{2/3} - 5x^{1/3} + 6 = 0$

87. $2x - 5\sqrt{x} + 3 = 0$

88. $(3x + 1) - 6\sqrt{3x + 1} + 8 = 0$

Getting Ready for the Next Section

Solve.

89. $5(2x + 1) = 12$

90. $4(3x - 2) = 21$

Use a calculator to evaluate. Give answers to 4 decimal places.

91. $\dfrac{100{,}000}{32{,}000}$

92. $\dfrac{1.4982}{3.5681} + 3$

93. $\dfrac{1}{2}\left(\dfrac{-0.6931}{1.4289} + 3\right)$

94. $1 + \dfrac{0.04}{52}$

Use the power rule to rewrite the following logarithms.

95. $\log 1.05^t$

96. $\log 1.033^t$

Use identities to simplify.

97. $\ln e^{0.05t}$

98. $\ln e^{-0.000121t}$

9.6 Exponential Equations and Change of Base

OBJECTIVES

A Solve exponential equations.

B Use the change-of-base property.

C Solve application problems whose solutions are found by solving logarithmic or exponential equations.

For items involved in exponential growth, the time it takes for a quantity to double is called the *doubling time.* For example, if you invest \$5,000 in an account that pays 5% annual interest, compounded quarterly, you may want to know how long it will take for your money to double in value. You can find this doubling time if you can solve the equation

$$10{,}000 = 5{,}000\,(1.0125)^{4t}$$

As you will see as you progress through this section, logarithms are the key to solving equations of this type.

Logarithms are very important in solving equations in which the variable appears as an exponent. The equation

$$5^x = 12$$

is an example of one such equation. Equations of this form are called *exponential equations.* Since the quantities 5^x and 12 are equal, so are their common logarithms. We begin our solution by taking the logarithm of both sides:

$$\log 5^x = \log 12$$

We now apply property 3 for logarithms, $\log x^r = r \log x$, to turn x from an exponent into a coefficient:

$$x \log 5 = \log 12$$

Dividing both sides by log 5 gives us

$$x = \frac{\log 12}{\log 5}$$

> **Note**
> Your answers to the problems in the problem sets may differ from those in the book slightly, depending on whether you round off intermediate answers or carry out the complete problem on a calculator. The differences will occur only in the right-most digit of your answers.

If we want a decimal approximation to the solution, we can find log 12 and log 5 on a calculator and divide:

$$x \approx \frac{1.0792}{0.6990}$$

$$\approx 1.5439$$

The complete problem looks like this:

$$5^x = 12$$

$$\log 5^x = \log 12$$

$$x \log 5 = \log 12$$

$$x = \frac{\log 12}{\log 5}$$

$$\approx \frac{1.0792}{0.6990}$$

$$\approx 1.5439$$

Here is another example of solving an exponential equation using logarithms.

EXAMPLE 1

Solve for x: $25^{2x+1} = 15$.

SOLUTION Taking the logarithm of both sides and then writing the exponent $(2x + 1)$ as a coefficient, we proceed as follows:

$$25^{2x+1} = 15$$

$$\log 25^{2x+1} = \log 15 \quad \textbf{Take the log of both sides}$$

$$(2x + 1) \log 25 = \log 15 \quad \textbf{Property 3}$$

$$2x + 1 = \frac{\log 15}{\log 25} \quad \textbf{Divide by log 25}$$

$$2x = \frac{\log 15}{\log 25} - 1 \quad \textbf{Add } -1 \textbf{ to both sides}$$

$$x = \frac{1}{2}\left(\frac{\log 15}{\log 25} - 1\right) \quad \textbf{Multiply both sides by } \tfrac{1}{2}$$

Using a calculator, we can write a decimal approximation to the answer:

$$x \approx \frac{1}{2}\left(\frac{1.1761}{1.3979} - 1\right)$$

$$\approx \frac{1}{2}(0.8413 - 1)$$

$$\approx \frac{1}{2}(-0.1587)$$

$$\approx -0.0794$$

If you invest P dollars in an account with an annual interest rate r that is compounded n times a year, then t years later the amount of money in that account will be

$$A = P\left(1 + \frac{r}{n}\right)^{nt}$$

EXAMPLE 2 How long does it take for \$5,000 to double if it is deposited in an account that yields 5% interest compounded once a year?

SOLUTION Substituting $P = 5{,}000$, $r = 0.05$, $n = 1$, and $A = 10{,}000$ into our formula, we have

$$10{,}000 = 5{,}000(1 + 0.05)^t$$
$$10{,}000 = 5{,}000(1.05)^t$$
$$2 = (1.05)^t \qquad \textbf{Divide by 5,000}$$

This is an exponential equation. We solve by taking the logarithm of both sides:

$$\log 2 = \log(1.05)^t$$
$$= t \log 1.05$$

Dividing both sides by log 1.05, we have

$$t = \frac{\log 2}{\log 1.05}$$
$$\approx 14.2$$

It takes a little over 14 years for \$5,000 to double if it earns 5% interest per year, compounded once a year.

There is a fourth property of logarithms we have not yet considered. This last property allows us to change from one base to another and is therefore called the *change-of-base property.*

Property 4 (Change of Base)

If a and b are both positive numbers other than 1, and if $x > 0$, then

$$\log_a x = \frac{\log_b x}{\log_b a}$$

↑ Base a ↑ Base b

The logarithm on the left side has a base of a, and both logarithms on the right side have a base of b. This allows us to change from base a to any other base b that is a positive number other than 1. Here is a proof of property 4 for logarithms.

Proof We begin by writing the identity

$$a^{\log_a x} = x$$

Taking the logarithm base b of both sides and writing the exponent $\log_a x$ as a coefficient, we have

$$\log_b a^{\log_a x} = \log_b x$$
$$\log_a x \log_b a = \log_b x$$

Dividing both sides by $\log_b a$, we have the desired result:

$$\frac{\log_a x \log_b a}{\log_b a} = \frac{\log_b x}{\log_b a}$$

$$\log_a x = \frac{\log_b x}{\log_b a}$$

We can use this property to find logarithms we could not otherwise compute on our calculators—that is, logarithms with bases other than 10 or e. The next example illustrates the use of this property.

EXAMPLE 3 Find $\log_8 24$.

SOLUTION Since we do not have base-8 logarithms on our calculators, we can change this expression to an equivalent expression that contains only base-10 logarithms:

$$\log_8 24 = \frac{\log 24}{\log 8} \qquad \textbf{Property 4}$$

Don't be confused. We did not just drop the base—we changed to base 10. We could have written the last line like this:

$$\log_8 24 = \frac{\log_{10} 24}{\log_{10} 8}$$

From our calculators, we write

$$\log_8 24 \approx \frac{1.3802}{0.9031}$$

$$\approx 1.5283$$

Application

EXAMPLE 4 Suppose that the population in a small city is 32,000 in the beginning of 1994 and that the city council assumes that the population size t years later can be estimated by the equation

$$P = 32{,}000e^{0.05t}$$

Approximately when will the city have a population of 50,000?

SOLUTION We substitute 50,000 for P in the equation and solve for t:

$$50{,}000 = 32{,}000e^{0.05t}$$

$$1.5625 = e^{0.05t} \qquad \frac{50{,}000}{32{,}000} = 1.5625$$

To solve this equation for t, we can take the natural logarithm of each side:

$$\ln 1.5625 = \ln e^{0.05t}$$

$$= 0.05t \ln e \qquad \textbf{Property 3 for logarithms}$$

$$= 0.05t \qquad \textbf{Because } \ln e = 1$$

$$t = \frac{\ln 1.5625}{0.05} \qquad \textbf{Divide each side by 0.05}$$

$$\approx 8.93 \text{ years}$$

We can estimate that the population will reach 50,000 toward the end of 2002.

USING TECHNOLOGY

Solving Exponential Equations

We can solve the equation $50{,}000 = 32{,}000e^{0.05t}$ from Example 4 with a graphing calculator by defining the expression on each side of the equation as a function. The solution to the equation will be the x-value of the point where the two graphs intersect.

First define functions $Y_1 = 50000$ and $Y_2 = 32{,}000e^{(0.05x)}$ as shown in Figure 1. Set your window variables so that

$$0 \le x \le 15;\ 0 \le y \le 70{,}000,\ \text{scale} = 10{,}000$$

Graph both functions, then use the appropriate command on your calculator to find the coordinates of the intersection point. From Figure 2 we see that the x-coordinate of this point is $x \approx 8.93$.

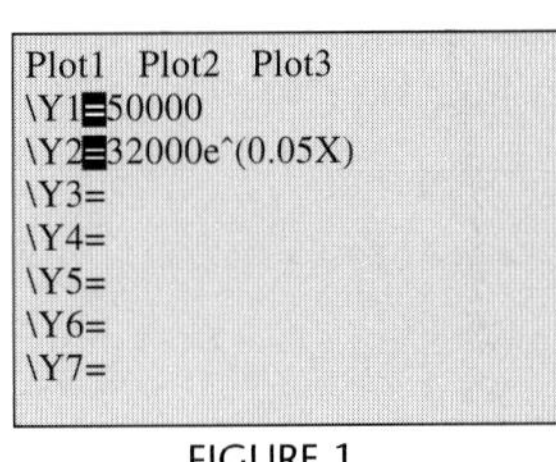

FIGURE 1

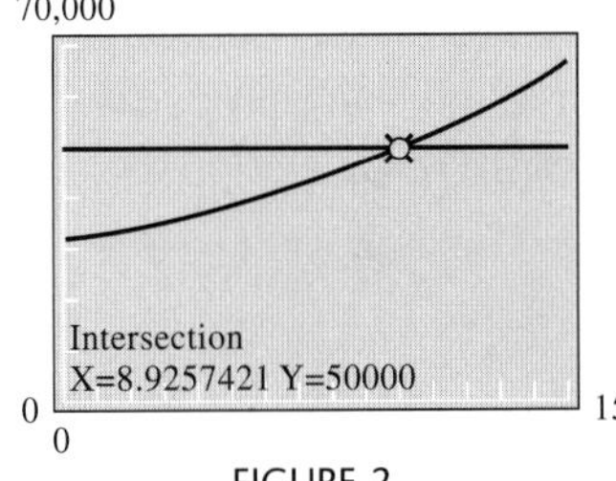

FIGURE 2

LINKING OBJECTIVES AND EXAMPLES

Next to each **objective** we have listed the examples that are best described by that objective.

A	1
B	3
C	2, 4

GETTING READY FOR CLASS

After reading through the preceding section, respond in your own words and in complete sentences.

1. What is an exponential equation?
2. How do logarithms help you solve exponential equations?
3. What is the change-of-base property?
4. Write an application modeled by the equation $A = 10{,}000\left(1 + \dfrac{0.08}{2}\right)^{2\cdot 5}$.

Problem Set 9.6

Online support materials can be found at www.thomsonedu.com/login

Solve each exponential equation. Use a calculator to write the answer to four decimal places.

1. $3^x = 5$
2. $4^x = 3$
3. $5^x = 3$
4. $3^x = 4$
5. $5^{-x} = 12$
6. $7^{-x} = 8$
7. $12^{-x} = 5$
8. $8^{-x} = 7$
9. $8^{x+1} = 4$
10. $9^{x+1} = 3$
11. $4^{x-1} = 4$
12. $3^{x-1} = 9$
13. $3^{2x+1} = 2$
14. $2^{2x+1} = 3$
15. $3^{1-2x} = 2$
16. $2^{1-2x} = 3$
17. $15^{3x-4} = 10$
18. $10^{3x-4} = 15$
19. $6^{5-2x} = 4$
20. $9^{7-3x} = 5$

Use the change-of-base property and a calculator to find a decimal approximation to each of the following logarithms.

21. $\log_8 16$
22. $\log_9 27$
23. $\log_{16} 8$
24. $\log_{27} 9$
25. $\log_7 15$
26. $\log_3 12$
27. $\log_{15} 7$
28. $\log_{12} 3$
29. $\log_8 240$
30. $\log_6 180$
31. $\log_4 321$
32. $\log_5 462$

Find a decimal approximation to each of the following natural logarithms.

33. ln 345
34. ln 3,450
35. ln 0.345
36. ln 0.0345
37. ln 10
38. ln 100
39. ln 45,000
40. ln 450,000

Applying the Concepts

41. **Compound Interest** How long will it take for \$500 to double if it is invested at 6% annual interest compounded twice a year?

42. **Compound Interest** How long will it take for \$500 to double if it is invested at 6% annual interest compounded 12 times a year?

43. **Compound Interest** How long will it take for \$1,000 to triple if it is invested at 12% annual interest compounded 6 times a year?

44. **Compound Interest** How long will it take for \$1,000 to become \$4,000 if it is invested at 12% annual interest compounded 6 times a year?

45. **Doubling Time** How long does it take for an amount of money P to double itself if it is invested at 8% interest compounded 4 times a year?

46. **Tripling Time** How long does it take for an amount of money P to triple itself if it is invested at 8% interest compounded 4 times a year?

47. **Tripling Time** If a \$25 investment is worth \$75 today, how long ago must that \$25 have been invested at 6% interest computed twice a year?

48. **Doubling Time** If a \$25 investment is worth \$50 today, how long ago must that \$25 have been invested at 6% interest computed twice a year?

Recall that if P dollars are invested in an account with annual interest rate r, compounded continuously, then the amount of money in the account after t years is given by the formula $A(t) = Pe^{rt}$.

49. **Continuously Compounded Interest** Repeat Problem 41 if the interest is compounded continuously.

50. **Continuously Compounded Interest** Repeat Problem 44 if the interest is compounded continuously.

51. **Continuously Compounded Interest** How long will it take \$500 to triple if it is invested at 6% annual interest, compounded continuously?

52. **Continuously Compounded Interest** How long will it take \$500 to triple if it is invested at 12% annual interest, compounded continuously?

= Videos available by instructor request
▸ = Online student support materials available at www.thomsonedu.com/login

Maintaining Your Skills

Find the vertex for each of the following parabolas, and then indicate if it is the highest or lowest point on the graph.

53. $y = 2x^2 + 8x - 15$

54. $y = 3x^2 - 9x - 10$

55. $y = 12x - 4x^2$

56. $y = 18x - 6x^2$

57. Maximum Height An object is projected into the air with an initial upward velocity of 64 feet per second. Its height h at any time t is given by the formula $h = 64t - 16t^2$. Find the time at which the object reaches its maximum height. Then, find the maximum height.

58. Maximum Height An object is projected into the air with an initial upward velocity of 64 feet per second from the top of a building 40 feet high. If the height h of the object t seconds after it is projected into the air is $h = 40 + 64t - 16t^2$, find the time at which the object reaches its maximum height. Then, find the maximum height it attains.

Extending the Concepts

59. Exponential Growth Suppose that the population in a small city is 32,000 at the beginning of 1994 and that the city council assumes that the population size t years later can be estimated by the equation

$$P(t) = 32{,}000e^{0.05t}$$

Approximately when will the city have a population of 64,000?

60. Exponential Growth Suppose the population of a city is given by the equation

$$P(t) = 100{,}000e^{0.05t}$$

where t is the number of years from the present time. How large is the population now? (Now corresponds to a certain value of t. Once you realize what that value of t is, the problem becomes very simple.)

61. Exponential Growth Suppose the population of a city is given by the equation

$$P(t) = 15{,}000e^{0.04t}$$

where t is the number of years from the present time. How long will it take for the population to reach 45,000?

62. Exponential Growth Suppose the population of a city is given by the equation

$$P(t) = 15{,}000e^{0.08t}$$

where t is the number of years from the present time. How long will it take for the population to reach 45,000?

63. Solve the formula $A = Pe^{rt}$ for t.

64. Solve the formula $A = Pe^{-rt}$ for t.

65. Solve the formula $A = P \cdot 2^{-kt}$ for t.

66. Solve the formula $A = P \cdot 2^{kt}$ for t.

67. Solve the formula $A = P(1 - r)^t$ for t.

68. Solve the formula $A = P(1 + r)^t$ for t.

Chapter 9 SUMMARY

Exponential Functions [9.1]

EXAMPLES

1. For the exponential function $f(x) = 2^x$,
$$f(0) = 2^0 = 1$$
$$f(1) = 2^1 = 2$$
$$f(2) = 2^2 = 4$$
$$f(3) = 2^3 = 8$$

Any function of the form

$$f(x) = b^x$$

where $b > 0$ and $b \neq 1$, is an *exponential function.*

One-to-One Functions [9.2]

2. The function $f(x) = x^2$ is not one-to-one because 9, which is in the range, comes from both 3 and -3 in the domain.

A function is a *one-to-one function* if every element in the range comes from exactly one element in the domain.

Inverse Functions [9.2]

3. The inverse of $f(x) = 2x - 3$ is
$$f^{-1}(x) = \frac{x+3}{2}$$

The *inverse* of a function is obtained by reversing the order of the coordinates of the ordered pairs belonging to the function. Only one-to-one functions have inverses that are also functions.

Definition of Logarithms [9.3]

4. The definition allows us to write expressions like
$$y = \log_3 27$$
equivalently in exponential form as
$$3^y = 27$$
which makes it apparent that y is 3.

If b is a positive number not equal to 1, then the expression

$$y = \log_b x$$

is equivalent to $x = b^y$; that is, in the expression $y = \log_b x$, y is the number to which we raise b in order to get x. Expressions written in the form $y = \log_b x$ are said to be in *logarithmic form.* Expressions like $x = b^y$ are in *exponential form.*

Two Special Identities [9.3]

5. Examples of the two special identities are
$$5^{\log_5 12} = 12$$
and
$$\log_8 8^3 = 3$$

For $b > 0$, $b \neq 1$, the following two expressions hold for all positive real numbers x:

(1) $b^{\log_b x} = x$

(2) $\log_b b^x = x$

Properties of Logarithms [9.4]

6. We can rewrite the expression

$$\log_{10} \frac{45^6}{273}$$

using the properties of logarithms, as

$$6 \log_{10} 45 - \log_{10} 273$$

If x, y, and b are positive real numbers, $b \neq 1$, and r is any real number, then:

1. $\log_b (xy) = \log_b x + \log_b y$
2. $\log_b \left(\frac{x}{y}\right) = \log_b x - \log_b y$

Common Logarithms [9.5]

7. $\log_{10} 10{,}000 = \log 10{,}000$
$= \log 10^4$
$= 4$

3. $\log_b x^r = r \log_b x$

Common logarithms are logarithms with a base of 10. To save time in writing, we omit the base when working with common logarithms; that is,

$$\log x = \log_{10} x$$

Natural Logarithms [9.5]

8. $\ln e = 1$
$\ln 1 = 0$

Natural logarithms, written *ln x*, are logarithms with a base of e, where the number e is an irrational number (like the number π). A decimal approximation for e is 2.7183. All the properties of exponents and logarithms hold when the base is e.

Change of Base [9.6]

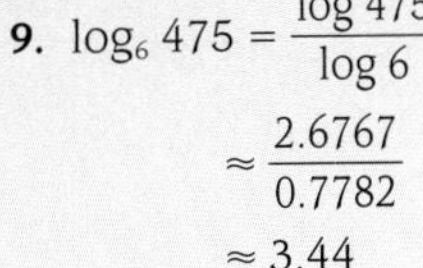

9. $\log_6 475 = \frac{\log 475}{\log 6}$
$\approx \frac{2.6767}{0.7782}$
≈ 3.44

If x, a, and b are positive real numbers, $a \neq 1$ and $b \neq 1$, then

$$\log_a x = \frac{\log_b x}{\log_b a}$$

! COMMON MISTAKES

The most common mistakes that occur with logarithms come from trying to apply the three properties of logarithms to situations in which they don't apply. For example, a very common mistake looks like this:

$$\frac{\log 3}{\log 2} = \log 3 - \log 2 \qquad \textbf{Mistake}$$

This is not a property of logarithms. To write the equation $\log 3 - \log 2$, we would have to start with

$$\log \frac{3}{2} \qquad \textit{NOT} \qquad \frac{\log 3}{\log 2}$$

There is a difference.

Chapter 9 Review Test

The problems below form a comprehensive review of the material in this chapter. They can be used to study for exams. If you would like to take a practice test on this chapter, you can use the odd-numbered problems. Give yourself an hour and work as many of the odd-numbered problems as possible. When you are finished, or when an hour has passed, check your answers with the answers in the back of the book. You can use the even-numbered problems for a second practice test.

Let $f(x) = 2^x$ and $g(x) = (\frac{1}{3})^x$, and find the following. [9.1]

1. $f(4)$ **2.** $f(-1)$

3. $g(2)$ **4.** $f(2) - g(-2)$

5. $f(-1) + g(1)$ **6.** $g(-1) + f(2)$

7. The graph of $y = f(x)$ **8.** The graph of $y = g(x)$

For each relation that follows, sketch the graph of the relation and its inverse, and write an equation for the inverse. [9.2]

9. $y = 2x + 1$ **10.** $y = x^2 - 4$

For each of the following functions, find the equation of the inverse. Write the inverse using the notation $f^{-1}(x)$ if the inverse is itself a function. [9.2]

11. $f(x) = 2x + 3$ **12.** $f(x) = x^2 - 1$

13. $f(x) = \frac{1}{2}x + 2$ **14.** $f(x) = 4 - 2x^2$

Write each equation in logarithmic form. [9.3]

15. $3^4 = 81$ **16.** $7^2 = 49$

17. $0.01 = 10^{-2}$ **18.** $2^{-3} = \frac{1}{8}$

Write each equation in exponential form. [9.3]

19. $\log_2 8 = 3$ **20.** $\log_3 9 = 2$

21. $\log_4 2 = \frac{1}{2}$ **22.** $\log_4 4 = 1$

Solve for x. [9.3]

23. $\log_5 x = 2$ **24.** $\log_{16} 8 = x$

25. $\log_x 0.01 = -2$

Graph each equation. [9.3]

26. $y = \log_2 x$ **27.** $y = \log_{1/2} x$

Simplify each expression. [9.3]

28. $\log_4 16$ **29.** $\log_{27} 9$

30. $\log_4(\log_3 3)$

Use the properties of logarithms to expand each expression. [9.4]

31. $\log_2 5x$ **32.** $\log_{10} \dfrac{2x}{y}$

33. $\log_a \dfrac{y^3\sqrt{x}}{z}$ **34.** $\log_{10} \dfrac{x^2}{y^3z^4}$

Write each expression as a single logarithm. [9.4]

35. $\log_2 x + \log_2 y$ **36.** $\log_3 x - \log_3 4$

37. $2 \log_a 5 - \frac{1}{2} \log_a 9$

38. $3 \log_2 x + 2 \log_2 y - 4 \log_2 z$

Solve each equation. [9.4]

39. $\log_2 x + \log_2 4 = 3$ **40.** $\log_2 x - \log_2 3 = 1$

41. $\log_3 x + \log_3(x - 2) = 1$

42. $\log_4(x + 1) - \log_4(x - 2) = 1$

43. $\log_6(x - 1) + \log_6 x = 1$ **44.** $\log_4(x - 3) + \log_4 x = 1$

Evaluate each expression. [9.5]

45. $\log 346$ **46.** $\log 0.713$

Find x. [9.5]

47. $\log x = 3.9652$ **48.** $\log x = -1.6003$

Simplify. [9.5]

49. $\ln e$ **50.** $\ln 1$

51. $\ln e^2$ **52.** $\ln e^{-4}$

Use the formula $\text{pH} = -\log[H^+]$ to find the pH of a solution with the given hydrogen ion concentration. [9.5]

53. $[H^+] = 7.9 \times 10^{-3}$ **54.** $[H^+] = 8.1 \times 10^{-6}$

Find $[H^+]$ for a solution when the given pH [9.5]

55. pH = 2.7 **56.** pH = 7.5

Solve each equation. [9.6]

57. $4^x = 8$ **58.** $4^{3x+2} = 5$

Use the change-of-base property and a calculator to evaluate each expression. Round your answers to the nearest hundredth. [9.6]

59. $\log_{16} 8$

60. $\log_{12} 421$

Use the formula $A = P\left(1 + \frac{r}{n}\right)^{nt}$ to solve each of the following problems. [9.6]

61. Investing How long does it take \$5,000 to double if it is deposited in an account that pays 16% annual interest compounded once a year?

62. Investing How long does it take \$10,000 to triple if it is deposited in an account that pays 12% annual interest compounded 6 times a year?

GROUP PROJECT Two Depreciation Models

Number of People 3

Time Needed 20 minutes

Equipment Paper, pencil, and graphing calculator

Background Recently, a consumer magazine contained an article on leasing a computer. The original price of the computer was \$2,500. The term of the lease was 24 months. At the end of the lease, the computer could be purchased for its residual value, \$188. This is enough information to find an equation that will give us the value of the computer t months after it has been purchased.

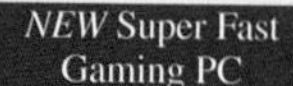

- **Only** \$99/**month** For 24 months
- **Processor** Dual Core 3.0 GHz
- **Monitor** 20-inch Widescreen
- **Hard Drive** 320 GB Ultra Fast
- **Memory** 2 GB Dual Channel
- **Video Card** VFast 9200XL
- **Optical Drive** Dual 24x DVD-ROM

Free Shipping
Buy today \$2,500

Procedure We will find models for two types of depreciation: linear depreciation and exponential depreciation. Here are the general equations:

Linear Depreciation	*Exponential Depreciation*
$V = mt + b$	$V = V_0e^{-kt}$

1. Let t represent time in months and V represent the value of the computer at time t; then find two ordered pairs (t, V) that correspond to the initial price of the computer and another ordered pair that corresponds to the residual value of \$188 when the computer is 2 years old.
2. Use the two ordered pairs to find m and b in the linear depreciation model; then write the equation that gives us linear depreciation.
3. Use the two ordered pairs to find V_0 and k in the exponential depreciation model; then write the equation that gives us exponential depreciation.
4. Graph each of the equations on your graphing calculator; then sketch the graphs on the following templates.
5. Find the value of the computer after 1 year, using both models.
6. Which of the two models do you think best describes the depreciation of a computer?

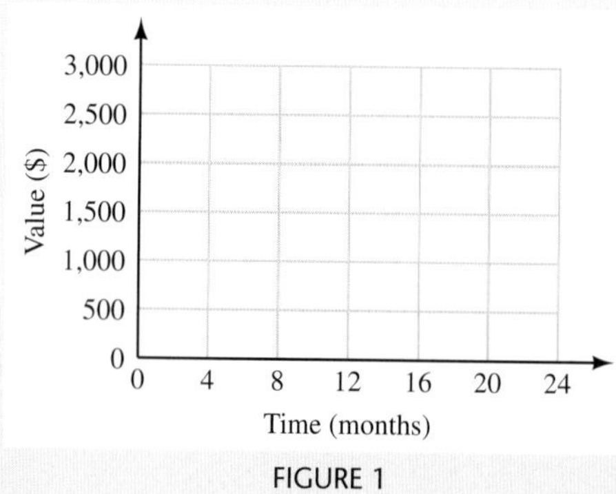

FIGURE 1

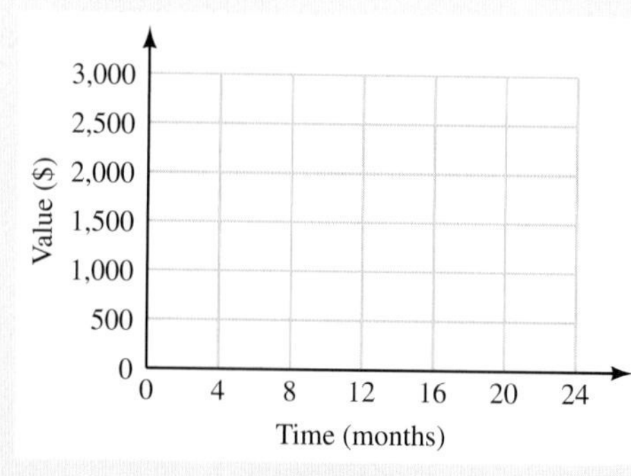

FIGURE 2

RESEARCH PROJECT

Drag Racing

20th Century Fox/The Kobal Collection

The movie *Heart Like a Wheel* is based on the racing career of drag racer Shirley Muldowney. The movie includes a number of races. Choose four races as the basis for your report. For each race you choose, give a written description of the events leading to the race, along with the details of the race itself. Then draw a graph that shows the speed of the dragster as a function of time. Do this for each race. One such graph from earlier in the chapter is shown here as an example. (For each graph, label the horizontal axis from 0 to 12 seconds, and label the vertical axis from 0 to 200 miles per hour.) Follow each graph with a description of any significant events that happen during the race (such as a motor malfunction or a crash) and their correlation to the graph.

Elapsed Time (sec)	Speed (mph)
0	0.0
1	72.7
2	129.9
3	162.8
4	192.2
5	212.4
6	228.1

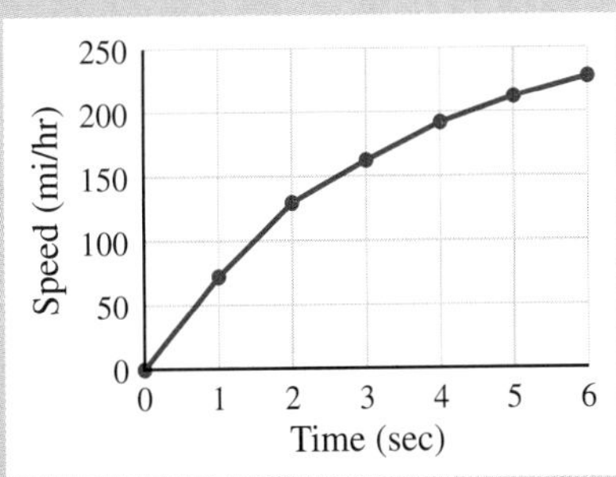

Conic Sections

10

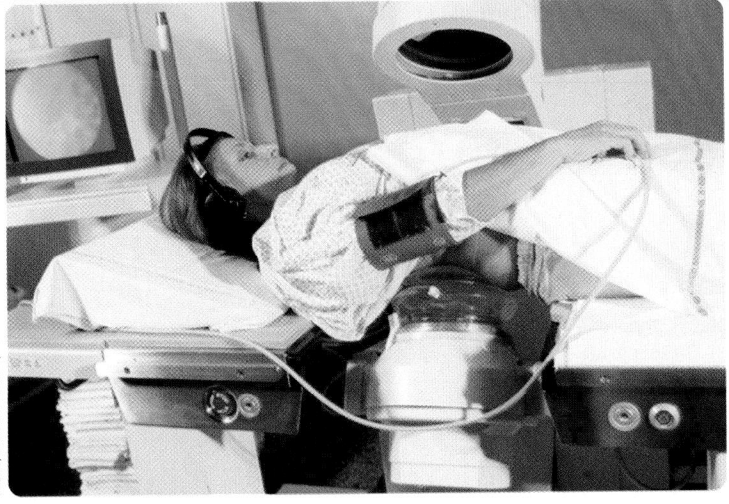

Phanie/Photo Researchers, Inc.

CHAPTER OUTLINE

One of the curves we will study in this chapter has interesting reflective properties. Figure 1(a) shows how you can draw one of these curves (an ellipse) using thumbtacks, string, pencil, and paper. Elliptical surfaces will reflect sound waves that originate at one focus through the other focus. This property of ellipses allows doctors to treat patients with kidney stones using a procedure called lithotripsy. A lithotripter is an elliptical device that creates sound waves that crush the kidney stone into small pieces, without surgery. The sound wave originates at one focus of the lithotripter. The energy from it reflects off the surface of the lithotripter and converges at the other focus, where the kidney stone is positioned. Figure 1(b) shows a cross-section of a lithotripter, with a patient positioned so the kidney stone is at the other focus.

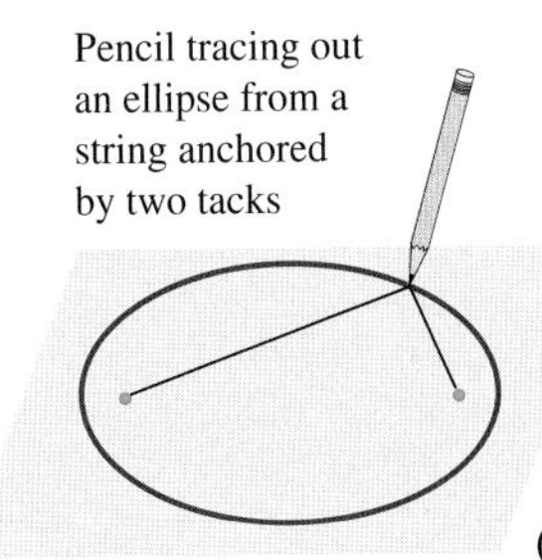

Pencil tracing out
an ellipse from a
string anchored
by two tacks

(a)

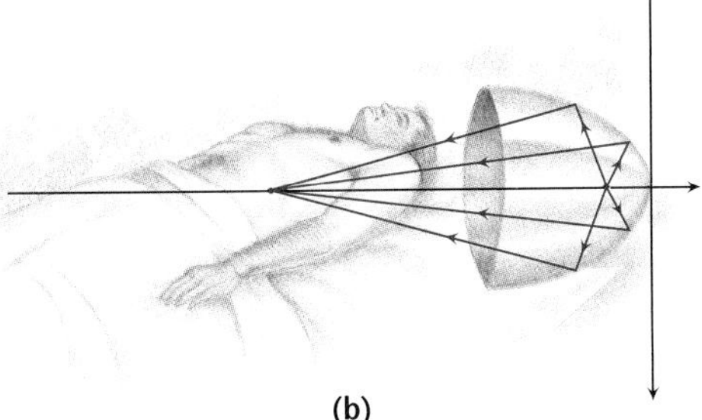

(b)

FIGURE 1

By studying the conic sections in this chapter, you will be better equipped to understand some of the more technical equipment that exists in the world outside of class.

10.1 The Circle

OBJECTIVES

A Use the distance formula.

B Write the equation of a circle, given its center and radius.

C Find the center and radius of a circle from its equation, and then sketch the graph.

Conic sections include ellipses, circles, hyperbolas, and parabolas. They are called conic sections because each can be found by slicing a cone with a plane as shown in Figure 1. We begin our work with conic sections by studying circles. Before we find the general equation of a circle, we must first derive what is known as the *distance formula.*

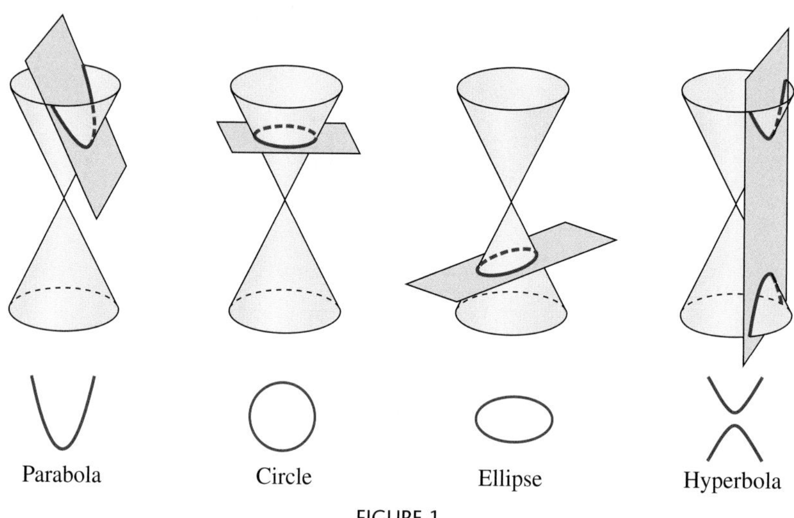

FIGURE 1

Suppose (x_1, y_1) and (x_2, y_2) are any two points in the first quadrant. (Actually, we could choose the two points to be anywhere on the coordinate plane. It is just more convenient to have them in the first quadrant.) We can name the points P_1 and P_2, respectively, and draw the diagram shown in Figure 2.

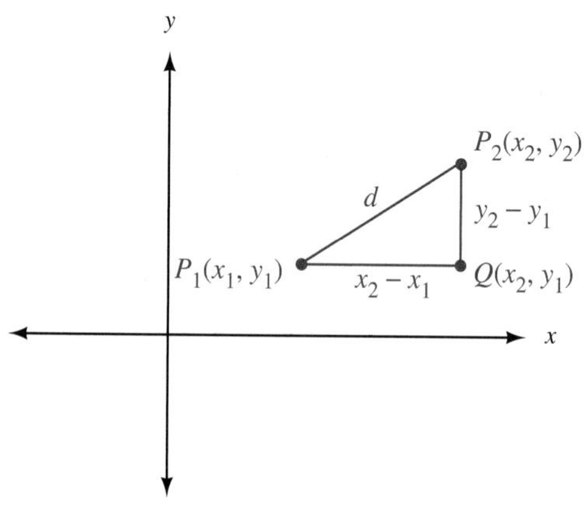

FIGURE 2

Notice the coordinates of point Q. The x-coordinate is x_2 since Q is directly below point P_2. The y-coordinate of Q is y_1 since Q is directly across from point P_1.

It is evident from the diagram that the length of P_2Q is $y_2 - y_1$ and the length of P_1Q is $x_2 - x_1$. Using the Pythagorean theorem, we have

$$(P_1P_2)^2 = (P_1Q)^2 + (P_2Q)^2$$

or

$$d^2 = (x_2 - x_1)^2 + (y_2 - y_1)^2$$

Taking the square root of both sides, we have

$$d = \sqrt{(x_2 - x_1)^2 + (y_2 - y_1)^2}$$

We know this is the positive square root since d is the distance from P_1 to P_2 and therefore must be positive. This formula is called the *distance formula.*

EXAMPLE 1 Find the distance between (3, 5) and (2, −1).

SOLUTION If we let (3, 5) be (x_1, y_1) and (2, −1) be (x_2, y_2) and apply the distance formula, we have

$$\begin{aligned} d &= \sqrt{(2-3)^2 + (-1-5)^2} \\ &= \sqrt{(-1)^2 + (-6)^2} \\ &= \sqrt{1 + 36} \\ &= \sqrt{37} \end{aligned}$$

EXAMPLE 2 Find x if the distance from $(x, 5)$ to (3, 4) is $\sqrt{2}$.

SOLUTION Using the distance formula, we have

$$\begin{aligned} \sqrt{2} &= \sqrt{(x-3)^2 + (5-4)^2} \\ 2 &= (x-3)^2 + 1^2 \\ 2 &= x^2 - 6x + 9 + 1 \\ 0 &= x^2 - 6x + 8 \\ 0 &= (x-4)(x-2) \\ x &= 4 \quad \text{or} \quad x = 2 \end{aligned}$$

The two solutions are 4 and 2, which indicates there are two points, (4, 5) and (2, 5), which are $\sqrt{2}$ units from (3, 4).

Circles

Because of their perfect symmetry, circles have been used for thousands of years in many disciplines, including art, science, and religion. The photograph shown in Figure 3 is of Stonehenge, a 4,500-year-old site in England. The arrangement of the stones is based on a circular plan that is thought to have both religious and astronomical significance. More recently, the design shown in Figure 4 began appearing in agricultural fields in England in the 1990s. Whoever made these designs chose the circle as their basic shape.

FIGURE 3

FIGURE 4

We can model circles very easily in algebra by using equations that are based on the distance formula.

Circle Theorem The equation of the circle with center at (a, b) and radius r is given by

$$(x - a)^2 + (y - b)^2 = r^2$$

Proof By definition, all points on the circle are a distance r from the center (a, b). If we let (x, y) represent any point on the circle, then (x, y) is r units from (a, b). Applying the distance formula, we have

$$r = \sqrt{(x - a)^2 + (y - b)^2}$$

Squaring both sides of this equation gives the equation of the circle:

$$(x - a)^2 + (y - b)^2 = r^2$$

We can use the circle theorem to find the equation of a circle, given its center and radius, or to find its center and radius, given the equation.

EXAMPLE 3

Find the equation of the circle with center at $(-3, 2)$ having a radius of 5.

SOLUTION We have $(a, b) = (-3, 2)$ and $r = 5$. Applying the circle theorem yields

$$[x - (-3)]^2 + (y - 2)^2 = 5^2$$

$$(x + 3)^2 + (y - 2)^2 = 25$$

EXAMPLE 4

Give the equation of the circle with radius 3 whose center is at the origin.

SOLUTION The coordinates of the center are $(0, 0)$, and the radius is 3. The equation must be

$$(x - 0)^2 + (y - 0)^2 = 3^2$$

$$x^2 + y^2 = 9$$

We can see from Example 4 that the equation of any circle with its center at the origin and radius r will be

$$x^2 + y^2 = r^2$$

EXAMPLE 5 Find the center and radius, and sketch the graph, of the circle whose equation is

$$(x - 1)^2 + (y + 3)^2 = 4$$

SOLUTION Writing the equation in the form

$$(x - a)^2 + (y - b)^2 = r^2$$

we have

$$(x - 1)^2 + [y - (-3)]^2 = 2^2$$

The center is at $(1, -3)$ and the radius is 2. The graph is shown in Figure 5.

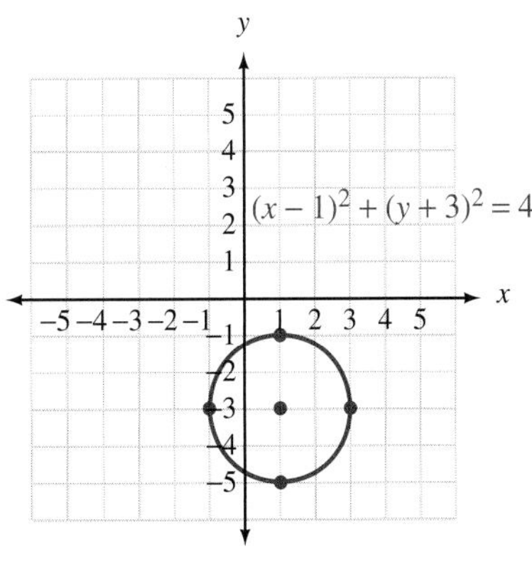

FIGURE 5

EXAMPLE 6 Sketch the graph of $x^2 + y^2 = 9$.

SOLUTION Since the equation can be written in the form

$$(x - 0)^2 + (y - 0)^2 = 3^2$$

it must have its center at $(0, 0)$ and a radius of 3. The graph is shown in Figure 6.

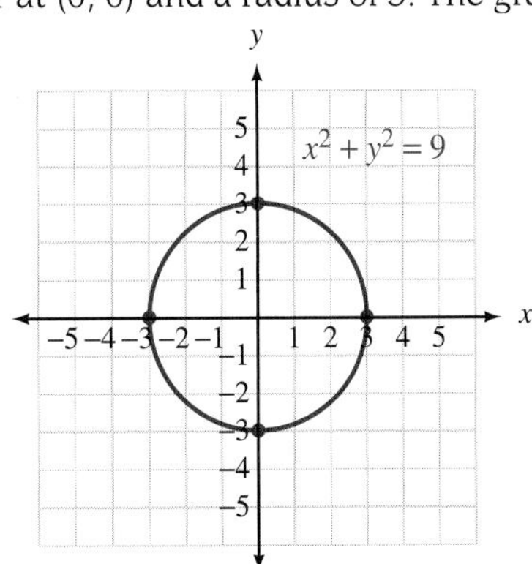

FIGURE 6

EXAMPLE 7

Sketch the graph of $x^2 + y^2 + 6x - 4y - 12 = 0$.

SOLUTION To sketch the graph, we must find the center and radius. The center and radius can be identified if the equation has the form

$$(x - a)^2 + (y - b)^2 = r^2$$

The original equation can be written in this form by completing the squares on x and y:

$$x^2 + y^2 + 6x - 4y - 12 = 0$$

$$x^2 + 6x + y^2 - 4y = 12$$

$$x^2 + 6x + \mathbf{9} + y^2 - 4y + \mathbf{4} = 12 + \mathbf{9} + \mathbf{4}$$

$$(x + 3)^2 + (y - 2)^2 = 25$$

$$(x + 3)^2 + (y - 2)^2 = 5^2$$

From the last line it is apparent that the center is at $(-3, 2)$ and the radius is 5. The graph is shown in Figure 7.

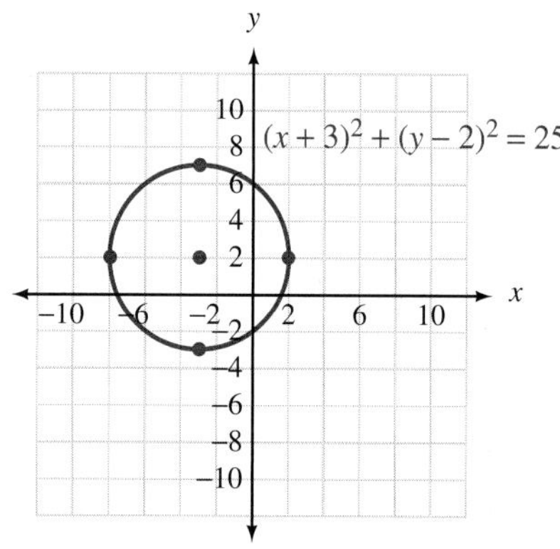

FIGURE 7

LINKING OBJECTIVES AND EXAMPLES

Next to each **objective** we have listed the examples that are best described by that objective.

A	1, 2
B	3, 4
C	5–7

GETTING READY FOR CLASS

After reading through the preceding section, respond in your own words and in complete sentences.

1. Describe the distance formula in words, as if you were explaining to someone how they should go about finding the distance between two points.
2. What is the mathematical definition of a circle?
3. How are the distance formula and the equation of a circle related?
4. When graphing a circle from its equation, why is completing the square sometimes useful?

Problem Set 10.1

Online support materials can be found at www.thomsonedu.com/login

Find the distance between the following points.

1. (3, 7) and (6, 3)
2. (4, 7) and (8, 1)
3. (0, 9) and (5, 0)
4. (−3, 0) and (0, 4)
5. (3, −5) and (−2, 1)
6. (−8, 9) and (−3, −2)
7. (−1, −2) and (−10, 5)
8. (−3, −8) and (−1, 6)
9. Find x so the distance between $(x, 2)$ and $(1, 5)$ is $\sqrt{13}$.
10. Find x so the distance between $(-2, 3)$ and $(x, 1)$ is 3.
11. Find y so the distance between $(7, y)$ and $(8, 3)$ is 1.
12. Find y so the distance between $(3, -5)$ and $(3, y)$ is 9.

Write the equation of the circle with the given center and radius.

13. Center (2, 3); $r = 4$
14. Center (3, −1); $r = 5$
15. Center (3, −2); $r = 3$
16. Center (−2, 4); $r = 1$
17. Center (−5, −1); $r = \sqrt{5}$
18. Center (−7, −6); $r = \sqrt{3}$
19. Center (0, −5); $r = 1$
20. Center (0, −1); $r = 7$
21. Center (0, 0); $r = 2$
22. Center (0, 0); $r = 5$

Give the center and radius, and sketch the graph of each of the following circles.

23. $x^2 + y^2 = 4$
24. $x^2 + y^2 = 16$
25. $(x - 1)^2 + (y - 3)^2 = 25$
26. $(x - 4)^2 + (y - 1)^2 = 36$
27. $(x + 2)^2 + (y - 4)^2 = 8$
28. $(x - 3)^2 + (y + 1)^2 = 12$
29. $(x + 1)^2 + (y + 1)^2 = 1$
30. $(x + 3)^2 + (y + 2)^2 = 9$
31. $x^2 + y^2 - 6y = 7$
32. $x^2 + y^2 + 10x = 0$
33. $x^2 + y^2 - 4x - 6y = -4$
34. $x^2 + y^2 - 4x + 2y = 4$
35. $x^2 + y^2 + 2x + y = \frac{11}{4}$
36. $x^2 + y^2 - 6x - y = -\frac{1}{4}$

Both of the following circles pass through the origin. In each case, find the equation.

37.

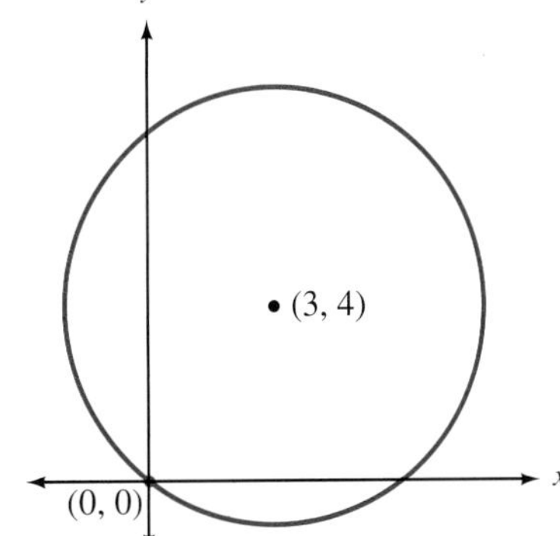

38.

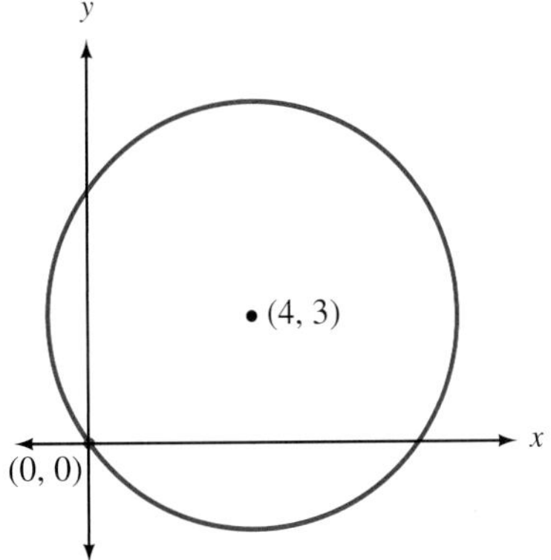

= Videos available by instructor request

▸ = Online student support materials available at www.thomsonedu.com/login

39. Find the equations of circles A, B, and C in the following diagram. The three points are the centers of the three circles.

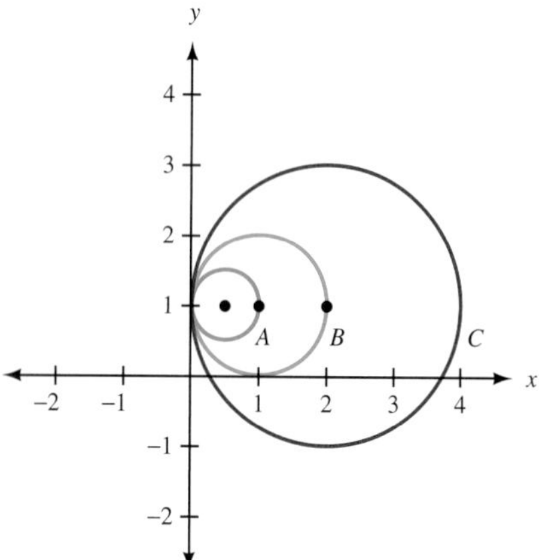

40. Each of the following circles passes through the origin. The centers are as shown. Find the equation of each circle.

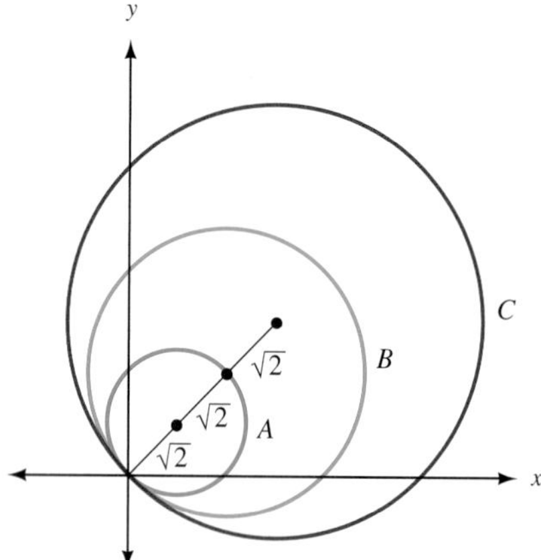

41. Find the equation of each of the three circles shown here.

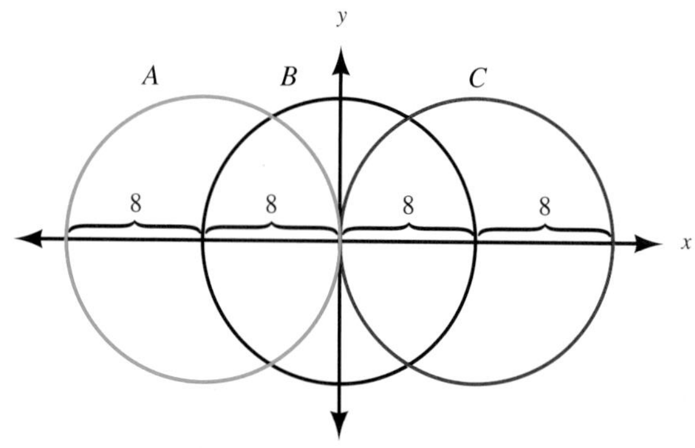

42. A parabola and a circle each contain the points $(-4, 0)$, $(0, 4)$, and $(4, 0)$. Sketch the graph of each curve on the same coordinate system, then write an equation for each curve.

43. Ferris Wheel A giant Ferris wheel has a diameter of 240 feet and sits 12 feet above the ground. As shown in the diagram below, the wheel is 500 feet from the entrance to the park. The xy-coordinate system containing the wheel has its origin on the ground at the center of the entrance. Write an equation that models the shape of the wheel.

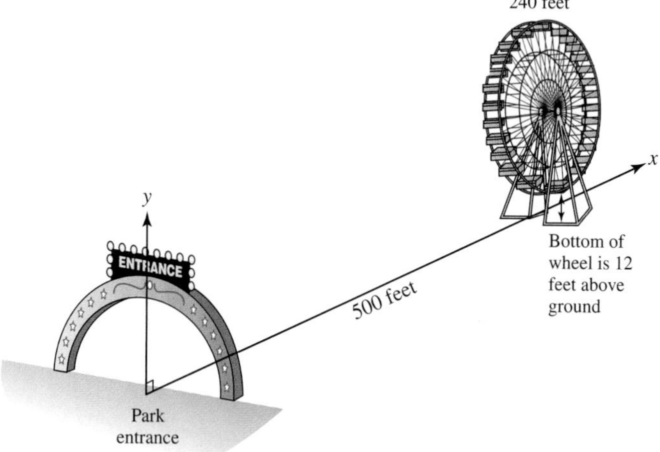

44. Magic Rings A magician is holding two rings that seem to lie in the same plane and intersect in two points. Each ring is 10 inches in diameter.

a. Find the equation of each ring if a coordinate system is placed with its origin at the center of the first ring and the x-axis contains the center of the second ring.

b. Find the equation of each ring if a coordinate system is placed with its origin at the center of the second ring and the x-axis contains the center of the first ring.

Maintaining Your Skills

Find the equation of the inverse of each of the following functions. Write the inverse using the notation $f^{-1}(x)$, if the inverse is itself a function.

45. $f(x) = 3^x$

46. $f(x) = 5^x$

47. $f(x) = 2x + 3$

48. $f(x) = 3x - 2$

49. $f(x) = \dfrac{x + 3}{5}$

50. $f(x) = \dfrac{x - 2}{6}$

Getting Ready for the Next Section

Solve.

51. $y^2 = 9$

52. $x^2 = 25$

53. $-y^2 = 4$

54. $-x^2 = 16$

55. $\dfrac{-x^2}{9} = 1$

56. $\dfrac{y^2}{100} = 1$

57. Divide $4x^2 + 9y^2$ by 36.

58. Divide $25x^2 + 4y^2$ by 100.

Find the x-intercepts and the y-intercepts.

59. $3x - 4y = 12$

60. $y = 3x^2 + 5x - 2$

61. If $\dfrac{x^2}{25} + \dfrac{y^2}{9} = 1$, find y when x is 3.

62. If $\dfrac{x^2}{25} + \dfrac{y^2}{9} = 1$, find y when x is -4.

Extending the Concepts

A circle is *tangent to* a line if it touches, but does not cross, the line.

63. Find the equation of the circle with center at (2, 3) if the circle is tangent to the y-axis.

64. Find the equation of the circle with center at (3, 2) if the circle is tangent to the x-axis.

65. Find the equation of the circle with center at (2, 3) if the circle is tangent to the vertical line $x = 4$.

66. Find the equation of the circle with center at (3, 2) if the circle is tangent to the horizontal line $y = 6$.

Find the distance from the origin to the center of each of the following circles.

67. $x^2 + y^2 - 6x + 8y = 144$

68. $x^2 + y^2 - 8x + 6y = 144$

69. $x^2 + y^2 - 6x - 8y = 144$

70. $x^2 + y^2 + 8x + 6y = 144$

71. If we were to solve the equation $x^2 + y^2 = 9$ for y, we would obtain the equation $y = \pm\sqrt{9 - x^2}$. This last equation is equivalent to the two equations $y = \sqrt{9 - x^2}$, in which y is always positive, and $y = -\sqrt{9 - x^2}$, in which y is always negative. Look at the graph of $x^2 + y^2 = 9$ in Example 6 of this section and indicate what part of the graph each of the two equations corresponds to.

72. Solve the equation $x^2 + y^2 = 9$ for x, and then indicate what part of the graph in Example 6 each of the two resulting equations corresponds to.

10.2 Ellipses and Hyperbolas

OBJECTIVES

A Graph an ellipse.

B Graph a hyperbola.

The photograph in Figure 1 shows Halley's comet as it passed close to earth in 1986. Like the planets in our solar system, it orbits the sun in an elliptical path (Figure 2). While it takes the earth 1 year to complete one orbit around the sun, it takes Halley's comet 76 years. The first known sighting of Halley's comet was in 239 B.C. Its most famous appearance occurred in 1066 A.D., when it was seen at the Battle of Hastings.

NASA

FIGURE 1

The Orbit of Halley's Comet

FIGURE 2

We begin this section with an introductory look at equations that produce ellipses. To simplify our work, we consider only ellipses that are centered about the origin.

Suppose we want to graph the equation

$$\frac{x^2}{25} + \frac{y^2}{9} = 1$$

We can find the y-intercepts by letting $x = 0$, and we can find the x-intercepts by letting $y = 0$:

When $x = 0$	When $y = 0$
$\frac{0^2}{25} + \frac{y^2}{9} = 1$	$\frac{x^2}{25} + \frac{0^2}{9} = 1$
$y^2 = 9$	$x^2 = 25$
$y = \pm 3$	$x = \pm 5$

The graph crosses the y-axis at (0, 3) and (0, −3) and the x-axis at (5, 0) and (−5, 0). Graphing these points and then connecting them with a smooth curve gives the graph shown in Figure 3.

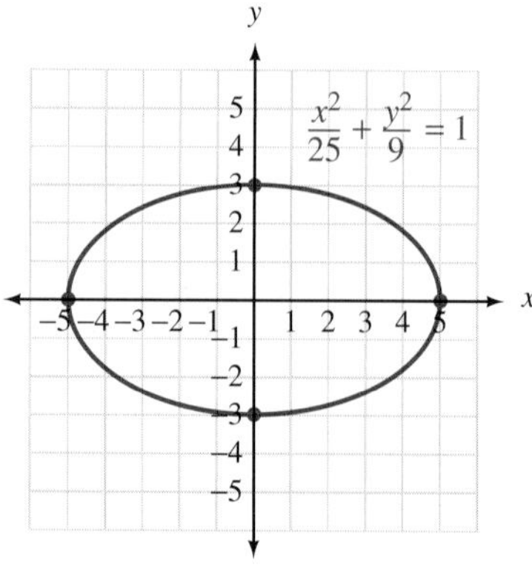

FIGURE 3

Note

We can find other ordered pairs on the graph by substituting values for x (or y) and then solving for y (or x). For example, if we let $x = 3$, then

$$\frac{3^2}{25} + \frac{y^2}{9} = 1$$

$$\frac{9}{25} + \frac{y^2}{9} = 1$$

$$0.36 + \frac{y^2}{9} = 1$$

$$\frac{y^2}{9} = 0.64$$

$$y^2 = 5.76$$

$$y = \pm 2.4$$

This would give us the two ordered pairs (3, −2.4) and (3, 2.4).

A graph of the type shown in Figure 3 is called an *ellipse.* If we were to find some other ordered pairs that satisfy our original equation, we would find that their graphs lie on the ellipse. Also, the coordinates of any point on the ellipse will satisfy the equation. We can generalize these results as follows.

The Ellipse The graph of any equation of the form

$$\frac{x^2}{a^2} + \frac{y^2}{b^2} = 1 \qquad \textbf{Standard form}$$

will be an *ellipse* centered at the origin. The ellipse will cross the x-axis at $(a, 0)$ and $(-a, 0)$. It will cross the y-axis at $(0, b)$ and $(0, -b)$. When a and b are equal, the ellipse will be a circle.

The most convenient way to graph an ellipse is to locate the intercepts.

EXAMPLE 1 Sketch the graph of $4x^2 + 9y^2 = 36$.

SOLUTION To write the equation in the form

$$\frac{x^2}{a^2} + \frac{y^2}{b^2} = 1$$

we must divide both sides by 36:

$$\frac{4x^2}{36} + \frac{9y^2}{36} = \frac{36}{36}$$

$$\frac{x^2}{9} + \frac{y^2}{4} = 1$$

The graph crosses the x-axis at $(3, 0)$, $(-3, 0)$ and the y-axis at $(0, 2)$, $(0, -2)$, as shown in Figure 4.

Note
When the equation is written in standard form, the x-intercepts are the positive and negative square roots of the number below x^2. The y-intercepts are the square roots of the number below y^2.

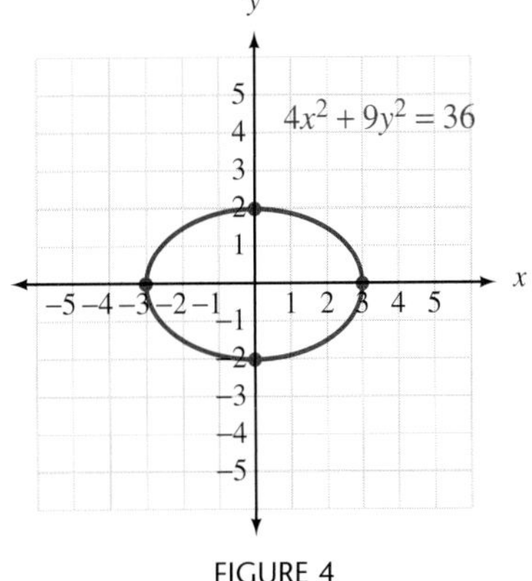

FIGURE 4

Hyperbolas

Figure 5 shows Europa, one of Jupiter's moons, as it was photographed by the Galileo space probe in the late 1990s. To speed up the trip from Earth to Jupiter—nearly a billion miles—Galileo made use of the *slingshot effect.* This involves flying a hyperbolic path very close to a planet, so that gravity can be used to gain velocity as the space probe hooks around the planet (Figure 6).

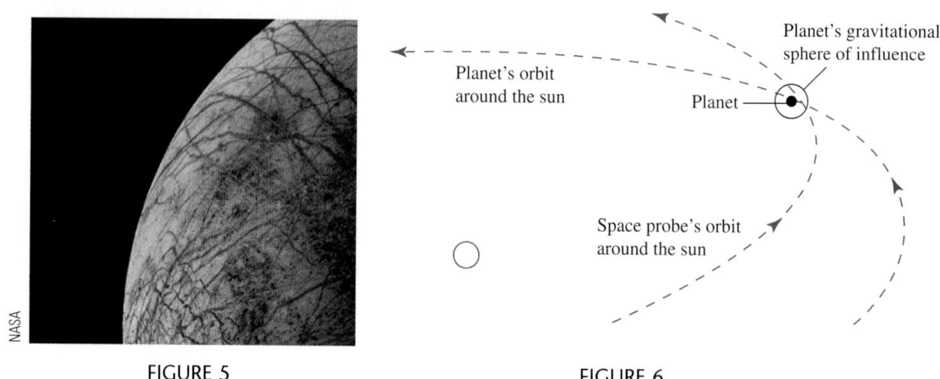

FIGURE 5

FIGURE 6

We use the rest of this section to consider equations that produce hyperbolas.

Consider the equation

$$\frac{x^2}{9} - \frac{y^2}{4} = 1$$

If we were to find a number of ordered pairs that are solutions to the equation and connect their graphs with a smooth curve, we would have Figure 7.

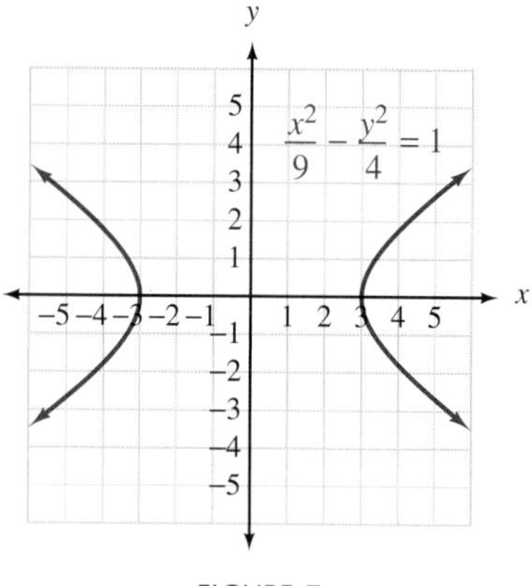

FIGURE 7

This graph is an example of a *hyperbola*. Notice that the graph has x-intercepts at $(3, 0)$ and $(-3, 0)$. The graph has no y-intercepts and hence does not cross the y-axis since substituting $x = 0$ into the equation yields

$$\frac{0^2}{9} - \frac{y^2}{4} = 1$$

$$-y^2 = 4$$

$$y^2 = -4$$

for which there is no real solution. We can, however, use the number below y^2 to help sketch the graph. If we draw a rectangle that has its sides parallel to the x- and y-axes and that passes through the x-intercepts and the points on the y-axis corresponding to the square roots of the number below y^2, $+2$ and -2, it looks like the rectangle in Figure 8. The lines that connect opposite corners of the rectangle are called *asymptotes.* The graph of the hyperbola

$$\frac{x^2}{9} - \frac{y^2}{4} = 1$$

will approach these lines. Figure 8 is the graph.

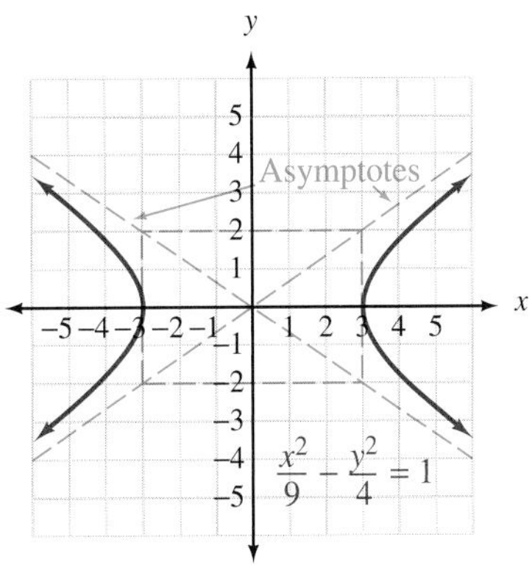

FIGURE 8

EXAMPLE 2

Graph the equation $\frac{y^2}{9} - \frac{x^2}{16} = 1$.

SOLUTION In this case the y-intercepts are 3 and -3, and the x-intercepts do not exist. We can use the square roots of the number below x^2, however, to find the asymptotes associated with the graph. The sides of the rectangle used to draw the asymptotes must pass through 3 and -3 on the y-axis and 4 and -4 on the x-axis. (See Figure 9.)

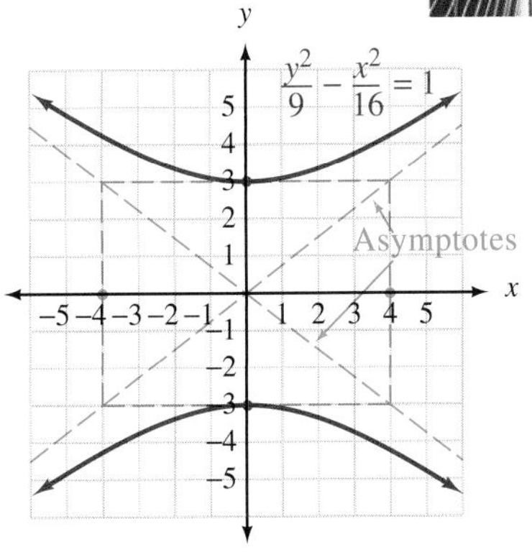

FIGURE 9

Here is a summary of what we have for hyperbolas.

The Hyperbola

The graph of the equation

$$\frac{x^2}{a^2} - \frac{y^2}{b^2} = 1$$

will be a hyperbola centered at the origin. The graph will have x-intercepts at $-a$ and a.

The graph of the equation

$$\frac{y^2}{a^2} - \frac{x^2}{b^2} = 1$$

will be a hyperbola centered at the origin. The graph will have y-intercepts at $-a$ and a.

As an aid in sketching either of the preceding equations, the asymptotes can be found by drawing lines through opposite corners of the rectangle whose sides pass through $-a$, a, $-b$, and b on the axes.

Ellipses and Hyperbolas Not Centered at the Origin

The following equation is that of an ellipse with its center at the point $(4, 1)$:

$$\frac{(x-4)^2}{9} + \frac{(y-1)^2}{4} = 1$$

To see why the center is at (4, 1) we substitute x' (read "x prime") for $x - 4$ and y' for $y - 1$ in the equation. That is,

If $\quad x' = x - 4$

and $\quad y' = y - 1$

the equation $\quad \dfrac{(x-4)^2}{9} + \dfrac{(y-1)^2}{4} = 1$

becomes $\quad \dfrac{(x')^2}{9} + \dfrac{(y')^2}{4} = 1$

This is the equation of an ellipse in a coordinate system with an x'-axis and a y'-axis. We call this new coordinate system the *$x'y'$-coordinate system*. The center of our ellipse is at the origin in the $x'y'$-coordinate system. The question is this: What are the coordinates of the center of this ellipse in the original xy-coordinate system? To answer this question, we go back to our original substitutions:

$$x' = x - 4$$

$$y' = y - 1$$

In the $x'y'$-coordinate system, the center of our ellipse is at $x' = 0, y' = 0$ (the origin of the $x'y'$ system). Substituting these numbers for x' and y', we have

$$0 = x - 4$$

$$0 = y - 1$$

Solving these equations for x and y will give us the coordinates of the center of our ellipse in the xy-coordinate system. As you can see, the solutions are $x = 4$ and $y = 1$ Therefore, in the xy-coordinate system, the center of our ellipse is at the point (4, 1). Figure 10 shows the graph.

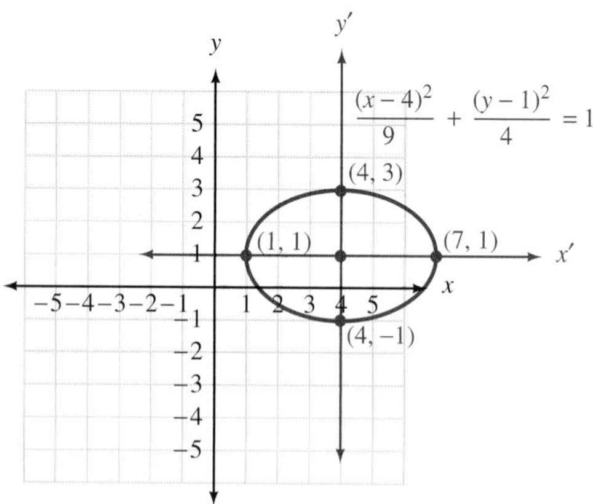

FIGURE 10

The coordinates of all points labeled in Figure 10 are given with respect to the xy-coordinate system. The x' and y' are shown simply for reference in our discussion. Note that the horizontal distance from the center to the vertices is 3—the square root of the denominator of the $(x-4)^2$ term. Likewise, the vertical distance from the center to the other vertices is 2—the square root of the denominator of the $(y-1)^2$ term.

We summarize the information above with the following:

An Ellipse with Center at (h, k) The graph of the equation

$$\frac{(x - h)^2}{a^2} + \frac{(y - k)^2}{b^2} = 1$$

will be an ellipse with center at (h, k). The vertices of the ellipse will be at the points, $(h + a, k)$, $(h - a, k)$, $(h, k + b)$, and $(h, k - b)$.

EXAMPLE 3

Graph the ellipse: $x^2 + 9y^2 + 4x - 54y + 76 = 0$.

SOLUTION To identify the coordinates of the center, we must complete the square on x and also on y. To begin, we rearrange the terms so that those containing x are together, those containing y are together, and the constant term is on the other side of the equal sign. Doing so gives us the following equation:

$$x^2 + 4x \qquad + 9y^2 - 54y \qquad = -76$$

Before we can complete the square on y, we must factor 9 from each term containing y:

$$x^2 + 4x \qquad + 9(y^2 - 6y \qquad) = -76$$

To complete the square on x, we add 4 to each side of the equation. To complete the square on y, we add 9 inside the parentheses. This increases the left side of the equation by 81 since each term within the parentheses is multiplied by 9. Therefore, we must add 81 to the right side of the equation also.

$$x^2 + 4x + \mathbf{4} + 9(y^2 - 6y + \mathbf{9}) = -76 + \mathbf{4} + \mathbf{81}$$

$$(x + 2)^2 + 9(y - 3)^2 = 9$$

To identify the distances to the vertices, we divide each term on both sides by 9:

$$\frac{(x + 2)^2}{9} + \frac{9(y - 3)^2}{9} = \frac{9}{9}$$

$$\frac{(x + 2)^2}{9} + \frac{(y - 3)^2}{1} = 1$$

The graph is an ellipse with center at $(-2, 3)$, as shown in Figure 11.

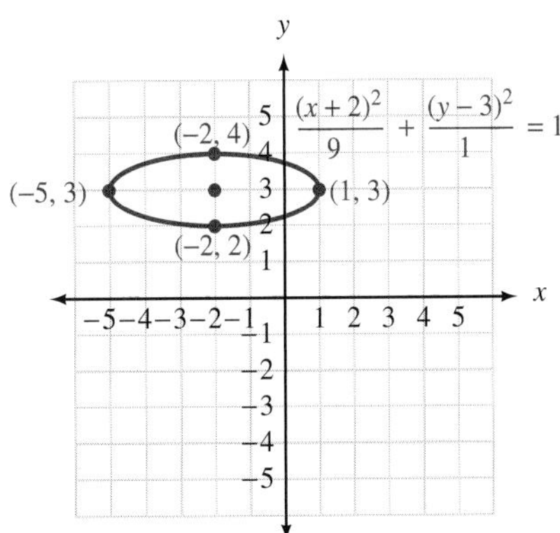

FIGURE 11

The ideas associated with graphing hyperbolas whose centers are not at the origin parallel the ideas just presented about graphing ellipses whose centers have been moved off the origin. Without showing the justification for doing so, we state the following guidelines for graphing hyperbolas:

> **Hyperbolas with Centers at (h, k)** The graphs of the equations
>
> $$\frac{(x-h)^2}{a^2} - \frac{(y-k)^2}{b^2} = 1 \quad \text{and} \quad \frac{(y-k)^2}{b^2} - \frac{(x-h)^2}{a^2} = 1$$
>
> will be hyperbolas with their centers at (h, k). The vertices of the graph of the first equation will be at the points $(h + a, k)$, and $(h - a, k)$, and the vertices for the graph of the second equation will be at $(h, k + b)$, $(h, k - b)$. In either case, the asymptotes can be found by connecting opposite corners of the rectangle that contains the four points $(h + a, k)$, $(h - a, k)$, $(h, k + b)$, and $(h, k - b)$.

EXAMPLE 4

Graph the hyperbola: $4x^2 - y^2 + 4y - 20 = 0$.

SOLUTION To identify the coordinates of the center of the hyperbola, we need to complete the square on y. (Because there is no linear term in x, we do not need to complete the square on x. The x-coordinate of the center will be $x = 0$.)

$$4x^2 - y^2 + 4y - 20 = 0$$

$$4x^2 - y^2 + 4y = 20 \qquad \textbf{Add 20 to each side}$$

$$4x^2 - 1(y^2 - 4y) = 20 \qquad \textbf{Factor } -1 \textbf{ from each term containing } y$$

To complete the square on y, we add 4 to the terms inside the parentheses. Doing so adds -4 to the left side of the equation because everything inside the parentheses is multiplied by -1. To keep from changing the equation we must add -4 to the right side also.

$$4x^2 - 1(y^2 - 4y + \mathbf{4}) = 20 - \mathbf{4} \qquad \textbf{Add } -4 \textbf{ to each side}$$

$$4x^2 - 1(y - 2)^2 = 16 \qquad y^2 - 4y + 4 = (y-2)^2$$

$$\frac{4x^2}{16} - \frac{(y-2)^2}{16} = \frac{16}{16} \qquad \textbf{Divide each side by 16}$$

$$\frac{x^2}{4} - \frac{(y-2)^2}{16} = 1 \qquad \textbf{Simplify each term}$$

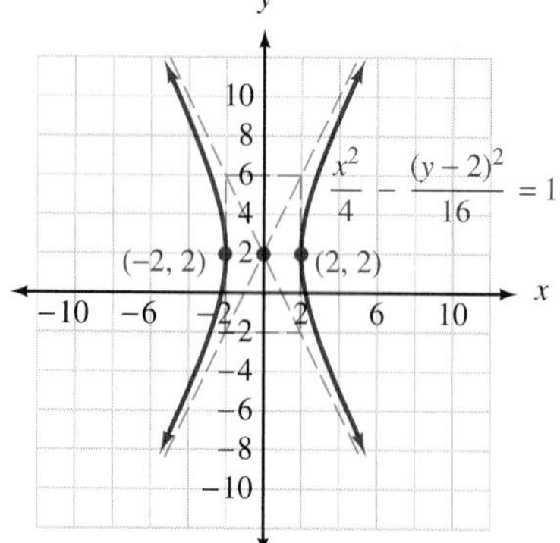

FIGURE 12

This is the equation of a hyperbola with center at $(0, 2)$. The graph opens to the right and left as shown in Figure 12.

LINKING OBJECTIVES AND EXAMPLES

Next to each **objective** we have listed the examples that are best described by that objective.

A	1, 3
B	2, 4

GETTING READY FOR CLASS

After reading through the preceding section, respond in your own words and in complete sentences.

1. How do we find the x-intercepts of a graph from the equation?
2. What is an ellipse?
3. How can you tell by looking at an equation if its graph will be an ellipse or a hyperbola?
4. Are the points on the asymptotes of a hyperbola in the solution set of the equation of the hyperbola? (That is, are the asymptotes actually part of the graph?) Explain.

Problem Set 10.2

Online support materials can be found at www.thomsonedu.com/login

Graph each of the following. Be sure to label both the x- and y-intercepts.

1. $\frac{x^2}{9} + \frac{y^2}{16} = 1$
2. $\frac{x^2}{25} + \frac{y^2}{4} = 1$
3. $\frac{x^2}{16} + \frac{y^2}{9} = 1$
4. $\frac{x^2}{4} + \frac{y^2}{25} = 1$
5. $\frac{x^2}{3} + \frac{y^2}{4} = 1$
6. $\frac{x^2}{4} + \frac{y^2}{3} = 1$
7. $4x^2 + 25y^2 = 100$
8. $4x^2 + 9y^2 = 36$
9. $x^2 + 8y^2 = 16$
10. $12x^2 + y^2 = 36$

Graph each of the following. Show the intercepts and the asymptotes in each case.

11. $\frac{x^2}{9} - \frac{y^2}{16} = 1$
12. $\frac{x^2}{25} - \frac{y^2}{4} = 1$
13. $\frac{x^2}{16} - \frac{y^2}{9} = 1$
14. $\frac{x^2}{4} - \frac{y^2}{25} = 1$
15. $\frac{y^2}{9} - \frac{x^2}{16} = 1$
16. $\frac{y^2}{25} - \frac{x^2}{4} = 1$
17. $\frac{y^2}{36} - \frac{x^2}{4} = 1$
18. $\frac{y^2}{4} - \frac{x^2}{36} = 1$
19. $x^2 - 4y^2 = 4$
20. $y^2 - 4x^2 = 4$
21. $16y^2 - 9x^2 = 144$
22. $4y^2 - 25x^2 = 100$

Find the x- and y-intercepts, if they exist, for each of the following. Do not graph.

23. $0.4x^2 + 0.9y^2 = 3.6$
24. $1.6x^2 + 0.9y^2 = 14.4$
25. $\frac{x^2}{0.04} - \frac{y^2}{0.09} = 1$
26. $\frac{y^2}{0.16} - \frac{x^2}{0.25} = 1$
27. $\frac{25x^2}{9} + \frac{25y^2}{4} = 1$
28. $\frac{16x^2}{9} + \frac{16y^2}{25} = 1$

Graph each of the following ellipses. In each case, label the coordinates of the center and the vertices.

29. $\frac{(x-4)^2}{4} + \frac{(y-2)^2}{9} = 1$
30. $\frac{(x-2)^2}{4} + \frac{(y-4)^2}{9} = 1$
31. $4x^2 + y^2 - 4y - 12 = 0$
32. $4x^2 + y^2 - 24x - 4y + 36 = 0$
33. $x^2 + 9y^2 + 4x - 54y + 76 = 0$
34. $4x^2 + y^2 - 16x + 2y + 13 = 0$

Graph each of the following hyperbolas. In each case, label the coordinates of the center and the vertices and show the asymptotes.

35. $\frac{(x-2)^2}{16} - \frac{y^2}{4} = 1$
36. $\frac{(y-2)^2}{16} - \frac{x^2}{4} = 1$
37. $9y^2 - x^2 - 4x + 54y + 68 = 0$
38. $4x^2 - y^2 - 24x + 4y + 28 = 0$

■ = Videos available by instructor request

▶ = Online student support materials available at www.thomsonedu.com/login

39. $4y^2 - 9x^2 - 16y + 72x - 164 = 0$

40. $4x^2 - y^2 - 16x - 2y + 11 = 0$

41. Find y when x is 4 in the equation $\frac{x^2}{25} + \frac{y^2}{9} = 1$.

42. Find x when y is 3 in the equation $\frac{x^2}{4} + \frac{y^2}{25} = 1$.

43. Find y when x is 1.8 in $16x^2 + 9y^2 = 144$.

44. Find y when x is 1.6 in $49x^2 + 4y^2 = 196$.

45. Give the equations of the two asymptotes in the graph you found in Problem 15.

46. Give the equations of the two asymptotes in the graph you found in Problem 16.

Find an equation for each graph.

47.

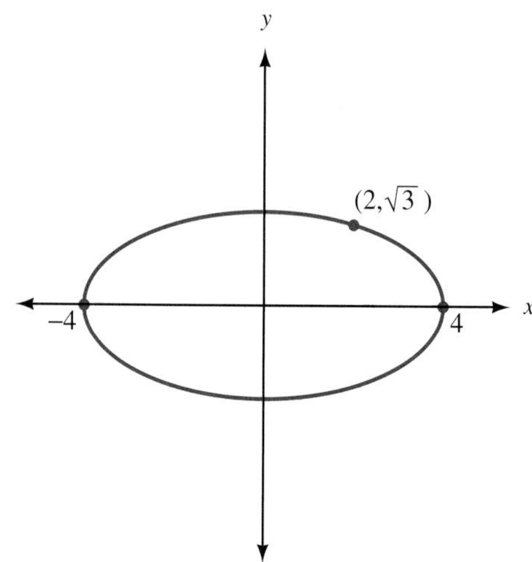

48.

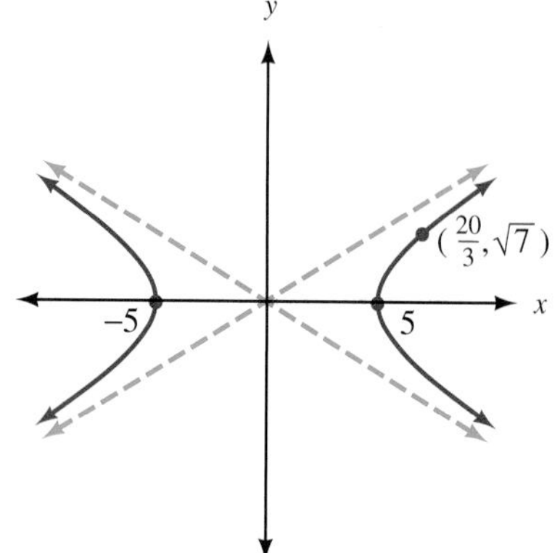

Applying the Concepts

The diagram shows the minor axis and the major axis for an ellipse.

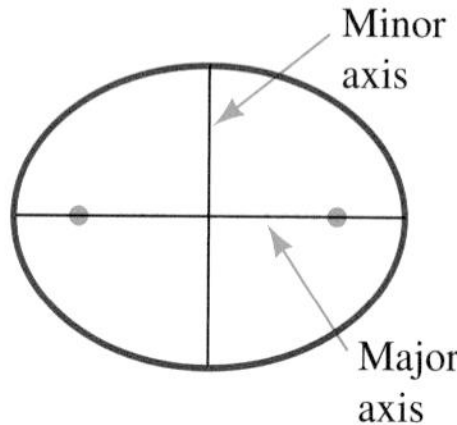

In any ellipse, the length of the major axis is $2a$, and the length of the minor axis is $2b$ (these are the same a and b that appear in the general equation of an ellipse). Each of the two points shown on the major axis is a focus of the ellipse. If the distance from the center of the ellipse to each focus is c, then it is always true that $a^2 = b^2 + c^2$. You will need this information for some of the problems that follow.

49. Archway A new theme park is planning an archway at its main entrance. The arch is to be in the form of a semi-ellipse with the major axis as the span. If the span is to be 40 feet and the height at the center is to be 10 feet, what is the equation of the ellipse? How far left and right of center could a 6-foot man walk upright under the arch?

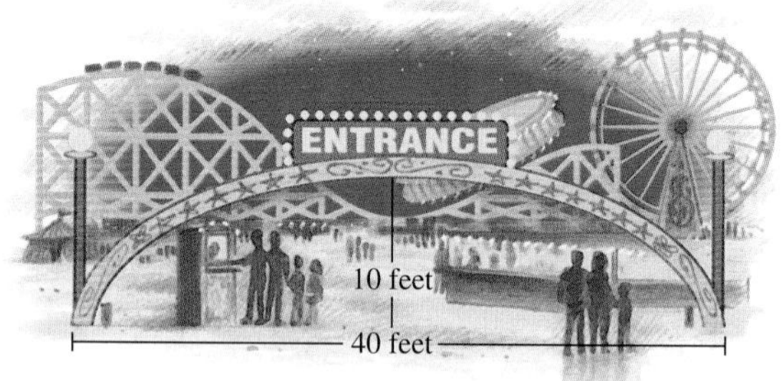

50. The Ellipse President's Park, located between the White House and the Washington Monument in Washington, DC, is also called the Ellipse. The park is enclosed by an elliptical path with major axis 458

meters and minor axis 390 meters. What is the equation for the path around the Ellipse?

National Park Service

51. **Elliptical Pool Table** A children's science museum plans to build an elliptical pool table to demonstrate that a ball rolled from a particular point (focus) will always go into a hole located at another particular point (the other focus). The focus needs to be 1 foot from the vertex of the ellipse. If the table is to be 8 feet long, how wide should it be? *Hint:* The distance from the center to each focus point is represented by c and is found by using the equation $a^2 = b^2 + c^2$.

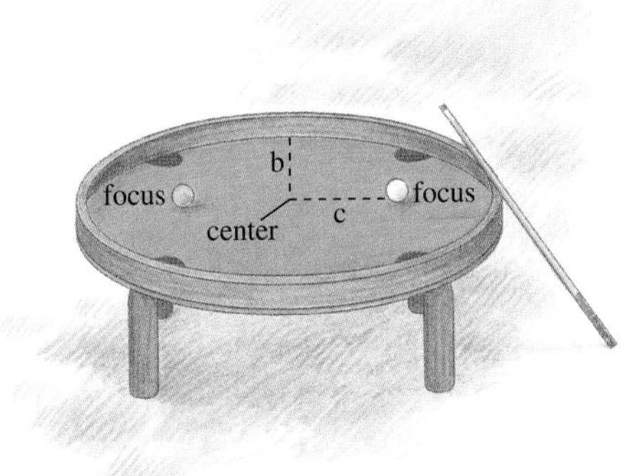

52. **Lithotripter** A lithotripter similar to the one mentioned in the introduction of the chapter is based on the ellipse $\frac{x^2}{36} + \frac{y^2}{25} = 1$. Determine how many units the kidney stone and the wave source (focus points) must be placed from the center of the ellipse. *Hint:* The distance from the center to each focus point is represented by c and is found by using the equation $a^2 = b^2 + c^2$.

Maintaining Your Skills

Let $f(x) = \frac{2}{x-2}$ and $g(x) = \frac{2}{x+2}$. Find the following.

53. $f(4)$
54. $g(4)$
55. $f[g(0)]$
56. $g[f(0)]$
57. $f(x) + g(x)$
58. $f(x) - g(x)$

Getting Ready for the Next Section

59. Which of the following are solutions to $x^2 + y^2 < 16$?
(0, 0) (4, 0) (0, 5)
60. Which of the following are solutions to $y \le x^2 - 2$?
(0, 0) (−2, 0) (0, −2)

Expand and multiply.

61. $(2y + 4)^2$
62. $(y + 3)^2$
63. Solve $x - 2y = 4$ for x.
64. Solve $2x + 3y = 6$ for y.

Simplify.

65. $x^2 - 2(x^2 - 3)$
66. $x^2 + (x^2 - 4)$

Factor.

67. $5y^2 + 16y + 12$
68. $3x^2 + 17x - 28$

Solve.

69. $y^2 = 4$
70. $x^2 = 25$
71. $-x^2 + 6 = 2$
72. $5y^2 + 16y + 12 = 0$

10.3 Second-Degree Inequalities and Nonlinear Systems

OBJECTIVES

A Graph second-degree inequalities.

B Solve systems of nonlinear equations.

C Graph the solution sets to systems of inequalities.

Previously we graphed linear inequalities by first graphing the boundary and then choosing a test point not on the boundary to indicate the region used for the solution set. The problems in this section are very similar. We will use the same general methods for graphing the inequalities in this section that we used when we graphed linear inequalities in two variables.

EXAMPLE 1 Graph $x^2 + y^2 < 16$.

SOLUTION The boundary is $x^2 + y^2 = 16$, which is a circle with center at the origin and a radius of 4. Since the inequality sign is $<$, the boundary is not included in the solution set and therefore must be represented with a broken line. The graph of the boundary is shown in Figure 1.

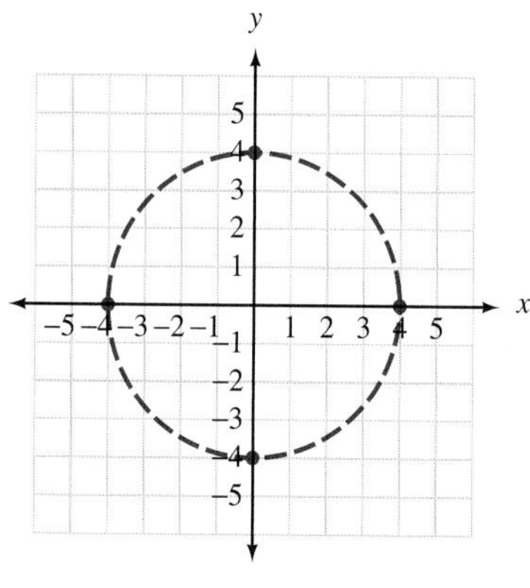

FIGURE 1

The solution set for $x^2 + y^2 < 16$ is either the region inside the circle or the region outside the circle. To see which region represents the solution set, we choose a convenient point not on the boundary and test it in the original inequality. The origin (0, 0) is a convenient point. Since the origin satisfies the inequality $x^2 + y^2 < 16$, all points in the same region will also satisfy the inequality. The graph of the solution set is shown in Figure 2.

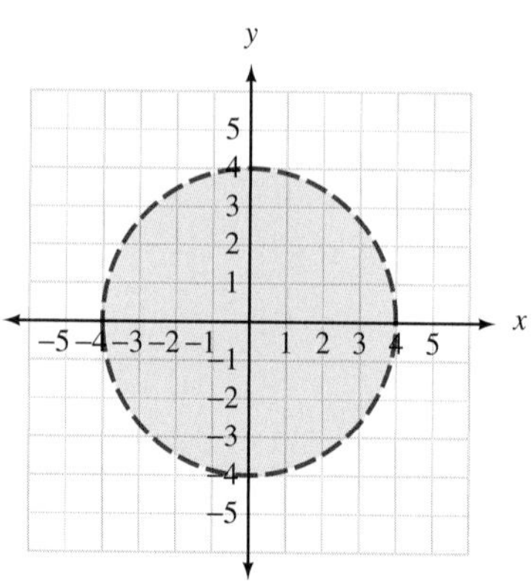

FIGURE 2

EXAMPLE 2

Graph the inequality $y \le x^2 - 2$.

SOLUTION The parabola $y = x^2 - 2$ is the boundary and is included in the solution set. Using (0, 0) as the test point, we see that $0 \le 0^2 - 2$ is a false statement, which means that the region containing (0, 0) is not in the solution set. (See Figure 3.)

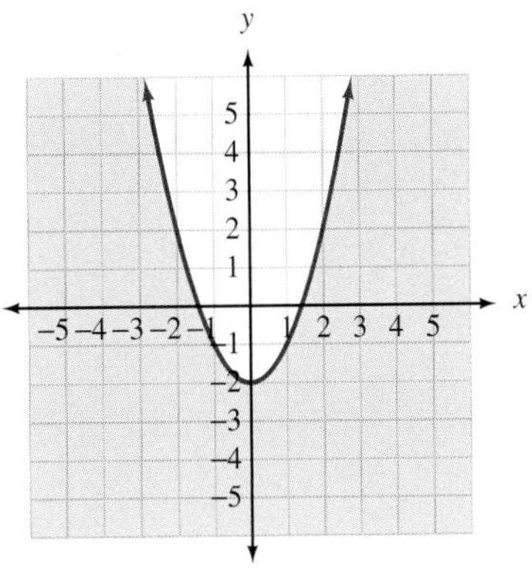

FIGURE 3

EXAMPLE 3

Graph $4y^2 - 9x^2 < 36$.

SOLUTION The boundary is the hyperbola $4y^2 - 9x^2 = 36$ and is not included in the solution set. Testing (0, 0) in the original inequality yields a true statement, which means that the region containing the origin is the solution set. (See Figure 4.)

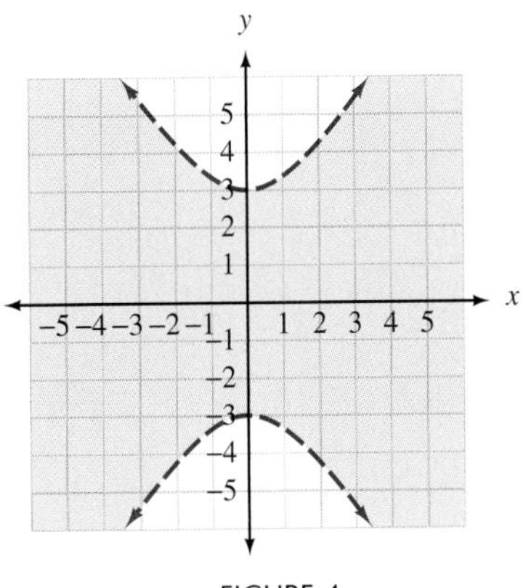

FIGURE 4

EXAMPLE 4

Solve the system.

$$x^2 + y^2 = 4$$

$$x - 2y = 4$$

SOLUTION In this case the substitution method is the most convenient. Solving the second equation for x in terms of y, we have

$$x - 2y = 4$$
$$x = 2y + 4$$

We now substitute $2y + 4$ for x in the first equation in our original system and proceed to solve for y:

$$(2y + 4)^2 + y^2 = 4$$
$$4y^2 + 16y + 16 + y^2 = 4 \quad \text{Expand } (2y + 4)^2$$
$$5y^2 + 16y + 16 = 4 \quad \text{Simplify left side}$$
$$5y^2 + 16y + 12 = 0 \quad \text{Add } -4 \text{ to each side}$$
$$(5y + 6)(y + 2) = 0 \quad \text{Factor}$$
$$5y + 6 = 0 \quad \text{or} \quad y + 2 = 0 \quad \text{Set factors equal to 0}$$
$$y = -\frac{6}{5} \quad \text{or} \quad y = -2 \quad \text{Solve}$$

These are the y-coordinates of the two solutions to the system. Substituting $y = -\frac{6}{5}$ into $x - 2y = 4$ and solving for x gives us $x = \frac{8}{5}$. Using $y = -2$ in the same equation yields $x = 0$. The two solutions to our system are $(\frac{8}{5}, -\frac{6}{5})$ and $(0, -2)$. Although graphing the system is not necessary, it does help us visualize the situation. (See Figure 5.)

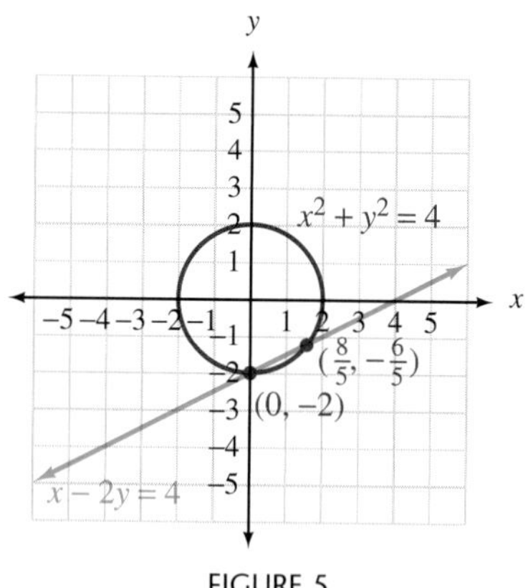

FIGURE 5

EXAMPLE 5

Solve the system.

$$16x^2 - 4y^2 = 64$$
$$x^2 + y^2 = 9$$

SOLUTION Since each equation is of the second degree in both x and y, it is easier to solve this system by eliminating one of the variables by addition. To eliminate y, we multiply the bottom equation by 4 and add the result to the top equation:

$$\begin{aligned} 16x^2 - 4y^2 &= 64 \\ 4x^2 + 4y^2 &= 36 \\ \hline 20x^2 \qquad &= 100 \\ x^2 &= 5 \\ x &= \pm\sqrt{5} \end{aligned}$$

The x-coordinates of the points of intersection are $\sqrt{5}$ and $-\sqrt{5}$. We substitute each back into the second equation in the original system and solve for y:

When $x = \sqrt{5}$

$$\begin{aligned} (\sqrt{5})^2 + y^2 &= 9 \\ 5 + y^2 &= 9 \\ y^2 &= 4 \\ y &= \pm 2 \end{aligned}$$

When $x = -\sqrt{5}$

$$\begin{aligned} (-\sqrt{5})^2 + y^2 &= 9 \\ 5 + y^2 &= 9 \\ y^2 &= 4 \\ y &= \pm 2 \end{aligned}$$

The four points of intersection are $(\sqrt{5}, 2)$, $(\sqrt{5}, -2)$, $(-\sqrt{5}, 2)$, and $(-\sqrt{5}, -2)$. Graphically the situation is as shown in Figure 6.

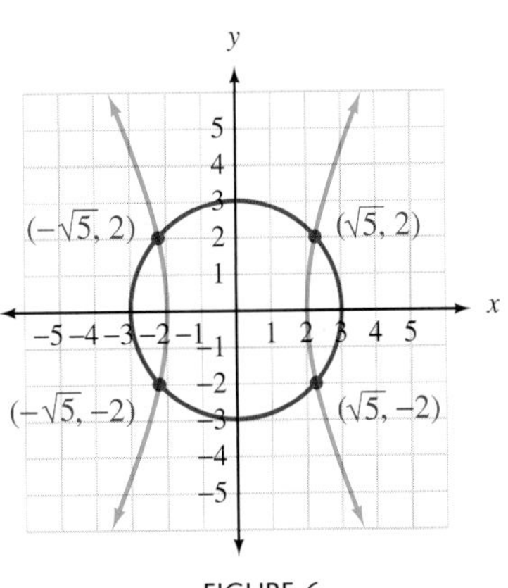

FIGURE 6

EXAMPLE 6 Solve the system.

$$x^2 - 2y = 2$$
$$y = x^2 - 3$$

SOLUTION We can solve this system using the substitution method. Replacing y in the first equation with $x^2 - 3$ from the second equation, we have

$$x^2 - 2(x^2 - 3) = 2$$
$$-x^2 + 6 = 2$$
$$x^2 = 4$$
$$x = \pm 2$$

Using either $+2$ or -2 in the equation $y = x^2 - 3$ gives us $y = 1$. The system has two solutions: $(2, 1)$ and $(-2, 1)$.

EXAMPLE 7 The sum of the squares of two numbers is 34. The difference of their squares is 16. Find the two numbers.

SOLUTION Let x and y be the two numbers. The sum of their squares is $x^2 + y^2$, and the difference of their squares is $x^2 - y^2$. (We can assume here that x^2 is the larger number.) The system of equations that describes the situation is

$$x^2 + y^2 = 34$$
$$x^2 - y^2 = 16$$

We can eliminate y by simply adding the two equations. The result of doing so is

$$2x^2 = 50$$
$$x^2 = 25$$
$$x = \pm 5$$

Substituting $x = 5$ into either equation in the system gives $y = \pm 3$. Using $x = -5$ gives the same results, $y = \pm 3$. The four pairs of numbers that are solutions to the original problem are 5 and 3, -5 and 3, 5 and -3, -5 and -3.

We now turn our attention to systems of inequalities. To solve a system of inequalities by graphing, we simply graph each inequality on the same set of axes. The solution set for the system is the region common to both graphs—the intersection of the individual solution sets.

EXAMPLE 8 Graph the solution set for the system.

$$x^2 + y^2 \le 9$$
$$\frac{x^2}{4} + \frac{y^2}{25} \ge 1$$

SOLUTION The boundary for the top inequality is a circle with center at the origin and a radius of 3. The solution set lies inside the boundary. The boundary for the second inequality is an ellipse. In this case the solution set lies outside the boundary. (See Figure 7.)

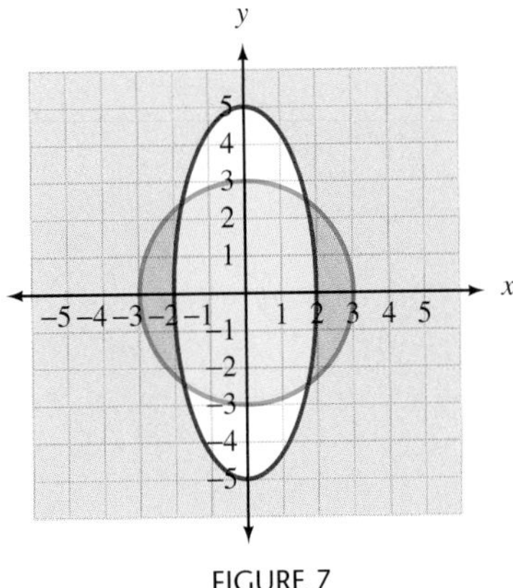

FIGURE 7

The solution set is the intersection of the two individual solution sets. 

LINKING OBJECTIVES AND EXAMPLES

Next to each **objective** we have listed the examples that are best described by that objective.

A	1–3
B	4–7
C	8

GETTING READY FOR CLASS

After reading through the preceding section, respond in your own words and in complete sentences.

1. What is the significance of a broken line when graphing inequalities?
2. Describe, in words, the set of points described by $(x - 3)^2 + (y - 2)^2 < 9$.
3. When solving the nonlinear systems whose graphs are a line and a circle, how many possible solutions can you expect?
4. When solving the nonlinear systems whose graphs are both circles, how many possible solutions can you expect?

Problem Set 10.3

Online support materials can be found at www.thomsonedu.com/login

Graph each of the following inequalities.

1. $x^2 + y^2 \le 49$
2. $x^2 + y^2 < 49$
3. $(x - 2)^2 + (y + 3)^2 < 16$
4. $(x + 3)^2 + (y - 2)^2 \ge 25$
5. $y < x^2 - 6x + 7$
6. $y \ge x^2 + 2x - 8$
7. $4x^2 + 25y^2 \le 100$
8. $25x^2 - 4y^2 > 100$

Solve each of the following systems of equations.

9. $x^2 + y^2 = 9$, $2x + y = 3$
10. $x^2 + y^2 = 9$, $x + 2y = 3$
11. $x^2 + y^2 = 16$, $x + 2y = 8$
12. $x^2 + y^2 = 16$, $x - 2y = 8$

= Videos available by instructor request
▶ = Online student support materials available at www.thomsonedu.com/login

13. $x^2 + y^2 = 25$
$x^2 - y^2 = 25$

14. $x^2 + y^2 = 4$
$2x^2 - y^2 = 5$

15. $x^2 + y^2 = 9$
$y = x^2 - 3$

16. $x^2 + y^2 = 4$
$y = x^2 - 2$

17. $x^2 + y^2 = 16$
$y = x^2 - 4$

18. $x^2 + y^2 = 1$
$y = x^2 - 1$

19. $3x + 2y = 10$
$y = x^2 - 5$

20. $4x + 2y = 10$
$y = x^2 - 10$

21. $y = x^2 + 2x - 3$
$y = -x + 1$

22. $y = -x^2 - 2x + 3$
$y = x - 1$

23. $y = x^2 - 6x + 5$
$y = x - 5$

24. $y = x^2 - 2x - 4$
$y = x - 4$

25. $4x^2 - 9y^2 = 36$
$4x^2 + 9y^2 = 36$

26. $4x^2 + 25y^2 = 100$
$4x^2 - 25y^2 = 100$

27. $x - y = 4$
$x^2 + y^2 = 16$

28. $x + y = 2$
$x^2 - y^2 = 4$

29. In a previous problem set, you found the equations for the three circles below. The equations are

Circle A	Circle B	Circle C
$(x + 8)^2 + y^2 = 64$	$x^2 + y^2 = 64$	$(x - 8)^2 + y^2 = 64$

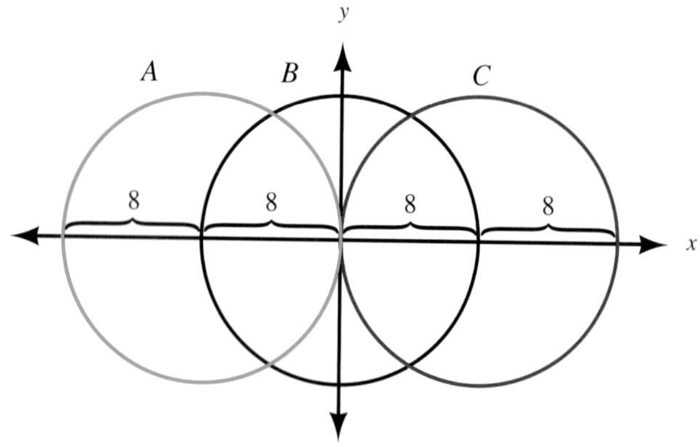

a. Find the points of intersection of circles A and B.

b. Find the points of intersection of circles B and C.

30. A magician is holding two rings that seem to lie in the same plane and intersect in two points. Each ring is 10 inches in diameter. If a coordinate system is placed with its origin at the center of the first ring and the x-axis contains the center of the second ring, then the equations are as follows:

First Ring	Second Ring
$x^2 + y^2 = 25$	$(x - 5)^2 + y^2 = 25$

Find the points of intersection of the two rings.

Graph the solution sets to the following systems.

31. $x^2 + y^2 < 9$
$y \geq x^2 - 1$

32. $x^2 + y^2 \leq 16$
$y < x^2 + 2$

33. $\dfrac{x^2}{9} + \dfrac{y^2}{25} \leq 1$
$\dfrac{x^2}{4} - \dfrac{y^2}{9} > 1$

34. $\dfrac{x^2}{4} + \dfrac{y^2}{16} \geq 1$
$\dfrac{x^2}{9} - \dfrac{y^2}{25} < 1$

35. $4x^2 + 9y^2 \leq 36$
$y > x^2 + 2$

36. $9x^2 + 4y^2 \geq 36$
$y < x^2 + 1$

37. A parabola and a circle each contain the points $(-4, 0)$, $(0, 4)$, and $(4, 0)$, as shown below. The equations for each of the curves are

Circle	Parabola
$x^2 + y^2 = 16$	$y = 4 - \frac{1}{4}x^2$

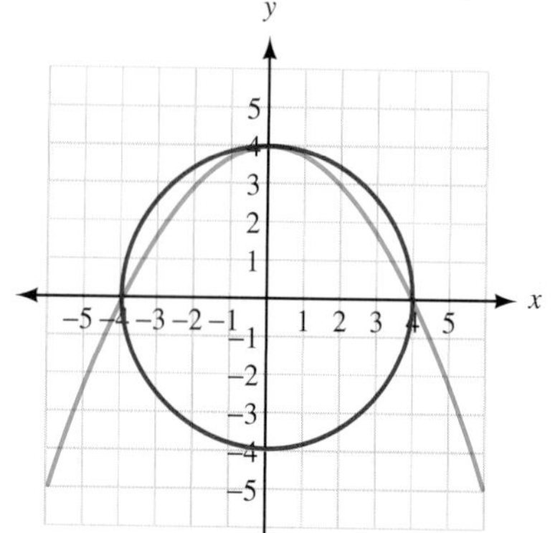

Write a system of inequalities that describes the regions in quadrants I and II that are between the two curves, if the boundaries are not included with the shaded region.

38. Find a system of inequalities that describes the shaded region in the figure below if the boundaries of the shaded region are included and equations of the circles are given by

Circle A	Circle B	Circle C
$(x + 8)^2 + y^2 = 64$	$x^2 + y^2 = 64$	$(x - 8)^2 + y^2 = 64$

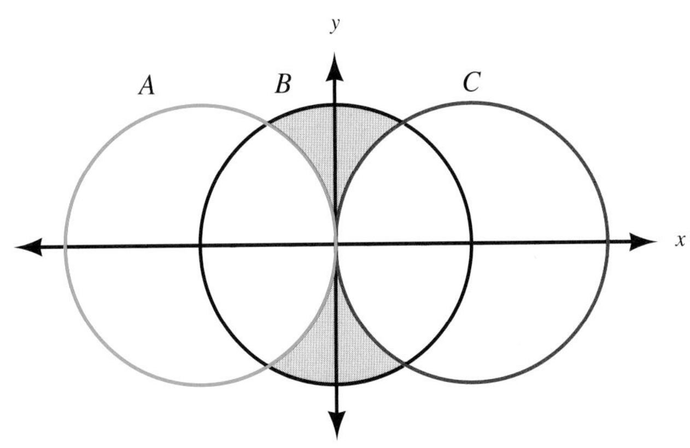

Applying the Concepts

39. Number Problems The sum of the squares of two numbers is 89. The difference of their squares is 39. Find the numbers.

40. Number Problems The difference of the squares of two numbers is 35. The sum of their squares is 37. Find the numbers.

41. Number Problems One number is 3 less than the square of another. Their sum is 9. Find the numbers.

42. Number Problems The square of one number is 2 less than twice the square of another. The sum of the squares of the two numbers is 25. Find the numbers.

Maintaining Your Skills

43. Let $g(x) = \left(x + \frac{2}{5}\right)^2$. Find all values for the variable x for which $g(x) = 0$.

44. Let $f(x) = (x + 2)^2 - 25$. Find all values for the variable x for which $f(x) = 0$.

For the problems below, let $f(x) = x^2 + 4x - 4$. Find all values for the variable x for which $f(x) = g(x)$.

45. $g(x) = 1$

46. $g(x) = -7$

47. $g(x) = x - 6$

48. $g(x) = x + 6$

Chapter 10 SUMMARY

Distance Formula [10.1]

EXAMPLES

1. The distance between (5, 2) and (−1, 1) is $d = \sqrt{(5 + 1)^2 + (2 - 1)^2} = \sqrt{37}$

The distance between the two points (x_1, y_1) and (x_2, y_2) is given by the formula

$$d = \sqrt{(x_2 - x_1)^2 + (y_2 - y_1)^2}$$

The Circle [10.1]

2. The graph of the circle $(x - 3)^2 + (y + 2)^2 = 25$ will have its center at (3, −2) and the radius will be 5.

The graph of any equation of the form

$$(x - a)^2 + (y - b)^2 = r^2$$

will be a circle having its center at (a, b) and a radius of r.

The Ellipse [10.2]

3. The ellipse $\frac{x^2}{9} + \frac{y^2}{4} = 1$ will cross the x-axis at 3 and −3 and will cross the y-axis at 2 and −2.

Any equation that can be put in the form

$$\frac{x^2}{a^2} + \frac{y^2}{b^2} = 1$$

will have an ellipse for its graph. The x-intercepts will be at a and $-a$, and the y-intercepts will be at b and $-b$.

The Hyperbola [10.2]

4. The hyperbola $\frac{x^2}{4} - \frac{y^2}{9} = 1$ will cross the x-axis at 2 and −2. It will not cross the y-axis.

The graph of an equation that can be put in either of the forms

$$\frac{x^2}{a^2} - \frac{y^2}{b^2} = 1 \quad \text{or} \quad \frac{y^2}{a^2} - \frac{x^2}{b^2} = 1$$

will be a hyperbola. The x-intercepts for the first equation will be at a and $-a$. The y-intercepts for the second equation will be at a and $-a$. Two straight lines, called *asymptotes*, are associated with the graph of every hyperbola. Although the asymptotes are not part of the hyperbola, they are useful in sketching the graph.

Second-Degree Inequalities in Two Variables [10.3]

5. The graph of the inequality
$$x^2 + y^2 < 9$$
is all points inside the circle with center at the origin and radius 3. The circle itself is not part of the solution and

We graph second-degree inequalities in two variables in much the same way that we graphed linear inequalities; that is, we begin by graphing the boundary, using a solid curve if the boundary is included in the solution (this happens when the inequality symbol is ≥ or ≤) or a broken curve if the boundary is not included in the solution (when the inequality symbol is > or <). After we have graphed the boundary, we choose a test point that is not on the boundary and

therefore is shown with a broken curve.

try it in the original inequality. A true statement indicates we are in the region of the solution. A false statement indicates we are not in the region of the solution.

Systems of Nonlinear Equations [10.3]

6. We can solve the system

$$x^2 + y^2 = 4$$
$$x = 2y + 4$$

by substituting $2y + 4$ from the second equation for x in the first equation:

$$(2y + 4)^2 + y^2 = 4$$
$$4y^2 + 16y + 16 + y^2 = 4$$
$$5y^2 + 16y + 12 = 0$$
$$(5y + 6)(y + 2) = 0$$
$$y = -\frac{6}{5} \quad \text{or} \quad y = -2$$

Substituting these values of y into the second equation in our system gives $x = \frac{8}{5}$ and $x = 0$. The solutions are $(\frac{8}{5}, -\frac{6}{5})$ and $(0, -2)$.

A system of nonlinear equations is two equations, at least one of which is not linear, considered at the same time. The solution set for the system consists of all ordered pairs that satisfy both equations. In most cases we use the substitution method to solve these systems; however, the addition method can be used if like variables are raised to the same power in both equations. It is sometimes helpful to graph each equation in the system on the same set of axes to anticipate the number and approximate positions of the solutions.

Chapter 10 Review Test

The problems below form a comprehensive review of the material in this chapter. They can be used to study for exams. If you would like to take a practice test on this chapter, you can use the odd-numbered problems. Give yourself an hour and work as many of the odd-numbered problems as possible. When you are finished, or when an hour has passed, check your answers with the answers in the back of the book. You can use the even-numbered problems for a second practice test.

Find the distance between the following points. [10.1]

1. (2, 6), (−1, 5) **2.** (3, −4), (1, −1)

3. (0, 3), (−4, 0) **4.** (−3, 7), (−3, −2)

5. Find x so that the distance between $(x, -1)$ and $(2, -4)$ is 5. [10.1]

6. Find y so that the distance between $(3, -4)$ and $(-3, y)$ is 10. [10.1]

Write the equation of the circle with the given center and radius. [10.1]

7. Center (3, 1), $r = 2$ **8.** Center (3, −1), $r = 4$

9. Center (−5, 0), $r = 3$ **10.** Center (−3, 4), $r = 3\sqrt{2}$

Find the equation of each circle. [10.1]

11. Center at the origin, x-intercepts ±5

12. Center at the origin, y-intercepts ±3

13. Center at (−2, 3) and passing through the point (2, 0)

14. Center at (−6, 8) and passing through the origin

Give the center and radius of each circle, and then sketch the graph. [10.1]

15. $x^2 + y^2 = 4$ **16.** $(x - 3)^2 + (y + 1)^2 = 16$

17. $x^2 + y^2 - 6x + 4y = -4$ **18.** $x^2 + y^2 + 4x - 2y = 4$

Graph each of the following. Label the x- and y-intercepts. [10.2]

19. $\frac{x^2}{4} + \frac{y^2}{9} = 1$ **20.** $4x^2 + y^2 = 16$

Graph the following. Show the asymptotes. [10.2]

21. $\frac{x^2}{4} - \frac{y^2}{9} = 1$ **22.** $4x^2 - y^2 = 16$

Graph each equation. [10.2]

23. $\frac{(x+2)^2}{9} + \frac{(y-3)^2}{1} = 1$ **24.** $\frac{(x-2)^2}{16} - \frac{y^2}{4} = 1$

25. $9y^2 - x^2 - 4x + 54y + 68 = 0$

26. $9x^2 + 4y^2 - 72x - 16y + 124 = 0$

Graph each of the following inequalities. [10.3]

27. $x^2 + y^2 < 9$ **28.** $(x + 2)^2 + (y - 1)^2 \le 4$

29. $y \ge x^2 - 1$ **30.** $9x^2 + 4y^2 \le 36$

Graph the solution set for each system. [10.3]

31. $x^2 + y^2 < 16$
$y > x^2 - 4$

32. $x + y \le 2$
$-x + y \le 2$
$y \ge -2$

Solve each system of equations. [10.3]

33. $x^2 + y^2 = 16$
$2x + y = 4$

34. $x^2 + y^2 = 4$
$y = x^2 - 2$

35. $9x^2 - 4y^2 = 36$
$9x^2 + 4y^2 = 36$

36. $2x^2 - 4y^2 = 8$
$x^2 + 2y^2 = 10$

Chapter 10 Projects

Conic Sections

GROUP PROJECT Constructing Ellipses

Number of People 4

Time Needed 20 minutes

Equipment Graph paper, pencils, string, and thumbtacks

Background The geometric definition for an ellipse is the set of points the sum of whose distances from two fixed points (called foci) is a constant. We can use this definition to draw an ellipse using thumbtacks, string, and a pencil.

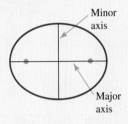

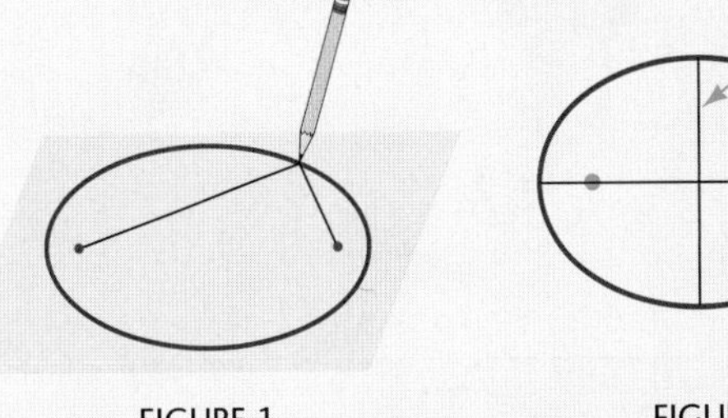

FIGURE 1 FIGURE 2

Procedure

1. Start with a piece of string 7 inches long. Place thumbtacks through the string $\frac{1}{2}$ inch from each end, then tack the string to a pad of graph paper so that the tacks are 4 inches apart. Pull the string tight with the tip of a pencil, then trace all the way around the two tacks. (See Figure 1.) The resulting diagram will be an ellipse.
2. The line segment that passes through the tacks (these are the foci) and connects opposite ends of the ellipse is called the major axis. The line segment perpendicular to the major axis that passes through the center of the ellipse and connects opposite ends of the ellipse is called the minor axis. (See Figure 2.) Measure the length of the major axis and the length of the minor axis. Record your results in Table 1.
3. Explain how drawing the ellipse as you have in step 2 shows that the geometric definition of an ellipse given at the beginning of this project is, in fact, correct.
4. Next, move the tacks so that they are 3 inches apart. Trace out that ellipse. Measure the length of the major axis and the length of the minor axis, and record your results in Table 1.
5. Repeat step 4 with the tacks 2 inches apart.
6. If the length of the string between the tacks stays at 6 inches, and the tacks were placed 6 inches apart, then the resulting ellipse would be a _____. If the tacks were placed 0 inches apart, then the resulting ellipse would be a _____.

TABLE 1
Ellipses (All Lengths Are Inches)

Length of String	Distance Between Foci	Length of Major Axis	Length of Minor Axis
6	4		
6	3		
6	2		

Hypatia of Alexandria

Bettmann/Corbis

The first woman mentioned in the history of mathematics is Hypatia of Alexandria. Research the life of Hypatia, and then write an essay that begins with a description of the time and place in which she lived and then goes on to give an indication of the type of person she was, her accomplishments in areas other than mathematics, and how she was viewed by her contemporaries.

Sequences and Series

David Young-Wolff/PhotoEdit

CHAPTER OUTLINE

Suppose you run up a balance of \$1,000 on a credit card that charges 1.65% interest each month (i.e., an annual rate of 19.8%). If you stop using the card and make the minimum payment of \$20 each month, how long will it take you to pay off the balance on the card? The answer can be found by using the formula

$$U_n = (1.0165)U_{n-1} - 20$$

where U_n stands for the current unpaid balance on the card, and U_{n-1} is the previous month's balance. The table below and Figure 1 were created from this formula and a graphing calculator. As you can see from the table, the balance on the credit card decreases very little each month.

Monthly Credit Card Balances

Previous Balance $U_{(n-1)}$	Monthly Interest Rate	Payment Number n	Monthly Payment	New Balance $U_{(n)}$
\$1,000.00	1.65%	1	\$20	\$996.50
\$996.50	1.65%	2	\$20	\$992.94
\$992.94	1.65%	3	\$20	\$989.32
\$989.32	1.65%	4	\$20	\$985.64
\$985.64	1.65%	5	\$20	\$981.90
.	.	.	.	.
.	.	.	.	.
.	.	.	.	.

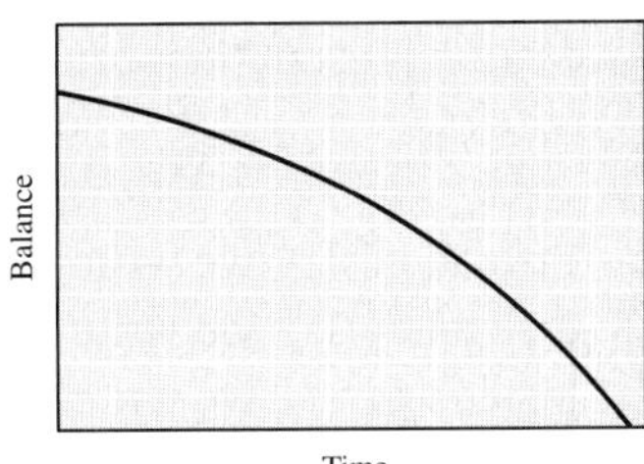

FIGURE 1

In the group project at the end of the chapter, you will use a graphing calculator to continue this table and in so doing find out just how many months it will take to pay off this credit card balance.

11.1 Sequences

OBJECTIVES

A Use a formula to find the terms of a sequence.

B Use a recursive formula.

C Find the general term of a sequence.

Many of the sequences in this chapter will be familiar to you on an intuitive level because you have worked with them for some time now. Here are some of those sequences:

The sequence of odd numbers

$$1, 3, 5, 7, \ldots$$

The sequence of even numbers

$$2, 4, 6, 8, \ldots$$

The sequence of squares

$$1^2, 2^2, 3^2, 4^2, \ldots = 1, 4, 9, 16, \ldots$$

The numbers in each of these sequences can be found from the formulas that define functions. For example, the sequence of even numbers can be found from the function

$$f(x) = 2x$$

by finding $f(1), f(2), f(3), f(4)$, and so forth. This gives us justification for the formal definition of a sequence.

DEFINITION A **sequence** is a function whose domain is the set of positive integers $\{1, 2, 3, 4, \ldots\}$.

As you can see, sequences are simply functions with a specific domain. If we want to form a sequence from the function $f(x) = 3x + 5$, we simply find $f(1)$, $f(2), f(3)$, and so on. Doing so gives us the sequence

$$8, 11, 14, 17, \ldots$$

because $f(1) = 3(1) + 5 = 8$, $f(2) = 3(2) + 5 = 11$, $f(3) = 3(3) + 5 = 14$, and $f(4) = 3(4) + 5 = 17$.

Notation Because the domain for a sequence is always the set $\{1, 2, 3, \ldots\}$, we can simplify the notation we use to represent the terms of a sequence. Using the letter a instead of f, and subscripts instead of numbers enclosed by parentheses, we can represent the sequence from the previous discussion as follows:

$$a_n = 3n + 5$$

Instead of $f(1)$, we write a_1 for the *first term* of the sequence.
Instead of $f(2)$, we write a_2 for the *second term* of the sequence.
Instead of $f(3)$, we write a_3 for the *third term* of the sequence.
Instead of $f(4)$, we write a_4 for the *fourth term* of the sequence.
Instead of $f(n)$, we write a_n for the *nth term* of the sequence.

The nth term is also called the *general term* of the sequence. The general term is used to define the other terms of the sequence. That is, if we are given the formula for the general term a_n, we can find any other term in the sequence. The following examples illustrate.

EXAMPLE 1 Find the first four terms of the sequence whose general term is given by $a_n = 2n - 1$.

SOLUTION The subscript notation a_n works the same way function notation works. To find the first, second, third, and fourth terms of this sequence, we simply substitute 1, 2, 3, and 4 for n in the formula $2n - 1$:

If the general term is $a_n = 2n - 1$
then the first term is $a_1 = 2(1) - 1 = 1$
the second term is $a_2 = 2(2) - 1 = 3$
the third term is $a_3 = 2(3) - 1 = 5$
the fourth term is $a_4 = 2(4) - 1 = 7$

The first four terms of the sequence in Example 1 are the odd numbers 1, 3, 5, and 7. The whole sequence can be written as

$$1, 3, 5, \ldots, 2n - 1, \ldots$$

Because each term in this sequence is larger than the preceding term, we say the sequence is an *increasing sequence.*

EXAMPLE 2 Write the first four terms of the sequence defined by

$$a_n = \frac{1}{n+1}$$

SOLUTION Replacing n with 1, 2, 3, and 4, we have, respectively, the first four terms:

$$\text{First term} = a_1 = \frac{1}{1+1} = \frac{1}{2}$$

$$\text{Second term} = a_2 = \frac{1}{2+1} = \frac{1}{3}$$

$$\text{Third term} = a_3 = \frac{1}{3+1} = \frac{1}{4}$$

$$\text{Fourth term} = a_4 = \frac{1}{4+1} = \frac{1}{5}$$

The sequence in Example 2 defined by

$$a_n = \frac{1}{n+1}$$

can be written as

$$\frac{1}{2}, \frac{1}{3}, \frac{1}{4}, \ldots, \frac{1}{n+1}, \ldots$$

Because each term in the sequence is smaller than the term preceding it, the sequence is said to be a *decreasing sequence.*

EXAMPLE 3 Find the fifth and sixth terms of the sequence whose general term is given by $a_n = \dfrac{(-1)^n}{n^2}$.

SOLUTION For the fifth term, we replace n with 5. For the sixth term, we replace n with 6:

$$\text{Fifth term} = a_5 = \frac{(-1)^5}{5^2} = \frac{-1}{25}$$

$$\text{Sixth term} = a_6 = \frac{(-1)^6}{6^2} = \frac{1}{36}$$

The sequence in Example 3 can be written as

$$-1, \frac{1}{4}, -\frac{1}{9}, \frac{1}{16}, \ldots, \frac{(-1)^n}{n^2}, \ldots$$

Because the terms alternate in sign—if one term is positive, then the next term is negative—we call this an *alternating sequence.* The first three examples all illustrate how we work with a sequence in which we are given a formula for the general term.

USING TECHNOLOGY

Finding Sequences on a Graphing Calculator

Method 1: Using a Table

We can use the table function on a graphing calculator to view the terms of a sequence. To view the terms of the sequence $a_n = 3n + 5$, we set $Y_1 = 3X + 5$. Then we use the table setup feature on the calculator to set the table minimum to 1, and the table increment to 1 also. Here is the setup and result for a TI-83.

Table Setup
Table minimum = 1
Table increment = 1
Independent variable: Auto
Dependent variable: Auto

Y Variables Setup
$Y_1 = 3x + 5$

The table will look like this:

X	Y
1	8
2	11
3	14
4	17
5	20

To find any particular term of a sequence, we change the independent variable setting to Ask, and then input the number of the term of the sequence we want to find. For example, if we want term a_{100}, then we input 100 for the independent variable, and the table gives us the value of 305 for that term.

Method 2: Using the Built-in seq(Function

Using this method, first find the seq(function. On a TI-83 it is found in the LIST OPS menu. To find terms a_1 through a_7 for $a_n = 3n + 5$, we first bring up the seq(function on our calculator, then we input the following four items, in order, separated by commas: 3X + 5, X, 1, 7. Then we close the parentheses. Our screen will look like this:

seq(3X + 5, X, 1, 7)

Pressing ENTER displays the first five terms of the sequence. Pressing the right arrow key repeatedly brings the remaining members of the sequence into view.

Method 3: Using the Built-in Seq Mode

Press the MODE key on your TI-83 and then select Seq (it's next to Func Par and Pol). Go to the Y variables list and set nMin = 1 and $u(n) = 3n + 5$. Then go to the TBLSET key to set up your table like the one shown in Method 1. Pressing TABLE will display the sequence you have defined.

Recursion Formulas

Let's go back to one of the first sequences we looked at in this section:

$$8, 11, 14, 17, \ldots$$

Each term in the sequence can be found by simply substituting positive integers for n in the formula $a_n = 3n + 5$. Another way to look at this sequence, however, is to notice that each term can be found by adding 3 to the preceding term; so, we could give all the terms of this sequence by simply saying

Start with 8, and then add 3 to each term to get the next term.

The same idea, expressed in symbols, looks like this:

$$a_1 = 8 \quad \text{and} \quad a_n = a_{n-1} + 3 \quad \text{for } n > 1$$

This formula is called a *recursion formula* because each term is written *recursively* in terms of the term or terms that precede it.

EXAMPLE 4 Write the first four terms of the sequence given recursively by

$$a_1 = 4 \quad \text{and} \quad a_n = 5a_{n-1} \quad \text{for } n > 1$$

SOLUTION The formula tells us to start the sequence with the number 4, and then multiply each term by 5 to get the next term. Therefore,

$$a_1 = 4$$

$$a_2 = 5a_1 = 5(4) = 20$$

$$a_3 = 5a_2 = 5(20) = 100$$

$$a_4 = 5a_3 = 5(100) = 500$$

The sequence is 4, 20, 100, 500, ...

USING TECHNOLOGY

Recursion Formulas on a Graphing Calculator

We can use a TI-83 graphing calculator to view the sequence defined recursively as

$$a_1 = 8, a_n = a_{n-1} + 3$$

First, put your TI-83 calculator in sequence mode by pressing the MODE key, and then selecting Seq (it's next to Func Par and Pol). Go to the Y variables list and set nMin = 1 and $u(n) = u(n - 1) + 3$. (The u is above the 7, and the n is on the X, T, θ, n and is automatically displayed if that key is pressed when the calculator is in the Seq mode.) Pressing TABLE will display the sequence you have defined.

Finding the General Term

In the first four examples, we found some terms of a sequence after being given the general term. In the next two examples, we will do the reverse. That is, given some terms of a sequence, we will find the formula for the general term.

EXAMPLE 5 Find a formula for the nth term of the sequence 2, 8, 18, 32, ...

SOLUTION Solving a problem like this involves some guessing. Looking over the first four terms, we see each is twice a perfect square:

$$2 = 2(1)$$

$$8 = 2(4)$$

$$18 = 2(9)$$

$$32 = 2(16)$$

If we write each square with an exponent of 2, the formula for the nth term becomes obvious:

$$a_1 = 2 = 2(1)^2$$

$$a_2 = 8 = 2(2)^2$$

$$a_3 = 18 = 2(3)^2$$

$$a_4 = 32 = 2(4)^2$$

$$\vdots$$

$$a_n = \quad 2(n)^2 = 2n^2$$

The general term of the sequence 2, 8, 18, 32, . . . is $a_n = 2n^2$.

EXAMPLE 6

Find the general term for the sequence $2, \frac{3}{8}, \frac{4}{27}, \frac{5}{64}, \ldots$.

SOLUTION The first term can be written as $\frac{2}{1}$. The denominators are all perfect cubes. The numerators are all 1 more than the base of the cubes in the denominators:

$$a_1 = \frac{2}{1} = \frac{1+1}{1^3}$$

$$a_2 = \frac{3}{8} = \frac{2+1}{2^3}$$

$$a_3 = \frac{4}{27} = \frac{3+1}{3^3}$$

$$a_4 = \frac{5}{64} = \frac{4+1}{4^3}$$

Observing this pattern, we recognize the general term to be

$$a_n = \frac{n+1}{n^3}$$

Note

Finding the nth term of a sequence from the first few terms is not always automatic. That is, it sometimes takes a while to recognize the pattern. Don't be afraid to guess at the formula for the general term. Many times, an incorrect guess leads to the correct formula.

LINKING OBJECTIVES AND EXAMPLES

Next to each **objective** we have listed the examples that are best described by that objective.

A	1–3
B	4
C	5, 6

GETTING READY FOR CLASS

After reading through the preceding section, respond in your own words and in complete sentences.

1. How are subscripts used to denote the terms of a sequence?
2. What is the relationship between the subscripts used to denote the terms of a sequence and function notation?
3. What is a decreasing sequence?
4. What is meant by a recursion formula for a sequence?

Problem Set 11.1

Online support materials can be found at www.thomsonedu.com/login

Write the first five terms of the sequences with the following general terms.

1. $a_n = 3n + 1$
2. $a_n = 2n + 3$
3. $a_n = 4n - 1$
4. $a_n = n + 4$
5. $a_n = n$
6. $a_n = -n$
7. $a_n = n^2 + 3$
8. $a_n = n^3 + 1$
9. $a_n = \frac{n}{n+3}$
10. $a_n = \frac{n+1}{n+2}$
11. $a_n = \frac{1}{n^2}$
12. $a_n = \frac{1}{n^3}$
13. $a_n = 2^n$
14. $a_n = 3^{-n}$
15. $a_n = 1 + \frac{1}{n}$

= Videos available by instructor request

▶ = Online student support materials available at www.thomsonedu.com/login

16. $a_n = n - \frac{1}{n}$

17. $a_n = (-2)^n$

18. $a_n = (-3)^n$

19. $a_n = 4 + (-1)^n$

20. $a_n = 10 + (-2)^n$

21. $a_n = (-1)^{n+1} \cdot \frac{n}{2n - 1}$

22. $a_n = (-1)^n \cdot \frac{2n + 1}{2n - 1}$

23. $a_n = n^2 \cdot 2^{-n}$

24. $a_n = n^n$

Write the first five terms of the sequences defined by the following recursion formulas.

25. $a_1 = 3 \quad a_n = -3a_{n-1} \quad n > 1$

26. $a_1 = 3 \quad a_n = a_{n-1} - 3 \quad n > 1$

27. $a_1 = 1 \quad a_n = 2a_{n-1} + 3 \quad n > 1$

28. $a_1 = 1 \quad a_n = a_{n-1} + n \quad n > 1$

29. $a_1 = 2 \quad a_n = 2a_{n-1} - 1 \quad n > 1$

30. $a_1 = -4 \quad a_n = -2a_{n-1} \quad n > 1$

31. $a_1 = 5 \quad a_n = 3a_{n-1} - 4 \quad n > 1$

32. $a_1 = -3 \quad a_n = -2a_{n-1} + 5 \quad n > 1$

33. $a_1 = 4 \quad a_n = 2a_{n-1} - a_1 \quad n > 1$

34. $a_1 = -3 \quad a_n = -2a_{n-1} - n \quad n > 1$

Determine the general term for each of the following sequences.

35. 4, 8, 12, 16, 20, . . .

36. 7, 10, 13, 16, . . .

37. 1, 4, 9, 16, . . .

38. 3, 12, 27, 48, . . .

39. 4, 8, 16, 32, . . .

40. −2, 4, −8, 16, . . .

41. $\frac{1}{4}, \frac{1}{8}, \frac{1}{16}, \frac{1}{32}, \ldots$

42. $\frac{1}{4}, \frac{2}{9}, \frac{3}{16}, \frac{4}{25}, \ldots$

43. 5, 8, 11, 14, . . .

44. 7, 5, 3, 1, . . .

45. −2, −6, −10, −14, . . .

46. −2, 2, −2, 2, . . .

47. 1, −2, 4, −8, . . .

48. −1, 3, −9, 27, . . .

49. $\log_2 3, \log_3 4, \log_4 5, \log_5 6, \ldots$

50. $0, \frac{3}{5}, \frac{8}{10}, \frac{15}{17}, \ldots$

Applying the Concepts

51. **Salary Increase** The entry level salary for a teacher is \$28,000 with 4% increases after every year of service.
   a. Write a sequence for this teacher's salary for the first 5 years.
   b. Find the general term of the sequence in part a.

52. **Holiday Account** To save money for holiday presents, a person deposits \$5 in a savings account on January 1, and then deposits an additional \$5 every week thereafter until Christmas.
   a. Write a sequence for the money in that savings account for the first 10 weeks of the year.
   b. Write the general term of the sequence in part a.
   c. If there are 50 weeks from January 1 to Christmas, how much money will be available for spending on Christmas presents?

53. **Akaka Falls** If a boulder fell from the top of Akaka Falls in Hawaii, the distance, in feet, the boulder would fall in each consecutive second would be modeled by a sequence whose general term is $a_n = 32n - 16$, where n represents the number of seconds.

George H. H. Huey/Corbis

   a. Write a sequence for the first 5 seconds the boulder falls.
   b. What is the total distance the boulder fell in 5 seconds?
   c. If Akaka Falls is approximately 420 feet high, will the boulder hit the ground within 5 seconds?

54. **Health Expenditures** The total national health expenditures, in billions of dollars, increased by approximately \$50 billion from 1995 to 1996 and from 1996 to 1997. The total national health expenditures for 1997 was approximately \$1,092 billion (U.S. Census Bureau, National Health Expenditures).
 a. If the same approximate increase is assumed to continue, what would be the estimated total expenditures for 1998, 1999, and 2000?
 b. Write the general term for the sequence, where n represents the number of years since 1997.
 c. If health expenditures continue to increase at this approximate rate, what would the expected expenditures be in 2010?

55. **Polygons** The formula for the sum of the interior angles of a polygon with n sides is $a_n = 180°(n - 2)$.
 a. Write a sequence to represent the sum of the interior angles of a polygon with 3, 4, 5, and 6 sides.
 b. What would be the sum of the interior angles of a polygon with 20 sides?
 c. What happens when $n = 2$ to indicate that a polygon cannot be formed with only two sides?

56. **Pendulum** A pendulum swings 10 feet left to right on its first swing. On each swing following the first, the pendulum swings $\frac{4}{5}$ of the previous swing.
 a. Write a sequence for the distance traveled by the pendulum on the first, second, and third swing.
 b. Write a general term for the sequence, where n represents the number of the swing.
 c. How far will the pendulum swing on its tenth swing? (Round to the nearest hundredth.)

Maintaining Your Skills

Find x in each of the following.

57. $\log_9 x = \dfrac{3}{2}$

58. $\log_x \dfrac{1}{4} = -2$

Simplify each expression.

59. $\log_2 32$

60. $\log_{10} 10{,}000$

61. $\log_3[\log_2 8]$

62. $\log_5[\log_6 6]$

Getting Ready for the Next Section

Simplify.

63. $-2 + 6 + 4 + 22$

64. $9 - 27 + 81 - 243$

65. $-8 + 16 - 32 + 64$

66. $-4 + 8 - 16 + 32 - 64$

67. $(1 - 3) + (4 - 3) + (9 - 3) + (16 - 3)$

68. $(1 - 3) + (9 + 1) + (16 + 1) + (25 + 1) + (36 + 1)$

69. $-\dfrac{1}{3} + \dfrac{1}{9} - \dfrac{1}{27} + \dfrac{1}{81}$

70. $\dfrac{1}{2} + \dfrac{2}{3} + \dfrac{3}{4} + \dfrac{4}{5}$

71. $\dfrac{1}{3} + \dfrac{1}{2} + \dfrac{3}{5} + \dfrac{2}{3}$

72. $\dfrac{1}{16} + \dfrac{1}{32} + \dfrac{1}{64}$

Extending the Concepts

73. As n increases, the terms in the sequence

$$a_n = \left(1 + \frac{1}{n}\right)^n$$

get closer and closer to the number e (that's the same e we used in defining natural logarithms). It takes some fairly large values of n, however, before we can see this happening. Use a calculator to find a_{100}, $a_{1,000}$, $a_{10,000}$, and $a_{100,000}$, and compare them to the decimal approximation we gave for the number e.

74. The sequence

$$a_n = \left(1 + \frac{1}{n}\right)^{-n}$$

gets close to the number $\frac{1}{e}$ as n becomes large. Use a calculator to find approximations for a_{100} and $a_{1,000}$, and then compare them to $\frac{1}{2.7183}$.

75. Write the first ten terms of the sequence defined by the recursion formula

$$a_1 = 1,\ a_2 = 1,\ a_n = a_{n-1} + a_{n-2} \qquad n > 2$$

76. Write the first ten terms of the sequence defined by the recursion formula

$$a_1 = 2,\ a_2 = 2,\ a_n = a_{n-1} + a_{n-2} \qquad n > 2$$

11.2 Series

OBJECTIVES

A Expand and simplify an expression written with summation notation.

B Write a series using summation notation.

There is an interesting relationship between the sequence of odd numbers and the sequence of squares that is found by adding the terms in the sequence of odd numbers.

$$\begin{aligned} 1 &= 1 \\ 1 + 3 &= 4 \\ 1 + 3 + 5 &= 9 \\ 1 + 3 + 5 + 7 &= 16 \end{aligned}$$

When we add the terms of a sequence the result is called a series.

> **DEFINITION** The sum of a number of terms in a sequence is called a **series.**

A sequence can be finite or infinite, depending on whether the sequence ends at the nth term. For example,

$$1, 3, 5, 7, 9$$

is a finite sequence, but

$$1, 3, 5, \ldots$$

is an infinite sequence. Associated with each of the preceding sequences is a series found by adding the terms of the sequence:

$$1 + 3 + 5 + 7 + 9 \qquad \text{Finite series}$$

$$1 + 3 + 5 + \ldots \qquad \text{Infinite series}$$

In this section, we will consider only finite series. We can introduce a new kind of notation here that is a compact way of indicating a finite series. The notation is called *summation notation,* or *sigma notation* because it is written using the Greek letter sigma. The expression

$$\sum_{i=1}^{4} (8i - 10)$$

is an example of an expression that uses summation notation. The summation notation in this expression is used to indicate the sum of all the expressions $8i - 10$ from $i = 1$ up to and including $i = 4$. That is,

$$\begin{aligned} \sum_{i=1}^{4} (8i - 10) &= (8 \cdot 1 - 10) + (8 \cdot 2 - 10) + (8 \cdot 3 - 10) + (8 \cdot 4 - 10) \\ &= -2 + 6 + 14 + 22 \\ &= 40 \end{aligned}$$

The letter i as used here is called the *index of summation,* or just *index* for short. Here are some examples illustrating the use of summation notation.

EXAMPLE 1 Expand and simplify $\sum_{i=1}^{5} (i^2 - 1)$.

SOLUTION We replace i in the expression $i^2 - 1$ with all consecutive integers from 1 up to 5, including 1 and 5:

$$\begin{aligned}\sum_{i=1}^{5} (i^2 - 1) &= (1^2 - 1) + (2^2 - 1) + (3^2 - 1) + (4^2 - 1) + (5^2 - 1)\\ &= 0 + 3 + 8 + 15 + 24\\ &= 50\end{aligned}$$

EXAMPLE 2 Expand and simplify $\sum_{i=3}^{6} (-2)^i$.

SOLUTION We replace i in the expression $(-2)^i$ with the consecutive integers beginning at 3 and ending at 6:

$$\begin{aligned}\sum_{i=3}^{6} (-2)^i &= (-2)^3 + (-2)^4 + (-2)^5 + (-2)^6\\ &= -8 + 16 + (-32) + 64\\ &= 40\end{aligned}$$

USING TECHNOLOGY

Summing Series on a Graphing Calculator

A TI-83 graphing calculator has a built-in sum(function that, when used with the seq(function, allows us to add the terms of a series. Let's repeat Example 1 using our graphing calculator. First, we go to LIST and select MATH. The fifth option in that list is sum(, which we select. Then we go to LIST again and select OPS. From that list we select seq(. Next we enter X^2 − 1, X, 1, 5, and then we close both sets of parentheses. Our screen shows the following:

sum(seq(X^2 − 1, X, 1, 5)) which will give us $\sum_{i=1}^{5} (i^2 - 1)$

When we press ENTER the calculator displays 50, which is the same result we obtained in Example 1.

EXAMPLE 3 Expand $\sum_{i=2}^{5} (x^i - 3)$.

SOLUTION We must be careful not to confuse the letter x with i. The index i is the quantity we replace by the consecutive integers from 2 to 5, not x:

$$\sum_{i=2}^{5} (x^i - 3) = (x^2 - 3) + (x^3 - 3) + (x^4 - 3) + (x^5 - 3)$$

In the first three examples, we were given an expression with summation notation and asked to expand it. The next examples in this section illustrate how we can write an expression in expanded form as an expression involving summation notation.

EXAMPLE 4

Write with summation notation: $1 + 3 + 5 + 7 + 9$.

SOLUTION A formula that gives us the terms of this sum is

$$a_i = 2i - 1$$

where i ranges from 1 up to and including 5. Notice we are using the subscript i in exactly the same way we used the subscript n in the previous section—to indicate the general term. Writing the sum

$$1 + 3 + 5 + 7 + 9$$

with summation notation looks like this:

$$\sum_{i=1}^{5} (2i - 1)$$

EXAMPLE 5

Write with summation notation: $3 + 12 + 27 + 48$.

SOLUTION We need a formula, in terms of i, that will give each term in the sum. Writing the sum as

$$3 \cdot 1^2 + 3 \cdot 2^2 + 3 \cdot 3^2 + 3 \cdot 4^2$$

we see the formula

$$a_i = 3 \cdot i^2$$

where i ranges from 1 up to and including 4. Using this formula and summation notation, we can represent the sum

$$3 + 12 + 27 + 48$$

as

$$\sum_{i=1}^{4} 3i^2$$

EXAMPLE 6

Write with summation notation:

$$\frac{x+3}{x^3} + \frac{x+4}{x^4} + \frac{x+5}{x^5} + \frac{x+6}{x^6}$$

SOLUTION A formula that gives each of these terms is

$$a_i = \frac{x+i}{x^i}$$

where i assumes all integer values between 3 and 6, including 3 and 6. The sum can be written as

$$\sum_{i=3}^{6} \frac{x+i}{x^i}$$

LINKING OBJECTIVES AND EXAMPLES

Next to each **objective** we have listed the examples that are best described by that objective.

A	1–3
B	4–6

GETTING READY FOR CLASS

After reading through the preceding section, respond in your own words and in complete sentences.

1. What is the difference between a sequence and a series?
2. Explain the summation notation $\sum_{i=1}^{4}$ in the series $\sum_{i=1}^{4}(2i+1)$.
3. When will a finite series result in a numerical value versus an algebraic expression?
4. Determine for what values of n the series $\sum_{i=1}^{n}(-1)^i$ will be equal to 0. Explain your answer.

Problem Set 11.2

Online support materials can be found at www.thomsonedu.com/login

Expand and simplify each of the following.

▶ **1.** $\sum_{i=1}^{4}(2i+4)$

2. $\sum_{i=1}^{5}(3i-1)$

3. $\sum_{i=2}^{3}(i^2-1)$

4. $\sum_{i=3}^{6}(i^2+1)$

5. $\sum_{i=1}^{4}(i^2-3)$

6. $\sum_{i=2}^{6}(2i^2+1)$

7. $\sum_{i=1}^{4}\frac{i}{1+i}$

8. $\sum_{i=1}^{3}\frac{i^2}{2i-1}$

9. $\sum_{i=1}^{4}(-3)^i$

10. $\sum_{i=1}^{4}\left(-\frac{1}{3}\right)^i$

▶ **11.** $\sum_{i=3}^{6}(-2)^i$

12. $\sum_{i=4}^{6}\left(-\frac{1}{2}\right)^i$

13. $\sum_{i=2}^{6}(-2)^i$

14. $\sum_{i=2}^{5}(-3)^i$

15. $\sum_{i=1}^{5}\left(-\frac{1}{2}\right)^i$

16. $\sum_{i=3}^{6}\left(-\frac{1}{3}\right)^i$

17. $\sum_{i=2}^{5}\frac{i-1}{i+1}$

18. $\sum_{i=2}^{4}\frac{i^2-1}{i^2+1}$

Expand the following.

19. $\sum_{i=1}^{5}(x+i)$

20. $\sum_{i=2}^{7}(x+1)^i$

21. $\sum_{i=1}^{4}(x-2)^i$

22. $\sum_{i=2}^{5}\left(x+\frac{1}{i}\right)^2$

23. $\sum_{i=1}^{5}\frac{x+i}{x-1}$

24. $\sum_{i=1}^{6}\frac{x-3i}{x+3i}$

▶ **25.** $\sum_{i=3}^{8}(x+i)^i$

26. $\sum_{i=1}^{5}(x+i)^{i+1}$

27. $\sum_{i=3}^{6}(x-2i)^{i+3}$

28. $\sum_{i=5}^{8}\left(\frac{x-i}{x+i}\right)^{2i}$

Write each of the following sums with summation notation. Do not calculate the sum.

29. $2+4+8+16$

30. $3+5+7+9+11$

▶ **31.** $4+8+16+32+64$

32. $3+8+15+24$

Write each of the following sums with summation notation. Do not calculate the sum. *Note:* More than one answer is possible.

33. $5+9+13+17+21$

34. $3-6+12-24+48$

35. $-4+8-16+32$

36. $15+24+35+48+63$

▶ **37.** $\frac{3}{4}+\frac{4}{5}+\frac{5}{6}+\frac{6}{7}+\frac{7}{8}$

▢ = Videos available by instructor request

▶ = Online student support materials available at www.thomsonedu.com/login

38. $\frac{1}{2} + \frac{2}{3} + \frac{3}{4} + \frac{4}{5}$

39. $\frac{1}{3} + \frac{2}{5} + \frac{3}{7} + \frac{4}{9}$

40. $\frac{3}{1} + \frac{5}{3} + \frac{7}{5} + \frac{9}{7}$

41. $(x - 2)^6 + (x - 2)^7 + (x - 2)^8 + (x - 2)^9$

42. $(x + 1)^3 + (x + 2)^4 + (x + 3)^5 + (x + 4)^6 + (x + 5)^7$

43. $\left(1 + \frac{1}{x}\right)^2 + \left(1 + \frac{2}{x}\right)^3 + \left(1 + \frac{3}{x}\right)^4 + \left(1 + \frac{4}{x}\right)^5$

44. $\frac{x - 1}{x + 2} + \frac{x - 2}{x + 4} + \frac{x - 3}{x + 6} + \frac{x - 4}{x + 8} + \frac{x - 5}{x + 10}$

45. $\frac{x}{x + 3} + \frac{x}{x + 4} + \frac{x}{x + 5}$

46. $\frac{x - 3}{x^3} + \frac{x - 4}{x^4} + \frac{x - 5}{x^5} + \frac{x - 6}{x^6}$

47. $x^2(x + 2) + x^3(x + 3) + x^4(x + 4)$

48. $x(x + 2)^2 + x(x + 3)^3 + x(x + 4)^4$

49. **Repeating Decimals** Any repeating, nonterminating decimal may be viewed as a series. For instance, $\frac{2}{3} = 0.6 + 0.06 + 0.006 + 0.0006 + \cdots$. Write the following fractions as series.

a. $\frac{1}{3}$ **b.** $\frac{2}{9}$ **c.** $\frac{3}{11}$

50. **Repeating Decimals** Refer to the previous exercise, and express the following repeating decimals as fractions.

a. $0.55555\cdots$ **b.** $1.33333\cdots$ **c.** $0.29292929\cdots$

Applying the Concepts

51. **Skydiving** A skydiver jumps from a plane and falls 16 feet the first second, 48 feet the second second, and 80 feet the third second. If he continues to fall in the same manner, how far will he fall the seventh second? What is the distance he falls in 7 seconds?

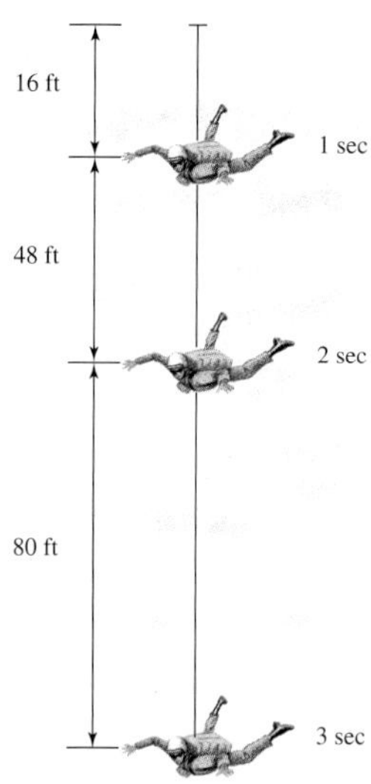

52. **Bacterial Growth** After 1 day, a colony of 50 bacteria reproduces to become 200 bacteria. After 2 days, they reproduce to become 800 bacteria. If they continue to reproduce at this rate, how many bacteria will be present after 4 days?

53. **Akaka Falls** In Section 11.1, when a boulder fell from the top of Akaka Falls in Hawaii, the sequence generated during the first 5 seconds the boulder fell was 16, 48, 80, 112, 144.

a. Write a finite series that represents the sum of this sequence.

b. The general term of this sequence was given as $a_n = 32n - 16$. Write the series produced in part a in summation notation.

54. **Pendulum** A pendulum swings 12 feet left to right on its first swing, and on each swing following the first, swings $\frac{3}{4}$ of the previous swing. The distance the pendulum traveled in 5 seconds can be expressed in summation notation $\sum_{i=1}^{5} 12(\frac{3}{4})^{i-1}$. Expand the summation notation and simplify. Round your final answer to the nearest tenth.

Maintaining Your Skills

Use the properties of logarithms to expand each of the following expressions.

55. $\log_2 x^3y$

56. $\log_7 \frac{x^2}{y^4}$

57. $\log_{10} \frac{\sqrt[3]{x}}{y^2}$

58. $\log_{10} \sqrt[3]{\frac{x}{y^2}}$

Write each expression as a single logarithm.

59. $\log_{10} x - \log_{10} y^2$

60. $\log_{10} x^2 + \log_{10} y^2$

61. $2 \log_3 x - 3 \log_3 y - 4 \log_3 z$

62. $\frac{1}{2} \log_6 x + \frac{1}{3} \log_6 y + \frac{1}{4} \log_6 z$

Solve each equation.

63. $\log_4 x - \log_4 5 = 2$

64. $\log_3 6 + \log_3 x = 4$

65. $\log_2 x + \log_2(x - 7) = 3$

66. $\log_5(x + 1) + \log_5(x - 3) = 1$

Getting Ready for the Next Section

Simplify.

67. $2 + 9(8)$

68. $\frac{1}{2} + 9\left(\frac{1}{2}\right)$

69. $\frac{10}{2}\left(\frac{1}{2} + 5\right)$

70. $\frac{10}{2}(2 + 74)$

71. $3 + (n - 1)2$

72. $7 + (n - 1)3$

Solve each system of equations.

73. $x + 2y = 7$
$x + 7y = 17$

74. $x + 3y = 14$
$x + 9y = 32$

Extending the Concepts

Solve each of the following equations for x.

75. $\sum_{i=1}^{4} (x - i) = 16$

76. $\sum_{i=3}^{8} (2x + i) = 30$

11.3 Arithmetic Sequences

OBJECTIVES

A Find the common difference for an arithmetic sequence.

B Find the general term for an arithmetic sequence.

C Find the sum of the first n terms of an arithmetic sequence.

In this and the following section, we will review and expand upon two major types of sequences, which we have worked with previously—arithmetic sequences and geometric sequences.

> **DEFINITION** An **arithmetic sequence** is a sequence of numbers in which each term is obtained from the preceding term by adding the same amount each time. An arithmetic sequence is also called an **arithmetic progression.**

The sequence

$$2, 6, 10, 14, \ldots$$

is an example of an arithmetic sequence, because each term is obtained from the preceding term by adding 4 each time. The amount we add each time—in this case, 4—is called the *common difference,* because it can be obtained by subtracting any two consecutive terms. (The term with the larger subscript must be written first.) The common difference is denoted by d.

EXAMPLE 1 Give the common difference d for the arithmetic sequence 4, 10, 16, 22, ...

SOLUTION Because each term can be obtained from the preceding term by adding 6, the common difference is 6. That is, $d = 6$.

EXAMPLE 2 Give the common difference for 100, 93, 86, 79, ...

SOLUTION The common difference in this case is $d = -7$, since adding -7 to any term always produces the next consecutive term.

EXAMPLE 3 Give the common difference for $\frac{1}{2}, 1, \frac{3}{2}, 2, \ldots$.

SOLUTION The common difference is $d = \frac{1}{2}$.

The General Term

The general term a_n of an arithmetic progression can always be written in terms of the first term a_1 and the common difference d. Consider the sequence from Example 1:

$$4, 10, 16, 22, \ldots$$

We can write each term in terms of the first term 4 and the common difference 6:

$$\begin{array}{llll} 4, & 4 + (1 \cdot 6), & 4 + (2 \cdot 6), & 4 + (3 \cdot 6), \ldots \\ a_1, & a_2, & a_3, & a_4, \ldots \end{array}$$

Observing the relationship between the subscript on the terms in the second line and the coefficients of the 6's in the first line, we write the general term for the sequence as

$$a_n = 4 + (n - 1)6$$

We generalize this result to include the general term of any arithmetic sequence.

> **Arithmetic Sequence** The **general term** of an arithmetic progression with first term a_1 and common difference d is given by
>
> $$a_n = a_1 + (n - 1)d$$

EXAMPLE 4 Find the general term for the sequence

$$7, 10, 13, 16, \ldots$$

SOLUTION The first term is $a_1 = 7$, and the common difference is $d = 3$. Substituting these numbers into the formula for the general term, we have

$$a_n = 7 + (n - 1)3$$

which we can simplify, if we choose, to

$$a_n = 7 + 3n - 3$$
$$= 3n + 4$$

EXAMPLE 5 Find the general term of the arithmetic progression whose third term a_3 is 7 and whose eighth term a_8 is 17.

SOLUTION According to the formula for the general term, the third term can be written as $a_3 = a_1 + 2d$, and the eighth term can be written as $a_8 = a_1 + 7d$. Because these terms are also equal to 7 and 17, respectively, we can write

$$a_3 = a_1 + 2d = 7$$
$$a_8 = a_1 + 7d = 17$$

To find a_1 and d, we simply solve the system:

$$a_1 + 2d = 7$$
$$a_1 + 7d = 17$$

We add the opposite of the top equation to the bottom equation. The result is

$$5d = 10$$
$$d = 2$$

To find a_1, we simply substitute 2 for d in either of the original equations and get

$$a_1 = 3$$

The general term for this progression is

$$a_n = 3 + (n - 1)2$$

which we can simplify to

$$a_n = 2n + 1$$

The sum of the first n terms of an arithmetic sequence is denoted by S_n. The following theorem gives the formula for finding S_n, which is sometimes called the nth *partial sum*.

Theorem 1 The sum of the first n terms of an arithmetic sequence whose first term is a_1 and whose nth term is a_n is given by

$$S_n = \frac{n}{2}(a_1 + a_n)$$

Proof We can write S_n in expanded form as

$$S_n = a_1 + [a_1 + d] + [a_1 + 2d] + \cdots + [a_1 + (n - 1)d]$$

We can arrive at this same series by starting with the last term a_n and subtracting d each time. Writing S_n this way, we have

$$S_n = a_n + [a_n - d] + [a_n - 2d] + \cdots + [a_n - (n - 1)d]$$

If we add the preceding two expressions term by term, we have

$$2S_n = (a_1 + a_n) + (a_1 + a_n) + (a_1 + a_n) + \cdots + (a_1 + a_n)$$

$$2S_n = n(a_1 + a_n)$$

$$S_n = \frac{n}{2}(a_1 + a_n)$$

EXAMPLE 6 Find the sum of the first 10 terms of the arithmetic progression 2, 10, 18, 26, ...

SOLUTION The first term is 2, and the common difference is 8. The tenth term is

$$\begin{aligned} a_{10} &= 2 + 9(8) \\ &= 2 + 72 \\ &= 74 \end{aligned}$$

Substituting $n = 10$, $a_1 = 2$, and $a_{10} = 74$ into the formula

$$S_n = \frac{n}{2}(a_1 + a_n)$$

we have

$$\begin{aligned} S_{10} &= \frac{10}{2}(2 + 74) \\ &= 5(76) \\ &= 380 \end{aligned}$$

The sum of the first 10 terms is 380.

LINKING OBJECTIVES AND EXAMPLES

Next to each **objective** we have listed the examples that are best described by that objective.

A	1–3
B	4, 5
C	6

GETTING READY FOR CLASS

After reading through the preceding section, respond in your own words and in complete sentences.

1. Explain how to determine if a sequence is arithmetic.
2. What is a common difference?
3. Suppose the value of a_5 is given. What other possible pieces of information could be given to have enough information to obtain the first 10 terms of the sequence?
4. Explain the formula $a_n = a_1 + (n - 1)d$ in words so that someone who wanted to find the nth term of an arithmetic sequence could do so from your description.

Problem Set 11.3

Online support materials can be found at www.thomsonedu.com/login

Determine which of the following sequences are arithmetic progressions. For those that are arithmetic progressions, identify the common difference d.

1. $1, 2, 3, 4, \ldots$
2. $4, 6, 8, 10, \ldots$
3. $1, 2, 4, 7, \ldots$
4. $1, 2, 4, 8, \ldots$
5. $50, 45, 40, \ldots$
6. $1, \frac{1}{2}, \frac{1}{4}, \frac{1}{8}, \ldots$
7. $1, 4, 9, 16, \ldots$
8. $5, 7, 9, 11, \ldots$
9. $\frac{1}{3}, 1, \frac{5}{3}, \frac{7}{3}, \ldots$
10. $5, 11, 17, \ldots$

Each of the following problems refers to arithmetic sequences.

11. If $a_1 = 3$ and $d = 4$, find a_n and a_{24}.
12. If $a_1 = 5$ and $d = 10$, find a_n and a_{100}.
13. If $a_1 = 6$ and $d = -2$, find a_{10} and S_{10}.
14. If $a_1 = 7$ and $d = -1$, find a_{24} and S_{24}.
15. If $a_6 = 17$ and $a_{12} = 29$, find the term a_1, the common difference d, and then find a_{30}.
16. If $a_5 = 23$ and $a_{10} = 48$, find the first term a_1, the common difference d, and then find a_{40}.
17. If the third term is 16 and the eighth term is 26, find the first term, the common difference, and then find a_{20} and S_{20}.
18. If the third term is 16 and the eighth term is 51, find the first term, the common difference, and then find a_{50} and S_{50}.
19. If $a_1 = 3$ and $d = 4$, find a_{20} and S_{20}.
20. If $a_1 = 40$ and $d = -5$, find a_{25} and S_{25}.
21. If $a_4 = 14$ and $a_{10} = 32$, find a_{40} and S_{40}.
22. If $a_7 = 0$ and $a_{11} = -\frac{8}{3}$, find a_{61} and S_{61}.
23. If $a_6 = -17$ and $S_6 = -12$, find a_1 and d.
24. If $a_{10} = 12$ and $S_{10} = 40$, find a_1 and d.
25. Find a_{85} for the sequence $14, 11, 8, 5, \ldots$
26. Find S_{100} for the sequence $-32, -25, -18, -11, \ldots$
27. If $S_{20} = 80$ and $a_1 = -4$, find d and a_{39}.
28. If $S_{24} = 60$ and $a_1 = 4$, find d and a_{116}.
29. Find the sum of the first 100 terms of the sequence $5, 9, 13, 17, \ldots$.
30. Find the sum of the first 50 terms of the sequence $8, 11, 14, 17, \ldots$.
31. Find a_{35} for the sequence $12, 7, 2, -3, \ldots$.
32. Find a_{45} for the sequence $25, 20, 15, 10, \ldots$.
33. Find the tenth term and the sum of the first 10 terms of the sequence $\frac{1}{2}, 1, \frac{3}{2}, 2, \ldots$.
34. Find the 15th term and the sum of the first 15 terms of the sequence $-\frac{1}{3}, 0, \frac{1}{3}, \frac{2}{3}, \ldots$.

Applying the Concepts

Straight-Line Depreciation Straight-line depreciation is an accounting method used to help spread the cost of new equipment over a number of years. The value at any time during the life of the machine can be found with a linear equation in two variables. For income tax purposes, however, it is the value at the end of the year that is most important, and for this reason sequences can be used.

35. **Value of a Copy Machine** A large copy machine sells for \$18,000 when it is new. Its value decreases

= Videos available by instructor request
= Online student support materials available at www.thomsonedu.com/login

$3,300 each year after that. We can use an arithmetic sequence to find the value of the machine at the end of each year. If we let a_0 represent the value when it is purchased, then a_1 is the value after 1 year, a_2 is the value after 2 years, and so on.

John Turner/Tony Stone Images

a. Write the first 5 terms of the sequence.

b. What is the common difference?

c. Construct a line graph for the first 5 terms of the sequence.

d. Use the line graph to estimate the value of the copy machine 2.5 years after it is purchased.

e. Write the sequence from part a using a recursive formula.

36. **Value of a Forklift** An electric forklift sells for $125,000 when new. Each year after that, it decreases $16,500 in value.

a. Write an arithmetic sequence that gives the value of the forklift at the end of each of the first 5 years after it is purchased.

b. What is the common difference for this sequence?

c. Construct a line graph for this sequence.

d. Use the line graph to estimate the value of the forklift 3.5 years after it is purchased.

e. Write the sequence from part (a) using a recursive formula.

Telegraph Color Library/FPG

37. **Distance** A rocket travels vertically 1,500 feet in its first second of flight, and then about 40 feet less each succeeding second. Use these estimates to answer the following questions.

a. Write a sequence of the vertical distance traveled by a rocket in each of its first 6 seconds.

b. Is the sequence in part (a) an arithmetic sequence? Explain why or why not.

c. What is the general term of the sequence in part (a)?

38. **Depreciation** Suppose an automobile sells for N dollars new, and then depreciates 40% each year.

a. Write a sequence for the value of this automobile (in terms of N) for each year.

b. What is the general term of the sequence in part (a)?

c. Is the sequence in part (a) an arithmetic sequence? Explain why it is or is not.

39. **Triangular Numbers** The first four triangular numbers are {1, 3, 6, 10, . . . }, and are illustrated in the following diagram.

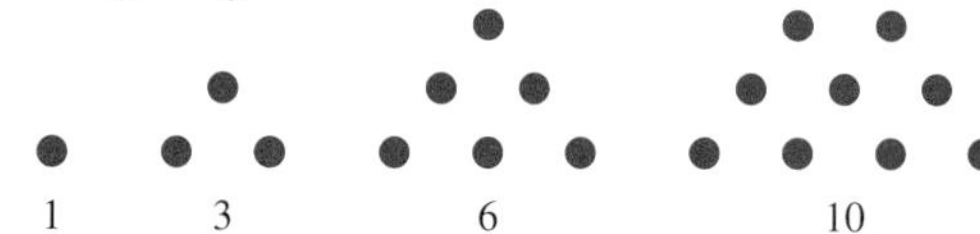

a. Write a sequence of the first 15 triangular numbers.

b. Write the recursive general term for the sequence of triangular numbers.

c. Is the sequence of triangular numbers an arithmetic sequence? Explain why it is or is not.

40. **Arithmetic Means** Three (or more) arithmetic means between two numbers may be found by forming an arithmetic sequence using the original two numbers and the arithmetic means. For example, three arithmetic means between 10 and 34 may be found by examining the sequence $\{10, a, b, c, 34\}$. For the sequence to be arithmetic, the common difference must be 6; therefore, $a = 16$, $b = 22$, and $c = 28$. Use this idea to answer the following questions.

a. Find four arithmetic means between 10 and 35.

b. Find three arithmetic means between 2 and 62.

c. Find five arithmetic means between 4 and 28.

Maintaining Your Skills

Find the following logarithms. Round to four decimal places.

41. $\log 576$

42. $\log 57{,}600$

43. $\log 0.0576$

44. $\log 0.000576$

Find x.

45. $\log x = 2.6484$

46. $\log x = 7.9832$

47. $\log x = -7.3516$

48. $\log x = -2.0168$

Getting Ready for the Next Section

Simplify.

49. $\frac{1}{8}\left(\frac{1}{2}\right)$

50. $\frac{1}{4}\left(\frac{1}{2}\right)$

51. $\frac{3\sqrt{3}}{3}$

52. $\frac{3}{\sqrt{3}}$

53. $2 \cdot 2^{n-1}$

54. $3 \cdot 3^{n-1}$

55. $\frac{ar^6}{ar^3}$

56. $\frac{ar^7}{ar^4}$

57. $\dfrac{\frac{1}{5}}{1 - \frac{1}{2}}$

58. $\dfrac{\frac{9}{10}}{1 - \frac{1}{10}}$

59. $\dfrac{-3[(-2)^8 - 1]}{-2 - 1}$

60. $\dfrac{4\left[\left(\frac{1}{2}\right)^6 - 1\right]}{\frac{1}{2} - 1}$

Extending the Concepts

61. Find a_1 and d for an arithmetic sequence with $S_7 = 147$ and $S_{13} = 429$.

62. Find a_1 and d for an arithmetic sequence with $S_{12} = 282$ and $S_{19} = 646$.

63. Find d for an arithmetic sequence in which $a_{15} - a_7 = -24$.

64. Find d for an arithmetic sequence in which $a_{25} - a_{15} = 120$.

65. Find the sum of the first 50 terms of the sequence $1, 1 + \sqrt{2}, 1 + 2\sqrt{2}, \ldots$.

66. Given $a_1 = -3$ and $d = 5$ in an arithmetic sequence, and $S_n = 890$, find n.

11.4 Geometric Sequences

OBJECTIVES

A Find the common ratio for a geometric sequence.

B Find the *n*th term for a geometric sequence.

C Find the general term for a geometric sequence.

D Find the sum of the first n terms of a geometric sequence.

E Find the sum of all terms of an infinite geometric sequence.

This section is concerned with the second major classification of sequences, called geometric sequences. The problems in this section are similar to the problems in the preceding section.

DEFINITION A sequence of numbers in which each term is obtained from the previous term by multiplying by the same amount each time is called a **geometric sequence.** Geometric sequences are also called **geometric progressions.**

The sequence

$$3, 6, 12, 24, \ldots$$

is an example of a geometric progression. Each term is obtained from the previous term by multiplying by 2. The amount by which we multiply each time—in this case, 2—is called the *common ratio.* The common ratio is denoted by r and can be found by taking the ratio of any two consecutive terms. (The term with the larger subscript must be in the numerator.)

EXAMPLE 1 Find the common ratio for the geometric progression.

$$\frac{1}{2}, \frac{1}{4}, \frac{1}{8}, \frac{1}{16}, \ldots$$

SOLUTION Because each term can be obtained from the term before it by multiplying by $\frac{1}{2}$, the common ratio is $\frac{1}{2}$. That is, $r = \frac{1}{2}$.

EXAMPLE 2 Find the common ratio for $\sqrt{3}, 3, 3\sqrt{3}, 9, \ldots$

SOLUTION If we take the ratio of the third term to the second term, we have

$$\frac{3\sqrt{3}}{3} = \sqrt{3}$$

The common ratio is $r = \sqrt{3}$.

Geometric Sequences The **general term** a_n of a geometric sequence with first term a_1 and common ratio r is given by

$$a_n = a_1 r^{n-1}$$

To see how we arrive at this formula, consider the following geometric progression whose common ratio is 3:

$$2, 6, 18, 54, \ldots$$

We can write each term of the sequence in terms of the first term 2 and the common ratio 3:

$$2 \cdot 3^0, \quad 2 \cdot 3^1, \quad 2 \cdot 3^2, \quad 2 \cdot 3^3, \ldots$$

$$a_1, \quad a_2, \quad a_3, \quad a_4, \ldots$$

Observing the relationship between the two preceding lines, we find we can write the general term of this progression as

$$a_n = 2 \cdot 3^{n-1}$$

Because the first term can be designated by a_1 and the common ratio by r, the formula $a_n = 2 \cdot 3^{n-1}$ coincides with the formula $a_n = a_1 r^{n-1}$

EXAMPLE 3 Find the general term for the geometric progression

$$5, 10, 20, \ldots$$

SOLUTION The first term is $a_1 = 5$, and the common ratio is $r = 2$. Using these values in the formula

$$a_n = a_1 r^{n-1}$$

we have

$$a_n = 5 \cdot 2^{n-1}$$

EXAMPLE 4 Find the tenth term of the sequence $3, \frac{3}{2}, \frac{3}{4}, \frac{3}{8}, \ldots$.

SOLUTION The sequence is a geometric progression with first term $a_1 = 3$ and common ratio $r = \frac{1}{2}$. The tenth term is

$$a_{10} = 3\left(\frac{1}{2}\right)^9 = \frac{3}{512}$$

EXAMPLE 5 Find the general term for the geometric progression whose fourth term is 16 and whose seventh term is 128.

SOLUTION The fourth term can be written as $a_4 = a_1r^3$, and the seventh term can be written as $a_7 = a_1r^6$.

$$a_4 = a_1r^3 = 16$$

$$a_7 = a_1r^6 = 128$$

We can solve for r by using the ratio $\frac{a_7}{a_4}$.

$$\frac{a_7}{a_4} = \frac{a_1r^6}{a_1r^3} = \frac{128}{16}$$

$$r^3 = 8$$

$$r = 2$$

The common ratio is 2. To find the first term, we substitute $r = 2$ into either of the original two equations. The result is

$$a_1 = 2$$

The general term for this progression is

$$a_n = 2 \cdot 2^{n-1}$$

which we can simplify by adding exponents, because the bases are equal:

$$a_n = 2^n$$

As was the case in the preceding section, the sum of the first n terms of a geometric progression is denoted by S_n, which is called the nth *partial sum* of the progression.

> **Theorem 2** The sum of the first n terms of a geometric progression with first term a_1 and common ratio r is given by the formula
>
> $$S_n = \frac{a_1(r^n - 1)}{r - 1}$$

Proof We can write the sum of the first n terms in expanded form:

$$S_n = a_1 + a_1r + a_1r^2 + \cdots + a_1r^{n-1} \qquad (1)$$

Then multiplying both sides by r, we have

$$rS_n = a_1r + a_1r^2 + a_1r^3 + \cdots + a_1r^n \qquad (2)$$

If we subtract the left side of equation (1) from the left side of equation (2) and do the same for the right sides, we end up with

$$rS_n - S_n = a_1r^n - a_1$$

We factor S_n from both terms on the left side and a_1 from both terms on the right side of this equation:

$$S_n(r - 1) = a_1(r^n - 1)$$

Dividing both sides by $r - 1$ gives the desired result:

$$S_n = \frac{a_1(r^n - 1)}{r - 1}$$

EXAMPLE 6 Find the sum of the first 10 terms of the geometric progression 5, 15, 45, 135, ...

SOLUTION The first term is $a_1 = 5$, and the common ratio is $r = 3$. Substituting these values into the formula for S_{10}, we have the sum of the first 10 terms of the sequence:

$$S_{10} = \frac{5(3^{10} - 1)}{3 - 1} = \frac{5(3^{10} - 1)}{2}$$

The answer can be left in this form. A calculator will give the result as 147,620.

Infinite Geometric Series

Suppose the common ratio for a geometric sequence is a number whose absolute value is less than 1—for instance, $\frac{1}{2}$. The sum of the first n terms is given by the formula

$$S_n = \frac{a_1\left[\left(\frac{1}{2}\right)^n - 1\right]}{\frac{1}{2} - 1}$$

As n becomes larger and larger, the term $\left(\frac{1}{2}\right)^n$ will become closer and closer to 0. That is, for $n = 10$, 20, and 30, we have the following approximations:

$$\left(\frac{1}{2}\right)^{10} \approx 0.001$$

$$\left(\frac{1}{2}\right)^{20} \approx 0.000001$$

$$\left(\frac{1}{2}\right)^{30} \approx 0.000000001$$

so that for large values of n, there is little difference between the expression

$$\frac{a_1(r^n - 1)}{r - 1}$$

and the expression

$$\frac{a_1(0 - 1)}{r - 1} = \frac{-a_1}{r - 1} = \frac{a_1}{1 - r} \quad \text{if} \quad |r| < 1$$

In fact, the sum of the terms of a geometric sequence in which $|r| < 1$ actually becomes the expression

$$\frac{a_1}{1 - r}$$

as n approaches infinity. To summarize, we have the following:

The Sum of an Infinite Geometric Series

If a geometric sequence has first term a_1 and common ratio r such that $|r| < 1$, then the following is called an **infinite geometric series:**

$$S = \sum_{i=0}^{\infty} a_1 r^i = a_1 + a_1 r + a_1 r^2 + a_1 r^3 + \cdots$$

Its sum is given by the formula

$$S = \frac{a_1}{1 - r}$$

EXAMPLE 7

Find the sum of the infinite geometric series

$$\frac{1}{5} + \frac{1}{10} + \frac{1}{20} + \frac{1}{40} + \cdots$$

SOLUTION The first term is $a_1 = \frac{1}{5}$, and the common ratio is $r = \frac{1}{2}$, which has an absolute value less than 1. Therefore, the sum of this series is

$$S = \frac{a_1}{1 - r} = \frac{\frac{1}{5}}{1 - \frac{1}{2}} = \frac{\frac{1}{5}}{\frac{1}{2}} = \frac{2}{5}$$

EXAMPLE 8

Show that 0.999 . . . is equal to 1.

SOLUTION We begin by writing 0.999 . . . as an infinite geometric series:

$$\begin{aligned} 0.999\ldots &= 0.9 + 0.09 + 0.009 + 0.0009 + \cdots \\ &= \frac{9}{10} + \frac{9}{100} + \frac{9}{1{,}000} + \frac{9}{10{,}000} + \cdots \\ &= \frac{9}{10} + \frac{9}{10}\left(\frac{1}{10}\right) + \frac{9}{10}\left(\frac{1}{10}\right)^2 + \frac{9}{10}\left(\frac{1}{10}\right)^3 + \cdots \end{aligned}$$

As the last line indicates, we have an infinite geometric series with $a_1 = \frac{9}{10}$ and $r = \frac{1}{10}$. The sum of this series is given by

$$S = \frac{a_1}{1-r} = \frac{\frac{9}{10}}{1-\frac{1}{10}} = \frac{\frac{9}{10}}{\frac{9}{10}} = 1$$

LINKING OBJECTIVES AND EXAMPLES

Next to each **objective** we have listed the examples that are best described by that objective.

A	1, 2
B	4
C	3, 5
D	6
E	7, 8

GETTING READY FOR CLASS

After reading through the preceding section, respond in your own words and in complete sentences.

1. What is a common ratio?
2. Explain the formula $a_n = a_1r^{n-1}$ in words so that someone who wanted to find the nth term of a geometric sequence could do so from your description.
3. When is the sum of an infinite geometric series a finite number?
4. Explain how a repeating decimal can be represented as an infinite geometric series.

Problem Set 11.4

Online support materials can be found at www.thomsonedu.com/login

Identify those sequences that are geometric progressions. For those that are geometric, give the common ratio r.

1. 1, 5, 25, 125, . . .
2. 6, 12, 24, 48, . . .
3. $\frac{1}{2}, \frac{1}{6}, \frac{1}{18}, \frac{1}{54}, \ldots$
4. 5, 10, 15, 20, . . .
5. 4, 9, 16, 25, . . .
6. $-1, \frac{1}{3}, -\frac{1}{9}, \frac{1}{27}, \ldots$
7. $-2, 4, -8, 16, \ldots$
8. 1, 8, 27, 64, . . .
9. 4, 6, 8, 10, . . .
10. $1, -3, 9, -27, \ldots$

Each of the following problems gives some information about a specific geometric progression.

11. If $a_1 = 4$ and $r = 3$, find a_n.
12. If $a_1 = 5$ and $r = 2$, find a_n.
13. If $a_1 = -2$ and $r = -\frac{1}{2}$, find a_6.
14. If $a_1 = 25$ and $r = -\frac{1}{5}$, find a_6.
15. If $a_1 = 3$ and $r = -1$, find a_{20}.
16. If $a_1 = -3$ and $r = -1$, find a_{20}.
17. If $a_1 = 10$ and $r = 2$, find S_{10}.
18. If $a_1 = 8$ and $r = 3$, find S_5.
19. If $a_1 = 1$ and $r = -1$, find S_{20}.
20. If $a_1 = 1$ and $r = -1$, find S_{21}.
21. Find a_8 for $\frac{1}{5}, \frac{1}{10}, \frac{1}{20}, \ldots$.
22. Find a_8 for $\frac{1}{2}, \frac{1}{10}, \frac{1}{50}, \ldots$.
23. Find S_5 for $-\frac{1}{2}, -\frac{1}{4}, -\frac{1}{8}, \ldots$.
24. Find S_6 for $-\frac{1}{2}, 1, -2, \ldots$.
25. Find a_{10} and S_{10} for $\sqrt{2}, 2, 2\sqrt{2}, \ldots$.

■ = Videos available by instructor request
▶ = Online student support materials available at www.thomsonedu.com/login

26. Find a_8 and S_8 for $\sqrt{3}, 3, 3\sqrt{3}, \ldots$.

27. Find a_6 and S_6 for $100, 10, 1, \ldots$.

28. Find a_6 and S_6 for $100, -10, 1, \ldots$.

29. If $a_4 = 40$ and $a_6 = 160$, find r.

30. If $a_5 = \frac{1}{8}$ and $a_8 = \frac{1}{64}$, find r.

31. Given the sequence $-3, 6, -12, 24, \ldots$, find a_8 and S_8.

32. Given the sequence $4, 2, 1, \frac{1}{2}, \ldots$, find a_9 and S_9.

33. Given $a_7 = 13$ and $a_{10} = 104$, find r.

34. Given $a_5 = -12$ and $a_8 = 324$, find r.

Find the sum of each geometric series.

35. $\frac{1}{2} + \frac{1}{4} + \frac{1}{8} + \cdots$

36. $\frac{1}{3} + \frac{1}{9} + \frac{1}{27} + \cdots$

37. $4 + 2 + 1 + \cdots$

38. $8 + 4 + 2 + \cdots$

39. $2 + 1 + \frac{1}{2} + \cdots$

40. $3 + 1 + \frac{1}{3} + \cdots$

41. $\frac{4}{3} - \frac{2}{3} + \frac{1}{3} + \cdots$

42. $6 - 4 + \frac{8}{3} + \cdots$

43. $\frac{2}{5} + \frac{4}{25} + \frac{8}{125} + \cdots$

44. $\frac{3}{4} + \frac{9}{16} + \frac{27}{64} + \cdots$

45. $\frac{3}{4} + \frac{1}{4} + \frac{1}{12} + \cdots$

46. $\frac{5}{3} + \frac{1}{3} + \frac{1}{15} + \cdots$

47. Show that $0.444\ldots$ is the same as $\frac{4}{9}$.

48. Show that $0.333\ldots$ is the same as $\frac{1}{3}$.

49. Show that $0.272727\ldots$ is the same as $\frac{3}{11}$.

50. Show that $0.545454\ldots$ is the same as $\frac{6}{11}$.

Applying the Concepts

Declining-Balance Depreciation The declining-balance method of depreciation is an accounting method businesses use to deduct most of the cost of new equipment during the first few years of purchase. The value at any time during the life of the machine can be found with a linear equation in two variables. For income tax purposes, however, it is the value at the end of the year that is most important, and for this reason sequences can be used.

51. **Value of a Crane** A construction crane sells for \$450,000 if purchased new. After that, the value decreases by 30% each year. We can use a geometric sequence to find the value of the crane at the end of each year. If we let a_0 represent the value when it is purchased, then a_1 is the value after 1 year, a_2 is the value after 2 years, and so on.
    a. Write the first five terms of the sequence.
    b. What is the common ratio?
    c. Construct a line graph for the first five terms of the sequence.
    d. Use the line graph to estimate the value of the crane 4.5 years after it is purchased.
    e. Write the sequence from part a using a recursive formula.

SuperStock

52. **Value of a Printing Press** A large printing press sells for \$375,000 when it is new. After that, its value decreases 25% each year.
    a. Write a geometric sequence that gives the value of the press at the end of each of the first 5 years after it is purchased.
    b. What is the common ratio for this sequence?
    c. Construct a line graph for this sequence.

d. Use the line graph to estimate the value of the printing press 1.5 years after it is purchased.

e. Write the sequence from part a using a recursive formula.

Jim McCrary/Tony Stone Images

53. Adding Terms Given the geometric series

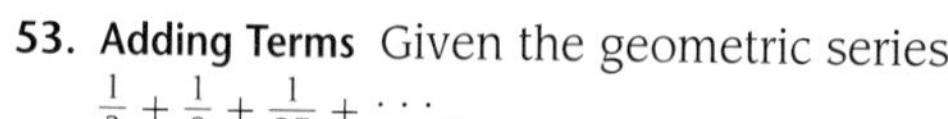

$\frac{1}{3} + \frac{1}{9} + \frac{1}{27} + \cdots,$

a. Find the sum of all the terms.

b. Find the sum of the first six terms.

c. Find the sum of all but the first six terms.

54. Perimeter Triangle *ABC* has a perimeter of 40 inches. A new triangle *XYZ* is formed by connecting the midpoints of the sides of the first triangle, as shown in the following figure. Because midpoints are joined, the perimeter of triangle *XYZ* will be one-half the perimeter of triangle *ABC*, or 20 inches.

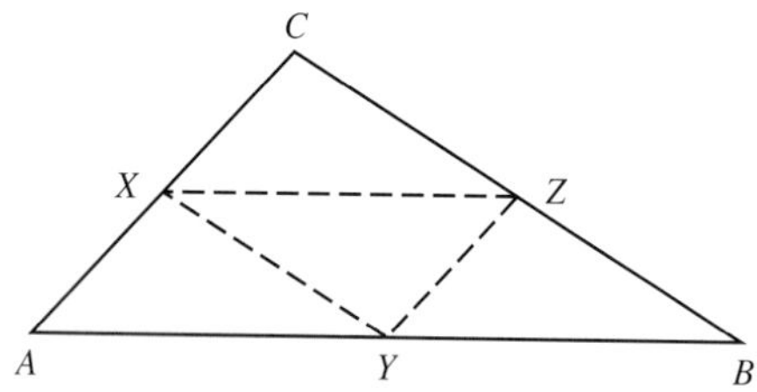

Midpoints of the sides of triangle *XYZ* are used to form a new triangle *RST*. If this pattern of using midpoints to draw triangles is continued seven more times, so there are a total of 10 triangles drawn, what will be the sum of the perimeters of these ten triangles?

55. Bouncing Ball A ball is dropped from a height of 20 feet. Each time it bounces it returns to $\frac{7}{8}$ of the height it fell from. If the ball is allowed to bounce an infinite number of times, find the total vertical distance that the ball travels.

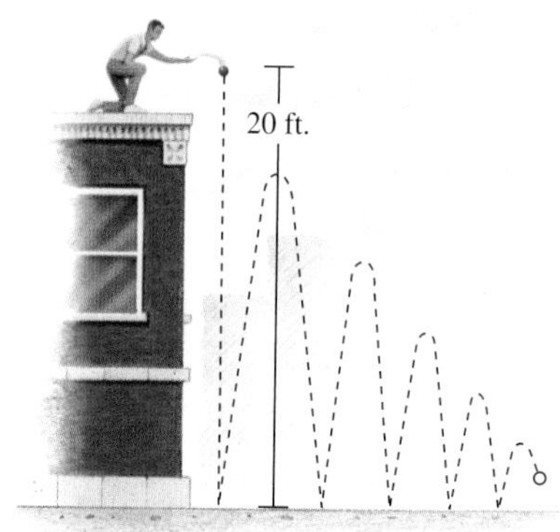

56. Stacking Paper Assume that a thin sheet of paper is 0.002 inch thick. The paper is torn in half, and the two halves placed together.

a. How thick is the pile of torn paper?

b. The pile of paper is torn in half again, and then the two halves placed together and torn in half again. The paper is large enough so this process may be performed a total of 5 times. How thick is the pile of torn paper?

c. Refer to the tearing and piling process described in part b. Assuming that somehow the original paper is large enough, how thick is the pile of torn paper if 25 tears are made?

57. Pendulum A pendulum swings 15 feet left to right on its first swing. On each swing following the first, the pendulum swings $\frac{4}{5}$ of the previous swing.

a. Write the general term for this geometric sequence.

b. If the pendulum is allowed to swing an infinite number of times, what is the total distance the pendulum will travel?

58. Bacteria Growth A bacteria colony of 50 bacteria doubles each day. The number of bacteria in the colony each day can be modeled as a geometric sequence.

a. What is the first term of the sequence? What is the common ratio?

b. Write the general term for this sequence.

c. What would be the total number of bacteria in the colony after 10 days?

Maintaining Your Skills

Find each of the following to the nearest hundredth.

59. $\log_4 20$

60. $\log_7 21$

61. $\ln 576$

62. $\ln 5{,}760$

63. Solve the formula $A = 10e^{5t}$ for t.

64. Solve the formula $A = P \cdot 2^{-5t}$ for t.

Getting Ready for the Next Section

65. $(x + y)^0$

66. $(x + y)^1$

Expand and multiply.

67. $(x + y)^2$

68. $(x + y)^3$

Simplify.

69. $\dfrac{6 \cdot 5 \cdot 4 \cdot 3 \cdot 2 \cdot 1}{(2 \cdot 1)(4 \cdot 3 \cdot 2 \cdot 1)}$

70. $\dfrac{7 \cdot 6 \cdot 5 \cdot 4 \cdot 3 \cdot 2 \cdot 1}{(5 \cdot 4 \cdot 3 \cdot 2 \cdot 1)(2 \cdot 1)}$

Extending the Concepts

71. Find the fraction form for 0.636363 . . . .

72. Find the fraction form for 0.123123123 . . . .

73. The sum of an infinite geometric series is 6 and the first term is 4. Find the common ratio.

74. The sum of an infinite geometric series is $\frac{5}{2}$ and the first term is 3. Find the common ratio.

75. **Sierpinski Triangle** In the sequence that follows, the figures are moving toward what is known as the Sierpinski triangle. To construct the figure in stage 2, we remove the triangle formed from the midpoints of the sides of the shaded region in stage 1. Likewise, the figure in stage 3 is found by removing the triangles formed by connecting the midponts of the sides of the shaded regions in stage 2. If we repeat this process infinitely many times, we arrive at the Sierpinski triangle.

   a. If the shaded region in stage 1 has an area of 1, find the area of the shaded regions in stages 2 through 4.

   b. Do the areas you found in part a form an arithmetic sequence or a geometric sequence?

Stage 1

Stage 2

Stage 3

Stage 4

   c. The Sierpinski triangle is the triangle that is formed after the process of forming the stages shown in the figure is repeated infinitely many times. What do you think the area of the shaded region of the Sierpinski triangle will be?

   d. Suppose the perimeter of the shaded region of the triangle in stage 1 is 1. If we were to find the perimeters of the shaded regions in the other stages, would we have an increasing sequence or a decreasing sequence?

11.5 The Binomial Expansion

OBJECTIVES

A Calculate binomial coefficients.

B Use the binomial formula to expand binomial powers.

C Find a particular term of a binomial expansion

The purpose of this section is to write and apply the formula for the expansion of expressions of the form $(x + y)^n$, where n is any positive integer. To write the formula, we must generalize the information in the following chart:

$$(x + y)^0 = 1$$
$$(x + y)^1 = x + y$$
$$(x + y)^2 = x^2 + 2xy + y^2$$
$$(x + y)^3 = x^3 + 3x^2y + 3xy^2 + y^3$$
$$(x + y)^4 = x^4 + 4x^3y + 6x^2y^2 + 4xy^3 + y^4$$
$$(x + y)^5 = x^5 + 5x^4y + 10x^3y^2 + 10x^2y^3 + 5xy^4 + y^5$$

Note

The polynomials to the right have been found by expanding the binomials on the left—we just haven't shown the work.

There are a number of similarities to notice among the polynomials on the right. Here is a list:

1. In each polynomial, the sequence of exponents on the variable x decreases to 0 from the exponent on the binomial at the left. (The exponent 0 is not shown, since $x^0 = 1$.)
2. In each polynomial, the exponents on the variable y increase from 0 to the exponent on the binomial at the left. (Because $y^0 = 1$, it is not shown in the first term.)
3. The sum of the exponents on the variables in any single term is equal to the exponent on the binomial at the left.

The pattern in the coefficients of the polynomials on the right can best be seen by writing the right side again without the variables. It looks like this:

Row 0						1					
Row 1					1		1				
Row 2				1		2		1			
Row 3			1		3		3		1		
Row 4		1		4		6		4		1	
Row 5	1		5		10		10		5		1

This triangle-shaped array of coefficients is called *Pascal's triangle.* Each entry in the triangular array is obtained by adding the two numbers above it. Each row begins and ends with the number 1. If we were to continue Pascal's triangle, the next two rows would be

Row 6		1		6		15		20		15		6		1	
Row 7	1		7		21		35		35		21		7		1

The coefficients for the terms in the expansion of $(x + y)^n$ are given in the nth row of Pascal's triangle.

There is an alternative method of finding these coefficients that does not involve Pascal's triangle. The alternative method involves *factorial notation.*

DEFINITION The expression **n!** is read "*n* factorial" and is the product of all the consecutive integers from *n* down to 1. For example,

$$1! = 1$$

$$2! = 2 \cdot 1 = 2$$

$$3! = 3 \cdot 2 \cdot 1 = 6$$

$$4! = 4 \cdot 3 \cdot 2 \cdot 1 = 24$$

$$5! = 5 \cdot 4 \cdot 3 \cdot 2 \cdot 1 = 120$$

The expression 0! is defined to be 1. We use factorial notation to define binomial coefficients as follows.

DEFINITION The expression $\binom{n}{r}$ is called a **binomial coefficient** and is defined by

$$\binom{n}{r} = \frac{n!}{r!(n-r)!}$$

EXAMPLE 1

Calculate the following binomial coefficients:

$$\binom{7}{5}, \binom{6}{2}, \binom{3}{0}$$

SOLUTION We simply apply the definition for binomial coefficients:

$$\binom{7}{5} = \frac{7!}{5!(7-5)!}$$

$$= \frac{7!}{5! \cdot 2!}$$

$$= \frac{7 \cdot 6 \cdot \cancel{5 \cdot 4 \cdot 3 \cdot 2 \cdot 1}}{(\cancel{5 \cdot 4 \cdot 3 \cdot 2 \cdot 1})(2 \cdot 1)}$$

$$= \frac{42}{2}$$

$$= 21$$

$$\binom{6}{2} = \frac{6!}{2!(6-2)!}$$

$$= \frac{6!}{2! \cdot 4!}$$

$$= \frac{6 \cdot 5 \cdot \cancel{4 \cdot 3 \cdot 2 \cdot 1}}{(2 \cdot 1)(\cancel{4 \cdot 3 \cdot 2 \cdot 1})}$$

$$= \frac{30}{2}$$

$$= 15$$

$$\binom{3}{0} = \frac{3!}{0!(3-0)!}$$

$$= \frac{3!}{0! \cdot 3!}$$

$$= \frac{\cancel{3 \cdot 2 \cdot 1}}{(1)(\cancel{3 \cdot 2 \cdot 1})}$$

$$= 1$$

If we were to calculate all the binomial coefficients in the following array, we would find they match exactly with the numbers in Pascal's triangle. That is why they are called binomial coefficients—because they are the coefficients of the expansion of $(x + y)^n$.

$$\binom{0}{0}$$

$$\binom{1}{0} \quad \binom{1}{1}$$

$$\binom{2}{0} \quad \binom{2}{1} \quad \binom{2}{2}$$

$$\binom{3}{0} \quad \binom{3}{1} \quad \binom{3}{2} \quad \binom{3}{3}$$

$$\binom{4}{0} \quad \binom{4}{1} \quad \binom{4}{2} \quad \binom{4}{3} \quad \binom{4}{4}$$

$$\binom{5}{0} \quad \binom{5}{1} \quad \binom{5}{2} \quad \binom{5}{3} \quad \binom{5}{4} \quad \binom{5}{5}$$

Using the new notation to represent the entries in Pascal's triangle, we can summarize everything we have noticed about the expansion of binomial powers of the form $(x + y)^n$.

The Binomial Expansion

If x and y represent real numbers and n is a positive integer, then the following formula is known as the *binomial expansion* or *binomial formula:*

$$(x + y)^n = \binom{n}{0} x^n y^0 + \binom{n}{1} x^{n-1} y^1 + \binom{n}{2} x^{n-2} y^2 + \cdots + \binom{n}{n} x^0 y^n$$

It does not make any difference, when expanding binomial powers of the form $(x + y)^n$, whether we use Pascal's triangle or the formula

$$\binom{n}{r} = \frac{n!}{r!(n - r)!}$$

to calculate the coefficients. We will show examples of both methods.

EXAMPLE 2 Expand $(x - 2)^3$.

SOLUTION Applying the binomial formula, we have

$$(x - 2)^3 = \binom{3}{0} x^3(-2)^0 + \binom{3}{1} x^2(-2)^1 + \binom{3}{2} x^1(-2)^2 + \binom{3}{3} x^0(-2)^3$$

The coefficients

$$\binom{3}{0}, \binom{3}{1}, \binom{3}{2}, \text{ and } \binom{3}{3}$$

can be found in the third row of Pascal's triangle. They are 1, 3, 3, and 1:

$$(x - 2)^3 = 1x^3(-2)^0 + 3x^2(-2)^1 + 3x^1(-2)^2 + 1x^0(-2)^3$$

$$= x^3 - 6x^2 + 12x - 8$$

EXAMPLE 3 Expand $(3x + 2y)^4$.

SOLUTION The coefficients can be found in the fourth row of Pascal's triangle.

1, 4, 6, 4, 1

Here is the expansion of $(3x + 2y)^4$:

$$(3x + 2y)^4 = 1(3x)^4 + 4(3x)^3(2y) + 6(3x)^2(2y)^2 + 4(3x)(2y)^3 + 1(2y)^4$$
$$= 81x^4 + 216x^3y + 216x^2y^2 + 96xy^3 + 16y^4$$

EXAMPLE 4 Write the first three terms in the expansion of $(x + 5)^9$.

SOLUTION The coefficients of the first three terms are

$$\binom{9}{0}, \binom{9}{1}, \text{ and } \binom{9}{2}$$

which we calculate as follows:

$$\binom{9}{0} = \frac{9!}{0! \cdot 9!} = \frac{\cancel{9 \cdot 8 \cdot 7 \cdot 6 \cdot 5 \cdot 4 \cdot 3 \cdot 2 \cdot 1}}{(1)(\cancel{9 \cdot 8 \cdot 7 \cdot 6 \cdot 5 \cdot 4 \cdot 3 \cdot 2 \cdot 1})} = \frac{1}{1} = 1$$

$$\binom{9}{1} = \frac{9!}{1! \cdot 8!} = \frac{9 \cdot \cancel{8 \cdot 7 \cdot 6 \cdot 5 \cdot 4 \cdot 3 \cdot 2 \cdot 1}}{(1)(\cancel{8 \cdot 7 \cdot 6 \cdot 5 \cdot 4 \cdot 3 \cdot 2 \cdot 1})} = \frac{9}{1} = 9$$

$$\binom{9}{2} = \frac{9!}{2! \cdot 7!} = \frac{9 \cdot 8 \cdot \cancel{7 \cdot 6 \cdot 5 \cdot 4 \cdot 3 \cdot 2 \cdot 1}}{(2 \cdot 1)(\cancel{7 \cdot 6 \cdot 5 \cdot 4 \cdot 3 \cdot 2 \cdot 1})} = \frac{72}{2} = 36$$

From the binomial formula, we write the first three terms:

$$(x + 5)^9 = 1 \cdot x^9 + 9 \cdot x^8(5) + 36x^7(5)^2 + \cdots$$
$$= x^9 + 45x^8 + 900x^7 + \cdots$$

The *k*th Term of a Binomial Expansion

If we look at each term in the expansion of $(x + y)^n$ as a term in a sequence, a_1, $a_2, a_3, \ldots,$ we can write

$$a_1 = \binom{n}{0} x^n y^0$$

$$a_2 = \binom{n}{1} x^{n-1} y^1$$

$$a_3 = \binom{n}{2} x^{n-2} y^2$$

$$a_4 = \binom{n}{3} x^{n-3} y^3 \qquad \text{and so on}$$

To write the formula for the general term, we simply notice that the exponent on y and the number below n in the coefficient are both 1 less than the term number. This observation allows us to write the following:

The General Term of a Binomial Expansion

The kth term in the expansion of $(x + y)^n$ is

$$a_k = \binom{n}{k-1} x^{n-(k-1)} y^{k-1}$$

EXAMPLE 5 Find the fifth term in the expansion of $(2x + 3y)^{12}$.

SOLUTION Applying the preceding formula, we have

$$a_5 = \binom{12}{4}(2x)^8(3y)^4$$

$$= \frac{12!}{4! \cdot 8!}(2x)^8(3y)^4$$

Notice that once we have one of the exponents, the other exponent and the denominator of the coefficient are determined: The two exponents add to 12 and match the numbers in the denominator of the coefficient.

Making the calculations from the preceding formula, we have

$$a_5 = 495(256x^8)(81y^4)$$

$$= 10{,}264{,}320x^8y^4$$

LINKING OBJECTIVES AND EXAMPLES

Next to each **objective** we have listed the examples that are best described by that objective.

A	1
B	2–4
C	5

GETTING READY FOR CLASS

After reading through the preceding section, respond in your own words and in complete sentences.

1. What is Pascal's triangle?
2. Why is $\binom{n}{0} = 1$ for any natural number?
3. State the binomial formula.
4. When is the binomial formula more efficient than multiplying to expand a binomial raised to a whole-number exponent?

Problem Set 11.5

Online support materials can be found at www.thomsonedu.com/login

Use the binomial formula to expand each of the following.

▶ **1.** $(x + 2)^4$

2. $(x - 2)^5$

3. $(x + y)^6$

4. $(x - 1)^6$

5. $(2x + 1)^5$

6. $(2x - 1)^4$

7. $(x - 2y)^5$

8. $(2x + y)^5$

9. $(3x - 2)^4$

10. $(2x - 3)^4$

▶ **11.** $(4x - 3y)^3$

12. $(3x - 4y)^3$

13. $(x^2 + 2)^4$

14. $(x^2 - 3)^3$

15. $(x^2 + y^2)^3$

16. $(x^2 - 3y)^4$

17. $(2x + 3y)^4$

18. $(2x - 1)^5$

19. $\left(\frac{x}{2} + \frac{y}{3}\right)^3$

20. $\left(\frac{x}{3} - \frac{y}{2}\right)^4$

21. $\left(\frac{x}{2} - 4\right)^3$

22. $\left(\frac{x}{3} + 6\right)^3$

23. $\left(\frac{x}{3} + \frac{y}{2}\right)^4$

24. $\left(\frac{x}{2} - \frac{y}{3}\right)^4$

Write the first four terms in the expansion of the following.

25. $(x + 2)^9$

26. $(x - 2)^9$

▶ **27.** $(x - y)^{10}$

28. $(x + 2y)^{10}$

29. $(x + 3)^{25}$

30. $(x - 1)^{40}$

31. $(x - 2)^{60}$

32. $\left(x + \frac{1}{2}\right)^{30}$

■ = Videos available by instructor request

▶ = Online student support materials available at www.thomsonedu.com/login

33. $(x - y)^{18}$

34. $(x - 2y)^{65}$

Write the first three terms in the expansion of each of the following.

35. $(x + 1)^{15}$

36. $(x - 1)^{15}$

37. $(x - y)^{12}$

38. $(x + y)^{12}$

39. $(x + 2)^{20}$

40. $(x - 2)^{20}$

Write the first two terms in the expansion of each of the following.

41. $(x + 2)^{100}$

42. $(x - 2)^{50}$

43. $(x + y)^{50}$

44. $(x - y)^{100}$

45. Find the ninth term in the expansion of $(2x + 3y)^{12}$.

46. Find the sixth term in the expansion of $(2x + 3y)^{12}$.

47. Find the fifth term of $(x - 2)^{10}$.

48. Find the fifth term of $(2x - 1)^{10}$.

49. Find the sixth term in the expansion of $(x - 2)^{12}$.

50. Find the ninth term in the expansion of $(7x - 1)^{10}$.

51. Find the third term in the expansion of $(x - 3y)^{25}$.

52. Find the 24th term in the expansion of $(2x - y)^{26}$.

53. Write the formula for the 12th term of $(2x + 5y)^{20}$. Do not simplify.

54. Write the formula for the eighth term of $(2x + 5y)^{20}$. Do not simplify.

55. Write the first three terms of the expansion of $(x^2y - 3)^{10}$.

56. Write the first three terms of the expansion of $(x - \frac{1}{x})^{50}$.

Applying the Concepts

57. **Probability** The third term in the expansion of $\left(\frac{1}{2} + \frac{1}{2}\right)^7$ will give the probability that in a family with 7 children, 5 will be boys and 2 will be girls. Find the third term.

58. **Probability** The fourth term in the expansion of $\left(\frac{1}{2} + \frac{1}{2}\right)^8$ will give the probability that in a family with 8 children, 3 will be boys and 5 will be girls. Find the fourth term.

Maintaining Your Skills

Solve each equation. Write your answers to the nearest hundreth.

59. $5^x = 7$

60. $10^x = 15$

61. $8^{2x+1} = 16$

62. $9^{3x-1} = 27$

63. **Compound Interest** How long will it take \$400 to double if it is invested in an account with an annual interest rate of 10% compounded four times a year?

64. **Compound Interest** How long will it take \$200 to become \$800 if it is invested in an account with an annual interest rate of 8% compounded four times a year?

Extending the Concepts

65. Calculate both $\binom{8}{5}$ and $\binom{8}{3}$ to show that they are equal.

66. Calculate both $\binom{10}{8}$ and $\binom{10}{2}$ to show that they are equal.

67. Simplify $\binom{20}{12}$ and $\binom{20}{8}$.

68. Simplify $\binom{15}{10}$ and $\binom{15}{5}$.

69. **Pascal's Triangle** Copy the first eight rows of Pascal's triangle into the eight rows of the triangular array to the right. (Each number in Pascal's triangle will go into one of the hexagons in the array.) Next, color in each hexagon that contains an odd number. What pattern begins to emerge from this coloring process?

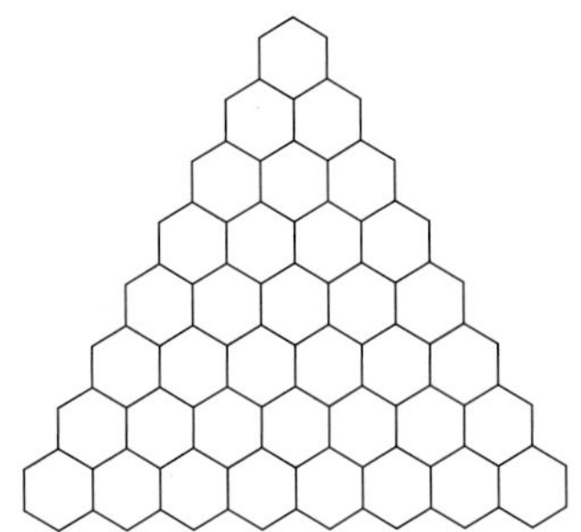

Chapter 11 SUMMARY

Sequences [11.1]

EXAMPLES

1. In the sequence 1, 3, 5, . . . , $2n - 1$, . . . , $a_1 = 1$, $a_2 = 3$, $a_3 = 5$, and $a_n = 2n - 1$.

A *sequence* is a function whose domain is the set of positive integers. The terms of a sequence are denoted by

$$a_1, a_2, a_3, \ldots, a_n, \ldots$$

where a_1 (read "a sub 1") is the first term, a_2 the second term, and a_n the nth or *general term.*

Summation Notation [11.2]

2. $\sum_{i=3}^{6} (-2)^i$
$= (-2)^3 + (-2)^4 + (-2)^5 + (-2)^6$
$= -8 + 16 + (-32) + 64$
$= 40$

The notation

$$\sum_{i=1}^{n} a_i = a_1 + a_2 + a_3 + \cdots + a_n$$

is called *summation notation* or *sigma notation.* The letter i as used here is called the *index of summation* or just *index.*

Arithmetic Sequences [11.3]

3. For the sequence 3, 7, 11, 15, . . . , $a_1 = 3$ and $d = 4$. The general term is

$$a_n = 3 + (n - 1)4$$
$$= 4n - 1$$

Using this formula to find the tenth term, we have

$$a_{10} = 4(10) - 1 = 39$$

The sum of the first 10 terms is

$$S_{10} = \frac{10}{2}(3 + 39) = 210$$

An *arithmetic sequence* is a sequence in which each term comes from the preceding term by adding a constant amount each time. If the first term of an arithmetic sequence is a_1 and the amount we add each time (called the *common difference*) is d, then the nth term of the progression is given by

$$a_n = a_1 + (n - 1)d$$

The sum of the first n terms of an arithmetic sequence is

$$S_n = \frac{n}{2}(a_1 + a_n)$$

S_n is called the nth *partial sum.*

Geometric Sequences [11.4]

4. For the geometric progression 3, 6, 12, 24, . . . , $a_1 = 3$ and $r = 2$. The general term is

$$a_n = 3 \cdot 2^{n-1}$$

The sum of the first 10 terms is

$$S_{10} = \frac{3(2^{10} - 1)}{2 - 1} = 3{,}069$$

A *geometric sequence* is a sequence of numbers in which each term comes from the previous term by multiplying by a constant amount each time. The constant by which we multiply each term to get the next term is called the *common ratio.* If the first term of a geometric sequence is a_1 and the common ratio is r, then the formula that gives the general term a_n is

$$a_n = a_1 r^{n-1}$$

The sum of the first n terms of a geometric sequence is given by the formula

$$S_n = \frac{a_1(r^n - 1)}{r - 1}$$

The Sum of an Infinite Geometric Series [11.4]

5. The sum of the series

$$\frac{1}{3} + \frac{1}{6} + \frac{1}{12} + \cdots$$

is

$$S = \frac{\frac{1}{3}}{1 - \frac{1}{2}} = \frac{\frac{1}{3}}{\frac{1}{2}} = \frac{2}{3}$$

If a geometric sequence has first term a_1 and common ratio r such that $|r| < 1$, then the following is called an *infinite geometric series:*

$$S = \sum_{i=0}^{\infty} a_1 r^i = a_1 + a_1 r + a_1 r^2 + a_1 r^3 + \cdots$$

Its sum is given by the formula

$$S = \frac{a_1}{1 - r}$$

Factorials [11.5]

The notation $n!$ is called *n factorial* and is defined to be the product of each consecutive integer from n down to 1. That is,

$0! = 1$ **(By definition)**

$1! = 1$

$2! = 2 \cdot 1$

$3! = 3 \cdot 2 \cdot 1$

$4! = 4 \cdot 3 \cdot 2 \cdot 1$

and so on.

Binomial Coefficients [11.5]

6.
$$\binom{7}{3} = \frac{7!}{3!(7-3)!} = \frac{7!}{3! \cdot 4!} = \frac{7 \cdot 6 \cdot 5 \cdot \cancel{4 \cdot 3 \cdot 2 \cdot 1}}{(3 \cdot 2 \cdot 1)(\cancel{4 \cdot 3 \cdot 2 \cdot 1})} = 35$$

The notation $\binom{n}{r}$ is called a *binomial coefficient* and is defined by

$$\binom{n}{r} = \frac{n!}{r!(n-r)!}$$

Binomial coefficients can be found by using the formula above or by *Pascal's triangle,* which is

```
                    1
                 1     1
              1     2     1
           1     3     3     1
        1     4     6     4     1
     1     5    10    10     5     1
```

and so on.

Binomial Expansion [11.5]

7. $(x + 2)^4$
$= x^4 + 4x^3 \cdot 2 + 6x^2 \cdot 2^2 + 4x \cdot 2^3 + 2^4$
$= x^4 + 8x^3 + 24x^2 + 32x + 16$

If n is a positive integer, then the formula for expanding $(x + y)^n$ is given by

$$(x + y)^n = \binom{n}{0} x^n y^0 + \binom{n}{1} x^{n-1} y^1 + \binom{n}{2} x^{n-2} y^2 + \cdots + \binom{n}{n} x^0 y^n$$

Chapter 11 Review Test

The problems below form a comprehensive review of the material in this chapter. They can be used to study for exams. If you would like to take a practice test on this chapter, you can use the odd-numbered problems. Give yourself an hour and work as many of the odd-numbered problems as possible. When you are finished, or when an hour has passed, check your answers with the answers in the back of the book. You can use the even-numbered problems for a second practice test.

Write the first four terms of the sequence with the following general terms. [11.1]

1. $a_n = 2n + 5$

2. $a_n = 3n - 2$

3. $a_n = n^2 - 1$

4. $a_n = \dfrac{n+3}{n+2}$

5. $a_1 = 4,\ a_n = 4a_{n-1},\ n > 1$

6. $a_1 = \dfrac{1}{4},\ a_n = \dfrac{1}{4}a_{n-1},\ n > 1$

Determine the general term for each of the following sequences. [11.1]

7. $2, 5, 8, 11, \ldots$

8. $-3, -1, 1, 3, 5, \ldots$

9. $1, 16, 81, 256, \ldots$

10. $2, 5, 10, 17, \ldots$

11. $\dfrac{1}{2}, \dfrac{1}{4}, \dfrac{1}{8}, \dfrac{1}{16}, \ldots$

12. $2, \dfrac{3}{4}, \dfrac{4}{9}, \dfrac{5}{16}, \dfrac{6}{25}, \ldots$

Expand and simplify each of the following. [11.2]

13. $\sum_{i=1}^{4} (2i + 3)$

14. $\sum_{i=1}^{3} (2i^2 - 1)$

15. $\sum_{i=2}^{3} \dfrac{i^2}{i+2}$

16. $\sum_{i=1}^{4} (-2)^{i-1}$

17. $\sum_{i=3}^{5} (4i + i^2)$

18. $\sum_{i=4}^{6} \dfrac{i+2}{i}$

Write each of the following sums with summation notation. [11.2]

19. $3 + 6 + 9 + 12$

20. $3 + 7 + 11 + 15$

21. $5 + 7 + 9 + 11 + 13$

22. $4 + 9 + 16$

23. $\dfrac{1}{3} + \dfrac{1}{4} + \dfrac{1}{5} + \dfrac{1}{6}$

24. $\dfrac{1}{3} + \dfrac{2}{9} + \dfrac{3}{27} + \dfrac{4}{81} + \dfrac{5}{243}$

25. $(x - 2) + (x - 4) + (x - 6)$

26. $\dfrac{x}{x+1} + \dfrac{x}{x+2} + \dfrac{x}{x+3} + \dfrac{x}{x+4}$

Determine which of the following sequences are arithmetic progressions, geometric progressions, or neither. [11.3, 11.4]

27. $1, -3, 9, -27, \ldots$

28. $7, 9, 11, 13, \ldots$

29. $5, 11, 17, 23, \ldots$

30. $\dfrac{1}{2}, \dfrac{1}{3}, \dfrac{1}{4}, \dfrac{1}{5}, \ldots$

31. $4, 8, 16, 32, \ldots$

32. $\dfrac{1}{2}, \dfrac{1}{4}, \dfrac{1}{8}, \dfrac{1}{16}, \ldots$

33. $12, 9, 6, 3, \ldots$

34. $2, 5, 9, 14, \ldots$

Each of the following problems refers to arithmetic progressions. [11.3]

35. If $a_1 = 2$ and $d = 3$, find a_n and a_{20}.

36. If $a_1 = 5$ and $d = -3$, find a_n and a_{16}.

37. If $a_1 = -2$ and $d = 4$, find a_{10} and S_{10}.

38. If $a_1 = 3$ and $d = 5$, find a_{16} and S_{16}.

39. If $a_5 = 21$ and $a_8 = 33$, find the first term a_1, the common difference d, and then find a_{10}.

40. If $a_3 = 14$ and $a_7 = 26$, find the first term a_1, the common difference d, and then find a_9 and S_9.

41. If $a_4 = -10$ and $a_8 = -18$, find the first term a_1, the common difference d, and then find a_{20} and S_{20}.

42. Find the sum of the first 100 terms of the sequence

$3, 7, 11, 15, 19, \ldots$

43. Find a_{40} for the sequence $100, 95, 90, 85, 80, \ldots$

Each of the following problems refers to infinite geometric progressions. [11.4]

44. If $a_1 = 3$ and $r = 2$, find a_n and a_{20}.

45. If $a_1 = 5$ and $r = -2$, find a_n and a_{16}.

46. If $a_1 = 4$ and $r = \dfrac{1}{2}$, find a_n and a_{10}.

47. If $a_1 = -2$ and $r = \dfrac{1}{3}$, find the sum.

48. If $a_1 = 4$ and $r = \dfrac{1}{2}$, find the sum.

49. If $a_3 = 12$ and $a_4 = 24$, find the first term a_1, the common ratio r, and then find a_6.

50. Find the tenth term of the sequence $3, 3\sqrt{3}, 9, 9\sqrt{3}, \ldots$

Use the binomial formula to expand each of the following. [11.5]

51. $(x - 2)^4$

52. $(2x + 3)^4$

53. $(3x + 2y)^3$

54. $(x^2 - 2)^5$

55. $\left(\frac{x}{2} + 3\right)^4$

56. $\left(\frac{x}{3} - \frac{y}{2}\right)^3$

Use the binomial formula to write the first three terms in the expansion of the following. [11.5]

57. $(x + 3y)^{10}$

58. $(x - 3y)^9$

59. $(x + y)^{11}$

60. $(x - 2y)^{12}$

Use the binomial formula to write the first two terms in the expansion of the following. [11.5]

61. $(x - 2y)^{16}$

62. $(x + 2y)^{32}$

63. $(x - 1)^{50}$

64. $(x + y)^{150}$

65. Find the sixth term in $(x - 3)^{10}$.

66. Find the fourth term in $(2x + 1)^9$.

Chapter 11 Projects

Sequences and Series

GROUP PROJECT Credit Card Payments

Number of People 2

Time Needed 20–30 minutes

Equipment Paper, pencil, and graphing calculator

Background In the beginning of this chapter, you were given a recursive function that you can use to find how long it will take to pay off a credit card. A graphing calculator can be used to model each payment by using the recall function on the graphing calculator. To set up this problem do the following:

(1) 1000 ENTER

(2) Round (1.0165ANS—20,2) ENTER

Note The *Round* function is under the MATH key in the NUM list.

Procedure Enter the preceding commands into your calculator. While one person in the group hits the ENTER key, another person counts, by keeping a tally of the "payments" made. The credit card is paid off when the calculator displays a negative number.

1. How many months did it take to pay off the credit card?
2. What was the amount of the last payment?
3. What was the total interest paid to the credit card company?
4. How much would you save if you paid \$25 per month instead of \$20?
5. If the credit card company raises the interest rate to 21.5% annual interest, how long will it take to pay off the balance? How much more would it cost in interest?
6. On larger balances, many credit card companies require a minimum payment of 2% of the outstanding balance. What is the recursion formula for this? How much would this save or cost you in interest?

Detach here and return with check or money order.

Summary of Corporate Card Account
Retain this portion for your files

Corporate Cardmember Name	Account Number	Closing Date
Leonardo Fibonacci	00000-1000-001	08-01-00

New Balance	Other Debits	Interest Rate	Minimum Payment
\$1,000.00	\$.00	19.8%	\$20.00

RESEARCH PROJECT

Building Squares from Odd Numbers

Corbis/Bettmann

A relationship exists between the sequence of squares and the sequence of odd numbers. In *The Book of Squares,* written in 1225, Leonardo Fibonacci has this to say about that relationship:

> I thought about the origin of all square numbers and discovered that they arise out of the increasing sequence of odd numbers.

Work with the sequence of odd numbers until you discover how it can be used to produce the sequence of squares. Then write an essay in which you give a written description of the relationship between the two sequences, along with a diagram that illustrates the relationship. Then see if you can use summation notation to write an equation that summarizes the whole relationship. Your essay should be clear and concise and written so that any of your classmates can read it and understand the relationship you are describing.

APPENDIX A
Resources

By gathering resources early in the term, before you need help, the information about these resources will be available to you when they are needed.

INSTRUCTOR

Knowing the contact information for your instructor is very important. You may already have this information from the course syllabus. It is a good idea to write it down again.

Name ______________________ Office Location ______________________
Available Hours: M ________ T ________ W ________ TH ________ F ________
Phone Number ____________ ext. ________ E-mail Address ______________________

TUTORING CENTER

Many schools offer tutoring, free of charge to their students. If this is the case at your school, find out when and where tutoring is offered.

Tutoring Location ______________________ Phone Number ____________ ext. ________
Available Hours: M ________ T ________ W ________ TH ________ F ________

COMPUTER LAB

Many schools offer a computer lab where students can use the online resources and software available with their textbook. Other students using the same software and websites as you can be very helpful. Find out where the computer lab at your school is located.

Computer Lab Location ______________________ Phone Number ____________ ext. ________
Available Hours: M ________ T ________ W ________ TH ________ F ________

VIDEO LESSONS

A complete set of video lessons is available to your school. These videos feature the author of your textbook presenting full-length, 15- to 20-minute lessons from every section of your textbook. If you miss class, or find yourself behind, these lessons will prove very useful.

Video Location ______________________ Phone Number ____________ ext. ________
Available Hours: M ________ T ________ W ________ TH ________ F ________

CLASSMATES

Form a study group and meet on a regular basis. When you meet try to speak to each other using proper mathematical language. That is, use the words that you see in the definition and property boxes in your textbook.

Name ______________________ Phone ____________ E-mail ______________________
Name ______________________ Phone ____________ E-mail ______________________
Name ______________________ Phone ____________ E-mail ______________________

Answers to Odd-Numbered Problems and Chapter Reviews

CHAPTER 1

Problem Set 1.1

1. $x + 5 = 2$ **3.** $6 - x = y$ **5.** $2t < y$ **7.** $x + y < x - y$ **9.** $3(x - 5) > y$ **11.** 36 **13.** 100 **15.** 8 **17.** 16 **19.** 10,000 **21.** 121 **23. a.** 19 **b.** 27 **c.** 27 **25. a.** 16 **b.** 12 **c.** 18 **27. a.** 33 **b.** 33 **29. a.** 144 **b.** 74 **c.** 144 **31. a.** 23 **b.** 41 **c.** 65 **33. a.** 39 **b.** 7 **c.** 5 **35. a.** 48 **b.** 24 **37. a.** 41 **b.** 95 **39.** 5,431 **41.** 138 **43.** 78 **45.** 152 **47.** 11 **49.** 87 **51.** 4 **53.** 22 **55.** 16 **57.** 625 **59.** 1 **61.** 0.6 **63.** 8,700 **65.** 5.046 **67.** 5 **69.** 128 **71.** 625 **73.** 31 **75.** 1,000 **77.** 750 **79.** 0.5 **81.** 1.6 **83.** 320 **85.** 1.7917 **87.** 2.0793 **89.** 785 **91.** 185.12 **93.** 12.3106 **95.** 0.1587 **97.** 196 **99.** 650 **101. a.** 7 **b.** 10 **c.** 25 **d.** 32 **103. a.** 121 **b.** 121 **c.** 101 **d.** 81 **105. a.** 1 **b.** 2,400 **c.** 0.52 **107.** {0, 1, 2, 3, 4, 5, 6} **109.** {2, 4} **111.** {1, 3, 5} **113.** {0, 1, 2, 3, 4, 5, 6} **115.** {0, 2} **117.** {0, 6} **119.** {0, 1, 2, 3, 4, 5, 6, 7} **121.** {1, 2, 4, 5} **123. a.** 0.03 seconds **b.** 0.37 seconds

Problem Set 1.2

1.

3.

5.
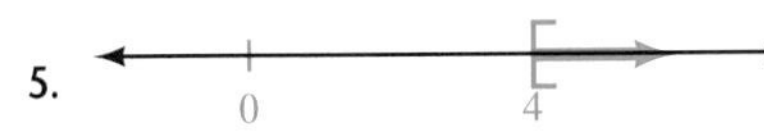

7.

9.

11.

13.

15.

17.

19.

21. ∅

23.
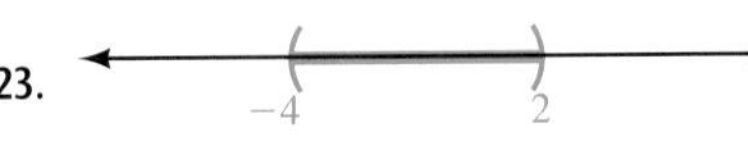

25.

27.

29.
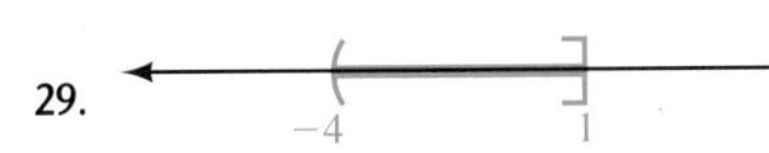

31.

33.
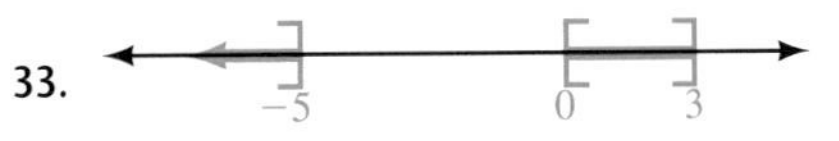

35.
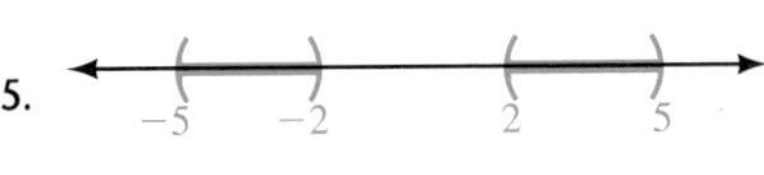

37. $x \geq 5$ **39.** $x \leq -3$ **41.** $x \leq 4$ **43.** $-4 < x < 4$ **45.** $-4 \leq x \leq 4$ **47.** 1, 2 **49.** $-6, -5.2, 0, 1, 2, 2.3, \frac{9}{2}$ **51.** $-\sqrt{7}, -\pi, \sqrt{17}$ **53.** 0, 1, 2 **55.** $2^2 \cdot 3 \cdot 5$ **57.** $2 \cdot 7 \cdot 19$ **59.** $3 \cdot 37$ **61.** $3^2 \cdot 41$ **63.** $\frac{3}{7}$ **65.** $\frac{11}{21}$ **67.** $\frac{3}{5}$ **69.** $\frac{5}{6}$ **71.** $\frac{5}{9}$ **73.** $\frac{3}{4}$ **75.** 40 **77.** $\frac{5}{11}$ **79.** -3 and 7 **81.** $3 < x < 13$ **83.** $x < 3$ or $x > 13$

85. a. $10 = 3 + 7$ Other answers are possible. **b.** $16 = 5 + 11$ Other answers are possible. **c.** $24 = 5 + 19$ Other answers are possible. **d.** $36 = 17 + 19$ Other answers are possible.

87. $50 \leq F \leq 270$ **89. a.** 98% **b.** 72% **c.** 54% **93.** $5 \cdot 4 \cdot 3 \cdot 2 \cdot 1 = 120$ **95.** $6! = 6 \cdot 5 \cdot 4 \cdot 3 \cdot 2 \cdot 1 = 6 \cdot 5!$

Problem Set 1.3

1. $4, -4, \frac{1}{4}$ **3.** $-\frac{1}{2}, \frac{1}{2}, -2$ **5.** $5, -5, \frac{1}{5}$ **7.** $\frac{3}{8}, -\frac{3}{8}, \frac{8}{3}$ **9.** $-\frac{1}{6}, \frac{1}{6}, -6$ **11.** $3, -3, \frac{1}{3}$ **13.** $-1, 1$ **15.** 0 **17.** 2 **19.** $\frac{3}{4}$ **21.** π **23.** -4 **25.** -2 **27.** $-\frac{3}{4}$ **29.** $\frac{21}{40}$ **31.** 2 **33.** $\frac{8}{27}$ **35.** $\frac{1}{10,000}$ **37.** $\frac{72}{385}$ **39.** 1 **41.** $6 + x$ **43.** $a + 8$ **45.** $15y$ **47.** x **49.** a **51.** x **53.** $3x + 18$ **55.** $12x + 8$ **57.** $15a + 10b$ **59.** $\frac{4}{3}x + 2$ **61.** $2 + y$ **63.** $40t + 8$ **65.** $6x - 3$ **67.** $10x - 15$ **69.** $9x + 3y - 6z$ **71.** $3x + 7y$ **73.** $6x + 7y$ **75.** $3x + 1$ **77.** $2x - 1$ **79.** $x + 2$ **81.** $a - 3$ **83.** $x + 24$ **85.** $3x - 2y$ **87.** $3x + 8y$ **89.** $8x + 5y$ **91.** $16x + 12$ **93.** $15x + 10$ **95.** $8y + 32$ **97.** $15t + 9$ **99.** $28x + 11$ **101.** $\frac{7}{15}$ **103.** $\frac{29}{35}$ **105.** $\frac{35}{144}$ **107.** $\frac{949}{1,260}$ **109.** $\frac{47}{105}$ **111.** 15 **113.** $14a + 7$ **115.** $6y + 6$

117. $12x + 2$ **119.** $8y + 11$ **121.** $24a + 15$ **123.** $11x + 20$ **125.** Commutative property of addition
127. Commutative property of multiplication **129.** Additive inverse **131.** Commutative property of addition
133. Associative and commutative properties of multiplication **135.** Commutative and associative properties of addition
137. Distributive property **139.** $12 + \frac{1}{4}(12) = 15$ **141. a.** Answers will vary. **b.** 2001 **c.** Negative

Problem Set 1.4

1. 4 **3.** −4 **5.** −10 **7.** −4 **9.** $\frac{19}{12}$ **11.** $-\frac{32}{105}$ **13.** −8 **15.** −12 **17.** $-7x$ **19.** 13 **21.** −14 **23.** $6a$
25. −15 **27.** 15 **29.** −24 **31.** $-10x$ **33.** x **35.** y **37.** $-8x + 6$ **39.** $-3a + 4$ **41.** −14 **43.** 18 **45.** 16
47. −19 **49.** 50 **51.** −2 **53.** 39 **55.** 11 **57.** −5 **59.** 11 **61.** 7 **63.** −44 **65.** −9 **67.** −9 **69.** $-3x$
71. $5y$ **73.** $x - 5$ **75.** $x - 7$ **77.** $-6x + 9y$ **79.** $3x - 18$ **81.** $3a + 6$ **83.** $3x + 60$ **85.** $x + 4y$ **87.** $-3x + 5y$
89. $11y$ **91.** $14x + 12$ **93.** $7m - 15$ **95.** $-2x + 9$ **97.** $7y + 10$ **99.** $-20x + 5$ **101.** $-11x + 10$ **103.** $0.01x + 500$
105. $0.02x + 1{,}500$ **107.** $-5a - 1$ **109.** $-\frac{2}{3}$ **111.** 32 **113.** $-\frac{1}{18}$ **115.** 2 **117.** $\frac{4}{3}$ **119.** 0 **121.** Undefined
123. $-\frac{2}{3}$ **125.** $\frac{5}{3}$ **127.** 11 **129.** 12

131.

a	b	Sum $a + b$	Difference $a - b$	Product ab	Quotient $\frac{a}{b}$
3	12	15	−9	36	$\frac{1}{4}$
−3	12	9	−15	−36	$-\frac{1}{4}$
3	−12	−9	15	−36	$-\frac{1}{4}$
−3	−12	−15	9	36	$\frac{1}{4}$

133.

x	$3(5x - 2)$	$15x - 6$	$15x - 2$
−2	−36	−36	−32
−1	−21	−21	−17
0	−6	−6	−2
1	9	9	13
2	24	24	28

135. a. 1 **b.** $\frac{3}{2}$ **c.** −1 **d.** 135 **137. a.** 5.25 **b.** −9.75 **c.** −16.875 **d.** −0.3 **139.** 1.5283 **141.** −0.0794 **143.** −0.0714
145. 3.4 **147.** 1.6 **149.** 1,200 **151.** 190 **153.** Santa Fe 10:10 P.M., Detroit 3:30 A.M.
155. a. 00:00:10 **b.** 00:02:20 **c.** 00:22:53

Problem Set 1.5

1. 5 **3.** 10 **5.** 25 **7.** 29 **9.** 23 **11.** △ **13.** ⊙ **15.** 17, 21 **17.** −2, −3 **19.** −4, −7 **21.** $-\frac{1}{2}, -\frac{3}{4}$ **23.** $\frac{5}{2}, 3$
25. 27 **27.** −270 **29.** $\frac{1}{8}$ **31.** $\frac{5}{2}$ **33.** −625 **35.** $-\frac{1}{125}$ **37. a.** 12 **b.** 16 **39.** 144 **41.** 2, 3, 5, among others
43. 2, 8, 34 **45.**

Two Numbers a and b	Their Product ab	Their Sum $a + b$
1, −24	−24	−23
−1, 24	−24	23
2, −12	−24	−10
−2, 12	−24	10
3, −8	−24	−5
−3, 8	−24	5
4, −6	−24	−2
−4, 6	−24	2

47. 41, 37.5, 34, 30.5, 27, 23.5; yes **49.** 41, 45.5, 50, 54.5, 59, 63.5; yes
51. The patient on Antidepressant 1 misses his morning dose; less than half of the antidepressant will remain in the body. The patient on Antidepressant 2 still has most of the medication in his body even after missing a dose because of the relatively long (5-day) half-life.

53.

Hours Since Discontinuing	Concentration (ng/mL)
0	60
4	30
8	15
12	7.5
16	3.75

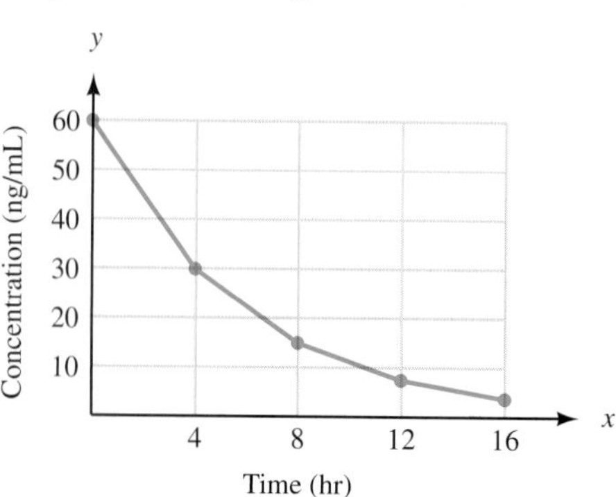

55.

Elevation (ft)	Boiling Point (°F)
−2,000	215.6
−1,000	213.8
0	212
1,000	210.2
2,000	208.4
3,000	206.6

57. a. The patient taking his medication
b. The patient stops taking his medication
c. 50 ng/mL
d. 4, 8, and 12 hours

59.

Temperature (Fahrenheit)	Shelf-Life (days)
32°	24
40°	10
50°	2
60°	1
70°	$\frac{1}{2}$

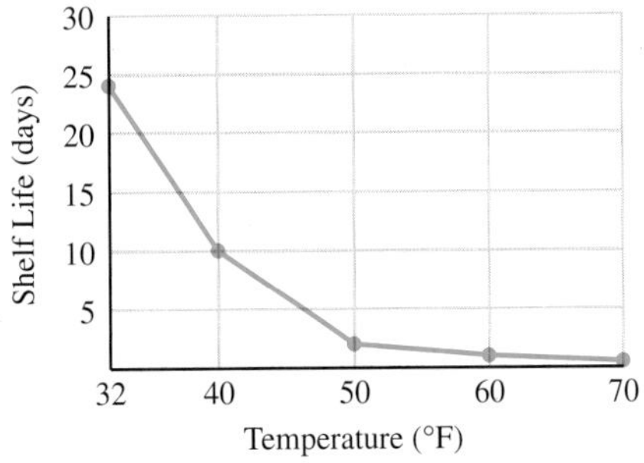

CHAPTER 1 REVIEW TEST

1. $x + 2$ **2.** $x - 2$ **3.** $\frac{x}{2}$ **4.** $2x$ **5.** $2(x + y)$ **6.** $2x + y$ **7.** 27 **8.** 125 **9.** 64 **10.** 1 **11.** 32 **12.** 81 **13.** 17 **14.** 4 **15.** 13 **16.** 7 **17.** 9 **18.** 32 **19.** 30 **20.** 43 **21.** {1, 2, 3, 4, 5, 6} **22.** {1, 3} **23.** {5} **24.** {6}

25.

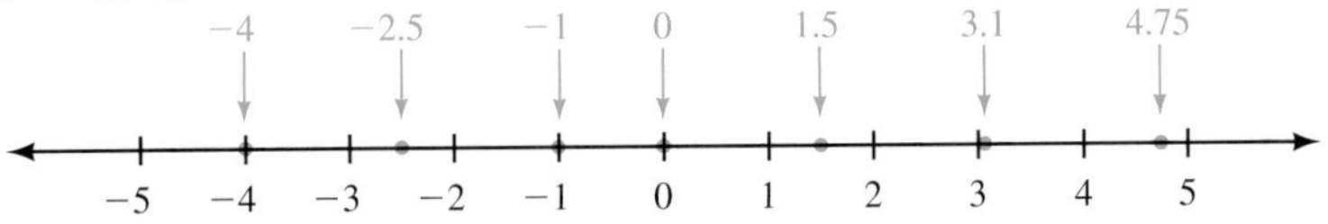

26.

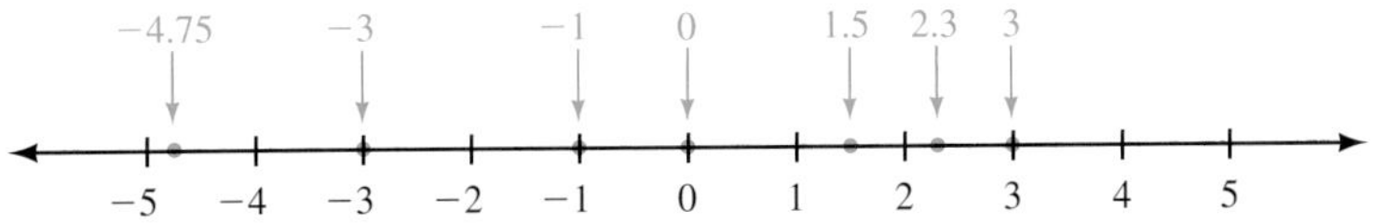

27. $-2, \frac{1}{2}$ **28.** $\frac{2}{5}, -\frac{5}{2}$ **29.** 3 **30.** −5

31. 4 **32.** 6 **33.** −7, 0, 5 **34.** $-7, -4.2, 0, \frac{3}{4}, 5$ **35.** $-\sqrt{3}, \pi$ **36.** $2^2 \cdot 3^2 \cdot 11^2$ **37.** $\frac{11}{13}$ **38.** 1 **39.** $\frac{27}{64}$ **40.** 2

41.

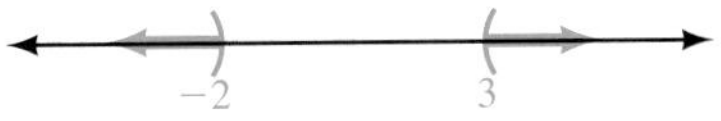

42.

43.

44.

45. $x \geq 4$ **46.** $x \leq 5$ **47.** $0 < x < 8$

48. $0 \leq x \leq 8$ **49.** $2y$ **50.** $3x$ **51.** $8x + 5$ **52.** $y + 2$ **53.** a **54.** c **55.** a **56.** b, d **57.** a, c **58.** f **59.** e **60.** g **61.** 2 **62.** −2 **63.** 1 **64.** 17 **65.** 3 **66.** −6 **67.** 2 **68.** 66 **69.** $-\frac{5}{6}$ **70.** $-\frac{5}{6}$ **71.** −42 **72.** 30 **73.** $21x$ **74.** $-6x$ **75.** $-6x + 10$ **76.** $-6x + 21$ **77.** $-x + 3$ **78.** $-15x + 3$ **79.** $-\frac{5}{6}$ **80.** −36 **81.** $\frac{1}{10}$ **82.** $-\frac{2}{7}$ **83.** −13 **84.** −17 **85.** 0 **86.** −1 **87.** − 36 **88.** − 34 **89.** 16 **90.** −24 **91.** 2 **92.** 39 **93.** 0 **94.** $-2y + 9$ **95.** $-18x - 14$ **96.** $5a - 22$ **97.** −1, arithmetic **98.** 8 **99.** −1, arithmetic **100.** $\frac{1}{16}$, geometric

CHAPTER 2

Problem Set 2.1

1. 8 **3.** 5 **5.** 2 **7.** −7 **9.** $-\frac{9}{2}$ **11.** −4 **13.** $-\frac{4}{3}$ **15.** −4 **17.** −2 **19.** $\frac{3}{4}$ **21.** 12 **23.** −10 **25.** 7 **27.** −7 **29.** 3 **31.** $\frac{4}{5}$ **33.** 1 **35.** 4 **37.** 17 **39.** 2 **41.** 6 **43.** $-\frac{4}{3}$ **45.** $-\frac{3}{2}$ **47.** $\frac{5}{3}$ **49.** $\frac{3}{2}$ **51.** 1 **53.** No solution

55. All real numbers are solutions. **57.** No solution **59.** $-\frac{2}{3}$ **61.** $-\frac{7}{640}$ **63.** 2 **65.** 20 **67.** 0.7 **69.** 2,400 **71.** 24 **73.** 6 **75.** 3 **77.** 2 **79.** 5 **81.** 8.5 **83.** 18 **85.** 36 **87.** 4 **89.** 30 **91.** 6,000 **93.** 5,000 **95.** $\frac{8}{15}$ **97.** $-\frac{5}{11}$ **99. a.** $\frac{5}{8}$ **b.** 0 **c.** $10x - 10$ **d.** 0 **e.** $8x - 40$ **f.** 5 **101. a.** $6.60 = $0.40n + $1.80 **b.** 12 miles

103. a. 3,937,000 = 1,125A **b.** 3,500 square miles **105.** Commutative **107.** Associative **109.** Commutative and associative **111.** Multiplicative identity **113.** Commutative **115.** Additive identity **117.** 0.5 **119.** 62.5 **121.** 0 **123.** 1.25 **125.** 13 **127. a.** 3 **b.** 5 **c.** 9 **129.** 2 **131.** $-\frac{3}{2}$ **133.** 6 **135.** −14

Problem Set 2.2

1. −3 **3.** 0 **5.** $\frac{3}{2}$ **7.** 4 **9.** $\frac{8}{5}$ **11.** 2 **13.** $-\frac{7}{640}$ **15.** 675 **17. a.** 0 **b.** 0 **19. a.** 23 **b.** 23 **21.** $-\frac{1}{2}$ **23.** 3 **25.** 2,400 **27.** $\ell = \frac{A}{w}$ **29.** $t = \frac{I}{pr}$ **31.** $T = \frac{PV}{nR}$ **33.** $x = \frac{y-b}{m}$ **35.** $F = \frac{9}{5}C + 32$ **37.** $v = \frac{h - 16t^2}{t}$ **39.** $d = \frac{A-a}{n-1}$ **41.** $y = -\frac{2}{3}x + 2$ **43.** $y = \frac{3}{5}x + 3$ **45.** $y = \frac{1}{3}x + 2$ **47.** $x = \frac{5}{a-b}$ **49.** $P = \frac{A}{1+rt}$ **51.** $y = -\frac{1}{4}x + 2$ **53.** $y = \frac{3}{5}x - 3$ **55.** $y = \frac{1}{2}x + \frac{3}{2}$ **57.** $y = -2x - 5$ **59.** $y = -\frac{2}{3}x + 1$ **61.** $y = -\frac{1}{2}x + \frac{7}{2}$ **63. a.** $y = 4x - 1$ **b.** $y = -\frac{1}{2}x$ **c.** $y = -3$ **65.** 20.52 **67.** 25% **69.** 925 **71. a.** $-\frac{15}{4} = -3.75$ **b.** −7 **c.** $y = \frac{4}{5}x + 4$ **d.** $x = \frac{5}{4}y - 5$ **73.** 16% **75.** \$3.25 **77.** 40% **79.** \$5.00 **81.** \$10.00 **83.** 6 miles per hour **85.** 42 miles per hour **87.** 55 miles per hour **89.** 84 kilometers per hour **91.** 6.8 feet per second **93.** 8 **95.** Shar = 128.4 beats per min, Sara = 140.4 beats per min **97.** 20 **99.** 31 **101.** 8 **103.** 1 **105.** $2x - 3$ **107.** $x + y = 180$ **109.** 30 **111.** 8.5 **113.** 6,000 **115.** $x = a - \frac{a}{b}y$ **117.** $a = \frac{bc}{b-c}$

Problem Set 2.3

1. 10 feet by 20 feet **3.** 7 feet **5.** 5 inches **7.** 4 meters **9.** \$92.00 **11.** \$200.00 **13.** \$3,260.66 **15.** \$99.6 million **17.** 20°, 160° **19. a.** 20.4°, 69.6° **b.** 38.4°, 141.6° **21.** 27°, 72°, 81° **23.** 34°, 44°, 102° **25.** 43°, 43°, 94° **27.** \$6,000 at 8%, \$3,000 at 9% **29.** \$5,000 at 12%, \$10,000 at 10% **31.** \$4,000 at 8%, \$2,000 at 9% **33.** 16 adults and 22 children's tickets **35.** \$54

37.

t	0	$\frac{1}{4}$	1	$\frac{7}{4}$	2
h	0	7	16	7	0

39.

Hot Coffee Sales

Year	Sales (billions of dollars)
2005	7
2006	7.5
2007	8
2008	8.6
2009	9.2

41.

Speed (miles per hour)	Distance (miles)
20	19
30	15
40	20
50	25
60	30
70	35

43.

Time (hours)	Distance Upstream (miles)	Distance Downstream (miles)
1	8	14
2	12	28
3	18	42
4	24	56
5	30	70
6	36	84

45.

Age (years)	Maximum Heart Rate (beats per minute)
18	202
19	201
20	200
21	199
22	198
23	197

47.

w	l	A
2	22	44
4	20	80
6	18	108
8	16	128
10	14	140
12	12	144

49. 

51. 

53. 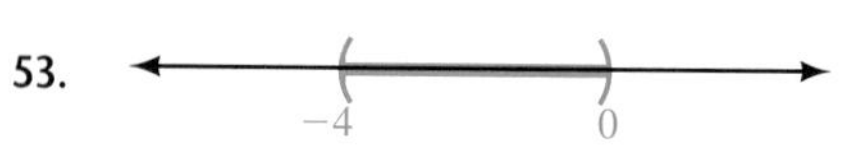

55.

57. 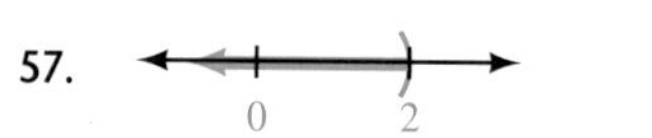

59.

61. −5 **63.** 6

Problem Set 2.4

1. $x \le \frac{3}{2}$ 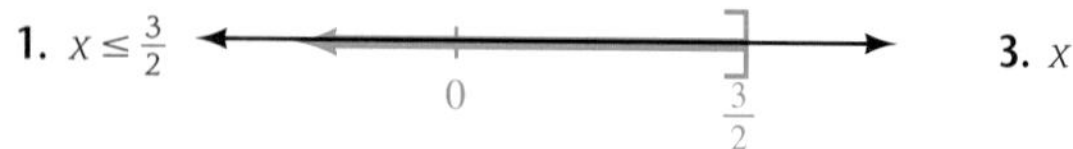

3. $x > 4$ 

5. $x \geq -5$  **7.** $x < 4$ 

9. $x \geq -6$ **11.** $x \geq 4$ 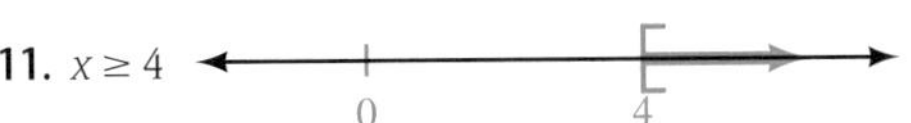

13. $x < -3$ **15.** $m \geq -1$

17. $x \geq -3$ **19.** $y \leq \frac{7}{2}$

21. $x < 6$ **23.** $y \geq -52$ **25.** $(-\infty, -2]$ **27.** $[1, \infty)$

29. $(-\infty, 3)$ **31.** $(-\infty, -1]$ **33.** $[-17, \infty)$ **35.** $x > 435$ **37.** $[3, 7]$

39. $(-4, 2)$ **41.** $[4, 6]$

43. $(-4, 2)$ **45.** $(-3, 3)$ 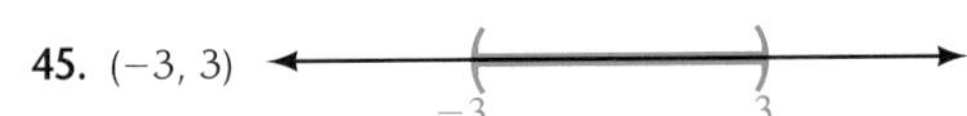

47. $(-\infty, -7] \cup [-3, \infty)$  **49.** $(-\infty, -1] \cup [\frac{3}{5}, \infty)$ 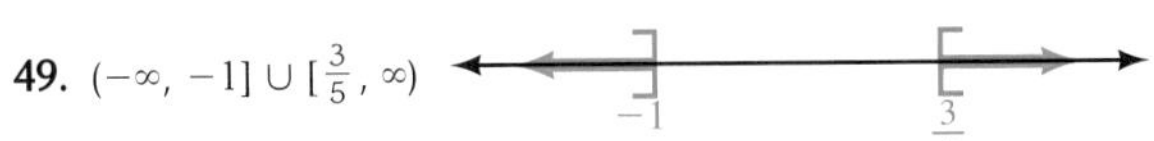

51. $(-\infty, -10) \cup (6, \infty)$ **53.** $-2 < x \leq 4$ **55.** $x < -4$ or $x \geq 1$

57. a. $x > 0$ **b.** $x \geq 0$ **c.** $x \geq 0$ **59.** $\varnothing$ **61.** all real numbers **63.** $\varnothing$ **65. a.** 1 **b.** 16 **c.** no **d.** $x > 16$
67. $p \leq 2$; set the price at \$2.00 or less per pad **69.** $p > 1.25$; charge more than \$1.25 per pad
71. During the years 1990 through 1994 **73.** Adults: $0.61 \leq r \leq 0.83$; 61% to 83%; Juveniles: $0.06 \leq r \leq 0.20$; 6% to 20%
75. a. 35° to 45° Celsius; $35° \leq C \leq 45°$ **b.** $20° \leq C \leq 30°$ **c.** −25° to −10° Celsius; $-25° \leq C \leq -10°$ **d.** $-20° \leq C \leq -5°$
77. 3 **79.** −3 **81.** The distance between x and 0 on the number line **83.** −3 **85.** 15 **87.** No solution **89.** −1
91. $x < \frac{c-b}{a}$ **93.** $\frac{-c-b}{a} < x < \frac{c-b}{a}$

Problem Set 2.5

1. −4, 4 **3.** −2, 2 **5.** $\varnothing$ **7.** −1, 1 **9.** $\varnothing$ **11.** −6, 6 **13.** −3, 7 **15.** $\frac{17}{3}, \frac{7}{3}$ **17.** 2, 4 **19.** $-\frac{5}{2}, \frac{5}{6}$ **21.** −1, 5
23. $\varnothing$ **25.** −4, 20 **27.** −4, 8 **29.** $-\frac{10}{3}, \frac{2}{3}$ **31.** $\varnothing$ **33.** $-1, \frac{3}{2}$ **35.** 5, 25 **37.** −30, 26 **39.** −12, 28 **41.** $-5, \frac{3}{5}$
43. $1, \frac{1}{9}$ **45.** $-\frac{1}{2}$ **47.** 0 **49.** $-\frac{1}{2}$ **51.** $-\frac{1}{6}, -\frac{7}{4}$ **53.** All real numbers **55.** All real numbers
57. a. $\frac{5}{4} = 1.25$ **b.** $\frac{5}{4} = 1.25$ **c.** 2 **d.** $\frac{1}{2}, 2$ **e.** $\frac{1}{3}, 4$ **59.** 1987 and 1995

61. **63.** 

65. $\frac{19}{15}$ **67.** 40 **69.** −2 **71.** 0 **73.** $x < 4$ **75.** $-\frac{11}{3} \leq a$ **77.** $t \leq -\frac{3}{2}$ **79.** $x = a - b$ or $x = a + b$

81. $x = \frac{-b-c}{a}$ or $x = \frac{-b+c}{a}$ **83.** $x = -\frac{a}{b}y - a$ or $x = -\frac{a}{b}y + a$

Problem Set 2.6

1. $-3 < x < 3$ **3.** $x \leq -2$ or $x \geq 2$ 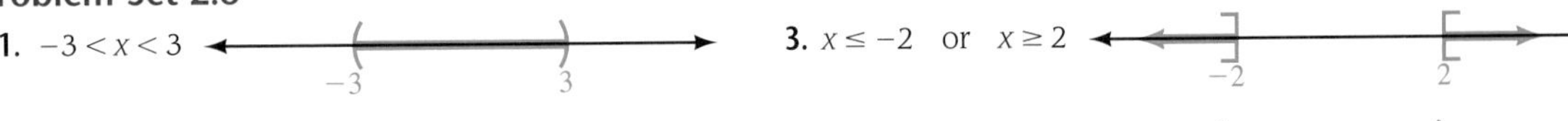

5. $-3 < x < 3$ **7.** $t < -7$ or $t > 7$ **9.** $\varnothing$ 

11. All real numbers **13.** $-4 < x < 10$ 

15. $a \le -9$ or $a \ge -1$ 

17. $\varnothing$ **19.** $-1 < x < 5$

21. $y \le -5$ or $y \ge -1$ **23.** $k \le -5$ or $k \ge 2$

25. $-1 < x < 7$ **27.** $a \le -2$ or $a \ge 1$

29. $-6 < x < \frac{8}{3}$ **31.** $x < 2$ or $x > 8$

33. $x \le -3$ or $x \ge 12$ **35.** $x < 2$ or $x > 6$

37. $0.99 < x < 1.01$ **39.** $x \le -\frac{3}{5}$ or $x \ge -\frac{2}{5}$ **41.** $-\frac{1}{6} \le x \le \frac{3}{2}$ **43.** $-0.05 < x < 0.25$ **45.** $|x| \le 4$ **47.** $|x - 5| \le 1$

49. a. 3 **b.** $\frac{4}{5}$ or -2 **c.** No **d.** $x < -2$ or $x > \frac{4}{5}$ **51.** $|x - 65| \le 10$ **53.** -6 **55.** 66 **57.** -13 **59.** 11 **61.** -4 **63.** 13

65. 16 **67.** -1 **69.** $a - b < x < a + b$ **71.** $x < \dfrac{b-c}{a}$ or $x > \dfrac{b+c}{a}$ **73.** $\dfrac{-c-b}{a} \le x \le \dfrac{c-b}{a}$

CHAPTER 2 REVIEW TEST

1. 10 **2.** 2 **3.** 2 **4.** -3 **5.** -3 **6.** 0 **7.** $\frac{2}{3}$ **8.** $\frac{10}{13}$ **9.** $\frac{5}{2}$ **10.** 1 **11.** $-\frac{5}{11}$ **12.** $-\frac{2}{9}$ **13.** $h = 17$

14. $t = 20$ **15.** $p = \dfrac{I}{rt}$ **16.** $x = \dfrac{y-b}{m}$ **17.** $y = \frac{4}{3}x - 4$ **18.** $v = \dfrac{d - 16t^2}{t}$ **19.** $y = -\frac{5}{3}x + 2$ **20.** $y = -\frac{2}{3}x + 2$

21. $y = -2x - 1$ **22.** $y = -3x + 1$ **23.** 4 feet by 12 feet **24.** 3 meters, 4 meters, 5 meters

25. a. 3,780 bricks **b.** 625 feet **26.** $24,875.24 **27.** $(-\infty, \frac{1}{2})$ **28.** $(-\infty, 8]$ **29.** $(-\infty, 12]$ **30.** $(-1, \infty)$ **31.** $[2, 6]$ **32.** $[0, 1]$

33. $(-\infty, -\frac{3}{2}] \cup [3, \infty)$ **34.** $(-\infty, 1) \cup [4, \infty)$ **35.** $-2, 2$ **36.** $-4, 4$ **37.** 2, 4 **38.** $-1, 4$ **39.** $-\frac{3}{2}, 3$ **40.** 5, 9 **41.** $-1, 1$

42. 0 **43.** **44.** **45.** $\varnothing$

46. 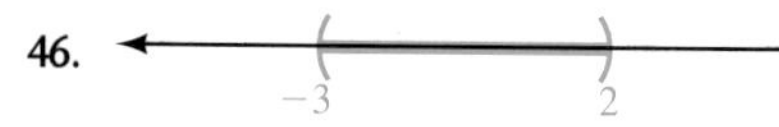

47. $\varnothing$ **48.** $\varnothing$ **49.** $\varnothing$ **50.** All real numbers except 0 **51.** $\varnothing$ **52.** All real numbers

CHAPTER 3

Problem Set 3.1

1.

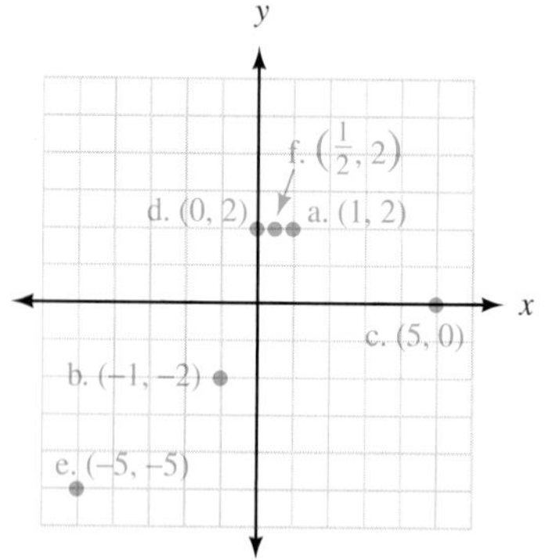

3. A (4, 1) B (−4, 3) C (−2, −5) D (2, −2) E (0, 5) F (−4, 0) G (1, 0)

5. x-intercept = 3, y-intercept = −2

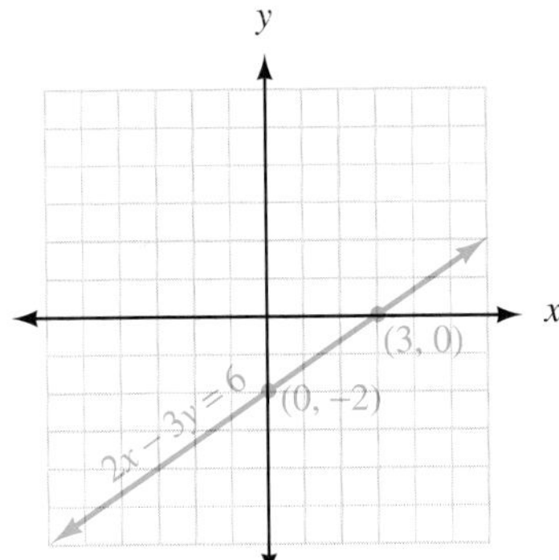

7. x-intercept = 5, y-intercept = −4

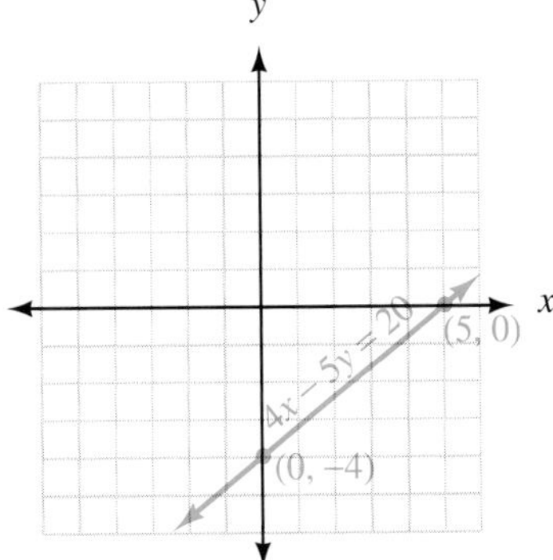

9. x-intercept = $-\frac{3}{2}$, y-intercept = 3

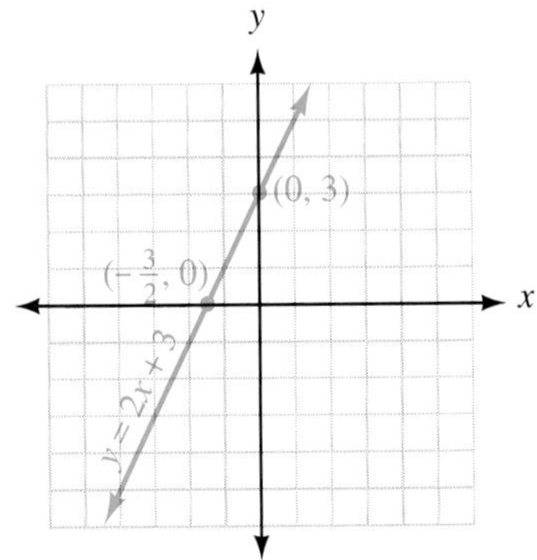

11. x-intercept = −4, y-intercept = 6

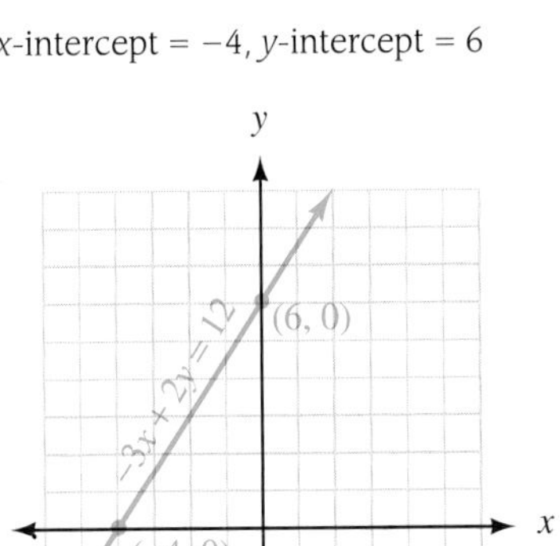

13. x-intercept = $\frac{10}{3}$, y-intercept = −4

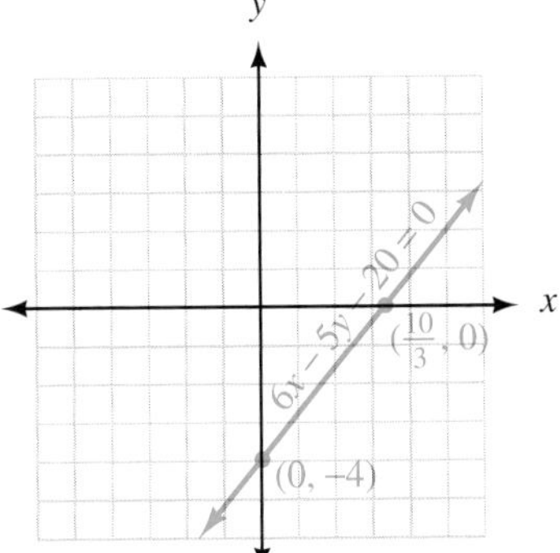

15. x-intercept = $\frac{5}{3}$, y-intercept = −5

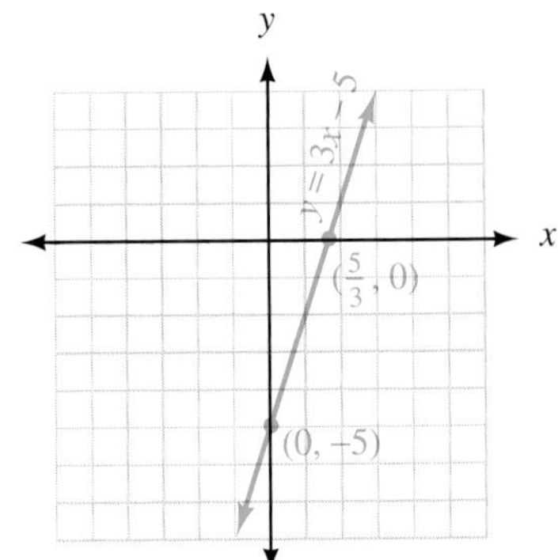

17. x-intercept = 2, y-intercept = 3

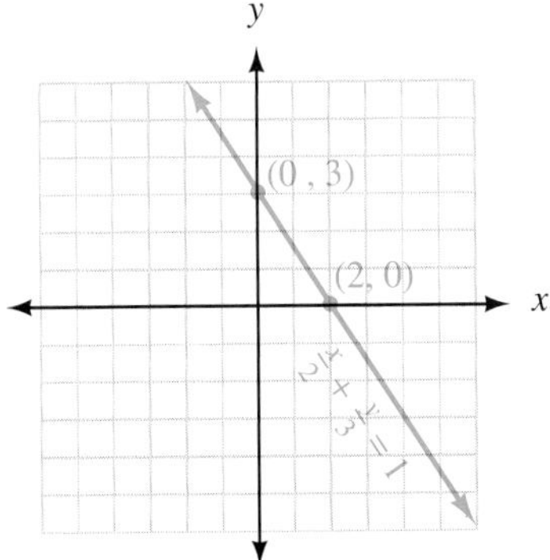

19. Table b

21.

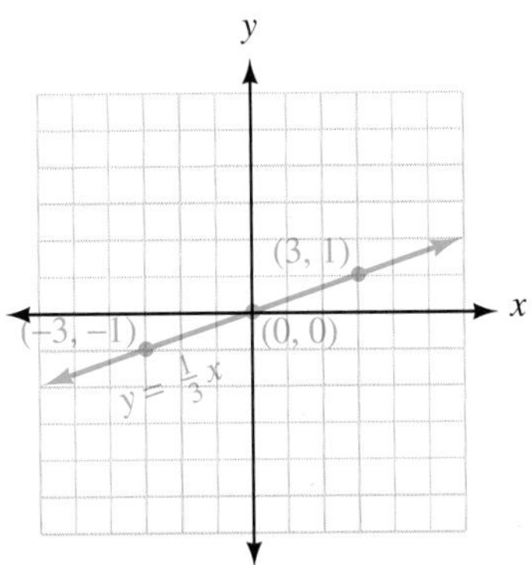

23.

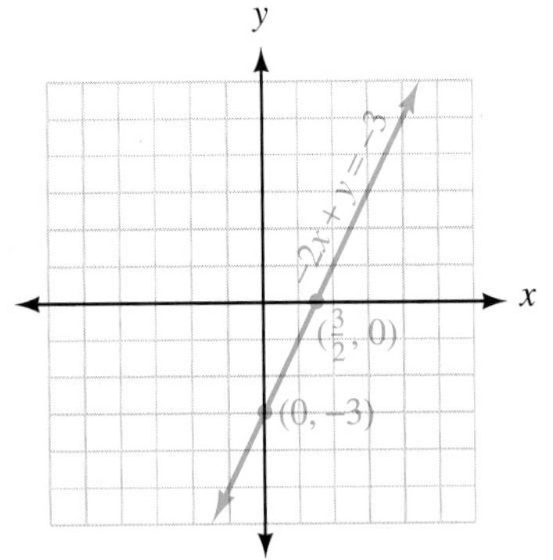

25.

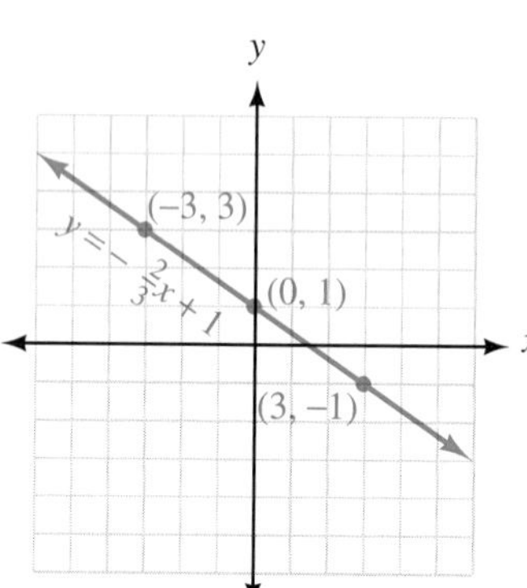

27.

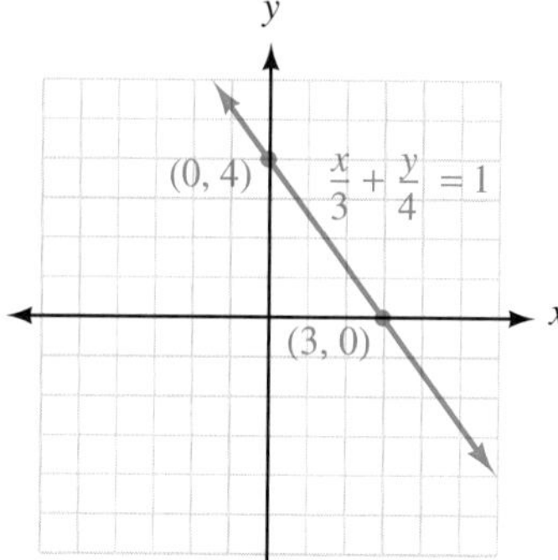

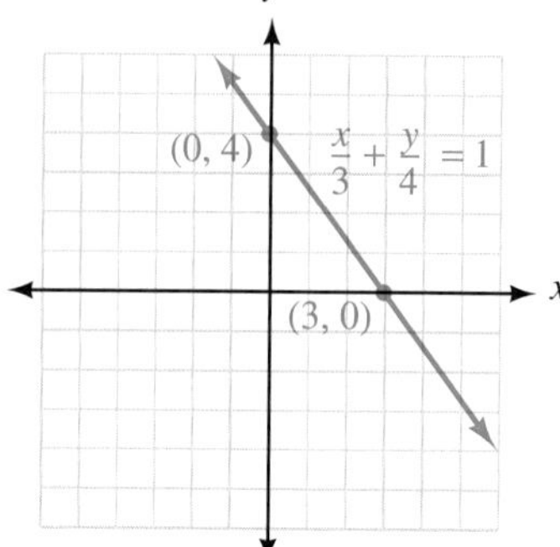

29. Equation b.

31. a. −7 **b.** −4 **c.** $-\frac{4}{3}$ **d.**

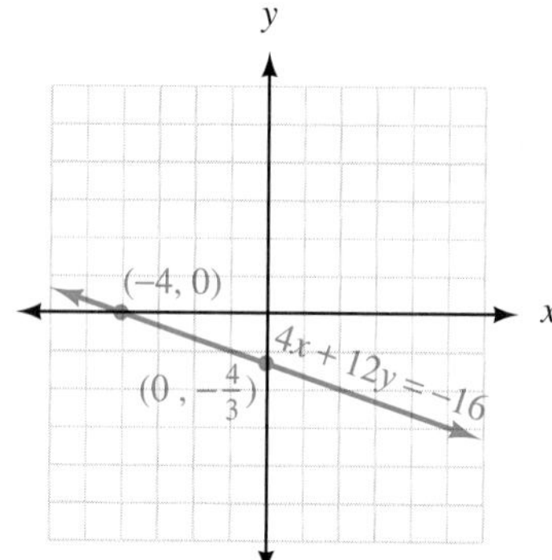

e. $y = -\frac{1}{3}x - \frac{4}{3}$

33. a.

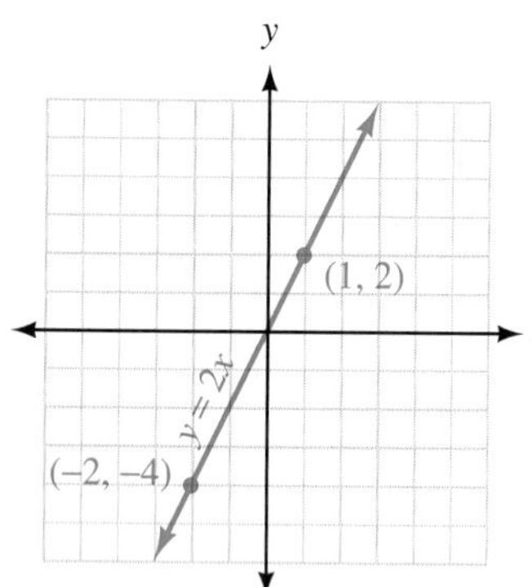

b.

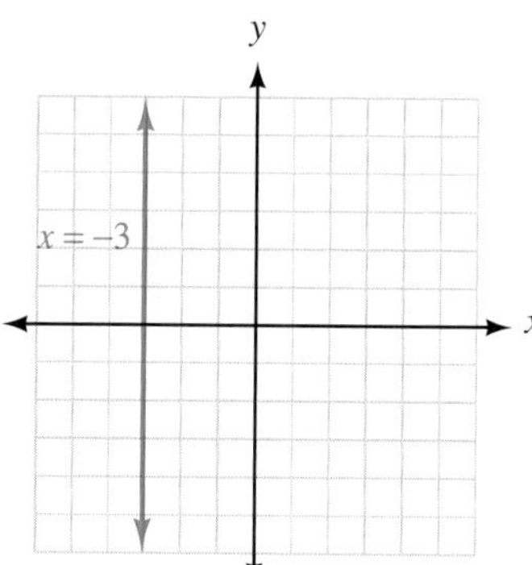

c.

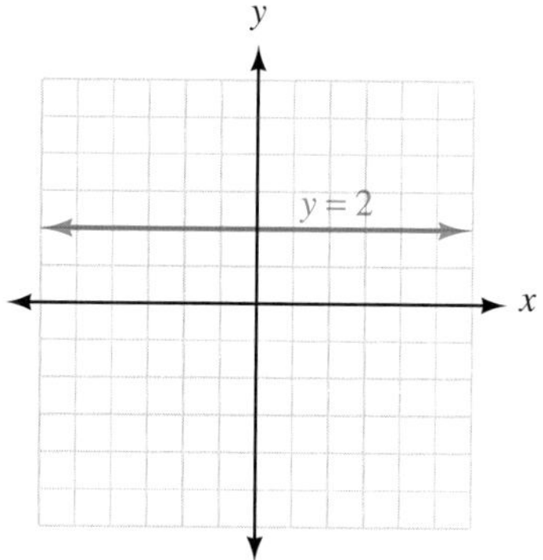

35. a.

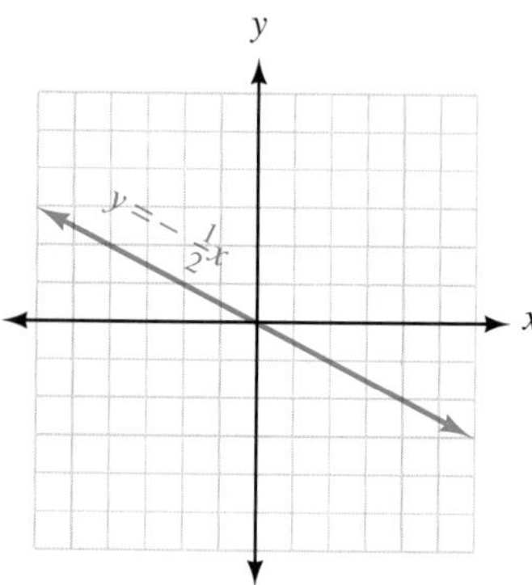

b.

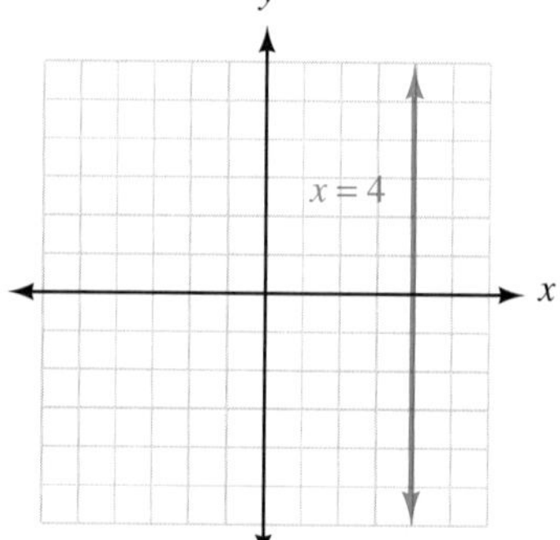

c.

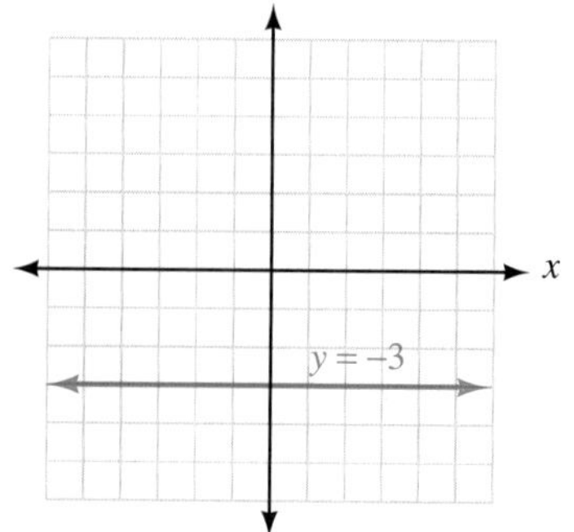

37.

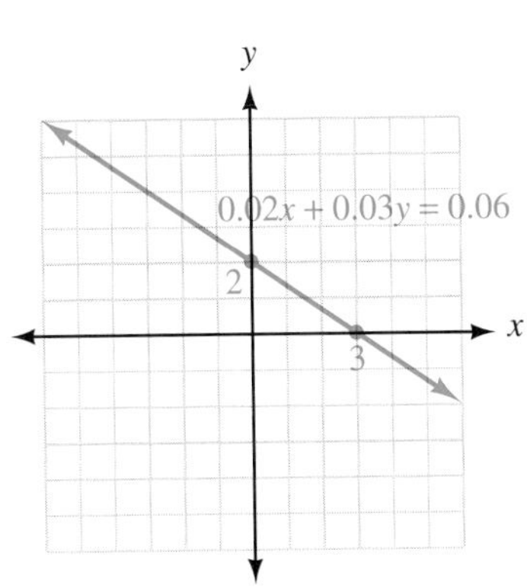

39.

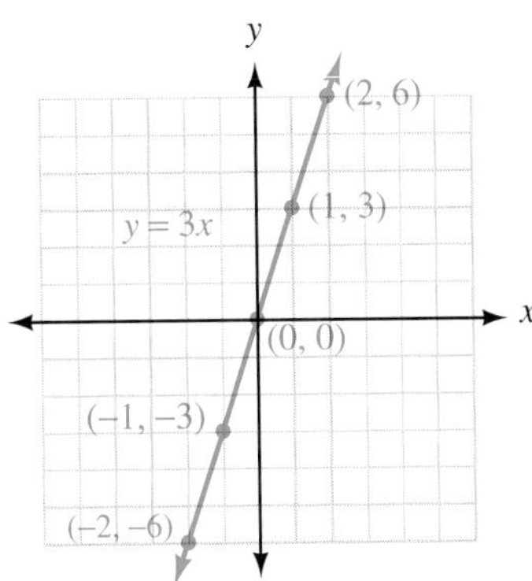

41. **a.** (5, 40), (10, 80), (20, 160) **b.** \$320 **c.** 30 hours **d.** No, if she works 35 hours, she should be paid \$280.

43.

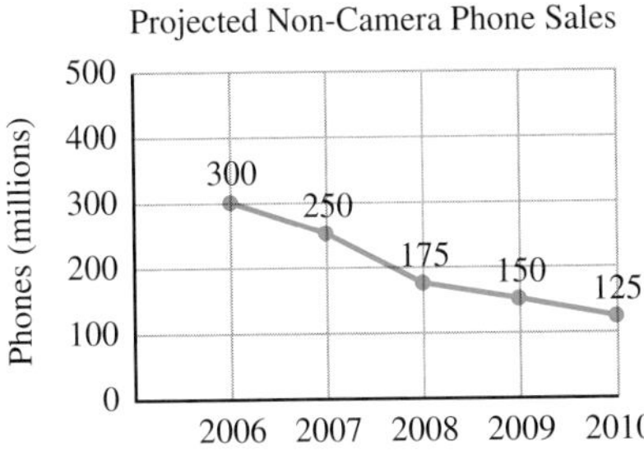

45. (1985, 20.2), (1990, 34.4), (1995, 44.8), (2000, 65.4), (2004, 99.4)

47. $A = (1, 2), B = (6, 7)$ 49. $A = (3, 3), B = (3, 6), C = (8, 6)$ 51. **a.** About 35% **b.** About 70% **c.** 2000 and 2001

53. 2 55. −2 57. 1,200 59. 2 61. $\frac{8}{15}$ 63. $-\frac{6}{100}$ 65. 1 67. $-\frac{4}{3}$ 69. Undefined 71. **a.** $\frac{3}{2}$ **b.** $-\frac{3}{2}$

73. x-intercept $= \frac{c}{a}$, y-intercept $= \frac{c}{b}$ 75. x-intercept $= a$, y-intercept $= b$

Problem Set 3.2

1. $\frac{3}{2}$ 3. No slope 5. $\frac{2}{3}$

7.

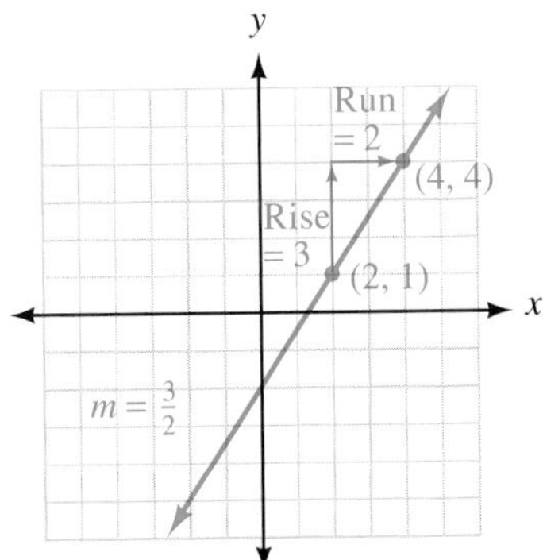

9.

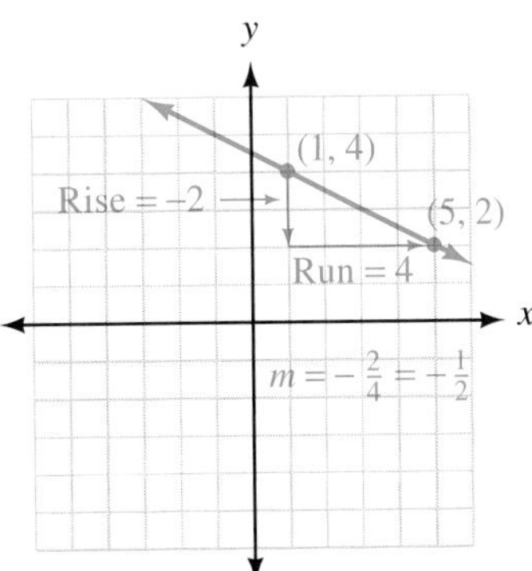

11.

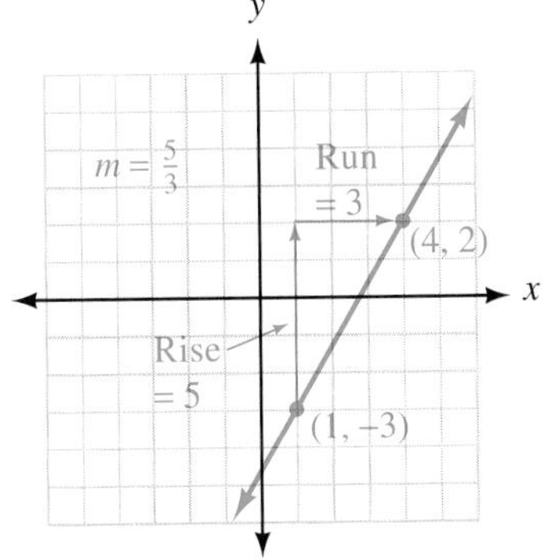

13.

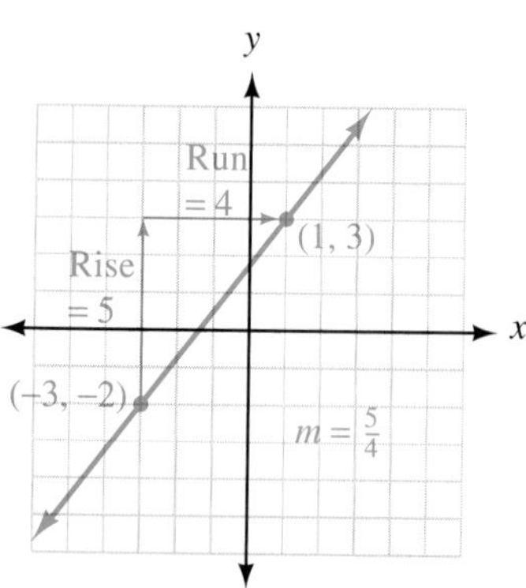

15.

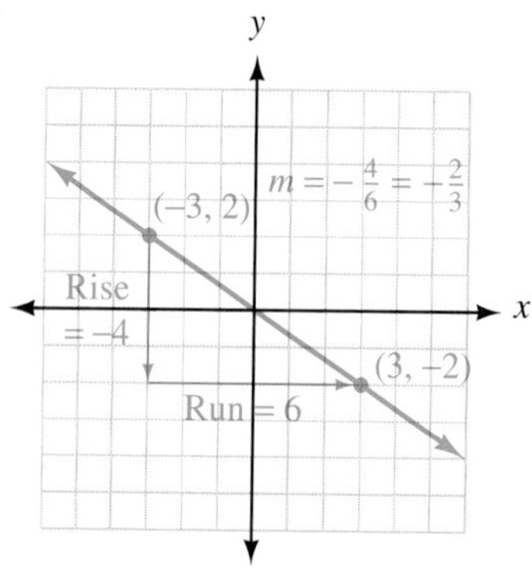

17.

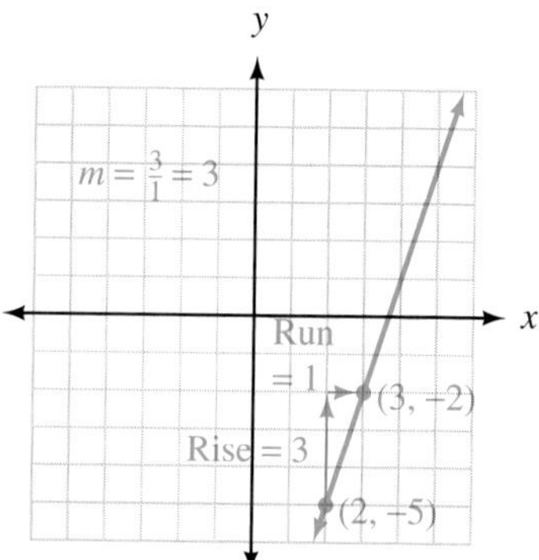

19. $x = 3$ 21. $a = 5$ 23. $b = 2$

25.

x	y
0	2
3	0

Slope $= -\frac{2}{3}$

27.

x	y
0	−5
3	−3

Slope $= \frac{2}{3}$

29.

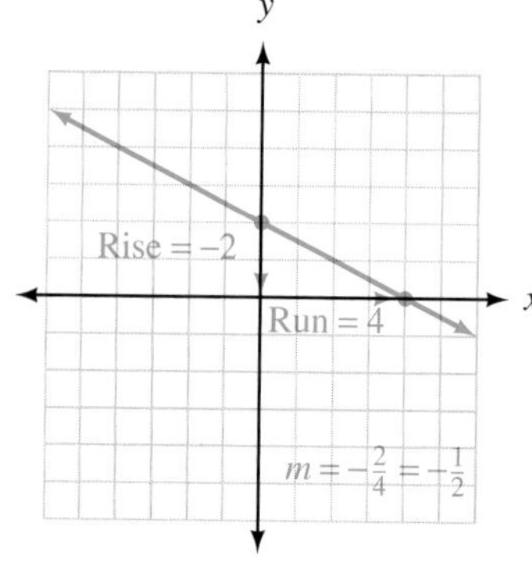

31. $\frac{1}{5}$ 33. 0 35. -1 37. $-\frac{3}{2}$ 39. a. yes b. no 41. a. 17.5 mph b. 40 km/h c. 120 ft/sec d. 28 m/min
43. Slopes: A, −50; B −75; C, −25 45. a. 15,800 feet b. $-\frac{7}{100}$
47. 6.5 percent/year. Over the 10 years from 1995 to 2005, the percent of adults that access the Internet has increased at an average rate of 6.5% each year.
49. a. −4,768.6 cases/year. New cases of rubella decreased at an average of 4,768.6 cases per year from 1969 to 1979.
b. −999.1 cases/year. New cases of rubella decreased at an average of 999.1 cases per year from 1979 to 1989.
c. The slope is 0, meaning that the number of new cases of German Measles is constant at 9 cases/year.
51. 0 53. $y = -\frac{3}{2}x + 6$ 55. $t = \frac{A - P}{Pr}$ 57. -1 59. -2 61. $y = mx + b$ 63. $y = -2x - 5$ 65. 5 67. $(\frac{5}{4}, \frac{7}{4})$

Problem Set 3.3

1. $y = 2x + 3$ 3. $y = x - 5$ 5. $y = \frac{1}{2}x + \frac{3}{2}$ 7. $y = 4$ 9. a. 3 b. $-\frac{1}{3}$ 11. a. -3 b. $\frac{1}{3}$ 13. a. $-\frac{2}{5}$ b. $\frac{5}{2}$

15. Slope = 3, y-intercept = −2, perpendicular slope $= -\frac{1}{3}$

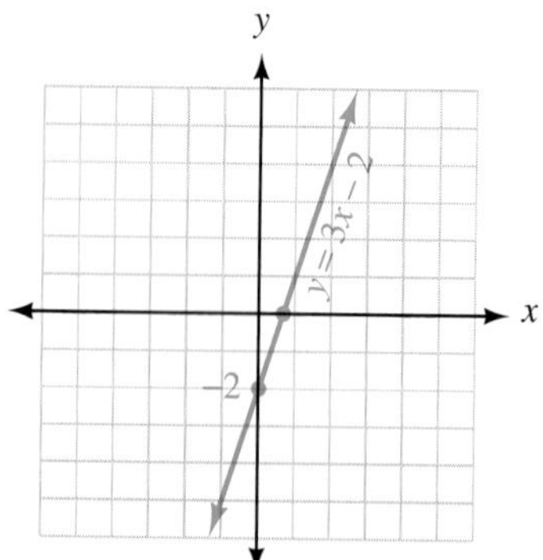

17. Slope $= \frac{2}{3}$, y-intercept = −4, perpendicular slope $= -\frac{3}{2}$

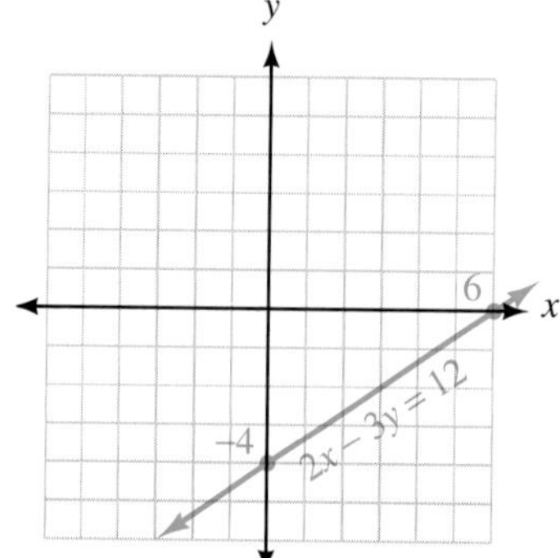

19. Slope $= -\frac{4}{5}$, y-intercept = 4, perpendicular slope $= \frac{5}{4}$

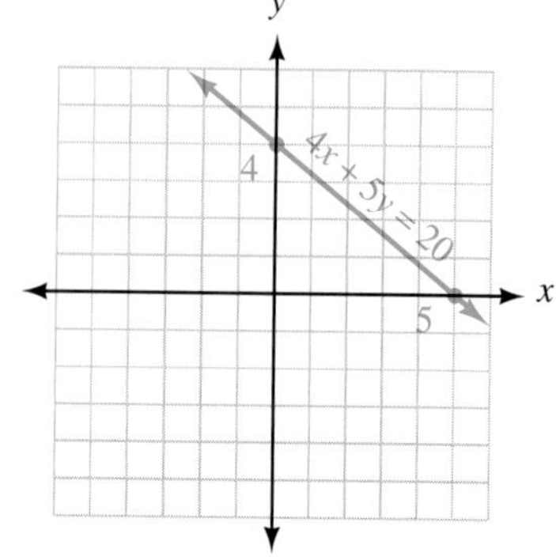

21. Slope = $\frac{1}{2}$, y-intercept = -4, $y = \frac{1}{2}x - 4$ **23.** Slope = $-\frac{2}{3}$, y-intercept = 3, $y = -\frac{2}{3}x + 3$ **25.** $y = 2x - 1$ **27.** $y = -\frac{1}{2}x - 1$

29. $y = -3x + 1$ **31.** $y = \frac{2}{3}x + \frac{2}{3}$ **33.** $y = -\frac{1}{4}x + \frac{3}{4}$ **35.** $x - y = 2$ **37.** $2x - y = 3$ **39.** $6x - 5y = 3$

41. **a.** x-intercept = $\frac{10}{3}$; y-intercept = -5 **b.** (4, 1) **c.** $y = \frac{3}{2}x - 5$ **d.** No

43. **a.** x-intercept = $\frac{4}{3}$; y-intercept = -2 **b.** (2, 1) **c.** $y = \frac{3}{2}x - 2$ **d.** No

45. **a.** 2 **b.** $y = 2x - 3$ **c.** $y = -3$ **d.** 2 **e.**

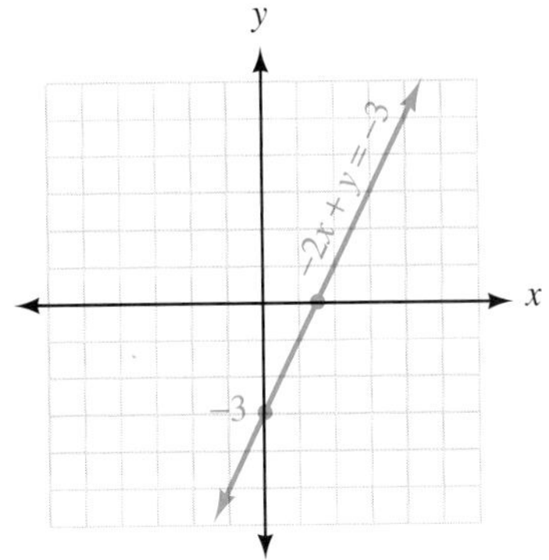

47. (0, −4), (2, 0); $y = 2x - 4$ **49.** (−2, 0) (0, 4); $y = 2x + 4$

51. Slope = 0,
y-intercept = -2

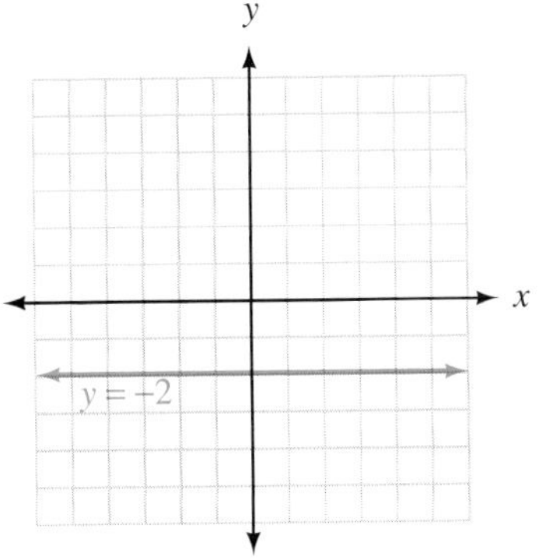

53. $y = 3x + 7$ **55.** $y = -\frac{5}{2}x - 13$ **57.** $y = \frac{1}{4}x + \frac{1}{4}$ **59.** $y = -\frac{2}{3}x + 2$ **61.** **b.** 86° F **63.** **a.** \$190,000 **b.** \$19 **c.** \$6.50

65. **a.** $y = \frac{69}{2}x - 68{,}994$ **b.** 213 books **67.** **a.** $y = 6.5x - 12{,}958.5$ **b.** In 2006 80.5% of adults will be online.
c. When we try to estimate the number of adults online in 2010 we get a number larger than 100%, which is impossible.

69. 23 inches, 5 inches **71.** \$46.50 **73.** (0, 0), (4, 0) **75.** (0, 0), (2, 0) **77.** $y = -\frac{3}{2}x + 12$ **79.** $y = -\frac{4}{3}x - \frac{3}{2}$

Problem Set 3.4

1.

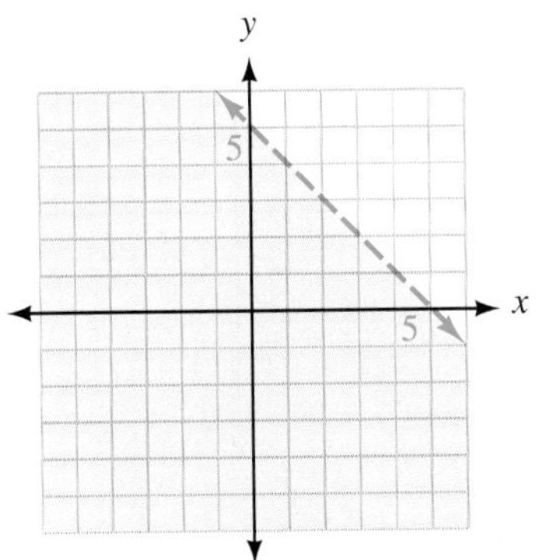

3.

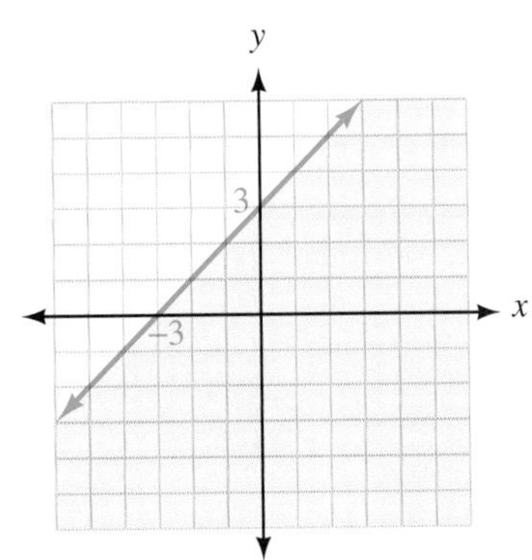

5.

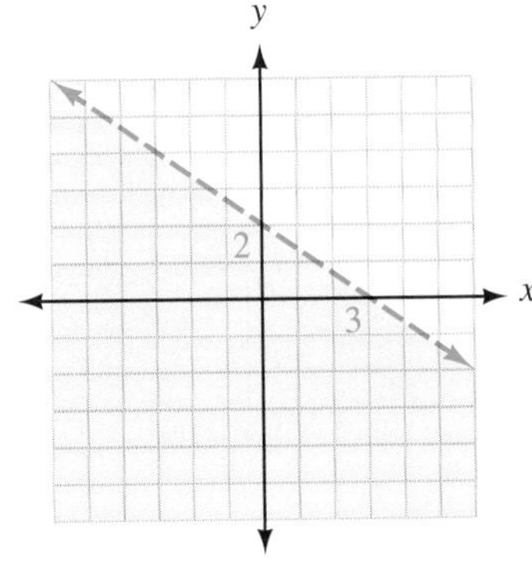

7.

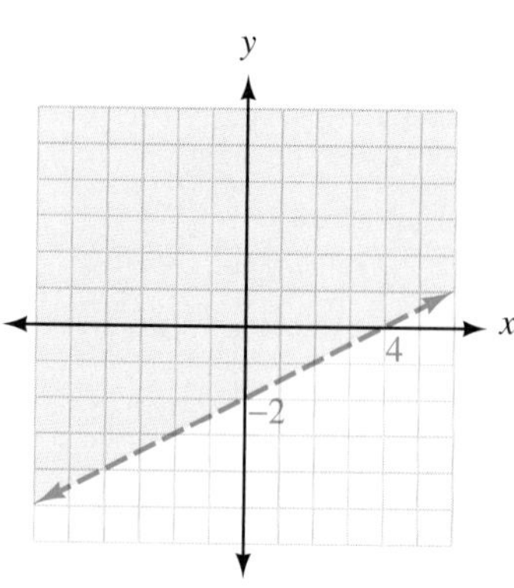

9.

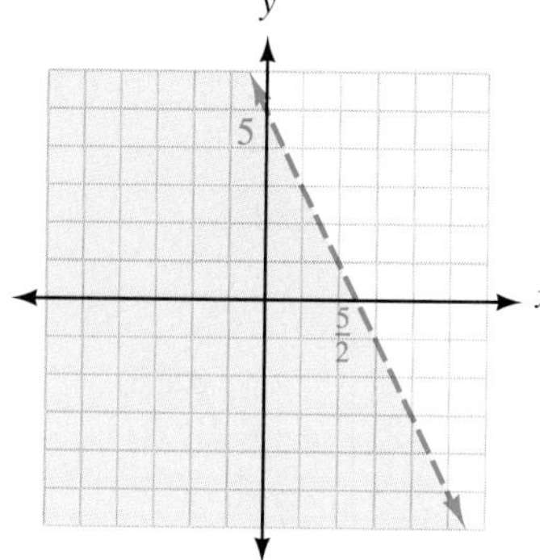

11.

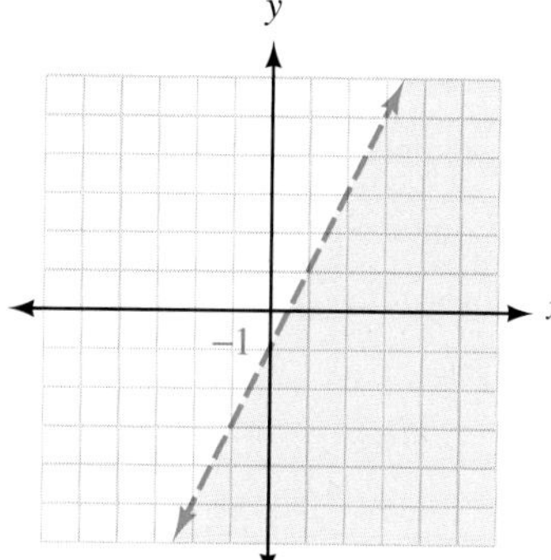

13.

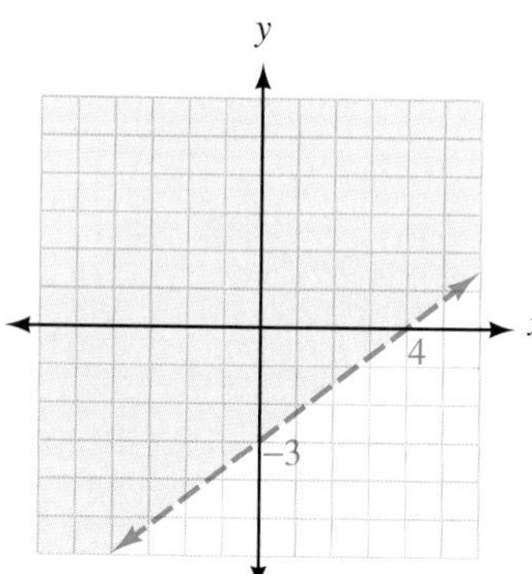

15.

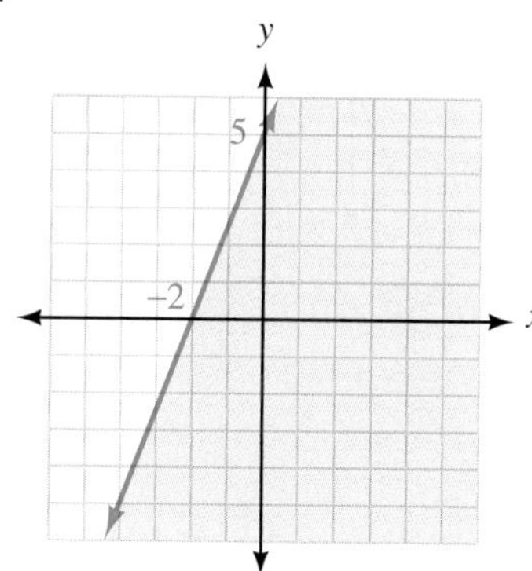

17.

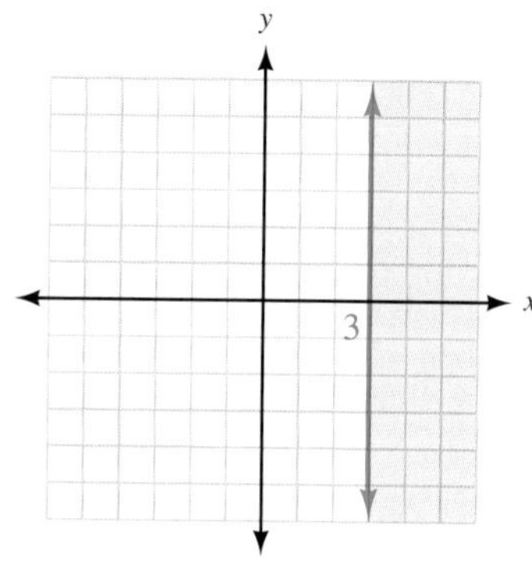

19.

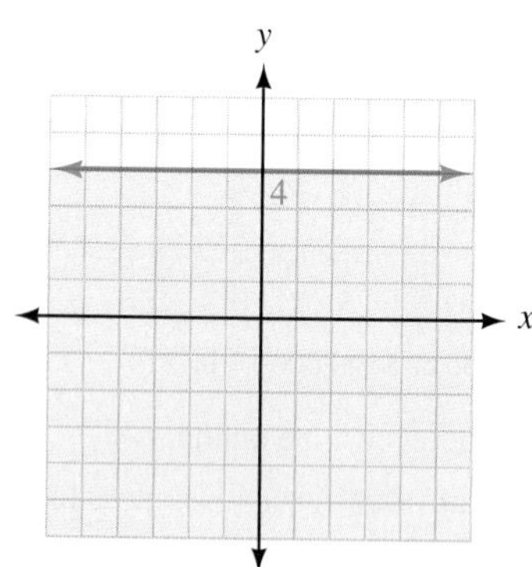

21.

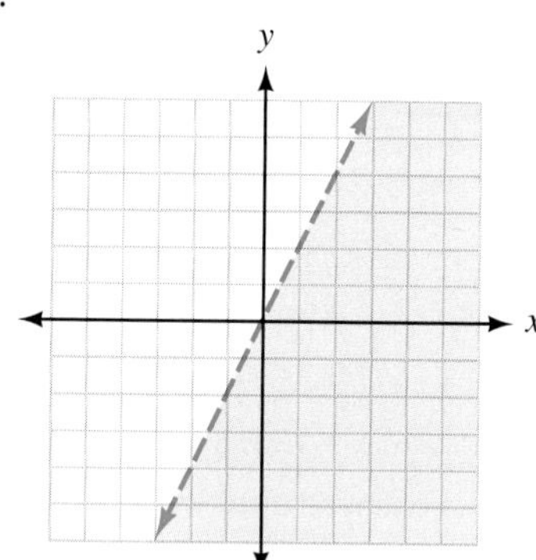

23.

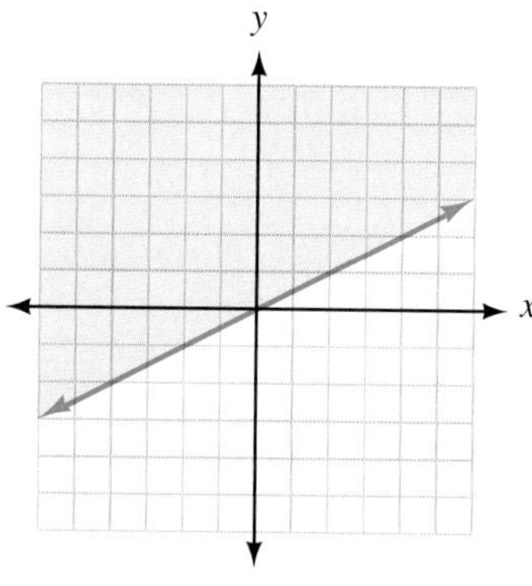

25.

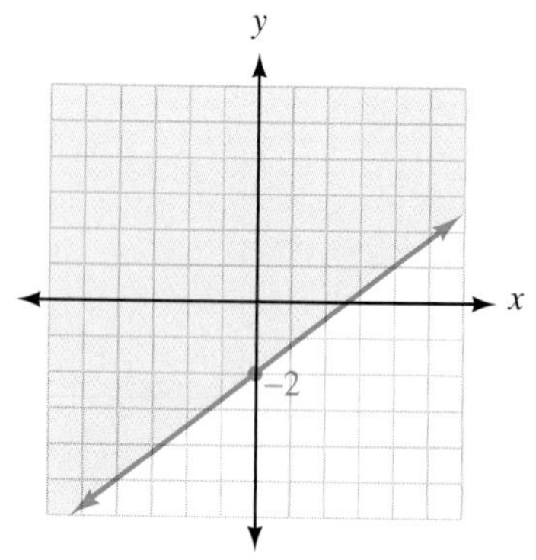

27.

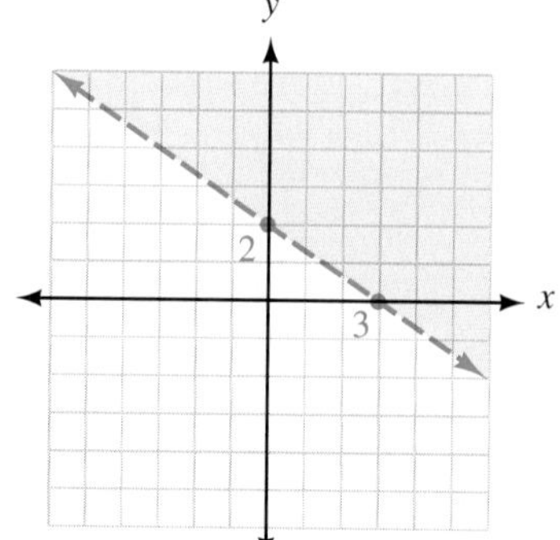

29.

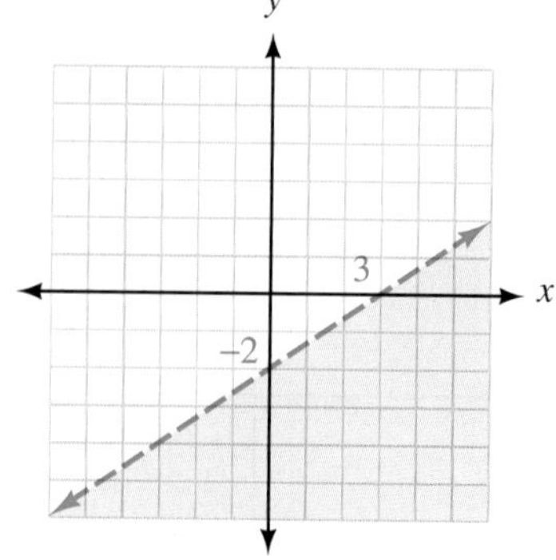

31.

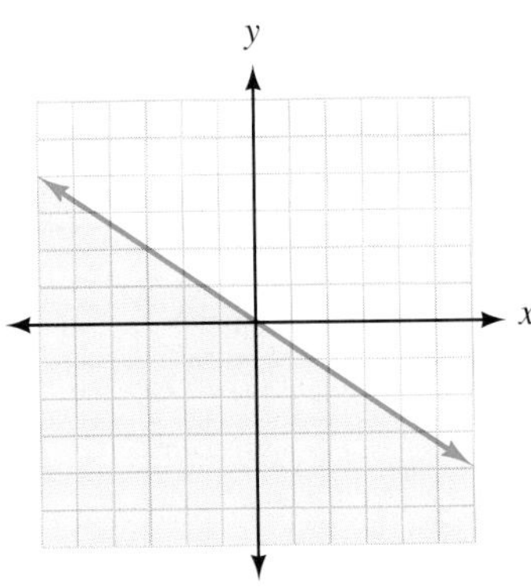

33.

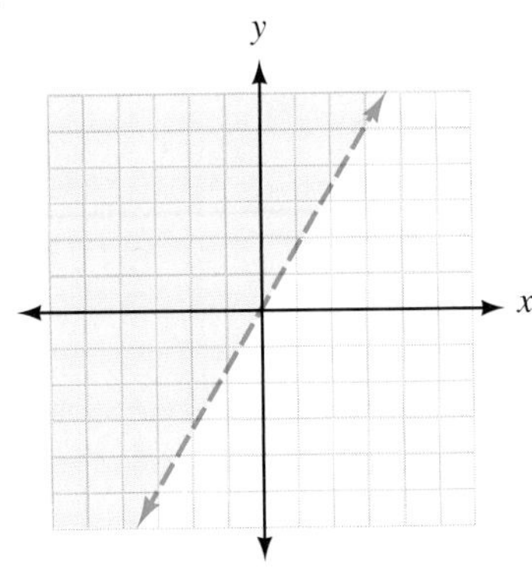

35.

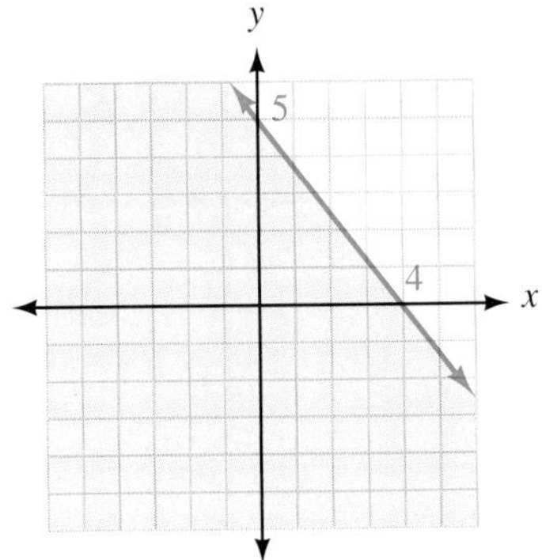

37. $x + y > 4$ **39.** $-x + 2y \leq 4$ **41. a.** $y < \frac{4}{3}$ **b.** $y > -\frac{4}{3}$ **c.** $y = -\frac{2}{3}x + 2$ **d.**

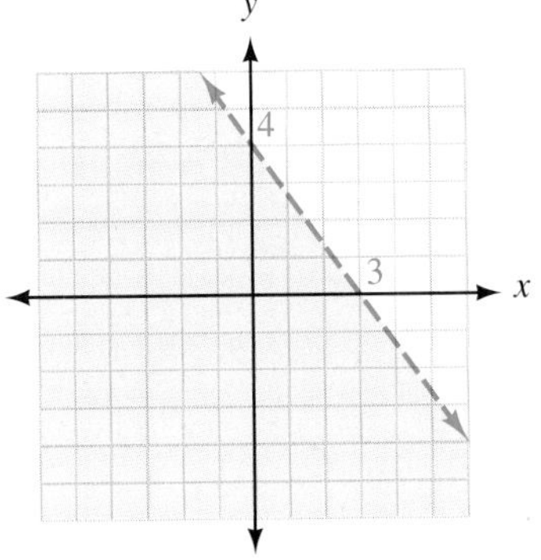

43.

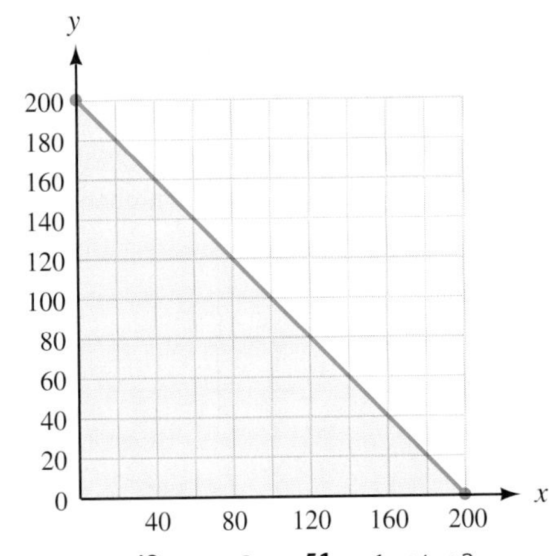

45.

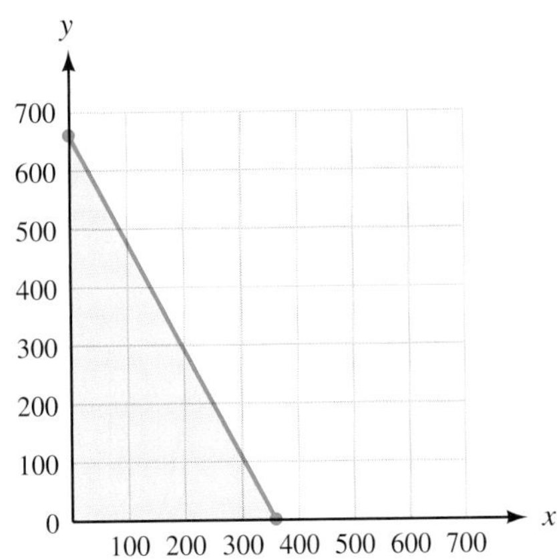

47. $y \leq 7$ **49.** $t > -2$ **51.** $-1 < t < 2$

53.

x	y
0	0
10	75
20	150

55.

x	y
0	0
1	1
1	−1

57.

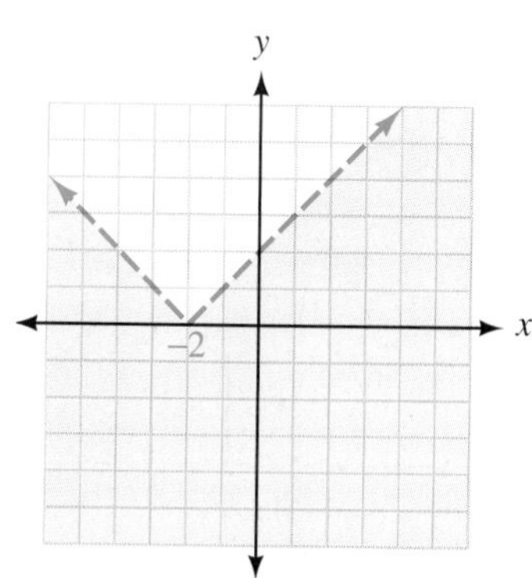

59.

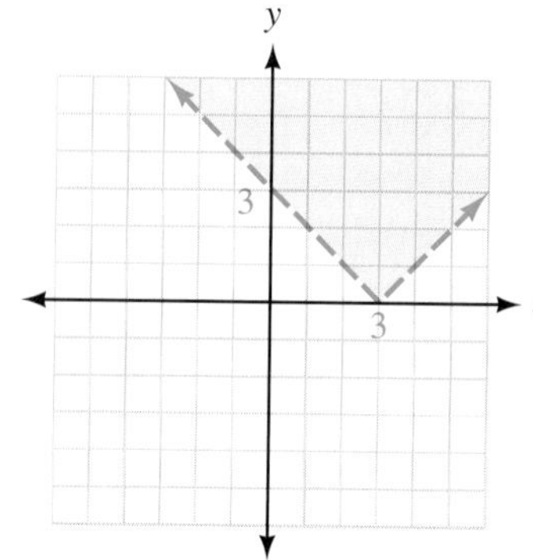

61.

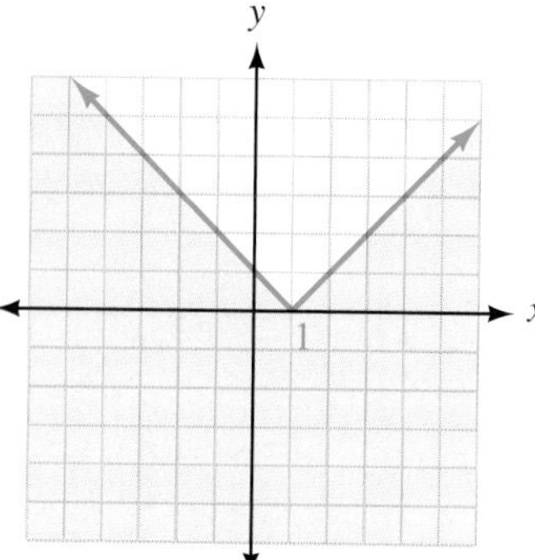

63.

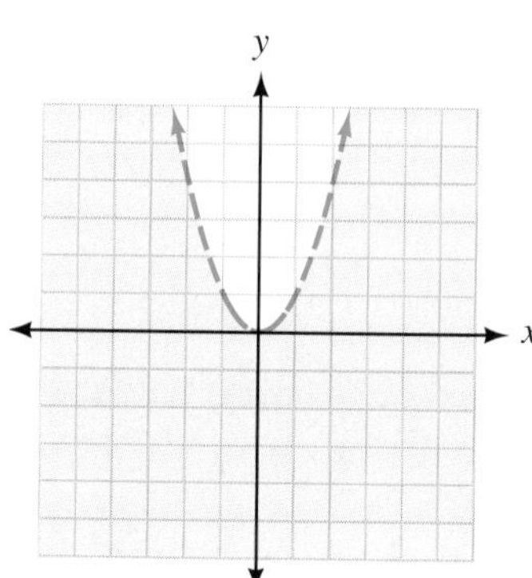

65.

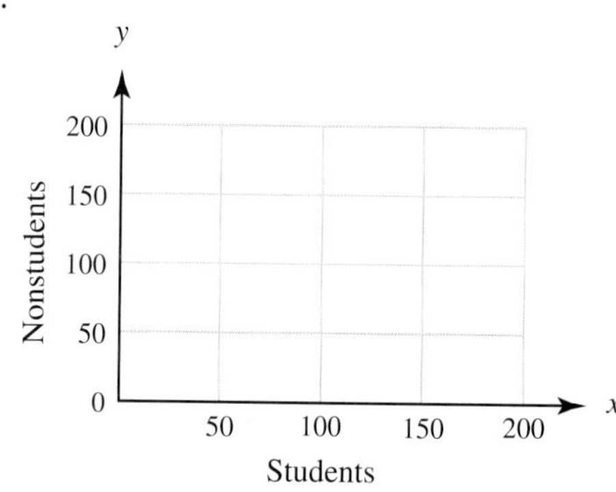

Problem Set 3.5

1. Domain = {1, 2, 4}; Range = {3, 5, 1}; a function
3. Domain = {−1, 1, 2}; range = {3, −5}; a function
5. Domain = {7, 3}; Range = {−1, 4}; not a function
7. Domain = {a, b, c, d}; range = {3, 4, 5}; a function
9. Domain = {a}; Range = {1, 2, 3, 4}; not a function
11. Yes
13. No
15. No
17. Yes
19. Yes
21. Domain = $\{x \mid -5 \le x \le 5\}$; Range = $\{y \mid 0 \le y \le 5\}$
23. Domain = $\{x \mid -5 \le x \le 3\}$; Range = $\{y \mid y = 3\}$
25. Domain = All real numbers; Range = $\{y \mid y \ge -1\}$; a function
27. Domain = All real numbers; Range = $\{y \mid y \ge 4\}$; a function
29. Domain = $\{x \mid x \ge -1\}$; Range = All real numbers; not a function

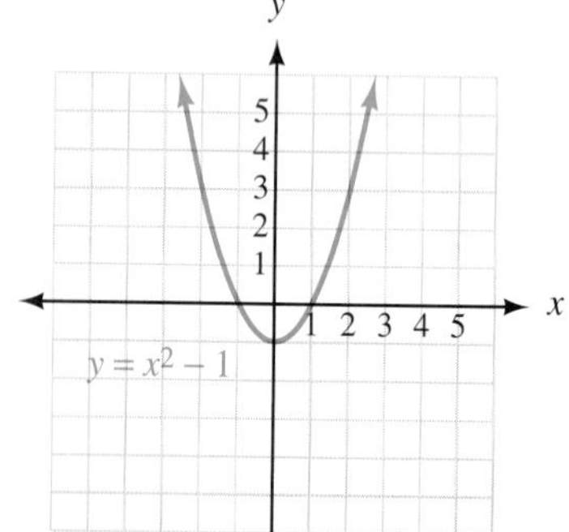

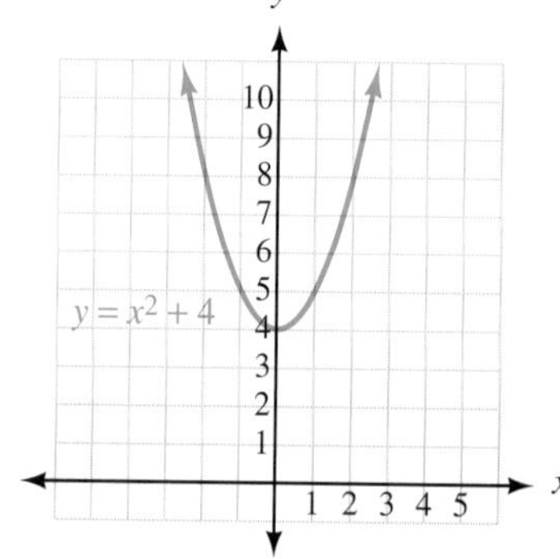

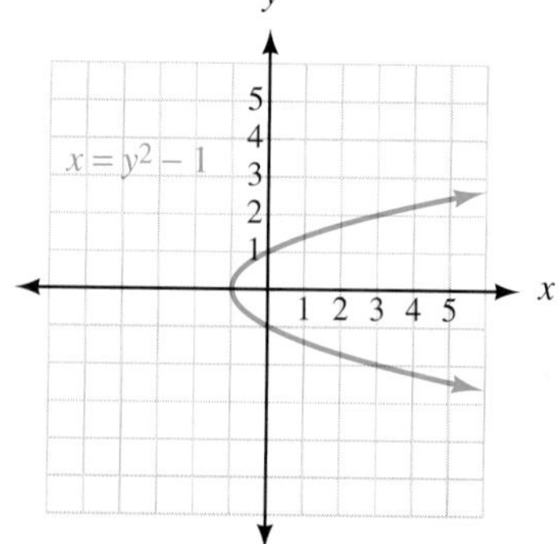

31. Domain = $\{x \mid x \geq 4\}$; Range = All real numbers; not a function

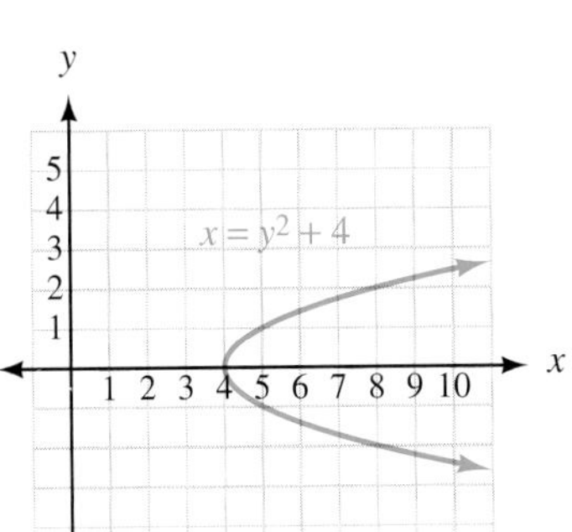

33. Domain = All real numbers; Range = $\{y \mid y \geq 0\}$; a function

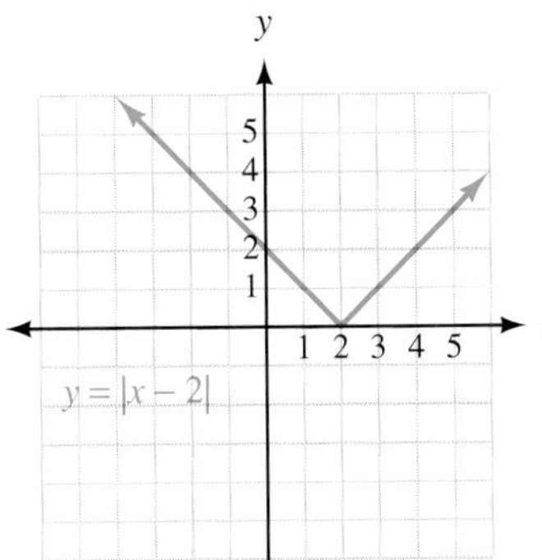

35. Domain = All real numbers; Range = $\{y \mid y \geq -2\}$; a function

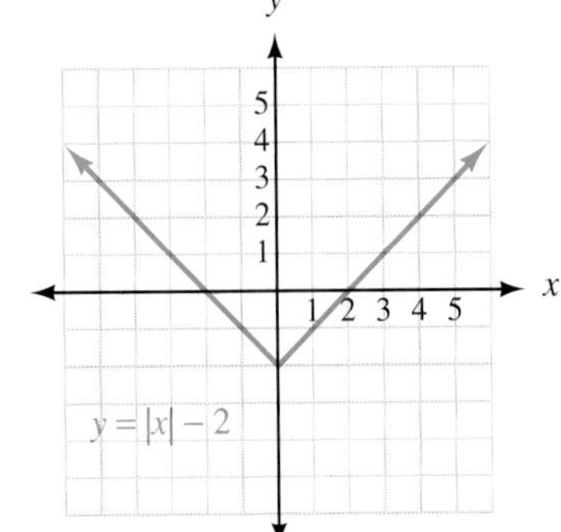

37. a. $y = 8.5x$ for $10 \leq x \leq 40$

b.

Hours Worked x	Function Rule $y = 8.5x$	Gross Pay ($) y
10	$y = 8.5(10) = 85$	85
20	$y = 8.5(20) = 170$	170
30	$y = 8.5(30) = 255$	255
40	$y = 8.5(40) = 340$	340

c.

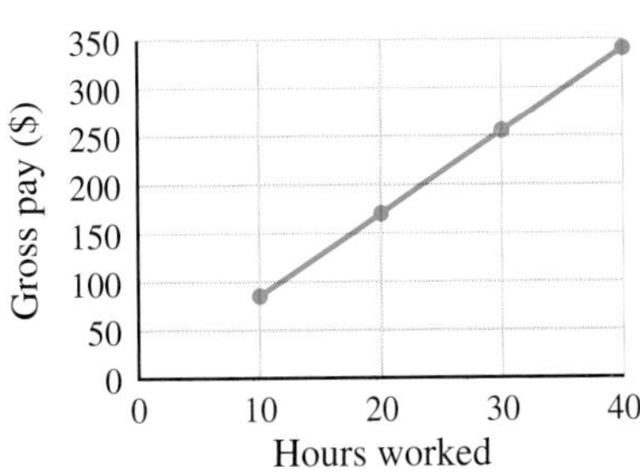

d. Domain = $\{x \mid 10 \leq x \leq 40\}$; Range = $\{y \mid 85 \leq y \leq 340\}$ **e.** Minimum = $85; Maximum = $340

39. a.

Time (sec) t	Function Rule $h = 16t - 16t^2$	Distance (ft) h
0	$h = 16(0) - 16(0)^2$	0
0.1	$h = 16(0.1) - 16(0.1)^2$	1.44
0.2	$h = 16(0.2) - 16(0.2)^2$	2.56
0.3	$h = 16(0.3) - 16(0.3)^2$	3.36
0.4	$h = 16(0.4) - 16(0.4)^2$	3.84
0.5	$h = 16(0.5) - 16(0.5)^2$	4
0.6	$h = 16(0.6) - 16(0.6)^2$	3.84
0.7	$h = 16(0.7) - 16(0.7)^2$	3.36
0.8	$h = 16(0.8) - 16(0.8)^2$	2.56
0.9	$h = 16(0.9) - 16(0.9)^2$	1.44
1	$h = 16(1) - 16(1)^2$	0

b. Domain = $\{t \mid 0 \leq t \leq 1\}$; Range = $\{h \mid 0 \leq h \leq 4\}$

c.

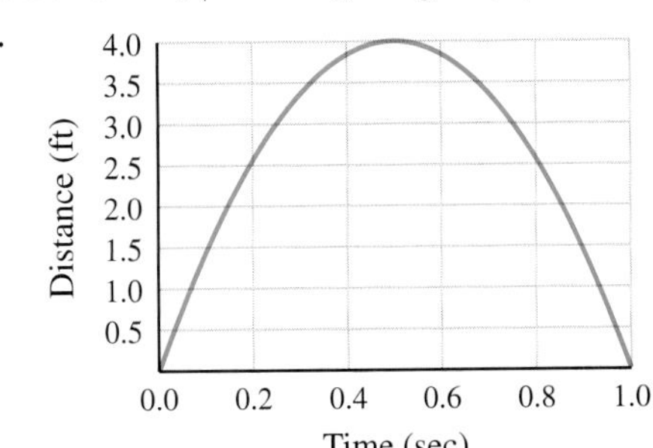

41. a.

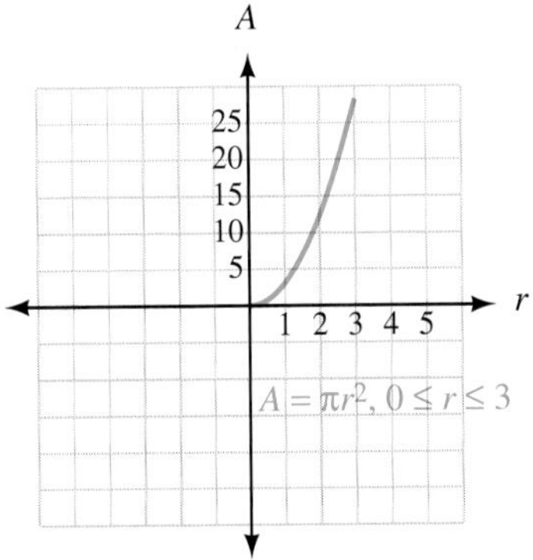

b. Domain = $\{r \mid 0 \leq r \leq 3\}$
Range = $\{A \mid 0 \leq A \leq 9\pi\}$

43. a. Yes **b.** Domain = $\{t \mid 0 \leq t \leq 6\}$; Range = $\{h \mid 0 \leq h \leq 60\}$ **c.** $t = 3$ **d.** $h = 60$ **e.** $t = 6$ **45. a.** III **b.** I **c.** II **d.** IV **47.** 10

49. -14 **51.** 1 **53.** -3 **55.** $-\frac{6}{5}$ **57.** $-\frac{7}{640}$ **59.** 150 **61.** 113 **63.** -9 **65.** **a.** 6 **b.** 7.5 **67.** **a.** 27 **b.** 6

69. Domain = All real numbers; Range = $\{y \mid y \le 5\}$; a function

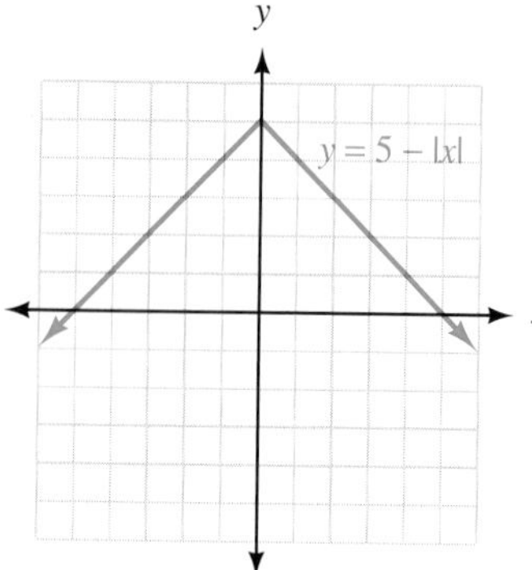

71. Domain = $\{x \mid x \ge 3\}$; Range = All real numbers; not a function

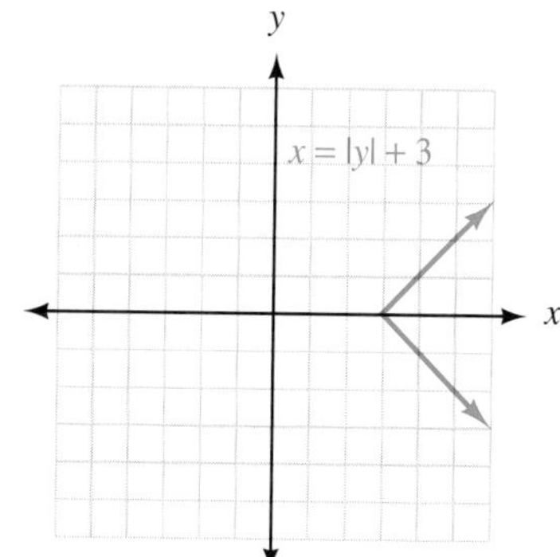

73. Domain = $\{x \mid -4 \le x \le 4\}$; Range = $\{y \mid -4 \le y \le 4\}$; not a function

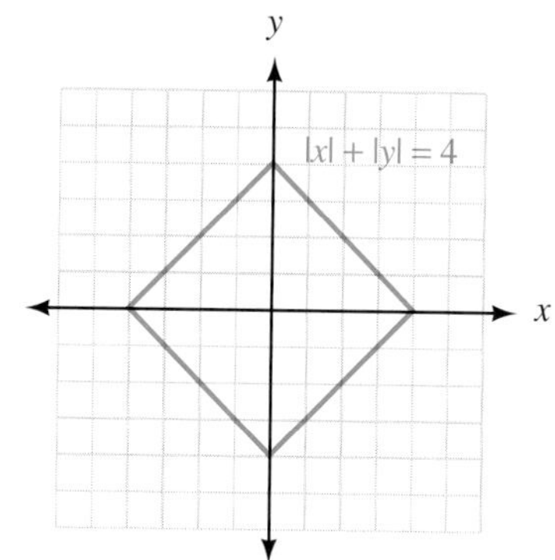

Problem Set 3.6

1. -1 **3.** -11 **5.** 2 **7.** 4 **9.** 35 **11.** -13 **13.** 1 **15.** -9 **17.** 8 **19.** 19 **21.** 16 **23.** 0 **25.** $3a^2 - 4a + 1$
27. 4 **29.** 0 **31.** 2 **33.** -8 **35.** -1 **37.** $2a^2 - 8$ **39.** $2b^2 - 8$ **41.** 0 **43.** -2 **45.** -3

47.

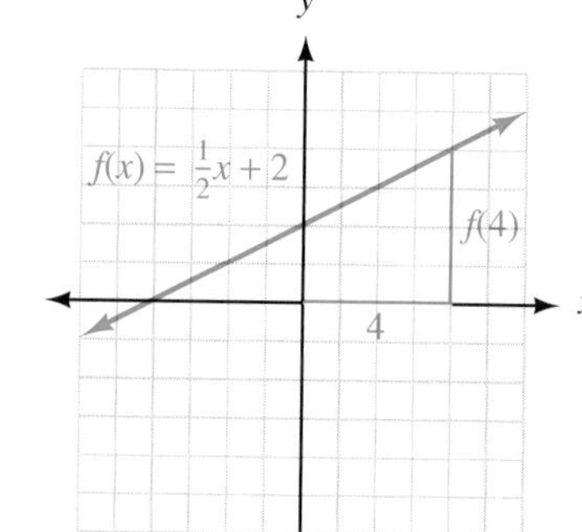

49. $x = 4$ **51.**

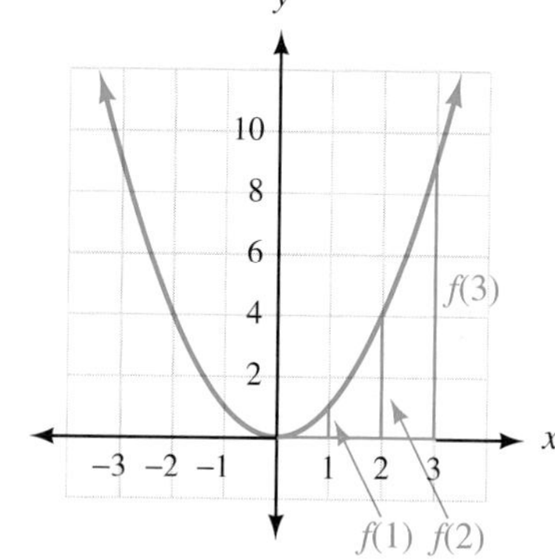

53. $V(3) = 300$, the painting is worth $300 in 3 years; $V(6) = 600$, the painting is worth $600 in 6 years.
55. **a.** True **b.** True **c.** True **d.** False **e.** True
57. **a.** $5,625 **b.** $1,500 **c.** $\{t \mid 0 \le t \le 5\}$ **d.**
e. $\{V(t) \mid 1{,}500 \le V(t) \le 18{,}000\}$ **f.** 2.42 years

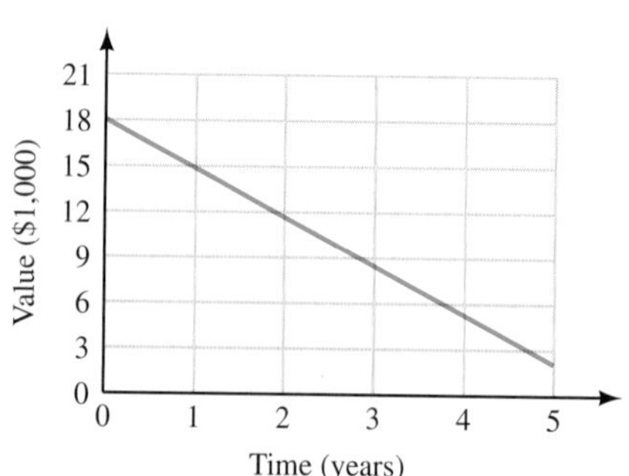

59. $-\frac{2}{3}, 4$ **61.** $-2, 1$ **63.** $\varnothing$ **65.** $-500 + 27x - 0.1x^2$ **67.** $6x^2 - 2x - 4$ **69.** $2x^2 + 8x + 8$ **71.** $0.6m - 42$
73. $4x^2 - 7x + 3$ **75.** **a.** 2 **b.** 0

77. **a.**

Weight (ounces)	0.6	1.0	1.1	2.5	3.0	4.8	5.0	5.3
Cost (cents)	39	39	63	87	87	135	135	159

b. More than 2 ounces, but not more than 3 ounces; $2 < x \le 3$

c. domain: $\{x \mid 0 < x \le 6\}$ **d.** range: $\{c \mid c = 39, 63, 87, 111, 135, 159\}$

Problem Set 3.7

1. $6x + 2$ **3.** $-2x + 8$ **5.** $4x - 7$ **7.** $3x^2 + x - 2$ **9.** $-2x + 3$ **11.** $9x^3 - 15x^2$ **13.** $\frac{3x^2}{3x - 5}$ **15.** $\frac{3x - 5}{3x^2}$ **17.** $3x^2 + 4x - 7$ **19.** 15 **21.** 98 **23.** $\frac{3}{2}$ **25.** 1 **27.** 40 **29.** 147 **31. a.** 81 **b.** 29 **c.** $(x + 4)^2$ **d.** $x^2 + 4$ **33. a.** -2 **b.** -1 **c.** $4x^2 + 12x - 1$ **37.** 6 **39.** 2 **41.** 3 **43.** -8 **45.** 6 **47.** 5 **49.** 3 **51.** -6 **53. a.** $R(x) = 11.5x - 0.05x^2$ **b.** $C(x) = 2x + 200$ **c.** $P(x) = -0.05x^2 + 9.5x - 200$ **d.** $\overline{C}(x) = 2 + \frac{200}{x}$ **55. a.** $M(x) = 220 - x$ **b.** 196 **c.** 142 **d.** 135 **e.** 128 **57.** $2 < x < 4$ **59.** $x < 4$ or $x > 8$ **61.** $-\frac{5}{7} \le x \le 1$ **63.** 196 **65.** 4 **67.** 1.6 **69.** 3 **71.** 2,400

Problem Set 3.8

1. Direct **3.** Direct **5.** Direct **7.** Direct **9.** Direct **11.** Inverse **13.** 30 **15.** 5 **17.** -6 **19.** $\frac{1}{2}$ **21.** 40 **23.** 225 **25.** 2 **27.** 10 **29.** $\frac{81}{5}$ **31.** 40.5 **33.** 64 **35.** 8 **37.** $\frac{50}{3}$ pounds **39. a.** $T = 4P$ **c.** 70 pounds per square inch **41.** 12 pounds per square inch **43. a.** $f = \frac{80}{d}$ **c.** An f-stop of 8

39. b.

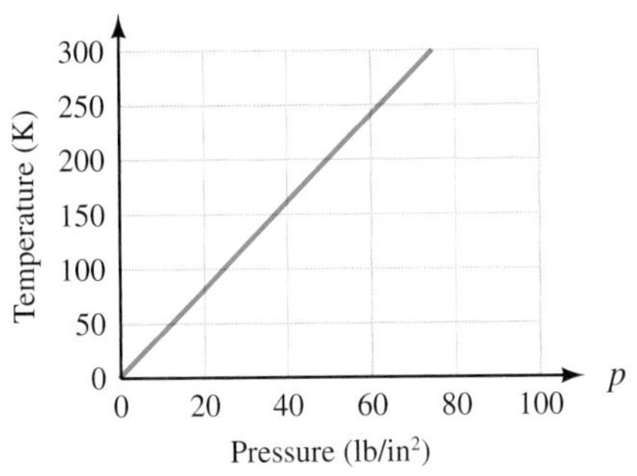

43. b.

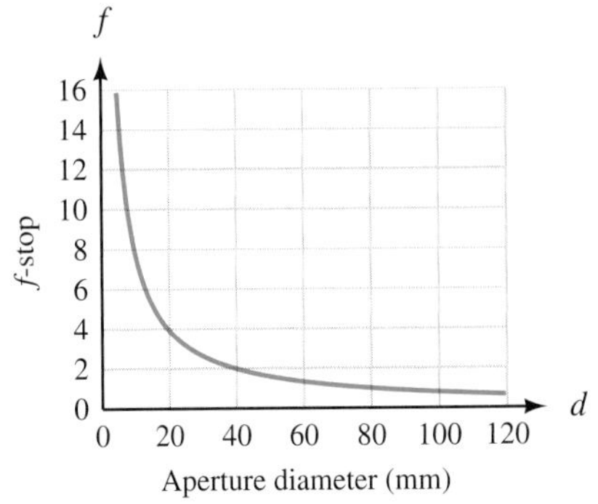

45. $\frac{1,504}{15}$ square inches **47.** 1.5 ohms **49.** 1.28 meters **51.** $F = G\frac{m_1m_2}{d^2}$ **53.** 12 **55.** 28 **57.** $-\frac{7}{4}$ **59.** $w = \frac{P - 21}{2}$ **61.** $t \ge -6$ **63.** $x < 6$ **65.** 6, 2 **67.** $\varnothing$ **69.** $x \le -9$ or $x \ge -1$

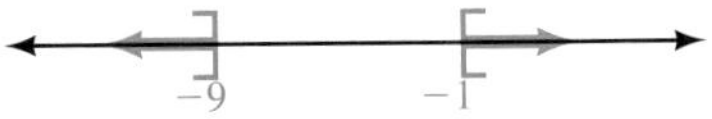

71. All real numbers **73. a.** Square of speed **b.** $d = 0.0675s^2$ **c.** About 204 feet from the cannon **d.** About 7.5 feet further away

CHAPTER 3 REVIEW TEST

1.

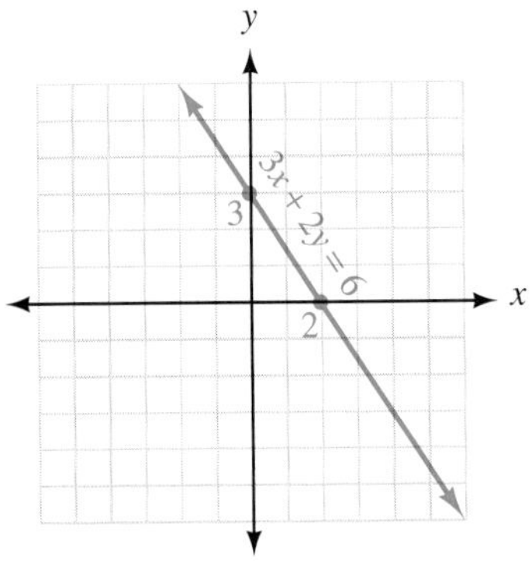

2.

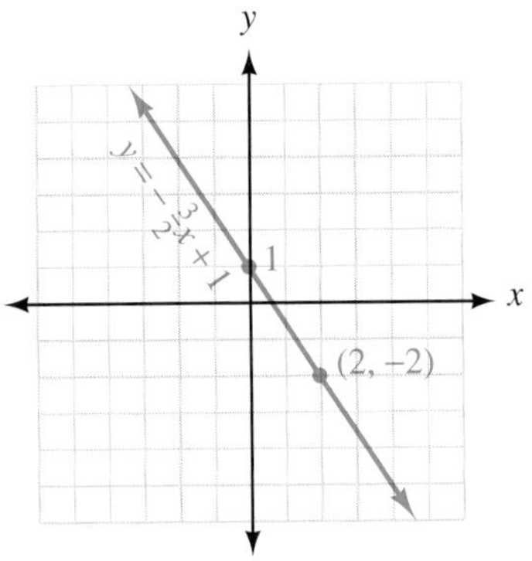

3.

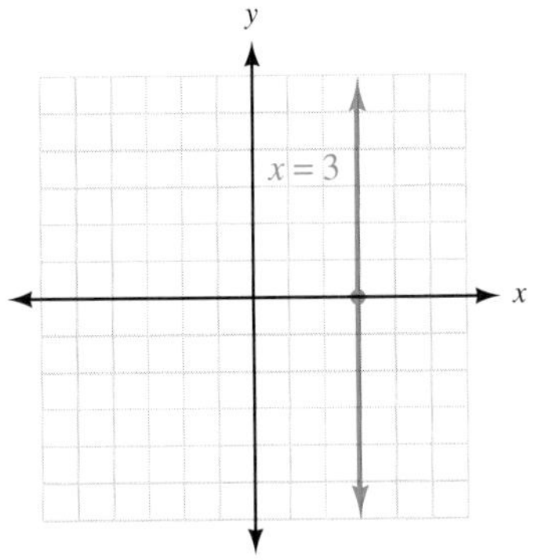

4. -2 **5.** 0 **6.** 3 **7.** 5 **8.** -5 **9.** 6 **10.** $y = 3x + 5$ **11.** $y = -2x$ **12.** $m = 3, b = -6$ **13.** $m = \frac{2}{3}, b = -3$ **14.** $y = 2x$ **15.** $y = -\frac{1}{3}x$ **16.** $y = 2x + 1$ **17.** $y = 7$ **18.** $y = -\frac{3}{2}x - \frac{17}{2}$ **19.** $y = 2x - 7$ **20.** $y = \frac{1}{3}x - \frac{2}{3}$

21.

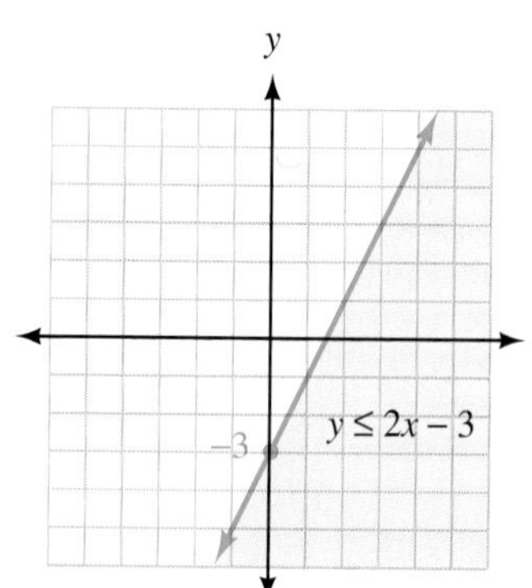

22.

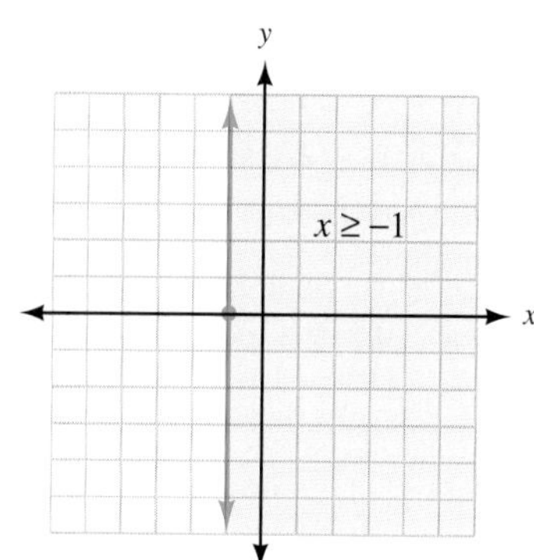

23. Domain = {2, 3, 4}; Range = {4, 3, 2}; a function

24. Domain = {6, −4, −2}; Range = {3, 0}; a function **25.** 0 **26.** 1 **27.** 1 **28.** $3a + 2$ **29.** 1 **30.** 31 **31.** 24 **32.** 6 **33.** 4 **34.** 25 **35.** 84 pounds **36.** 16 footcandles

CHAPTER 4

Problem Set 4.1

1. (4, 3)

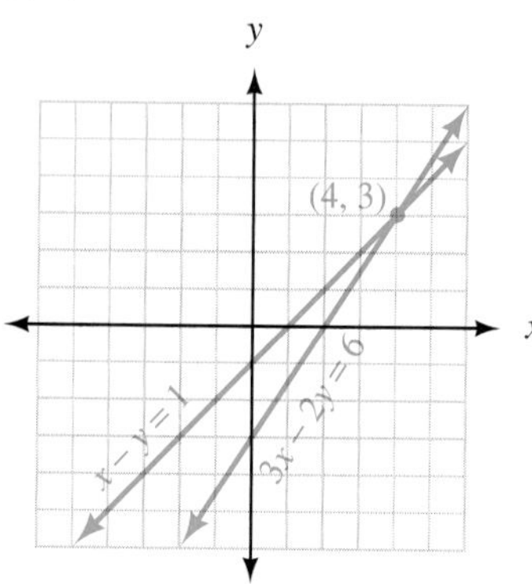

3. (−5, −6)

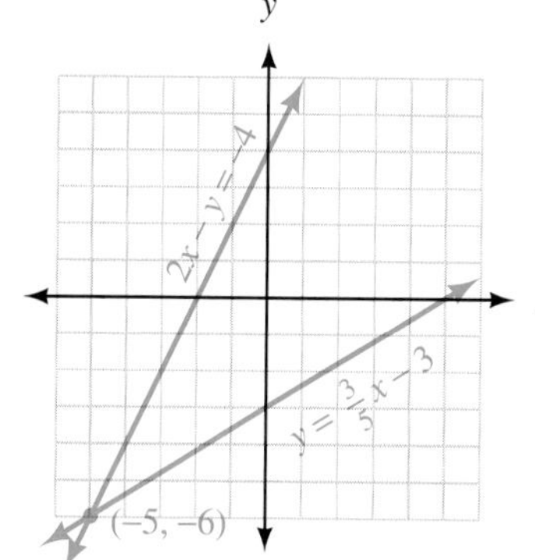

5. (4, 2)

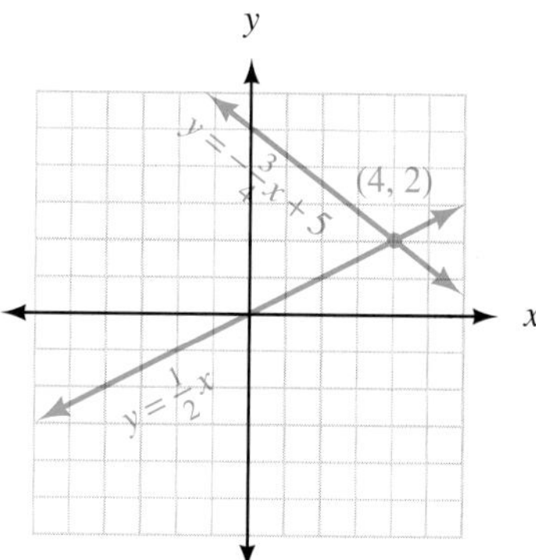

7. Lines are parallel; there is no solution.

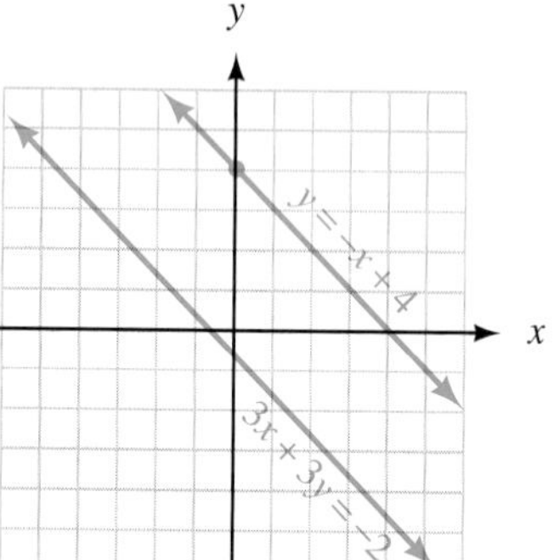

9. Lines coincide; any solution to one of the equations is a solution to the other

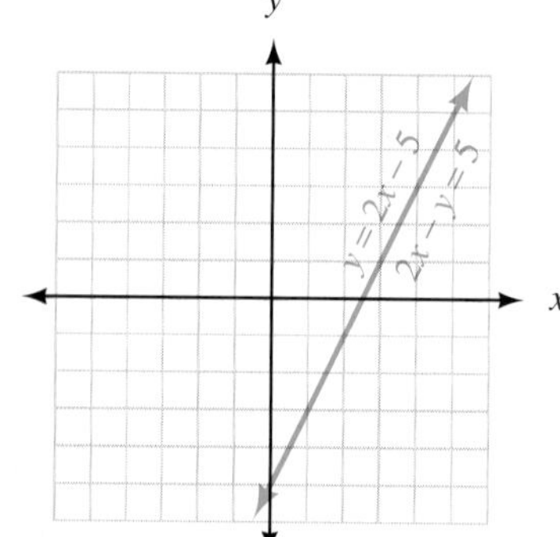

11. (2, 3) **13.** (1, 1) **15.** Lines coincide: $\{(x, y) \mid 3x - 2y = 6\}$ **17.** $(1, -\frac{1}{2})$ **19.** $(\frac{1}{2}, -3)$ **21.** $(-\frac{8}{3}, 5)$ **23.** (2, 2) **25.** Parallel lines; ∅ **27.** (12, 30) **29.** (10, 24) **31.** (3, −3) **33.** $(\frac{4}{3}, -2)$ **35.** $y = 5,\ z = 2$ **37.** (2, 4) **39.** Lines coincide: $\{(x, y) \mid 2x - y = 5\}$ **41.** Lines coincide: $\{(x, y) \mid x = \frac{3}{2}y\}$ **43.** $(-\frac{15}{43}, -\frac{27}{43})$ **45.** $(\frac{60}{43}, \frac{46}{43})$ **47.** $(\frac{9}{41}, -\frac{11}{41})$ **49.** $(-\frac{11}{7}, -\frac{20}{7})$ **51.** $(2, \frac{4}{3})$ **53.** (−12, −12) **55.** Parallel, ∅ **57.** $y = 5,\ z = 2$ **59.** $(\frac{3}{2}, \frac{3}{8})$ **61.** $(-4, -\frac{8}{3})$

63. a. $-y$ **b.** -2 **c.** -2 **d.**

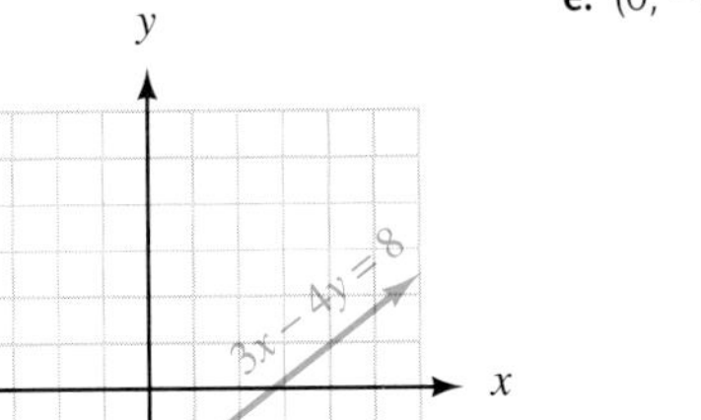

e. $(0, -2)$ **65.** $(6{,}000, 4{,}000)$ **67.** 2 **69.** $(4, 0)$ **71.** 3

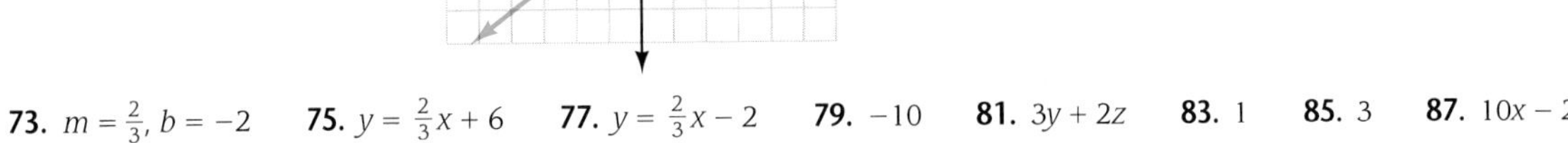

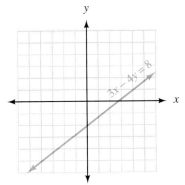

73. $m = \frac{2}{3}, b = -2$ **75.** $y = \frac{2}{3}x + 6$ **77.** $y = \frac{2}{3}x - 2$ **79.** -10 **81.** $3y + 2z$ **83.** 1 **85.** 3 **87.** $10x - 2z$

89. $9x + 3y - 6z$ **91.** $a = 3, b = -2$ **93.** $a = -\frac{8}{3}, b = 16$

Problem Set 4.2

1. $(1, 2, 1)$ **3.** $(2, 1, 3)$ **5.** $(2, 0, 1)$ **7.** $(\frac{1}{2}, \frac{2}{3}, -\frac{1}{2})$ **9.** No solution, inconsistent system **11.** $(4, -3, -5)$ **13.** No unique solution **15.** $(4, -5, -3)$ **17.** No unique solution **19.** $(\frac{1}{2}, 1, 2)$ **21.** $(\frac{1}{2}, \frac{1}{3}, \frac{1}{4})$ **23.** $(\frac{10}{3}, -\frac{5}{3}, -\frac{1}{3})$ **25.** $(\frac{1}{4}, -\frac{1}{3}, \frac{1}{8})$ **27.** $(6, 8, 12)$ **29.** $(3, 6, 4)$ **31.** $(1, 3, 1)$ **33.** $(-1, 2, -2)$ **35.** 4 amp, 3 amp, 1 amp **37.** 147 **39.** 50 **41.** 100 **43.** -2 **45.** 11 **47.** -14 **49.** -4 **51.** $(1, 2, 3, 4)$

Problem Set 4.3

1. 3 **3.** 5 **5.** -1 **7.** 0 **9.** 2 **11.** -3 **13.** -2 **15.** -3 **17.** 3 **19.** 0 **21.** 3 **23.** 8 **25.** 6 **27.** -228

29. $\begin{vmatrix} y & x \\ m & 1 \end{vmatrix} = y - mx = b; y = mx + b$ **31. a.** $y = 0.3x + 3.4$ **b.** $y = 0.3(2) + 3.4$; \$4 billion

33. domain $= \{1, 3, 4\}$; range $= \{2, 4\}$; is a function **35.** domain $= \{1, 2, 3\}$; range $= \{1, 2, 3\}$; is a function **37.** is a function **39.** not a function **41.** -10 **43.** 3 **45.** $-\frac{5}{11}$ **47.** 13 **49.** 22 **51.** 13 **53.** 4 **55.** 4

Problem Set 4.4

1. $(3, 1)$ **3.** Lines are parallel; $\varnothing$ **5.** $(-\frac{15}{43}, -\frac{27}{43})$ **7.** $(\frac{60}{43}, \frac{46}{43})$ **9.** $(3, -1, 2)$ **11.** $(\frac{1}{2}, \frac{5}{2}, 1)$ **13.** No unique solution **15.** $(-\frac{10}{91}, -\frac{9}{13}, \frac{107}{91})$ **17.** $(\frac{71}{13}, -\frac{12}{13}, \frac{24}{13})$ **19.** $(3, 1, 2)$ **21.** $x = 50$ items **23.** 3 **25.** 0 **27.** 1 **29.** 3 **31.** $3x + 2$

33. $-\frac{160}{9}$ **35.** 320 **37.** $2x + 5y$ **39.** 6 **41.** $y = -1, z = 5$ **43.** $\begin{aligned} x + 2y &= 1 \\ 3x + 4y &= 0 \end{aligned}$

Problem Set 4.5

1. $y = 2x + 3, x + y = 18$; the two numbers are 5 and 13. **3.** 10, 16 **5.** 1, 3, 4

7. Let $x =$ the number of adult tickets and $y =$ the number of children's tickets.

$x + y = 925$ 225 adult and 700 children's tickets

$2x + y = 1{,}150$

9. Let $x =$ the amount invested at 6% and $y =$ the amount invested at 7%.

$x + y = 20{,}000$ He has \$12,000 at 6% and \$8,000 at 7%.

$0.06x + 0.07y = 1{,}280$

11. \$4,000 at 6%, \$8,000 at 7.5% **13.** \$200 at 6%, \$1,400 at 8%, \$600 at 9% **15.** 3 gallons of 50%, 6 gallons of 20% **17.** 5 gallons of 20%, 10 gallons of 14% **19.** 12.5 lbs of 40%, 37.5 lbs of 60% **21.** boat; 9 mph; current: 3 mph **23.** airplane: 270 mph; wind: 30 mph **25.** 12 nickels, 8 dimes **27.** 3 nickels, 3 dimes, 3 quarters **29.** 110 nickels **31.** $x = -200p + 700$; 100 items **33.** $h = -16t^2 + 64t + 80$

35.

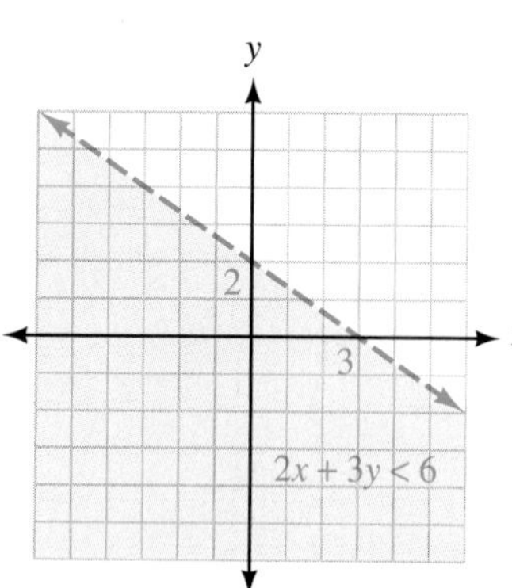

37.

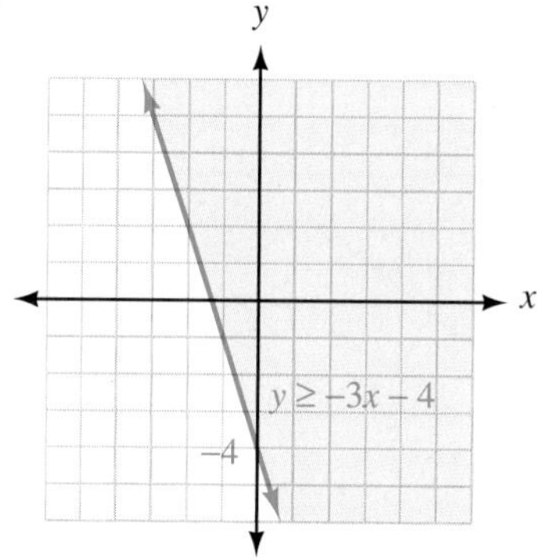

39.

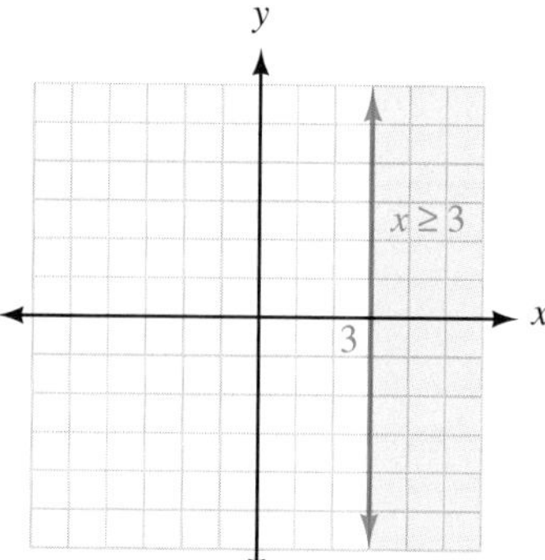

41. No **43.** (4, 0) **45.** $x > 435$

47.

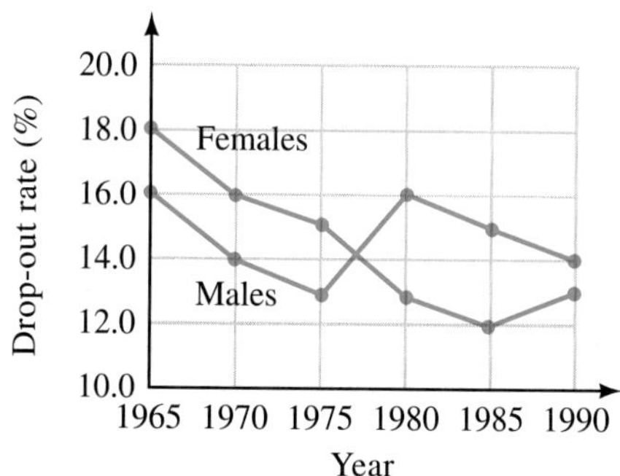

49. **a.** $M = 0.6x - 1172$ **b.** $F = -0.4x + 805$ **c.** year 1977

Problem Set 4.6

1.

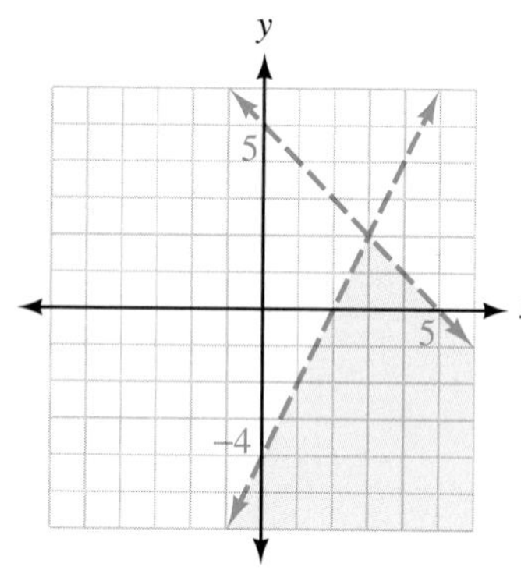

3.

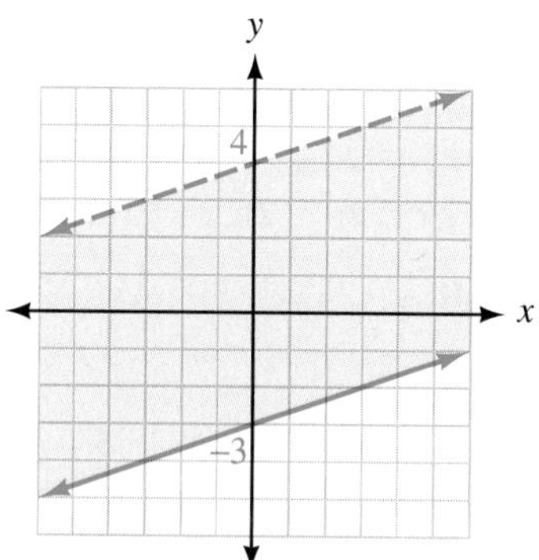

5.

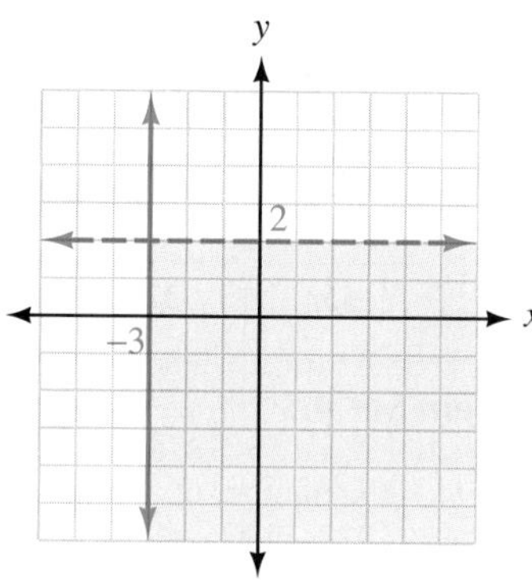

7.

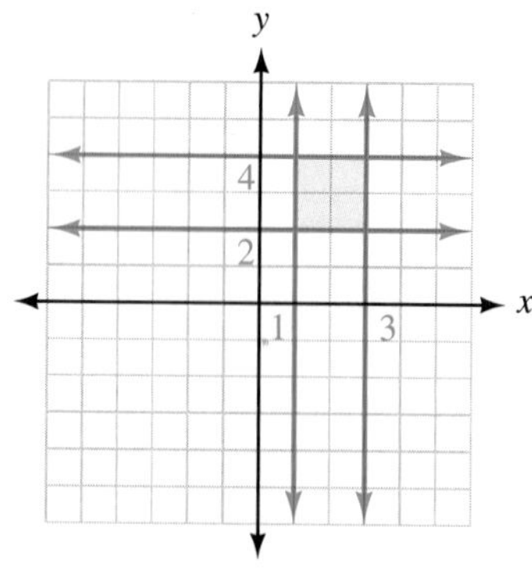

9.

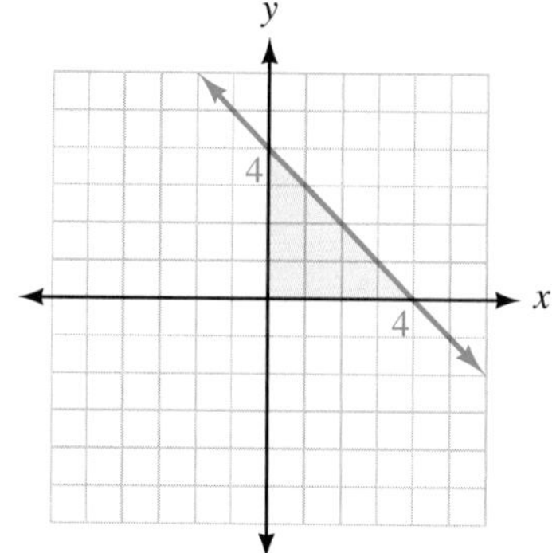

11.

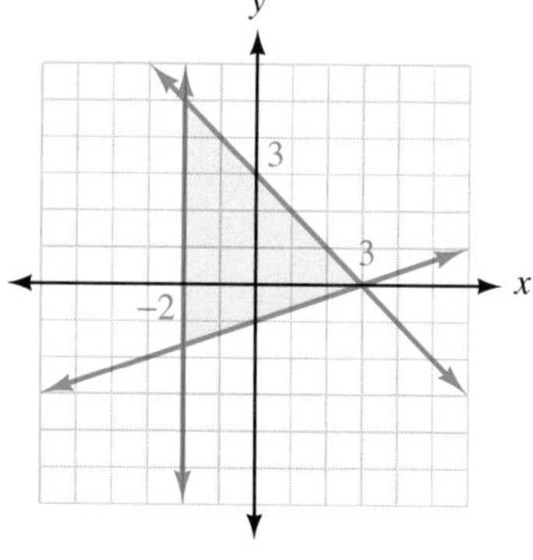

13.

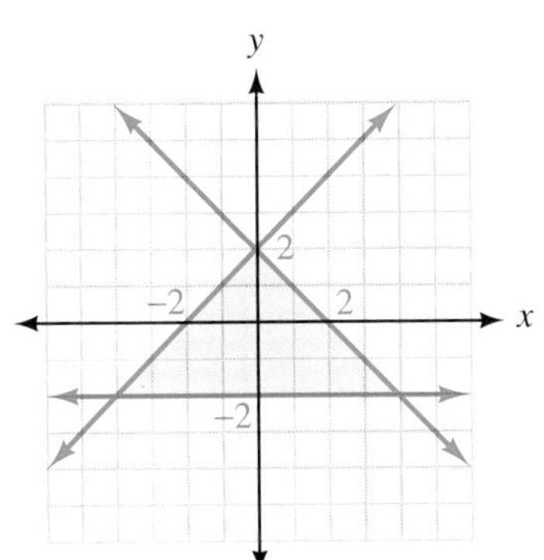

15.

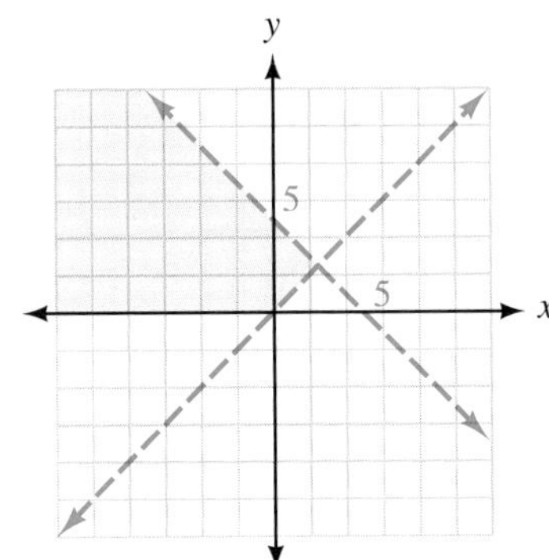

17.

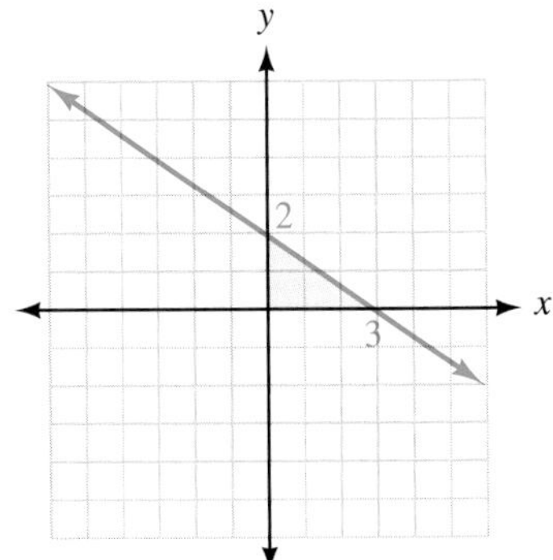

19. $x + y \leq 4$
$-x + y < 4$

21. $x + y \geq 4$
$-x + y < 4$

23. a. $0.55x + 0.65y \leq 40$
$x \geq 2y$
$x > 15$
$y \geq 0$

b. 10 65-cent stamps

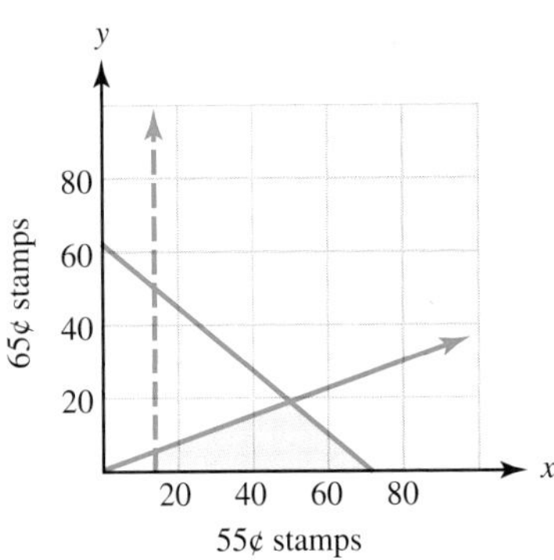
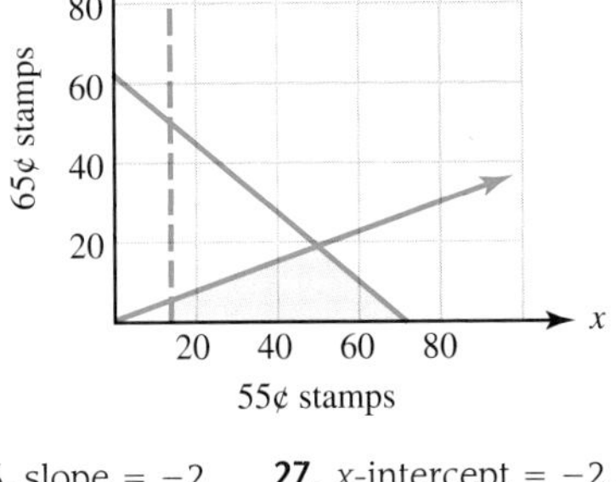

25. x-intercept $= 3$, y-intercept $= 6$, slope $= -2$ **27.** x-intercept $= -2$, no y-intercept, no slope

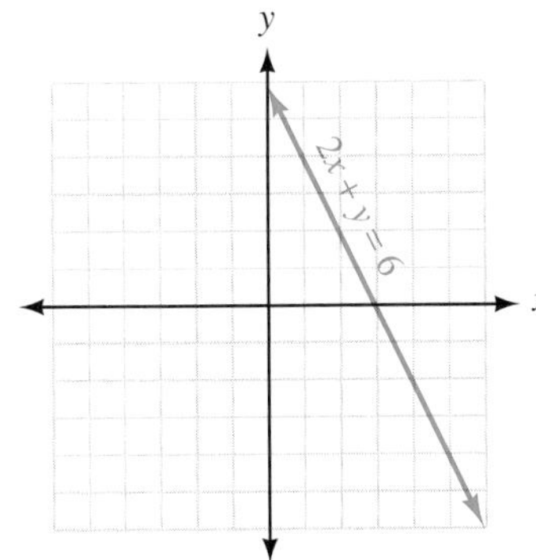

29. $y = -\frac{3}{7}x + \frac{5}{7}$ **31.** $x = 4$ **33.** Domain = All real numbers; Range = $\{y \mid y \geq -9\}$; a function **35.** -4 **37.** 4 **39.** $\frac{81}{4}$

CHAPTER 4 REVIEW TEST

1. $(6, -2)$ **2.** $(2, -4)$ **3.** No solution (parallel lines) **4.** $(1, -1)$ **5.** $(0, 1)$ **6.** No unique solution (lines coincide) **7.** $(3, 1)$ **8.** $(\frac{3}{2}, 0)$ **9.** $(3, 5)$ **10.** $(4, 8)$ **11.** $(3, 12)$ **12.** $(3, -2)$ **13.** $(\frac{3}{2}, \frac{1}{2})$ **14.** $(4, 1)$ **15.** $(6, -2)$ **16.** $(7, -4)$ **17.** $(-5, 3)$ **18.** No solution (parallel lines) **19.** $(3, -1, 4)$ **20.** $(2, \frac{1}{2}, -3)$ **21.** $(-1, \frac{1}{2}, \frac{3}{2})$ **22.** $(2, \frac{1}{3}, \frac{2}{3})$ **23.** No unique solution (dependent system) **24.** No unique solution (dependent system) **25.** $(2, -1, 4)$ **26.** No unique solution (dependent system) **27.** 23 **28.** -3 **29.** -3 **30.** -16 **31.** 36 **32.** -2 **33.** $\frac{4}{7}$ **34.** -4 **35.** $(\frac{7}{29}, -\frac{19}{29})$ **36.** $(\frac{34}{41}, -\frac{18}{41})$ **37.** No unique solution (lines coincide) **38.** $(\frac{17}{27}, -\frac{11}{81})$ **39.** $(-\frac{50}{33}, -\frac{23}{33})$ **40.** $(\frac{11}{14}, -\frac{1}{14}, \frac{11}{14})$ **41.** $(\frac{113}{46}, \frac{59}{23}, -\frac{7}{23})$ **42.** No solution (inconsistent) **43.** 47 adult, 80 children **44.** 12 dimes, 8 quarters **45.** \$5,000 at 12%, \$7,000 at 15% **46.** Boat: 12 mph; current: 2 mph

47.

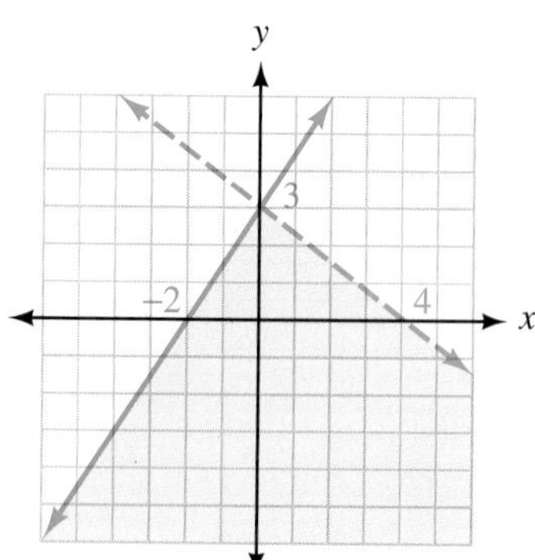

48.

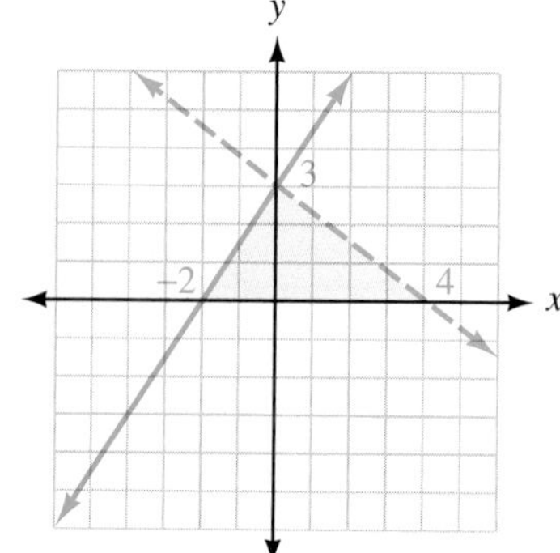

49.

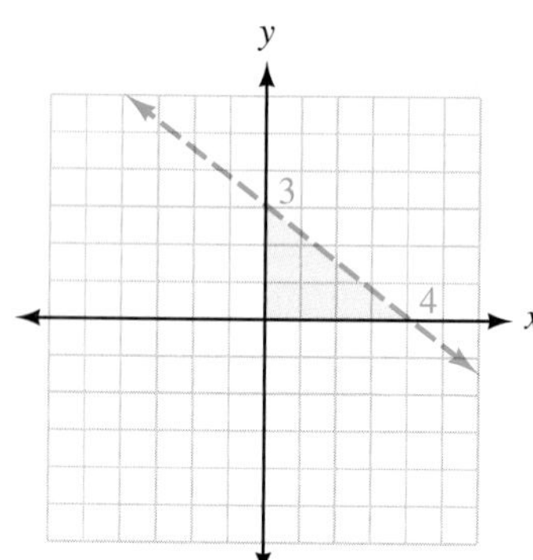

50.

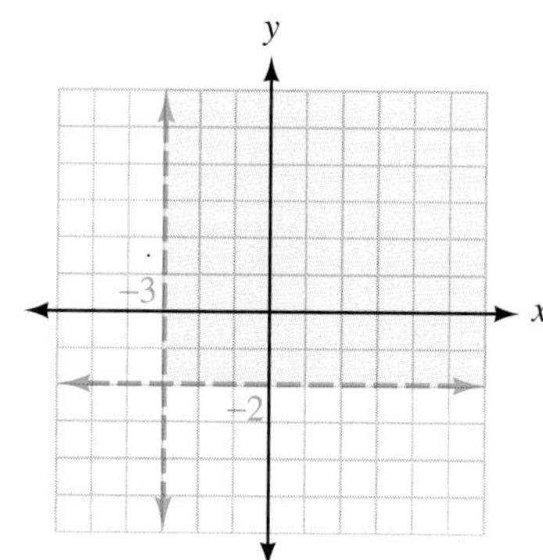

CHAPTER 5

Problem Set 5.1

1. 16 **3.** −16 **5.** −0.027 **7.** 32 **9.** $\frac{1}{8}$ **11.** $\frac{25}{36}$ **13.** x^9 **15.** 64 **17.** $-\frac{8}{27}x^6$ **19.** $-6a^6$ **21.** $\frac{1}{9}$ **23.** $-\frac{1}{32}$ **25.** $\frac{16}{9}$ **27.** 17 **29.** x^3 **31.** $\frac{a^6}{b^{15}}$ **33.** $\frac{8}{125y^{18}}$ **35.** $\frac{1}{5}$ **37.** $\frac{24a^{12}c^6}{b^3}$ **39.** $\frac{8x^{22}}{81y^{23}}$ **41.** $\frac{1}{x^{10}}$ **43.** a^{10} **45.** $\frac{1}{t^6}$ **47.** x^{12} **49.** x^{18} **51.** $\frac{1}{x^{22}}$ **53.** $\frac{a^3b^7}{4}$ **55.** $\frac{y^{38}}{x^{16}}$ **57.** $\frac{16y^{16}}{x^8}$ **59.** x^4y^6 **61.** x^2y **63.** $3ab^2$ **65.** $2a$ **67.** $4xy^4$ **69. a.** 32 **b.** 64 **c.** 32 **d.** 64 **71. a.** 8 **b.** 8 **c.** $\frac{1}{16}$ **d.** $\frac{1}{16}$ **73. a.** $\frac{1}{7}$ **b.** $\frac{1}{11}$ **c.** $\frac{1}{2x}$ **d.** $\frac{1}{8x^2}$ **75.** 2^n **77.** r^3 **79.** 3.78×10^5 **81.** 4.9×10^3 **83.** 3.7×10^{-4} **85.** 4.95×10^{-3} **87.** 5,340 **89.** 7,800,000 **91.** 0.00344 **93.** 0.49 **95.** 8×10^4 **97.** 2×10^9 **99.** 2.5×10^{-6} **101.** $80x^3y^6$ **103.** $72x^3y^3$ **105.** $15xy$ **107.** $12x^7y^6$ **109.** $54a^6b^2c^4$ **111.** $2x^3$ **113.** $4x$ **115.** $-4xy^4$ **117.** $-2x^2y^2$ **119.** 50 **121.** 7.9×10^{-5} **123.** 6.3×10^8 **125. a.** 7.3×10^4 **b.** 1.94×10^5 **c.** 1.9×10^6 **127.** 1.003×10^{19} miles **129.** 4.22×10^{11} **131.** 1.5×10^{13} **133.** (1, 2) **135.** $(\frac{5}{2}, -1)$ **137.** (0, 3) **139.** (3, 4) **141.** $5x$ **143.** $8x^2$ **145.** $2x^3$ **147.** −5 **149.** $-2x + 3$ **151.** $9x + 6$ **153.** 1,200 **155.** $\frac{1}{x^3}$ **157.** y^3 **159.** x^5

Problem Set 5.2

1. Trinomial, 2, 5 **3.** Binomial, 1, 3 **5.** Trinomial, 2, 8 **7.** Polynomial, 3, 4 **9.** Monomial, 0, $-\frac{3}{4}$ **11.** Trinomial, 3, 6 **13.** $7x + 1$ **15.** $2x^2 + 7x - 15$ **17.** $12a^2 - 7ab - 10b^2$ **19.** $x^2 - 13x + 3$ **21.** $\frac{1}{4}x^2 - \frac{7}{12}x - \frac{1}{4}$ **23.** $-y^3 - y^2 - 4y + 7$ **25.** $2x^3 + x^2 - 3x - 17$ **27.** $\frac{1}{14}x^2 + \frac{1}{7}xy + \frac{5}{7}y^2$ **29.** $-3a^3 + 6a^2b - 5ab^2$ **31.** $-3x$ **33.** $3x^2 - 12xy$ **35.** $17x^5 - 12$ **37.** $14a^2 - 2ab + 8b^2$ **39.** $2 - x$ **41.** $10x - 5$ **43.** $9x - 35$ **45.** $9y - 4x$ **47.** $9a + 2$ **49.** −2 **51. a.** 208 **b.** 103 **53. a.** 104 **b.** −15 **55. a.** 110 **b.** −120 **57. a.** −5,000 **b.** 3,000 **59. a.** $3x - 3a$ **b.** $2x - 2a$ **61.** 240 feet for both **63.** $P(x) = -300 + 40x - 0.5x^2$; $P(60) = \$300$ **65.** $P(x) = -800 + 3.5x - 0.002x^2$; $P(1{,}000) = \$700$ **67.** $x - 5$ **69.** $x - 7$ **71.** $5x - 1$ **73.** $x - 1$ **75.** $3x + 8y$ **77.** $2x^2 + 7x - 15$ **79.** $6x^3 - 11x^2y + 11xy^2 - 12y^3$ **81.** $-12x^4$ **83.** $20x^5$ **85.** a^6 **87.** 650 **89. a.** 5 **b.** −4 **c.** −3 **d.** 3 **e.** 4 **f.** 0 **g.** −4 **h.** 4

Problem Set 5.3

1. $12x^3 - 10x^2 + 8x$ **3.** $-3a^5 + 18a^4 - 21a^2$ **5.** $2a^5b - 2a^3b^2 + 2a^2b^4$ **7.** $x^2 - 2x - 15$ **9.** $6x^4 - 19x^2 + 15$ **11.** $x^3 + 9x^2 + 23x + 15$ **13.** $a^3 - b^3$ **15.** $8x^3 + y^3$ **17.** $2a^3 - a^2b - ab^2 - 3b^3$ **19.** $x^2 + x - 6$ **21.** $6a^2 + 13a + 6$ **23.** $20 - 2t - 6t^2$ **25.** $x^6 - 2x^3 - 15$ **27.** $20x^2 - 9xy - 18y^2$ **29.** $18t^2 - \frac{2}{9}$ **31.** $25x^2 + 20xy + 4y^2$ **33.** $25 - 30t^3 + 9t^6$ **35.** $4a^2 - 9b^2$ **37.** $9r^4 - 49s^2$ **39.** $y^2 + 3y + \frac{9}{4}$ **41.** $a^2 - a + \frac{1}{4}$ **43.** $x^2 + \frac{1}{2}x + \frac{1}{16}$ **45.** $t^2 + \frac{2}{3}t + \frac{1}{9}$ **47.** $\frac{1}{9}x^2 - \frac{4}{25}$ **49.** $x^3 - 6x^2 + 12x - 8$ **51.** $x^3 - \frac{3}{2}x^2 + \frac{3}{4}x - \frac{1}{8}$ **53.** $3x^3 - 18x^2 + 33x - 18$ **55.** $a^2b^2 + b^2 + 8a^2 + 8$ **57.** $3x^2 + 12x + 14$ **59.** $24x$ **61.** $x^2 + 4x - 5$ **63.** $4a^2 - 30a + 56$ **65.** $32a^2 + 20a - 18$ **67.** $(x + y)^2 + (x + y) - 20 = x^2 + 2xy + y^2 + x + y - 20$ **69.** $2^4 - 3^4 = -65$; $(2 - 3)^4 = 1$; $(2^2 + 3^2)(2 + 3)(2 - 3) = -65$ **71.** $R(p) = 900p - 300p^2$; $R(x) = 3x - \frac{x^2}{300}$; \$672 **73.** $R(p) = 350p - 10p^2$; $R(x) = 35x - \frac{x^2}{10}$; \$1,852.50 **75.** $P(x) = -500 + 30x - \frac{x^2}{10}$; $P(60) = \$940$ **77.** $A = 100 + 400r + 600r^2 + 400r^3 + 100r^4$ **79.** (1, 2, 3) **81.** (1, 3, 1) **83.** $8a^2$ **85.** -48 **87.** $2a^3b$ **89.** $-3b^2$ **91.** $-y^4$ **93.** $x^{2n} - 5x^n + 6$ **95.** $10x^{2n} + 13x^n - 3$ **97.** $x^{2n} + 10x^n + 25$ **99.** $x^{3n} + 1$

Problem Set 5.4

1. $5x^2(2x - 3)$ **3.** $9y^3(y^3 + 2)$ **5.** $3ab(3a - 2b)$ **7.** $7xy^2(3y^2 + x)$ **9.** $3(a^2 - 7a + 10)$ **11.** $4x(x^2 - 4x - 5)$ **13.** $10x^2y^2(x^2 + 2xy - 3y^2)$ **15.** $xy(-x + y - xy)$ **17.** $2xy^2z(2x^2 - 4xz + 3z^2)$ **19.** $5abc(4abc - 6b + 5ac)$ **21.** $(a - 2b)(5x - 3y)$ **23.** $3(x + y)^2(x^2 - 2y^2)$ **25.** $(x + 5)(2x^2 + 7x + 6)$ **27.** $(x + 1)(3y + 2a)$ **29.** $(xy + 1)(x + 3)$ **31.** $(x - 2)(3y^2 + 4)$ **33.** $(x - a)(x - b)$ **35.** $(b + 5)(a - 1)$ **37.** $(b^2 + 1)(a^4 - 5)$ **39.** $(x + 3)(x^2 - 4)$ **41.** $(x + 2)(x^2 - 25)$ **43.** $(2x + 3)(x^2 - 4)$ **45.** $(x + 3)(4x^2 - 9)$ **47.** 6 **49.** $P(1 + r) + P(1 + r)r = (1 + r)(P + Pr) = (1 + r)P(1 + r) = P(1 + r)^2$ **51.** $p = 11.5 - 0.05x$; \$5.25 **53.** \$28.50 **55.** 36 **57.** 0 **59.** $3x^2(x^2 - 3xy - 6y^2)$ **61.** $(x - 3)(2x^2 - 4x - 3)$ **63.** $3x^2 + 5x - 2$ **65.** $3x^2 - 5x + 2$ **67.** $x^2 + 5x + 6$ **69.** $6y^2 + y - 35$ **71.** $20 - 19a + 3a^2$ **73.**

Two Numbers a and b	Their Product ab	Their Sum $a + b$
1, −24	−24	−23
−1, 24	−24	23
2, −12	−24	−10
−2, 12	−24	10
3, −8	−24	−5
−3, 8	−24	5
4, −6	−24	−2
−4, 6	−24	2

Problem Set 5.5

1. $(x + 3)(x + 4)$ **3.** $(x + 3)(x - 4)$ **5.** $(y + 3)(y - 2)$ **7.** $(2 - x)(8 + x)$ **9.** $(2 + x)(6 + x)$ **11.** $3(a - 2)(a - 5)$ **13.** $4x(x - 5)(x + 1)$ **15.** $(x + 2y)(x + y)$ **17.** $(a + 6b)(a - 3b)$ **19.** $(x - 8a)(x + 6a)$ **21.** $(x - 6b)^2$ **23.** $3(x - 3y)(x + y)$ **25.** $2a^3(a^2 + 2ab + 2b^2)$ **27.** $10x^2y^2(x + 3y)(x - y)$ **29.** $(2x - 3)(x + 5)$ **31.** $(2x - 5)(x + 3)$ **33.** $(2x - 3)(x - 5)$ **35.** $(2x - 5)(x - 3)$ **37.** Prime **39.** $(2 + 3a)(1 + 2a)$ **41.** $15(4y + 3)(y - 1)$ **43.** $x^2(3x - 2)(2x + 1)$ **45.** $10r(2r - 3)^2$ **47.** $(4x + y)(x - 3y)$ **49.** $(2x - 3a)(5x + 6a)$ **51.** $(3a + 4b)(6a - 7b)$ **53.** $200(1 + 2t)(3 - 2t)$ **55.** $y^2(3y - 2)(3y + 5)$ **57.** $2a^2(3 + 2a)(4 - 3a)$ **59.** $2x^2y^2(4x + 3y)(x - y)$ **61.** $100(3x^2 + 1)(x^2 + 3)$ **63.** $(5a^2 + 3)(4a^2 + 5)$ **65.** $3(3 + 4r^2)(1 - r^2)$ **67.** $(x + 5)(2x + 3)(x + 2)$ **69.** $(2x + 3)(x + 5)(x + 2)$ **71.** $(3x - 5)(x + 4)(x - 3)$ **73.** $(2x - 3)(3x - 4)(x - 2)$ **75.** $(3x - 5)(4x + 9)(x + 3)$ **77.** $(3x + 2)(2x - 5)(5x - 2)$ **79.** $(5x + 3)(4x + 7)(2x + 3)$ **81.** $9x^2 - 25y^2$ **83.** $a + 250$ **85.** $12x - 35$ **87.** $9x + 8$ **89.** $7x + 8$ **91.** $y = 2(2x - 1)(x + 5)$; $y = 0$ when $x = \frac{1}{2}$ or $x = -5$, $y = 42$ when $x = 2$ **93.** $4x^2 - 9$ **95.** $4x^2 - 12x + 9$ **97.** $8x^3 - 27$ **99.** $\frac{5}{8}$ **101.** x^3 **103.** $4x^2$ **105.** $\frac{1}{2}$ **107.** x^2 **109.** $3x$ **111.** $2y$ **113.** $(2x^3 + 5y^2)(4x^3 + 3y^2)$ **115.** $(3x - 5)(x + 100)$ **117.** $(\frac{1}{4}x + 1)(\frac{1}{2}x + 2)$ **119.** $(2x + 0.5)(x + 0.5)$

Problem Set 5.6

1. $(x - 3)^2$ **3.** $(a - 6)^2$ **5.** $(5 - t)^2$ **7.** $(\frac{1}{3}x + 3)^2$ **9.** $(2y^2 - 3)^2$ **11.** $(4a + 5b)^2$ **13.** $(\frac{1}{5} + \frac{1}{4}t^2)^2$ **15.** $(y + \frac{3}{2})^2$ **17.** $(a - \frac{1}{2})^2$ **19.** $(x - \frac{1}{4})^2$ **21.** $(t + \frac{1}{3})^2$ **23.** $4(2x - 3)^2$ **25.** $3a(5a + 1)^2$ **27.** $(x + 2 + 3)^2 = (x + 5)^2$ **29.** $(x + 3)(x - 3)$ **31.** $(7x + 8y)(7x - 8y)$ **33.** $(2a + \frac{1}{2})(2a - \frac{1}{2})$ **35.** $(x + \frac{3}{5})(x - \frac{3}{5})$ **37.** $(3x + 4y)(3x - 4y)$ **39.** $10(5 + t)(5 - t)$ **41.** $(x^2 + 9)(x + 3)(x - 3)$ **43.** $(3x^3 + 1)(3x^3 - 1)$ **45.** $(4a^2 + 9)(2a + 3)(2a - 3)$ **47.** $(\frac{1}{9} + \frac{y^2}{4})(\frac{1}{3} + \frac{y}{2})(\frac{1}{3} - \frac{y}{2})$ **49.** $(x - y)(x + y)(x^2 + xy + y^2)(x^2 - xy + y^2)$ **51.** $2a(a - 2)(a + 2)(a^2 + 2a + 4)(a^2 - 2a + 4)$ **53.** $(x + 1)(x - 5)$ **55.** $y(y + 8)$

57. $(x-5+y)(x-5-y)$ **59.** $(a+4+b)(a+4-b)$ **61.** $(x+y+a)(x+y-a)$ **63.** $(x+3)(x+2)(x-2)$ **65.** $(x+2)(x+5)(x-5)$ **67.** $(2x+3)(x+2)(x-2)$ **69.** $(x+3)(2x+3)(2x-3)$ **71.** $(2x-15)(2x+5)$ **73.** $(a-3-4b)(a-3+4b)$ **75.** $(a-3-4b)(a-3+4b)$ **77.** $(x-3)^2(x+4)$ **79.** $(x-y)(x^2+xy+y^2)$ **81.** $(a+2)(a^2-2a+4)$ **83.** $(3+x)(9-3x+x^2)$ **85.** $(y-1)(y^2+y+1)$ **87.** $10(r-5)(r^2+5r+25)$ **89.** $(4+3a)(16-12a+9a^2)$ **91.** $(2x-3y)(4x^2+6xy+9y^2)$ **93.** $(t+\frac{1}{3})(t^2-\frac{1}{3}t+\frac{1}{9})$ **95.** $(3x-\frac{1}{3})(9x^2+x+\frac{1}{9})$ **97.** $(4a+5b)(16a^2-20ab+25b^2)$ **99.** 30 and −30 **101.** $(-\frac{15}{43}, -\frac{27}{43})$ **103.** (1, 3, 1) **105.** $y(y^2+25)$ **107.** $2ab^3(b^2+4b+1)$ **109.** $(2x-3)(2x+a)$ **111.** $(x+2)(x-2)$ **113.** $(x-3)^2$ **115.** $(3a-4)(2a-1)$ **117.** $(x+2)(x^2-2x+4)$ **119.** $(a-b+3)(a+b-3)$ **121.** $(x-y-8)(x+y+2)$ **123.** $k=144$ **125.** $k=\pm 126$

Problem Set 5.7

1. $(x+9)(x-9)$ **3.** $(x-3)(x+5)$ **5.** $(x+2)(x+3)^2$ **7.** $(x^2+2)(y^2+1)$ **9.** $2ab(a^2+3a+1)$ **11.** Does not factor: prime **13.** $3(2a+5)(2a-5)$ **15.** $(3x-2y)^2$ **17.** $(5-t)^2$ **19.** $4x(x^2+4y^2)$ **21.** $2y(y+5)^2$ **23.** $a^4(a+2b)(a^2-2ab+4b^2)$ **25.** $(t+3+x)(t+3-x)$ **27.** $(x+5)(x+3)(x-3)$ **29.** $5(a+b)^2$ **31.** Does not factor: prime **33.** $3(x+2y)(x+3y)$ **35.** $(3a+\frac{1}{3})^2$ **37.** $(x-3)(x-7)^2$ **39.** $(x+8)(x-8)$ **41.** $(2-5x)(4+3x)$ **43.** $a^5(7a+3)(7a-3)$ **45.** $(r+\frac{1}{5})(r-\frac{1}{5})$ **47.** Does not factor: prime **49.** $100(x-3)(x+2)$ **51.** $a(5a+3)(5a+1)$ **53.** $(3x^2+1)(x^2-5)$ **55.** $3a^2b(2a-1)(4a^2+2a+1)$ **57.** $(4-r)(16+4r+r^2)$ **59.** $5x^2(2x+3)(2x-3)$ **61.** $100(2t+3)(2t-3)$ **63.** $2x^3(4x-5)(2x-3)$ **65.** $(y+1)(y-1)(y^2-y+1)(y^2+y+1)$ **67.** $2(5+a)(5-a)$ **69.** $3x^2y^2(2x+3y)^2$ **71.** $(x-2+y)(x-2-y)$ **73.** $(a-\frac{2}{3}b)^2$ **75.** $(x-\frac{2}{5}y)^2$ **77.** $(a-\frac{5}{6}b)^2$ **79.** $(x-\frac{4}{5}y)^2$ **81.** $(2x-3)(x-5)(x+2)$ **83.** $(x-4)^3(x-3)$ **85.** $2(y-3)(y^2+3y+9)$ **87.** $2(a-4b)(a^2+4ab+16b^2)$ **89.** $2(x+6y)(x^2-6xy+36y^2)$ **91.** 60 geese; 48 ducks **93.** 150 oranges; 144 apples **95.** $2x^2+2x+1$ **97.** t^2-4t+3 **99.** $(x-6)(x+4)$ **101.** $x(2x+1)(x-3)$ **103.** $(x+2)(x-3)(x+3)$ **105.** 6 **107.** $-\frac{1}{2}$

Problem Set 5.8

1. 6, −1 **3.** 0, 2, 3 **5.** $\frac{1}{3}, -4$ **7.** $\frac{2}{3}, \frac{3}{2}$ $\frac{4}{3}$ **9.** 5, −5 **11.** 0, −3, 7 **13.** $-4, \frac{5}{2}$ **15.** $0, \frac{4}{3}$ **17.** $-\frac{1}{5}, \frac{1}{3}$ **19.** $0, -\frac{4}{3},$ **21.** −10, 0 **23.** −5, 1 **25.** 1, 2 **27.** −2, 3 **29.** $-2, \frac{1}{4}$ **31.** −3, −2, 2 **33.** −2, −5, 5 **35.** $-\frac{3}{2}, -2, 2$ **37.** $-3, -\frac{3}{2}, \frac{3}{2}$ **39.** $-2, \frac{5}{3}$ **41.** −1, 9 **43.** 0, −3 **45.** −4, −2 **47.** −4, 2 **49.** $-\frac{3}{2}$ **51.** −2, 8 **53.** −3 **55.** −7, 1 **57.** 0, 5 **59.** −1, 8 **61. a.** $\frac{25}{9}$ **b.** $-\frac{5}{3}, \frac{5}{3}$ **c.** −3, 3 **d.** $\frac{5}{3}$ **63.** 3 hours **65.** −5, −4 or 4, 5 **67.** 24 feet **69.** 6, 8, 10 **71.** 2 feet, 8 feet **73.** 18 inches, 4 inches **75.** 0 and 2 seconds **77.** 1 and 2 seconds **79.** 0 and $\frac{3}{2}$ seconds **81.** 2 and 3 seconds **83.** \$4 or \$8 **85.** \$7 or \$10 **87.** (1, 2) **89.** (15, 12) **91.** (3, −2, 1) **93.** \$4,000 at 5%, \$8,000 at 6%

95.

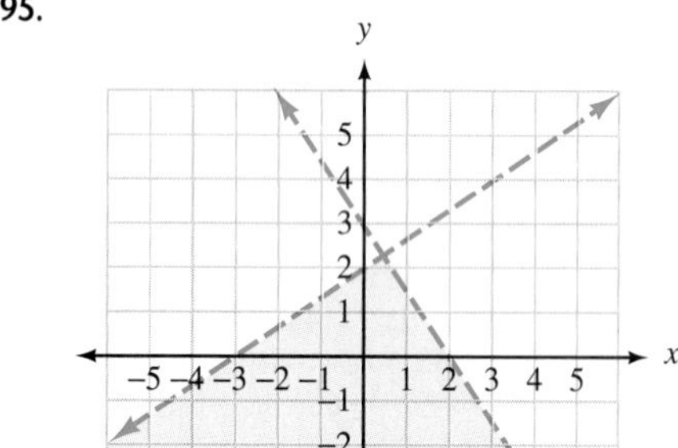

97.

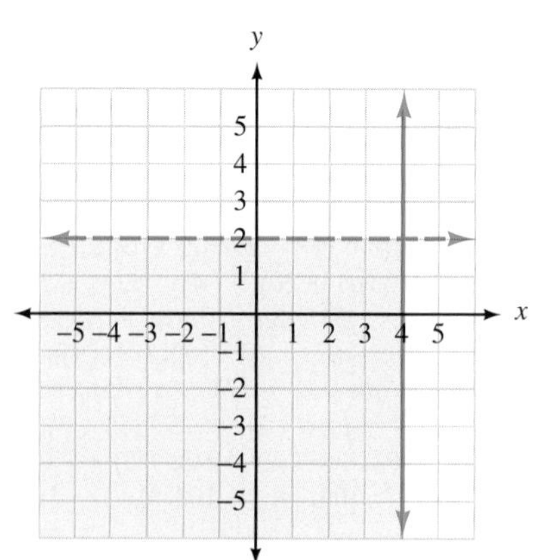

CHAPTER 5 REVIEW TEST

1. x^{10} **2.** $25x^6$ **3.** $-32x^{18}y^8$ **4.** $\frac{1}{8}$ **5.** $\frac{9}{4}$ **6.** $\frac{1}{2}$ **7.** 3.45×10^7 **8.** 3.57×10^{-3} **9.** 44,500 **10.** 0.000445 **11.** $\frac{1}{a^9}$ **12.** $\frac{x^{12}}{4}$ **13.** x^2 **14.** 8×10^{-2} **15.** 4×10^{-10} **16.** $2x^2-5x+7$ **17.** $2x^3-2x^2-2x-4$ **18.** x^2-2x-3 **19.** $30x+12$ **20.** 15 **21.** $12x^3-6x^2+3x$ **22.** $2a^4b^3+4a^3b^4+2a^2b^5$ **23.** $18-9y+y^2$ **24.** $6x^4+5x^2-4$ **25.** $2t^3-4t^2-6t$ **26.** x^3+27 **27.** $8x^3-27$ **28.** a^4-4a^2+4 **29.** $9x^2+30x+25$ **30.** $16x^2-24xy+9y^2$ **31.** $x^2-\frac{1}{9}$ **32.** $4a^2-b^2$ **33.** x^3-3x^2+3x-1

34. $x^{2m} - 4$ **35.** $3xy(2x^3 - 3y^3 + 6x^2y^2)$ **36.** $4(x + y)^2(x^2 - 2y^2)$ **37.** $(4x^2 + 5)(2 - y)$ **38.** $(1 - y)(1 + y)(x^3 + 8b^2)$ **39.** $(x - 2)(x - 3)$ **40.** $2x(x + 5)(x - 3)$ **41.** $(5a - 4b)(4a - 5b)$ **42.** $x^2(3x + 2)(2x - 5)$ **43.** $3y(4x + 5)(2x - 3)$ **44.** $(x^2 + 4)(x + 2)(x - 2)$ **45.** $3(a^2 + 3)^2$ **46.** $(a - 2)(a^2 + 2a + 4)$ **47.** $5x(x + 3y)^2$ **48.** $3ab(a - 3b)(a + 3b)$ **49.** $(x - 5 + y)(x - 5 - y)$ **50.** $(6 - 5a)(6 + 5a)$ **51.** $(x + 3)(x - 3)(x + 4)$ **52.** $-3, -2$ **53.** $-\frac{1}{2}, \frac{4}{5}$ **54.** $-\frac{5}{3}, \frac{5}{3}$ **55.** $0, -2$ **56.** $-3, 6$ **57.** $-4, 3, -3$ **58.** $-10, -8$ or $8, 10$ **59.** $-5, -4$ or $4, 5$ **60.** $3, 4, 5$ **61.** $6, 8, 10$

CHAPTER 6

Problem Set 6.1

1. a. $\frac{1}{5}$ **b.** $\frac{1}{x - 3}, x \neq \pm 3$ **c.** $\frac{x}{x + 3}, x \neq \pm 3$ **d.** $\frac{x^2 + 3x + 9}{x + 3}, x \neq \pm 3$ **e.** $\frac{x^2 + 3x + 9}{x^2}, x \neq 0, x \neq 3$

3. $h(0) = -3, h(-3) = 3, h(3) = 0, h(-1)$ is undefined, $h(1) = -1$ **5.** $\{x \mid x \neq 1\}$ **7.** $\{t \mid t \neq 4, t \neq -4\}$ **9.** $\{x \mid x \neq 5\}$ **11.** All real numbers **13.** $\{x \mid x \neq 0\}$ **15.** $\{x \mid x \neq -4, x \neq 5\}$ **17.** $\frac{x - 4}{6}$ **19.** $\frac{4x - 3y}{x(x + y)}$ **21.** $(a^2 + 9)(a + 3)$ **23.** $\frac{a - 6}{a + 6}$ **25.** $\frac{2y + 3}{y + 1}$ **27.** $\frac{a^2 - ab + b^2}{a - b}$ **29.** $\frac{2(x - 1)}{x}$ **31.** $\frac{2x + 3y}{2x + y}$ **33.** $\frac{x + 3}{y - 4}$ **35.** $\frac{x + b}{x - 2b}$ **37.** $x + 2$ **39.** $\frac{2x^2 - 5}{3x - 2}$ **41.** $\frac{x^2 + 2x + 4}{x + 2}$ **43.** $4 + t$ **45.** $4x^2 + 6x +$ 9 **47.** -1 **49.** $-(y + 6)$ **51.** $\frac{-(3a + 1)}{3a - 1}$ or $-\frac{3a + 1}{3a - 1}$ **53.** 3 **55.** $x + a$ **57.** 2; 2 **59.** Undefined; 4 **61.** 1; 1 **63.** 3; 3

65. The graph of $y = x + 2$ includes the point (2, 4) while the other graph does not.

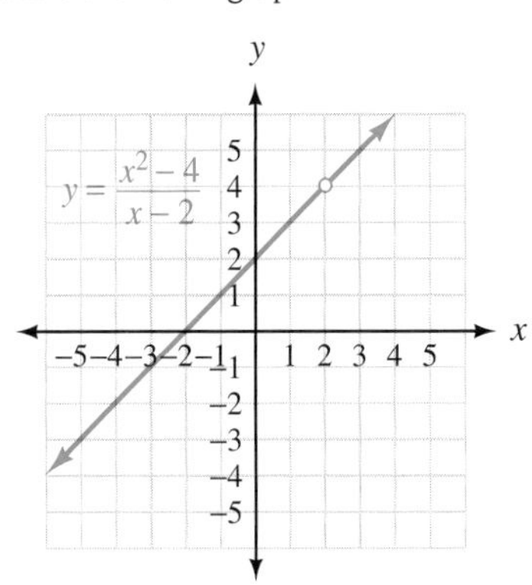

67. a. $x^2 - 49$ **b.** $x^2 - 14x + 49$ **c.** $7x^3 - 98x^2 + 343x$ **d.** $\frac{x}{7}$

69.

Weeks x	Weight (pounds) $W(x)$
0	200
1	194
4	184
12	173
24	168

71. 7.9 feet per second **73. a.** Domain $= \{t \mid 20 \leq t \leq 50\}$

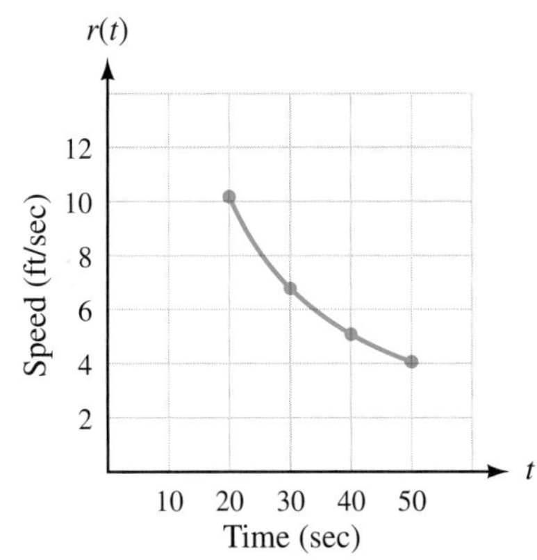

75. a. Domain $= \{d \mid 1 \leq d \leq 6\}$

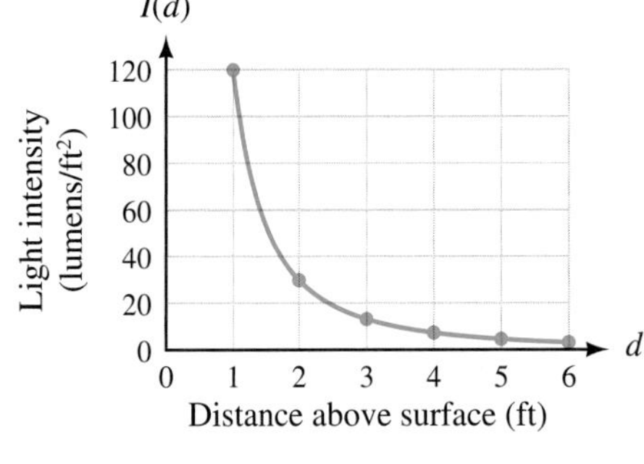

77. $3x^2 - 7x + 4$ **79.** 9 **81.** $8x^2$ **83.** $2x^3$ **85.** $-2x^2y^2$ **87.** 185.12 **89.** $4x^3 - 8x^2$ **91.** $4x^3 - 6x - 20$ **93.** $-3x + 9$ **95.** $(x + a)(x - a)$ **97.** $(x + y)(x - 7y)$ **99. a.** 2 **b.** -4 **c.** Undefined **d.** 2 **e.** 1 **f.** -6 **g.** 4 **h.** -3

Problem Set 6.2

1. $2x^2 - 4x + 3$ **3.** $-2x^2 - 3x + 4$ **5.** $2y^2 + \frac{5}{2} - \frac{3}{2y^2}$ **7.** $-\frac{5}{2}x + 4 + \frac{3}{x}$ **9.** $4ab^3 + 6a^2b$ **11.** $-xy + 2y^2 + 3xy^2$ **13.** $x + 2$ **15.** $a - 3$ **17.** $5x + 6y$ **19.** $x^2 + xy + y^2$ **21.** $(y^2 + 4)(y + 2)$ **23.** $(x + 2)(x + 5)$ **25.** $x - 7 + \frac{7}{x + 2}$ **27.** $2x^2 - 5x + 1 + \frac{4}{x + 1}$ **29.** $y^2 - 3y - 13$ **31.** $3y^2 + 6y + 8 + \frac{37}{2y - 4}$ **33.** $a^3 + 2a^2 + 4a + 6 + \frac{17}{a - 2}$ **35.** $y^3 + 2y^2 + 4y + 8$ **37.** $h(x) = \frac{x + 6}{4}; \{x \mid x \neq 6\}$

39. $h(x) = \frac{x-8}{x+4}$; $\{x \mid x \neq -4, x \neq 8\}$ **41.** $h(x) = x^2 + 3x + 9$; $\{x \mid x \neq 3\}$ **43. a.** 4 **b.** 4 **45. a.** 5 **b.** 5 **47. a.** $2x + h$ **b.** $x + a$

49. a. $2x + h$ **b.** $x + a$ **51. a.** $2x + h - 3$ **b.** $x + a - 3$ **53. a.** $4x + 2h + 3$ **b.** $2x + 2a + 3$
57. a. $(x + 4)(x + 5)(x + 1)$ **b.** $(x + 5)(x + 1)$ **59. a.** $(x + 2)(x + 4)(x - 3)$ **b.** $(x + 4)(x - 3)$ **61.** 13 is the same as the remainder

63. a.

x	1	5	10	15	20
$C(x)$	2.15	2.75	3.50	4.25	5.00

b. $\overline{C}(x) = \frac{2}{x} + 0.15$

c.

x	1	5	10	15	20
$\overline{C}(x)$	2.15	0.55	0.35	0.28	0.25

d. It decreases.

65. $\frac{21}{10}$ **67.** $\frac{11}{8}$ **69.** $\frac{1}{18}$ **71.** 32 **73.** 17 **75.** $\frac{x^{16}}{y^{22}}$ **77.** $\frac{2}{3}$ **79.** $20x^2y^2$ **81.** $72x^4y^5$ **83.** $(x + 2)(x - 2)$ **85.** $x^2(x - y)$

87. $2(y - 1)(y + 1)$ **89.** $4x^3 - x^2 + 3$ **91.** $0.5x^2 - 0.4x + 0.3$ **93.** $\frac{3}{2}x - \frac{5}{2} + \frac{1}{2x + 4}$ **95.** $\frac{2}{3}x + \frac{1}{3} + \frac{2}{3x - 1}$

Problem Set 6.3

1. $\frac{1}{6}$ **3.** $\frac{9}{4}$ **5.** $\frac{1}{2}$ **7.** $\frac{15y}{x^2}$ **9.** $\frac{b}{a}$ **11.** $\frac{2y^5}{z^3}$ **13.** $\frac{x + 3}{x + 2}$ **15.** $y + 1$ **17.** $\frac{3(x + 4)}{x - 2}$ **19.** 1 **21.** $\frac{(a - 2)(a + 2)}{a - 5}$

23. $\frac{9t^2 - 6t + 4}{4t^2 - 2t + 1}$ **25.** $\frac{x + 3}{x + 4}$ **27.** $\frac{5a - b}{9a^2 + 15ab + 25b^2}$ **29.** 2 **31.** $\frac{x(x - 1)}{x^2 + 1}$ **33.** $\frac{(a + 4b)(a - 3b)}{(a - 4b)(a + 5b)}$ **35.** $\frac{2y - 1}{2y - 3}$

37. $\frac{(y - 2)(y + 1)}{(y + 2)(y - 1)}$ **39.** $\frac{x - 1}{x + 1}$ **41.** $\frac{x - 2}{x + 3}$ **43. a.** $\frac{5}{21}$ **b.** $\frac{5x + 3}{25x^2 + 15x + 9}$ **c.** $\frac{5x - 3}{25x^2 + 15x + 9}$ **d.** $\frac{5x + 3}{5x - 3}$

45. a. $\frac{(x + 2)^2}{(x - 1)^2}$ **b.** $(x - 3)^2$ **47.** $\frac{(x + 2)(x - 4)}{4(x + 9)}$ **49.** $(x + 3)(x + 4)$ **51.** $3x$ **53.** $2(x + 5)$ **55.** $x - 2$ **57.** $-(y - 4)$

59. $(a - 5)(a + 1)$ **61.** $\frac{(x - 4)^2}{x - 3}$ **63.** $-\frac{x - 1}{x + 3}$ **65.** $(y - 2)(x - 7)$ **67.** $\frac{(x + 3)^2}{x - 4}$ **69.** $\frac{(2x - 3)(x - 1)}{x - 2}$

71.

Number of Copies	Price per Copy ($)
1	20.33
10	9.33
20	6.40
50	4.00
100	3.05

73. $305.00 **75.** $x^2 - 6x + 9$

77. $10x^5 + 8x^3 - 6x^2$ **79.** $12a^2 + 11a - 5$ **81.** $12xy - 6x + 28y - 14$ **83.** $9 - 6t^2 + t^4$ **85.** $3x^3 + 18x^2 + 33x + 18$
87. $\frac{2}{3}$ **89.** $\frac{47}{105}$ **91.** $x - 7$ **93.** $(x + 1)(x - 1)$ **95.** $2(x + 5)$ **97.** $(a - b)(a^2 + ab + b^2)$ **99.** $\frac{x^4 - x^2y^2 + y^4}{x^2 + y^2}$

101. $\frac{(a + 5)(a - 1)}{3a^2 - 2a + 1}$ **103.** $\frac{a(c - 1)}{a - b}$

Problem Set 6.4

1. $\frac{5}{4}$ **3.** $\frac{1}{3}$ **5.** $\frac{41}{24}$ **7.** $\frac{19}{144}$ **9.** $\frac{31}{24}$ **11.** 1 **13.** -1 **15.** $\frac{1}{x + y}$ **17.** 1 **19.** $\frac{a^2 + 2a - 3}{a^3}$ **21.** 1

23. a. $\frac{1}{16}$ **b.** $\frac{9}{4}$ **c.** $\frac{13}{24}$ **d.** $\frac{5(x + 3)}{(x - 3)^2}$ **e.** $\frac{x + 3}{5}$ **f.** $\frac{x - 2}{x - 3}$ **25.** $\frac{4 - 3t}{2t^2}$ **27.** $\frac{1}{2}$ **29.** $\frac{x + 3}{2(x + 1)}$ **31.** $\frac{a - b}{a^2 + ab + b^2}$ **33.** $\frac{2y - 3}{4y^2 + 6y + 9}$

35. $\frac{2(2x - 3)}{(x - 3)(x - 2)}$ **37.** $\frac{1}{2t - 7}$ **39.** $\frac{4}{(a - 3)(a + 1)}$ **41.** $\frac{-4x^2}{(2x + 1)(2x - 1)(4x^2 + 2x + 1)}$ **43.** $\frac{2}{(2x + 3)(4x + 3)}$

45. $\frac{a}{(a + 4)(a + 5)}$ **47.** $\frac{x + 1}{(x - 2)(x + 3)}$ **49.** $\frac{x - 1}{(x + 1)(x + 2)}$ **51.** $\frac{1}{(x + 2)(x + 1)}$ **53.** $\frac{1}{(x + 2)(x + 3)}$ **55.** $\frac{4x + 5}{2x + 1}$ **57.** $\frac{22 - 5t}{4 - t}$

59. $\frac{2x^2 + 3x - 4}{2x + 3}$ **61.** $\frac{2x - 3}{2x}$ **63.** $\frac{1}{2}$ **65.** $-\frac{2x - 11}{4(x - 4)}$ **67.** $\frac{3}{x + 4}$ **69.** $\frac{(2x + 1)(x + 5)}{(x - 2)(x + 1)(x + 3)}$ **71.** $\frac{(x + 2)(x - 1)}{(3x - 2)(3x + 2)(x - 2)}$

73. $\frac{3(x - 2)(x + 1)}{(2x + 1)(x - 1)(x - 3)}$ **75.** $\frac{2x}{(x - 9)(x - 7)}$ **77.** $x + \frac{4}{x} = \frac{x^2 + 4}{x}$ **79.** $\frac{51}{10} = 5.1$
81. a. $T = 120$ months **b.** The two objects will never meet. **83.** 5.4×10^4 **85.** 3.4×10^{-4} **87.** 6,440 **89.** 0.00644

91. 1.2×10^4 **93.** $\frac{6}{5}$ **95.** $x + 2$ **97.** $3 - x$ **99.** $(x + 2)(x - 2)$ **101.** $\frac{x-1}{x+3}$ **103.** $\frac{a^2 + ab + a - v}{u + v}$ **105.** $\frac{6x + 5}{(4x - 1)(3x - 4)}$
107. $\frac{y(y^2 + 1)}{(y + 1)^2(y - 1)^2}$

Problem Set 6.5

1. $\frac{9}{8}$ **3.** $\frac{2}{15}$ **5.** $\frac{119}{20}$ **7.** $\frac{1}{x+1}$ **9.** $\frac{a+1}{a-1}$ **11.** $\frac{y-x}{y+x}$ **13.** $\frac{1}{(x+5)(x-2)}$ **15.** $\frac{1}{a^2-a+1}$ **17.** $\frac{x+3}{x+2}$ **19.** $\frac{a+3}{a-2}$
21. $\frac{9x^2+6x+4}{x(x+1)}$ **23.** $\frac{x}{x-2}$ **25.** $\frac{x-3}{x}$ **27.** $\frac{x+4}{x+2}$ **29.** $\frac{a-1}{a+1}$ **31.** $-\frac{x}{3}$ **33.** $\frac{y^2+1}{2y}$ **35.** $\frac{-x^2+x-1}{x-1}$ **37.** $\frac{5}{3}$
39. $\frac{2x-1}{2x+3}$ **41.** $-\frac{1}{x(x+h)}$ **43.** $\frac{3c+4a-2b}{5}$ **45.** $\frac{(t-4)(t+1)}{(t+6)(t-3)}$ **47.** $\frac{(5b-1)(b+5)}{2(2b-11)}$ **49.** $-\frac{3}{2x+14}$ **51.** $2m - 9$
53. a. $-\frac{4}{ax}$ **b.** $-\frac{1}{(x+1)(a+1)}$ **c.** $-\frac{a+x}{a^2x^2}$ **55.** $(a^{-1} + b^{-1})^{-1} = \left(\frac{1}{a} + \frac{1}{b}\right)^{-1} = \left(\frac{a+b}{ab}\right)^{-1} = \frac{ab}{a+b}$
57. a. As v approaches 0, the denominator approaches 1. **b.** $v = \frac{fs}{h} - s$ **59.** −15 **61.** 5 **63.** 1 **65.** −3, 4 **67.** 2, 3
69. $xy - 2x$ **71.** $3x - 18$ **73.** ab **75.** $(y + 5)(y - 5)$ **77.** $x(a + b)$ **79.** 2 **81.** $\frac{1}{3}$ **83.** $\frac{40}{243}$ **85.** $\frac{a-2b}{a+2b}$
87. $a + b$ **89.** $-\frac{qt}{q+t}$

Problem Set 6.6

1. $-\frac{35}{3}$ **3.** $-\frac{18}{5}$ **5.** $\frac{36}{11}$ **7.** 2 **9.** 5 **11.** 2 **13.** −3, 4 **15.** $1, -\frac{4}{3}$ **17.** $-\frac{9}{5}, 5$ **19.** $\frac{9}{2}$ **21.** ∅ **23.** $3, -\frac{4}{3}$
25. $-\frac{4}{3}$ **27.** ∅ **29. a.** $\frac{1}{3}$ **b.** 3 **c.** 9 **d.** 4 **e.** $\frac{1}{3}, 3$ **31. a.** $\frac{6}{(x-4)(x+3)}$ **b.** $\frac{x-3}{x-4}$ **c.** Possible solutions −2 and 5; only 5 checks; 5
33. Possible solution −1, which does not check; ∅ **35.** 5 **37.** $-\frac{1}{2}, \frac{5}{3}$ **39.** $\frac{2}{3}$ **41.** 18
43. Possible solution 4, which does not check; ∅ **45.** Possible solutions 3 and −4; only −4 checks; −4 **47.** −6 **49.** −5 **51.** $\frac{53}{17}$
53. Possible solutions 1 and 2; only 2 checks; 2 **55.** Possible solution 3 , which does not check; ∅ **57.** $\frac{22}{3}$ **59.** 2 **61.** 1, 5
63. $x = \frac{ab}{a-b}$ **65.** $R = \frac{R_1R_2}{R_1 + R_2}$ **67.** $y = \frac{x-3}{x-1}$ **69.** $y = \frac{1-x}{3x-2}$ **73.**

Time t (sec)	Speed of Kayak Relative to the Water v (m/sec)	Current of the River c (m/sec)
240	4	1
300	4	2
514	4	3
338	3	1
540	3	2
N/A	3	3

75. 5 **77.** 8 meters, 13 meters **79.** −6, −5 or 5, 6 **81.** 3, 4, 5 **83.** 2,358 **85.** 12.3 **87.** 3 **89.** 9, −1 **91.** 60
93. $-\frac{2}{3}, -\frac{5}{2}, \frac{5}{2}$ **95.** −2, 2, 3 **97.** $x = -\frac{a}{5}$ **99.** $v = \frac{16t^2 + s}{t}$ **101.** $f = \frac{pg}{g - p}$

Problem Set 6.7

As you can see, in addition to the answers to the problems we have included some of the equations used to solve the problems. Remember, you should attempt the problems on your own before looking here to check your answers or equations.

1. $\frac{1}{x} + \frac{1}{3x} = \frac{20}{3}$; $\frac{1}{5}$ and $\frac{3}{5}$ **3.** $x + \frac{1}{x} = \frac{10}{3}$; 3 or $\frac{1}{3}$ **5.** $\frac{1}{x} + \frac{1}{x+1} = \frac{7}{12}$; 3, 4 **7.** $\frac{7+x}{9+x} = \frac{5}{6}$; 3
9. Let x = speed of current; $\frac{1.5}{5-x} = \frac{3}{5+x}$; $\frac{5}{3}$ miles per hour **11.** $\frac{8}{x+2} + \frac{8}{x-2} = 3$; 6 miles per hour
13. Train A: 75 miles per hour, Train B: 60 miles per hour; let x = speed of B; $\frac{150}{x+15} = \frac{120}{x}$
15. 540 miles per hour; let x = speed of 747; $\frac{810}{270} - 1\frac{1}{2} = \frac{810}{x}$ **17.** 54 miles per hour; let x = usual speed; $\frac{270}{x+6} + \frac{1}{2} = \frac{270}{x}$
19. Let x = time to fill the tank with both open; $\frac{1}{8} - \frac{1}{16} = \frac{1}{x}$; 16 hours **21.** $7\frac{1}{2}$ minutes; $\frac{1}{3} - \frac{1}{5} = \frac{1}{x}$

23. Let x = time to fill with hot water; $\frac{1}{x} + \frac{1}{3.5} = \frac{1}{2.1}$; $5\frac{1}{4}$ minutes **25.** 51.1 acres **27.** 5.9 miles per hour
29. 20.7 miles per hour **31.** 4.6 miles per hour **33.** 3.6 miles per hour **35.** $\frac{1}{3}\left[\left(x + \frac{2}{3}x\right) + \frac{1}{3}\left(x + \frac{2}{3}x\right)\right] = 10, x = \frac{27}{2}$ **37.** $\dfrac{2}{3a}$
39. $(x - 3)(x + 2)$ **41.** 1 **43.** $\dfrac{3 - x}{3 + x}$ **45.** Possible solution 3, which does not check; $\varnothing$

CHAPTER 6 REVIEW TEST

1. $\frac{25x^2}{7y^3}$ **2.** $\dfrac{a(a - b)}{4}$ **3.** $\dfrac{x - 5}{x + 5}$ **4.** $\dfrac{a + 1}{a - 1}$ **5.** $3x + 2 + \frac{4}{x}$ **6.** $-9b + 5a - 7a^2b^2$ **7.** $x^{3n} - x^{2n}$ **8.** $x + 2$ **9.** $5x + 6y$
10. $y^3 + 2y^2 + 4y + 8$ **11.** $4x + 1 - \dfrac{2}{2x - 7}$ **12.** $y^2 - 3y - 13$ **13.** $\frac{9}{5}$ **14.** $\frac{3x}{4y^2}$ **15.** $\dfrac{x - 1}{x^2 + 1}$ **16.** 1 **17.** $\dfrac{x + 2}{x - 2}$
18. $(2x - 3)(x + 3)$ **19.** $\frac{31}{30}$ **20.** -1 **21.** $\dfrac{x^2 + x + 1}{x^3}$ **22.** $\dfrac{1}{(y + 4)(y + 3)}$ **23.** $\dfrac{x - 1}{2(x + 1)(x + 2)}$ **24.** $\dfrac{15x - 2}{5x - 2}$ **25.** 5
26. $\dfrac{1}{a^2 - a + 1}$ **27.** $\dfrac{x^2 + x + 1}{x^2 + 1}$ **28.** $\dfrac{x + 3}{x + 2}$ **29.** 6 **30.** 1 **31.** -6 **32.** Possible solution -3, which does not check; $\varnothing$
33. $\frac{22}{3}$ **34.** Possible solutions 4 and -5; only 4 checks **35.** Car: 30 miles per hour; truck: 20 miles per hour
36. 7.5 miles per hour **37.** 742 miles per hour

CHAPTER 7

Problem Set 7.1

1. 12 **3.** Not a real number **5.** -7 **7.** -3 **9.** 2 **11.** Not a real number **13.** 0.2 **15.** 0.2 **17.** 5 **19.** -6
21. $\frac{1}{6}$ **23.** $\frac{2}{5}$ **25.** $6a^4$ **27.** $3a^4$ **29.** $2x^2y$ **31.** $2a^3b^5$ **33.** 6 **35.** -3 **37.** 2 **39.** -2 **41.** 2 **43.** $\frac{9}{5}$ **45.** 9
47. 125 **49.** $\frac{1}{3}$ **51.** $\frac{1}{27}$ **53.** $\frac{6}{5}$ **55.** $\frac{8}{27}$ **57.** 7 **59.** $\frac{3}{4}$ **61.** $x^{4/5}$ **63.** a **65.** $\dfrac{1}{x^{2/5}}$ **67.** $x^{1/6}$ **69.** $x^{9/25}y^{1/2}z^{1/5}$
71. $\dfrac{b^{7/4}}{a^{1/8}}$ **73.** $y^{3/10}$ **75.** $\dfrac{1}{a^2b^4}$ **77.** $\dfrac{s^{1/2}}{r^{20}}$ **79.** $10b^3$ **81.** 25 mph
83. **a.** 424 picometers **b.** 600 picometers **c.** 6×10^{-10} meters **85.** **a.** B **b.** A **c.** C **d.** (0, 0) and (1, 1) **87.** $x^6 - x^3$
89. $x^2 + 2x - 15$ **91.** $x^4 - 10x^2 + 25$ **93.** $x^3 - 27$ **95.** $x^6 - x^5$ **97.** $12a^2 - 11ab + 2b^2$ **99.** $x^6 - 4$ **101.** $3x - 4x^3y$
103. $(x - 5)(x + 2)$ **105.** $(3x - 2)(2x + 5)$ **107.** x^2 **109.** t **111.** $x^{1/3}$ **113.** $(9^{1/2} + 4^{1/2})^2 = (3 + 2)^2 = 5^2 = 25 \neq 9 + 4$
115. $\sqrt{\sqrt{a}} = (a^{1/2})^{1/2} = a^{1/4} = \sqrt[4]{a}$

Problem Set 7.2

1. $x + x^2$ **3.** $a^2 - a$ **5.** $6x^3 - 8x^2 + 10x$ **7.** $12x^2 - 36y^2$ **9.** $x^{4/3} - 2x^{2/3} - 8$ **11.** $a - 10a^{1/2} + 21$ **13.** $20y^{2/3} - 7y^{1/3} - 6$
15. $10x^{4/3} + 21x^{2/3}y^{1/2} + 9y$ **17.** $t + 10t^{1/2} + 25$ **19.** $x^3 + 8x^{3/2} + 16$ **21.** $a - 2a^{1/2}b^{1/2} + b$ **23.** $4x - 12x^{1/2}y^{1/2} + 9y$ **25.** $a - 3$
27. $x^3 - y^3$ **29.** $t - 8$ **31.** $4x^3 - 3$ **33.** $x + y$ **35.** $a - 8$ **37.** $8x + 1$ **39.** $t - 1$ **41.** $2x^{1/2} + 3$ **43.** $3x^{1/3} - 4y^{1/3}$
45. $3a - 2b$ **47.** $3(x - 2)^{1/2}(4x - 11)$ **49.** $5(x - 3)^{7/5}(x - 6)$ **51.** $3(x + 1)^{1/2}(3x^2 + 3x + 2)$ **53.** $(x^{1/3} - 2)(x^{1/3} - 3)$
55. $(a^{1/5} - 4)(a^{1/5} + 2)$ **57.** $(2y^{1/3} + 1)(y^{1/3} - 3)$ **59.** $(3t^{1/5} + 5)(3t^{1/5} - 5)$ **61.** $(2x^{1/7} + 5)^2$ **63.** -8 **65.** 90 **67.** -3 **69.** 0
71. $\dfrac{3 + x}{x^{1/2}}$ **73.** $\dfrac{x + 5}{x^{1/3}}$ **75.** $\dfrac{x^3 + 3x^2 + 1}{(x^3 + 1)^{1/2}}$ **77.** $\dfrac{-4}{(x^2 + 4)^{1/2}}$ **79.** 15.8% **81.** $\dfrac{1}{x^2 + 9}$ **83.** $3x - 4x^3y$ **85.** $5x - 4$
87. $x^2 + 5x + 25$ **89.** 5 **91.** 6 **93.** $4x^2y$ **95.** $5y$ **97.** 3 **99.** 2 **101.** $2ab$ **103.** 25 **105.** $48x^4y^2$ **107.** $4x^6y^6$

Problem Set 7.3

1. $2\sqrt{2}$ **3.** $7\sqrt{2}$ **5.** $12\sqrt{2}$ **7.** $4\sqrt{5}$ **9.** $4\sqrt{3}$ **11.** $15\sqrt{3}$ **13.** $3\sqrt[3]{2}$ **15.** $4\sqrt[3]{2}$ **17.** $6\sqrt[3]{2}$ **19.** $2\sqrt[5]{2}$ **21.** $3x\sqrt{2x}$
23. $2y\sqrt[4]{2y^3}$ **25.** $2xy^2\sqrt[3]{5xy}$ **27.** $4abc^2\sqrt{3b}$ **29.** $2bc\sqrt[3]{6a^2c}$ **31.** $2xy^2\sqrt[5]{2x^3y^2}$ **33.** $3xy^2z\sqrt[5]{x^2}$ **35.** $2\sqrt{3}$
37. $\sqrt{-20}$, which is not a real number **39.** $\dfrac{\sqrt{11}}{2}$ **41.** $\dfrac{2\sqrt{3}}{3}$ **43.** $\dfrac{5\sqrt{6}}{6}$ **45.** $\dfrac{\sqrt{2}}{2}$ **47.** $\dfrac{\sqrt{5}}{5}$ **49.** $2\sqrt[3]{4}$ **51.** $\dfrac{2\sqrt[3]{3}}{3}$
53. $\dfrac{\sqrt[4]{24x^2}}{2x}$ **55.** $\dfrac{\sqrt[4]{8y^3}}{y}$ **57.** $\dfrac{\sqrt[3]{36xy^2}}{3y}$ **59.** $\dfrac{\sqrt[3]{6xy^2}}{3y}$ **61.** $\dfrac{\sqrt[4]{2x}}{2x}$ **63.** $\dfrac{3x\sqrt{15xy}}{5y}$ **65.** $\dfrac{5xy\sqrt{6xz}}{2z}$ **67.** $\dfrac{2ab\sqrt[3]{6ac^2}}{3c}$
69. $\dfrac{2xy^2\sqrt[3]{3z^2}}{3z}$ **71.** $\sqrt{x}$ **73.** $\sqrt[6]{xy}$ **75.** $\sqrt[12]{a}$ **77.** $x\sqrt[9]{6x}$ **79.** $ab^2c\sqrt[6]{c}$ **81.** $abc^2\sqrt[15]{3a^2b}$ **83.** $2b^2\sqrt[3]{a}$ **85.** $5|x|$
87. $3|xy|\sqrt{3x}$ **89.** $|x - 5|$ **91.** $|2x + 3|$ **93.** $2|a(a + 2)|$ **95.** $2|x|\sqrt{x - 2}$ **97.** $\sqrt{9 + 16} = \sqrt{25} = 5$; $\sqrt{9} + \sqrt{16} = 3 + 4 = 7$

99. $5\sqrt{13}$ feet **101. a.** 13 feet **b.** $2\sqrt{14} \approx 7.5$ feet **105.** $\frac{y^3}{x^2}$ **107.** 1 **109.** $\frac{4x^2 - 6x + 9}{9x^2 - 3x + 1}$ **111.** $7x$ **113.** $27xy^2$ **115.** $\frac{5}{6}x$

117. $3\sqrt{2}$ **119.** $5y\sqrt{3xy}$ **121.** $2a\sqrt[3]{ab^2}$ **123.** $12\sqrt[3]{5}$ **125.** $6\sqrt[3]{49}$ **127.** $\frac{\sqrt[10]{a^7}}{a}$ **129.** $\frac{\sqrt[20]{a^9}}{a}$

131.

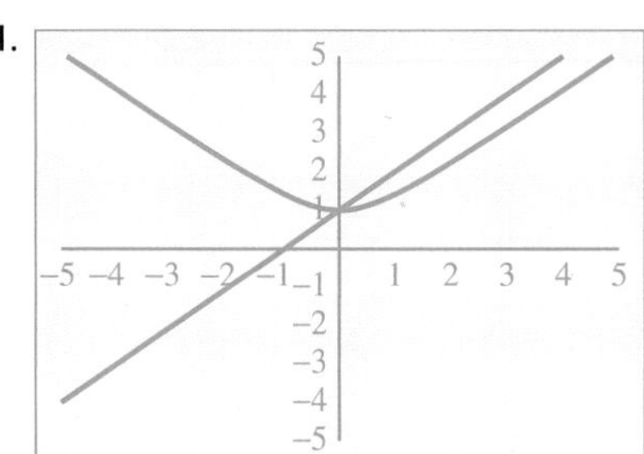

133. About $\frac{3}{4}$ of a unit when $x = 2$ **135.** $x = 0$

Problem Set 7.4

1. $7\sqrt{5}$ **3.** $-x\sqrt{7}$ **5.** $\sqrt[3]{10}$ **7.** $9\sqrt[5]{6}$ **9.** 0 **11.** $\sqrt{5}$ **13.** $-32\sqrt{2}$ **15.** $-3x\sqrt{2}$ **17.** $-2\sqrt[3]{2}$ **19.** $8x\sqrt[3]{xy^2}$

21. $3a^2b\sqrt{3ab}$ **23.** $11ab\sqrt[3]{3a^2b}$ **25.** $10xy\sqrt[4]{3y}$ **27.** $\sqrt{2}$ **29.** $\frac{8\sqrt{5}}{15}$ **31.** $\frac{(x-1)\sqrt{x}}{x}$ **33.** $\frac{3\sqrt{2}}{2}$ **35.** $\frac{5\sqrt{6}}{6}$ **37.** $\frac{8\sqrt[3]{25}}{5}$

39. a. $8\sqrt{2x}$ **b.** $-4\sqrt{2x}$ **41. a.** $4\sqrt{2x}$ **b.** $2\sqrt{2x}$ **43. a.** $3x\sqrt{2}$ **b.** $-x\sqrt{2}$ **45. a.** $3\sqrt{2x} + 3$ **b.** $-\sqrt{2x} - 7$

47. $\sqrt{12} \approx 3.464$; $2\sqrt{3} = 2(1.732) \approx 3.464$ **49.** $\sqrt{8} + \sqrt{18} \approx 2.828 + 4.243 = 7.071$; $\sqrt{50} \approx 7.071$; $\sqrt{26} \approx 5.099$ **55.** $\frac{\sqrt{3}}{2}$

57. 1 **59.** $\frac{13 - 3t}{3 - t}$ **61.** $\frac{6}{4x + 3}$ **63.** $\frac{x - y}{x^2 + xy + y^2}$ **65.** 6 **67.** $4x^2 + 3xy - y^2$ **69.** $x^2 + 6x + 9$ **71.** $x^2 - 4$ **73.** $6\sqrt{2}$

75. 6 **77.** $9x$ **79.** $\frac{\sqrt{6}}{2}$ **81.** $-xy$ **83.** $-xy^2z^3$ **85.** $5b^2c\sqrt[4]{2a^3b}$ **87.** $-3ab\sqrt[3]{2a}$

Problem Set 7.5

1. $3\sqrt{2}$ **3.** $10\sqrt{21}$ **5.** 720 **7.** 54 **9.** $\sqrt{6} - 9$ **11.** $24 + 6\sqrt[3]{4}$ **13.** $2 + 2\sqrt[3]{3}$ **15.** $xy\sqrt[3]{y} + x^2\sqrt[3]{y}$ **17.** $2x^2\sqrt[4]{x} + 2x^3\sqrt[4]{2}$
19. $7 + 2\sqrt{6}$ **21.** $x + 2\sqrt{x} - 15$ **23.** $34 + 20\sqrt{3}$ **25.** $19 + 8\sqrt{3}$ **27.** $x - 6\sqrt{x} + 9$ **29.** $4a - 12\sqrt{ab} + 9b$ **31.** $x + 4\sqrt{x} - 4$
33. $x - 6\sqrt{x - 5} + 4$ **35.** 1 **37.** $a - 49$ **39.** $25 - x$ **41.** $x - 8$ **43.** $10 + 6\sqrt{3}$ **45.** $5 + \sqrt[3]{12} + \sqrt[3]{18}$

47. $x^2 + x\sqrt[3]{x^2y^2} + \sqrt[3]{xy} + y$ **49.** $\frac{\sqrt{2}}{2}$ **51.** $\frac{\sqrt{x}}{x}$ **53.** $\frac{4\sqrt{3}}{3}$ **55.** $\frac{x\sqrt{6}}{3}$ **57.** $\frac{2\sqrt{10x}}{5x}$ **59.** $\frac{\sqrt{2}}{2}$ **61.** $\frac{2x\sqrt{2y}}{y}$

63. $\frac{a\sqrt{6c}}{2bc}$ **65.** $\frac{2ac\sqrt[3]{b^2}}{b}$ **67.** $\frac{\sqrt{3} + 1}{2}$ **69.** $\frac{5 - \sqrt{5}}{4}$ **71.** $\frac{x + 3\sqrt{x}}{x - 9}$ **73.** $\frac{10 + 3\sqrt{5}}{11}$ **75.** $\frac{3\sqrt{x} + 3\sqrt{y}}{x - y}$

77. $2 + \sqrt{3}$ **79.** $\frac{11 - 4\sqrt{7}}{3}$ **81.** $\frac{a + 2\sqrt{ab} + b}{a - b}$ **83.** $\frac{x + 4\sqrt{x} + 4}{x - 4}$ **85.** $\frac{5 - \sqrt{21}}{4}$ **87.** $\frac{\sqrt{x} - 3x + 2}{1 - x}$ **89.** $\frac{8\sqrt{3}}{3}$

91. $\frac{11\sqrt{5}}{5}$ **93.** $-4\sqrt{3}$ **95.** $5\sqrt{3}$

97. $(\sqrt[3]{2} + \sqrt[3]{3})(\sqrt[3]{4} - \sqrt[3]{6} + \sqrt[3]{9})$
$= \sqrt[3]{8} - \sqrt[3]{12} + \sqrt[3]{18} + \sqrt[3]{12} - \sqrt[3]{18} + \sqrt[3]{27}$
$= 2 + 3 = 5$

99. $10\sqrt{3}$ **101.** $x + 6\sqrt{x} + 9$ **103.** 75 **105.** $\frac{5\sqrt{2}}{4}$ seconds; $\frac{5}{2}$ seconds **111.** $-\frac{1}{8}$ **113.** $\frac{y - 2}{y + 2}$ **115.** $\frac{2x + 1}{2x - 1}$

117. $t^2 + 10t + 25$ **119.** x **121.** 7 **123.** $-4, -3$ **125.** $-6, -3$ **127.** $-5, -2$ **129.** $\frac{x(\sqrt{x - 2} - 4)}{x - 18}$

131. $\frac{x(\sqrt{x + 5} + 5)}{x - 20}$ **133.** $\frac{3(\sqrt{5x} - x)}{5 - x}$

Problem Set 7.6

1. 4 **3.** ∅ **5.** 5 **7.** ∅ **9.** $\frac{39}{2}$ **11.** ∅ **13.** 5 **15.** 3 **17.** $-\frac{32}{3}$ **19.** 3, 4 **21.** $-1, -2$
23. Possible solutions $\frac{1}{2}$ and 11; only 11 checks; 11 **25.** 3, 7 **27.** $-\frac{1}{4}$, 14 **29.** -1 **31.** ∅ **33.** 7 **35.** 0, 3 **37.** -4
39. 8 **41.** 0 **43.** 9 **45.** 0 **47.** 8 **49.** Possible solution 9, which does not check; ∅
51. Possible solutions 0 and 32; only 0 checks; 0 **53.** Possible solutions -2 and 6; only 6 checks; 6 **55.** $\frac{1}{2}$ **57.** $\frac{1}{2}$, 1

59. 5, 13 **61.** $-\frac{3}{2}$ **63.** Possible solution 1, which does not check; ∅ **65.** $\frac{1}{2}$
67. Possible solution 2, which does not check; ∅ **69.** 1 **71.** −11

73.

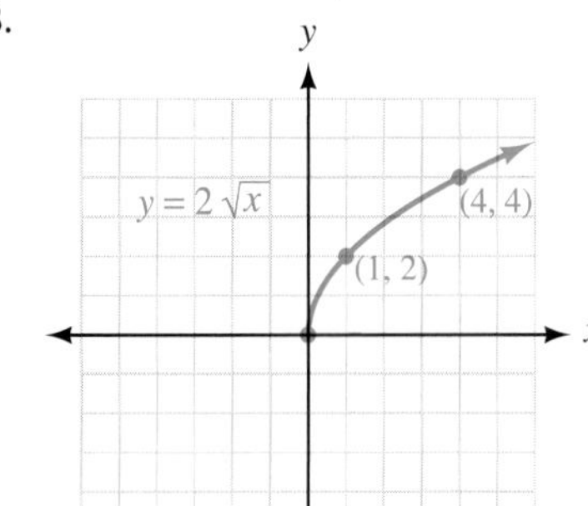

75.

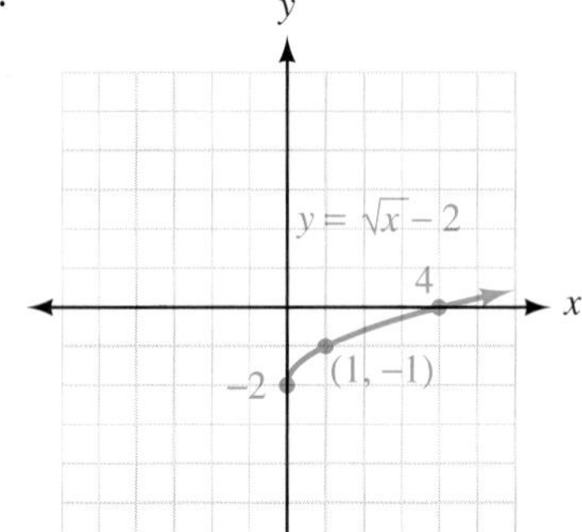

77.

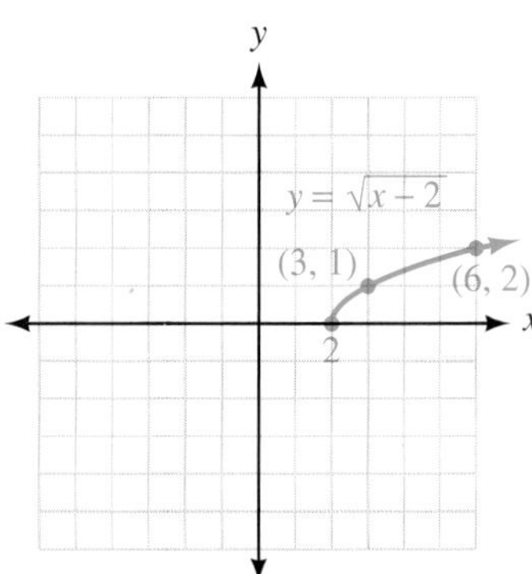

79.

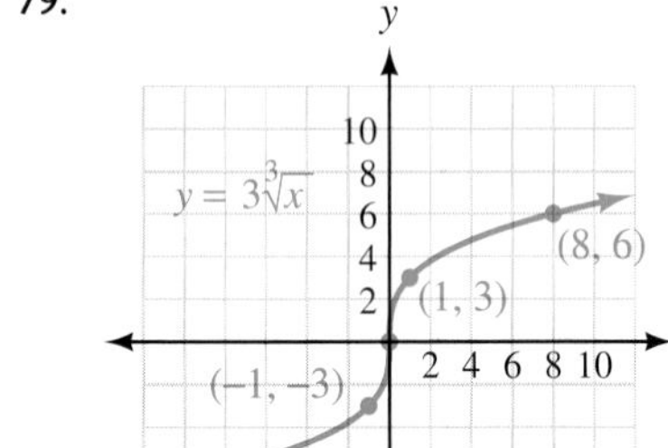

81.

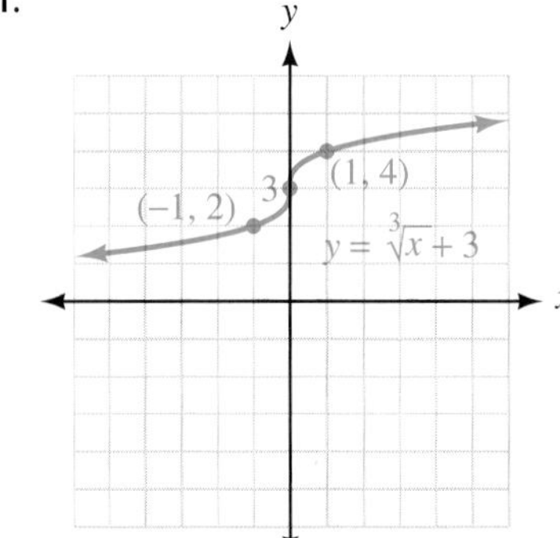

83.

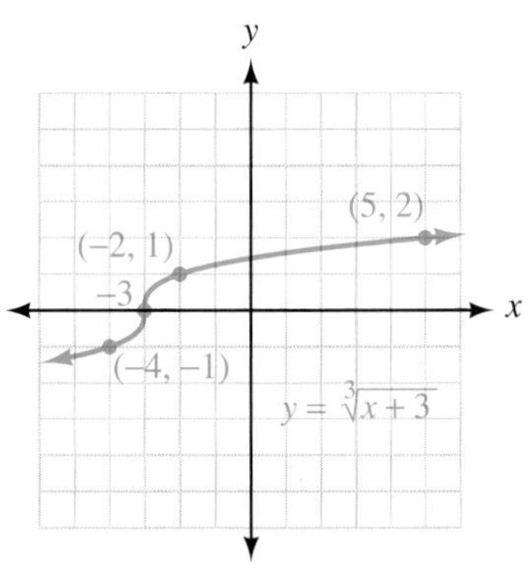

85. $h = 100 - 16t^2$ **87.** $\frac{392}{121} \approx 3.24$ feet **89.** 5 meters **91.** 2,500 meters **93.** $\sqrt{6} - 2$ **95.** $x + 10\sqrt{x} + 25$ **97.** $\dfrac{x - 3\sqrt{x}}{x - 9}$
99. 5 **101.** $2\sqrt{3}$ **103.** −1 **105.** 1 **107.** 4 **109.** 2 **111.** $10 - 2x$ **113.** $2 - 3x$ **115.** $6 + 7x - 20x^2$ **117.** $8x - 12x^2$
119. $4 + 12x + 9x^2$ **121.** $4 - 9x^2$

Problem Set 7.7

1. $6i$ **3.** $-5i$ **5.** $6i\sqrt{2}$ **7.** $-2i\sqrt{3}$ **9.** 1 **11.** −1 **13.** $-i$ **15.** $x = 3, y = -1$ **17.** $x = -2, y = -\frac{1}{2}$ **19.** $x = -8, y = -5$
21. $x = 7, y = \frac{1}{2}$ **23.** $x = \frac{3}{7}, y = \frac{2}{5}$ **25.** $5 + 9i$ **27.** $5 - i$ **29.** $2 - 4i$ **31.** $1 - 6i$ **33.** $2 + 2i$ **35.** $-1 - 7i$ **37.** $6 + 8i$
39. $2 - 24i$ **41.** $-15 + 12i$ **43.** $18 + 24i$ **45.** $10 + 11i$ **47.** $21 + 23i$ **49.** $-2 + 2i$ **51.** $2 - 11i$ **53.** $-21 + 20i$ **55.** $-2i$
57. $-7 - 24i$ **59.** 5 **61.** 40 **63.** 13 **65.** 164 **67.** $-3 - 2i$ **69.** $-2 + 5i$ **71.** $\frac{8}{13} + \frac{12}{13}i$ **73.** $-\frac{18}{13} - \frac{12}{13}i$ **75.** $-\frac{5}{13} + \frac{12}{13}i$
77. $\frac{13}{15} - \frac{2}{5}i$ **79.** $R = -11 - 7i$ ohms **81.** $-\frac{3}{2}$ **83.** $-3, \frac{1}{2}$ **85.** $\frac{5}{4}$ or $\frac{4}{5}$

CHAPTER 7 REVIEW TEST

1. 7 **2.** −3 **3.** 2 **4.** 27 **5.** $2x^3y^2$ **6.** $\frac{1}{16}$ **7.** x^2 **8.** a^2b^4 **9.** $a^{7/20}$ **10.** $a^{5/12}b^{8/3}$ **11.** $12x + 11x^{1/2}y^{1/2} - 15y$
12. $a^{2/3} - 10a^{1/3} + 25$ **13.** $4x^{1/2} + 2x^{5/6}$ **14.** $2(x-3)^{1/4}(4x - 13)$ **15.** $\dfrac{x+5}{x^{1/4}}$ **16.** $2\sqrt{3}$ **17.** $5\sqrt{2}$ **18.** $2\sqrt[3]{2}$ **19.** $3x\sqrt{2}$
20. $4ab^2c\sqrt{5a}$ **21.** $2abc\sqrt[4]{2bc^2}$ **22.** $\dfrac{3\sqrt{2}}{2}$ **23.** $3\sqrt[3]{4}$ **24.** $\dfrac{4x\sqrt{21xy}}{7y}$ **25.** $\dfrac{2y\sqrt[3]{45x^2z^2}}{3z}$ **26.** $-2x\sqrt{6}$ **27.** $3\sqrt{3}$
28. $\dfrac{8\sqrt{5}}{5}$ **29.** $7\sqrt{2}$ **30.** $11a^2b\sqrt{3ab}$ **31.** $-4xy\sqrt[3]{xz^2}$ **32.** $\sqrt{6} - 4$ **33.** $x - 5\sqrt{x} + 6$ **34.** $3\sqrt{5} + 6$ **35.** $6 + \sqrt{35}$

36. $\frac{63 + 12\sqrt{7}}{47}$ **37.** 0 **38.** 3 **39.** 5 **40.** $\varnothing$

41.

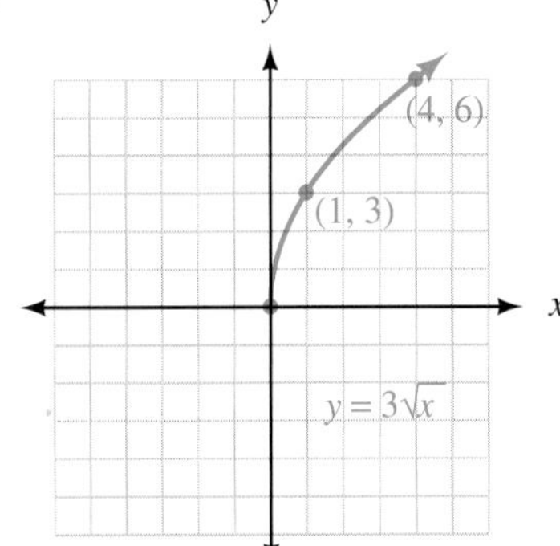

42.

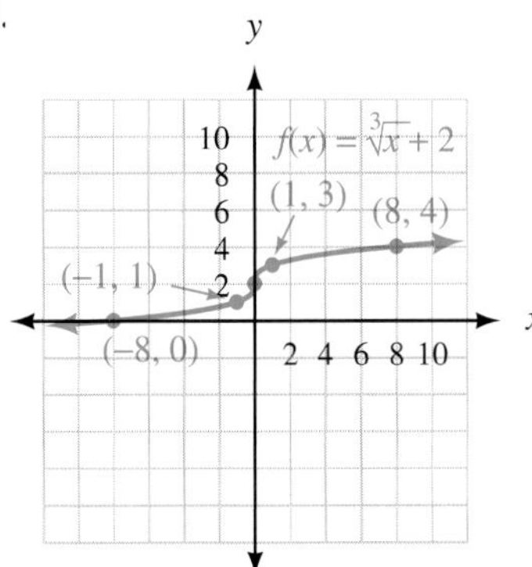

43. 1 **44.** $-i$ **45.** $x = -\frac{3}{2}, y = -\frac{1}{2}$ **46.** $x = -2, y = -4$ **47.** $9 + 3i$ **48.** $-7 + 4i$ **49.** $-6 + 12i$ **50.** $5 + 14i$
51. $12 + 16i$ **52.** 25 **53.** $1 - 3i$ **54.** $-\frac{6}{5} + \frac{3}{5}i$ **55.** 22.5 feet **56.** 65 yards

CHAPTER 8

Problem Set 8.1

1. ± 5 **3.** $\pm\frac{\sqrt{3}}{2}$ **5.** $\pm 2i\sqrt{3}$ **7.** $\pm\frac{3\sqrt{5}}{2}$ **9.** $-2, 3$ **11.** $\frac{-3 \pm 3i}{2}$ **13.** $-4 \pm 3i\sqrt{3}$ **15.** $\frac{3 \pm 2i}{2}$
17. $x^2 + 12x + 36 = (x + 6)^2$ **19.** $x^2 - 4x + 4 = (x - 2)^2$ **21.** $a^2 - 10a + 25 = (a - 5)^2$ **23.** $x^2 + 5x + \frac{25}{4} = (x + \frac{5}{2})^2$
25. $y^2 - 7y + \frac{49}{4} = (y - \frac{7}{2})^2$ **27.** $x^2 + \frac{1}{2}x + \frac{1}{16} = (x + \frac{1}{4})^2$ **29.** $x^2 + \frac{2}{3}x + \frac{1}{9} = (x + \frac{1}{3})^2$ **31.** $-3, -9$ **33.** $1 \pm 2i$
35. $4 \pm \sqrt{15}$ **37.** $\frac{5 \pm \sqrt{37}}{2}$ **39.** $1 \pm \sqrt{5}$ **41.** $\frac{4 \pm \sqrt{13}}{3}$ **43.** $\frac{3 \pm i\sqrt{71}}{8}$ **45. a.** No. **b.** $\pm 3i$ **47. a.** 0, 6 **b.** 0, 6
49. a. $-7, 5$ **b.** $-7, 5$ **51.** No **53. a.** $\frac{7}{5}$ **b.** 3 **c.** $\frac{7 \pm 2\sqrt{2}}{5}$ **d.** $\frac{71}{5}$ **e.** 3 **55.** ± 1 **57.** ± 1 **59.** ± 1

61.

x	$f(x)$	$g(x)$	$h(x)$
-2	49	49	25
-1	25	25	13
0	9	9	9
1	1	1	13
2	1	1	25

63. 3

65. a. $-1, 6$ **b.** $-1, 6$ **c.** -6 **d.** -10 **67.** $\frac{\sqrt{3}}{2}$ inch, 1 inch **69.** $\sqrt{2}$ inches **71.** 781 feet **73.** 7.3% to the nearest tenth
75. $20\sqrt{2} \approx 28$ feet **77.** $3\sqrt{5}$ **79.** $3y^2\sqrt{3y}$ **81.** $3x^2y\sqrt[3]{2y^2}$ **83.** $\frac{3\sqrt{2}}{2}$ **85.** $\sqrt[3]{2}$ **87.** 13 **89.** 7 **91.** $\frac{1}{2}$ **93.** 13
95. $(3t - 2)(9t^2 + 6t + 4)$ **97.** $x = \pm 2a$ **99.** $x = \frac{-p \pm \sqrt{p^2 - 4q}}{2}$ **101.** $x = \frac{-p \pm \sqrt{p^2 - 12q}}{6}$ **103.** $(x - 5)^2 + (y - 3)^2 = 2^2$

Problem Set 8.2

1. a. $\frac{-2 \pm \sqrt{10}}{3}$ **b.** $\frac{2 \pm \sqrt{10}}{3}$ **c.** $\frac{-2 \pm i\sqrt{2}}{3}$ **d.** $\frac{-2 \pm \sqrt{10}}{2}$ **e.** $\frac{2 \pm i\sqrt{2}}{2}$ **3. a.** $1 \pm i$ **b.** $1 \pm 2i$ **c.** $-1 \pm i$ **5.** 1, 2
7. $\frac{2 \pm i\sqrt{14}}{3}$ **9.** 0, 5 **11.** $0, -\frac{4}{3}$ **13.** $\frac{3 \pm \sqrt{5}}{4}$ **15.** $-3 \pm \sqrt{17}$ **17.** $\frac{-1 \pm i\sqrt{5}}{2}$ **19.** 1 **21.** $\frac{1 \pm i\sqrt{47}}{6}$ **23.** $4 \pm \sqrt{2}$
25. $\frac{1}{2}, 1$ **27.** $-\frac{1}{2}, 3$ **29.** $\frac{-1 \pm i\sqrt{7}}{2}$ **31.** $1 \pm \sqrt{2}$ **33.** $\frac{-3 \pm \sqrt{5}}{2}$ **35.** $-5, 3$ **37.** $2, -1 \pm i\sqrt{3}$ **39.** $-\frac{3}{2}, \frac{3 \pm 3i\sqrt{3}}{4}$
41. $\frac{1}{5}, \frac{-1 \pm i\sqrt{3}}{10}$ **43.** $0, \frac{-1 \pm i\sqrt{5}}{2}$ **45.** $0, 1 \pm i$ **47.** $0, \frac{-1 \pm i\sqrt{2}}{3}$ **49.** a and b **51. a.** $\frac{5}{3}, 0$ **b.** $\frac{5}{3}, 0$ **53.** No, $2 \pm i\sqrt{3}$
55. Yes **57. a.** $-1, 3$ **b.** $1 \pm i\sqrt{7}$ **c.** $-2, 2$ **d.** $2 \pm 2\sqrt{2}$ **59. a.** $-2, \frac{5}{3}$ **b.** $\frac{3}{2}, \frac{-3 \pm 3i\sqrt{3}}{4}$ **c.** $\varnothing$ **d.** $1, -1$ **61.** 2 seconds

63. $40 \pm 20 = 20$ or 60 items **65.** 0.49 centimeter (8.86 cm is impossible)

67. a. $\ell + w = 10$, $\ell w = 15$
b. 8.16 yards, 1.84 yards
c. Two answers are possible because either dimension (long or short) may be considered the length.

69. $4y + 1 + \frac{-2}{2y - 7}$ **71.** $x^2 + 7x + 12$ **73.** 5 **75.** $\frac{27}{125}$ **77.** $\frac{1}{4}$ **79.** $21x^3y$ **81.** 169 **83.** 0 **85.** ± 12 **87.** $x^2 - x - 6$

89. $x^3 - 4x^2 - 3x + 18$ **91.** $-2\sqrt{3}, \sqrt{3}$ **93.** $\frac{-1 \pm \sqrt{3}}{\sqrt{2}} = \frac{-\sqrt{2} \pm \sqrt{6}}{2}$ **95.** $-2i, i$

Problem Set 8.3

1. $D = 16$, two rational **3.** $D = 0$, one rational **5.** $D = 5$, two irrational **7.** $D = 17$, two irrational **9.** $D = 36$, two rational **11.** $D = 116$, two irrational **13.** ± 10 **15.** ± 12 **17.** 9 **19.** -16 **21.** $\pm 2\sqrt{6}$ **23.** $x^2 - 7x + 10 = 0$ **25.** $t^2 - 3t - 18 = 0$ **27.** $y^3 - 4y^2 - 4y + 16 = 0$ **29.** $2x^2 - 7x + 3 = 0$ **31.** $4t^2 - 9t - 9 = 0$ **33.** $6x^3 - 5x^2 - 54x + 45 = 0$ **35.** $10a^2 - a - 3 = 0$ **37.** $9x^3 - 9x^2 - 4x + 4 = 0$ **39.** $x^4 - 13x^2 + 36 = 0$ **41.** $f(x) = x^2 + x - 2$ **43.** $f(x) = x^2 - 6x + 7$ **45.** $f(x) = x^2 - 4x + 5$ **47.** $f(x) = 4x^2 - 20x + 18$ or $2x^2 - 10x + 9$ **49.** $f(x) = 4x^2 - 12x + 14$ or $2x^2 - 6x + 7$ **51.** $f(x) = x^3 - 4x^2 + 7x$ **53. a.** $y = g(x)$ **b.** $y = h(x)$ **c.** $y = f(x)$ **55.** $a - 2a^{1/2} - 15$ **57.** $x - 64$ **59.** $5xy^2 - 4x^2y$ **61.** $2(x + 1)^{1/3}(4x + 3)$ **63.** $(3x^{1/3} + 2)^2$ **65.** $x^2 - 7x$ **67.** $9a^2 - 6a - 3$ **69.** $24a^2 + 94a + 90$ **71.** $\pm i\sqrt{2}$ **73.** 4 **75.** -5 **77.** $-3, 2$ **79.** $-\frac{1}{3}, \frac{5}{2}$ **81.** $-\frac{1}{2}, \frac{5}{2}$ **83.** $\frac{1 + \sqrt{2}}{2} \approx 1.21$, $\frac{1 - \sqrt{2}}{2} \approx -0.21$ **85.** $\frac{2}{3}, 1, -1$ **87.** $\frac{2}{3}, \sqrt{3}, \approx 1.73, -\sqrt{3}, \approx -1.73$

Problem Set 8.4

1. 1, 2 **3.** $-8, -\frac{5}{2}$ **5.** $\pm 3, \pm 1$ **7.** $\pm 2, \pm\sqrt{3}$ **9.** $\pm 3, \pm i\sqrt{3}$ **11.** $\pm 2i, \pm i\sqrt{5}$ **13.** $\frac{7}{2}, 4$ **15.** $-\frac{9}{8}, \frac{1}{2}$ **17.** $\pm\frac{\sqrt{30}}{6}, \pm i$

19. $\pm\frac{\sqrt{21}}{3}, \pm\frac{i\sqrt{21}}{3}$ **21.** 4, 25 **23.** Possible solutions 25 and 9; only 25 checks; 25

25. Possible solutions $\frac{25}{9}$ and $\frac{49}{4}$; only $\frac{25}{9}$ checks; $\frac{25}{9}$ **27.** Possible solutions 4 and 16; only 16 checks; 16

29. Possible solutions 9 and 36; only 9 checks; 9 **31.** Possible solutions $\frac{1}{4}$ and 25; only $\frac{1}{4}$ checks; $\frac{1}{4}$ **33.** 27, 38 **35.** 4, 12

37. $\pm 2, \pm 2i$ **39.** $\pm 2, \pm 2i\sqrt{2}$ **41.** $\frac{1}{2}, 1$ **43.** $\pm\frac{\sqrt{5}}{5}$ **45.** $-216, 8$ **47.** $\pm 3i\sqrt{3}$ **49.** $t = \frac{v \pm \sqrt{v^2 + 64h}}{32}$

51. $x = \frac{-4 \pm 2\sqrt{4 - k}}{k}$ **53.** $x = -y$ **55.** $t = \frac{1 \pm \sqrt{1 + h}}{4}$

57. a.

y: 0, 100, 200, 300, 400, 500, 600, 700
x: 0, 100, 200, 300, 400, 500, 600

b. 630 feet

59. Let $x = BC$. Then $\frac{4}{x} = \frac{x}{x + 4}$, or $4(x + 4) = x^2$.

$$x = \frac{4 + 4\sqrt{5}}{2} = 4\left(\frac{1 + \sqrt{5}}{2}\right)$$

61. $3\sqrt{7}$ **63.** $5\sqrt{2}$ **65.** $39x^2y\sqrt{5x}$ **67.** $-11 + 6\sqrt{5}$ **69.** $x + 4\sqrt{x} + 4$ **71.** $\frac{7 + 2\sqrt{7}}{3}$ **73.** -2 **75.** 1,322.5

77. $-\frac{7}{640}$ **79.** 1, 5 **81.** $-3, 1$ **83.** $\frac{3}{2} \pm \frac{1}{2}i$ **85.** 9; 3 **87.** 1; 1 **89.** x-intercepts $= -2, 0, 2$; y-intercept $= 0$

91. x-intercepts $= -3, -\frac{1}{3}, 3$; y-intercept $= -9$ **93.** $\frac{1}{2}$ and -1

Problem Set 8.5

1. x-intercepts $= -3, 1$;
Vertex $= (-1, -4)$

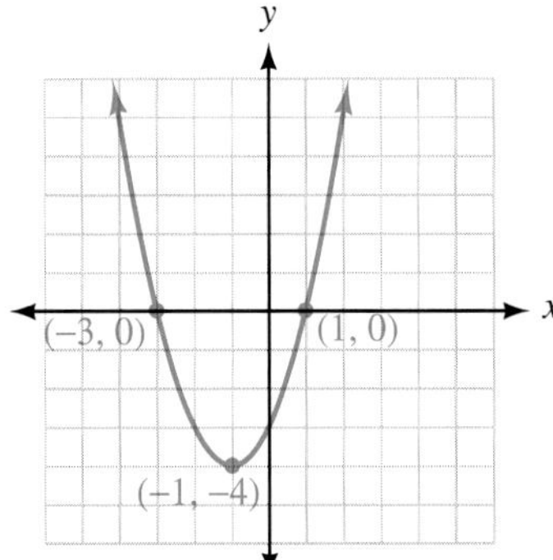

3. x-intercepts $= -5, 1$;
Vertex $= (-2, 9)$

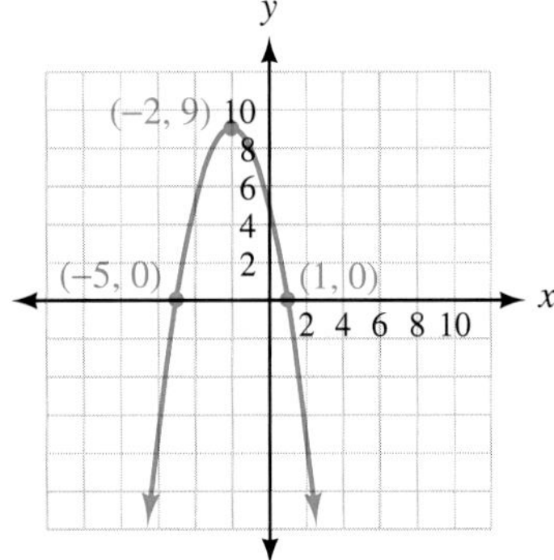

5. x-intercepts $= -1, 1$;
Vertex $= (0, -1)$

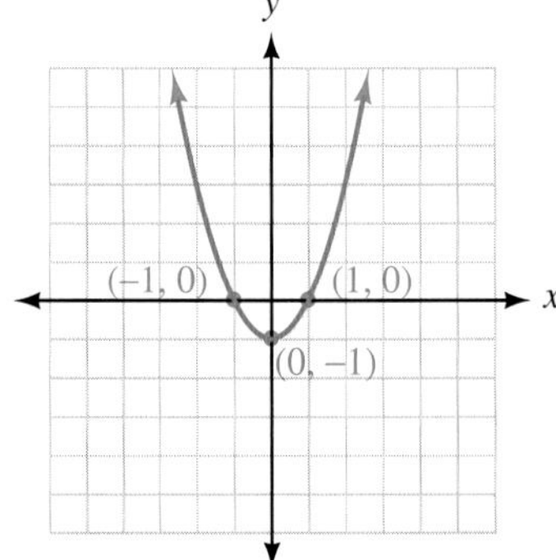

7. x-intercepts $= 3, -3$;
Vertex $= (0, 9)$

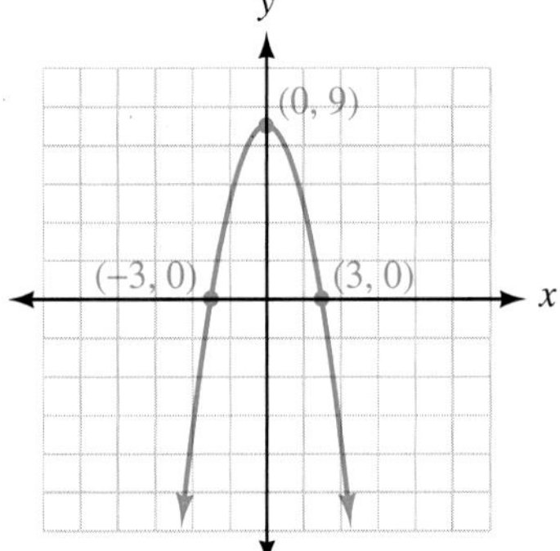

9. x-intercepts $= -1, 3$;
Vertex $= (1, -8)$

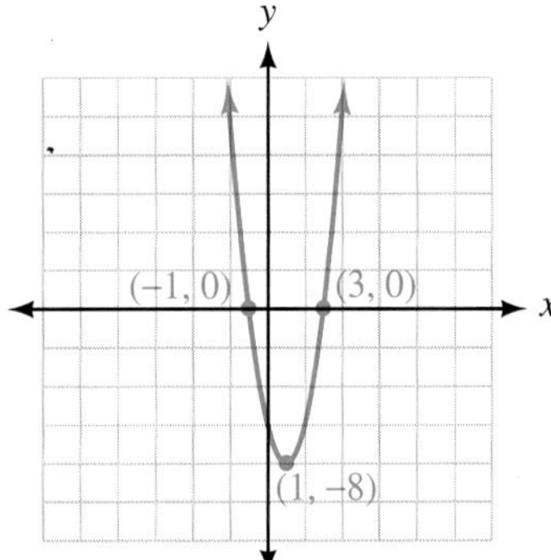

11. x-intercepts $= 1 + \sqrt{5}, 1 - \sqrt{5}$;
Vertex $= (1, -5)$

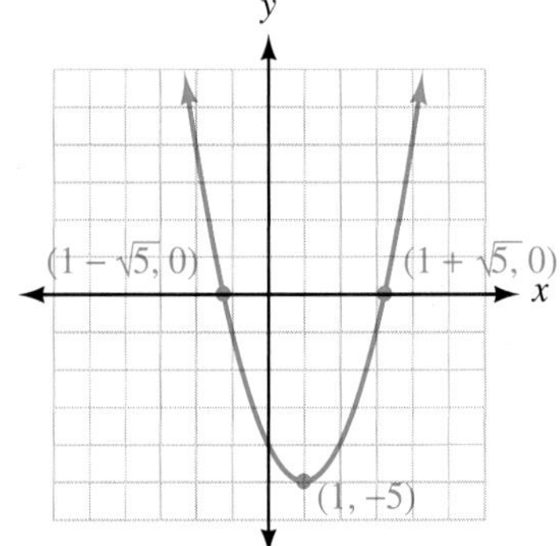

13. Vertex $= (2, -8)$

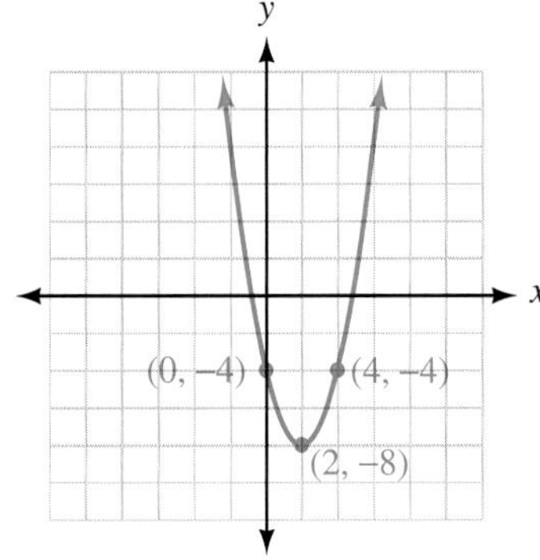

15. Vertex $= (1, -4)$

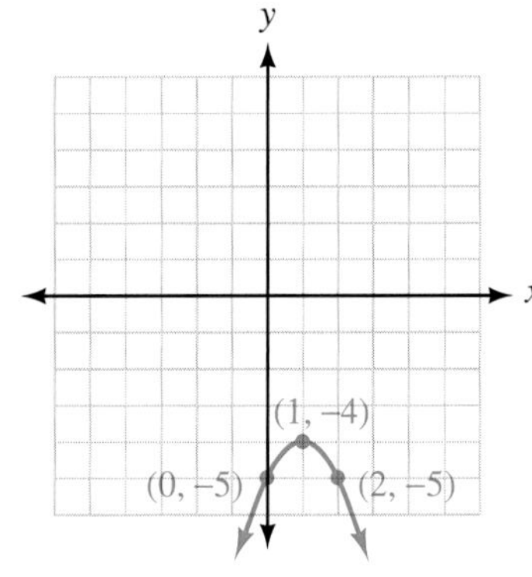

17. Vertex $= (0, 1)$

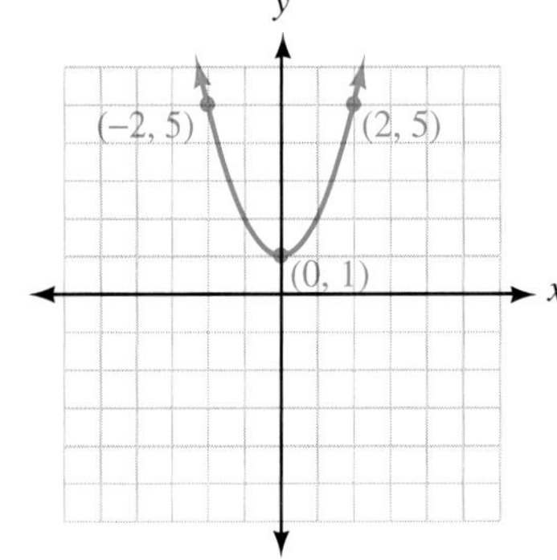

19. Vertex = (0, −3)

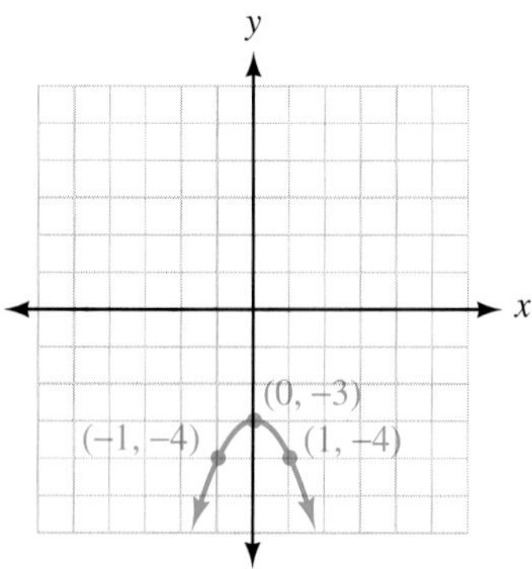

21. (3, −4) lowest **23.** (1, 9) highest

25. (2, 16) highest **27.** (−4, 16) highest **29.** 875 patterns; maximum profit $731.25

31. The ball is in her hand when $h(t) = 0$, which means $t = 0$ or $t = 2$ seconds. Maximum height is $h(1) = 16$ feet.

33. Maximum $R = \$3{,}600$ when $p = \$6.00$ **35.** Maximum $R = \$7{,}225$ when $p = \$8.50$

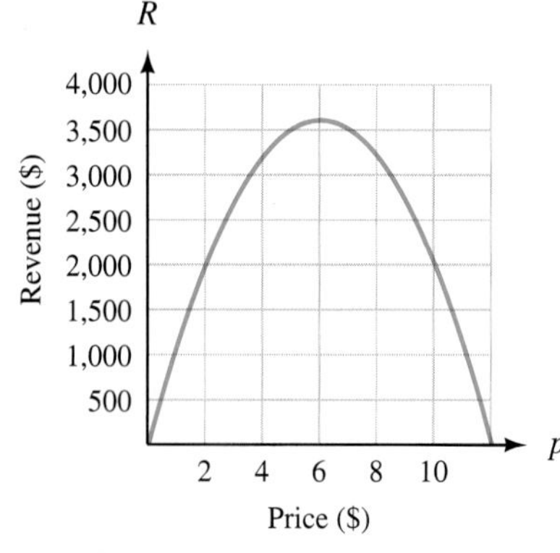

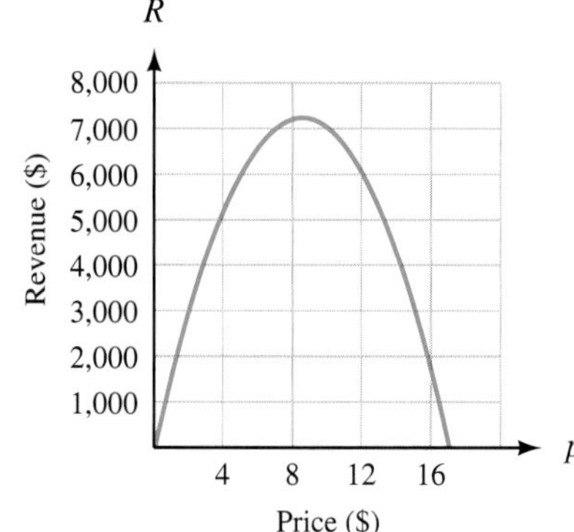

37. $y = -\frac{1}{135}(x - 90)^2 + 60$

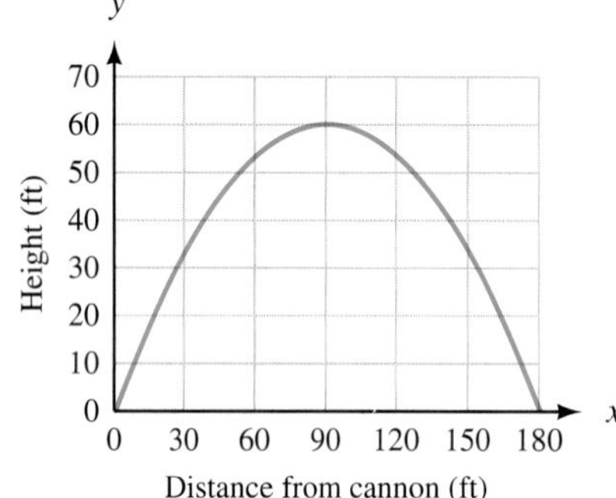

39. $1 - i$ **41.** $27 + 5i$ **43.** $\frac{1}{10} + \frac{3}{10}i$ **45.** −2, 4 **47.** $-\frac{1}{2}, \frac{2}{3}$ **49.** 3 **51.** $y = (x - 2)^2 - 4$

Problem Set 8.6

1. 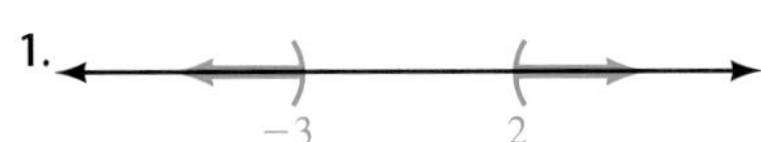

3. 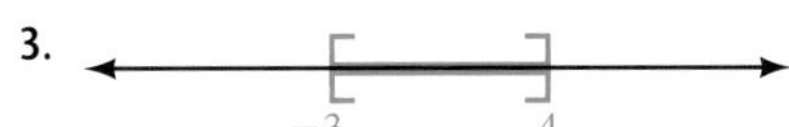

5. 

7. 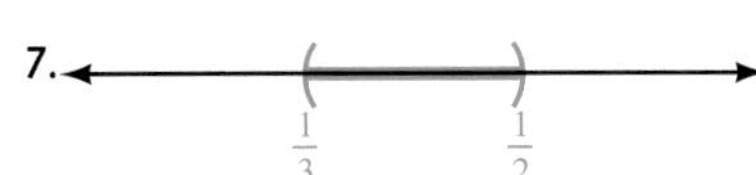

9. 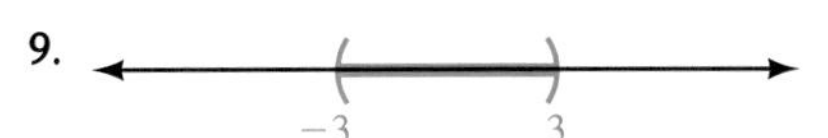

11. 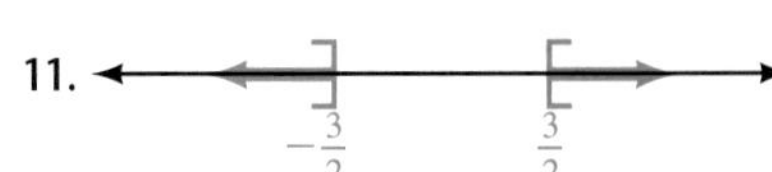

13. 

15. All real numbers **17.** No solution; ∅

19. 

21. 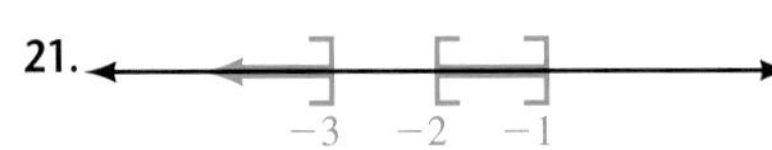

23. 

25. 

27. 

29.
−3 2 4

31. 

33. a. $x - 1 > 0$ **b.** $x - 1 \geq 0$ **c.** $x - 1 \geq 0$ **35.** ∅ **37.** $x = 1$ **39.** all real numbers **41.** $x > 1$ or $x < 1$ **43.** ∅
45. $x > 3$ or $x < 3$; all real numbers except 3 **47. a.** $-2 < x < 2$ **b.** $x < -2$ or $x > 2$ **c.** $x = -2$ or $x = 2$
49. a. $-2 < x < 5$ **b.** $x < -2$ or $x > 5$ **c.** $x = -2$ or $x = 5$
51. a. $x < -1$ or $1 < x < 3$ **b.** $-1 < x < 1$ or $x > 3$ **c.** $x = -1$ or $x = 1$ or $x = 3$ **53.** $x \geq 4$; the width is at least 4 inches
55. $5 \leq p \leq 8$; charge at least \$5 but no more than \$8 for each radio **57.** 1.5625 **59.** 0.6549 **61.** $\frac{5}{3}$
63. Possible solutions 1 and 6; only 6 checks; 6

65.

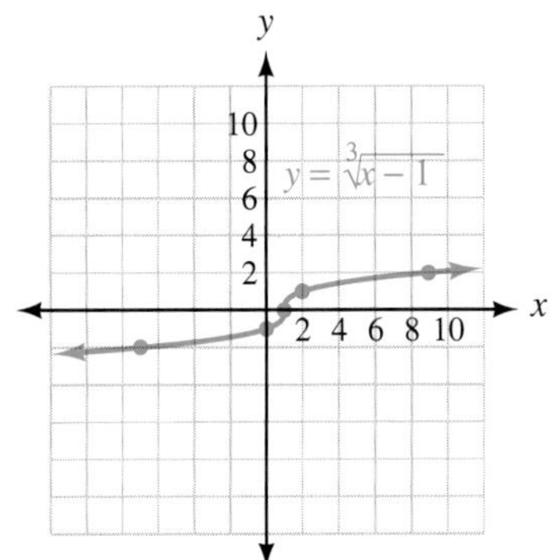

67. 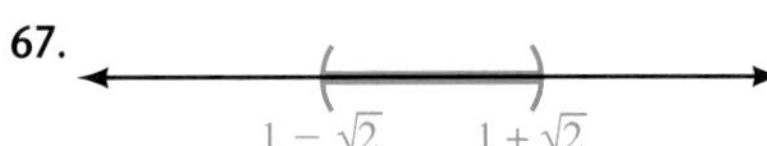

69. 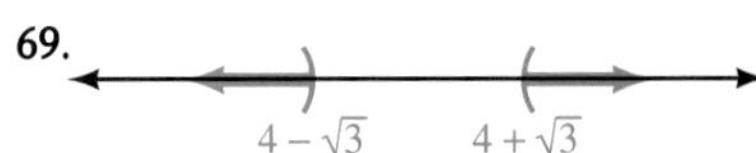

CHAPTER 8 REVIEW TEST

1. 0, 5 **2.** 0, $\frac{4}{3}$ **3.** $\frac{4 \pm 7i}{3}$ **4.** $-3 \pm \sqrt{3}$ **5.** −5, 2 **6.** −3, −2 **7.** 3 **8.** 2 **9.** $\frac{-3 \pm \sqrt{3}}{2}$ **10.** $\frac{3 \pm \sqrt{5}}{2}$ **11.** −5, 2

12. 0, $\frac{9}{4}$ **13.** $\frac{2 \pm i\sqrt{15}}{2}$ **14.** $1 \pm \sqrt{2}$ **15.** $-\frac{4}{3}, \frac{1}{2}$ **16.** 3 **17.** $5 \pm \sqrt{7}$ **18.** 0, $\frac{5 \pm \sqrt{21}}{2}$ **19.** −1, $\frac{3}{5}$ **20.** 3, $\frac{-3 \pm 3i\sqrt{3}}{2}$

21. $\frac{1 \pm i\sqrt{2}}{3}$ **22.** $\frac{3 \pm \sqrt{29}}{2}$ **23.** 100 or 170 items **24.** 20 or 60 items **25.** $D = 0$; 1 rational **26.** $D = 0$; 1 rational

27. $D = 25$; 2 rational **28.** $D = 361$; 2 rational **29.** $D = 5$; 2 irrational **30.** $D = 5$; 2 irrational **31.** $D = -23$; 2 complex
32. $D = -87$; 2 complex **33.** ±20 **34.** ±20 **35.** 4 **36.** 4 **37.** 25 **38.** 49 **39.** $x^2 - 8x + 15 = 0$ **40.** $x^2 - 2x - 8 = 0$
41. $2y^2 + 7y - 4 = 0$ **42.** $t^3 - 5t^2 - 9t + 45 = 0$ **43.** −4, 12 **44.** $-\frac{3}{4}, -\frac{1}{6}$ **45.** $\pm 2, \pm i\sqrt{3}$

46. Possible solutions 4 and 1, only 4 checks; 4 **47.** $\frac{9}{4}$, 16 **48.** 4 **49.** 4 **50.** 7 **51.** $t = \frac{5 \pm \sqrt{25 + 16h}}{16}$

52. $t = \frac{v \pm \sqrt{v^2 + 640}}{32}$ **53.** −1 2 **54.** $\frac{2}{3}$ 4

55.

56.

57.
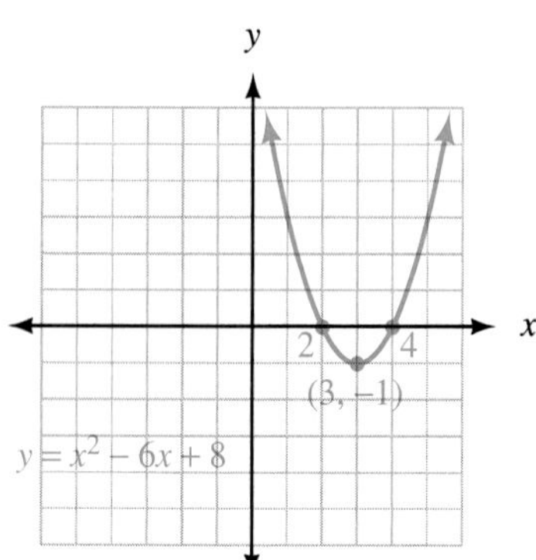

58.
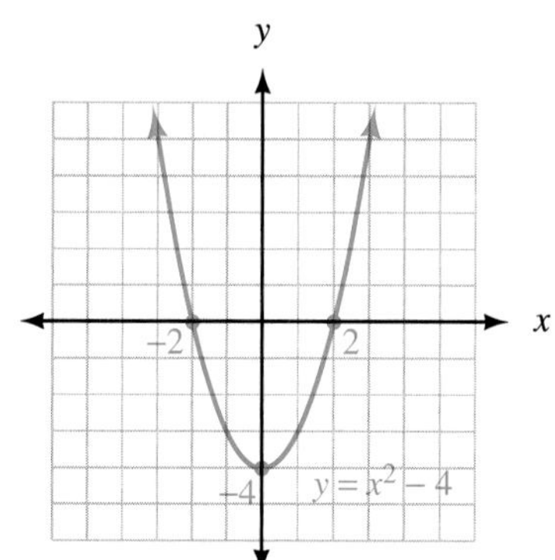

CHAPTER 9

Problem Set 9.1

1. 1 **3.** 2 **5.** $\frac{1}{27}$ **7.** 13

9.
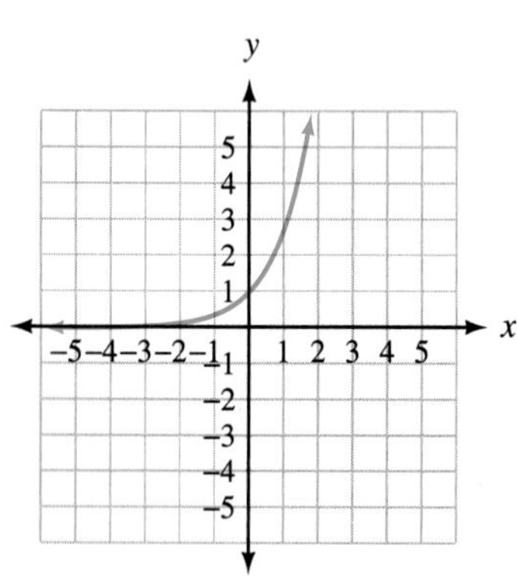

11.
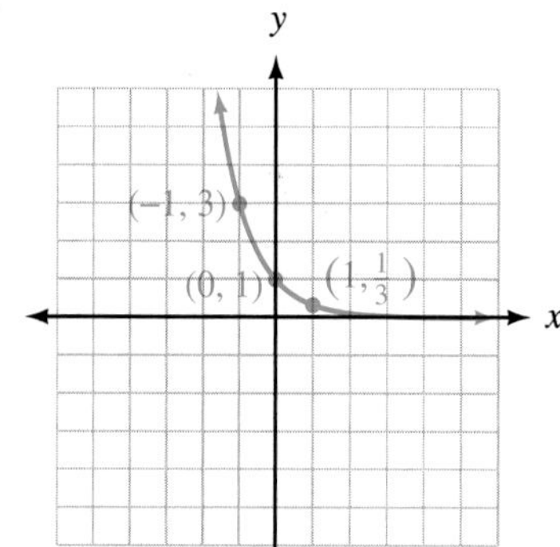

13.
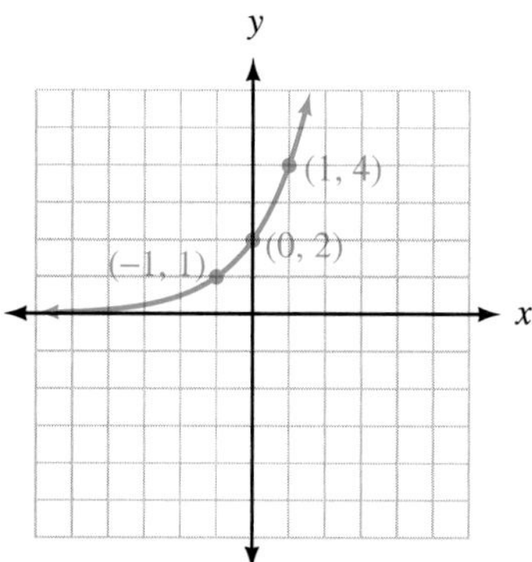

15.
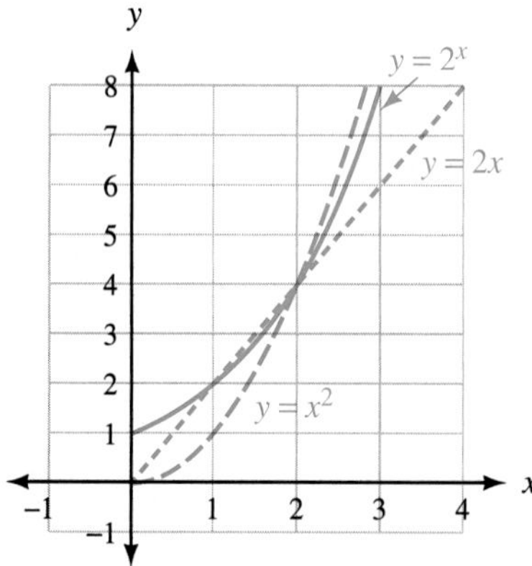

17.
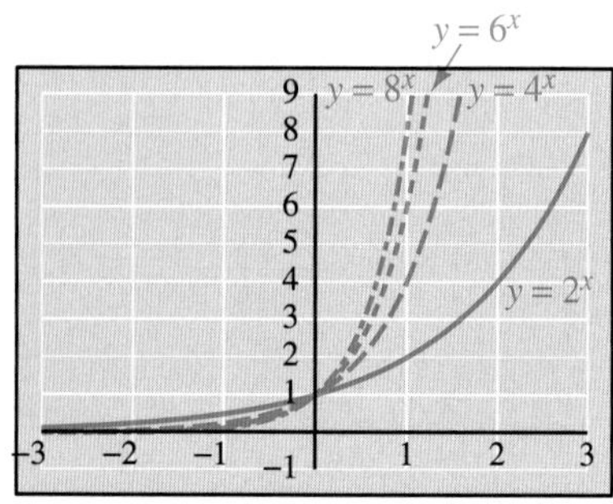

19.

21. $h = 6 \cdot (\frac{2}{3})^n$; 5th bounce: $6(\frac{2}{3})^5 \approx 0.79$ feet **23.** After 8 days, 700 micrograms; After 11 days, $1{,}400 \cdot 2^{-11/8} \approx 539.8$ micrograms

25. a. $A(t) = 1{,}200\left(1 + \frac{.06}{4}\right)^{4t}$ **b.** \$1,932.39 **c.** About 11.64 years **d.** \$1,939.29

27. a. The function underestimated the expenditures by \$69 billion.
b. \$4,123 billion
\$4,577 billion
\$5,080 billion

29.

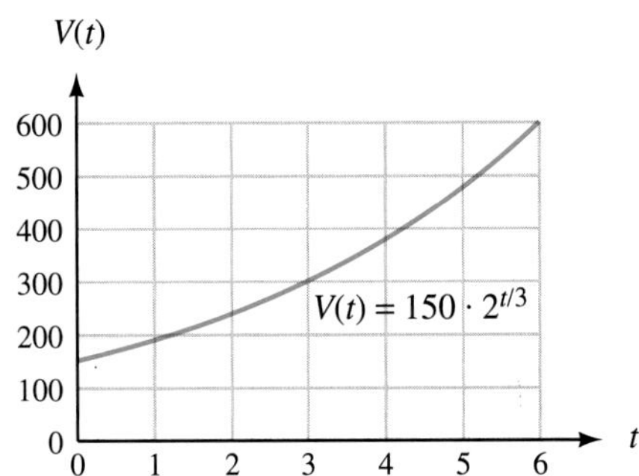

31. After 6 years

33. **a.** \$129,138.48 **b.** $\{t \mid 0 \le t \le 6\}$

c.

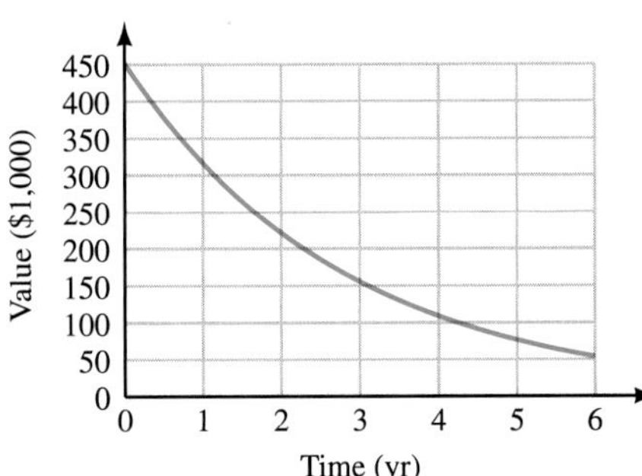

d. $\{V(t) \mid 52{,}942.05 \le V(t) \le 450{,}000\}$

e. After approximately 4 years and 8 months

35. $D = \{1, 3, 4\}$, $R = \{2, 4, 1\}$, a function **37.** $\{x \mid x \ge -\frac{1}{3}\}$ **39.** -18 **41.** 2

43. $y = \dfrac{x+3}{2}$ **45.** $y = \pm\sqrt{x+2}$ **47.** $y = \dfrac{2x-4}{x-1}$ **49.** $y = x^2 + 3$

51.

t	$s(t)$
0	0
1	83.3
2	138.9
3	175.9
4	200.6
5	217.1
6	228.1

250
200
150
100
50
0
Speed (mi/hr)
0 1 2 3 4 5 6
Time (sec)

Problem Set 9.2

1. $f^{-1}(x) = \dfrac{x+1}{3}$ **3.** $f^{-1}(x) = \sqrt[3]{x}$ **5.** $f^{-1}(x) = \dfrac{x-3}{x-1}$ **7.** $f^{-1}(x) = 4x + 3$ **9.** $f^{-1}(x) = 2(x+3) = 2x + 6$

11. $f^{-1}(x) = \dfrac{1-x}{3x-2}$ 

13.

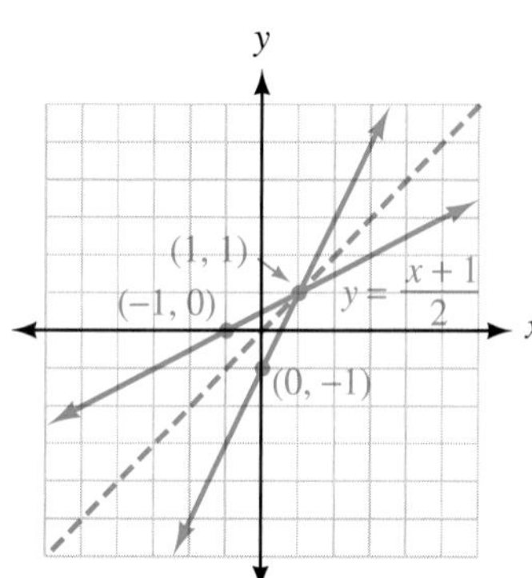

15.

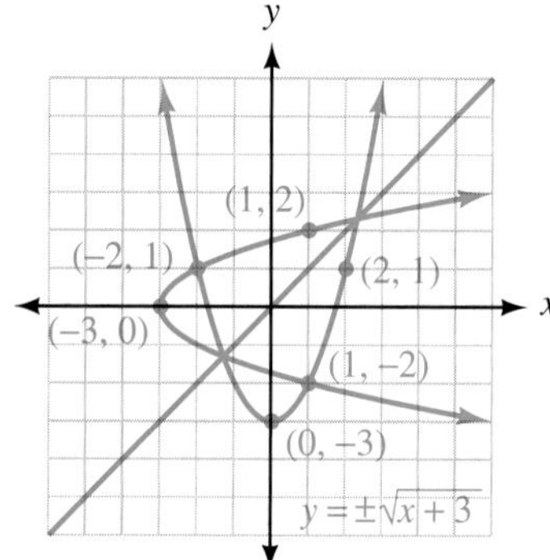

17.

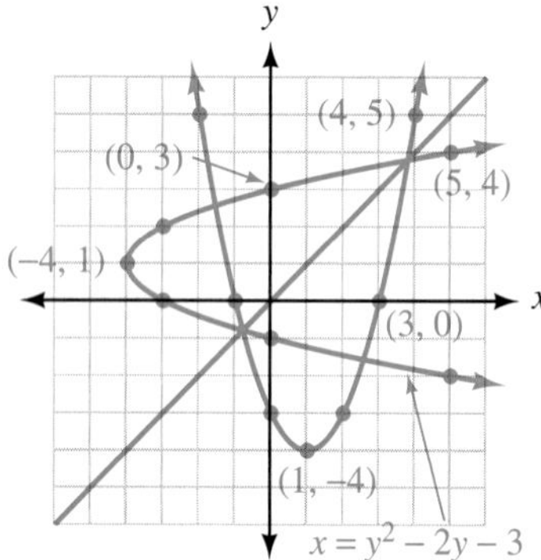

19.

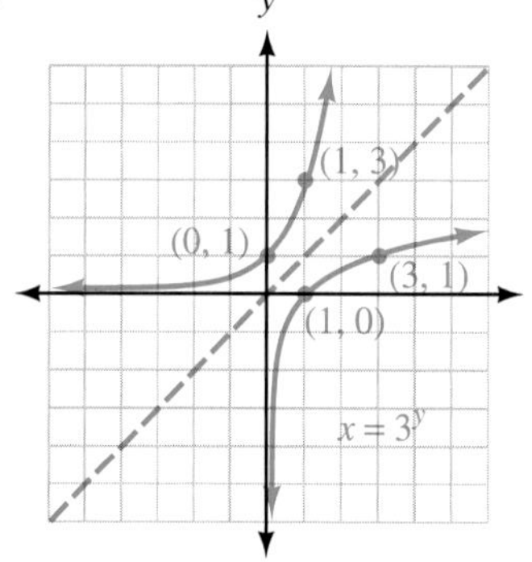

21.

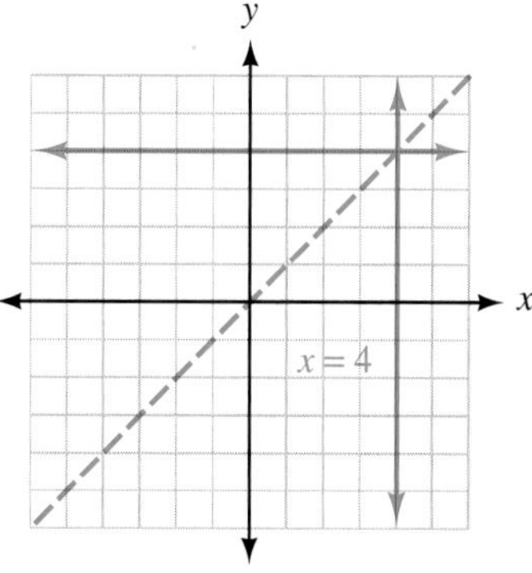

23.

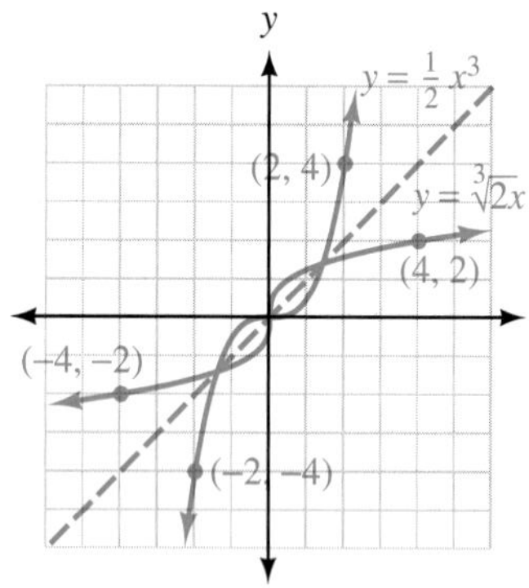

25.

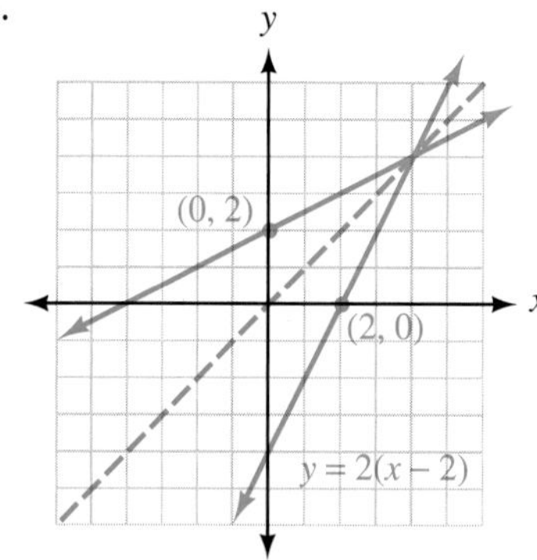

27.

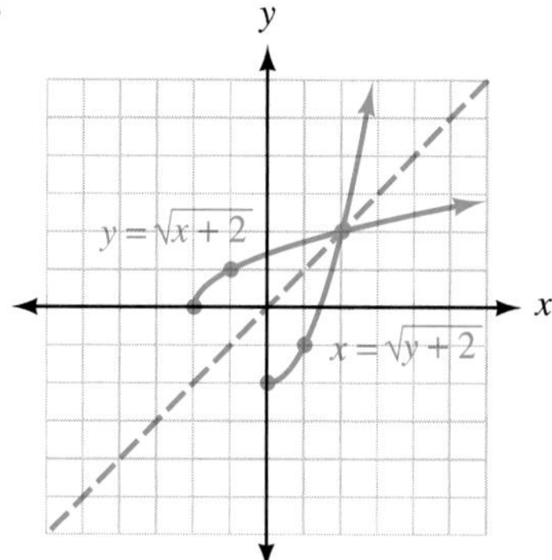

29. a. Yes **b.** No **c.** Yes **31. a.** 4 **b.** $\frac{4}{3}$ **c.** 2 **d.** 2 **33.** $f^{-1}(x) = \frac{1}{x}$
35. a. −3 **b.** −6 **c.** 2 **d.** 3 **e.** −2 **f.** 3 **g.** They are inverses of each other. **37.** −2, 3 **39.** 25 **41.** $5 \pm \sqrt{17}$ **43.** $1 \pm i$
45. $\frac{1}{9}$ **47.** $\frac{2}{3}$ **49.** $\sqrt[3]{4}$ **51.** 3 **53.** 4 **55.** 4 **57.** 1 **59.** $f^{-1}(x) = \dfrac{x-5}{3}$ **61.** $f^{-1}(x) = \sqrt[3]{x-1}$ **63.** $f^{-1}(x) = \dfrac{2x-4}{x-1}$
65. a. 1 **b.** 2 **c.** 5 **d.** 0 **e.** 1 **f.** 2 **g.** 2 **h.** 5

Problem Set 9.3

1. $\log_2 16 = 4$ **3.** $\log_5 125 = 3$ **5.** $\log_{10} 0.01 = -2$ **7.** $\log_2 \frac{1}{32} = -5$ **9.** $\log_{1/2} 8 = -3$ **11.** $\log_3 27 = 3$ **13.** $10^2 = 100$
15. $2^6 = 64$ **17.** $8^0 = 1$ **19.** $10^{-3} = 0.001$ **21.** $6^2 = 36$ **23.** $5^{-2} = \frac{1}{25}$ **25.** 9 **27.** $\frac{1}{125}$ **29.** 4 **31.** $\frac{1}{3}$ **33.** 2 **35.** $\sqrt[3]{5}$

37.

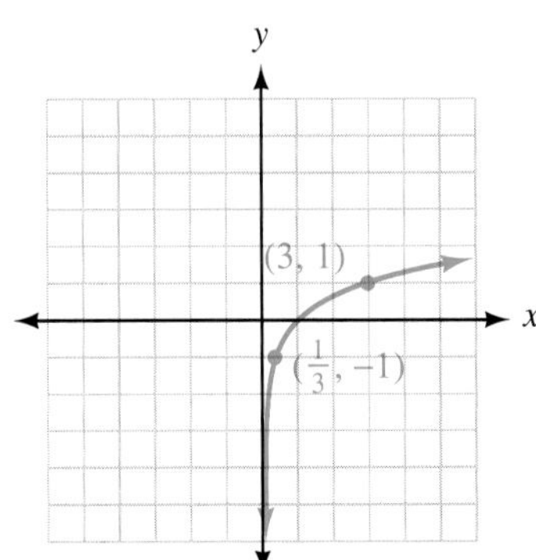

39.

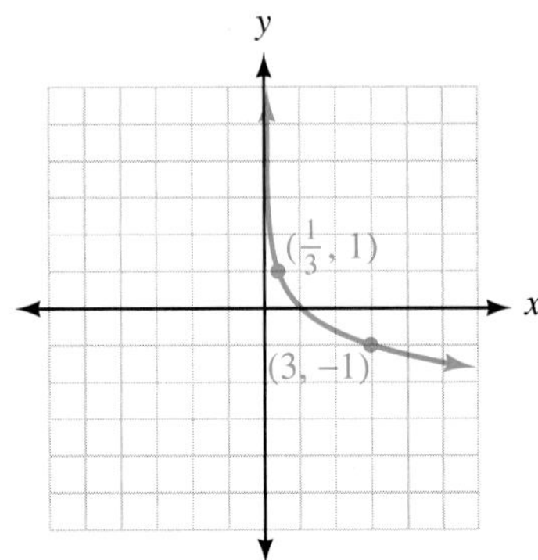

41.

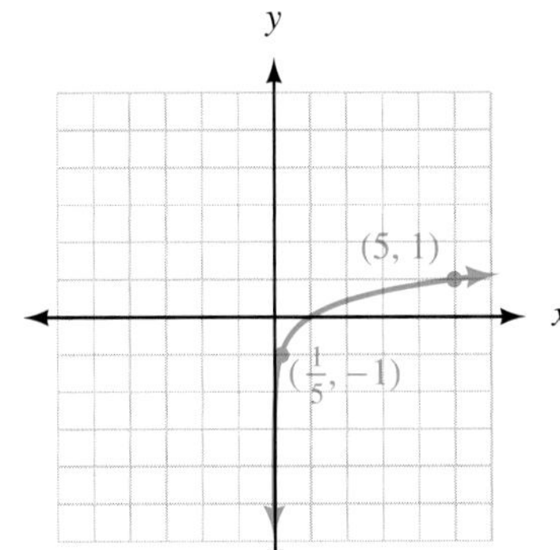

43.

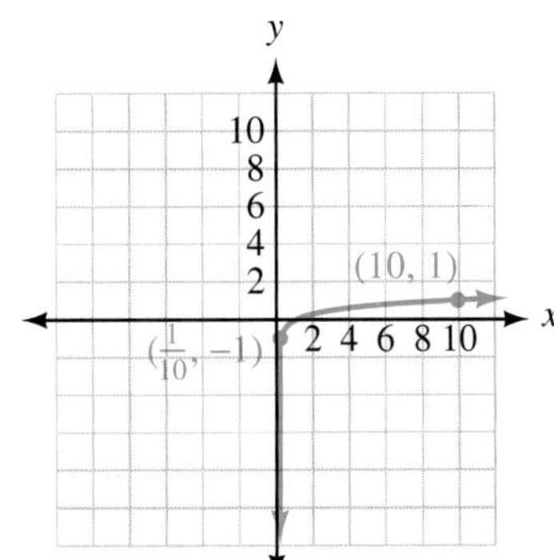

45. 4 **47.** $\frac{3}{2}$ **49.** 3 **51.** 1 **53.** 0 **55.** 0 **57.** $\frac{1}{2}$ **59.** 7 **61.** 10^{-6} **63.** 2 **65.** 10^8 times as large **67.** 25; 5 **69.** 1; 1 **71.** $\pm 3i$ **73.** $\pm 2i$ **75.** $\frac{-2 \pm \sqrt{10}}{2}$ **77.** $\frac{2 \pm i\sqrt{3}}{2}$ **79.** 5, $\frac{-5 \pm 5i\sqrt{3}}{2}$ **81.** 0, 5, −1 **83.** $\frac{3 \pm \sqrt{29}}{2}$ **85.** 4 **87.** −4, 2 **89.** $-\frac{11}{8}$

91. $2^3 = (x + 2)(x)$ **93.** $3^4 = \frac{x - 2}{x + 1}$ **95. a.**

x	$f(x)$
−1	$\frac{1}{8}$
0	1
1	8
2	64

b.

x	$f^{-1}(x)$
$\frac{1}{8}$	−1
1	0
8	1
64	2

c. $f(x) = 8^x$ **d.** $f^{-1}(x) = \log_8 x$

Problem Set 9.4

1. $\log_3 4 + \log_3 x$ **3.** $\log_6 5 - \log_6 x$ **5.** $5 \log_2 y$ **7.** $\frac{1}{3} \log_9 z$ **9.** $2 \log_6 x + 4 \log_6 y$ **11.** $\frac{1}{2} \log_5 x + 4 \log_5 y$ **13.** $\log_b x + \log_b y - \log_b z$ **15.** $\log_{10} 4 - \log_{10} x - \log_{10} y$ **17.** $2 \log_{10} x + \log_{10} y - \frac{1}{2} \log_{10} z$ **19.** $3 \log_{10} x + \frac{1}{2} \log_{10} y - 4 \log_{10} z$ **21.** $\frac{2}{3} \log_b x + \frac{1}{3} \log_b y - \frac{4}{3} \log_b z$ **23.** $\log_b xz$ **25.** $\log_3 \frac{x^2}{y^3}$ **27.** $\log_{10} \sqrt{x}\sqrt[3]{y}$ **29.** $\log_2 \frac{x^3\sqrt{y}}{z}$ **31.** $\log_2 \frac{\sqrt{x}}{y^3 z^4}$ **33.** $\log_{10} \frac{x^{3/2}}{y^{3/4} z^{4/5}}$ **35.** $\frac{2}{3}$ **37.** 18 **39.** Possible solutions −1 and 3; only 3 checks; 3 **41.** 3 **43.** Possible solutions −2 and 4; only 4 checks; 4 **45.** Possible solutions −1 and 4; only 4 checks; 4 **47.** Possible solutions $-\frac{5}{2}$ and $\frac{5}{3}$; only $\frac{5}{3}$ checks; $\frac{5}{3}$ **51.** N is 2 in both cases **53.** $\frac{x^2}{3} + \frac{y^2}{36}$ **55.** $\frac{x^2}{4} + \frac{y^2}{25}$ **57.** $D = -7$; two complex **59.** $x^2 - 2x - 15 = 0$ **61.** $3y^2 - 11y + 6 = 0$ **63.** 1 **65.** 1 **67.** 4 **69.** 2.5×10^{-6} **71.** 51

Problem Set 9.5

1. 2.5775 **3.** 1.5775 **5.** 3.5775 **7.** −1.4225 **9.** 4.5775 **11.** 2.7782 **13.** 3.3032 **15.** −2.0128 **17.** −1.5031 **19.** −0.3990 **21.** 759 **23.** 0.00759 **25.** 1,430 **27.** 0.00000447 **29.** 0.0000000918 **31.** 10^{10} **33.** 10^{-10} **35.** 10^{20} **37.** $\frac{1}{100}$ **39.** 1,000 **41.** 1 **43.** 5 **45.** x **47.** $\ln 10 + 3t$ **49.** $\ln A - 2t$ **51.** 2.7080 **53.** −1.0986 **55.** 2.1972 **57.** 2.7724 **59. a.** 1.1886 **b.** 0.2218 **61. a.** 0.6667 **b.** −0.3010 **63.** 3.19 **65.** 1.78×10^{-5} **67.** 3.16×10^5

69. 2.00×10^8 **71.** 10 times larger **73.**

Location	Date	Magnitude, M	Shock Wave, T
Moresby Island	January 23	4.0	1.00×10^4
Vancouver Island	April 30	5.3	1.99×10^5
Quebec City	June 29	3.2	1.58×10^3
Mould Bay	November 13	5.2	1.58×10^5
St. Lawrence	December 14	3.7	5.01×10^3

75. 12.9% **77.** 5.3% **79.**

x	$(1 + x)^{1/x}$
1	2
0.5	2.25
0.1	2.5937
0.01	2.7048
0.001	2.7169
0.0001	2.7181
0.00001	2.7183

e **81.** 0, −3 **83.** −4, −2 **85.** $\pm 2, \pm i\sqrt{2}$ **87.** $1, \frac{9}{4}$ **89.** 0.7

91. 3.125 **93.** 1.2575 **95.** $t \cdot \log(1.05)$ **97.** $0.05t$

Problem Set 9.6

1. 1.4650 **3.** 0.6826 **5.** −1.5440 **7.** −0.6477 **9.** $-\frac{1}{3}$ **11.** 2 **13.** −0.1845 **15.** 0.1845 **17.** 1.6168 **19.** 2.1131 **21.** 1.3333 **23.** 0.7500 **25.** 1.3917 **27.** 0.7186 **29.** 2.6356 **31.** 4.1632 **33.** 5.8435 **35.** −1.0642 **37.** 2.3026 **39.** 10.7144 **41.** 11.7 years **43.** 9.25 years **45.** 8.75 years **47.** 18.58 years **49.** 11.55 years **51.** 18.31 years **53.** lowest point: (−2, −23) **55.** highest point: $(\frac{3}{2}, 9)$ **57.** 2 seconds, 64 feet **59.** 13.9 years later or toward the end of 2007 **61.** 27.5 years **63.** $t = \frac{1}{r} \ln \frac{A}{P}$ **65.** $t = \frac{1}{k} \frac{\log P - \log A}{\log 2}$ **67.** $t = \frac{\log A - \log P}{\log(1 - r)}$

CHAPTER 9 REVIEW TEST

1. 16 **2.** $\frac{1}{2}$ **3.** $\frac{1}{9}$ **4.** −5 **5.** $\frac{5}{6}$ **6.** 7 **7.**

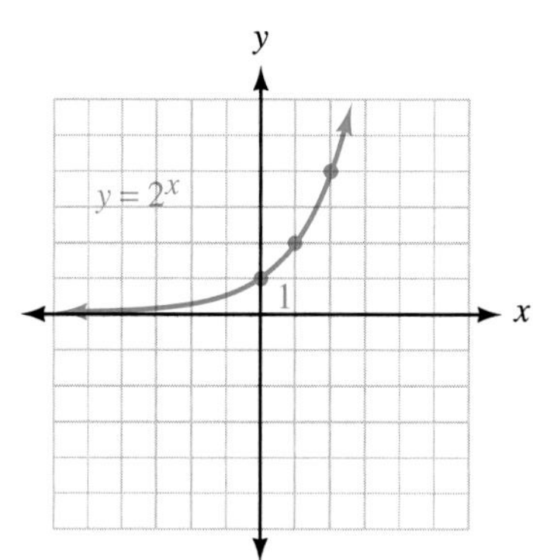

8.

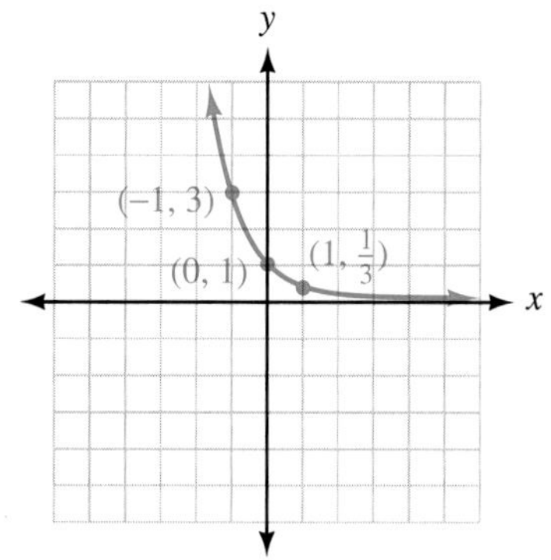

9.

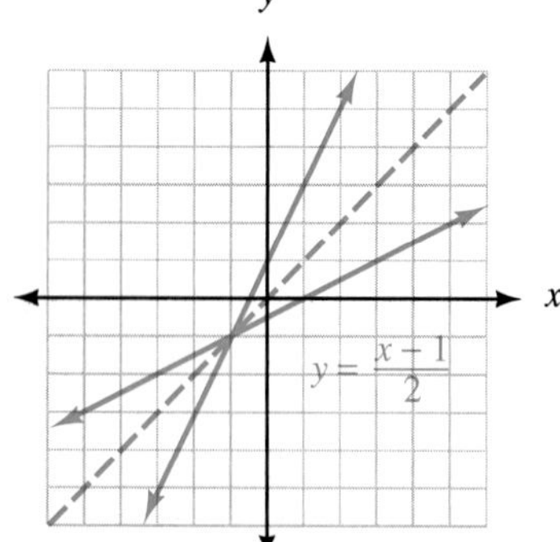

10.

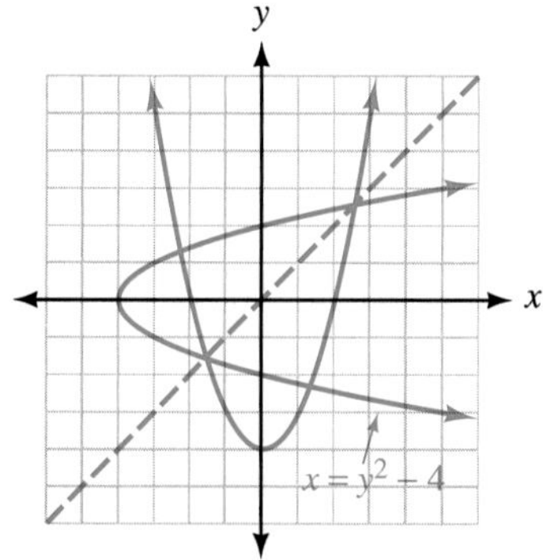

11. $f^{-1}(x) = \frac{x - 3}{2}$

12. $y = \pm\sqrt{x + 1}$ **13.** $f^{-1}(x) = 2x - 4$ **14.** $y = \pm\sqrt{\frac{4 - x}{2}}$ **15.** $\log_3 81 = 4$ **16.** $\log_7 49 = 2$ **17.** $\log_{10} 0.01 = -2$ **18.** $\log_2 \frac{1}{8} = -3$ **19.** $2^3 = 8$ **20.** $3^2 = 9$ **21.** $4^{1/2} = 2$ **22.** $4^1 = 4$ **23.** 25 **24.** $\frac{3}{4}$ **25.** 10

26.

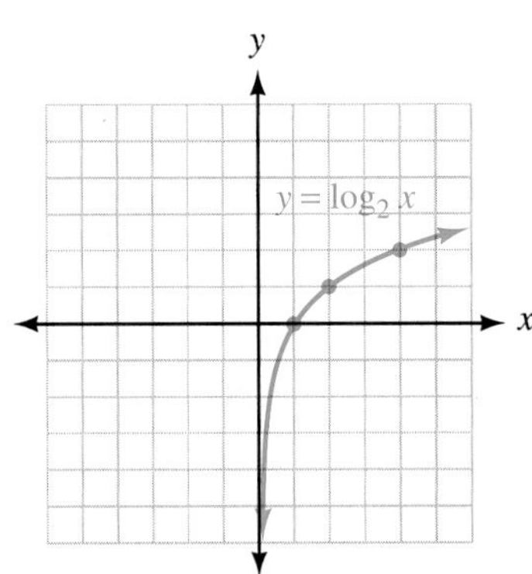

27.

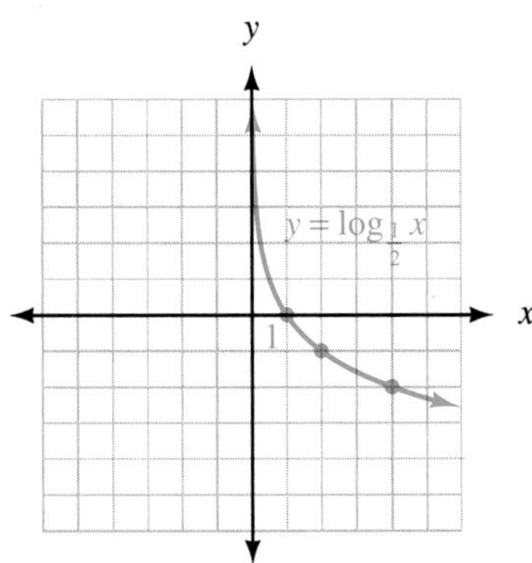

28. 2 **29.** $\frac{2}{3}$ **30.** 0 **31.** $\log_2 5 + \log_2 x$

32. $\log_{10} 2 + \log_{10} x - \log_{10} y$ **33.** $\frac{1}{2}\log_a x + 3\log_a y - \log_a z$ **34.** $2\log_{10} x - 3\log_{10} y - 4\log_{10} z$ **35.** $\log_2 xy$ **36.** $\log_3 \frac{x}{4}$

37. $\log_a \frac{25}{3}$ **38.** $\log_2 \frac{x^3y^2}{z^4}$ **39.** 2 **40.** 6 **41.** Possible solutions -1 and 3; only 3 checks; 3 **42.** 3

43. Possible solutions -2 and 3; only 3 checks; 3 **44.** Possible solutions -1 and 4; only 4 checks; 4 **45.** 2.5391 **46.** -0.1469
47. 9,230 **48.** 0.0251 **49.** 1 **50.** 0 **51.** 2 **52.** -4 **53.** 2.1 **54.** 5.1 **55.** 2.0×10^{-3} **56.** 3.2×10^{-8} **57.** $\frac{3}{2}$

58. $x = \frac{1}{3}\left(\frac{\log 5}{\log 4} - 2\right) \approx -0.28$ **59.** 0.75 **60.** 2.43 **61.** About 4.67 years **62.** About 9.25 years

CHAPTER 10

Problem Set 10.1

1. 5 **3.** $\sqrt{106}$ **5.** $\sqrt{61}$ **7.** $\sqrt{130}$ **9.** 3 or -1 **11.** 3 **13.** $(x-2)^2 + (y-3)^2 = 16$ **15.** $(x-3)^2 + (y+2)^2 = 9$
17. $(x+5)^2 + (y+1)^2 = 5$ **19.** $x^2 + (y+5)^2 = 1$ **21.** $x^2 + y^2 = 4$

23. Center = (0, 0), Radius = 2

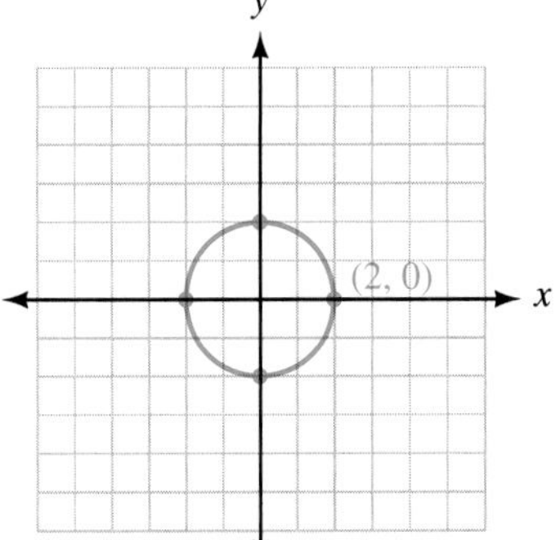

25. Center = (1, 3), Radius = 5

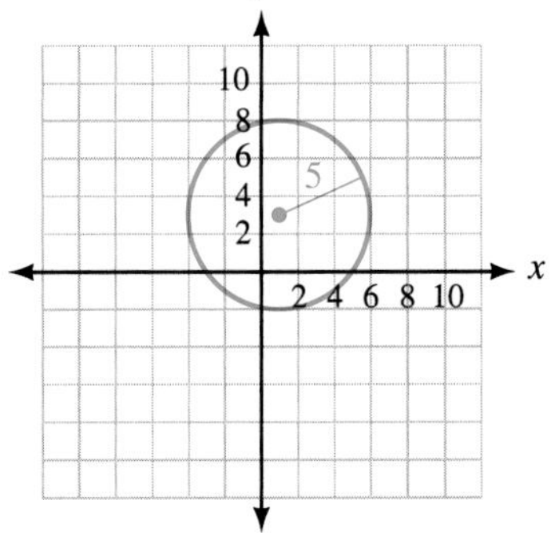

27. Center = (−2, 4), Radius = $2\sqrt{2}$

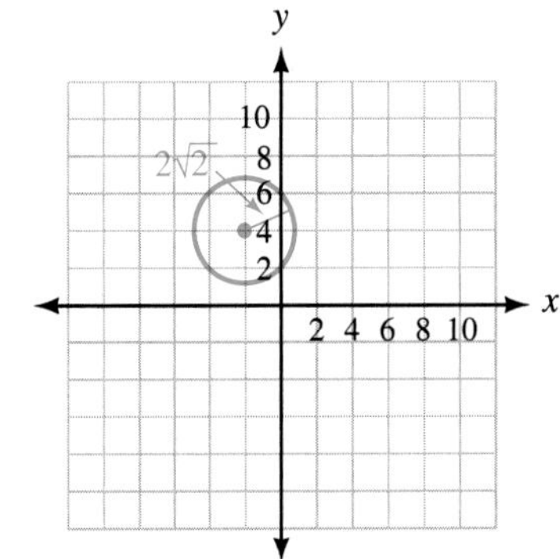

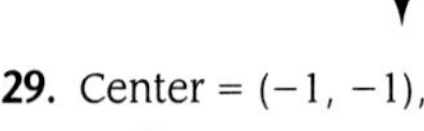
29. Center = (−1, −1), Radius = 1

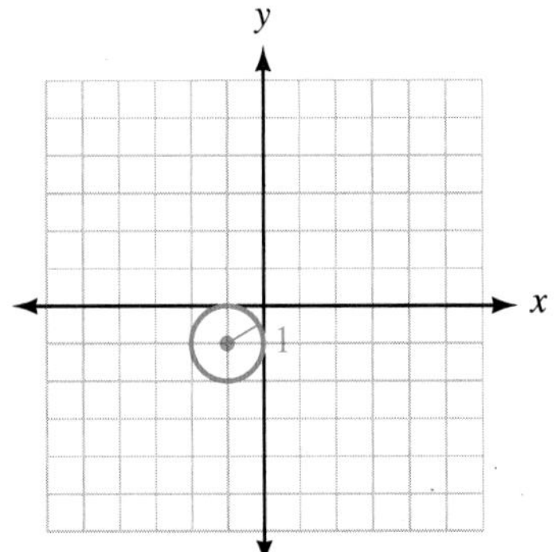

31. Center = (0, 3),
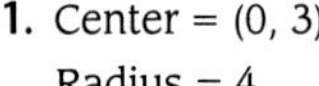
Radius = 4

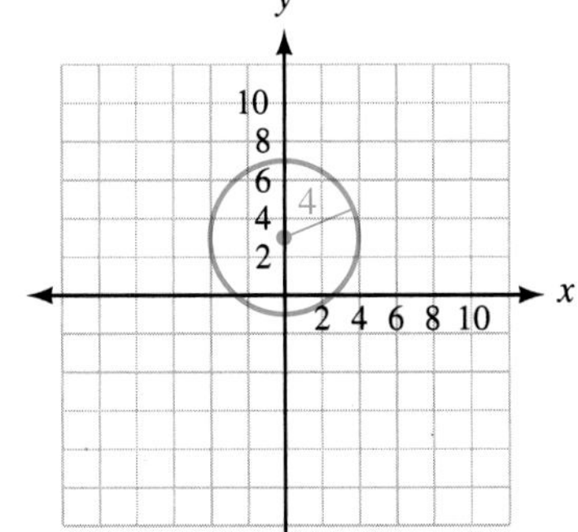

33. Center = (2, 3), Radius = 3

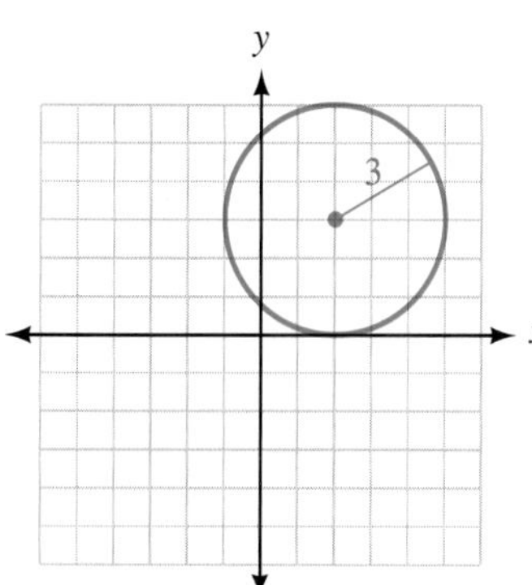

35. Center = $(-1, -\frac{1}{2})$, Radius = 2

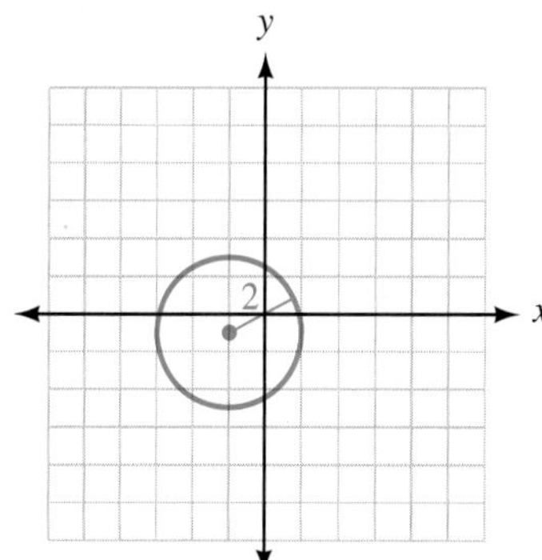

37. $(x - 3)^2 + (y - 4)^2 = 25$

39. A: $(x - \frac{1}{2})^2 + (y - 1)^2 = \frac{1}{4}$
B: $(x - 1)^2 + (y - 1)^2 = 1$
C: $(x - 2)^2 + (y - 1)^2 = 4$

41. A: $(x + 8)^2 + y^2 = 64$
B: $x^2 + y^2 = 64$
C: $(x - 8)^2 + y^2 = 64$

43. $(x - 500)^2 + (y - 132)^2 = 120^2$ **45.** $f^{-1}(x) = \log_3 x$ **47.** $f^{-1}(x) = \frac{x - 3}{2}$ **49.** $f^{-1}(x) = 5x - 3$ **51.** $y = \pm 3$

53. $y = \pm 2i$, no real solution **55.** $x = \pm 3i$, no real solution **57.** $\frac{x^2}{9} + \frac{y^2}{4}$ **59.** x-intercept 4, y-intercept -3 **61.** ± 2.4

63. $(x - 2)^2 + (y - 3)^2 = 4$ **65.** $(x - 2)^2 + (y - 3)^2 = 4$ **67.** 5 **69.** 5

71. $y = \sqrt{9 - x^2}$ corresponds to the top half; $y = -\sqrt{9 - x^2}$ to the bottom half.

Problem Set 10.2

1.

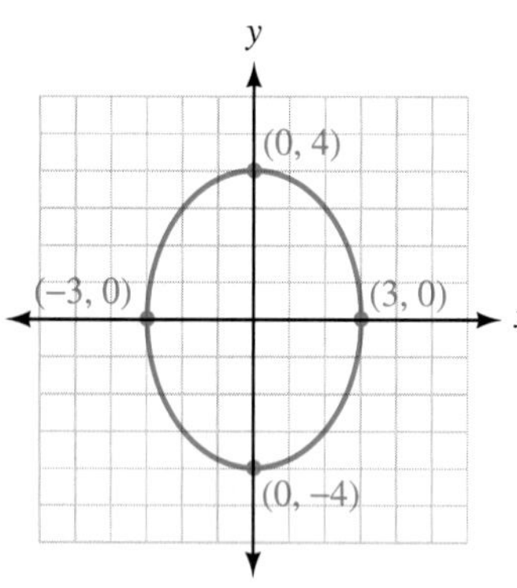

3.

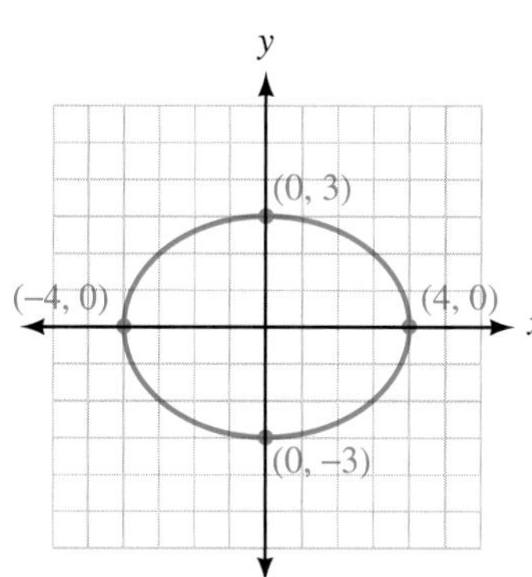

5.

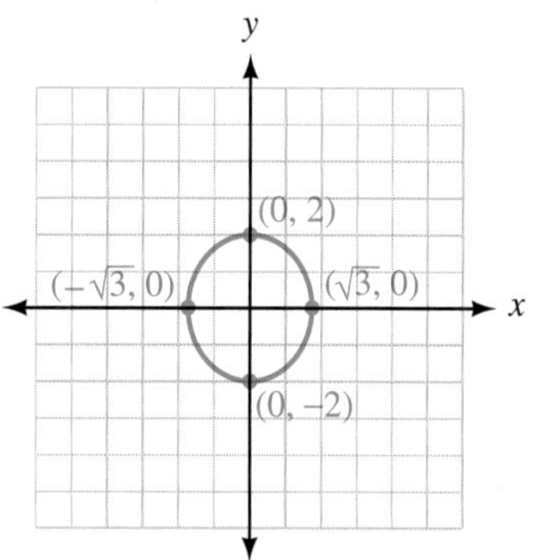

7.

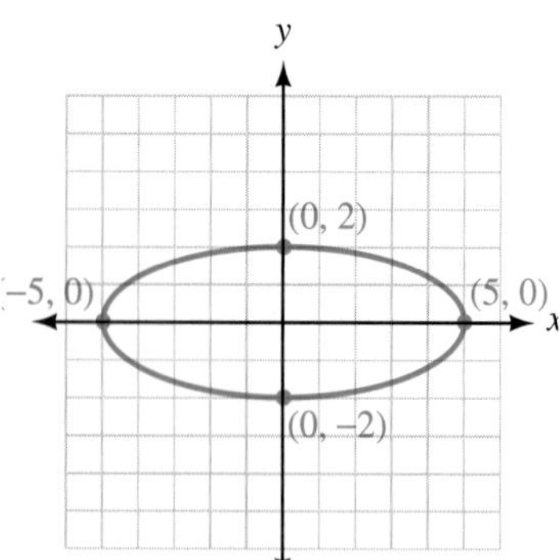

9.

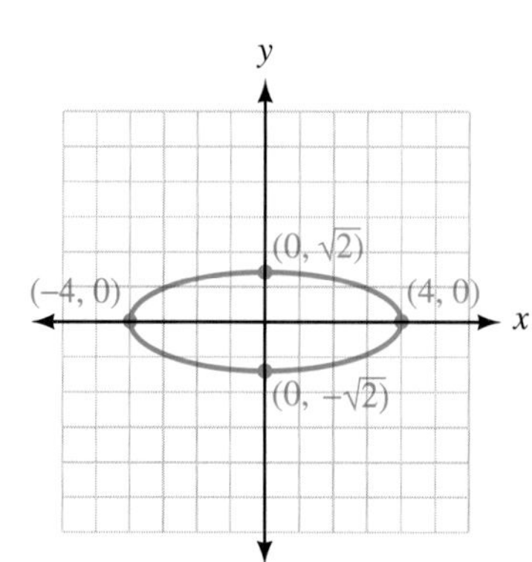

11.

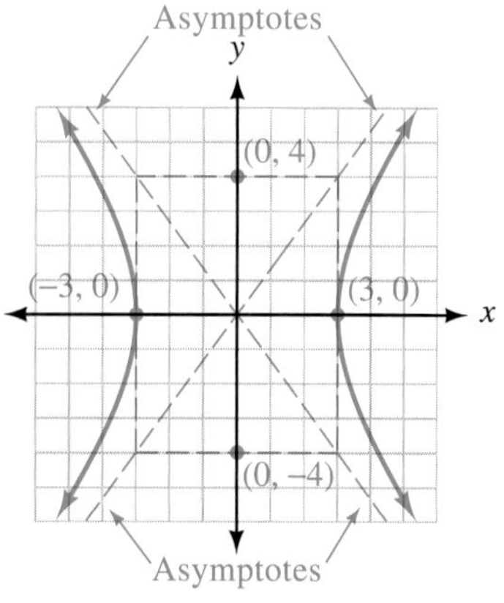

13.

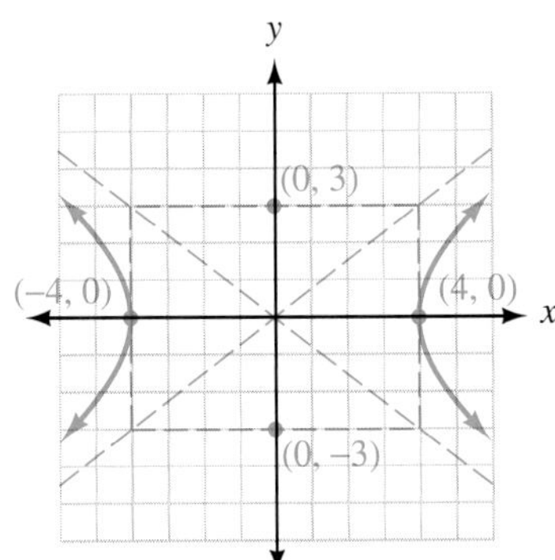

15.

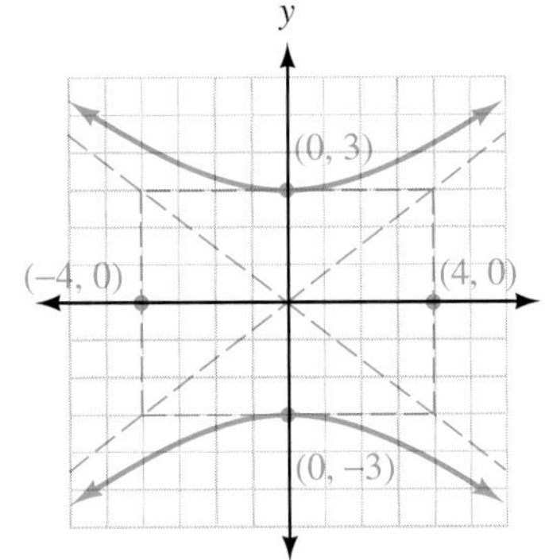

17.

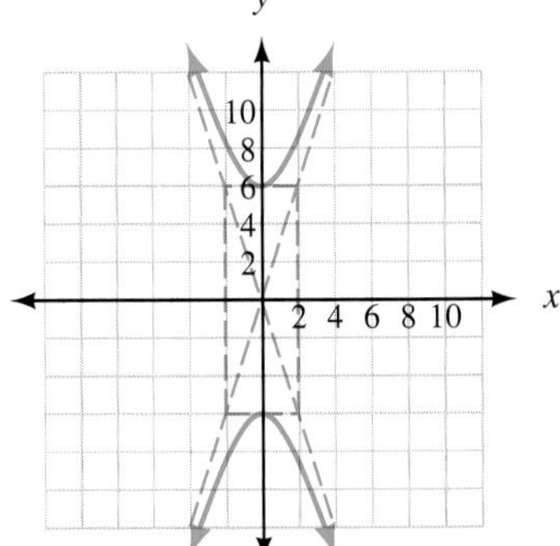

19.

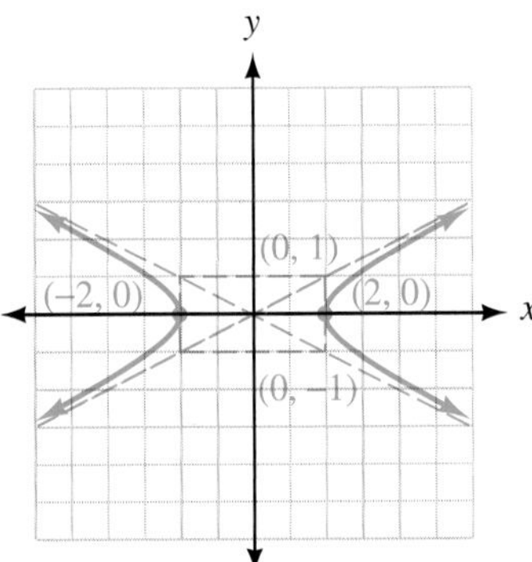

21.

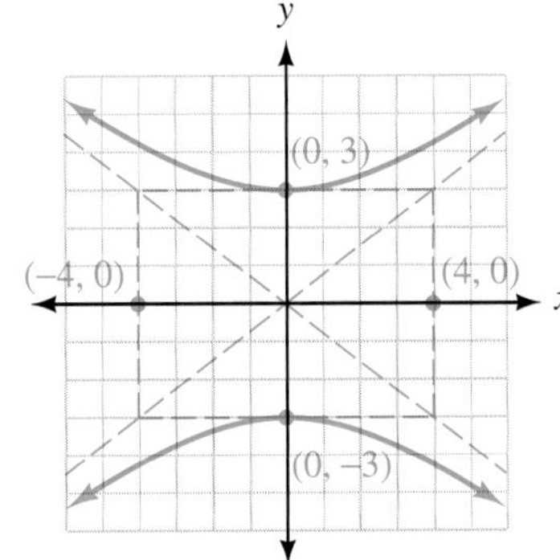

23. x-intercepts = ± 3, y-intercepts = ± 2

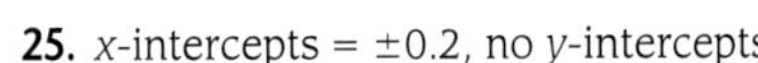

25. x-intercepts = ± 0.2, no y-intercepts

27. x-intercepts = $\pm\frac{3}{5}$, y-intercepts = $\pm\frac{2}{5}$

29.

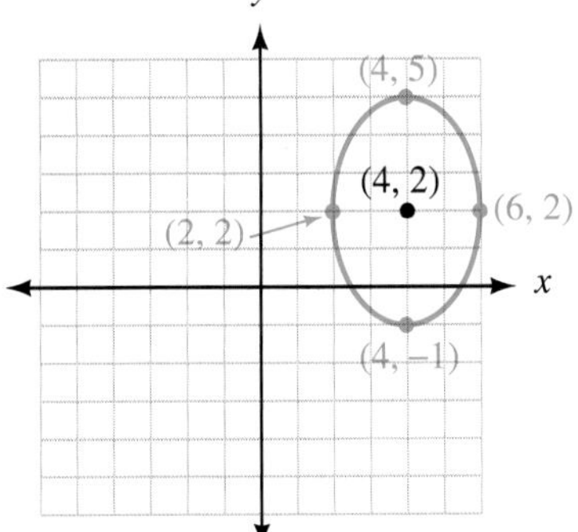

31.

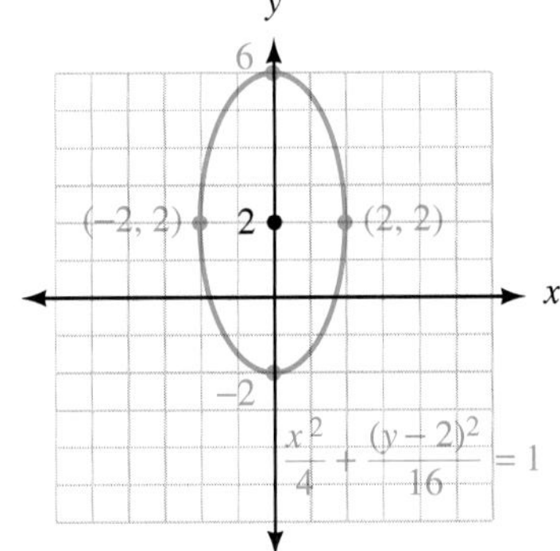

33.

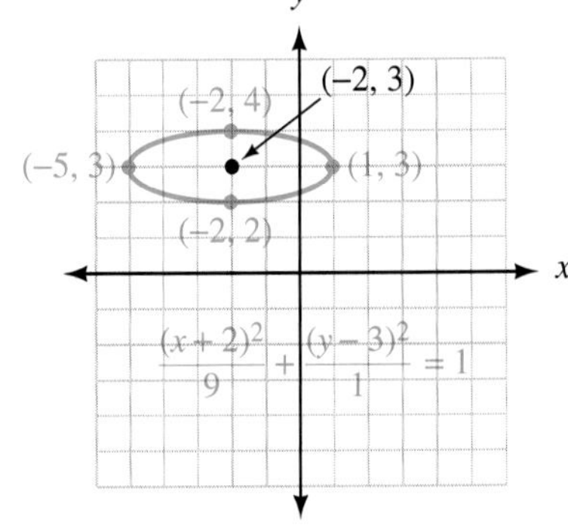

35.

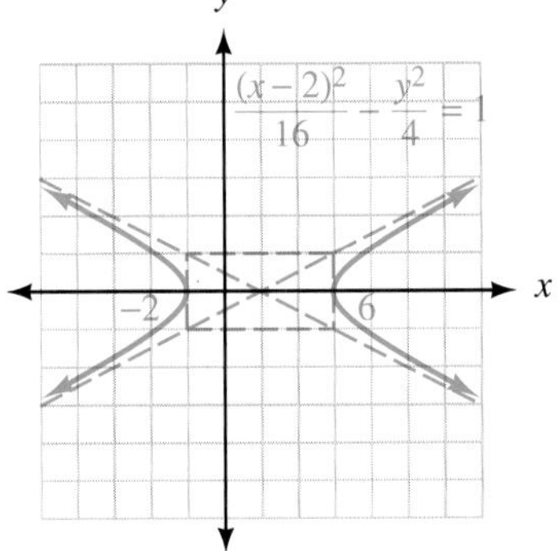

37.

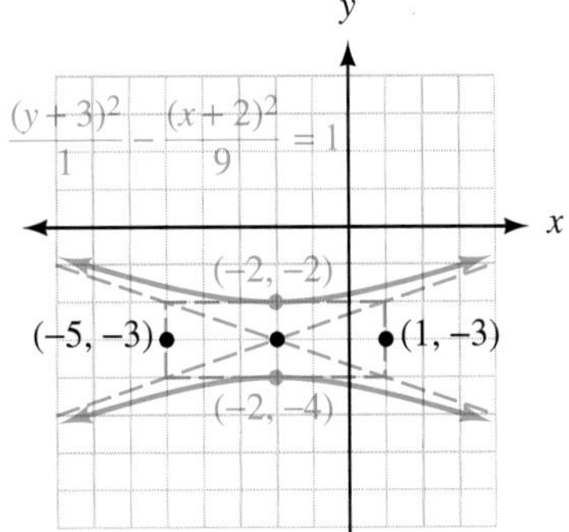

39.

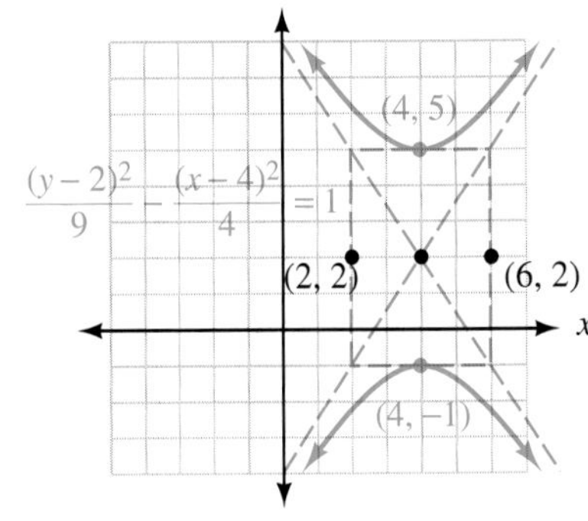

41. $\pm\frac{9}{5}$ **43.** ± 3.2 **45.** $y = \frac{3}{4}x, y = -\frac{3}{4}x$ **47.** $\frac{x^2}{16} + \frac{y^2}{4} = 1$
49. The equation is $\frac{x^2}{20^2} + \frac{y^2}{10^2} = 1$. A 6-foot man could walk upright under the arch anywhere between 16 feet to the left and 16 feet the right of the center. **51.** About 5.3 feet wide **53.** 1 **55.** -2 **57.** $\frac{4x}{(x-2)(x+2)}$ **59.** (0, 0) **61.** $4y^2 + 16y + 16$
63. $x = 2y + 4$ **65.** $-x^2 + 6$ **67.** $(5y + 6)(y + 2)$ **69.** ± 2 **71.** ± 2

Problem Set 10.3

1.

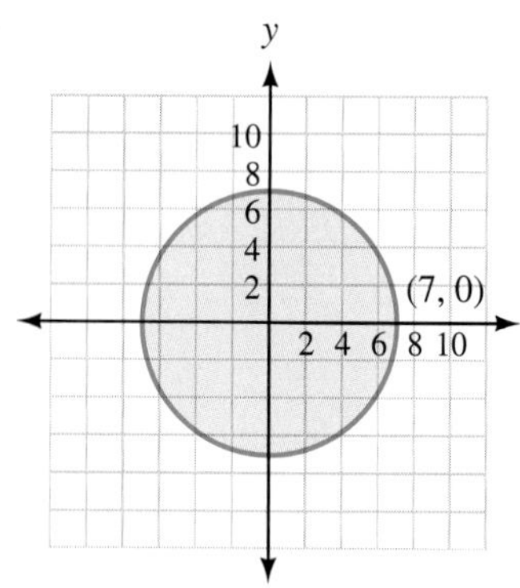

3.

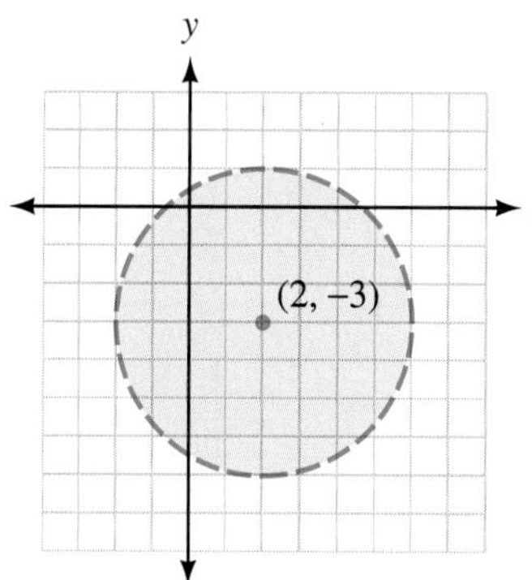

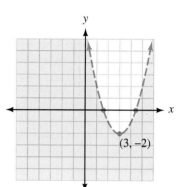

7.

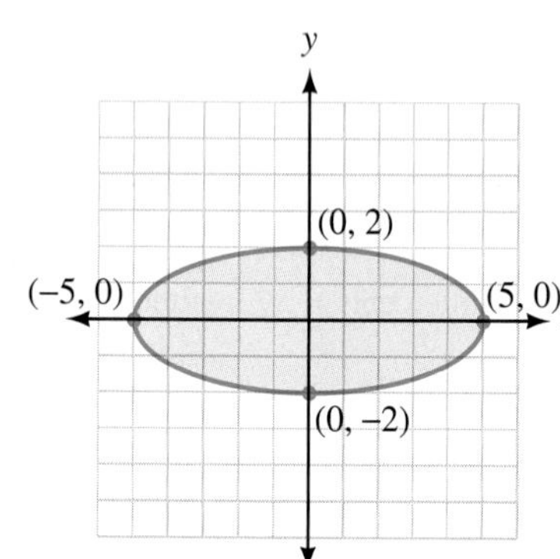

9. $(0, 3), (\frac{12}{5}, -\frac{9}{5})$ **11.** $(0, 4), (\frac{16}{5}, \frac{12}{5})$ **13.** $(5, 0), (-5, 0)$ **15.** $(0, -3), (\sqrt{5}, 2), (-\sqrt{5}, 2)$ **17.** $(0, -4), (\sqrt{7}, 3), (-\sqrt{7}, 3)$
19. $(-4, 11), (\frac{5}{2}, \frac{5}{4})$ **21.** $(-4, 5), (1, 0)$ **23.** $(2, -3), (5, 0)$ **25.** $(3, 0), (-3, 0)$ **27.** $(4, 0), (0, -4)$
29. a. $(-4, 4\sqrt{3})$ and $(-4, -4\sqrt{3})$ **b.** $(4, 4\sqrt{3})$ and $(4, -4\sqrt{3})$

31.

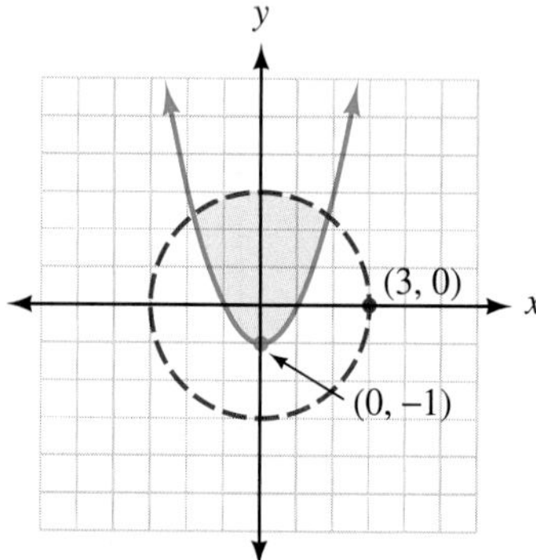

33.

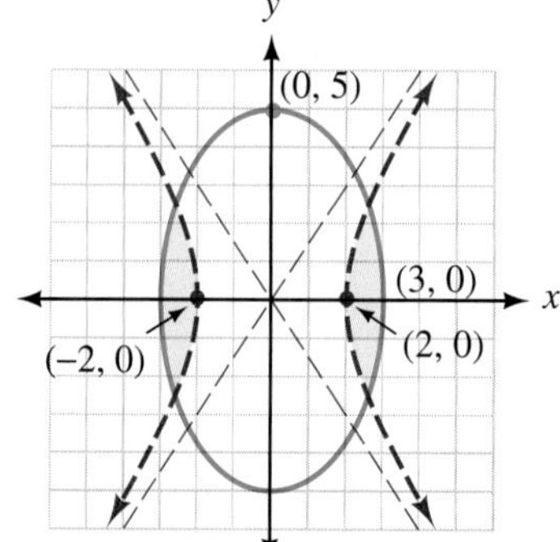

35. No intersection

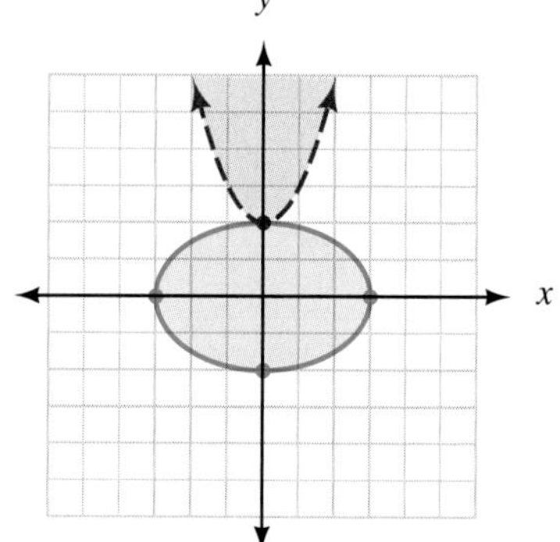

37. $x^2 + y^2 < 16$
$y > 4 - \frac{1}{4}x^2$
39. 8, 5 or −8, −5 or 8, −5 or −8, 5 **41.** 6, 3 or 13, −4 **43.** $-\frac{2}{5}$ **45.** −5, 1 **47.** −2, −1

CHAPTER 10 REVIEW TEST

1. $\sqrt{10}$ **2.** $\sqrt{13}$ **3.** 5 **4.** 9 **5.** −2, 6 **6.** −12, 4 **7.** $(x-3)^2+(y-1)^2=4$ **8.** $(x-3)^2+(y+1)^2=16$ **9.** $(x+5)^2+y^2=9$ **10.** $(x+3)^2+(y-4)^2=18$ **11.** $x^2+y^2=25$ **12.** $x^2+y^2=9$ **13.** $(x+2)^2+(y-3)^2=25$ **14.** $(x+6)^2+(y-8)^2=100$

15. (0, 0); $r=2$

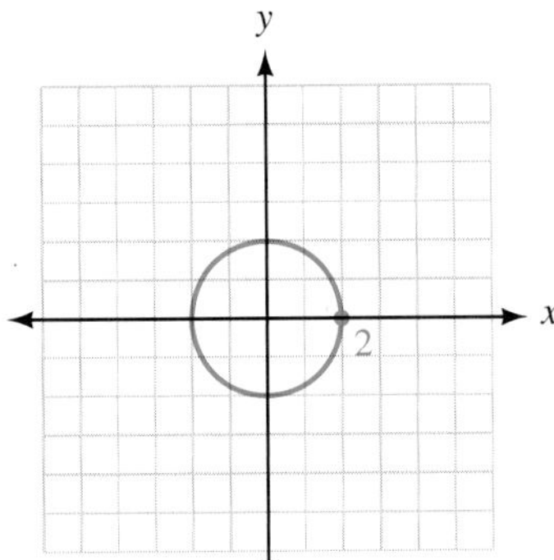

16. (3, −1); $r=4$

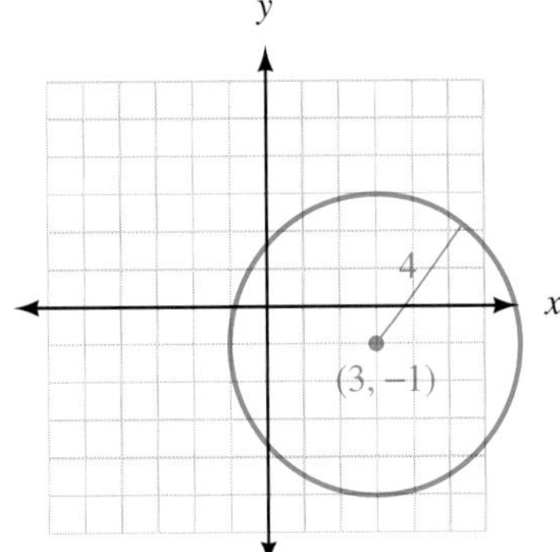

17. (3, −2); $r=3$

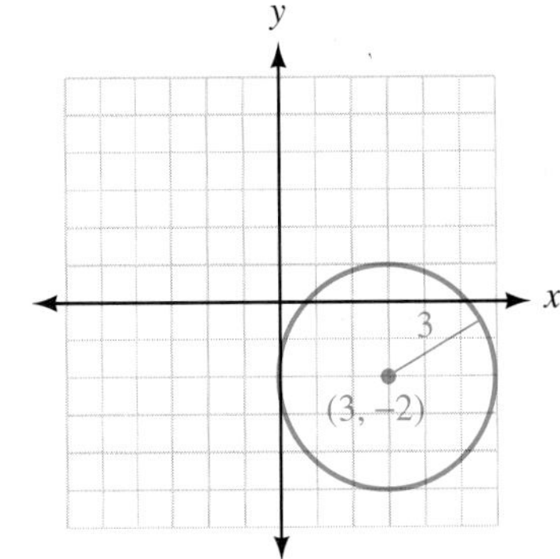

18. (−2, 1); $r=3$

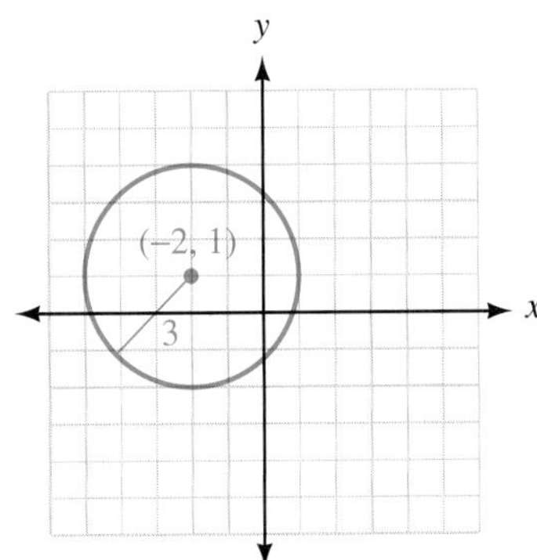

19.

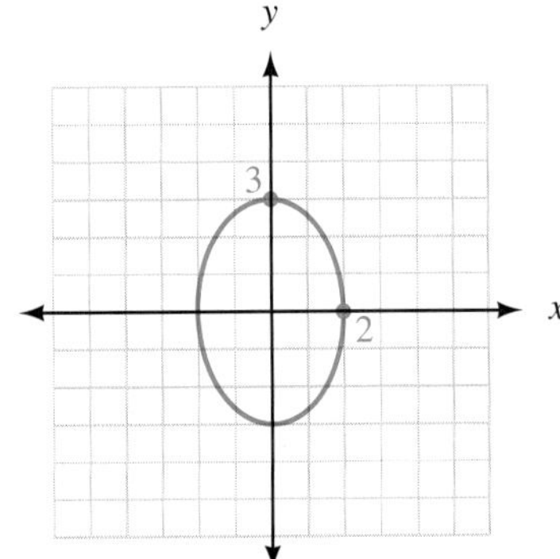

20.

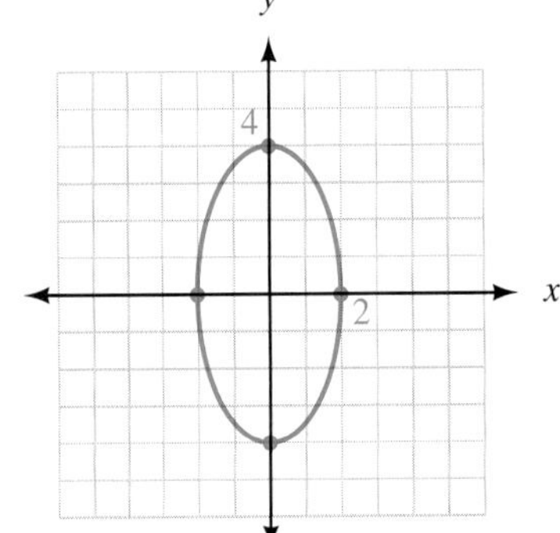

21.

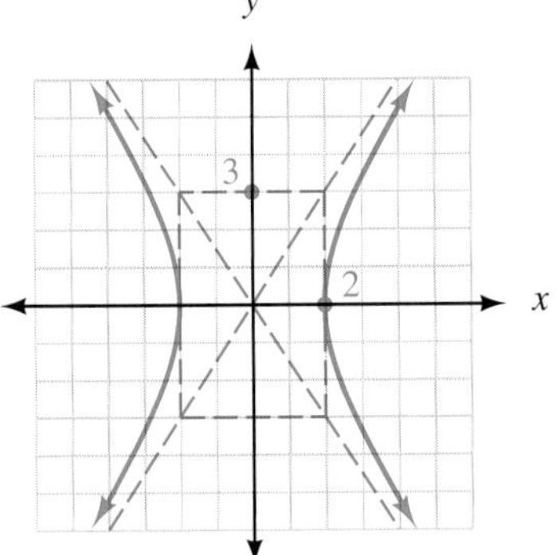

22.

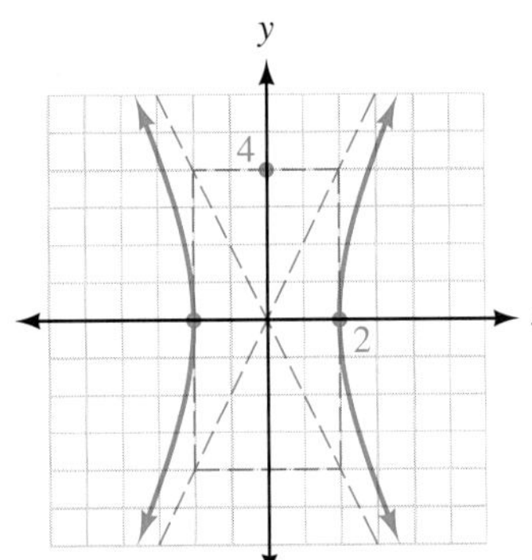

23.

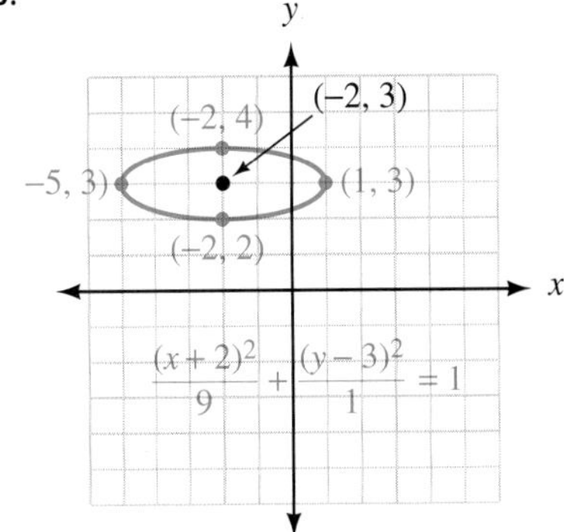

24.

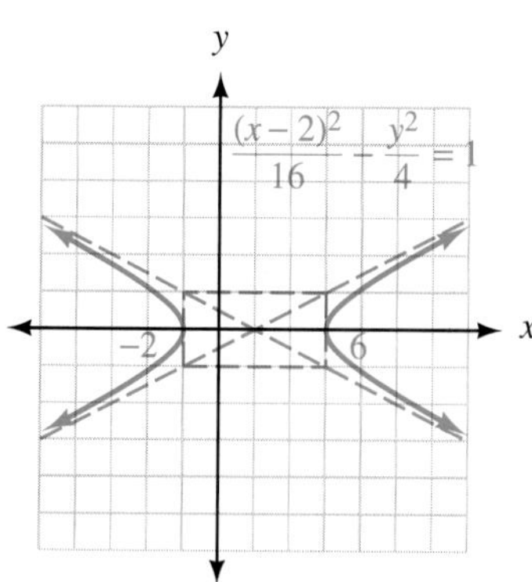

25.

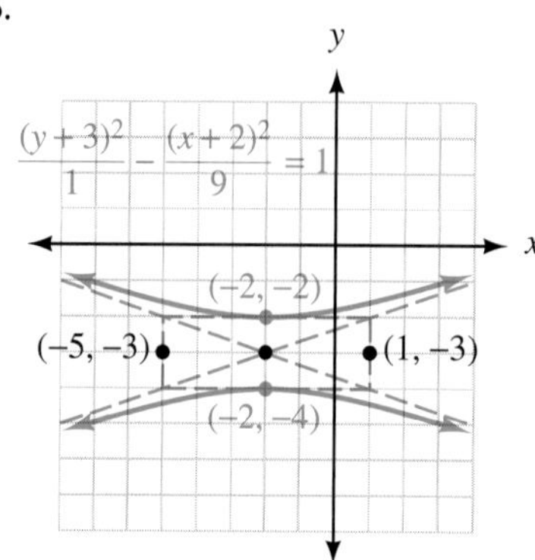

26.

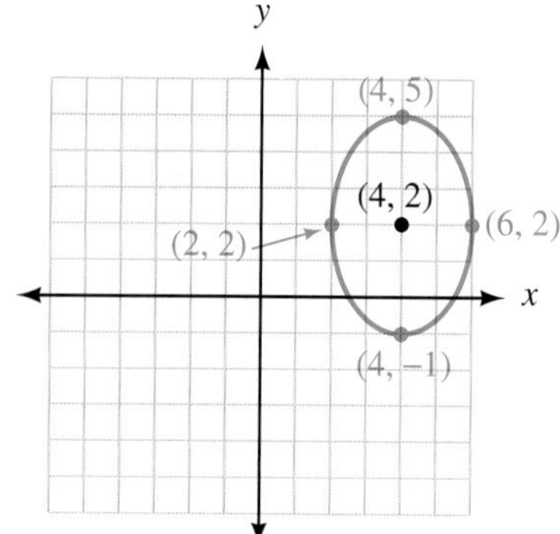

27.

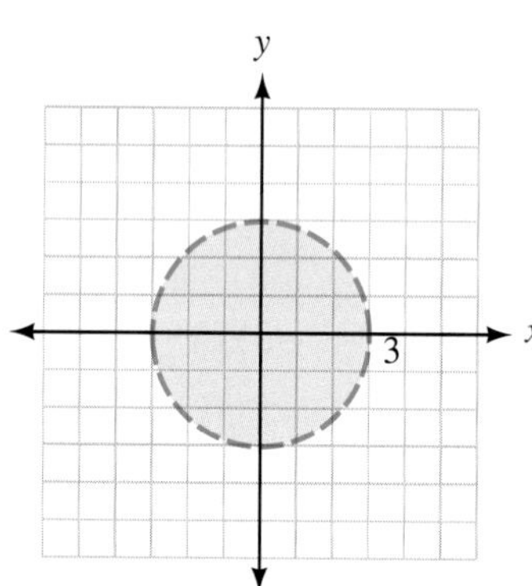

28.

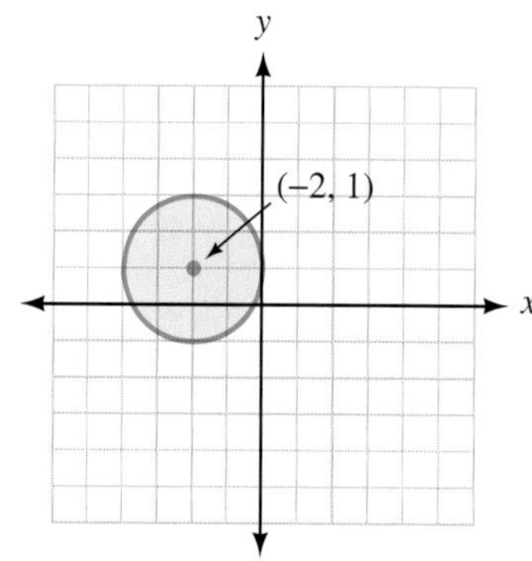

29.

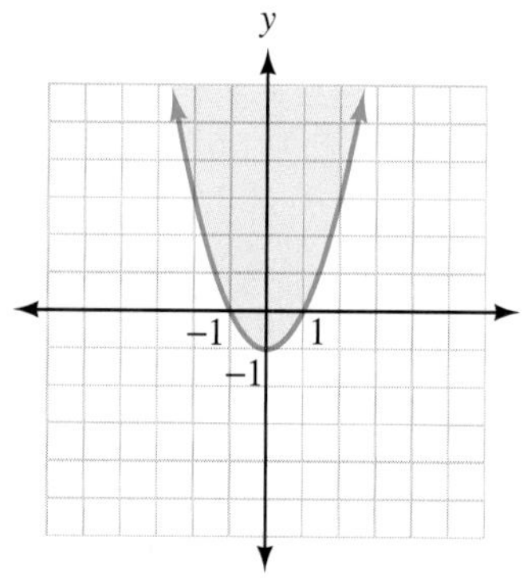

30.

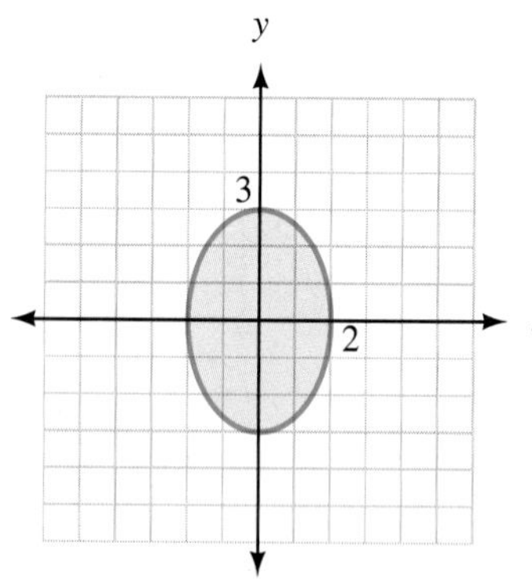

31.

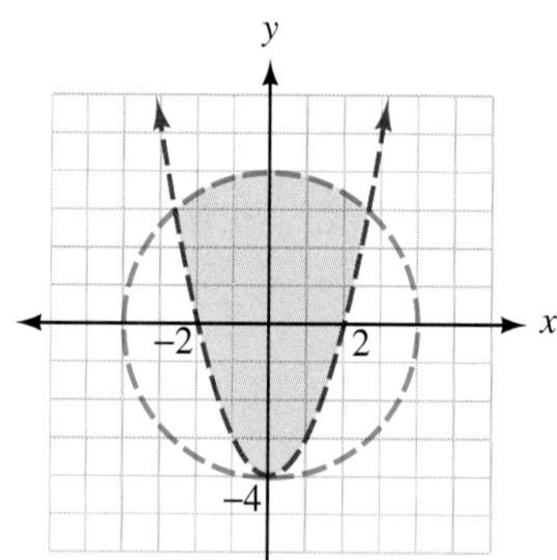

32.

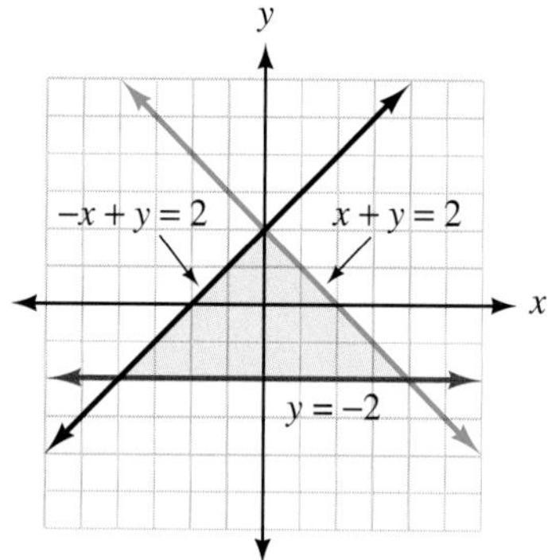

33. $(0, 4), \left(\frac{16}{5}, -\frac{12}{5}\right)$ **34.** $(0, -2), (\sqrt{3}, 1), (-\sqrt{3}, 1)$ **35.** $(-2, 0), (2, 0)$ **36.** $\left(-\sqrt{7}, -\frac{\sqrt{6}}{2}\right), \left(-\sqrt{7}, \frac{\sqrt{6}}{2}\right), \left(\sqrt{7}, -\frac{\sqrt{6}}{2}\right), \left(\sqrt{7}, \frac{\sqrt{6}}{2}\right)$

CHAPTER 11

Problem Set 11.1

1. 4, 7, 10, 13, 16 **3.** 3, 7, 11, 15, 19 **5.** 1, 2, 3, 4, 5 **7.** 4, 7, 12, 19, 28 **9.** $\frac{1}{4}, \frac{2}{5}, \frac{1}{2}, \frac{4}{7}, \frac{5}{8}$ **11.** $1, \frac{1}{4}, \frac{1}{9}, \frac{1}{16}, \frac{1}{25}$
13. 2, 4, 8, 16, 32 **15.** $2, \frac{3}{2}, \frac{4}{3}, \frac{5}{4}, \frac{6}{5}$ **17.** −2, 4, −8, 16, −32 **19.** 3, 5, 3, 5, 3 **21.** $1, -\frac{2}{3}, \frac{3}{5}, -\frac{4}{7}, \frac{5}{9}$ **23.** $\frac{1}{2}, 1, \frac{9}{8}, 1, \frac{25}{32}$
25. 3, −9, 27, −81, 243 **27.** 1, 5, 13, 29, 61 **29.** 2, 3, 5, 9, 17 **31.** 5, 11, 29, 83, 245 **33.** 4, 4, 4, 4, 4 **35.** $a_n = 4n$
37. $a_n = n^2$ **39.** $a_n = 2^{n+1}$ **41.** $a_n = \frac{1}{2^{n+1}}$ **43.** $a_n = 3n + 2$ **45.** $a_n = -4n + 2$ **47.** $a_n = (-2)^{n-1}$
49. $a_n = \log_{n+1}(n + 2)$ **51.** **a.** \$28000, \$29120, \$30284.80, \$31,496.19, \$32756.04 **b.** $a_n = 28000(1.04)^{n-1}$
53. **a.** 16 ft, 48 ft, 80 ft, 112 ft, 144 ft **b.** 400 ft **c.** No **55.** **a.** 180°, 360°, 540°, 720° **b.** 3,240° **c.** Sum of interior angles would be 0°
57. 27 **59.** 5 **61.** 1 **63.** 30 **65.** 40 **67.** 18 **69.** $-\frac{20}{81}$ **71.** $\frac{21}{10}$
73. $a_{100} \approx 2.7048$, $a_{1,000} \approx 2.7169$, $a_{10,000} \approx 2.7181$, $a_{100,000} \approx 2.7183$ **75.** 1, 1, 2, 3, 5, 8, 13, 21, 34, 55

Problem Set 11.2

1. 36 **3.** 11 **5.** 18 **7.** $\frac{163}{60}$ **9.** 60 **11.** 40 **13.** 44 **15.** $-\frac{11}{32}$ **17.** $\frac{21}{10}$ **19.** $5x + 15$

21. $(x-2) + (x-2)^2 + (x-2)^3 + (x-2)^4$ **23.** $\frac{x+1}{x-1} + \frac{x+2}{x-1} + \frac{x+3}{x-1} + \frac{x+4}{x-1} + \frac{x+5}{x-1}$

25. $(x+3)^3 + (x+4)^4 + (x+5)^5 + (x+6)^6 + (x+7)^7 + (x+8)^8$ **27.** $(x-6)^6 + (x-8)^7 + (x-10)^8 + (x-12)^9$

29. $\sum_{i=1}^{4} 2^i$ **31.** $\sum_{i=2}^{6} 2^i$ **33.** $\sum_{i=1}^{5} (4i+1)$ **35.** $\sum_{i=2}^{5} -(-2)^i$ **37.** $\sum_{i=3}^{7} \frac{i}{i+1}$ **39.** $\sum_{i=1}^{4} \frac{i}{2i+1}$ **41.** $\sum_{i=6}^{9} (x-2)^i$ **43.** $\sum_{i=1}^{4} (1 + \frac{i}{x})^{i+1}$

45. $\sum_{i=3}^{5} \frac{x}{x+i}$ **47.** $\sum_{i=2}^{4} x^i(x+i)$

49. a. $0.3 + 0.03 + 0.003 + 0.0003 + \ldots$ **b.** $0.2 + 0.02 + 0.002 + 0.0002 + \ldots$ **c.** $0.27 + 0.0027 + 0.000027 + \ldots$

51. seventh second: 208 feet; total: 784 feet **53. a.** $16 + 48 + 80 + 112 + 144$ **b.** $\sum_{i=1}^{5} (32i - 16)$ **55.** $3\log_2 x + \log_2 y$

57. $\frac{1}{3}\log_{10} x - 2\log_{10} y$ **59.** $\log_{10} \frac{x}{y^2}$ **61.** $\log_3 \frac{x^2}{y^3 z^4}$ **63.** 80 **65.** Possible solutions -1 and 8; only 8. **67.** 74 **69.** $\frac{55}{2}$

71. $2n + 1$ **73.** $(3, 2)$ **75.** $\frac{13}{2}$

Problem Set 11.3

1. arithmetic; $d = 1$ **3.** not arithmetic **5.** arithmetic; $d = -5$ **7.** not arithmetic **9.** arithmetic; $d = \frac{2}{3}$

11. $a_n = 4n - 1$; $a_{24} = 95$ **13.** $a_{10} = -12$; $S_{10} = -30$ **15.** $a_1 = 7$; $d = 2$; $a_{30} = 65$ **17.** $a_1 = 12$; $d = 2$; $a_{20} = 50$; $S_{20} = 620$

19. $a_{20} = 79$; $S_{20} = 820$ **21.** $a_{40} = 122$; $S_{40} = 2{,}540$ **23.** $a_1 = 13$; $d = -6$ **25.** $a_{85} = -238$ **27.** $d = \frac{16}{19}$; $a_{85} = 28$

29. 20,300 **31.** $a_{35} = -158$ **33.** $a_{10} = 5$; $S_{10} = \frac{55}{2}$

35. a. \$18,000, \$14,700, \$11,400, \$8,100, \$4,800 **b.** −\$3,300 **c.**

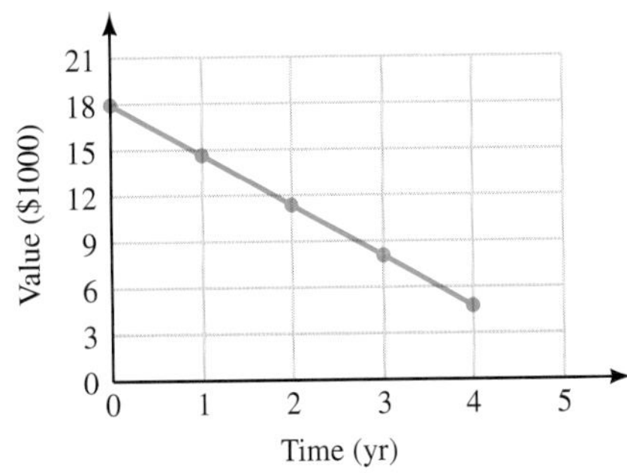

d. \$9,750 **e.** $a_0 = 18000$; $a_n = a_{n-1} - 3300$ for $n \geq 1$ **37. a.** 1500 ft, 1460 ft, 1420 ft, 1380 ft, 1340 ft, 1300 ft **b.** It is arithmetic because the same amount is subtracted from each succeeding term. **c.** $a_n = 1{,}540 - 40n$

39. a. 1, 3, 6, 10, 15, 21, 28, 36, 45, 55, 66, 78, 91, 105, 120 **b.** $a_1 = 1$; $a_n = n + a_{n-1}$ for $n \geq 2$ **c.** It is not arithmetic because the same amount is not added to each term.

41. 2.7604 **43.** -1.2396 **45.** 445 **47.** 4.45×10^{-8} **49.** $\frac{1}{16}$ **51.** $\sqrt{3}$ **53.** 2^n **55.** r^3 **57.** $\frac{2}{5}$

59. 255 **61.** $a_1 = 9, d = 4$ **63.** $d = -3$ **65.** $S_{50} = 50 + 1225\sqrt{2}$

Problem Set 11.4

1. geometric; $r = 5$ **3.** geometric; $r = \frac{1}{3}$ **5.** not geometric **7.** geometric; $r = -2$ **9.** not geometric

11. $a_n = 4 \cdot 3^{n-1}$ **13.** $a_6 = \frac{1}{16}$ **15.** $a_{20} = -3$ **17.** $S_{10} = 10{,}230$ **19.** $S_{20} = 0$ **21.** $a_8 = \frac{1}{640}$ **23.** $S_5 = -\frac{31}{32}$

25. $a_{10} = 32$; $S_{10} = 62 + 31\sqrt{2}$ **27.** $a_6 = \frac{1}{1000}$; $S_6 = 111.111$ **29.** ± 2 **31.** $a_8 = 384$; $S_8 = 255$ **33.** $r = 2$

35. 1 **37.** 8 **39.** 4 **41.** $\frac{8}{9}$ **43.** $\frac{2}{3}$ **45.** $\frac{9}{8}$ **51. a.** \$450,000, \$315,000, \$220,500, \$154,350, \$108,045 **b.** 0.7

c.

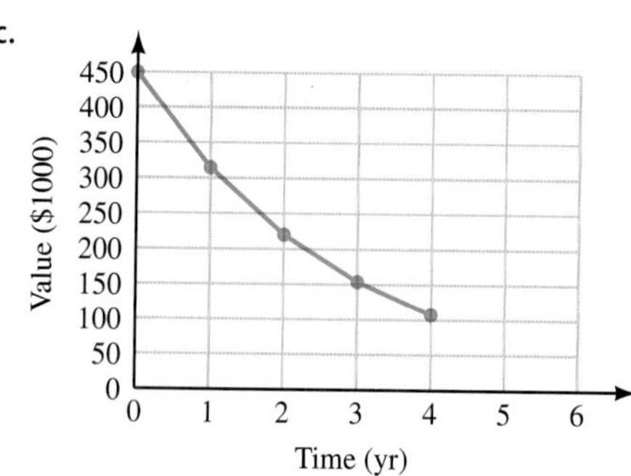

d. Approximately \$90,000 **e.** $a_0 = 450000$; $a_n = 0.7a_{n-1}$ for $n \geq 1$

53. a. $\frac{1}{2}$ **b.** $\frac{364}{729}$ **c.** $\frac{1}{1458}$ **55.** 300 feet **57. a.** $a_n = 15(\frac{4}{5})^{n-1}$ **b.** 75 feet **59.** 2.16 **61.** 6.36 **63.** $t = \frac{1}{5}\ln\frac{A}{10}$ **65.** 1
67. $x^3 + 2xy + y^2$ **69.** 15 **71.** $\frac{7}{11}$ **73.** $r = \frac{1}{3}$
75. a. stage 1:1; stage 2: $\frac{3}{4}$; stage 3: $\frac{9}{16}$; stage 4: $\frac{27}{64}$ **b.** geometric sequence **c.** 0 **d.** increasing sequence

Problem Set 11.5

1. $x^4 + 8x^3 + 24x^2 + 32x + 16$ **3.** $x^6 + 6x^5y + 15x^4y^2 + 20x^3y^3 + 15x^2y^4 + 6xy^5 + y^6$
5. $32x^5 + 80x^4 + 80x^3 + 40x^2 + 10x + 1$ **7.** $x^5 - 10x^4y + 40x^3y^2 - 80x^2y^3 + 80xy^4 - 32y^5$
9. $81x^4 - 216x^3 + 216x^2 - 96x + 16$ **11.** $64x^3 - 144x^2y + 108xy^2 - 27y^3$ **13.** $x^8 + 8x^6 + 24x^4 + 32x^2 + 16$
15. $x^6 + 3x^4y^2 + 3x^2y^4 + y^6$ **17.** $16x^4 + 96x^3y + 216x^2y^2 + 216xy^3 + 81y^4$ **19.** $\frac{x^3}{8} + \frac{x^2y}{4} + \frac{xy^2}{6} + \frac{y^3}{27}$
21. $\frac{x^3}{8} - 3x^2 + 24x - 64$ **23.** $\frac{x^4}{81} + \frac{2x^3y}{27} + \frac{x^2y^2}{6} + \frac{xy^3}{6} + \frac{y^4}{16}$ **25.** $x^9 + 18x^8 + 144x^7 + 672x^6$
27. $x^{10} - 10x^9y + 45x^8y^2 - 120x^7y^3$ **29.** $x^{25} + 75x^{24} + 2{,}700x^{23} + 62{,}100x^{22}$ **31.** $x^{60} - 120x^{59} + 7{,}080x^{58} - 273{,}760x^{57}$
33. $x^{18} - 18x^{17}y + 153x^{16}y^2 - 816x^{15}y^3$ **35.** $x^{15} + 15x^{14} + 105x^{13}$ **37.** $x^{12} - 12x^{11}y + 66x^{10}y^2$ **39.** $x^{20} + 40x^{19} + 760x^{18}$
41. $x^{100} + 200x^{99}$ **43.** $x^{50} + 50x^{49}y$ **45.** $51{,}963{,}120x^4y^8$ **47.** $3{,}360x^6$ **49.** $-25{,}344x^7$ **51.** $2{,}700x^{23}y^2$
53. $\binom{20}{11}(2x)^9(5y)^{11} = \frac{20!}{11!9!}(2x)^9(5y)^{11}$ **55.** $x^{20}y^{10} - 30x^{18}y^9 + 405x^{16}y^8$ **57.** $\frac{21}{128}$ **59.** $x \approx 1.21$ **61.** $\frac{1}{6}$ **63.** ≈ 17.5 years
65. 56 **67.** 125,970

CHAPTER 11 REVIEW TEST

1. 7, 9, 11, 13 **2.** 1, 4, 7, 10 **3.** 0, 3, 8, 15 **4.** $\frac{4}{3}, \frac{5}{4}, \frac{6}{5}, \frac{7}{6}$ **5.** 4, 16, 64, 256 **6.** $\frac{1}{4}, \frac{1}{16}, \frac{1}{64}, \frac{1}{256}$ **7.** $a_n = 3n - 1$
8. $a_n = 2n - 5$ **9.** $a_n = n^4$ **10.** $a_n = n^2 + 1$ **11.** $a_n = (\frac{1}{2})^n = 2^{-n}$ **12.** $a_n = \frac{n+1}{n^2}$ **13.** 32 **14.** 25 **15.** $\frac{14}{5}$
16. −5 **17.** 98 **18.** $\frac{127}{30}$ **19.** $\sum_{i=1}^{4} 3i$ **20.** $\sum_{i=1}^{4} (4i - 1)$ **21.** $\sum_{i=1}^{5} (2i + 3)$ **22.** $\sum_{i=2}^{4} i^2$ **23.** $\sum_{i=1}^{4} \frac{1}{i+2}$ **24.** $\sum_{i=1}^{5} \frac{i}{3^i}$
25. $\sum_{i=1}^{3} (x - 2i)$ **26.** $\sum_{i=1}^{4} \frac{x}{x+i}$ **27.** geometric **28.** arithmetic **29.** arithmetic **30.** neither **31.** geometric
32. geometric **33.** arithmetic **34.** neither **35.** $a_n = 3n - 1$; $a_{20} = 59$ **36.** $a_n = 8 - 3n$; $a_{16} = -40$
37. $a_{10} = 34$; $S_{10} = 160$ **38.** $a_{16} = 78$; $S_{16} = 648$ **39.** $a_1 = 5$; $d = 4$; $a_{10} = 41$ **40.** $a_1 = 8$; $d = 3$; $a_9 = 32$; $S_9 = 180$
41. $a_1 = -4$; $d = -2$; $a_{20} = -42$; $S_{20} = -460$ **42.** 20,100 **43.** $a_{40} = -95$ **44.** $a_n = 3(2)^{n-1}$; $a_{20} = 1{,}572{,}864$
45. $a_n = 5(-2)^{n-1}$; $a_{16} = -163{,}840$ **46.** $a_n = 4(\frac{1}{2})^{n-1}$; $a_{10} = \frac{1}{128}$ **47.** −3 **48.** 8 **49.** $a_1 = 3$; $r = 2$; $a_6 = 96$ **50.** $243\sqrt{3}$
51. $x^4 - 8x^3 + 24x^2 - 32x + 16$ **52.** $16x^4 + 96x^3 + 216x^2 + 216x + 81$ **53.** $27x^3 + 54x^2y + 36xy^2 + 8y^3$
54. $x^{10} - 10x^8 + 40x^6 - 80x^4 + 80x^2 - 32$ **55.** $\frac{1}{16}x^4 + \frac{3}{2}x^3 + \frac{27}{2}x^2 + 54x + 81$ **56.** $\frac{1}{27}x^3 - \frac{1}{6}x^2y + \frac{1}{4}xy^2 - \frac{1}{8}y^3$
57. $x^{10} + 30x^9y + 405x^8y^2$ **58.** $x^9 - 27x^8y + 324x^7y^2$ **59.** $x^{11} + 11x^{10}y + 55x^9y^2$ **60.** $x^{12} - 24x^{11}y + 264x^{10}y^2$
61. $x^{16} - 32x^{15}y$ **62.** $x^{32} + 64x^{31}y$ **63.** $x^{50} - 50x^{49}$ **64.** $x^{150} + 150x^{149}y$ **65.** $-61{,}236x^5$ **66.** $5376x^6$

Index

Functions: A Summary

More about Functions [Section 3.5]

Since functions are rules that assign values from one set (the domain) to a second set (the range), the result is a set of ordered pairs, where the first coordinate in each ordered pair comes from the domain and the second coordinate of each ordered pair comes from the range. This fact allows us to write an alternate definition for functions.

Alternate Definition A *function* is a set of ordered pairs in which no two different ordered pairs have the same first coordinate. The set of all first coordinates is called the *domain* of the function. The set of all second coordinates is called the *range* of the function.

In order to classify together all sets of ordered pairs, whether they are functions or not, we include the following definitions.

Relations [Section 3.5]

A *relation* is a rule that pairs each element in one set, called the *domain,* with one or more elements from a second set, called the *range.*

Alternate Definition A *relation* is a set of ordered pairs. The set of all first coordinates is the *domain* of the relation. The set of all second coordinates is the *range* of the relation.

Example Figure 9 is a scatter diagram showing the advertised prices of used Ford Mustangs appearing in a Los Angeles newspaper one day in 1995. The set of data from which the scatter diagram was constructed is not a function because some values of x are paired with more than one value of y.

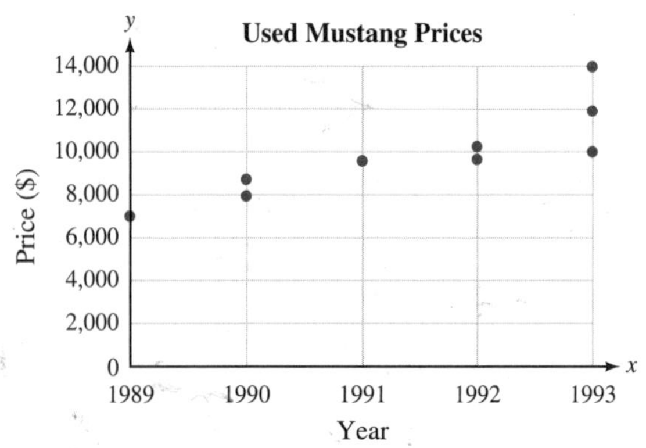

FIGURE 9 **Scatter diagram of a relation that is not also a function.**

As a consequence of the definitions for functions and relations, we have the following two facts: Every function is also a relation, but not every relation is a function.

Vertical Line Test [Section 3.5]

If a vertical line crosses the graph of a relation in more than one place, the relation cannot be a function. If no vertical line can be found that crosses a graph in more than one place, then the graph is the graph of a function.

Examples The curve shown in Figure 10 does not represent a function since there are many vertical lines that cross this graph in more than one place.

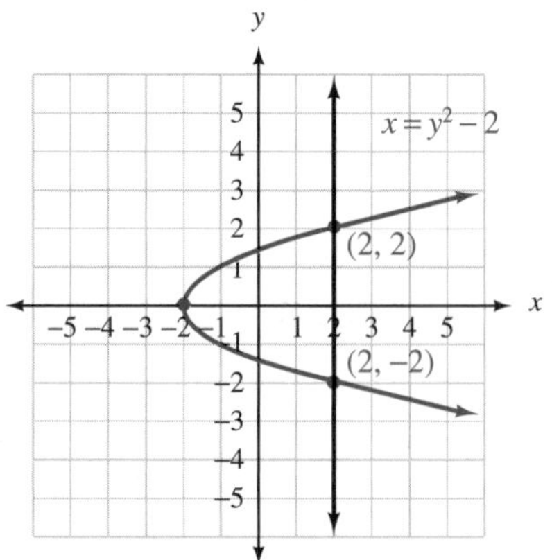

FIGURE 10 **The graph of the relation $x = y^2 - 2$.**

Algebra with Functions [Section 3.7]

If we are given functions f and g with a common domain, we can define four other functions as follows.

$(f+g)(x) = f(x) + g(x)$	The function $f + g$ is the sum of the functions f and g.
$(f-g)(x) = f(x) - g(x)$	The function $f - g$ is the difference of the functions f and g.
$(fg)(x) = f(x)g(x)$	The function fg is the product of the functions f and g.
$\left(\frac{f}{g}\right)(x) = \frac{f(x)}{g(x)}$	The function f/g is the quotient of the functions f and g, where $g(x) \neq 0$.

Composition of Functions [Section 3.7]

To find the composition of two functions f and g, we first require that the range of g have numbers in com-